RSMeans®

Residential Cost Data

19th Annual Edition

- *Square Foot Costs*
- *Systems Costs*
- *Unit Costs*

2000

Senior Editors

Howard M Chandler
Robert W. Mewis

Contributing Editors

Barbara Balboni
John H. Chiang, PE
Paul C. Crosscup
Jennifer L. Curran
Stephen E. Donnelly
J. Robert Lang
Robert C. McNichols
Melville J. Mossman, PE
John J. Moylan
Jeannene D. Murphy
Peter T. Nightingale
Stephen C. Plotner
Michael J. Regan
William R. Tennyson, II
Phillip R. Waier, PE

Manager, Engineering Operations

John H. Ferguson, PE

President

Durwood S. Snead

Vice President and General Manager

Roger J. Grant

Vice President, Sales and Marketing

John M. Shea

Production Manager

Michael Kokernak

Production Coordinator

Marion E. Schofield

Technical Support

Wayne D. Anderson
Thomas J. Dion
Michael H. Donelan
Jonathan Forgit
Gary L. Hoitt
Paula Reale-Camelio
Kathryn S. Rodriguez
Sheryl A. Rose
James N. Wills

Art Director

Helen A. Marcella

Book & Cover Design

Norman R. Forgit

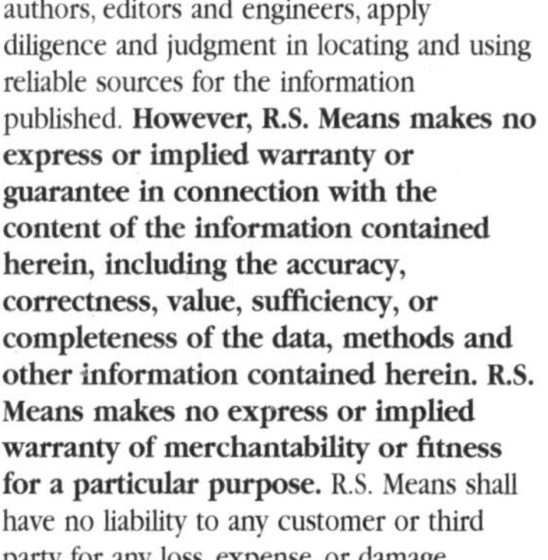

First Printing

Foreword

R.S. Means Co., Inc. is owned by CMD Group, a leading worldwide provider of proprietary construction information. CMD Group is comprised of three synergistic product groups crafted to be the complete resource for reliable, timely and actionable construction market data. In North America, CMD Group encompasses: Architects' First Source, an innovative product selection and specification solution in print and on the Internet; Construction Market Data (CMD), the source for construction activity information, as well as early planning reports for the design community; Associated Construction Publications, with 14 magazines, one of the largest editorial networks dedicated to U.S. highway and heavy construction coverage; Manufacturer's Survey Associates (MSA), a leading estimating and quantity survey firm in the U.S.; R.S. Means, the authority on construction cost data in North America; CMD Canada, the leading supplier of project information, industry news and forecasting data products for the Canadian construction industry; and BIMSA/Mexico, the dominant distributor of information on building projects and construction throughout Mexico. Worldwide, CMD Group includes Byggfakta Scandinavia, providing construction market data to Denmark, Estonia, Finland, Norway and Sweden; and Cordell Building Information Services, the market leader for construction and cost information in Australia.

Our Mission

Since 1942, R.S. Means Company, Inc. has been actively engaged in construction cost publishing and consulting throughout North America.

Today, over fifty years after the company began, our primary objective remains the same: to provide you, the construction and facilities professional, with the most current and comprehensive construction cost data possible.

Whether you are a contractor, an owner, an architect, an engineer, a facilities manager, or anyone else who needs a fast and reliable construction cost estimate, you'll find this publication to be a highly useful and necessary tool.

Today, with the constant flow of new construction methods and materials, it's difficult to find the time to look at and evaluate all the different construction cost possibilities. In addition, because labor and material costs keep changing, last year's cost information is not a reliable basis for today's estimate or budget.

That's why so many construction professionals turn to R.S. Means. We keep track of the costs for you, along with a wide range of other key information, from city cost indexes . . . to productivity rates . . . to crew composition . . . to contractor's overhead and profit rates.

R.S. Means performs these functions by collecting data from all facets of the industry, and organizing it in a format that is instantly accessible to you. From the preliminary budget to the detailed unit price estimate, you'll find the data in this book useful for all phases of construction cost determination.

The Staff, the Organization, and Our Services

When you purchase one of R.S. Means' publications, you are in effect hiring the services of a full-time staff of construction and engineering professionals.

Our thoroughly experienced and highly qualified staff works daily at collecting, analyzing, and disseminating comprehensive cost information for your needs. These staff members have years of practical construction experience and engineering training prior to joining the firm. As a result, you can count on them not only for the cost figures, but also for additional background reference information that will help you create a realistic estimate.

The Means organization is always prepared to help you solve construction problems through its five major divisions: Construction and Cost Data Publishing, Electronic Products and Services, Consulting Services, Insurance Division, and Educational Services.

Besides a full array of construction cost estimating books, Means also publishes a number of other reference works for the construction industry. Subjects include construction estimating and project and business management; special topics such as HVAC, roofing, plumbing, and hazardous waste remediation; and a library of facility management references.

In addition, you can access all of our construction cost data through your computer with Means CostWorks 2000 CD-ROM, an electronic tool that offers over 50,000 lines of Means construction cost data.

What's more, you can increase your knowledge and improve your construction estimating and management performance with a Means Construction Seminar or In-House Training Program. These two-day seminar programs offer unparalleled opportunities for everyone in your organization to get updated on a wide variety of construction-related issues.

Means also is a worldwide provider of construction cost management and analysis services for commercial and government owners and of claims and valuation services for insurers.

In short, R.S. Means can provide you with the tools and expertise for constructing accurate and dependable construction estimates and budgets in a variety of ways.

Robert Snow Means Established a Tradition of Quality That Continues Today

Robert Snow Means spent years building his company, making certain he always delivered a quality product.

Today, at R.S. Means, we do more than talk about the quality of our data and the usefulness of our books. We stand behind all of our data, from historical cost indexes... to construction materials and techniques... to current costs.

If you have any questions about our products or services, please call us toll-free at 1-800-334-3509. Our customer service representatives will be happy to assist you.

Table of Contents

Foreword	ii
How the Book Is Built: An Overview	iv
How To Use the Book: The Details	v
Square Foot Cost Section	1
Assemblies Cost Section	97
Unit Price Section	287
Reference Section	511
Reference Numbers	512
Crew Listings	543
Location Factors	566
Abbreviations	572
Index	577
Other Means Publications and Reference Works	Yellow Pages
Installing Contractor's Overhead & Profit	Inside Back Cover

RESIDENTIAL

SQUARE FOOT COSTS

UNIT PRICES

GENERAL REQUIREMENTS 1
SITE CONSTRUCTION 2
CONCRETE 3
MASONRY 4
METALS 5
WOOD & PLASTICS 6
THERMAL & MOISTURE PROTECTION 7
DOORS & WINDOWS 8
FINISHES 9
SPECIALTIES 10
EQUIPMENT 11
FURNISHINGS 12
SPECIAL CONSTRUCTION 13
CONVEYING SYSTEMS 14
MECHANICAL 15
ELECTRICAL 16

ASSEMBLIES

SITE WORK 1
FOUNDATIONS 2
FRAMING 3
EXTERIOR WALLS 4
ROOFING 5
INTERIOR CONSTRUCTION 6
SPECIALTIES 7
MECHANICAL 8
ELECTRICAL 9

REFERENCE INFORMATION

REFERENCE NUMBERS
CREWS
LOCATION FACTORS
INDEX

How the Book Is Built: An Overview

A Powerful Construction Tool

You have in your hands one of the most powerful construction tools available today. A successful project is built on the foundation of an accurate and dependable estimate. This book will enable you to construct just such an estimate.

For the casual user the book is designed to be:

- quickly and easily understood so you can get right to your estimate
- filled with valuable information so you can understand the necessary factors that go into the cost estimate

For the regular user, the book is designed to be:

- a handy desk reference that can be quickly referred to for key costs
- a comprehensive, fully reliable source of current construction costs and productivity rates, so you'll be prepared to estimate any project
- a source book for preliminary project cost, product selections, and alternate materials and methods

To meet all of these requirements we have organized the book into the following clearly defined sections.

Square Foot Cost Section

This section lists Square Foot costs for typical residential construction projects. The organizational format used divides the projects into basic building classes. These classes are defined at the beginning of the section. The individual projects are further divided into ten common components of construction. An outline of a typical page layout, an explanation of Square Foot prices, and a Table of Contents are located at the beginning of the section.

Assemblies Cost Section

This section uses an "Assemblies" (sometimes referred to as "systems") format grouping all the functional elements of a building into 9 construction divisions.

At the top of each "Assembly" cost table is an illustration, a brief description, and the design criteria used to develop the cost. Each of the components and its contributing cost to the system is shown.

Material: These cost figures include a standard 10% markup for "handling". They are national average material costs as of January of the current year and include delivery to the job site.

Installation: The installation costs include labor and equipment, plus a markup for the installing contractor's overhead and profit.

For a complete breakdown and explanation of a typical "Assemblies" page, see "How To Use Assemblies Cost Tables" at the beginning of the Assembly Section.

Unit Price Section

All cost data has been divided into the 16 divisions according to the MasterFormat system of classification and numbering as developed by the Construction Specifications Institute (CSI) and Construction Specifications Canada (CSC). For a listing of these divisions and an outline of their subdivisions, see the Unit Price Section Table of Contents.

Estimating tips are included at the beginning of each division.

Reference Section

This section includes information on Reference Numbers, Crew Listings, Location Factors, and a listing of Abbreviations. It is visually identified by a vertical gray bar on the edge of pages.

Reference Numbers: At the beginning of selected major classifications in the Unit Price Section are "reference numbers" shown in bold squares. These numbers refer you to related information in the Reference Section.

In this section, you'll find reference tables, explanations, and estimating information that support how we develop the unit price data. Also included are alternate pricing methods, technical data, and estimating procedures, along with information on design and economy in construction. You'll also find helpful tips on what to expect and what to avoid when estimating and constructing your project.

It is recommended that you refer to the Reference Section if a "reference number" appears within the section you are estimating.

Crew Listings: This section lists all the crews referenced in the book. For the purposes of this book, a crew is composed of more than one trade classification and/or the addition of power equipment to any trade classification. Power equipment is included in the cost of the crew. Costs are shown both with bare labor rates and with the installing contractor's overhead and profit added. For each, the total crew cost per eight-hour day and the composite cost per labor-hour are listed.

Location Factors: Costs vary depending upon regional economy. You can adjust the "national average" costs in this book to over 930 major cities throughout the U.S. and Canada by using the data in this section.

Abbreviations: A listing of the abbreviations used throughout this book, along with the terms they represent, is included.

Index

A comprehensive listing of all terms and subjects in this book to help you find what you need quickly when you are not sure where it falls in MasterFormat.

The Scope of This Book

This book is designed to be as comprehensive and as easy to use as possible. To that end we have made certain assumptions and limited its scope in three key ways:

1. We have established material prices based on a "national average."
2. We have computed labor costs based on a 7 major region average of open shop wage rates.
3. We have targeted the data for projects of a certain size range.

Project Size

This book is intended for use by those involved primarily in Residential construction costing less than $750,000. This includes the construction of homes, row houses, townhouses, condominiums and apartments.

With reasonable exercise of judgment the figures can be used for any building work. For other types of projects, such as repair and remodeling or commercial buildings, consult the appropriate MEANS publication for more information.

How to Use the Book: The Details

What's Behind the Numbers? The Development of Cost Data

The staff at R.S. Means continuously monitors developments in the construction industry in order to ensure reliable, thorough and up-to-date cost information.

While *overall* construction costs may vary relative to general economic conditions, price fluctuations within the industry are dependent upon many factors. Individual price variations may, in fact, be opposite to overall economic trends. Therefore, costs are continually monitored and complete updates are published yearly. Also, new items are frequently added in response to changes in materials and methods.

Costs—$ (U.S.)

All costs represent U.S. national averages and are given in U.S. dollars. The Means Location Factors can be used to adjust costs to a particular location. The Location Factors for Canada can be used to adjust U.S. national averages to local costs in Canadian dollars.

Material Costs

The R.S. Means staff contacts manufacturers, dealers, distributors, and contractors all across the U.S. and Canada to determine national average material costs. If you have access to current material costs for your specific location, you may wish to make adjustments to reflect differences from the national average. Included within material costs are fasteners for a normal installation. R.S. Means engineers use manufacturers' recommendations, written specifications and/or standard construction practice for size and spacing of fasteners. Adjustments to material costs may be required for your specific application or location. Material costs do not include sales tax.

Labor Costs

Labor costs are based on the average of open shop wages from across the U.S. for the current year. Rates along with overhead and profit markups are listed on the inside back cover of this book.

- If wage rates in your area vary from those used in this book, or if rate increases are expected within a given year, labor costs should be adjusted accordingly.

Labor costs reflect productivity based on actual working conditions. These figures include time spent during a normal workday on tasks other than actual installation, such as material receiving and handling, mobilization at site, site movement, breaks, and cleanup.

Productivity data is developed over an extended period so as not to be influenced by abnormal variations and reflects a typical average.

Equipment Costs

Equipment costs include not only rental, but also operating costs for equipment under normal use. The operating costs include parts and labor for routine servicing such as repair and replacement of pumps, filters and worn lines. Normal operating expendables such as fuel, lubricants, tires and electricity (where applicable) are also included. Extaordinary operating expendables with highly variable wear patterns such as diamond bits and blades are excluded. These costs are included under materials. Equipment rental rates are obtained from industry sources throughout North America—contractors, suppliers, dealers, manufacturers, and distributors.

Crew Equipment Cost/Day—The power equipment required for each crew is included in the crew cost. The daily cost for crew equipment is based on dividing the weekly bare rental rate by 5 (number of working days per week), and then adding the hourly operating cost times 8 (hours per day). This "Crew Equipment Cost/Day" is listed in Subdivision 01590.

Factors Affecting Costs

Costs can vary depending upon a number of variables. Here's how we have handled the main factors affecting costs.

Quality—The prices for materials and the workmanship upon which productivity is based represent sound construction work. They are also in line with U.S. government specifications.

Overtime—We have made no allowance for overtime. If you anticipate premium time or work beyond normal working hours, be sure to make an appropriate adjustment to your labor costs.

Productivity—The productivity, daily output, and labor-hour figures for each line item are based on working an eight-hour day in daylight hours in moderate temperatures. For work that extends beyond normal work hours or is performed under adverse conditions, productivity may decrease. (See the section in "How To Use the Unit Price Pages" for more on productivity.)

Size of Project—The size, scope of work, and type of construction project will have a significant impact on cost. Economies of scale can reduce costs for large projects. Unit costs can often run higher for small projects. Costs in this book are intended for the size and type of project as previously described in "How the Book Is Built: An Overview." Costs for projects of a significantly different size or type should be adjusted accordingly.

Location—Material prices in this book are for metropolitan areas. However, in dense urban areas, traffic and site storage limitations may increase costs. Beyond a 20-mile radius of large cities, extra trucking or transportation charges may also increase the material costs slightly. On the other hand, lower wage rates may be in effect. Be sure to consider both these factors when preparing an estimate, particularly if the job site is located in a central city or remote rural location.

In addition, highly specialized subcontract items may require travel and per diem expenses for mechanics.

Other factors—

- season of year
- contractor management
- weather conditions
- local union restrictions
- building code requirements
- availability of:
    - adequate energy
    - skilled labor
    - building materials
- owner's special requirements/restrictions
- safety requirements
- environmental considerations

General Conditions—The "Square Foot" and "Assemblies" sections of this book use costs that include the installing contractor's overhead and profit (O&P). The Unit Price Section presents cost data in two ways: Bare Costs and Total Cost including O&P (Overhead and Profit). General Conditions, when applicable, should also be added to the Total Cost including O&P. The costs for General Conditions are listed in Division 1 of the Unit Price Section and the Reference Section of this book. General Conditions for the *Installing Contractor* may range from 0% to 10% of the Total Cost including O&P. For the *General* or *Prime Contractor*, costs for General Conditions may range from 5% to 15% of the Total Cost including O&P, with a figure of 10% as the most typical allowance.

Overhead & Profit—Total Cost including O&P for the *Installing Contractor* is shown in the last column on both the Unit Price and the Assemblies pages of this book. This figure is the sum of the bare material cost plus 10% for profit, the base labor cost plus overhead and profit, and the bare equipment cost plus 10% for profit. Details for the calculation of Overhead and Profit on labor are shown on the inside back cover and in the Reference Section of this book. (See the "How to Use the Unit Price Pages" for an example of this calculation.)

Unpredictable Factors—General business conditions influence "in-place" costs of all items. Substitute materials and construction methods may have to be employed. These may affect the installed cost and/or life cycle costs. Such factors may be difficult to evaluate and cannot necessarily be predicted on the basis of the job's location in a particular section of the country. Thus, where these factors apply, you may find significant, but unavoidable cost variations for which you will have to apply a measure of judgment to your estimate.

Rounding of Costs

In general, all unit prices in excess of $5.00 have been rounded to make them easier to use and still maintain adequate precision of the results. The rounding rules we have chosen are in the following table.

Prices from . . .	Rounded to the nearest . . .
$.01 to $5.00	$.01
$5.01 to $20.00	$.05
$20.01 to $100.00	$.50
$100.01 to $300.00	$1.00
$300.01 to $1,000.00	$5.00
$1,000.01 to $10,000.00	$25.00
$10,000.01 to $50,000.00	$100.00
$50,000.01 and above	$500.00

Final Checklist

Estimating can be a straightforward process provided you remember the basics. Here's a checklist of some of the items you should remember to do before completing your estimate.

Did you remember to . . .

- factor in the Location Factor for your locale
- take into consideration which items have been marked up and by how much
- mark up the entire estimate sufficiently for your purposes
- read the background information on techniques and technical matters that could impact your project time span and cost
- include all components of your project in the final estimate
- double check your figures to be sure of your accuracy
- call R.S. Means if you have any questions about your estimate or the data you've found in our publications

Remember, R.S. Means stands behind its publications. If you have any questions about your estimate . . . about the costs you've used from our books . . . or even about the technical aspects of the job that may affect your estimate, feel free to call the R.S. Means editors at 1-800-334-3509.

Square Foot Cost Section

Table of Contents

Page

Introduction
General 3
How to Use 4
Building Classes 8
Configurations 9
Building Types 10
Garage Types 12
Building Components 13
Exterior Wall Construction 14

Residential Cost Estimate Worksheets
Instructions 15
Model Residence Example 17

Page

Economy 21
Illustrations 23
1 Story 24
1-1/2 Story 26
2 Story 28
Bi-Level 30
Tri-Level 32
Wings & Ells 34

Average 35
Illustrations 37
1 Story 38
1-1/2 Story 40
2 Story 42
2-1/2 Story 44
3 Story 46
Bi-Level 48
Tri-Level 50
Solid Wall (log home) 1 Story 52
Solid Wall (log home) 2 Story 54
Wings & Ells 56

Custom 57
Illustrations 59
1 Story 60
1-1/2 Story 62
2 Story 64
2-1/2 Story 66
3 Story 68
Bi-Level 70
Tri-Level 72
Wings & Ells 74

Page

Luxury 75
Illustrations 77
1 Story 78
1-1/2 Story 80
2 Story 82
2-1/2 Story 84
3 Story 86
Bi-Level 88
Tri-Level 90
Wings & Ells 92

Residential Modifications
Kitchen Cabinets 93
Kitchen Counter Tops 93
Vanity Bases 94
Solid Surface Vanity Tops 94
Fireplaces & Chimneys 94
Windows & Skylights 94
Dormers 94
Appliances 95
Breezeway 95
Porches 95
Finished Attic 95
Alarm System 95
Sauna, Prefabricated 95
Garages 96
Swimming Pools 96
Wood & Coal Stoves 96
Sidewalks 96
Fencing 96
Carport 96

Introduction to Square Foot Cost Section

The Square Foot Cost Section of this manual contains costs per square foot for four classes of construction in seven building types. Costs are listed for various exterior wall materials which are typical of the class and building type. There are cost tables for Wings and Ells with modification tables to adjust the base cost of each class of building. Non-standard items can easily be added to the standard structures.

Cost estimating for a residence is a three-step process:
(1) Identification
(2) Listing dimensions
(3) Calculations

Guidelines and a sample cost estimating procedure are shown on the following pages.

Identification

To properly identify a residential building, the class of construction, type, and exterior wall material must be determined. Page 8 has drawings and guidelines for determining the class of construction. There are also detailed specifications and additional drawings at the beginning of each set of tables to further aid in proper building class and type identification.

Sketches for eight types of residential buildings and their configurations are shown on pages 10 and 11. Definitions of living area are next to each sketch. Sketches and definitions of garage types are on page 12.

Living Area

Base cost tables are prepared as costs per square foot of living area. The living area of a residence is that area which is suitable and normally designed for full time living. It does not include basement recreation rooms or finished attics, although these areas are often considered full time living areas by the owners.

Living area is calculated from the exterior dimensions without the need to adjust for exterior wall thickness. When calculating the living area of a 1-1/2 story, two story, three story or tri-level residence, overhangs and other differences in size and shape between floors must be considered.

Only the floor area with a ceiling height of six feet or more in a 1-1/2 story residence is considered living area. In bi-levels and tri-levels, the areas that are below grade are considered living area, even when these areas may not be completely finished.

Base Tables and Modifications

Base cost tables show the base cost per square foot without a basement, with one full bath and one full kitchen for economy and average homes and an additional half bath for custom and luxury models. Adjustments for finished and unfinished basements are part of the base cost tables. Adjustments for multi family residences, additional bathrooms, townhouses, alternative roofs, and air conditioning and heating systems are listed in Modifications, Adjustments and Alternatives tables below the base cost tables.

Costs for other modifications, adjustments and alternatives, including garages, breezeways and site improvements, are on pages 93 to 96.

Listing of Dimensions

To use this section of the manual only the dimensions used to calculate the horizontal area of the building and additions, modifications, adjustments and alternatives are needed. The dimensions, normally the length and width, can come from drawings or field measurements. For ease in calculation, consider measuring in tenths of feet, i.e., 9 ft. 6 in. = 9.5 ft., 9 ft. 4 in. = 9.3 ft.

In all cases, make a sketch of the building. Any protrusions or other variations in shape should be noted on the sketch with dimensions.

Calculations

The calculations portion of the estimate is a two-step activity:
(1) The selection of appropriate costs from the tables
(2) Computations

Selection of Appropriate Costs

To select the appropriate cost from the base tables, the following information is needed:

(1) Class of construction
    (a) Economy
    (b) Average
    (c) Custom
    (d) Luxury
(2) Type of residence
    (a) One story
    (b) 1-1/2 story
    (c) 2 story
    (d) 3 story
    (e) Bi-level
    (f) Tri-level
(3) Occupancy
    (a) One family
    (b) Two family
    (c) Three family
(4) Building configuration
    (a) Detached
    (b) Town/Row house
    (c) Semi-detached
(5) Exterior wall construction
    (a) Wood frame
    (b) Brick veneer
    (c) Solid masonry
(6) Living areas

Modifications, adjustments and alternatives are classified by class, type and size.

Computations

The computation process should take the following sequence:
(1) Multiply the base cost by the area
(2) Add or subtract the modifications, adjustments, alternatives
(3) Apply the location modifier

When selecting costs, interpolate or use the cost that most nearly matches the structure under study. This applies to size, exterior wall construction and class.

How to Use the Residential Square Foot Cost Pages

The following is a detailed explanation of a sample entry in the Residential Square Foot Cost Section. Each bold number below corresponds to the item being described on the facing page with the appropriate component of the sample entry following in parenthesis.

Prices listed are costs that include overhead and profit of the installing contractor. Total model costs include an additional mark-up for General Contractors' overhead and profit, and fees specific to class of construction.

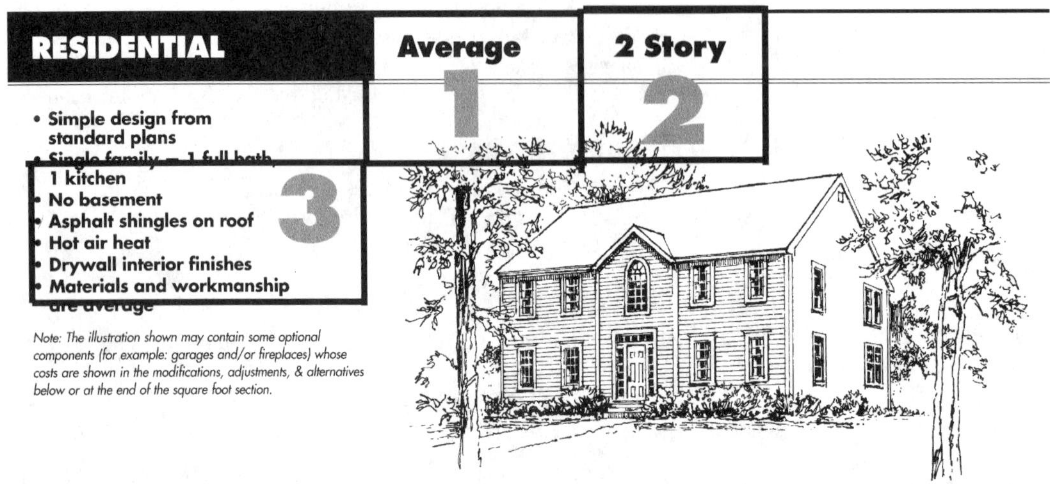

RESIDENTIAL | **Average** 1 | **2 Story** 2

- **Simple design from standard plans**
- **Single family — 1 full bath, 1 kitchen**
- **No basement** 3
- **Asphalt shingles on roof**
- **Hot air heat**
- **Drywall interior finishes**
- **Materials and workmanship are average**

Note: The illustration shown may contain some optional components (for example: garages and/or fireplaces) whose costs are shown in the modifications, adjustments, & alternatives below or at the end of the square foot section.

Base cost per square foot of living area

Exterior Wall (4)	1000	1200	1400	1600	1800	Living Area (5) 2000	2200	2600	3000	3400	3800
Wood Siding - Wood Frame	91.60	82.20	78.85	76.45	73.25	70.55	[illegible]	64.75	60.80	59.30	57.55
Brick Veneer - Wood Frame	99.10	89.20	85.35	82.65	79.10	76.20	[illegible]	69.55	65.25	63.55	61.60
Stucco on Wood Frame	92.20	82.85	79.40	77.00	73.75	71.05	[illegible]	65.15	61.15	59.70	57.90
Solid Masonry	109.40	98.70	94.20	91.15	87.10	83.95	81.50	76.20	71.40	69.35	67.15
Finished Basement, Add (7)	13.10	12.60	12.20	11.90	11.60	11.40	[illegible]	10.75	10.45	10.30	10.10
Unfinished Basement, Add	5.45	5.05	4.80	4.60	4.40	4.30	[illegible]	3.85	3.65	3.55	3.45

Modifications

Add to the total cost (9)

Upgrade Kitchen Cabinets	$ + 2282
Solid Surface Countertops	+ 826
Full Bath - including plumbing, wall and floor finishes	+ 3531
Half Bath - including plumbing, wall and floor finishes	+ 2247
One Car Attached Garage	+ 8054
One Car Detached Garage	+ 10,547
Fireplace & Chimney	+ 3965

Adjustments

For multi family - add to total cost

Additional Kitchen	$ + 3707
Additional Bath	+ 3531
Additional Entry & Exit	+ 1085
Separate Heating	+ 1192
Separate Electric	+ 1342

For Townhouse/Rowhouse - Multiply cost per square foot by (10)

Inner Unit	.90
End Unit	.95

Alternatives

Add to or deduct from the cost per square foot of living area (11)

Cedar Shake Roof	+ 1.10
Clay Tile Roof	+ 2.30
Slate Roof	+ 3.85
Upgrade Walls to Skim Coat Plaster	+ .36
Upgrade Ceilings to Textured Finish	+ .43
Air Conditioning, in Heating Ductwork	+ 1.48
In Separate Ductwork	+ 3.77
Heating Systems, Hot Water	+ 1.60
Heat Pump	+ 1.90
Electric Heat	− 1.01
Not Heated	− 2.67

Additional upgrades or components

Kitchen Cabinets & Countertops	Page 93
Bathroom Vanities	94
Fireplaces & Chimneys	94
Windows, Skylights & Dormers	94
Appliances	95
Breezeways & Porches	95
Finished Attic	95
Garages	96
Site Improvements	96
Wings & Ells	56

(12)

Important: See the Reference Section for Location Factors (to adjust for your city) and Estimating Forms

1

Class of Construction (Average)

The class of construction depends upon the design and specifications of the plan. The four classes are economy, average, custom and luxury.

2

Type of Residence (2 Story)

The building type describes the number of stories or levels in the model. The seven building types are 1 story, 1-1/2 story, 2 story, 2-1/2 story, 3 story, bi-level and tri-level.

3

Specification Highlights (Hot Air Heat)

These specifications include information concerning the components of the model, including the number of baths, roofing types, HVAC systems, and materials and workmanship. If the components listed are not appropriate, modifications can be made by consulting the information shown lower on the page or the Assemblies Section.

4

Exterior Wall System (Wood Siding - Wood Frame)

This section includes the types of exterior wall systems and the structural frame used. The exterior wall systems shown are typical of the class of construction and the building type shown.

5

Living Areas (2000 SF)

The living area is that area of the residence which is suitable and normally designed for full time living. It does not include basement recreation rooms or finished attics. Living area is calculated from the exterior dimensions without the need to adjust for exterior wall thickness. When calculating the living area of a 1-1/2 story, 2 story, 3 story or tri-level residence, overhangs and other differences in size and shape between floors must be considered. Only the floor area with a ceiling height of six feet or more in a 1-1/2 story residence is considered living area. In bi-levels and tri-levels, the areas that are below grade are considered living area, even when these areas may not be completely finished. A range of various living areas for the residential model are shown to aid in selection of values from the matrix.

6

Base Costs per Square Foot of Living Area ($70.55)

Base cost tables show the cost per square foot of living area without a basement, with one full bath and one full kitchen for economy and average homes and an additional half bath for custom and luxury models. When selecting costs, interpolate or use the cost that most nearly matches the residence under consideration for size, exterior wall system and class of construction. Prices listed are costs that include overhead and profit of the installing contractor, a general contractor markup and an allowance for plans that vary by class of construction. For additional information on contractor overhead and architectural fees, see the Reference Section.

7

Basement Types (Finished)

The two types of basements are finished or unfinished. The specifications and components for both are shown on the Building Classes page in the Introduction to this section.

8

Additional Costs for Basements ($11.40)

These values indicate the additional cost per square foot of living area for either a finished or an unfinished basement.

9

Modifications and Adjustments (Solid Surface Countertops $826)

Modifications and Adjustments are costs added to or subtracted from the total cost of the residence. The total cost of the residence is equal to the cost per square foot of living area times the living area. Typical modifications and adjustments include kitchens, baths, garages and fireplaces.

10

Multiplier for Townhouse/ Rowhouse (Inner Unit .90)

The multipliers shown adjust the base costs per square foot of living area for the common wall condition encountered in townhouses or rowhouses.

11

Alternatives (Skim Coat Plaster $.36)

Alternatives are costs added to or subtracted from the base cost per square foot of living area. Typical alternatives include variations in kitchens, baths, roofing, air conditioning and heating systems.

12

Additional Upgrades or Components (Wings & Ells page 37)

Costs for additional upgrades or components, including wings or ells, breezeways, porches, finished attics and site improvements, are shown in other locations in the Residential Square Foot Cost Section.

How to Use the Square Foot Cost Pages *(Continued)*

2

1 **Average 2 Story**
Living Area - 2000 S.F.
Perimeter - 135 L.F.

		Labor-Hours	Cost Per Square Foot Of Living Area		
			Mat.	Labor	Total
1 Site Work	Site preparation for slab; 4' deep trench excavation for foundation wall.	.034		.51	.51
2 Foundation	Continuous reinforced concrete footing 8" deep x 18" wide; dampproofed and insulated reinforced concrete foundation wall, 8" thick, 4' deep, 4" concrete slab on 4" crushed stone base and polyethylene vapor barrier, trowel finish.	.066	1.93	2.61	4.54
3 Framing	Exterior walls - 2" x 4" wood studs, 16" O.C.; 1/2" plywood sheathing; 2" x 6" rafters 16" O.C. with 1/2" plywood sheathing, 4 in 12 pitch; 2" x 6" ceiling joists 16" O.C.; 2" x 8" floor joists 16" O.C. with 5/8" plywood subfloor; 1/2" plywood subfloor on 1" x 2" wood sleepers 16" O.C.	.131	[illegible]	5.54	[illegible]
4 Exterior Walls	Beveled wood siding and #15 felt building paper on insulated wood frame walls; 6" attic insulation; double hung windows; 3 flush solid core wood exterior doors with storms.	.111	10.27	5.33	15.60
5 Roofing	25 year asphalt shingles; #15 felt building paper; aluminum gutters, downspouts, drip edge and flashings.	[illegible]	.42	[illegible]	1.20
6 Interiors	Walls and ceilings, 1/2" taped and finished drywall, primed and painted with 2 coats; painted baseboard and trim, finished hardwood floor 40%, carpet with 1/2" underlayment 40%, vinyl tile with 1/2" underlayment 15%, ceramic tile with 1/2" underlayment 5%; hollow core and louvered interior doors.	.232	13.12	13.06	26.18
7 Specialties	Average grade kitchen cabinets - 14 L.F. wall and base with plastic laminate counter top and kitchen sink; 40 gallon electric water heater.	.021	1.08	.45	1.53
8 Mechanical	1 lavatory, white, wall hung; 1 water closet, white; 1 bathtub with shower; enameled steel, white; gas fired warm air heat.	.060	1.97	1.82	3.79
9 Electrical	200 Amp. service; romex wiring; incandescent lighting fixtures, switches, receptacles.	.039	.65	1.06	1.71
10 Overhead	Contractor's overhead and profit and plans.		2.43	2.17	4.60
		Total	37.22	33.33	**70.55**

3 4 5 6 7 8 9

1 Specifications

The parameters for an example dwelling from the previous pages are listed here. Included are the square foot dimensions of the proposed building. Living Area takes into account the number of floors and other factors needed to define a building's total square footage. Perimeter dimensions are defined in terms of linear feet.

2 Building Type

This is a sketch of a cross section view through the dwelling. It is shown to help define the living area for the building type. For more information, see the Building Types pages in the Introduction.

3 Components (4 Exterior Walls)

This page contains the ten components needed to develop the complete square foot cost of the typical dwelling specified. All components are defined with a description of the materials and/or task involved. Use cost figures from each component to estimate the cost per square foot of that section of the project. The components listed on this page are typical of all sizes of residences from the facing page. Specific quantities of components required would vary with the size of the dwelling and the exterior wall system.

4 Labor-Hours (.111)

Use this column to determine the number of labor-hours needed to perform a task. This figure will give the builder labor-hours per square foot of building. The total labor-hours per component is determined by multiplying the living area times the labor-hours listed on that line.

5 Materials (10.27)

This column gives the amount needed to develop the cost of materials. The figures given here are not bare costs. Ten percent has been added to bare material cost for profit.

6 Installation (5.33)

Installation includes labor and equipment costs. The labor rates included here incorporate the total overhead and profit costs for the installing contractor. The average mark-up used to create these figures is 72.0% over and above bare labor costs. The equipment rates include 10% for profit.

7 Line Totals (15.60)

The extreme right-hand column lists the sum of two figures. Use this total to determine the sum of material cost plus installation cost. The result is a convenient total cost for each of the ten components.

8 Overhead

The costs in components 1 through 9 include overhead and profit for the installing contractor. Item 10 is overhead and profit for the general contractor. This is typically a percentage mark-up of all other costs. The amount depends on size and type of dwelling, building class and economic conditions. An allowance for plans or design has been included where appropriate.

9 Bottom Line Total (70.55)

This figure is the complete square foot cost for the construction project and equals the sum of total material and total labor costs. To determine total project cost, multiply the bottom line total times the living area.

Building Classes

Economy Class

An economy class residence is usually built from stock plans. The materials and workmanship are sufficient to satisfy building codes. Low construction cost is more important than distinctive features. The overall shape of the foundation and structure is seldom other than square or rectangular.

An unfinished basement includes 7′ high 8″ thick foundation wall composed of either concrete block or cast-in-place concrete.

Included in the finished basement cost are inexpensive paneling or drywall as the interior finish on the foundation walls, a low cost sponge backed carpeting adhered to the concrete floor, a drywall ceiling, and overhead lighting.

Custom Class

A custom class residence is usually built from plans and specifications with enough features to give the building a distinction of design. Materials and workmanship are generally above average with obvious attention given to construction details. Construction normally exceeds building code requirements.

An unfinished basement includes a 7′-6″ high 10″ thick cast-in-place concrete foundation wall or a 7′-6″ high 12″ thick concrete block foundation wall.

A finished basement includes painted drywall on insulated 2″ x 4″ wood furring as the interior finish to the concrete walls, a suspended ceiling, carpeting adhered to the concrete floor, overhead lighting and heating.

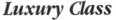

Average Class

An average class residence is a simple design and built from standard plans. Materials and workmanship are average, but often exceed minimum building codes. There are frequently special features that give the residence some distinctive characteristics.

An unfinished basement includes 7′-6″ high 8″ thick foundation wall composed of either cast-in-place concrete or concrete block.

Included in the finished basement are plywood paneling or drywall on furring that is fastened to the foundation walls, sponge backed carpeting adhered to the concrete floor, a suspended ceiling, overhead lighting and heating.

Luxury Class

A luxury class residence is built from an architect's plan for a specific owner. It is unique both in design and workmanship. There are many special features, and construction usually exceeds all building codes. It is obvious that primary attention is placed on the owner's comfort and pleasure. Construction is supervised by an architect.

An unfinished basement includes 8′ high 12″ thick foundation wall that is composed of cast-in-place concrete or concrete block.

A finished basement includes painted drywall on 2″ x 4″ wood furring as the interior finish, suspended ceiling, tackless carpet on wood subfloor with sleepers, overhead lighting and heating.

Configurations

Detached House

This category of residence is a free-standing separate building with or without an attached garage. It has four complete walls.

Semi-Detached House

This category of residence has two living units side-by-side. The common wall is a fireproof wall. Semi-detached residences can be treated as a row house with two end units. Semi-detached residences can be any of the building types.

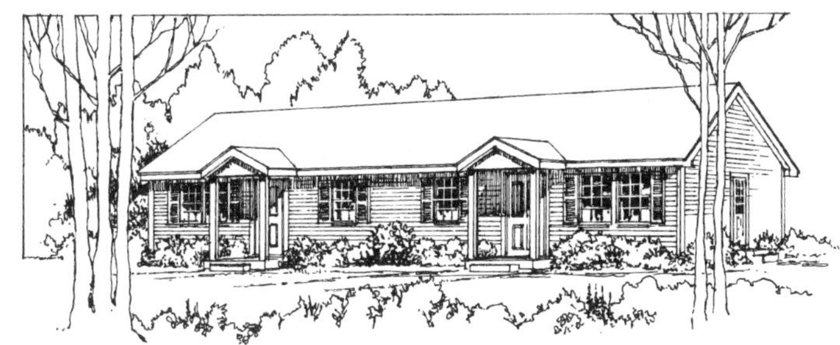

Town/Row House

This category of residence has a number of attached units made up of inner units and end units. The units are joined by common walls. The inner units have only two exterior walls. The common walls are fireproof. The end units have three walls and a common wall. Town houses/row houses can be any of the building types.

Building Types

One Story

This is an example of a one-story dwelling. The living area of this type of residence is confined to the ground floor. The headroom in the attic is usually too low for use as a living area.

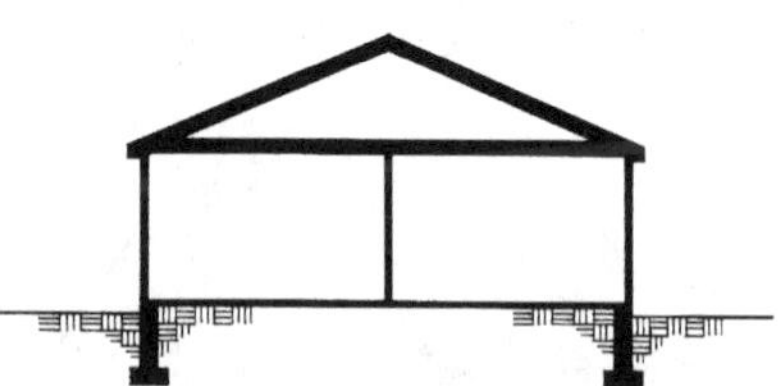

One-and-a-half Story

The living area in the upper level of this type of residence is 50% to 90% of the ground floor. This is made possible by a combination of this design's high-peaked roof and/or dormers. Only the upper level area with a ceiling height of 6 feet or more is considered living area. The living area of this residence is the sum of the ground floor area plus the area on the second level with a ceiling height of 6 feet or more.

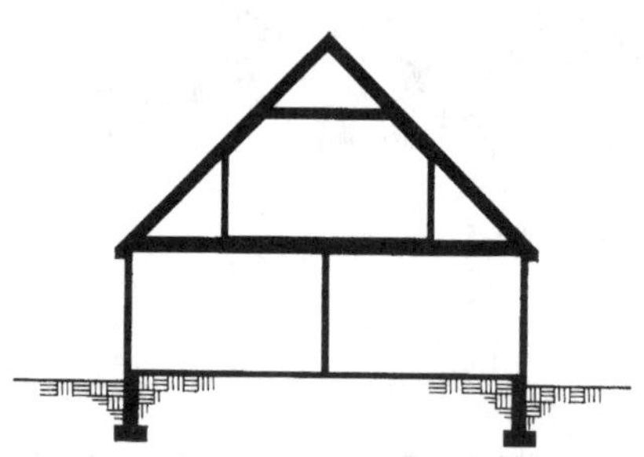

One Story with Finished Attic

The main living area in this type of residence is the ground floor. The upper level or attic area has sufficient headroom for comfortable use as a living area. This is made possible by a high peaked roof. The living area in the attic is less than 50% of the ground floor. The living area of this type of residence is the ground floor area only. The finished attic is considered an adjustment.

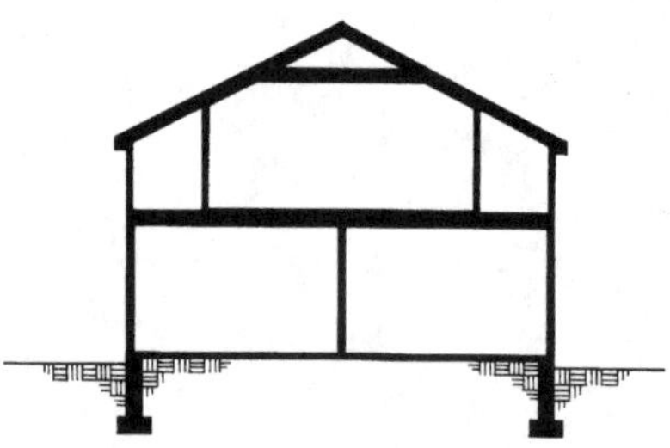

Two Story

This type of residence has a second floor or upper level area which is equal or nearly equal to the ground floor area. The upper level of this type of residence can range from 90% to 110% of the ground floor area, depending on setbacks or overhangs. The living area is the sum of the ground floor area and the upper level floor area.

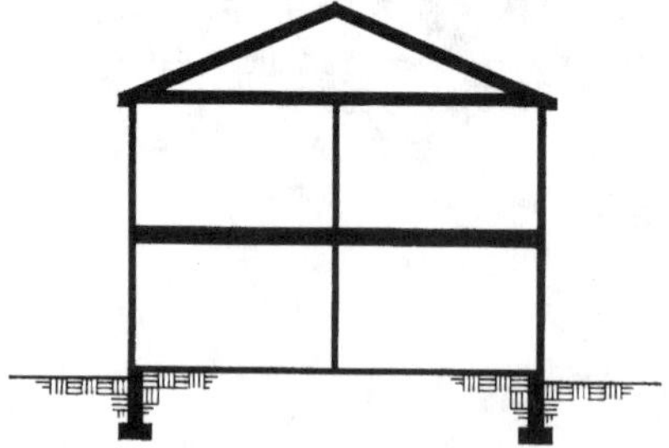

Two-and-one-half Story

This type of residence has two levels of equal or nearly equal area and a third level which has a living area that is 50% to 90% of the ground floor. This is made possible by a high peaked roof, extended wall heights and/or dormers. Only the upper level area with a ceiling height of 6 feet or more is considered living area. The living area of this residence is the sum of the ground floor area, the second floor area and the area on the third level with a ceiling height of 6 feet or more.

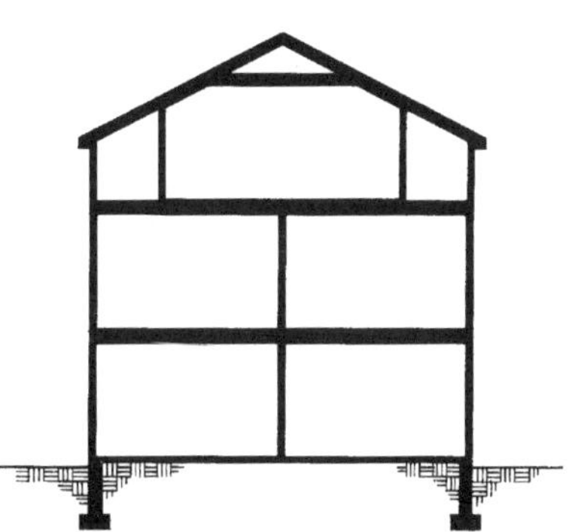

Bi-level

This type of residence has two living areas, one above the other. One area is about 4 feet below grade and the second is about 4 feet above grade. Both areas are equal in size. The lower level in this type of residence is originally designed and built to serve as a living area and not as a basement. Both levels have full ceiling heights. The living area is the sum of the lower level area and the upper level area.

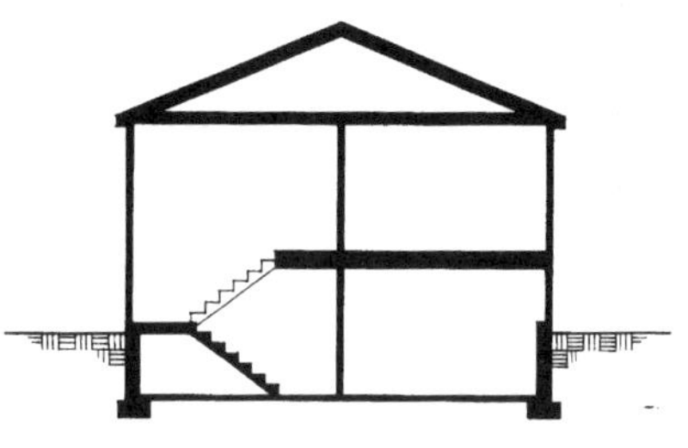

Three Story

This type of residence has three levels which are equal or nearly equal. As in the 2 story residence, the second and third floor areas may vary slightly depending on setbacks or overhangs. The living area is the sum of the ground floor area and the two upper level floor areas.

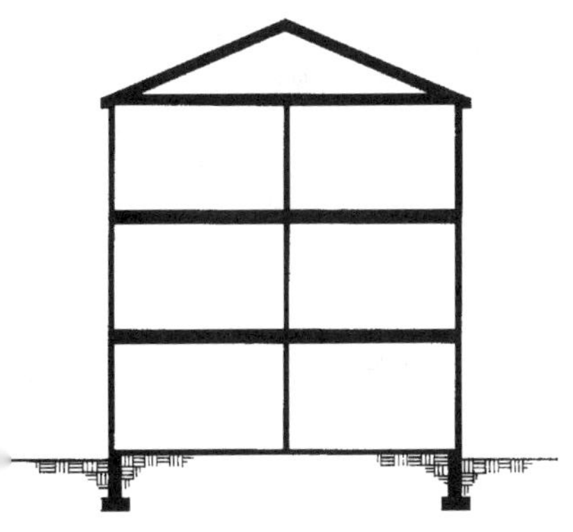

Tri-level

This type of residence has three levels of living area, one at grade level, one about 4 feet below grade and one about 4 feet above grade. All levels are originally designed to serve as living areas. All levels have full ceiling heights. The living area is a sum of the areas of each of the three levels.

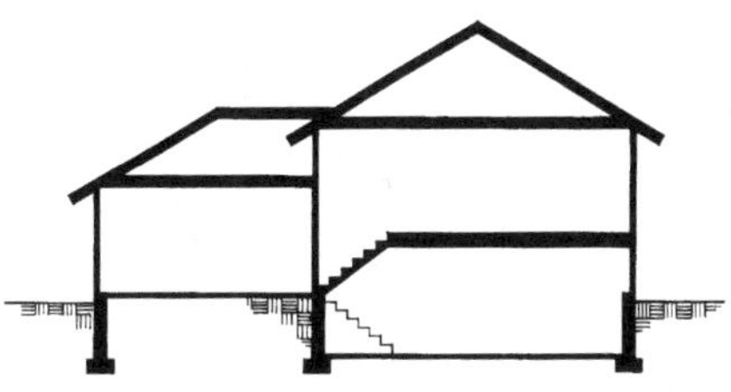

Garage Types

Attached Garage

Shares a common wall with the dwelling. Access is typically through a door between dwelling and garage.

Basement Garage

Constructed under the roof of the dwelling but below the living area.

Built-In Garage

Constructed under the second floor living space and above basement level of dwelling. Reduces gross square feet of living area.

Detached Garage

Constructed apart from the main dwelling. Shares no common area or wall with the dwelling.

Building Components

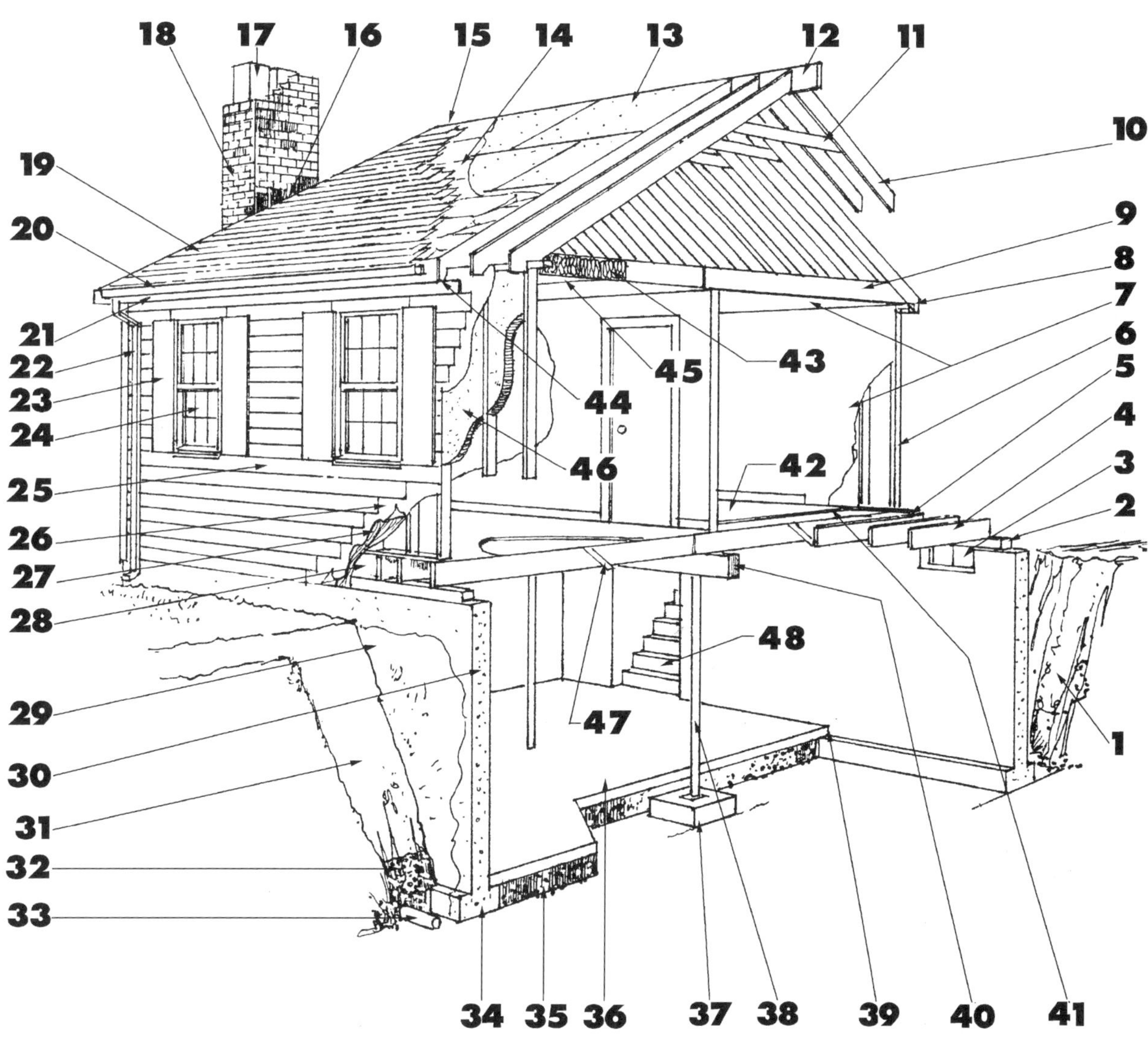

1. Excavation
2. Sill Plate
3. Basement Window
4. Floor Joist
5. Shoe Plate
6. Studs
7. Drywall
8. Plate
9. Ceiling Joists
10. Rafters
11. Collar Ties
12. Ridge Rafter
13. Roof Sheathing
14. Roof Felt
15. Roof Shingles
16. Flashing
17. Flue Lining
18. Chimney
19. Roof Shingles
20. Gutter
21. Fascia
22. Downspout
23. Shutter
24. Window
25. Wall Shingles
26. Building Paper
27. Wall Sheathing
28. Fire Stop
29. Dampproofing
30. Foundation Wall
31. Backfill
32. Drainage Stone
33. Drainage Tile
34. Wall Footing
35. Gravel
36. Concrete Slab
37. Column Footing
38. Pipe Column
39. Expansion Joint
40. Girder
41. Sub-floor
42. Finish Floor
43. Attic Insulation
44. Soffit
45. Ceiling Strapping
46. Wall Insulation
47. Cross Bridging
48. Bulkhead Stairs

Exterior Wall Construction

Typical Frame Construction

Typical wood frame construction consists of wood studs with insulation between them. A typical exterior surface is made up of sheathing, building paper and exterior siding consisting of wood, vinyl, aluminum or stucco over the wood sheathing.

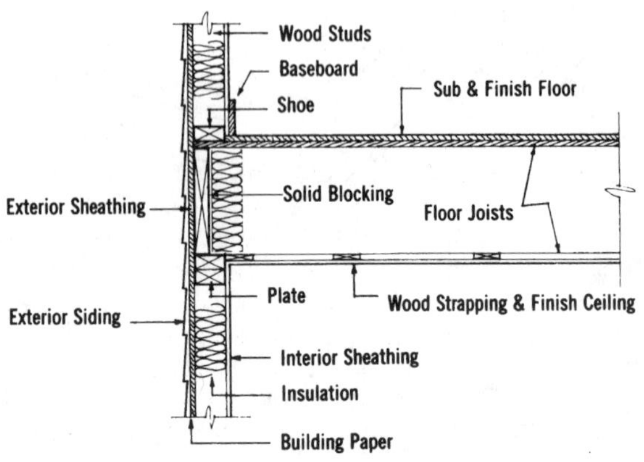

Brick Veneer

Typical brick veneer construction consists of wood studs with insulation between them. A typical exterior surface is sheathing, building paper and an exterior of brick tied to the sheathing, with metal strips.

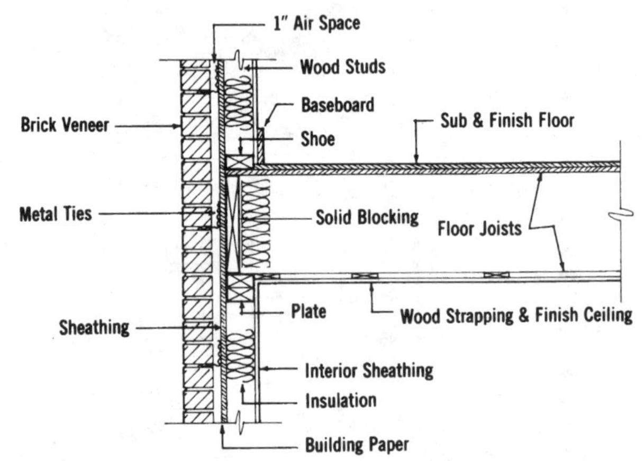

Stone

Typical solid masonry construction consists of a stone or block wall covered on the exterior with brick, stone or other masonry.

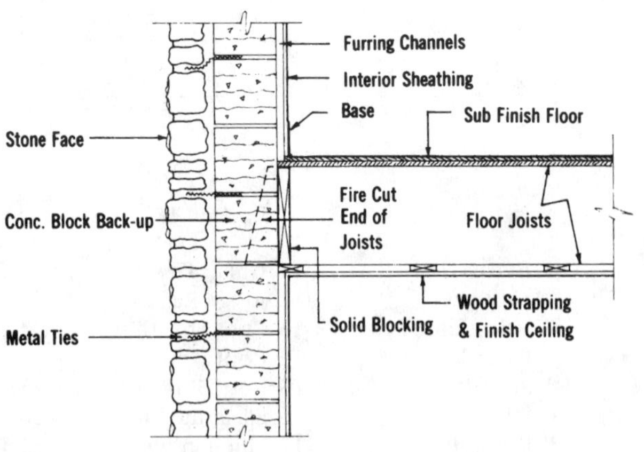

Residential Cost Estimate Worksheets

Worksheet Instructions

The residential cost estimate worksheet can be used as an outline for developing a residential construction or replacement cost. It is also useful for insurance appraisals. The design of the worksheet helps eliminate errors and omissions. To use the worksheet, follow the example below.

1. Fill out the owner's name, residence address, the estimator or appraiser's name, some type of project identifying number or code, and the date.
2. Determine from the plans, specifications, owner's description, photographs or any other means possible the class of construction. The models in this book use economy, average, custom and luxury as classes. Fill in the appropriate box.
3. Fill in the appropriate box for the residence type, configuration, occupancy, and exterior wall. If you require clarification, the pages preceding this worksheet describe each of these.
4. Next, the living area of the residence must be established. The heated or air conditioned space of the residence, not including the basement, should be measured. It is easiest to break the structure up into separate components as shown in the example. The main house (A), a one-and-one-half story wing (B), and a one story wing (C). The breezeway (D), garage (E) and open covered porch (F) will be treated differently. Data entry blocks for the living area are included on the worksheet for your use. Keep each level of each component separate, and fill out the blocks as shown.
5. By using the information on the worksheet, find the model, wing or ell in the following square foot cost pages that best matches the class, type, exterior finish and size of the residence being estimated. Use the *modifications, adjustments, and alternatives* to determine the adjusted cost per square foot of living area for each component.
6. For each component, multiply the cost per square foot by the living area square footage. If the residence is a townhouse/rowhouse, a multiplier should be applied based upon the configuration.
7. The second page of the residential cost estimate worksheet has space for the additional components of a house. The cost for additional bathrooms, finished attic space, breezeways, porches, fireplaces, appliance or cabinet upgrades, and garages should be added on this page. The information for each of these components is found with the model being used, or in the *modifications, adjustments, and alternatives* pages.
8. Add the total from page one of the estimate worksheet and the items listed on page two. The sum is the adjusted total building cost.
9. Depending on the use of the final estimated cost, one of the remaining two boxes should be filled out. Any additional items or exclusions should be added or subtracted at this time. The data contained in this book is a national average. Construction costs are different throughout the country. To allow for this difference, a location factor based upon the first three digits of the residence's zip code must be applied. The location factor is a multiplier that increases or decreases the adjusted total building cost. Find the appropriate location factor and calculate the local cost. If depreciation is a concern, a dollar figure should be subtracted at this point.
10. No residence will match a model exactly. Many differences will be found. At this level of estimating, a variation of plus or minus 10% should be expected.

Adjustments Instructions

No residence matches a model exactly in shape, material, or specifications.

The common differences are:

1. Two or more exterior wall systems
   a. Partial basement
   b. Partly finished basement
2. Specifications or features that are between two classes
3. Crawl space instead of a basement

EXAMPLES

Below are quick examples. See pages 17-19 for complete examples of cost adjustments for these differences:

1. Residence "A" is an average one-story structure with 1,600 S.F. of living area and no basement. Three walls are wood siding on wood frame, and the fourth wall is brick veneer on wood frame. The brick veneer wall is 35% of the exterior wall area.

   Use page 38 to calculate the Base Cost per S.F. of Living Area.
   Wood Siding for 1,600 S.F. = $70.75 per S.F.
   Brick Veneer for 1,600 S.F. = $80.95 per S.F.
   .65 ($70.75) + .35 ($80.95) = $74.32 per S.F. of Living Area.

2. a. Residence "B" is the same as Residence "A"; However, it has an unfinished basement under 50% of the building. To adjust the $74.32 per S.F. of living area for this partial basement, use page 38.
   $74.32 + .50 ($6.95) = $77.80 per S.F. of Living Area.

   b. Residence "C" is the same as Residence "A"; However, it has a full basement under the entire building. 640 S.F. or 40% of the basement area is finished.
   Using Page 42:
   $74.32 + .40 ($18.10) + .60 ($6.95) = $85.73 per S.F. of Living Area.

3. When specifications or features of a building are between classes, estimate the percent deviation, and use two tables to calculate the cost per S.F.

   A two-story residence with wood siding and 1,800 S.F. of living area has features 30% better than Average, but 70% less than Custom.

   From pages 42 and 64:
   Custom 1,800 S.F. Base Cost = $94.55 per S.F.
   Average 1,800 S.F. Base Cost = $73.25 per S.F.
   DIFFERENCE = $21.30 per S.F.

   Cost is $73.25 + .30 ($21.30) = $79.64 per S.F. of Living Area.

4. To add the cost of a crawl space, use the cost of an unfinished basement as a maximum. For specific costs of components to be added or deducted, such as vapor barrier, underdrain, and floor, see the "Assemblies" section, pages 101 to 281.

Model Residence Example

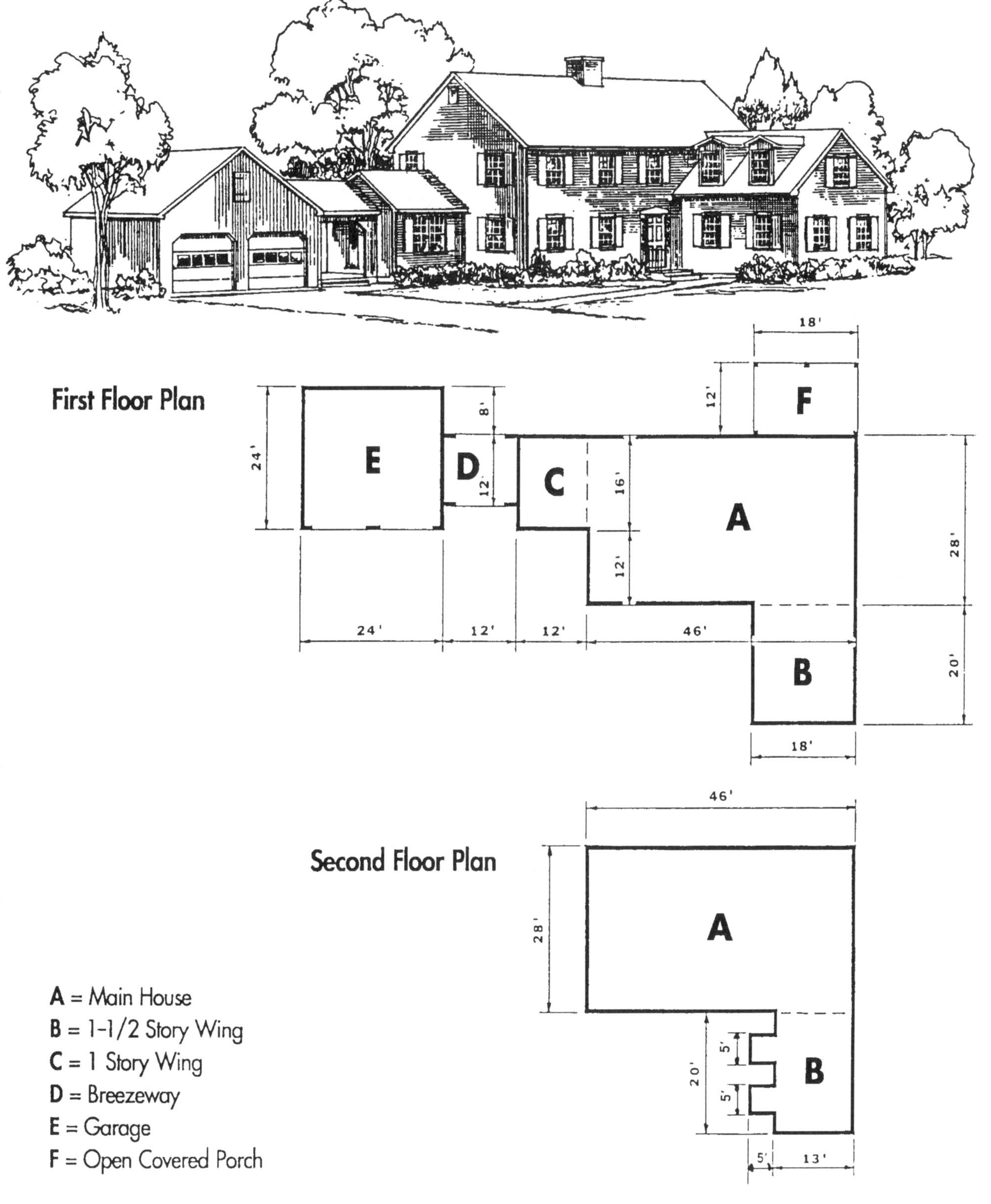

A = Main House
B = 1-1/2 Story Wing
C = 1 Story Wing
D = Breezeway
E = Garage
F = Open Covered Porch

RESIDENTIAL
COST ESTIMATE

OWNERS NAME:	Albert Westenberg	APPRAISER:	Nicole Wojtowicz
RESIDENCE ADDRESS:	300 Sygiel Road	PROJECT:	# 55
CITY, STATE, ZIP CODE:	Three Rivers, MA 01080	DATE:	Jan. 3, 2000

CLASS OF CONSTRUCTION
- ☐ ECONOMY
- ☑ AVERAGE
- ☐ CUSTOM
- ☐ LUXURY

RESIDENCE TYPE
- ☐ 1 STORY
- ☐ 11/2 STORY
- ☑ 2 STORY
- ☐ 2 1/2 STORY
- ☐ 3 STORY
- ☐ BI-LEVEL
- ☐ TRI-LEVEL

CONFIGURATION
- ☑ DETACHED
- ☐ TOWN/ROW HOUSE
- ☐ SEMI-DETACHED

OCCUPANCY
- ☑ ONE FAMILY
- ☐ TWO FAMILY
- ☐ THREE FAMILY
- ☐ OTHER

EXTERIOR WALL SYSTEM
- ☑ WOOD SIDING - WOOD FRAME
- ☐ BRICK VENEER - WOOD FRAME
- ☐ STUCCO ON WOOD FRAME
- ☐ PAINTED CONCRETE BLOCK
- ☐ SOLID MASONRY (AVERAGE & CUSTOM)
- ☐ STONE VENEER - WOOD FRAME
- ☐ SOLID BRICK (LUXURY)
- ☐ SOLID STONE (LUXURY)

* LIVING AREA (Main Building)		
First Level	1288	S.F.
Second level	1288	S.F.
Third Level		S.F.
Total	2576	S.F.

* LIVING AREA ~~(Wing~~ or Ell) (B)		
First Level	360	S.F.
Second level	310	S.F.
Third Level		S.F.
Total	670	S.F.

* LIVING AREA (Wing or Ell) (C)		
First Level	192	S.F.
Second level		S.F.
Third Level		S.F.
Total	192	S.F.

* Basement Area is not part of living area.

MAIN BUILDING		COSTS PER S.F. LIVING AREA
Cost per Square Foot of Living Area, from Page 42	$	64.75
Basement Addition: ______ % Finished, 100 % Unfinished	+	3.85
Roof Cover Adjustment: Cedar Shake Type, Page 42 (Add or Deduct)	( )	1.00
Central Air Conditioning: ☐ Separate Ducts ☑ Heating Ducts, Page 42	+	1.48
Heating System Adjustment: ______ Type, Page ______ (Add or Deduct)	( )	______
Main Building: Adjusted Cost per S.F. of Living Area	$	71.08

MAIN BUILDING TOTAL COST: $ 71.08 /S.F. (Cost per S.F. Living Area) x 2,576 S.F. (Living Area) x —— (Town/Row House Multiplier (Use 1 for Detached)) = $ 183,102 TOTAL COST

~~WING~~ OR ELL (B) 1-1/2 STORY		COSTS PER S.F. LIVING AREA
Cost per Square Foot of Living Area, from Page 56	$	57.20
Basement Addition: 100 % Finished, ______ % Unfinished	+	15.95
Roof Cover Adjustment: ______ Type, Page ______ (Add or Deduct)	( )	-
Central Air Conditioning: ☐ Separate Ducts ☑ Heating Ducts, Page 40	+	1.85
Heating System Adjustment: ______ Type, Page ______ (Add or Deduct)	( )	—
~~Wing~~ or Ell (B): Adjusted Cost per S.F. of Living Area	$	75.00

WING OR ELL (B) TOTAL COST: $ 75.00 /S.F. (Cost per S.F. Living Area) x 670 S.F. (Living Area) = $ 50,250 TOTAL COST

WING OR ~~ELL~~ (C) 1 STORY		COSTS PER S.F. LIVING AREA
Cost per Square Foot of Living Area, from Page 56 (WOOD SIDING)	$	85.40
Basement Addition: ______ % Finished, ______ % Unfinished	+	—
Roof Cover Adjustment: ______ Type, Page (Add or Deduct)	( )	—
Central Air Conditioning: ☐ Separate Ducts ☐ Heating Ducts, Page ______	+	—
Heating System Adjustment: ______ Type, Page ______ (Add or Deduct)	( )	—
Wing or Ell (C) Adjusted Cost per S.F. of Living Area	$	85.40

WING OR ELL (C) TOTAL COST: $ 85.40 /S.F. (Cost per S.F. Living Area) x 192 S.F. (Living Area) = $ 16,397 TOTAL COST

TOTAL THIS PAGE: 249,749

Page 1 of 2

RESIDENTIAL
COST ESTIMATE

Total Page 1			$ 249,749
	QUANTITY	UNIT COST	
Additional Bathrooms: 2 Full, 1 Half 2 @ 3531 1 @ 2247			+ 9,309
Finished Attic: N/A Ft. x ____ Ft.	S.F.		
Breezeway: ☑ Open ☐ closed 12 Ft. x 12 Ft.	144 S.F.	15.00	+ 2,160
Covered Porch: ☑ Open ☐ Enclosed 18 Ft. x 12 Ft.	216 S.F.	25.20	+ 5,443
Fireplace: ☑ Interior Chimney ☐ Exterior Chimney ☑ No. of Flues (2) ☑ Additional Fireplaces 1 - 2nd Story			+ 6,690
Appliances:			+ —
Kitchen Cabinets Adjustments: (±)			—
☑ Garage ☐ Carport: 2 Car(s) Description Wood, Attached (±)			13,676
Miscellaneous:			+

ADJUSTED TOTAL BUILDING COST $ 287,027

REPLACEMENT COST		
ADJUSTED TOTAL BUILDING COST	$	287,027
Site Improvements		
(A) Paving & Sidewalks	$	
(B) Landscaping	$	
(C) Fences	$	
(D) Swimming Pools	$	
(E) Miscellaneous	$	
TOTAL	$	287,027
Location Factor	x	1.05
Location Replacement Cost	$	301,378
Depreciation	-$	30,138
LOCAL DEPRECIATED COST	$	271,240

INSURANCE COST		
ADJUSTED TOTAL BUILDING COST	$	
Insurance Exclusions		
(A) Footings, sitework, Underground Piping	-$	
(B) Architects Fees	-$	
Total Building Cost Less Exclusion	$	
Location Factor	x	
LOCAL INSURABLE REPLACEMENT COST	$	

SKETCH AND ADDITIONAL CALCULATIONS

Economy Class

1 Story

1-1/2 Story

2 Story

Bi-Level

Tri-Level

- **Mass produced from stock plans**
- **Single family — 1 full bath, 1 kitchen**
- **No basement**
- **Asphalt shingles on roof**
- **Hot air heat**
- **Drywall interior finishes**
- **Materials and workmanship are sufficient to meet codes**

Note: The illustration shown may contain some optional components (for example: garages and/or fireplaces) whose costs are shown in the modifications, adjustments, & alternatives below or at the end of the square foot section.

©Home Planners, Inc.

Base cost per square foot of living area

	Living Area										
Exterior Wall	**600**	**800**	**1000**	**1200**	**1400**	**1600**	**1800**	**2000**	**2400**	**2800**	**3200**
Wood Siding - Wood Frame	77.05	70.35	64.80	60.25	56.50	53.80	52.50	50.85	47.25	44.65	42.90
Brick Veneer - Wood Frame	83.35	76.00	69.95	64.85	60.70	57.75	56.30	54.45	50.50	47.65	45.70
Stucco on Wood Frame	75.75	69.20	63.75	59.25	55.60	53.00	51.70	50.15	46.55	44.05	42.35
Painted Concrete Block	78.10	71.35	65.70	61.00	57.20	54.50	53.15	51.45	47.80	45.20	43.40
Finished Basement, Add	17.65	16.60	15.80	15.15	14.60	14.25	14.05	13.75	13.35	13.00	12.75
Unfinished Basement, Add	9.00	8.05	7.35	6.80	6.30	6.00	5.85	5.60	5.25	4.95	4.75

Modifications

Add to the total cost

Upgrade Kitchen Cabinets	$ + 242
Solid Surface Countertops	+ 469
Full Bath - including plumbing, wall and floor finishes	+ 2814
Half Bath - including plumbing, wall and floor finishes	+ 1791
One Car Attached Garage	+ 7575
One Car Detached Garage	+ 9783
Fireplace & Chimney	+ 3480

Adjustments

For multi family - add to total cost

Additional Kitchen	$ + 1992
Additional Bath	+ 2800
Additional Entry & Exit	+ 1085
Separate Heating	+ 1192
Separate Electric	+ 702

For Townhouse/Rowhouse - Multiply cost per square foot by

Inner Unit	.95
End Unit	.97

Alternatives

Add to or deduct from the cost per square foot of living area

Composition Roll Roofing	– .5
Cedar Shake Roof	+ 2.4
Upgrade Walls and Ceilings to Skim Coat Plaster	+ .4
Upgrade Ceilings to Textured Finish	+ .4
Air Conditioning, in Heating Ductwork	+ 2.4
In Separate Ductwork	+ 4.6
Heating Systems, Hot Water	+ 1.7
Heat Pump	+ 1.5
Electric Heat	– 1.4
Not Heated	– 3.0

Additional upgrades or components

Kitchen Cabinets & Countertops	Page 9
Bathroom Vanities	9
Fireplaces & Chimneys	9
Windows, Skylights & Dormers	9
Appliances	9
Breezeways & Porches	9
Finished Attic	9
Garages	9
Site Improvements	9
Wings & Ells	3

Important: See the Reference Section for Location Factors (to adjust for your city) and Estimating Form

Economy 1 Story

Living Area - 1200 S.F.
Perimeter - 146 L.F.

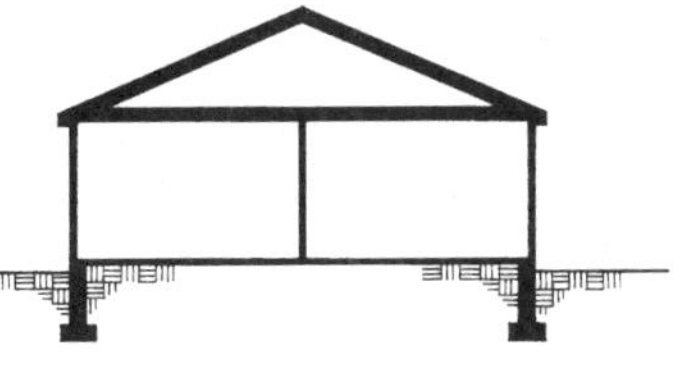

		Labor-Hours	Cost Per Square Foot Of Living Area		
			Mat.	Labor	Total
1 Site Work	Site preparation for slab; 4' deep trench excavation for foundation wall.	.060		.84	.84
2 Foundation	Continuous reinforced concrete footing, 8" deep x 18" wide; dampproofed and insulated 8" thick reinforced concrete block foundation wall, 4' deep; 4" concrete slab on 4" crushed stone base and polyethylene vapor barrier, trowel finish.	.131	3.49	4.67	8.16
3 Framing	Exterior walls - 2" x 4" wood studs, 16" O.C.; 1/2" insulation board sheathing; wood truss roof frame, 24" O.C. with 1/2" plywood sheathing, 4 in 12 pitch.	.098	4.13	3.94	8.07
4 Exterior Walls	Metal lath reinforced stucco exterior on insulated wood frame walls; 6" attic insulation; sliding sash wood windows; 2 flush solid core wood exterior doors with storms.	.110	6.55	5.32	11.87
5 Roofing	20 year asphalt shingles; #15 felt building paper; aluminum gutters, downspouts, drip edge and flashings.	.047	.82	1.51	2.33
6 Interiors	Walls and ceilings, 1/2" taped and finished drywall, primed and painted with 2 coats; painted baseboard and trim; rubber backed carpeting 80%, asphalt tile 20%; hollow core wood interior doors.	.243	7.97	9.52	17.49
7 Specialties	Economy grade kitchen cabinets - 6 L.F. wall and base with plastic laminate counter top and kitchen sink; 30 gallon electric water heater.	.004	1.07	.58	1.65
8 Mechanical	1 lavatory, white, wall hung; 1 water closet, white; 1 bathtub, enameled steel, white; gas fired warm air heat.	.086	2.52	2.12	4.64
9 Electrical	100 Amp. service; romex wiring; incandescent lighting fixtures, switches, receptacles.	.036	.47	.93	1.40
10 Overhead	Contractor's overhead and profit		1.35	1.45	2.80
		Total	28.37	30.88	**59.25**

- **Mass produced from stock plans**
- **Single family — 1 full bath, 1 kitchen**
- **No basement**
- **Asphalt shingles on roof**
- **Hot air heat**
- **Drywall interior finishes**
- **Materials and workmanship are sufficient to meet codes**

Note: The illustration shown may contain some optional components (for example: garages and/or fireplaces) whose costs are shown in the modifications, adjustments, & alternatives below or at the end of the square foot section.

Base cost per square foot of living area

Exterior Wall	Living Area										
	600	800	1000	1200	1400	1600	1800	2000	2400	2800	3200
Wood Siding - Wood Frame	88.65	74.35	66.00	62.70	60.40	56.05	54.30	52.05	47.70	46.15	44.30
Brick Veneer - Wood Frame	97.45	80.75	71.95	68.25	65.70	60.90	58.85	56.35	51.50	49.75	47.65
Stucco on Wood Frame	86.80	73.05	64.80	61.55	59.30	55.05	53.35	51.15	46.90	45.40	43.60
Painted Concrete Block	90.20	75.45	67.05	63.65	61.30	56.90	55.10	52.80	48.35	46.75	44.85
Finished Basement, Add	13.60	11.55	11.00	10.60	10.30	9.90	9.65	9.45	9.05	8.85	8.60
Unfinished Basement, Add	7.85	6.05	5.60	5.20	4.95	4.60	4.40	4.25	3.85	3.70	3.50

Modifications

Add to the total cost

Upgrade Kitchen Cabinets	$ + 242
Solid Surface Countertops	+ 469
Full Bath - including plumbing, wall and floor finishes	+ 2814
Half Bath - including plumbing, wall and floor finishes	+ 1791
One Car Attached Garage	+ 7575
One Car Detached Garage	+ 9783
Fireplace & Chimney	+ 3480

Adjustments

For multi family - add to total cost

Additional Kitchen	$ + 1992
Additional Bath	+ 2800
Additional Entry & Exit	+ 1085
Separate Heating	+ 1192
Separate Electric	+ 702

For Townhouse/Rowhouse - Multiply cost per square foot by

Inner Unit	.95
End Unit	.97

Alternatives

Add to or deduct from the cost per square foot of living area

Composition Roll Roofing	− .35
Cedar Shake Roof	+ 1.75
Upgrade Walls and Ceilings to Skim Coat Plaster	+ .46
Upgrade Ceilings to Textured Finish	+ .43
Air Conditioning, in Heating Ductwork	+ 1.80
In Separate Ductwork	+ 4.02
Heating Systems, Hot Water	+ 1.62
Heat Pump	+ 1.74
Electric Heat	− 1.16
Not Heated	− 2.82

Additional upgrades or components

Kitchen Cabinets & Countertops	Page 93
Bathroom Vanities	94
Fireplaces & Chimneys	94
Windows, Skylights & Dormers	94
Appliances	93
Breezeways & Porches	95
Finished Attic	93
Garages	96
Site Improvements	96
Wings & Ells	34

Important: See the Reference Section for Location Factors (to adjust for your city) and Estimating Form

Economy 1-1/2 Story

Living Area - 1600 S.F.
Perimeter - 135 L.F.

		Labor-Hours	Cost Per Square Foot Of Living Area		
			Mat.	Labor	Total
1 Site Work	Site preparation for slab; 4' deep trench excavation for foundation wall.	.041		.63	.63
2 Foundation	Continuous reinforced concrete footing, 8" deep x 18" wide; dampproofed and insulated 8" thick reinforced concrete block foundation wall, 4' deep; 4" concrete slab on 4" crushed stone base and polyethylene vapor barrier, trowel finish.	.073	2.34	3.17	5.51
3 Framing	Exterior walls - 2" x 4" wood studs, 16" O.C.; 1/2" insulation board sheathing; 2" x 6" rafters, 16" O.C. with 1/2" plywood sheathing, 8 in 12 pitch; 2" x 8" floor joists 16" O.C. with bridging and 5/8" plywood subfloor.	.090	4.27	4.43	8.70
4 Exterior Walls	Beveled wood siding and #15 felt building paper on insulated wood frame walls; 6" attic insulation; double hung windows; 2 flush solid core wood exterior doors with storms.	.077	7.84	4.65	12.49
5 Roofing	20 year asphalt shingles; #15 felt building paper; aluminum gutters, downspouts, drip edge and flashings.	.029	.51	.94	1.45
6 Interiors	Walls and ceilings, 1/2" taped and finished drywall, primed and painted with 2 coats; painted baseboard and trim; rubber backed carpeting 80%, asphalt tile 20%; hollow core wood interior doors.	.204	8.17	9.97	18.14
7 Specialties	Economy grade kitchen cabinets - 6 L.F. wall and base with plastic laminate counter top and kitchen sink; 30 gallon electric water heater.	.020	.80	.43	1.23
8 Mechanical	1 lavatory, white, wall hung; 1 water closet, white; 1 bathtub, enameled steel, white; gas fired warm air heat.	.079	2.10	1.89	3.99
9 Electrical	100 Amp. service; romex wiring; incandescent lighting fixtures, switches, receptacles.	.033	.42	.84	1.26
10 Overhead	Contractor's overhead and profit.		1.32	1.33	2.65
		Total	27.77	28.28	**56.05**

RESIDENTIAL | Economy | 2 Story

- **Mass produced from stock plans**
- **Single family – 1 full bath, 1 kitchen**
- **No basement**
- **Asphalt shingles on roof**
- **Hot air heat**
- **Drywall interior finishes**
- **Materials and workmanship are sufficient to meet codes**

Note: The illustration shown may contain some optional components (for example: garages and/or fireplaces) whose costs are shown in the modifications, adjustments, & alternatives below or at the end of the square foot section.

Base cost per square foot of living area

Exterior Wall	Living Area 1000	1200	1400	1600	1800	2000	2200	2600	3000	3400	3800
Wood Siding - Wood Frame	72.15	64.65	61.85	59.85	57.40	55.05	53.55	50.30	47.05	45.80	44.45
Brick Veneer - Wood Frame	79.00	71.00	67.75	65.55	62.75	60.20	58.45	54.75	51.10	49.65	48.15
Stucco on Wood Frame	70.75	63.35	60.60	58.70	56.35	54.00	52.55	49.40	46.20	45.00	43.70
Painted Concrete Block	73.35	65.75	62.90	60.85	58.35	55.95	54.40	51.05	47.75	46.45	45.10
Finished Basement, Add	9.30	8.85	8.55	8.35	8.10	7.95	7.80	7.50	7.25	7.10	7.00
Unfinished Basement, Add	4.90	4.50	4.25	4.05	3.85	3.75	3.55	3.30	3.10	3.00	2.90

Modifications

Add to the total cost

Upgrade Kitchen Cabinets	$ + 242
Solid Surface Countertops	+ 469
Full Bath - including plumbing, wall and floor finishes	+ 2814
Half Bath - including plumbing, wall and floor finishes	+ 1791
One Car Attached Garage	+ 7575
One Car Detached Garage	+ 9783
Fireplace & Chimney	+ 3845

Adjustments

For multi family - add to total cost

Additional Kitchen	$ + 1992
Additional Bath	+ 2800
Additional Entry & Exit	+ 1085
Separate Heating	+ 1192
Separate Electric	+ 702

For Townhouse/Rowhouse - Multiply cost per square foot by

Inner Unit	.93
End Unit	.96

Alternatives

Add to or deduct from the cost per square foot of living area

Composition Roll Roofing	– .2
Cedar Shake Roof	+ 1.2
Upgrade Walls and Ceilings to Skim Coat Plaster	+ .4
Upgrade Ceilings to Textured Finish	+ .4
Air Conditioning, in Heating Ductwork	+ 1.4
In Separate Ductwork	+ 3.6
Heating Systems, Hot Water	+ 1.5
Heat Pump	+ 1.8
Electric Heat	– 1.0
Not Heated	– 2.6

Additional upgrades or components

Kitchen Cabinets & Countertops	Page 9
Bathroom Vanities	9
Fireplaces & Chimneys	9
Windows, Skylights & Dormers	9
Appliances	9
Breezeways & Porches	9
Finished Attic	9
Garages	9
Site Improvements	9
Wings & Ells	3

Economy 2 Story

Living Area - 2000 S.F.
Perimeter - 135 L.F.

		Labor-Hours	Cost Per Square Foot Of Living Area		
			Mat.	Labor	Total
1 Site Work	Site preparation for slab; 4′ deep trench excavation for foundation wall.	.034		.50	.50
2 Foundation	Continuous reinforced concrete footing, 8″ deep x 18″ wide; dampproofed and insulated 8″ thick reinforced concrete block foundation wall, 4′ deep; 4″ concrete slab on 4″ crushed stone base and polyethylene vapor barrier, trowel finish.	.069	1.88	2.54	4.42
3 Framing	Exterior walls - 2″ x 4″ wood studs, 16″ O.C.; 1/2″ insulation board sheathing; wood truss roof frame, 24″ O.C. with 1/2″ plywood sheathing, 4 in 12 pitch; 2″ x 8″ floor joists 16″ O.C. with bridging and 5/8″ plywood subfloor.	.112	4.37	4.66	9.03
4 Exterior Walls	Beveled wood siding and #15 felt building paper on insulated wood frame walls; 6″ attic insulation; double hung windows; 2 flush solid core wood exterior doors with storms.	.107	8.07	4.81	12.88
5 Roofing	20 year asphalt shingles; #15 felt building paper; aluminum gutters, downspouts, drip edge and flashings.	.024	.41	.76	1.17
6 Interiors	Walls and ceilings, 1/2″ taped and finished drywall, primed and painted with 2 coats; painted baseboard and trim; rubber backed carpeting 80%, asphalt tile 20%; hollow core wood interior doors.	.219	8.40	10.27	18.67
7 Specialties	Economy grade kitchen cabinets - 6 L.F. wall and base with plastic laminate counter top and kitchen sink; 30 gallon electric water heater.	.017	.64	.36	1.00
8 Mechanical	1 lavatory, white, wall hung; 1 water closet, white; 1 bathtub, enameled steel, white; gas fired warm air heat.	.061	1.84	1.77	3.61
9 Electrical	100 Amp. service; romex wiring; incandescent lighting fixtures; switches, receptacles.	.030	.39	.78	1.17
10 Overhead	Contractor's overhead and profit		1.30	1.30	2.60
		Total	27.30	27.75	**55.05**

- **Mass produced from stock plans**
- **Single family — 1 full bath, 1 kitchen**
- **No basement**
- **Asphalt shingles on roof**
- **Hot air heat**
- **Drywall interior finishes**
- **Materials and workmanship are sufficient to meet codes**

Note: The illustration shown may contain some optional components (for example: garages and/or fireplaces) whose costs are shown in the modifications, adjustments, & alternatives below or at the end of the square foot section.

Base cost per square foot of living area

Exterior Wall	Living Area 1000	1200	1400	1600	1800	2000	2200	2600	3000	3400	3800
Wood Siding - Wood Frame	66.75	59.65	57.15	55.40	53.25	**51.05**	49.70	46.80	43.80	42.75	41.55
Brick Veneer - Wood Frame	71.90	64.40	61.60	59.60	57.25	54.85	53.35	50.15	46.85	45.65	44.30
Stucco on Wood Frame	65.70	58.70	56.25	54.50	52.40	50.25	48.95	46.15	43.20	42.15	40.95
Painted Concrete Block	67.65	60.50	57.90	56.10	53.95	51.70	50.35	47.35	44.35	43.25	42.00
Finished Basement, Add	9.30	8.85	8.55	8.35	8.10	7.95	7.80	7.50	7.25	7.10	7.00
Unfinished Basement, Add	4.90	4.50	4.25	4.05	3.85	3.75	3.55	3.30	3.10	3.00	2.90

Modifications

Add to the total cost

Upgrade Kitchen Cabinets	$ + 242
Solid Surface Countertops	+ 469
Full Bath - including plumbing, wall and floor finishes	+ 2814
Half Bath - including plumbing, wall and floor finishes	+ 1791
One Car Attached Garage	+ 7575
One Car Detached Garage	+ 9783
Fireplace & Chimney	+ 3845

Adjustments

For multi family - add to total cost

Additional Kitchen	$ + 1992
Additional Bath	+ 2800
Additional Entry & Exit	+ 1085
Separate Heating	+ 1192
Separate Electric	+ 702

For Townhouse/Rowhouse - Multiply cost per square foot by

Inner Unit	.94
End Unit	.97

Alternatives

Add to or deduct from the cost per square foot of living area

Composition Roll Roofing	– .25
Cedar Shake Roof	+ 1.20
Upgrade Walls and Ceilings to Skim Coat Plaster	+ .44
Upgrade Ceilings to Textured Finish	+ .43
Air Conditioning, in Heating Ductwork	+ 1.44
In Separate Ductwork	+ 3.66
Heating Systems, Hot Water	+ 1.55
Heat Pump	+ 1.85
Electric Heat	– 1.01
Not Heated	– 2.67

Additional upgrades or components

Kitchen Cabinets & Countertops	Page 93
Bathroom Vanities	94
Fireplaces & Chimneys	94
Windows, Skylights & Dormers	94
Appliances	95
Breezeways & Porches	95
Finished Attic	95
Garages	96
Site Improvements	96
Wings & Ells	34

Economy Bi-Level

Living Area - 2000 S.F.
Perimeter - 135 L.F.

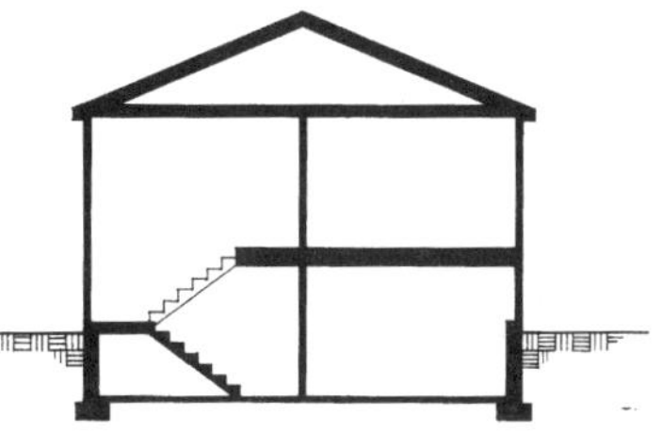

		Labor-Hours	Cost Per Square Foot Of Living Area		
			Mat.	Labor	Total
1 Site Work	Excavation for lower level, 4′ deep. Site preparation for slab.	.029		.50	.50
2 Foundation	Continuous reinforced concrete footing, 8″ deep x 18″ wide; dampproofed and insulated 8″ thick reinforced concrete block foundation wall, 4′ deep; 4″ concrete slab on 4″ crushed stone base and polyethylene vapor barrier, trowel finish.	.069	1.88	2.54	4.42
3 Framing	Exterior walls - 2″ x 4″ wood studs, 16″ O.C.; 1/2″ insulation board sheathing; wood truss roof frame, 24″ O.C. with 1/2″ plywood sheathing, 4 in 12 pitch; 2″ x 8″ floor joists 16″ O.C. with bridging and 5/8″ plywood subfloor.	.107	4.09	4.38	8.47
4 Exterior Walls	Beveled wood siding and #15 felt building paper on insulated wood frame walls; 6″ attic insulation; double hung windows; 2 flush solid core wood exterior doors with storms.	.089	6.29	3.73	10.02
5 Roofing	20 year asphalt shingles; #15 felt building paper; aluminum gutters, downspouts, drip edge and flashings.	.024	.41	.76	1.17
6 Interiors	Walls and ceilings, 1/2″ taped and finished drywall, primed and painted with 2 coats; painted baseboard and trim, rubber backed carpeting 80%, asphalt tile 20%; hollow core wood interior doors.	.213	8.23	10.01	18.24
7 Specialties	Economy grade kitchen cabinets - 6 L.F. wall and base with plastic laminate counter top and kitchen sink; 30 gallon electric water heater.	.018	.64	.36	1.00
8 Mechanical	1 lavatory, white, wall hung; 1 water closet, white; 1 bathtub, enameled steel, white; gas fired warm air heat.	.061	1.84	1.77	3.61
9 Electrical	100 Amp. service; romex wiring; incandescent lighting fixtures; switches, receptacles.	.030	.39	.78	1.17
10 Overhead	Contractor's overhead and profit.		1.19	1.26	2.45
		Total	24.96	26.09	**51.05**

- **Mass produced from stock plans**
- **Single family — 1 full bath, 1 kitchen**
- **No basement**
- **Asphalt shingles on roof**
- **Hot air heat**
- **Drywall interior finishes**
- **Materials and workmanship are sufficient to meet codes**

Note: The illustration shown may contain some optional components (for example: garages and/or fireplaces) whose costs are shown in the modifications, adjustments, & alternatives below or at the end of the square foot section.

©Design Basics, Inc.

Base cost per square foot of living area

Exterior Wall	Living Area 1200	1500	1800	2000	2200	2400	2800	3200	3600	4000	4400
Wood Siding - Wood Frame	60.75	55.85	52.05	50.70	48.65	**46.75**	45.25	43.30	41.00	40.30	38.55
Brick Veneer - Wood Frame	65.45	60.10	55.90	54.40	52.15	50.00	48.40	46.20	43.70	42.95	41.00
Stucco on Wood Frame	59.80	54.95	51.25	49.95	47.90	46.05	44.55	42.65	40.40	39.75	38.05
Solid Masonry	61.55	56.60	52.70	51.35	49.25	47.30	45.80	43.80	41.45	40.75	38.95
Finished Basement, Add*	11.05	10.55	10.10	9.90	9.70	9.50	9.30	9.10	8.90	8.80	8.60
Unfinished Basement, Add*	5.35	4.90	4.55	4.35	4.20	4.00	3.85	3.65	3.50	3.40	3.25

*Basement under middle level only.

Modifications

Add to the total cost

Upgrade Kitchen Cabinets	$ + 242
Solid Surface Countertops	+ 469
Full Bath - including plumbing, wall and floor finishes	+ 2814
Half Bath - including plumbing, wall and floor finishes	+ 1791
One Car Attached Garage	+ 7575
One Car Detached Garage	+ 9783
Fireplace & Chimney	+ 3845

Adjustments

For multi family - add to total cost

Additional Kitchen	$ + 1992
Additional Bath	+ 2800
Additional Entry & Exit	+ 1085
Separate Heating	+ 1192
Separate Electric	+ 702

For Townhouse/Rowhouse - Multiply cost per square foot by

Inner Unit	.93
End Unit	.96

Alternatives

Add to or deduct from the cost per square foot of living area

Composition Roll Roofing	− .35
Cedar Shake Roof	+ 1.75
Upgrade Walls and Ceilings to Skim Coat Plaster	+ .40
Upgrade Ceilings to Textured Finish	+ .43
Air Conditioning, in Heating Ductwork	+ 1.20
In Separate Ductwork	+ 3.42
Heating Systems, Hot Water	+ 1.50
Heat Pump	+ 1.92
Electric Heat	− .92
Not Heated	− 2.58

Additional upgrades or components

Kitchen Cabinets & Countertops	Page 93
Bathroom Vanities	94
Fireplaces & Chimneys	94
Windows, Skylights & Dormers	94
Appliances	95
Breezeways & Porches	95
Finished Attic	95
Garages	96
Site Improvements	96
Wings & Ells	34

Important: See the Reference Section for Location Factors (to adjust for your city) and Estimating Form

Economy Tri-Level

Living Area - 2400 S.F.
Perimeter - 163 L.F.

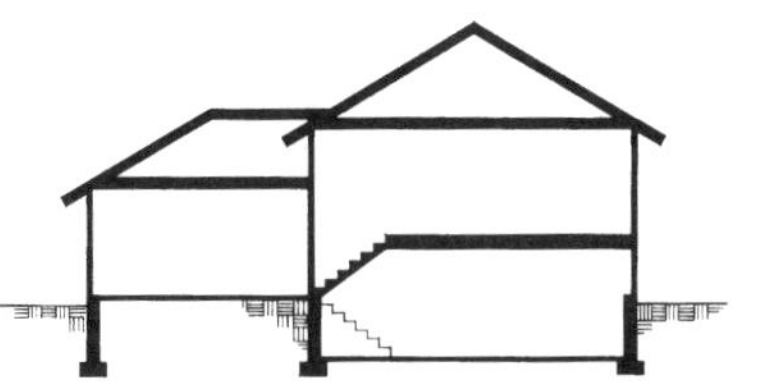

		Labor-Hours	Cost Per Square Foot Of Living Area		
			Mat.	Labor	Total
1 Site Work	Site preparation for slab; 4′ deep trench excavation for foundation wall, excavation for lower level, 4′ deep.	.027		.42	.42
2 Foundation	Continuous reinforced concrete footing, 8″ deep x 18″ wide; dampproofed and insulated 8″ thick reinforced concrete block foundation wall, 4′ deep; 4″ concrete slab on 4″ crushed stone base and polyethylene vapor barrier, trowel finish.	.071	2.11	2.76	4.87
3 Framing	Exterior walls - 2″ x 4″ wood studs, 16″ O.C.; 1/2″ insulation board sheathing; wood truss roof frame, 24″ O.C. with 1/2″ plywood sheathing, 4 in 12 pitch; 2″ x 8″ floor joists 16″ O.C. with bridging and 5/8″ plywood subfloor.	.094	3.83	3.93	7.76
4 Exterior Walls	Beveled wood siding and #15 felt building paper on insulated wood frame walls; 6″ attic insulation; double hung windows; 2 flush solid core wood exterior doors with storms.	.081	5.44	3.22	8.66
5 Roofing	20 year asphalt shingles; #15 felt building paper; aluminum gutters, downspouts, drip edge and flashings.	.032	.55	1.01	1.56
6 Interiors	Walls and ceilings, 1/2″ taped and finished drywall, primed and painted with 2 coats; painted baseboard and trim, rubber backed carpeting 80%, asphalt tile 20%; hollow core wood interior doors.	.177	7.15	8.77	15.92
7 Specialties	Economy grade kitchen cabinets - 6 L.F. wall and base with plastic laminate counter top and kitchen sink; 30 gallon electric water heater.	.014	.54	.30	.84
8 Mechanical	1 lavatory, white, wall hung; 1 water closet, white; 1 bathtub, enameled steel, white; gas fired warm air heat.	.057	1.68	1.68	3.36
9 Electrical	100 Amp. service; romex wiring; incandescent lighting fixtures, switches, receptacles.	.029	.37	.75	1.12
10 Overhead	Contractor's overhead and profit		1.08	1.16	2.24
		Total	22.75	24.00	**46.75**

1 Story — Base cost per square foot of living area

Exterior Wall	Living Area 50	100	200	300	400	500	600	700
Wood Siding - Wood Frame	105.60	80.05	69.00	56.75	53.15	50.95	49.50	49.80
Brick Veneer - Wood Frame	121.60	91.45	78.50	63.10	58.85	56.30	54.60	54.70
Stucco on Wood Frame	102.35	77.70	67.00	55.45	52.00	49.85	48.50	48.80
Painted Concrete Block	108.40	82.05	70.65	57.85	54.15	51.90	50.40	50.65
Finished Basement, Add	26.65	21.75	19.70	16.30	15.60	15.20	14.95	14.75
Unfinished Basement, Add	16.40	12.30	10.55	7.70	7.10	6.75	6.55	6.40

1-1/2 Story — Base cost per square foot of living area

Exterior Wall	Living Area 100	200	300	400	500	600	700	800
Wood Siding - Wood Frame	83.65	67.05	56.85	50.35	47.30	45.75	43.85	43.35
Brick Veneer - Wood Frame	97.90	78.45	66.35	57.75	54.10	52.20	49.95	49.40
Stucco on Wood Frame	80.70	64.70	54.85	48.80	45.85	44.40	42.55	42.10
Painted Concrete Block	86.15	69.00	58.50	51.65	48.50	46.85	44.90	44.40
Finished Basement, Add	17.85	15.80	14.45	12.90	12.50	12.25	12.00	11.95
Unfinished Basement, Add	10.20	8.45	7.30	6.05	5.70	5.50	5.25	5.25

2 Story — Base cost per square foot of living area

Exterior Wall	Living Area 100	200	400	600	800	1000	1200	1400
Wood Siding - Wood Frame	86.30	63.75	54.00	43.80	40.65	38.70	37.50	37.85
Brick Veneer - Wood Frame	102.25	75.15	63.50	50.15	46.35	44.05	42.55	42.75
Stucco on Wood Frame	83.00	61.45	52.00	42.50	39.50	37.65	36.45	36.85
Painted Concrete Block	89.05	65.75	55.65	44.90	41.65	39.65	38.35	38.70
Finished Basement, Add	13.40	10.90	9.90	8.20	7.85	7.65	7.50	7.45
Unfinished Basement, Add	8.25	6.15	5.30	3.85	3.60	3.40	3.30	3.25

Base costs do not include bathroom or kitchen facilities. Use Modifications/Adjustments/Alternatives on pages 93-96 where appropriate.

Average Class

1 Story

1-1/2 Story

2 Story

2-1/2 Story

Bi-Level

Tri-Level

- **Simple design from standard plans**
- **Single family — 1 full bath, 1 kitchen**
- **No basement**
- **Asphalt shingles on roof**
- **Hot air heat**
- **Drywall interior finishes**
- **Materials and workmanship are average**

Note: The illustration shown may contain some optional components (for example: garages and/or fireplaces) whose costs are shown in the modifications, adjustments, & alternatives below or at the end of the square foot section.

Base cost per square foot of living area

Exterior Wall	Living Area 600	800	1000	1200	1400	1600	1800	2000	2400	2800	3200
Wood Siding - Wood Frame	99.15	91.05	84.10	78.50	74.10	**70.75**	68.95	67.05	62.60	59.50	57.35
Brick Veneer - Wood Frame	112.00	103.15	95.60	89.40	84.60	80.95	79.00	76.85	72.05	68.70	66.30
Stucco on Wood Frame	105.65	97.45	90.45	84.75	80.35	77.00	75.20	73.30	68.75	65.70	63.50
Solid Masonry	120.70	110.90	102.65	95.75	90.35	86.40	84.25	81.80	76.50	72.85	70.15
Finished Basement, Add	22.65	21.20	20.20	19.30	18.55	18.10	17.80	17.40	16.90	16.45	16.05
Unfinished Basement, Add	9.95	9.00	8.35	7.75	7.30	6.95	6.80	6.55	6.20	5.90	5.70

Modifications

Add to the total cost

Upgrade Kitchen Cabinets	$ + 2282
Solid Surface Countertops	+ 826
Full Bath - including plumbing, wall and floor finishes	+ 3531
Half Bath - including plumbing, wall and floor finishes	+ 2247
One Car Attached Garage	+ 8054
One Car Detached Garage	+ 10,547
Fireplace & Chimney	+ 3555

Adjustments

For multi family - add to total cost

Additional Kitchen	$ + 3707
Additional Bath	+ 3531
Additional Entry & Exit	+ 1085
Separate Heating	+ 1192
Separate Electric	+ 1342

For Townhouse/Rowhouse - Multiply cost per square foot by

Inner Unit	.92
End Unit	.96

Alternatives

Add to or deduct from the cost per square foot of living area

Cedar Shake Roof	+ 2.15
Clay Tile Roof	+ 4.65
Slate Roof	+ 7.75
Upgrade Walls to Skim Coat Plaster	+ .30
Upgrade Ceilings to Textured Finish	+ .43
Air Conditioning, in Heating Ductwork	+ 2.47
In Separate Ductwork	+ 4.75
Heating Systems, Hot Water	+ 1.78
Heat Pump	+ 1.60
Electric Heat	– 1.16
Not Heated	– 2.82

Additional upgrades or components

Kitchen Cabinets & Countertops	Page 93
Bathroom Vanities	94
Fireplaces & Chimneys	94
Windows, Skylights & Dormers	94
Appliances	95
Breezeways & Porches	95
Finished Attic	95
Garages	96
Site Improvements	96
Wings & Ells	56

Important: See the Reference Section for Location Factors (to adjust for your city) and Estimating Form

Average 1 Story

Living Area - 1600 S.F.
Perimeter - 163 L.F.

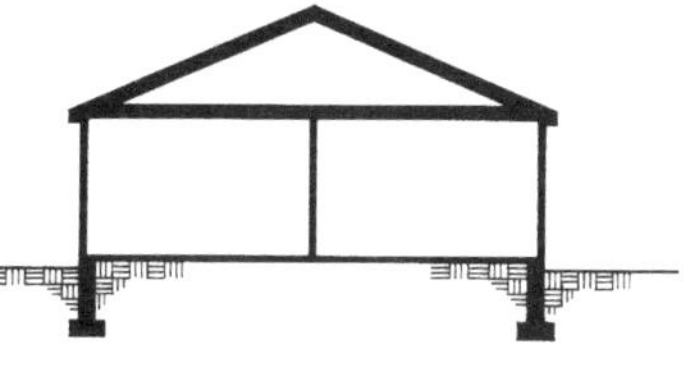

		Labor-Hours	Cost Per Square Foot Of Living Area		
			Mat.	Labor	Total
1 Site Work	Site preparation for slab; 4′ deep trench excavation for foundation wall.	.048		.65	.65
2 Foundation	Continuous reinforced concrete footing 8″ deep x 18″ wide; dampproofed and insulated reinforced concrete foundation wall, 8″ thick, 4′ deep; 4″ concrete slab on 4″ crushed stone base and polyethylene vapor barrier, trowel finish.	.113	3.26	4.26	7.52
3 Framing	Exterior walls - 2″ x 4″ wood studs, 16″ O.C.; 1/2″ plywood sheathing; 2″ x 6″ rafters 16″ O.C. with 1/2″ plywood sheathing, 4 in 12 pitch; 2″ x 6″ ceiling joists 16″ O.C.; 1/2″ plywood subfloor on 1″ x 2″ wood sleepers 16″ O.C.	.136	5.59	5.90	11.49
4 Exterior Walls	Beveled wood siding and #15 felt building paper on insulated wood frame walls; 6″ attic insulation; double hung windows; 3 flush solid core wood exterior doors with storms.	.098	8.48	4.39	12.87
5 Roofing	25 year asphalt shingles; #15 felt building paper; aluminum gutters, downspouts, drip edge and flashings.	.047	.84	1.55	2.39
6 Interiors	Walls and ceilings, 1/2″ taped and finished drywall, primed and painted with 2 coats; painted baseboard and trim, finished hardwood floor 40%, carpet with 1/2″ underlayment 40%, vinyl tile with 1/2″ underlayment 15%, ceramic tile with 1/2″ underlayment 5%; hollow core and louvered interior doors.	.251	11.82	11.39	23.21
7 Specialties	Average grade kitchen cabinets - 14 L.F. wall and base with plastic laminate counter top and kitchen sink; 40 gallon electric water heater.	.009	1.36	.56	1.92
8 Mechanical	1 lavatory, white, wall hung; 1 water closet, white; 1 bathtub with shower, enameled steel, white; gas fired warm air heat.	.098	2.26	1.94	4.20
9 Electrical	200 Amp. service; romex wiring; incandescent lighting fixtures, switches, receptacles.	.041	.73	1.15	1.88
10 Overhead	Contractor's overhead and profit and plans.		2.41	2.21	4.62
		Total	36.75	34.00	**70.75**

RESIDENTIAL	**Average**	**1-1/2 Story**

SQUARE FOOT COSTS

- **Simple design from standard plans**
- **Single family — 1 full bath, 1 kitchen**
- **No basement**
- **Asphalt shingles on roof**
- **Hot air heat**
- **Drywall interior finishes**
- **Materials and workmanship are average**

Note: The illustration shown may contain some optional components (for example: garages and/or fireplaces) whose costs are shown in the modifications, adjustments, & alternatives below or at the end of the square foot section.

©By Designer

Base cost per square foot of living area

Exterior Wall	Living Area 600	800	1000	1200	1400	1600	1800	2000	2400	2800	3200
Wood Siding - Wood Frame	112.35	95.40	85.00	81.05	78.20	72.95	70.80	67.95	62.75	60.90	58.55
Brick Veneer - Wood Frame	122.05	102.45	91.50	87.15	84.00	78.25	75.85	72.70	66.95	64.90	62.25
Stucco on Wood Frame	113.15	96.00	85.55	81.55	78.70	73.40	71.25	68.35	63.10	61.25	58.90
Solid Masonry	134.20	111.25	99.65	94.75	91.30	84.85	82.10	78.65	72.20	69.85	66.85
Finished Basement, Add	18.55	15.75	15.05	14.50	14.15	13.60	13.30	13.00	12.40	12.15	11.85
Unfinished Basement, Add	8.50	6.70	6.25	5.90	5.65	5.25	5.05	4.90	4.50	4.35	4.15

Modifications

Add to the total cost

Upgrade Kitchen Cabinets	$ + 2282
Solid Surface Countertops	+ 826
Full Bath - including plumbing, wall and floor finishes	+ 3531
Half Bath - including plumbing, wall and floor finishes	+ 2247
One Car Attached Garage	+ 8054
One Car Detached Garage	+ 10,547
Fireplace & Chimney	+ 3555

Adjustments

For multi family - add to total cost

Additional Kitchen	$ + 3707
Additional Bath	+ 3531
Additional Entry & Exit	+ 1085
Separate Heating	+ 1192
Separate Electric	+ 1342

For Townhouse/Rowhouse - Multiply cost per square foot by

Inner Unit	.92
End Unit	.96

Alternatives

Add to or deduct from the cost per square foot of living area

Cedar Shake Roof	+ 1.5
Clay Tile Roof	+ 3.3
Slate Roof	+ 5.6
Upgrade Walls to Skim Coat Plaster	+ .3
Upgrade Ceilings to Textured Finish	+ .4
Air Conditioning, in Heating Ductwork	+ 1.8
In Separate Ductwork	+ 4.1
Heating Systems, Hot Water	+ 1.6
Heat Pump	+ 1.7
Electric Heat	– 1.0
Not Heated	– 2.7

Additional upgrades or components

Kitchen Cabinets & Countertops	Page 9
Bathroom Vanities	9
Fireplaces & Chimneys	9
Windows, Skylights & Dormers	9
Appliances	9
Breezeways & Porches	9
Finished Attic	9
Garages	9
Site Improvements	9
Wings & Ells	5

Important: See the Reference Section for Location Factors (to adjust for your city) and Estimating Form

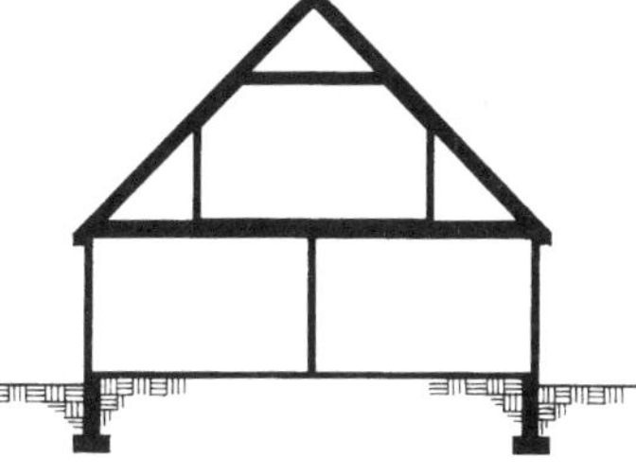

Average 1-1/2 Story

Living Area - 1800 S.F.
Perimeter - 144 L.F.

		Labor-Hours	Cost Per Square Foot Of Living Area		
			Mat.	Labor	Total
1 Site Work	Site preparation for slab; 4′ deep trench excavation for foundation wall.	.037		.58	.58
2 Foundation	Continuous reinforced concrete footing 8″ deep x 18″ wide; dampproofed and insulated reinforced concrete foundation wall, 8″ thick, 4′ deep; 4″ concrete slab on 4″ crushed stone base and polyethylene vapor barrier, trowel finish.	.073	2.32	3.13	5.45
3 Framing	Exterior walls - 2″ x 4″ wood studs, 16″ O.C.; 1/2″ plywood sheathing; 2″ x 6″ rafters 16″ O.C. with 1/2″ plywood sheathing, 8 in 12 pitch; 2″ x 8″ floor joists 16″ O.C. with 5/8″ plywood subfloor; 1/2″ plywood subfloor on 1″ x 2″ wood sleepers 16″ O.C.	.098	5.31	5.40	10.71
4 Exterior Walls	Beveled wood siding and #15 felt building paper on insulated wood frame walls; 6″ attic insulation; double hung windows; 3 flush solid core wood exterior doors with storms.	.078	9.35	4.86	14.21
5 Roofing	25 year asphalt shingles; #15 felt building paper; aluminum gutters, downspouts, drip edge and flashings.	.029	.52	.97	1.49
6 Interiors	Walls and ceilings, 1/2″ taped and finished drywall, primed and painted with 2 coats; painted baseboard and trim, finished hardwood floor 40%, carpet with 1/2″ underlayment 40%, vinyl tile with 1/2″ underlayment 15%, ceramic tile with 1/2″ underlayment 5%; hollow core and louvered interior doors.	.225	13.20	13.08	26.28
7 Specialties	Average grade kitchen cabinets - 14 L.F. wall and base with plastic laminate counter top and kitchen sink; 40 gallon electric water heater.	.022	1.20	.50	1.70
8 Mechanical	1 lavatory, white, wall hung; 1 water closet, white; 1 bathtub with shower, enameled steel, white; gas fired warm air heat.	.049	2.11	1.87	3.98
9 Electrical	200 Amp. service; romex wiring; incandescent lighting fixtures, switches, receptacles.	.039	.68	1.10	1.78
10 Overhead	Contractor's overhead and profit and plans.		2.45	2.17	4.62
		Total	37.14	33.66	**70.80**

- **Simple design from standard plans**
- **Single family — 1 full bath, 1 kitchen**
- **No basement**
- **Asphalt shingles on roof**
- **Hot air heat**
- **Drywall interior finishes**
- **Materials and workmanship are average**

Note: The illustration shown may contain some optional components (for example: garages and/or fireplaces) whose costs are shown in the modifications, adjustments, & alternatives below or at the end of the square foot section.

Base cost per square foot of living area

Exterior Wall	Living Area 1000	1200	1400	1600	1800	2000	2200	2600	3000	3400	3800
Wood Siding - Wood Frame	91.60	82.20	78.85	76.45	73.25	70.55	68.80	64.75	60.80	59.30	57.55
Brick Veneer - Wood Frame	99.10	89.20	85.35	82.65	79.10	76.20	74.15	69.55	65.25	63.55	61.60
Stucco on Wood Frame	92.20	82.85	79.40	77.00	73.75	71.05	69.25	65.15	61.15	59.70	57.90
Solid Masonry	109.40	98.70	94.20	91.15	87.10	83.95	81.50	76.20	71.40	69.35	67.15
Finished Basement, Add	13.10	12.60	12.20	11.90	11.60	11.40	11.15	10.75	10.45	10.30	10.10
Unfinished Basement, Add	5.45	5.05	4.80	4.60	4.40	4.30	4.10	3.85	3.65	3.55	3.45

Modifications

Add to the total cost

Upgrade Kitchen Cabinets	$ + 2282
Solid Surface Countertops	+ 826
Full Bath - including plumbing, wall and floor finishes	+ 3531
Half Bath - including plumbing, wall and floor finishes	+ 2247
One Car Attached Garage	+ 8054
One Car Detached Garage	+ 10,547
Fireplace & Chimney	+ 3965

Adjustments

For multi family - add to total cost

Additional Kitchen	$ + 3707
Additional Bath	+ 3531
Additional Entry & Exit	+ 1085
Separate Heating	+ 1192
Separate Electric	+ 1342

For Townhouse/Rowhouse - Multiply cost per square foot by

Inner Unit	.90
End Unit	.95

Alternatives

Add to or deduct from the cost per square foot of living area

Cedar Shake Roof	+ 1.10
Clay Tile Roof	+ 2.30
Slate Roof	+ 3.85
Upgrade Walls to Skim Coat Plaster	+ .36
Upgrade Ceilings to Textured Finish	+ .43
Air Conditioning, in Heating Ductwork	+ 1.48
In Separate Ductwork	+ 3.77
Heating Systems, Hot Water	+ 1.60
Heat Pump	+ 1.90
Electric Heat	– 1.01
Not Heated	– 2.67

Additional upgrades or components

Kitchen Cabinets & Countertops	Page 93
Bathroom Vanities	94
Fireplaces & Chimneys	94
Windows, Skylights & Dormers	94
Appliances	95
Breezeways & Porches	95
Finished Attic	95
Garages	96
Site Improvements	96
Wings & Ells	56

Average 2 Story

Living Area - 2000 S.F.
Perimeter - 135 L.F.

		Labor-Hours	Cost Per Square Foot Of Living Area		
			Mat.	Labor	Total
1 Site Work	Site preparation for slab; 4' deep trench excavation for foundation wall.	.034		.51	.51
2 Foundation	Continuous reinforced concrete footing 8" deep x 18" wide; dampproofed and insulated reinforced concrete foundation wall, 8" thick, 4' deep, 4" concrete slab on 4" crushed stone base and polyethylene vapor barrier, trowel finish.	.066	1.93	2.61	4.54
3 Framing	Exterior walls - 2" x 4" wood studs, 16" O.C.; 1/2" plywood sheathing; 2" x 6" rafters 16" O.C. with 1/2" plywood sheathing, 4 in 12 pitch; 2" x 6" ceiling joists 16" O.C.; 2" x 8" floor joists 16" O.C. with 5/8" plywood subfloor; 1/2" plywood subfloor on 1" x 2" wood sleepers 16" O.C.	.131	5.35	5.54	10.89
4 Exterior Walls	Beveled wood siding and #15 felt building paper on insulated wood frame walls; 6" attic insulation; double hung windows; 3 flush solid core wood exterior doors with storms.	.111	10.27	5.33	15.60
5 Roofing	25 year asphalt shingles; #15 felt building paper; aluminum gutters, downspouts, drip edge and flashings.	.024	.42	.78	1.20
6 Interiors	Walls and ceilings, 1/2" taped and finished drywall, primed and painted with 2 coats; painted baseboard and trim, finished hardwood floor 40%, carpet with 1/2" underlayment 40%, vinyl tile with 1/2" underlayment 15%, ceramic tile with 1/2" underlayment 5%; hollow core and louvered interior doors.	.232	13.12	13.06	26.18
7 Specialties	Average grade kitchen cabinets - 14 L.F. wall and base with plastic laminate counter top and kitchen sink; 40 gallon electric water heater.	.021	1.08	.45	1.53
8 Mechanical	1 lavatory, white, wall hung; 1 water closet, white; 1 bathtub with shower; enameled steel, white; gas fired warm air heat.	.060	1.97	1.82	3.79
9 Electrical	200 Amp. service; romex wiring; incandescent lighting fixtures, switches, receptacles.	.039	.65	1.06	1.71
10 Overhead	Contractor's overhead and profit and plans.		2.43	2.17	4.60
		Total	37.22	33.33	**70.55**

- **Simple design from standard plans**
- **Single family — 1 full bath, 1 kitchen**
- **No basement**
- **Asphalt shingles on roof**
- **Hot air heat**
- **Drywall interior finishes**
- **Materials and workmanship are average**

Note: The illustration shown may contain some optional components (for example: garages and/or fireplaces) whose costs are shown in the modifications, adjustments, & alternatives below or at the end of the square foot section.

Base cost per square foot of living area

	Living Area										
Exterior Wall	**1200**	**1400**	**1600**	**1800**	**2000**	**2400**	**2800**	**3200**	**3600**	**4000**	**4400**
Wood Siding - Wood Frame	91.10	86.20	77.70	76.70	74.25	69.70	66.65	62.95	61.35	57.95	57.05
Brick Veneer - Wood Frame	99.10	93.45	84.35	83.40	80.50	75.35	72.05	67.85	66.00	62.25	61.20
Stucco on Wood Frame	91.80	86.85	78.25	77.30	74.80	70.15	67.10	63.40	61.80	58.30	57.40
Solid Masonry	109.10	102.45	92.70	91.70	88.35	82.40	78.80	73.95	71.80	67.60	66.50
Finished Basement, Add	11.15	10.60	10.20	10.15	9.90	9.45	9.30	9.00	8.80	8.60	8.50
Unfinished Basement, Add	4.50	4.15	3.90	3.85	3.70	3.40	3.30	3.10	3.00	2.85	2.80

Modifications

Add to the total cost

Upgrade Kitchen Cabinets	$ + 2282
Solid Surface Countertops	+ 826
Full Bath - including plumbing, wall and floor finishes	+ 3531
Half Bath - including plumbing, wall and floor finishes	+ 2247
One Car Attached Garage	+ 8054
One Car Detached Garage	+ 10,547
Fireplace & Chimney	+ 4510

Adjustments

For multi family - add to total cost

Additional Kitchen	$ + 3707
Additional Bath	+ 3531
Additional Entry & Exit	+ 1085
Separate Heating	+ 1192
Separate Electric	+ 1342

For Townhouse/Rowhouse - Multiply cost per square foot by

Inner Unit	.90
End Unit	.95

Alternatives

Add to or deduct from the cost per square foot of living area

Cedar Shake Roof	+ .9
Clay Tile Roof	+ 2.0
Slate Roof	+ 3.3
Upgrade Walls to Skim Coat Plaster	+ .3
Upgrade Ceilings to Textured Finish	+ .4
Air Conditioning, in Heating Ductwork	+ 1.2
In Separate Ductwork	+ 3.3
Heating Systems, Hot Water	+ 2.0
Heat Pump	+ 1.8
Electric Heat	– 1.4
Not Heated	– 3.0

Additional upgrades or components

Kitchen Cabinets & Countertops — Page 9
Bathroom Vanities
Fireplaces & Chimneys
Windows, Skylights & Dormers
Appliances
Breezeways & Porches
Finished Attic
Garages
Site Improvements
Wings & Ells

Average 2-1/2 Story

Living Area - 3200 S.F.
Perimeter - 150 L.F.

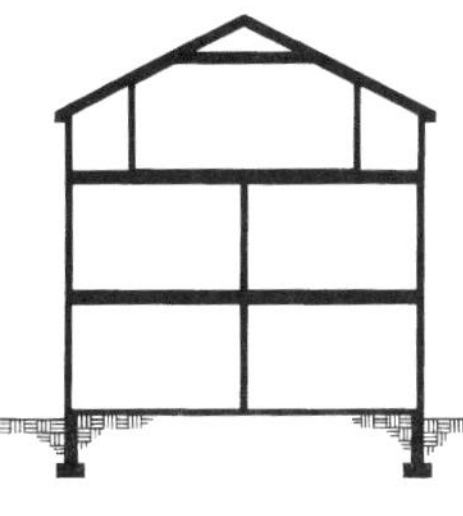

		Labor-Hours	Cost Per Square Foot Of Living Area		
			Mat.	Labor	Total
1 Site Work	Site preparation for slab; 4′ deep trench excavation for foundation wall.	.046		.32	.32
2 Foundation	Continuous reinforced concrete footing 8″ deep x 18″ wide, dampproofed and insulated reinforced concrete foundation wall, 8″ thick, 4′ deep; 4″ concrete slab on 4″ crushed stone base and polyethylene vapor barrier, trowel finish.	.061	1.38	1.85	3.23
3 Framing	Exterior walls - 2″ x 4″ wood studs, 16″ O.C.; 1/2″ plywood sheathing; 2″ x 6″ rafters 16″ O.C. with 1/2″ plywood sheathing, 4 in 12 pitch; 2″ x 6″ ceiling joists 16″ O.C.; 2″ x 8″ floor joists 16″ O.C. with 5/8″ plywood subfloor; 1/2″ plywood subfloor on 1″ x 2″ wood sleepers 16″ O.C.	.127	5.09	5.38	10.47
4 Exterior Walls	Beveled wood siding and #15 felt building paper on insulated wood frame walls; 6″ attic insulation; double hung windows; 3 flush solid core wood exterior doors with storms.	.136	8.70	4.52	13.22
5 Roofing	25 year asphalt shingles; #15 felt building paper; aluminum gutters, downspouts, drip edge and flashings.	.018	.33	.60	.93
6 Interiors	Walls and ceilings, 1/2″ taped and finished drywall, primed and painted with 2 coats; painted baseboard and trim, finished hardwood floor 40%, carpet with 1/2″ underlayment 40%, vinyl tile with 1/2″ underlayment 15%, ceramic tile with 1/2″ underlayment 5%; hollow core and louvered interior doors.	.286	12.60	12.47	25.07
7 Specialties	Average grade kitchen cabinets - 14 L.F. wall and base with plastic laminate counter top and kitchen sink; 40 gallon electric water heater.	.030	.68	.29	.97
8 Mechanical	1 lavatory, white, wall hung; 1 water closet, white; 1 bathtub with shower, enameled steel, white; gas fired warm air heat.	.072	1.57	1.61	3.18
9 Electrical	200 Amp. service; romex wiring; incandescent lighting fixtures, switches, receptacles.	.046	.52	.93	1.45
10 Overhead	Contractor's overhead and profit and plans.		2.16	1.95	4.11
		Total	33.03	29.92	**62.95**

RESIDENTIAL	Average	3 Story

- **Simple design from standard plans**
- **Single family — 1 full bath, 1 kitchen**
- **No basement**
- **Asphalt shingles on roof**
- **Hot air heat**
- **Drywall interior finishes**
- **Materials and workmanship are average**

Note: The illustration shown may contain some optional components (for example: garages and/or fireplaces) whose costs are shown in the modifications, adjustments, & alternatives below or at the end of the square foot section.

Base cost per square foot of living area

	Living Area										
Exterior Wall	**1500**	**1800**	**2100**	**2500**	**3000**	**3500**	**4000**	**4500**	**5000**	**5500**	**6000**
Wood Siding - Wood Frame	85.60	76.85	73.90	71.60	66.30	64.40	60.95	57.40	56.45	54.95	53.75
Brick Veneer - Wood Frame	93.10	83.85	80.40	77.80	71.95	69.75	65.80	61.90	60.75	59.10	57.65
Stucco on Wood Frame	86.25	77.45	74.50	72.15	66.75	64.85	61.35	57.75	56.80	55.25	54.10
Solid Masonry	102.50	92.50	88.55	85.55	79.00	76.40	71.85	67.50	66.20	64.30	62.60
Finished Basement, Add	9.55	9.20	8.90	8.70	8.40	8.20	8.00	7.80	7.70	7.60	7.50
Unfinished Basement, Add	3.65	3.45	3.25	3.10	2.90	2.75	2.60	2.50	2.45	2.35	2.30

Modifications

Add to the total cost

Upgrade Kitchen Cabinets	$ + 2282
Solid Surface Countertops	+ 826
Full Bath - including plumbing, wall and floor finishes	+ 3531
Half Bath - including plumbing, wall and floor finishes	+ 2247
One Car Attached Garage	+ 8054
One Car Detached Garage	+ 10,547
Fireplace & Chimney	+ 4510

Adjustments

For multi family - add to total cost

Additional Kitchen	$ + 3707
Additional Bath	+ 3531
Additional Entry & Exit	+ 1085
Separate Heating	+ 1192
Separate Electric	+ 1342

For Townhouse/Rowhouse - Multiply cost per square foot by

Inner Unit	.88
End Unit	.94

Alternatives

Add to or deduct from the cost per square foot of living area

Cedar Shake Roof	+ .7
Clay Tile Roof	+ 1.5
Slate Roof	+ 2.6
Upgrade Walls to Skim Coat Plaster	+ .3
Upgrade Ceilings to Textured Finish	+ .4
Air Conditioning, in Heating Ductwork	+ 1.2
In Separate Ductwork	+ 3.3
Heating Systems, Hot Water	+ 2.0
Heat Pump	+ 1.8
Electric Heat	- 1.2
Not Heated	- 2.8

Additional upgrades or components

Kitchen Cabinets & Countertops	Page 9
Bathroom Vanities	9
Fireplaces & Chimneys	9
Windows, Skylights & Dormers	9
Appliances	9
Breezeways & Porches	9
Finished Attic	9
Garages	9
Site Improvements	9
Wings & Ells	5

Important: See the Reference Section for Location Factors (to adjust for your city) and Estimating Form

Average 3 Story

Living Area - 3000 S.F.
Perimeter - 135 L.F.

		Labor-Hours	Cost Per Square Foot Of Living Area		
			Mat.	Labor	Total
1 Site Work	Site preparation for slab; 4′ deep trench excavation for foundation wall.	.038		.34	.34
2 Foundation	Continuous reinforced concrete footing 8″ deep x 18″ wide, dampproofed and insulated reinforced concrete foundation wall, 8″ thick, 4′ deep; 4″ concrete slab on 4″ crushed stone base and polyethylene vapor barrier, trowel finish.	.053	1.27	1.75	3.02
3 Framing	Exterior walls - 2″ x 4″ wood studs, 16″ O.C.; 1/2″ plywood sheathing; 2″ x 6″ rafters 16″ O.C. with 1/2″ plywood sheathing, 4 in 12 pitch; 2″ x 6″ ceiling joists 16″ O.C.; 2″ x 8″ floor joists 16″ O.C. with 5/8″ plywood subfloor; 1/2″ plywood subfloor on 1″ x 2″ wood sleepers 16″ O.C.	.128	5.20	5.53	10.73
4 Exterior Walls	Horizontal beveled wood siding; #15 felt building paper; 3-1/2″ batt insulation; wood double hung windows; 3 flush solid core wood exterior doors; storms and screens.	.139	9.94	5.16	15.10
5 Roofing	25 year asphalt shingles; #15 felt building paper; aluminum gutters, downspouts, drip edge and flashings.	.014	.28	.51	.79
6 Interiors	Walls and ceilings, 1/2″ taped and finished drywall, primed and painted with 2 coats; painted baseboard and trim, finished hardwood floor 40%, carpet with 1/2″ underlayment 40%, vinyl tile with 1/2″ underlayment 15%, ceramic tile with 1/2″ underlayment 5%; hollow core and louvered interior doors.	.280	13.15	13.10	26.25
7 Specialties	Average grade kitchen cabinets - 14 L.F. wall and base with plastic laminate counter top and kitchen sink; 40 gallon electric water heater.	.025	.72	.30	1.02
8 Mechanical	1 lavatory, white, wall hung; 1 water closet, white; 1 bathtub with shower, enameled steel, white; gas fired warm air heat.	.065	1.60	1.62	3.22
9 Electrical	200 Amp. service; romex wiring; incandescent lighting fixtures, switches, receptacles.	.042	.53	.94	1.47
10 Overhead	Contractor's overhead and profit and plans.		2.30	2.06	4.36
		Total	34.99	31.31	**66.30**

RESIDENTIAL	Average	Bi-Level

- **Simple design from standard plans**
- **Single family — 1 full bath, 1 kitchen**
- **No basement**
- **Asphalt shingles on roof**
- **Hot air heat**
- **Drywall interior finishes**
- **Materials and workmanship are average**

Note: The illustration shown may contain some optional components (for example: garages and/or fireplaces) whose costs are shown in the modifications, adjustments, & alternatives below or at the end of the square foot section.

Base cost per square foot of living area

Exterior Wall	Living Area 1000	1200	1400	1600	1800	2000	2200	2600	3000	3400	3800
Wood Siding - Wood Frame	85.05	76.25	73.20	71.05	68.15	65.70	64.15	60.55	56.90	55.65	54.05
Brick Veneer - Wood Frame	90.70	81.45	78.05	75.70	72.55	69.95	68.15	64.20	60.25	58.80	57.10
Stucco on Wood Frame	85.55	76.70	73.60	71.50	68.55	66.05	64.50	60.85	57.20	55.90	54.35
Solid Masonry	97.75	87.95	84.20	81.55	78.05	75.20	73.20	68.70	64.45	62.80	60.90
Finished Basement, Add	13.10	12.60	12.20	11.90	11.60	11.40	11.15	10.75	10.45	10.30	10.10
Unfinished Basement, Add	5.45	5.05	4.80	4.60	4.40	4.30	4.10	3.85	3.65	3.55	3.45

Modifications

Add to the total cost

Upgrade Kitchen Cabinets	$ + 2282
Solid Surface Countertops	+ 826
Full Bath - including plumbing, wall and floor finishes	+ 3531
Half Bath - including plumbing, wall and floor finishes	+ 2247
One Car Attached Garage	+ 8054
One Car Detached Garage	+ 10,547
Fireplace & Chimney	+ 3555

Adjustments

For multi family - add to total cost

Additional Kitchen	$ + 3707
Additional Bath	+ 3531
Additional Entry & Exit	+ 1085
Separate Heating	+ 1192
Separate Electric	+ 1342

For Townhouse/Rowhouse - Multiply cost per square foot by

Inner Unit	.91
End Unit	.96

Alternatives

Add to or deduct from the cost per square foot of living area

Cedar Shake Roof	+ 1.10
Clay Tile Roof	+ 2.30
Slate Roof	+ 3.85
Upgrade Walls to Skim Coat Plaster	+ .33
Upgrade Ceilings to Textured Finish	+ .43
Air Conditioning, in Heating Ductwork	+ 1.48
In Separate Ductwork	+ 3.77
Heating Systems, Hot Water	+ 1.60
Heat Pump	+ 1.90
Electric Heat	– 1.01
Not Heated	– 2.67

Additional upgrades or components

Kitchen Cabinets & Countertops	Page 93
Bathroom Vanities	94
Fireplaces & Chimneys	94
Windows, Skylights & Dormers	94
Appliances	95
Breezeways & Porches	95
Finished Attic	95
Garages	96
Site Improvements	96
Wings & Ells	56

Important: See the Reference Section for Location Factors (to adjust for your city) and Estimating Forms

Average Bi-Level

Living Area - 2000 S.F.
Perimeter - 135 L.F.

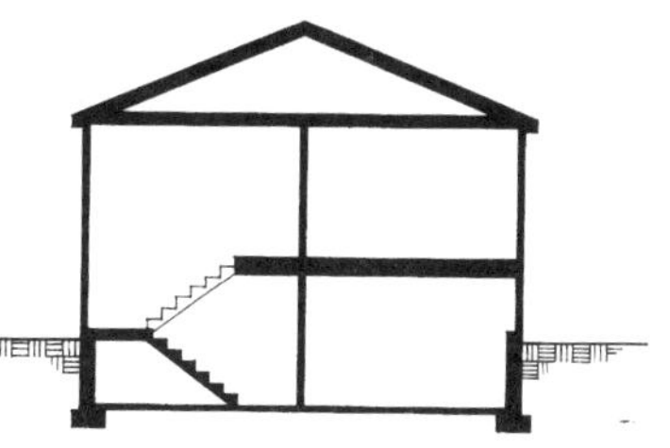

		Labor-Hours	Cost Per Square Foot Of Living Area		
			Mat.	Labor	Total
1 Site Work	Excavation for lower level, 4′ deep. Site preparation for slab.	.029		.51	.51
2 Foundation	Continuous reinforced concrete footing 8″ deep x 18″ wide, dampproofed and insulated reinforced concrete foundation wall, 8″ thick, 4′ deep; 4″ concrete slab on 4″ crushed stone base and polyethylene vapor barrier, trowel finish.	.066	1.93	2.61	4.54
3 Framing	Exterior walls - 2″ x 4″ wood studs, 16″ O.C.; 1/2″ plywood sheathing; 2″ x 6″ rafters 16″ O.C. with 1/2″ plywood sheathing, 4 in 12 pitch; 2″ x 6″ ceiling joists 16″ O.C.; 2″ x 8″ floor joists 16″ O.C. with 5/8″ plywood subfloor; 1/2″ plywood subfloor on 1″ x 2″ wood sleepers 16″ O.C.	.118	5.06	5.25	10.31
4 Exterior Walls	Horizontal beveled wood siding; #15 felt building paper; 3-1/2″ batt insulation; wood double hung windows; 3 flush solid core wood exterior doors; storms and screens.	.091	7.95	4.12	12.07
5 Roofing	25 year asphalt shingles; #15 felt building paper; aluminum gutters, downspouts, drip edge and flashings.	.024	.42	.78	1.20
6 Interiors	Walls and ceilings, 1/2″ taped and finished drywall, primed and painted with 2 coats; painted baseboard and trim, finished hardwood floor 40%, carpet with 1/2″ underlayment 40%, vinyl tile with 1/2″ underlayment 15%, ceramic tile with 1/2″ underlayment 5%; hollow core and louvered interior doors.	.217	12.94	12.79	25.73
7 Specialties	Average grade kitchen cabinets - 14 L.F. wall and base with plastic laminate counter top and kitchen sink; 40 gallon electric water heater.	.021	1.08	.45	1.53
8 Mechanical	1 lavatory, white, wall hung; 1 water closet, white; 1 bathtub with shower, enameled steel, white; gas fired warm air heat.	.061	1.97	1.82	3.79
9 Electrical	200 Amp. service; romex wiring; incandescent lighting fixtures, switches, receptacles.	.039	.65	1.06	1.71
10 Overhead	Contractor's overhead and profit and plans.		2.24	2.07	4.31
		Total	34.24	31.46	**65.70**

- **Simple design from standard plans**
- **Single family — 1 full bath, 1 kitchen**
- **No basement**
- **Asphalt shingles on roof**
- **Hot air heat**
- **Drywall interior finishes**
- **Materials and workmanship are average**

Note: The illustration shown may contain some optional components (for example: garages and/or fireplaces) whose costs are shown in the modifications, adjustments, & alternatives below or at the end of the square foot section.

Base cost per square foot of living area

Exterior Wall	Living Area 1200	1500	1800	2100	2400	2700	3000	3400	3800	4200	4600
Wood Siding - Wood Frame	79.45	73.45	68.80	65.20	62.45	60.75	59.25	57.55	54.95	52.80	51.60
Brick Veneer - Wood Frame	84.65	78.15	73.00	69.05	66.05	64.25	62.50	60.75	57.90	55.55	54.25
Stucco on Wood Frame	79.95	73.85	69.15	65.50	62.75	61.05	59.50	57.85	55.20	53.00	51.85
Solid Masonry	91.10	84.00	78.30	73.90	70.55	68.60	66.60	64.70	61.55	59.00	57.60
Finished Basement, Add*	14.95	14.30	13.70	13.20	12.90	12.70	12.45	12.30	12.00	11.75	11.65
Unfinished Basement, Add*	6.05	5.60	5.20	4.90	4.70	4.60	4.40	4.30	4.15	4.00	3.95

*Basement under middle level only.

Modifications

Add to the total cost

Upgrade Kitchen Cabinets	$ + 2282
Solid Surface Countertops	+ 826
Full Bath - including plumbing, wall and floor finishes	+ 3531
Half Bath - including plumbing, wall and floor finishes	+ 2247
One Car Attached Garage	+ 8054
One Car Detached Garage	+ 10,547
Fireplace & Chimney	+ 3965

Adjustments

For multi family - add to total cost

Additional Kitchen	$ + 3707
Additional Bath	+ 3531
Additional Entry & Exit	+ 1085
Separate Heating	+ 1192
Separate Electric	+ 1342

For Townhouse/Rowhouse - Multiply cost per square foot by

Inner Unit	.90
End Unit	.95

Alternatives

Add to or deduct from the cost per square foot of living area

Cedar Shake Roof	+ 1.5
Clay Tile Roof	+ 3.3
Slate Roof	+ 5.6
Upgrade Walls to Skim Coat Plaster	+ .2
Upgrade Ceilings to Textured Finish	+ .4
Air Conditioning, in Heating Ductwork	+ 1.2
In Separate Ductwork	+ 3.5
Heating Systems, Hot Water	+ 1.5
Heat Pump	+ 1.9
Electric Heat	– .8
Not Heated	– 2.5

Additional upgrades or components

Kitchen Cabinets & Countertops	Page 9
Bathroom Vanities	9
Fireplaces & Chimneys	9
Windows, Skylights & Dormers	9
Appliances	9
Breezeways & Porches	9
Finished Attic	9
Garages	9
Site Improvements	9
Wings & Ells	5

Important: See the Reference Section for Location Factors (to adjust for your city) and Estimating Forn

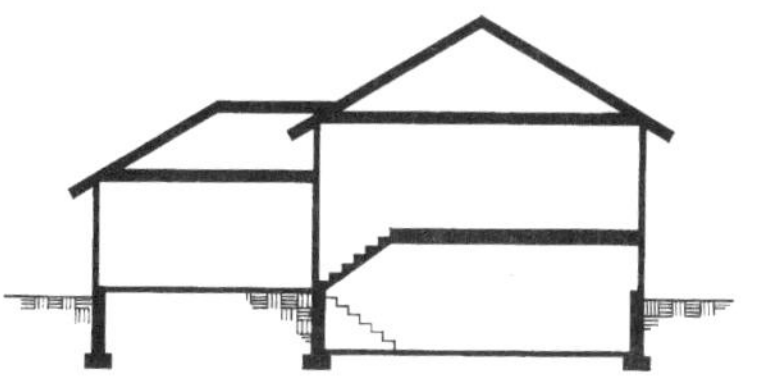

Average Tri-Level

Living Area - 2400 S.F.
Perimeter - 163 L.F.

		Labor-Hours	Cost Per Square Foot Of Living Area		
			Mat.	Labor	Total
1 Site Work	Site preparation for slab; 4′ deep trench excavation for foundation wall, excavation for lower level, 4′ deep.	.029		.43	.43
2 Foundation	Continuous reinforced concrete footing 8″ deep x 18″ wide; dampproofed and insulated reinforced concrete foundation wall, 8″ thick, 4′ deep; 4″ concrete slab on 4″ crushed stone base and polyethylene vapor barrier, trowel finish.	.080	2.17	2.84	5.01
3 Framing	Exterior walls - 2″ x 4″ wood studs, 16″ O.C.; 1/2″ plywood sheathing; 2″ x 6″ rafters 16″ O.C. with 1/2″ plywood sheathing, 4 in 12 pitch; 2″ x 6″ ceiling joists 16″ O.C.; 2″ x 8″ floor joists 16″ O.C. with 5/8″ plywood subfloor; 1/2″ plywood subfloor on 1″ x 2″ wood sleepers 16″ O.C.	.124	4.97	4.99	9.96
4 Exterior Walls	Horizontal beveled wood siding: #15 felt building paper; 3-1/2″ batt insulation; wood double hung windows; 3 flush solid core wood exterior doors; storms and screens.	.083	6.87	3.56	10.43
5 Roofing	25 year asphalt shingles; #15 felt building paper; aluminum gutters, downspouts, drip edge and flashings.	.032	.57	1.04	1.61
6 Interiors	Walls and ceilings, 1/2″ taped and finished drywall, primed and painted with 2 coats; painted baseboard and trim, finished hardwood floor 40%, carpet with 1/2″ underlayment 40%, vinyl tile with 1/2″ underlayment 15%, ceramic tile with 1/2″ underlayment 5%; hollow core and louvered interior doors.	.186	12.58	12.02	24.60
7 Specialties	Average grade kitchen cabinets - 14 L.F. wall and base with plastic laminate counter top and kitchen sink; 40 gallon electric water heater.	.012	.90	.37	1.27
8 Mechanical	1 lavatory, white, wall hung; 1 water closet, white; 1 bathtub with shower, enameled steel, white; gas fired warm air heat.	.059	1.74	1.73	3.47
9 Electrical	200 Amp. service; romex wiring; incandescent lighting fixtures, switches, receptacles.	.036	.60	1.00	1.60
10 Overhead	Contractor's overhead and profit and plans.		2.12	1.95	4.07
		Total	32.52	29.93	**62.45**

RESIDENTIAL | Solid Wall | 1 Story

- **Post and beam frame**
- **Log exterior walls**
- **Simple design from standard plans**
- **Single family — 1 full bath, 1 kitchen**
- **No basement**
- **Asphalt shingles on roof**
- **Hot air heat**
- **Drywall interior finishes**
- **Materials and workmanship are average**

Note: The illustration shown may contain some optional components (for example: garages and/or fireplaces) whose costs are shown in the modifications, adjustments, & alternatives below or at the end of the square foot section.

Base cost per square foot of living area

	Living Area										
Exterior Wall	**600**	**800**	**1000**	**1200**	**1400**	**1600**	**1800**	**2000**	**2400**	**2800**	**3200**
6" Log - Solid Wall	98.50	89.10	81.60	75.80	71.10	**67.10**	65.20	63.80	59.40	56.00	54.20
8" Log - Solid Wall	97.80	88.30	81.00	75.20	70.60	66.60	64.70	63.40	59.10	55.50	53.70
Finished Basement, Add	22.60	21.30	20.30	19.40	18.70	18.30	18.00	17.60	17.10	16.70	16.30
Unfinished Basement, Add	9.60	8.80	8.20	7.60	7.20	6.80	6.70	6.50	6.10	5.80	5.60

Modifications

Add to the total cost

Upgrade Kitchen Cabinets	$ + 2282
Solid Surface Countertops	+ 826
Full Bath - including plumbing, wall and floor finishes	+ 3531
Half Bath - including plumbing, wall and floor finishes	+ 2247
One Car Attached Garage	+ 8054
One Car Detached Garage	+ 10,547
Fireplace & Chimney	+ 3555

Adjustments

For multi family - add to total cost

Additional Kitchen	$ + 3707
Additional Bath	+ 3531
Additional Entry & Exit	+ 1085
Separate Heating	+ 1192
Separate Electric	+ 1342

For Townhouse/Rowhouse - Multiply cost per square foot by

Inner Unit	.92
End Unit	.96

Alternatives

Add to or deduct from the cost per square foot of living area

Cedar Shake Roof	+ 2.15
Air Conditioning, in Heating Ductwork	+ 2.47
In Separate Ductwork	+ 4.75
Heating Systems, Hot Water	+ 1.78
Heat Pump	+ 1.60
Electric Heat	– 1.16
Not Heated	– 2.82

Additional upgrades or components

Kitchen Cabinets & Countertops	Page 93
Bathroom Vanities	94
Fireplaces & Chimneys	94
Windows, Skylights & Dormers	94
Appliances	95
Breezeways & Porches	95
Finished Attic	95
Garages	96
Site Improvements	96
Wings & Ells	56

Important: See the Reference Section for Location Factors (to adjust for your city) and Estimating Forms

Solid Wall 1 Story

Living Area - 1600 S.F.
Perimeter - 163 L.F.

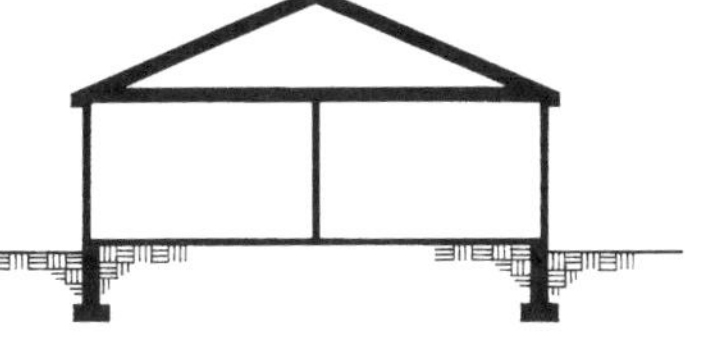

		Labor-Hours	Cost Per Square Foot Of Living Area		
			Mat.	Labor	Total
1 Site Work	Site preparation for slab; 4' deep trench excavation for foundation wall.	.048		.79	.79
2 Foundation	Continuous reinforced concrete footing 8" deep x 18" wide; dampproofed and insulated reinforced concrete foundation wall, 8" thick, 4' deep; 4" concrete slab on 4" crushed stone base and polyethylene vapor barrier, trowel finish.	.113	3.70	4.63	8.33
3 Framing / **4 Exterior Walls**	Exterior walls - Precut traditional log home. Handicrafted white cedar or pine logs. Delivery included.	.201	14.08	8.99	23.07
5 Roofing	25 year asphalt shingles; #15 felt building paper; aluminum gutters, downspouts, drip edge and flashings.	.047	.92	1.61	2.53
6 Interiors	Walls and ceilings, 1/2" taped and finished drywall, primed and painted with 2 coats; painted baseboard and trim, finished hardwood floor 40%, carpet with 1/2" underlayment 40%, vinyl tile with 1/2" underlayment 15%, ceramic tile with 1/2" underlayment 5%; hollow core and louvered interior doors.	.232	10.05	9.19	19.24
7 Specialties	Average grade kitchen cabinets - 14 L.F. wall and base with plastic laminate counter top and kitchen sink; 40 gallon electric water heater.	.009	1.68	.57	2.25
8 Mechanical	1 lavatory, white, wall hung; 1 water closet, white; 1 bathtub with shower, enameled steel, white; gas fired warm air heat.	.098	2.65	2.05	4.70
9 Electrical	200 Amp. service; romex wiring; incandescent lighting fixtures, switches, receptacles.	.041	.84	1.23	2.07
10 Overhead	Contractor's overhead and profit and plans.		2.12	2.00	4.12
		Total	36.04	31.06	**67.10**

- **Post and beam frame**
- **Log exterior walls**
- **Simple design from standard plans**
- **Single family — 1 full bath, 1 kitchen**
- **No basement**
- **Asphalt shingles on roof**
- **Hot air heat**
- **Drywall interior finishes**
- **Materials and workmanship are average**

Note: The illustration shown may contain some optional components (for example: garages and/or fireplaces) whose costs are shown in the modifications, adjustments, & alternatives below or at the end of the square foot section.

Base cost per square foot of living area

Exterior Wall	Living Area 1000	1200	1400	1600	1800	2000	2200	2600	3000	3400	3800
6" Log-Solid	98.60	81.30	77.60	74.80	71.80	68.90	66.90	63.00	58.40	56.80	55.30
8" Log-Solid	94.70	84.90	80.80	78.20	75.10	71.80	69.80	65.50	61.10	59.50	57.60
Finished Basement, Add	13.10	12.70	12.20	12.00	11.70	11.50	11.30	10.90	10.60	10.40	10.30
Unfinished Basement, Add	5.30	4.90	4.70	4.50	4.30	4.00	4.00	3.80	3.60	3.50	3.40

Modifications

Add to the total cost

Upgrade Kitchen Cabinets	$ + 2282
Solid Surface Countertops	+ 826
Full Bath - including plumbing, wall and floor finishes	+ 3531
Half Bath - including plumbing, wall and floor finishes	+ 2247
One Car Attached Garage	+ 8054
One Car Detached Garage	+ 10,547
Fireplace & Chimney	+ 3965

Adjustments

For multi family - add to total cost

Additional Kitchen	$ + 3707
Additional Bath	+ 3531
Additional Entry & Exit	+ 1085
Separate Heating	+ 1192
Separate Electric	+ 1342

For Townhouse/Rowhouse - Multiply cost per square foot by

Inner Unit	.92
End Unit	.96

Alternatives

Add to or deduct from the cost per square foot of living area

Cedar Shake Roof	+ 1.55
Air Conditioning, in Heating Ductwork	+ 1.48
In Separate Ductwork	+ 3.77
Heating Systems, Hot Water	+ 1.60
Heat Pump	+ 1.90
Electric Heat	– 1.01
Not Heated	– 2.67

Additional upgrades or components

Kitchen Cabinets & Countertops	Page 93
Bathroom Vanities	94
Fireplaces & Chimneys	94
Windows, Skylights & Dormers	94
Appliances	95
Breezeways & Porches	95
Finished Attic	95
Garages	96
Site Improvements	96
Wings & Ells	56

Important: See the Reference Section for Location Factors (to adjust for your city) and Estimating Forms

Solid Wall 2 Story

Living Area - 2000 S.F.
Perimeter - 135 L.F.

		Labor-Hours	Cost Per Square Foot Of Living Area		
			Mat.	Labor	Total
1 Site Work	Site preparation for slab; 4′ deep trench excavation for foundation wall.	.034		.59	.59
2 Foundation	Continuous reinforced concrete footing 8″ deep x 18″ wide; dampproofed and insulated reinforced concrete foundation wall, 8″ thick, 4′ deep; 4″ concrete slab on 4″ crushed stone base and polyethylene vapor barrier, trowel finish.	.066	2.22	2.81	5.03
3 Framing **4 Exterior Walls**	Exterior walls - Precut traditional log home. Handicrafted white cedar or pine logs. Delivery included.	.232	16.59	10.78	27.37
5 Roofing	25 year asphalt shingles; #15 felt building paper; aluminum gutters, downspouts, drip edge and flashings.	.024	.46	.83	1.29
6 Interiors	Walls and ceilings, 1/2″ taped and finished drywall, primed and painted with 2 coats; painted baseboard and trim, finished hardwood floor 40%, carpet with 1/2″ underlayment 40%, vinyl tile with 1/2″ underlayment 15%, ceramic tile with 1/2″ underlayment 5%; hollow core and louvered interior doors.	.225	11.25	11.17	22.42
7 Specialties	Average grade kitchen cabinets - 14 L.F. wall and base with plastic laminate counter top and kitchen sink; 40 gallon electric water heater.	.021	1.32	.47	1.79
8 Mechanical	1 lavatory, white, wall hung; 1 water closet, white; 1 bathtub with shower, enameled steel, white; gas fired warm air heat.	.060	2.32	1.90	4.22
9 Electrical	200 Amp. service; romex wiring; incandescent lighting fixtures, switches, receptacles.	.039	.76	1.11	1.87
10 Overhead	Contractor's overhead and profit and plans.		2.42	1.90	4.32
		Total	37.34	31.56	**68.90**

1 Story — Base cost per square foot of living area

Exterior Wall	Living Area 50	100	200	300	400	500	600	700
Wood Siding - Wood Frame	129.70	98.85	85.40	71.30	66.90	64.25	62.45	63.15
Brick Veneer - Wood Frame	141.40	105.50	89.95	72.45	67.30	64.25	62.15	62.65
Stucco on Wood Frame	131.10	99.80	86.15	71.80	67.30	64.65	62.80	63.45
Solid Masonry	169.20	127.05	108.85	86.95	81.00	77.40	75.95	76.05
Finished Basement, Add	35.95	28.95	26.05	21.25	20.25	19.65	19.30	19.00
Unfinished Basement, Add	17.35	13.25	11.50	8.65	8.10	7.75	7.50	7.35

1-1/2 Story — Base cost per square foot of living area

Exterior Wall	Living Area 100	200	300	400	500	600	700	800
Wood Siding - Wood Frame	107.25	85.80	73.00	65.35	61.45	59.60	57.20	56.50
Brick Veneer - Wood Frame	142.45	108.10	89.95	78.35	72.85	69.95	66.70	65.60
Stucco on Wood Frame	128.15	96.65	80.35	70.90	66.00	63.45	60.55	59.50
Solid Masonry	162.05	123.75	102.95	88.55	82.25	78.80	75.10	73.90
Finished Basement, Add	24.25	21.35	19.40	17.30	16.70	16.30	15.95	15.90
Unfinished Basement, Add	10.95	9.25	8.10	6.85	6.50	6.25	6.05	6.00

2 Story — Base cost per square foot of living area

Exterior Wall	Living Area 100	200	400	600	800	1000	1200	1400
Wood Siding - Wood Frame	107.65	79.95	67.85	55.95	52.00	49.60	48.00	48.80
Brick Veneer - Wood Frame	146.35	103.05	83.55	66.45	60.90	57.55	55.35	55.70
Stucco on Wood Frame	130.30	91.60	74.00	60.10	55.20	52.25	50.25	50.80
Solid Masonry	168.25	118.70	96.60	75.15	68.75	64.90	62.30	62.40
Finished Basement, Add	19.25	15.80	14.30	11.90	11.40	11.10	10.95	10.80
Unfinished Basement, Add	8.80	6.75	5.85	4.45	4.15	4.00	3.85	3.80

Base costs do not include bathroom or kitchen facilities. Use Modifications/Adjustments/Alternatives on pages 93-96 where appropriate.

Custom Class

1 Story

1-1/2 Story

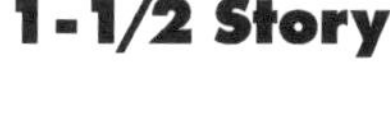

2 Story

2-1/2 Story

Bi-Level

Tri-Level

- **A distinct residence from designer's plans**
- **Single family — 1 full bath, 1 half bath, 1 kitchen**
- **No basement**
- **Asphalt shingles on roof**
- **Forced hot air heat/air conditioning**
- **Drywall interior finishes**
- **Materials and workmanship are above average**

Note: The illustration shown may contain some optional components (for example: garages and/or fireplaces) whose costs are shown in the modifications, adjustments, & alternatives below or at the end of the square foot section.

Base cost per square foot of living area

Exterior Wall	Living Area 800	1000	1200	1400	1600	1800	2000	2400	2800	3200	3600
Wood Siding - Wood Frame	122.75	111.90	103.10	96.40	91.25	88.40	85.40	**79.00**	74.65	71.50	67.85
Brick Veneer - Wood Frame	132.80	121.55	112.40	105.40	100.10	97.15	94.00	87.40	82.85	79.55	75.80
Stone Veneer - Wood Frame	137.45	125.75	116.25	108.90	103.35	100.30	96.95	90.10	85.35	81.85	77.95
Solid Masonry	137.50	125.85	116.30	108.95	103.40	100.35	96.95	90.10	85.35	81.90	77.95
Finished Basement, Add	30.70	29.20	27.90	26.85	26.10	25.75	25.15	24.35	23.70	23.20	22.75
Unfinished Basement, Add	12.95	12.15	11.45	10.90	10.55	10.35	10.05	9.65	9.30	9.00	8.80

Modifications

Add to the total cost

Upgrade Kitchen Cabinets	$ + 631
Solid Surface Countertops	+ 1180
Full Bath - including plumbing, wall and floor finishes	+ 4184
Half Bath - including plumbing, wall and floor finishes	+ 2663
Two Car Attached Garage	+ 16,313
Two Car Detached Garage	+ 18,591
Fireplace & Chimney	+ 3740

Adjustments

For multi family - add to total cost

Additional Kitchen	$ + 8100
Additional Full Bath & Half Bath	+ 6847
Additional Entry & Exit	+ 1085
Separate Heating & Air Conditioning	+ 4115
Separate Electric	+ 1342

For Townhouse/Rowhouse - Multiply cost per square foot by

Inner Unit	.90
End Unit	.95

Alternatives

Add to or deduct from the cost per square foot of living area

Cedar Shake Roof	+ 1.8
Clay Tile Roof	+ 4.3
Slate Roof	+ 7.4
Upgrade Ceilings to Textured Finish	+ .4
Air Conditioning, in Heating Ductwork	Base Syster
Heating Systems, Hot Water	+ 1.8
Heat Pump	+ 1.6
Electric Heat	– 2.1
Not Heated	– 3.2

Additional upgrades or components

Kitchen Cabinets & Countertops	Page 9
Bathroom Vanities	9
Fireplaces & Chimneys	9
Windows, Skylights & Dormers	9
Appliances	9
Breezeways & Porches	9
Finished Attic	9
Garages	9
Site Improvements	9
Wings & Ells	7

Important: See the Reference Section for Location Factors (to adjust for your city) and Estimating Form

Custom 1 Story

Living Area - 2400 S.F.
Perimeter - 207 L.F.

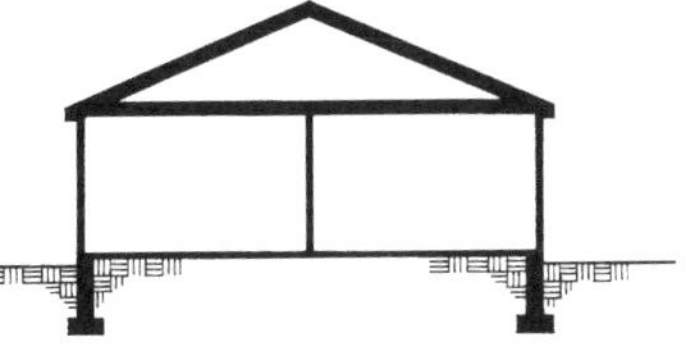

		Labor-Hours	Cost Per Square Foot Of Living Area		
			Mat.	Labor	Total
1 Site Work	Site preparation for slab; 4′ deep trench excavation for foundation wall.	.028		.51	.51
2 Foundation	Continuous reinforced concrete footing 8″ deep x 18″ wide; dampproofed and insulated reinforced concrete foundation wall, 8″ thick, 4′ deep; 4″ concrete slab on 4″ crushed stone base and polyethylene vapor barrier, trowel finish.	.113	3.67	4.29	7.96
3 Framing	Exterior walls - 2″ x 6″ wood studs, 16″ O.C.; 1/2″ plywood sheathing; 2″ x 8″ rafters 16″ O.C. with 1/2″ plywood sheathing, 4 in 12 pitch; 2″ x 6″ ceiling joists 16″ O.C.; 5/8″ plywood subfloor on 1″ x 3″ wood sleepers 16″ O.C.	.190	3.79	4.27	8.06
4 Exterior Walls	Horizontal beveled wood siding; #15 felt building paper; 6″ batt insulation; wood double hung windows; 3 solid core wood exterior doors; storms and screens.	.085	10.57	3.05	13.62
5 Roofing	30 year asphalt shingles; #15 felt building paper; aluminum gutters, downspouts and drip edge; copper flashings.	.082	2.85	2.25	5.10
6 Interiors	Walls and ceilings - 5/8″ drywall, skim coat plaster, painted with primer and 2 coats; hardwood baseboard and trim, sanded and finished; hardwood floor 70%, ceramic tile with underlayment 20%, vinyl tile with underlayment 10%; wood panel interior doors, primed and painted with 2 coats.	.292	14.13	11.30	25.43
7 Specialties	Custom grade kitchen cabinets - 20 L.F. wall and base with plastic laminate counter top and kitchen sink; 4 L.F. bathroom vanity; 75 gallon electric water heater, medicine cabinet.	.019	3.04	.56	3.60
8 Mechanical	Gas fired warm air heat/air conditioning; one full bath including: bathtub, corner shower, built in lavatory and water closet; one 1/2 bath including: built in lavatory and water closet.	.092	3.74	1.99	5.73
9 Electrical	200 Amp. service; romex wiring; fluorescent and incandescent lighting fixtures, switches, receptacles.	.039	.67	1.13	1.80
10 Overhead	Contractor's overhead and profit and design.		4.26	2.93	7.19
		Total	46.72	32.28	**79.00**

- **A distinct residence from designer's plans**
- **Single family – 1 full bath, 1 half bath, 1 kitchen**
- **No basement**
- **Asphalt shingles on roof**
- **Forced hot air heat/air conditioning**
- **Drywall interior finishes**
- **Materials and workmanship are above average**

Note: The illustration shown may contain some optional components (for example: garages and/or fireplaces) whose costs are shown in the modifications, adjustments, & alternatives below or at the end of the square foot section.

Base cost per square foot of living area

Exterior Wall	Living Area										
	1000	1200	1400	1600	1800	2000	2400	2800	3200	3600	4000
Wood Siding - Wood Frame	113.05	106.50	101.80	94.70	91.40	87.30	79.95	77.15	73.80	71.30	67.85
Brick Veneer - Wood Frame	117.20	110.40	105.50	98.05	94.60	90.30	82.65	79.65	76.10	73.55	69.95
Stone Veneer - Wood Frame	122.10	114.95	109.85	102.05	98.35	93.85	85.80	82.65	78.90	76.20	72.45
Solid Masonry	122.15	115.05	109.90	102.10	98.40	93.90	85.80	82.70	78.90	76.25	72.40
Finished Basement, Add	20.40	19.60	19.10	18.25	17.80	17.40	16.55	16.15	15.70	15.50	15.15
Unfinished Basement, Add	8.75	8.30	8.05	7.60	7.35	7.15	6.70	6.50	6.30	6.15	6.00

Modifications

Add to the total cost

Upgrade Kitchen Cabinets	$ + 631
Solid Surface Countertops	+ 1180
Full Bath - including plumbing, wall and floor finishes	+ 4184
Half Bath - including plumbing, wall and floor finishes	+ 2663
Two Car Attached Garage	+ 16,313
Two Car Detached Garage	+ 18,591
Fireplace & Chimney	+ 3740

Adjustments

For multi family - add to total cost

Additional Kitchen	$ + 8100
Additional Full Bath & Half Bath	+ 6847
Additional Entry & Exit	+ 1085
Separate Heating & Air Conditioning	+ 4115
Separate Electric	+ 1342

For Townhouse/Rowhouse - Multiply cost per square foot by

Inner Unit	.90
End Unit	.95

Alternatives

Add to or deduct from the cost per square foot of living area

Cedar Shake Roof	+ 1.3
Clay Tile Roof	+ 3.1
Slate Roof	+ 5.3
Upgrade Ceilings to Textured Finish	+ .4
Air Conditioning, in Heating Ductwork	Base System
Heating Systems, Hot Water	+ 1.7
Heat Pump	+ 1.8
Electric Heat	– 1.9
Not Heated	– 2.9

Additional upgrades or components

Kitchen Cabinets & Countertops	Page 9
Bathroom Vanities	9
Fireplaces & Chimneys	9
Windows, Skylights & Dormers	9
Appliances	9
Breezeways & Porches	9
Finished Attic	9
Garages	9
Site Improvements	9
Wings & Ells	7

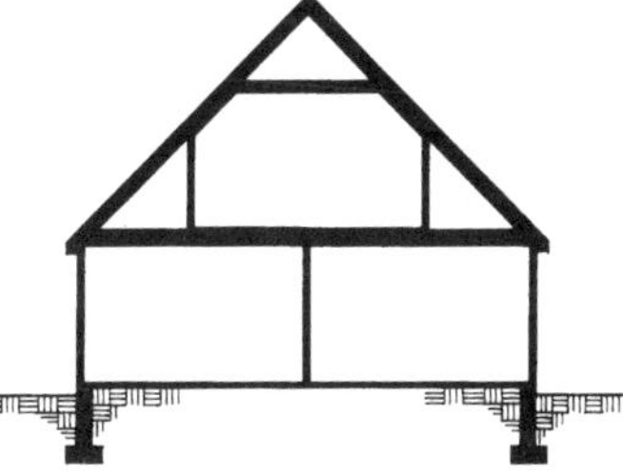

Custom 1-1/2 Story

Living Area - 2800 S.F.
Perimeter - 175 L.F.

SQUARE FOOT COSTS

		Labor-Hours	Cost Per Square Foot Of Living Area		
			Mat.	Labor	Total
1 Site Work	Site preparation for slab; 4′ deep trench excavation for foundation wall.	.028		.44	.44
2 Foundation	Continuous reinforced concrete footing 8″ deep x 18″ wide; dampproofed and insulated reinforced concrete foundation wall, 8″ thick, 4′ deep; 4″ concrete slab on 4″ crushed stone base and polyethylene vapor barrier, trowel finish.	.065	2.57	3.04	5.61
3 Framing	Exterior walls - 2″ x 6″ wood studs, 16″ O.C.; 1/2″ plywood sheathing; 2″ x 8″ rafters 16″ O.C. with 1/2″ plywood sheathing, 8 in 12 pitch; 2″ x 10″ floor joists 16″ O.C. with 5/8″ plywood subfloor; 5/8″ plywood subfloor on 1″ x 3″ wood sleepers 16″ O.C.	.192	4.15	4.38	8.53
4 Exterior Walls	Horizontal beveled wood siding; #15 felt building paper; 6″ batt insulation; wood double hung windows; 3 solid core wood exterior doors; storms and screens.	.064	11.06	3.18	14.24
5 Roofing	30 year asphalt shingles; #15 felt building paper; aluminum gutters, downspouts and drip edge; copper flashings.	.048	1.79	1.40	3.19
6 Interiors	Walls and ceilings - 5/8″ drywall, skim coat plaster, painted with primer and 2 coats; hardwood baseboard and trim, sanded and finished; hardwood floor 70%, ceramic tile with underlayment 20%, vinyl tile with underlayment 10%; wood panel interior doors, primed and painted with 2 coats.	.259	15.38	12.83	28.21
7 Specialties	Custom grade kitchen cabinets - 20 L.F. wall and base with plastic laminate counter top and kitchen sink; 4 L.F. bathroom vanity; 75 gallon electric water heater, medicine cabinet.	.030	2.61	.49	3.10
8 Mechanical	Gas fired warm air heat/air conditioning; one full bath including: bathtub, corner shower, built in lavatory and water closet; one 1/2 bath including: built in lavatory and water closet.	.084	3.25	1.84	5.09
9 Electrical	200 Amp. service; romex wiring; fluorescent and incandescent lighting fixtures, switches, receptacles.	.038	.63	1.09	1.72
10 Overhead	Contractor's overhead and profit and design.		4.15	2.87	7.02
		Total	45.59	31.56	**77.15**

- **A distinct residence from designer's plans**
- **Single family — 1 full bath, 1 half bath, 1 kitchen**
- **No basement**
- **Asphalt shingles on roof**
- **Forced hot air heat/air conditioning**
- **Drywall interior finishes**
- **Materials and workmanship are above average**

Note: The illustration shown may contain some optional components (for example: garages and/or fireplaces) whose costs are shown in the modifications, adjustments, & alternatives below or at the end of the square foot section.

Base cost per square foot of living area

Exterior Wall	Living Area 1200	1400	1600	1800	2000	2400	2800	3200	3600	4000	4400
Wood Siding - Wood Frame	108.50	102.90	99.05	94.55	90.75	84.35	79.40	75.70	73.30	71.20	69.10
Brick Veneer - Wood Frame	112.85	107.00	102.95	98.20	94.30	87.55	82.35	78.40	75.95	73.70	71.50
Stone Veneer - Wood Frame	118.10	111.90	107.65	102.65	98.55	91.40	85.80	81.65	79.05	76.65	74.35
Solid Masonry	118.15	111.95	107.70	102.65	98.60	91.40	85.85	81.70	79.15	76.65	74.35
Finished Basement, Add	16.45	15.80	15.35	14.90	14.65	13.95	13.40	13.05	12.85	12.60	12.40
Unfinished Basement, Add	7.00	6.70	6.45	6.25	6.10	5.75	5.45	5.30	5.20	5.00	4.95

Modifications

Add to the total cost

Upgrade Kitchen Cabinets	$ + 631
Solid Surface Countertops	+ 1180
Full Bath - including plumbing, wall and floor finishes	+ 4184
Half Bath - including plumbing, wall and floor finishes	+ 2663
Two Car Attached Garage	+ 16,313
Two Car Detached Garage	+ 18,591
Fireplace & Chimney	+ 4220

Adjustments

For multi family - add to total cost

Additional Kitchen	$ + 8100
Additional Full Bath & Half Bath	+ 6847
Additional Entry & Exit	+ 1085
Separate Heating & Air Conditioning	+ 4115
Separate Electric	+ 1342

For Townhouse/Rowhouse - Multiply cost per square foot by

Inner Unit	.87
End Unit	.93

Alternatives

Add to or deduct from the cost per square foot of living area

Cedar Shake Roof	+ .9
Clay Tile Roof	+ 2.1
Slate Roof	+ 3.7
Upgrade Ceilings to Textured Finish	+ .4
Air Conditioning, in Heating Ductwork	Base System
Heating Systems, Hot Water	+ 1.6
Heat Pump	+ 1.9
Electric Heat	– 1.9
Not Heated	– 2.7

Additional upgrades or components

Kitchen Cabinets & Countertops	Page 9
Bathroom Vanities	9
Fireplaces & Chimneys	9
Windows, Skylights & Dormers	9
Appliances	9
Breezeways & Porches	9
Finished Attic	9
Garages	9
Site Improvements	9
Wings & Ells	7

Important: See the Reference Section for Location Factors (to adjust for your city) and Estimating Form

Custom 2 Story

Living Area - 2800 S.F.
Perimeter - 156 L.F.

		Labor-Hours	Cost Per Square Foot Of Living Area		
			Mat.	Labor	Total
1 Site Work	Site preparation for slab; 4′ deep trench excavation for foundation wall.	.024		.44	.44
2 Foundation	Continuous reinforced concrete footing 8″ deep x 18″ wide; dampproofed and insulated reinforced concrete foundation wall, 8″ thick, 4′ deep; 4″ concrete slab on 4″ crushed stone base and polyethylene vapor barrier, trowel finish.	.058	2.20	2.60	4.80
3 Framing	Exterior walls - 2″ x 6″ wood studs, 16″ O.C.; 1/2″ plywood sheathing; 2″ x 8″ rafters 16″ O.C. with 1/2″ plywood sheathing, 6 in 12 pitch; 2″ x 8″ ceiling joists 16″ O.C.; 2″ x 10″ floor joists 16″ O.C. with 5/8″ plywood subfloor; 5/8″ plywood subfloor on 1″ x 3″ wood sleepers 16″ O.C.	.159	4.36	4.50	8.86
4 Exterior Walls	Horizontal beveled wood siding; #15 felt building paper; 6″ batt insulation; wood double hung windows; 3 solid core wood exterior doors; storms and screens.	.091	12.68	3.66	16.34
5 Roofing	30 year asphalt shingles; #15 felt building paper; aluminum gutters, downspouts and drip edge; copper flashings.	.042	1.43	1.13	2.56
6 Interiors	Walls and ceilings - 5/8″ drywall, skim coat plaster, painted with primer and 2 coats; hardwood baseboard and trim, sanded and finished; hardwood floor 70%, ceramic tile with underlayment 20%, vinyl tile with underlayment 10%; wood panel interior doors, primed and painted with 2 coats.	.271	15.79	13.36	29.15
7 Specialties	Custom grade kitchen cabinets - 20 L.F. wall and base with plastic laminate counter top and kitchen sink; 4 L.F. bathroom vanity; 75 gallon electric water heater, medicine cabinet.	.028	2.61	.49	3.10
8 Mechanical	Gas fired warm air heat/air conditioning; one full bath including: bathtub, corner shower; built in lavatory and water closet; one 1/2 bath including: built in lavatory and water closet.	.078	3.36	1.87	5.23
9 Electrical	200 Amp. service; romex wiring; fluorescent and incandescent lighting fixtures, switches, receptacles.	.038	.63	1.09	1.72
10 Overhead	Contractor's overhead and profit and design.		4.30	2.90	7.20
		Total	47.36	32.04	**79.40**

- **A distinct residence from designer's plans**
- **Single family — 1 full bath, 1 half bath, 1 kitchen**
- **No basement**
- **Asphalt shingles on roof**
- **Forced hot air heat/air conditioning**
- **Drywall interior finishes**
- **Materials and workmanship are above average**

Note: The illustration shown may contain some optional components (for example: garages and/or fireplaces) whose costs are shown in the modifications, adjustments, & alternatives below or at the end of the square foot section.

Base cost per square foot of living area

	Living Area										
Exterior Wall	1500	1800	2100	2400	2800	3200	3600	4000	4500	5000	5500
Wood Siding - Wood Frame	110.25	98.60	92.70	88.50	84.35	79.25	76.85	72.50	70.45	68.10	65.90
Brick Veneer - Wood Frame	114.80	102.80	96.50	92.05	87.75	82.35	79.80	75.20	73.05	70.55	68.25
Stone Veneer - Wood Frame	120.30	107.80	100.95	96.35	91.80	86.00	83.30	78.45	76.10	73.50	71.00
Solid Masonry	120.35	107.85	101.00	96.40	91.85	86.05	83.30	78.50	76.15	73.50	71.05
Finished Basement, Add	13.05	12.45	11.75	11.45	11.15	10.70	10.45	10.15	9.95	9.75	9.60
Unfinished Basement, Add	5.60	5.35	4.95	4.80	4.65	4.40	4.30	4.15	4.00	3.95	3.80

Modifications

Add to the total cost

Upgrade Kitchen Cabinets	$ + 631
Solid Surface Countertops	+ 1180
Full Bath - including plumbing, wall and floor finishes	+ 4184
Half Bath - including plumbing, wall and floor finishes	+ 2663
Two Car Attached Garage	+ 16,313
Two Car Detached Garage	+ 18,591
Fireplace & Chimney	+ 4220

Adjustments

For multi family - add to total cost

Additional Kitchen	$ + 8100
Additional Full Bath & Half Bath	+ 6847
Additional Entry & Exit	+ 1085
Separate Heating & Air Conditioning	+ 4115
Separate Electric	+ 1342

For Townhouse/Rowhouse - Multiply cost per square foot by

Inner Unit	.87
End Unit	.94

Alternatives

Add to or deduct from the cost per square foot of living area

Cedar Shake Roof	+ .8
Clay Tile Roof	+ 1.9
Slate Roof	+ 3.2
Upgrade Ceilings to Textured Finish	+ .4
Air Conditioning, in Heating Ductwork	Base Syster
Heating Systems, Hot Water	+ 2.0
Heat Pump	+ 1.8
Electric Heat	– 3.1
Not Heated	– 2.7

Additional upgrades or components

Kitchen Cabinets & Countertops	Page 9
Bathroom Vanities	9
Fireplaces & Chimneys	9
Windows, Skylights & Dormers	9
Appliances	9
Breezeways & Porches	9
Finished Attic	9
Garages	9
Site Improvements	9
Wings & Ells	7

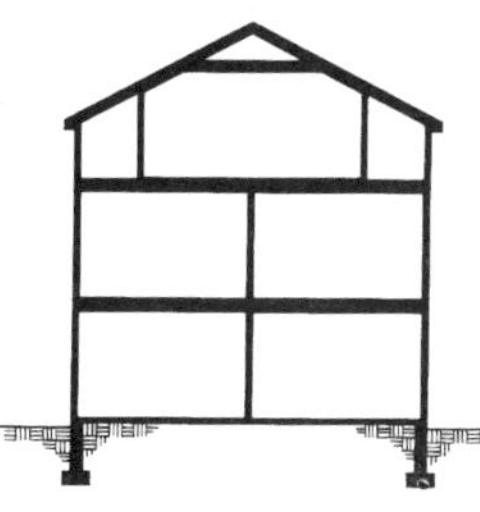

Custom 2-1/2 Story

Living Area - 3200 S.F.
Perimeter - 150 L.F.

		Labor-Hours	Cost Per Square Foot Of Living Area		
			Mat.	Labor	Total
1 Site Work	Site preparation for slab; 4′ deep trench excavation for foundation wall.	.048		.39	.39
2 Foundation	Continuous reinforced concrete footing 8″ deep x 18″ wide; dampproofed and insulated reinforced concrete foundation wall, 8″ thick, 4′ deep; 4″ concrete slab on 4″ crushed stone base and polyethylene vapor barrier, trowel finish.	.063	1.80	2.15	3.95
3 Framing	Exterior walls - 2″ x 6″ wood studs, 16″ O.C.; 1/2″ plywood sheathing; 2″ x 8″ rafters 16″ O.C. with 1/2″ plywood sheathing, 6 in 12 pitch; 2″ x 8″ ceiling joists 16″ O.C.; 2″ x 10″ floor joists 16″ O.C. with 5/8″ plywood subfloor; 5/8″ plywood subfloor on 1″ x 3″ wood sleepers 16″ O.C.	.177	4.59	4.64	9.23
4 Exterior Walls	Horizontal beveled wood siding; #15 felt building paper; 6″ batt insulation; wood double hung windows; 3 solid core wood exterior doors; storms and screens.	.134	13.25	3.84	17.09
5 Roofing	30 year asphalt shingles; #15 felt building paper; aluminum gutters, downspouts and drip edge; copper flashings.	.032	1.10	.87	1.97
6 Interiors	Walls and ceilings - 5/8″ drywall, skim coat plaster, painted with primer and 2 coats; hardwood baseboard and trim, sanded and finished; hardwood floor 70%, ceramic tile with underlayment 20%, vinyl tile with underlayment 10%; wood panel interior doors, primed and painted with 2 coats.	.354	16.19	13.95	30.14
7 Specialties	Custom grade kitchen cabinets - 20 L.F. wall and base with plastic laminate counter top and kitchen sink; 4 L.F. bathroom vanity; 75 gallon electric water heater, medicine cabinet.	.053	2.30	.43	2.73
8 Mechanical	Gas fired warm air heat/air conditioning; one full bath including: bathtub, corner shower; built in lavatory and water closet; one 1/2 bath including: built in lavatory and water closet.	.104	3.08	1.81	4.89
9 Electrical	200 Amp. service; romex wiring; fluorescent and incandescent lighting fixtures, switches, receptacles.	.048	.60	1.06	1.66
10 Overhead	Contractor's overhead and profit and design.		4.29	2.91	7.20
		Total	47.20	32.05	**79.25**

RESIDENTIAL	Custom	3 Story

- **A distinct residence from designer's plans**
- **Single family — 1 full bath, 1 half bath, 1 kitchen**
- **No basement**
- **Asphalt shingles on roof**
- **Forced hot air heat/air conditioning**
- **Drywall interior finishes**
- **Materials and workmanship are above average**

Note: The illustration shown may contain some optional components (for example: garages and/or fireplaces) whose costs are shown in the modifications, adjustments, & alternatives below or at the end of the square foot section.

Base cost per square foot of living area

Exterior Wall	Living Area 1500	1800	2100	2500	3000	3500	4000	4500	5000	5500	6000
Wood Siding - Wood Frame	110.35	98.80	94.20	90.65	**83.65**	80.75	76.10	71.50	70.10	68.10	66.40
Brick Veneer - Wood Frame	115.10	103.20	98.35	94.60	87.20	84.15	79.15	74.35	72.80	70.75	68.90
Stone Veneer - Wood Frame	120.80	108.40	103.20	99.25	91.45	88.15	82.80	77.70	76.10	73.85	71.85
Solid Masonry	120.85	108.45	103.30	99.30	91.50	88.20	82.85	77.75	76.10	73.90	71.85
Finished Basement, Add	11.45	10.95	10.50	10.20	9.70	9.45	9.10	8.80	8.65	8.55	8.40
Unfinished Basement, Add	4.95	4.65	4.45	4.30	4.05	3.90	3.70	3.55	3.50	3.40	3.35

Modifications

Add to the total cost

Upgrade Kitchen Cabinets	$ + 631
Solid Surface Countertops	+ 1180
Full Bath - including plumbing, wall and floor finishes	+ 4184
Half Bath - including plumbing, wall and floor finishes	+ 2663
Two Car Attached Garage	+ 16,313
Two Car Detached Garage	+ 18,591
Fireplace & Chimney	+ 4770

Adjustments

For multi family - add to total cost

Additional Kitchen	$ + 8100
Additional Full Bath & Half Bath	+ 6847
Additional Entry & Exit	+ 1085
Separate Heating & Air Conditioning	+ 4115
Separate Electric	+ 1342

For Townhouse/Rowhouse - Multiply cost per square foot by

Inner Unit	.85
End Unit	.93

Alternatives

Add to or deduct from the cost per square foot of living area

Cedar Shake Roof	+ .6
Clay Tile Roof	+ 1.4
Slate Roof	+ 2.5
Upgrade Ceilings to Textured Finish	+ .4
Air Conditioning, in Heating Ductwork	Base Syster
Heating Systems, Hot Water	+ 2.0
Heat Pump	+ 1.8
Electric Heat	– 3.1
Not Heated	– 2.7

Additional upgrades or components

Kitchen Cabinets & Countertops	Page 9
Bathroom Vanities	9
Fireplaces & Chimneys	9
Windows, Skylights & Dormers	9
Appliances	9
Breezeways & Porches	9
Finished Attic	9
Garages	9
Site Improvements	9
Wings & Ells	7

Custom 3 Story

Living Area - 3000 S.F.
Perimeter - 135 L.F.

		Labor-Hours	Cost Per Square Foot Of Living Area Mat.	Labor	Total
1 Site Work	Site preparation for slab; 4′ deep trench excavation for foundation wall.	.048		.41	.41
2 Foundation	Continuous reinforced concrete footing 8″ deep x 18″ wide; dampproofed and insulated reinforced concrete foundation wall, 8″ thick, 4′ deep; 4″ concrete slab on 4″ crushed stone base and polyethylene vapor barrier, trowel finish.	.060	1.68	2.03	3.71
3 Framing	Exterior walls - 2″ x 6″ wood studs, 16″ O.C.; 1/2″ plywood sheathing; 2″ x 8″ rafters 16″ O.C. with 1/2″ plywood sheathing, 6 in 12 pitch; 2″ x 8″ ceiling joists 16″ O.C.; 2″ x 10″ floor joists 16″ O.C. with 5/8″ plywood subfloor; 5/8″ plywood subfloor on 1″ x 3″ wood sleepers 16″ O.C.	.191	4.78	4.79	9.57
4 Exterior Walls	Horizontal beveled wood siding; #15 felt building paper; 6″ batt insulation; wood double hung windows; 3 solid core wood exterior doors; storms and screens.	.150	15.18	4.36	19.54
5 Roofing	30 year asphalt shingles; #15 felt building paper; aluminum gutters, downspouts and drip edge; copper flashings.	.028	.95	.75	1.70
6 Interiors	Walls and ceilings - 5/8″ drywall, skim coat plaster, painted with primer and 2 coats; hardwood baseboard and trim, sanded and finished; hardwood floor 70%, ceramic tile with underlayment 20%, vinyl tile with underlayment 10%; wood panel interior doors, primed and painted with 2 coats.	.409	16.81	14.68	31.49
7 Specialties	Custon grade kitchen cabinets - 20 L.F. wall and base with plastic laminate counter top and kitchen sink; 4 L.F. bathroom vanity; 75 gallon electric water heater, medicine cabinet.	.053	2.44	.45	2.89
8 Mechanical	Gas fired warm air heat/air conditioning; one full bath including: bathtub, corner shower; built in lavatory and water closet; one 1/2 bath including: built in lavatory and water closet.	.105	3.21	1.84	5.05
9 Electrical	200 Amp. service; romex wiring; fluorescent and incandescent lighting fixtures, switches, receptacles.	.048	.61	1.07	1.68
10 Overhead	Contractor's overhead and profit and design.		4.56	3.05	7.61
		Total	50.22	33.43	**83.65**

RESIDENTIAL | Custom | Bi-Level

- A distinct residence from designer's plans
- Single family – 1 full bath, 1 half bath, 1 kitchen
- No basement
- Asphalt shingles on roof
- Forced hot air heat/air conditioning
- Drywall interior finishes
- Materials and workmanship are above average

Note: The illustration shown may contain some optional components (for example: garages and/or fireplaces) whose costs are shown in the modifications, adjustments, & alternatives below or at the end of the square foot section.

Base cost per square foot of living area

Exterior Wall	Living Area 1200	1400	1600	1800	2000	2400	2800	3200	3600	4000	4400
Wood Siding - Wood Frame	100.80	95.70	92.15	88.10	84.50	78.70	74.30	70.90	68.70	66.85	64.85
Brick Veneer - Wood Frame	104.10	98.80	95.10	90.85	87.20	81.10	76.50	72.95	70.70	68.70	66.65
Stone Veneer - Wood Frame	108.00	102.45	98.60	94.15	90.35	84.00	79.10	75.40	73.00	70.95	68.80
Solid Masonry	108.05	102.50	98.60	94.20	90.40	84.00	79.10	75.40	73.05	70.95	68.80
Finished Basement, Add	16.45	15.80	15.35	14.90	14.65	13.95	13.40	13.05	12.85	12.60	12.40
Unfinished Basement, Add	7.00	6.70	6.45	6.25	6.10	5.75	5.45	5.30	5.20	5.00	4.95

Modifications

Add to the total cost

Upgrade Kitchen Cabinets	$ + 631
Solid Surface Countertops	+ 1180
Full Bath - including plumbing, wall and floor finishes	+ 4184
Half Bath - including plumbing, wall and floor finishes	+ 2663
Two Car Attached Garage	+ 16,313
Two Car Detached Garage	+ 18,591
Fireplace & Chimney	+ 3740

Adjustments

For multi family - add to total cost

Additional Kitchen	$ + 8100
Additional Full Bath & Half Bath	+ 6847
Additional Entry & Exit	+ 1085
Separate Heating & Air Conditioning	+ 4115
Separate Electric	+ 1342

For Townhouse/Rowhouse - Multiply cost per square foot by

Inner Unit	.89
End Unit	.95

Alternatives

Add to or deduct from the cost per square foot of living area

Cedar Shake Roof	+ .95
Clay Tile Roof	+ 2.15
Slate Roof	+ 3.70
Upgrade Ceilings to Textured Finish	+ .44
Air Conditioning, in Heating Ductwork	Base System
Heating Systems, Hot Water	+ 1.62
Heat Pump	+ 1.93
Electric Heat	– 1.95
Not Heated	– 2.79

Additional upgrades or components

Kitchen Cabinets & Countertops	Page 93
Bathroom Vanities	94
Fireplaces & Chimneys	94
Windows, Skylights & Dormers	94
Appliances	95
Breezeways & Porches	95
Finished Attic	95
Garages	96
Site Improvements	96
Wings & Ells	74

Important: See the Reference Section for Location Factors (to adjust for your city) and Estimating Forms

Custom Bi-Level

Living Area - 2800 S.F.
Perimeter - 156 L.F.

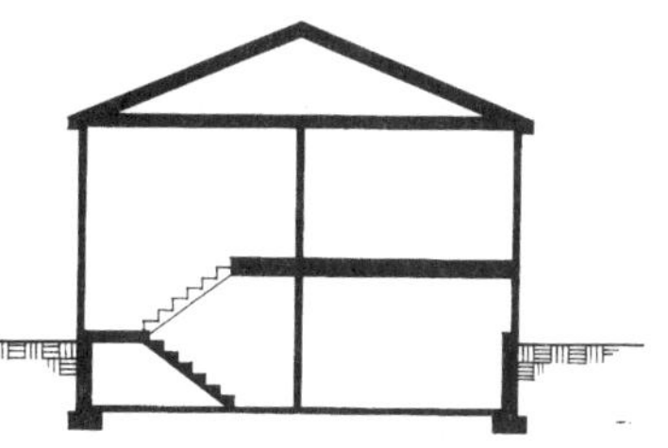

		Labor-Hours	Cost Per Square Foot Of Living Area Mat.	Labor	Total
1 Site Work	Excavation for lower level, 4′ deep. Site preparation for slab.	.024		.44	.44
2 Foundation	Continuous reinforced concrete footing 8″ deep x 18″ wide; dampproofed and insulated reinforced concrete foundation wall, 8″ thick, 4′ deep; 4″ concrete slab on 4″ crushed stone base and polyethylene vapor barrier, trowel finish.	.058	2.20	2.60	4.80
3 Framing	Exterior walls - 2″ x 6″ wood studs, 16″ O.C.; 1/2″ plywood sheathing; 2″ x 8″ rafters 16″ O.C. with 1/2″ plywood sheathing, 6 in 12 pitch; 2″ x 8″ ceiling joists 16″ O.C.; 2″ x 10″ floor joists 16″ O.C. with 5/8″ plywood subfloor; 5/8″ plywood subfloor on 1″ x 3″ wood sleepers 16″ O.C.	.147	4.09	4.29	8.38
4 Exterior Walls	Horizontal beveled wood siding; #15 felt building paper; 6″ batt insulation; wood double hung windows; 3 solid core wood exterior doors; storms and screens.	.079	9.74	2.82	12.56
5 Roofing	30 year asphalt shingles; #15 felt building paper; aluminum gutters, downspouts and drip edge; copper flashings.	.033	1.43	1.13	2.56
6 Interiors	Walls and ceilings - 5/8″ drywall, skim coat plaster, painted with primer and 2 coats; hardwood baseboard and trim, sanded and finished; hardwood floor 70%, ceramic tile with underlayment 20%, vinyl tile with underlayment 10%; wood panel interior doors, primed and painted with 2 coats.	.257	15.64	13.12	28.76
7 Specialties	Custom grade kitchen cabinets - 20 L.F. wall and base with plastic laminate counter top and kitchen sink; 4 L.F. bathroom vanity; 75 gallon electric water heater, medicine cabinet.	.028	2.61	.49	3.10
8 Mechanical	Gas fired warm air heat/air conditioning; one full bath including: bathtub, corner shower, built in lavatory and water closet; one 1/2 bath including: built in lavatory and water closet.	.078	3.36	1.87	5.23
9 Electrical	200 Amp. service; romex wiring; fluorescent and incandescent lighting fixtures, switches, receptacles.	.038	.63	1.09	1.72
10 Overhead	Contractor's overhead and profit and design.		3.97	2.78	6.75
		Total	43.67	30.63	**74.30**

- **A distinct residence from designer's plans**
- **Single family — 1 full bath, 1 half bath, 1 kitchen**
- **No basement**
- **Asphalt shingles on roof**
- **Forced hot air heat/air conditioning**
- **Drywall interior finishes**
- **Materials and workmanship are above average**

Note: The illustration shown may contain some optional components (for example: garages and/or fireplaces) whose costs are shown in the modifications, adjustments, & alternatives below or at the end of the square foot section.

Base cost per square foot of living area

Exterior Wall	Living Area 1200	1500	1800	2100	2400	2800	3200	3600	4000	4500	5000
Wood Siding - Wood Frame	103.30	94.35	87.40	82.15	78.10	75.20	71.75	68.05	66.70	62.90	60.95
Brick Veneer - Wood Frame	106.55	97.35	90.05	84.60	80.40	77.40	73.80	69.95	68.55	64.60	62.55
Stone Veneer - Wood Frame	110.45	100.85	93.25	87.45	83.10	80.00	76.20	72.20	70.70	66.60	64.40
Solid Masonry	110.50	100.90	93.30	87.50	83.15	80.00	76.20	72.25	70.75	66.65	64.45
Finished Basement, Add*	20.50	19.45	18.60	17.85	17.40	17.05	16.65	16.25	16.05	15.65	15.40
Unfinished Basement, Add*	8.65	8.10	7.65	7.30	7.05	6.90	6.65	6.45	6.35	6.15	6.00

*Basement under middle level only.

Modifications

Add to the total cost

Upgrade Kitchen Cabinets	$ + 631
Solid Surface Countertops	+ 1180
Full Bath - including plumbing, wall and floor finishes	+ 4184
Half Bath - including plumbing, wall and floor finishes	+ 2663
Two Car Attached Garage	+ 16,313
Two Car Detached Garage	+ 18,591
Fireplace & Chimney	+ 4220

Adjustments

For multi family - add to total cost

Additional Kitchen	$ + 8100
Additional Full Bath & Half Bath	+ 6847
Additional Entry & Exit	+ 1085
Separate Heating & Air Conditioning	+ 4115
Separate Electric	+ 1342

For Townhouse/Rowhouse - Multiply cost per square foot by

Inner Unit	.87
End Unit	.94

Alternatives

Add to or deduct from the cost per square foot of living area

Cedar Shake Roof	+ 1.3
Clay Tile Roof	+ 3.1
Slate Roof	+ 5.3
Upgrade Ceilings to Textured Finish	+ .4
Air Conditioning, in Heating Ductwork	Base System
Heating Systems, Hot Water	+ 1.5
Heat Pump	+ 2.0
Electric Heat	– 1.7
Not Heated	– 2.7

Additional upgrades or components

Kitchen Cabinets & Countertops	Page 9
Bathroom Vanities	9
Fireplaces & Chimneys	9
Windows, Skylights & Dormers	9
Appliances	9
Breezeways & Porches	9
Finished Attic	9
Garages	9
Site Improvements	9
Wings & Ells	7

Important: See the Reference Section for Location Factors (to adjust for your city) and Estimating Form

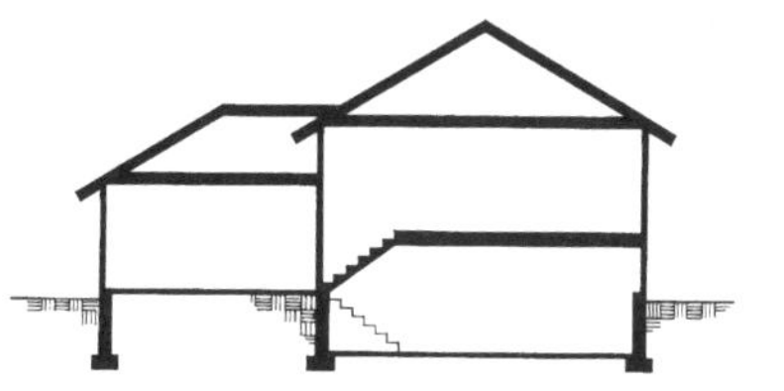

Custom Tri-Level

Living Area - 3200 S.F.
Perimeter - 198 L.F.

		Labor-Hours	Cost Per Square Foot Of Living Area		
			Mat.	Labor	Total
1 Site Work	Site preparation for slab; 4′ deep trench excavation for foundation wall, excavation for lower level, 4′ deep.	.023		.39	.39
2 Foundation	Continuous reinforced concrete footing 8″ deep x 18″ wide; dampproofed and insulated reinforced concrete foundation wall, 8″ thick, 4′ deep; 4″ concrete slab on 4″ crushed stone base and polyethylene vapor barrier, trowel finish.	.073	2.57	3.01	5.58
3 Framing	Exterior walls - 2″ x 6″ wood studs, 16″ O.C.; 1/2″ plywood sheathing; 2″ x 8″ rafters 16″ O.C. with 1/2″ plywood sheathing, 6 in 12 pitch; 2″ x 8″ ceiling joists 16″ O.C.; 2″ x 10″ floor joists 16″ O.C. with 5/8″ plywood subfloor; 5/8″ plywood subfloor on 1″ x 3″ wood sleepers 16″ O.C.	.162	3.95	4.24	8.19
4 Exterior Walls	Horizontal beveled wood siding; #15 felt building paper; 6″ batt insulation; wood double hung windows; 3 solid core wood exterior doors; storms and screens.	.076	9.24	2.70	11.94
5 Roofing	30 year asphalt shingles; #15 felt building paper; aluminum gutters, downspouts and drip edge; copper flashings.	.045	1.90	1.49	3.39
6 Interiors	Walls and ceilings - 5/8″ drywall, skim coat plaster, painted with primer and 2 coats; hardwood baseboard and trim, sanded and finished; hardwood floor 70%, ceramic tile with underlayment 20%, vinyl tile with underlayment 10%; wood panel interior doors, primed and painted with 2 coats.	.242	14.56	11.89	26.45
7 Specialties	Custom grade kitchen cabinets - 20 L.F. wall and base with plastic laminate counter top and kitchen sink; 4 L.F. bathroom vanity; 75 gallon electric water heater, medicine cabinet.	.026	2.30	.43	2.73
8 Mechanical	Gas fired warm air heat/air conditioning; one full bath including: bathtub, corner shower, built in lavatory and water closet; one 1/2 bath including: built in lavatory and water closet.	.073	3.08	1.81	4.89
9 Electrical	200 Amp. service; romex wiring; fluorescent and incandescent lighting fixtures, switches, receptacles.	.036	.60	1.06	1.66
10 Overhead	Contractor's overhead and profit and design.		3.83	2.70	6.53
		Total	42.03	29.72	**71.75**

to adjust fc

1 Story — Base cost per square foot of living area

Exterior Wall	Living Area 50	100	200	300	400	500	600	700
Wood Siding - Wood Frame	160.20	120.95	103.85	85.20	79.60	76.20	74.00	74.70
Brick Veneer - Wood Frame	171.30	128.85	110.45	89.60	83.55	79.90	77.55	78.10
Stone Veneer - Wood Frame	184.50	138.30	118.30	94.85	88.30	84.30	81.70	82.15
Solid Masonry	184.65	138.40	118.40	94.90	88.30	84.35	81.80	82.20
Finished Basement, Add	50.05	40.60	36.65	30.05	28.75	27.95	27.40	27.05
Unfinished Basement, Add	38.80	26.05	20.10	15.30	13.95	13.15	12.60	12.20

1-1/2 Story — Base cost per square foot of living area

Exterior Wall	Living Area 100	200	300	400	500	600	700	800
Wood Siding - Wood Frame	131.70	105.75	89.85	79.80	74.95	72.65	69.70	68.85
Brick Veneer - Wood Frame	141.60	113.65	96.45	84.90	79.75	77.15	73.90	73.05
Stone Veneer - Wood Frame	153.40	123.10	104.30	91.05	85.40	82.50	79.00	78.05
Solid Masonry	153.50	123.20	104.40	91.10	85.45	82.55	79.00	78.15
Finished Basement, Add	33.25	29.30	26.65	23.75	23.00	22.45	22.00	21.90
Unfinished Basement, Add	22.80	16.85	14.15	11.95	11.15	10.60	10.15	10.00

2 Story — Base cost per square foot of living area

Exterior Wall	Living Area 100	200	400	600	800	1000	1200	1400
Wood Siding - Wood Frame	134.35	99.30	84.05	68.35	63.35	60.30	58.35	59.20
Brick Veneer - Wood Frame	145.45	107.25	90.65	72.75	67.30	64.00	61.85	62.60
Stone Veneer - Wood Frame	158.60	116.65	98.50	78.00	72.05	68.40	66.05	66.65
Solid Masonry	158.75	116.80	98.60	78.05	72.05	68.45	66.10	66.70
Finished Basement, Add	25.05	20.30	18.35	15.05	14.40	14.00	13.75	13.55
Unfinished Basement, Add	19.40	13.05	10.05	7.65	7.00	6.60	6.30	6.10

Base costs do not include bathroom or kitchen facilities. Use Modifications/Adjustments/Alternatives on pages 93-96 where appropriate.

Luxury Class

1 Story

1-1/2 Story

2 Story

2-1/2 Story

Bi-Level

Tri-Level

- **Unique residence built from an architect's plan**
- **Single family — 1 full bath, 1 half bath, 1 kitchen**
- **No basement**
- **Cedar shakes on roof**
- **Forced hot air heat/air conditioning**
- **Double drywall interior**
- **Many special features**
- **Extraordinary materials and workmanship**

Note: The illustration shown may contain some optional components (for example: garages and/or fireplaces) whose costs are shown in the modifications, adjustments, & alternatives below or at the end of the square foot section.

Base cost per square foot of living area

Exterior Wall	Living Area 1000	1200	1400	1600	1800	2000	2400	2800	3200	3600	4000
Wood Siding - Wood Frame	138.70	128.40	120.65	114.70	111.35	107.80	100.50	**95.45**	91.85	87.70	84.30
Brick Veneer - Wood Frame	142.30	131.65	123.60	117.50	114.05	110.35	102.80	97.55	93.80	89.50	86.00
Solid Brick	151.05	139.55	130.80	124.20	120.55	116.45	108.30	102.70	98.55	93.95	90.20
Solid Stone	151.50	140.00	131.20	124.60	120.85	116.75	108.65	103.00	98.80	94.20	90.45
Finished Basement, Add	36.20	34.55	33.25	32.30	31.80	31.10	30.10	29.30	28.65	28.05	27.55
Unfinished Basement, Add	13.80	13.05	12.40	12.00	11.75	11.45	10.95	10.60	10.30	10.05	9.80

Modifications

Add to the total cost

Upgrade Kitchen Cabinets	$ + 832
Solid Surface Countertops	+ 1326
Full Bath - including plumbing, wall and floor finishes	+ 4854
Half Bath - including plumbing, wall and floor finishes	+ 3089
Two Car Attached Garage	+ 19,703
Two Car Detached Garage	+ 22,432
Fireplace & Chimney	+ 5285

Adjustments

For multi family - add to total cost

Additional Kitchen	$ + 10,340
Additional Full Bath & Half Bath	+ 7953
Additional Entry & Exit	+ 1629
Separate Heating & Air Conditioning	+ 4115
Separate Electric	+ 1342

For Townhouse/Rowhouse - Multiply cost per square foot by

Inner Unit	.90
End Unit	.95

Alternatives

Add to or deduct from the cost per square foot of living area

Heavyweight Asphalt Shingles	– 1.8
Clay Tile Roof	+ 2.5
Slate Roof	+ 5.7
Upgrade Ceilings to Textured Finish	+ .4
Air Conditioning, in Heating Ductwork	Base Syste
Heating Systems, Hot Water	+ 2.9
Heat Pump	+ 1.7
Electric Heat	– 1.9
Not Heated	– 3.3

Additional upgrades or components

Kitchen Cabinets & Countertops	Page 9
Bathroom Vanities	9
Fireplaces & Chimneys	9
Windows, Skylights & Dormers	9
Appliances	9
Breezeways & Porches	9
Finished Attic	9
Garages	9
Site Improvements	9
Wings & Ells	9

Luxury 1 Story

Living Area - 2800 S.F.
Perimeter - 219 L.F.

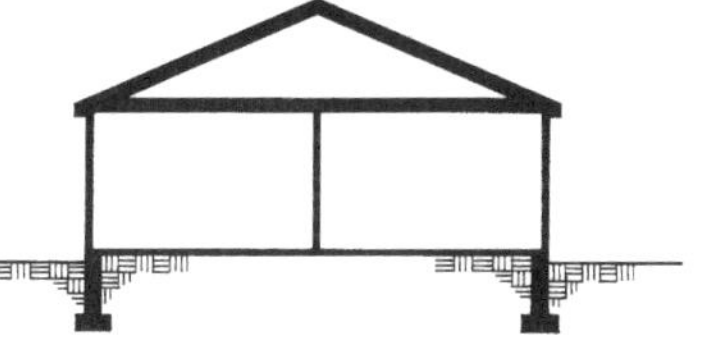

		Labor-Hours	Cost Per Square Foot Of Living Area		
			Mat.	Labor	Total
1 Site Work	Site preparation for slab; 4′ deep trench excavation for foundation wall.	.028		.47	.47
2 Foundation	Continuous reinforced concrete footing 8″ deep x 18″ wide; dampproofed and insulated reinforced concrete foundation wall, 12″ thick, 4′ deep; 4″ concrete slab on 4″ crushed stone base and polyethylene vapor barrier, trowel finish.	.098	4.32	4.45	8.77
3 Framing	Exterior walls - 2″ x 6″ wood studs, 16″ O.C.; 5/8″ plywood sheathing; 2″ x 10″ rafters 16″ O.C. with 5/8″ plywood sheathing, 6 in 12 pitch; 2″ x 8″ ceiling joists 16″ O.C.; 5/8″ plywood subfloor on 1″ x 3″ wood sleepers 16″ O.C.	.260	9.99	8.74	18.73
4 Exterior Walls	Horizontal beveled wood siding; #15 felt building paper; 6″ batt insulation; wood double hung windows; 3 solid core wood exterior doors; storms and screens.	.204	9.33	3.23	12.56
5 Roofing	Red cedar shingles; #15 felt building paper; aluminum gutters, downspouts and drip edge; copper flashings.	.082	3.33	2.62	5.95
6 Interiors	Walls and ceilings - 5/8″ drywall, skim coat plaster, painted with primer and 2 coats; hardwood baseboard and trim, sanded and finished; hardwood floor 70%, ceramic tile with underlayment 20%, vinyl tile with underlayment 10%; wood panel interior doors, primed and painted with 2 coats.	.287	12.47	12.16	24.63
7 Specialties	Luxury grade kitchen cabinets - 25 L.F. wall and base with plastic laminate counter top and kitchen sink; 6 L.F. bathroom vanity; 75 gallon electric water heater; medicine cabinet.	.052	3.39	.60	3.99
8 Mechanical	Gas fired warm air heat/air conditioning; one full bath including: bathtub, corner shower; built in lavatory and water closet; one 1/2 bath including: built in lavatory and water closet.	.078	3.76	2.07	5.83
9 Electrical	200 Amp. service; romex wiring; fluorescent and incandescent lighting fixtures; intercom, switches, receptacles.	.044	.73	1.30	2.03
10 Overhead	Contractor's overhead and profit and architect's fees.		7.11	5.38	12.49
		Total	54.43	41.02	**95.45**

- **Unique residence built from an architect's plan**
- **Single family — 1 full bath, 1 half bath, 1 kitchen**
- **No basement**
- **Cedar shakes on roof**
- **Forced hot air heat/air conditioning**
- **Double drywall interior**
- **Many special features**
- **Extraordinary materials and workmanship**

Note: The illustration shown may contain some optional components (for example: garages and/or fireplaces) whose costs are shown in the modifications, adjustments, & alternatives below or at the end of the square foot section.

Base cost per square foot of living area

Exterior Wall	Living Area 1000	1200	1400	1600	1800	2000	2400	2800	3200	3600	4000
Wood Siding - Wood Frame	130.95	123.10	117.50	109.35	105.45	100.65	92.25	**88.90**	85.10	82.20	78.25
Brick Veneer - Wood Frame	135.10	127.00	121.25	112.75	108.65	103.70	94.95	91.50	87.45	84.45	80.40
Solid Brick	145.25	136.55	130.25	120.95	116.40	111.10	101.50	97.65	93.15	90.00	85.50
Solid Stone	145.80	137.00	130.80	121.40	116.85	111.50	101.85	98.00	93.45	90.25	85.80
Finished Basement, Add	25.30	24.35	23.65	22.65	22.05	21.55	20.45	20.00	19.40	19.15	18.70
Unfinished Basement, Add	9.85	9.40	9.10	8.60	8.35	8.15	7.65	7.40	7.15	7.00	6.80

Modifications

Add to the total cost

Upgrade Kitchen Cabinets	$ + 832
Solid Surface Countertops	+ 1326
Full Bath - including plumbing, wall and floor finishes	+ 4854
Half Bath - including plumbing, wall and floor finishes	+ 3089
Two Car Attached Garage	+ 19,703
Two Car Detached Garage	+ 22,432
Fireplace & Chimney	+ 5285

Adjustments

For multi family - add to total cost

Additional Kitchen	$ + 10,340
Additional Full Bath & Half Bath	+ 7953
Additional Entry & Exit	+ 1629
Separate Heating & Air Conditioning	+ 4115
Separate Electric	+ 1342

For Townhouse/Rowhouse - Multiply cost per square foot by

Inner Unit	.90
End Unit	.95

Alternatives

Add to or deduct from the cost per square foot of living area

Heavyweight Asphalt Shingles	– 1.35
Clay Tile Roof	+ 1.80
Slate Roof	+ 4.15
Upgrade Ceilings to Textured Finish	+ .44
Air Conditioning, in Heating Ductwork	Base System
Heating Systems, Hot Water	+ 2.57
Heat Pump	+ 1.95
Electric Heat	– 1.95
Not Heated	– 3.02

Additional upgrades or components

Kitchen Cabinets & Countertops	Page 93
Bathroom Vanities	94
Fireplaces & Chimneys	94
Windows, Skylights & Dormers	94
Appliances	95
Breezeways & Porches	95
Finished Attic	95
Garages	96
Site Improvements	96
Wings & Ells	97

Luxury 1-1/2 Story

Living Area - 2800 S.F.
Perimeter - 175 L.F.

		Labor-Hours	Cost Per Square Foot Of Living Area		
			Mat.	Labor	Total
1 Site Work	Site preparation for slab; 4' deep trench excavation for foundation wall.	.025		.47	.47
2 Foundation	Continuous reinforced concrete footing 8" deep x 18" wide; dampproofed and insulated reinforced concrete foundation wall, 12" thick, 4' deep; 4" concrete slab on 4" crushed stone base and polyethylene vapor barrier, trowel finish.	.066	3.12	3.35	6.47
3 Framing	Exterior walls - 2" x 6" wood studs, 16" O.C.; 5/8" plywood sheathing; 2" x 10" rafters 16" O.C. with 5/8" plywood sheathing, 8 in 12 pitch; 2" x 8" ceiling joists 16" O.C.; 2" x 12" floor joists 16" O.C. with 5/8" plywood subfloor; 5/8" plywood subfloor on 1" x 3" wood sleepers 16" O.C.	.189	6.19	5.63	11.82
4 Exterior Walls	Horizontal beveled wood siding; #15 felt building paper; 6" batt insulation; wood double hung windows; 3 solid core wood exterior doors; storms and screens.	.174	10.63	3.69	14.32
5 Roofing	Red cedar shingles; #15 felt building paper; aluminum gutters, downspouts and drip edge; copper flashings.	.065	2.08	1.64	3.72
6 Interiors	Walls and ceilings - 5/8" drywall, skim coat plaster, painted with primer and 2 coats; hardwood baseboard and trim, sanded and finished; hardwood floor 70%, ceramic tile with underlayment 20%, vinyl tile with underlayment 10%; wood panel interior doors, primed and painted with 2 coats.	.260	14.27	14.40	28.67
7 Specialties	Luxury grade kitchen cabinets - 25 L.F. wall and base with plastic laminate counter top and kitchen sink; 6 L.F. bathroom vanity; 75 gallon electric water heater; medicine cabinet.	.062	3.39	.60	3.99
8 Mechanical	Gas fired warm air heat/air conditioning; one full bath including: bathtub, corner shower; built in lavatory and water closet; one 1/2 bath including: built in lavatory and water closet.	.080	3.76	2.07	5.83
9 Electrical	200 Amp. service; romex wiring; fluorescent and incandescent lighting fixtures; intercom, switches, receptacles.	.044	.73	1.30	2.03
10 Overhead	Contractor's overhead and profit and architect's fees.		6.63	4.95	11.58
		Total	50.80	38.10	**88.90**

RESIDENTIAL | Luxury | 2 Story

- **Unique residence built from an architect's plan**
- **Single family — 1 full bath, 1 half bath, 1 kitchen**
- **No basement**
- **Cedar shakes on roof**
- **Forced hot air heat/air conditioning**
- **Double drywall interior**
- **Many special features**
- **Extraordinary materials and workmanship**

Note: The illustration shown may contain some optional components (for example: garages and/or fireplaces) whose costs are shown in the modifications, adjustments, & alternatives below or at the end of the square foot section.

Base cost per square foot of living area

Exterior Wall	Living Area										
	1200	1400	1600	1800	2000	2400	2800	3200	3600	4000	4400
Wood Siding - Wood Frame	124.25	117.70	113.25	108.00	103.65	96.30	90.70	86.40	83.70	81.25	78.85
Brick Veneer - Wood Frame	128.70	121.90	117.20	111.75	107.25	99.55	93.65	89.20	86.35	83.80	81.30
Solid Brick	139.50	132.00	126.85	120.85	116.00	107.45	100.80	95.90	92.85	89.90	87.20
Solid Stone	140.10	132.55	127.40	121.35	116.50	107.90	101.20	96.25	93.20	90.20	87.50
Finished Basement, Add	20.40	19.60	19.05	18.50	18.15	17.30	16.60	16.20	15.90	15.55	15.40
Unfinished Basement, Add	7.95	7.55	7.35	7.05	6.90	6.50	6.20	6.00	5.90	5.70	5.60

Modifications

Add to the total cost

Upgrade Kitchen Cabinets	$ + 832
Solid Surface Countertops	+ 1326
Full Bath - including plumbing, wall and floor finishes	+ 4854
Half Bath - including plumbing, wall and floor finishes	+ 3089
Two Car Attached Garage	+ 19,703
Two Car Detached Garage	+ 22,432
Fireplace & Chimney	+ 5785

Adjustments

For multi family - add to total cost

Additional Kitchen	$ + 10,340
Additional Full Bath & Half Bath	+ 7953
Additional Entry & Exit	+ 1629
Separate Heating & Air Conditioning	+ 4115
Separate Electric	+ 1342

For Townhouse/Rowhouse - Multiply cost per square foot by

Inner Unit	.86
End Unit	.93

Alternatives

Add to or deduct from the cost per square foot of living area

Heavyweight Asphalt Shingles	– .9
Clay Tile Roof	+ 1.2
Slate Roof	+ 2.8
Upgrade Ceilings to Textured Finish	+ .4
Air Conditioning, in Heating Ductwork	Base Syste
Heating Systems, Hot Water	+ 2.3
Heat Pump	+ 2.0
Electric Heat	– 1.7
Not Heated	– 2.8

Additional upgrades or components

Kitchen Cabinets & Countertops Page 9
Bathroom Vanities 9
Fireplaces & Chimneys 9
Windows, Skylights & Dormers 9
Appliances 9
Breezeways & Porches 9
Finished Attic 9
Garages 9
Site Improvements 9
Wings & Ells 9

Important: See the Reference Section for Location Factors (to adjust for your city) and Estimating Form

Luxury 2 Story

Living Area - 3200 S.F.
Perimeter - 163 L.F.

		Labor-Hours	Cost Per Square Foot Of Living Area		
			Mat.	Labor	Total
1 Site Work	Site preparation for slab; 4′ deep trench excavation for foundation wall.	.024		.42	.42
2 Foundation	Continuous reinforced concrete footing 8″ deep x 18″ wide; dampproofed and insulated reinforced concrete foundation wall, 12″ thick, 4′ deep; 4″ concrete slab on 4″ crushed stone base and polyethylene vapor barrier, trowel finish.	.058	2.53	2.72	5.25
3 Framing	Exterior walls - 2″ x 6″ wood studs, 16″ O.C.; 5/8″ plywood sheathing; 2″ x 10″ rafters 16″ O.C. with 5/8″ plywood sheathing, 6 in 12 pitch; 2″ x 8″ ceiling joists 16″ O.C.; 2″ x 12″ floor joists 16″ O.C. with 5/8″ plywood subfloor; 5/8″ plywood subfloor on 1″ x 3″ wood sleepers 16″ O.C.	.193	6.22	5.58	11.80
4 Exterior Walls	Horizontal beveled wood siding, #15 felt building paper; 6″ batt insulation; wood double hung windows; 3 solid core wood exterior doors; storms and screens.	.247	11.34	3.97	15.31
5 Roofing	Red cedar shingles; #15 felt building paper; aluminum gutters, downspouts and drip edge; copper flashings.	.049	1.66	1.31	2.97
6 Interiors	Walls and ceilings - 5/8″ drywall, skim coat plaster, painted with primer and 2 coats; hardwood baseboard and trim, sanded and finished; hardwood floor 70%, ceramic tile with underlayment 20%, vinyl tile with underlayment 10%; wood panel interior doors, primed and painted with 2 coats.	.252	14.13	14.34	28.47
7 Specialties	Luxury grade kitchen cabinets - 25 L.F. wall and base with plastic laminate counter top and kitchen sink; 6 L.F. bathroom vanity; 75 gallon electric water heater; medicine cabinet.	.057	2.97	.52	3.49
8 Mechanical	Gas fired warm air heat/air conditioning; one full bath including: bathtub, corner shower; built in lavatory and water closet; one 1/2 bath including: built in lavatory and water closet.	.071	3.45	2.00	5.45
9 Electrical	200 Amp. service; romex wiring; fluorescent and incandescent lighting fixtures; intercom, switches, receptacles.	.042	.70	1.27	1.97
10 Overhead	Contractor's overhead and profit and architect's fee.		6.45	4.82	11.27
		Total	49.45	36.95	**86.40**

- **Unique residence built from an architect's plan**
- **Single family — 1 full bath, 1 half bath, 1 kitchen**
- **No basement**
- **Cedar shakes on roof**
- **Forced hot air heat/air conditioning**
- **Double drywall interior**
- **Many special features**
- **Extraordinary materials and workmanship**

Note: The illustration shown may contain some optional components (for example: garages and/or fireplaces) whose costs are shown in the modifications, adjustments, & alternatives below or at the end of the square foot section.

Base cost per square foot of living area

Exterior Wall	Living Area										
	1500	1800	2100	2500	3000	3500	4000	4500	5000	5500	6000
Wood Siding - Wood Frame	125.30	111.95	105.30	99.70	**93.15**	87.50	82.20	79.90	77.20	74.70	72.55
Brick Veneer - Wood Frame	129.90	116.20	109.10	103.35	96.40	90.45	84.95	82.50	79.70	77.05	74.85
Solid Brick	141.20	126.60	118.40	112.10	104.40	97.65	91.65	88.85	85.75	82.85	80.35
Solid Stone	141.80	127.10	118.85	112.55	104.80	98.05	92.00	89.20	86.05	83.15	80.65
Finished Basement, Add	16.25	15.50	14.65	14.15	13.55	13.00	12.60	12.35	12.10	11.90	11.75
Unfinished Basement, Add	6.35	6.05	5.60	5.40	5.15	4.90	4.70	4.55	4.45	4.35	4.30

Modifications

Add to the total cost

Upgrade Kitchen Cabinets	$ + 832
Solid Surface Countertops	+ 1326
Full Bath - including plumbing, wall and floor finishes	+ 4854
Half Bath - including plumbing, wall and floor finishes	+ 3089
Two Car Attached Garage	+ 19,703
Two Car Detached Garage	+ 22,432
Fireplace & Chimney	+ 6335

Adjustments

For multi family - add to total cost

Additional Kitchen	$ + 10,340
Additional Full Bath & Half Bath	+ 7953
Additional Entry & Exit	+ 1629
Separate Heating & Air Conditioning	+ 4115
Separate Electric	+ 1342

For Townhouse/Rowhouse - Multiply cost per square foot by

Inner Unit	.86
End Unit	.93

Alternatives

Add to or deduct from the cost per square foot of living area

Heavyweight Asphalt Shingles	– .8
Clay Tile Roof	+ 1.1
Slate Roof	+ 2.5
Upgrade Ceilings to Textured Finish	+ .4
Air Conditioning, in Heating Ductwork	Base System
Heating Systems, Hot Water	+ 2.2
Heat Pump	+ 2.1
Electric Heat	– 3.1
Not Heated	– 2.8

Additional upgrades or components

Kitchen Cabinets & Countertops	Page 9
Bathroom Vanities	9
Fireplaces & Chimneys	9
Windows, Skylights & Dormers	9
Appliances	9
Breezeways & Porches	9
Finished Attic	9
Garages	9
Site Improvements	9
Wings & Ells	9

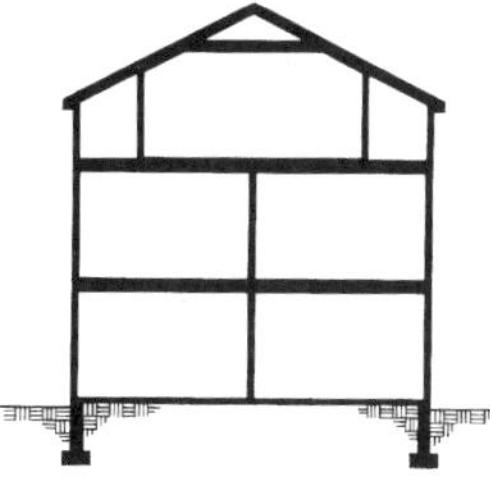

Luxury 2-1/2 Story

Living Area - 3000 S.F.
Perimeter - 148 L.F.

		Labor-Hours	Cost Per Square Foot Of Living Area		
			Mat.	Labor	Total
1 Site Work	Site preparation for slab; 4′ deep trench excavation for foundation wall.	.055		.44	.44
2 Foundation	Continuous reinforced concrete footing 8″ deep x 18″ wide; dampproofed and insulated reinforced concrete foundation wall, 12″ thick, 4′ deep; 4″ concrete slab on 4″ crushed stone base and polyethylene vapor barrier, trowel finish.	.067	2.22	2.45	4.67
3 Framing	Exterior walls - 2″ x 6″ wood studs, 16″ O.C.; 5/8″ plywood sheathing; 2″ x 10″ rafters 16″ O.C. with 5/8″ plywood sheathing, 6 in 12 pitch; 2″ x 8″ ceiling joists 16″ O.C.; 2″ x 12″ floor joists 16″ O.C. with 5/8″ plywood subfloor; 5/8″ plywood subfloor on 1″ x 3″ wood sleepers 16″ O.C.	.209	6.44	5.75	12.19
4 Exterior Walls	Horizontal beveled wood siding; #15 felt building paper; 6″ batt insulation; wood double hung windows; 3 solid core wood exterior doors; storms and screens.	.405	13.28	4.66	17.94
5 Roofing	Red cedar shingles; #15 felt building paper; aluminum gutters, downspouts and drip edge; copper flashings.	.039	1.28	1.01	2.29
6 Interiors	Walls and ceilings - 5/8″ drywall, skim coat plaster, painted with primer and 2 coats; hardwood baseboard and trim, sanded and finished; hardwood floor 70%, ceramic tile with underlayment 20%, vinyl tile with underlayment 10%; wood panel interior doors, primed and painted with 2 coats.	.341	15.80	16.32	32.12
7 Specialties	Luxury grade kitchen cabinets - 25 L.F. wall and base with plastic laminate counter top and kitchen sink; 6 L.F. bathroom vanity; 75 gallon electric water heater; medicine cabinet.	.119	3.17	.55	3.72
8 Mechanical	Gas fired warm air heat/air conditioning; one full bath including: bathtub, corner shower; built in lavatory and water closet; one 1/2 bath including: built in lavatory and water closet.	.103	3.59	2.03	5.62
9 Electrical	200 Amp. service; romex wiring; fluorescent and incandescent lighting fixtures; intercom, switches, receptacles.	.054	.71	1.28	1.99
10 Overhead	Contractor's overhead and profit and architect's fee.		6.98	5.19	12.17
		Total	53.47	39.68	**93.15**

- **Unique residence built from an architect's plan**
- **Single family — 1 full bath, 1 half bath, 1 kitchen**
- **No basement**
- **Cedar shakes on roof**
- **Forced hot air heat/air conditioning**
- **Double drywall interior**
- **Many special features**
- **Extraordinary materials and workmanship**

Note: The illustration shown may contain some optional components (for example: garages and/or fireplaces) whose costs are shown in the modifications, adjustments, & alternatives below or at the end of the square foot section.

Base cost per square foot of living area

	Living Area										
Exterior Wall	**1500**	**1800**	**2100**	**2500**	**3000**	**3500**	**4000**	**4500**	**5000**	**5500**	**6000**
Wood Siding - Wood Frame	125.10	111.85	106.60	102.40	**94.50**	91.15	85.90	80.75	79.15	76.85	74.95
Brick Veneer - Wood Frame	129.95	116.30	110.80	106.40	98.10	94.60	89.00	83.60	81.90	79.50	77.45
Solid Brick	141.60	127.10	120.90	116.05	106.85	102.85	96.55	90.60	88.65	86.00	83.55
Solid Stone	142.25	127.70	121.45	116.55	107.30	103.30	96.95	90.95	89.00	86.35	83.85
Finished Basement, Add	14.20	13.55	13.00	12.65	12.05	11.70	11.30	10.90	10.75	10.60	10.35
Unfinished Basement, Add	5.60	5.30	5.05	4.85	4.60	4.45	4.20	4.05	4.00	3.90	3.80

Modifications

Add to the total cost

Upgrade Kitchen Cabinets	$ + 832
Solid Surface Countertops	+ 1326
Full Bath - including plumbing, wall and floor finishes	+ 4854
Half Bath - including plumbing, wall and floor finishes	+ 3089
Two Car Attached Garage	+ 19,703
Two Car Detached Garage	+ 22,432
Fireplace & Chimney	+ 6335

Adjustments

For multi family - add to total cost

Additional Kitchen	$ + 10,340
Additional Full Bath & Half Bath	+ 7953
Additional Entry & Exit	+ 1629
Separate Heating & Air Conditioning	+ 4115
Separate Electric	+ 1342

For Townhouse/Rowhouse - Multiply cost per square foot by

Inner Unit	.84
End Unit	.92

Alternatives

Add to or deduct from the cost per square foot of living area

Heavyweight Asphalt Shingles	– .60
Clay Tile Roof	+ .85
Slate Roof	+ 1.90
Upgrade Ceilings to Textured Finish	+ .44
Air Conditioning, in Heating Ductwork	Base System
Heating Systems, Hot Water	+ 2.28
Heat Pump	+ 2.12
Electric Heat	– 3.17
Not Heated	– 2.73

Additional upgrades or components

Kitchen Cabinets & Countertops	Page 93
Bathroom Vanities	94
Fireplaces & Chimneys	94
Windows, Skylights & Dormers	94
Appliances	95
Breezeways & Porches	95
Finished Attic	95
Garages	96
Site Improvements	96
Wings & Ells	92

Important: See the Reference Section for Location Factors (to adjust for your city) and Estimating Forms

Luxury 3 Story

Living Area - 3000 S.F.
Perimeter - 135 L.F.

		Labor-Hours	Cost Per Square Foot Of Living Area		
			Mat.	Labor	Total
1 Site Work	Site preparation for slab; 4′ deep trench excavation for foundation wall.	.055		.44	.44
2 Foundation	Continuous reinforced concrete footing 8″ deep x 18″ wide; dampproofed and insulated reinforced concrete foundation wall, 12″ thick, 4′ deep; 4″ concrete slab on 4″ crushed stone base and polyethylene vapor barrier, trowel finish.	.063	2.00	2.22	4.22
3 Framing	Exterior walls - 2″ x 6″ wood studs, 16″ O.C.; 5/8″ plywood sheathing; 2″ x 10″ rafters 16″ O.C. with 5/8″ plywood sheathing, 6 in 12 pitch; 2″ x 8″ ceiling joists 16″ O.C.; 2″ x 12″ floor joists 16″ O.C. with 5/8″ plywood subfloor; 5/8″ plywood subfloor on 1″ x 3″ wood sleepers 16″ O.C.	.225	6.57	5.78	12.35
4 Exterior Walls	Horizontal beveled wood siding; #15 felt building paper; 6″ batt insulation; wood double hung windows; 3 solid core wood exterior doors; storms and screens.	.454	14.54	5.08	19.62
5 Roofing	Red cedar shingles; #15 felt building paper; aluminum gutters, downspouts and drip edge; copper flashings.	.034	1.11	.87	1.98
6 Interiors	Walls and ceilings - 5/8″ drywall, skim coat plaster, painted with primer and 2 coats; hardwood baseboard and trim, sanded and finished; hardwood floor 70%, ceramic tile with underlayment 20%, vinyl tile with underlayment 10%; wood panel interior doors, primed and painted with 2 coats.	.390	15.81	16.40	32.21
7 Specialties	Luxury grade kitchen cabinets - 25 L.F. wall and base with plastic laminate counter top and kitchen sink; 6 L.F. bathroom vanity; 75 gallon electric water heater; medicine cabinet.	.119	3.17	.55	3.72
8 Mechanical	Gas fired warm air heat/air conditioning; one full bath including: bathtub, corner shower; built in lavatory and water closet; one 1/2 bath including: built in lavatory and water closet.	.103	3.59	2.03	5.62
9 Electrical	200 Amp. service; romex wiring; fluorescent and incandescent lighting fixtures; intercom, switches, receptacles.	.053	.71	1.28	1.99
10 Overhead	Contractor's overhead and profit and architect's fees.		7.13	5.22	12.35
		Total	54.63	39.87	**94.50**

- **Unique residence built from an architect's plan**
- **Single family — 1 full bath, 1 half bath, 1 kitchen**
- **No basement**
- **Cedar shakes on roof**
- **Forced hot air heat/air conditioning**
- **Double drywall interior**
- **Many special features**
- **Extraordinary materials and workmanship**

Note: The illustration shown may contain some optional components (for example: garages and/or fireplaces) whose costs are shown in the modifications, adjustments, & alternatives below or at the end of the square foot section.

Base cost per square foot of living area

Exterior Wall	Living Area 1200	1400	1600	1800	2000	2400	2800	3200	3600	4000	4400
Wood Siding - Wood Frame	115.95	110.00	105.80	101.10	96.95	90.25	85.20	**81.25**	78.70	76.60	74.30
Brick Veneer - Wood Frame	119.30	113.10	108.80	103.85	99.65	92.70	87.40	83.35	80.70	78.50	76.15
Solid Brick	127.40	120.70	116.00	110.65	106.25	98.60	92.80	88.40	85.60	83.05	80.55
Solid Stone	127.85	121.10	116.40	111.05	106.55	98.90	93.05	88.65	85.85	83.35	80.80
Finished Basement, Add	20.40	19.60	19.05	18.50	18.15	17.30	16.60	16.20	15.90	15.55	15.40
Unfinished Basement, Add	7.95	7.55	7.35	7.05	6.90	6.50	6.20	6.00	5.90	5.70	5.60

Modifications

Add to the total cost

Upgrade Kitchen Cabinets	$ + 832
Solid Surface Countertops	+ 1326
Full Bath - including plumbing, wall and floor finishes	+ 4854
Half Bath - including plumbing, wall and floor finishes	+ 3089
Two Car Attached Garage	+ 19,703
Two Car Detached Garage	+ 22,432
Fireplace & Chimney	+ 5285

Adjustments

For multi family - add to total cost

Additional Kitchen	$ + 10,340
Additional Full Bath & Half Bath	+ 7953
Additional Entry & Exit	+ 1629
Separate Heating & Air Conditioning	+ 4115
Separate Electric	+ 1342

For Townhouse/Rowhouse - Multiply cost per square foot by

Inner Unit	.89
End Unit	.94

Alternatives

Add to or deduct from the cost per square foot of living area

Heavyweight Asphalt Shingles	– .9
Clay Tile Roof	+ 1.2
Slate Roof	+ 2.8
Upgrade Ceilings to Textured Finish	+ .4
Air Conditioning, in Heating Ductwork	Base Syste
Heating Systems, Hot Water	+ 2.3
Heat Pump	+ 2.0
Electric Heat	– 1.7
Not Heated	– 2.8

Additional upgrades or components

Kitchen Cabinets & Countertops	Page 9
Bathroom Vanities	9
Fireplaces & Chimneys	9
Windows, Skylights & Dormers	9
Appliances	9
Breezeways & Porches	9
Finished Attic	9
Garages	9
Site Improvements	9
Wings & Ells	9

Important: See the Reference Section for Location Factors (to adjust for your city) and Estimating Forn

Luxury Bi-Level

Living Area - 3200 S.F.
Perimeter - 163 L.F.

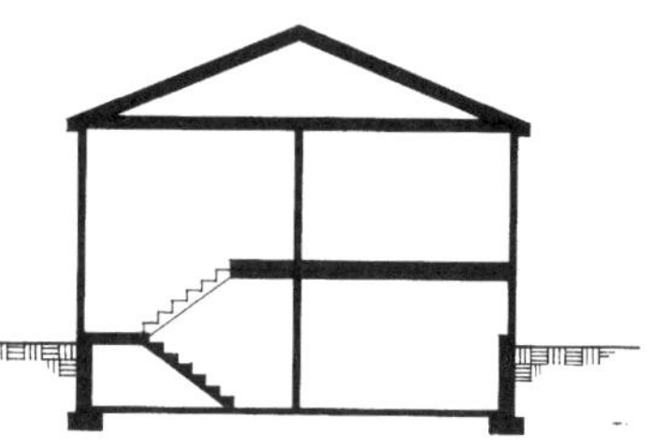

		Labor-Hours	Cost Per Square Foot Of Living Area		
			Mat.	Labor	Total
1 Site Work	Excavation for lower level, 4' deep. Site preparation for slab.	.024		.42	.42
2 Foundation	Continuous reinforced concrete footing 8" deep x 18" wide; dampproofed and insulated reinforced concrete foundation wall, 12" thick, 4' deep; 4" concrete slab on 4" crushed stone base and polyethylene vapor barrier, trowel finish.	.058	2.53	2.72	5.25
3 Framing	Exterior walls - 2" x 6" wood studs, 16" O.C.; 5/8" plywood sheathing; 2" x 10" rafters 16" O.C. with 5/8" plywood sheathing, 6 in 12 pitch; 2" x 8" ceiling joists 16" O.C.; 2" x 12" floor joists 16" O.C. with 5/8" plywood subfloor; 5/8" plywood subfloor on 1" x 3" wood sleepers 16" O.C.	.232	5.90	5.32	11.22
4 Exterior Walls	Horizontal beveled wood siding; #15 felt building paper; 6" batt insulation; wood double hung windows; 3 solid core wood exterior doors; storms and screens.	.185	8.74	3.05	11.79
5 Roofing	Red cedar shingles: #15 felt building paper; aluminum gutters, downspouts and drip edge; copper flashings.	.042	1.66	1.31	2.97
6 Interiors	Walls and ceilings - 5/8" drywall, skim coat plaster, painted with primer and 2 coats; hardwood baseboard and trim, sanded and finished; hardwood floor 70%, ceramic tile with underlayment 20%; vinyl tile with underlayment 10%; wood panel interior doors, primed and painted with 2 coats.	.238	13.98	14.11	28.09
7 Specialties	Luxury grade kitchen cabinets - 25 L.F. wall and base with plastic laminate counter top and kitchen sink; 6 L.F. bathroom vanity; 75 gallon electric water heater; medicine cabinet.	.056	2.97	.52	3.49
8 Mechanical	Gas fired warm air heat/air conditioning; one full bath including: bathtub, corner shower; built in lavatory and water closet; one 1/2 bath including: built in lavatory and water closet.	.071	3.45	2.00	5.45
9 Electrical	200 Amp. service; romex wiring; fluorescent and incandescent lighting fixtures; intercom, switches, receptacles.	.042	.70	1.27	1.97
10 Overhead	Contractor's overhead and profit and architect's fees.		6.00	4.60	10.60
	Total		45.93	35.32	**81.25**

- **Unique residence built from an architect's plan**
- **Single family — 1 full bath, 1 half bath, 1 kitchen**
- **No basement**
- **Cedar shakes on roof**
- **Forced hot air heat/air conditioning**
- **Double drywall interior**
- **Many special features**
- **Extraordinary materials and workmanship**

Note: The illustration shown may contain some optional components (for example: garages and/or fireplaces) whose costs are shown in the modifications, adjustments, & alternatives below or at the end of the square foot section.

©Home Planners, Inc.

Base cost per square foot of living area

Exterior Wall	Living Area 1500	1800	2100	2400	2800	3200	3600	4000	4500	5000	5500
Wood Siding - Wood Frame	109.05	100.95	94.95	90.25	86.85	82.85	78.70	77.15	72.80	70.60	68.10
Brick Veneer - Wood Frame	112.05	103.70	97.40	92.55	89.05	84.90	80.60	79.00	74.50	72.20	69.60
Solid Brick	119.30	110.25	103.40	98.20	94.45	89.90	85.30	83.50	78.70	76.10	73.30
Solid Stone	119.75	110.60	103.70	98.50	94.70	90.20	85.50	83.75	78.90	76.30	73.50
Finished Basement, Add*	24.15	23.05	22.10	21.55	21.15	20.55	20.10	19.85	19.35	19.00	18.70
Unfinished Basement, Add*	9.20	8.70	8.25	8.00	7.85	7.55	7.35	7.20	6.95	6.85	6.65

*Basement under middle level only.

Modifications

Add to the total cost

Upgrade Kitchen Cabinets	$ + 832
Solid Surface Countertops	+ 1326
Full Bath - including plumbing, wall and floor finishes	+ 4854
Half Bath - including plumbing, wall and floor finishes	+ 3089
Two Car Attached Garage	+ 19,703
Two Car Detached Garage	+ 22,432
Fireplace & Chimney	+ 5785

Adjustments

For multi family - add to total cost

Additional Kitchen	$ + 10,340
Additional Full Bath & Half Bath	+ 7953
Additional Entry & Exit	+ 1629
Separate Heating & Air Conditioning	+ 4115
Separate Electric	+ 1342

For Townhouse/Rowhouse - Multiply cost per square foot by

Inner Unit	.86
End Unit	.93

Alternatives

Add to or deduct from the cost per square foot of living area

Heavyweight Asphalt Shingles	– 1.3
Clay Tile Roof	+ 1.8
Slate Roof	+ 4.1
Upgrade Ceilings to Textured Finish	+ .4
Air Conditioning, in Heating Ductwork	Base Syster
Heating Systems, Hot Water	+ 2.2
Heat Pump	+ 2.1
Electric Heat	– 1.6
Not Heated	– 2.7

Additional upgrades or components

Kitchen Cabinets & Countertops	Page 9
Bathroom Vanities	9
Fireplaces & Chimneys	9
Windows, Skylights & Dormers	9
Appliances	9
Breezeways & Porches	9
Finished Attic	9
Garages	9
Site Improvements	9
Wings & Ells	9

Luxury Tri-Level

Living Area - 3600 S.F.
Perimeter - 207 L.F.

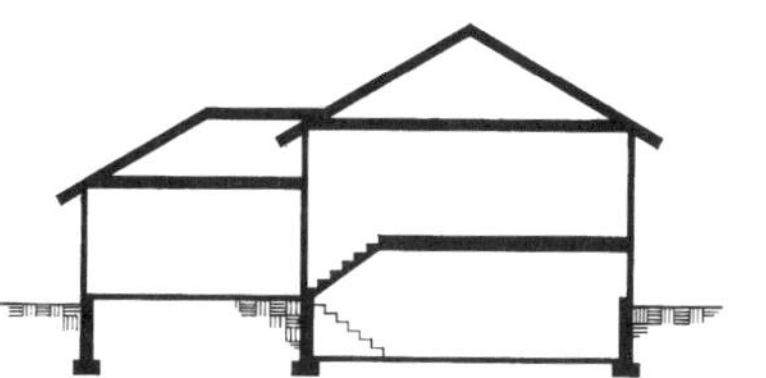

		Labor-Hours	Cost Per Square Foot Of Living Area		
			Mat.	Labor	Total
1 Site Work	Site preparation for slab; 4' deep trench excavation for foundation wall, excavation for lower level, 4' deep.	.021		.37	.37
2 Foundation	Continuous reinforced concrete footing 8" deep x 18" wide; dampproofed and insulated reinforced concrete foundation wall, 12" thick, 4' deep; 4" concrete slab on 4" crushed stone base and polyethylene vapor barrier, trowel finish.	.109	3.02	3.15	6.17
3 Framing	Exterior walls - 2" x 6" wood studs, 16" O.C.; 5/8" plywood sheathing; 2" x 10" rafters 16" O.C. with 5/8" plywood sheathing, 6 in 12 pitch; 2" x 8" ceiling joists 16" O.C.; 2" x 12" floor joists 16" O.C. with 5/8" plywood subfloor; 5/8" plywood subfloor on 1" x 3" wood sleepers 16" O.C.	.204	5.83	5.35	11.18
4 Exterior Walls	Horizontal beveled wood siding; #15 felt building paper; 6" batt insulation; wood double hung windows; 3 solid core wood exterior doors; storms and screens.	.181	8.26	2.89	11.15
5 Roofing	Red cedar shingles; #15 felt building paper; aluminum gutters, downspouts and drip edge; copper flashings.	.056	2.23	1.75	3.98
6 Interiors	Walls and ceilings - 5/8" drywall, skim coat plaster, painted with primer and 2 coats; hardwood baseboard and trim, sanded and finished; hardwood floor 70%, ceramic tile with underlayment 20%, vinyl tile with underlayment 10%; wood panel interior doors, primed and painted with 2 coats.	.217	12.80	12.66	25.46
7 Specialties	Luxury grade kitchen cabinets - 25 L.F. wall and base with plastic laminate counter top and kitchen sink; 6 L.F. bathroom vanity; 75 gallon electric water heater; medicine cabinet.	.048	2.63	.46	3.09
8 Mechanical	Gas fired warm air heat/air conditioning; one full bath including: bathtub, corner shower; built in lavatory and water closet; one 1/2 bath including: built in lavatory and water closet.	.057	3.18	1.92	5.10
9 Electrical	200 Amp. service; romex wiring; fluorescent and incandescent lighting fixtures; intercom, switches, receptacles.	.039	.68	1.24	1.92
10 Overhead	Contractor's overhead and profit and architect's fees.		5.79	4.49	10.28
		Total	44.42	34.28	**78.70**

1 Story — Base cost per square foot of living area

Exterior Wall	Living Area 50	100	200	300	400	500	600	700
Wood Siding - Wood Frame	182.00	138.80	119.95	99.70	93.50	89.80	87.35	88.20
Brick Veneer - Wood Frame	193.20	146.80	126.65	104.15	97.55	93.55	90.95	91.65
Solid Brick	220.45	166.25	142.85	114.95	107.30	102.65	99.60	99.95
Solid Stone	221.95	167.30	143.75	115.55	107.80	103.10	100.05	100.40
Finished Basement, Add	63.40	51.15	46.05	37.55	35.85	34.85	34.15	33.65
Unfinished Basement, Add	26.90	21.05	18.65	14.60	13.75	13.30	12.95	12.75

1-1/2 Story — Base cost per square foot of living area

Exterior Wall	Living Area 100	200	300	400	500	600	700	800
Wood Siding - Wood Frame	148.45	119.85	102.35	91.45	86.20	83.70	80.40	79.50
Brick Veneer - Wood Frame	158.45	127.85	109.05	96.70	91.05	88.25	84.70	83.75
Solid Brick	182.80	147.30	125.25	109.35	102.70	99.25	95.15	94.10
Solid Stone	184.10	148.35	126.15	110.00	103.35	99.85	95.70	94.70
Finished Basement, Add	41.95	36.85	33.45	29.70	28.65	28.00	27.40	27.30
Unfinished Basement, Add	17.35	14.90	13.30	11.50	11.00	10.70	10.40	10.35

2 Story — Base cost per square foot of living area

Exterior Wall	Living Area 100	200	400	600	800	1000	1200	1400
Wood Siding - Wood Frame	149.80	111.40	94.60	77.70	72.20	68.85	66.70	67.75
Brick Veneer - Wood Frame	161.05	119.45	101.30	82.15	76.20	72.60	70.30	71.20
Solid Brick	188.30	138.90	117.50	92.95	85.95	81.70	78.90	79.55
Solid Stone	189.75	139.95	118.40	93.55	86.50	82.20	79.40	80.00
Finished Basement, Add	31.75	25.60	23.05	18.80	18.00	17.45	17.10	16.90
Unfinished Basement, Add	13.45	10.55	9.30	7.30	6.90	6.65	6.50	6.35

Base costs do not include bathroom or kitchen facilities. Use Modifications/Adjustments/Alternatives on pages 93-96 where appropriate.

 Important: See the Reference Section for Location Factors (to adjust for your city) and Estimating Forms

Kitchen cabinets - Base units, hardwood *(Cost per Unit)*

	Economy	Average	Custom	Luxury
24" deep, 35" high,				
One top drawer,				
One door below				
12" wide	$120	$160	$215	$280
15" wide	163	217	290	380
18" wide	173	230	305	405
21" wide	182	242	320	425
24" wide	202	269	360	470
Four drawers				
12" wide	266	355	470	620
15" wide	274	365	485	640
18" wide	289	385	510	675
24" wide	311	415	550	725
Two top drawers,				
Two doors below				
27" wide	236	315	420	550
30" wide	251	335	445	585
33" wide	263	350	465	615
36" wide	270	360	480	630
42" wide	289	385	510	675
48" wide	304	405	540	710
Range or sink base				
(Cost per unit)				
Two doors below				
30" wide	203	270	360	475
33" wide	217	289	385	505
36" wide	225	300	400	525
42" wide	240	320	425	560
48" wide	255	340	450	595
Corner Base Cabinet				
(Cost per unit)				
36" wide	184	245	325	430
Lazy Susan *(Cost per unit)*				
With revolving door	289	385	510	675

Kitchen cabinets - Wall cabinets, hardwood *(Cost per Unit)*

	Economy	Average	Custom	Luxury
12" deep, 2 doors				
12" high				
30" wide	$118	$157	$210	$ 275
36" wide	137	182	240	320
15" high				
30" wide	136	181	240	315
33" wide	144	192	255	335
36" wide	148	197	260	345
24" high				
30" wide	155	206	275	360
36" wide	169	225	300	395
42" wide	187	249	330	430
30" high, 1 door				
12" wide	107	143	190	250
15" wide	122	163	215	285
18" wide	134	179	240	315
24" wide	146	194	260	340
30" high, 2 doors				
27" wide	161	214	285	370
30" wide	173	231	305	405
36" wide	197	262	350	460
42" wide	212	283	375	495
48" wide	236	315	420	550
Corner wall, 30" high				
24" wide	131	175	235	305
30" wide	161	214	285	375
36" wide	170	227	300	395
Broom closet				
84" high, 24" deep				
18" wide	330	440	585	770
Oven Cabinet				
84" high, 24" deep				
27" wide	491	655	870	1145

Kitchen countertops *(Cost per L.F.)*

	Economy	Average	Custom	Luxury
Solid Surface				
24" wide, no backsplash	80	106	140	185
with backsplash	86	115	155	200
Stock plastic laminate, 24" wide				
with backsplash	11	14	20	25
Custom plastic laminate, no splash				
7/8" thick, alum. molding	20	27	35	45
1-1/4" thick, no splash	22	30	40	50
Marble				
1/2" - 3/4" thick w/splash	37	49	65	85
Maple, laminated				
1-1/2" thick w/splash	36	48	65	85
Stainless steel				
(per S.F.)	69	92	120	160
Cutting blocks, recessed				
16" x 20" x 1" (each)	60	80	105	140

Vanity bases *(Cost per Unit)*

	Economy	Average	Custom	Luxury
2 door, 30" high, 21" deep				
24" wide	122	163	215	285
30" wide	133	177	235	310
36" wide	151	201	265	350
48" wide	187	249	330	435

Solid surface vanity tops *(Cost Each)*

	Economy	Average	Custom	Luxury
Center bowl				
22" x 25"	$278	$375	$525	$567
22" x 31"	320	432	605	653
22" x 37"	365	493	690	745
22" x 49"	455	614	860	929

Fireplaces & Chimneys *(Cost per Unit)*

	1-1/2 Story	2 Story	3 Story
Economy (prefab metal)			
Exterior chimney & 1 fireplace	$3480	$3845	$4210
Interior chimney & 1 fireplace	3330	3705	3885
Average (masonry)			
Exterior chimney & 1 fireplace	3555	3965	4510
Interior chimney & 1 fireplace	3330	3735	4070
For more than 1 flue, add	265	435	730
For more than 1 fireplace, add	2520	2520	2520
Custom (masonry)			
Exterior chimney & 1 fireplace	3740	4220	4770
Interior chimney & 1 fireplace	3510	3970	4280
For more than 1 flue, add	295	505	690
For more than 1 fireplace, add	2685	2685	2685
Luxury (masonry)			
Exterior chimney & 1 fireplace	5285	5785	6335
Interior chimney & 1 fireplace	5045	5520	5830
For more than 1 flue, add	435	730	1020
For more than 1 fireplace, add	4160	4160	4160

Windows and Skylights *(Cost Each)*

	Economy	Average	Custom	Luxury
Fixed Picture Windows				
3'-6" x 4'-0"	$ 244	$ 305	$ 320	$ 477
4'-0" x 6'-0"	351	438	460	685
5'-0" x 6'-0"	385	481	505	880
6'-0" x 6'-0"	435	543	570	895
Bay/Bow Windows				
8'-0" x 5'-0"	990	1250	1275	1325
10'-0" x 5'-0"	1350	1425	1700	1800
10'-0" x 6'-0"	1485	1568	1850	2350
12'-0" x 6'-0"	2075	2190	2375	3000
Palladian Windows				
3'-2" x 6'-4"		1260	1400	1631
4'-0" x 6'-0"		1283	1425	1660
5'-5" x 6'-10"		1620	1800	2097
8'-0" x 6'-0"		1980	2200	2563
Skylights				
46" x 21-1/2"	325	362	488	707
46" x 28"	433	481	586	850
57" x 44"	548	609	743	1077

Dormers *(Cost/S.F. of plan area)*

	Economy	Average	Custom	Luxury
Framing and Roofing Only				
Gable dormer, 2" x 6" roof frame	$18	$21	$24	$37
2" x 8" roof frame	19	22	25	39
Shed dormer, 2" x 6" roof frame	12	13	15	24
2" x 8" roof frame	13	14	16	25
2" x 10" roof frame	14	15	17	26

Appliances *(Cost per Unit)*

	Economy	Average	Custom	Luxury
Range				
30″ free standing, 1 oven	$ 325	$ 863	$1131	$1400
2 oven	1275	1488	1594	1700
30″ built-in, 1 oven	630	743	946	1150
2 oven	1100	1538	1756	1975
21″ free standing				
1 oven	350	483	549	615
Counter Top Ranges				
4 burner standard	237	404	487	570
As above with griddle	490	575	618	660
Microwave Oven	192	225	367	390
Combination Range, Refrigerator, Sink				
30″ wide	880	1328	1551	1775
60″ wide	2575	2962	3155	3348
72″ wide	2925	3364	3584	3803
Comb. Range, Refrig., Sink, Microwave Oven & Ice Maker	4567	5252	5595	5937
Compactor				
4 to 1 compaction	425	460	478	495
Deep Freeze				
15 to 23 C.F.	445	543	591	640
30 C.F.	855	965	1020	1075
Dehumidifier, portable, auto.				
15 pint	204	235	250	265
30 pint	242	279	297	315
Washing Machine, automatic	445	835	1030	1225
Water Heater				
Electric, glass lined				
30 gal.	310	380	415	450
80 gal.	635	815	905	995
Water Heater, Gas, glass lined				
30 gal.	425	528	579	630
50 gal.	800	988	1081	1175
Water Softener, automatic				
30 grains/gal.	575	586	592	597
100 grains/gal.	710	817	870	923
Dishwasher, built-in				
2 cycles	360	413	439	465
4 or more cycles	355	540	633	725
Dryer, automatic	435	635	735	835
Garage Door Opener	245	298	324	350
Garbage Disposal	78	153	191	228
Heater, Electric, built-in				
1250 watt ceiling type	148	184	201	219
1250 watt wall type	179	209	223	238
Wall type w/blower				
1500 watt	204	226	237	248
3000 watt	365	411	434	457
Hood For Range, 2 speed				
30″ wide	107	349	469	590
42″ wide	250	475	588	700
Humidifier, portable				
7 gal. per day	102	111	116	120
15 gal. per day	164	181	189	197
Ice Maker, automatic				
13 lb. per day	595	685	729	774
51 lb. per day	1000	1150	1225	1300
Refrigerator, no frost				
10-12 C.F.	505	638	704	770
14-16 C.F.	530	590	620	650
18-20 C.F.	585	785	885	985
21-29 C.F.	925	1950	2463	2975
Sump Pump, 1/3 H.P.	192	266	303	340

Breezeway *(Cost per S.F.)*

Class	Type	Area (S.F.)			
		50	100	150	200
Economy	Open	$ 15.35	$13.10	$10.95	$10.75
	Enclosed	73.70	56.95	47.30	41.40
Average	Open	19.50	17.20	15.00	13.65
	Enclosed	82.80	61.65	50.45	44.40
Custom	Open	28.50	25.05	21.85	20.00
	Enclosed	117.05	87.05	71.20	62.55
Luxury	Open	29.50	25.85	23.40	22.60
	Enclosed	118.85	88.15	71.25	63.85

Porches *(Cost per S.F.)*

Class	Type	Area (S.F.)				
		25	50	100	200	300
Economy	Open	$ 45.35	$ 30.40	$23.70	$20.10	$17.15
	Enclosed	90.70	63.20	47.80	37.25	31.95
Average	Open	56.45	37.63	28.95	25.20	22.95
	Enclosed	111.55	75.55	57.20	44.30	37.50
Custom	Open	73.90	49.05	36.95	32.25	28.90
	Enclosed	146.95	100.40	76.45	59.45	51.30
Luxury	Open	79.70	52.10	38.50	34.60	30.75
	Enclosed	156.05	109.60	81.35	63.15	54.40

Finished attic *(Cost per S.F.)*

Class	Area (S.F.)				
	400	500	600	800	1000
Economy	$12.15	$11.70	$11.20	$11.00	$10.70
Average	19.05	18.50	18.15	17.85	17.40
Custom	23.20	22.65	22.20	21.85	21.40
Luxury	30.00	29.35	28.65	28.00	27.50

Alarm system *(Cost per System)*

	Burglar Alarm	Smoke Detector
Economy	$ 345	$ 50
Average	400	63
Custom	673	129
Luxury	1025	160

Sauna, prefabricated
(Cost per unit, including heater and controls—7′ high)

Size	Cost
6′ x 4′	$3875
6′ x 5′	3900
6′ x 6′	4550
6′ x 9′	5625
8′ x 10′	6575
8′ x 12′	7600
10′ x 12′	8675

Garages *

(Costs include exterior wall systems comparable with the quality of the residence. Included in the cost is an allowance for one personnel door, manual overhead door(s) and electrical fixture.)

Class	Type									
	Detached			Attached			Built-in		Basement	
Economy	One Car	Two Car	Three Car	One Car	Two Car	Three Car	One Car	Two Car	One Car	Two Car
Wood	$ 9783	$14,853	$19,923	$ 7575	$13,005	$18,075	$-1313	$-2626	$1080	$1381
Masonry	13,415	19,398	25,381	9848	16,191	22,175	-1759	-3518		
Average										
Wood	10,547	15,809	21,072	8054	13,676	18,938	-1407	-2814	1226	1673
Masonry	13,543	19,559	25,574	9928	16,304	22,319	-1775	-3086		
Custom										
Wood	12,308	18,591	24,873	9519	16,313	22,595	-3646	-4081	1707	2635
Masonry	15,012	21,974	28,936	11,210	18,685	25,647	-3978	-4745		
Luxury										
Wood	14,680	22,432	30,184	11,439	19,703	27,455	-3794	-4377	2403	3672
Masonry	18,169	26,798	35,427	13,622	22,763	31,393	-4223	-5234		

*See the Introduction to this section for definitions of garage types.

Swimming pools *(Cost per S.F.)*

Residential (includes equipment)	
In-ground	$ 18-44
Deck equipment	1.30
Paint pool, preparation & 3 coats (epoxy)	2.63
Rubber base paint	2.44
Pool Cover	.72
Swimming Pool Heaters *(Cost per unit)*	
(not including wiring, external piping, base or pad)	
Gas	
155 MBH	$ 2450
190 MBH	3300
500 MBH	8050
Electric	
15 KW 7200 gallon pool	1925
24 KW 9600 gallon pool	2525
54 KW 24,000 gallon pool	3775

Wood and coal stoves

Wood Only	
Free Standing (minimum)	$ 1250
Fireplace Insert (minimum)	1260
Coal Only	
Free Standing	$ 1424
Fireplace Insert	1559
Wood and Coal	
Free Standing	$ 2928
Fireplace Insert	3000

Sidewalks *(Cost per S.F.)*

Concrete, 3000 psi with wire mesh	4" thick	$ 2.34
	5" thick	2.85
	6" thick	3.19
Precast concrete patio blocks (natural)	2" thick	7.99
Precast concrete patio blocks (colors)	2" thick	8.34
Flagstone, bluestone	1" thick	10.35
Flagstone, bluestone	1-1/2" thick	13.70
Slate (natural, irregular)	3/4" thick	7.23
Slate (random rectangular)	1/2" thick	8.90
Seeding		
Fine grading & seeding includes lime, fertilizer & seed	per S.Y.	1.67
Lawn Sprinkler System	per S.F.	.66

Fencing *(Cost per L.F.)*

Chain Link, 4' high, galvanized	$ 10.15
Gate, 4' high (each)	119.00
Cedar Picket, 3' high, 2 rail	8.80
Gate (each)	112.00
3 Rail, 4' high	9.80
Gate (each)	120.00
Cedar Stockade, 3 Rail, 6' high	10.05
Gate (each)	120.00
Board & Battens, 2 sides 6' high, pine	14.70
6' high, cedar	20.50
No. 1 Cedar, basketweave, 6' high	11.75
Gate, 6' high (each)	138.00

Carport *(Cost per S.F.)*

Economy	$ 5.71
Average	8.87
Custom	13.37
Luxury	15.30

Assemblies Section

Table of Contents

Table No.		Page
1 Site Work		**101**
1-04	Footing Excavation	102
1-08	Foundation Excavation	104
1-12	Utility Trenching	106
1-16	Sidewalk	108
1-20	Driveway	110
1-24	Septic	112
1-60	Chain Link Fence	114
1-64	Wood Fence	115
2 Foundations		**117**
2-04	Footing	118
2-08	Block Wall	120
2-12	Concrete Wall	122
2-16	Wood Wall Foundation	124
2-20	Floor Slab	126
3 Framing		**129**
3-02	Floor (Wood)	130
3-04	Floor (Wood)	132
3-06	Floor (Wood)	134
3-08	Exterior Wall	136
3-12	Gable End Roof	138
3-16	Truss Roof	140
3-20	Hip Roof	142
3-24	Gambrel Roof	144
3-28	Mansard Roof	146
3-32	Shed/Flat Roof	148
3-40	Gable Dormer	150
3-44	Shed Dormer	152
3-48	Partition	154
4 Exterior Walls		**157**
4-02	Masonry Block Wall	158
4-04	Brick/Stone Veneer	160
4-08	Wood Siding	162
4-12	Shingle Siding	164
4-16	Metal & Plastic Siding	166
4-20	Insulation	168
4-28	Double Hung Window	170
4-32	Casement Window	172
4-36	Awning Window	174
4-40	Sliding Window	176
4-44	Bow/Bay Window	178
4-48	Fixed Window	180
4-52	Entrance Door	182
4-53	Sliding Door	184
4-56	Residential Overhead Door	186
4-58	Aluminum Window	188
4-60	Storm Door & Window	190
4-64	Shutters/Blinds	191
5 Roofing		**193**
5-04	Gable End Roofing	194
5-08	Hip Roof Roofing	196
5-12	Gambrel Roofing	198
5-16	Mansard Roofing	200
5-20	Shed Roofing	202
5-24	Gable Dormer Roofing	204
5-28	Shed Dormer Roofing	206
5-32	Skylight/Skywindow	208
5-34	Built-up Roofing	210
6 Interiors		**213**
6-04	Drywall & Thincoat Wall	214
6-08	Drywall & Thincoat Ceiling	216
6-12	Plaster & Stucco Wall	218
6-16	Plaster & Stucco Ceiling	220
6-18	Suspended Ceiling	222
6-20	Interior Door	224
6-24	Closet Door	226
6-60	Carpet	228
6-64	Flooring	229
6-90	Stairways	230
7 Specialties		**233**
7-08	Kitchen	234
7-12	Appliances	236
7-16	Bath Accessories	237
7-24	Masonry Fireplace	238
7-30	Prefabricated Fireplace	240
7-32	Greenhouse	242
7-36	Swimming Pool	243
7-40	Wood Deck	244
8 Mechanical		**247**
8-04	Two Fixture Lavatory	248
8-12	Three Fixture Bathroom	250
8-16	Three Fixture Bathroom	252
8-20	Three Fixture Bathroom	254
8-24	Three Fixture Bathroom	256
8-28	Three Fixture Bathroom	258
8-32	Three Fixture Bathroom	260
8-36	Four Fixture Bathroom	262
8-40	Four Fixture Bathroom	264
8-44	Five Fixture Bathroom	266
8-60	Gas Fired Heating/Cooling	268
8-64	Oil Fired Heating/Cooling	270
8-68	Hot Water Heating	272
8-80	Rooftop Heating/Cooling	274
9 Electrical		**277**
9-10	Electric Service	278
9-20	Electric Heating	279
9-30	Wiring Devices	280
9-40	Light Fixtures	281

How to Use the Assemblies Cost Tables

The following is a detailed explanation of a sample Assemblies Cost Table. Included are an illustration and accompanying system descriptions. Additionally, related systems and price sheets may be included. Next to each bold number below is the item being described with the appropriate component of the sample entry following in parenthesis. General contractors should add an additional markup to the figures shown in the Assemblies section. Note: Throughout this section, the words assembly and system are used interchangeably.

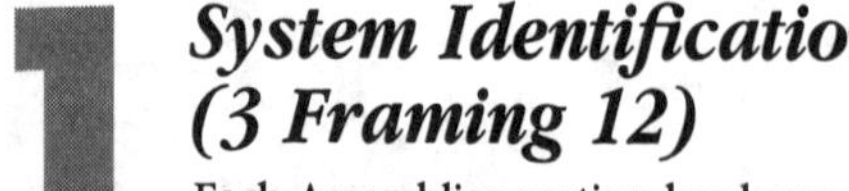

System Identification (3 Framing 12)

Each Assemblies section has been assigned a unique identification number, component category, system number and system description.

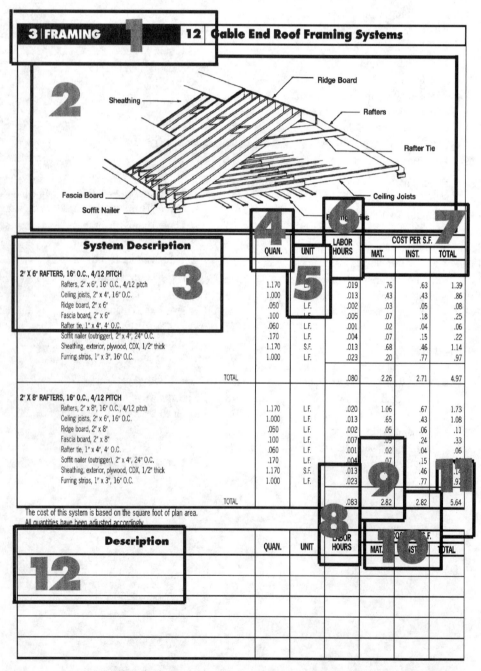

3 | FRAMING | **12 | Gable End Roof Framing Systems**

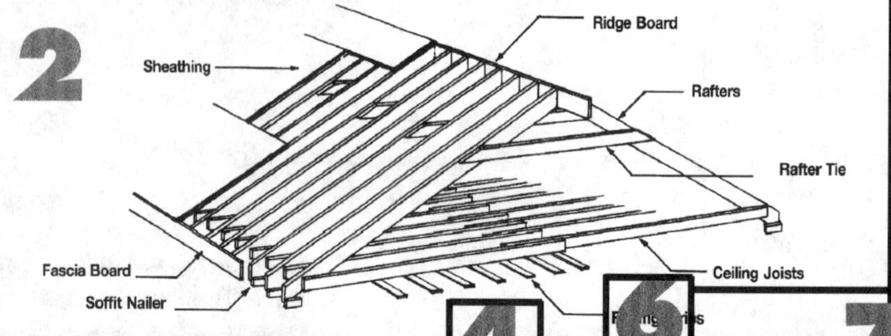

Ridge Board
Sheathing
Rafters
Rafter Tie
Fascia Board
Soffit Nailer
Ceiling Joists
Furring Strips

System Description		QUAN.	UNIT	LABOR HOURS	COST PER S.F. MAT.	COST PER S.F. INST.	COST PER S.F. TOTAL
2" X 6" RAFTERS, 16" O.C., 4/12 PITCH							
Rafters, 2" x 6", 16" O.C., 4/12 pitch		1.170	L.F.	.019	.76	.63	1.39
Ceiling joists, 2" x 4", 16" O.C.		1.000	[illegible]	.013	.43	.43	.86
Ridge board, 2" x 6"		.050	L.F.	.002	.03	.05	.08
Fascia board, 2" x 6"		.100	L.F.	.005	.07	.18	.25
Rafter tie, 1" x 4", 4' O.C.		.060	L.F.	.001	.02	.04	.06
Soffit nailer (outrigger), 2" x 4", 24" O.C.		.170	L.F.	.004	.07	.15	.22
Sheathing, exterior, plywood, CDX, 1/2" thick		1.170	S.F.	.013	.68	.46	1.14
Furring strips, 1" x 3", 16" O.C.		1.000	L.F.	.023	.20	.77	.97
	TOTAL			.080	2.26	2.71	4.97
2" X 8" RAFTERS, 16" O.C., 4/12 PITCH							
Rafters, 2" x 8", 16" O.C., 4/12 pitch		1.170	L.F.	.020	1.06	.67	1.73
Ceiling joists, 2" x 6", 16" O.C.		1.000	L.F.	.013	.65	.43	1.08
Ridge board, 2" x 8"		.050	L.F.	.002	.05	.06	.11
Fascia board, 2" x 8"		.100	L.F.	.007	.09	.24	.33
Rafter tie, 1" x 4", 4' O.C.		.060	L.F.	.001	.02	.04	.06
Soffit nailer (outrigger), 2" x 4", 24" O.C.		.170	L.F.	.004	.07	.15	[illegible]
Sheathing, exterior, plywood, CDX, 1/2" thick		1.170	S.F.	.013	[illegible]	.46	[illegible]
Furring strips, 1" x 3", 16" O.C.		1.000	L.F.	.023	[illegible]	.77	[illegible]
	TOTAL			.083	2.82	2.82	5.64

The cost of this system is based on the square foot of plan area.
All quantities have been adjusted accordingly.

Description	QUAN.	UNIT	LABOR HOURS	COST PER S.F. MAT.	COST PER S.F. INST.	COST PER S.F. TOTAL

Important: See the Reference Section for critical supporting data - Reference Nos., Crews & Location Factors

2 *Illustration*

At the top of most assembly pages is an illustration with individual components labeled. Elements involved in the total system function are shown.

3 *System Description (2" x 6" Rafters, 16" O.C., 4/12 Pitch)*

The components of a typical system are listed separately to show what has been included in the development of the total system price. Each page includes a brief outline of any special conditions to be used when pricing a system. Alternative components can be found on the opposite page. Simply insert any chosen new element into the chart to develop a custom system.

4 *Quantities for Each Component*

Each material in a system is shown with the quantity required for the system unit. For example, there are 1.170 L.F. of rafter per S.F. of plan area.

5 *Unit of Measure for Each Component*

The abbreviated designation indicates the unit of measure, as defined by industry standards, upon which the individual component has been priced. In this example, items are priced by the linear foot (L.F.) or the square foot (S.F.).

6 *Labor Hours*

This is the amount of time it takes to install the quantity of the individual component.

7 *Unit of Measure (Cost per S.F.)*

In the three right-hand columns, each cost figure is adjusted to agree with the unit of measure for the entire system. In this case, cost per S.F. is the common unit of measure.

8 *Labor Hours (.083)*

The labor hours column shows the amount of time necessary to install the system per the unit of measure. For example, it takes .083 labor hours to install one square foot (plan area) of this roof framing system.

9 *Materials (2.82)*

This column contains the material cost of each element. These cost figures include 10% for profit.

10 *Installation (2.82)*

This column contains labor and equipment costs. Labor rates include bare cost and the installing contractor's overhead and profit. On the average, the labor cost will be 71.8% over the bare labor cost. Equipment costs include 10% for profit.

11 *Totals (5.64)*

The figure in this column is the sum of the material and installation costs.

12 *Work Sheet*

Using the selective price sheet on the page opposite each system, it is possible to create estimates with alternative items for any number of systems.

Division 1
Site Work

Backfill

Excavate

System Description	QUAN.	UNIT	LABOR HOURS	COST EACH MAT.	COST EACH INST.	COST EACH TOTAL
BUILDING, 24′ X 38′, 4′ DEEP						
Clear and strip, dozer, light trees, 30′ from building	.190	Acre	9.120		470.25	470.25
Excavate, backhoe	174.000	C.Y.	2.320		252.30	252.30
Backfill, dozer, 4″ lifts, no compaction	87.000	C.Y.	.580		83.52	83.52
Rough grade, dozer, 30′ from building	87.000	C.Y.	.580		83.52	83.52
TOTAL			12.600		889.59	889.59
BUILDING, 26′ X 46′, 4′ DEEP						
Clear and strip, dozer, light trees, 30′ from building	.210	Acre	10.080		519.75	519.75
Excavate, backhoe	201.000	C.Y.	2.680		291.45	291.45
Backfill, dozer, 4″ lifts, no compaction	100.000	C.Y.	.667		96.00	96.00
Rough grade, dozer, 30′ from building	100.000	C.Y.	.667		96.00	96.00
TOTAL			14.094		1003.20	1003.20
BUILDING, 26′ X 60′, 4′ DEEP						
Clear and strip, dozer, light trees, 30′ from building	.240	Acre	11.520		594.00	594.00
Excavate, backhoe	240.000	C.Y.	3.200		348.00	348.00
Backfill, dozer, 4″ lifts, no compaction	120.000	C.Y.	.800		115.20	115.20
Rough grade, dozer, 30′ from building	120.000	C.Y.	.800		115.20	115.20
TOTAL			16.320		1172.40	1172.40
BUILDING, 30′ X 66′, 4′ DEEP						
Clear and strip, dozer, light trees, 30′ from building	.260	Acre	12.480		643.50	643.50
Excavate, backhoe	268.000	C.Y.	3.573		388.60	388.60
Backfill, dozer, 4″ lifts, no compaction	134.000	C.Y.	.893		128.64	128.64
Rough grade, dozer, 30′ from building	134.000	C.Y.	.893		128.64	128.64
TOTAL			17.839		1289.38	1289.38

The costs in this system are on a cost each basis.
Quantities are based on 1′-0″ clearance on each side of footing.

Description	QUAN.	UNIT	LABOR HOURS	COST EACH MAT.	COST EACH INST.	COST EACH TOTAL

Footing Excavation Price Sheet	QUAN.	UNIT	LABOR HOURS	COST EACH		
				MAT.	INST.	TOTAL
Clear and grub, medium brush, 30′ from building, 24′ x 38′	.190	Acre	9.120		470.00	470.00
26′ x 46′	.210	Acre	10.080		520.00	520.00
26′ x 60′	.240	Acre	11.520		595.00	595.00
30′ x 66′	.260	Acre	12.480		640.00	640.00
Light trees, to 6″ dia. cut & chip, 24′ x 38′	.190	Acre	9.120		470.00	470.00
26′ x 46′	.210	Acre	10.080		520.00	520.00
26′ x 60′	.240	Acre	11.520		595.00	595.00
30′ x 66′	.260	Acre	12.480		640.00	640.00
Medium trees, to 10″ dia. cut & chip, 24′ x 38′	.190	Acre	13.029		675.00	675.00
26′ x 46′	.210	Acre	14.400		745.00	745.00
26′ x 60′	.240	Acre	16.457		855.00	855.00
30′ x 66′	.260	Acre	17.829		925.00	925.00
Excavation, footing, 24′ x 38′, 2′ deep	68.000	C.Y.	.907		99.00	99.00
4′ deep	174.000	C.Y.	2.320		252.00	252.00
8′ deep	384.000	C.Y.	5.120		555.00	555.00
26′ x 46′, 2′ deep	79.000	C.Y.	1.053		115.00	115.00
4′ deep	201.000	C.Y.	2.680		292.00	292.00
8′ deep	404.000	C.Y.	5.387		585.00	585.00
26′ x 60′, 2′ deep	94.000	C.Y.	1.253		136.00	136.00
4′ deep	240.000	C.Y.	3.200		350.00	350.00
8′ deep	483.000	C.Y.	6.440		700.00	700.00
30′ x 66′, 2′ deep	105.000	C.Y.	1.400		153.00	153.00
4′ deep	268.000	C.Y.	3.573		390.00	390.00
8′ deep	539.000	C.Y.	7.187		785.00	785.00
Backfill, 24′ x 38′, 2″ lifts, no compaction	34.000	C.Y.	.227		32.50	32.50
Compaction, air tamped	34.000	C.Y.	2.267		224.00	224.00
4″ lifts, no compaction	87.000	C.Y.	.580		83.50	83.50
Compaction, air tamped	87.000	C.Y.	5.800		575.00	575.00
8″ lifts, no compaction	192.000	C.Y.	1.280		184.00	184.00
Compaction, air tamped	192.000	C.Y.	12.800		1250.00	1250.00
26′ x 46′, 2″ lifts, no compaction	40.000	C.Y.	.267		38.50	38.50
Compaction, air tamped	40.000	C.Y.	2.667		263.00	263.00
4″ lifts, no compaction	100.000	C.Y.	.667		96.00	96.00
Compaction, air tamped	100.000	C.Y.	6.667		660.00	660.00
8″ lifts, no compaction	202.000	C.Y.	1.347		194.00	194.00
Compaction, air tamped	202.000	C.Y.	13.467		1325.00	1325.00
26′ x 60′, 2″ lifts, no compaction	47.000	C.Y.	.313		45.50	45.50
Compaction, air tamped	47.000	C.Y.	3.133		310.00	310.00
4″ lifts, no compaction	120.000	C.Y.	.800		116.00	116.00
Compaction, air tamped	120.000	C.Y.	8.000		785.00	785.00
8″ lifts, no compaction	242.000	C.Y.	1.613		232.00	232.00
Compaction, air tamped	242.000	C.Y.	16.133		1600.00	1600.00
30′ x 66′, 2″ lifts, no compaction	53.000	C.Y.	.353		50.50	50.50
Compaction, air tamped	53.000	C.Y.	3.533		350.00	350.00
4″ lifts, no compaction	134.000	C.Y.	.893		129.00	129.00
Compaction, air tamped	134.000	C.Y.	8.933		880.00	880.00
8″ lifts, no compaction	269.000	C.Y.	1.793		258.00	258.00
Compaction, air tamped	269.000	C.Y.	17.933		1775.00	1775.00
Rough grade, 30′ from building, 24′ x 38′	87.000	C.Y.	.580		83.50	83.50
26′ x 46′	100.000	C.Y.	.667		96.00	96.00
26′ x 60′	120.000	C.Y.	.800		116.00	116.00
30′ x 66′	134.000	C.Y.	.893		129.00	129.00

08 Foundation Excavation Systems

Backfill

Excavate

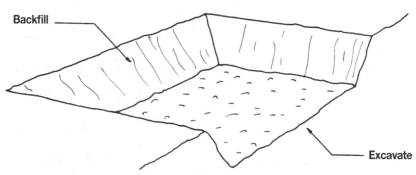

System Description	QUAN.	UNIT	LABOR HOURS	COST EACH		
				MAT.	INST.	TOTAL
BUILDING, 24′ X 38′, 8′ DEEP						
Clear & grub, dozer, medium brush, 30′ from building	.190	Acre	2.027		171.00	171.00
Excavate, track loader, 1-1/2 C.Y. bucket	550.000	C.Y.	7.857		616.00	616.00
Backfill, dozer, 8″ lifts, no compaction	180.000	C.Y.	1.200		172.80	172.80
Rough grade, dozer, 30′ from building	280.000	C.Y.	1.867		268.80	268.80
TOTAL			12.951		1228.60	1228.60
BUILDING, 26′ X 46′, 8′ DEEP						
Clear & grub, dozer, medium brush, 30′ from building	.210	Acre	2.240		189.00	189.00
Excavate, track loader, 1-1/2 C.Y. bucket	672.000	C.Y.	9.600		752.64	752.64
Backfill, dozer, 8″ lifts, no compaction	220.000	C.Y.	1.467		211.20	211.20
Rough grade, dozer, 30′ from building	340.000	C.Y.	2.267		326.40	326.40
TOTAL			15.574		1479.24	1479.24
BUILDING, 26′ X 60′, 8′ DEEP						
Clear & grub, dozer, medium brush, 30′ from building	.240	Acre	2.560		216.00	216.00
Excavate, track loader, 1-1/2 C.Y. bucket	829.000	C.Y.	11.843		928.48	928.48
Backfill, dozer, 8″ lifts, no compaction	270.000	C.Y.	1.800		259.20	259.20
Rough grade, dozer, 30′ from building	420.000	C.Y.	2.800		403.20	403.20
TOTAL			19.003		1806.88	1806.88
BUILDING, 30′ X 66′, 8′ DEEP						
Clear & grub, dozer, medium brush, 30′ from building	.260	Acre	2.773		234.00	234.00
Excavate, track loader, 1-1/2 C.Y. bucket	990.000	C.Y.	14.143		1108.80	1108.80
Backfill dozer, 8″ lifts, no compaction	320.000	C.Y.	2.133		307.20	307.20
Rough grade, dozer, 30′ from building	500.000	C.Y.	3.333		480.00	480.00
TOTAL			22.382		2130.00	2130.00

The costs in this system are on a cost each basis.
Quantities are based on 1′-0″ clearance beyond footing projection.

Description	QUAN.	UNIT	LABOR HOURS	COST EACH		
				MAT.	INST.	TOTAL

Important: See the Reference Section for critical supporting data - Reference Nos., Crews & Location Factors

Foundation Excavation Price Sheet	QUAN.	UNIT	LABOR HOURS	COST EACH		
				MAT.	INST.	TOTAL
Clear & grub, medium brush, 30′ from building, 24′ x 38′	.190	Acre	2.027		171.00	171.00
26′ x 46′	.210	Acre	2.240		189.00	189.00
26′ x 60′	.240	Acre	2.560		217.00	217.00
30′ x 66′	.260	Acre	2.773		234.00	234.00
Light trees, to 6″ dia. cut & chip, 24′ x 38′	.190	Acre	9.120		470.00	470.00
26′ x 46′	.210	Acre	10.080		520.00	520.00
26′ x 60′	.240	Acre	11.520		595.00	595.00
30′ x 66′	.260	Acre	12.480		640.00	640.00
Medium trees, to 10″ dia. cut & chip, 24′ x 38′	.190	Acre	13.029		675.00	675.00
26′ x 46′	.210	Acre	14.400		745.00	745.00
26′ x 60′	.240	Acre	16.457		855.00	855.00
30′ x 66′	.260	Acre	17.829		925.00	925.00
Excavation, basement, 24′ x 38′, 2′ deep	98.000	C.Y.	1.400		110.00	110.00
4′ deep	220.000	C.Y.	3.143		246.00	246.00
8′ deep	550.000	C.Y.	7.857		620.00	620.00
26′ x 46′, 2′ deep	123.000	C.Y.	1.757		138.00	138.00
4′ deep	274.000	C.Y.	3.914		305.00	305.00
8′ deep	672.000	C.Y.	9.600		750.00	750.00
26′ x 60′, 2′ deep	157.000	C.Y.	2.243		176.00	176.00
4′ deep	345.000	C.Y.	4.929		385.00	385.00
8′ deep	829.000	C.Y.	11.843		930.00	930.00
30′ x 66′, 2′ deep	192.000	C.Y.	2.743		215.00	215.00
4′ deep	419.000	C.Y.	5.986		470.00	470.00
8′ deep	990.000	C.Y.	14.143		1100.00	1100.00
Backfill, 24′ x 38′, 2″ lifts, no compaction	32.000	C.Y.	.213		30.50	30.50
Compaction, air tamped	32.000	C.Y.	2.133		210.00	210.00
4″ lifts, no compaction	72.000	C.Y.	.480		69.50	69.50
Compaction, air tamped	72.000	C.Y.	4.800		475.00	475.00
8″ lifts, no compaction	180.000	C.Y.	1.200		173.00	173.00
Compaction, air tamped	180.000	C.Y.	12.000		1175.00	1175.00
26′ x 46′, 2″ lifts, no compaction	40.000	C.Y.	.267		38.50	38.50
Compaction, air tamped	40.000	C.Y.	2.667		263.00	263.00
4″ lifts, no compaction	90.000	C.Y.	.600		86.50	86.50
Compaction, air tamped	90.000	C.Y.	6.000		595.00	595.00
8″ lifts, no compaction	220.000	C.Y.	1.467		212.00	212.00
Compacton, air tamped	220.000	C.Y.	14.667		1450.00	1450.00
26′ x 60′, 2″ lifts, no compaction	50.000	C.Y.	.333		48.00	48.00
Compaction, air tamped	50.000	C.Y.	3.333		330.00	330.00
4″ lifts, no compaction	110.000	C.Y.	.733		106.00	106.00
Compaction, air tamped	110.000	C.Y.	7.333		720.00	720.00
8″ lifts, no compaction	270.000	C.Y.	1.800		260.00	260.00
Compaction, air tamped	270.000	C.Y.	18.000		1775.00	1775.00
30′ x 66′, 2″ lifts, no compaction	60.000	C.Y.	.400		57.50	57.50
Compaction, air tamped	60.000	C.Y.	4.000		395.00	395.00
4″ lifts, no compaction	130.000	C.Y.	.867		125.00	125.00
Compaction, air tamped	130.000	C.Y.	8.667		855.00	855.00
8″ lifts, no compaction	320.000	C.Y.	2.133		310.00	310.00
Compaction, air tamped	320.000	C.Y.	21.333		2100.00	2100.00
Rough grade, 30′ from building, 24′ x 38′	280.000	C.Y.	1.867		269.00	269.00
26′ x 46′	340.000	C.Y.	2.267		325.00	325.00
26′ x 60′	420.000	C.Y.	2.800		405.00	405.00
30′ x 66′	500.000	C.Y.	3.333		480.00	480.00

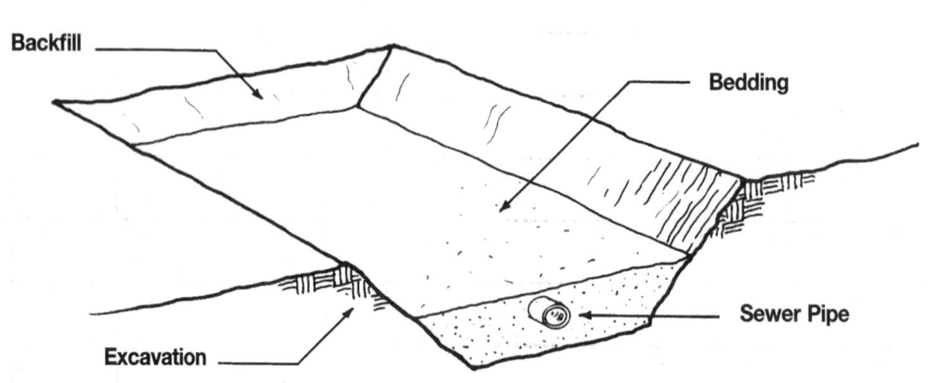

System Description	QUAN.	UNIT	LABOR HOURS	COST PER L.F. MAT.	INST.	TOTAL
2′ DEEP						
Excavation, backhoe	.296	C.Y.	.032		1.35	1.35
Bedding, sand	.111	C.Y.	.044	1.38	1.32	2.70
Utility, sewer, 6″ cast iron	1.000	L.F.	.283	12.21	8.90	21.11
Backfill, incl. compaction	.185	C.Y.	.044		1.07	1.07
TOTAL			.403	13.59	12.64	26.23
4′ DEEP						
Excavation, backhoe	.889	C.Y.	.095		4.06	4.06
Bedding, sand	.111	C.Y.	.044	1.38	1.32	2.70
Utility, sewer, 6″ cast iron	1.000	L.F.	.283	12.21	8.90	21.11
Backfill, incl. compaction	.778	C.Y.	.183		4.51	4.51
TOTAL			.605	13.59	18.79	32.38
6′ DEEP						
Excavation, backhoe	1.770	C.Y.	.189		8.09	8.09
Bedding, sand	.111	C.Y.	.044	1.38	1.32	2.70
Utility, sewer, 6″ cast iron	1.000	L.F.	.283	12.21	8.90	21.11
Backfill, incl. compaction	1.660	C.Y.	.391		9.63	9.63
TOTAL			.907	13.59	27.94	41.53
8′ DEEP						
Excavation, backhoe	2.960	C.Y.	.316		13.53	13.53
Bedding, sand	.111	C.Y.	.044	1.38	1.32	2.70
Utility, sewer, 6″ cast iron	1.000	L.F.	.283	12.21	8.90	21.11
Backfill, incl. compaction	2.850	C.Y.	.671		16.53	16.53
TOTAL			1.314	13.59	40.28	53.87

The costs in this system are based on a cost per linear foot of trench, and based on 2′ wide at bottom of trench up to 6′ deep.

Description	QUAN.	UNIT	LABOR HOURS	COST PER L.F. MAT.	INST.	TOTAL

Important: See the Reference Section for critical supporting data - Reference Nos., Crews & Location Factors

Utility Trenching Price Sheet	QUAN.	UNIT	LABOR HOURS	COST PER L.F.		
				MAT.	INST.	TOTAL
Excavation, bottom of trench 2′ wide, 2′ deep	.296	C.Y.	.032		1.35	1.35
4′ deep	.889	C.Y.	.095		4.06	4.06
6′ deep	1.770	C.Y.	.142		6.75	6.75
8′ deep	2.960	C.Y.	.105		10.20	10.20
Bedding, sand, bottom of trench 2′ wide, no compaction, pipe, 2″ diameter	.070	C.Y.	.028	.87	.83	1.70
4″ diameter	.084	C.Y.	.034	1.04	.99	2.03
6″ diameter	.105	C.Y.	.042	1.30	1.24	2.54
8″ diameter	.122	C.Y.	.049	1.51	1.44	2.95
Compacted, pipe, 2″ diameter	.074	C.Y.	.030	.92	.87	1.79
4″ diameter	.092	C.Y.	.037	1.14	1.09	2.23
6″ diameter	.111	C.Y.	.044	1.38	1.32	2.70
8″ diameter	.129	C.Y.	.052	1.60	1.53	3.13
3/4″ stone, bottom of trench 2′ wide, pipe, 4″ diameter	.082	C.Y.	.033	1.02	.97	1.99
6″ diameter	.099	C.Y.	.040	1.23	1.17	2.40
3/8″ stone, bottom of trench 2′ wide, pipe, 4″ diameter	.084	C.Y.	.034	1.04	.99	2.03
6″ diameter	.102	C.Y.	.041	1.26	1.20	2.46
Utilities, drainage & sewerage, corrugated plastic, 6″ diameter	1.000	L.F.	.069	3.16	1.78	4.94
8″ diameter	1.000	L.F.	.072	4.84	1.86	6.70
Bituminous fiber, 4″ diameter	1.000	L.F.	.064	2.43	1.66	4.09
6″ diameter	1.000	L.F.	.069	3.16	1.78	4.94
8″ diameter	1.000	L.F.	.072	4.84	1.86	6.70
Concrete, non-reinforced, 6″ diameter	1.000	L.F.	.181	3.82	5.65	9.47
8″ diameter	1.000	L.F.	.214	4.20	6.65	10.85
PVC, SDR 35, 4″ diameter	1.000	L.F.	.064	2.43	1.66	4.09
6″ diameter	1.000	L.F.	.069	3.16	1.78	4.94
8″ diameter	1.000	L.F.	.072	4.84	1.86	6.70
Vitrified clay, 4″ diameter	1.000	L.F.	.091	1.97	2.35	4.32
6″ diameter	1.000	L.F.	.120	3.22	3.11	6.33
8″ diameter	1.000	L.F.	.140	4.55	4.55	9.10
Gas & service, polyethylene, 1-1/4″ diameter	1.000	L.F.	.059	.64	1.74	2.38
Steel sched.40, 1″ diameter	1.000	L.F.	.107	2.63	3.86	6.49
2″ diameter	1.000	L.F.	.114	4.13	4.13	8.26
Sub-drainage, PVC, perforated, 3″ diameter	1.000	L.F.	.064	2.43	1.66	4.09
4″ diameter	1.000	L.F.	.064	2.43	1.66	4.09
5″ diameter	1.000	L.F.	.069	3.16	1.78	4.94
6″ diameter	1.000	L.F.	.069	3.16	1.78	4.94
Porous wall concrete, 4″ diameter	1.000	L.F.	.072	1.96	1.86	3.82
Vitrified clay, perforated, 4″ diameter	1.000	L.F.	.120	2.12	3.74	5.86
6″ diameter	1.000	L.F.	.152	3.30	4.75	8.05
Water service, copper, type K, 3/4″	1.000	L.F.	.083	1.56	3.02	4.58
1″ diameter	1.000	L.F.	.093	2.01	3.39	5.40
PVC, 3/4″	1.000	L.F.	.121	1.74	4.39	6.13
1″ diameter	1.000	L.F.	.134	2.18	4.85	7.03
Backfill, bottom of trench 2′ wide no compact, 2′ deep, pipe, 2″ diameter	.226	C.Y.	.053		1.31	1.31
4″ diameter	.212	C.Y.	.050		1.23	1.23
6″ diameter	.185	C.Y.	.044		1.07	1.07
4′ deep, pipe, 2″ diameter	.819	C.Y.	.193		4.75	4.75
4″ diameter	.805	C.Y.	.189		4.67	4.67
6″ diameter	.778	C.Y.	.183		4.51	4.51
6′ deep, pipe, 2″ diameter	1.700	C.Y.	.400		9.85	9.85
4″ diameter	1.690	C.Y.	.398		9.80	9.80
6″ diameter	1.660	C.Y.	.391		9.65	9.65
8′ deep, pipe, 2″ diameter	2.890	C.Y.	.680		16.75	16.75
4″ diameter	2.870	C.Y.	.675		16.65	16.65
6″ diameter	2.850	C.Y.	.671		16.55	16.55

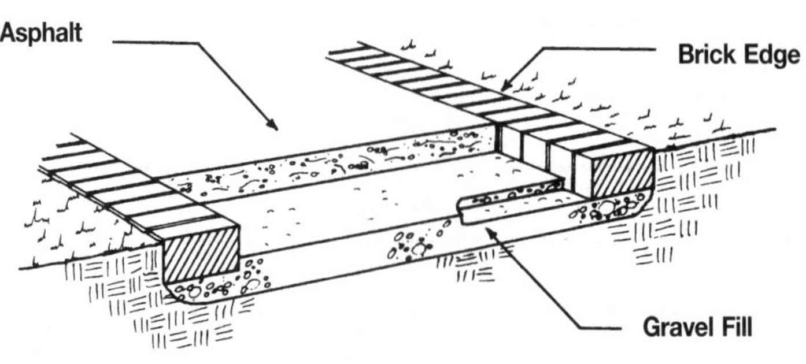

System Description	QUAN.	UNIT	LABOR HOURS	COST PER S.F.		
				MAT.	INST.	TOTAL
ASPHALT SIDEWALK SYSTEM, 3′ WIDE WALK						
Gravel fill, 4″ deep	1.000	S.F.	.001	.21	.03	.24
Compact fill	.012	C.Y.	.001		.01	.01
Handgrade	1.000	S.F.	.004		.11	.11
Walking surface, bituminous paving, 2″ thick	1.000	S.F.	.007	.39	.22	.61
Edging, brick, laid on edge	.670	L.F.	.079	.93	2.40	3.33
TOTAL			.092	1.53	2.77	4.30
CONCRETE SIDEWALK SYSTEM, 3′ WIDE WALK						
Gravel fill, 4″ deep	1.000	S.F.	.001	.21	.03	.24
Compact fill	.012	C.Y.	.001		.01	.01
Handgrade	1.000	S.F.	.004		.11	.11
Walking surface, concrete, 4″ thick	1.000	S.F.	.040	1.15	1.19	2.34
Edging, brick, laid on edge	.670	L.F.	.079	.93	2.40	3.33
TOTAL			.125	2.29	3.74	6.03
PAVERS, BRICK SIDEWALK SYSTEM, 3′ WIDE WALK						
Sand base fill, 4″ deep	1.000	S.F.	.001	.14	.06	.20
Compact fill	.012	C.Y.	.001		.01	.01
Handgrade	1.000	S.F.	.004		.11	.11
Walking surface, brick pavers	1.000	S.F.	.160	3.32	4.84	8.16
Edging, redwood, untreated, 1″ x 4″	.670	L.F.	.032	1.95	1.09	3.04
TOTAL			.198	5.41	6.11	11.52

The costs in this system are based on a cost per square foot of sidewalk area. Concrete used is 3000 p.s.i.

Description	QUAN.	UNIT	LABOR HOURS	COST PER S.F.		
				MAT.	INST.	TOTAL

Important: See the Reference Section for critical supporting data - Reference Nos., Crews & Location Factors

Sidewalk Price Sheet	QUAN.	UNIT	LABOR HOURS	COST PER S.F.		
				MAT.	INST.	TOTAL
Base, crushed stone, 3" deep	1.000	S.F.	.001	.52	.06	.58
6" deep	1.000	S.F.	.001	1.04	.06	1.10
9" deep	1.000	S.F.	.002	1.56	.09	1.65
12" deep	1.000	S.F.	.002	1.66	.11	1.77
Bank run gravel, 6" deep	1.000	S.F.	.001	.32	.05	.37
9" deep	1.000	S.F.	.001	.47	.07	.54
12" deep	1.000	S.F.	.001	.63	.08	.71
Compact base, 3" deep	.009	C.Y.	.001		.01	.01
6" deep	.019	C.Y.	.001		.03	.03
9" deep	.028	C.Y.	.001		.04	.04
Handgrade	1.000	S.F.	.004		.11	.11
Surface, brick, pavers dry joints, laid flat, running bond	1.000	S.F.	.160	3.32	4.84	8.16
Basket weave	1.000	S.F.	.168	3.04	5.10	8.14
Herringbone	1.000	S.F.	.174	3.04	5.25	8.29
Laid on edge, running bond	1.000	S.F.	.229	2.19	6.90	9.09
Mortar jts. laid flat, running bond	1.000	S.F.	.192	3.98	5.80	9.78
Basket weave	1.000	S.F.	.202	3.65	6.10	9.75
Herringbone	1.000	S.F.	.209	3.65	6.30	9.95
Laid on edge, running bond	1.000	S.F.	.274	2.63	8.30	10.93
Bituminous paving, 1-1/2" thick	1.000	S.F.	.006	.29	.17	.46
2" thick	1.000	S.F.	.007	.39	.22	.61
2-1/2" thick	1.000	S.F.	.008	.49	.23	.72
Sand finish, 3/4" thick	1.000	S.F.	.001	.17	.07	.24
1" thick	1.000	S.F.	.001	.21	.09	.30
Concrete, reinforced, broom finish, 4" thick	1.000	S.F.	.040	1.15	1.19	2.34
5" thick	1.000	S.F.	.044	1.54	1.31	2.85
6" thick	1.000	S.F.	.047	1.79	1.40	3.19
Crushed stone, white marble, 3" thick	1.000	S.F.	.009	.25	.23	.48
Bluestone, 3" thick	1.000	S.F.	.009	.20	.23	.43
Flagging, bluestone, 1"	1.000	S.F.	.198	4.40	5.95	10.35
1-1/2"	1.000	S.F.	.188	8.00	5.70	13.70
Slate, natural cleft, 3/4"	1.000	S.F.	.174	1.98	5.25	7.23
Random rect., 1/2"	1.000	S.F.	.152	4.29	4.61	8.90
Granite blocks	1.000	S.F.	.174	5.80	5.25	11.05
Edging, corrugated aluminum, 4", 3' wide walk	.666	L.F.	.008	.21	.28	.49
4' wide walk	.500	L.F.	.006	.16	.21	.37
6", 3' wide walk	.666	L.F.	.010	.26	.33	.59
4' wide walk	.500	L.F.	.007	.20	.25	.45
Redwood-cedar-cypress, 1" x 4", 3' wide walk	.666	L.F.	.021	.98	.72	1.70
4' wide walk	.500	L.F.	.016	.74	.54	1.28
2" x 4", 3' wide walk	.666	L.F.	.032	1.95	1.09	3.04
4' wide walk	.500	L.F.	.024	1.47	.82	2.29
Brick, dry joints, 3' wide walk	.666	L.F.	.079	.93	2.40	3.33
4' wide walk	.500	L.F.	.059	.70	1.79	2.49
Mortar joints, 3' wide walk	.666	L.F.	.095	1.12	2.88	4.00
4' wide walk	.500	L.F.	.071	.83	2.15	2.98

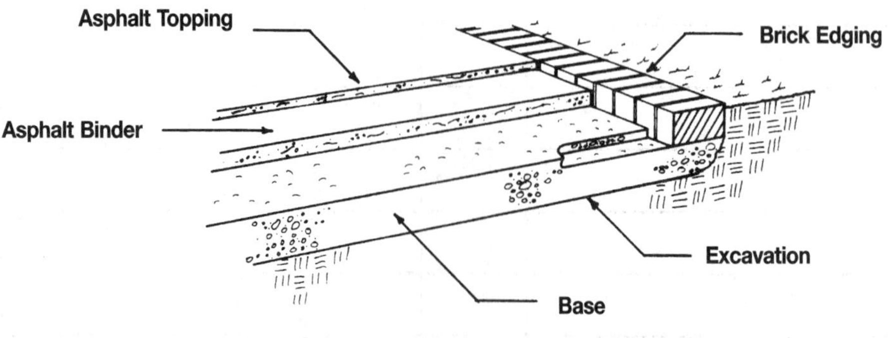

System Description	QUAN.	UNIT	LABOR HOURS	COST PER S.F.		
				MAT.	INST.	TOTAL
ASPHALT DRIVEWAY TO 10′ WIDE						
Excavation, driveway to 10′ wide, 6″ deep	.019	C.Y.	.001		.02	.02
Base, 6″ crushed stone	1.000	S.F.	.001	1.04	.06	1.10
Handgrade base	1.000	S.F.	.004		.11	.11
Surface, asphalt, 2″ thick base, 1″ topping	1.000	S.F.	.003	.60	.21	.81
Edging, brick pavers	.200	L.F.	.024	.28	.72	1.00
TOTAL			.033	1.92	1.12	3.04
CONCRETE DRIVEWAY TO 10′ WIDE						
Excavation, driveway to 10′ wide, 6″ deep	.019	C.Y.	.001		.02	.02
Base, 6″ crushed stone	1.000	S.F.	.001	1.04	.06	1.10
Handgrade base	1.000	S.F.	.004		.11	.11
Surface, concrete, 4″ thick	1.000	S.F.	.040	1.15	1.19	2.34
Edging, brick pavers	.200	L.F.	.024	.28	.72	1.00
TOTAL			.070	2.47	2.10	4.57
PAVERS, BRICK DRIVEWAY TO 10′ WIDE						
Excavation, driveway to 10′ wide, 6″ deep	.019	C.Y.	.001		.02	.02
Base, 6″ sand	1.000	S.F.	.001	.22	.09	.31
Handgrade base	1.000	S.F.	.004		.11	.11
Surface, pavers, brick laid flat, running bond	1.000	S.F.	.160	3.32	4.84	8.16
Edging, redwood, untreated, 2″ x 4″	.200	L.F.	.010	.59	.33	.92
TOTAL			.176	4.13	5.39	9.52

Description	QUAN.	UNIT	LABOR HOURS	COST PER S.F.		
				MAT.	INST.	TOTAL

Important: See the Reference Section for critical supporting data - Reference Nos., Crews & Location Factors

Driveway Price Sheet	QUAN.	UNIT	LABOR HOURS	COST PER S.F.		
				MAT.	INST.	TOTAL
Excavation, by machine, 10′ wide, 6″ deep	.019	C.Y.	.001		.02	.02
12″ deep	.037	C.Y.	.001		.06	.06
18″ deep	.055	C.Y.	.001		.08	.08
20′ wide, 6″ deep	.019	C.Y.	.001		.02	.02
12″ deep	.037	C.Y.	.001		.06	.06
18″ deep	.055	C.Y.	.001		.08	.08
Base, crushed stone, 10′ wide, 3″ deep	1.000	S.F.	.001	.52	.04	.56
6″ deep	1.000	S.F.	.001	1.04	.06	1.10
9″ deep	1.000	S.F.	.002	1.56	.09	1.65
20′ wide, 3″ deep	1.000	S.F.	.001	.52	.04	.56
6″ deep	1.000	S.F.	.001	1.04	.06	1.10
9″ deep	1.000	S.F.	.002	1.56	.09	1.65
Bank run gravel, 10′ wide, 3″ deep	1.000	S.F.	.001	.16	.03	.19
6″ deep	1.000	S.F.	.001	.32	.05	.37
9″ deep	1.000	S.F.	.001	.47	.07	.54
20′ wide, 3″ deep	1.000	S.F.	.001	.16	.03	.19
6″ deep	1.000	S.F.	.001	.32	.05	.37
9″ deep	1.000	S.F.	.001	.47	.07	.54
Handgrade, 10′ wide	1.000	S.F.	.004		.11	.11
20′ wide	1.000	S.F.	.004		.11	.11
Surface, asphalt, 10′ wide, 3/4″ topping, 1″ base	1.000	S.F.	.002	.47	.16	.63
2″ base	1.000	S.F.	.003	.56	.19	.75
1″ topping, 1″ base	1.000	S.F.	.002	.51	.18	.69
2″ base	1.000	S.F.	.003	.60	.21	.81
20′ wide, 3/4″ topping, 1″ base	1.000	S.F.	.002	.47	.16	.63
2″ base	1.000	S.F.	.003	.56	.19	.75
1″ topping, 1″ base	1.000	S.F.	.002	.51	.18	.69
2″ base	1.000	S.F.	.003	.60	.21	.81
Concrete, 10′ wide, 4″ thick	1.000	S.F.	.040	1.15	1.19	2.34
6″ thick	1.000	S.F.	.047	1.79	1.40	3.19
20′ wide, 4″ thick	1.000	S.F.	.040	1.15	1.19	2.34
6″ thick	1.000	S.F.	.047	1.79	1.40	3.19
Paver, brick 10′ wide dry joints, running bond, laid flat	1.000	S.F.	.160	3.32	4.84	8.16
Laid on edge	1.000	S.F.	.229	2.19	6.90	9.09
Mortar joints, laid flat	1.000	S.F.	.192	3.98	5.80	9.78
Laid on edge	1.000	S.F.	.274	2.63	8.30	10.93
20′ wide, running bond, dry jts., laid flat	1.000	S.F.	.160	3.32	4.84	8.16
Laid on edge	1.000	S.F.	.229	2.19	6.90	9.09
Mortar joints, laid flat	1.000	S.F.	.192	3.98	5.80	9.78
Laid on edge	1.000	S.F.	.274	2.63	8.30	10.93
Crushed stone, 10′ wide, white marble, 3″	1.000	S.F.	.009	.25	.23	.48
Bluestone, 3″	1.000	S.F.	.009	.20	.23	.43
20′ wide, white marble, 3″	1.000	S.F.	.009	.25	.23	.48
Bluestone, 3″	1.000	S.F.	.009	.20	.23	.43
Soil cement, 10′ wide	1.000	S.F.	.007	.21	.57	.78
20′ wide	1.000	S.F.	.007	.21	.57	.78
Granite blocks, 10′ wide	1.000	S.F.	.174	5.80	5.25	11.05
20′ wide	1.000	S.F.	.174	5.80	5.25	11.05
Asphalt block, solid 1-1/4″ thick	1.000	S.F.	.119	3.12	3.58	6.70
Solid 3″ thick	1.000	S.F.	.123	4.37	3.72	8.09
Edging, brick, 10′ wide	.200	L.F.	.024	.28	.72	1.00
20′ wide	.100	L.F.	.012	.14	.36	.50
Redwood, untreated 2″ x 4″, 10′ wide	.200	L.F.	.010	.59	.33	.92
20′ wide	.100	L.F.	.005	.29	.16	.45
Granite, 4 1/2″ x 12″ straight, 10′ wide	.200	L.F.	.032	1.55	1.31	2.86
20′ wide	.100	L.F.	.016	.78	.66	1.44
Finishes, asphalt sealer, 10′ wide	1.000	S.F.	.023	.47	.59	1.06
20′ wide	1.000	S.F.	.023	.47	.59	1.06
Concrete, exposed aggregate 10′ wide	1.000	S.F.	.013	.41	.39	.80
20′ wide	1.000	S.F.	.013	.41	.39	.80

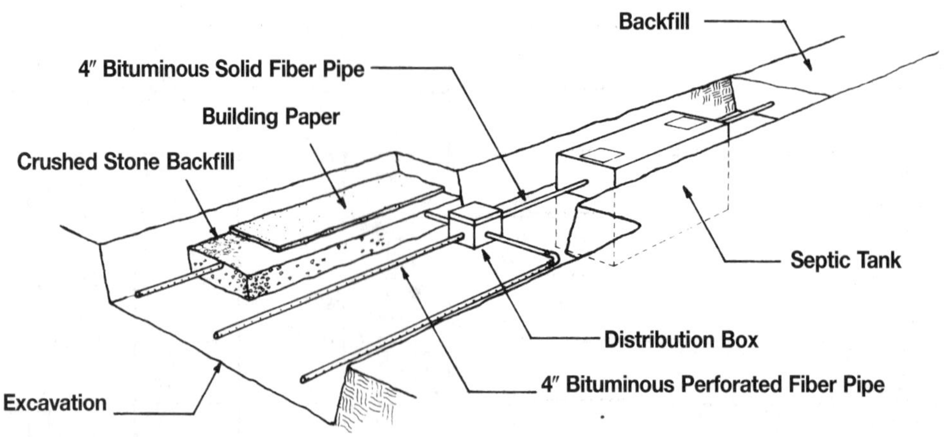

System Description	QUAN.	UNIT	LABOR HOURS	COST EACH		
				MAT.	INST.	TOTAL
SEPTIC SYSTEM WITH 1000 S.F. LEACHING FIELD, 1000 GALLON TANK						
Tank, 1000 gallon, concrete	1.000	Ea.	3.500	535.00	114.00	649.00
Distribution box, concrete	1.000	Ea.	1.000	95.00	25.00	120.00
4″ PVC pipe	25.000	L.F.	1.600	60.75	41.50	102.25
Tank and field excavation	119.000	C.Y.	6.566		636.65	636.65
Crushed stone backfill	76.000	C.Y.	12.160	1786.00	442.32	2228.32
Backfill with excavated material	36.000	C.Y.	.240		34.56	34.56
Building paper	125.000	S.Y.	2.432	33.75	78.75	112.50
4″ PVC perforated pipe	145.000	L.F.	9.280	352.35	240.70	593.05
4″ pipe fittings	2.000	Ea.	2.286	6.20	75.00	81.20
TOTAL			39.064	2869.05	1688.48	4557.53
SEPTIC SYSTEM WITH 2 LEACHING PITS, 1000 GALLON TANK						
Tank, 1000 gallon, concrete	1.000	Ea.	3.500	535.00	114.00	649.00
Distribution box, concrete	1.000	Ea.	1.000	95.00	25.00	120.00
4″ PVC pipe	75.000	L.F.	4.800	182.25	124.50	306.75
Excavation for tank only	20.000	C.Y.	1.103		107.00	107.00
Crushed stone backfill	10.000	C.Y.	1.600	235.00	58.20	293.20
Backfill with excavated material	55.000	C.Y.	.367		52.80	52.80
Pits, 6′ diameter, including excavation and stone backfill	2.000	Ea.		1100.00		1100.00
TOTAL			12.370	2147.25	481.50	2628.75

The costs in this system include all necessary piping and excavation.

Description	QUAN.	UNIT	LABOR HOURS	COST EACH		
				MAT.	INST.	TOTAL

Septic Systems Price Sheet	QUAN.	UNIT	LABOR HOURS	COST EACH		
				MAT.	INST.	TOTAL
Tank, precast concrete, 1000 gallon	1.000	Ea.	3.500	535.00	114.00	649.00
2000 gallon	1.000	Ea.	5.600	1100.00	183.00	1283.00
Distribution box, concrete, 5 outlets	1.000	Ea.	1.000	95.00	25.00	120.00
12 outlets	1.000	Ea.	2.000	248.00	49.50	297.50
4″ pipe, PVC, solid	25.000	L.F.	1.600	61.00	41.50	102.50
Tank and field excavation, 1000 S.F. field	119.000	C.Y.	6.566		635.00	635.00
2000 S.F. field	190.000	C.Y.	10.483		1025.00	1025.00
Tank excavation only, 1000 gallon tank	20.000	C.Y.	1.103		107.00	107.00
2000 gallon tank	32.000	C.Y.	1.766		172.00	172.00
Backfill, crushed stone 1000 S.F. field	76.000	C.Y.	12.160	1775.00	445.00	2220.00
2000 S.F. field	140.000	C.Y.	22.400	3300.00	815.00	4115.00
Backfill with excavated material, 1000 S.F. field	36.000	C.Y.	.240		34.50	34.50
2000 S.F. field	60.000	C.Y.	.400		57.50	57.50
6′ diameter pits	55.000	C.Y.	.367		52.50	52.50
3′ diameter pits	42.000	C.Y.	.280		40.50	40.50
Building paper, 1000 S.F. field	125.000	S.Y.	2.378	33.00	77.00	110.00
2000 S.F. field	250.000	S.Y.	4.865	67.50	158.00	225.50
4″ pipe, PVC, perforated, 1000 S.F. field	145.000	L.F.	9.280	350.00	241.00	591.00
2000 S.F. field	265.000	L.F.	16.960	645.00	440.00	1085.00
Pipe fittings, bituminous fiber, 1000 S.F. field	2.000	Ea.	2.286	6.20	75.00	81.20
2000 S.F. field	4.000	Ea.	4.571	12.40	150.00	162.40
Leaching pit, including excavation and stone backfill, 3′ diameter	1.000	Ea.		415.00		415.00
6′ diameter	1.000	Ea.		550.00		550.00

60 Chain Link Fence Price Sheet

System Description	QUAN.	UNIT	LABOR HOURS	COST PER UNIT MAT.	INST.	TOTAL
Chain link fence						
Galv.9ga. wire, 1-5/8"post 10'O.C., 1-3/8"top rail, 2"corner post, 3'hi	1.000	L.F.	.130	4.25	3.36	7.61
4' high	1.000	L.F.	.141	6.50	3.66	10.16
6' high	1.000	L.F.	.209	7.35	5.40	12.75
Add for gate 3' wide 1-3/8" frame 3' high	1.000	Ea.	2.000	43.00	52.00	95.00
4' high	1.000	Ea.	2.400	56.50	62.00	118.50
6' high	1.000	Ea.	2.400	85.00	62.00	147.00
Add for gate 4' wide 1-3/8" frame 3' high	1.000	Ea.	2.667	54.50	69.00	123.50
4' high	1.000	Ea.	2.667	73.50	69.00	142.50
6' high	1.000	Ea.	3.000	108.00	77.50	185.50
Alum.9ga. wire, 1-5/8"post, 10'O.C., 1-3/8"top rail, 2"corner post,3'hi	1.000	L.F.	.130	5.10	3.36	8.46
4' high	1.000	L.F.	.141	5.85	3.66	9.51
6' high	1.000	L.F.	.209	7.50	5.40	12.90
Add for gate 3' wide 1-3/8" frame 3' high	1.000	Ea.	2.000	50.00	52.00	102.00
4' high	1.000	Ea.	2.400	68.00	62.00	130.00
6' high	1.000	Ea.	2.400	102.00	62.00	164.00
Add for gate 4' wide 1-3/8" frame 3' high	1.000	Ea.	2.400	68.00	62.00	130.00
4' high	1.000	Ea.	2.667	90.50	69.00	159.50
6' high	1.000	Ea.	3.000	142.00	77.50	219.50
Vinyl 9ga. wire, 1-5/8"post 10'O.C., 1-3/8"top rail, 2"corner post,3'hi	1.000	L.F.	.130	4.53	3.36	7.89
4' high	1.000	L.F.	.141	7.50	3.66	11.16
6' high	1.000	L.F.	.209	8.65	5.40	14.05
Add for gate 3' wide 1-3/8" frame 3' high	1.000	Ea.	2.000	56.50	52.00	108.50
4' high	1.000	Ea.	2.400	73.50	62.00	135.50
6' high	1.000	Ea.	2.400	113.00	62.00	175.00
Add for gate 4' wide 1-3/8" frame 3' high	1.000	Ea.	2.400	77.00	62.00	139.00
4' high	1.000	Ea.	2.667	102.00	69.00	171.00
6' high	1.000	Ea.	3.000	147.00	77.50	224.50
Tennis court, chain link fence, 10' high						
Galv.11ga.wire, 2"post 10'O.C., 1-3/8"top rail, 2-1/2"corner post	1.000	L.F.	.253	11.35	6.55	17.90
Add for gate 3' wide 1-3/8" frame	1.000	Ea.	2.400	142.00	62.00	204.00
Alum.11ga.wire, 2"post 10'O.C., 1-3/8"top rail, 2-1/2"corner post	1.000	L.F.	.253	15.85	6.55	22.40
Add for gate 3' wide 1-3/8" frame	1.000	Ea.	2.400	181.00	62.00	243.00
Vinyl 11ga.wire,2"post 10' O.C.,1-3/8"top rail,2-1/2"corner post	1.000	L.F.	.253	13.60	6.55	20.15
Add for gate 3' wide 1-3/8" frame	1.000	Ea.	2.400	204.00	62.00	266.00
Railings, commercial						
Aluminum balcony rail, 1-1/2" posts with pickets	1.000	L.F.	.164	40.50	7.52	48.02
With expanded metal panels	1.000	L.F.	.164	51.50	7.52	59.02
With porcelain enamel panel inserts	1.000	L.F.	.164	44.00	7.52	51.52
Mild steel, ornamental rounded top rail	1.000	L.F.	.164	38.50	7.52	46.02
As above, but pitch down stairs	1.000	L.F.	.183	42.00	8.38	50.38
Steel pipe, welded, 1-1/2" round, painted	1.000	L.F.	.160	11.30	7.31	18.61
Galvanized	1.000	L.F.	.160	16.00	7.31	23.31
Residential, stock units, mild steel, deluxe	1.000	L.F.	.102	8.80	4.65	13.45
Economy	1.000	L.F.	.102	6.60	4.65	11.25

64 Wood Fence Price Sheet

System Description	QUAN.	UNIT	LABOR HOURS	COST PER UNIT MAT.	INST.	TOTAL
Basketweave, 3/8"x4" boards, 2"x4" stringers on spreaders, 4"x4" posts						
No. 1 cedar, 6' high	1.000	L.F.	.150	7.85	3.89	11.74
Treated pine, 6' high	1.000	L.F.	.160	9.45	4.14	13.59
Board fence, 1"x4" boards, 2"x4" rails, 4"x4" posts						
Preservative treated, 2 rail, 3' high	1.000	L.F.	.166	5.75	4.29	10.04
4' high	1.000	L.F.	.178	6.30	4.60	10.90
3 rail, 5' high	1.000	L.F.	.185	7.15	4.78	11.93
6' high	1.000	L.F.	.192	8.15	4.97	13.12
Western cedar, No. 1, 2 rail, 3' high	1.000	L.F.	.166	6.30	4.29	10.59
3 rail, 4' high	1.000	L.F.	.178	7.45	4.60	12.05
5' high	1.000	L.F.	.185	8.55	4.78	13.33
6' high	1.000	L.F.	.192	9.40	4.97	14.37
No. 1 cedar, 2 rail, 3' high	1.000	L.F.	.166	9.25	4.29	13.54
4' high	1.000	L.F.	.178	10.50	4.60	15.10
3 rail, 5' high	1.000	L.F.	.185	12.20	4.78	16.98
6' high	1.000	L.F.	.192	13.60	4.97	18.57
Shadow box, 1"x6" boards, 2"x4" rails, 4"x4" posts						
Fir, pine or spruce, treated, 3 rail, 6' high	1.000	L.F.	.160	10.55	4.14	14.69
No. 1 cedar, 3 rail, 4' high	1.000	L.F.	.185	12.75	4.78	17.53
6' high	1.000	L.F.	.192	15.75	4.97	20.72
Open rail, split rails, No. 1 cedar, 2 rail, 3' high	1.000	L.F.	.150	4.76	3.89	8.65
3 rail, 4' high	1.000	L.F.	.160	6.60	4.14	10.74
No. 2 cedar, 2 rail, 3' high	1.000	L.F.	.150	3.89	3.89	7.78
3 rail, 4' high	1.000	L.F.	.160	4.44	4.14	8.58
Open rail, rustic rails, No. 1 cedar, 2 rail, 3' high	1.000	L.F.	.150	3.21	3.89	7.10
3 rail, 4' high	1.000	L.F.	.160	4.27	4.14	8.41
No. 2 cedar, 2 rail, 3' high	1.000	L.F.	.150	3.12	3.89	7.01
3 rail, 4' high	1.000	L.F.	.160	3.30	4.14	7.44
Rustic picket, molded pine pickets, 2 rail, 3' high	1.000	L.F.	.171	4.60	4.44	9.04
3 rail, 4' high	1.000	L.F.	.197	5.29	5.11	10.40
No. 1 cedar, 2 rail, 3' high	1.000	L.F.	.171	6.10	4.44	10.54
3 rail, 4' high	1.000	L.F.	.197	7.02	5.11	12.13
Picket fence, fir, pine or spruce, preserved, treated						
2 rail, 3' high	1.000	L.F.	.171	3.96	4.44	8.40
3 rail, 4' high	1.000	L.F.	.185	4.68	4.78	9.46
Western cedar, 2 rail, 3' high	1.000	L.F.	.171	4.95	4.44	9.39
3 rail, 4' high	1.000	L.F.	.185	5.05	4.78	9.83
No. 1 cedar, 2 rail, 3' high	1.000	L.F.	.171	9.90	4.44	14.34
3 rail, 4' high	1.000	L.F.	.185	11.55	4.78	16.33
Stockade, No. 1 cedar, 3-1/4" rails, 6' high	1.000	L.F.	.150	9.35	3.89	13.24
8' high	1.000	L.F.	.155	12.10	4.01	16.11
No. 2 cedar, treated rails, 6' high	1.000	L.F.	.150	9.35	3.89	13.24
Treated pine, treated rails, 6' high	1.000	L.F.	.150	9.15	3.89	13.04
Gates, No. 2 cedar, picket, 3'-6" wide 4' high	1.000	Ea.	2.667	49.50	69.00	118.50
No. 2 cedar, rustic round, 3' wide, 3' high	1.000	Ea.	2.667	63.00	69.00	132.00
No. 2 cedar, stockade screen, 3'-6" wide, 6' high	1.000	Ea.	3.000	55.00	77.50	132.50
General, wood, 3'-6" wide, 4' high	1.000	Ea.	2.400	48.40	62.00	110.40
6' high	1.000	Ea.	3.000	60.50	77.50	138.00

Division 2 Foundations

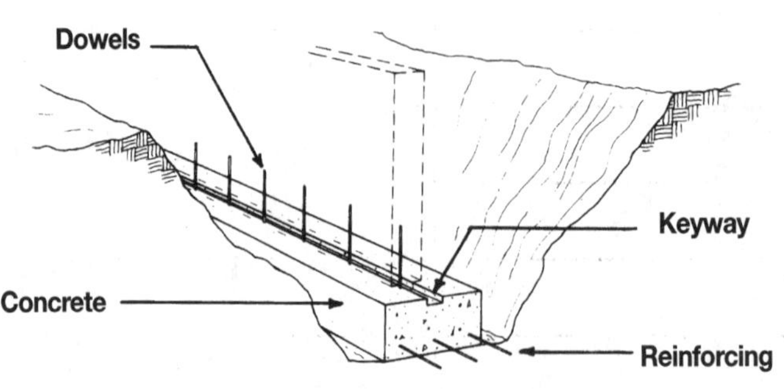

System Description	QUAN.	UNIT	LABOR HOURS	COST PER L.F.		
				MAT.	INST.	TOTAL
8″ THICK BY 18″ WIDE FOOTING						
Concrete, 3000 psi	.040	C.Y.		2.76		2.76
Place concrete, direct chute	.040	C.Y.	.016		.45	.45
Forms, footing, 4 uses	1.330	SFCA	.103	.81	3.02	3.83
Reinforcing, 1/2″ diameter bars, 2 each	1.380	Lb.	.011	.44	.41	.85
Keyway, 2″ x 4″, beveled, 4 uses	1.000	L.F.	.015	.23	.51	.74
Dowels, 1/2″ diameter bars, 2′ long, 6′ O.C.	.166	Ea.	.006	.07	.22	.29
TOTAL			.151	4.31	4.61	8.92
12″ THICK BY 24″ WIDE FOOTING						
Concrete, 3000 psi	.070	C.Y.		4.83		4.83
Place concrete, direct chute	.070	C.Y.	.028		.79	.79
Forms, footing, 4 uses	2.000	SFCA	.155	1.22	4.54	5.76
Reinforcing, 1/2″ diameter bars, 2 each	1.380	Lb.	.011	.44	.41	.85
Keyway, 2″ x 4″, beveled, 4 uses	1.000	L.F.	.015	.23	.51	.74
Dowels, 1/2″ diameter bars, 2′ long, 6′ O.C.	.166	Ea.	.006	.07	.22	.29
TOTAL			.215	6.79	6.47	13.26
12″ THICK BY 36″ WIDE FOOTING						
Concrete, 3000 psi	.110	C.Y.		7.59		7.59
Place concrete, direct chute	.110	C.Y.	.044		1.24	1.24
Forms, footing, 4 uses	2.000	SFCA	.155	1.22	4.54	5.76
Reinforcing, 1/2″ diameter bars, 2 each	1.380	Lb.	.011	.44	.41	.85
Keyway, 2″ x 4″, beveled, 4 uses	1.000	L.F.	.015	.23	.51	.74
Dowels, 1/2″ diameter bars, 2′ long, 6′ O.C.	.166	Ea.	.006	.07	.22	.29
TOTAL			.231	9.55	6.92	16.47

The footing costs in this system are on a cost per linear foot basis

Description	QUAN.	UNIT	LABOR HOURS	COST PER S.F.		
				MAT.	INST.	TOTAL

Important: See the Reference Section for critical supporting data - Reference Nos., Crews & Location Factors

Footing Price Sheet	QUAN.	UNIT	LABOR HOURS	COST PER L.F. MAT.	INST.	TOTAL
Concrete, 8″ thick by 18″ wide footing						
2000 psi concrete	.040	C.Y.		2.66		2.66
2500 psi concrete	.040	C.Y.		2.68		2.68
3000 psi concrete	.040	C.Y.		2.76		2.76
3500 psi concrete	.040	C.Y.		2.82		2.82
4000 psi concrete	.040	C.Y.		2.94		2.94
12″ thick by 24″ wide footing						
2000 psi concrete	.070	C.Y.		4.66		4.66
2500 psi concrete	.070	C.Y.		4.69		4.69
3000 psi concrete	.070	C.Y.		4.83		4.83
3500 psi concrete	.070	C.Y.		4.94		4.94
4000 psi concrete	.070	C.Y.		5.15		5.15
12″ thick by 36″ wide footing						
2000 psi concrete	.110	C.Y.		7.30		7.30
2500 psi concrete	.110	C.Y.		7.35		7.35
3000 psi concrete	.110	C.Y.		7.60		7.60
3500 psi concrete	.110	C.Y.		7.75		7.75
4000 psi concrete	.110	C.Y.		8.10		8.10
Place concrete, 8″ thick by 18″ wide footing, direct chute	.040	C.Y.	.016		.45	.45
Pumped concrete	.040	C.Y.	.017		.67	.67
Crane & bucket	.040	C.Y.	.032		1.31	1.31
12″ thick by 24″ wide footing, direct chute	.070	C.Y.	.028		.79	.79
Pumped concrete	.070	C.Y.	.030		1.18	1.18
Crane & bucket	.070	C.Y.	.056		2.29	2.29
12″ thick by 36″ wide footing, direct chute	.110	C.Y.	.044		1.24	1.24
Pumped concrete	.110	C.Y.	.047		1.86	1.86
Crane & bucket	.110	C.Y.	.088		3.60	3.60
Forms, 8″ thick footing, 1 use	1.330	SFCA	.140	.17	4.10	4.27
4 uses	1.330	SFCA	.103	.81	3.02	3.83
12″ thick footing, 1 use	2.000	SFCA	.210	.26	6.15	6.41
4 uses	2.000	SFCA	.155	1.22	4.54	5.76
Reinforcing, 3/8″ diameter bar, 1 each	.400	Lb.	.003	.13	.12	.25
2 each	.800	Lb.	.006	.26	.24	.50
3 each	1.200	Lb.	.009	.38	.36	.74
1/2″ diameter bar, 1 each	.700	Lb.	.005	.22	.21	.43
2 each	1.380	Lb.	.011	.44	.41	.85
3 each	2.100	Lb.	.016	.67	.63	1.30
5/8″ diameter bar, 1 each	1.040	Lb.	.008	.33	.31	.64
2 each	2.080	Lb.	.016	.67	.62	1.29
Keyway, beveled, 2″ x 4″, 1 use	1.000	L.F.	.030	.46	1.02	1.48
2 uses	1.000	L.F.	.023	.35	.77	1.12
2″ x 6″, 1 use	1.000	L.F.	.032	.64	1.08	1.72
2 uses	1.000	L.F.	.024	.48	.81	1.29
Dowels, 2 feet long, 6′ O.C., 3/8″ bar	.166	Ea.	.005	.04	.20	.24
1/2″ bar	.166	Ea.	.006	.07	.22	.29
5/8″ bar	.166	Ea.	.006	.12	.24	.36
3/4″ bar	.166	Ea.	.006	.12	.24	.36

08 Block Wall Systems

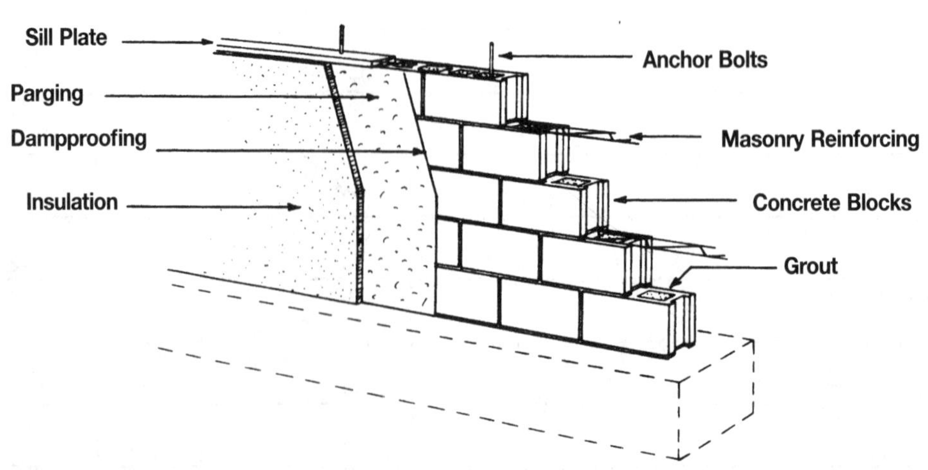

System Description	QUAN.	UNIT	LABOR HOURS	COST PER S.F. MAT.	COST PER S.F. INST.	COST PER S.F. TOTAL
8″ WALL, GROUTED, FULL HEIGHT						
Concrete block, 8″ x 16″ x 8″	1.000	S.F.	.093	1.65	2.88	4.53
Masonry reinforcing, every second course	.750	L.F.	.002	.09	.07	.16
Parging, plastering with portland cement plaster, 1 coat	1.000	S.F.	.014	.23	.44	.67
Dampproofing, bituminous coating, 1 coat	1.000	S.F.	.012	.07	.38	.45
Insulation, 1″ rigid polystyrene	1.000	S.F.	.010	.33	.34	.67
Grout, solid, pumped	1.000	S.F.	.059	1.00	1.81	2.81
Anchor bolts, 1/2″ diameter, 8″ long, 4′ O.C.	.060	Ea.	.002	.03	.08	.11
Sill plate, 2″ x 4″, treated	.250	L.F.	.007	.18	.25	.43
TOTAL			.199	3.58	6.25	9.83
12″ WALL, GROUTED, FULL HEIGHT						
Concrete block, 8″ x 16″ x 12″	1.000	S.F.	.122	2.39	3.67	6.06
Masonry reinforcing, every second course	.750	L.F.	.003	.11	.11	.22
Parging, plastering with portland cement plaster, 1 coat	1.000	S.F.	.014	.23	.44	.67
Dampproofing, bituminous coating, 1 coat	1.000	S.F.	.012	.07	.38	.45
Insulation, 1″ rigid polystyrene	1.000	S.F.	.010	.33	.34	.67
Grout, solid, pumped	1.000	S.F.	.063	1.62	1.92	3.54
Anchor bolts, 1/2″ diameter, 8″ long, 4′ O.C.	.060	Ea.	.002	.03	.08	.11
Sill plate, 2″ x 4″, treated	.250	L.F.	.007	.18	.25	.43
TOTAL			.233	4.96	7.19	12.15

The costs in this system are based on a square foot of wall. Do not subtract for window or door openings.

Description	QUAN.	UNIT	LABOR HOURS	COST PER S.F. MAT.	COST PER S.F. INST.	COST PER S.F. TOTAL

Block Wall Systems	QUAN.	UNIT	LABOR HOURS	COST PER S.F.		
				MAT.	INST.	TOTAL
Concrete, block, 8″ x 16″ x, 6″ thick	1.000	S.F.	.089	1.43	2.75	4.18
8″ thick	1.000	S.F.	.093	1.65	2.88	4.53
10″ thick	1.000	S.F.	.095	2.25	2.95	5.20
12″ thick	1.000	S.F.	.122	2.39	3.67	6.06
Solid block, 8″ x 16″ x, 6″ thick	1.000	S.F.	.091	1.81	2.81	4.62
8″ thick	1.000	S.F.	.096	2.28	2.98	5.26
10″ thick	1.000	S.F.	.096	2.28	2.98	5.26
12″ thick	1.000	S.F.	.126	3.56	3.82	7.38
Masonry reinforcing, wire strips, to 8″ wide, every course	1.500	L.F.	.004	.18	.14	.32
Every 2nd course	.750	L.F.	.002	.09	.07	.16
Every 3rd course	.500	L.F.	.001	.06	.05	.11
Every 4th course	.400	L.F.	.001	.05	.04	.09
Wire strips to 12″ wide, every course	1.500	L.F.	.006	.21	.21	.42
Every 2nd course	.750	L.F.	.003	.11	.11	.22
Every 3rd course	.500	L.F.	.002	.07	.07	.14
Every 4th course	.400	L.F.	.002	.06	.06	.12
Parging, plastering with portland cement plaster, 1 coat	1.000	S.F.	.014	.23	.44	.67
2 coats	1.000	S.F.	.022	.36	.68	1.04
Dampproofing, bituminous, brushed on, 1 coat	1.000	S.F.	.012	.07	.38	.45
2 coats	1.000	S.F.	.016	.11	.51	.62
Sprayed on, 1 coat	1.000	S.F.	.010	.08	.31	.39
2 coats	1.000	S.F.	.016	.16	.51	.67
Troweled on, 1/16″ thick	1.000	S.F.	.016	.17	.51	.68
1/8″ thick	1.000	S.F.	.020	.31	.63	.94
1/2″ thick	1.000	S.F.	.023	1.01	.73	1.74
Insulation, rigid, fiberglass, 1.5#/C.F., unfaced						
1-1/2″ thick R 6.2	1.000	S.F.	.008	.37	.27	.64
2″ thick R 8.5	1.000	S.F.	.008	.45	.27	.72
3″ thick R 13	1.000	S.F.	.010	.54	.34	.88
Foamglass, 1-1/2″ thick R 2.64	1.000	S.F.	.010	1.60	.34	1.94
2″ thick R 5.26	1.000	S.F.	.011	2.98	.37	3.35
Perlite, 1″ thick R 2.77	1.000	S.F.	.010	.26	.34	.60
2″ thick R 5.55	1.000	S.F.	.011	.53	.37	.90
Polystyrene, extruded, 1″ thick R 5.4	1.000	S.F.	.010	.33	.34	.67
2″ thick R 10.8	1.000	S.F.	.011	.95	.37	1.32
Molded 1″ thick R 3.85	1.000	S.F.	.010	.13	.34	.47
2″ thick R 7.7	1.000	S.F.	.011	.54	.37	.91
Grout, concrete block cores, 6″ thick	1.000	S.F.	.044	.75	1.36	2.11
8″ thick	1.000	S.F.	.059	1.00	1.81	2.81
10″ thick	1.000	S.F.	.061	1.31	1.86	3.17
12″ thick	1.000	S.F.	.063	1.62	1.92	3.54
Anchor bolts, 2′ on center, 1/2″ diameter, 8″ long	.120	Ea.	.005	.06	.16	.22
12″ long	.120	Ea.	.005	.16	.17	.33
3/4″ diameter, 8″ long	.120	Ea.	.006	.23	.20	.43
12″ long	.120	Ea.	.006	.28	.22	.50
4′ on center, 1/2″ diameter, 8″ long	.060	Ea.	.002	.03	.08	.11
12″ long	.060	Ea.	.003	.08	.09	.17
3/4″ diameter, 8″ long	.060	Ea.	.003	.11	.10	.21
12″ long	.060	Ea.	.003	.14	.11	.25
Sill plates, treated, 2″ x 4″	.250	L.F.	.007	.18	.25	.43
4″ x 4″	.250	L.F.	.007	.49	.23	.72

12 Concrete Wall Systems

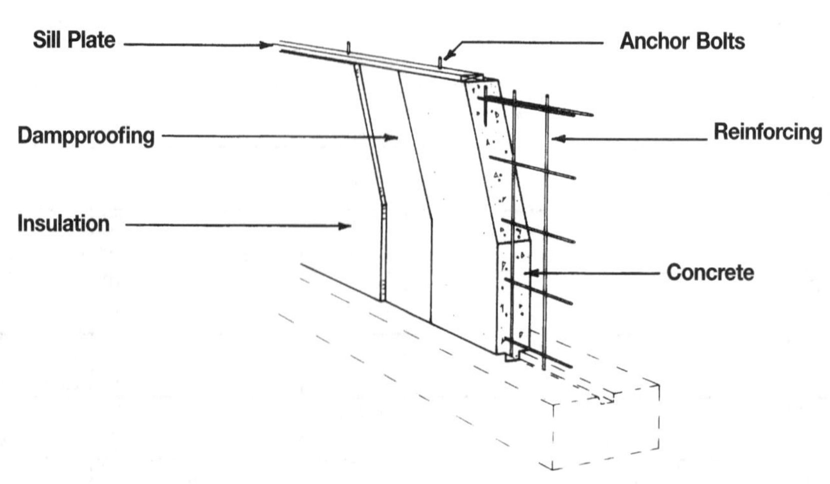

System Description	QUAN.	UNIT	LABOR HOURS	COST PER S.F. MAT.	COST PER S.F. INST.	COST PER S.F. TOTAL
8″ THICK, POURED CONCRETE WALL						
Concrete, 8″ thick , 3000 psi	.025	C.Y.		1.73		1.73
Forms, prefabricated plywood, 4 uses per month	2.000	SFCA	.076	.82	2.28	3.10
Reinforcing, light	.670	Lb.	.004	.21	.14	.35
Placing concrete, direct chute	.025	C.Y.	.013		.37	.37
Dampproofing, brushed on, 2 coats	1.000	S.F.	.016	.11	.51	.62
Rigid insulation, 1″ polystyrene	1.000	S.F.	.010	.33	.34	.67
Anchor bolts, 1/2″ diameter, 12″ long, 4′ O.C.	.060	Ea.	.003	.08	.09	.17
Sill plates, 2″ x 4″, treated	.250	L.F.	.007	.18	.25	.43
TOTAL			.129	3.46	3.98	7.44
12″ THICK, POURED CONCRETE WALL						
Concrete, 12″ thick, 3000 psi	.040	C.Y.		2.76		2.76
Forms, prefabricated plywood, 4 uses per month	2.000	SFCA	.076	.82	2.28	3.10
Reinforcing, light	1.000	Lb.	.005	.32	.21	.53
Placing concrete, direct chute	.040	C.Y.	.019		.54	.54
Dampproofing, brushed on, 2 coats	1.000	S.F.	.016	.11	.51	.62
Rigid insulation, 1″ polystyrene	1.000	S.F.	.010	.33	.34	.67
Anchor bolts, 1/2″ diameter, 12″ long, 4′ O.C.	.060	Ea.	.003	.08	.09	.17
Sill plates, 2″ x 4″ treated	.250	L.F.	.007	.18	.25	.43
TOTAL			.136	4.60	4.22	8.82

The costs in this system are based on sq. ft. of wall. Do not subtract for window and door openings. The costs assume a 4′ high wall.

Description	QUAN.	UNIT	LABOR HOURS	COST PER S.F. MAT.	COST PER S.F. INST.	COST PER S.F. TOTAL

Important: See the Reference Section for critical supporting data - Reference Nos., Crews & Location Factors

Concrete Wall Price Sheet	QUAN.	UNIT	LABOR HOURS	COST PER S.F. MAT.	INST.	TOTAL
Formwork, prefabricated plywood, 1 use per month	2.000	SFCA	.081	2.48	2.44	4.92
4 uses per month	2.000	SFCA	.076	.82	2.28	3.10
Job built forms, 1 use per month	2.000	SFCA	.320	4.18	9.60	13.78
4 uses per month	2.000	SFCA	.221	1.38	6.60	7.98
Reinforcing, 8″ wall, light reinforcing	.670	Lb.	.004	.21	.14	.35
Heavy reinforcing	1.500	Lb.	.008	.48	.32	.80
10″ wall, light reinforcing	.850	Lb.	.005	.27	.18	.45
Heavy reinforcing	2.000	Lb.	.011	.64	.42	1.06
12″ wall light reinforcing	1.000	Lb.	.005	.32	.21	.53
Heavy reinforcing	2.250	Lb.	.012	.72	.47	1.19
Placing concrete, 8″ wall, direct chute	.025	C.Y.	.013		.37	.37
Pumped concrete	.025	C.Y.	.016		.63	.63
Crane & bucket	.025	C.Y.	.023		.92	.92
10″ wall, direct chute	.030	C.Y.	.016		.45	.45
Pumped concrete	.030	C.Y.	.019		.76	.76
Crane & bucket	.030	C.Y.	.027		1.12	1.12
12″ wall, direct chute	.040	C.Y.	.019		.54	.54
Pumped concrete	.040	C.Y.	.023		.92	.92
Crane & bucket	.040	C.Y.	.032		1.31	1.31
Dampproofing, bituminous, brushed on, 1 coat	1.000	S.F.	.012	.07	.38	.45
2 coats	1.000	S.F.	.016	.11	.51	.62
Sprayed on, 1 coat	1.000	S.F.	.010	.08	.31	.39
2 coats	1.000	S.F.	.016	.16	.51	.67
Troweled on, 1/16″ thick	1.000	S.F.	.016	.17	.51	.68
1/8″ thick	1.000	S.F.	.020	.31	.63	.94
1/2″ thick	1.000	S.F.	.023	1.01	.73	1.74
Insulation rigid, fiberglass, 1.5#/C.F., unfaced						
1-1/2″ thick, R 6.2	1.000	S.F.	.008	.37	.27	.64
2″ thick, R 8.3	1.000	S.F.	.008	.45	.27	.72
3″ thick, R 12.4	1.000	S.F.	.010	.54	.34	.88
Foamglass, 1-1/2″ thick R 2.64	1.000	S.F.	.010	1.60	.34	1.94
2″ thick R 5.26	1.000	S.F.	.011	2.98	.37	3.35
Perlite, 1″ thick R 2.77	1.000	S.F.	.010	.26	.34	.60
2″ thick R 5.55	1.000	S.F.	.011	.53	.37	.90
Polystyrene, extruded, 1″ thick R 5.40	1.000	S.F.	.010	.33	.34	.67
2″ thick R 10.8	1.000	S.F.	.011	.95	.37	1.32
Molded, 1″ thick R 3.85	1.000	S.F.	.010	.13	.34	.47
2″ thick R 7.70	1.000	S.F.	.011	.54	.37	.91
Anchor bolts, 2′ on center, 1/2″ diameter, 8″ long	.120	Ea.	.005	.06	.16	.22
12″ long	.120	Ea.	.005	.16	.17	.33
3/4″ diameter, 8″ long	.120	Ea.	.006	.23	.20	.43
12″ long	.120	Ea.	.006	.28	.22	.50
Sill plates, treated lumber, 2″ x 4″	.250	L.F.	.007	.18	.25	.43
4″ x 4″	.250	L.F.	.007	.49	.23	.72

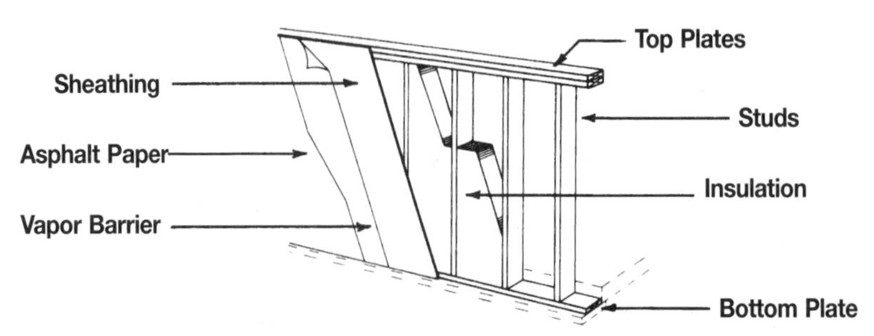

System Description	QUAN.	UNIT	LABOR HOURS	COST PER S.F.		
				MAT.	INST.	TOTAL
2″ X 4″ STUDS, 16″ O.C., WALL						
Studs, 2″ x 4″, 16″ O.C., treated	1.000	L.F.	.015	.72	.49	1.21
Plates, double top plate, single bottom plate, treated, 2″ x 4″	.750	L.F.	.011	.54	.37	.91
Sheathing, 1/2″, exterior grade, CDX, treated	1.000	S.F.	.014	.78	.48	1.26
Asphalt paper, 15# roll	1.100	S.F.	.002	.03	.08	.11
Vapor barrier, 4 mil polyethylene	1.000	S.F.	.002	.03	.07	.10
Insulation, batts, fiberglass, 3-1/2″ thick, R 11	1.000	S.F.	.005	.24	.17	.41
TOTAL			.049	2.34	1.66	4.00
2″ X 6″ STUDS, 16″ O.C., WALL						
Studs, 2″ x 6″, 16″ O.C., treated	1.000	L.F.	.016	1.16	.54	1.70
Plates, double top plate, single bottom plate, treated, 2″ x 6″	.750	L.F.	.012	.87	.41	1.28
Sheathing, 5/8″ exterior grade, CDX, treated	1.000	S.F.	.015	.91	.51	1.42
Asphalt paper, 15# roll	1.100	S.F.	.002	.03	.08	.11
Vapor barrier, 4 mil polyethylene	1.000	S.F.	.002	.03	.07	.10
Insulation, batts, fiberglass, 6″ thick, R 19	1.000	S.F.	.006	.34	.20	.54
TOTAL			.053	3.34	1.81	5.15
2″ X 8″ STUDS, 16″ O.C., WALL						
Studs, 2″ x 8″, 16″ O.C. treated	1.000	L.F.	.018	1.62	.60	2.22
Plates, double top plate, single bottom plate, treated, 2″ x 8″	.750	L.F.	.013	1.22	.45	1.67
Sheathing, 3/4″ exterior grade, CDX, treated	1.000	S.F.	.016	1.13	.55	1.68
Asphalt paper, 15# roll	1.100	S.F.	.002	.03	.08	.11
Vapor barrier, 4 mil polyethylene	1.000	S.F.	.002	.03	.07	.10
Insulation, batts, fiberglass, 9″ thick, R 30	1.000	S.F.	.006	.62	.20	.82
TOTAL			.057	4.65	1.95	6.60

The costs in this system are based on a sq. ft. of wall area. Do not Subtract for window or door openings. The costs assume a 4′ high wall.

Description	QUAN.	UNIT	LABOR HOURS	COST PER S.F.		
				MAT.	INST.	TOTAL

Important: See the Reference Section for critical supporting data - Reference Nos., Crews & Location Factors

Wood Wall Foundation Price Sheet	QUAN.	UNIT	LABOR HOURS	COST PER S.F.		
				MAT.	INST.	TOTAL
Studs, treated, 2″ x 4″, 12″ O.C.	1.250	L.F.	.018	.90	.61	1.51
16″ O.C.	1.000	L.F.	.015	.72	.49	1.21
2″ x 6″, 12″ O.C.	1.250	L.F.	.020	1.45	.68	2.13
16″ O.C.	1.000	L.F.	.016	1.16	.54	1.70
2″ x 8″, 12″ O.C.	1.250	L.F.	.022	2.03	.75	2.78
16″ O.C.	1.000	L.F.	.018	1.62	.60	2.22
Plates, treated double top single bottom, 2″ x 4″	.750	L.F.	.011	.54	.37	.91
2″ x 6″	.750	L.F.	.012	.87	.41	1.28
2″ x 8″	.750	L.F.	.013	1.22	.45	1.67
Sheathing, treated exterior grade CDX, 1/2″ thick	1.000	S.F.	.014	.78	.48	1.26
5/8″ thick	1.000	S.F.	.015	.91	.51	1.42
3/4″ thick	1.000	S.F.	.016	1.13	.55	1.68
Asphalt paper, 15# roll	1.100	S.F.	.002	.03	.08	.11
Vapor barrier, polyethylene, 4 mil	1.000	S.F.	.002	.03	.07	.10
10 mil	1.000	S.F.	.002	.07	.07	.14
Insulation, rigid, fiberglass, 1.5#/C.F., unfaced	1.000	S.F.	.008	.28	.27	.55
1-1/2″ thick, R 6.2	1.000	S.F.	.008	.37	.27	.64
2″ thick, R 8.3	1.000	S.F.	.008	.45	.27	.72
3″ thick, R 12.4	1.000	S.F.	.010	.55	.35	.90
Foamglass 1 1/2″ thick, R 2.64	1.000	S.F.	.010	1.60	.34	1.94
2″ thick, R 5.26	1.000	S.F.	.011	2.98	.37	3.35
Perlite 1″ thick, R 2.77	1.000	S.F.	.010	.26	.34	.60
2″ thick, R 5.55	1.000	S.F.	.011	.53	.37	.90
Polystyrene, extruded, 1″ thick, R 5.40	1.000	S.F.	.010	.33	.34	.67
2″ thick, R 10.8	1.000	S.F.	.011	.95	.37	1.32
Molded 1″ thick, R 3.85	1.000	S.F.	.010	.13	.34	.47
2″ thick, R 7.7	1.000	S.F.	.011	.54	.37	.91
Non rigid, batts, fiberglass, paper backed, 3-1/2″ thick roll, R 11	1.000	S.F.	.005	.24	.17	.41
6″, R 19	1.000	S.F.	.006	.34	.20	.54
9″, R 30	1.000	S.F.	.006	.62	.20	.82
12″, R 38	1.000	S.F.	.006	.79	.20	.99
Mineral fiber, paper backed, 3-1/2″, R 13	1.000	S.F.	.005	.26	.17	.43
6″, R 19	1.000	S.F.	.005	.41	.17	.58
10″, R 30	1.000	S.F.	.006	.64	.20	.84

20 Floor Slab Systems

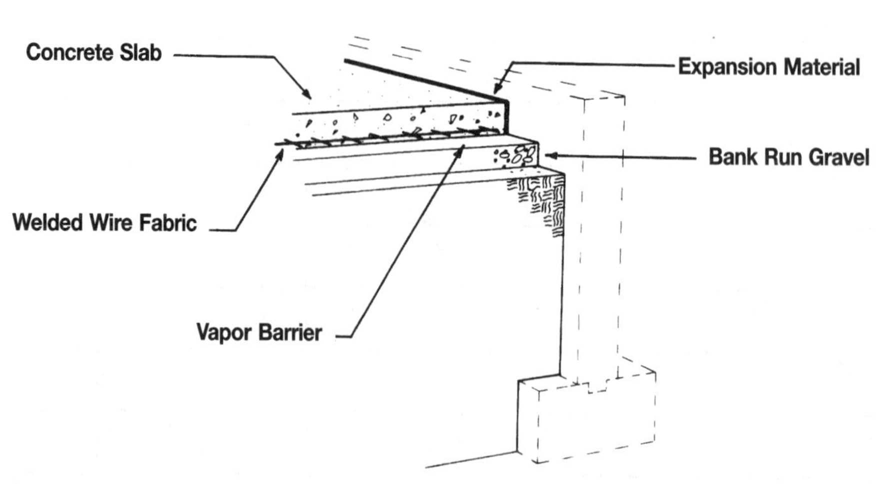

System Description	QUAN.	UNIT	LABOR HOURS	COST PER S.F. MAT.	COST PER S.F. INST.	COST PER S.F. TOTAL
4″ THICK SLAB						
Concrete, 4″ thick, 3000 psi concrete	.012	C.Y.		.83		.83
Place concrete, direct chute	.012	C.Y.	.005		.15	.15
Bank run gravel, 4″ deep	1.000	S.F.	.001	.24	.04	.28
Polyethylene vapor barrier, .006″ thick	1.000	S.F.	.002	.03	.07	.10
Edge forms, expansion material	.100	L.F.	.005	.04	.16	.20
Welded wire fabric, 6 x 6, 10/10 (W1.4/W1.4)	1.100	S.F.	.005	.09	.20	.29
Steel trowel finish	1.000	S.F.	.015		.45	.45
TOTAL			.033	1.23	1.07	2.30
6″ THICK SLAB						
Concrete, 6″ thick, 3000 psi concrete	.019	C.Y.		1.31		1.31
Place concrete, direct chute	.019	C.Y.	.008		.23	.23
Bank run gravel, 4″ deep	1.000	S.F.	.001	.24	.04	.28
Polyethylene vapor barrier, .006″ thick	1.000	S.F.	.002	.03	.07	.10
Edge forms, expansion material	.100	L.F.	.005	.04	.16	.20
Welded wire fabric, 6 x 6, 10/10 (W1.4/W1.4)	1.100	S.F.	.005	.09	.20	.29
Steel trowel finish	1.000	S.F.	.015		.45	.45
TOTAL			.036	1.71	1.15	2.86

The slab costs in this section are based on a cost per square foot of floor area.

Description	QUAN.	UNIT	LABOR HOURS	COST PER S.F. MAT.	COST PER S.F. INST.	COST PER S.F. TOTAL

Important: See the Reference Section for critical supporting data - Reference Nos., Crews & Location Factors

Floor Slab Price Sheet	QUAN.	UNIT	LABOR HOURS	COST PER S.F. MAT.	INST.	TOTAL
Concrete, 4″ thick slab, 2000 psi concrete	.012	C.Y.		.80		.80
2500 psi concrete	.012	C.Y.		.80		.80
3000 psi concrete	.012	C.Y.		.83		.83
3500 psi concrete	.012	C.Y.		.85		.85
4000 psi concrete	.012	C.Y.		.88		.88
4500 psi concrete	.012	C.Y.		.91		.91
5″ thick slab, 2000 psi concrete	.015	C.Y.		1.00		1.00
2500 psi concrete	.015	C.Y.		1.01		1.01
3000 psi concrete	.015	C.Y.		1.04		1.04
3500 psi concrete	.015	C.Y.		1.06		1.06
4000 psi concrete	.015	C.Y.		1.10		1.10
4500 psi concrete	.015	C.Y.		1.14		1.14
6″ thick slab, 2000 psi concrete	.019	C.Y.		1.26		1.26
2500 psi concrete	.019	C.Y.		1.27		1.27
3000 psi concrete	.019	C.Y.		1.31		1.31
3500 psi concrete	.019	C.Y.		1.34		1.34
4000 psi concrete	.019	C.Y.		1.40		1.40
4500 psi concrete	.019	C.Y.		1.44		1.44
Place concrete, 4″ slab, direct chute	.012	C.Y.	.005		.15	.15
Pumped concrete	.012	C.Y.	.006		.23	.23
Crane & bucket	.012	C.Y.	.008		.32	.32
5″ slab, direct chute	.015	C.Y.	.007		.18	.18
Pumped concrete	.015	C.Y.	.007		.29	.29
Crane & bucket	.015	C.Y.	.010		.41	.41
6″ slab, direct chute	.019	C.Y.	.008		.23	.23
Pumped concrete	.019	C.Y.	.009		.37	.37
Crane & bucket	.019	C.Y.	.012		.51	.51
Gravel, bank run, 4″ deep	1.000	S.F.	.001	.24	.04	.28
6″ deep	1.000	S.F.	.001	.32	.05	.37
9″ deep	1.000	S.F.	.001	.47	.07	.54
12″ deep	1.000	S.F.	.001	.63	.08	.71
3/4″ crushed stone, 3″ deep	1.000	S.F.	.001	.52	.04	.56
6″ deep	1.000	S.F.	.001	1.04	.06	1.10
9″ deep	1.000	S.F.	.002	1.56	.09	1.65
12″ deep	1.000	S.F.	.002	1.66	.11	1.77
Vapor barrier polyethylene, .004″ thick	1.000	S.F.	.002	.03	.07	.10
.006″ thick	1.000	S.F.	.002	.03	.07	.10
Edge forms, expansion material, 4″ thick slab	.100	L.F.	.004	.02	.10	.12
6″ thick slab	.100	L.F.	.005	.04	.16	.20
Welded wire fabric 6 x 6, 10/10 (W1.4/W1.4)	1.100	S.F.	.005	.09	.20	.29
6 x 6, 6/6 (W2.9/W2.9)	1.100	S.F.	.006	.19	.24	.43
4 x 4, 10/10 (W1.4/W1.4)	1.100	S.F.	.006	.13	.22	.35
Finish concrete, screed finish	1.000	S.F.	.009		.27	.27
Float finish	1.000	S.F.	.011		.34	.34
Steel trowel, for resilient floor	1.000	S.F.	.013		.41	.41
For finished floor	1.000	S.F.	.015		.45	.45

For information about Means Estimating Seminars, see yellow pages 11 and 12 in back of book

Division 3 Framing

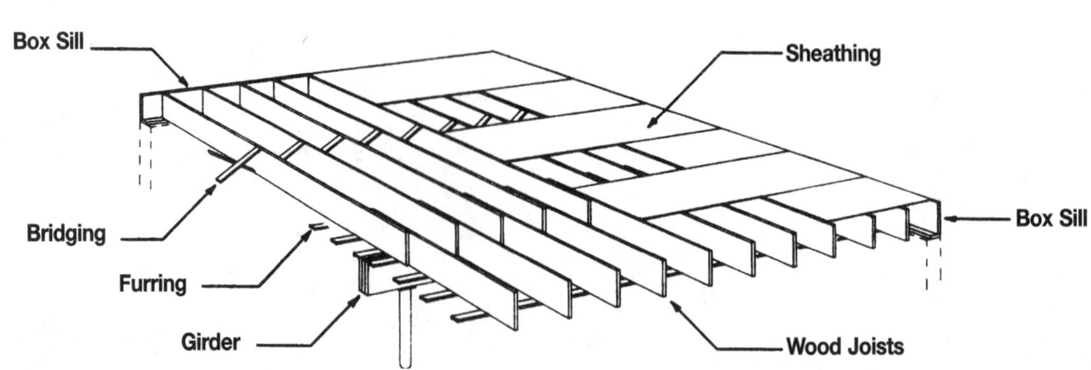

System Description	QUAN.	UNIT	LABOR HOURS	COST PER S.F.		
				MAT.	INST.	TOTAL
2″ X 8″, 16″ O.C.						
Wood joists, 2″ x 8″, 16″ O.C.	1.000	L.F.	.015	.91	.49	1.40
Bridging, 1″ x 3″, 6′ O.C.	.080	Pr.	.005	.02	.17	.19
Box sills, 2″ x 8″	.150	L.F.	.002	.14	.07	.21
Girder, including lally columns, 3- 2″ x 8″	.125	L.F.	.015	.43	.53	.96
Sheathing, plywood, subfloor, 5/8″ CDX	1.000	S.F.	.012	.73	.40	1.13
Furring, 1″ x 3″, 16″ O.C.	1.000	L.F.	.023	.20	.77	.97
TOTAL			.072	2.43	2.43	4.86
2″ X 10″, 16″ O.C.						
Wood joists, 2″ x 10″, 16″ O.C.	1.000	L.F.	.018	1.34	.60	1.94
Bridging, 1″ x 3″, 6′ O.C.	.080	Pr.	.005	.02	.17	.19
Box sills, 2″ x 10″	.150	L.F.	.003	.20	.09	.29
Girder, including lally columns, 3-2″ x 10″	.125	L.F.	.016	.59	.56	1.15
Sheathing, plywood, subfloor, 5/8″ CDX	1.000	S.F.	.012	.73	.40	1.13
Furring, 1″ x 3″,16″ O.C.	1.000	L.F.	.023	.20	.77	.97
TOTAL			.077	3.08	2.59	5.67
2″ X 12″, 16″ O.C.						
Wood joists, 2″ x 12″, 16″ O.C.	1.000	L.F.	.018	1.62	.62	2.24
Bridging, 1″ x 3″, 6′ O.C.	.080	Pr.	.005	.02	.17	.19
Box sills, 2″ x 12″	.150	L.F.	.003	.24	.09	.33
Girder, including lally columns, 3-2″ x 12″	.125	L.F.	.017	.70	.59	1.29
Sheathing, plywood, subfloor, 5/8″ CDX	1.000	S.F.	.012	.73	.40	1.13
Furring, 1″ x 3″, 16″ O.C.	1.000	L.F.	.023	.20	.77	.97
TOTAL			.078	3.51	2.64	6.15

Floor costs on this page are given on a cost per square foot basis.

Description	QUAN.	UNIT	LABOR HOURS	COST PER S.F.		
				MAT.	INST.	TOTAL

Floor Framing Price Sheet (Wood)	QUAN.	UNIT	LABOR HOURS	COST PER S.F.		
				MAT.	INST.	TOTAL
Joists, #2 or better, pine, 2″ x 4″, 12″ O.C.	1.250	L.F.	.016	.54	.54	1.08
16″ O.C.	1.000	L.F.	.013	.43	.43	.86
2″ x 6″, 12″ O.C.	1.250	L.F.	.016	.81	.54	1.35
16″ O.C.	1.000	L.F.	.013	.65	.43	1.08
2″ x 8″, 12″ O.C.	1.250	L.F.	.018	1.14	.61	1.75
16″ O.C.	1.000	L.F.	.015	.91	.49	1.40
2″ x 10″, 12″ O.C.	1.250	L.F.	.022	1.68	.75	2.43
16″ O.C.	1.000	L.F.	.018	1.34	.60	1.94
2″x 12″, 12″ O.C.	1.250	L.F.	.023	2.03	.78	2.81
16″ O.C.	1.000	L.F.	.018	1.62	.62	2.24
Bridging, wood 1″ x 3″, joists 12″ O.C.	.100	Pr.	.006	.03	.21	.24
16″ O.C.	.080	Pr.	.005	.02	.17	.19
Metal, galvanized, joists 12″ O.C.	.100	Pr.	.006	.06	.21	.27
16″ O.C.	.080	Pr.	.005	.05	.17	.22
Compression type, joists 12″ O.C.	.100	Pr.	.004	.14	.14	.28
16″ O.C.	.080	Pr.	.003	.11	.11	.22
Box sills, #2 or better pine, 2″ x 4″	.150	L.F.	.002	.06	.06	.12
2″ x 6″	.150	L.F.	.002	.10	.06	.16
2″ x 8″	.150	L.F.	.002	.14	.07	.21
2″ x 10″	.150	L.F.	.003	.20	.09	.29
2″ x 12″	.150	L.F.	.003	.24	.09	.33
Girders, including lally columns, 3 pieces spiked together, 2″ x 8″	.125	L.F.	.015	.43	.53	.96
2″ x 10″	.125	L.F.	.016	.59	.56	1.15
2″ x 12″	.125	L.F.	.017	.70	.59	1.29
Solid girders, 3″ x 8″	.040	L.F.	.004	.20	.14	.34
3″ x 10″	.040	L.F.	.004	.22	.14	.36
3″ x 12″	.040	L.F.	.004	.25	.15	.40
4″ x 8″	.040	L.F.	.004	.25	.15	.40
4″ x 10″	.040	L.F.	.004	.29	.15	.44
4″ x 12″	.040	L.F.	.004	.33	.16	.49
Steel girders, bolted & including fabrication, wide flange shapes						
12″ deep, 14#/l.f.	1.000	L.F.	.061	10.90	3.95	14.85
10″ deep, 15#/l.f.	1.000	L.F.	.067	11.90	4.31	16.21
8″ deep, 10#/l.f.	1.000	L.F.	.067	7.90	4.31	12.21
6″ deep, 9#/l.f.	1.000	L.F.	.067	7.15	4.31	11.46
5″ deep, 16#/l.f.	1.000	L.F.	.064	11.50	4.17	15.67
Sheathing, plywood exterior grade CDX, 1/2″ thick	1.000	S.F.	.011	.58	.39	.97
5/8″ thick	1.000	S.F.	.012	.73	.40	1.13
3/4″ thick	1.000	S.F.	.013	.89	.43	1.32
Boards, 1″ x 8″ laid regular	1.000	S.F.	.016	1.00	.54	1.54
Laid diagonal	1.000	S.F.	.019	1.00	.64	1.64
1″ x 10″ laid regular	1.000	S.F.	.015	1.11	.49	1.60
Laid diagonal	1.000	S.F.	.018	1.12	.60	1.72
Furring, 1″ x 3″, 12″ O.C.	1.250	L.F.	.029	.25	.96	1.21
16″ O.C.	1.000	L.F.	.023	.20	.77	.97
24″ O.C.	.750	L.F.	.017	.15	.58	.73

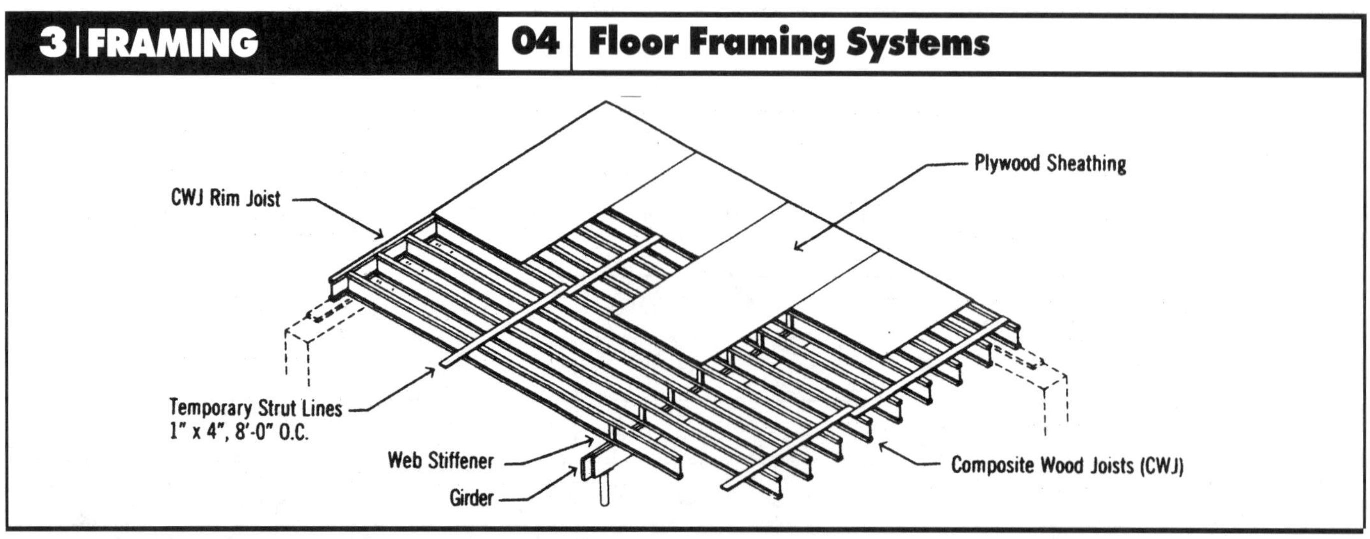

System Description	QUAN.	UNIT	LABOR HOURS	COST PER S.F.		
				MAT.	INST.	TOTAL
9-1/2" COMPOSITE WOOD JOISTS, 16" O.C.						
CWJ, 9-1/2", 16" O.C., 15' span	1.000	L.F.	.018	1.55	.60	2.15
Temp. strut line, 1" x 4", 8' O.C.	.160	L.F.	.003	.06	.11	.17
CWJ rim joist, 9-1/2"	.150	L.F.	.003	.23	.09	.32
Girder, including lally columns, 3- 2" x 8"	.125	L.F.	.015	.43	.53	.96
Sheathing, plywood, subfloor, 5/8" CDX	1.000	S.F.	.012	.73	.40	1.13
TOTAL			.051	3.00	1.73	4.73
11-1/2" COMPOSITE WOOD JOISTS, 16" O.C.						
CWJ, 11-1/2", 16" O.C., 18' span	1.000	L.F.	.018	1.68	.62	2.30
Temp. strut line, 1" x 4", 8' O.C.	.160	L.F.	.003	.06	.11	.17
CWJ rim joist, 11-1/2"	.150	L.F.	.003	.25	.09	.34
Girder, including lally columns, 3-2" x 10"	.125	L.F.	.016	.59	.56	1.15
Sheathing, plywood, subfloor, 5/8" CDX	1.000	S.F.	.012	.73	.40	1.13
Furring, 1" x 3",16" O.C.	1.000	L.F.	.023			
TOTAL			.052	3.31	1.78	5.09
14" COMPOSITE WOOD JOISTS, 16" O.C.						
CWJ, 14", 16" O.C., 22' span	1.000	L.F.	.020	1.88	.66	2.54
Temp. strut line, 1" x 4", 8' O.C.	.160	L.F.	.003	.06	.11	.17
CWJ rim joist, 14"	.150	L.F.	.003	.28	.10	.38
Girder, including lally columns, 3-2" x 12"	.600	L.F.	.017	.70	.59	1.29
Sheathing, plywood, subfloor, 5/8" CDX	1.000	S.F.	.012	.73	.40	1.13
Furring, 1" x 3", 16" O.C.	1.000	L.F.	.023			
TOTAL			.055	3.65	1.86	5.51

Floor costs on this page are given on a cost per square foot basis.

Description	QUAN.	UNIT	LABOR HOURS	COST PER S.F.		
				MAT.	INST.	TOTAL

FRAMING 3

Floor Framing Price Sheet (Wood)	QUAN.	UNIT	LABOR HOURS	COST PER S.F.		
				MAT.	INST.	TOTAL
Composite wood joist 9-1/2" deep, 12" O.C.	1.250	L.F.	.022	1.94	.75	2.69
16" O.C.	1.000	L.F.	.018	1.55	.60	2.15
11-1/2" deep, 12" O.C.	1.250	L.F.	.023	2.09	.77	2.86
16" O.C.	1.000	L.F.	.018	1.68	.62	2.30
14" deep, 12" O.C.	1.250	L.F.	.024	2.34	.83	3.17
16" O.C.	1.000	L.F.	.020	1.88	.66	2.54
16 " deep, 12" O.C.	1.250	L.F.	.026	3.06	.86	3.92
16" O.C.	1.000	L.F.	.021	2.45	.69	3.14
CWJ rim joist, 9-1/2"	.150	L.F.	.003	.23	.09	.32
11-1/2"	.150	L.F.	.003	.25	.09	.34
14"	.150	L.F.	.003	.28	.10	.38
16"	.150	L.F.	.003	.37	.10	.47
Girders, including lally columns, 3 pieces spiked together, 2" x 8"	.125	L.F.	.015	.43	.53	.96
2" x 10"	.125	L.F.	.016	.59	.56	1.15
2" x 12"	.125	L.F.	.017	.70	.59	1.29
Solid girders, 3" x 8"	.040	L.F.	.004	.20	.14	.34
3" x 10"	.040	L.F.	.004	.22	.14	.36
3" x 12"	.040	L.F.	.004	.25	.15	.40
4" x 8"	.040	L.F.	.004	.25	.15	.40
4" x 10"	.040	L.F.	.004	.29	.15	.44
4" x 12"	.040	L.F.	.004	.33	.16	.49
Steel girders, bolted & including fabrication, wide flange shapes						
12" deep, 14#/l.f.	1.000	L.F.	.061	10.90	3.95	14.85
10" deep, 15#/l.f.	1.000	L.F.	.067	11.90	4.31	16.21
8" deep, 10#/l.f.	1.000	L.F.	.067	7.90	4.31	12.21
6" deep, 9#/l.f.	1.000	L.F.	.067	7.15	4.31	11.46
5" deep, 16#/l.f.	1.000	L.F.	.064	11.50	4.17	15.67
Sheathing, plywood exterior grade CDX, 1/2" thick	1.000	S.F.	.011	.58	.39	.97
5/8" thick	1.000	S.F.	.012	.73	.40	1.13
3/4" thick	1.000	S.F.	.013	.89	.43	1.32
Boards, 1" x 8" laid regular	1.000	S.F.	.016	1.00	.54	1.54
Laid diagonal	1.000	S.F.	.019	1.00	.64	1.64
1" x 10" laid regular	1.000	S.F.	.015	1.11	.49	1.60
Laid diagonal	1.000	S.F.	.018	1.12	.60	1.72
Furring, 1" x 3", 12" O.C.	1.250	L.F.	.029	.25	.96	1.21
16" O.C.	1.000	L.F.	.023	.20	.77	.97
24" O.C.	.750	L.F.	.017	.15	.58	.73

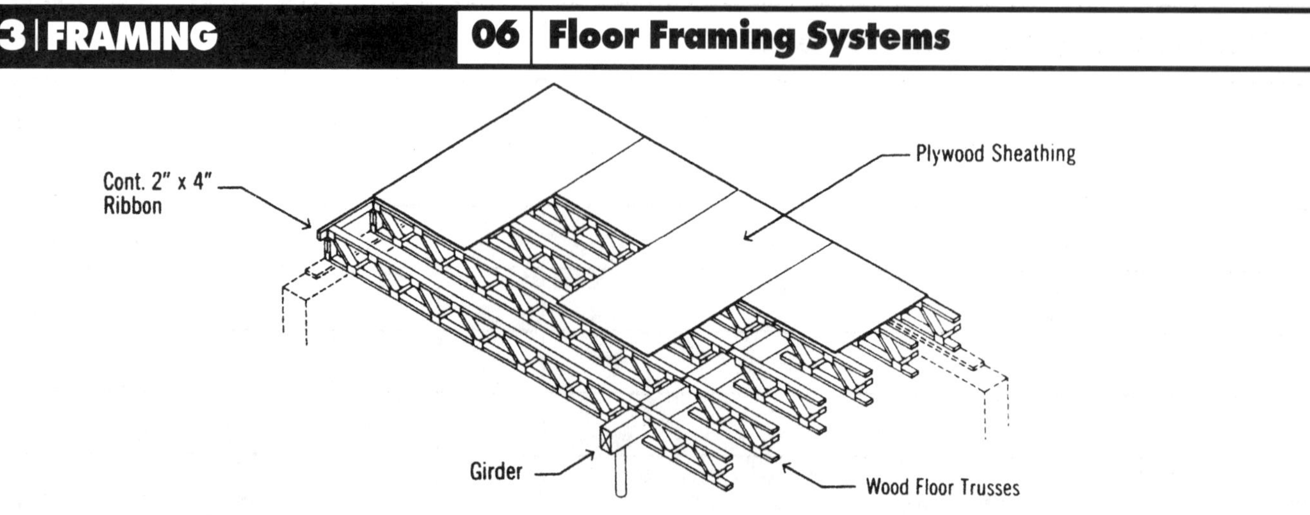

System Description	QUAN.	UNIT	LABOR HOURS	COST PER S.F. MAT.	INST.	TOTAL
12" OPEN WEB JOISTS, 16" O.C.						
OWJ 12", 16" O.C., 21' span	1.000	L.F.	.018	1.73	.62	2.35
Continuous ribbing, 2" x 4"	.150	L.F.	.002	.06	.06	.12
Girder, including lally columns, 3- 2" x 8"	.125	L.F.	.015	.43	.53	.96
Sheathing, plywood, subfloor, 5/8" CDX	1.000	S.F.	.012	.73	.40	1.13
Furring, 1" x 3", 16" O.C.	1.000	L.F.	.023	.20	.77	.97
TOTAL			.070	3.15	2.38	5.53
14" OPEN WEB WOOD JOISTS, 16" O.C.						
OWJ 14", 16" O.C., 22' span	1.000	L.F.	.020	2.00	.66	2.66
Continuous ribbing, 2" x 4"	.150	L.F.	.002	.06	.06	.12
Girder, including lally columns, 3-2" x 10"	.125	L.F.	.016	.59	.56	1.15
Sheathing, plywood, subfloor, 5/8" CDX	1.000	S.F.	.012	.73	.40	1.13
Furring, 1" x 3",16" O.C.	1.000	L.F.	.023	.20	.77	.97
TOTAL			.073	3.58	2.45	6.03
16" OPEN WEB WOOD JOISTS, 16" O.C.						
OWJ 16", 16" O.C., 24' span	1.000	L.F.	.021	2.10	.69	2.79
Continuous ribbing, 2" x 4"	.150	L.F.	.002	.06	.06	.12
Girder, including lally columns, 3-2" x 12"	.125	L.F.	.017	.70	.59	1.29
Sheathing, plywood, subfloor, 5/8" CDX	1.000	S.F.	.012	.73	.40	1.13
Furring, 1" x 3", 16" O.C.	1.000	L.F.	.023	.20	.77	.97
TOTAL			.075	3.79	2.51	6.30

Floor costs on this page are given on a cost per square foot basis.

Description	QUAN.	UNIT	LABOR HOURS	COST PER S.F. MAT.	INST.	TOTAL

Floor Framing Price Sheet (Wood)	QUAN.	UNIT	LABOR HOURS	COST PER S.F.		
				MAT.	INST.	TOTAL
Open web joists, 12" deep, 12" O.C.	1.250	L.F.	.023	2.16	.77	2.93
16" O.C.	1.000	L.F.	.018	1.73	.62	2.35
14" deep, 12" O.C.	1.250	L.F.	.024	2.50	.83	3.33
16" O.C.	1.000	L.F.	.020	2.00	.66	2.66
16 " deep, 12" O.C.	1.250	L.F.	.026	2.63	.86	3.49
16" O.C.	1.000	L.F.	.021	2.10	.69	2.79
18" deep, 12" O.C.	1.250	L.F.	.027	2.72	.91	3.63
16" O.C.	1.000	L.F.	.022	2.18	.73	2.91
Continuous ribbing, 2" x 4"	.150	L.F.	.002	.06	.06	.12
2" x 6"	.150	L.F.	.002	.10	.06	.16
2" x 8"	.150	L.F.	.002	.14	.07	.21
2" x 10"	.150	L.F.	.003	.20	.09	.29
2" x 12"	.150	L.F.	.003	.24	.09	.33
Girders, including lally columns, 3 pieces spiked together, 2" x 8"	.125	L.F.	.015	.43	.53	.96
2" x 10"	.125	L.F.	.016	.59	.56	1.15
2" x 12"	.125	L.F.	.017	.70	.59	1.29
Solid girders, 3" x 8"	.040	L.F.	.004	.20	.14	.34
3" x 10"	.040	L.F.	.004	.22	.14	.36
3" x 12"	.040	L.F.	.004	.25	.15	.40
4" x 8"	.040	L.F.	.004	.25	.15	.40
4" x 10"	.040	L.F.	.004	.29	.15	.44
4" x 12"	.040	L.F.	.004	.33	.16	.49
Steel girders, bolted & including fabrication, wide flange shapes						
12" deep, 14#/l.f.	1.000	L.F.	.061	10.90	3.95	14.85
10" deep, 15#/l.f.	1.000	L.F.	.067	11.90	4.31	16.21
8" deep, 10#/l.f.	1.000	L.F.	.067	7.90	4.31	12.21
6" deep, 9#/l.f.	1.000	L.F.	.067	7.15	4.31	11.46
5" deep, 16#/l.f.	1.000	L.F.	.064	11.50	4.17	15.67
Sheathing, plywood exterior grade CDX, 1/2" thick	1.000	S.F.	.011	.58	.39	.97
5/8" thick	1.000	S.F.	.012	.73	.40	1.13
3/4" thick	1.000	S.F.	.013	.89	.43	1.32
Boards, 1" x 8" laid regular	1.000	S.F.	.016	1.00	.54	1.54
Laid diagonal	1.000	S.F.	.019	1.00	.64	1.64
1" x 10" laid regular	1.000	S.F.	.015	1.11	.49	1.60
Laid diagonal	1.000	S.F.	.018	1.12	.60	1.72
Furring, 1" x 3", 12" O.C.	1.250	L.F.	.029	.25	.96	1.21
16" O.C.	1.000	L.F.	.023	.20	.77	.97
24" O.C.	.750	L.F.	.017	.15	.58	.73

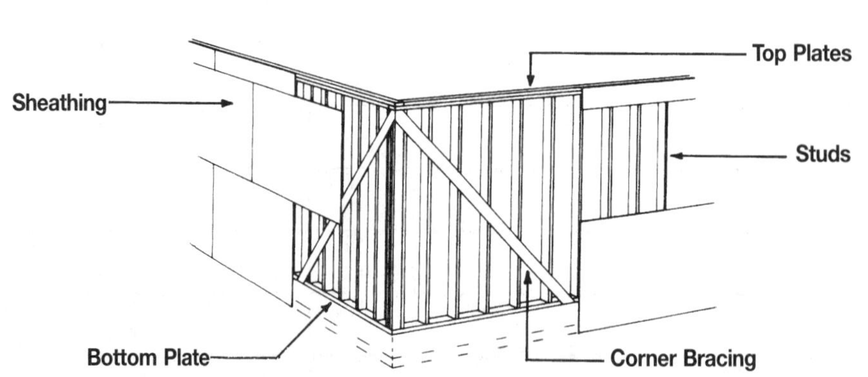

System Description	QUAN.	UNIT	LABOR HOURS	COST PER S.F. MAT.	INST.	TOTAL
2″ X 4″, 16″ O.C.						
2″ x 4″ studs, 16″ O.C.	1.000	L.F.	.015	.43	.49	.92
Plates, 2″ x 4″, double top, single bottom	.375	L.F.	.005	.16	.18	.34
Corner bracing, let-in, 1″ x 6″	.063	L.F.	.003	.03	.11	.14
Sheathing, 1/2″ plywood, CDX	1.000	S.F.	.011	.58	.39	.97
TOTAL			.034	1.20	1.17	2.37
2″ X 4″, 24″ O.C.						
2″ x 4″ studs, 24″ O.C.	.750	L.F.	.011	.32	.37	.69
Plates, 2″ x 4″, double top, single bottom	.375	L.F.	.005	.16	.18	.34
Corner bracing, let-in, 1″ x 6″	.063	L.F.	.002	.03	.07	.10
Sheathing, 1/2″ plywood, CDX	1.000	S.F.	.011	.58	.39	.97
TOTAL			.029	1.09	1.01	2.10
2″ X 6″, 16″ O.C.						
2″ x 6″ studs, 16″ O.C.	1.000	L.F.	.016	.65	.54	1.19
Plates, 2″ x 6″, double top, single bottom	.375	L.F.	.006	.24	.20	.44
Corner bracing, let-in, 1″ x 6″	.063	L.F.	.003	.03	.11	.14
Sheathing, 1/2″ plywood, CDX	1.000	S.F.	.011	.58	.39	.97
TOTAL			.036	1.50	1.24	2.74
2″ X 6″, 24″ O.C.						
2″ x 6″ studs, 24″ O.C.	.750	L.F.	.012	.49	.41	.90
Plates, 2″ x 6″, double top, single bottom	.375	L.F.	.006	.24	.20	.44
Corner bracing, let-in, 1″ x 6″	.063	L.F.	.002	.03	.07	.10
Sheathing, 1/2″ plywood, CDX	1.000	S.F.	.011	.58	.39	.97
TOTAL			.031	1.34	1.07	2.41

The wall costs on this page are given in cost per square foot of wall.
For window and door openings see below.

Description	QUAN.	UNIT	LABOR HOURS	COST PER S.F. MAT.	INST.	TOTAL

Exterior Wall Framing Price Sheet	QUAN.	UNIT	LABOR HOURS	COST PER S.F.		
				MAT.	INST.	TOTAL
Studs, #2 or better, 2" x 4", 12" O.C.	1.250	L.F.	.018	.54	.61	1.15
16" O.C.	1.000	L.F.	.015	.43	.49	.92
24" O.C.	.750	L.F.	.011	.32	.37	.69
32" O.C.	.600	L.F.	.009	.26	.29	.55
2" x 6", 12" O.C.	1.250	L.F.	.020	.81	.68	1.49
16" O.C.	1.000	L.F.	.016	.65	.54	1.19
24" O.C.	.750	L.F.	.012	.49	.41	.90
32" O.C.	.600	L.F.	.010	.39	.32	.71
2" x 8", 12" O.C.	1.250	L.F.	.025	1.48	.85	2.33
16" O.C.	1.000	L.F.	.020	1.18	.68	1.86
24" O.C.	.750	L.F.	.015	.89	.51	1.40
32" O.C.	.600	L.F.	.012	.71	.41	1.12
Plates, #2 or better, double top, single bottom, 2" x 4"	.375	L.F.	.005	.16	.18	.34
2" x 6"	.375	L.F.	.006	.24	.20	.44
2" x 8"	.375	L.F.	.008	.44	.26	.70
Corner bracing, let-in 1" x 6" boards, studs, 12" O.C.	.070	L.F.	.004	.04	.13	.17
16" O.C.	.063	L.F.	.003	.03	.11	.14
24" O.C.	.063	L.F.	.002	.03	.07	.10
32" O.C.	.057	L.F.	.002	.03	.07	.10
Let-in steel ("T" shape), studs, 12" O.C.	.070	L.F.	.001	.03	.03	.06
16" O.C.	.063	L.F.	.001	.03	.03	.06
24" O.C.	.063	L.F.	.001	.03	.03	.06
32" O.C.	.057	L.F.	.001	.02	.03	.05
Sheathing, plywood CDX, 3/8" thick	1.000	S.F.	.010	.45	.35	.80
1/2" thick	1.000	S.F.	.011	.58	.39	.97
5/8" thick	1.000	S.F.	.012	.73	.42	1.15
3/4" thick	1.000	S.F.	.013	.89	.45	1.34
Boards, 1" x 6", laid regular	1.000	S.F.	.025	1.43	.83	2.26
Laid diagonal	1.000	S.F.	.027	1.43	.92	2.35
1" x 8", laid regular	1.000	S.F.	.021	1.43	.71	2.14
Laid diagonal	1.000	S.F.	.025	1.43	.83	2.26
Wood fiber, regular, no vapor barrier, 1/2" thick	1.000	S.F.	.013	.58	.45	1.03
5/8" thick	1.000	S.F.	.013	.78	.45	1.23
Asphalt impregnated 25/32" thick	1.000	S.F.	.013	.39	.45	.84
1/2" thick	1.000	S.F.	.013	.32	.45	.77
Polystyrene, regular, 3/4" thick	1.000	S.F.	.010	.33	.34	.67
2" thick	1.000	S.F.	.011	.95	.37	1.32
Fiberglass, foil faced, 1" thick	1.000	S.F.	.008	.84	.27	1.11
2" thick	1.000	S.F.	.009	1.40	.30	1.70

Window & Door Openings	QUAN.	UNIT	LABOR HOURS	COST EACH		
				MAT.	INST.	TOTAL
The following costs are to be added to the total costs of the wall for each opening. Do not subtract the area of the openings.						
Headers, 2" x 6" double, 2' long	4.000	L.F.	.178	2.60	6.00	8.60
3' long	6.000	L.F.	.267	3.90	9.00	12.90
4' long	8.000	L.F.	.356	5.20	12.00	17.20
5' long	10.000	L.F.	.444	6.50	15.00	21.50
2" x 8" double, 4' long	8.000	L.F.	.376	7.30	12.70	20.00
5' long	10.000	L.F.	.471	9.10	15.90	25.00
6' long	12.000	L.F.	.565	10.90	19.10	30.00
8' long	16.000	L.F.	.753	14.55	25.50	40.05
2" x 10" double, 4' long	8.000	L.F.	.400	10.70	13.50	24.20
6' long	12.000	L.F.	.600	16.10	20.50	36.60
8' long	16.000	L.F.	.800	21.50	27.00	48.50
10' long	20.000	L.F.	1.000	27.00	34.00	61.00
2" x 12" double, 8' long	16.000	L.F.	.853	26.00	29.00	55.00
12' long	24.000	L.F.	1.280	39.00	43.00	82.00

12 Gable End Roof Framing Systems

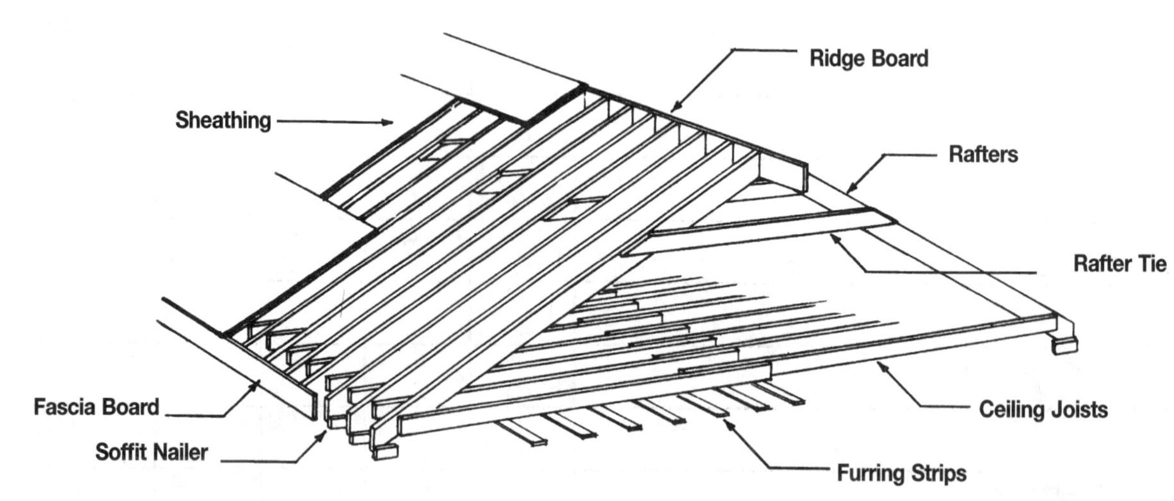

System Description	QUAN.	UNIT	LABOR HOURS	COST PER S.F. MAT.	COST PER S.F. INST.	COST PER S.F. TOTAL
2″ X 6″ RAFTERS, 16″ O.C., 4/12 PITCH						
Rafters, 2″ x 6″, 16″ O.C., 4/12 pitch	1.170	L.F.	.019	.76	.63	1.39
Ceiling joists, 2″ x 4″, 16″ O.C.	1.000	L.F.	.013	.43	.43	.86
Ridge board, 2″ x 6″	.050	L.F.	.002	.03	.05	.08
Fascia board, 2″ x 6″	.100	L.F.	.005	.07	.18	.25
Rafter tie, 1″ x 4″, 4′ O.C.	.060	L.F.	.001	.02	.04	.06
Soffit nailer (outrigger), 2″ x 4″, 24″ O.C.	.170	L.F.	.004	.07	.15	.22
Sheathing, exterior, plywood, CDX, 1/2″ thick	1.170	S.F.	.013	.68	.46	1.14
Furring strips, 1″ x 3″, 16″ O.C.	1.000	L.F.	.023	.20	.77	.97
TOTAL			.080	2.26	2.71	4.97
2″ X 8″ RAFTERS, 16″ O.C., 4/12 PITCH						
Rafters, 2″ x 8″, 16″ O.C., 4/12 pitch	1.170	L.F.	.020	1.06	.67	1.73
Ceiling joists, 2″ x 6″, 16″ O.C.	1.000	L.F.	.013	.65	.43	1.08
Ridge board, 2″ x 8″	.050	L.F.	.002	.05	.06	.11
Fascia board, 2″ x 8″	.100	L.F.	.007	.09	.24	.33
Rafter tie, 1″ x 4″, 4′ O.C.	.060	L.F.	.001	.02	.04	.06
Soffit nailer (outrigger), 2″ x 4″, 24″ O.C.	.170	L.F.	.004	.07	.15	.22
Sheathing, exterior, plywood, CDX, 1/2″ thick	1.170	S.F.	.013	.68	.46	1.14
Furring strips, 1″ x 3″, 16″ O.C.	1.000	L.F.	.023	.20	.77	.97
TOTAL			.083	2.82	2.82	5.64

The cost of this system is based on the square foot of plan area.
All quantities have been adjusted accordingly.

Description	QUAN.	UNIT	LABOR HOURS	COST PER S.F. MAT.	COST PER S.F. INST.	COST PER S.F. TOTAL

Gable End Roof Framing Price Sheet	QUAN.	UNIT	LABOR HOURS	COST PER S.F. MAT.	INST.	TOTAL
Rafters, #2 or better, 16″ O.C., 2″ x 6″, 4/12 pitch	1.170	L.F.	.019	.76	.63	1.39
8/12 pitch	1.330	L.F.	.027	.86	.90	1.76
2″ x 8″, 4/12 pitch	1.170	L.F.	.020	1.06	.67	1.73
8/12 pitch	1.330	L.F.	.028	1.21	.96	2.17
2″ x 10″, 4/12 pitch	1.170	L.F.	.030	1.57	1.01	2.58
8/12 pitch	1.330	L.F.	.043	1.78	1.45	3.23
24″ O.C., 2″ x 6″, 4/12 pitch	.940	L.F.	.015	.61	.51	1.12
8/12 pitch	1.060	L.F.	.021	.69	.72	1.41
2″ x 8″, 4/12 pitch	.940	L.F.	.016	.86	.54	1.40
8/12 pitch	1.060	L.F.	.023	.96	.76	1.72
2″ x 10″, 4/12 pitch	.940	L.F.	.024	1.26	.81	2.07
8/12 pitch	1.060	L.F.	.034	1.42	1.16	2.58
Ceiling joist, #2 or better, 2″ x 4″, 16″ O.C.	1.000	L.F.	.013	.43	.43	.86
24″ O.C.	.750	L.F.	.010	.32	.32	.64
2″ x 6″, 16″ O.C.	1.000	L.F.	.013	.65	.43	1.08
24″ O.C.	.750	L.F.	.010	.49	.32	.81
2″ x 8″, 16″ O.C.	1.000	L.F.	.015	.91	.49	1.40
24″ O.C.	.750	L.F.	.011	.68	.37	1.05
2″ x 10″, 16″ O.C.	1.000	L.F.	.018	1.34	.60	1.94
24″ O.C.	.750	L.F.	.013	1.01	.45	1.46
Ridge board, #2 or better, 1″ x 6″	.050	L.F.	.001	.05	.05	.10
1″ x 8″	.050	L.F.	.001	.07	.05	.12
1″ x 10″	.050	L.F.	.002	.08	.05	.13
2″ x 6″	.050	L.F.	.002	.03	.05	.08
2″ x 8″	.050	L.F.	.002	.05	.06	.11
2″ x 10″	.050	L.F.	.002	.07	.07	.14
Fascia board, #2 or better, 1″ x 6″	.100	L.F.	.004	.05	.13	.18
1″ x 8″	.100	L.F.	.005	.06	.15	.21
1″ x 10″	.100	L.F.	.005	.07	.17	.24
2″ x 6″	.100	L.F.	.006	.07	.19	.26
2″ x 8″	.100	L.F.	.007	.09	.24	.33
2″ x 10″	.100	L.F.	.004	.27	.12	.39
Rafter tie, #2 or better, 4′ O.C., 1″ x 4″	.060	L.F.	.001	.02	.04	.06
1″ x 6″	.060	L.F.	.001	.03	.04	.07
2″ x 4″	.060	L.F.	.002	.03	.05	.08
2″ x 6″	.060	L.F.	.002	.04	.07	.11
Soffit nailer (outrigger), 2″ x 4″, 16″ O.C.	.220	L.F.	.006	.09	.19	.28
24″ O.C.	.170	L.F.	.004	.07	.15	.22
2″ x 6″, 16″ O.C.	.220	L.F.	.006	.11	.22	.33
24″ O.C.	.170	L.F.	.005	.09	.17	.26
Sheathing, plywood CDX, 4/12 pitch, 3/8″ thick.	1.170	S.F.	.012	.53	.41	.94
1/2″ thick	1.170	S.F.	.013	.68	.46	1.14
5/8″ thick	1.170	S.F.	.014	.85	.49	1.34
8/12 pitch, 3/8″	1.330	S.F.	.014	.60	.47	1.07
1/2″ thick	1.330	S.F.	.015	.77	.52	1.29
5/8″ thick	1.330	S.F.	.016	.97	.56	1.53
Boards, 4/12 pitch roof, 1″ x 6″	1.170	S.F.	.026	1.67	.87	2.54
1″ x 8″	1.170	S.F.	.021	1.67	.73	2.40
8/12 pitch roof, 1″ x 6″	1.330	S.F.	.029	1.90	.98	2.88
1″ x 8″	1.330	S.F.	.024	1.90	.82	2.72
Furring, 1″ x 3″, 12″ O.C.	1.200	L.F.	.027	.24	.92	1.16
16″ O.C.	1.000	L.F.	.023	.20	.77	.97
24″ O.C.	.800	L.F.	.018	.16	.62	.78

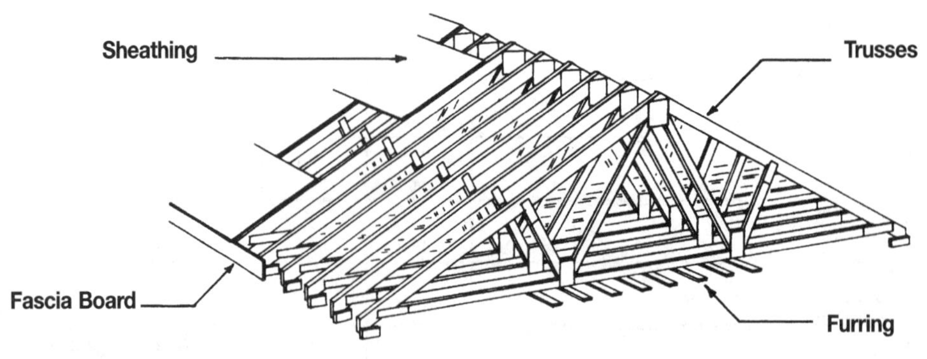

System Description	QUAN.	UNIT	LABOR HOURS	COST PER S.F.		
				MAT.	INST.	TOTAL
TRUSS, 16" O.C., 4/12 PITCH, 1' OVERHANG, 26' SPAN						
Truss, 40# loading, 16" O.C., 4/12 pitch, 26' span	.030	Ea.	.021	2.15	.89	3.04
Fascia board, 2" x 6"	.100	L.F.	.005	.07	.18	.25
Sheathing, exterior, plywood, CDX, 1/2" thick	1.170	S.F.	.013	.68	.46	1.14
Furring, 1" x 3", 16" O.C.	1.000	L.F.	.023	.20	.77	.97
TOTAL			.062	3.10	2.30	5.40
TRUSS, 16" O.C., 8/12 PITCH, 1' OVERHANG, 26' SPAN						
Truss, 40# loading, 16" O.C., 8/12 pitch, 26' span	.030	Ea.	.023	2.54	1.00	3.54
Fascia board, 2" x 6"	.100	L.F.	.005	.07	.18	.25
Sheathing, exterior, plywood, CDX, 1/2" thick	1.330	S.F.	.015	.77	.52	1.29
Furring, 1" x 3", 16" O.C.	1.000	L.F.	.023	.20	.77	.97
TOTAL			.066	3.58	2.47	6.05
TRUSS, 24" O.C., 4/12 PITCH, 1' OVERHANG, 26' SPAN						
Truss, 40# loading, 24" O.C., 4/12 pitch, 26' span	.020	Ea.	.014	1.43	.60	2.03
Fascia board, 2" x 6"	.100	L.F.	.005	.07	.18	.25
Sheathing, exterior, plywood, CDX, 1/2" thick	1.170	S.F.	.013	.68	.46	1.14
Furring, 1" x 3", 16" O.C.	1.000	L.F.	.023	.20	.77	.97
TOTAL			.055	2.38	2.01	4.39
TRUSS, 24" O.C., 8/12 PITCH, 1' OVERHANG, 26' SPAN						
Truss, 40# loading, 24" O.C., 8/12 pitch, 26' span	.020	Ea.	.015	1.69	.66	2.35
Fascia board, 2" x 6"	.100	L.F.	.005	.07	.18	.25
Sheathing, exterior, plywood, CDX, 1/2" thick	1.330	S.F.	.015	.77	.52	1.29
Furring, 1" x 3", 16" O.C.	1.000	L.F.	.023	.20	.77	.97
TOTAL			.058	2.73	2.13	4.86

The cost of this system is based on the square foot of plan area.
A one foot overhang is included.

Description	QUAN.	UNIT	LABOR HOURS	COST PER S.F.		
				MAT.	INST.	TOTAL

Truss Roof Framing Price Sheet

Truss Roof Framing Price Sheet	QUAN.	UNIT	LABOR HOURS	COST PER S.F. MAT.	INST.	TOTAL
Truss, 40# loading, including 1′ overhang, 4/12 pitch, 24′ span, 16″ O.C.	.033	Ea.	.022	1.55	.93	2.48
24″ O.C.	.022	Ea.	.015	1.03	.62	1.65
26′ span, 16″ O.C.	.030	Ea.	.021	2.15	.89	3.04
24″ O.C.	.020	Ea.	.014	1.43	.60	2.03
28′ span, 16″ O.C.	.027	Ea.	.020	1.50	.87	2.37
24″ O.C.	.019	Ea.	.014	1.05	.62	1.67
32′ span, 16″ O.C.	.024	Ea.	.019	1.88	.82	2.70
24″ O.C.	.016	Ea.	.013	1.26	.54	1.80
36′ span, 16″ O.C.	.022	Ea.	.019	2.06	.82	2.88
24″ O.C.	.015	Ea.	.013	1.40	.56	1.96
8/12 pitch, 24′ span, 16″ O.C.	.033	Ea.	.024	2.57	1.03	3.60
24″ O.C.	.022	Ea.	.016	1.72	.68	2.40
26′ span, 16″ O.C.	.030	Ea.	.023	2.54	1.00	3.54
24″ O.C.	.020	Ea.	.015	1.69	.66	2.35
28′ span, 16″ O.C.	.027	Ea.	.022	2.47	.94	3.41
24″ O.C.	.019	Ea.	.016	1.74	.66	2.40
32′ span, 16″ O.C.	.024	Ea.	.021	2.45	.92	3.37
24″ O.C.	.016	Ea.	.014	1.63	.61	2.24
36′ span, 16″ O.C.	.022	Ea.	.021	2.55	.92	3.47
24″ O.C.	.015	Ea.	.015	1.74	.62	2.36
Fascia board, #2 or better, 1″ x 6″	.100	L.F.	.004	.05	.13	.18
1″ x 8″	.100	L.F.	.005	.06	.15	.21
1″ x 10″	.100	L.F.	.005	.07	.17	.24
2″ x 6″	.100	L.F.	.006	.07	.19	.26
2″ x 8″	.100	L.F.	.007	.09	.24	.33
2″ x 10″	.100	L.F.	.009	.13	.30	.43
Sheathing, plywood CDX, 4/12 pitch, 3/8″ thick	1.170	S.F.	.012	.53	.41	.94
1/2″ thick	1.170	S.F.	.013	.68	.46	1.14
5/8″ thick	1.170	S.F.	.014	.85	.49	1.34
8/12 pitch, 3/8″ thick	1.330	S.F.	.014	.60	.47	1.07
1/2″ thick	1.330	S.F.	.015	.77	.52	1.29
5/8″ thick	1.330	S.F.	.016	.97	.56	1.53
Boards, 4/12 pitch, 1″ x 6″	1.170	S.F.	.026	1.67	.87	2.54
1″ x 8″	1.170	S.F.	.021	1.67	.73	2.40
8/12 pitch, 1″ x 6″	1.330	S.F.	.029	1.90	.98	2.88
1″ x 8″	1.330	S.F.	.024	1.90	.82	2.72
Furring, 1″ x 3″, 12″ O.C.	1.200	L.F.	.027	.24	.92	1.16
16″ O.C.	1.000	L.F.	.023	.20	.77	.97
24″ O.C.	.800	L.F.	.018	.16	.62	.78

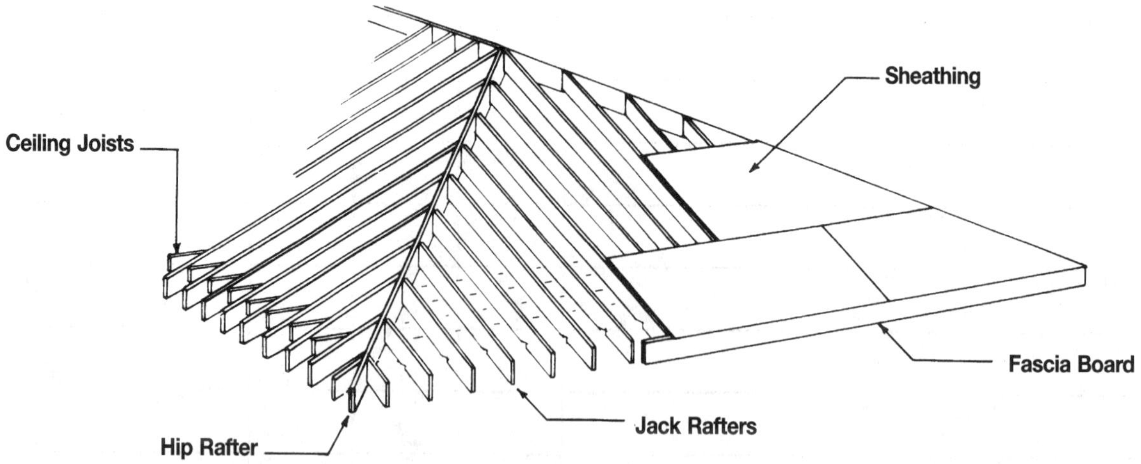

20 Hip Roof Framing Systems

Ceiling Joists
Sheathing
Fascia Board
Jack Rafters
Hip Rafter

System Description	QUAN.	UNIT	LABOR HOURS	COST PER S.F. MAT.	COST PER S.F. INST.	COST PER S.F. TOTAL
2″ X 6″, 16″ O.C., 4/12 PITCH						
Hip rafters, 2″ x 6″, 4/12 pitch	.160	L.F.	.003	.10	.11	.21
Jack rafters, 2″ x 6″, 16″ O.C., 4/12 pitch	1.430	L.F.	.038	.93	1.29	2.22
Ceiling joists, 2″ x 4″, 16″ O.C.	1.000	L.F.	.013	.43	.43	.86
Fascia board, 2″ x 6″	.220	L.F.	.012	.15	.41	.56
Soffit nailer (outrigger), 2″ x 4″, 24″ O.C.	.220	L.F.	.006	.09	.19	.28
Sheathing, 1/2″ exterior plywood, CDX	1.570	S.F.	.018	.91	.61	1.52
Furring strips, 1″ x 3″, 16″ O.C.	1.000	L.F.	.023	.20	.77	.97
TOTAL			.113	2.81	3.81	6.62
2″ X 8″, 16″ O.C., 4/12 PITCH						
Hip rafters, 2″ x 8″, 4/12 pitch	.160	L.F.	.004	.15	.12	.27
Jack rafters, 2″ x 8″, 16″ O.C., 4/12 pitch	1.430	L.F.	.047	1.30	1.57	2.87
Ceiling joists, 2″ x 6″, 16″ O.C.	1.000	L.F.	.013	.65	.43	1.08
Fascia board, 2″ x 8″	.220	L.F.	.016	.20	.53	.73
Soffit nailer (outrigger), 2″ x 4″, 24″ O.C.	.220	L.F.	.006	.09	.19	.28
Sheathing, 1/2″ exterior plywood, CDX	1.570	S.F.	.018	.91	.61	1.52
Furring strips, 1″ x 3″, 16″ O.C.	1.000	L.F.	.023	.20	.77	.97
TOTAL			.127	3.50	4.22	7.72

The cost of this system is based on S.F. of plan area. Measurement is area under the hip roof only. See gable roof system for added costs.

Description	QUAN.	UNIT	LABOR HOURS	COST PER S.F. MAT.	COST PER S.F. INST.	COST PER S.F. TOTAL

Important: See the Reference Section for critical supporting data - Reference Nos., Crews & Location Factors

Hip Roof Framing Price Sheet

	QUAN.	UNIT	LABOR HOURS	COST PER S.F.		
				MAT.	INST.	TOTAL
Hip rafters, #2 or better, 2″ x 6″, 4/12 pitch	.160	L.F.	.003	.10	.11	.21
8/12 pitch	.210	L.F.	.006	.14	.19	.33
2″ x 8″, 4/12 pitch	.160	L.F.	.004	.15	.12	.27
8/12 pitch	.210	L.F.	.006	.19	.21	.40
2″ x 10″, 4/12 pitch	.160	L.F.	.004	.21	.15	.36
8/12 pitch roof	.210	L.F.	.008	.28	.26	.54
Jack rafters, #2 or better, 16″ O.C., 2″ x 6″, 4/12 pitch	1.430	L.F.	.038	.93	1.29	2.22
8/12 pitch	1.800	L.F.	.061	1.17	2.05	3.22
2″ x 8″, 4/12 pitch	1.430	L.F.	.047	1.30	1.57	2.87
8/12 pitch	1.800	L.F.	.075	1.64	2.52	4.16
2″ x 10″, 4/12 pitch	1.430	L.F.	.051	1.92	1.72	3.64
8/12 pitch	1.800	L.F.	.082	2.41	2.77	5.18
24″ O.C., 2″ x 6″, 4/12 pitch	1.150	L.F.	.031	.75	1.04	1.79
8/12 pitch	1.440	L.F.	.049	.94	1.64	2.58
2″ x 8″, 4/12 pitch	1.150	L.F.	.038	1.05	1.27	2.32
8/12 pitch	1.440	L.F.	.060	1.31	2.02	3.33
2″ x 10″, 4/12 pitch	1.150	L.F.	.041	1.54	1.38	2.92
8/12 pitch	1.440	L.F.	.066	1.93	2.22	4.15
Ceiling joists, #2 or better, 2″ x 4″, 16″ O.C.	1.000	L.F.	.013	.43	.43	.86
24″ O.C.	.750	L.F.	.010	.32	.32	.64
2″ x 6″, 16″ O.C.	1.000	L.F.	.013	.65	.43	1.08
24″ O.C.	.750	L.F.	.010	.49	.32	.81
2″ x 8″, 16″ O.C.	1.000	L.F.	.015	.91	.49	1.40
24″ O.C.	.750	L.F.	.011	.68	.37	1.05
2″ x 10″, 16″ O.C.	1.000	L.F.	.018	1.34	.60	1.94
24″ O.C.	.750	L.F.	.013	1.01	.45	1.46
Fascia board, #2 or better, 1″ x 6″	.220	L.F.	.009	.11	.29	.40
1″ x 8″	.220	L.F.	.010	.13	.34	.47
1″ x 10″	.220	L.F.	.011	.14	.38	.52
2″ x 6″	.220	L.F.	.013	.16	.42	.58
2″ x 8″	.220	L.F.	.016	.20	.53	.73
2″ x 10″	.220	L.F.	.020	.29	.66	.95
Soffit nailer (outrigger), 2″ x 4″, 16″ O.C.	.280	L.F.	.007	.12	.24	.36
24″ O.C.	.220	L.F.	.006	.09	.19	.28
2″ x 8″, 16″ O.C.	.280	L.F.	.007	.19	.23	.42
24″ O.C.	.220	L.F.	.005	.15	.18	.33
Sheathing, plywood CDX, 4/12 pitch, 3/8″ thick	1.570	S.F.	.016	.71	.55	1.26
1/2″ thick	1.570	S.F.	.018	.91	.61	1.52
5/8″ thick	1.570	S.F.	.019	1.15	.66	1.81
8/12 pitch, 3/8″ thick	1.900	S.F.	.020	.86	.67	1.53
1/2″ thick	1.900	S.F.	.022	1.10	.74	1.84
5/8″ thick	1.900	S.F.	.023	1.39	.80	2.19
Boards, 4/12 pitch, 1″ x 6″ boards	1.450	S.F.	.032	2.07	1.07	3.14
1″ x 8″ boards	1.450	S.F.	.027	2.07	.90	2.97
8/12 pitch, 1″ x 6″ boards	1.750	S.F.	.039	2.50	1.30	3.80
1″ x 8″ boards	1.750	S.F.	.032	2.50	1.09	3.59
Furring, 1″ x 3″, 12″ O.C.	1.200	L.F.	.027	.24	.92	1.16
16″ O.C.	1.000	L.F.	.023	.20	.77	.97
24″ O.C.	.800	L.F.	.018	.16	.62	.78

24 Gambrel Roof Framing Systems

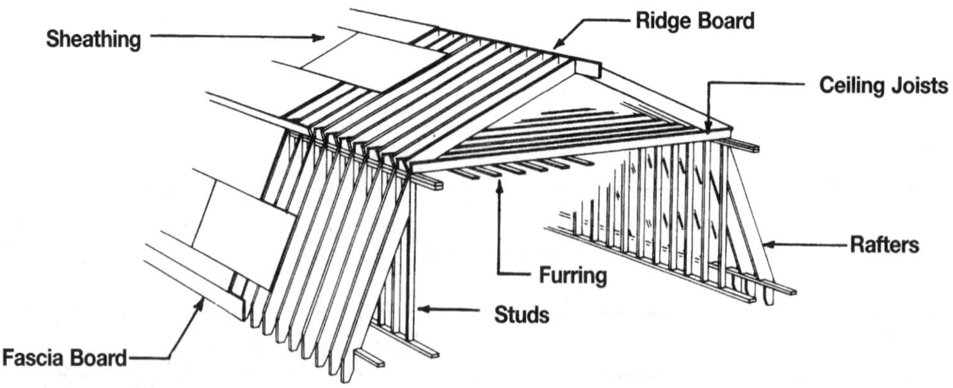

System Description	QUAN.	UNIT	LABOR HOURS	COST PER S.F. MAT.	COST PER S.F. INST.	COST PER S.F. TOTAL
2" X 6" RAFTERS, 16" O.C.						
Roof rafters, 2" x 6", 16" O.C.	1.430	L.F.	.029	.93	.97	1.90
Ceiling joists, 2" x 6", 16" O.C.	.710	L.F.	.009	.46	.31	.77
Stud wall, 2" x 4", 16" O.C., including plates	.790	L.F.	.012	.34	.42	.76
Furring strips, 1" x 3", 16" O.C.	.710	L.F.	.016	.14	.55	.69
Ridge board, 2" x 8"	.050	L.F.	.002	.05	.06	.11
Fascia board, 2" x 6"	.100	L.F.	.006	.07	.19	.26
Sheathing, exterior grade plywood, 1/2" thick	1.450	S.F.	.017	.84	.57	1.41
TOTAL			.091	2.83	3.07	5.90
2" X 8" RAFTERS, 16" O.C.						
Roof rafters, 2" x 8", 16" O.C.	1.430	L.F.	.031	1.30	1.03	2.33
Ceiling joists, 2" x 6", 16" O.C.	.710	L.F.	.009	.46	.31	.77
Stud wall, 2" x 4", 16" O.C., including plates	.790	L.F.	.012	.34	.42	.76
Furring strips, 1" x 3", 16" O.C.	.710	L.F.	.016	.14	.55	.69
Ridge board, 2" x 8"	.050	L.F.	.002	.05	.06	.11
Fascia board, 2" x 8"	.100	L.F.	.007	.09	.24	.33
Sheathing, exterior grade plywood, 1/2" thick	1.450	S.F.	.017	.84	.57	1.41
TOTAL			.094	3.22	3.18	6.40

The cost of this system is based on the square foot of plan area on the first floor.

Description	QUAN.	UNIT	LABOR HOURS	COST PER S.F. MAT.	COST PER S.F. INST.	COST PER S.F. TOTAL

Gambrel Roof Framing Price Sheet	QUAN.	UNIT	LABOR HOURS	COST PER S.F.		
				MAT.	INST.	TOTAL
Roof rafters, #2 or better, 2″ x 6″, 16″ O.C.	1.430	L.F.	.029	.93	.97	1.90
24″ O.C.	1.140	L.F.	.023	.74	.78	1.52
2″ x 8″, 16″ O.C.	1.430	L.F.	.031	1.30	1.03	2.33
24″ O.C.	1.140	L.F.	.024	1.04	.82	1.86
2″ x 10″, 16″ O.C.	1.430	L.F.	.046	1.92	1.56	3.48
24″ O.C.	1.140	L.F.	.037	1.53	1.24	2.77
Ceiling joist, #2 or better, 2″ x 4″, 16″ O.C.	.710	L.F.	.009	.31	.31	.62
24″ O.C.	.570	L.F.	.007	.25	.25	.50
2″ x 6″, 16″ O.C.	.710	L.F.	.009	.46	.31	.77
24″ O.C.	.570	L.F.	.007	.37	.25	.62
2″ x 8″, 16″ O.C.	.710	L.F.	.010	.65	.35	1.00
24″ O.C.	.570	L.F.	.008	.52	.28	.80
Stud wall, #2 or better, 2″ x 4″, 16″ O.C.	.790	L.F.	.012	.34	.42	.76
24″ O.C.	.630	L.F.	.010	.27	.33	.60
2″ x 6″, 16″ O.C.	.790	L.F.	.014	.51	.48	.99
24″ O.C.	.630	L.F.	.011	.41	.38	.79
Furring, 1″ x 3″, 16″ O.C.	.710	L.F.	.016	.14	.55	.69
24″ O.C.	.590	L.F.	.013	.12	.45	.57
Ridge board, #2 or better, 1″ x 6″	.050	L.F.	.001	.05	.05	.10
1″ x 8″	.050	L.F.	.001	.07	.05	.12
1″ x 10″	.050	L.F.	.002	.08	.05	.13
2″ x 6″	.050	L.F.	.002	.03	.05	.08
2″ x 8″	.050	L.F.	.002	.05	.06	.11
2″ x 10″	.050	L.F.	.002	.07	.07	.14
Fascia board, #2 or better, 1″ x 6″	.100	L.F.	.004	.05	.13	.18
1″ x 8″	.100	L.F.	.005	.06	.15	.21
1″ x 10″	.100	L.F.	.005	.07	.17	.24
2″ x 6″	.100	L.F.	.006	.07	.19	.26
2″ x 8″	.100	L.F.	.007	.09	.24	.33
2″ x 10″	.100	L.F.	.009	.13	.30	.43
Sheathing, plywood, exterior grade CDX, 3/8″ thick	1.450	S.F.	.015	.65	.51	1.16
1/2″ thick	1.450	S.F.	.017	.84	.57	1.41
5/8″ thick	1.450	S.F.	.018	1.06	.61	1.67
3/4″ thick	1.450	S.F.	.019	1.29	.65	1.94
Boards, 1″ x 6″, laid regular	1.450	S.F.	.032	2.07	1.07	3.14
Laid diagonal	1.450	S.F.	.036	2.07	1.20	3.27
1″ x 8″, laid regular	1.450	S.F.	.027	2.07	.90	2.97
Laid diagonal	1.450	S.F.	.032	2.07	1.07	3.14

28 Mansard Roof Framing Systems

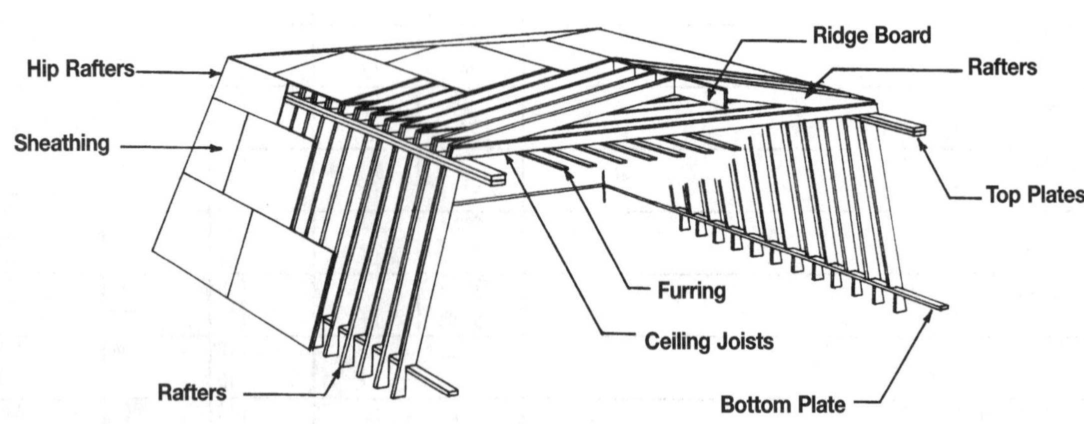

System Description	QUAN.	UNIT	LABOR HOURS	COST PER S.F.		
				MAT.	INST.	TOTAL
2" X 6" RAFTERS, 16" O.C.						
Roof rafters, 2" x 6", 16" O.C.	1.210	L.F.	.033	.79	1.11	1.90
Rafter plates, 2" x 6", double top, single bottom	.364	L.F.	.010	.24	.33	.57
Ceiling joists, 2" x 4", 16" O.C.	.920	L.F.	.012	.40	.40	.80
Hip rafter, 2" x 6"	.070	L.F.	.002	.05	.07	.12
Jack rafter, 2" x 6", 16" O.C.	1.000	L.F.	.039	.65	1.32	1.97
Ridge board, 2" x 6"	.018	L.F.	.001	.01	.02	.03
Sheathing, exterior grade plywood, 1/2" thick	2.210	S.F.	.025	1.28	.86	2.14
Furring strips, 1" x 3", 16" O.C.	.920	L.F.	.021	.18	.71	.89
TOTAL			.143	3.60	4.82	8.42
2" X 8" RAFTERS, 16" O.C.						
Roof rafters, 2" x 8", 16" O.C.	1.210	L.F.	.036	1.10	1.21	2.31
Rafter plates, 2" x 8", double top, single bottom	.364	L.F.	.011	.33	.36	.69
Ceiling joists, 2" x 6", 16" O.C.	.920	L.F.	.012	.60	.40	1.00
Hip rafter, 2" x 8"	.070	L.F.	.002	.06	.08	.14
Jack rafter, 2" x 8", 16" O.C.	1.000	L.F.	.048	.91	1.61	2.52
Ridge board, 2" x 8"	.018	L.F.	.001	.02	.02	.04
Sheathing, exterior grade plywood, 1/2" thick	2.210	S.F.	.025	1.28	.86	2.14
Furring strips, 1" x 3", 16" O.C.	.920	L.F.	.021	.18	.71	.89
TOTAL			.156	4.48	5.25	9.73

The cost of this system is based on the square foot of plan area.

Description	QUAN.	UNIT	LABOR HOURS	COST PER S.F.		
				MAT.	INST.	TOTAL

Mansard Roof Framing Price Sheet	QUAN.	UNIT	LABOR HOURS	COST PER S.F.		
				MAT.	INST.	TOTAL
Roof rafters, #2 or better, 2″ x 6″, 16″ O.C.	1.210	L.F.	.033	.79	1.11	1.90
24″ O.C.	.970	L.F.	.026	.63	.89	1.52
2″ x 8″, 16″ O.C.	1.210	L.F.	.036	1.10	1.21	2.31
24″ O.C.	.970	L.F.	.029	.88	.97	1.85
2″ x 10″, 16″ O.C.	1.210	L.F.	.046	1.62	1.54	3.16
24″ O.C.	.970	L.F.	.037	1.30	1.23	2.53
Rafter plates, #2 or better double top single bottom, 2″ x 6″	.364	L.F.	.010	.24	.33	.57
2″ x 8″	.364	L.F.	.011	.33	.36	.69
2″ x 10″	.364	L.F.	.014	.49	.46	.95
Ceiling joist, #2 or better, 2″ x 4″, 16″ O.C.	.920	L.F.	.012	.40	.40	.80
24″ O.C.	.740	L.F.	.009	.32	.32	.64
2″ x 6″, 16″ O.C.	.920	L.F.	.012	.60	.40	1.00
24″ O.C.	.740	L.F.	.009	.48	.32	.80
2″ x 8″, 16″ O.C.	.920	L.F.	.013	.84	.45	1.29
24″ O.C.	.740	L.F.	.011	.67	.36	1.03
Hip rafter, #2 or better, 2″ x 6″	.070	L.F.	.002	.05	.07	.12
2″ x 8″	.070	L.F.	.002	.06	.08	.14
2″ x 10″	.070	L.F.	.003	.09	.10	.19
Jack rafter, #2 or better, 2″ x 6″, 16″ O.C.	1.000	L.F.	.039	.65	1.32	1.97
24″ O.C.	.800	L.F.	.031	.52	1.06	1.58
2″ x 8″, 16″ O.C.	1.000	L.F.	.048	.91	1.61	2.52
24″ O.C.	.800	L.F.	.038	.73	1.29	2.02
Ridge board, #2 or better, 1″ x 6″	.018	L.F.	.001	.02	.02	.04
1″ x 8″	.018	L.F.	.001	.02	.02	.04
1″ x 10″	.018	L.F.	.001	.03	.02	.05
2″ x 6″	.018	L.F.	.001	.01	.02	.03
2″ x 8″	.018	L.F.	.001	.02	.02	.04
2″ x 10″	.018	L.F.	.001	.02	.02	.04
Sheathing, plywood exterior grade CDX, 3/8″ thick	2.210	S.F.	.023	.99	.77	1.76
1/2″ thick	2.210	S.F.	.025	1.28	.86	2.14
5/8″ thick	2.210	S.F.	.027	1.61	.93	2.54
3/4″ thick	2.210	S.F.	.029	1.97	.99	2.96
Boards, 1″ x 6″, laid regular	2.210	S.F.	.049	3.16	1.64	4.80
Laid diagonal	2.210	S.F.	.054	3.16	1.83	4.99
1″ x 8″, laid regular	2.210	S.F.	.040	3.16	1.37	4.53
Laid diagonal	2.210	S.F.	.049	3.16	1.64	4.80
Furring, 1″ x 3″, 12″ O.C.	1.150	L.F.	.026	.23	.89	1.12
24″ O.C.	.740	L.F.	.017	.15	.57	.72

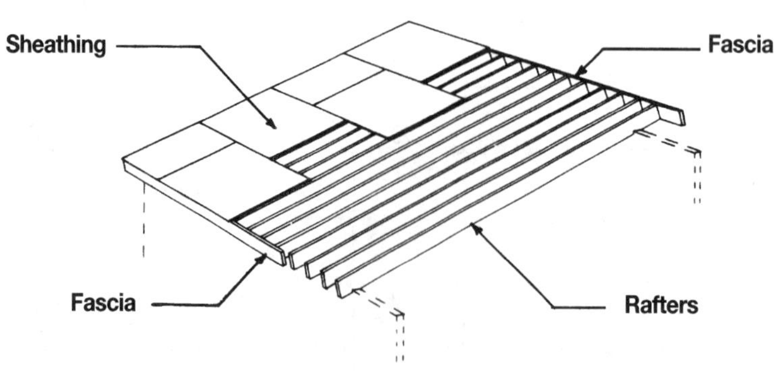

System Description	QUAN.	UNIT	LABOR HOURS	COST PER S.F. MAT.	COST PER S.F. INST.	COST PER S.F. TOTAL
2" X 6",16" O.C., 4/12 PITCH						
Rafters, 2" x 6", 16" O.C., 4/12 pitch	1.170	L.F.	.019	.76	.63	1.39
Fascia, 2" x 6"	.100	L.F.	.006	.07	.19	.26
Bridging, 1" x 3", 6' O.C.	.080	Pr.	.005	.02	.17	.19
Sheathing, exterior grade plywood, 1/2" thick	1.230	S.F.	.014	.71	.48	1.19
TOTAL			.044	1.56	1.47	3.03
2" X 6", 24" O.C., 4/12 PITCH						
Rafters, 2" x 6", 24" O.C., 4/12 pitch	.940	L.F.	.015	.61	.51	1.12
Fascia, 2" x 6"	.100	L.F.	.006	.07	.19	.26
Bridging, 1" x 3", 6' O.C.	.060	Pr.	.004	.02	.12	.14
Sheathing, exterior grade plywood, 1/2" thick	1.230	S.F.	.014	.71	.48	1.19
TOTAL			.039	1.41	1.30	2.71
2" X 8", 16" O.C., 4/12 PITCH						
Rafters, 2" x 8", 16" O.C., 4/12 pitch	1.170	L.F.	.020	1.06	.67	1.73
Fascia, 2" x 8"	.100	L.F.	.007	.09	.24	.33
Bridging, 1" x 3", 6' O.C.	.080	Pr.	.005	.02	.17	.19
Sheathing, exterior grade plywood, 1/2" thick	1.230	S.F.	.014	.71	.48	1.19
TOTAL			.046	1.88	1.56	3.44
2" X 8", 24" O.C., 4/12 PITCH						
Rafters, 2" x 8", 24" O.C., 4/12 pitch	.940	L.F.	.016	.86	.54	1.40
Fascia, 2" x 8"	.100	L.F.	.007	.09	.24	.33
Bridging, 1" x 3", 6' O.C.	.060	Pr.	.004	.02	.12	.14
Sheathing, exterior grade plywood, 1/2" thick	1.230	S.F.	.014	.71	.48	1.19
TOTAL			.041	1.68	1.38	3.06

The cost of this system is based on the square foot of plan area.
A 1' overhang is assumed. No ceiling joists or furring are included.

Description	QUAN.	UNIT	LABOR HOURS	COST PER S.F. MAT.	COST PER S.F. INST.	COST PER S.F. TOTAL

Important: See the Reference Section for critical supporting data - Reference Nos., Crews & Location Factors

Shed/Flat Roof Framing Price Sheet	QUAN.	UNIT	LABOR HOURS	COST PER S.F.		
				MAT.	INST.	TOTAL
Rafters, #2 or better, 16″ O.C., 2″ x 4″, 0 - 4/12 pitch	1.170	L.F.	.014	.57	.47	1.04
5/12 - 8/12 pitch	1.330	L.F.	.020	.65	.68	1.33
2″ x 6″, 0 - 4/12 pitch	1.170	L.F.	.019	.76	.63	1.39
5/12 - 8/12 pitch	1.330	L.F.	.027	.86	.90	1.76
2″ x 8″, 0 - 4/12 pitch	1.170	L.F.	.020	1.06	.67	1.73
5/12 - 8/12 pitch	1.330	L.F.	.028	1.21	.96	2.17
2″ x 10″, 0 - 4/12 pitch	1.170	L.F.	.030	1.57	1.01	2.58
5/12 - 8/12 pitch	1.330	L.F.	.043	1.78	1.45	3.23
24″ O.C., 2″ x 4″, 0 - 4/12 pitch	.940	L.F.	.011	.46	.38	.84
5/12 - 8/12 pitch	1.060	L.F.	.021	.69	.72	1.41
2″ x 6″, 0 - 4/12 pitch	.940	L.F.	.015	.61	.51	1.12
5/12 - 8/12 pitch	1.060	L.F.	.021	.69	.72	1.41
2″ x 8″, 0 - 4/12 pitch	.940	L.F.	.016	.86	.54	1.40
5/12 - 8/12 pitch	1.060	L.F.	.023	.96	.76	1.72
2″ x 10″, 0 - 4/12 pitch	.940	L.F.	.024	1.26	.81	2.07
5/12 - 8/12 pitch	1.060	L.F.	.034	1.42	1.16	2.58
Fascia, #2 or better,, 1″ x 4″	.100	L.F.	.003	.04	.10	.14
1″ x 6″	.100	L.F.	.004	.05	.13	.18
1″ x 8″	.100	L.F.	.005	.06	.15	.21
1″ x 10″	.100	L.F.	.005	.07	.17	.24
2″ x 4″	.100	L.F.	.005	.06	.16	.22
2″ x 6″	.100	L.F.	.006	.07	.19	.26
2″ x 8″	.100	L.F.	.007	.09	.24	.33
2″ x 10″	.100	L.F.	.009	.13	.30	.43
Bridging, wood 6′ O.C., 1″ x 3″, rafters, 16″ O.C.	.080	Pr.	.005	.02	.17	.19
24″ O.C.	.060	Pr.	.004	.02	.12	.14
Metal, galvanized, rafters, 16″ O.C.	.080	Pr.	.005	.05	.17	.22
24″ O.C.	.060	Pr.	.003	.15	.12	.27
Compression type, rafters, 16″ O.C.	.080	Pr.	.003	.11	.11	.22
24″ O.C.	.060	Pr.	.002	.08	.08	.16
Sheathing, plywood, exterior grade, 3/8″ thick, flat 0 - 4/12 pitch	1.230	S.F.	.013	.55	.43	.98
5/12 - 8/12 pitch	1.330	S.F.	.014	.60	.47	1.07
1/2″ thick, flat 0 - 4/12 pitch	1.230	S.F.	.014	.71	.48	1.19
5/12 - 8/12 pitch	1.330	S.F.	.015	.77	.52	1.29
5/8″ thick, flat 0 - 4/12 pitch	1.230	S.F.	.015	.90	.52	1.42
5/12 - 8/12 pitch	1.330	S.F.	.016	.97	.56	1.53
3/4″ thick, flat 0 - 4/12 pitch	1.230	S.F.	.016	1.09	.55	1.64
5/12 - 8/12 pitch	1.330	S.F.	.018	1.18	.60	1.78
Boards, 1″ x 6″, laid regular, flat 0 - 4/12 pitch	1.230	S.F.	.027	1.76	.91	2.67
5/12 - 8/12 pitch	1.330	S.F.	.041	1.90	1.38	3.28
Laid diagonal, flat 0 - 4/12 pitch	1.230	S.F.	.030	1.76	1.02	2.78
5/12 - 8/12 pitch	1.330	S.F.	.044	1.90	1.50	3.40
1″ x 8″, laid regular, flat 0 - 4/12 pitch	1.230	S.F.	.022	1.76	.76	2.52
5/12 - 8/12 pitch	1.330	S.F.	.034	1.90	1.13	3.03
Laid diagonal, flat 0 - 4/12 pitch	1.230	S.F.	.027	1.76	.91	2.67
5/12 - 8/12 pitch	1.330	S.F.	.044	1.90	1.50	3.40

40 Gable Dormer Framing Systems

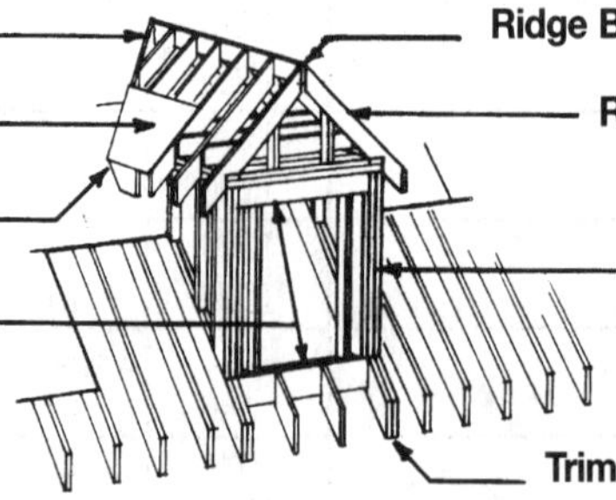

Valley Rafter
Ridge Board
Sheathing
Rafters
Fascia Board
Studs & Plates
Headers
Trimmer Rafters

System Description	QUAN.	UNIT	LABOR HOURS	COST PER S.F.		
				MAT.	INST.	TOTAL
2" X 6", 16" O.C.						
Dormer rafter, 2" x 6", 16" O.C.	1.330	L.F.	.036	.86	1.22	2.08
Ridge board, 2" x 6"	.280	L.F.	.009	.18	.30	.48
Trimmer rafters, 2" x 6"	.880	L.F.	.014	.57	.48	1.05
Wall studs & plates, 2" x 4", 16" O.C.	3.160	L.F.	.056	1.36	1.90	3.26
Fascia, 2" x 6"	.220	L.F.	.012	.15	.41	.56
Valley rafter, 2" x 6", 16" O.C.	.280	L.F.	.009	.18	.30	.48
Cripple rafter, 2" x 6", 16" O.C.	.560	L.F.	.022	.36	.74	1.10
Headers, 2" x 6", doubled	.670	L.F.	.030	.44	1.01	1.45
Ceiling joist, 2" x 4", 16" O.C.	1.000	L.F.	.013	.43	.43	.86
Sheathing, exterior grade plywood, 1/2" thick	3.610	S.F.	.041	2.09	1.41	3.50
TOTAL			.242	6.62	8.20	14.82
2" X 8", 16" O.C.						
Dormer rafter, 2" x 8", 16" O.C.	1.330	L.F.	.039	1.21	1.33	2.54
Ridge board, 2" x 8"	.280	L.F.	.010	.25	.34	.59
Trimmer rafter, 2" x 8"	.880	L.F.	.015	.80	.50	1.30
Wall studs & plates, 2" x 4", 16" O.C.	3.160	L.F.	.056	1.36	1.90	3.26
Fascia, 2" x 8"	.220	L.F.	.016	.20	.53	.73
Valley rafter, 2" x 8", 16" O.C.	.280	L.F.	.010	.25	.32	.57
Cripple rafter, 2" x 8", 16" O.C.	.560	L.F.	.027	.51	.90	1.41
Headers, 2" x 8", doubled	.670	L.F.	.032	.61	1.07	1.68
Ceiling joist, 2" x 4", 16" O.C.	1.000	L.F.	.013	.43	.43	.86
Sheathing,, exterior grade plywood, 1/2" thick	3.610	S.F.	.041	2.09	1.41	3.50
TOTAL			.259	7.71	8.73	16.44

The cost in this system is based on the square foot of plan area.
The measurement being the plan area of the dormer only.

Description	QUAN.	UNIT	LABOR HOURS	COST PER S.F.		
				MAT.	INST.	TOTAL

Important: See the Reference Section for critical supporting data - Reference Nos., Crews & Location Factors

Gable Dormer Framing Price Sheet	QUAN.	UNIT	LABOR HOURS	COST PER S.F.		
				MAT.	INST.	TOTAL
Dormer rafters, #2 or better, 2" x 4", 16" O.C.	1.330	L.F.	.029	.69	.98	1.67
24" O.C.	1.060	L.F.	.023	.55	.78	1.33
2" x 6", 16" O.C.	1.330	L.F.	.036	.86	1.22	2.08
24" O.C.	1.060	L.F.	.029	.69	.98	1.67
2" x 8", 16" O.C.	1.330	L.F.	.039	1.21	1.33	2.54
24" O.C.	1.060	L.F.	.031	.96	1.06	2.02
Ridge board, #2 or better, 1" x 4"	.280	L.F.	.006	.22	.20	.42
1" x 6"	.280	L.F.	.007	.28	.25	.53
1" x 8"	.280	L.F.	.008	.37	.27	.64
2" x 4"	.280	L.F.	.007	.15	.24	.39
2" x 6"	.280	L.F.	.009	.18	.30	.48
2" x 8"	.280	L.F.	.010	.25	.34	.59
Trimmer rafters, #2 or better, 2" x 4"	.880	L.F.	.011	.46	.38	.84
2" x 6"	.880	L.F.	.014	.57	.48	1.05
2" x 8"	.880	L.F.	.015	.80	.50	1.30
2" x 10"	.880	L.F.	.022	1.18	.76	1.94
Wall studs & plates, #2 or better, 2" x 4" studs, 16" O.C.	3.160	L.F.	.056	1.36	1.90	3.26
24" O.C.	2.800	L.F.	.050	1.20	1.68	2.88
2" x 6" studs, 16" O.C.	3.160	L.F.	.063	2.05	2.15	4.20
24" O.C.	2.800	L.F.	.056	1.82	1.90	3.72
Fascia, #2 or better, 1" x 4"	.220	L.F.	.006	.08	.22	.30
1" x 6"	.220	L.F.	.008	.10	.26	.36
1" x 8"	.220	L.F.	.009	.12	.31	.43
2" x 4"	.220	L.F.	.011	.14	.36	.50
2" x 6"	.220	L.F.	.014	.17	.46	.63
2" x 8"	.220	L.F.	.016	.20	.53	.73
Valley rafter, #2 or better, 2" x 4"	.280	L.F.	.007	.15	.24	.39
2" x 6"	.280	L.F.	.009	.18	.30	.48
2" x 8"	.280	L.F.	.010	.25	.32	.57
2" x 10"	.280	L.F.	.012	.38	.40	.78
Cripple rafter, #2 or better, 2" x 4", 16" O.C.	.560	L.F.	.018	.29	.59	.88
24" O.C.	.450	L.F.	.014	.23	.48	.71
2" x 6", 16" O.C.	.560	L.F.	.022	.36	.74	1.10
24" O.C.	.450	L.F.	.018	.29	.59	.88
2" x 8", 16" O.C.	.560	L.F.	.027	.51	.90	1.41
24" O.C.	.450	L.F.	.021	.41	.72	1.13
Headers, #2 or better double header, 2" x 4"	.670	L.F.	.024	.35	.81	1.16
2" x 6"	.670	L.F.	.030	.44	1.01	1.45
2" x 8"	.670	L.F.	.032	.61	1.07	1.68
2" x 10"	.670	L.F.	.034	.90	1.13	2.03
Ceiling joist, #2 or better, 2" x 4", 16" O.C.	1.000	L.F.	.013	.43	.43	.86
24" O.C.	.800	L.F.	.010	.34	.34	.68
2" x 6", 16" O.C.	1.000	L.F.	.013	.65	.43	1.08
24" O.C.	.800	L.F.	.010	.52	.34	.86
Sheathing, plywood exterior grade, 3/8" thick	3.610	S.F.	.038	1.62	1.26	2.88
1/2" thick	3.610	S.F.	.041	2.09	1.41	3.50
5/8" thick	3.610	S.F.	.044	2.64	1.52	4.16
3/4" thick	3.610	S.F.	.048	3.21	1.62	4.83
Boards, 1" x 6", laid regular	3.610	S.F.	.089	5.15	3.00	8.15
Laid diagonal	3.610	S.F.	.099	5.15	3.32	8.47
1" x 8", laid regular	3.610	S.F.	.076	5.15	2.56	7.71
Laid diagonal	3.610	S.F.	.089	5.15	3.00	8.15

44 Shed Dormer Framing Systems

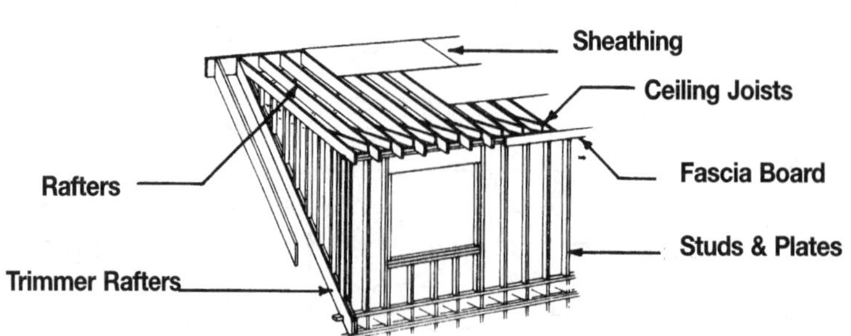

System Description	QUAN.	UNIT	LABOR HOURS	COST PER S.F.		
				MAT.	INST.	TOTAL
2″ X 6″ RAFTERS, 16″ O.C.						
Dormer rafter, 2″ x 6″, 16″ O.C.	1.080	L.F.	.029	.70	.99	1.69
Trimmer rafter, 2″ x 6″	.400	L.F.	.006	.26	.22	.48
Studs & plates, 2″ x 4″, 16″ O.C.	2.750	L.F.	.049	1.18	1.65	2.83
Fascia, 2″ x 6″	.250	L.F.	.014	.17	.46	.63
Ceiling joist, 2″ x 4″, 16″ O.C.	1.000	L.F.	.013	.43	.43	.86
Sheathing, exterior grade plywood, CDX, 1/2″ thick	2.940	S.F.	.034	1.71	1.15	2.86
TOTAL			.145	4.45	4.90	9.35
2″ X 8″ RAFTERS, 16″ O.C.						
Dormer rafter, 2″ x 8″, 16″ O.C.	1.080	L.F.	.032	.98	1.08	2.06
Trimmer rafter, 2″ x 8″	.400	L.F.	.007	.36	.23	.59
Studs & plates, 2″ x 4″, 16″ O.C.	2.750	L.F.	.049	1.18	1.65	2.83
Fascia, 2″ x 8″	.250	L.F.	.018	.23	.60	.83
Ceiling joist, 2″ x 6″, 16″ O.C.	1.000	L.F.	.013	.65	.43	1.08
Sheathing, exterior grade plywood, CDX, 1/2″ thick	2.940	S.F.	.034	1.71	1.15	2.86
TOTAL			.153	5.11	5.14	10.25
2″ X 10″ RAFTERS, 16″ O.C.						
Dormer rafter, 2″ x 10″, 16″ O.C.	1.080	L.F.	.041	1.45	1.37	2.82
Trimmer rafter, 2″ x 10″	.400	L.F.	.010	.54	.34	.88
Studs & plates, 2″ x 4″, 16″ O.C.	2.750	L.F.	.049	1.18	1.65	2.83
Fascia, 2″ x 10″	.250	L.F.	.022	.34	.75	1.09
Ceiling joist, 2″ x 6″, 16″ O.C.	1.000	L.F.	.013	.65	.43	1.08
Sheathing, exterior grade plywood, CDX, 1/2″ thick	2.940	S.F.	.034	1.71	1.15	2.86
TOTAL			.169	5.87	5.69	11.56

The cost in this system is based on the square foot of plan area.
The measurement is the plan area of the dormer only.

Description	QUAN.	UNIT	LABOR HOURS	COST PER S.F.		
				MAT.	INST.	TOTAL

FRAMING 3

Shed Dormer Framing Price Sheet	QUAN.	UNIT	LABOR HOURS	COST PER S.F.		
				MAT.	INST.	TOTAL
Dormer rafters, #2 or better, 2″ x 4″, 16″ O.C.	1.080	L.F.	.023	.56	.79	1.35
24″ O.C.	.860	L.F.	.019	.45	.63	1.08
2″ x 6″, 16″ O.C.	1.080	L.F.	.029	.70	.99	1.69
24″ O.C.	.860	L.F.	.023	.56	.79	1.35
2″ x 8″, 16″ O.C.	1.080	L.F.	.032	.98	1.08	2.06
24″ O.C.	.860	L.F.	.025	.78	.86	1.64
2″ x 10″, 16″ O.C.	1.080	L.F.	.041	1.45	1.37	2.82
24″ O.C.	.860	L.F.	.032	1.15	1.09	2.24
Trimmer rafter, #2 or better, 2″ x 4″	.400	L.F.	.005	.21	.17	.38
2″ x 6″	.400	L.F.	.006	.26	.22	.48
2″ x 8″	.400	L.F.	.007	.36	.23	.59
2″ x 10″	.400	L.F.	.010	.54	.34	.88
Studs & plates, #2 or better, 2″ x 4″, 16″ O.C.	2.750	L.F.	.049	1.18	1.65	2.83
24″ O.C.	2.200	L.F.	.039	.95	1.32	2.27
2″ x 6″, 16″ O.C.	2.750	L.F.	.055	1.79	1.87	3.66
24″ O.C.	2.200	L.F.	.044	1.43	1.50	2.93
Fascia, #2 or better, 1″ x 4″	.250	L.F.	.006	.08	.22	.30
1″ x 6″	.250	L.F.	.008	.10	.26	.36
1″ x 8″	.250	L.F.	.009	.12	.31	.43
2″ x 4″	.250	L.F.	.011	.14	.36	.50
2″ x 6″	.250	L.F.	.014	.17	.46	.63
2″ x 8″	.250	L.F.	.018	.23	.60	.83
Ceiling joist, #2 or better, 2″ x 4″, 16″ O.C.	1.000	L.F.	.013	.43	.43	.86
24″ O.C.	.800	L.F.	.010	.34	.34	.68
2″ x 6″, 16″ O.C.	1.000	L.F.	.013	.65	.43	1.08
24″ O.C.	.800	L.F.	.010	.52	.34	.86
2″ x 8″, 16″ O.C.	1.000	L.F.	.015	.91	.49	1.40
24″ O.C.	.800	L.F.	.012	.73	.39	1.12
Sheathing, plywood exterior grade, 3/8″ thick	2.940	S.F.	.031	1.32	1.03	2.35
1/2″ thick	2.940	S.F.	.034	1.71	1.15	2.86
5/8″ thick	2.940	S.F.	.036	2.15	1.23	3.38
3/4″ thick	2.940	S.F.	.039	2.62	1.32	3.94
Boards, 1″ x 6″, laid regular	2.940	S.F.	.072	4.20	2.44	6.64
Laid diagonal	2.940	S.F.	.080	4.20	2.70	6.90
1″ x 8″, laid regular	2.940	S.F.	.061	4.20	2.09	6.29
Laid diagonal	2.940	S.F.	.072	4.20	2.44	6.64

Window Openings	QUAN.	UNIT	LABOR HOURS	COST EACH		
				MAT.	INST.	TOTAL
The following are to be added to the total cost of the dormers for window openings. Do not subtract window area from the stud wall quantities.						
Headers, 2″ x 6″ doubled, 2′ long	4.000	L.F.	.178	2.60	6.00	8.60
3′ long	6.000	L.F.	.267	3.90	9.00	12.90
4′ long	8.000	L.F.	.356	5.20	12.00	17.20
5′ long	10.000	L.F.	.444	6.50	15.00	21.50
2″ x 8″ doubled, 4′ long	8.000	L.F.	.376	7.30	12.70	20.00
5′ long	10.000	L.F.	.471	9.10	15.90	25.00
6′ long	12.000	L.F.	.565	10.90	19.10	30.00
8′ long	16.000	L.F.	.753	14.55	25.50	40.05
2″ x 10″ doubled, 4′ long	8.000	L.F.	.400	10.70	13.50	24.20
6′ long	12.000	L.F.	.600	16.10	20.50	36.60
8′ long	16.000	L.F.	.800	21.50	27.00	48.50
10′ long	20.000	L.F.	1.000	27.00	34.00	61.00

48 Partition Framing Systems

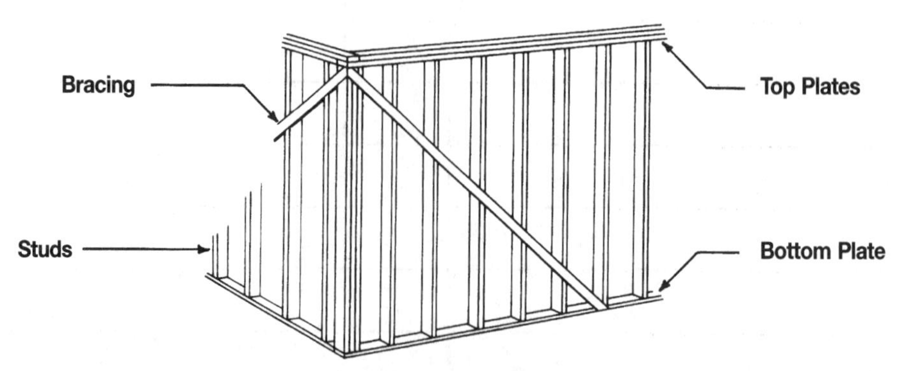

System Description	QUAN.	UNIT	LABOR HOURS	COST PER S.F.		
				MAT.	INST.	TOTAL
2" X 4", 16" O.C.						
2" x 4" studs, #2 or better, 16" O.C.	1.000	L.F.	.015	.43	.49	.92
Plates, double top, single bottom	.375	L.F.	.005	.16	.18	.34
Cross bracing, let-in, 1" x 6"	.080	L.F.	.004	.04	.14	.18
TOTAL			.024	.63	.81	1.44
2" X 4", 24" O.C.						
2" x 4" studs, #2 or better, 24" O.C.	.800	L.F.	.012	.34	.39	.73
Plates, double top, single bottom	.375	L.F.	.005	.16	.18	.34
Cross bracing, let-in, 1" x 6"	.080	L.F.	.003	.04	.09	.13
TOTAL			.020	.54	.66	1.20
2" X 6", 16" O.C.						
2" x 6" studs, #2 or better, 16" O.C.	1.000	L.F.	.016	.65	.54	1.19
Plates, double top, single bottom	.375	L.F.	.006	.24	.20	.44
Cross bracing, let-in, 1" x 6"	.080	L.F.	.004	.04	.14	.18
TOTAL			.026	.93	.88	1.81
2" X 6", 24" O.C.						
2" x 6" studs, #2 or better, 24" O.C.	.800	L.F.	.013	.52	.43	.95
Plates, double top, single bottom	.375	L.F.	.006	.24	.20	.44
Cross bracing, let-in, 1" x 6"	.080	L.F.	.003	.04	.09	.13
TOTAL			.022	.80	.72	1.52

The costs in this system are based on a square foot of wall area. Do not subtract for door or window openings.

Description	QUAN.	UNIT	LABOR HOURS	COST PER S.F.		
				MAT.	INST.	TOTAL

Important: See the Reference Section for critical supporting data - Reference Nos., Crews & Location Factors

Partition Framing Price Sheet	QUAN.	UNIT	LABOR HOURS	COST PER S.F.		
				MAT.	INST.	TOTAL
Wood studs, #2 or better, 2″ x 4″, 12″ O.C.	1.250	L.F.	.018	.54	.61	1.15
16″ O.C.	1.000	L.F.	.015	.43	.49	.92
24″ O.C.	.800	L.F.	.012	.34	.39	.73
32″ O.C.	.650	L.F.	.009	.28	.32	.60
2″ x 6″, 12″ O.C.	1.250	L.F.	.020	.81	.68	1.49
16″ O.C.	1.000	L.F.	.016	.65	.54	1.19
24″ O.C.	.800	L.F.	.013	.52	.43	.95
32″ O.C.	.650	L.F.	.010	.42	.35	.77
Plates, #2 or better double top single bottom, 2″ x 4″	.375	L.F.	.005	.16	.18	.34
2″ x 6″	.375	L.F.	.006	.24	.20	.44
2″ x 8″	.375	L.F.	.005	.34	.18	.52
Cross bracing, let-in, 1″ x 6″ boards studs, 12″ O.C.	.080	L.F.	.005	.05	.18	.23
16″ O.C.	.080	L.F.	.004	.04	.14	.18
24″ O.C.	.080	L.F.	.003	.04	.09	.13
32″ O.C.	.080	L.F.	.002	.03	.07	.10
Let-in steel (T shaped) studs, 12″ O.C.	.080	L.F.	.001	.04	.05	.09
16″ O.C.	.080	L.F.	.001	.03	.04	.07
24″ O.C.	.080	L.F.	.001	.03	.04	.07
32″ O.C.	.080	L.F.	.001	.03	.03	.06
Steel straps studs, 12″ O.C.	.080	L.F.	.001	.05	.04	.09
16″ O.C.	.080	L.F.	.001	.04	.04	.08
24″ O.C.	.080	L.F.	.001	.04	.04	.08
32″ O.C.	.080	L.F.	.001	.04	.03	.07
Metal studs, load bearing 24″ O.C., 20 ga. galv., 2-1/2″ wide	1.000	S.F.	.015	.41	.51	.92
3-5/8″ wide	1.000	S.F.	.015	.46	.52	.98
4″ wide	1.000	S.F.	.016	.51	.53	1.04
6″ wide	1.000	S.F.	.016	.63	.54	1.17
16 ga., 2-1/2″ wide	1.000	S.F.	.017	.47	.58	1.05
3-5/8″ wide	1.000	S.F.	.017	.54	.59	1.13
4″ wide	1.000	S.F.	.018	.59	.60	1.19
6″ wide	1.000	S.F.	.018	.73	.62	1.35
Non-load bearing 24″ O.C., 25 ga. galv., 1-5/8″ wide	1.000	S.F.	.015	.12	.52	.64
2-1/2″ wide	1.000	S.F.	.016	.13	.53	.66
3-5/8″ wide	1.000	S.F.	.016	.15	.54	.69
4″ wide	1.000	S.F.	.016	.18	.55	.73
6″ wide	1.000	S.F.	.017	.24	.56	.80
20 ga., 2-1/2″ wide	1.000	S.F.	.016	.22	.53	.75
3-5/8″ wide	1.000	S.F.	.016	.24	.54	.78
4″ wide	1.000	S.F.	.016	.30	.55	.85
6″ wide	1.000	S.F.	.017	.35	.56	.91

Window & Door Openings	QUAN.	UNIT	LABOR HOURS	COST EACH		
				MAT.	INST.	TOTAL
The following costs are to be added to the total costs of the walls. Do not subtract openings from total wall area.						
Headers, 2″ x 6″ double, 2′ long	4.000	L.F.	.178	2.60	6.00	8.60
3′ long	6.000	L.F.	.267	3.90	9.00	12.90
4′ long	8.000	L.F.	.356	5.20	12.00	17.20
5′ long	10.000	L.F.	.444	6.50	15.00	21.50
2″ x 8″ double, 4′ long	8.000	L.F.	.376	7.30	12.70	20.00
5′ long	10.000	L.F.	.471	9.10	15.90	25.00
6′ long	12.000	L.F.	.565	10.90	19.10	30.00
8′ long	16.000	L.F.	.753	14.55	25.50	40.05
2″ x 10″ double, 4′ long	8.000	L.F.	.400	10.70	13.50	24.20
6′ long	12.000	L.F.	.600	16.10	20.50	36.60
8′ long	16.000	L.F.	.800	21.50	27.00	48.50
10′ long	20.000	L.F.	1.000	27.00	34.00	61.00
2″ x 12″ double, 8′ long	16.000	L.F.	.853	26.00	29.00	55.00
12′ long	24.000	L.F.	1.280	39.00	43.00	82.00

Division 4
Exterior Walls

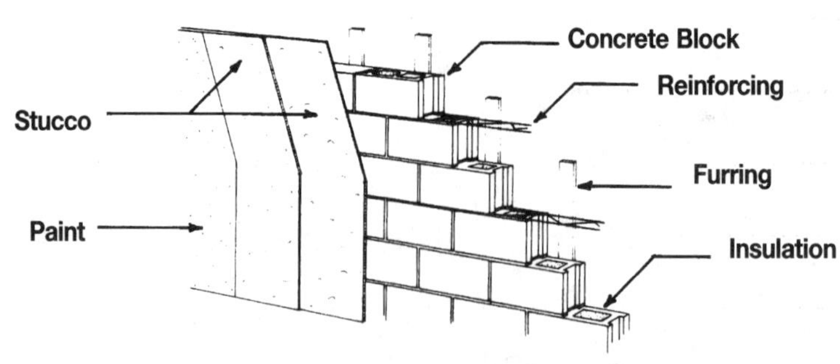

System Description	QUAN.	UNIT	LABOR HOURS	COST PER S.F.		
				MAT.	INST.	TOTAL
6″ THICK CONCRETE BLOCK WALL						
6″ thick concrete block, 6″ x 8″ x 16″	1.000	S.F.	.100	1.17	3.10	4.27
Masonry reinforcing, truss strips every other course	.625	L.F.	.002	.08	.06	.14
Furring, 1″ x 3″, 16″ O.C.	1.000	L.F.	.016	.20	.55	.75
Masonry insulation, poured vermiculite	1.000	S.F.	.013	.53	.45	.98
Stucco, 2 coats	1.000	S.F.	.069	.19	2.11	2.30
Masonry paint, 2 coats	1.000	S.F.	.016	.17	.48	.65
TOTAL			.216	2.34	6.75	9.09
8″ THICK CONCRETE BLOCK WALL						
8″ thick concrete block, 8″ x 8″ x 16″	1.000	S.F.	.107	1.39	3.30	4.69
Masonry reinforcing, truss strips every other course	.625	L.F.	.002	.08	.06	.14
Furring, 1″ x 3″, 16″ O.C.	1.000	L.F.	.016	.20	.55	.75
Masonry insulation, poured vermiculite	1.000	S.F.	.018	.70	.59	1.29
Stucco, 2 coats	1.000	S.F.	.069	.19	2.11	2.30
Masonry paint, 2 coats	1.000	S.F.	.016	.17	.48	.65
TOTAL			.228	2.73	7.09	9.82
12″ THICK CONCRETE BLOCK WALL						
12″ thick concrete block, 12″ x 8″ x 16″	1.000	S.F.	.141	2.12	4.27	6.39
Masonry reinforcing, truss strips every other course	.625	L.F.	.003	.09	.09	.18
Furring, 1″ x 3″, 16″ O.C.	1.000	L.F.	.016	.20	.55	.75
Masonry insulation, poured vermiculite	1.000	S.F.	.026	1.04	.88	1.92
Stucco, 2 coats	1.000	S.F.	.069	.19	2.11	2.30
Masonry paint, 2 coats	1.000	S.F.	.016	.17	.48	.65
TOTAL			.271	3.81	8.38	12.19

Costs for this system are based on a square foot of wall area. Do not subtract for window openings.

Description	QUAN.	UNIT	LABOR HOURS	COST PER S.F.		
				MAT.	INST.	TOTAL

Masonry Block Price Sheet	QUAN.	UNIT	LABOR HOURS	COST PER S.F.		
				MAT.	INST.	TOTAL
Block concrete, 8″ x 16″ regular, 4″ thick	1.000	S.F.	.093	.84	2.88	3.72
6″ thick	1.000	S.F.	.100	1.17	3.10	4.27
8″ thick	1.000	S.F.	.107	1.39	3.30	4.69
10″ thick	1.000	S.F.	.111	1.98	3.44	5.42
12″ thick	1.000	S.F.	.141	2.12	4.27	6.39
Solid block, 4″ thick	1.000	S.F.	.096	1.27	2.98	4.25
6″ thick	1.000	S.F.	.104	1.56	3.22	4.78
8″ thick	1.000	S.F.	.111	2.02	3.44	5.46
10″ thick	1.000	S.F.	.133	2.96	4.01	6.97
12″ thick	1.000	S.F.	.148	3.29	4.46	7.75
Lightweight, 4″ thick	1.000	S.F.	.093	.84	2.88	3.72
6″ thick	1.000	S.F.	.100	1.17	3.10	4.27
8″ thick	1.000	S.F.	.107	1.39	3.30	4.69
10″ thick	1.000	S.F.	.111	1.98	3.44	5.42
12″ thick	1.000	S.F.	.141	2.12	4.27	6.39
Split rib profile, 4″ thick	1.000	S.F.	.116	2.08	3.59	5.67
6″ thick	1.000	S.F.	.123	2.51	3.81	6.32
8″ thick	1.000	S.F.	.131	3.02	4.06	7.08
10″ thick	1.000	S.F.	.157	3.32	4.77	8.09
12″ thick	1.000	S.F.	.175	3.69	5.30	8.99
Masonry reinforcing, wire truss strips, every course, 8″ block	1.375	L.F.	.004	.17	.12	.29
12″ block	1.375	L.F.	.006	.19	.19	.38
Every other course, 8″ block	.625	L.F.	.002	.08	.06	.14
12″ block	.625	L.F.	.003	.09	.09	.18
Furring, wood, 1″ x 3″, 12″ O.C.	1.250	L.F.	.020	.25	.69	.94
16″ O.C.	1.000	L.F.	.016	.20	.55	.75
24″ O.C.	.800	L.F.	.013	.16	.44	.60
32″ O.C.	.640	L.F.	.010	.13	.35	.48
Steel, 3/4″ channels, 12″ O.C.	1.250	L.F.	.034	.20	1.07	1.27
16″ O.C.	1.000	L.F.	.030	.18	.95	1.13
24″ O.C.	.800	L.F.	.023	.12	.72	.84
32″ O.C.	.640	L.F.	.018	.10	.58	.68
Masonry insulation, vermiculite or perlite poured 4″ thick	1.000	S.F.	.009	.34	.29	.63
6″ thick	1.000	S.F.	.013	.53	.45	.98
8″ thick	1.000	S.F.	.018	.70	.59	1.29
10″ thick	1.000	S.F.	.021	.85	.72	1.57
12″ thick	1.000	S.F.	.026	1.04	.88	1.92
Block inserts polystyrene, 6″ thick	1.000	S.F.		.83		.83
8″ thick	1.000	S.F.		.83		.83
10″ thick	1.000	S.F.		.99		.99
12″ thick	1.000	S.F.		1.05		1.05
Stucco, 1 coat	1.000	S.F.	.057	.15	1.73	1.88
2 coats	1.000	S.F.	.069	.19	2.11	2.30
3 coats	1.000	S.F.	.081	.22	2.48	2.70
Painting, 1 coat	1.000	S.F.	.011	.12	.33	.45
2 coats	1.000	S.F.	.016	.17	.48	.65
Primer & 1 coat	1.000	S.F.	.013	.19	.39	.58
2 coats	1.000	S.F.	.018	.25	.54	.79
Lath, metal lath expanded 2.5 lb/S.Y., painted	1.000	S.F.	.010	.16	.33	.49
Galvanized	1.000	S.F.	.012	.18	.36	.54

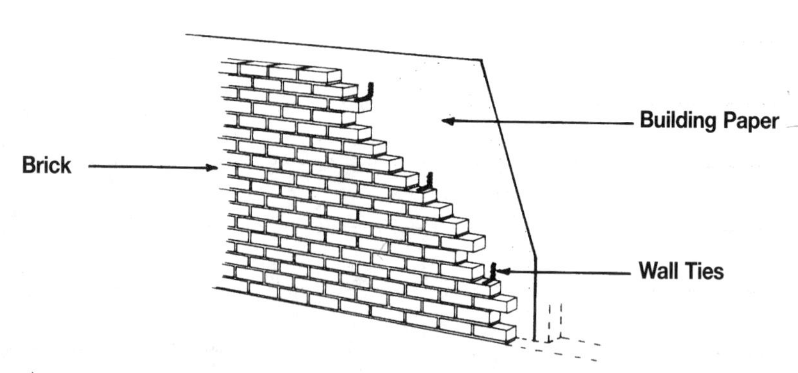

System Description	QUAN.	UNIT	LABOR HOURS	COST PER S.F.		
				MAT.	INST.	TOTAL
SELECT COMMON BRICK						
Brick, select common, running bond	1.000	S.F.	.174	2.95	5.40	8.35
Wall ties, 7/8" x 7", 22 gauge	1.000	Ea.	.008	.05	.26	.31
Building paper, #15 asphalt	1.100	S.F.	.002	.03	.08	.11
Trim, pine, painted	.125	L.F.	.004	.08	.14	.22
TOTAL			.188	3.11	5.88	8.99
RED FACED COMMON BRICK						
Brick, common, red faced, running bond	1.000	S.F.	.182	2.95	5.65	8.60
Wall ties, 7/8" x 7", 22 gauge	1.000	Ea.	.008	.05	.26	.31
Building paper, #15 asphalt	1.100	S.F.	.002	.03	.08	.11
Trim, pine, painted	.125	L.F.	.004	.08	.14	.22
TOTAL			.196	3.11	6.13	9.24
BUFF OR GREY FACE BRICK						
Brick, buff or grey	1.000	S.F.	.182	3.13	5.65	8.78
Wall ties, 7/8" x 7", 22 gauge	1.000	Ea.	.008	.05	.26	.31
Building paper, #15 asphalt	1.100	S.F.	.002	.03	.08	.11
Trim, pine, painted	.125	L.F.	.004	.08	.14	.22
TOTAL			.196	3.29	6.13	9.42
STONE WORK, ROUGH STONE, AVERAGE						
Stone work, rough stone, average	1.000	S.F.	.179	7.27	5.39	12.66
Wall ties, 7/8" x 7", 22 gauge	1.000	Ea.	.008	.05	.26	.31
Building paper, #15 asphalt	1.000	S.F.	.002	.03	.08	.11
Trim, pine, painted	.125	L.F.	.004	.08	.14	.22
TOTAL			.193	7.43	5.87	13.30

The costs in this system are based on a square foot of wall area. Do not subtract area for window & door openings.

Description	QUAN.	UNIT	LABOR HOURS	COST PER S.F.		
				MAT.	INST.	TOTAL

Brick/Stone Veneer Price Sheet	QUAN.	UNIT	LABOR HOURS	COST PER S.F. MAT.	INST.	TOTAL
Brick						
Select common, running bond	1.000	S.F.	.174	2.95	5.40	8.35
Red faced, running bond	1.000	S.F.	.182	2.95	5.65	8.60
Buff or grey faced, running bond	1.000	S.F.	.182	3.13	5.65	8.78
Header every 6th course	1.000	S.F.	.216	3.44	6.70	10.14
English bond	1.000	S.F.	.286	4.41	8.85	13.26
Flemish bond	1.000	S.F.	.195	3.11	6.05	9.16
Common bond	1.000	S.F.	.267	3.92	8.25	12.17
Stack bond	1.000	S.F.	.182	3.13	5.65	8.78
Jumbo, running bond	1.000	S.F.	.092	3.88	2.85	6.73
Norman, running bond	1.000	S.F.	.125	3.51	3.87	7.38
Norwegian, running bond	1.000	S.F.	.107	2.29	3.30	5.59
Economy, running bond	1.000	S.F.	.129	2.34	3.99	6.33
Engineer, running bond	1.000	S.F.	.154	2.07	4.76	6.83
Roman, running bond	1.000	S.F.	.160	4.97	4.95	9.92
Utility, running bond	1.000	S.F.	.089	3.26	2.75	6.01
Glazed, running bond	1.000	S.F.	.190	11.00	5.90	16.90
Stone work, rough stone, average	1.000	S.F.	.179	7.25	5.40	12.65
Maximum	1.000	S.F.	.267	10.85	8.05	18.90
Wall ties, galvanized, corrugated 7/8″ x 7″, 22 gauge	1.000	Ea.	.008	.05	.26	.31
16 gauge	1.000	Ea.	.008	.15	.26	.41
Cavity wall, every 3rd course 6″ long Z type, 1/4″ diameter	1.330	L.F.	.010	.29	.34	.63
3/16″ diameter	1.330	L.F.	.010	.14	.34	.48
8″ long, Z type, 1/4″ diameter	1.330	L.F.	.010	.34	.34	.68
3/16″ diameter	1.330	L.F.	.010	.15	.34	.49
Building paper, aluminum and kraft laminated foil, 1 side	1.000	S.F.	.002	.04	.07	.11
2 sides	1.000	S.F.	.002	.06	.07	.13
#15 asphalt paper	1.100	S.F.	.002	.03	.08	.11
Polyethylene, .002″ thick	1.000	S.F.	.002	.01	.07	.08
.004″ thick	1.000	S.F.	.002	.03	.07	.10
.006″ thick	1.000	S.F.	.002	.03	.07	.10
.008″ thick	1.000	S.F.	.002			
.010″ thick	1.000	S.F.	.002	.07	.07	.14
Trim, 1″ x 4″, cedar	.125	L.F.	.005	.18	.17	.35
Fir	.125	L.F.	.005	.07	.17	.24
Redwood	.125	L.F.	.005	.18	.17	.35
White pine	.125	L.F.	.005	.07	.17	.24

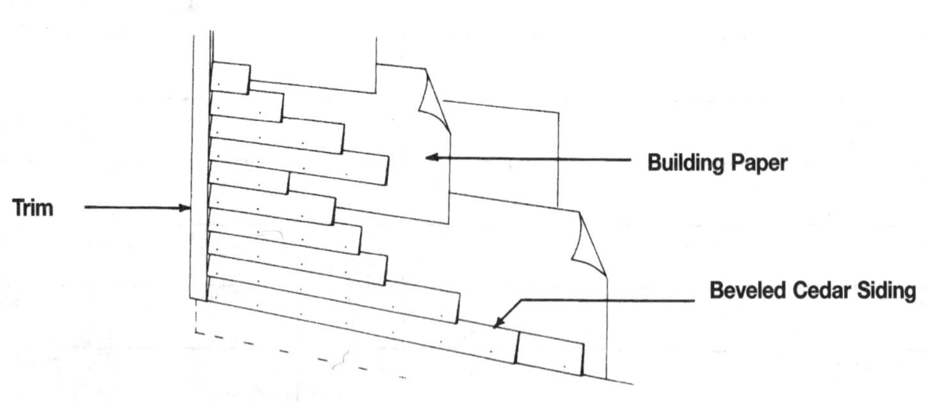

System Description	QUAN.	UNIT	LABOR HOURS	COST PER S.F.		
				MAT.	INST.	TOTAL
1/2" X 6" BEVELED CEDAR SIDING, "A" GRADE						
1/2" x 6" beveled cedar siding	1.000	S.F.	.032	2.09	1.08	3.17
#15 asphalt felt paper	1.100	S.F.	.002	.03	.08	.11
Trim, cedar	.125	L.F.	.005	.18	.17	.35
Paint, primer & 2 coats	1.000	S.F.	.017	.17	.52	.69
TOTAL			.056	2.47	1.85	4.32
1/2" X 8" BEVELED CEDAR SIDING, "A" GRADE						
1/2" x 8" beveled cedar siding	1.000	S.F.	.029	1.72	.98	2.70
#15 asphalt felt paper	1.100	S.F.	.002	.03	.08	.11
Trim, cedar	.125	L.F.	.005	.18	.17	.35
Paint, primer & 2 coats	1.000	S.F.	.017	.17	.52	.69
TOTAL			.053	2.10	1.75	3.85
1" X 4" TONGUE & GROOVE, REDWOOD, VERTICAL GRAIN						
1" x 4" tongue & groove, vertical, redwood	1.000	S.F.	.033	4.40	1.13	5.53
#15 asphalt felt paper	1.100	S.F.	.002	.03	.08	.11
Trim, redwood	.125	L.F.	.005	.18	.17	.35
Sealer, 1 coat, stain, 1 coat	1.000	S.F.	.013	.11	.40	.51
TOTAL			.053	4.72	1.78	6.50
1" X 6" TONGUE & GROOVE, REDWOOD, VERTICAL GRAIN						
1" x 6" tongue & groove, vertical, redwood	1.000	S.F.	.024	4.40	.79	5.19
#15 asphalt felt paper	1.100	S.F.	.002	.03	.08	.11
Trim, redwood	.125	L.F.	.005	.18	.17	.35
Sealer, 1 coat, stain, 1 coat	1.000	S.F.	.013	.11	.40	.51
TOTAL			.044	4.72	1.44	6.16

The costs in this system are based on a square foot of wall area.
Do not subtract area for door or window openings.

Description	QUAN.	UNIT	LABOR HOURS	COST PER S.F.		
				MAT.	INST.	TOTAL

Important: See the Reference Section for critical supporting data - Reference Nos., Crews & Location Factors

Wood Siding Price Sheet

Wood Siding Price Sheet	QUAN.	UNIT	LABOR HOURS	COST PER S.F.		
				MAT.	INST.	TOTAL
Siding, beveled cedar, "B" grade, 1/2" x 6"	1.000	S.F.	.028	1.38	.93	2.31
1/2" x 8"	1.000	S.F.	.023	1.38	.77	2.15
"A" grade, 1/2" x 6"	1.000	S.F.	.032	2.09	1.08	3.17
1/2" x 8"	1.000	S.F.	.029	1.72	.98	2.70
Clear grade, 1/2" x 6"	1.000	S.F.	.028	1.49	.93	2.42
1/2" x 8"	1.000	S.F.	.023	1.49	.77	2.26
Redwood, clear vertical grain, 1/2" x 6"	1.000	S.F.	.036	2.97	1.20	4.17
1/2" x 8"	1.000	S.F.	.032	2.41	1.08	3.49
Clear all heart vertical grain, 1/2" x 6"	1.000	S.F.	.028	2.09	.93	3.02
1/2" x 8"	1.000	S.F.	.023	2.10	.77	2.87
Siding board & batten, cedar, "B" grade, 1" x 10"	1.000	S.F.	.031	2.29	1.04	3.33
1" x 12"	1.000	S.F.	.031	2.29	1.04	3.33
Redwood, clear vertical grain, 1" x 10"	1.000	S.F.	.043	2.60	1.44	4.04
1" x 12"	1.000	S.F.	.018	4.40	.60	5.00
White pine, #2 & better, 1" x 10"	1.000	S.F.	.029	.76	.98	1.74
1" x 12"	1.000	S.F.	.029	.76	.98	1.74
Siding vertical, tongue & groove, cedar "B" grade, 1" x 4"	1.000	S.F.	.033	3.06	1.13	4.19
1" x 6"	1.000	S.F.	.024	2.33	.79	3.12
1" x 8"	1.000	S.F.	.024	2.29	.79	3.08
1" x 10"	1.000	S.F.	.021	2.12	.71	2.83
"A" grade, 1" x 4"	1.000	S.F.	.033	4.55	1.13	5.68
1" x 6"	1.000	S.F.	.024	3.50	.79	4.29
1" x 8"	1.000	S.F.	.024	3.43	.79	4.22
1" x 10"	1.000	S.F.	.021	3.18	.71	3.89
Clear vertical grain, 1" x 4"	1.000	S.F.	.033	5.10	1.13	6.23
1" x 6"	1.000	S.F.	.024	3.94	.79	4.73
1" x 8"	1.000	S.F.	.024	4.13	.79	4.92
1" x 10"	1.000	S.F.	.021	3.87	.71	4.58
Redwood, clear vertical grain, 1" x 4"	1.000	S.F.	.033	4.40	1.13	5.53
1" x 6"	1.000	S.F.	.024	4.40	.79	5.19
1" x 8"	1.000	S.F.	.024	4.40	.79	5.19
1" x 10"	1.000	S.F.	.021	3.81	.71	4.52
Clear all heart vertical grain, 1" x 4"	1.000	S.F.	.033	5.40	1.13	6.53
1" x 6"	1.000	S.F.	.024	4.19	.79	4.98
1" x 8"	1.000	S.F.	.024	4.13	.79	4.92
1" x 10"	1.000	S.F.	.021	3.81	.71	4.52
White pine, 1" x 8"	1.000	S.F.	.024	.87	.79	1.66
Siding plywood, texture 1-11 cedar, 3/8" thick	1.000	S.F.	.024	1.25	.80	2.05
5/8" thick	1.000	S.F.	.024	1.20	.80	2.00
Redwood, 3/8" thick	1.000	S.F.	.024	1.25	.80	2.05
5/8" thick	1.000	S.F.	.024	2.08	.80	2.88
Fir, 3/8" thick	1.000	S.F.	.024	.67	.80	1.47
5/8" thick	1.000	S.F.	.024	1.11	.80	1.91
Southern yellow pine, 3/8" thick	1.000	S.F.	.024	.67	.80	1.47
5/8" thick	1.000	S.F.	.024	.94	.80	1.74
Hard board, 7/16" thick primed, plain finish	1.000	S.F.	.025	1.14	.83	1.97
Board finish	1.000	S.F.	.023	.81	.77	1.58
Polyvinyl coated, 3/8" thick	1.000	S.F.	.021	1.02	.72	1.74
5/8" thick	1.000	S.F.	.024	.94	.80	1.74
Paper, #15 asphalt felt	1.100	S.F.	.002	.03	.08	.11
Trim, cedar	.125	L.F.	.005	.18	.17	.35
Fir	.125	L.F.	.005	.07	.17	.24
Redwood	.125	L.F.	.005	.18	.17	.35
White pine	.125	L.F.	.005	.07	.17	.24
Painting, primer, & 1 coat	1.000	S.F.	.013	.11	.40	.51
2 coats	1.000	S.F.	.017	.17	.52	.69
Stain, sealer, & 1 coat	1.000	S.F.	.017	.08	.52	.60
2 coats	1.000	S.F.	.019	.14	.56	.70

12 Shingle Siding Systems

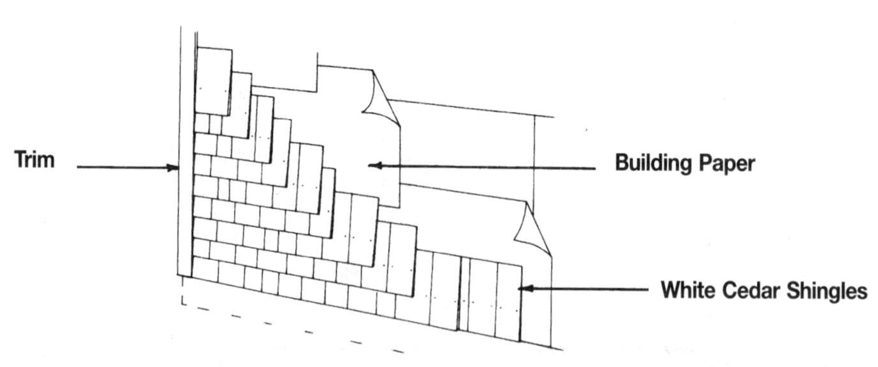

System Description	QUAN.	UNIT	LABOR HOURS	COST PER S.F.		
				MAT.	INST.	TOTAL
WHITE CEDAR SHINGLES, 5″ EXPOSURE						
White cedar shingles, 16″ long, grade "A", 5″ exposure	1.000	S.F.	.033	1.24	1.13	2.37
#15 asphalt felt paper	1.100	S.F.	.002	.03	.08	.11
Trim, cedar	.125	S.F.	.005	.18	.17	.35
Paint, primer & 1 coat	1.000	S.F.	.017	.08	.52	.60
TOTAL			.057	1.53	1.90	3.43
NO. 1 PERFECTIONS, 5-1/2″ EXPOSURE						
No. 1 perfections, red cedar, 5-1/2″ exposure	1.000	S.F.	.029	1.81	.98	2.79
#15 asphalt felt paper	1.100	S.F.	.002	.03	.08	.11
Trim, cedar	.125	S.F.	.005	.18	.17	.35
Stain, sealer & 1 coat	1.000	S.F.	.017	.08	.52	.60
TOTAL			.053	2.10	1.75	3.85
RESQUARED & REBUTTED PERFECTIONS, 5-1/2″ EXPOSURE						
Resquared & rebutted perfections, 5-1/2″ exposure	1.000	S.F.	.027	2.19	.90	3.09
#15 asphalt felt paper	1.100	S.F.	.002	.03	.08	.11
Trim, cedar	.125	S.F.	.005	.18	.17	.35
Stain, sealer & 1 coat	1.000	S.F.	.017	.08	.52	.60
TOTAL			.051	2.48	1.67	4.15
HAND-SPLIT SHAKES, 8-1/2″ EXPOSURE						
Hand-split red cedar shakes, 18″ long, 8-1/2″ exposure	1.000	S.F.	.040	1.07	1.35	2.42
#15 asphalt felt paper	1.100	S.F.	.002	.03	.08	.11
Trim, cedar	.125	S.F.	.005	.18	.17	.35
Stain, sealer & 1 coat	1.000	S.F.	.017	.08	.52	.60
TOTAL			.064	1.36	2.12	3.48

The costs in this system are based on a square foot of wall area.
Do not subtract area for door or window openings.

Description	QUAN.	UNIT	LABOR HOURS	COST PER S.F.		
				MAT.	INST.	TOTAL

Shingle Siding Price Sheet	QUAN.	UNIT	LABOR HOURS	COST PER S.F.		
				MAT.	INST.	TOTAL
Shingles wood, white cedar 16" long, "A" grade, 5" exposure	1.000	S.F.	.033	1.24	1.13	2.37
7" exposure	1.000	S.F.	.030	1.12	1.02	2.14
8-1/2" exposure	1.000	S.F.	.032	.71	1.08	1.79
10" exposure	1.000	S.F.	.028	.62	.95	1.57
"B" grade, 5" exposure	1.000	S.F.	.040	1.29	1.35	2.64
7" exposure	1.000	S.F.	.028	.90	.95	1.85
8-1/2" exposure	1.000	S.F.	.024	.77	.81	1.58
10" exposure	1.000	S.F.	.020	.65	.68	1.33
Fire retardant, "A" grade, 5" exposure	1.000	S.F.	.033	1.57	1.13	2.70
7" exposure	1.000	S.F.	.030	1.42	1.02	2.44
8-1/2" exposure	1.000	S.F.	.027	1.25	.90	2.15
10" exposure	1.000	S.F.	.023	1.10	.79	1.89
"B" grade, 5" exposure	1.000	S.F.	.040	1.62	1.35	2.97
7" exposure	1.000	S.F.	.028	1.13	.95	2.08
8-1/2" exposure	1.000	S.F.	.024	.97	.81	1.78
10" exposure	1.000	S.F.	.020	.82	.68	1.50
No. 1 perfections red cedar, 18" long, 5-1/2" exposure	1.000	S.F.	.029	1.81	.98	2.79
7" exposure	1.000	S.F.	.036	1.33	1.20	2.53
8-1/2" exposure	1.000	S.F.	.032	1.20	1.08	2.28
10" exposure	1.000	S.F.	.025	.93	.84	1.77
Fire retardant, 5" exposure	1.000	S.F.	.029	2.13	.98	3.11
7" exposure	1.000	S.F.	.036	1.65	1.20	2.85
8-1/2" exposure	1.000	S.F.	.032	1.48	1.08	2.56
10" exposure	1.000	S.F.	.025	1.15	.84	1.99
Resquared & rebutted, 5-1/2" exposure	1.000	S.F.	.027	2.19	.90	3.09
7" exposure	1.000	S.F.	.024	1.97	.81	2.78
8-1/2" exposure	1.000	S.F.	.021	1.75	.72	2.47
10" exposure	1.000	S.F.	.019	1.53	.63	2.16
Fire retardant, 5" exposure	1.000	S.F.	.027	2.51	.90	3.41
7" exposure	1.000	S.F.	.024	2.25	.81	3.06
8-1/2" exposure	1.000	S.F.	.021	2.00	.72	2.72
10" exposure	1.000	S.F.	.023	1.35	.77	2.12
Hand-split, red cedar, 24" long, 7" exposure	1.000	S.F.	.045	2.13	1.51	3.64
8-1/2" exposure	1.000	S.F.	.038	1.82	1.30	3.12
10" exposure	1.000	S.F.	.032	1.52	1.08	2.60
12" exposure	1.000	S.F.	.026	1.22	.86	2.08
Fire retardant, 7" exposure	1.000	S.F.	.045	2.59	1.51	4.10
8-1/2" exposure	1.000	S.F.	.038	2.22	1.30	3.52
10" exposure	1.000	S.F.	.032	1.85	1.08	2.93
12" exposure	1.000	S.F.	.026	1.48	.86	2.34
18" long, 5" exposure	1.000	S.F.	.068	1.82	2.30	4.12
7" exposure	1.000	S.F.	.048	1.28	1.62	2.90
8-1/2" exposure	1.000	S.F.	.040	1.07	1.35	2.42
10" exposure	1.000	S.F.	.036	.96	1.22	2.18
Fire retardant, 5" exposure	1.000	S.F.	.068	2.38	2.30	4.68
7" exposure	1.000	S.F.	.048	1.68	1.62	3.30
8-1/2" exposure	1.000	S.F.	.040	1.40	1.35	2.75
10" exposure	1.000	S.F.	.036	1.26	1.22	2.48
Paper, #15 asphalt felt	1.100	S.F.	.002	.03	.07	.10
Trim, cedar	.125	S.F.	.005	.18	.17	.35
Fir	.125	S.F.	.005	.07	.17	.24
Redwood	.125	S.F.	.005	.18	.17	.35
White pine	.125	S.F.	.005	.07	.17	.24
Painting, primer, & 1 coat	1.000	S.F.	.013	.11	.40	.51
2 coats	1.000	S.F.	.017	.17	.52	.69
Staining, sealer, & 1 coat	1.000	S.F.	.017	.08	.52	.60
2 coats	1.000	S.F.	.019	.14	.56	.70

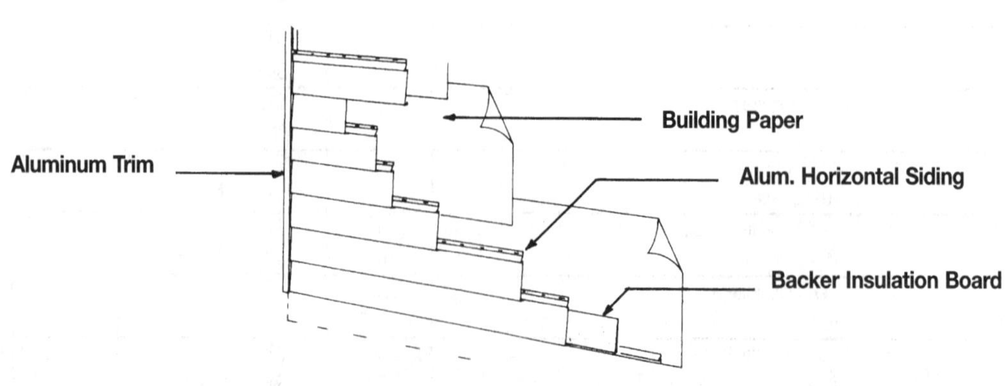

System Description	QUAN.	UNIT	LABOR HOURS	COST PER S.F. MAT.	INST.	TOTAL
ALUMINUM CLAPBOARD SIDING, 8″ WIDE, WHITE						
Aluminum horizontal siding, 8″ clapboard	1.000	S.F.	.031	1.35	1.05	2.40
Backer, insulation board	1.000	S.F.	.008	.37	.27	.64
Trim, aluminum	.600	L.F.	.016	.61	.53	1.14
Paper, #15 asphalt felt	1.100	S.F.	.002	.03	.08	.11
TOTAL			.057	2.36	1.93	4.29
ALUMINUM VERTICAL BOARD & BATTEN, WHITE						
Aluminum vertical board & batten	1.000	S.F.	.027	1.51	.92	2.43
Backer insulation board	1.000	S.F.	.008	.37	.27	.64
Trim, aluminum	.600	L.F.	.016	.61	.53	1.14
Paper, #15 asphalt felt	1.100	S.F.	.002	.03	.08	.11
TOTAL			.053	2.52	1.80	4.32
VINYL CLAPBOARD SIDING, 8″ WIDE, WHITE						
PVC vinyl horizontal siding, 8″ clapboard	1.000	S.F.	.032	.68	1.09	1.77
Backer, insulation board	1.000	S.F.	.008	.37	.27	.64
Trim, vinyl	.600	L.F.	.014	.41	.47	.88
Paper, #15 asphalt felt	1.100	S.F.	.002	.03	.08	.11
TOTAL			.056	1.49	1.91	3.40
VINYL VERTICAL BOARD & BATTEN, WHITE						
PVC vinyl vertical board & batten	1.000	S.F.	.029	1.43	.98	2.41
Backer, insulation board	1.000	S.F.	.008	.37	.27	.64
Trim, vinyl	.600	L.F.	.014	.41	.47	.88
Paper, #15 asphalt felt	1.100	S.F.	.002	.03	.08	.11
TOTAL			.053	2.24	1.80	4.04

The costs in this system are on a square foot of wall basis.
Do not subtract openings from wall area.

Description	QUAN.	UNIT	LABOR HOURS	COST PER S.F. MAT.	INST.	TOTAL

 Important: See the Reference Section for critical supporting data - Reference Nos., Crews & Location Factors

Metal & Plastic Siding Price Sheet

Metal & Plastic Siding Price Sheet	QUAN.	UNIT	LABOR HOURS	COST PER S.F. MAT.	COST PER S.F. INST.	COST PER S.F. TOTAL
Siding, aluminum, .024″ thick, smooth, 8″ wide, white	1.000	S.F.	.031	1.35	1.05	2.40
Color	1.000	S.F.	.031	1.44	1.05	2.49
Double 4″ pattern, 8″ wide, white	1.000	S.F.	.031	1.29	1.05	2.34
Color	1.000	S.F.	.031	1.38	1.05	2.43
Double 5″ pattern, 10″ wide, white	1.000	S.F.	.029	1.29	.98	2.27
Color	1.000	S.F.	.029	1.38	.98	2.36
Embossed, single, 8″ wide, white	1.000	S.F.	.031	1.60	1.05	2.65
Color	1.000	S.F.	.031	1.69	1.05	2.74
Double 4″ pattern, 8″ wide, white	1.000	S.F.	.031	1.45	1.05	2.50
Color	1.000	S.F.	.031	1.54	1.05	2.59
Double 5″ pattern, 10″ wide, white	1.000	S.F.	.029	1.45	.98	2.43
Color	1.000	S.F.	.029	1.54	.98	2.52
Alum siding with insulation board, smooth, 8″ wide, white	1.000	S.F.	.031	1.30	1.05	2.35
Color	1.000	S.F.	.031	1.39	1.05	2.44
Double 4″ pattern, 8″ wide, white	1.000	S.F.	.031	1.28	1.05	2.33
Color	1.000	S.F.	.031	1.37	1.05	2.42
Double 5″ pattern, 10″ wide, white	1.000	S.F.	.029	1.28	.98	2.26
Color	1.000	S.F.	.029	1.37	.98	2.35
Embossed, single, 8″ wide, white	1.000	S.F.	.031	1.49	1.05	2.54
Color	1.000	S.F.	.031	1.58	1.05	2.63
Double 4″ pattern, 8″ wide, white	1.000	S.F.	.031	1.51	1.05	2.56
Color	1.000	S.F.	.031	1.60	1.05	2.65
Double 5″ pattern, 10″ wide, white	1.000	S.F.	.029	1.51	.98	2.49
Color	1.000	S.F.	.029	1.60	.98	2.58
Aluminum, shake finish, 10″ wide, white	1.000	S.F.	.029	1.60	.98	2.58
Color	1.000	S.F.	.029	1.69	.98	2.67
Aluminum, vertical, 12″ wide, white	1.000	S.F.	.027	1.51	.92	2.43
Color	1.000	S.F.	.027	1.60	.92	2.52
Vinyl siding, 8″ wide, smooth, white	1.000	S.F.	.032	.68	1.09	1.77
Color	1.000	S.F.	.032	.77	1.09	1.86
10″ wide, Dutch lap, smooth, white	1.000	S.F.	.029	.70	.98	1.68
Color	1.000	S.F.	.029	.79	.98	1.77
Double 4″ pattern, 8″ wide, white	1.000	S.F.	.032	.61	1.09	1.70
Color	1.000	S.F.	.032	.70	1.09	1.79
Double 5″ pattern, 10″ wide, white	1.000	S.F.	.029	.57	.98	1.55
Color	1.000	S.F.	.029	.66	.98	1.64
Embossed, single, 8″ wide, white	1.000	S.F.	.032	.73	1.09	1.82
Color	1.000	S.F.	.032	.82	1.09	1.91
10″ wide, white	1.000	S.F.	.029	.74	.98	1.72
Color	1.000	S.F.	.029	.83	.98	1.81
Double 4″ pattern, 8″ wide, white	1.000	S.F.	.032	.63	1.09	1.72
Color	1.000	S.F.	.032	.72	1.09	1.81
Double 5″ pattern, 10″ wide, white	1.000	S.F.	.029	.64	.98	1.62
Color	1.000	S.F.	.029	.73	.98	1.71
Vinyl, shake finish, 10″ wide, white	1.000	S.F.	.029	2.09	.98	3.07
Color	1.000	S.F.	.029	2.18	.98	3.16
Vinyl, vertical, double 5″ pattern, 10″ wide, white	1.000	S.F.	.029	1.43	.98	2.41
Color	1.000	S.F.	.029	1.52	.98	2.50
Backer board, installed in siding panels 8″ or 10″ wide	1.000	S.F.	.008	.37	.27	.64
4′ x 8′ sheets, polystyrene, 3/4″ thick	1.000	S.F.	.010	.33	.34	.67
4′ x 8′ fiberboard, plain	1.000	S.F.	.008	.37	.27	.64
Trim, aluminum, white	.600	L.F.	.016	.61	.53	1.14
Color	.600	L.F.	.016	.66	.53	1.19
Vinyl, white	.600	L.F.	.014	.41	.47	.88
Color	.600	L.F.	.014	.44	.47	.91
Paper, #15 asphalt felt	1.100	S.F.	.002	.03	.08	.11
Kraft paper, plain	1.100	S.F.	.002	.04	.08	.12
Foil backed	1.100	S.F.	.002	.07	.08	.15

DESCRIPTION	QUAN.	UNIT	LABOR HOURS	COST PER S.F.		
				MAT.	INST.	TOTAL
Poured insulation, cellulose fiber, R3.8 per inch (1" thick)	1.000	S.F.	.003	.04	.11	.15
Fiberglass , R4.0 per inch (1" thick)	1.000	S.F.	.003	.03	.11	.14
Mineral wool, R3.0 per inch (1" thick)	1.000	S.F.	.003	.03	.11	.14
Polystyrene, R4.0 per inch (1" thick)	1.000	S.F.	.003	.18	.11	.29
Vermiculite, R2.7 per inch (1" thick)	1.000	S.F.	.003	.13	.11	.24
Perlite, R2.7 per inch (1" thick)	1.000	S.F.	.003	.13	.11	.24
Reflective insulation, aluminum foil reinforced with scrim	1.000	S.F.	.004	.15	.14	.29
Reinforced with woven polyolefin	1.000	S.F.	.004	.19	.14	.33
With single bubble air space, R8.8	1.000	S.F.	.005	.29	.18	.47
With double bubble air space, R9.8	1.000	S.F.	.005	.31	.18	.49
Rigid insulation, fiberglass, unfaced,						
1-1/2" thick, R6.2	1.000	S.F.	.008	.37	.27	.64
2" thick, R8.3	1.000	S.F.	.008	.45	.27	.72
2-1/2" thick, R10.3	1.000	S.F.	.010	.54	.34	.88
3" thick, R12.4	1.000	S.F.	.010	.54	.34	.88
Foil faced, 1" thick, R4.3	1.000	S.F.	.008	.84	.27	1.11
1-1/2" thick, R6.2	1.000	S.F.	.008	1.12	.27	1.39
2" thick, R8.7	1.000	S.F.	.009	1.40	.30	1.70
2-1/2" thick, R10.9	1.000	S.F.	.010	1.65	.34	1.99
3" thick, R13.0	1.000	S.F.	.010	1.80	.34	2.14
Foam glass, 1-1/2" thick R2.64	1.000	S.F.	.010	1.60	.34	1.94
2" thick R5.26	1.000	S.F.	.011	2.98	.37	3.35
Perlite, 1" thick R2.77	1.000	S.F.	.010	.26	.34	.60
2" thick R5.55	1.000	S.F.	.011	.53	.37	.90
Polystyrene, extruded, blue, 2.2#/C.F., 3/4" thick R4	1.000	S.F.	.010	.33	.34	.67
1-1/2" thick R8.1	1.000	S.F.	.011	.65	.37	1.02
2" thick R10.8	1.000	S.F.	.011	.95	.37	1.32
Molded bead board, white, 1" thick R3.85	1.000	S.F.	.010	.13	.34	.47
1-1/2" thick, R5.6	1.000	S.F.	.011	.35	.37	.72
2" thick, R7.7	1.000	S.F.	.011	.54	.37	.91
Non-rigid insulation, batts						
Fiberglass, kraft faced, 3-1/2" thick, R11, 11" wide	1.000	S.F.	.005	.24	.17	.41
15" wide	1.000	S.F.	.005	.24	.17	.41
23" wide	1.000	S.F.	.005	.24	.17	.41
6" thick, R19, 11" wide	1.000	S.F.	.006	.34	.20	.54
15" wide	1.000	S.F.	.006	.34	.20	.54
23" wide	1.000	S.F.	.006	.34	.20	.54
9" thick, R30, 15" wide	1.000	S.F.	.006	.62	.20	.82
23" wide	1.000	S.F.	.006	.62	.20	.82
12" thick, R38, 15" wide	1.000	S.F.	.006	.79	.20	.99
23" wide	1.000	S.F.	.006	.79	.20	.99
Fiberglass, foil faced, 3-1/2" thick, R11, 15" wide	1.000	S.F.	.005	.35	.17	.52
23" wide	1.000	S.F.	.005	.35	.17	.52
6" thick, R19, 15" thick	1.000	S.F.	.005	.43	.17	.60
23" wide	1.000	S.F.	.005	.43	.17	.60
9" thick, R30, 15" wide	1.000	S.F.	.006	.74	.20	.94
23" wide	1.000	S.F.	.006	.74	.20	.94

Insulation Systems	QUAN.	UNIT	LABOR HOURS	COST PER S.F. MAT.	INST.	TOTAL
Non-rigid insulation batts						
Fiberglass unfaced, 3-1/2" thick, R11, 15" wide	1.000	S.F.	.005	.21	.17	.38
23" wide	1.000	S.F.	.005	.21	.17	.38
6" thick, R19, 15" wide	1.000	S.F.	.006	.35	.20	.55
23" wide	1.000	S.F.	.006	.35	.20	.55
9" thick, R19, 15" wide	1.000	S.F.	.007	.62	.23	.85
23" wide	1.000	S.F.	.007	.62	.23	.85
12" thick, R38, 15" wide	1.000	S.F.	.007	.79	.23	1.02
23" wide	1.000	S.F.	.007	.79	.23	1.02
Mineral fiber batts, 3" thick, R11	1.000	S.F.	.005	.26	.17	.43
3-1/2" thick, R13	1.000	S.F.	.005	.26	.17	.43
6" thick, R19	1.000	S.F.	.005	.41	.17	.58
6-1/2" thick, R22	1.000	S.F.	.005	.41	.17	.58
10" thick, R30	1.000	S.F.	.006	.64	.20	.84

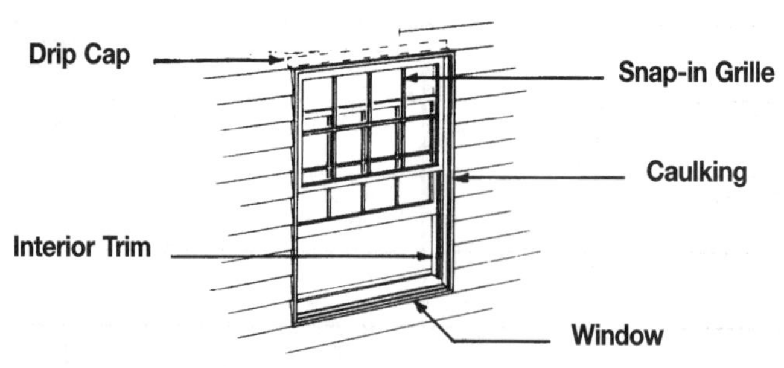

System Description	QUAN.	UNIT	LABOR HOURS	COST EACH MAT.	COST EACH INST.	COST EACH TOTAL
BUILDER'S QUALITY WOOD WINDOW 2′ X 3′, DOUBLE HUNG						
Window, primed, builder's quality, 2′ x 3′, insulating glass	1.000	Ea.	.800	167.00	27.00	194.00
Trim, interior casing	11.000	L.F.	.367	9.35	12.43	21.78
Paint, interior & exterior, primer & 2 coats	2.000	Face	1.778	1.98	53.00	54.98
Caulking	10.000	L.F.	.323	1.80	10.90	12.70
Snap-in grille	1.000	Set	.333	63.50	11.25	74.75
Drip cap, metal	2.000	L.F.	.040	.44	1.36	1.80
TOTAL			3.641	244.07	115.94	360.01
PLASTIC CLAD WOOD WINDOW 3′ X 4′, DOUBLE HUNG						
Window, plastic clad, premium, 3′ x 4′, insulating glass	1.000	Ea.	.889	345.00	30.00	375.00
Trim, interior casing	15.000	L.F.	.500	12.75	16.95	29.70
Paint, interior, primer & 2 coats	1.000	Face	.889	.99	26.50	27.49
Caulking	14.000	L.F.	.452	2.52	15.26	17.78
Snap-in grille	1.000	Set	.333	63.50	11.25	74.75
TOTAL			3.063	424.76	99.96	524.72
METAL CLAD WOOD WINDOW, 3′ X 5′, DOUBLE HUNG						
Window, metal clad, deluxe, 3′ x 5′, insulating glass	1.000	Ea.	1.000	272.00	34.00	306.00
Trim, interior casing	17.000	L.F.	.567	14.45	19.21	33.66
Paint, interior, primer & 2 coats	1.000	Face	.889	.99	26.50	27.49
Caulking	16.000	L.F.	.516	2.88	17.44	20.32
Snap-in grille	1.000	Set	.235	114.00	7.95	121.95
Drip cap, metal	3.000	L.F.	.060	.66	2.04	2.70
TOTAL			3.267	404.98	107.14	512.12

The cost of this system is on a cost per each window basis.

Description	QUAN.	UNIT	LABOR HOURS	COST EACH MAT.	COST EACH INST.	COST EACH TOTAL

Important: See the Reference Section for critical supporting data - Reference Nos., Crews & Location Factors

Double Hung Window Price Sheet	QUAN.	UNIT	LABOR HOURS	COST EACH		
				MAT.	INST.	TOTAL
Windows, double-hung, builder's quality, 2′ x 3′, single glass	1.000	Ea.	.800	117.00	27.00	144.00
Insulating glass	1.000	Ea.	.800	167.00	27.00	194.00
3′ x 4′, single glass	1.000	Ea.	.889	154.00	30.00	184.00
Insulating glass	1.000	Ea.	.889	203.00	30.00	233.00
4′ x 4′-6″, single glass	1.000	Ea.	1.000	189.00	34.00	223.00
Insulating glass	1.000	Ea.	1.000	250.00	34.00	284.00
Plastic clad premium insulating glass, 2′-6″ x 3′	1.000	Ea.	.800	264.00	27.00	291.00
3′ x 3′-6″	1.000	Ea.	.800	320.00	27.00	347.00
3′ x 4′	1.000	Ea.	.889	345.00	30.00	375.00
3′ x 4′-6″	1.000	Ea.	.889	360.00	30.00	390.00
3′ x 5′	1.000	Ea.	1.000	375.00	34.00	409.00
3′-6″ x 6′	1.000	Ea.	1.000	625.00	34.00	659.00
Metal clad deluxe insulating glass, 2′-6″ x 3′	1.000	Ea.	.800	187.00	27.00	214.00
3′ x 3′-6″	1.000	Ea.	.800	221.00	27.00	248.00
3′ x 4′	1.000	Ea.	.889	235.00	30.00	265.00
3′ x 4′-6″	1.000	Ea.	.889	254.00	30.00	284.00
3′ x 5′	1.000	Ea.	1.000	272.00	34.00	306.00
3′-6″ x 6′	1.000	Ea.	1.000	330.00	34.00	364.00
Trim, interior casing, window 2′ x 3′	11.000	L.F.	.367	9.35	12.45	21.80
2′-6″ x 3′	12.000	L.F.	.400	10.20	13.55	23.75
3′ x 3′-6″	14.000	L.F.	.467	11.90	15.80	27.70
3′ x 4′	15.000	L.F.	.500	12.75	16.95	29.70
3′ x 4′-6″	16.000	L.F.	.533	13.60	18.10	31.70
3′ x 5′	17.000	L.F.	.567	14.45	19.20	33.65
3′-6″ x 6′	20.000	L.F.	.667	17.00	22.50	39.50
4′ x 4′-6″	18.000	L.F.	.600	15.30	20.50	35.80
Paint or stain, interior or exterior, 2′ x 3′ window, 1 coat	1.000	Face	.444	.35	13.30	13.65
2 coats	1.000	Face	.727	.70	22.00	22.70
Primer & 1 coat	1.000	Face	.727	.67	22.00	22.67
Primer & 2 coats	1.000	Face	.889	.99	26.50	27.49
3′ x 4′ window, 1 coat	1.000	Face	.667	.71	19.95	20.66
2 coats	1.000	Face	.667	.81	19.95	20.76
Primer & 1 coat	1.000	Face	.727	1.01	22.00	23.01
Primer & 2 coats	1.000	Face	.889	.99	26.50	27.49
4′ x 4′-6″ window, 1 coat	1.000	Face	.667	.71	19.95	20.66
2 coats	1.000	Face	.667	.81	19.95	20.76
Primer & 1 coat	1.000	Face	.727	1.01	22.00	23.01
Primer & 2 coats	1.000	Face	.889	.99	26.50	27.49
Caulking, window, 2′ x 3′	10.000	L.F.	.323	1.80	10.90	12.70
2′-6″ x 3′	11.000	L.F.	.355	1.98	12.00	13.98
3′ x 3′-6″	13.000	L.F.	.419	2.34	14.15	16.49
3′ x 4′	14.000	L.F.	.452	2.52	15.25	17.77
3′ x 4′-6″	15.000	L.F.	.484	2.70	16.35	19.05
3′ x 5′	16.000	L.F.	.516	2.88	17.45	20.33
3′-6″ x 6′	19.000	L.F.	.613	3.42	20.50	23.92
4′ x 4′-6″	17.000	L.F.	.548	3.06	18.55	21.61
Grilles, glass size to, 16″ x 24″ per sash	1.000	Set	.333	63.50	11.25	74.75
32″ x 32″ per sash	1.000	Set	.235	114.00	7.95	121.95
Drip cap, aluminum, 2′ long	2.000	L.F.	.040	.44	1.36	1.80
3′ long	3.000	L.F.	.060	.66	2.04	2.70
4′ long	4.000	L.F.	.080	.88	2.72	3.60
Wood, 2′ long	2.000	L.F.	.067	1.70	2.26	3.96
3′ long	3.000	L.F.	.100	2.55	3.39	5.94
4′ long	4.000	L.F.	.133	3.40	4.52	7.92

32 Casement Window Systems

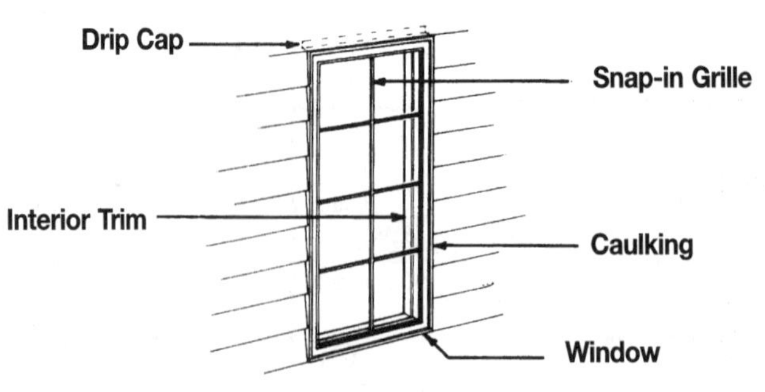

System Description	QUAN.	UNIT	LABOR HOURS	COST EACH		
				MAT.	INST.	TOTAL
BUILDER'S QUALITY WINDOW, WOOD, 2′ BY 3′, CASEMENT						
Window, primed, builder's quality, 2′ x 3′, insulating glass	1.000	Ea.	.800	215.00	27.00	242.00
Trim, interior casing	11.000	L.F.	.367	9.35	12.43	21.78
Paint, interior & exterior, primer & 2 coats	2.000	Face	1.778	1.98	53.00	54.98
Caulking	10.000	L.F.	.323	1.80	10.90	12.70
Snap-in grille	1.000	Ea.	.267	30.00	9.00	39.00
Drip cap, metal	2.000	L.F.	.040	.44	1.36	1.80
TOTAL			3.575	258.57	113.69	372.26
PLASTIC CLAD WOOD WINDOW, 2′ X 4′, CASEMENT						
Window, plastic clad, premium, 2′ x 4′, insulating glass	1.000	Ea.	.889	254.00	30.00	284.00
Trim, interior casing	13.000	L.F.	.433	11.05	14.69	25.74
Paint, interior, primer & 2 coats	1.000	Ea.	.889	.99	26.50	27.49
Caulking	12.000	L.F.	.387	2.16	13.08	15.24
Snap-in grille	1.000	Ea.	.267	30.00	9.00	39.00
TOTAL			2.865	298.20	93.27	391.47
METAL CLAD WOOD WINDOW, 2′ X 5′, CASEMENT						
Window, metal clad, deluxe, 2′ x 5′, insulating glass	1.000	Ea.	1.000	241.00	34.00	275.00
Trim, interior casing	15.000	L.F.	.500	12.75	16.95	29.70
Paint, interior, primer & 2 coats	1.000	Ea.	.889	.99	26.50	27.49
Caulking	14.000	L.F.	.452	2.52	15.26	17.78
Snap-in grille	1.000	Ea.	.250	40.50	8.45	48.95
Drip cap, metal	12.000	L.F.	.040	.44	1.36	1.80
TOTAL			3.131	298.20	102.52	400.72

The cost of this system is on a cost per each window basis.

Description	QUAN.	UNIT	LABOR HOURS	COST EACH		
				MAT.	INST.	TOTAL

 Important: See the Reference Section for critical supporting data - Reference Nos., Crews & Location Factors

Casement Window Price Sheet	QUAN.	UNIT	LABOR HOURS	COST EACH		
				MAT.	INST.	TOTAL
Window, casement, builders quality, 2′ x 3′, single glass	1.000	Ea.	.800	175.00	27.00	202.00
Insulating glass	1.000	Ea.	.800	215.00	27.00	242.00
2′ x 4′-6″, single glass	1.000	Ea.	.727	660.00	24.50	684.50
Insulating glass	1.000	Ea.	.727	570.00	24.50	594.50
2′ x 6′, single glass	1.000	Ea.	.889	880.00	30.00	910.00
Insulating glass	1.000	Ea.	.889	850.00	30.00	880.00
Plastic clad premium insulating glass, 2′ x 3′	1.000	Ea.	.800	179.00	27.00	206.00
2′ x 4′	1.000	Ea.	.889	188.00	30.00	218.00
2′ x 5′	1.000	Ea.	1.000	257.00	34.00	291.00
2′ x 6′	1.000	Ea.	1.000	325.00	34.00	359.00
Metal clad deluxe insulating glass, 2′ x 3′	1.000	Ea.	.800	176.00	27.00	203.00
2′ x 4′	1.000	Ea.	.889	212.00	30.00	242.00
2′ x 5′	1.000	Ea.	1.000	241.00	34.00	275.00
2′ x 6′	1.000	Ea.	1.000	277.00	34.00	311.00
Trim, interior casing, window 2′ x 3′	11.000	L.F.	.367	9.35	12.45	21.80
2′ x 4′	13.000	L.F.	.433	11.05	14.70	25.75
2′ x 4′-6″	14.000	L.F.	.467	11.90	15.80	27.70
2′ x 5′	15.000	L.F.	.500	12.75	16.95	29.70
2′ x 6′	17.000	L.F.	.567	14.45	19.20	33.65
Paint or stain, interior or exterior, 2′ x 3′ window, 1 coat	1.000	Face	.444	.35	13.30	13.65
2 coats	1.000	Face	.727	.70	22.00	22.70
Primer & 1 coat	1.000	Face	.727	.67	22.00	22.67
Primer & 2 coats	1.000	Face	.889	.99	26.50	27.49
2′ x 4′ window, 1 coat	1.000	Face	.444	.35	13.30	13.65
2 coats	1.000	Face	.727	.70	22.00	22.70
Primer & 1 coat	1.000	Face	.727	.67	22.00	22.67
Primer & 2 coats	1.000	Face	.889	.99	26.50	27.49
2′ x 6′ window, 1 coat	1.000	Face	.667	.71	19.95	20.66
2 coats	1.000	Face	.667	.81	19.95	20.76
Primer & 1 coat	1.000	Face	.727	1.01	22.00	23.01
Primer & 2 coats	1.000	Face	.889	.99	26.50	27.49
Caulking, window, 2′ x 3′	10.000	L.F.	.323	1.80	10.90	12.70
2′ x 4′	12.000	L.F.	.387	2.16	13.10	15.26
2′ x 4′-6″	13.000	L.F.	.419	2.34	14.15	16.49
2′ x 5′	14.000	L.F.	.452	2.52	15.25	17.77
2′ x 6′	16.000	L.F.	.516	2.88	17.45	20.33
Grilles, glass size, to 20″ x 36″	1.000	Ea.	.267	30.00	9.00	39.00
To 20″ x 56″	1.000	Ea.	.250	40.50	8.45	48.95
Drip cap, metal, 2′ long	2.000	L.F.	.040	.44	1.36	1.80
Wood, 2′ long	2.000	L.F.	.067	1.70	2.26	3.96

36 Awning Window Systems

Drip Cap
Interior Trim
Snap-in Grille
Caulking
Window

System Description	QUAN.	UNIT	LABOR HOURS	COST EACH		
				MAT.	INST.	TOTAL
BUILDER'S QUALITY WINDOW, WOOD, 34" X 22", AWNING						
Window, builder quality, 34" x 22", insulating glass	1.000	Ea.	.800	242.00	27.00	269.00
Trim, interior casing	10.500	L.F.	.350	8.93	11.87	20.80
Paint, interior & exterior, primer & 2 coats	2.000	Face	1.778	1.98	53.00	54.98
Caulking	9.500	L.F.	.306	1.71	10.36	12.07
Snap-in grille	1.000	Ea.	.267	24.00	9.00	33.00
Drip cap, metal	3.000	L.F.	.060	.66	2.04	2.70
TOTAL			3.561	279.28	113.27	392.55
PLASTIC CLAD WOOD WINDOW, 40" X 28", AWNING						
Window, plastic clad, premium, 40" x 28", insulating glass	1.000	Ea.	.889	315.00	30.00	345.00
Trim interior casing	13.500	L.F.	.450	11.48	15.26	26.74
Paint, interior, primer & 2 coats	1.000	Face	.889	.99	26.50	27.49
Caulking	12.500	L.F.	.403	2.25	13.63	15.88
Snap-in grille	1.000	Ea.	.267	24.00	9.00	33.00
TOTAL			2.898	353.72	94.39	448.11
METAL CLAD WOOD WINDOW, 48" X 36", AWNING						
Window, metal clad, deluxe, 48" x 36", insulating glass	1.000	Ea.	1.000	261.00	34.00	295.00
Trim, interior casing	15.000	L.F.	.500	12.75	16.95	29.70
Paint, interior, primer & 2 coats	1.000	Face	.889	.99	26.50	27.49
Caulking	14.000	L.F.	.452	2.52	15.26	17.78
Snap-in grille	1.000	Ea.	.250	31.00	8.45	39.45
Drip cap, metal	4.000	L.F.	.080	.88	2.72	3.60
TOTAL			3.171	309.14	103.88	413.02

The cost of this system is on a cost per each window basis.

Description	QUAN.	UNIT	LABOR HOURS	COST EACH		
				MAT.	INST.	TOTAL

Awning Window Price Sheet	QUAN.	UNIT	LABOR HOURS	COST EACH		
				MAT.	INST.	TOTAL
Windows, awning, builder's quality, 34″ x 22″, insulated glass	1.000	Ea.	.800	242.00	27.00	269.00
Low E glass	1.000	Ea.	.800	242.00	27.00	269.00
40″ x 28″, insulated glass	1.000	Ea.	.889	410.00	30.00	440.00
Low E glass	1.000	Ea.	.889	410.00	30.00	440.00
48″ x 36″, insulated glass	1.000	Ea.	1.000	755.00	34.00	789.00
Low E glass	1.000	Ea.	1.000	755.00	34.00	789.00
Plastic clad premium insulating glass, 34″ x 22″	1.000	Ea.	.800	212.00	27.00	239.00
40″ x 22″	1.000	Ea.	.800	234.00	27.00	261.00
36″ x 28″	1.000	Ea.	.889	250.00	30.00	280.00
36″ x 36″	1.000	Ea.	.889	315.00	30.00	345.00
48″ x 28″	1.000	Ea.	1.000	297.00	34.00	331.00
60″ x 36″	1.000	Ea.	1.000	410.00	34.00	444.00
Metal clad deluxe insulating glass, 34″ x 22″	1.000	Ea.	.800	203.00	27.00	230.00
40″ x 22″	1.000	Ea.	.800	222.00	27.00	249.00
36″ x 25″	1.000	Ea.	.889	226.00	30.00	256.00
40″ x 30″	1.000	Ea.	.889	294.00	30.00	324.00
48″ x 28″	1.000	Ea.	1.000	265.00	34.00	299.00
60″ x 36″	1.000	Ea.	1.000	261.00	34.00	295.00
Trim, interior casing window, 34″ x 22″	10.500	L.F.	.350	8.95	11.85	20.80
40″ x 22″	11.500	L.F.	.383	9.80	13.00	22.80
36″ x 28″	12.500	L.F.	.417	10.65	14.15	24.80
40″ x 28″	13.500	L.F.	.450	11.50	15.25	26.75
48″ x 28″	14.500	L.F.	.483	12.35	16.40	28.75
48″ x 36″	15.000	L.F.	.500	12.75	16.95	29.70
Paint or stain, interior or exterior, 34″ x 22″, 1 coat	1.000	Face	.444	.35	13.30	13.65
2 coats	1.000	Face	.727	.70	22.00	22.70
Primer & 1 coat	1.000	Face	.727	.67	22.00	22.67
Primer & 2 coats	1.000	Face	.889	.99	26.50	27.49
36″ x 28″, 1 coat	1.000	Face	.444	.35	13.30	13.65
2 coats	1.000	Face	.727	.70	22.00	22.70
Primer & 1 coat	1.000	Face	.727	.67	22.00	22.67
Primer & 2 coats	1.000	Face	.889	.99	26.50	27.49
48″ x 36″, 1 coat	1.000	Face	.667	.71	19.95	20.66
2 coats	1.000	Face	.667	.81	19.95	20.76
Primer & 1 coat	1.000	Face	.727	1.01	22.00	23.01
Primer & 2 coats	1.000	Face	.889	.99	26.50	27.49
Caulking, window, 34″ x 22″	9.500	L.F.	.306	1.71	10.35	12.06
40″ x 22″	10.500	L.F.	.339	1.89	11.45	13.34
36″ x 28″	11.500	L.F.	.371	2.07	12.55	14.62
40″ x 28″	12.500	L.F.	.403	2.25	13.65	15.90
48″ x 28″	13.500	L.F.	.435	2.43	14.70	17.13
48″ x 36″	14.000	L.F.	.452	2.52	15.25	17.77
Grilles, glass size, to 28″ by 16″	1.000	Ea.	.267	24.00	9.00	33.00
To 44″ by 24″	1.000	Ea.	.250	31.00	8.45	39.45
Drip cap, aluminum, 3′ long	3.000	L.F.	.060	.66	2.04	2.70
3′-6″ long	3.500	L.F.	.070	.77	2.38	3.15
4′ long	4.000	L.F.	.080	.88	2.72	3.60
Wood, 3′ long	3.000	L.F.	.100	2.55	3.39	5.94
3′-6″ long	3.500	L.F.	.117	2.98	3.96	6.94
4′ long	4.000	L.F.	.133	3.40	4.52	7.92

40 Sliding Window Systems

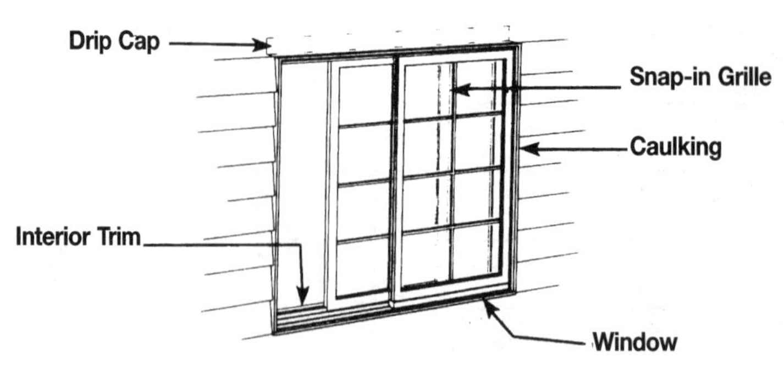

System Description	QUAN.	UNIT	LABOR HOURS	COST EACH		
				MAT.	INST.	TOTAL
BUILDER'S QUALITY WOOD WINDOW, 3′ X 2′, SLIDING						
Window, primed, builder's quality, 3′ x 2′, insul. glass	1.000	Ea.	.800	178.00	27.00	205.00
Trim, interior casing	11.000	L.F.	.367	9.35	12.43	21.78
Paint, interior & exterior, primer & 2 coats	2.000	Face	1.778	1.98	53.00	54.98
Caulking	10.000	L.F.	.323	1.80	10.90	12.70
Snap-in grille	1.000	Set	.333	34.50	11.25	45.75
Drip cap, metal	3.000	L.F.	.060	.66	2.04	2.70
TOTAL			3.661	226.29	116.62	342.91
PLASTIC CLAD WOOD WINDOW, 4′ X 3′-6″, SLIDING						
Window, plastic clad, premium, 4′ x 3′-6″, insulating glass	1.000	Ea.	.889	615.00	30.00	645.00
Trim, interior casing	16.000	L.F.	.533	13.60	18.08	31.68
Paint, interior, primer & 2 coats	1.000	Face	.889	.99	26.50	27.49
Caulking	17.000	L.F.	.548	3.06	18.53	21.59
Snap-in grille	1.000	Set	.333	34.50	11.25	45.75
TOTAL			3.192	667.15	104.36	771.51
METAL CLAD WOOD WINDOW, 6′ X 5′, SLIDING						
Window, metal clad, deluxe, 6′ x 5′, insulating glass	1.000	Ea.	1.000	515.00	34.00	549.00
Trim, interior casing	23.000	L.F.	.767	19.55	25.99	45.54
Paint, interior, primer & 2 coats	1.000	Face	.889	.99	26.50	27.49
Caulking	22.000	L.F.	.710	3.96	23.98	27.94
Snap-in grille	1.000	Set	.364	54.50	12.25	66.75
Drip cap, metal	6.000	L.F.	.120	1.32	4.08	5.40
TOTAL			3.850	595.32	126.80	722.12

The cost of this system is on a cost per each window basis.

Description	QUAN.	UNIT	LABOR HOURS	COST EACH		
				MAT.	INST.	TOTAL

Important: See the Reference Section for critical supporting data - Reference Nos., Crews & Location Factors

Sliding Window Price Sheet	QUAN.	UNIT	LABOR HOURS	COST EACH		
				MAT.	INST.	TOTAL
Windows, sliding, builder's quality, 3′ x 3′, single glass	1.000	Ea.	.800	141.00	27.00	168.00
Insulating glass	1.000	Ea.	.800	178.00	27.00	205.00
4′ x 3′-6″, single glass	1.000	Ea.	.889	167.00	30.00	197.00
Insulating glass	1.000	Ea.	.889	209.00	30.00	239.00
6′ x 5′, single glass	1.000	Ea.	1.000	305.00	34.00	339.00
Insulating glass	1.000	Ea.	1.000	365.00	34.00	399.00
Plastic clad premium insulating glass, 3′ x 3′	1.000	Ea.	.800	515.00	27.00	542.00
4′ x 3′-6″	1.000	Ea.	.889	615.00	30.00	645.00
5′ x 4′	1.000	Ea.	.889	710.00	30.00	740.00
6′ x 5′	1.000	Ea.	1.000	885.00	34.00	919.00
Metal clad deluxe insulating glass, 3′ x 3′	1.000	Ea.	.800	286.00	27.00	313.00
4′ x 3′-6″	1.000	Ea.	.889	350.00	30.00	380.00
5′ x 4′	1.000	Ea.	.889	420.00	30.00	450.00
6′ x 5′	1.000	Ea.	1.000	515.00	34.00	549.00
Trim, interior casing, window 3′ x 2′	11.000	L.F.	.367	9.35	12.45	21.80
3′ x 3′	13.000	L.F.	.433	11.05	14.70	25.75
4′ x 3′-6″	16.000	L.F.	.533	13.60	18.10	31.70
5′ x 4′	19.000	L.F.	.633	16.15	21.50	37.65
6′ x 5′	23.000	L.F.	.767	19.55	26.00	45.55
Paint or stain, interior or exterior, 3′ x 2′ window, 1 coat	1.000	Face	.444	.35	13.30	13.65
2 coats	1.000	Face	.727	.70	22.00	22.70
Primer & 1 coat	1.000	Face	.727	.67	22.00	22.67
Primer & 2 coats	1.000	Face	.889	.99	26.50	27.49
4′ x 3′-6″ window, 1 coat	1.000	Face	.667	.71	19.95	20.66
2 coats	1.000	Face	.667	.81	19.95	20.76
Primer & 1 coat	1.000	Face	.727	1.01	22.00	23.01
Primer & 2 coats	1.000	Face	.889	.99	26.50	27.49
6′ x 5′ window, 1 coat	1.000	Face	.889	1.97	26.50	28.47
2 coats	1.000	Face	1.333	3.60	40.00	43.60
Primer & 1 coat	1.000	Face	1.333	3.57	40.00	43.57
Primer & 2 coats	1.000	Face	1.600	5.25	48.00	53.25
Caulking, window, 3′ x 2′	10.000	L.F.	.323	1.80	10.90	12.70
3′ x 3′	12.000	L.F.	.387	2.16	13.10	15.26
4′ x 3′-6″	15.000	L.F.	.484	2.70	16.35	19.05
5′ x 4′	18.000	L.F.	.581	3.24	19.60	22.84
6′ x 5′	22.000	L.F.	.710	3.96	24.00	27.96
Grilles, glass size, to 14″ x 36″	1.000	Set	.333	34.50	11.25	45.75
To 36″ x 36″	1.000	Set	.364	54.50	12.25	66.75
Drip cap, aluminum, 3′ long	3.000	L.F.	.060	.66	2.04	2.70
4′ long	4.000	L.F.	.080	.88	2.72	3.60
5′ long	5.000	L.F.	.100	1.10	3.40	4.50
6′ long	6.000	L.F.	.120	1.32	4.08	5.40
Wood, 3′ long	3.000	L.F.	.100	2.55	3.39	5.94
4′ long	4.000	L.F.	.133	3.40	4.52	7.92
5′ long	5.000	L.F.	.167	4.25	5.65	9.90
6′ long	6.000	L.F.	.200	5.10	6.80	11.90

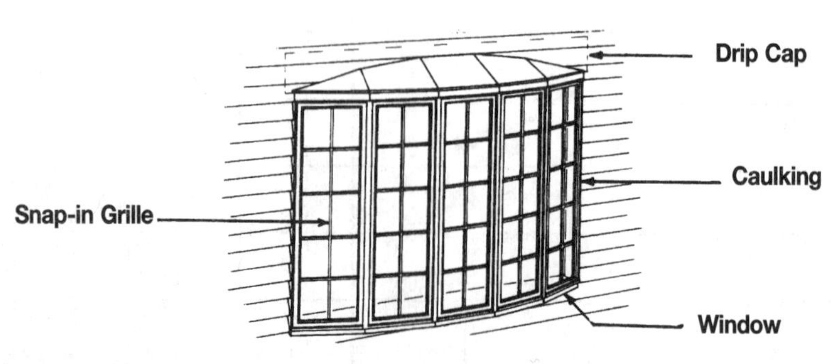

System Description	QUAN.	UNIT	LABOR HOURS	COST EACH		
				MAT.	INST.	TOTAL
AWNING TYPE BOW WINDOW, BUILDER'S QUALITY, 8′ X 5′						
Window, primed, builder's quality, 8′ x 5′, insulating glass	1.000	Ea.	1.600	1200.00	54.00	1254.00
Trim, interior casing	27.000	L.F.	.900	22.95	30.51	53.46
Paint, interior & exterior, primer & 1 coat	2.000	Face	3.200	10.50	96.00	106.50
Drip cap, vinyl	1.000	Ea.	.533	74.00	18.00	92.00
Caulking	26.000	L.F.	.839	4.68	28.34	33.02
Snap-in grilles	1.000	Set	1.067	120.00	36.00	156.00
TOTAL			8.139	1432.13	262.85	1694.98
CASEMENT TYPE BOW WINDOW, PLASTIC CLAD, 10′ X 6′						
Window, plastic clad, premium, 10′ x 6′, insulating glass	1.000	Ea.	2.286	1825.00	77.00	1902.00
Trim, interior casing	33.000	L.F.	1.100	28.05	37.29	65.34
Paint, interior, primer & 1 coat	1.000	Face	1.778	1.98	53.00	54.98
Drip cap, vinyl	1.000	Ea.	.615	80.50	21.00	101.50
Caulking	32.000	L.F.	1.032	5.76	34.88	40.64
Snap-in grilles	1.000	Set	1.333	150.00	45.00	195.00
TOTAL			8.144	2091.29	268.17	2359.46
DOUBLE HUNG TYPE, METAL CLAD, 9′ X 5′						
Window, metal clad, deluxe, 9′ x 5′, insulating glass	1.000	Ea.	2.667	1100.00	90.00	1190.00
Trim, interior casing	29.000	L.F.	.967	24.65	32.77	57.42
Paint, interior, primer & 1 coat	1.000	Face	1.778	1.98	53.00	54.98
Drip cap, vinyl	1.000	Set	.615	80.50	21.00	101.50
Caulking	28.000	L.F.	.903	5.04	30.52	35.56
Snap-in grilles	1.000	Set	1.067	120.00	36.00	156.00
TOTAL			7.997	1332.17	263.29	1595.46

The cost of this system is on a cost per each window basis.

Description	QUAN.	UNIT	LABOR HOURS	COST EACH		
				MAT.	INST.	TOTAL

Important: See the Reference Section for critical supporting data - Reference Nos., Crews & Location Factors

Bow/Bay Window Price Sheet	QUAN.	UNIT	LABOR HOURS	COST EACH		
				MAT.	INST.	TOTAL
Windows, bow awning type, builder's quality, 8′ x 5′, insulating glass	1.000	Ea.	1.600	935.00	54.00	989.00
Low E glass	1.000	Ea.	1.600	1200.00	54.00	1254.00
12′ x 6′, insulating glass	1.000	Ea.	2.667	1250.00	90.00	1340.00
Low E glass	1.000	Ea.	2.667	1325.00	90.00	1415.00
Plastic clad premium insulating glass, 6′ x 4′	1.000	Ea.	1.600	1025.00	54.00	1079.00
9′ x 4′	1.000	Ea.	2.000	1350.00	67.50	1417.50
10′ x 5′	1.000	Ea.	2.286	2275.00	77.00	2352.00
12′ x 6′	1.000	Ea.	2.667	2900.00	90.00	2990.00
Metal clad deluxe insulating glass, 6′ x 4′	1.000	Ea.	1.600	845.00	54.00	899.00
9′ x 4′	1.000	Ea.	2.000	1200.00	67.50	1267.50
10′ x 5′	1.000	Ea.	2.286	1625.00	77.00	1702.00
12′ x 6′	1.000	Ea.	2.667	2275.00	90.00	2365.00
Bow casement type, builder's quality, 8′ x 5′, single glass	1.000	Ea.	1.600	1425.00	54.00	1479.00
Insulating glass	1.000	Ea.	1.600	1725.00	54.00	1779.00
12′ x 6′, single glass	1.000	Ea.	2.667	1775.00	90.00	1865.00
Insulating glass	1.000	Ea.	2.667	1850.00	90.00	1940.00
Plastic clad premium insulating glass, 8′ x 5′	1.000	Ea.	1.600	1225.00	54.00	1279.00
10′ x 5′	1.000	Ea.	2.000	1750.00	67.50	1817.50
10′ x 6′	1.000	Ea.	2.286	1825.00	77.00	1902.00
12′ x 6′	1.000	Ea.	2.667	2175.00	90.00	2265.00
Metal clad deluxe insulating glass, 8′ x 5′	1.000	Ea.	1.600	1250.00	54.00	1304.00
10′ x 5′	1.000	Ea.	2.000	1350.00	67.50	1417.50
10′ x 6′	1.000	Ea.	2.286	1600.00	77.00	1677.00
12′ x 6′	1.000	Ea.	2.667	2225.00	90.00	2315.00
Bow, double hung type, builder's quality, 8′ x 4′, single glass	1.000	Ea.	1.600	1000.00	54.00	1054.00
Insulating glass	1.000	Ea.	1.600	1075.00	54.00	1129.00
9′ x 5′, single glass	1.000	Ea.	2.667	1075.00	90.00	1165.00
Insulating glass	1.000	Ea.	2.667	1125.00	90.00	1215.00
Plastic clad premium insulating glass, 7′ x 4′	1.000	Ea.	1.600	1025.00	54.00	1079.00
8′ x 4′	1.000	Ea.	2.000	1050.00	67.50	1117.50
8′ x 5′	1.000	Ea.	2.286	1100.00	77.00	1177.00
9′ x 5′	1.000	Ea.	2.667	1125.00	90.00	1215.00
Metal clad deluxe insulating glass, 7′ x 4′	1.000	Ea.	1.600	960.00	54.00	1014.00
8′ x 4′	1.000	Ea.	2.000	995.00	67.50	1062.50
8′ x 5′	1.000	Ea.	2.286	1025.00	77.00	1102.00
9′ x 5′	1.000	Ea.	2.667	1100.00	90.00	1190.00
Trim, interior casing, window 7′ x 4′	1.000	Ea.	.767	19.55	26.00	45.55
8′ x 5′	1.000	Ea.	.900	23.00	30.50	53.50
10′ x 6′	1.000	Ea.	1.100	28.00	37.50	65.50
12′ x 6′	1.000	Ea.	1.233	31.50	42.00	73.50
Paint or stain, interior, or exterior, 7′ x 4′ window, 1 coat	1.000	Face	.889	1.97	26.50	28.47
Primer & 1 coat	1.000	Face	1.333	3.57	40.00	43.57
8′ x 5′ window, 1 coat	1.000	Face	.889	1.97	26.50	28.47
Primer & 1 coat	1.000	Face	1.333	3.57	40.00	43.57
10′ x 6′ window, 1 coat	1.000	Face	1.333	1.42	40.00	41.42
Primer & 1 coat	1.000	Face	1.778	1.98	53.00	54.98
12′ x 6′ window, 1 coat	1.000	Face	1.778	3.94	53.00	56.94
Primer & 1 coat	1.000	Face	2.667	7.15	80.00	87.15
Drip cap, vinyl moulded window, 7′ long	1.000	Ea.	.533	74.00	18.00	92.00
8′ long	1.000	Ea.	.533	74.00	18.00	92.00
10′ long	1.000	Ea.	.615	80.50	21.00	101.50
12′ long	1.000	Ea.	.615	80.50	21.00	101.50
Caulking, window, 7′ x 4′	1.000	Ea.	.710	3.96	24.00	27.96
8′ x 5′	1.000	Ea.	.839	4.68	28.50	33.18
10′ x 6′	1.000	Ea.	1.032	5.75	35.00	40.75
12′ x 6′	1.000	Ea.	1.161	6.50	39.00	45.50
Grilles, window, 7′ x 4′	1.000	Set	.800	90.00	27.00	117.00
8′ x 5′	1.000	Set	1.067	120.00	36.00	156.00
10′ x 6′	1.000	Set	1.333	150.00	45.00	195.00
12′ x 6′	1.000	Set	1.600	180.00	54.00	234.00

48 Fixed Window Systems

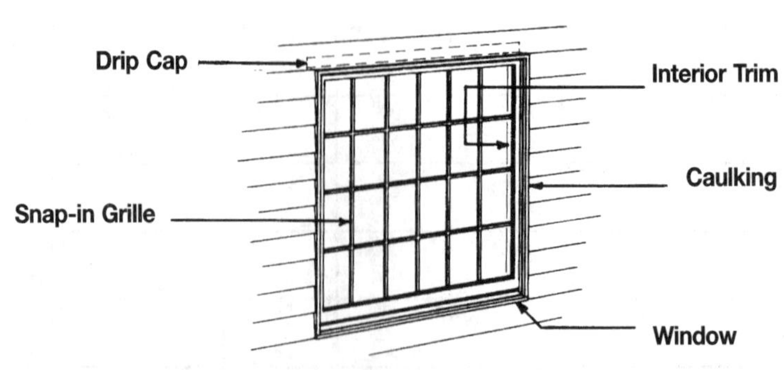

System Description	QUAN.	UNIT	LABOR HOURS	COST EACH		
				MAT.	INST.	TOTAL
BUILDER'S QUALITY PICTURE WINDOW, 4′ X 4′						
Window, primed, builder's quality, 4′ x 4′, insulating glass	1.000	Ea.	1.333	220.00	45.00	265.00
Trim, interior casing	17.000	L.F.	.567	14.45	19.21	33.66
Paint, interior & exterior, primer & 2 coats	2.000	Face	1.778	1.98	53.00	54.98
Caulking	16.000	L.F.	.516	2.88	17.44	20.32
Snap-in grille	1.000	Ea.	.267	121.00	9.00	130.00
Drip cap, metal	4.000	L.F.	.080	.88	2.72	3.60
TOTAL			4.541	361.19	146.37	507.56
PLASTIC CLAD WOOD WINDOW, 4′-6″ X 6′-6″						
Window, plastic clad, prem., 4′-6″ x 6′-6″, insul. glass	1.000	Ea.	1.455	635.00	49.00	684.00
Trim, interior casing	23.000	L.F.	.767	19.55	25.99	45.54
Paint, interior, primer & 2 coats	1.000	Face	.889	.99	26.50	27.49
Caulking	22.000	L.F.	.710	3.96	23.98	27.94
Snap-in grille	1.000	Ea.	.267	121.00	9.00	130.00
TOTAL			4.088	780.50	134.47	914.97
METAL CLAD WOOD WINDOW, 6′-6″ X 6′-6″						
Window, metal clad, deluxe, 6′-6″ x 6′-6″, insulating glass	1.000	Ea.	1.600	515.00	54.00	569.00
Trim interior casing	27.000	L.F.	.900	22.95	30.51	53.46
Paint, interior, primer & 2 coats	1.000	Face	1.600	5.25	48.00	53.25
Caulking	26.000	L.F.	.839	4.68	28.34	33.02
Snap-in grille	1.000	Ea.	.267	121.00	9.00	130.00
Drip cap, metal	6.500	L.F.	.130	1.43	4.42	5.85
TOTAL			5.336	670.31	174.27	844.58

The cost of this system is on a cost per each window basis.

Description	QUAN.	UNIT	LABOR HOURS	COST EACH		
				MAT.	INST.	TOTAL

Fixed Window Price Sheet	QUAN.	UNIT	LABOR HOURS	COST EACH		
				MAT.	INST.	TOTAL
Window-picture, builder's quality, 4′ x 4′, single glass	1.000	Ea.	1.333	262.00	45.00	307.00
Insulating glass	1.000	Ea.	1.333	220.00	45.00	265.00
4′ x 4′-6″, single glass	1.000	Ea.	1.455	218.00	49.00	267.00
Insulating glass	1.000	Ea.	1.455	222.00	49.00	271.00
5′ x 4′, single glass	1.000	Ea.	1.455	220.00	49.00	269.00
Insulating glass	1.000	Ea.	1.455	276.00	49.00	325.00
6′ x 4′-6″, single glass	1.000	Ea.	1.600	310.00	54.00	364.00
Insulating glass	1.000	Ea.	1.600	380.00	54.00	434.00
Plastic clad premium insulating glass, 4′ x 4′	1.000	Ea.	1.333	465.00	45.00	510.00
4′-6″ x 6′-6″	1.000	Ea.	1.455	635.00	49.00	684.00
5′-6″ x 6′-6″	1.000	Ea.	1.600	825.00	54.00	879.00
6′-6″ x 6′-6″	1.000	Ea.	1.600	840.00	54.00	894.00
Metal clad deluxe insulating glass, 4′ x 4′	1.000	Ea.	1.333	277.00	45.00	322.00
4′-6″ x 6′-6″	1.000	Ea.	1.455	410.00	49.00	459.00
5′-6″ x 6′-6″	1.000	Ea.	1.600	450.00	54.00	504.00
6′-6″ x 6′-6″	1.000	Ea.	1.600	515.00	54.00	569.00
Trim, interior casing, window 4′ x 4′	17.000	L.F.	.567	14.45	19.20	33.65
4′-6″ x 4′-6″	19.000	L.F.	.633	16.15	21.50	37.65
5′-0″ x 4′-0″	19.000	L.F.	.633	16.15	21.50	37.65
4′-6″ x 6′-6″	23.000	L.F.	.767	19.55	26.00	45.55
5′-6″ x 6′-6″	25.000	L.F.	.833	21.50	28.50	50.00
6′-6″ x 6′-6″	27.000	L.F.	.900	23.00	30.50	53.50
Paint or stain, interior or exterior, 4′ x 4′ window, 1 coat	1.000	Face	.667	.71	19.95	20.66
2 coats	1.000	Face	.667	.81	19.95	20.76
Primer & 1 coat	1.000	Face	.727	1.01	22.00	23.01
Primer & 2 coats	1.000	Face	.889	.99	26.50	27.49
4′-6″ x 6′-6″ window, 1 coat	1.000	Face	.667	.71	19.95	20.66
2 coats	1.000	Face	.667	.81	19.95	20.76
Primer & 1 coat	1.000	Face	.727	1.01	22.00	23.01
Primer & 2 coats	1.000	Face	.889	.99	26.50	27.49
6′-6″ x 6′-6″ window, 1 coat	1.000	Face	.889	1.97	26.50	28.47
2 coats	1.000	Face	1.333	3.60	40.00	43.60
Primer & 1 coat	1.000	Face	1.333	3.57	40.00	43.57
Primer & 2 coats	1.000	Face	1.600	5.25	48.00	53.25
Caulking, window, 4′ x 4′	1.000	Ea.	.516	2.88	17.45	20.33
4′-6″ x 4′-6″	1.000	Ea.	.581	3.24	19.60	22.84
5′-0″ x 4′-0″	1.000	Ea.	.581	3.24	19.60	22.84
4′-6″ x 6′-6″	1.000	Ea.	.710	3.96	24.00	27.96
5′-6″ x 6′-6″	1.000	Ea.	.774	4.32	26.00	30.32
6′-6″ x 6′-6″	1.000	Ea.	.839	4.68	28.50	33.18
Grilles, glass size, to 48″ x 48″	1.000	Ea.	.267	121.00	9.00	130.00
To 60″ x 68″	1.000	Ea.	.286	134.00	9.65	143.65
Drip cap, aluminum, 4′ long	4.000	L.F.	.080	.88	2.72	3.60
4′-6″ long	4.500	L.F.	.090	.99	3.06	4.05
5′ long	5.000	L.F.	.100	1.10	3.40	4.50
6′ long	6.000	L.F.	.120	1.32	4.08	5.40
Wood, 4′ long	4.000	L.F.	.133	3.40	4.52	7.92
4′-6″ long	4.500	L.F.	.150	3.83	5.10	8.93
5′ long	5.000	L.F.	.167	4.25	5.65	9.90
6′ long	6.000	L.F.	.200	5.10	6.80	11.90

52 Entrance Door Systems

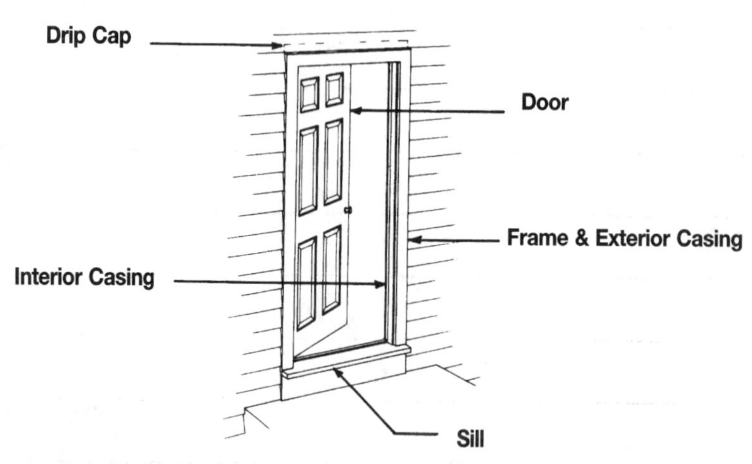

System Description	QUAN.	UNIT	LABOR HOURS	COST EACH MAT.	COST EACH INST.	COST EACH TOTAL
COLONIAL, 6 PANEL, 3′ X 6′-8″, WOOD						
Door, 3′ x 6′-8″ x 1-3/4″ thick, pine, 6 panel colonial	1.000	Ea.	1.067	355.00	36.00	391.00
Frame, 5-13/16″ deep, incl. exterior casing & drip cap	17.000	L.F.	.725	124.10	24.48	148.58
Interior casing, 2-1/2″ wide	18.000	L.F.	.600	15.30	20.34	35.64
Sill, 8/4 x 8″ deep	3.000	L.F.	.480	43.50	16.20	59.70
Butt hinges, brass, 4-1/2″ x 4-1/2″	1.500	Pr.		13.43		13.43
Lockset	1.000	Ea.	.571	31.00	19.30	50.30
Weatherstripping, metal, spring type, bronze	1.000	Set	1.053	17.00	35.50	52.50
Paint, interior & exterior, primer & 2 coats	2.000	Face	1.778	10.40	53.00	63.40
TOTAL			6.274	609.73	204.82	814.55
SOLID CORE BIRCH, FLUSH, 3′ X 6′-8″						
Door, 3′ x 6′-8″, 1-3/4″ thick, birch, flush solid core	1.000	Ea.	1.067	83.50	36.00	119.50
Frame, 5-13/16″ deep, incl. exterior casing & drip cap	17.000	L.F.	.725	124.10	24.48	148.58
Interior casing, 2-1/2″ wide	18.000	L.F.	.600	15.30	20.34	35.64
Sill, 8/4 x 8″ deep	3.000	L.F.	.480	43.50	16.20	59.70
Butt hinges, brass, 4-1/2″ x 4-1/2″	1.500	Pr.		13.43		13.43
Lockset	1.000	Ea.	.571	31.00	19.30	50.30
Weatherstripping, metal, spring type, bronze	1.000	Set	1.053	17.00	35.50	52.50
Paint, interior & exterior, primer & 2 coats	2.000	Face	1.778	9.74	53.00	62.74
TOTAL			6.274	337.57	204.82	542.39

These systems are on a cost per each door basis.

Description	QUAN.	UNIT	LABOR HOURS	COST EACH MAT.	COST EACH INST.	COST EACH TOTAL

Important: See the Reference Section for critical supporting data - Reference Nos., Crews & Location Factors

Entrance Door Price Sheet	QUAN.	UNIT	LABOR HOURS	COST EACH		
				MAT.	INST.	TOTAL
Door exterior wood 1-3/4" thick, pine, dutch door, 2'-8" x 6'-8" minimum	1.000	Ea.	1.333	605.00	45.00	650.00
Maximum	1.000	Ea.	1.600	640.00	54.00	694.00
3'-0" x 6'-8", minimum	1.000	Ea.	1.333	635.00	45.00	680.00
Maximum	1.000	Ea.	1.600	685.00	54.00	739.00
Colonial, 6 panel, 2'-8" x 6'-8"	1.000	Ea.	1.000	330.00	34.00	364.00
3'-0" x 6'-8"	1.000	Ea.	1.067	355.00	36.00	391.00
8 panel, 2'-6" x 6'-8"	1.000	Ea.	1.000	490.00	34.00	524.00
3'-0" x 6'-8"	1.000	Ea.	1.067	530.00	36.00	566.00
Flush, birch, solid core, 2'-8" x 6'-8"	1.000	Ea.	1.000	88.50	34.00	122.50
3'-0" x 6'-8"	1.000	Ea.	1.067	83.50	36.00	119.50
Porch door, 2'-8" x 6'-8"	1.000	Ea.	1.000	207.00	34.00	241.00
3'-0" x 6'-8"	1.000	Ea.	1.067	238.00	36.00	274.00
Hand carved mahogany, 2'-8" x 6'-8"	1.000	Ea.	1.067	475.00	36.00	511.00
3'-0" x 6'-8"	1.000	Ea.	1.067	510.00	36.00	546.00
Rosewood, 2'-8" x 6'-8"	1.000	Ea.	1.067	740.00	36.00	776.00
3'-0" x 6-8"	1.000	Ea.	1.067	770.00	36.00	806.00
Door, metal clad wood 1-3/8" thick raised panel, 2'-8" x 6'-8"	1.000	Ea.	1.067	264.00	36.00	300.00
3'-0" x 6'-8"	1.000	Ea.	1.067	262.00	36.00	298.00
Deluxe metal door, 2'-8" x 6'-8"	1.000	Ea.	1.231	410.00	41.50	451.50
3'-0" x 6'-8"	1.000	Ea.	1.231	405.00	41.50	446.50
Frame, pine, including exterior trim & drip cap, 5/4, x 4-9/16" deep	17.000	L.F.	.725	73.00	24.50	97.50
5-13/16" deep	17.000	L.F.	.725	124.00	24.50	148.50
6-9/16" deep	17.000	L.F.	.725	106.00	24.50	130.50
Safety glass lites, add	1.000	Ea.		26.00		26.00
Interior casing, 2'-8" x 6'-8" door	18.000	L.F.	.600	15.30	20.50	35.80
3'-0" x 6'-8" door	19.000	L.F.	.633	16.15	21.50	37.65
Sill, oak, 8/4 x 8" deep	3.000	L.F.	.480	43.50	16.20	59.70
8/4 x 10" deep	3.000	L.F.	.533	58.50	18.00	76.50
Butt hinges, steel plated, 4-1/2" x 4-1/2", plain	1.500	Pr.		13.45		13.45
Ball bearing	1.500	Pr.		30.00		30.00
Bronze, 4-1/2" x 4-1/2", plain	1.500	Pr.		15.30		15.30
Ball bearing	1.500	Pr.		31.50		31.50
Lockset, minimum	1.000	Ea.	.571	31.00	19.30	50.30
Maximum	1.000	Ea.	1.000	130.00	34.00	164.00
Weatherstripping, metal, interlocking, zinc	1.000	Set	2.667	12.80	90.00	102.80
Bronze	1.000	Set	2.667	20.00	90.00	110.00
Spring type, bronze	1.000	Set	1.053	17.00	35.50	52.50
Rubber, minimum	1.000	Set	1.053	4.55	35.50	40.05
Maximum	1.000	Set	1.143	5.20	38.50	43.70
Felt minimum	1.000	Set	.571	2.10	19.30	21.40
Maximum	1.000	Set	.615	2.28	21.00	23.28
Paint or stain, flush door, interior or exterior, 1 coat	2.000	Face	.941	3.46	28.00	31.46
2 coats	2.000	Face	1.455	6.90	44.00	50.90
Primer & 1 coat	2.000	Face	1.455	6.50	44.00	50.50
Primer & 2 coats	2.000	Face	1.778	9.75	53.00	62.75
Paneled door, interior & exterior, 1 coat	2.000	Face	1.143	3.70	34.00	37.70
2 coats	2.000	Face	2.000	7.40	60.00	67.40
Primer & 1 coat	2.000	Face	1.455	6.90	44.00	50.90
Primer & 2 coats	2.000	Face	1.778	10.40	53.00	63.40

53 Sliding Door Systems

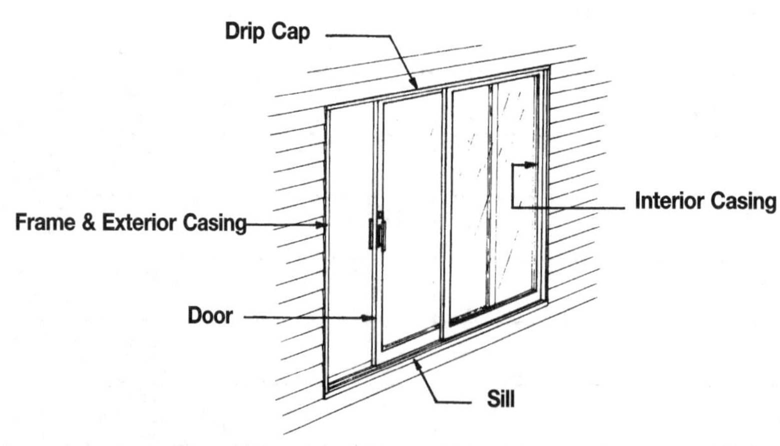

System Description	QUAN.	UNIT	LABOR HOURS	COST EACH MAT.	COST EACH INST.	COST EACH TOTAL
WOOD SLIDING DOOR, 8′ WIDE, PREMIUM						
Wood, 5/8″ thick tempered insul. glass, 8′ wide, premium	1.000	Ea.	5.333	1175.00	180.00	1355.00
Interior casing	22.000	L.F.	.733	18.70	24.86	43.56
Exterior casing	22.000	L.F.	.733	18.70	24.86	43.56
Sill, oak, 8/4 x 8″ deep	8.000	L.F.	1.280	116.00	43.20	159.20
Drip cap	8.000	L.F.	.160	1.76	5.44	7.20
Paint, interior & exterior, primer & 2 coats	2.000	Face	2.816	14.96	84.48	99.44
TOTAL			11.055	1345.12	362.84	1707.96
ALUMINUM SLIDING DOOR, 8′ WIDE, PREMIUM						
Aluminum, 5/8″ tempered insul. glass, 8′ wide, premium	1.000	Ea.	5.333	1350.00	180.00	1530.00
Interior casing	22.000	L.F.	.733	18.70	24.86	43.56
Exterior casing	22.000	L.F.	.733	18.70	24.86	43.56
Sill, oak, 8/4 x 8″ deep	8.000	L.F.	1.280	116.00	43.20	159.20
Drip cap	8.000	L.F.	.160	1.76	5.44	7.20
Paint, interior & exterior, primer & 2 coats	2.000	Face	2.816	14.96	84.48	99.44
TOTAL			11.055	1520.12	362.84	1882.96

The cost of this system is on a cost per each door basis.

Description	QUAN.	UNIT	LABOR HOURS	COST EACH MAT.	COST EACH INST.	COST EACH TOTAL

Important: See the Reference Section for critical supporting data - Reference Nos., Crews & Location Factors

Sliding Door Price Sheet	QUAN.	UNIT	LABOR HOURS	MAT.	INST.	TOTAL
Sliding door, wood, 5/8" thick, tempered insul. glass, 6' wide, premium	1.000	Ea.	4.000	1025.00	135.00	1160.00
Economy	1.000	Ea.	4.000	745.00	135.00	880.00
8'wide, wood premium	1.000	Ea.	5.333	1175.00	180.00	1355.00
Economy	1.000	Ea.	5.333	875.00	180.00	1055.00
12' wide, wood premium	1.000	Ea.	6.400	1900.00	216.00	2116.00
Economy	1.000	Ea.	6.400	1250.00	216.00	1466.00
Aluminum, 5/8" thick, tempered insul. glass, 6'wide, premium	1.000	Ea.	4.000	1350.00	135.00	1485.00
Economy	1.000	Ea.	4.000	560.00	135.00	695.00
8'wide, premium	1.000	Ea.	5.333	1350.00	180.00	1530.00
Economy	1.000	Ea.	5.333	1125.00	180.00	1305.00
12' wide, premium	1.000	Ea.	6.400	2250.00	216.00	2466.00
Economy	1.000	Ea.	6.400	1400.00	216.00	1616.00
Interior casing, 6' wide door	20.000	L.F.	.667	17.00	22.50	39.50
8' wide door	22.000	L.F.	.733	18.70	25.00	43.70
12' wide door	26.000	L.F.	.867	22.00	29.50	51.50
Exterior casing, 6' wide door	20.000	L.F.	.667	17.00	22.50	39.50
8' wide door	22.000	L.F.	.733	18.70	25.00	43.70
12' wide door	26.000	L.F.	.867	22.00	29.50	51.50
Sill, oak, 8/4 x 8" deep, 6' wide door	6.000	L.F.	.960	87.00	32.50	119.50
8' wide door	8.000	L.F.	1.280	116.00	43.00	159.00
12' wide door	12.000	L.F.	1.920	174.00	65.00	239.00
8/4 x 10" deep, 6' wide door	6.000	L.F.	1.067	117.00	36.00	153.00
8' wide door	8.000	L.F.	1.422	156.00	48.00	204.00
12' wide door	12.000	L.F.	2.133	235.00	72.00	307.00
Drip cap, 6' wide door	6.000	L.F.	.120	1.32	4.08	5.40
8' wide door	8.000	L.F.	.160	1.76	5.45	7.21
12' wide door	12.000	L.F.	.240	2.64	8.15	10.79
Paint or stain, interior & exterior, 6' wide door, 1 coat	2.000	Face	1.600	4.80	48.00	52.80
2 coats	2.000	Face	1.600	4.80	48.00	52.80
Primer & 1 coat	2.000	Face	1.778	9.60	53.50	63.10
Primer & 2 coats	2.000	Face	2.560	13.60	77.00	90.60
8' wide door, 1 coat	2.000	Face	1.760	5.30	53.00	58.30
2 coats	2.000	Face	1.760	5.30	53.00	58.30
Primer & 1 coat	2.000	Face	1.956	10.55	59.00	69.55
Primer & 2 coats	2.000	Face	2.816	14.95	84.50	99.45
12' wide door, 1 coat	2.000	Face	2.080	6.25	62.50	68.75
2 coats	2.000	Face	2.080	6.25	62.50	68.75
Primer & 1 coat	2.000	Face	2.311	12.50	69.50	82.00
Primer & 2 coats	2.000	Face	3.328	17.70	100.00	117.70
Aluminum door, trim only, interior & exterior, 6' door, 1 coat	2.000	Face	.800	2.40	24.00	26.40
2 coats	2.000	Face	.800	2.40	24.00	26.40
Primer & 1 coat	2.000	Face	.889	4.80	27.00	31.80
Primer & 2 coats	2.000	Face	1.280	6.80	38.50	45.30
8' wide door, 1 coat	2.000	Face	.880	2.64	26.50	29.14
2 coats	2.000	Face	.880	2.64	26.50	29.14
Primer & 1 coat	2.000	Face	.978	5.30	29.50	34.80
Primer & 2 coats	2.000	Face	1.408	7.50	42.00	49.50
12' wide door, 1 coat	2.000	Face	1.040	3.12	31.00	34.12
2 coats	2.000	Face	1.040	3.12	31.00	34.12
Primer & 1 coat	2.000	Face	1.156	6.25	35.00	41.25
Primer & 2 coats	2.000	Face	1.664	8.85	50.00	58.85

56 Residential Garage Door Systems

Drip Cap
Exterior Trim
Door
Weatherstripping
Jamb

System Description	QUAN.	UNIT	LABOR HOURS	COST EACH		
				MAT.	INST.	TOTAL
OVERHEAD, SECTIONAL GARAGE DOOR, 9′ X 7′						
Wood, overhead sectional door, std., incl. hardware, 9′ x 7′	1.000	Ea.	2.000	400.00	67.50	467.50
Jamb & header blocking, 2″ x 6″	25.000	L.F.	.901	16.25	30.50	46.75
Exterior trim	25.000	L.F.	.833	21.25	28.25	49.50
Paint, interior & exterior, primer & 2 coats	2.000	Face	3.556	20.80	106.00	126.80
Weatherstripping, molding type	1.000	Set	.767	19.55	25.99	45.54
Drip cap	9.000	L.F.	.180	1.98	6.12	8.10
TOTAL			8.237	479.83	264.36	744.19
OVERHEAD, SECTIONAL GARAGE DOOR, 16′ X 7′						
Wood, overhead sectional, std., incl. hardware, 16′ x 7′	1.000	Ea.	2.667	800.00	90.00	890.00
Jamb & header blocking, 2″ x 6″	30.000	L.F.	1.081	19.50	36.60	56.10
Exterior trim	30.000	L.F.	1.000	25.50	33.90	59.40
Paint, interior & exterior, primer & 2 coats	2.000	Face	5.333	31.20	159.00	190.20
Weatherstripping, molding type	1.000	Set	1.000	25.50	33.90	59.40
Drip cap	16.000	L.F.	.320	3.52	10.88	14.40
TOTAL			11.401	905.22	364.28	1269.50
OVERHEAD, SWING-UP TYPE, GARAGE DOOR, 16′ X 7′						
Wood, overhead, swing-up, std., incl. hardware, 16′ x 7′	1.000	Ea.	2.667	590.00	90.00	680.00
Jamb & header blocking, 2″ x 6″	30.000	L.F.	1.081	19.50	36.60	56.10
Exterior trim	30.000	L.F.	1.000	25.50	33.90	59.40
Paint, interior & exterior, primer & 2 coats	2.000	Face	5.333	31.20	159.00	190.20
Weatherstripping, molding type	1.000	Set	1.000	25.50	33.90	59.40
Drip cap	16.000	L.F.	.320	3.52	10.88	14.40
TOTAL			11.401	695.22	364.28	1059.50

This system is on a cost per each door basis.

Description	QUAN.	UNIT	LABOR HOURS	COST EACH		
				MAT.	INST.	TOTAL

Important: See the Reference Section for critical supporting data - Reference Nos., Crews & Location Factors

Resi Garage Door Price Sheet

Resi Garage Door Price Sheet	QUAN.	UNIT	LABOR HOURS	COST EACH		
				MAT.	INST.	TOTAL
Overhead, sectional, including hardware, fiberglass, 9′ x 7′, standard	1.000	Ea.	3.030	560.00	102.00	662.00
Deluxe	1.000	Ea.	3.030	655.00	102.00	757.00
16′ x 7′, standard	1.000	Ea.	2.667	865.00	90.00	955.00
Deluxe	1.000	Ea.	2.667	1075.00	90.00	1165.00
Hardboard, 9′ x 7′, standard	1.000	Ea.	2.000	355.00	67.50	422.50
Deluxe	1.000	Ea.	2.000	430.00	67.50	497.50
16′ x 7′, standard	1.000	Ea.	2.667	630.00	90.00	720.00
Deluxe	1.000	Ea.	2.667	735.00	90.00	825.00
Metal, 9′ x 7′, standard	1.000	Ea.	3.030	282.00	102.00	384.00
Deluxe	1.000	Ea.	2.000	560.00	67.50	627.50
16′ x 7′, standard	1.000	Ea.	5.333	555.00	180.00	735.00
Deluxe	1.000	Ea.	2.667	770.00	90.00	860.00
Wood, 9′ x 7′, standard	1.000	Ea.	2.000	400.00	67.50	467.50
Deluxe	1.000	Ea.	2.000	1125.00	67.50	1192.50
16′ x 7′, standard	1.000	Ea.	2.667	800.00	90.00	890.00
Deluxe	1.000	Ea.	2.667	1675.00	90.00	1765.00
Overhead swing-up type including hardware, fiberglass, 9′ x 7′, standard	1.000	Ea.	2.000	550.00	67.50	617.50
Deluxe	1.000	Ea.	2.000	650.00	67.50	717.50
16′ x 7′, standard	1.000	Ea.	2.667	690.00	90.00	780.00
Deluxe	1.000	Ea.	2.667	805.00	90.00	895.00
Hardboard, 9′ x 7′, standard	1.000	Ea.	2.000	284.00	67.50	351.50
Deluxe	1.000	Ea.	2.000	375.00	67.50	442.50
16′ x 7′, standard	1.000	Ea.	2.667	395.00	90.00	485.00
Deluxe	1.000	Ea.	2.667	590.00	90.00	680.00
Metal, 9′ x 7′, standard	1.000	Ea.	2.000	310.00	67.50	377.50
Deluxe	1.000	Ea.	2.000	490.00	67.50	557.50
16′ x 7′, standard	1.000	Ea.	2.667	490.00	90.00	580.00
Deluxe	1.000	Ea.	2.667	785.00	90.00	875.00
Wood, 9′ x 7′, standard	1.000	Ea.	2.000	340.00	67.50	407.50
Deluxe	1.000	Ea.	2.000	545.00	67.50	612.50
16′ x 7′, standard	1.000	Ea.	2.667	590.00	90.00	680.00
Deluxe	1.000	Ea.	2.667	835.00	90.00	925.00
Jamb & header blocking, 2″ x 6″, 9′ x 7′ door	25.000	L.F.	.901	16.25	30.50	46.75
16′ x 7′ door	30.000	L.F.	1.081	19.50	36.50	56.00
2″ x 8″, 9′ x 7′ door	25.000	L.F.	1.000	23.00	34.00	57.00
16′ x 7′ door	30.000	L.F.	1.200	27.50	40.50	68.00
Exterior trim, 9′ x 7′ door	25.000	L.F.	.833	21.50	28.50	50.00
16′ x 7′ door	30.000	L.F.	1.000	25.50	34.00	59.50
Paint or stain, interior & exterior, 9′ x 7′ door, 1 coat	1.000	Face	2.286	7.40	68.50	75.90
2 coats	1.000	Face	4.000	14.80	120.00	134.80
Primer & 1 coat	1.000	Face	2.909	13.85	88.00	101.85
Primer & 2 coats	1.000	Face	3.556	21.00	106.00	127.00
16′ x 7′ door, 1 coat	1.000	Face	3.429	11.10	103.00	114.10
2 coats	1.000	Face	6.000	22.00	180.00	202.00
Primer & 1 coat	1.000	Face	4.364	21.00	132.00	153.00
Primer & 2 coats	1.000	Face	5.333	31.00	159.00	190.00
Weatherstripping, molding type, 9′ x 7′ door	1.000	Set	.767	19.55	26.00	45.55
16′ x 7′ door	1.000	Set	1.000	25.50	34.00	59.50
Drip cap, 9′ door	9.000	L.F.	.180	1.98	6.10	8.08
16′ door	16.000	L.F.	.320	3.52	10.90	14.42
Garage door opener, economy	1.000	Ea.	1.000	228.00	34.00	262.00
Deluxe, including remote control	1.000	Ea.	1.000	330.00	34.00	364.00

58 Aluminum Window Systems

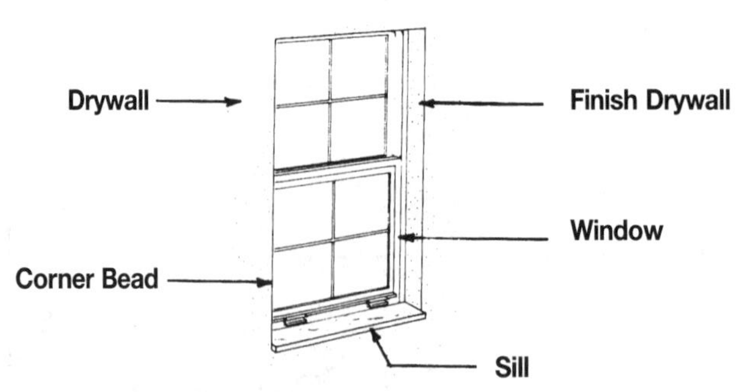

System Description	QUAN.	UNIT	LABOR HOURS	COST EACH MAT.	COST EACH INST.	COST EACH TOTAL
SINGLE HUNG, 2′ X 3′ OPENING						
Window, 2′ x 3′ opening, enameled, insulating glass	1.000	Ea.	1.600	165.00	67.00	232.00
Blocking, 1″ x 3″ furring strip nailers	10.000	L.F.	.145	2.00	4.90	6.90
Drywall, 1/2″ thick, standard	5.000	S.F.	.040	1.45	1.35	2.80
Corner bead, 1″ x 1″, galvanized steel	8.000	L.F.	.160	.80	5.44	6.24
Finish drywall, tape and finish corners inside and outside	16.000	L.F.	.233	1.12	7.84	8.96
Sill, slate	2.000	L.F.	.400	15.50	11.90	27.40
TOTAL			2.578	185.87	98.43	284.30
SLIDING, 3′ X 2′ OPENING						
Window, 3′ x 2′ opening, enameled, insulating glass	1.000	Ea.	1.600	170.00	67.00	237.00
Blocking, 1″ x 3″ furring strip nailers	10.000	L.F.	.145	2.00	4.90	6.90
Drywall, 1/2″ thick, standard	5.000	S.F.	.040	1.45	1.35	2.80
Corner bead, 1″ x 1″, galvanized steel	7.000	L.F.	.140	.70	4.76	5.46
Finish drywall, tape and finish corners inside and outside	14.000	L.F.	.204	.98	6.86	7.84
Sill, slate	3.000	L.F.	.600	23.25	17.85	41.10
TOTAL			2.729	198.38	102.72	301.10
AWNING, 3′-1″ X 3′-2″						
Window, 3′-1″ x 3′-2″ opening, enameled, insul. glass	1.000	Ea.	1.600	190.00	67.00	257.00
Blocking, 1″ x 3″ furring strip, nailers	12.500	L.F.	.182	2.50	6.13	8.63
Drywall, 1/2″ thick, standard	4.500	S.F.	.036	1.31	1.22	2.53
Corner bead, 1″ x 1″, galvanized steel	9.250	L.F.	.185	.93	6.29	7.22
Finish drywall, tape and finish corners, inside and outside	18.500	L.F.	.269	1.30	9.07	10.37
Sill, slate	3.250	L.F.	.650	25.19	19.34	44.53
TOTAL			2.922	221.23	109.05	330.28

Description	QUAN.	UNIT	LABOR HOURS	COST PER S.F. MAT.	COST PER S.F. INST.	COST PER S.F. TOTAL

Important: See the Reference Section for critical supporting data - Reference Nos., Crews & Location Factors

Aluminum Window Price Sheet	QUAN.	UNIT	LABOR HOURS	COST EACH		
				MAT.	INST.	TOTAL
Window, aluminum, awning, 3′-1″ x 3′-2″, standard glass	1.000	Ea.	1.600	203.00	67.00	270.00
Insulating glass	1.000	Ea.	1.600	190.00	67.00	257.00
4′-5″ x 5′-3″, standard glass	1.000	Ea.	2.000	286.00	84.00	370.00
Insulating glass	1.000	Ea.	2.000	330.00	84.00	414.00
Casement, 3′-1″ x 3′-2″, standard glass	1.000	Ea.	1.600	284.00	67.00	351.00
Insulating glass	1.000	Ea.	1.600	284.00	67.00	351.00
Single hung, 2′ x 3′, standard glass	1.000	Ea.	1.600	136.00	67.00	203.00
Insulating glass	1.000	Ea.	1.600	165.00	67.00	232.00
2′-8″ x 6′-8″, standard glass	1.000	Ea.	2.000	290.00	84.00	374.00
Insulating glass	1.000	Ea.	2.000	370.00	84.00	454.00
3′-4″ x 5′-0″, standard glass	1.000	Ea.	1.778	187.00	74.50	261.50
Insulating glass	1.000	Ea.	1.778	262.00	74.50	336.50
Sliding, 3′ x 2′, standard glass	1.000	Ea.	1.600	153.00	67.00	220.00
Insulating glass	1.000	Ea.	1.600	170.00	67.00	237.00
5′ x 3′, standard glass	1.000	Ea.	1.778	195.00	74.50	269.50
Insulating glass	1.000	Ea.	1.778	272.00	74.50	346.50
8′ x 4′, standard glass	1.000	Ea.	2.667	280.00	112.00	392.00
Insulating glass	1.000	Ea.	2.667	450.00	112.00	562.00
Blocking, 1″ x 3″ furring, opening 3′ x 2′	10.000	L.F.	.145	2.00	4.90	6.90
3′ x 3′	12.500	L.F.	.182	2.50	6.15	8.65
3′ x 5′	16.000	L.F.	.233	3.20	7.85	11.05
4′ x 4′	16.000	L.F.	.233	3.20	7.85	11.05
4′ x 5′	18.000	L.F.	.262	3.60	8.80	12.40
4′ x 6′	20.000	L.F.	.291	4.00	9.80	13.80
4′ x 8′	24.000	L.F.	.349	4.80	11.75	16.55
6′-8″ x 2′-8″	19.000	L.F.	.276	3.80	9.30	13.10
Drywall, 1/2″ thick, standard, opening 3′ x 2′	5.000	S.F.	.040	1.45	1.35	2.80
3′ x 3′	6.000	S.F.	.048	1.74	1.62	3.36
3′ x 5′	8.000	S.F.	.064	2.32	2.16	4.48
4′ x 4′	8.000	S.F.	.064	2.32	2.16	4.48
4′ x 5′	9.000	S.F.	.072	2.61	2.43	5.04
4′ x 6′	10.000	S.F.	.080	2.90	2.70	5.60
4′ x 8′	12.000	S.F.	.096	3.48	3.24	6.72
6′-8″ x 2′	9.500	S.F.	.076	2.76	2.57	5.33
Corner bead, 1″ x 1″, galvanized steel, opening 3′ x 2′	7.000	L.F.	.140	.70	4.76	5.46
3′ x 3′	9.000	L.F.	.180	.90	6.10	7.00
3′ x 5′	11.000	L.F.	.220	1.10	7.50	8.60
4′ x 4′	12.000	L.F.	.240	1.20	8.15	9.35
4′ x 5′	13.000	L.F.	.260	1.30	8.85	10.15
4′ x 6′	14.000	L.F.	.280	1.40	9.50	10.90
4′ x 8′	16.000	L.F.	.320	1.60	10.90	12.50
6′-8″ x 2′	15.000	L.F.	.300	1.50	10.20	11.70
Tape and finish corners, inside and outside, opening 3′ x 2′	14.000	L.F.	.204	.98	6.85	7.83
3′ x 3′	18.000	L.F.	.262	1.26	8.80	10.06
3′ x 5′	22.000	L.F.	.320	1.54	10.80	12.34
4′ x 4′	24.000	L.F.	.349	1.68	11.75	13.43
4′ x 5′	26.000	L.F.	.378	1.82	12.75	14.57
4′ x 6′	28.000	L.F.	.407	1.96	13.70	15.66
4′ x 8′	32.000	L.F.	.465	2.24	15.70	17.94
6′-8″ x 2′	30.000	L.F.	.436	2.10	14.70	16.80
Sill, slate, 2′ long	2.000	L.F.	.400	15.50	11.90	27.40
3′ long	3.000	L.F.	.600	23.50	17.85	41.35
4′ long	4.000	L.F.	.800	31.00	24.00	55.00
Wood, 1-5/8″ x 5-1/8″, 2′ long	2.000	L.F.	.128	5.15	4.32	9.47
3′ long	3.000	L.F.	.192	7.70	6.50	14.20
4′ long	4.000	L.F.	.256	10.30	8.65	18.95

System Description	QUAN.	UNIT	LABOR HOURS	COST EACH MAT.	COST EACH INST.	COST EACH TOTAL
Storm door, aluminum, combination, storm & screen, anodized, 2′-6″ x 6′-8″	1.000	Ea.	1.067	151.00	36.00	187.00
2′-8″ x 6′-8″	1.000	Ea.	1.143	181.00	38.50	219.50
3′-0″ x 6′-8″	1.000	Ea.	1.143	181.00	38.50	219.50
Mill finish, 2′-6″ x 6′-8″	1.000	Ea.	1.067	213.00	36.00	249.00
2′-8″ x 6′-8″	1.000	Ea.	1.143	213.00	38.50	251.50
3′-0″ x 6′-8″	1.000	Ea.	1.143	231.00	38.50	269.50
Painted, 2′-6″ x 6′-8″	1.000	Ea.	1.067	208.00	36.00	244.00
2′-8″ x 6′-8″	1.000	Ea.	1.143	194.00	38.50	232.50
3′-0″ x 6′-8″	1.000	Ea.	1.143	216.00	38.50	254.50
Wood, combination, storm & screen, crossbuck, 2′-6″ x 6′-9″	1.000	Ea.	1.455	219.00	49.00	268.00
2′-8″ x 6′-9″	1.000	Ea.	1.600	219.00	54.00	273.00
3′-0″ x 6′-9″	1.000	Ea.	1.778	226.00	60.00	286.00
Full lite, 2′-6″ x 6′-9″	1.000	Ea.	1.455	225.00	49.00	274.00
2′-8″ x 6′-9″	1.000	Ea.	1.600	225.00	54.00	279.00
3′-0″ x 6′-9″	1.000	Ea.	1.778	232.00	60.00	292.00
Windows, aluminum, combination storm & screen, basement, 1′-10″ x 1′-0″	1.000	Ea.	.533	27.00	18.00	45.00
2′-9″ x 1′-6″	1.000	Ea.	.533	29.50	18.00	47.50
3′-4″ x 2′-0″	1.000	Ea.	.533	36.00	18.00	54.00
Double hung, anodized, 2′-0″ x 3′-5″	1.000	Ea.	.533	70.50	18.00	88.50
2′-6″ x 5′-0″	1.000	Ea.	.571	94.50	19.30	113.80
4′-0″ x 6′-0″	1.000	Ea.	.640	200.00	21.50	221.50
Painted, 2′-0″ x 3′-5″	1.000	Ea.	.533	84.00	18.00	102.00
2′-6″ x 5′-0″	1.000	Ea.	.571	135.00	19.30	154.30
4′-0″ x 6′-0″	1.000	Ea.	.640	242.00	21.50	263.50
Fixed window, anodized, 4′-6″ x 4′-6″	1.000	Ea.	.640	108.00	21.50	129.50
5′-8″ x 4′-6″	1.000	Ea.	.800	122.00	27.00	149.00
Painted, 4′-6″ x 4′-6″	1.000	Ea.	.640	108.00	21.50	129.50
5′-8″ x 4′-6″	1.000	Ea.	.800	122.00	27.00	149.00

 Important: See the Reference Section for critical supporting data - Reference Nos., Crews & Location Factors

64 Shutters/Blinds Systems

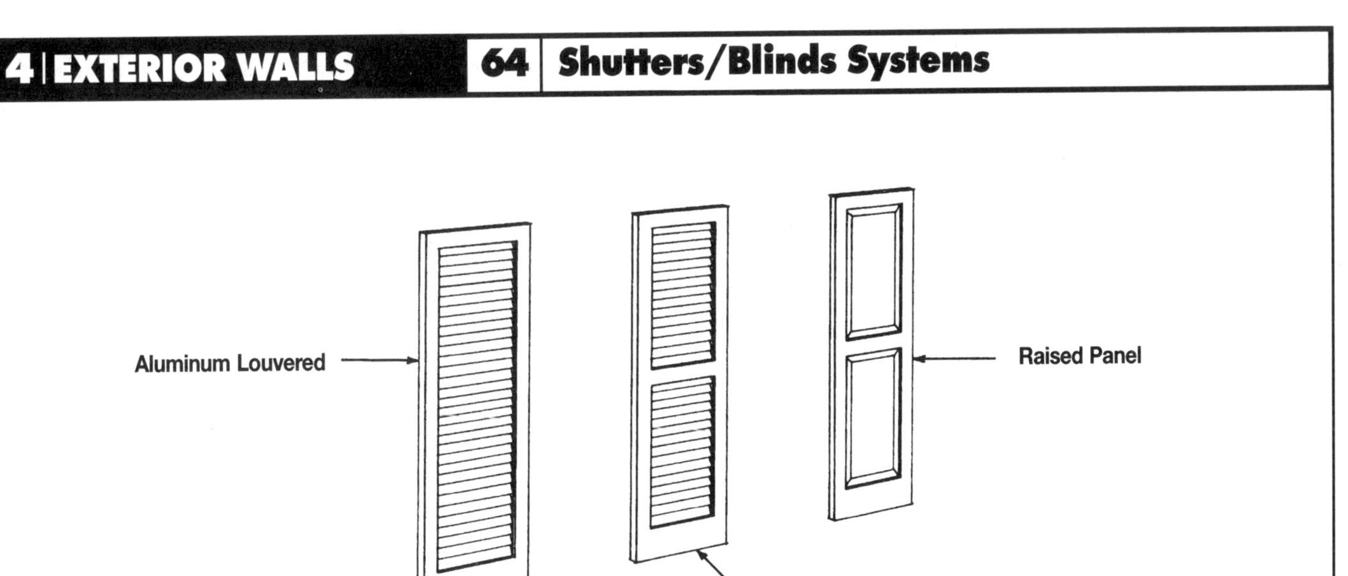

System Description	QUAN.	UNIT	LABOR HOURS	COST PER PAIR MAT.	INST.	TOTAL
Shutters, exterior blinds, aluminum, louvered, 1'-4" wide, 3"-0" long	1.000	Set	.800	32.50	27.00	59.50
4'-0" long	1.000	Set	.800	36.00	27.00	63.00
5'-4" long	1.000	Set	.800	41.50	27.00	68.50
6'-8" long	1.000	Set	.889	52.50	30.00	82.50
Wood, louvered, 1'-2" wide, 3'-3" long	1.000	Set	.800	45.00	27.00	72.00
4'-7" long	1.000	Set	.800	54.00	27.00	81.00
5'-3" long	1.000	Set	.800	58.50	27.00	85.50
1'-6" wide, 3'-3" long	1.000	Set	.800	49.50	27.00	76.50
4'-7" long	1.000	Set	.800	67.00	27.00	94.00
Polystyrene, solid raised panel, 3'-3" wide, 3'-0" long	1.000	Set	.800	138.00	27.00	165.00
3'-11" long	1.000	Set	.800	172.00	27.00	199.00
5'-3" long	1.000	Set	.800	217.00	27.00	244.00
6'-8" long	1.000	Set	.889	245.00	30.00	275.00
Polystyrene, louvered, 1'-2" wide, 3'-3" long	1.000	Set	.800	44.00	27.00	71.00
4'-7" long	1.000	Set	.800	54.50	27.00	81.50
5'-3" long	1.000	Set	.800	58.50	27.00	85.50
6'-8" long	1.000	Set	.889	95.50	30.00	125.50
Vinyl, louvered, 1'-2" wide, 4'-7" long	1.000	Set	.720	51.50	24.50	76.00
1'-4" x 6'-8" long	1.000	Set	.889	87.00	30.00	117.00

Division 5
Roofing

04 Gable End Roofing Systems

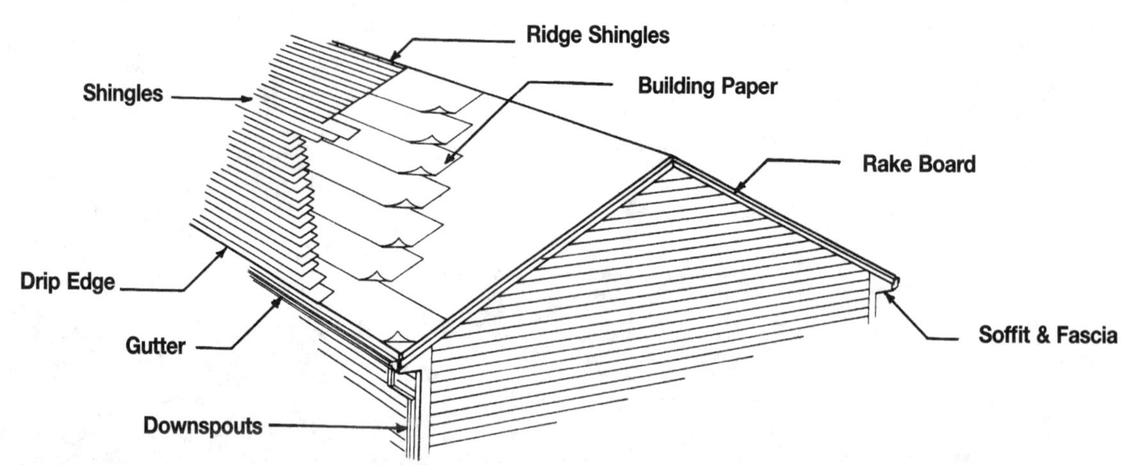

System Description	QUAN.	UNIT	LABOR HOURS	COST PER S.F. MAT.	INST.	TOTAL
ASPHALT, ROOF SHINGLES, CLASS A						
Shingles, inorganic class A, 210-235 lb./sq., 4/12 pitch	1.160	S.F.	.017	.34	.55	.89
Drip edge, metal, 5" wide	.150	L.F.	.003	.04	.10	.14
Building paper, #15 felt	1.300	S.F.	.002	.04	.06	.10
Ridge shingles, asphalt	.042	L.F.	.001	.03	.03	.06
Soffit & fascia, white painted aluminum, 1′ overhang	.083	L.F.	.012	.19	.41	.60
Rake trim, painted, 1" x 6"	.040	L.F.	.004	.04	.10	.14
Gutter, seamless, aluminum painted	.083	L.F.	.006	.10	.20	.30
Downspouts, aluminum painted	.035	L.F.	.002	.04	.06	.10
TOTAL			.047	.82	1.51	2.33
WOOD, CEDAR SHINGLES NO. 1 PERFECTIONS, 18" LONG						
Shingles, wood, cedar, No. 1 perfections, 4/12 pitch	1.160	S.F.	.035	2.17	1.18	3.35
Drip edge, metal, 5" wide	.150	L.F.	.003	.04	.10	.14
Building paper, #15 felt	1.300	S.F.	.002	.04	.06	.10
Ridge shingles, cedar	.042	L.F.	.001	.11	.04	.15
Soffit & fascia, white painted aluminum, 1′ overhang	.083	L.F.	.012	.19	.41	.60
Rake trim, painted, 1" x 6"	.040	L.F.	.004	.04	.10	.14
Gutter, seamless, aluminum, painted	.083	L.F.	.006	.10	.20	.30
Downspouts, aluminum, painted	.035	L.F.	.002	.04	.06	.10
TOTAL			.065	2.73	2.15	4.88

The prices in these systems are based on a square foot of plan area.
All quantities have been adjusted accordingly.

Description	QUAN.	UNIT	LABOR HOURS	COST PER S.F. MAT.	INST.	TOTAL

Important: See the Reference Section for critical supporting data - Reference Nos., Crews & Location Factors

Gable End Roofing Price Sheet	QUAN.	UNIT	LABOR HOURS	COST PER S.F.		
				MAT.	INST.	TOTAL
Shingles, asphalt, inorganic, class A, 210-235 lb./sq., 4/12 pitch	1.160	S.F.	.017	.34	.55	.89
8/12 pitch	1.330	S.F.	.019	.37	.60	.97
Laminated, multi-layered, 240-260 lb./sq., 4/12 pitch	1.160	S.F.	.021	.45	.68	1.13
8/12 pitch	1.330	S.F.	.023	.49	.73	1.22
Premium laminated, multi-layered, 260-300 lb./sq., 4/12 pitch	1.160	S.F.	.027	.57	.87	1.44
8/12 pitch	1.330	S.F.	.030	.62	.94	1.56
Clay tile, Spanish tile, red, 4/12 pitch	1.160	S.F.	.053	4.08	1.69	5.77
8/12 pitch	1.330	S.F.	.058	4.42	1.83	6.25
Mission tile, red, 4/12 pitch	1.160	S.F.	.083	8.35	2.65	11.00
8/12 pitch	1.330	S.F.	.090	9.05	2.87	11.92
French tile, red, 4/12 pitch	1.160	S.F.	.071	7.55	2.26	9.81
8/12 pitch	1.330	S.F.	.077	8.20	2.44	10.64
Slate, Buckingham, Virginia, black, 4/12 pitch	1.160	S.F.	.055	7.15	1.74	8.89
8/12 pitch	1.330	S.F.	.059	7.75	1.89	9.64
Vermont, black or grey, 4/12 pitch	1.160	S.F.	.055	4.68	1.74	6.42
8/12 pitch	1.330	S.F.	.059	5.05	1.89	6.94
Wood, No. 1 red cedar, 5X, 16″ long, 5″ exposure, 4/12 pitch	1.160	S.F.	.038	1.99	1.30	3.29
8/12 pitch	1.330	S.F.	.042	2.16	1.40	3.56
Fire retardant, 4/12 pitch	1.160	S.F.	.038	2.39	1.30	3.69
8/12 pitch	1.330	S.F.	.042	2.59	1.40	3.99
18″ long, No.1 perfections, 5″ exposure, 4/12 pitch	1.160	S.F.	.035	2.17	1.18	3.35
8/12 pitch	1.330	S.F.	.038	2.35	1.27	3.62
Fire retardant, 4/12 pitch	1.160	S.F.	.035	2.55	1.18	3.73
8/12 pitch	1.330	S.F.	.038	2.76	1.27	4.03
Resquared & rebutted, 18″ long, 6″ exposure, 4/12 pitch	1.160	S.F.	.032	2.63	1.08	3.71
8/12 pitch	1.330	S.F.	.035	2.85	1.17	4.02
Fire retardant, 4/12 pitch	1.160	S.F.	.032	3.01	1.08	4.09
8/12 pitch	1.330	S.F.	.035	3.26	1.17	4.43
Wood shakes hand split, 24″ long, 10″ exposure, 4/12 pitch	1.160	S.F.	.038	1.82	1.30	3.12
8/12 pitch	1.330	S.F.	.042	1.98	1.40	3.38
Fire retardant, 4/12 pitch	1.160	S.F.	.038	2.22	1.30	3.52
8/12 pitch	1.330	S.F.	.042	2.41	1.40	3.81
18″ long, 8″ exposure, 4/12 pitch	1.160	S.F.	.048	1.28	1.62	2.90
8/12 pitch	1.330	S.F.	.052	1.39	1.76	3.15
Fire retardant, 4/12 pitch	1.160	S.F.	.048	1.68	1.62	3.30
8/12 pitch	1.330	S.F.	.052	1.82	1.76	3.58
Drip edge, metal, 5″ wide	.150	L.F.	.003	.04	.10	.14
8″ wide	.150	L.F.	.003	.05	.10	.15
Building paper, #15 asphalt felt	1.300	S.F.	.002	.04	.06	.10
Ridge shingles, asphalt	.042	L.F.	.001	.03	.03	.06
Clay	.042	L.F.	.002	.40	.05	.45
Slate	.042	L.F.	.002	.39	.05	.44
Wood, shingles	.042	L.F.	.001	.11	.04	.15
Shakes	.042	L.F.	.001	.11	.04	.15
Soffit & fascia, aluminum, vented, 1′ overhang	.083	L.F.	.012	.19	.41	.60
2′ overhang	.083	L.F.	.013	.28	.45	.73
Vinyl, vented, 1′ overhang	.083	L.F.	.011	.12	.37	.49
2′ overhang	.083	L.F.	.012	.18	.41	.59
Wood, board fascia, plywood soffit, 1′ overhang	.083	L.F.	.004	.02	.11	.13
2′ overhang	.083	L.F.	.006	.03	.17	.20
Rake trim, painted, 1″ x 6″	.040	L.F.	.004	.04	.10	.14
1″ x 8″	.040	L.F.	.004	.07	.12	.19
Gutter, 5″ box, aluminum, seamless, painted	.083	L.F.	.006	.10	.20	.30
Vinyl	.083	L.F.	.006	.09	.20	.29
Downspout, 2″ x 3″, aluminum, one story house	.035	L.F.	.001	.03	.05	.08
Two story house	.060	L.F.	.003	.05	.09	.14
Vinyl, one story house	.035	L.F.	.002	.04	.06	.10
Two story house	.060	L.F.	.003	.05	.09	.14

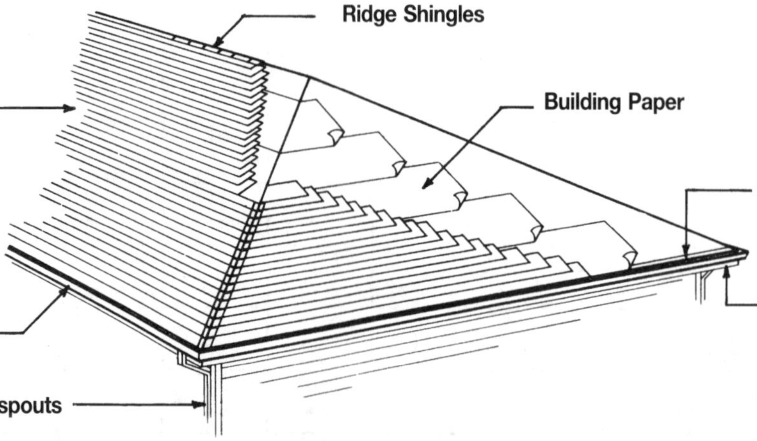

08 Hip Roof Roofing Systems

Ridge Shingles
Shingles
Building Paper
Drip Edge
Soffit & Fascia
Gutter
Downspouts

System Description	QUAN.	UNIT	LABOR HOURS	COST PER S.F.		
				MAT.	INST.	TOTAL
ASPHALT, ROOF SHINGLES, CLASS A						
Shingles, inorganic, class A, 210-235 lb./sq. 4/12 pitch	1.570	S.F.	.023	.46	.74	1.20
Drip edge, metal, 5″ wide	.122	L.F.	.002	.03	.08	.11
Building paper, #15 asphalt felt	1.800	S.F.	.002	.05	.08	.13
Ridge shingles, asphalt	.075	L.F.	.002	.06	.06	.12
Soffit & fascia, white painted aluminum, 1′ overhang	.120	L.F.	.017	.28	.59	.87
Gutter, seamless, aluminum, painted	.120	L.F.	.008	.14	.29	.43
Downspouts, aluminum, painted	.035	L.F.	.002	.04	.06	.10
TOTAL			.056	1.06	1.90	2.96
WOOD, CEDAR SHINGLES, NO. 1 PERFECTIONS, 18″ LONG						
Shingles, red cedar, No. 1 perfections, 5″ exp., 4/12 pitch	1.570	S.F.	.047	2.90	1.57	4.47
Drip edge, metal, 5″ wide	.122	L.F.	.002	.03	.08	.11
Building paper, #15 asphalt felt	1.800	S.F.	.002	.05	.08	.13
Ridge shingles, wood, cedar	.075	L.F.	.002	.20	.07	.27
Soffit & fascia, white painted aluminum, 1′ overhang	.120	L.F.	.017	.28	.59	.87
Gutter, seamless, aluminum, painted	.120	L.F.	.008	.14	.29	.43
Downspouts, aluminum, painted	.035	L.F.	.002	.04	.06	.10
TOTAL			.080	3.64	2.74	6.38

The prices in these systems are based on a square foot of plan area.
All quantities have been adjusted accordingly.

Description	QUAN.	UNIT	LABOR HOURS	COST PER S.F.		
				MAT.	INST.	TOTAL

 Important: See the Reference Section for critical supporting data - Reference Nos., Crews & Location Factors

Hip Roof - Roofing Price Sheet	QUAN.	UNIT	LABOR HOURS	COST PER S.F.		
				MAT.	INST.	TOTAL
Shingles, asphalt, inorganic, class A, 210-235 lb./sq., 4/12 pitch	1.570	S.F.	.023	.46	.74	1.20
8/12 pitch	1.850	S.F.	.028	.54	.87	1.41
Laminated, multi-layered, 240-260 lb./sq., 4/12 pitch	1.570	S.F.	.028	.60	.90	1.50
8/12 pitch	1.850	S.F.	.034	.71	1.07	1.78
Prem. laminated, multi-layered, 260-300 lb./sq., 4/12 pitch	1.570	S.F.	.037	.76	1.16	1.92
8/12 pitch	1.850	S.F.	.043	.90	1.38	2.28
Clay tile, Spanish tile, red, 4/12 pitch	1.570	S.F.	.071	5.45	2.26	7.71
8/12 pitch	1.850	S.F.	.084	6.45	2.68	9.13
Mission tile, red, 4/12 pitch	1.570	S.F.	.111	11.10	3.54	14.64
8/12 pitch	1.850	S.F.	.132	13.20	4.20	17.40
French tile, red, 4/12 pitch	1.570	S.F.	.095	10.10	3.01	13.11
8/12 pitch	1.850	S.F.	.113	11.95	3.57	15.52
Slate, Buckingham, Virginia, black, 4/12 pitch	1.570	S.F.	.073	9.50	2.32	11.82
8/12 pitch	1.850	S.F.	.087	11.30	2.76	14.06
Vermont, black or grey, 4/12 pitch	1.570	S.F.	.073	6.25	2.32	8.57
8/12 pitch	1.850	S.F.	.087	7.40	2.76	10.16
Wood, red cedar, No.1 5X, 16″ long, 5″ exposure, 4/12 pitch	1.570	S.F.	.051	2.66	1.73	4.39
8/12 pitch	1.850	S.F.	.061	3.15	2.05	5.20
Fire retardant, 4/12 pitch	1.570	S.F.	.051	3.19	1.73	4.92
8/12 pitch	1.850	S.F.	.061	3.78	2.05	5.83
18″ long, No.1 perfections, 5″ exposure, 4/12 pitch	1.570	S.F.	.047	2.90	1.57	4.47
8/12 pitch	1.850	S.F.	.055	3.44	1.86	5.30
Fire retardant, 4/12 pitch	1.570	S.F.	.047	3.40	1.57	4.97
8/12 pitch	1.850	S.F.	.055	4.04	1.86	5.90
Resquared & rebutted, 18″ long, 6″ exposure, 4/12 pitch	1.570	S.F.	.043	3.50	1.44	4.94
8/12 pitch	1.850	S.F.	.051	4.16	1.71	5.87
Fire retardant, 4/12 pitch	1.570	S.F.	.043	4.00	1.44	5.44
8/12 pitch	1.850	S.F.	.051	4.76	1.71	6.47
Wood shakes hand split, 24″ long, 10″ exposure, 4/12 pitch	1.570	S.F.	.051	2.43	1.73	4.16
8/12 pitch	1.850	S.F.	.061	2.89	2.05	4.94
Fire retardant, 4/12 pitch	1.570	S.F.	.051	2.96	1.73	4.69
8/12 pitch	1.850	S.F.	.061	3.52	2.05	5.57
18″ long, 8″ exposure, 4/12 pitch	1.570	S.F.	.064	1.71	2.16	3.87
8/12 pitch	1.850	S.F.	.076	2.03	2.57	4.60
Fire retardant, 4/12 pitch	1.570	S.F.	.064	2.24	2.16	4.40
8/12 pitch	1.850	S.F.	.076	2.66	2.57	5.23
Drip edge, metal, 5″ wide	.122	L.F.	.002	.03	.08	.11
8″ wide	.122	L.F.	.002	.04	.08	.12
Building paper, #15 asphalt felt	1.800	S.F.	.002	.05	.08	.13
Ridge shingles, asphalt	.075	L.F.	.002	.06	.06	.12
Clay	.075	L.F.	.003	.71	.10	.81
Slate	.075	L.F.	.003	.70	.10	.80
Wood, shingles	.075	L.F.	.002	.20	.07	.27
Shakes	.075	L.F.	.002	.20	.07	.27
Soffit & fascia, aluminum, vented, 1′ overhang	.120	L.F.	.017	.28	.59	.87
2′ overhang	.120	L.F.	.019	.41	.65	1.06
Vinyl, vented, 1′ overhang	.120	L.F.	.016	.18	.54	.72
2′ overhang	.120	L.F.	.017	.26	.59	.85
Wood, board fascia, plywood soffit, 1′ overhang	.120	L.F.	.004	.02	.11	.13
2′ overhang	.120	L.F.	.006	.03	.17	.20
Gutter, 5″ box, aluminum, seamless, painted	.120	L.F.	.008	.14	.29	.43
Vinyl	.120	L.F.	.009	.13	.29	.42
Downspout, 2″ x 3″, aluminum, one story house	.035	L.F.	.002	.04	.06	.10
Two story house	.060	L.F.	.003	.05	.09	.14
Vinyl, one story house	.035	L.F.	.001	.03	.05	.08
Two story house	.060	L.F.	.003	.05	.09	.14

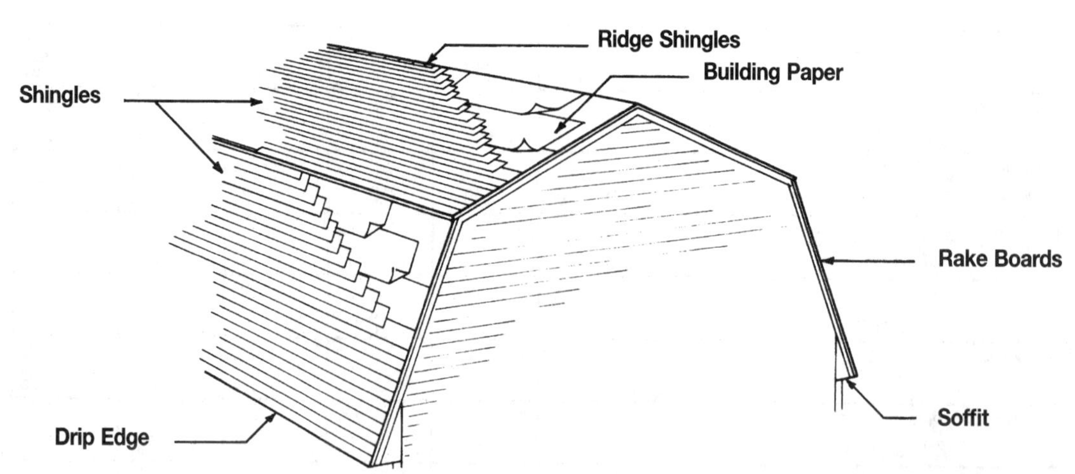

System Description	QUAN.	UNIT	LABOR HOURS	COST PER S.F.		
				MAT.	INST.	TOTAL
ASPHALT, ROOF SHINGLES, CLASS A						
Shingles, asphalt, inorganic, class A, 210-235 lb./sq.	1.450	S.F.	.022	.43	.69	1.12
Drip edge, metal, 5″ wide	.146	L.F.	.003	.04	.10	.14
Building paper, #15 asphalt felt	1.500	S.F.	.002	.04	.07	.11
Ridge shingles, asphalt	.042	L.F.	.001	.03	.03	.06
Soffit & fascia, painted aluminum, 1′ overhang	.083	L.F.	.012	.19	.41	.60
Rake, trim, painted, 1″ x 6″	.063	L.F.	.006	.07	.17	.24
Gutter, seamless, alumunum, painted	.083	L.F.	.006	.10	.20	.30
Downspouts, aluminum, painted	.042	L.F.	.002	.05	.07	.12
TOTAL			.054	.95	1.74	2.69
WOOD, CEDAR SHINGLES, NO. 1 PERFECTIONS, 18″ LONG						
Shingles, wood, red cedar, No. 1 perfections, 5″ exposure	1.450	S.F.	.044	2.72	1.47	4.19
Drip edge, metal, 5″ wide	.146	L.F.	.003	.04	.10	.14
Building paper, #15 asphalt felt	1.500	S.F.	.002	.04	.07	.11
Ridge shingles, wood	.042	L.F.	.001	.11	.04	.15
Soffit & fascia, white painted aluminum, 1′ overhang	.083	L.F.	.012	.19	.41	.60
Rake, trim, painted, 1″ x 6″	.063	L.F.	.004	.06	.13	.19
Gutter, seamless, aluminum, painted	.083	L.F.	.006	.10	.20	.30
Downspouts, aluminum, painted	.042	L.F.	.002	.05	.07	.12
TOTAL			.074	3.31	2.49	5.80

The prices in this system are based on a square foot of plan area.
All quantities have been adjusted accordingly.

Description	QUAN.	UNIT	LABOR HOURS	COST PER S.F.		
				MAT.	INST.	TOTAL

Gambrel Roofing Price Sheet	QUAN.	UNIT	LABOR HOURS	COST PER S.F.		
				MAT.	INST.	TOTAL
Shingles, asphalt, standard, inorganic, class A, 210-235 lb./sq.	1.450	S.F.	.022	.43	.69	1.12
Laminated, multi-layered, 240-260 lb./sq.	1.450	S.F.	.027	.56	.85	1.41
Premium laminated, multi-layered, 260-300 lb./sq.	1.450	S.F.	.034	.71	1.09	1.80
Slate, Buckingham, Virginia, black	1.450	S.F.	.069	8.95	2.18	11.13
Vermont, black or grey	1.450	S.F.	.069	5.85	2.18	8.03
Wood, red cedar, No.1 5X, 16″ long, 5″ exposure, plain	1.450	S.F.	.048	2.49	1.62	4.11
Fire retardant	1.450	S.F.	.048	2.99	1.62	4.61
18″ long, No.1 perfections, 6″ exposure, plain	1.450	S.F.	.044	2.72	1.47	4.19
Fire retardant	1.450	S.F.	.044	3.19	1.47	4.66
Resquared & rebutted, 18″ long, 6″ exposure, plain	1.450	S.F.	.040	3.29	1.35	4.64
Fire retardant	1.450	S.F.	.040	3.75	1.35	5.10
Shakes, hand split, 24″ long, 10″ exposure, plain	1.450	S.F.	.048	2.28	1.62	3.90
Fire retardant	1.450	S.F.	.048	2.78	1.62	4.40
18″ long, 8″ exposure, plain	1.450	S.F.	.060	1.61	2.03	3.64
Fire retardant	1.450	S.F.	.060	2.11	2.03	4.14
Drip edge, metal, 5″ wide	.146	L.F.	.003	.04	.10	.14
8″ wide	.146	L.F.	.003	.05	.10	.15
Building paper, #15 asphalt felt	1.500	S.F.	.002	.04	.07	.11
Ridge shingles, asphalt	.042	L.F.	.001	.03	.03	.06
Slate	.042	L.F.	.002	.39	.05	.44
Wood, shingles	.042	L.F.	.001	.11	.04	.15
Shakes	.042	L.F.	.001	.11	.04	.15
Soffit & fascia, aluminum, vented, 1′ overhang	.083	L.F.	.012	.19	.41	.60
2′ overhang	.083	L.F.	.013	.28	.45	.73
Vinyl vented, 1′ overhang	.083	L.F.	.011	.12	.37	.49
2′ overhang	.083	L.F.	.012	.18	.41	.59
Wood board fascia, plywood soffit, 1′ overhang	.083	L.F.	.004	.02	.11	.13
2′ overhang	.083	L.F.	.006	.03	.17	.20
Rake trim, painted, 1″ x 6″	.063	L.F.	.006	.07	.17	.24
1″ x 8″	.063	L.F.	.007	.09	.22	.31
Gutter, 5″ box, aluminum, seamless, painted	.083	L.F.	.006	.10	.20	.30
Vinyl	.083	L.F.	.006	.09	.20	.29
Downspout 2″ x 3″, aluminum, one story house	.042	L.F.	.002	.03	.06	.09
Two story house	.070	L.F.	.003	.05	.11	.16
Vinyl, one story house	.042	L.F.	.002	.03	.06	.09
Two story house	.070	L.F.	.003	.05	.11	.16

16 Mansard Roofing Systems

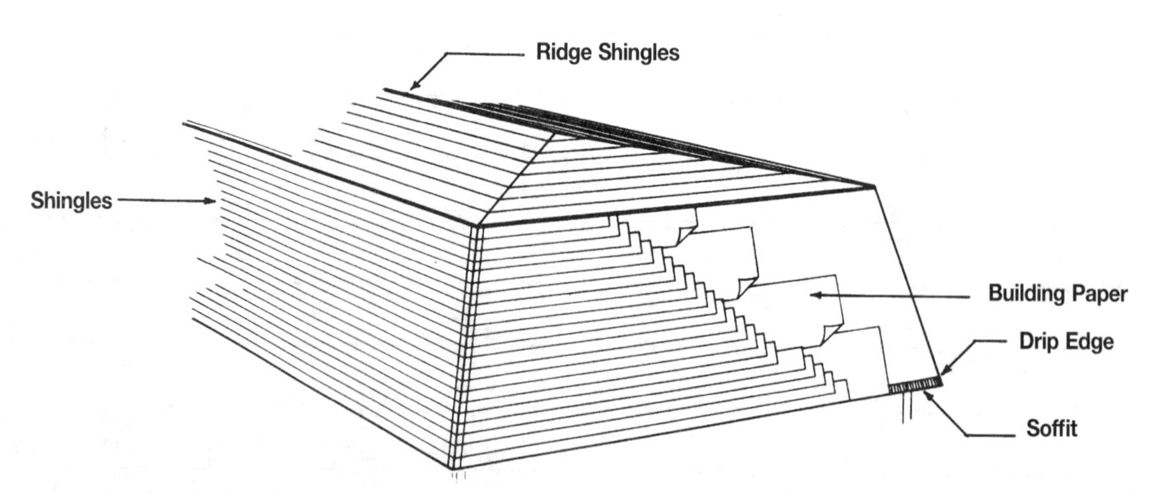

System Description	QUAN.	UNIT	LABOR HOURS	COST PER S.F.		
				MAT.	INST.	TOTAL
ASPHALT, ROOF SHINGLES, CLASS A						
Shingles, standard inorganic class A 210-235 lb./sq.	2.210	S.F.	.032	.63	1.01	1.64
Drip edge, metal, 5″ wide	.122	L.F.	.002	.03	.08	.11
Building paper, #15 asphalt felt	2.300	S.F.	.003	.07	.10	.17
Ridge shingles, asphalt	.090	L.F.	.002	.07	.07	.14
Soffit & fascia, white painted aluminum, 1′ overhang	.122	L.F.	.018	.28	.60	.88
Gutter, seamless, aluminum, painted	.122	L.F.	.008	.14	.30	.44
Downspouts, aluminum, painted	.042	L.F.	.002	.05	.07	.12
TOTAL			.067	1.27	2.23	3.50
WOOD, CEDAR SHINGLES, NO. 1 PERFECTIONS, 18″ LONG						
Shingles, wood, red cedar, No. 1 perfections, 5″ exposure	2.210	S.F.	.064	3.98	2.16	6.14
Drip edge, metal, 5″ wide	.122	L.F.	.002	.03	.08	.11
Building paper, #15 asphalt felt	2.300	S.F.	.003	.07	.10	.17
Ridge shingles, wood	.090	L.F.	.003	.24	.09	.33
Soffit & fascia, white painted aluminum, 1′ overhang	.122	L.F.	.018	.28	.60	.88
Gutter, seamless, aluminum, painted	.122	L.F.	.008	.14	.30	.44
Downspouts, aluminum, painted	.042	L.F.	.002	.05	.07	.12
TOTAL			.100	4.79	3.40	8.19

The prices in these systems are based on a square foot of plan area.
All quantities have been adjusted accordingly.

Description	QUAN.	UNIT	LABOR HOURS	COST PER S.F.		
				MAT.	INST.	TOTAL

Mansard Roofing Price Sheet	QUAN.	UNIT	LABOR HOURS	COST PER S.F.		
				MAT.	INST.	TOTAL
Shingles, asphalt, standard, inorganic, class A, 210-235 lb./sq.	2.210	S.F.	.032	.63	1.01	1.64
Laminated, multi-layered, 240-260 lb./sq.	2.210	S.F.	.039	.83	1.24	2.07
Premium laminated, multi-layered, 260-300 lb./sq.	2.210	S.F.	.050	1.05	1.60	2.65
Slate Buckingham, Virginia, black	2.210	S.F.	.101	13.10	3.19	16.29
Vermont, black or grey	2.210	S.F.	.101	8.60	3.19	11.79
Wood, red cedar, No.1 5X, 16″ long, 5″ exposure, plain	2.210	S.F.	.070	3.65	2.38	6.03
Fire retardant	2.210	S.F.	.070	4.38	2.38	6.76
18″ long, No.1 perfections 6″ exposure, plain	2.210	S.F.	.064	3.98	2.16	6.14
Fire retardant	2.210	S.F.	.064	4.67	2.16	6.83
Resquared & rebutted, 18″ long, 6″ exposure, plain	2.210	S.F.	.059	4.82	1.98	6.80
Fire retardant	2.210	S.F.	.059	5.50	1.98	7.48
Shakes, hand split, 24″ long 10″ exposure, plain	2.210	S.F.	.070	3.34	2.38	5.72
Fire retardant	2.210	S.F.	.070	4.07	2.38	6.45
18″ long, 8″ exposure, plain	2.210	S.F.	.088	2.35	2.97	5.32
Fire retardant	2.210	S.F.	.088	3.08	2.97	6.05
Drip edge, metal, 5″ wide	.122	S.F.	.002	.03	.08	.11
8″ wide	.122	S.F.	.002	.04	.08	.12
Building paper, #15 asphalt felt	2.300	S.F.	.003	.07	.10	.17
Ridge shingles, asphalt	.090	L.F.	.002	.07	.07	.14
Slate	.090	L.F.	.004	.84	.11	.95
Wood, shingles	.090	L.F.	.003	.24	.09	.33
Shakes	.090	L.F.	.003	.24	.09	.33
Soffit & fascia, aluminum vented, 1′ overhang	.122	L.F.	.018	.28	.60	.88
2′ overhang	.122	L.F.	.020	.42	.66	1.08
Vinyl vented, 1′ overhang	.122	L.F.	.016	.18	.55	.73
2′ overhang	.122	L.F.	.018	.26	.60	.86
Wood board fascia, plywood soffit, 1′ overhang	.122	L.F.	.013	.31	.42	.73
2′ overhang	.122	L.F.	.019	.42	.63	1.05
Gutter, 5″ box, aluminum, seamless, painted	.122	L.F.	.008	.14	.30	.44
Vinyl	.122	L.F.	.009	.13	.30	.43
Downspout 2″ x 3″, aluminum, one story house	.042	L.F.	.002	.03	.06	.09
Two story house	.070	L.F.	.003	.05	.11	.16
Vinyl, one story house	.042	L.F.	.002	.03	.06	.09
Two story house	.070	L.F.	.003	.05	.11	.16

20 Shed Roofing Systems

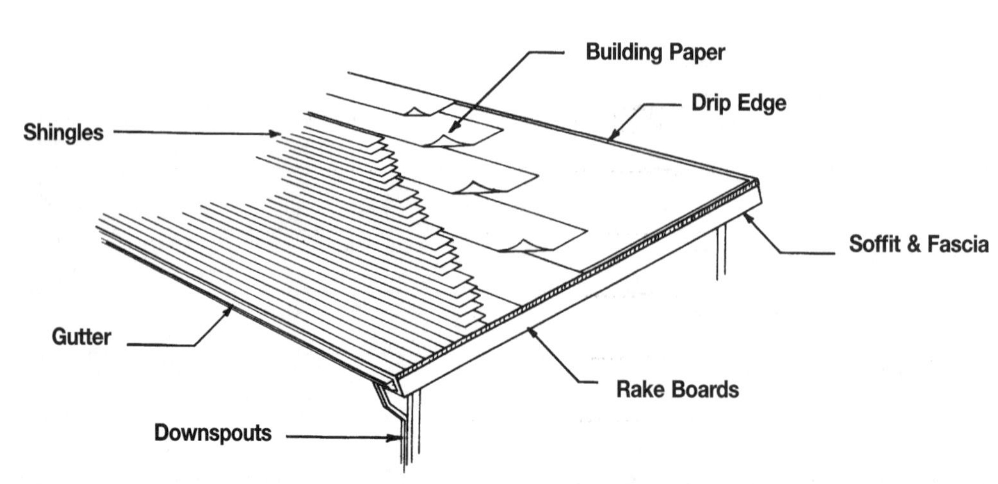

System Description	QUAN.	UNIT	LABOR HOURS	COST PER S.F. MAT.	COST PER S.F. INST.	COST PER S.F. TOTAL
ASPHALT, ROOF SHINGLES, CLASS A						
Shingles, inorganic class A 210-235 lb./sq. 4/12 pitch	1.230	S.F.	.019	.37	.60	.97
Drip edge, metal, 5″ wide	.100	L.F.	.002	.02	.07	.09
Building paper, #15 asphalt felt	1.300	S.F.	.002	.04	.06	.10
Soffit & fascia, white painted aluminum, 1′ overhang	.080	L.F.	.012	.19	.39	.58
Rake trim, painted, 1″ x 6″	.043	L.F.	.004	.04	.12	.16
Gutter, seamless, aluminum, painted	.040	L.F.	.003	.05	.10	.15
Downspouts, painted aluminum	.020	L.F.	.001	.02	.03	.05
TOTAL			.043	.73	1.37	2.10
WOOD, CEDAR SHINGLES, NO. 1 PERFECTIONS, 18″ LONG						
Shingles, red cedar, No. 1 perfections, 5″ exp., 4/12 pitch	1.230	S.F.	.035	2.17	1.18	3.35
Drip edge, metal, 5″ wide	.100	L.F.	.002	.02	.07	.09
Building paper, #15 asphalt felt	1.300	S.F.	.002	.04	.06	.10
Soffit & fascia, white painted aluminum, 1′ overhang	.080	L.F.	.012	.19	.39	.58
Rake trim, painted, 1″ x 6″	.043	L.F.	.003	.04	.09	.13
Gutter, seamless, aluminum, painted	.040	L.F.	.003	.05	.10	.15
Downspouts, painted aluminum	.020	L.F.	.001	.02	.03	.05
TOTAL			.058	2.53	1.92	4.45

The prices in these systems are based on a square foot of plan area.
All quantities have been adjusted accordingly.

Description	QUAN.	UNIT	LABOR HOURS	COST PER S.F. MAT.	COST PER S.F. INST.	COST PER S.F. TOTAL

Shed Roofing Price Sheet	QUAN.	UNIT	LABOR HOURS	COST PER S.F. MAT.	INST.	TOTAL
Shingles, asphalt, inorganic, class A, 210-235 lb./sq., 4/12 pitch	1.230	S.F.	.017	.34	.55	.89
8/12 pitch	1.330	S.F.	.019	.37	.60	.97
Laminated, multi-layered, 240-260 lb./sq. 4/12 pitch	1.230	S.F.	.021	.45	.68	1.13
8/12 pitch	1.330	S.F.	.023	.49	.73	1.22
Premium laminated, multi-layered, 260-300 lb./sq. 4/12 pitch	1.230	S.F.	.027	.57	.87	1.44
8/12 pitch	1.330	S.F.	.030	.62	.94	1.56
Clay tile, Spanish tile, red, 4/12 pitch	1.230	S.F.	.053	4.08	1.69	5.77
8/12 pitch	1.330	S.F.	.058	4.42	1.83	6.25
Mission tile, red, 4/12 pitch	1.230	S.F.	.083	8.35	2.65	11.00
8/12 pitch	1.330	S.F.	.090	9.05	2.87	11.92
French tile, red, 4/12 pitch	1.230	S.F.	.071	7.55	2.26	9.81
8/12 pitch	1.330	S.F.	.077	8.20	2.44	10.64
Slate, Buckingham, Virginia, black, 4/12 pitch	1.230	S.F.	.055	7.15	1.74	8.89
8/12 pitch	1.330	S.F.	.059	7.75	1.89	9.64
Vermont, black or grey, 4/12 pitch	1.230	S.F.	.055	4.68	1.74	6.42
8/12 pitch	1.330	S.F.	.059	5.05	1.89	6.94
Wood, red cedar, No.1 5X, 16″ long, 5″ exposure, 4/12 pitch	1.230	S.F.	.038	1.99	1.30	3.29
8/12 pitch	1.330	S.F.	.042	2.16	1.40	3.56
Fire retardant, 4/12 pitch	1.230	S.F.	.038	2.39	1.30	3.69
8/12 pitch	1.330	S.F.	.042	2.59	1.40	3.99
18″ long, 6″ exposure, 4/12 pitch	1.230	S.F.	.035	2.17	1.18	3.35
8/12 pitch	1.330	S.F.	.038	2.35	1.27	3.62
Fire retardant, 4/12 pitch	1.230	S.F.	.035	2.55	1.18	3.73
8/12 pitch	1.330	S.F.	.038	2.76	1.27	4.03
Resquared & rebutted, 18″ long, 6″ exposure, 4/12 pitch	1.230	S.F.	.032	2.63	1.08	3.71
8/12 pitch	1.330	S.F.	.035	2.85	1.17	4.02
Fire retardant, 4/12 pitch	1.230	S.F.	.032	3.01	1.08	4.09
8/12 pitch	1.330	S.F.	.035	3.26	1.17	4.43
Wood shakes, hand split, 24″ long, 10″ exposure, 4/12 pitch	1.230	S.F.	.038	1.82	1.30	3.12
8/12 pitch	1.330	S.F.	.042	1.98	1.40	3.38
Fire retardant, 4/12 pitch	1.230	S.F.	.038	2.22	1.30	3.52
8/12 pitch	1.330	S.F.	.042	2.41	1.40	3.81
18″ long, 8″ exposure, 4/12 pitch	1.230	S.F.	.048	1.28	1.62	2.90
8/12 pitch	1.330	S.F.	.052	1.39	1.76	3.15
Fire retardant, 4/12 pitch	1.230	S.F.	.048	1.68	1.62	3.30
8/12 pitch	1.330	S.F.	.052	1.82	1.76	3.58
Drip edge, metal, 5″ wide	.100	L.F.	.002	.02	.07	.09
8″ wide	.100	L.F.	.002	.04	.07	.11
Building paper, #15 asphalt felt	1.300	S.F.	.002	.04	.06	.10
Soffit & fascia, aluminum vented, 1′ overhang	.080	L.F.	.012	.19	.39	.58
2′ overhang	.080	L.F.	.013	.27	.43	.70
Vinyl vented, 1′ overhang	.080	L.F.	.011	.12	.36	.48
2′ overhang	.080	L.F.	.012	.17	.39	.56
Wood board fascia, plywood soffit, 1′ overhang	.080	L.F.	.010	.21	.31	.52
2′ overhang	.080	L.F.	.014	.28	.47	.75
Rake, trim, painted, 1″ x 6″	.043	L.F.	.004	.04	.12	.16
1″ x 8″	.043	L.F.	.004	.04	.12	.16
Gutter, 5″ box, aluminum, seamless, painted	.040	L.F.	.003	.05	.10	.15
Vinyl	.040	L.F.	.003	.04	.10	.14
Downspout 2″ x 3″, aluminum, one story house	.020	L.F.	.001	.02	.03	.05
Two story house	.020	L.F.	.001	.03	.05	.08
Vinyl, one story house	.020	L.F.	.001	.02	.03	.05
Two story house	.020	L.F.	.001	.03	.05	.08

24 Gable Dormer Roofing Systems

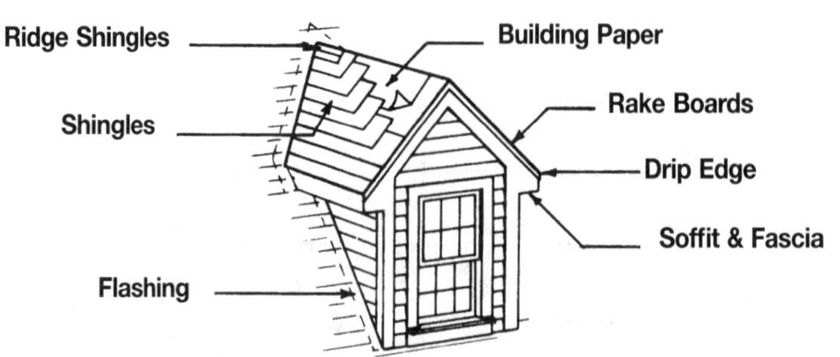

System Description	QUAN.	UNIT	LABOR HOURS	COST PER S.F.		
				MAT.	INST.	TOTAL
ASPHALT, ROOF SHINGLES, CLASS A						
Shingles, standard inorganic class A 210-235 lb./sq	1.400	S.F.	.020	.40	.64	1.04
Drip edge, metal, 5" wide	.220	L.F.	.004	.05	.15	.20
Building paper, #15 asphalt felt	1.500	S.F.	.002	.04	.07	.11
Ridge shingles, asphalt	.280	L.F.	.007	.22	.22	.44
Soffit & fascia, aluminum, vented	.220	L.F.	.032	.51	1.08	1.59
Flashing, aluminum, mill finish, .013" thick	1.500	S.F.	.083	.56	3.00	3.56
TOTAL			.148	1.78	5.16	6.94
WOOD, CEDAR, NO. 1 PERFECTIONS						
Shingles, red cedar, No.1 perfections, 18" long, 5" exp.	1.400	S.F.	.041	2.53	1.37	3.90
Drip edge, metal, 5" wide	.220	L.F.	.004	.05	.15	.20
Building paper, #15 asphalt felt	1.500	S.F.	.002	.04	.07	.11
Ridge shingles, wood	.280	L.F.	.008	.74	.27	1.01
Soffit & fascia, aluminum, vented	.220	L.F.	.032	.51	1.08	1.59
Flashing, aluminum, mill finish, .013" thick	1.500	S.F.	.083	.56	3.00	3.56
TOTAL			.170	4.43	5.94	10.37
SLATE, BUCKINGHAM, BLACK						
Shingles, Buckingham, Virginia, black	1.400	S.F.	.064	8.33	2.03	10.36
Drip edge, metal, 5" wide	.220	L.F.	.004	.05	.15	.20
Building paper, #15 asphalt felt	1.500	S.F.	.002	.04	.07	.11
Ridge shingles, slate	.280	L.F.	.011	2.62	.36	2.98
Soffit & fascia, aluminum, vented	.220	L.F.	.032	.51	1.08	1.59
Flashing, copper, 16 oz.	1.500	S.F.	.104	5.03	3.80	8.83
TOTAL			.217	16.58	7.49	24.07

The prices in these systems are based on a square foot of plan area under the dormer roof.

Description	QUAN.	UNIT	LABOR HOURS	COST PER S.F.		
				MAT.	INST.	TOTAL

Important: See the Reference Section for critical supporting data - Reference Nos., Crews & Location Factors

Gable Dormer Roofing Price Sheet	QUAN.	UNIT	LABOR HOURS	COST PER S.F. MAT.	INST.	TOTAL
Shingles, asphalt, standard, inorganic, class A, 210-235 lb./sq.	1.400	S.F.	.020	.40	.64	1.04
Laminated, multi-layered, 240-260 lb./sq.	1.400	S.F.	.025	.53	.79	1.32
Premium laminated, multi-layered, 260-300 lb./sq.	1.400	S.F.	.032	.67	1.02	1.69
Clay tile, Spanish tile, red	1.400	S.F.	.062	4.76	1.97	6.73
Mission tile, red	1.400	S.F.	.097	9.75	3.09	12.84
French tile, red	1.400	S.F.	.083	8.80	2.63	11.43
Slate Buckingham, Virginia, black	1.400	S.F.	.064	8.35	2.03	10.38
Vermont, black or grey	1.400	S.F.	.064	5.45	2.03	7.48
Wood, red cedar, No.1 5X, 16″ long, 5″ exposure	1.400	S.F.	.045	2.32	1.51	3.83
Fire retardant	1.400	S.F.	.045	2.78	1.51	4.29
18″ long, No.1 perfections, 5″ exposure	1.400	S.F.	.041	2.53	1.37	3.90
Fire retardant	1.400	S.F.	.041	2.97	1.37	4.34
Resquared & rebutted, 18″ long, 5″ exposure	1.400	S.F.	.037	3.07	1.26	4.33
Fire retardant	1.400	S.F.	.037	3.51	1.26	4.77
Shakes hand split, 24″ long, 10″ exposure	1.400	S.F.	.045	2.13	1.51	3.64
Fire retardant	1.400	S.F.	.045	2.59	1.51	4.10
18″ long, 8″ exposure	1.400	S.F.	.056	1.50	1.89	3.39
Fire retardant	1.400	S.F.	.056	1.96	1.89	3.85
Drip edge, metal, 5″ wide	.220	L.F.	.004	.05	.15	.20
8″ wide	.220	L.F.	.004	.08	.15	.23
Building paper, #15 asphalt felt	1.500	S.F.	.002	.04	.07	.11
Clay	.280	L.F.	.011	2.66	.36	3.02
Slate	.280	L.F.	.011	2.62	.36	2.98
Wood	.280	L.F.	.008	.74	.27	1.01
Soffit & fascia, aluminum, vented	.220	L.F.	.032	.51	1.08	1.59
Vinyl, vented	.220	L.F.	.029	.33	.99	1.32
Wood, board fascia, plywood soffit	.220	L.F.	.026	.58	.83	1.41
Flashing, aluminum, .013″ thick	1.500	S.F.	.083	.56	3.00	3.56
.032″ thick	1.500	S.F.	.083	1.64	3.00	4.64
.040″ thick	1.500	S.F.	.083	2.25	3.00	5.25
.050″ thick	1.500	S.F.	.083	2.97	3.00	5.97
Copper, 16 oz.	1.500	S.F.	.104	5.05	3.80	8.85
20 oz.	1.500	S.F.	.109	6.30	3.96	10.26
24 oz.	1.500	S.F.	.114	7.60	4.16	11.76
32 oz.	1.500	S.F.	.120	10.05	4.35	14.40

28 Shed Dormer Roofing Systems

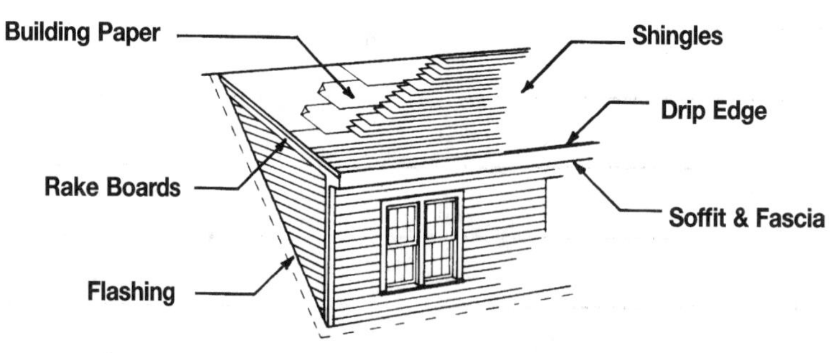

System Description	QUAN.	UNIT	LABOR HOURS	COST PER S.F. MAT.	COST PER S.F. INST.	COST PER S.F. TOTAL
ASPHALT, ROOF SHINGLES, CLASS A						
Shingles, standard inorganic class A 210-235 lb./sq.	1.100	S.F.	.016	.31	.51	.82
Drip edge, aluminum, 5″ wide	.250	L.F.	.005	.06	.17	.23
Building paper, #15 asphalt felt	1.200	S.F.	.002	.03	.05	.08
Soffit & fascia, aluminum, vented, 1′ overhang	.250	L.F.	.036	.58	1.23	1.81
Flashing, aluminum, mill finish, 0.013″ thick	.800	L.F.	.044	.30	1.60	1.90
TOTAL			.103	1.28	3.56	4.84
WOOD, CEDAR, NO. 1 PERFECTIONS, 18″ LONG						
Shingles, wood, red cedar, #1 perfections, 5″ exposure	1.100	S.F.	.032	1.99	1.08	3.07
Drip edge, aluminum, 5″ wide	.250	L.F.	.005	.06	.17	.23
Building paper, #15 asphalt felt	1.200	S.F.	.002	.03	.05	.08
Soffit & fascia, aluminum, vented, 1′ overhang	.250	L.F.	.036	.58	1.23	1.81
Flashing, aluminum, mill finish, 0.013″ thick	.800	L.F.	.044	.30	1.60	1.90
TOTAL			.119	2.96	4.13	7.09
SLATE, BUCKINGHAM, BLACK						
Shingles, slate, Buckingham, black	1.100	S.F.	.050	6.55	1.60	8.15
Drip edge, aluminum, 5″ wide	.250	L.F.	.005	.06	.17	.23
Building paper, #15 asphalt felt	1.200	S.F.	.002	.03	.05	.08
Soffit & fascia, aluminum, vented, 1′ overhang	.250	L.F.	.036	.58	1.23	1.81
Flashing, copper, 16 oz.	.800	L.F.	.056	2.68	2.02	4.70
TOTAL			.149	9.90	5.07	14.97

The prices in this system are based on a square foot of plan area under the dormer roof.

Description	QUAN.	UNIT	LABOR HOURS	COST PER S.F. MAT.	COST PER S.F. INST.	COST PER S.F. TOTAL

Important: See the Reference Section for critical supporting data - Reference Nos., Crews & Location Factors

Shed Dormer Roofing Price Sheet

	QUAN.	UNIT	LABOR HOURS	COST PER S.F. MAT.	INST.	TOTAL
Shingles, asphalt, standard, inorganic, class A, 210-235 lb./sq.	1.100	S.F.	.016	.31	.51	.82
Laminated, multi-layered, 240-260 lb./sq.	1.100	S.F.	.020	.41	.62	1.03
Premium laminated, multi-layered, 260-300 lb./sq.	1.100	S.F.	.025	.52	.80	1.32
Clay tile, Spanish tile, red	1.100	S.F.	.049	3.74	1.55	5.29
Mission tile, red	1.100	S.F.	.077	7.65	2.43	10.08
French tile, red	1.100	S.F.	.065	6.95	2.07	9.02
Slate Buckingham, Virginia, black	1.100	S.F.	.050	6.55	1.60	8.15
Vermont, black or grey	1.100	S.F.	.050	4.29	1.60	5.89
Wood, red cedar, No. 1 5X, 16″ long, 5″ exposure	1.100	S.F.	.035	1.83	1.19	3.02
Fire retardant	1.100	S.F.	.035	2.19	1.19	3.38
18″ long, No.1 perfections, 5″ exposure	1.100	S.F.	.032	1.99	1.08	3.07
Fire retardant	1.100	S.F.	.032	2.34	1.08	3.42
Resquared & rebutted, 18″ long, 5″ exposure	1.100	S.F.	.029	2.41	.99	3.40
Fire retardant	1.100	S.F.	.029	2.76	.99	3.75
Shakes hand split, 24″ long, 10″ exposure	1.100	S.F.	.035	1.67	1.19	2.86
Fire retardant	1.100	S.F.	.035	2.03	1.19	3.22
18″ long, 8″ exposure	1.100	S.F.	.044	1.18	1.49	2.67
Fire retardant	1.100	S.F.	.044	1.54	1.49	3.03
Drip edge, metal, 5″ wide	.250	L.F.	.005	.06	.17	.23
8″ wide	.250	L.F.	.005	.09	.17	.26
Building paper, #15 asphalt felt	1.200	S.F.	.002	.03	.05	.08
Soffit & fascia, aluminum, vented	.250	L.F.	.036	.58	1.23	1.81
Vinyl, vented	.250	L.F.	.033	.37	1.13	1.50
Wood, board fascia, plywood soffit	.250	L.F.	.030	.66	.96	1.62
Flashing, aluminum, .013″ thick	.800	L.F.	.044	.30	1.60	1.90
.032″ thick	.800	L.F.	.044	.87	1.60	2.47
.040″ thick	.800	L.F.	.044	1.20	1.60	2.80
.050″ thick	.800	L.F.	.044	1.58	1.60	3.18
Copper, 16 oz.	.800	L.F.	.056	2.68	2.02	4.70
20 oz.	.800	L.F.	.058	3.36	2.11	5.47
24 oz.	.800	L.F.	.061	4.04	2.22	6.26
32 oz.	.800	L.F.	.064	5.35	2.32	7.67

32 Skylight/Skywindow Systems

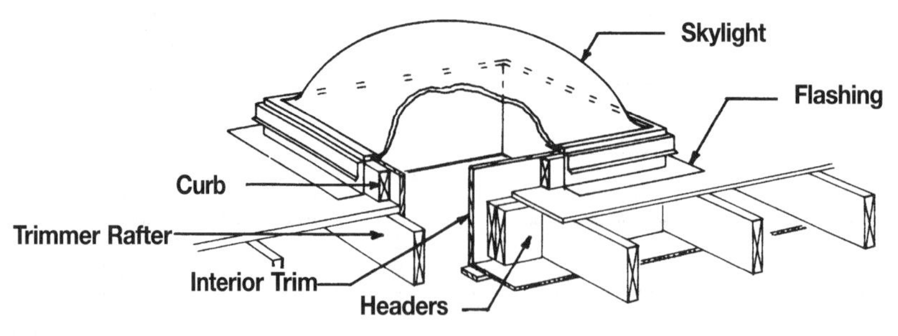

System Description	QUAN.	UNIT	LABOR HOURS	COST EACH		
				MAT.	INST.	TOTAL
SKYLIGHT, FIXED, 32" X 32"						
Skylight, fixed bubble, insulating, 32" x 32"	1.000	Ea.	1.422	62.93	43.38	106.31
Trimmer rafters, 2" x 6"	28.000	L.F.	.448	18.20	15.12	33.32
Headers, 2" x 6"	6.000	L.F.	.267	3.90	9.00	12.90
Curb, 2" x 4"	12.000	L.F.	.154	5.16	5.16	10.32
Flashing, aluminum, .013" thick	13.500	S.F.	.745	5.00	27.00	32.00
Trim, interior casing, painted	12.000	L.F.	.696	10.80	22.44	33.24
TOTAL			3.732	105.99	122.10	228.09
SKYLIGHT, FIXED, 48" X 48"						
Skylight, fixed bubble, insulating, 48" x 48"	1.000	Ea.	1.296	130.40	39.52	169.92
Trimmer rafters, 2" x 6"	28.000	L.F.	.448	18.20	15.12	33.32
Headers, 2" x 6"	8.000	L.F.	.356	5.20	12.00	17.20
Curb, 2" x 4"	16.000	L.F.	.205	6.88	6.88	13.76
Flashing, aluminum, .013" thick	16.000	S.F.	.883	5.92	32.00	37.92
Trim, interior casing, painted	16.000	L.F.	.927	14.40	29.92	44.32
TOTAL			4.115	181.00	135.44	316.44
SKYWINDOW, OPERATING, 24" X 48"						
Skywindow, operating, thermopane glass, 24" x 48"	1.000	Ea.	3.200	550.00	97.50	647.50
Trimmer rafters, 2" x 6"	28.000	L.F.	.448	18.20	15.12	33.32
Headers, 2" x 6"	8.000	L.F.	.267	3.90	9.00	12.90
Curb, 2" x 4"	14.000	L.F.	.179	6.02	6.02	12.04
Flashing, aluminum, .013" thick	14.000	S.F.	.772	5.18	28.00	33.18
Trim, interior casing, painted	14.000	L.F.	.811	12.60	26.18	38.78
TOTAL			5.677	595.90	181.82	777.72

The prices in these systems are on a cost each basis.

Description	QUAN.	UNIT	LABOR HOURS	COST EACH		
				MAT.	INST.	TOTAL

Important: See the Reference Section for critical supporting data - Reference Nos., Crews & Location Factors

Skylight/Skywindow Price Sheet	QUAN.	UNIT	LABOR HOURS	COST EACH		
				MAT.	INST.	TOTAL
Skylight, fixed bubble insulating, 24" x 24"	1.000	Ea.	.800	35.50	24.50	60.00
32" x 32"	1.000	Ea.	1.422	63.00	43.50	106.50
32" x 48"	1.000	Ea.	.864	87.00	26.50	113.50
48" x 48"	1.000	Ea.	1.296	130.00	39.50	169.50
Ventilating bubble insulating, 36" x 36"	1.000	Ea.	2.667	375.00	81.50	456.50
52" x 52"	1.000	Ea.	2.667	565.00	81.50	646.50
28" x 52"	1.000	Ea.	3.200	440.00	97.50	537.50
36" x 52"	1.000	Ea.	3.200	475.00	97.50	572.50
Skywindow, operating, thermopane glass, 24" x 48"	1.000	Ea.	3.200	550.00	97.50	647.50
32" x 48"	1.000	Ea.	3.556	575.00	109.00	684.00
Trimmer rafters, 2" x 6"	28.000	L.F.	.448	18.20	15.10	33.30
2" x 8"	28.000	L.F.	.472	25.50	15.95	41.45
2" x 10"	28.000	L.F.	.711	37.50	24.00	61.50
Headers, 24" window, 2" x 6"	4.000	L.F.	.178	2.60	6.00	8.60
2" x 8"	4.000	L.F.	.188	3.64	6.35	9.99
2" x 10"	4.000	L.F.	.200	5.35	6.75	12.10
32" window, 2" x 6"	6.000	L.F.	.267	3.90	9.00	12.90
2" x 8"	6.000	L.F.	.282	5.45	9.55	15.00
2" x 10"	6.000	L.F.	.300	8.05	10.15	18.20
48" window, 2" x 6"	8.000	L.F.	.356	5.20	12.00	17.20
2" x 8"	8.000	L.F.	.376	7.30	12.70	20.00
2" x 10"	8.000	L.F.	.400	10.70	13.50	24.20
Curb, 2" x 4", skylight, 24" x 24"	8.000	L.F.	.102	3.44	3.44	6.88
32" x 32"	12.000	L.F.	.154	5.15	5.15	10.30
32" x 48"	14.000	L.F.	.179	6.00	6.00	12.00
48" x 48"	16.000	L.F.	.205	6.90	6.90	13.80
Flashing, aluminum .013" thick, skylight, 24" x 24"	9.000	S.F.	.497	3.33	18.00	21.33
32" x 32"	13.500	S.F.	.745	5.00	27.00	32.00
32" x 48"	14.000	S.F.	.772	5.20	28.00	33.20
48" x 48"	16.000	S.F.	.883	5.90	32.00	37.90
Copper 16 oz., skylight, 24" x 24"	9.000	S.F.	.626	30.00	23.00	53.00
32" x 32"	13.500	S.F.	.939	45.00	34.00	79.00
32" x 48"	14.000	S.F.	.974	47.00	35.50	82.50
48" x 48"	16.000	S.F.	1.113	53.50	40.50	94.00
Trim, interior casing painted, 24" x 24"	8.000	L.F.	.347	7.60	11.45	19.05
32" x 32"	12.000	L.F.	.520	11.40	17.15	28.55
32" x 48"	14.000	L.F.	.607	13.30	20.00	33.30
48" x 48"	16.000	L.F.	.693	15.20	23.00	38.20

34 Built-up Roofing Systems

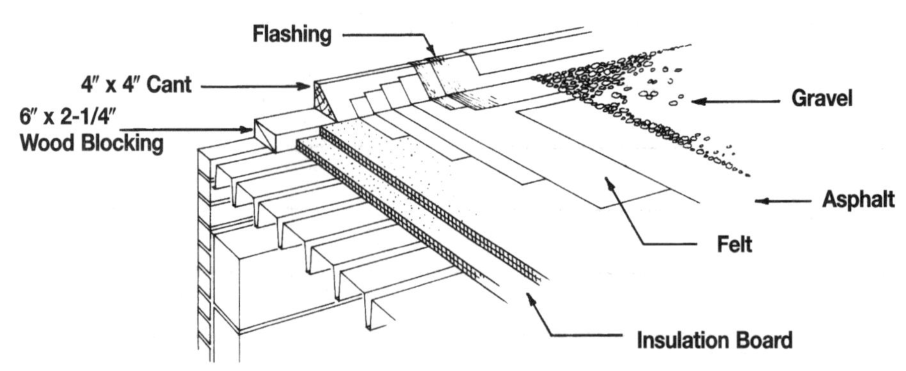

System Description	QUAN.	UNIT	LABOR HOURS	COST PER S.F. MAT.	COST PER S.F. INST.	COST PER S.F. TOTAL
ASPHALT, ORGANIC, 4-PLY, INSULATED DECK						
Membrane, asphalt, 4-plies #15 felt, gravel surfacing	1.000	S.F.	.025	.53	.98	1.51
Insulation board, 2-layers of 1-1/16" glass fiber	2.000	S.F.	.016	1.58	.50	2.08
Wood blocking, treated, 6" x 2-1/4" & 4" x 4" cant	.040	L.F.	.005	.15	.18	.33
Flashing, aluminum, 0.040" thick	.050	S.F.	.003	.08	.10	.18
TOTAL			.049	2.34	1.76	4.10
ASPHALT, INORGANIC, 3-PLY, INSULATED DECK						
Membrane, asphalt, 3-plies type IV glass felt, gravel surfacing	1.000	S.F.	.028	.49	1.07	1.56
Insulation board, 2-layers of 1-1/16" glass fiber	2.000	S.F.	.016	1.58	.50	2.08
Wood blocking, treated, 6" x 2-1/4" & 4" x 4" cant	.040	L.F.	.005	.15	.18	.33
Flashing, aluminum, 0.040" thick	.050	S.F.	.003	.08	.10	.18
TOTAL			.052	2.30	1.85	4.15
COAL TAR, ORGANIC, 4-PLY, INSULATED DECK						
Membrane, coal tar, 4-plies #15 felt, gravel surfacing	1.000	S.F.	.027	1.11	1.02	2.13
Insulation board, 2-layers of 1-1/16" glass fiber	2.000	S.F.	.016	1.58	.50	2.08
Wood blocking, treated, 6" x 2-1/4" & 4" x 4" cant	.040	L.F.	.005	.15	.18	.33
Flashing, aluminum, 0.040" thick	.050	S.F.	.003	.08	.10	.18
TOTAL			.051	2.92	1.80	4.72
COAL TAR, INORGANIC, 3-PLY, INSULATED DECK						
Membrane, coal tar, 3-plies type IV glass felt, gravel surfacing	1.000	S.F.	.029	.91	1.13	2.04
Insulation board, 2-layers of 1-1/16" glass fiber	2.000	S.F.	.016	1.58	.50	2.08
Wood blocking, treated, 6" x 2-1/4" & 4" x 4" cant	.040	L.F.	.005	.15	.18	.33
Flashing, aluminum, 0.040" thick	.050	S.F.	.003	.08	.10	.18
TOTAL			.053	2.72	1.91	4.63

Description	QUAN.	UNIT	LABOR HOURS	COST PER S.F. MAT.	COST PER S.F. INST.	COST PER S.F. TOTAL

Important: See the Reference Section for critical supporting data - Reference Nos., Crews & Location Factors

Built-Up Roofing Price Sheet	QUAN.	UNIT	LABOR HOURS	COST PER S.F.		
				MAT.	INST.	TOTAL
Membrane, asphalt, 4-plies #15 organic felt, gravel surfacing	1.000	S.F.	.025	.53	.98	1.51
Asphalt base sheet & 3-plies #15 asphalt felt	1.000	S.F.	.025	.39	.98	1.37
3-plies type IV glass fiber felt	1.000	S.F.	.028	.49	1.07	1.56
4-plies type IV glass fiber felt	1.000	S.F.	.028	.59	1.07	1.66
Coal tar, 4-plies #15 organic felt, gravel surfacing	1.000	S.F.	.027			
4-plies tarred felt	1.000	S.F.	.027	1.11	1.02	2.13
3-plies type IV glass fiber felt	1.000	S.F.	.029	.91	1.13	2.04
4-plies type IV glass fiber felt	1.000	S.F.	.027	1.24	1.02	2.26
Roll, asphalt, 1-ply #15 organic felt, 2-plies mineral surfaced	1.000	S.F.	.021	.37	.79	1.16
3-plies type IV glass fiber, 1-ply mineral surfaced	1.000	S.F.	.022	.54	.85	1.39
Insulation boards, glass fiber, 1-1/16″ thick	1.000	S.F.	.008	.79	.25	1.04
2-1/16″ thick	1.000	S.F.	.010	1.17	.32	1.49
2-7/16″ thick	1.000	S.F.	.010	1.32	.32	1.64
Expanded perlite, 1″ thick	1.000	S.F.	.010	.26	.32	.58
1-1/2″ thick	1.000	S.F.	.010	.40	.32	.72
2″ thick	1.000	S.F.	.011	.53	.36	.89
Fiberboard, 1″ thick	1.000	S.F.	.010	.35	.32	.67
1-1/2″ thick	1.000	S.F.	.010	.53	.32	.85
2″ thick	1.000	S.F.	.010	.70	.32	1.02
Extruded polystyrene, 15 PSI compressive strength, 2″ thick R10	1.000	S.F.	.006	.36	.20	.56
3″ thick R15	1.000	S.F.	.008	.80	.25	1.05
4″ thick R20	1.000	S.F.	.008	1.11	.25	1.36
Tapered for drainage	1.000	S.F.	.005	.36	.17	.53
40 PSI compressive strength, 1″ thick R5	1.000	S.F.	.005	.36	.17	.53
2″ thick R10	1.000	S.F.	.006	.71	.20	.91
3″ thick R15	1.000	S.F.	.008	1.05	.25	1.30
4″ thick R20	1.000	S.F.	.008	1.40	.25	1.65
Fiberboard high density, 1/2″ thick R1.3	1.000	S.F.	.008	.20	.25	.45
1″ thick R2.5	1.000	S.F.	.010	.37	.32	.69
1 1/2″ thick R3.8	1.000	S.F.	.010	.61	.32	.93
Polyisocyanurate, 1 1/2″ thick R10.87	1.000	S.F.	.006	.37	.20	.57
2″ thick R14.29	1.000	S.F.	.007	.46	.23	.69
3 1/2″ thick R25	1.000	S.F.	.008	.77	.25	1.02
Tapered for drainage	1.000	S.F.	.006	.40	.18	.58
Expanded polystyrene, 1″ thick	1.000	S.F.	.005	.19	.17	.36
2″ thick R10	1.000	S.F.	.006	.35	.20	.55
3″ thick	1.000	S.F.	.006	.53	.20	.73
Wood blocking, treated, 6″ x 2″ & 4″ x 4″ cant	.040	L.F.	.002	.09	.08	.17
6″ x 4-1/2″ & 4″ x 4″ cant	.040	L.F.	.005	.15	.18	.33
6″ x 5″ & 4″ x 4″ cant	.040	L.F.	.007	.18	.23	.41
Flashing, aluminum, 0.019″ thick	.050	S.F.	.003	.04	.10	.14
0.032″ thick	.050	S.F.	.003	.05	.10	.15
0.040″ thick	.050	S.F.	.003	.08	.10	.18
Copper sheets, 16 oz., under 500 lbs.	.050	S.F.	.003	.17	.13	.30
Over 500 lbs.	.050	S.F.	.003	.16	.09	.25
20 oz., under 500 lbs.	.050	S.F.	.004	.21	.13	.34
Over 500 lbs.	.050	S.F.	.003	.20	.10	.30
Stainless steel, 32 gauge	.050	S.F.	.003	.12	.09	.21
28 gauge	.050	S.F.	.003	.14	.09	.23
26 gauge	.050	S.F.	.003	.17	.09	.26
24 gauge	.050	S.F.	.003	.23	.09	.32

Division 6
Interiors

04 Drywall & Thincoat Wall Systems

Corners
Paint
Trim
Finish
Drywall

System Description	QUAN.	UNIT	LABOR HOURS	COST PER S.F.		
				MAT.	INST.	TOTAL
1/2" SHEETROCK, TAPED & FINISHED						
Drywall, 1/2" thick, standard	1.000	S.F.	.008	.29	.27	.56
Finish, taped & finished joints	1.000	S.F.	.008	.04	.27	.31
Corners, taped & finished, 32 L.F. per 12' x 12' room	.083	L.F.	.001	.01	.04	.05
Painting, primer & 2 coats	1.000	S.F.	.011	.16	.32	.48
Trim, baseboard, painted	.125	L.F.	.006	.19	.21	.40
TOTAL			.034	.69	1.11	1.80
THINCOAT, SKIM-COAT, ON 1/2" BACKER DRYWALL						
Drywall, 1/2" thick, thincoat backer	1.000	S.F.	.008	.29	.27	.56
Thincoat plaster	1.000	S.F.	.011	.08	.34	.42
Corners, taped & finished, 32 L.F. per 12' x 12' room	.083	L.F.	.001	.01	.04	.05
Painting, primer & 2 coats	1.000	S.F.	.011	.16	.32	.48
Trim, baseboard, painted	.125	L.F.	.006	.19	.21	.40
TOTAL			.037	.73	1.18	1.91
5/8" SHEETROCK, TAPED & FINISHED						
Drywall, 5/8" thick, standard	1.000	S.F.	.008	.32	.27	.59
Finish, taped & finished joints	1.000	S.F.	.008	.04	.27	.31
Corners, taped & finished, 32 L.F. per 12' x 12' room	.083	L.F.	.001	.01	.04	.05
Painting, primer & 2 coats	1.000	S.F.	.011	.16	.32	.48
Trim, baseboard, painted	.125	L.F.	.006	.19	.21	.40
TOTAL			.034	.72	1.11	1.83

The costs in this system are based on a square foot of wall.
Do not deduct for openings.

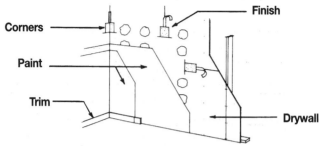

Description	QUAN.	UNIT	LABOR HOURS	COST PER S.F.		
				MAT.	INST.	TOTAL

Drywall & Thincoat Wall Price Sheet	QUAN.	UNIT	LABOR HOURS	COST PER S.F. MAT.	INST.	TOTAL
Drywall-sheetrock, 1/2" thick, standard	1.000	S.F.	.008	.29	.27	.56
Fire resistant	1.000	S.F.	.008	.28	.27	.55
Water resistant	1.000	S.F.	.008	.35	.27	.62
5/8" thick, standard	1.000	S.F.	.008	.32	.27	.59
Fire resistant	1.000	S.F.	.008	.32	.27	.59
Water resistant	1.000	S.F.	.008	.39	.27	.66
Drywall backer for thincoat system, 1/2" thick	1.000	S.F.	.008	.29	.27	.56
5/8" thick	1.000	S.F.	.008	.32	.27	.59
Finish drywall, taped & finished	1.000	S.F.	.008	.04	.27	.31
Texture spray	1.000	S.F.	.010	.06	.31	.37
Thincoat plaster, including tape	1.000	S.F.	.011	.08	.34	.42
Corners drywall, taped & finished, 32 L.F. per 4' x 4' room	.250	L.F.	.004	.02	.12	.14
6' x 6' room	.110	L.F.	.002	.01	.05	.06
10' x 10' room	.100	L.F.	.001	.01	.05	.06
12' x 12' room	.083	L.F.	.001	.01	.04	.05
16' x 16' room	.063	L.F.	.001		.03	.03
Thincoat system, 32 L.F. per 4' x 4' room	.250	L.F.	.003	.02	.09	.11
6' x 6' room	.110	L.F.	.001	.01	.04	.05
10' x 10' room	.100	L.F.	.001	.01	.03	.04
12' x 12' room	.083	L.F.	.001	.01	.03	.04
16' x 16' room	.063	L.F.	.001	.01	.02	.03
Painting, primer, & 1 coat	1.000	S.F.	.008	.10	.25	.35
& 2 coats	1.000	S.F.	.011	.16	.32	.48
Wallpaper, $7/double roll	1.000	S.F.	.013	.30	.38	.68
$17/double roll	1.000	S.F.	.015	.54	.45	.99
$40/double roll	1.000	S.F.	.018	1.01	.55	1.56
Tile, ceramic adhesive thin set, 4 1/4" x 4 1/4" tiles	1.000	S.F.	.084	2.20	2.36	4.56
6" x 6" tiles	1.000	S.F.	.080	2.89	2.25	5.14
Pregrouted sheets	1.000	S.F.	.067	4.48	1.87	6.35
Trim, painted or stained, baseboard	.125	L.F.	.006	.19	.21	.40
Base shoe	.125	L.F.	.005	.12	.18	.30
Chair rail	.125	L.F.	.005	.13	.17	.30
Cornice molding	.125	L.F.	.004	.10	.14	.24
Cove base, vinyl	.125	L.F.	.003	.06	.10	.16
Paneling, not including furring or trim						
Plywood, prefinished, 1/4" thick, 4' x 8' sheets, vert. grooves						
Birch faced, minimum	1.000	S.F.	.032	.79	1.08	1.87
Average	1.000	S.F.	.038	1.21	1.29	2.50
Maximum	1.000	S.F.	.046	1.78	1.54	3.32
Mahogany, African	1.000	S.F.	.040	2.29	1.35	3.64
Philippine (lauan)	1.000	S.F.	.032	.97	1.08	2.05
Oak or cherry, minimum	1.000	S.F.	.032	1.91	1.08	2.99
Maximum	1.000	S.F.	.040	2.95	1.35	4.30
Rosewood	1.000	S.F.	.050	4.19	1.69	5.88
Teak	1.000	S.F.	.040	2.95	1.35	4.30
Chestnut	1.000	S.F.	.043	4.37	1.44	5.81
Pecan	1.000	S.F.	.040	1.87	1.35	3.22
Walnut, minimum	1.000	S.F.	.032	2.50	1.08	3.58
Maximum	1.000	S.F.	.040	4.76	1.35	6.11

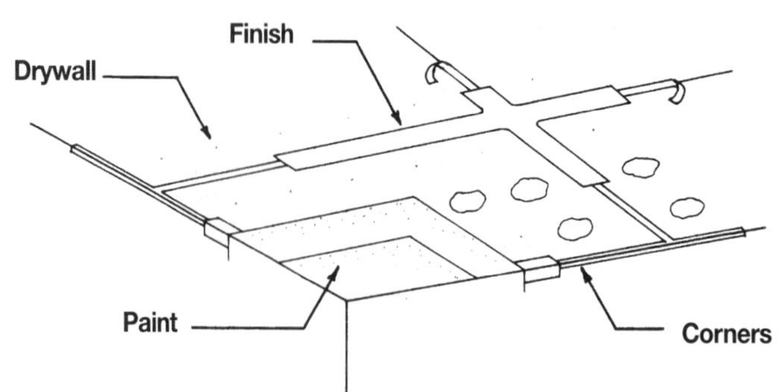

System Description	QUAN.	UNIT	LABOR HOURS	COST PER S.F. MAT.	COST PER S.F. INST.	COST PER S.F. TOTAL
1/2" SHEETROCK, TAPED & FINISHED						
Drywall, 1/2" thick, standard	1.000	S.F.	.008	.29	.27	.56
Finish, taped & finished	1.000	S.F.	.008	.04	.27	.31
Corners, taped & finished, 12' x 12' room	.333	L.F.	.005	.02	.16	.18
Paint, primer & 2 coats	1.000	S.F.	.011	.16	.32	.48
TOTAL			.032	.51	1.02	1.53
THINCOAT, SKIM COAT ON 1/2" BACKER DRYWALL						
Drywall, 1/2" thick, thincoat backer	1.000	S.F.	.008	.29	.27	.56
Thincoat plaster	1.000	S.F.	.011	.08	.34	.42
Corners, taped & finished, 12' x 12' room	.333	L.F.	.005	.02	.16	.18
Paint, primer & 2 coats	1.000	S.F.	.011	.16	.32	.48
TOTAL			.035	.55	1.09	1.64
WATER-RESISTANT SHEETROCK, 1/2" THICK, TAPED & FINISHED						
Drywall, 1/2" thick, water-resistant	1.000	S.F.	.008	.35	.27	.62
Finish, taped & finished	1.000	S.F.	.008	.04	.27	.31
Corners, taped & finished, 12' x 12' room	.333	L.F.	.005	.02	.16	.18
Paint, primer & 2 coats	1.000	S.F.	.011	.16	.32	.48
TOTAL			.032	.57	1.02	1.59
5/8" SHEETROCK, TAPED & FINISHED						
Drywall, 5/8" thick, standard	1.000	S.F.	.008	.32	.27	.59
Finish, taped & finished	1.000	S.F.	.008	.04	.27	.31
Corners, taped & finished, 12' x 12' room	.333	L.F.	.005	.02	.16	.18
Paint, primer & 2 coats	1.000	S.F.	.011	.16	.32	.48
TOTAL			.032	.54	1.02	1.56

The costs in this system are based on a square foot of ceiling.

Description	QUAN.	UNIT	LABOR HOURS	COST PER S.F. MAT.	COST PER S.F. INST.	COST PER S.F. TOTAL

Drywall & Thincoat Ceilings	QUAN.	UNIT	LABOR HOURS	COST PER S.F.		
				MAT.	INST.	TOTAL
Drywall-sheetrock, 1/2" thick, standard	1.000	S.F.	.008	.29	.27	.56
Fire resistant	1.000	S.F.	.008	.28	.27	.55
Water resistant	1.000	S.F.	.008	.35	.27	.62
5/8" thick, standard	1.000	S.F.	.008	.32	.27	.59
Fire resistant	1.000	S.F.	.008	.32	.27	.59
Water resistant	1.000	S.F.	.008	.39	.27	.66
Drywall backer for thincoat system, 1/2" thick	1.000	S.F.	.016	.57	.54	1.11
5/8" thick	1.000	S.F.	.016	.60	.54	1.14
Finish drywall, taped & finished	1.000	S.F.	.008	.04	.27	.31
Texture spray	1.000	S.F.	.010	.06	.31	.37
Thincoat plaster	1.000	S.F.	.011	.08	.34	.42
Corners taped & finished, 4' x 4' room	1.000	L.F.	.015	.07	.49	.56
6' x 6' room	.667	L.F.	.010	.05	.33	.38
10' x 10' room	.400	L.F.	.006	.03	.20	.23
12' x 12' room	.333	L.F.	.005	.02	.16	.18
16' x 16' room	.250	L.F.	.003	.01	.09	.10
Thincoat system, 4' x 4' room	1.000	L.F.	.011	.08	.34	.42
6' x 6' room	.667	L.F.	.007	.05	.22	.27
10' x 10' room	.400	L.F.	.004	.03	.14	.17
12' x 12' room	.333	L.F.	.004	.03	.12	.15
16' x 16' room	.250	L.F.	.002	.01	.06	.07
Painting, primer & 1 coat	1.000	S.F.	.008	.10	.25	.35
& 2 coats	1.000	S.F.	.011	.16	.32	.48
Wallpaper, double roll, solid pattern, avg. workmanship	1.000	S.F.	.013	.30	.38	.68
Basic pattern, avg. workmanship	1.000	S.F.	.015	.54	.45	.99
Basic pattern, quality workmanship	1.000	S.F.	.018	1.01	.55	1.56
Tile, ceramic adhesive thin set, 4 1/4" x 4 1/4" tiles	1.000	S.F.	.084	2.20	2.36	4.56
6" x 6" tiles	1.000	S.F.	.080	2.89	2.25	5.14
Pregrouted sheets	1.000	S.F.	.067	4.48	1.87	6.35

12 Plaster & Stucco Wall Systems

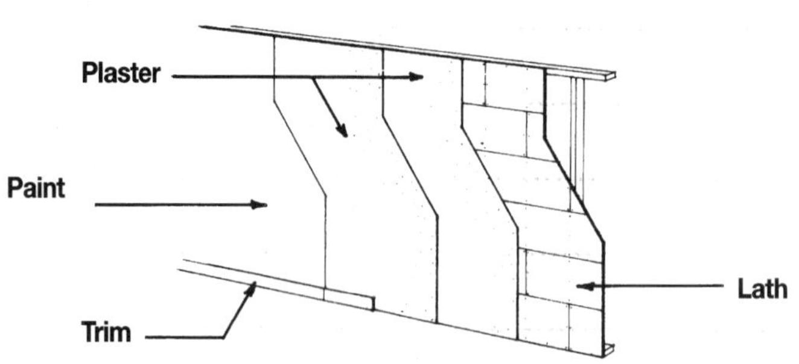

System Description	QUAN.	UNIT	LABOR HOURS	COST PER S.F.		
				MAT.	INST.	TOTAL
PLASTER ON GYPSUM LATH						
Plaster, gypsum or perlite, 2 coats	1.000	S.F.	.053	.44	1.64	2.08
Lath, 3/8″ gypsum	1.000	S.F.	.010	.42	.33	.75
Corners, expanded metal, 32 L.F. per 12′ x 12′ room	.083	L.F.	.002	.01	.06	.07
Painting, primer & 2 coats	1.000	S.F.	.011	.16	.32	.48
Trim, baseboard, painted	.125	L.F.	.006	.19	.21	.40
TOTAL			.082	1.22	2.56	3.78
PLASTER ON METAL LATH						
Plaster, gypsum or perlite, 2 coats	1.000	S.F.	.053	.44	1.64	2.08
Lath, 2.5 Lb. diamond, metal	1.000	S.F.	.010	.16	.33	.49
Corners, expanded metal, 32 L.F. per 12′ x 12′ room	.083	L.F.	.002	.01	.06	.07
Painting, primer & 2 coats	1.000	S.F.	.011	.16	.32	.48
Trim, baseboard, painted	.125	L.F.	.006	.19	.21	.40
TOTAL			.082	.96	2.56	3.52
STUCCO ON METAL LATH						
Stucco, 2 coats	1.000	S.F.	.041	.23	1.25	1.48
Lath, 2.5 Lb. diamond, metal	1.000	S.F.	.010	.16	.33	.49
Corners, expanded metal, 32 L.F. per 12′ x 12′ room	.083	L.F.	.002	.01	.06	.07
Painting, primer & 2 coats	1.000	S.F.	.011	.16	.32	.48
Trim, baseboard, painted	.125	L.F.	.006	.19	.21	.40
TOTAL			.070	.75	2.17	2.92

The costs in these systems are based on a per square foot of wall area.
Do not deduct for openings.

Description	QUAN.	UNIT	LABOR HOURS	COST PER S.F.		
				MAT.	INST.	TOTAL

Plaster & Stucco Wall Price Sheet	QUAN.	UNIT	LABOR HOURS	COST PER S.F.		
				MAT.	INST.	TOTAL
Plaster, gypsum or perlite, 2 coats	1.000	S.F.	.053	.44	1.64	2.08
3 coats	1.000	S.F.	.065	.62	1.98	2.60
Lath, gypsum, standard, 3/8" thick	1.000	S.F.	.010	.42	.33	.75
1/2" thick	1.000	S.F.	.013	.44	.40	.84
Fire resistant, 3/8" thick	1.000	S.F.	.013	.43	.40	.83
1/2" thick	1.000	S.F.	.014	.48	.43	.91
Metal, diamond, 2.5 Lb.	1.000	S.F.	.010	.16	.33	.49
3.4 Lb.	1.000	S.F.	.012	.25	.37	.62
Rib, 2.75 Lb.	1.000	S.F.	.012	.25	.37	.62
3.4 Lb.	1.000	S.F.	.013	.35	.40	.75
Corners, expanded metal, 32 L.F. per 4' x 4' room	.250	L.F.	.005	.03	.17	.20
6' x 6' room	.110	L.F.	.002	.01	.07	.08
10' x 10' room	.100	L.F.	.002	.01	.07	.08
12' x 12' room	.083	L.F.	.002	.01	.06	.07
16' x 16' room	.063	L.F.	.001	.01	.04	.05
Painting, primer & 1 coats	1.000	S.F.	.008	.10	.25	.35
Primer & 2 coats	1.000	S.F.	.011	.16	.32	.48
Wallpaper, low price double roll	1.000	S.F.	.013	.30	.38	.68
Medium price double roll	1.000	S.F.	.015	.54	.45	.99
High price double roll	1.000	S.F.	.018	1.01	.55	1.56
Tile, ceramic thin set, 4-1/4" x 4-1/4" tiles	1.000	S.F.	.084	2.20	2.36	4.56
6" x 6" tiles	1.000	S.F.	.080	2.89	2.25	5.14
Pregrouted sheets	1.000	S.F.	.067	4.48	1.87	6.35
Trim, painted or stained, baseboard	.125	L.F.	.006	.19	.21	.40
Base shoe	.125	L.F.	.005	.12	.18	.30
Chair rail	.125	L.F.	.005	.13	.17	.30
Cornice molding	.125	L.F.	.004	.10	.14	.24
Cove base, vinyl	.125	L.F.	.003	.06	.10	.16
Paneling not including furring or trim						
Plywood, prefinished, 1/4" thick, 4' x 8' sheets, vert. grooves						
Birch faced, minimum	1.000	S.F.	.032	.79	1.08	1.87
Average	1.000	S.F.	.038	1.21	1.29	2.50
Maximum	1.000	S.F.	.046	1.78	1.54	3.32
Mahogany, African	1.000	S.F.	.040	2.29	1.35	3.64
Philippine (lauan)	1.000	S.F.	.032	.97	1.08	2.05
Oak or cherry, minimum	1.000	S.F.	.032	1.91	1.08	2.99
Maximum	1.000	S.F.	.040	2.95	1.35	4.30
Rosewood	1.000	S.F.	.050	4.19	1.69	5.88
Teak	1.000	S.F.	.040	2.95	1.35	4.30
Chestnut	1.000	S.F.	.043	4.37	1.44	5.81
Pecan	1.000	S.F.	.040	1.87	1.35	3.22
Walnut, minimum	1.000	S.F.	.032	2.50	1.08	3.58
Maximum	1.000	S.F.	.040	4.76	1.35	6.11

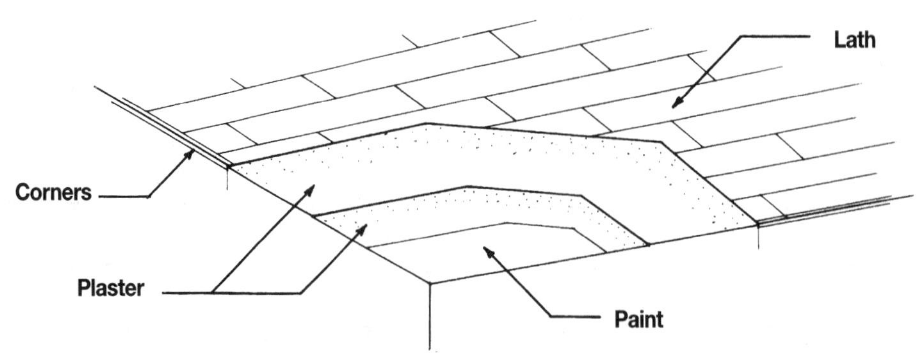

System Description	QUAN.	UNIT	LABOR HOURS	COST PER S.F.		
				MAT.	INST.	TOTAL
PLASTER ON GYPSUM LATH						
Plaster, gypsum or perlite, 2 coats	1.000	S.F.	.061	.42	1.86	2.28
Lath, 3/8" gypsum	1.000	S.F.	.014	.42	.46	.88
Corners, expanded metal, 12' x 12' room	.330	L.F.	.007	.03	.22	.25
Painting, primer & 2 coats	1.000	S.F.	.011	.16	.32	.48
TOTAL			.093	1.03	2.86	3.89
PLASTER ON METAL LATH						
Plaster, gypsum or perlite, 2 coats	1.000	S.F.	.061	.42	1.86	2.28
Lath, 2.5 Lb. diamond, metal	1.000	S.F.	.012	.16	.37	.53
Corners, expanded metal, 12' x 12' room	.330	L.F.	.007	.03	.22	.25
Painting, primer & 2 coats	1.000	S.F.	.011	.16	.32	.48
TOTAL			.091	.77	2.77	3.54
STUCCO ON GYPSUM LATH						
Stucco, 2 coats	1.000	S.F.	.041	.23	1.25	1.48
Lath, 3/8" gypsum	1.000	S.F.	.014	.42	.46	.88
Corners, expanded metal, 12' x 12' room	.330	L.F.	.007	.03	.22	.25
Painting, primer & 2 coats	1.000	S.F.	.011	.16	.32	.48
TOTAL			.073	.84	2.25	3.09
STUCCO ON METAL LATH						
Stucco, 2 coats	1.000	S.F.	.041	.23	1.25	1.48
Lath, 2.5 Lb. diamond, metal	1.000	S.F.	.012	.16	.37	.53
Corners, expanded metal, 12' x 12' room	.330	L.F.	.007	.03	.22	.25
Painting, primer & 2 coats	1.000	S.F.	.011	.16	.32	.48
TOTAL			.071	.58	2.16	2.74

The costs in these systems are based on a square foot of ceiling area.

Description	QUAN.	UNIT	LABOR HOURS	COST PER S.F.		
				MAT.	INST.	TOTAL

Plaster & Stucco Ceiling Price Sheet	QUAN.	UNIT	LABOR HOURS	COST PER S.F.		
				MAT.	INST.	TOTAL
Plaster, gypsum or perlite, 2 coats	1.000	S.F.	.061	.42	1.86	2.28
3 coats	1.000	S.F.	.065	.62	1.98	2.60
Lath, gypsum, standard, 3/8" thick	1.000	S.F.	.014	.42	.46	.88
1/2" thick	1.000	S.F.	.015	.43	.48	.91
Fire resistant, 3/8" thick	1.000	S.F.	.017	.43	.53	.96
1/2" thick	1.000	S.F.	.018	.48	.56	1.04
Metal, diamond, 2.5 Lb.	1.000	S.F.	.012	.16	.37	.53
3.4 Lb.	1.000	S.F.	.015	.25	.47	.72
Rib, 2.75 Lb.	1.000	S.F.	.012	.25	.37	.62
3.4 Lb.	1.000	S.F.	.013	.35	.40	.75
Corners expanded metal, 4' x 4' room	1.000	L.F.	.020	.10	.68	.78
6' x 6' room	.667	L.F.	.013	.07	.46	.53
10' x 10' room	.400	L.F.	.008	.04	.27	.31
12' x 12' room	.333	L.F.	.007	.03	.22	.25
16' x 16' room	.250	L.F.	.004	.02	.13	.15
Painting, primer & 1 coat	1.000	S.F.	.008	.10	.25	.35
Primer & 2 coats	1.000	S.F.	.011	.16	.32	.48

6 INTERIORS

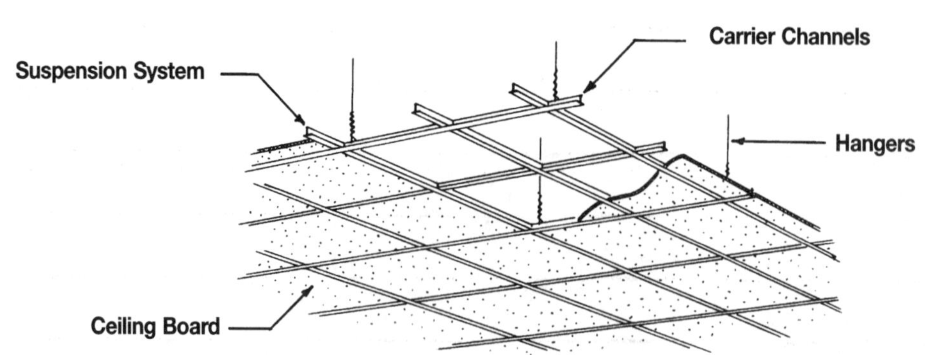

System Description	QUAN.	UNIT	LABOR HOURS	COST PER S.F.		
				MAT.	INST.	TOTAL
2′ X 2′ GRID, FILM FACED FIBERGLASS, 5/8″ THICK						
Suspension system, 2′ x 2′ grid, T bar	1.000	S.F.	.012	.44	.42	.86
Ceiling board, film faced fiberglass, 5/8″ thick	1.000	S.F.	.013	.56	.43	.99
Carrier channels, 1-1/2″ x 3/4″	1.000	S.F.	.017	.23	.57	.80
Hangers, #12 wire	1.000	S.F.	.002	.03	.06	.09
TOTAL			.044	1.26	1.48	2.74
2′ X 4′ GRID, FILM FACED FIBERGLASS, 5/8″ THICK						
Suspension system, 2′ x 4′ grid, T bar	1.000	S.F.	.010	.35	.34	.69
Ceiling board, film faced fiberglass, 5/8″ thick	1.000	S.F.	.013	.56	.43	.99
Carrier channels, 1-1/2″ x 3/4″	1.000	S.F.	.017	.23	.57	.80
Hangers, #12 wire	1.000	S.F.	.002	.03	.06	.09
TOTAL			.042	1.17	1.40	2.57
2′ X 2′ GRID, MINERAL FIBER, REVEAL EDGE, 1″ THICK						
Suspension system, 2′ x 2′ grid, T bar	1.000	S.F.	.012	.44	.42	.86
Ceiling board, mineral fiber, reveal edge, 1″ thick	1.000	S.F.	.013	1.25	.45	1.70
Carrier channels, 1-1/2″ x 3/4″	1.000	S.F.	.017	.23	.57	.80
Hangers, #12 wire	1.000	S.F.	.002	.03	.06	.09
TOTAL			.044	1.95	1.50	3.45
2′ X 4′ GRID, MINERAL FIBER, REVEAL EDGE, 1″ THICK						
Suspension system, 2′ x 4′ grid, T bar	1.000	S.F.	.010	.35	.34	.69
Ceiling board, mineral fiber, reveal edge, 1″ thick	1.000	S.F.	.013	1.25	.45	1.70
Carrier channels, 1-1/2″ x 3/4″	1.000	S.F.	.017	.23	.57	.80
Hangers, #12 wire	1.000	S.F.	.002	.03	.06	.09
TOTAL			.042	1.86	1.42	3.28

Description	QUAN.	UNIT	LABOR HOURS	COST PER S.F.		
				MAT.	INST.	TOTAL

 Important: See the Reference Section for critical supporting data - Reference Nos., Crews & Location Factors

Suspended Ceiling Price Sheet	QUAN.	UNIT	LABOR HOURS	COST PER S.F.		
				MAT.	INST.	TOTAL
Suspension systems, T bar, 2′ x 2′ grid	1.000	S.F.	.012	.44	.42	.86
2′ x 4′ grid	1.000	S.F.	.010	.35	.34	.69
Concealed Z bar, 12″ module	1.000	S.F.	.015	.39	.52	.91
Ceiling boards, fiberglass, film faced, 2′ x 2′ or 2′ x 4′, 5/8″ thick	1.000	S.F.	.013	.56	.43	.99
3/4″ thick	1.000	S.F.	.013	1.17	.45	1.62
3″ thick thermal R11	1.000	S.F.	.018	1.38	.60	1.98
Glass cloth faced, 3/4″ thick	1.000	S.F.	.016	1.82	.54	2.36
1″ thick	1.000	S.F.	.016	2.05	.56	2.61
1-1/2″ thick, nubby face	1.000	S.F.	.017	2.43	.57	3.00
Mineral fiber boards, 5/8″ thick, aluminum face 2′ x 2′	1.000	S.F.	.013	1.40	.45	1.85
2′ x 4′	1.000	S.F.	.012	.92	.42	1.34
Standard faced, 2′ x 2′ or 2′ x 4′	1.000	S.F.	.012	.65	.40	1.05
Plastic coated face, 2′ x 2′ or 2′ x 4′	1.000	S.F.	.020	1.00	.68	1.68
Fire rated, 2 hour rating, 5/8″ thick	1.000	S.F.	.012	.88	.40	1.28
Reveal edge, 2′ x 2′ or 2′ x 4′, painted, 1″ thick	1.000	S.F.	.013	1.11	.40	1.51
2″ thick	1.000	S.F.	.015	1.83	.44	2.27
2-1/2″ thick	1.000	S.F.	.016	2.50	.48	2.98
3″ thick	1.000	S.F.	.018	2.80	.53	3.33
Luminous panels, prismatic, acrylic	1.000	S.F.	.020	1.80	.68	2.48
Polystyrene	1.000	S.F.	.020	.92	.68	1.60
Flat or ribbed, acrylic	1.000	S.F.	.020	3.16	.68	3.84
Polystyrene	1.000	S.F.	.020	2.16	.68	2.84
Drop pan, white, acrylic	1.000	S.F.	.020	4.62	.68	5.30
Polystyrene	1.000	S.F.	.020	3.86	.68	4.54
Carrier channels, 4′-0″ on center, 3/4″ x 1-1/2″	1.000	S.F.	.017	.23	.57	.80
1-1/2″ x 3-1/2″	1.000	S.F.	.017	.43	.57	1.00
Hangers, #12 wire	1.000	S.F.	.002	.03	.06	.09

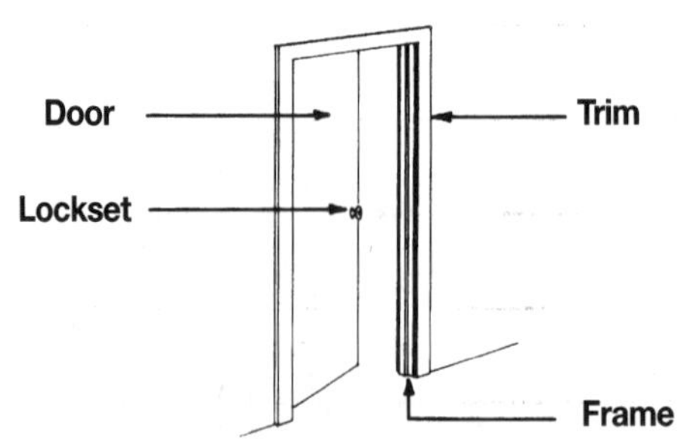

System Description	QUAN.	UNIT	LABOR HOURS	COST EACH MAT.	COST EACH INST.	COST EACH TOTAL
LAUAN, FLUSH DOOR, HOLLOW CORE						
Door, flush, lauan, hollow core, 2′-8″ wide x 6′-8″ high	1.000	Ea.	.889	30.00	30.00	60.00
Frame, pine, 4-5/8″ jamb	17.000	L.F.	.725	96.05	24.48	120.53
Trim, casing, painted	34.000	L.F.	1.473	32.30	48.62	80.92
Butt hinges, chrome, 3-1/2″ x 3-1/2″	1.500	Pr.		24.08		24.08
Lockset, passage	1.000	Ea.	.500	13.85	16.90	30.75
Paint, door & frame, primer & 2 coats	2.000	Face	3.886	11.02	116.00	127.02
TOTAL			7.473	207.30	236.00	443.30
BIRCH, FLUSH DOOR, HOLLOW CORE						
Door, flush, birch, hollow core, 2′-8″ wide x 6′-8″ high	1.000	Ea.	.889	45.00	30.00	75.00
Frame, pine, 4-5/8″ jamb	17.000	L.F.	.725	96.05	24.48	120.53
Trim, casing, painted	34.000	L.F.	1.473	32.30	48.62	80.92
Butt hinges, chrome, 3-1/2″ x 3-1/2″	1.500	Pr.		24.08		24.08
Lockset, passage	1.000	Ea.	.500	13.85	16.90	30.75
Paint, door & frame, primer & 2 coats	2.000	Face	3.886	11.02	116.00	127.02
TOTAL			7.473	222.30	236.00	458.30
RAISED PANEL, SOLID, PINE DOOR						
Door, pine, raised panel, 2′-8″ wide x 6′-8″ high	1.000	Ea.	.889	153.00	30.00	183.00
Frame, pine, 4-5/8″ jamb	17.000	L.F.	.725	96.05	24.48	120.53
Trim, casing, painted	34.000	L.F.	1.473	32.30	48.62	80.92
Butt hinges, bronze, 3-1/2″ x 3-1/2″	1.500	Pr.		27.38		27.38
Lockset, passage	1.000	Ea.	.500	13.85	16.90	30.75
Paint, door & frame, primer & 2 coats	2.000	Face	8.000	15.38	240.00	255.38
TOTAL			11.587	337.96	360.00	697.96

The costs in these systems are based on a cost per each door.

Description	QUAN.	UNIT	LABOR HOURS	COST EACH MAT.	COST EACH INST.	COST EACH TOTAL

 Important: See the Reference Section for critical supporting data - Reference Nos., Crews & Location Factors

Interior Door Price Sheet	QUAN.	UNIT	LABOR HOURS	COST EACH		
				MAT.	INST.	TOTAL
Door, hollow core, lauan 1-3/8″ thick, 6′-8″ high x 1′-6″ wide	1.000	Ea.	.889	24.00	30.00	54.00
2′-0″ wide	1.000	Ea.	.889	25.50	30.00	55.50
2′-6″ wide	1.000	Ea.	.889	29.00	30.00	59.00
2′-8″ wide	1.000	Ea.	.889	30.00	30.00	60.00
3′-0″ wide	1.000	Ea.	.941	32.50	32.00	64.50
Birch 1-3/8″ thick, 6′-8″ high x 1′-6″ wide	1.000	Ea.	.889	34.00	30.00	64.00
2′-0″ wide	1.000	Ea.	.889	37.00	30.00	67.00
2′-6″ wide	1.000	Ea.	.889	43.00	30.00	73.00
2′-8″ wide	1.000	Ea.	.889	45.00	30.00	75.00
3′-0″ wide	1.000	Ea.	.941	47.00	32.00	79.00
Louvered pine 1-3/8″ thick, 6′-8″ high x 1′-6″ wide	1.000	Ea.	.842	109.00	28.50	137.50
2′-0″ wide	1.000	Ea.	.889	127.00	30.00	157.00
2′-6″ wide	1.000	Ea.	.889	141.00	30.00	171.00
2′-8″ wide	1.000	Ea.	.889	153.00	30.00	183.00
3′-0″ wide	1.000	Ea.	.941	162.00	32.00	194.00
Paneled pine 1-3/8″ thick, 6′-8″ high x 1′-6″ wide	1.000	Ea.	.842	109.00	28.50	137.50
2′-0″ wide	1.000	Ea.	.889	127.00	30.00	157.00
2′-6″ wide	1.000	Ea.	.889	141.00	30.00	171.00
2′-8″ wide	1.000	Ea.	.889	153.00	30.00	183.00
3′-0″ wide	1.000	Ea.	.941	162.00	32.00	194.00
Frame, pine, 1′-6″ thru 2′-0″ wide door, 3-5/8″ deep	16.000	L.F.	.683	67.00	23.00	90.00
4-5/8″ deep	16.000	L.F.	.683	90.50	23.00	113.50
5-5/8″ deep	16.000	L.F.	.683	85.00	23.00	108.00
2′-6″ thru 3′0″ wide door, 3-5/8″ deep	17.000	L.F.	.725	71.00	24.50	95.50
4-5/8″ deep	17.000	L.F.	.725	96.00	24.50	120.50
5-5/8″ deep	17.000	L.F.	.725	90.00	24.50	114.50
Trim, casing, painted, both sides, 1′-6″ thru 2′-6″ wide door	32.000	L.F.	1.855	29.00	60.00	89.00
2′-6″ thru 3′-0″ wide door	34.000	L.F.	1.969	30.50	63.50	94.00
Butt hinges 3-1/2″ x 3-1/2″, steel plated, chrome	1.500	Pr.		24.00		24.00
Bronze	1.500	Pr.		27.50		27.50
Locksets, passage, minimum	1.000	Ea.	.500	13.85	16.90	30.75
Maximum	1.000	Ea.	.575	15.95	19.45	35.40
Privacy, miniumum	1.000	Ea.	.625	17.30	21.00	38.30
Maximum	1.000	Ea.	.675	18.70	23.00	41.70
Paint 2 sides, primer & 2 cts., flush door, 1′-6″ to 2′-0″ wide	2.000	Face	5.547	12.85	166.00	178.85
2′-6″ thru 3′-0″ wide	2.000	Face	6.933	16.05	208.00	224.05
Louvered door, 1′-6″ thru 2′-0″ wide	2.000	Face	6.400	12.30	192.00	204.30
2′-6″ thru 3′-0″ wide	2.000	Face	8.000	15.40	240.00	255.40
Paneled door, 1′-6″ thru 2′-0″ wide	2.000	Face	6.400	12.30	192.00	204.30
2′-6″ thru 3′-0″ wide	2.000	Face	8.000	15.40	240.00	255.40

24 Closet Door Systems

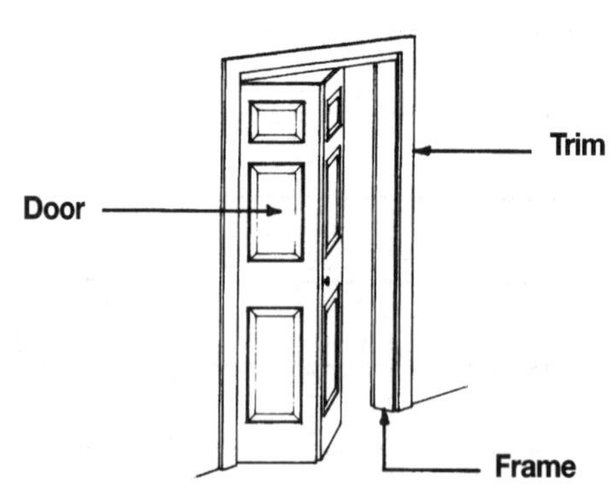

System Description	QUAN.	UNIT	LABOR HOURS	COST EACH		
				MAT.	INST.	TOTAL
BI-PASSING, FLUSH, LAUAN, HOLLOW CORE, 4′-0″ X 6′-8″						
Door, flush, lauan, hollow core, 4′-0″ x 6′-8″ opening	1.000	Ea.	1.333	165.00	45.00	210.00
Frame, pine, 4-5/8″ jamb	18.000	L.F.	.768	101.70	25.92	127.62
Trim, both sides, casing, painted	36.000	L.F.	2.086	32.40	67.32	99.72
Paint, door & frame, primer & 2 coats	2.000	Face	3.886	11.02	116.00	127.02
TOTAL			8.073	310.12	254.24	564.36
BI-PASSING, FLUSH, BIRCH, HOLLOW CORE, 6′-0″ X 6′-8″						
Door, flush, birch, hollow core, 6′-0″ x 6′-8″ opening	1.000	Ea.	1.600	218.00	54.00	272.00
Frame, pine, 4-5/8″ jamb	19.000	L.F.	.811	107.35	27.36	134.71
Trim, both sides, casing, painted	38.000	L.F.	2.203	34.20	71.06	105.26
Paint, door & frame, primer & 2 coats	2.000	Face	4.857	13.78	145.00	158.78
TOTAL			9.471	373.33	297.42	670.75
BI-FOLD, PINE, PANELED, 3′-0″ X 6′-8″						
Door, pine, paneled, 3′-0″ x 6′-8″ opening	1.000	Ea.	1.231	118.00	41.50	159.50
Frame, pine, 4-5/8″ jamb	17.000	L.F.	.725	96.05	24.48	120.53
Trim, both sides, casing, painted	34.000	L.F.	1.969	30.60	63.58	94.18
Paint, door & frame, primer & 2 coats	2.000	Face	8.000	15.38	240.00	255.38
TOTAL			11.925	260.03	369.56	629.59
BI-FOLD, PINE, LOUVERED, 6′-0″ X 6′-8″						
Door, pine, louvered, 6′-0″ x 6′-8″ opening	1.000	Ea.	1.600	229.00	54.00	283.00
Frame, pine, 4-5/8″ jamb	19.000	L.F.	.811	107.35	27.36	134.71
Trim, both sides, casing, painted	38.000	L.F.	2.203	34.20	71.06	105.26
Paint, door & frame, primer & 2 coats	2.000	Face	10.000	19.23	300.00	319.23
TOTAL			14.614	389.78	452.42	842.20

The costs in this system are based on a cost per each door.

Description	QUAN.	UNIT	LABOR HOURS	COST EACH		
				MAT.	INST.	TOTAL

Closet Door Price Sheet	QUAN.	UNIT	LABOR HOURS	COST EACH		
				MAT.	INST.	TOTAL
Doors, bi-passing, pine, louvered, 4'-0" x 6'-8" opening	1.000	Ea.	1.333	365.00	45.00	410.00
6'-0" x 6'-8" opening	1.000	Ea.	1.600	445.00	54.00	499.00
Paneled, 4'-0" x 6'-8" opening	1.000	Ea.	1.333	350.00	45.00	395.00
6'-0" x 6'-8" opening	1.000	Ea.	1.600	430.00	54.00	484.00
Flush, birch, hollow core, 4'-0" x 6'-8" opening	1.000	Ea.	1.333	182.00	45.00	227.00
6'-0" x 6'-8" opening	1.000	Ea.	1.600	218.00	54.00	272.00
Flush, lauan, hollow core, 4'-0" x 6'-8" opening	1.000	Ea.	1.333	165.00	45.00	210.00
6'-0" x 6'-8" opening	1.000	Ea.	1.600	179.00	54.00	233.00
Bi-fold, pine, louvered, 3'-0" x 6'-8" opening	1.000	Ea.	1.231	118.00	41.50	159.50
6'-0" x 6'-8" opening	1.000	Ea.	1.600	229.00	54.00	283.00
Paneled, 3'-0" x 6'-8" opening	1.000	Ea.	1.231	118.00	41.50	159.50
6'-0" x 6'-8" opening	1.000	Ea.	1.600	229.00	54.00	283.00
Flush, birch, hollow core, 3'-0" x 6'-8" opening	1.000	Ea.	1.231	51.50	41.50	93.00
6'-0" x 6'-8" opening	1.000	Ea.	1.600	102.00	54.00	156.00
Flush, lauan, hollow core, 3'-0" x 6'8" opening	1.000	Ea.	1.231	147.00	41.50	188.50
6'-0" x 6'-8" opening	1.000	Ea.	1.600	289.00	54.00	343.00
Frame pine, 3'-0" door, 3-5/8" deep	17.000	L.F.	.725	71.00	24.50	95.50
4-5/8" deep	17.000	L.F.	.725	96.00	24.50	120.50
5-5/8" deep	17.000	L.F.	.725	90.00	24.50	114.50
4'-0" door, 3-5/8" deep	18.000	L.F.	.768	75.50	26.00	101.50
4-5/8" deep	18.000	L.F.	.768	102.00	26.00	128.00
5-5/8" deep	18.000	L.F.	.768	95.50	26.00	121.50
6'-0" door, 3-5/8" deep	19.000	L.F.	.811	79.50	27.50	107.00
4-5/8" deep	19.000	L.F.	.811	107.00	27.50	134.50
5-5/8" deep	19.000	L.F.	.811	101.00	27.50	128.50
Trim both sides, painted 3'-0" x 6'-8" door	34.000	L.F.	1.969	30.50	63.50	94.00
4'-0" x 6'-8" door	36.000	L.F.	2.086	32.50	67.50	100.00
6'-0" x 6'-8" door	38.000	L.F.	2.203	34.00	71.00	105.00
Paint 2 sides, primer & 2 cts., flush door & frame, 3' x 6'-8" opng	2.000	Face	2.914	8.25	87.00	95.25
4'-0" x 6'-8" opening	2.000	Face	3.886	11.00	116.00	127.00
6'-0" x 6'-8" opening	2.000	Face	4.857	13.80	145.00	158.80
Paneled door & frame, 3'-0" x 6'-8" opening	2.000	Face	6.000	11.55	180.00	191.55
4'-0" x 6'-8" opening	2.000	Face	8.000	15.40	240.00	255.40
6'-0" x 6'-8" opening	2.000	Face	10.000	19.25	300.00	319.25
Louvered door & frame, 3'-0" x 6'-8" opening	2.000	Face	6.000	11.55	180.00	191.55
4'-0" x 6'-8" opening	2.000	Face	8.000	15.40	240.00	255.40
6'-0" x 6'-8" opening	2.000	Face	10.000	19.25	300.00	319.25

System Description	QUAN.	UNIT	LABOR HOURS	COST PER S.F. MAT.	INST.	TOTAL
Carpet, direct glue-down, nylon, level loop, 26 oz.	1.000	S.F.	.018	1.62	.56	2.18
32 oz.	1.000	S.F.	.018	2.15	.56	2.71
40 oz.	1.000	S.F.	.018	2.63	.56	3.19
Nylon, plush, 20 oz.	1.000	S.F.	.018	1.03	.56	1.59
24 oz.	1.000	S.F.	.018	1.12	.56	1.68
30 oz.	1.000	S.F.	.018	1.54	.56	2.10
36 oz.	1.000	S.F.	.018	1.98	.56	2.54
42 oz.	1.000	S.F.	.022	2.26	.67	2.93
48 oz.	1.000	S.F.	.022	2.70	.67	3.37
54 oz.	1.000	S.F.	.022	3.09	.67	3.76
Olefin, 15 oz.	1.000	S.F.	.018	.56	.56	1.12
22 oz.	1.000	S.F.	.018	.67	.56	1.23
Tile, foam backed, needle punch	1.000	S.F.	.014	2.45	.44	2.89
Tufted loop or shag	1.000	S.F.	.014	1.03	.44	1.47
Wool, 36 oz., level loop	1.000	S.F.	.018	4.61	.56	5.17
32 oz., patterned	1.000	S.F.	.020	4.54	.62	5.16
48 oz., patterned	1.000	S.F.	.020	5.85	.62	6.47
Padding, sponge rubber cushion, minimum	1.000	S.F.	.006	.31	.18	.49
Maximum	1.000	S.F.	.006	.85	.18	1.03
Felt, 32 oz. to 56 oz., minimum	1.000	S.F.	.006	.39	.18	.57
Maximum	1.000	S.F.	.006	.64	.18	.82
Bonded urethane, 3/8″ thick, minimum	1.000	S.F.	.006	.40	.18	.58
Maximum	1.000	S.F.	.006	.68	.18	.86
Prime urethane, 1/4″ thick, minimum	1.000	S.F.	.006	.23	.18	.41
Maximum	1.000	S.F.	.006	.41	.18	.59
Stairs, for stairs, add to above carpet prices	1.000	Riser	.267		8.30	8.30
Underlayment plywood, 3/8″ thick	1.000	S.F.	.011	.68	.36	1.04
1/2″ thick	1.000	S.F.	.011	.88	.37	1.25
5/8″ thick	1.000	S.F.	.011	.84	.39	1.23
3/4″ thick	1.000	S.F.	.012	1.07	.42	1.49
Particle board, 3/8″ thick	1.000	S.F.	.011	.42	.36	.78
1/2″ thick	1.000	S.F.	.011	.44	.37	.81
5/8″ thick	1.000	S.F.	.011	.51	.39	.90
3/4″ thick	1.000	S.F.	.012	.63	.42	1.05
Hardboard, 4′ x 4′, 0.215″ thick	1.000	S.F.	.011	.45	.36	.81

System Description	QUAN.	UNIT	LABOR HOURS	COST PER S.F. MAT.	INST.	TOTAL
Resilient flooring, asphalt tile on concrete, 1/8" thick						
Color group B	1.000	S.F.	.020	1.06	.62	1.68
Color group C & D	1.000	S.F.	.020	1.17	.62	1.79
Asphalt tile on wood subfloor, 1/8" thick						
Color group B	1.000	S.F.	.020	1.26	.62	1.88
Color group C & D	1.000	S.F.	.020	1.37	.62	1.99
Vinyl composition tile, 12" x 12", 1/16" thick	1.000	S.F.	.016	.76	.50	1.26
Embossed	1.000	S.F.	.016	.95	.50	1.45
Marbleized	1.000	S.F.	.016	.95	.50	1.45
Plain	1.000	S.F.	.016	1.07	.50	1.57
.080" thick, embossed	1.000	S.F.	.016	.94	.50	1.44
Marbleized	1.000	S.F.	.016	1.08	.50	1.58
Plain	1.000	S.F.	.016	1.57	.50	2.07
1/8" thick, marbleized	1.000	S.F.	.016	1.03	.50	1.53
Plain	1.000	S.F.	.016	2.21	.50	2.71
Vinyl tile, 12" x 12", .050" thick, minimum	1.000	S.F.	.016	1.69	.50	2.19
Maximum	1.000	S.F.	.016	3.31	.50	3.81
1/8" thick, minimum	1.000	S.F.	.016	2.13	.50	2.63
Maximum	1.000	S.F.	.016	4.79	.50	5.29
1/8" thick, solid colors	1.000	S.F.	.016	3.42	.50	3.92
Florentine pattern	1.000	S.F.	.016	3.93	.50	4.43
Marbleized or travertine pattern	1.000	S.F.	.016	8.05	.50	8.55
Vinyl sheet goods, backed, .070" thick, minimum	1.000	S.F.	.032	1.38	1.00	2.38
Maximum	1.000	S.F.	.040	2.55	1.24	3.79
.093" thick, minimum	1.000	S.F.	.035	1.65	1.08	2.73
Maximum	1.000	S.F.	.040	2.92	1.24	4.16
.125" thick, minimum	1.000	S.F.	.035	1.79	1.08	2.87
Maximum	1.000	S.F.	.040	3.93	1.24	5.17
Wood, oak, finished in place, 25/32" x 2-1/2" clear	1.000	S.F.	.074	3.68	2.26	5.94
Select	1.000	S.F.	.074	4.03	2.26	6.29
No. 1 common	1.000	S.F.	.074	4.29	2.26	6.55
Prefinished, oak, 2-1/2" wide	1.000	S.F.	.047	6.45	1.59	8.04
3-1/4" wide	1.000	S.F.	.043	8.25	1.46	9.71
Ranch plank, oak, random width	1.000	S.F.	.055	8.00	1.86	9.86
Parquet, 5/16" thick, finished in place, oak, minimum	1.000	S.F.	.077	4.05	2.36	6.41
Maximum	1.000	S.F.	.107	6.35	3.37	9.72
Teak, minimum	1.000	S.F.	.077	5.30	2.36	7.66
Maximum	1.000	S.F.	.107	8.70	3.37	12.07
Sleepers, treated, 16" O.C., 1" x 2"	1.000	S.F.	.007	.22	.23	.45
1" x 3"	1.000	S.F.	.008	.22	.27	.49
2" x 4"	1.000	S.F.	.011	.72	.36	1.08
2" x 6"	1.000	S.F.	.012	1.16	.42	1.58
Subfloor, plywood, 1/2" thick	1.000	S.F.	.011	.58	.36	.94
5/8" thick	1.000	S.F.	.012	.73	.40	1.13
3/4" thick	1.000	S.F.	.013	.89	.43	1.32
Ceramic tile, color group 2, 1" x 1"	1.000	S.F.	.087	4.52	2.46	6.98
2" x 2" or 2" x 1"	1.000	S.F.	.084	4.37	2.36	6.73
Color group 1, 8" x 8"		S.F.	.064	3.32	1.80	5.12
12" x 12"		S.F.	.049	4.16	1.38	5.54
16" x 16"		S.F.	.029	6.35	.82	7.17

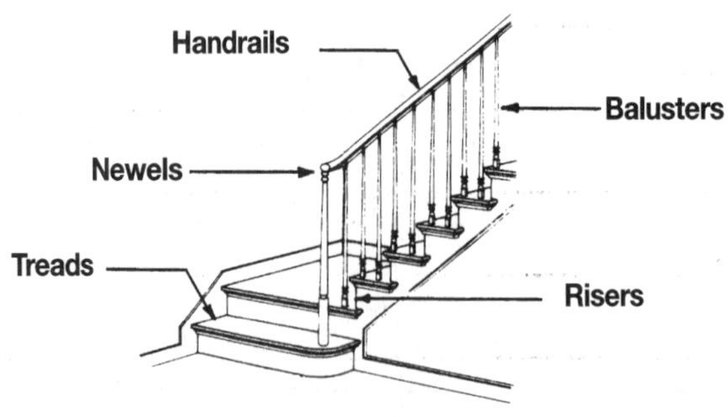

System Description	QUAN.	UNIT	LABOR HOURS	COST EACH		
				MAT.	INST.	TOTAL
7 RISERS, OAK TREADS, BOX STAIRS						
Treads, oak, 9-1/2" x 1-1/16" thick	6.000	Ea.	2.667	153.00	90.00	243.00
Risers, 3/4" thick, beech	7.000	Ea.	2.625	133.35	88.62	221.97
Balusters, birch, 30" high	12.000	Ea.	3.429	83.40	115.80	199.20
Newels, 3-1/4" wide	2.000	Ea.	2.286	74.00	77.00	151.00
Handrails, oak laminated	7.000	L.F.	.933	42.70	31.50	74.20
Stringers, 2" x 10", 3 each	21.000	L.F.	.305	9.03	10.29	19.32
TOTAL			12.245	495.48	413.21	908.69
14 RISERS, OAK TREADS, BOX STAIRS						
Treads, oak, 9-1/2" x 1-1/16" thick	13.000	Ea.	5.778	331.50	195.00	526.50
Risers, 3/4" thick, beech	14.000	Ea.	5.250	266.70	177.24	443.94
Balusters, birch, 30" high	26.000	Ea.	7.429	180.70	250.90	431.60
Newels, 3-1/4" wide	2.000	Ea.	2.286	74.00	77.00	151.00
Handrails, oak, laminated	14.000	L.F.	1.867	85.40	63.00	148.40
Stringers, 2" x 10", 3 each	42.000	L.F.	5.169	56.28	174.30	230.58
TOTAL			27.779	994.58	937.44	1932.02
14 RISERS, PINE TREADS, BOX STAIRS						
Treads, pine, 9-1/2" x 3/4" thick	13.000	Ea.	5.778	117.65	195.00	312.65
Risers, 3/4" thick, pine	14.000	Ea.	5.091	72.66	171.78	244.44
Balusters, pine, 30" high	26.000	Ea.	7.429	117.26	250.90	368.16
Newels, 3-1/4" wide	2.000	Ea.	2.286	74.00	77.00	151.00
Handrails, oak, laminated	14.000	L.F.	1.867	85.40	63.00	148.40
Stringers, 2" x 10", 3 each	42.000	L.F.	5.169	56.28	174.30	230.58
TOTAL			27.620	523.25	931.98	1455.23

Description	QUAN.	UNIT	LABOR HOURS	COST EACH		
				MAT.	INST.	TOTAL

Important: See the Reference Section for critical supporting data - Reference Nos., Crews & Location Factors

Stairway Price Sheet

Stairway Price Sheet	QUAN.	UNIT	LABOR HOURS	COST EACH MAT.	COST EACH INST.	COST EACH TOTAL
Treads, oak, 1-1/16" x 9-1/2", 3' long, 7 riser stair	6.000	Ea.	2.667	153.00	90.00	243.00
14 riser stair	13.000	Ea.	5.778	330.00	195.00	525.00
1-1/16" x 11-1/2", 3' long, 7 riser stair	6.000	Ea.	2.667	153.00	90.00	243.00
14 riser stair	13.000	Ea.	5.778	330.00	195.00	525.00
Pine, 3/4" x 9-1/2", 3' long, 7 riser stair	6.000	Ea.	2.667	54.50	90.00	144.50
14 riser stair	13.000	Ea.	5.778	118.00	195.00	313.00
3/4" x 11-1/4", 3' long, 7 riser stair	6.000	Ea.	2.667	61.50	90.00	151.50
14 riser stair	13.000	Ea.	5.778	133.00	195.00	328.00
Risers, oak, 3/4" x 7-1/2" high, 7 riser stair	7.000	Ea.	2.625	116.00	88.50	204.50
14 riser stair	14.000	Ea.	5.250	231.00	177.00	408.00
Beech, 3/4" x 7-1/2" high, 7 riser stair	7.000	Ea.	2.625	133.00	88.50	221.50
14 riser stair	14.000	Ea.	5.250	267.00	177.00	444.00
Baluster, turned, 30" high, pine, 7 riser stair	12.000	Ea.	3.429	54.00	116.00	170.00
14 riser stair	26.000	Ea.	7.429	117.00	251.00	368.00
30" birch, 7 riser stair	12.000	Ea.	3.429	83.50	116.00	199.50
14 riser stair	26.000	Ea.	7.429	181.00	251.00	432.00
42" pine, 7 riser stair	12.000	Ea.	3.556	79.00	120.00	199.00
14 riser stair	26.000	Ea.	7.704	172.00	260.00	432.00
42" birch, 7 riser stair	12.000	Ea.	3.556	101.00	120.00	221.00
14 riser stair	26.000	Ea.	7.704	218.00	260.00	478.00
Newels, 3-1/4" wide, starting, 7 riser stair	2.000	Ea.	2.286	74.00	77.00	151.00
14 riser stair	2.000	Ea.	2.286	74.00	77.00	151.00
Landing, 7 riser stair	2.000	Ea.	3.200	162.00	108.00	270.00
14 riser stair	2.000	Ea.	3.200	162.00	108.00	270.00
Handrails, oak, laminated, 7 riser stair	7.000	L.F.	.933	42.50	31.50	74.00
14 riser stair	14.000	L.F.	1.867	85.50	63.00	148.50
Stringers, fir, 2" x 10" 7 riser stair	21.000	L.F.	2.585	28.00	87.00	115.00
14 riser stair	42.000	L.F.	5.169	56.50	174.00	230.50
2" x 12", 7 riser stair	21.000	L.F.	2.585	34.00	87.00	121.00
14 riser stair	42.000	L.F.	5.169	68.00	174.00	242.00

Special Stairways

Special Stairways	QUAN.	UNIT	LABOR HOURS	COST EACH MAT.	COST EACH INST.	COST EACH TOTAL
Basement stairs, prefabricated, open risers	1.000	Flight	4.000	630.00	135.00	765.00
Curved stairways, 3'-3" wide, prefabricated oak, 9' high	1.000	Flight	22.857	7025.00	770.00	7795.00
10' high	1.000	Flight	22.857	7925.00	770.00	8695.00
Open two sides, 9' high	1.000	Flight	32.000	11000.00	1075.00	12075.00
10' high	1.000	Flight	32.000	11900.00	1075.00	12975.00
Spiral stairs, oak, 4'-6" diameter, prefabricated, 9' high	1.000	Flight	10.667	4400.00	360.00	4760.00
Aluminum, 5'-0" diameter stock unit	1.000	Flight	9.956	2850.00	455.00	3305.00
Custom unit	1.000	Flight	9.956	5400.00	455.00	5855.00
Cast iron, 4'-0" diameter, minimum	1.000	Flight	9.956	2550.00	455.00	3005.00
Maximum	1.000	Flight	17.920	3475.00	820.00	4295.00
Steel, industrial, pre-erected, 3'-6" wide, bar rail	1.000	Flight	7.724	2550.00	500.00	3050.00
Picket rail	1.000	Flight	7.724	2850.00	500.00	3350.00

Division 7
Specialties

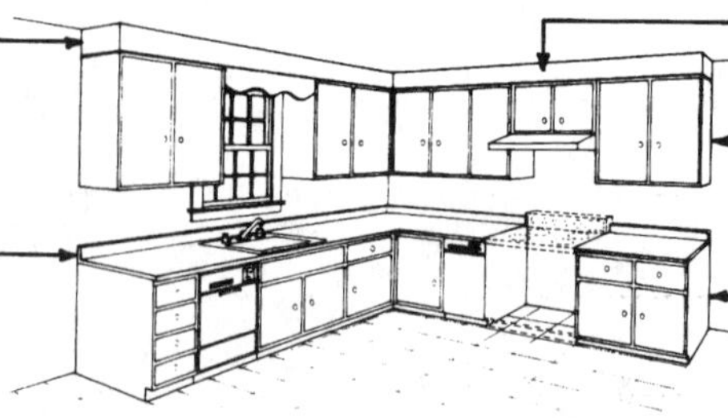

Soffit Drywall
Soffit Framing
Top Cabinets
Counter Top
Bottom Cabinets

System Description	QUAN.	UNIT	LABOR HOURS	COST PER L.F. MAT.	INST.	TOTAL
KITCHEN, ECONOMY GRADE						
Top cabinets, economy grade	1.000	L.F.	.171	28.32	5.76	34.08
Bottom cabinets, economy grade	1.000	L.F.	.256	42.48	8.64	51.12
Counter top, laminated plastic, post formed	1.000	L.F.	.267	9.40	9.00	18.40
Blocking, wood, 2" x 4"	1.000	L.F.	.032	.43	1.08	1.51
Soffit, framing, wood, 2" x 4"	4.000	L.F.	.071	1.72	2.40	4.12
Soffit drywall, painted	2.000	S.F.	.060	.84	2.04	2.88
TOTAL			.857	83.19	28.92	112.11
AVERAGE GRADE						
Top cabinets, average grade	1.000	L.F.	.213	35.40	7.20	42.60
Bottom cabinets, average grade	1.000	L.F.	.320	53.10	10.80	63.90
Counter top, laminated plastic, square edge, incl. backsplash	1.000	L.F.	.267	28.50	9.00	37.50
Blocking, wood, 2" x 4"	1.000	L.F.	.032	.43	1.08	1.51
Soffit framing, wood, 2" x 4"	4.000	L.F.	.071	1.72	2.40	4.12
Soffit drywall, painted	2.000	S.F.	.060	.84	2.04	2.88
TOTAL			.963	119.99	32.52	152.51
CUSTOM GRADE						
Top cabinets, custom grade	1.000	L.F.	.256	99.20	8.60	107.80
Bottom cabinets, custom grade	1.000	L.F.	.384	148.80	12.90	161.70
Counter top, laminated plastic, square edge, incl. backsplash	1.000	L.F.	.267	28.50	9.00	37.50
Blocking, wood, 2" x 4"	1.000	L.F.	.032	.43	1.08	1.51
Soffit framing, wood, 2" x 4"	4.000	L.F.	.071	1.72	2.40	4.12
Soffit drywall, painted	2.000	S.F.	.060	.84	2.04	2.88
TOTAL			1.070	279.49	36.02	315.51

Description	QUAN.	UNIT	LABOR HOURS	COST PER L.F. MAT.	INST.	TOTAL

Kitchen Price Sheet	QUAN.	UNIT	LABOR HOURS	COST PER L.F.		
				MAT.	INST.	TOTAL
Top cabinets, economy grade	1.000	L.F.	.171	28.50	5.75	34.25
Average grade	1.000	L.F.	.213	35.50	7.20	42.70
Custom grade	1.000	L.F.	.256	99.00	8.60	107.60
Bottom cabinets, economy grade	1.000	L.F.	.256	42.50	8.65	51.15
Average grade	1.000	L.F.	.320	53.00	10.80	63.80
Custom grade	1.000	L.F.	.384	149.00	12.90	161.90
Counter top, laminated plastic, 7/8″ thick, no splash	1.000	L.F.	.267	17.30	9.00	26.30
With backsplash	1.000	L.F.	.267	22.50	9.00	31.50
1-1/4″ thick, no splash	1.000	L.F.	.286	20.50	9.65	30.15
With backsplash	1.000	L.F.	.286	25.50	9.65	35.15
Post formed, laminated plastic	1.000	L.F.	.267	9.40	9.00	18.40
Marble, with backsplash, minimum	1.000	L.F.	.471	33.00	15.95	48.95
Maximum	1.000	L.F.	.615	84.00	21.00	105.00
Maple, solid laminated, no backsplash	1.000	L.F.	.286	34.00	9.65	43.65
With backsplash	1.000	L.F.	.286	38.50	9.65	48.15
Blocking, wood, 2″ x 4″	1.000	L.F.	.032	.43	1.08	1.51
2″ x 6″	1.000	L.F.	.036	.65	1.22	1.87
2″ x 8″	1.000	L.F.	.040	.91	1.35	2.26
Soffit framing, wood, 2″ x 3″	4.000	L.F.	.064	1.36	2.16	3.52
2″ x 4″	4.000	L.F.	.071	1.72	2.40	4.12
Soffit, drywall, painted	2.000	S.F.	.060	.84	2.04	2.88
Paneling, standard	2.000	S.F.	.064	1.58	2.16	3.74
Deluxe	2.000	S.F.	.091	3.56	3.08	6.64
Sinks, porcelain on cast iron, single bowl, 21″ x 24″	1.000	Ea.	10.334	270.00	340.00	610.00
21″ x 30″	1.000	Ea.	10.334	410.00	340.00	750.00
Double bowl, 20″ x 32″	1.000	Ea.	10.810	370.00	355.00	725.00
Stainless steel, single bowl, 16″ x 20″	1.000	Ea.	10.334	355.00	340.00	695.00
22″ x 25″	1.000	Ea.	10.334	385.00	340.00	725.00
Double bowl, 20″ x 32″	1.000	Ea.	10.810	194.00	355.00	549.00

System Description	QUAN.	UNIT	LABOR HOURS	COST EACH MAT.	COST EACH INST.	COST EACH TOTAL
All appliances include plumbing and electrical rough-in & hook-ups						
Range, free standing, minimum	1.000	Ea.	3.600	350.00	112.00	462.00
Maximum	1.000	Ea.	6.000	1375.00	172.00	1547.00
Built-in, minimum	1.000	Ea.	3.333	350.00	121.00	471.00
Maximum	1.000	Ea.	10.000	940.00	345.00	1285.00
Counter top range, 4-burner, minimum	1.000	Ea.	3.333	254.00	121.00	375.00
Maximum	1.000	Ea.	4.667	540.00	169.00	709.00
Compactor, built-in, minimum	1.000	Ea.	2.215	380.00	76.50	456.50
Maximum	1.000	Ea.	3.282	415.00	113.00	528.00
Dishwasher, built-in, minimum	1.000	Ea.	6.735	315.00	244.00	559.00
Maximum	1.000	Ea.	9.235	330.00	335.00	665.00
Garbage disposer, minimum	1.000	Ea.	2.810	59.00	102.00	161.00
Maximum	1.000	Ea.	2.810	209.00	102.00	311.00
Microwave oven, minimum	1.000	Ea.	2.615	107.00	95.00	202.00
Maximum	1.000	Ea.	4.615	415.00	168.00	583.00
Range hood, ducted, minimum	1.000	Ea.	4.658	72.00	160.00	232.00
Maximum	1.000	Ea.	5.991	510.00	206.00	716.00
Ductless, minimum	1.000	Ea.	2.615	63.50	91.00	154.50
Maximum	1.000	Ea.	3.948	500.00	137.00	637.00
Refrigerator, 16 cu.ft., minimum	1.000	Ea.	2.000	580.00	49.50	629.50
Maximum	1.000	Ea.	3.200	930.00	79.00	1009.00
16 cu.ft. with icemaker, minimum	1.000	Ea.	4.210	715.00	125.00	840.00
Maximum	1.000	Ea.	5.410	1075.00	154.00	1229.00
19 cu.ft., minimum	1.000	Ea.	2.667	705.00	66.00	771.00
Maximum	1.000	Ea.	4.667	1225.00	116.00	1341.00
19 cu.ft. with icemaker, minimum	1.000	Ea.	5.143	900.00	148.00	1048.00
Maximum	1.000	Ea.	7.143	1425.00	198.00	1623.00
Sinks, porcelain on cast iron single bowl, 21″ x 24″	1.000	Ea.	10.334	270.00	340.00	610.00
21″ x 30″	1.000	Ea.	10.334	410.00	340.00	750.00
Double bowl, 20″ x 32″	1.000	Ea.	10.810	370.00	355.00	725.00
Stainless steel, single bowl 16″ x 20″	1.000	Ea.	10.334	355.00	340.00	695.00
22″ x 25″	1.000	Ea.	10.334	385.00	340.00	725.00
Double bowl, 20″ x 32″	1.000	Ea.	10.810	194.00	355.00	549.00
Water heater, electric, 30 gallon	1.000	Ea.	3.636	263.00	132.00	395.00
40 gallon	1.000	Ea.	4.000	284.00	145.00	429.00
Gas, 30 gallon	1.000	Ea.	4.000	345.00	145.00	490.00
75 gallon	1.000	Ea.	5.333	885.00	194.00	1079.00
Wall, packaged terminal heater/air conditioner cabinet, wall sleeve, louver, electric heat, thermostat, manual changeover, 208V						
6000 BTUH cooling, 8800 BTU heating	1.000	Ea.	2.667	1000.00	87.50	1087.50
9000 BTUH cooling, 13,900 BTU heating	1.000	Ea.	3.200	1025.00	105.00	1130.00
12,000 BTUH cooling, 13,900 BTU heating	1.000	Ea.	4.000	1075.00	132.00	1207.00
15,000 BTUH cooling, 13,900 BTU heating	1.000	Ea.	5.333	1150.00	175.00	1325.00

Important: See the Reference Section for critical supporting data - Reference Nos., Crews & Location Factors

System Description	QUAN.	UNIT	LABOR HOURS	COST EACH MAT.	INST.	TOTAL
Curtain rods, stainless, 1″ diameter, 3′ long	1.000	Ea.	.615	22.50	21.00	43.50
5′ long	1.000	Ea.	.615	22.50	21.00	43.50
Grab bar, 1″ diameter, 12″ long	1.000	Ea.	.283	33.00	9.55	42.55
36″ long	1.000	Ea.	.340	42.00	11.50	53.50
1-1/4″ diameter, 12″ long	1.000	Ea.	.333	39.00	11.25	50.25
36″ long	1.000	Ea.	.400	49.50	13.50	63.00
1-1/2″ diameter, 12″ long	1.000	Ea.	.383	45.00	12.95	57.95
36″ long	1.000	Ea.	.460	57.00	15.55	72.55
Mirror, 18″ x 24″	1.000	Ea.	.400	67.00	13.50	80.50
72″ x 24″	1.000	Ea.	1.333	216.00	45.00	261.00
Medicine chest with mirror, 18″ x 24″	1.000	Ea.	.400	100.00	13.50	113.50
36″ x 24″	1.000	Ea.	.600	150.00	20.50	170.50
Toilet tissue dispenser, surface mounted, minimum	1.000	Ea.	.267	11.65	9.00	20.65
Maximum	1.000	Ea.	.400	17.50	13.50	31.00
Flush mounted, minimum	1.000	Ea.	.293	12.80	9.90	22.70
Maximum	1.000	Ea.	.427	18.65	14.40	33.05
Towel bar, 18″ long, minimum	1.000	Ea.	.278	26.50	9.40	35.90
Maximum	1.000	Ea.	.348	33.00	11.75	44.75
24″ long, minimum	1.000	Ea.	.313	29.50	10.60	40.10
Maximum	1.000	Ea.	.383	36.50	12.95	49.45
36″ long, minimum	1.000	Ea.	.381	38.00	12.85	50.85
Maximum	1.000	Ea.	.419	42.00	14.15	56.15

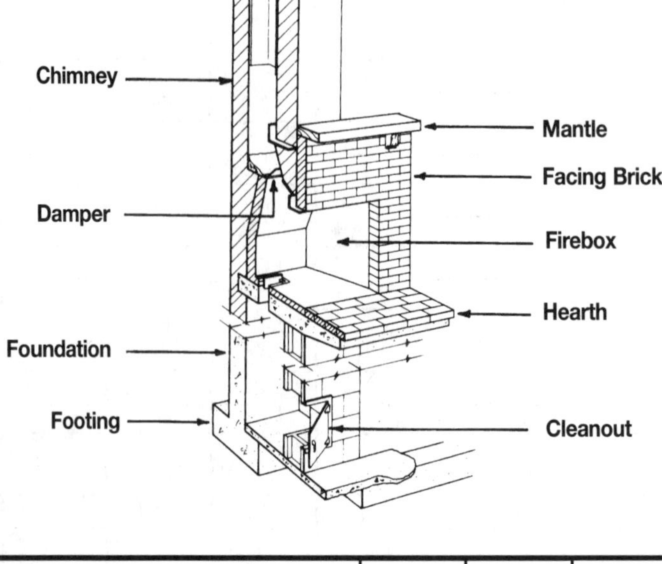

System Description	QUAN.	UNIT	LABOR HOURS	COST EACH		
				MAT.	INST.	TOTAL
MASONRY FIREPLACE						
Footing, 8″ thick, concrete, 4′ x 7′	.700	C.Y.	2.110	70.70	63.07	133.77
Foundation, concrete block, 32″ x 60″ x 4′ deep	1.000	Ea.	5.333	85.80	165.00	250.80
Fireplace, brick firebox, 30″ x 29″ opening	1.000	Ea.	40.000	390.00	1200.00	1590.00
Damper, cast iron, 30″ opening	1.000	Ea.	1.333	78.00	45.00	123.00
Facing brick, standard size brick, 6′ x 5′	30.000	S.F.	5.217	88.50	162.00	250.50
Hearth, standard size brick, 3′ x 6′	1.000	Ea.	8.000	145.00	242.00	387.00
Chimney, standard size brick, 8″ x 12″ flue, one story house	12.000	V.L.F.	12.000	258.00	360.00	618.00
Mantle, 4″ x 8″, wood	6.000	L.F.	1.333	27.96	45.00	72.96
Cleanout, cast iron, 8″ x 8″	1.000	Ea.	.667	19.25	22.50	41.75
TOTAL			75.993	1163.21	2304.57	3467.78

The costs in this system are on a cost each basis.

Description	QUAN.	UNIT	LABOR HOURS	COST EACH		
				MAT.	INST.	TOTAL

 Important: See the Reference Section for critical supporting data - Reference Nos., Crews & Location Factors

Masonry Fireplace Price Sheet

Masonry Fireplace Price Sheet	QUAN.	UNIT	LABOR HOURS	COST EACH MAT.	INST.	TOTAL
Footing 8″ thick, 3′ x 6′	.440	C.Y.	1.326	44.50	39.50	84.00
4′ x 7′	.700	C.Y.	2.110	70.50	63.00	133.50
5′ x 8′	1.000	C.Y.	3.014	101.00	90.00	191.00
1′ thick, 3′ x 6′	.670	C.Y.	2.020	67.50	60.50	128.00
4′ x 7′	1.030	C.Y.	3.105	104.00	92.50	196.50
5′ x 8′	1.480	C.Y.	4.461	149.00	133.00	282.00
Foundation-concrete block, 24″ x 48″, 4′ deep	1.000	Ea.	4.267	68.50	132.00	200.50
8′ deep	1.000	Ea.	8.533	137.00	264.00	401.00
24″ x 60″, 4′ deep	1.000	Ea.	4.978	80.00	154.00	234.00
8′ deep	1.000	Ea.	9.956	160.00	310.00	470.00
32″ x 48″, 4′ deep	1.000	Ea.	4.711	76.00	146.00	222.00
8′ deep	1.000	Ea.	9.422	152.00	292.00	444.00
32″ x 60″, 4′ deep	1.000	Ea.	5.333	86.00	165.00	251.00
8′ deep	1.000	Ea.	10.844	174.00	335.00	509.00
32″ x 72″, 4′ deep	1.000	Ea.	6.133	98.50	190.00	288.50
8′ deep	1.000	Ea.	12.267	197.00	380.00	577.00
Fireplace, brick firebox 30″ x 29″ opening	1.000	Ea.	40.000	390.00	1200.00	1590.00
48″ x 30″ opening	1.000	Ea.	60.000	585.00	1800.00	2385.00
Steel fire box with registers, 25″ opening	1.000	Ea.	26.667	750.00	810.00	1560.00
48″ opening	1.000	Ea.	44.000	1125.00	1325.00	2450.00
Damper, cast iron, 30″ opening	1.000	Ea.	1.333	78.00	45.00	123.00
36″ opening	1.000	Ea.	1.556	91.00	52.50	143.50
Steel, 30″ opening	1.000	Ea.	1.333	69.50	45.00	114.50
36″ opening	1.000	Ea.	1.556	81.00	52.50	133.50
Facing for fireplace, standard size brick, 6′ x 5′	30.000	S.F.	5.217	88.50	162.00	250.50
7′ x 5′	35.000	S.F.	6.087	103.00	189.00	292.00
8′ x 6′	48.000	S.F.	8.348	142.00	259.00	401.00
Fieldstone, 6′ x 5′	30.000	S.F.	5.217	400.00	162.00	562.00
7′ x 5′	35.000	S.F.	6.087	470.00	189.00	659.00
8′ x 6′	48.000	S.F.	8.348	645.00	259.00	904.00
Sheetrock on metal, studs, 6′ x 5′	30.000	S.F.	1.038	15.90	35.00	50.90
7′ x 5′	35.000	S.F.	1.211	18.55	41.00	59.55
8′ x 6′	48.000	S.F.	1.661	25.50	56.00	81.50
Hearth, standard size brick, 3′ x 6′	1.000	Ea.	8.000	145.00	242.00	387.00
3′ x 7′	1.000	Ea.	9.280	168.00	281.00	449.00
3′ x 8′	1.000	Ea.	10.640	193.00	320.00	513.00
Stone, 3′ x 6′	1.000	Ea.	8.000	156.00	242.00	398.00
3′ x 7′	1.000	Ea.	9.280	181.00	281.00	462.00
3′ x 8′	1.000	Ea.	10.640	207.00	320.00	527.00
Chimney, standard size brick , 8″ x 12″ flue, one story house	12.000	V.L.F.	12.000	258.00	360.00	618.00
Two story house	20.000	V.L.F.	20.000	430.00	600.00	1030.00
Mantle wood, beams, 4″ x 8″	6.000	L.F.	1.333	28.00	45.00	73.00
4″ x 10″	6.000	L.F.	1.371	35.00	46.00	81.00
Ornate, prefabricated, 6′ x 3′-6″ opening, minimum	1.000	Ea.	1.600	142.00	54.00	196.00
Maximum	1.000	Ea.	1.600	173.00	54.00	227.00
Cleanout, door and frame, cast iron, 8″ x 8″	1.000	Ea.	.667	19.25	22.50	41.75
12″ x 12″	1.000	Ea.	.800	36.50	27.00	63.50

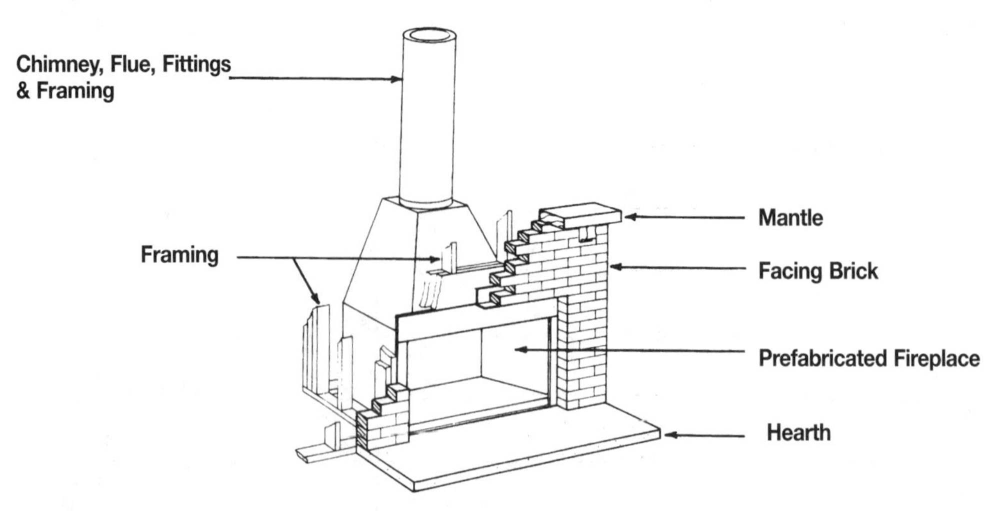

System Description	QUAN.	UNIT	LABOR HOURS	COST EACH		
				MAT.	INST.	TOTAL
PREFABRICATED FIREPLACE						
Prefabricated fireplace, metal, minimum	1.000	Ea.	6.154	1025.00	208.00	1233.00
Framing, 2″ x 4″ studs, 6′ x 5′	35.000	L.F.	.509	15.05	17.15	32.20
Sheetrock, 1/2″ fire resistant, 6′ x 5′	40.000	S.F.	.640	12.80	21.60	34.40
Facing, brick, standard size brick, 6′ x 5′	30.000	S.F.	5.217	88.50	162.00	250.50
Hearth, standard size brick, 3′ x 6′	1.000	Ea.	8.000	145.00	242.00	387.00
Chimney, one story house, framing, 2″ x 4″ studs	80.000	L.F.	1.164	34.40	39.20	73.60
Sheathing, plywood, 5/8″ thick	32.000	S.F.	.759	38.40	25.60	64.00
Flue, 10″ metal, insulated pipe	12.000	V.L.F.	4.000	258.00	130.80	388.80
Fittings, ceiling support	1.000	Ea.	.667	127.00	22.00	149.00
Fittings, joist shield	1.000	Ea.	.667	71.00	22.00	93.00
Fittings, roof flashing	1.000	Ea.	.667	144.00	22.00	166.00
Mantle beam, wood, 4″ x 8″	6.000	L.F.	1.333	27.96	45.00	72.96
TOTAL			29.777	1987.11	957.35	2944.46

The costs in this system are on a cost each basis.

Description	QUAN.	UNIT	LABOR HOURS	COST EACH		
				MAT.	INST.	TOTAL

Prefabricated Fireplace Price Sheet	QUAN.	UNIT	LABOR HOURS	COST EACH		
				MAT.	INST.	TOTAL
Prefabricated fireplace, minimum	1.000	Ea.	6.154	1025.00	208.00	1233.00
Average	1.000	Ea.	8.000	1225.00	270.00	1495.00
Maximum	1.000	Ea.	8.889	2950.00	300.00	3250.00
Framing, 2″ x 4″ studs, fireplace, 6′ x 5′	35.000	L.F.	.509	15.05	17.15	32.20
7′ x 5′	40.000	L.F.	.582	17.20	19.60	36.80
8′ x 6′	45.000	L.F.	.655	19.35	22.00	41.35
Sheetrock, 1/2″ thick, fireplace, 6′ x 5′	40.000	S.F.	.640	12.80	21.50	34.30
7′ x 5′	45.000	S.F.	.720	14.40	24.50	38.90
8′ x 6′	50.000	S.F.	.800	16.00	27.00	43.00
Facing for fireplace, brick, 6′ x 5′	30.000	S.F.	5.217	88.50	162.00	250.50
7′ x 5′	35.000	S.F.	6.087	103.00	189.00	292.00
8′ x 6′	48.000	S.F.	8.348	142.00	259.00	401.00
Fieldstone, 6′ x 5′	30.000	S.F.	5.217	420.00	162.00	582.00
7′ x 5′	35.000	S.F.	6.087	490.00	189.00	679.00
8′ x 6′	48.000	S.F.	8.348	670.00	259.00	929.00
Hearth, standard size brick, 3′ x 6′	1.000	Ea.	8.000	145.00	242.00	387.00
3′ x 7′	1.000	Ea.	9.280	168.00	281.00	449.00
3′ x 8′	1.000	Ea.	10.640	193.00	320.00	513.00
Stone, 3′ x 6′	1.000	Ea.	8.000	156.00	242.00	398.00
3′ x 7′	1.000	Ea.	9.280	181.00	281.00	462.00
3′ x 8′	1.000	Ea.	10.640	207.00	320.00	527.00
Chimney, framing, 2″ x 4″, one story house	80.000	L.F.	1.164	34.50	39.00	73.50
Two story house	120.000	L.F.	1.745	51.50	59.00	110.50
Sheathing, plywood, 5/8″ thick	32.000	S.F.	.759	38.50	25.50	64.00
Stucco on plywood	32.000	S.F.	1.124	43.50	37.00	80.50
Flue, 10″ metal pipe, insulated, one story house	12.000	V.L.F.	4.000	258.00	131.00	389.00
Two story house	20.000	V.L.F.	6.667	430.00	218.00	648.00
Fittings, ceiling support	1.000	Ea.	.667	127.00	22.00	149.00
Fittings joist sheild, one story house	1.000	Ea.	.667	71.00	22.00	93.00
Two story house	2.000	Ea.	1.333	142.00	44.00	186.00
Fittings roof flashing	1.000	Ea.	.667	144.00	22.00	166.00
Mantle, wood beam, 4″ x 8″	6.000	L.F.	1.333	28.00	45.00	73.00
4″ x 10″	6.000	L.F.	1.371	35.00	46.00	81.00
Ornate prefabricated, 6′ x 3′-6″ opening, minimum	1.000	Ea.	1.600	142.00	54.00	196.00
Maximum	1.000	Ea.	1.600	173.00	54.00	227.00

System Description	QUAN.	UNIT	LABOR HOURS	COST EACH		
				MAT.	INST.	TOTAL
Economy, lean to, shell only, not including 2′ stub wall, fndtn, flrs, heat			18.824	1350.00	565.00	1915.00
4′ x 16′	1.000	Ea.	26.212	1900.00	885.00	2785.00
4′ x 24′	1.000	Ea.	30.259	2175.00	1025.00	3200.00
6′ x 10′	1.000	Ea.	16.552	1600.00	560.00	2160.00
6′ x 16′	1.000	Ea.	23.034	2225.00	775.00	3000.00
6′ x 24′	1.000	Ea.	29.793	2850.00	1000.00	3850.00
8′ x 10′	1.000	Ea.	22.069	2125.00	745.00	2870.00
8′ x 16′	1.000	Ea.	38.400	3700.00	1300.00	5000.00
8′ x 24′	1.000	Ea.	49.655	4775.00	1675.00	6450.00
Free standing, 8′ x 8′	1.000	Ea.	17.356	2475.00	585.00	3060.00
8′ x 16′	1.000	Ea.	30.210	4300.00	1025.00	5325.00
8′ x 24′	1.000	Ea.	39.051	5550.00	1325.00	6875.00
10′ x 10′	1.000	Ea.	18.824	2950.00	635.00	3585.00
10′ x 16′	1.000	Ea.	24.094	3775.00	815.00	4590.00
10′ x 24′	1.000	Ea.	31.624	4950.00	1075.00	6025.00
14′ x 10′	1.000	Ea.	20.741	3700.00	700.00	4400.00
14′ x 16′	1.000	Ea.	24.889	4450.00	840.00	5290.00
14′ x 24′	1.000	Ea.	33.348	5975.00	1125.00	7100.00
Standard,lean to,shell only,not incl.2′stub wall, fndtn,flrs, heat 4′x10′	1.000	Ea.	28.235	2050.00	955.00	3005.00
4′ x 16′	1.000	Ea.	39.341	2850.00	1325.00	4175.00
4′ x 24′	1.000	Ea.	45.412	3275.00	1525.00	4800.00
6′ x 10′	1.000	Ea.	24.828	2375.00	835.00	3210.00
6′ x 16′	1.000	Ea.	34.538	3325.00	1175.00	4500.00
6′ x 24′	1.000	Ea.	44.690	4300.00	1500.00	5800.00
8′ x 10′	1.000	Ea.	33.103	3175.00	1125.00	4300.00
8′ x 16′	1.000	Ea.	57.600	5525.00	1950.00	7475.00
8′ x 24′	1.000	Ea.	74.483	7150.00	2500.00	9650.00
Free standing, 8′ x 8′	1.000	Ea.	26.034	3700.00	880.00	4580.00
8′ x 16′	1.000	Ea.	45.315	6425.00	1525.00	7950.00
8′ x 24′	1.000	Ea.	58.576	8325.00	1975.00	10300.00
10′ x 10′	1.000	Ea.	28.235	4425.00	955.00	5380.00
10′ x 16′	1.000	Ea.	36.141	5675.00	1225.00	6900.00
10′ x 24′	1.000	Ea.	47.435	7425.00	1600.00	9025.00
14′ x 10′	1.000	Ea.	31.111	5575.00	1050.00	6625.00
14′ x 16′	1.000	Ea.	37.333	6675.00	1250.00	7925.00
14′ x 24′	1.000	Ea.	50.030	8950.00	1700.00	10650.00
Deluxe,lean to,shell only,not incl.2′stub wall, fndtn, flrs or heat, 4′x10′	1.000	Ea.	20.645	3525.00	695.00	4220.00
4′ x 16′	1.000	Ea.	33.032	5625.00	1125.00	6750.00
4′ x 24′	1.000	Ea.	49.548	8450.00	1675.00	10125.00
6′ x 10′	1.000	Ea.	30.968	5275.00	1050.00	6325.00
6′ x 16′	1.000	Ea.	49.548	8450.00	1675.00	10125.00
6′ x 24′	1.000	Ea.	74.323	12700.00	2500.00	15200.00
8′ x 10′	1.000	Ea.	41.290	7050.00	1400.00	8450.00
8′ x 16′	1.000	Ea.	66.065	11300.00	2225.00	13525.00
8′ x 24′	1.000	Ea.	99.097	16900.00	3350.00	20250.00
Freestanding, 8′ x 8′	1.000	Ea.	18.618	4875.00	625.00	5500.00
8′ x 16′	1.000	Ea.	37.236	9725.00	1250.00	10975.00
8′ x 24′	1.000	Ea.	55.855	14600.00	1875.00	16475.00
10′ x 10′	1.000	Ea.	29.091	7600.00	980.00	8580.00
10′ x 16′	1.000	Ea.	46.545	12200.00	1575.00	13775.00
10′ x 24′	1.000	Ea.	69.818	18200.00	2350.00	20550.00
14′ x 10′	1.000	Ea.	40.727	10600.00	1375.00	11975.00
14′ x 16′	1.000	Ea.	65.164	17000.00	2200.00	19200.00
14′ x 24′	1.000	Ea.	97.745	25500.00	3300.00	28800.00

36 Swimming Pool Systems

System Description	QUAN.	UNIT	LABOR HOURS	COST EACH		
				MAT.	INST.	TOTAL
Swimming pools, vinyl lined, metal sides, sand bottom, 12′ x 28′	1.000	Ea.	50.176	3075.00	1775.00	4850.00
12′ x 32′	1.000	Ea.	55.365	3400.00	1975.00	5375.00
12′ x 36′	1.000	Ea.	60.060	3675.00	2175.00	5850.00
16′ x 32′	1.000	Ea.	66.797	4100.00	2400.00	6500.00
16′ x 36′	1.000	Ea.	71.189	4375.00	2550.00	6925.00
16′ x 40′	1.000	Ea.	74.702	4575.00	2675.00	7250.00
20′ x 36′	1.000	Ea.	77.859	4775.00	2800.00	7575.00
20′ x 40′	1.000	Ea.	82.133	5050.00	2925.00	7975.00
20′ x 44′	1.000	Ea.	90.347	5550.00	3250.00	8800.00
24′ x 40′	1.000	Ea.	98.560	6050.00	3525.00	9575.00
24′ x 44′	1.000	Ea.	108.416	6650.00	3875.00	10525.00
24′ x 48′	1.000	Ea.	118.272	7250.00	4225.00	11475.00
Vinyl lined, concrete sides, 12′ x 28′	1.000	Ea.	79.445	4875.00	2850.00	7725.00
12′ x 32′	1.000	Ea.	88.816	5450.00	3175.00	8625.00
12′ x 36′	1.000	Ea.	97.655	6000.00	3500.00	9500.00
16′ x 32′	1.000	Ea.	111.391	6825.00	3975.00	10800.00
16′ x 36′	1.000	Ea.	121.352	7450.00	4350.00	11800.00
16′ x 40′	1.000	Ea.	130.443	8000.00	4675.00	12675.00
28′ x 36′	1.000	Ea.	140.582	8625.00	5025.00	13650.00
20′ x 40′	1.000	Ea.	149.333	9150.00	5350.00	14500.00
20′ x 44′	1.000	Ea.	164.267	10100.00	5900.00	16000.00
24′ x 40′	1.000	Ea.	179.200	11000.00	6425.00	17425.00
24′ x 44′	1.000	Ea.	197.120	12100.00	7075.00	19175.00
24′ x 48′	1.000	Ea.	215.040	13200.00	7700.00	20900.00
Gunite, bottom and sides, 12′ x 28′	1.000	Ea.	129.766	6375.00	4650.00	11025.00
12′ x 32′	1.000	Ea.	142.163	6975.00	5100.00	12075.00
12′ x 36′	1.000	Ea.	153.027	7500.00	5475.00	12975.00
16′ x 32′	1.000	Ea.	167.741	8225.00	6000.00	14225.00
16′ x 36′	1.000	Ea.	176.419	8650.00	6300.00	14950.00
16′ x 40′	1.000	Ea.	182.367	8950.00	6525.00	15475.00
20′ x 36′	1.000	Ea.	187.948	9225.00	6725.00	15950.00
20′ x 40′	1.000	Ea.	179.200	12200.00	6425.00	18625.00
20′ x 44′	1.000	Ea.	197.120	13400.00	7075.00	20475.00
24′ x 40′	1.000	Ea.	215.040	14600.00	7700.00	22300.00
24′ x 44′	1.000	Ea.	273.241	13400.00	9775.00	23175.00
24′ x 48′	1.000	Ea.	298.074	14600.00	10700.00	25300.00

40 Wood Deck Systems

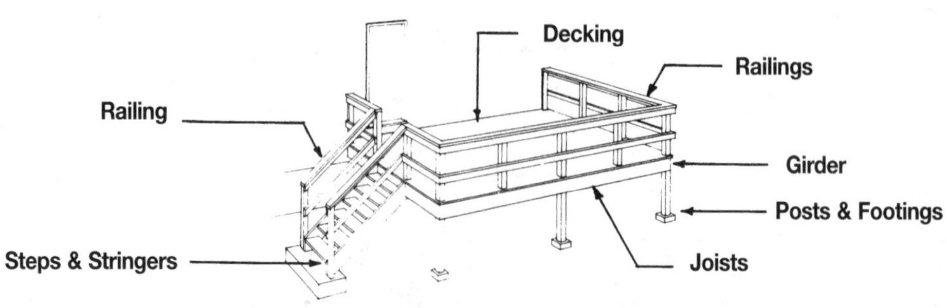

System Description	QUAN.	UNIT	LABOR HOURS	COST PER S.F. MAT.	COST PER S.F. INST.	COST PER S.F. TOTAL
8′ X 12′ DECK, PRESSURE TREATED LUMBER, JOISTS 16″ O.C.						
Decking, 2″ x 6″ lumber	2.080	L.F.	.027	1.62	.89	2.51
Joists, 2″ x 8″, 16″ O.C.	1.000	L.F.	.015	1.08	.49	1.57
Girder, 2″ x 10″	.125	L.F.	.002	.20	.08	.28
Posts, 4″ x 4″, including concrete footing	.250	L.F.	.022	.55	.68	1.23
Stairs, 2″ x 10″ stringers, 2″ x 10″ steps	1.000	Set	.020	3.15	.68	3.83
Railings, 2″ x 4″	1.000	L.F.	.026	.52	.87	1.39
TOTAL			.112	7.12	3.69	10.81
12′ X 16′ DECK, PRESSURE TREATED LUMBER, JOISTS 24″ O.C.						
Decking, 2″ x 6″	2.080	L.F.	.027	1.62	.89	2.51
Joists, 2″ x 10″, 24″ O.C.	.800	L.F.	.014	1.24	.48	1.72
Girder, 2″ x 10″	.083	L.F.	.001	.13	.05	.18
Posts, 4″ x 4″, including concrete footing	.122	L.F.	.017	.37	.50	.87
Stairs, 2″ x 10″ stringers, 2″ x 10″ steps	1.000	Set	.012	1.89	.41	2.30
Railings, 2″ x 4″	.670	L.F.	.017	.35	.58	.93
TOTAL			.088	5.60	2.91	8.51
12′ X 24′ DECK, REDWOOD OR CEDAR, JOISTS 16″ O.C.						
Decking, 2″ x 6″ redwood	2.080	L.F.	.027	9.15	.89	10.04
Joists, 2″ x 10″, 16″ O.C.	1.000	L.F.	.018	7.35	.60	7.95
Girder, 2″ x 10″	.083	L.F.	.001	.61	.05	.66
Post, 4″ x 4″, including concrete footing	.111	L.F.	.021	.90	.63	1.53
Stairs, 2″ x 10″ stringers, 2″ x 10″ steps	1.000	Set	.012	1.61	.41	2.02
Railings, 2″ x 4″	.540	L.F.	.005	1.58	.15	1.73
TOTAL			.084	21.20	2.73	23.93

The costs in this system are on a square foot basis.

Description	QUAN.	UNIT	LABOR HOURS	COST PER S.F. MAT.	COST PER S.F. INST.	COST PER S.F. TOTAL

Important: See the Reference Section for critical supporting data - Reference Nos., Crews & Location Factors

Wood Deck Price Sheet	QUAN.	UNIT	LABOR HOURS	COST PER S.F. MAT.	INST.	TOTAL
Decking, treated lumber, 1″ x 4″	3.430	L.F.	.031	2.38	1.06	3.44
1″ x 6″	2.180	L.F.	.033	2.50	1.11	3.61
2″ x 4″	3.200	L.F.	.041	1.67	1.38	3.05
2″ x 6″	2.080	L.F.	.027	1.62	.89	2.51
Redwood or cedar,, 1″ x 4″	3.430	L.F.	.035	2.78	1.17	3.95
1″ x 6″	2.180	L.F.	.036	2.91	1.22	4.13
2″ x 4″	3.200	L.F.	.028	9.70	.95	10.65
2″ x 6″	2.080	L.F.	.027	9.15	.89	10.04
Joists for deck, treated lumber, 2″ x 8″, 16″ O.C.	1.000	L.F.	.015	1.08	.49	1.57
24″ O.C.	.800	L.F.	.012	.87	.39	1.26
2″ x 10″, 16″ O.C.	1.000	L.F.	.018	1.56	.60	2.16
24″ O.C.	.800	L.F.	.014	1.24	.48	1.72
Redwood or cedar, 2″ x 8″, 16″ O.C.	1.000	L.F.	.015	5.85	.49	6.34
24″ O.C.	.800	L.F.	.012	4.68	.39	5.07
2″ x 10″, 16″ O.C.	1.000	L.F.	.018	7.35	.60	7.95
24″ O.C.	.800	L.F.	.014	5.90	.48	6.38
Girder for joists, treated lumber, 2″ x 10″, 8′ x 12′ deck	.125	L.F.	.002	.20	.08	.28
12′ x 16′ deck	.083	L.F.	.001	.13	.05	.18
12′ x 24′ deck	.083	L.F.	.001	.13	.05	.18
Redwood or cedar, 2″ x 10″, 8′ x 12′ deck	.125	L.F.	.002	.92	.08	1.00
12′ x 16′ deck	.083	L.F.	.001	.61	.05	.66
12′ x 24′ deck	.083	L.F.	.001	.61	.05	.66
Posts, 4″ x 4″, including concrete footing, 8′ x 12′ deck	.250	L.F.	.022	.55	.68	1.23
12′ x 16′ deck	.122	L.F.	.017	.37	.50	.87
12′ x 24′ deck	.111	L.F.	.017	.36	.48	.84
Stairs 2″ x 10″ stringers, treated lumber, 8′ x 12′ deck	1.000	Set	.020	3.15	.68	3.83
12′ x 16′ deck	1.000	Set	.012	1.89	.41	2.30
12′ x 24′ deck	1.000	Set	.008	1.26	.27	1.53
Redwood or cedar, 8′ x 12′ deck	1.000	Set	.040	5.35	1.35	6.70
12′ x 16′ deck	1.000	Set	.020	2.68	.68	3.36
12′ x 24′ deck	1.000	Set	.012	1.61	.41	2.02
Railings 2″ x 4″, treated lumber, 8′ x 12′ deck	1.000	L.F.	.026	.52	.87	1.39
12′ x 16′ deck	.670	L.F.	.017	.35	.58	.93
12′ x 24′ deck	.540	L.F.	.014	.28	.47	.75
Redwood or cedar, 8′ x 12′ deck	1.000	L.F.	.009	2.95	.29	3.24
12′ x 16′ deck	.670	L.F.	.006	1.94	.19	2.13
12′ x 24′ deck	.540	L.F.	.005	1.58	.15	1.73

Division 8
Mechanical

04 Two Fixture Lavatory Systems

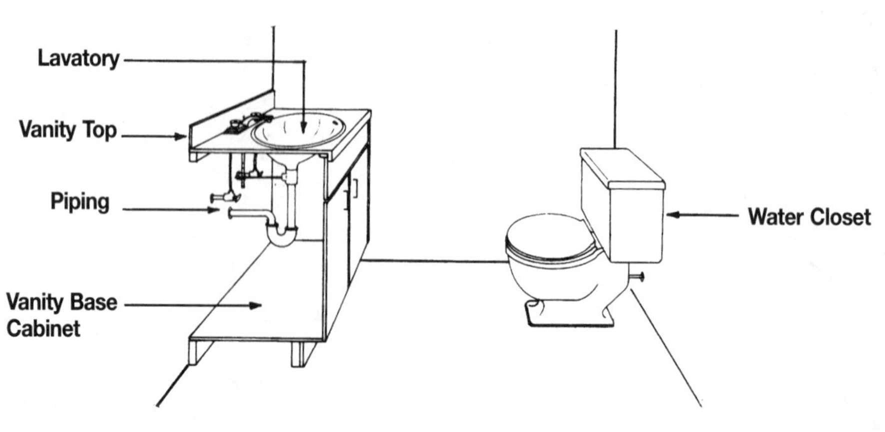

System Description	QUAN.	UNIT	LABOR HOURS	COST EACH		
				MAT.	INST.	TOTAL
LAVATORY INSTALLED WITH VANITY, PLUMBING IN 2 WALLS						
Water closet, floor mounted, 2 piece, close coupled, white	1.000	Ea.	3.019	139.00	98.50	237.50
Rough-in supply, waste and vent for water closet	1.000	Ea.	2.376	63.30	79.74	143.04
Lavatory, 20" x 18", P.E. cast iron white	1.000	Ea.	2.500	219.00	81.50	300.50
Rough-in supply, waste and vent for lavatory	1.000	Ea.	2.844	62.30	96.50	158.80
Piping, supply, copper 1/2" diameter, type "L"	10.000	L.F.	.988	12.50	35.90	48.40
Waste, cast iron, 4" diameter, no-hub	7.000	L.F.	1.931	71.40	63.00	134.40
Vent, cast iron, 2" diameter, no-hub	12.000	L.F.	2.866	75.60	93.60	169.20
Vanity base cabinet, 2 door, 30" wide	1.000	Ea.	1.000	143.00	34.00	177.00
Vanity top, plastic & laminated, square edge	2.670	L.F.	.712	76.10	24.03	100.13
TOTAL			18.236	862.20	606.77	1468.97
LAVATORY WITH WALL-HUNG LAVATORY, PLUMBING IN 2 WALLS						
Water closet, floor mounted, 2 piece close coupled, white	1.000	Ea.	3.019	139.00	98.50	237.50
Rough-in supply, waste and vent for water closet	1.000	Ea.	2.376	63.30	79.74	143.04
Lavatory, 20" x 18", P.E. cast iron, wall hung, white	1.000	Ea.	2.000	220.00	65.50	285.50
Rough-in supply, waste and vent for lavatory	1.000	Ea.	2.844	62.30	96.50	158.80
Piping, supply, copper 1/2" diameter, type "L"	10.000	L.F.	.988	12.50	35.90	48.40
Waste, cast iron, 4" diameter, no-hub	7.000	L.F.	1.931	71.40	63.00	134.40
Vent, cast iron, 2" diameter, no hub	12.000	L.F.	2.866	75.60	93.60	169.20
Carrier, steel for studs, no arms	1.000	Ea.	1.143	29.00	41.50	70.50
TOTAL			17.167	673.10	574.24	1247.34

Description	QUAN.	UNIT	LABOR HOURS	COST EACH		
				MAT.	INST.	TOTAL

Important: See the Reference Section for critical supporting data - Reference Nos., Crews & Location Factors

Two Fixture Lavatory Price Sheet	QUAN.	UNIT	LABOR HOURS	COST EACH		
				MAT.	INST.	TOTAL
Water closet, close coupled standard 2 piece, white	1.000	Ea.	3.019	139.00	98.50	237.50
Color	1.000	Ea.	3.019	165.00	98.50	263.50
One piece elongated bowl, white	1.000	Ea.	3.019	480.00	98.50	578.50
Color	1.000	Ea.	3.019	660.00	98.50	758.50
Low profile, one piece elongated bowl, white	1.000	Ea.	3.019	610.00	98.50	708.50
Color	1.000	Ea.	3.019	550.00	98.50	648.50
Rough-in for water closet						
1/2" copper supply, 4" cast iron waste, 2" cast iron vent	1.000	Ea.	2.376	63.50	79.50	143.00
4" PVC waste, 2" PVC vent	1.000	Ea.	2.678	37.50	89.50	127.00
4" copper waste , 2" copper vent	1.000	Ea.	2.520	48.00	87.00	135.00
3" cast iron waste, 1-1/2" cast iron vent	1.000	Ea.	2.244	56.00	75.50	131.50
3" PVC waste, 1-1/2" PVC vent	1.000	Ea.	2.388	32.00	83.50	115.50
3" copper waste, 1-1/2" copper vent	1.000	Ea.	2.524	44.00	84.50	128.50
1/2" PVC supply, 4" PVC waste, 2" PVC vent	1.000	Ea.	2.974	47.00	101.00	148.00
3" PVC waste, 1-1/2" PVC vent	1.000	Ea.	2.684	42.00	94.00	136.00
1/2" steel supply, 4" cast iron waste, 2" cast iron vent	1.000	Ea.	2.545	66.50	86.00	152.50
4" cast iron waste, 2" steel vent	1.000	Ea.	2.590	57.00	87.50	144.50
4" PVC waste, 2" PVC vent	1.000	Ea.	2.847	40.50	96.00	136.50
Lavatory, vanity top mounted, P.E. on cast iron 20" x 18" white	1.000	Ea.	2.500	219.00	81.50	300.50
Color	1.000	Ea.	2.500	249.00	81.50	330.50
Steel, enameled 10" x 17" white	1.000	Ea.	2.759	111.00	90.00	201.00
Color	1.000	Ea.	2.500	117.00	81.50	198.50
Vitreous china 20" x 16", white	1.000	Ea.	2.963	202.00	96.50	298.50
Color	1.000	Ea.	2.963	202.00	96.50	298.50
Wall hung, P.E. on cast iron, 20" x 18", white	1.000	Ea.	2.000	220.00	65.50	285.50
Color	1.000	Ea.	2.000	250.00	65.50	315.50
Vitreous china 19" x 17", white	1.000	Ea.	2.286	160.00	74.50	234.50
Color	1.000	Ea.	2.286	178.00	74.50	252.50
Rough-in supply waste and vent for lavatory						
1/2" copper supply, 2" cast iron waste, 1-1/2" cast iron vent	1.000	Ea.	2.844	62.50	96.50	159.00
2" PVC waste, 1-1/2" PVC vent	1.000	Ea.	2.962	37.50	104.00	141.50
2" copper waste, 1-1/2" copper vent	1.000	Ea.	2.308	36.50	84.00	120.50
1-1/2" PVC waste, 1-1/4" PVC vent	1.000	Ea.	2.639	36.00	95.50	131.50
1-1/2" copper waste, 1-1/4" copper vent	1.000	Ea.	2.114	32.50	77.00	109.50
1/2" PVC supply, 2" PVC waste, 1-1/2" PVC vent	1.000	Ea.	3.455	53.50	122.00	175.50
1-1/2" PVC waste, 1-1/4" PVC vent	1.000	Ea.	3.132	52.50	114.00	166.50
1/2" steel supply, 2" cast iron waste, 1-1/2" cast iron vent	1.000	Ea.	3.126	67.50	107.00	174.50
2" cast iron waste, 2" steel vent	1.000	Ea.	3.225	58.50	110.00	168.50
2" PVC waste, 1-1/2" PVC vent	1.000	Ea.	3.244	42.50	114.00	156.50
1-1/2" PVC waste, 1-1/4" PVC vent	1.000	Ea.	2.921	41.00	106.00	147.00
Piping, supply, 1/2" copper, type "L"	10.000	L.F.	.988	12.50	36.00	48.50
1/2" steel	10.000	L.F.	1.270	17.50	46.00	63.50
1/2" PVC	10.000	L.F.	1.481	29.00	54.00	83.00
Waste, 4" cast iron	7.000	L.F.	1.931	71.50	63.00	134.50
4" copper	7.000	L.F.	2.800	63.50	91.50	155.00
4" PVC	7.000	L.F.	2.333	39.50	76.50	116.00
Vent, 2" cast iron	12.000	L.F.	2.866	75.50	93.50	169.00
2" copper	12.000	L.F.	2.182	40.00	79.00	119.00
2" PVC	12.000	L.F.	3.254	38.50	106.00	144.50
2" steel	12.000	Ea.	3.000	47.00	98.00	145.00
Vanity base cabinet, 2 door, 24" x 30"	1.000	Ea.	1.000	143.00	34.00	177.00
24" x 36"	1.000	Ea.	1.200	160.00	40.50	200.50
Vanity top, laminated plastic, square edge 25" x 32"	2.670	L.F.	.712	76.00	24.00	100.00
25" x 38"	3.170	L.F.	.845	90.50	28.50	119.00
Post formed, laminated plastic, 25" x 32"	2.670	L.F.	.712	25.00	24.00	49.00
25" x 38"	3.170	L.F.	.845	30.00	28.50	58.50
Cultured marble, 25" x 32" with bowl	1.000	Ea.	2.500	169.00	81.50	250.50
25" x 38" with bowl	1.000	Ea.	2.500	202.00	81.50	283.50
Carrier for lavatory, steel for studs	1.000	Ea.	1.143	29.00	41.50	70.50
Wood 2" x 8" blocking	1.330	L.F.	.053	1.21	1.80	3.01

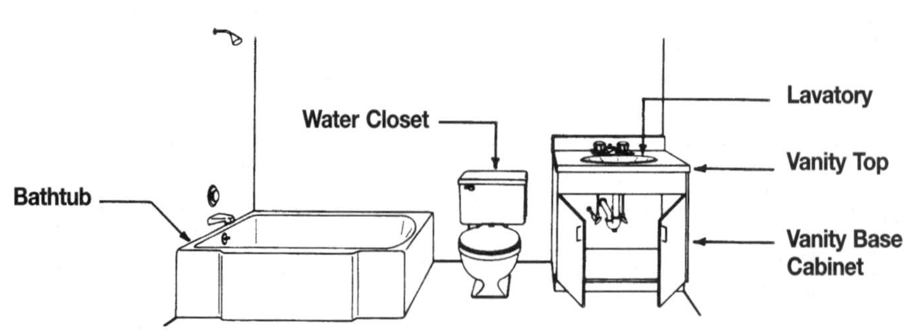

System Description	QUAN.	UNIT	LABOR HOURS	COST EACH MAT.	COST EACH INST.	COST EACH TOTAL
BATHROOM INSTALLED WITH VANITY						
Water closet, floor mounted, 2 piece, close coupled, white	1.000	Ea.	3.019	139.00	98.50	237.50
Rough-in supply, waste and vent for water closet	1.000	Ea.	2.376	63.30	79.74	143.04
Lavatory, 20″ x 18″, P.E. cast iron with accessories, white	1.000	Ea.	2.500	219.00	81.50	300.50
Rough-in supply, waste and vent for lavatory	1.000	Ea.	2.791	61.70	94.70	156.40
Bathtub, P.E. cast iron, 5′ long with accessories, white	1.000	Ea.	3.636	335.00	119.00	454.00
Rough-in supply, waste and vent for bathtub	1.000	Ea.	2.409	53.86	84.50	138.36
Piping, supply, 1/2″ copper	20.000	L.F.	1.975	25.00	71.80	96.80
Waste, 4″ cast iron, no hub	9.000	L.F.	2.483	91.80	81.00	172.80
Vent, 2″ galvanized steel	6.000	L.F.	1.500	23.58	48.90	72.48
Vanity base cabinet, 2 door, 30″ wide	1.000	Ea.	1.000	143.00	34.00	177.00
Vanity top, plastic laminated square edge	2.670	L.F.	.712	58.74	24.03	82.77
TOTAL			24.401	1213.98	817.67	2031.65
BATHROOM WITH WALL HUNG LAVATORY						
Water closet, floor mounted, 2 piece, close coupled, white	1.000	Ea.	3.019	139.00	98.50	237.50
Rough-in supply, waste and vent for water closet	1.000	Ea.	2.376	63.30	79.74	143.04
Lavatory, 20″ x 18″ P.E. cast iron, wall hung, white	1.000	Ea.	2.000	220.00	65.50	285.50
Rough-in supply, waste and vent for lavatory	1.000	Ea.	2.791	61.70	94.70	156.40
Bathtub, P.E. cast iron, 5′ long with accessories, white	1.000	Ea.	3.636	335.00	119.00	454.00
Rough-in supply, waste and vent for bathtub	1.000	Ea.	3.297	70.00	116.90	186.90
Piping, supply, 1/2″ copper	20.000	L.F.	1.975	25.00	71.80	96.80
Waste, 4″ cast iron, no hub	9.000	L.F.	2.483	91.80	81.00	172.80
Vent, 2″ galvanized steel	6.000	L.F.	1.500	23.58	48.90	72.48
Carrier, steel, for studs, no arms	1.000	Ea.	1.143	29.00	41.50	70.50
TOTAL			24.220	1058.38	817.54	1875.92

The costs in this system are a cost each basis. All necessary piping is included.

Description	QUAN.	UNIT	LABOR HOURS	COST EACH MAT.	COST EACH INST.	COST EACH TOTAL

 Important: See the Reference Section for critical supporting data - Reference Nos., Crews & Location Factors

Three Fixture Bathroom Price Sheet	QUAN.	UNIT	LABOR HOURS	COST EACH		
				MAT.	INST.	TOTAL
ater closet, close coupled standard 2 piece, white	1.000	Ea.	3.019	139.00	98.50	237.50
Color	1.000	Ea.	3.019	165.00	98.50	263.50
One piece, elongated bowl, white	1.000	Ea.	3.019	480.00	98.50	578.50
Color	1.000	Ea.	3.019	660.00	98.50	758.50
Low profile, one piece elongated bowl, white	1.000	Ea.	3.019	610.00	98.50	708.50
Color	1.000	Ea.	3.019	550.00	98.50	648.50
ough-in, for water closet						
1/2″ copper supply, 4″ cast iron waste, 2″ cast iron vent	1.000	Ea.	2.376	63.50	79.50	143.00
4″ PVC/DWV waste, 2″ PVC vent	1.000	Ea.	2.678	37.50	89.50	127.00
4″ copper waste, 2″ copper vent	1.000	Ea.	2.520	48.00	87.00	135.00
3″ cast iron waste, 1-1/2″ cast iron vent	1.000	Ea.	2.244	56.00	75.50	131.50
3″ PVC waste, 1-1/2″ PVC vent	1.000	Ea.	2.388	32.00	83.50	115.50
3″ copper waste, 1-1/2″ copper vent	1.000	Ea.	2.014	34.00	70.00	104.00
1/2″ PVC supply, 4″ PVC waste, 2″ PVC vent	1.000	Ea.	2.974	47.00	101.00	148.00
3″ PVC waste, 1-1/2″ PVC supply	1.000	Ea.	2.684	42.00	94.00	136.00
1/2″ steel supply, 4″ cast iron waste, 2″ cast iron vent	1.000	Ea.	2.545	66.50	86.00	152.50
4″ cast iron waste, 2″ steel vent	1.000	Ea.	2.590	57.00	87.50	144.50
4″ PVC waste, 2″ PVC vent	1.000	Ea.	2.847	40.50	96.00	136.50
avatory, wall hung, P.E. cast iron 20″ x 18″, white	1.000	Ea.	2.000	220.00	65.50	285.50
Color	1.000	Ea.	2.000	250.00	65.50	315.50
Vitreous china 19″ x 17″, white	1.000	Ea.	2.286	160.00	74.50	234.50
Color	1.000	Ea.	2.286	178.00	74.50	252.50
Lavatory, for vanity top, P.E. cast iron 20″ x 18″″, white	1.000	Ea.	2.500	219.00	81.50	300.50
Color	1.000	Ea.	2.500	249.00	81.50	330.50
Steel, enameled 20″ x 17″, white	1.000	Ea.	2.759	111.00	90.00	201.00
Color	1.000	Ea.	2.500	117.00	81.50	198.50
Vitreous china 20″ x 16″, white	1.000	Ea.	2.963	202.00	96.50	298.50
Color	1.000	Ea.	2.963	202.00	96.50	298.50
ough-in, for lavatory						
1/2″ copper supply, 1-1/2″ C.I. waste, 1-1/2″ C.I. vent	1.000	Ea.	2.791	61.50	94.50	156.00
1-1/2″ PVC waste, 1-1/4″ PVC vent	1.000	Ea.	2.639	36.00	95.50	131.50
1/2″ steel supply, 1-1/4″ cast iron waste, 1-1/4″ steel vent	1.000	Ea.	2.890	52.50	99.00	151.50
1-1/4″ PVC@ waste, 1-1/4″ PVC vent	1.000	Ea.	2.794	41.00	101.00	142.00
1/2″ PVC supply, 1-1/2″ PVC waste, 1-1/2″ PVC vent	1.000	Ea.	3.259	53.00	118.00	171.00
athtub, P.E. cast iron, 5′ long corner with fittings, white	1.000	Ea.	3.636	335.00	119.00	454.00
Color	1.000	Ea.	3.636	385.00	119.00	504.00
ough-in, for bathtub						
1/2″ copper supply, 4″ cast iron waste, 1-1/2″ copper vent	1.000	Ea.	2.409	54.00	84.50	138.50
4″ PVC waste, 1-1/2″ PVC vent	1.000	Ea.	2.877	41.50	101.00	142.50
1/2″ steel supply, 4″ cast iron waste, 1-1/2″ steel vent	1.000	Ea.	2.898	60.00	99.50	159.50
4″ PVC waste, 1-1/2″ PVC vent	1.000	Ea.	3.159	46.50	111.00	157.50
1/2″ PVC supply, 4″ PVC waste, 1-1/2″ PVC vent	1.000	Ea.	3.370	58.00	119.00	177.00
ping, supply 1/2″ copper	20.000	L.F.	1.975	25.00	72.00	97.00
1/2″ steel	20.000	L.F.	2.540	35.00	92.00	127.00
1/2″ PVC	20.000	L.F.	2.963	58.00	108.00	166.00
ping, waste, 4″ cast iron no hub	9.000	L.F.	2.483	92.00	81.00	173.00
4″ PVC/DWV	9.000	L.F.	3.000	51.00	98.00	149.00
4″ copper/DWV	9.000	L.F.	3.600	81.50	117.00	198.50
ping, vent 2″ cast iron no hub	6.000	L.F.	1.433	38.00	47.00	85.00
2″ copper/DWV	6.000	L.F.	1.091	20.00	39.50	59.50
2″ PVC/DWV	6.000	L.F.	1.627	19.20	53.00	72.20
2″ steel, galvanized	6.000	L.F.	1.500	23.50	49.00	72.50
anity base cabinet, 2 door, 24″ x 30″	1.000	Ea.	1.000	143.00	34.00	177.00
24″ x 36″	1.000	Ea.	1.200	160.00	40.50	200.50
anity top, laminated plastic square edge 25″ x 32″	2.670	L.F.	.712	58.50	24.00	82.50
25″ x 38″	3.160	L.F.	.843	69.50	28.50	98.00
Cultured marble, 25″ x 32″, with bowl	1.000	Ea.	2.500	169.00	81.50	250.50
25″ x 38″, with bowl	1.000	Ea.	2.500	202.00	81.50	283.50
arrier, for lavatory, steel for studs, no arms	1.000	Ea.	1.143	29.00	41.50	70.50
Wood, 2″ x 8″ blocking	1.300	L.F.	.052	1.18	1.76	2.94

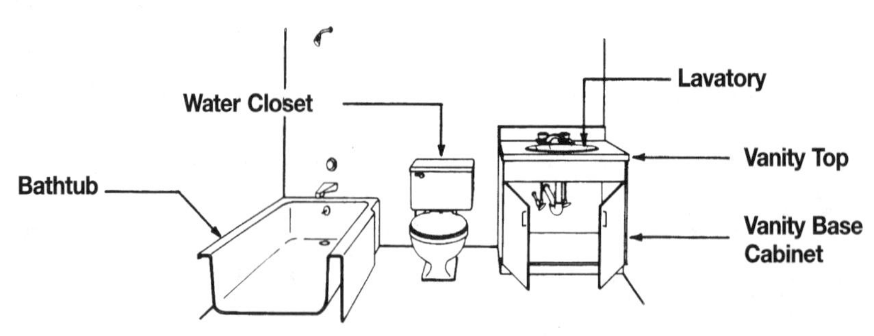

System Description	QUAN.	UNIT	LABOR HOURS	COST EACH		
				MAT.	INST.	TOTAL
BATHROOM WITH LAVATORY INSTALLED IN VANITY						
Water closet, floor mounted, 2 piece, close coupled, white	1.000	Ea.	3.019	139.00	98.50	237.50
Rough-in supply, waste and vent for water closet	1.000	Ea.	2.376	63.30	79.74	143.04
Lavatory, 20" x 18", P.E. cast iron with accessories, white	1.000	Ea.	2.500	219.00	81.50	300.50
Rough-in supply, waste and vent for lavatory	1.000	Ea.	2.791	61.70	94.70	156.40
Bathtub, P.E. cast iron 5′ long with accessories, white	1.000	Ea.	3.636	335.00	119.00	454.00
Rough-in supply, waste and vent for bathtub	1.000	Ea.	2.409	53.86	84.50	138.36
Piping, supply, 1/2" copper	10.000	L.F.	.988	12.50	35.90	48.40
Waste, 4" cast iron, no hub	6.000	L.F.	1.655	61.20	54.00	115.20
Vent, 2" galvanized steel	6.000	L.F.	1.500	23.58	48.90	72.48
Vanity base cabinet, 2 door, 30" wide	1.000	Ea.	1.000	143.00	34.00	177.00
Vanity top, plastic laminated square edge	2.670	L.F.	.712	58.74	24.03	82.77
TOTAL			22.586	1170.88	754.77	1925.65
BATHROOM WITH WALL HUNG LAVATORY						
Water closet, floor mounted, 2 piece, close coupled, white	1.000	Ea.	3.019	139.00	98.50	237.50
Rough-in supply, waste and vent for water closet	1.000	Ea.	2.376	63.30	79.74	143.04
Lavatory, 20" x 18" P.E. cast iron, wall hung, white	1.000	Ea.	2.000	220.00	65.50	285.50
Rough-in supply, waste and vent for lavatory	1.000	Ea.	2.791	61.70	94.70	156.40
Bathtub, P.E. cast iron, 5′ long with accessories, white	1.000	Ea.	3.636	335.00	119.00	454.00
Rough-in supply, waste and vent for bathtub	1.000	Ea.	2.409	53.86	84.50	138.36
Piping, supply, 1/2" copper	10.000	L.F.	.988	12.50	35.90	48.40
Waste, 4" cast iron, no hub	6.000	L.F.	1.655	61.20	54.00	115.20
Vent, 2" galvanized steel	6.000	L.F.	1.500	23.58	48.90	72.48
Carrier, steel, for studs, no arms	1.000	Ea.	1.143	29.00	41.50	70.50
TOTAL			21.517	999.14	722.24	1721.38

The costs in this system are on a cost each basis. All necessary piping is included.

Description	QUAN.	UNIT	LABOR HOURS	COST EACH		
				MAT.	INST.	TOTAL

Important: See the Reference Section for critical supporting data - Reference Nos., Crews & Location Factors

Three Fixture Bathroom Price Sheet	QUAN.	UNIT	LABOR HOURS	COST EACH MAT.	INST.	TOTAL
Water closet, close coupled standard 2 piece, white	1.000	Ea.	3.019	139.00	98.50	237.50
Color	1.000	Ea.	3.019	165.00	98.50	263.50
One piece elongated bowl, white	1.000	Ea.	3.019	480.00	98.50	578.50
Color	1.000	Ea.	3.019	660.00	98.50	758.50
Low profile, one piece elongated bowl, white	1.000	Ea.	3.019	610.00	98.50	708.50
Color	1.000	Ea.	3.019	550.00	98.50	648.50
Rough-in for water closet						
1/2" copper supply, 4" cast iron waste, 2" cast iron vent	1.000	Ea.	2.376	63.50	79.50	143.00
4" PVC/DWV waste, 2" PVC vent	1.000	Ea.	2.678	37.50	89.50	127.00
4" carrier waste, 2" copper vent	1.000	Ea.	2.520	48.00	87.00	135.00
3" cast iron waste, 1-1/2" cast iron vent	1.000	Ea.	2.244	56.00	75.50	131.50
3" PVC waste, 1-1/2" PVC vent	1.000	Ea.	2.388	32.00	83.50	115.50
3" copper waste, 1-1/2" copper vent	1.000	Ea.	2.014	34.00	70.00	104.00
1/2" PVC supply, 4" PVC waste, 2" PVC vent	1.000	Ea.	2.974	47.00	101.00	148.00
3" PVC waste, 1-1/2" PVC supply	1.000	Ea.	2.684	42.00	94.00	136.00
1/2" steel supply, 4" cast iron waste, 2" cast iron vent	1.000	Ea.	2.545	66.50	86.00	152.50
4" cast iron waste, 2" steel vent	1.000	Ea.	2.590	57.00	87.50	144.50
4" PVC waste, 2" PVC vent	1.000	Ea.	2.847	40.50	96.00	136.50
Lavatory, wall hung, PE cast iron 20" x 18", white	1.000	Ea.	2.000	220.00	65.50	285.50
Color	1.000	Ea.	2.000	250.00	65.50	315.50
Vitreous china 19" x 17", white	1.000	Ea.	2.286	160.00	74.50	234.50
Color	1.000	Ea.	2.286	178.00	74.50	252.50
Lavatory, for vanity top, PE cast iron 20" x 18", white	1.000	Ea.	2.500	219.00	81.50	300.50
Color	1.000	Ea.	2.500	249.00	81.50	330.50
Steel enameled 20" x 17", white	1.000	Ea.	2.759	111.00	90.00	201.00
Color	1.000	Ea.	2.500	117.00	81.50	198.50
Vitreous china 20" x 16", white	1.000	Ea.	2.963	202.00	96.50	298.50
Color	1.000	Ea.	2.963	202.00	96.50	298.50
Rough-in for lavatory						
1/2" copper supply, 1-1/2" cast iron waste, 1-1/2" cast iron vent	1.000	Ea.	2.791	61.50	94.50	156.00
1-1/2" PVC waste, 1-1/4" PVC vent	1.000	Ea.	2.639	36.00	95.50	131.50
1/2" steel supply, 1-1/4" cast iron waste, 1-1/4" steel vent	1.000	Ea.	2.890	52.50	99.00	151.50
1-1/4" PVC waste, 1-1/4" PVC vent	1.000	Ea.	2.794	41.00	101.00	142.00
1/2" PVC supply, 1-1/2" PVC waste, 1-1/2" PVC vent	1.000	Ea.	3.259	53.00	118.00	171.00
Bathtub, PE cast iron, 5' long corner with fittings, white	1.000	Ea.	3.636	335.00	119.00	454.00
Color	1.000	Ea.	3.636	385.00	119.00	504.00
Rough-in for bathtub						
1/2" copper supply, 4" cast iron waste, 1-1/2" copper vent	1.000	Ea.	2.409	54.00	84.50	138.50
4" PVC waste, 1/2" PVC vent	1.000	Ea.	2.877	41.50	101.00	142.50
1/2" steel supply, 4" cast iron waste, 1-1/2" steel vent	1.000	Ea.	2.898	60.00	99.50	159.50
4" PVC waste, 1-1/2" PVC vent	1.000	Ea.	3.159	46.50	111.00	157.50
1/2" PVC supply, 4" PVC waste, 1-1/2" PVC vent	1.000	Ea.	3.370	58.00	119.00	177.00
Piping supply, 1/2" copper	10.000	L.F.	.988	12.50	36.00	48.50
1/2" steel	10.000	L.F.	1.270	17.50	46.00	63.50
1/2" PVC	10.000	L.F.	1.481	29.00	54.00	83.00
Piping waste, 4" cast iron no hub	6.000	L.F.	1.655	61.00	54.00	115.00
4" PVC/DWV	6.000	L.F.	2.000	34.00	65.50	99.50
4" copper/DWV	6.000	L.F.	2.400	54.50	78.50	133.00
Piping vent 2" cast iron no hub	6.000	L.F.	1.433	38.00	47.00	85.00
2" copper/DWV	6.000	L.F.	1.091	20.00	39.50	59.50
2" PVC/DWV	6.000	L.F.	1.627	19.20	53.00	72.20
2" steel, galvanized	6.000	L.F.	1.500	23.50	49.00	72.50
Vanity base cabinet, 2 door, 24" x 30"	1.000	Ea.	1.000	143.00	34.00	177.00
24" x 36"	1.000	Ea.	1.200	160.00	40.50	200.50
Vanity top, laminated plastic square edge 25" x 32"	2.670	L.F.	.712	58.50	24.00	82.50
25" x 38"	3.160	L.F.	.843	69.50	28.50	98.00
Cultured marble, 25" x 32", with bowl	1.000	Ea.	2.500	169.00	81.50	250.50
25" x 38", with bowl	1.000	Ea.	2.500	202.00	81.50	283.50
Carrier, for lavatory, steel for studs, no arms	1.000	Ea.	1.143	29.00	41.50	70.50
Wood, 2" x 8" blocking	1.300	L.F.	.052	1.18	1.76	2.94

Three Fixture Bathroom Systems

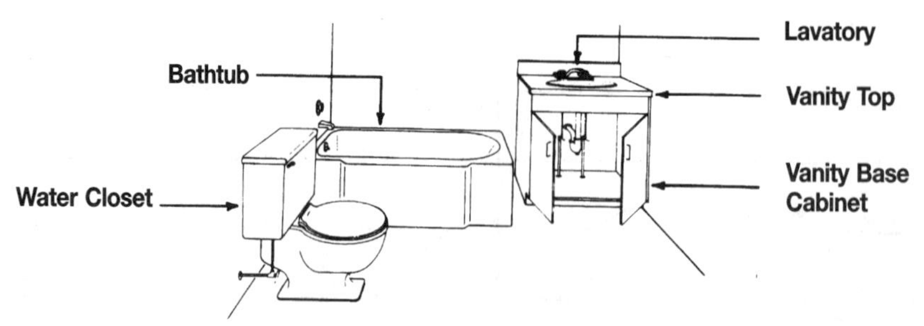

System Description	QUAN.	UNIT	LABOR HOURS	COST EACH		
				MAT.	INST.	TOTAL
BATHROOM WITH LAVATORY INSTALLED IN VANITY						
Water closet, floor mounted, 2 piece, close coupled, white	1.000	Ea.	3.019	139.00	98.50	237.50
Rough-in supply, waste and vent for water closet	1.000	Ea.	2.376	63.30	79.74	143.04
Lavatory, 20″ x 18″, PE cast iron with accessories, white	1.000	Ea.	2.500	219.00	81.50	300.50
Rough-in supply, waste and vent for lavatory	1.000	Ea.	2.791	61.70	94.70	156.40
Bathtub, P.E. cast iron, 5′ long with accessories, white	1.000	Ea.	3.636	335.00	119.00	454.00
Rough-in supply, waste and vent for bathtub	1.000	Ea.	2.409	53.86	84.50	138.36
Piping, supply, 1/2″ copper	32.000	L.F.	3.160	40.00	114.88	154.88
Waste, 4″ cast iron, no hub	12.000	L.F.	3.310	122.40	108.00	230.40
Vent, 2″ galvanized steel	6.000	L.F.	1.500	23.58	48.90	72.48
Vanity base cabinet, 2 door, 30″ wide	1.000	Ea.	1.000	143.00	34.00	177.00
Vanity top, plastic laminated square edge	2.670	L.F.	.712	58.74	24.03	82.77
TOTAL			26.413	1259.58	887.75	2147.33
BATHROOM WITH WALL HUNG LAVATORY						
Water closet, floor mounted, 2 piece, close coupled, white	1.000	Ea.	3.019	139.00	98.50	237.50
Rough-in supply, waste and vent for water closet	1.000	Ea.	2.376	63.30	79.74	143.04
Lavatory, 20″ x 18″ P.E. cast iron, wall hung, white	1.000	Ea.	2.000	220.00	65.50	285.50
Rough-in supply, waste and vent for lavatory	1.000	Ea.	2.791	61.70	94.70	156.40
Bathtub, P.E. cast iron, 5′ long with accessories, white	1.000	Ea.	3.636	335.00	119.00	454.00
Rough-in supply, waste and vent for bathtub	1.000	Ea.	2.409	53.86	84.50	138.36
Piping, supply, 1/2″ copper	32.000	L.F.	3.160	40.00	114.88	154.88
Waste, 4″ cast iron, no hub	12.000	L.F.	3.310	122.40	108.00	230.40
Vent, 2″ galvanized steel	6.000	L.F.	1.500	23.58	48.90	72.48
Carrier steel, for studs, no arms	1.000	Ea.	1.143	29.00	41.50	70.50
TOTAL			25.344	1087.84	855.22	1943.06

The costs in this system are on a cost each basis. All necessary piping is included.

Description	QUAN.	UNIT	LABOR HOURS	COST EACH		
				MAT.	INST.	TOTAL

Important: See the Reference Section for critical supporting data - Reference Nos., Crews & Location Factors

Three Fixture Bathroom Price Sheet	QUAN.	UNIT	LABOR HOURS	COST EACH		
				MAT.	INST.	TOTAL
Water closet, close coupled, standard 2 piece, white	1.000	Ea.	3.019	139.00	98.50	237.50
Color	1.000	Ea.	3.019	165.00	98.50	263.50
One piece, elongated bowl, white	1.000	Ea.	3.019	480.00	98.50	578.50
Color	1.000	Ea.	3.019	660.00	98.50	758.50
Low profile, one piece, elongated bowl, white	1.000	Ea.	3.019	610.00	98.50	708.50
Color	1.000	Ea.	3.019	550.00	98.50	648.50
Rough-in, for water closet						
1/2″ copper supply, 4″ cast iron waste, 2″ cast iron vent	1.000	Ea.	2.376	63.50	79.50	143.00
4″ PVC/DWV waste, 2″ PVC vent	1.000	Ea.	2.678	37.50	89.50	127.00
4″ copper waste, 2″ copper vent	1.000	Ea.	2.520	48.00	87.00	135.00
3″ cast iron waste, 1-1/2″ cast iron vent	1.000	Ea.	2.244	56.00	75.50	131.50
3″ PVC waste, 1-1/2″ PVC vent	1.000	Ea.	2.388	32.00	83.50	115.50
3″ copper waste, 1-1/2″ copper vent	1.000	Ea.	2.014	34.00	70.00	104.00
1/2″ PVC supply, 4″ PVC waste, 2″ PVC vent	1.000	Ea.	2.974	47.00	101.00	148.00
3″ PVC waste, 1-1/2″ PVC supply	1.000	Ea.	2.684	42.00	94.00	136.00
1/2″ steel supply, 4″ cast iron waste, 2″ cast iron vent	1.000	Ea.	2.545	66.50	86.00	152.50
4″ cast iron waste, 2″ steel vent	1.000	Ea.	2.590	57.00	87.50	144.50
4″ PVC waste, 2″ PVC vent	1.000	Ea.	2.847	40.50	96.00	136.50
Lavatory wall hung, P.E. cast iron, 20″ x 18″, white	1.000	Ea.	2.000	220.00	65.50	285.50
Color	1.000	Ea.	2.000	250.00	65.50	315.50
Vitreous china, 19″ x 17″, white	1.000	Ea.	2.286	160.00	74.50	234.50
Color	1.000	Ea.	2.286	178.00	74.50	252.50
Lavatory, for vanity top, P.E., cast iron, 20″ x 18″, white	1.000	Ea.	2.500	219.00	81.50	300.50
Color	1.000	Ea.	2.500	249.00	81.50	330.50
Steel, enameled, 20″ x 17″, white	1.000	Ea.	2.759	111.00	90.00	201.00
Color	1.000	Ea.	2.500	117.00	81.50	198.50
Vitreous china, 20″ x 16″, white	1.000	Ea.	2.963	202.00	96.50	298.50
Color	1.000	Ea.	2.963	202.00	96.50	298.50
Rough-in, for lavatory						
1/2″ copper supply, 1-1/2″ C.I. waste, 1-1/2″ C.I. vent	1.000	Ea.	2.791	61.50	94.50	156.00
1-1/2″ PVC waste, 1-1/4″ PVC vent	1.000	Ea.	2.639	36.00	95.50	131.50
1/2″ steel supply, 1-1/4″ cast iron waste, 1-1/4″ steel vent	1.000	Ea.	2.890	52.50	99.00	151.50
1-1/4″ PVC waste, 1-1/4″ PVC vent	1.000	Ea.	2.794	41.00	101.00	142.00
1/2″ PVC supply, 1-1/2″ PVC waste, 1-1/2″ PVC vent	1.000	Ea.	3.259	53.00	118.00	171.00
Bathtub, P.E. cast iron, 5′ long corner with fittings, white	1.000	Ea.	3.636	335.00	119.00	454.00
Color	1.000	Ea.	3.636	385.00	119.00	504.00
Rough-in, for bathtub						
1/2″ copper supply, 4″ cast iron waste, 1-1/2″ copper vent	1.000	Ea.	2.409	54.00	84.50	138.50
4″ PVC waste, 1/2″ PVC vent	1.000	Ea.	2.877	41.50	101.00	142.50
1/2″ steel supply, 4″ cast iron waste, 1-1/2″ steel vent	1.000	Ea.	2.898	60.00	99.50	159.50
4″ PVC waste, 1-1/2″ PVC vent	1.000	Ea.	3.159	46.50	111.00	157.50
1/2″ PVC supply, 4″ PVC waste, 1-1/2″ PVC vent	1.000	Ea.	3.370	58.00	119.00	177.00
Piping, supply, 1/2″ copper	32.000	L.F.	3.160	40.00	115.00	155.00
1/2″ steel	32.000	L.F.	4.063	56.00	148.00	204.00
1/2″ PVC	32.000	L.F.	4.741	92.50	173.00	265.50
Piping, waste, 4″ cast iron no hub	12.000	L.F.	3.310	122.00	108.00	230.00
4″ PVC/DWV	12.000	L.F.	4.000	68.00	131.00	199.00
4″ copper/DWV	12.000	L.F.	4.800	109.00	157.00	266.00
Piping, vent, 2″ cast iron no hub	6.000	L.F.	1.433	38.00	47.00	85.00
2″ copper/DWV	6.000	L.F.	1.091	20.00	39.50	59.50
2″ PVC/DWV	6.000	L.F.	1.627	19.20	53.00	72.20
2″ steel, galvanized	6.000	L.F.	1.500	23.50	49.00	72.50
Vanity base cabinet, 2 door, 24″ x 30″	1.000	Ea.	1.000	143.00	34.00	177.00
24″ x 36″	1.000	Ea.	1.200	160.00	40.50	200.50
Vanity top, laminated plastic square edge, 25″ x 32″	2.670	L.F.	.712	58.50	24.00	82.50
25″ x 38″	3.160	L.F.	.843	69.50	28.50	98.00
Cultured marble, 25″ x 32″, with bowl	1.000	Ea.	2.500	169.00	81.50	250.50
25″ x 38″, with bowl	1.000	Ea.	2.500	202.00	81.50	283.50
Carrier, for lavatory, steel for studs, no arms	1.000	Ea.	1.143	29.00	41.50	70.50
Wood, 2″ x 8″ blocking	1.300	L.F.	.052	1.18	1.76	2.94

24 Three Fixture Bathroom Systems

Corner Bathtub
Water Closet
Lavatory
Vanity Top
Vanity Base Cabinet

System Description	QUAN.	UNIT	LABOR HOURS	COST EACH MAT.	COST EACH INST.	COST EACH TOTAL
BATHROOM WITH LAVATORY INSTALLED IN VANITY						
Water closet, floor mounted, 2 piece, close coupled, white	1.000	Ea.	3.019	139.00	98.50	237.50
Rough-in supply waste and vent for water closet	1.000	Ea.	2.376	63.30	79.74	143.04
Lavatory, 20" x 18", P.E. cast iron with fittings, white	1.000	Ea.	2.500	219.00	81.50	300.50
Rough-in supply waste and vent for lavatory	1.000	Ea.	2.791	61.70	94.70	156.40
Bathtub, P.E. cast iron, corner with fittings, white	1.000	Ea.	3.636	1325.00	119.00	1444.00
Rough-in supply waste and vent for bathtub	1.000	Ea.	2.409	53.86	84.50	138.36
Piping supply, 1/2" copper	32.000	L.F.	3.160	40.00	114.88	154.88
Waste, 4" cast iron, no hub	12.000	L.F.	3.310	122.40	108.00	230.40
Vent, 2" steel, galvanized	6.000	L.F.	1.500	23.58	48.90	72.48
Vanity base cabinet, 2 door, 30" wide	1.000	Ea.	1.000	143.00	34.00	177.00
Vanity top, plastic laminated, square edge	2.670	L.F.	.712	76.10	24.03	100.13
TOTAL			26.413	2266.94	887.75	3154.69
BATHROOM WITH WALL HUNG LAVATORY						
Water closet, floor mounted, 2 piece, close coupled, white	1.000	Ea.	3.019	139.00	98.50	237.50
Rough-in supply waste and vent for water closet	1.000	Ea.	2.376	63.30	79.74	143.04
Lavatory, 20" x 18", P.E. cast iron, with fittings, white	1.000	Ea.	2.000	220.00	65.50	285.50
Rough-in supply, waste and vent, lavatory	1.000	Ea.	2.791	61.70	94.70	156.40
Bathtub, P.E. cast iron, corner, with fittings, white	1.000	Ea.	3.636	1325.00	119.00	1444.00
Rough-in supply, waste and vent, bathtub	1.000	Ea.	2.409	53.86	84.50	138.36
Piping, supply, 1/2" copper	32.000	L.F.	3.160	40.00	114.88	154.88
Waste, 4" cast iron, no hub	12.000	L.F.	3.310	122.40	108.00	230.40
Vent, 2" steel, galvanized	6.000	L.F.	1.500	23.58	48.90	72.48
Carrier, steel, for studs, no arms	1.000	Ea.	1.143	29.00	41.50	70.50
TOTAL			25.344	2077.84	855.22	2933.06

The costs in this system are on a cost each basis. All necessary piping is included.

Description	QUAN.	UNIT	LABOR HOURS	COST EACH MAT.	COST EACH INST.	COST EACH TOTAL

Important: See the Reference Section for critical supporting data - Reference Nos., Crews & Location Factors

Three Fixture Bathroom Price Sheet	QUAN.	UNIT	LABOR HOURS	COST EACH		
				MAT.	INST.	TOTAL
Water closet, close coupled, standard 2 piece, white	1.000	Ea.	3.019	139.00	98.50	237.50
Color	1.000	Ea.	3.019	165.00	98.50	263.50
One piece elongated bowl, white	1.000	Ea.	3.019	480.00	98.50	578.50
Color	1.000	Ea.	3.019	660.00	98.50	758.50
Low profile, one piece elongated bowl, white	1.000	Ea.	3.019	610.00	98.50	708.50
Color	1.000	Ea.	3.019	550.00	98.50	648.50
Rough-in, for water closet						
1/2" copper supply, 4" cast iron waste, 2" cast iron vent	1.000	Ea.	2.376	63.50	79.50	143.00
4" PVC/DWV waste, 2" PVC vent	1.000	Ea.	2.678	37.50	89.50	127.00
4" copper waste, 2" copper vent	1.000	Ea.	2.520	48.00	87.00	135.00
3" cast iron waste, 1-1/2" cast iron vent	1.000	Ea.	2.244	56.00	75.50	131.50
3" PVC waste, 1-1/2" PVC vent	1.000	Ea.	2.388	32.00	83.50	115.50
3" copper waste, 1-1/2" copper vent	1.000	Ea.	2.014	34.00	70.00	104.00
1/2" PVC supply, 4" PVC waste, 2" PVC vent	1.000	Ea.	2.974	47.00	101.00	148.00
3" PVC waste, 1-1/2" PVC supply	1.000	Ea.	2.684	42.00	94.00	136.00
1/2" steel supply, 4" cast iron waste, 2" cast iron vent	1.000	Ea.	2.545	66.50	86.00	152.50
4" cast iron waste, 2" steel vent	1.000	Ea.	2.590	57.00	87.50	144.50
4" PVC waste, 2" PVC vent	1.000	Ea.	2.847	40.50	96.00	136.50
Lavatory, wall hung P.E. cast iron 20" x 18", white	1.000	Ea.	2.000	220.00	65.50	285.50
Color	1.000	Ea.	2.000	250.00	65.50	315.50
Vitreous china 19" x 17", white	1.000	Ea.	2.286	160.00	74.50	234.50
Color	1.000	Ea.	2.286	178.00	74.50	252.50
Lavatory, for vanity top, P.E., cast iron, 20" x 18", white	1.000	Ea.	2.500	219.00	81.50	300.50
Color	1.000	Ea.	2.500	249.00	81.50	330.50
Steel enameled 20" x 17", white	1.000	Ea.	2.759	111.00	90.00	201.00
Color	1.000	Ea.	2.500	117.00	81.50	198.50
Vitreous china 20" x 16", white	1.000	Ea.	2.963	202.00	96.50	298.50
Color	1.000	Ea.	2.963	202.00	96.50	298.50
Rough-in, for lavatory						
1/2" copper supply, 1-1/2" cast iron waste, 1-1/2" cast iron vent	1.000	Ea.	2.791	61.50	94.50	156.00
1-1/2" PVC waste, 1-1/4" PVC vent	1.000	Ea.	2.639	36.00	95.50	131.50
1/2" steel supply, 1-1/4" cast iron waste, 1-1/4" steel vent	1.000	Ea.	2.890	52.50	99.00	151.50
1-1/4" PVC waste, 1-1/4" PVC vent	1.000	Ea.	2.794	41.00	101.00	142.00
1/2" PVC supply, 1-1/2" PVC waste, 1-1/2" PVC vent	1.000	Ea.	3.259	53.00	118.00	171.00
Bathtub, P.E. cast iron, corner with fittings, white	1.000	Ea.	3.636	1325.00	119.00	1444.00
Color	1.000	Ea.	4.000	1500.00	131.00	1631.00
Rough-in, for bathtub						
1/2" copper supply, 4" cast iron waste, 1-1/2" copper vent	1.000	Ea.	2.409	54.00	84.50	138.50
4" PVC waste, 1-1/2" PVC vent	1.000	Ea.	2.877	41.50	101.00	142.50
1/2" steel supply, 4" cast iron waste, 1-1/2" steel vent	1.000	Ea.	2.898	60.00	99.50	159.50
4" PVC waste, 1-1/2" PVC vent	1.000	Ea.	3.159	46.50	111.00	157.50
1/2" PVC supply, 4" PVC waste, 1-1/2" PVC vent	1.000	Ea.	3.370	58.00	119.00	177.00
Piping, supply, 1/2" copper	32.000	L.F.	3.160	40.00	115.00	155.00
1/2" steel	32.000	L.F.	4.063	56.00	148.00	204.00
1/2" PVC	32.000	L.F.	4.741	92.50	173.00	265.50
Piping, waste, 4" cast iron, no hub	12.000	L.F.	3.310	122.00	108.00	230.00
4" PVC/DWV	12.000	L.F.	4.000	68.00	131.00	199.00
4" copper/DWV	12.000	L.F.	4.800	109.00	157.00	266.00
Piping, vent 2" cast iron, no hub	6.000	L.F.	1.433	38.00	47.00	85.00
2" copper/DWV	6.000	L.F.	1.091	20.00	39.50	59.50
2" PVC/DWV	6.000	L.F.	1.627	19.20	53.00	72.20
2" steel, galvanized	6.000	L.F.	1.500	23.50	49.00	72.50
Vanity base cabinet, 2 door, 24" x 30"	1.000	Ea.	1.000	143.00	34.00	177.00
24" x 36"	1.000	Ea.	1.200	160.00	40.50	200.50
Vanity top, laminated plastic square edge 25" x 32"	2.670	L.F.	.712	76.00	24.00	100.00
25" x 38"	3.160	L.F.	.843	90.00	28.50	118.50
Cultured marble, 25" x 32", with bowl	1.000	Ea.	2.500	169.00	81.50	250.50
25" x 38", with bowl	1.000	Ea.	2.500	202.00	81.50	283.50
Carrier, for lavatory, steel for studs, no arms	1.000	Ea.	1.143	29.00	41.50	70.50
Wood, 2" x 8" blocking	1.300	L.F.	.053	1.21	1.80	3.01

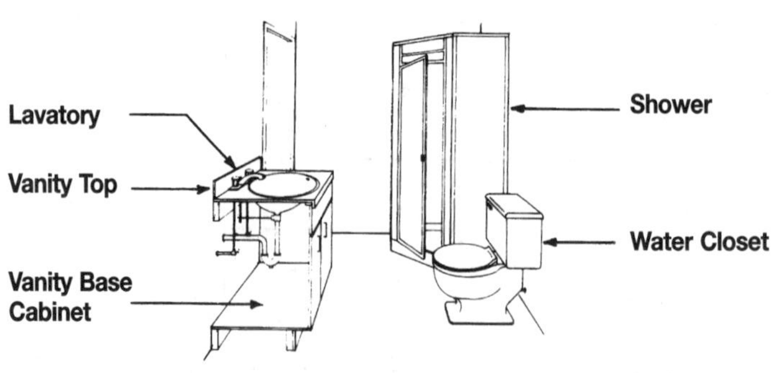

System Description	QUAN.	UNIT	LABOR HOURS	COST EACH		
				MAT.	INST.	TOTAL
BATHROOM WITH SHOWER, LAVATORY INSTALLED IN VANITY						
Water closet, floor mounted, 2 piece, close coupled, white	1.000	Ea.	3.019	139.00	98.50	237.50
Rough-in supply, waste and vent for water closet	1.000	Ea.	2.376	63.30	79.74	143.04
Lavatory, 20" x 18" P.E. cast iron with fittings, white	1.000	Ea.	2.500	219.00	81.50	300.50
Rough-in supply, waste and vent	1.000	Ea.	2.791	61.70	94.70	156.40
Shower, steel enameled, stone base, corner, white	1.000	Ea.	8.000	345.00	261.00	606.00
Rough-in supply, waste and vent	1.000	Ea.	3.238	63.95	111.59	175.54
Piping supply, 1/2" copper	36.000	L.F.	4.148	52.50	150.78	203.28
Waste 4" cast iron, no hub	7.000	L.F.	2.759	102.00	90.00	192.00
Vent 2" steel galvanized	6.000	L.F.	2.250	35.37	73.35	108.72
Vanity base 2 door, 30" wide	1.000	Ea.	1.000	143.00	34.00	177.00
Vanity top, plastic laminated, square edge	2.170	L.F.	.712	60.08	24.03	84.11
TOTAL			32.793	1284.90	1099.19	2384.09
BATHROOM WITH SHOWER, WALL HUNG LAVATORY						
Water closet, floor mounted, close coupled	1.000	Ea.	3.019	139.00	98.50	237.50
Rough-in supply, waste and vent for water closet	1.000	Ea.	2.376	63.30	79.74	143.04
Lavatory, 20" x 18" P.E. cast iron with fittings, white	1.000	Ea.	2.000	220.00	65.50	285.50
Rough-in supply, waste and vent for lavatory	1.000	Ea.	2.791	61.70	94.70	156.40
Shower, steel enameled, stone base, white	1.000	Ea.	8.000	345.00	261.00	606.00
Rough-in supply, waste and vent for shower	1.000	Ea.	3.238	63.95	111.59	175.54
Piping supply, 1/2" copper	36.000	L.F.	4.148	52.50	150.78	203.28
Waste, 4" cast iron, no hub	7.000	L.F.	2.759	102.00	90.00	192.00
Vent, 2" steel, galvanized	6.000	L.F.	2.250	35.37	73.35	108.72
Carrier, steel, for studs, no arms	1.000	Ea.	1.143	29.00	41.50	70.50
TOTAL			31.724	1111.82	1066.66	2178.48

The costs in this system are on a cost each basis. All necessary piping is included.

Description	QUAN.	UNIT	LABOR HOURS	COST EACH		
				MAT.	INST.	TOTAL

Three Fixture Bathroom Price Sheet	QUAN.	UNIT	LABOR HOURS	COST EACH MAT.	INST.	TOTAL
Water closet, close coupled, standard 2 piece, white	1.000	Ea.	3.019	139.00	98.50	237.50
Color	1.000	Ea.	3.019	165.00	98.50	263.50
One piece elongated bowl, white	1.000	Ea.	3.019	480.00	98.50	578.50
Color	1.000	Ea.	3.019	660.00	98.50	758.50
Low profile, one piece elongated bowl, white	1.000	Ea.	3.019	610.00	98.50	708.50
Color	1.000	Ea.	3.019	550.00	98.50	648.50
Rough-in, for water closet						
1/2" copper supply, 4" cast iron waste, 2" cast iron vent	1.000	Ea.	2.376	63.50	79.50	143.00
4" PVC/DWV waste, 2" PVC vent	1.000	Ea.	2.678	37.50	89.50	127.00
4" copper waste, 2" copper vent	1.000	Ea.	2.520	48.00	87.00	135.00
3" cast iron waste, 1-1/2" cast iron vent	1.000	Ea.	2.244	56.00	75.50	131.50
3" PVC waste, 1-1/2" PVC vent	1.000	Ea.	2.388	32.00	83.50	115.50
3" copper waste, 1-1/2" copper vent	1.000	Ea.	2.014	34.00	70.00	104.00
1/2" PVC supply, 4" PVC waste, 2" PVC vent	1.000	Ea.	2.974	47.00	101.00	148.00
3" PVC waste, 1-1/2" PVC supply	1.000	Ea.	2.684	42.00	94.00	136.00
1/2" steel supply, 4" cast iron waste, 2" cast iron vent	1.000	Ea.	2.545	66.50	86.00	152.50
4" cast iron waste, 2" steel vent	1.000	Ea.	2.590	57.00	87.50	144.50
4" PVC waste, 2" PVC vent	1.000	Ea.	2.847	40.50	96.00	136.50
Lavatory, wall hung, P.E. cast iron 20" x 18", white	1.000	Ea.	2.000	220.00	65.50	285.50
Color	1.000	Ea.	2.000	250.00	65.50	315.50
Vitreous china 19" x 17", white	1.000	Ea.	2.286	160.00	74.50	234.50
Color	1.000	Ea.	2.286	178.00	74.50	252.50
Lavatory, for vanity top, P.E. cast iron 20" x 18", white	1.000	Ea.	2.500	219.00	81.50	300.50
Color	1.000	Ea.	2.500	249.00	81.50	330.50
Steel enameled 20" x 17", white	1.000	Ea.	2.759	111.00	90.00	201.00
Color	1.000	Ea.	2.500	117.00	81.50	198.50
Vitreous china 20" x 16", white	1.000	Ea.	2.963	202.00	96.50	298.50
Color	1.000	Ea.	2.963	202.00	96.50	298.50
Rough-in, for lavatory						
1/2" copper supply, 1-1/2" cast iron waste, 1-1/2" cast iron vent	1.000	Ea.	2.791	61.50	94.50	156.00
1-1/2" PVC waste, 1-1/2" PVC vent	1.000	Ea.	2.639	36.00	95.50	131.50
1/2" steel supply, 1-1/4" cast iron waste, 1-1/4" steel vent	1.000	Ea.	2.890	52.50	99.00	151.50
1-1/4" PVC waste, 1-1/4" PVC vent	1.000	Ea.	2.921	41.00	106.00	147.00
1/2" PVC supply, 1-1/2" PVC waste, 1-1/2" PVC vent	1.000	Ea.	3.259	53.00	118.00	171.00
Shower, steel enameled stone base, 32" x 32", white	1.000	Ea.	8.000	345.00	261.00	606.00
Color	1.000	Ea.	7.822	840.00	255.00	1095.00
36" x 36" white	1.000	Ea.	8.889	905.00	290.00	1195.00
Color	1.000	Ea.	8.889	955.00	290.00	1245.00
Rough-in, for shower						
1/2" copper supply, 4" cast iron waste, 1-1/2" copper vent	1.000	Ea.	3.238	64.00	112.00	176.00
4" PVC waste, 1-1/2" PVC vent	1.000	Ea.	3.429	42.00	119.00	161.00
1/2" steel supply, 4" cast iron waste, 1-1/2" steel vent	1.000	Ea.	3.665	68.50	127.00	195.50
4" PVC waste, 1-1/2" PVC vent	1.000	Ea.	3.881	50.00	135.00	185.00
1/2" PVC supply, 4" PVC waste, 1-1/2" PVC vent	1.000	Ea.	4.219	68.50	148.00	216.50
Piping, supply, 1/2" copper	36.000	L.F.	4.148	52.50	151.00	203.50
1/2" steel	36.000	L.F.	5.333	73.50	194.00	267.50
1/2" PVC	36.000	L.F.	6.222	121.00	227.00	348.00
Piping, waste, 4" cast iron no hub	7.000	L.F.	2.759	102.00	90.00	192.00
4" PVC/DWV	7.000	L.F.	3.333	56.50	109.00	165.50
4" copper/DWV	7.000	L.F.	4.000	90.50	131.00	221.50
Piping, vent, 2" cast iron no hub	6.000	L.F.	2.149	56.50	70.00	126.50
2" copper/DWV	6.000	L.F.	1.636	30.00	59.50	89.50
2" PVC/DWV	6.000	L.F.	2.441	29.00	79.50	108.50
2" steel, galvanized	6.000	L.F.	2.250	35.50	73.50	109.00
Vanity base cabinet, 2 door, 24" x 30"	1.000	Ea.	1.000	143.00	34.00	177.00
24" x 36"	1.000	Ea.	1.200	160.00	40.50	200.50
Vanity top, laminated plastic square edge, 25" x 32"	2.170	L.F.	.712	60.00	24.00	84.00
25" x 38"	2.670	L.F.	.845	71.50	28.50	100.00
Carrier, for lavatory, steel for studs, no arms	1.000	Ea.	1.143	29.00	41.50	70.50
Wood, 2" x 8" blocking	1.300	L.F.	.052	1.18	1.76	2.94

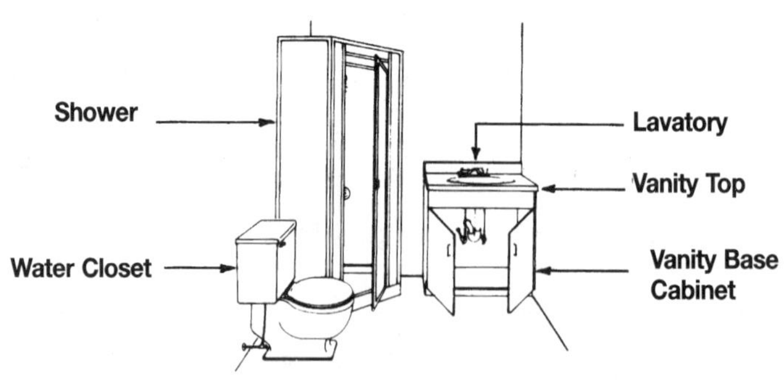

System Description	QUAN.	UNIT	LABOR HOURS	COST EACH		
				MAT.	INST.	TOTAL
BATHROOM WITH LAVATORY INSTALLED IN VANITY						
Water closet, floor mounted, 2 piece, close coupled, white	1.000	Ea.	3.019	139.00	98.50	237.50
Rough-in supply, waste and vent for water closet	1.000	Ea.	2.376	63.30	79.74	143.04
Lavatory, 20" x 18", P.E. cast iron with fittings, white	1.000	Ea.	2.500	219.00	81.50	300.50
Rough-in supply, waste and vent for lavatory	1.000	Ea.	2.791	61.70	94.70	156.40
Shower, steel enameled, stone base, corner, white	1.000	Ea.	8.000	345.00	261.00	606.00
Rough-in supply, waste and vent for shower	1.000	Ea.	3.238	63.95	111.59	175.54
Piping, supply, 1/2" copper	36.000	L.F.	3.556	45.00	129.24	174.24
Waste, 4" cast iron, no hub	7.000	L.F.	1.931	71.40	63.00	134.40
Vent, 2" steel, galvanized	6.000	L.F.	1.500	23.58	48.90	72.48
Vanity base, 2 door, 30" wide	1.000	Ea.	1.000	143.00	34.00	177.00
Vanity top, plastic laminated, square edge	2.670	L.F.	.712	58.74	24.03	82.77
TOTAL			30.623	1233.67	1026.20	2259.87
BATHROOM, WITH WALL HUNG LAVATORY						
Water closet, floor mounted, 2 piece, close coupled, white	1.000	Ea.	3.019	139.00	98.50	237.50
Rough-in supply, waste and vent for water closet	1.000	Ea.	2.376	63.30	79.74	143.04
Lavatory, wall hung, 20" x 18" P.E. cast iron with fittings, white	1.000	Ea.	2.000	220.00	65.50	285.50
Rough-in supply, waste and vent for lavatory	1.000	Ea.	2.791	61.70	94.70	156.40
Shower, steel enameled, stone base, corner, white	1.000	Ea.	8.000	345.00	261.00	606.00
Rough-in supply, waste and vent for shower	1.000	Ea.	3.238	63.95	111.59	175.54
Piping, supply, 1/2" copper	36.000	L.F.	3.556	45.00	129.24	174.24
Waste, 4" cast iron, no hub	7.000	L.F.	1.931	71.40	63.00	134.40
Vent, 2" steel, galvanized	6.000	L.F.	1.500	23.58	48.90	72.48
Carrier, steel, for studs, no arms	1.000	Ea.	1.143	29.00	41.50	70.50
TOTAL			29.554	1061.93	993.67	2055.60

The costs in this system are on a cost each basis. All necessary piping is included.

Description	QUAN.	UNIT	LABOR HOURS	COST EACH		
				MAT.	INST.	TOTAL

Important: See the Reference Section for critical supporting data - Reference Nos., Crews & Location Factors

Three Fixture Bathroom Price Sheet	QUAN.	UNIT	LABOR HOURS	COST EACH		
				MAT.	INST.	TOTAL
Water closet, close coupled, standard 2 piece, white	1.000	Ea.	3.019	139.00	98.50	237.50
Color	1.000	Ea.	3.019	165.00	98.50	263.50
One piece elongated bowl, white	1.000	Ea.	3.019	480.00	98.50	578.50
Color	1.000	Ea.	3.019	660.00	98.50	758.50
Low profile one piece elongated bowl, white	1.000	Ea.	3.019	610.00	98.50	708.50
Color	1.000	Ea.	3.623	660.00	118.00	778.00
Rough-in, for water closet						
1/2" copper supply, 4" cast iron waste, 2" cast iron vent	1.000	Ea.	2.376	63.50	79.50	143.00
4" P.V.C./DWV waste, 2" PVC vent	1.000	Ea.	2.678	37.50	89.50	127.00
4" copper waste, 2" copper vent	1.000	Ea.	2.520	48.00	87.00	135.00
3" cast iron waste, 1-1/2" cast iron vent	1.000	Ea.	2.244	56.00	75.50	131.50
3" PVC waste, 1-1/2" PVC vent	1.000	Ea.	2.388	32.00	83.50	115.50
3" copper waste, 1-1/2" copper vent	1.000	Ea.	2.014	34.00	70.00	104.00
1/2" P.V.C. supply, 4" P.V.C. waste, 2" P.V.C. vent	1.000	Ea.	2.974	47.00	101.00	148.00
3" P.V.C. waste, 1-1/2" P.V.C. vent	1.000	Ea.	2.684	42.00	94.00	136.00
1/2" steel supply, 4" cast iron waste, 2" cast iron vent	1.000	Ea.	2.545	66.50	86.00	152.50
4" cast iron waste, 2" steel vent	1.000	Ea.	2.590	57.00	87.50	144.50
4" P.V.C. waste, 2" P.V.C. vent	1.000	Ea.	2.847	40.50	96.00	136.50
Lavatory, wall hung P.E. cast iron 20" x 18", white	1.000	Ea.	2.000	220.00	65.50	285.50
Color	1.000	Ea.	2.000	250.00	65.50	315.50
Vitreous china 19" x 17", white	1.000	Ea.	2.286	160.00	74.50	234.50
Color	1.000	Ea.	2.286	178.00	74.50	252.50
Lavatory, for vanity top P.E. cast iron 20" x 18", white	1.000	Ea.	2.500	219.00	81.50	300.50
Color	1.000	Ea.	2.500	249.00	81.50	330.50
Steel enameled 20" x 17", white	1.000	Ea.	2.759	111.00	90.00	201.00
Color	1.000	Ea.	2.500	117.00	81.50	198.50
Vitreous china 20" x 16", white	1.000	Ea.	2.963	202.00	96.50	298.50
Color	1.000	Ea.	2.963	202.00	96.50	298.50
Rough-in, for lavatory						
1/2" copper supply, 1-1/2" cast iron waste, 1-1/2" cast iron vent	1.000	Ea.	2.791	61.50	94.50	156.00
1-1/2" P.V.C. waste, 1-1/2" P.V.C. vent	1.000	Ea.	2.639	36.00	95.50	131.50
1/2" steel supply, 1-1/2" cast iron waste, 1-1/4" steel vent	1.000	Ea.	2.890	52.50	99.00	151.50
1-1/2" P.V.C. waste, 1-1/4" P.V.C. vent	1.000	Ea.	2.921	41.00	106.00	147.00
1/2" P.V.C. supply, 1-1/2" P.V.C. waste, 1-1/2" P.V.C. vent	1.000	Ea.	3.259	53.00	118.00	171.00
Shower, steel enameled stone base, 32" x 32", white	1.000	Ea.	8.000	345.00	261.00	606.00
Color	1.000	Ea.	7.822	840.00	255.00	1095.00
36" x 36", white	1.000	Ea.	8.889	905.00	290.00	1195.00
Color	1.000	Ea.	8.889	955.00	290.00	1245.00
Rough-in, for shower						
1/2" copper supply, 2" cast iron waste, 1-1/2" copper vent	1.000	Ea.	3.161	60.50	110.00	170.50
2" P.V.C. waste, 1-1/2" P.V.C. vent	1.000	Ea.	3.429	42.00	119.00	161.00
1/2" steel supply, 2" cast iron waste, 1-1/2" steel vent	1.000	Ea.	3.887	92.00	134.00	226.00
2" P.V.C. waste, 1-1/2" P.V.C. vent	1.000	Ea.	3.881	50.00	135.00	185.00
1/2" P.V.C. supply, 2" P.V.C. waste, 1-1/2" P.V.C. vent	1.000	Ea.	4.219	68.50	148.00	216.50
Piping, supply, 1/2" copper	36.000	L.F.	3.556	45.00	129.00	174.00
1/2" steel	36.000	L.F.	4.571	63.00	166.00	229.00
1/2" P.V.C.	36.000	L.F.	5.333	104.00	194.00	298.00
Waste, 4" cast iron, no hub	7.000	L.F.	1.931	71.50	63.00	134.50
4" P.V.C./DWV	7.000	L.F.	2.333	39.50	76.50	116.00
4" copper/DWV	7.000	L.F.	2.800	63.50	91.50	155.00
Vent, 2" cast iron, no hub	6.000	L.F.	1.091	20.00	39.50	59.50
2" copper/DWV	6.000	L.F.	1.091	20.00	39.50	59.50
2" P.V.C./DWV	6.000	L.F.	1.627	19.20	53.00	72.20
2" steel, galvanized	6.000	L.F.	1.500	23.50	49.00	72.50
Vanity base cabinet, 2 door, 24" x 30"	1.000	Ea.	1.000	143.00	34.00	177.00
24" x 36"	1.000	Ea.	1.200	160.00	40.50	200.50
Vanity top, laminated plastic square edge, 25" x 32"	2.670	L.F.	.712	58.50	24.00	82.50
25" x 38"	3.170	L.F.	.845	69.50	28.50	98.00
Carrier , for lavatory, steel, for studs, no arms	1.000	Ea.	1.143	29.00	41.50	70.50
Wood, 2" x 8" blocking	1.300	L.F.	.052	1.18	1.76	2.94

36 Four Fixture Bathroom Systems

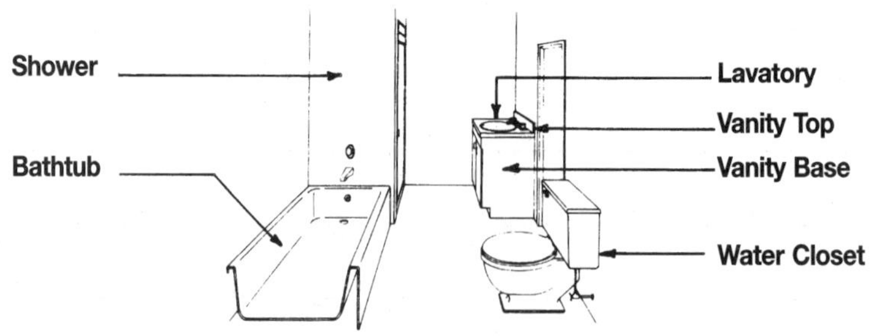

System Description	QUAN.	UNIT	LABOR HOURS	COST EACH		
				MAT.	INST.	TOTAL
BATHROOM WITH LAVATORY INSTALLED IN VANITY						
Water closet, floor mounted, 2 piece, close coupled, white	1.000	Ea.	3.019	139.00	98.50	237.50
Rough-in supply, waste and vent for water closet	1.000	Ea.	2.376	63.30	79.74	143.04
Lavatory, 20" x 18" P.E. cast iron with fittings, white	1.000	Ea.	2.500	219.00	81.50	300.50
Shower, steel, enameled, stone base, corner, white	1.000	Ea.	8.889	825.00	290.00	1115.00
Rough-in supply, waste and vent for lavatory and shower	2.000	Ea.	7.667	163.00	261.88	424.88
Bathtub, P.E. cast iron, 5' long with fittings, white	1.000	Ea.	3.636	335.00	119.00	454.00
Rough-in supply, waste and vent for bathtub	1.000	Ea.	2.409	53.86	84.50	138.36
Piping, supply, 1/2" copper	42.000	L.F.	4.148	52.50	150.78	203.28
Waste, 4" cast iron, no hub	10.000	L.F.	2.759	102.00	90.00	192.00
Vent, 2" steel galvanized	13.000	L.F.	3.250	51.09	105.95	157.04
Vanity base, 2 doors, 30" wide	1.000	Ea.	1.000	143.00	34.00	177.00
Vanity top, plastic laminated, square edge	2.670	L.F.	.712	58.74	24.03	82.77
TOTAL			42.365	2205.49	1419.88	3625.37
BATHROOM WITH WALL HUNG LAVATORY						
Water closet, floor mounted, 2 piece, close coupled, white	1.000	Ea.	3.019	139.00	98.50	237.50
Rough-in supply, waste and vent for water closet	1.000	Ea.	2.376	63.30	79.74	143.04
Lavatory, 20" x 18" P.E. cast iron with fittings, white	1.000	Ea.	2.000	220.00	65.50	285.50
Shower, steel enameled, stone base, corner , white	1.000	Ea.	8.889	825.00	290.00	1115.00
Rough-in supply, waste and vent for lavatory and shower	2.000	Ea.	7.667	163.00	261.88	424.88
Bathtub, P.E. cast iron, 5' long with fittings, white	1.000	Ea.	3.636	335.00	119.00	454.00
Rough-in supply, waste and vent for bathtub	1.000	Ea.	2.409	53.86	84.50	138.36
Piping, supply, 1/2" copper	42.000	L.F.	4.148	52.50	150.78	203.28
Waste, 4" cast iron, no hub	10.000	L.F.	2.759	102.00	90.00	192.00
Vent, 2" steel galvanized	13.000	L.F.	3.250	51.09	105.95	157.04
Carrier, steel, for studs, no arms	1.000	Ea.	1.143	29.00	41.50	70.50
TOTAL			41.296	2033.75	1387.35	3421.10

The costs in this system are on a cost each basis. All necessary piping is included.

Description	QUAN.	UNIT	LABOR HOURS	COST EACH		
				MAT.	INST.	TOTAL

Four Fixture Bathroom Price Sheet	QUAN.	UNIT	LABOR HOURS	COST EACH		
				MAT.	INST.	TOTAL
Water closet, close coupled, standard 2 piece, white	1.000	Ea.	3.019	139.00	98.50	237.50
Color	1.000	Ea.	3.019	165.00	98.50	263.50
One piece elongated bowl, white	1.000	Ea.	3.019	480.00	98.50	578.50
Color	1.000	Ea.	3.019	660.00	98.50	758.50
Low profile, one piece elongated bowl, white	1.000	Ea.	3.019	610.00	98.50	708.50
Color	1.000	Ea.	3.019	550.00	98.50	648.50
Rough-in, for water closet						
1/2" copper supply, 4" cast iron waste, 2" cast iron vent	1.000	Ea.	2.376	63.50	79.50	143.00
4" PVC/DWV waste, 2" PVC vent	1.000	Ea.	2.678	37.50	89.50	127.00
4" copper waste, 2" copper vent	1.000	Ea.	2.520	48.00	87.00	135.00
3" cast iron waste, 1-1/2" cast iron vent	1.000	Ea.	2.244	56.00	75.50	131.50
3" P.V.C. waste, 1-1/2" P.V.C. vent	1.000	Ea.	2.388	32.00	83.50	115.50
3" copper waste, 1-1/2" copper vent	1.000	Ea.	2.014	34.00	70.00	104.00
1/2" P.V.C. supply, 4" P.V.C. waste, 2" P.V.C. vent	1.000	Ea.	2.974	47.00	101.00	148.00
3" P.V.C. waste, 1-1/2" P.V.C. vent	1.000	Ea.	2.684	42.00	94.00	136.00
1/2" steel supply, 4" cast iron waste, 2" cast iron vent	1.000	Ea.	2.545	66.50	86.00	152.50
4" cast iron waste, 2" steel vent	1.000	Ea.	2.590	57.00	87.50	144.50
4" P.V.C. waste, 2" P.V.C. vent	1.000	Ea.	2.847	40.50	96.00	136.50
Lavatory, wall hung P.E. cast iron 20" x 18", white	1.000	Ea.	2.000	220.00	65.50	285.50
Color	1.000	Ea.	2.000	250.00	65.50	315.50
Vitreous china 19" x 17", white	1.000	Ea.	2.286	160.00	74.50	234.50
Color	1.000	Ea.	2.286	178.00	74.50	252.50
Lavatory for vanity top, P.E. cast iron 20" x 18", white	1.000	Ea.	2.500	219.00	81.50	300.50
Color	1.000	Ea.	2.500	249.00	81.50	330.50
Steel enameled, 20" x 17", white	1.000	Ea.	2.759	111.00	90.00	201.00
Color	1.000	Ea.	2.500	117.00	81.50	198.50
Vitreous china 20" x 16", white	1.000	Ea.	2.963	202.00	96.50	298.50
Color	1.000	Ea.	2.963	202.00	96.50	298.50
Shower, steel enameled stone base, 36" square, white	1.000	Ea.	8.889	825.00	290.00	1115.00
Color	1.000	Ea.	8.889	890.00	290.00	1180.00
Rough-in, for lavatory or shower						
1/2" copper supply, 1-1/2" cast iron waste, 1-1/2" cast iron vent	1.000	Ea.	3.834	81.50	131.00	212.50
1-1/2" P.V.C. waste, 1-1/4" P.V.C. vent	1.000	Ea.	3.675	49.50	133.00	182.50
1/2" steel supply, 1-1/4" cast iron waste, 1-1/4" steel vent	1.000	Ea.	4.103	75.50	141.00	216.50
1-1/4" P.V.C. waste, 1-1/4" P.V.C. vent	1.000	Ea.	3.937	57.00	143.00	200.00
1/2" P.V.C. supply, 1-1/2" P.V.C. waste, 1-1/2" P.V.C. vent	1.000	Ea.	4.592	76.50	167.00	243.50
Bathtub, P.E. cast iron, 5' long with fittings, white	1.000	Ea.	3.636	335.00	119.00	454.00
Color	1.000	Ea.	3.636	385.00	119.00	504.00
Steel, enameled 5' long with fittings, white	1.000	Ea.	2.909	256.00	95.00	351.00
Color	1.000	Ea.	2.909	256.00	95.00	351.00
Rough-in, for bathtub						
1/2" copper supply, 4" cast iron waste, 1-1/2" copper vent	1.000	Ea.	2.409	54.00	84.50	138.50
4" P.V.C. waste, 1-1/2" P.V.C. vent	1.000	Ea.	2.877	41.50	101.00	142.50
1/2" steel supply, 4" cast iron waste, 1-1/2" steel vent	1.000	Ea.	2.898	60.00	99.50	159.50
4" P.V.C. waste, 1-1/2" P.V.C. vent	1.000	Ea.	3.159	46.50	111.00	157.50
1/2" P.V.C. supply, 4" P.V.C. waste, 1-1/2" P.V.C. vent	1.000	Ea.	3.370	58.00	119.00	177.00
Piping, supply, 1/2" copper	42.000	L.F.	4.148	52.50	151.00	203.50
1/2" steel	42.000	L.F.	5.333	73.50	194.00	267.50
1/2" P.V.C.	42.000	L.F.	6.222	121.00	227.00	348.00
Waste, 4" cast iron, no hub	10.000	L.F.	2.759	102.00	90.00	192.00
4" P.V.C./DWV	10.000	L.F.	3.333	56.50	109.00	165.50
4" copper/DWV	10.000	Ea.	4.000	90.50	131.00	221.50
Vent 2" cast iron, no hub	13.000	L.F.	3.104	82.00	101.00	183.00
2" copper/DWV	13.000	L.F.	2.364	43.50	86.00	129.50
2" P.V.C./DWV	13.000	L.F.	3.525	41.50	115.00	156.50
2" steel, galvanized	13.000	L.F.	3.250	51.00	106.00	157.00
Vanity base cabinet, 2 doors, 30" wide	1.000	Ea.	1.000	143.00	34.00	177.00
Vanity top, plastic laminated, square edge	2.670	L.F.	.712	58.50	24.00	82.50
Carrier, steel for studs, no arms	1.000	Ea.	1.143	29.00	41.50	70.50
Wood, 2" x 8" blocking	1.300	L.F.	.052	1.18	1.76	2.94

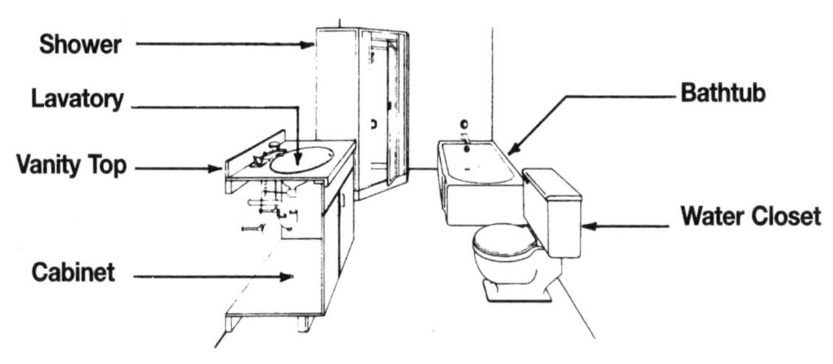

System Description	QUAN.	UNIT	LABOR HOURS	COST EACH		
				MAT.	INST.	TOTAL
BATHROOM WITH LAVATORY INSTALLED IN VANITY						
Water closet, floor mounted, 2 piece, close coupled, white	1.000	Ea.	3.019	139.00	98.50	237.50
Rough-in supply, waste and vent for water closet	1.000	Ea.	2.376	63.30	79.74	143.04
Lavatory, 20″ x 18″ P.E. cast iron with fittings, white	1.000	Ea.	2.500	219.00	81.50	300.50
Shower, steel, enameled, stone base, corner, white	1.000	Ea.	8.889	825.00	290.00	1115.00
Rough-in supply waste and vent for lavatory and shower	2.000	Ea.	7.667	163.00	261.88	424.88
Bathtub, P.E. cast iron, 5′ long with fittings, white	1.000	Ea.	3.636	335.00	119.00	454.00
Rough-in supply waste and vent for bathtub	1.000	Ea.	2.409	53.86	84.50	138.36
Piping supply, 1/2″ copper	42.000	L.F.	4.938	62.50	179.50	242.00
Waste, 4″ cast iron, no hub	10.000	L.F.	4.138	153.00	135.00	288.00
Vent, 2″ steel galvanized	13.000	L.F.	4.500	70.74	146.70	217.44
Vanity base, 2 doors, 30″ wide	1.000	Ea.	1.000	143.00	34.00	177.00
Vanity top, plastic laminated, square edge	2.670	L.F.	.712	60.08	24.03	84.11
TOTAL			45.784	2287.48	1534.35	3821.83
BATHROOM WITH WALL HUNG LAVATORY						
Water closet, floor mounted, 2 piece, close coupled, white	1.000	Ea.	3.019	139.00	98.50	237.50
Rough-in supply, waste and vent for water closet	1.000	Ea.	2.376	63.30	79.74	143.04
Lavatory, 20″ x 18″ P.E. cast iron with fittings, white	1.000	Ea.	2.000	220.00	65.50	285.50
Shower, steel enameled, stone base, corner, white	1.000	Ea.	8.889	825.00	290.00	1115.00
Rough-in supply, waste and vent for lavatory and shower	2.000	Ea.	7.667	163.00	261.88	424.88
Bathtub, P.E. cast iron, 5″ long with fittings, white	1.000	Ea.	3.636	335.00	119.00	454.00
Rough-in supply, waste and vent for bathtub	1.000	Ea.	2.409	53.86	84.50	138.36
Piping, supply, 1/2″ copper	42.000	L.F.	4.938	62.50	179.50	242.00
Waste, 4″ cast iron, no hub	10.000	L.F.	4.138	153.00	135.00	288.00
Vent, 2″ steel galvanized	13.000	L.F.	4.500	70.74	146.70	217.44
Carrier, steel for studs, no arms	1.000	Ea.	1.143	29.00	41.50	70.50
TOTAL			44.715	2114.40	1501.82	3616.22

The costs in this system are on a cost each basis. All necessary piping is included.

Description	QUAN.	UNIT	LABOR HOURS	COST EACH		
				MAT.	INST.	TOTAL

Important: See the Reference Section for critical supporting data - Reference Nos., Crews & Location Factors

Four Fixture Bathroom Price Sheet	QUAN.	UNIT	LABOR HOURS	COST EACH		
				MAT.	INST.	TOTAL
Water closet, close coupled, standard 2 piece, white	1.000	Ea.	3.019	139.00	98.50	237.50
Color	1.000	Ea.	3.019	165.00	98.50	263.50
One piece, elongated bowl, white	1.000	Ea.	3.019	480.00	98.50	578.50
Color	1.000	Ea.	3.019	660.00	98.50	758.50
Low profile, one piece elongated bowl, white	1.000	Ea.	3.019	610.00	98.50	708.50
Color	1.000	Ea.	3.019	550.00	98.50	648.50
Rough-in, for water closet						
1/2″ copper supply, 4″ cast iron waste, 2″ cast iron vent	1.000	Ea.	2.376	63.50	79.50	143.00
4″ PVC/DWV waste, 2″ PVC vent	1.000	Ea.	2.678	37.50	89.50	127.00
4″ copper waste, 2″ copper vent	1.000	Ea.	2.520	48.00	87.00	135.00
3″ cast iron waste, 1-1/2″ cast iron vent	1.000	Ea.	2.244	56.00	75.50	131.50
3″ PVC waste, 1-1/2″ PVC vent	1.000	Ea.	2.388	32.00	83.50	115.50
3″ PVC waste, 1-1/2″ PVC vent	1.000	Ea.	2.014	34.00	70.00	104.00
1/2″ PVC supply, 4″ PVC waste, 2″ PVC vent	1.000	Ea.	2.974	47.00	101.00	148.00
3″ PVC waste, 1-1/2″ PVC vent	1.000	Ea.	2.684	42.00	94.00	136.00
1/2″ steel supply, 4″ cast iron waste, 2″ cast iron vent	1.000	Ea.	2.545	66.50	86.00	152.50
4″ cast iron waste, 2″ steel vent	1.000	Ea.	2.590	57.00	87.50	144.50
4″ PVC waste, 2″ PVC vent	1.000	Ea.	2.847	40.50	96.00	136.50
Lavatory wall hung, P.E. cast iron 20″ x 18″, white	1.000	Ea.	2.000	220.00	65.50	285.50
Color	1.000	Ea.	2.000	250.00	65.50	315.50
Vitreous china 19″ x 17″, white	1.000	Ea.	2.286	160.00	74.50	234.50
Color	1.000	Ea.	2.286	178.00	74.50	252.50
Lavatory for vanity top, P.E. cast iron, 20″ x 18″, white	1.000	Ea.	2.500	219.00	81.50	300.50
Color	1.000	Ea.	2.500	249.00	81.50	330.50
Steel, enameled 20″ x 17″, white	1.000	Ea.	2.759	111.00	90.00	201.00
Color	1.000	Ea.	2.500	117.00	81.50	198.50
Vitreous china 20″ x 16″, white	1.000	Ea.	2.963	202.00	96.50	298.50
Color	1.000	Ea.	2.963	202.00	96.50	298.50
Shower, steel enameled, stone base 36″ square, white	1.000	Ea.	8.889	825.00	290.00	1115.00
Color	1.000	Ea.	8.889	890.00	290.00	1180.00
Rough-in, for lavatory and shower						
1/2″ copper supply, 1-1/2″ cast iron waste, 1-1/2″ cast iron vent	1.000	Ea.	7.667	163.00	262.00	425.00
1-1/2″ PVC waste, 1-1/4″ PVC vent	1.000	Ea.	7.351	99.50	267.00	366.50
1/2″ steel supply, 1-1/4″ cast iron waste, 1-1/4″ steel vent	1.000	Ea.	8.205	151.00	283.00	434.00
1-1/4″ PVC waste, 1-1/4″ PVC vent	1.000	Ea.	7.873	114.00	286.00	400.00
1/2″ PVC supply, 1-1/2″ PVC waste, 1-1/2″ PVC vent	1.000	Ea.	9.185	153.00	335.00	488.00
Bathtub, P.E. cast iron, 5′ long with fittings, white	1.000	Ea.	3.636	335.00	119.00	454.00
Color	1.000	Ea.	3.636	385.00	119.00	504.00
Steel enameled, 5′ long with fittings, white	1.000	Ea.	2.909	256.00	95.00	351.00
Color	1.000	Ea.	2.909	256.00	95.00	351.00
Rough-in, for bathtub						
1/2″ copper supply, 4″ cast iron waste, 1-1/2″ copper vent	1.000	Ea.	2.409	54.00	84.50	138.50
4″ PVC waste, 1-1/2″ PVC vent	1.000	Ea.	2.877	41.50	101.00	142.50
1/2″ steel supply, 4″ cast iron waste, 1-1/2″ steel vent	1.000	Ea.	2.898	60.00	99.50	159.50
4″ PVC waste, 1-1/2″ PVC vent	1.000	Ea.	3.159	46.50	111.00	157.50
1/2″ PVC supply, 4″ PVC waste, 1-1/2″ PVC vent	1.000	Ea.	3.370	58.00	119.00	177.00
Piping supply, 1/2″ copper	42.000	L.F.	4.148	52.50	151.00	203.50
1/2″ steel	42.000	L.F.	5.333	73.50	194.00	267.50
1/2″ PVC	42.000	L.F.	6.222	121.00	227.00	348.00
Piping, waste, 4″ cast iron, no hub	10.000	L.F.	3.586	133.00	117.00	250.00
4″ PVC/DWV	10.000	L.F.	4.333	73.50	142.00	215.50
4″ copper/DWV	10.000	L.F.	5.200	118.00	170.00	288.00
Piping, vent, 2″ cast iron, no hub	13.000	L.F.	3.104	82.00	101.00	183.00
2″ copper/DWV	13.000	L.F.	2.364	43.50	86.00	129.50
2″ PVC/DWV	13.000	L.F.	3.525	41.50	115.00	156.50
2″ steel, galvanized	13.000	L.F.	3.250	51.00	106.00	157.00
Vanity base cabinet, 2 doors, 30″ wide	1.000	Ea.	1.000	143.00	34.00	177.00
Vanity top, plastic laminated, square edge	3.160	L.F.	.843	69.50	28.50	98.00
Carrier, steel, for studs, no arms	1.000	Ea.	1.143	29.00	41.50	70.50
Wood, 2″ x 8″ blocking	1.300	L.F.	.052	1.18	1.76	2.94

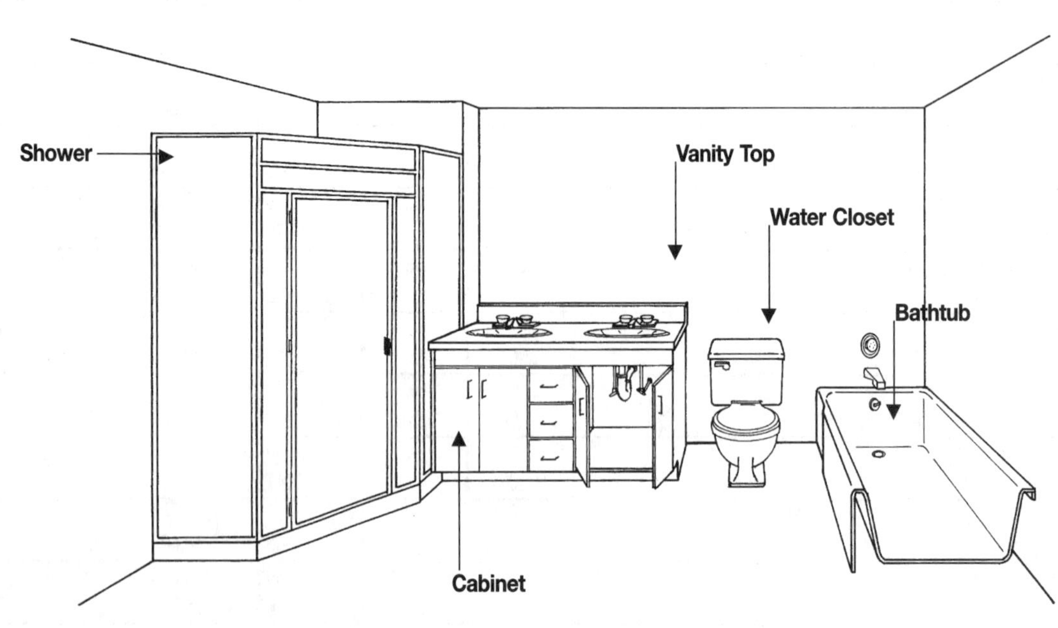

System Description	QUAN.	UNIT	LABOR HOURS	COST EACH		
				MAT.	INST.	TOTAL
BATHROOM WITH SHOWER, BATHTUB, LAVATORIES IN VANITY						
Water closet, floor mounted, 1 piece combination, white	1.000	Ea.	3.019	610.00	98.50	708.50
Rough-in supply, waste and vent for water closet	1.000	Ea.	2.376	63.30	79.74	143.04
Lavatory, 20″ x 16″, vitreous china oval, with fittings, white	2.000	Ea.	5.926	404.00	193.00	597.00
Shower, steel enameled, stone base, corner, white	1.000	Ea.	8.889	825.00	290.00	1115.00
Rough-in supply waste and vent for lavatory and shower	3.000	Ea.	8.371	185.10	284.10	469.20
Bathtub, P.E. cast iron, 5′ long with fittings, white	1.000	Ea.	3.636	335.00	119.00	454.00
Rough-in supply, waste and vent for bathtub	1.000	Ea.	2.684	64.06	93.50	157.56
Piping, supply, 1/2″ copper	42.000	L.F.	4.148	52.50	150.78	203.28
Waste, 4″ cast iron, no hub	10.000	L.F.	2.759	102.00	90.00	192.00
Vent, 2″ steel galvanized	13.000	L.F.	3.250	51.09	105.95	157.04
Vanity base, 2 door, 24″ x 48″	1.000	Ea.	1.400	202.00	47.00	249.00
Vanity top, plastic laminated, square edge	4.170	L.F.	1.112	91.74	37.53	129.27
TOTAL			47.570	2985.79	1589.10	4574.89

The costs in this system are on a cost each basis. All necessary piping is included.

Description	QUAN.	UNIT	LABOR HOURS	COST EACH		
				MAT.	INST.	TOTAL

 Important: See the Reference Section for critical supporting data - Reference Nos., Crews & Location Factors

Five Fixture Bathroom Price Sheet	QUAN.	UNIT	LABOR HOURS	COST EACH		
				MAT.	INST.	TOTAL
Water closet, close coupled, standard 2 piece, white	1.000	Ea.	3.019	139.00	98.50	237.50
Color	1.000	Ea.	3.019	165.00	98.50	263.50
One piece elongated bowl, white	1.000	Ea.	3.019	480.00	98.50	578.50
Color	1.000	Ea.	3.019	660.00	98.50	758.50
Low profile, one piece elongated bowl, white	1.000	Ea.	3.019	610.00	98.50	708.50
Color	1.000	Ea.	3.019	550.00	98.50	648.50
Rough-in, supply, waste and vent for water closet						
1/2" copper supply, 4" cast iron waste, 2" cast iron vent	1.000	Ea.	2.376	63.50	79.50	143.00
4" P.V.C./DWV waste, 2" P.V.C. vent	1.000	Ea.	2.678	37.50	89.50	127.00
4" copper waste, 2" copper vent	1.000	Ea.	2.520	48.00	87.00	135.00
3" cast iron waste, 1-1/2" cast iron vent	1.000	Ea.	2.244	56.00	75.50	131.50
3" P.V.C. waste, 1-1/2" P.V.C. vent	1.000	Ea.	2.388	32.00	83.50	115.50
3" copper waste, 1-1/2" copper vent	1.000	Ea.	2.014	34.00	70.00	104.00
1/2" P.V.C. supply, 4" P.V.C. waste, 2" P.V.C. vent	1.000	Ea.	2.974	47.00	101.00	148.00
3" P.V.C. waste, 1-1/2" P.V.C. supply	1.000	Ea.	2.684	42.00	94.00	136.00
1/2" steel supply, 4" cast iron waste, 2" cast iron vent	1.000	Ea.	2.545	66.50	86.00	152.50
4" cast iron waste, 2" steel vent	1.000	Ea.	2.590	57.00	87.50	144.50
4" P.V.C. waste, 2" P.V.C. vent	1.000	Ea.	2.847	40.50	96.00	136.50
Lavatory, wall hung, P.E. cast iron 20" x 18", white	2.000	Ea.	4.000	440.00	131.00	571.00
Color	2.000	Ea.	4.000	500.00	131.00	631.00
Vitreous china, 19" x 17", white	2.000	Ea.	4.571	320.00	149.00	469.00
Color	2.000	Ea.	4.571	355.00	149.00	504.00
Lavatory, for vanity top, P.E. cast iron, 20" x 18", white	2.000	Ea.	5.000	440.00	163.00	603.00
Color	2.000	Ea.	5.000	500.00	163.00	663.00
Steel enameled 20" x 17", white	2.000	Ea.	5.517	222.00	180.00	402.00
Color	2.000	Ea.	5.000	234.00	163.00	397.00
Vitreous china 20" x 16", white	2.000	Ea.	5.926	405.00	193.00	598.00
Color	2.000	Ea.	5.926	405.00	193.00	598.00
Shower, steel enameled, stone base 36" square, white	1.000	Ea.	8.889	825.00	290.00	1115.00
Color	1.000	Ea.	8.889	890.00	290.00	1180.00
Rough-in, for lavatory or shower						
1/2" copper supply, 1-1/2" cast iron waste, 1-1/2" cast iron vent	3.000	Ea.	8.371	185.00	284.00	469.00
1-1/2" P.V.C. waste, 1-1/4" P.V.C. vent	3.000	Ea.	7.916	109.00	287.00	396.00
1/2" steel supply, 1-1/4" cast iron waste, 1-1/4" steel vent	3.000	Ea.	8.671	158.00	297.00	455.00
1-1/4" P.V.C. waste, 1-1/4" P.V.C. vent	3.000	Ea.	8.381	123.00	305.00	428.00
1/2" P.V.C. supply, 1-1/2" P.V.C. waste, 1-1/2" P.V.C. vent	3.000	Ea.	9.777	159.00	355.00	514.00
Bathtub, P.E. cast iron 5' long with fittings, white	1.000	Ea.	3.636	335.00	119.00	454.00
Color	1.000	Ea.	3.636	385.00	119.00	504.00
Steel, enameled 5' long with fittings, white	1.000	Ea.	2.909	256.00	95.00	351.00
Color	1.000	Ea.	2.909	256.00	95.00	351.00
Rough-in, for bathtub						
1/2" copper supply, 4" cast iron waste, 1-1/2" copper vent	1.000	Ea.	2.684	64.00	93.50	157.50
4" P.V.C. waste, 1-1/2" P.V.C. vent	1.000	Ea.	3.210	47.00	112.00	159.00
1/2" steel supply, 4" cast iron waste, 1-1/2" steel vent	1.000	Ea.	3.173	70.00	108.00	178.00
4" P.V.C. waste, 1-1/2" P.V.C. vent	1.000	Ea.	3.492	52.00	122.00	174.00
1/2" P.V.C. supply, 4" P.V.C. waste, 1-1/2" P.V.C. vent	1.000	Ea.	3.703	63.50	130.00	193.50
Piping, supply, 1/2" copper	42.000	L.F.	4.148	52.50	151.00	203.50
1/2" steel	42.000	L.F.	5.333	73.50	194.00	267.50
1/2" P.V.C.	42.000	L.F.	6.222	121.00	227.00	348.00
Piping, waste, 4" cast iron, no hub	10.000	L.F.	2.759	102.00	90.00	192.00
4" P.V.C./DWV	10.000	L.F.	3.333	56.50	109.00	165.50
4" copper/DWV	10.000	L.F.	4.000	90.50	131.00	221.50
Piping, vent, 2" cast iron, no hub	13.000	L.F.	3.104	82.00	101.00	183.00
2" copper/DWV	13.000	L.F.	2.364	43.50	86.00	129.50
2" P.V.C./DWV	13.000	L.F.	3.525	41.50	115.00	156.50
2" steel, galvanized	13.000	L.F.	3.250	51.00	106.00	157.00
Vanity base cabinet, 2 doors, 24" x 48"	1.000	Ea.	1.400	202.00	47.00	249.00
Vanity top, plastic laminated, square edge	4.170	L.F.	1.112	91.50	37.50	129.00
Carrier, steel, for studs, no arms	1.000	Ea.	1.143	29.00	41.50	70.50
Wood, 2" x 8" blocking	1.300	L.F.	.052	1.18	1.76	2.94

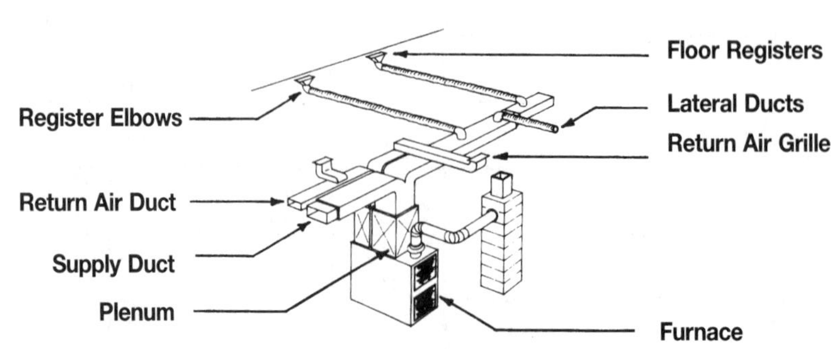

System Description	QUAN.	UNIT	LABOR HOURS	COST PER SYSTEM		
				MAT.	INST.	TOTAL
HEATING ONLY, GAS FIRED HOT AIR, ONE ZONE, 1200 S.F. BUILDING						
Furnace, gas, up flow	1.000	Ea.	5.000	720.00	163.00	883.00
Intermittent pilot	1.000	Ea.		145.00		145.00
Supply duct, rigid fiberglass	176.000	S.F.	12.069	156.64	408.32	564.96
Return duct, sheet metal, galvanized	158.000	Lb.	16.136	181.70	546.68	728.38
Lateral ducts, 6″ flexible fiberglass	144.000	L.F.	8.862	341.28	289.44	630.72
Register, elbows	12.000	Ea.	3.200	85.80	104.40	190.20
Floor registers, enameled steel	12.000	Ea.	3.000	209.40	108.60	318.00
Floor grille, return air	2.000	Ea.	.727	47.00	26.40	73.40
Thermostat	1.000	Ea.	1.000	27.00	36.50	63.50
Plenum	1.000	Ea.	1.000	67.50	32.50	100.00
TOTAL			50.994	1981.32	1715.84	3697.16
HEATING/COOLING, GAS FIRED FORCED AIR, ONE ZONE, 1200 S.F. BUILDING						
Furnace, including plenum, compressor, coil	1.000	Ea.	14.720	3220.00	483.00	3703.00
Intermittent pilot	1.000	Ea.		145.00		145.00
Supply duct, rigid fiberglass	176.000	S.F.	12.069	156.64	408.32	564.96
Return duct, sheet metal, galvanized	158.000	Lb.	16.136	181.70	546.68	728.38
Lateral duct, 6″ flexible fiberglass	144.000	L.F.	8.862	341.28	289.44	630.72
Register elbows	12.000	Ea.	3.200	85.80	104.40	190.20
Floor registers, enameled steel	12.000	Ea.	3.000	209.40	108.60	318.00
Floor grille return air	2.000	Ea.	.727	47.00	26.40	73.40
Thermostat	1.000	Ea.	1.000	27.00	36.50	63.50
Refrigeration piping (precharged)	25.000	L.F.		165.00		165.00
TOTAL			59.714	4575.00	2000.00	6575.00

The costs in these systems are based on complete system basis. For larger buildings use the price sheet on the opposite page.

Description	QUAN.	UNIT	LABOR HOURS	COST PER SYSTEM		
				MAT.	INST.	TOTAL

Important: See the Reference Section for critical supporting data - Reference Nos., Crews & Location Factors

Gas Heating/Cooling Price Sheet	QUAN.	UNIT	LABOR HOURS	COST EACH		
				MAT.	INST.	TOTAL
Furnace, heating only, 100 MBH, area to 1200 S.F.	1.000	Ea.	5.000	720.00	163.00	883.00
120 MBH, area to 1500 S.F.	1.000	Ea.	5.000	720.00	163.00	883.00
160 MBH, area to 2000 S.F.	1.000	Ea.	5.714	965.00	187.00	1152.00
200 MBH, area to 2400 S.F.	1.000	Ea.	6.154	2025.00	201.00	2226.00
Heating/cooling, 100 MBH heat, 36 MBH cool, to 1200 S.F.	1.000	Ea.	16.000	3500.00	525.00	4025.00
120 MBH heat, 42 MBH cool, to 1500 S.F.	1.000	Ea.	18.462	3725.00	625.00	4350.00
144 MBH heat, 47 MBH cool, to 2000 S.F.	1.000	Ea.	20.000	4325.00	680.00	5005.00
200 MBH heat, 60 MBH cool, to 2400 S.F.	1.000	Ea.	34.286	4525.00	1150.00	5675.00
Intermittent pilot, 100 MBH furnace	1.000	Ea.		145.00		145.00
200 MBH furnace	1.000	Ea.		145.00		145.00
Supply duct, rectangular, area to 1200 S.F., rigid fiberglass	176.000	S.F.	12.069	157.00	410.00	567.00
Sheet metal insulated	228.000	Lb.	31.331	340.00	1050.00	1390.00
Area to 1500 S.F., rigid fiberglass	176.000	S.F.	12.069	157.00	410.00	567.00
Sheet metal insulated	228.000	Lb.	31.331	340.00	1050.00	1390.00
Area to 2400 S.F., rigid fiberglass	205.000	S.F.	14.057	182.00	475.00	657.00
Sheet metal insulated	271.000	Lb.	37.048	400.00	1250.00	1650.00
Round flexibie, insulated 6″ diameter, to 1200 S.F.	156.000	L.F.	9.600	370.00	315.00	685.00
To 1500 S.F.	184.000	L.F.	11.323	435.00	370.00	805.00
8″ diameter, to 2000 S.F.	269.000	L.F.	23.911	775.00	780.00	1555.00
To 2400 S.F.	248.000	L.F.	22.044	715.00	720.00	1435.00
Return duct, sheet metal galvanized, to 1500 S.F.	158.000	Lb.	16.136	182.00	545.00	727.00
To 2400 S.F.	191.000	Lb.	19.506	220.00	660.00	880.00
Lateral ducts, flexible round 6″ insulated, to 1200 S.F.	144.000	L.F.	8.862	340.00	289.00	629.00
To 1500 S.F.	172.000	L.F.	10.585	410.00	345.00	755.00
To 2000 S.F.	261.000	L.F.	16.062	620.00	525.00	1145.00
To 2400 S.F.	300.000	L.F.	18.462	710.00	605.00	1315.00
Spiral steel insulated, to 1200 S.F.	144.000	L.F.	20.069	305.00	655.00	960.00
To 1500 S.F.	172.000	L.F.	23.955	365.00	780.00	1145.00
To 2000 S.F.	261.000	L.F.	36.354	555.00	1175.00	1730.00
To 2400 S.F.	300.000	L.F.	41.829	635.00	1350.00	1985.00
Rectangular sheet metal galvanized insulated, to 1200 S.F.	228.000	Lb.	39.056	410.00	1300.00	1710.00
To 1500 S.F.	344.000	Lb.	53.966	575.00	1800.00	2375.00
To 2000 S.F.	522.000	Lb.	81.928	870.00	2725.00	3595.00
To 2400 S.F.	600.000	Lb.	94.191	1000.00	3150.00	4150.00
Register elbows, to 1500 S.F.	12.000	Ea.	3.200	86.00	104.00	190.00
To 2400 S.F.	14.000	Ea.	3.733	100.00	122.00	222.00
Floor registers, enameled steel w/damper, to 1500 S.F.	12.000	Ea.	3.000	209.00	109.00	318.00
To 2400 S.F.	14.000	Ea.	4.308	279.00	156.00	435.00
Return air grille, area to 1500 S.F. 12″ x 12″	2.000	Ea.	.727	47.00	26.50	73.50
Area to 2400 S.F. 8″ x 16″	2.000	Ea.	.444	42.00	16.15	58.15
Area to 2400 S.F. 8″ x 16″	2.000	Ea.	.727	47.00	26.50	73.50
16″ x 16″	1.000	Ea.	.364	33.50	13.20	46.70
Thermostat, manual, 1 set back	1.000	Ea.	1.000	27.00	36.50	63.50
Electric, timed, 1 set back	1.000	Ea.	1.000	87.00	36.50	123.50
2 set back	1.000	Ea.	1.000	116.00	36.50	152.50
Plenum, heating only, 100 M.B.H.	1.000	Ea.	1.000	67.50	32.50	100.00
120 MBH	1.000	Ea.	1.000	67.50	32.50	100.00
160 MBH	1.000	Ea.	1.000	67.50	32.50	100.00
200 MBH	1.000	Ea.	1.000	67.50	32.50	100.00
Refrigeration piping, 3/8″	25.000	L.F.		10.50		10.50
3/4″	25.000	L.F.		21.00		21.00
7/8″	25.000	L.F.		24.50		24.50
Diffusers, ceiling, 6″ diameter, to 1500 S.F.	10.000	Ea.	4.444	170.00	162.00	332.00
To 2400 S.F.	12.000	Ea.	6.000	216.00	218.00	434.00
Floor, aluminum, adjustable, 2-1/4″ x 12″ to 1500 S.F.	12.000	Ea.	3.000	148.00	109.00	257.00
To 2400 S.F.	14.000	Ea.	3.500	172.00	127.00	299.00
Side wall, aluminum, adjustable, 8″ x 4″, to 1500 S.F.	12.000	Ea.	3.000	330.00	109.00	439.00
5″ x 10″ to 2400 S.F.	12.000	Ea.	3.692	445.00	134.00	579.00

64 Oil Fired Heating/Cooling Systems

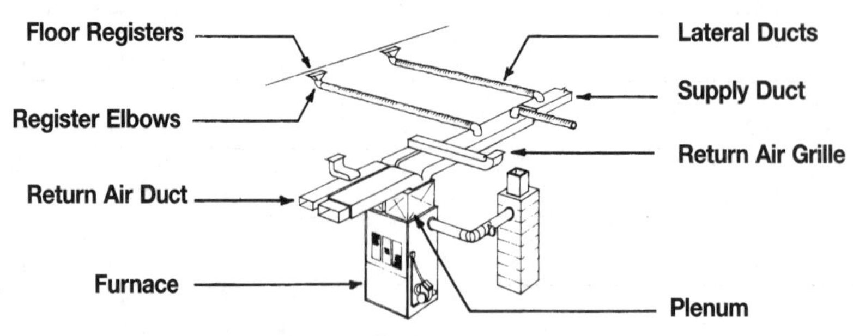

System Description	QUAN.	UNIT	LABOR HOURS	COST PER SYSTEM MAT.	COST PER SYSTEM INST.	COST PER SYSTEM TOTAL
HEATING ONLY, OIL FIRED HOT AIR, ONE ZONE, 1200 S.F. BUILDING						
Furnace, oil fired, atomizing gun type burner	1.000	Ea.	4.571	850.00	149.00	999.00
Oil piping to furnace	1.000	Ea.	3.181	50.95	115.20	166.15
Oil tank, 275 gallon, on legs	1.000	Ea.	3.200	254.00	105.00	359.00
Supply duct, rigid fiberglass	176.000	S.F.	12.069	156.64	408.32	564.96
Return duct, sheet metal, galvanized	158.000	Lb.	16.136	181.70	546.68	728.38
Lateral ducts, 6″ flexible fiberglass	144.000	L.F.	8.862	341.28	289.44	630.72
Register elbows	12.000	Ea.	3.200	85.80	104.40	190.20
Floor register, enameled steel	12.000	Ea.	3.000	209.40	108.60	318.00
Floor grille, return air	2.000	Ea.	.727	47.00	26.40	73.40
Thermostat	1.000	Ea.	1.000	27.00	36.50	63.50
TOTAL			55.946	2203.77	1889.54	4093.31
HEATING/COOLING, OIL FIRED, FORCED AIR, ONE ZONE, 1200 S.F. BUILDING						
Furnace, including plenum, compressor, coil	1.000	Ea.	16.000	3725.00	525.00	4250.00
Oil piping to furnace	1.000	Ea.	3.412	111.85	122.80	234.65
Oil tank, 275 gallon on legs	1.000	Ea.	3.200	254.00	105.00	359.00
Supply duct, rigid fiberglass	176.000	S.F.	12.069	156.64	408.32	564.96
Return duct, sheet metal, galvanized	158.000	Lb.	16.136	181.70	546.68	728.38
Lateral ducts, 6″ flexible fiberglass	144.000	L.F.	8.862	341.28	289.44	630.72
Register elbows	12.000	Ea.	3.200	85.80	104.40	190.20
Floor registers, enameled steel	12.000	Ea.	3.000	209.40	108.60	318.00
Floor grille, return air	2.000	Ea.	.727	47.00	26.40	73.40
Refrigeration piping (precharged)	25.000	L.F.		165.00		165.00
TOTAL			66.606	5277.67	2236.64	7514.31

Description	QUAN.	UNIT	LABOR HOURS	COST PER SYSTEM MAT.	COST PER SYSTEM INST.	COST PER SYSTEM TOTAL

Important: See the Reference Section for critical supporting data - Reference Nos., Crews & Location Factors

Oil Fired Heating/Cooling	QUAN.	UNIT	LABOR HOURS	COST EACH MAT.	INST.	TOTAL
Furnace, heating, 95.2 MBH, area to 1200 S.F.	1.000	Ea.	4.571	850.00	149.00	999.00
123.2 MBH, area to 1500 S.F.	1.000	Ea.	5.000	1200.00	163.00	1363.00
151.2 MBH, area to 2000 S.F.	1.000	Ea.	5.333	1325.00	174.00	1499.00
200 MBH, area to 2400 S.F.	1.000	Ea.	6.154	2150.00	201.00	2351.00
Heating/cooling, 95.2 MBH heat, 36 MBH cool, to 1200 S.F.	1.000	Ea.	16.000	3725.00	525.00	4250.00
112 MBH heat, 42 MBH cool, to 1500 S.F.	1.000	Ea.	24.000	5600.00	790.00	6390.00
151 MBH heat, 47 MBH cool, to 2000 S.F.	1.000	Ea.	20.800	4850.00	685.00	5535.00
184.8 MBH heat, 60 MBH cool, to 2400 S.F.	1.000	Ea.	24.000	5125.00	815.00	5940.00
Oil piping to furnace, 3/8" dia., copper	1.000	Ea.	3.412	112.00	123.00	235.00
Oil tank, on legs above ground, 275 gallons	1.000	Ea.	3.200	254.00	105.00	359.00
550 gallons	1.000	Ea.	5.926	900.00	195.00	1095.00
Below ground, 275 gallons	1.000	Ea.	3.200	254.00	105.00	359.00
550 gallons	1.000	Ea.	5.926	900.00	195.00	1095.00
1000 gallons	1.000	Ea.	6.400	1125.00	211.00	1336.00
Supply duct, rectangular, area to 1200 S.F., rigid fiberglass	176.000	S.F.	12.069	157.00	410.00	567.00
Sheet metal, insulated	228.000	Lb.	31.331	340.00	1050.00	1390.00
Area to 1500 S.F., rigid fiberglass	176.000	S.F.	12.069	157.00	410.00	567.00
Sheet metal, insulated	228.000	Lb.	31.331	340.00	1050.00	1390.00
Area to 2400 S.F., rigid fiberglass	205.000	S.F.	14.057	182.00	475.00	657.00
Sheet metal, insulated	271.000	Lb.	37.048	400.00	1250.00	1650.00
Round flexible, insulated, 6" diameter to 1200 S.F.	156.000	L.F.	9.600	370.00	315.00	685.00
To 1500 S.F.	184.000	L.F.	11.323	435.00	370.00	805.00
8" diameter to 2000 S.F.	269.000	L.F.	23.911	775.00	780.00	1555.00
To 2400 S.F.	269.000	L.F.	22.044	715.00	720.00	1435.00
Return duct, sheet metal galvanized, to 1500 S.F.	158.000	Lb.	16.136	182.00	545.00	727.00
To 2400 S.F.	191.000	Lb.	19.506	220.00	660.00	880.00
Lateral ducts, flexible round, 6", insulated to 1200 S.F.	144.000	L.F.	8.862	340.00	289.00	629.00
To 1500 S.F.	172.000	L.F.	10.585	410.00	345.00	755.00
To 2000 S.F.	261.000	L.F.	16.062	620.00	525.00	1145.00
To 2400 S.F.	300.000	L.F.	18.462	710.00	605.00	1315.00
Spiral steel, insulated to 1200 S.F.	144.000	L.F.	20.069	305.00	655.00	960.00
To 1500 S.F.	172.000	L.F.	23.955	365.00	780.00	1145.00
To 2000 S.F.	261.000	L.F.	36.354	555.00	1175.00	1730.00
To 2400 S.F.	300.000	L.F.	41.829	635.00	1350.00	1985.00
Rectangular sheet metal galvanized insulated, to 1200 S.F.	288.000	Lb.	45.184	480.00	1500.00	1980.00
To 1500 S.F.	344.000	Lb.	53.966	575.00	1800.00	2375.00
To 2000 S.F.	522.000	Lb.	81.928	870.00	2725.00	3595.00
To 2400 S.F.	600.000	Lb.	94.191	1000.00	3150.00	4150.00
Register elbows, to 1500 S.F.	12.000	Ea.	3.200	86.00	104.00	190.00
To 2400 S.F.	14.000	Ea.	3.733	100.00	122.00	222.00
Floor registers, enameled steel w/damper, to 1500 S.F.	12.000	Ea.	3.000	209.00	109.00	318.00
To 2400 S.F.	14.000	Ea.	4.308	279.00	156.00	435.00
Return air grille, area to 1500 S.F., 12" x 12"	2.000	Ea.	.727	47.00	26.50	73.50
12" x 24"	1.000	Ea.	.444	42.00	16.15	58.15
Area to 2400 S.F., 8" x 16"	2.000	Ea.	.727	47.00	26.50	73.50
16" x 16"	1.000	Ea.	.364	33.50	13.20	46.70
Thermostat, manual, 1 set back	1.000	Ea.	1.000	27.00	36.50	63.50
Electric, timed, 1 set back	1.000	Ea.	1.000	87.00	36.50	123.50
2 set back	1.000	Ea.	1.000	116.00	36.50	152.50
Refrigeration piping, 3/8"	25.000	L.F.		10.50		10.50
3/4"	25.000	L.F.		21.00		21.00
Diffusers, ceiling, 6" diameter, to 1500 S.F.	10.000	Ea.	4.444	170.00	162.00	332.00
To 2400 S.F.	12.000	Ea.	6.000	216.00	218.00	434.00
Floor, aluminum, adjustable, 2-1/4" x 12" to 1500 S.F.	12.000	Ea.	3.000	148.00	109.00	257.00
To 2400 S.F.	14.000	Ea.	3.500	172.00	127.00	299.00
Side wall, aluminum, adjustable, 8" x 4", to 1500 S.F.	12.000	Ea.	3.000	330.00	109.00	439.00
5" x 10" to 2400 S.F.	12.000	Ea.	3.692	445.00	134.00	579.00

68 Hot Water Heating Systems

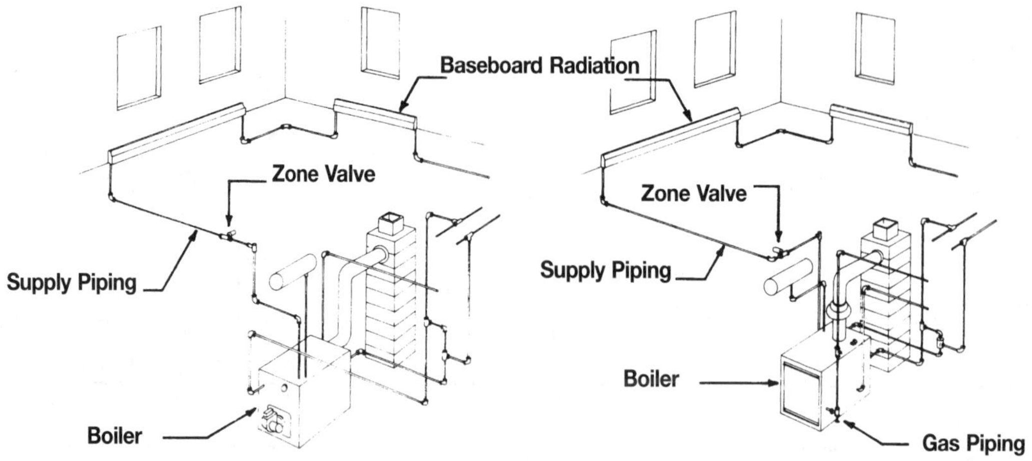

System Description	QUAN.	UNIT	LABOR HOURS	COST EACH		
				MAT.	INST.	TOTAL
OIL FIRED HOT WATER HEATING SYSTEM, AREA TO 1200 S.F.						
Boiler package, oil fired, 97 MBH, area to 1200 S.F. building	1.000	Ea.	15.000	1425.00	475.00	1900.00
Oil piping, 1/4" flexible copper tubing	1.000	Ea.	3.242	51.60	117.05	168.65
Oil tank, 275 gallon, with black iron filler pipe	1.000	Ea.	3.200	254.00	105.00	359.00
Supply piping, 3/4" copper tubing	176.000	L.F.	18.526	290.40	672.32	962.72
Supply fittings, copper 3/4"	36.000	Ea.	15.158	15.48	550.80	566.28
Supply valves, 3/4"	2.000	Ea.	.800	89.00	29.00	118.00
Baseboard radiation, 3/4"	106.000	L.F.	35.333	457.92	1160.70	1618.62
Zone valve	1.000	Ea.	.400	66.00	14.60	80.60
TOTAL			91.659	2649.40	3124.47	5773.87
OIL FIRED HOT WATER HEATING SYSTEM, AREA TO 2400 S.F.						
Boiler package, oil fired, 225 MBH, area to 2400 S.F. building	1.000	Ea.	19.704	4525.00	625.00	5150.00
Oil piping, 3/8" flexible copper tubing	1.000	Ea.	3.289	82.60	118.55	201.15
Oil tank, 550 gallon, with black iron pipe filler pipe	1.000	Ea.	5.926	900.00	195.00	1095.00
Supply piping, 3/4" copper tubing	228.000	L.F.	24.000	376.20	870.96	1247.16
Supply fittings, copper	46.000	Ea.	19.368	19.78	703.80	723.58
Supply valves	2.000	Ea.	.800	89.00	29.00	118.00
Baseboard radiation	212.000	L.F.	70.667	915.84	2321.40	3237.24
Zone valve	1.000	Ea.	.400	66.00	14.60	80.60
TOTAL			144.154	6974.42	4878.31	11852.73

The costs in this system are on a cost each basis. The costs represent total cost for the system based on a gross square foot of plan area.

Description	QUAN.	UNIT	LABOR HOURS	COST EACH		
				MAT.	INST.	TOTAL

Important: See the Reference Section for critical supporting data - Reference Nos., Crews & Location Factors

Hot Water Heating Price Sheet	QUAN.	UNIT	LABOR HOURS	COST EACH		
				MAT.	INST.	TOTAL
Boiler, oil fired, 97 MBH, area to 1200 S.F.	1.000	Ea.	15.000	1425.00	475.00	1900.00
118 MBH, area to 1500 S.F.	1.000	Ea.	16.506	2300.00	525.00	2825.00
161 MBH, area to 2000 S.F.	1.000	Ea.	18.405	4000.00	585.00	4585.00
215 MBH, area to 2400 S.F.	1.000	Ea.	19.704	4525.00	625.00	5150.00
Oil piping, (valve & filter), 3/8″ copper	1.000	Ea.	3.289	82.50	119.00	201.50
1/4″ copper	1.000	Ea.	3.242	51.50	117.00	168.50
Oil tank, filler pipe and cap on legs, 275 gallon	1.000	Ea.	3.200	254.00	105.00	359.00
550 gallon	1.000	Ea.	5.926	900.00	195.00	1095.00
Buried underground, 275 gallon	1.000	Ea.	3.200	254.00	105.00	359.00
550 gallon	1.000	Ea.	5.926	900.00	195.00	1095.00
1000 gallon	1.000	Ea.	6.400	1125.00	211.00	1336.00
Supply piping copper, area to 1200 S.F., 1/2″ tubing	176.000	L.F.	17.383	220.00	630.00	850.00
3/4″ tubing	176.000	L.F.	18.526	290.00	670.00	960.00
Area to 1500 S.F., 1/2″ tubing	186.000	L.F.	18.370	233.00	670.00	903.00
3/4″ tubing	186.000	L.F.	19.579	305.00	710.00	1015.00
Area to 2000 S.F., 1/2″ tubing	204.000	L.F.	20.148	255.00	730.00	985.00
3/4″ tubing	204.000	L.F.	21.474	335.00	780.00	1115.00
Area to 2400 S.F., 1/2″ tubing	228.000	L.F.	22.519	285.00	820.00	1105.00
3/4″ tubing	228.000	L.F.	24.000	375.00	870.00	1245.00
Supply pipe fittings copper, area to 1200 S.F., 1/2″	36.000	Ea.	14.400	7.20	520.00	527.20
3/4″	36.000	Ea.	15.158	15.50	550.00	565.50
Area to 1500 S.F., 1/2″	40.000	Ea.	16.000	8.00	580.00	588.00
3/4″	40.000	Ea.	16.842	17.20	610.00	627.20
Area to 2000 S.F., 1/2″	44.000	Ea.	17.600	8.80	640.00	648.80
3/4″	44.000	Ea.	18.526	18.90	675.00	693.90
Area to 2400, S.F., 1/2″	46.000	Ea.	18.400	9.20	665.00	674.20
3/4″	46.000	Ea.	19.368	19.80	705.00	724.80
Supply valves, 1/2″ pipe size	2.000	Ea.	.667	64.00	24.00	88.00
3/4″	2.000	Ea.	.800	89.00	29.00	118.00
Baseboard radiation, area to 1200 S.F., 1/2″ tubing	106.000	L.F.	28.267	790.00	930.00	1720.00
3/4″ tubing	106.000	L.F.	35.333	460.00	1150.00	1610.00
Area to 1500 S.F., 1/2″ tubing	134.000	L.F.	35.733	1000.00	1175.00	2175.00
3/4″ tubing	134.000	L.F.	44.667	580.00	1475.00	2055.00
Area to 2000 S.F., 1/2″ tubing	178.000	L.F.	47.467	1325.00	1550.00	2875.00
3/4″ tubing	178.000	L.F.	59.333	770.00	1950.00	2720.00
Area to 2400 S.F., 1/2″ tubing	212.000	L.F.	56.533	1575.00	1850.00	3425.00
3/4″ tubing	212.000	L.F.	70.667	915.00	2325.00	3240.00
Zone valves, 1/2″ tubing	1.000	Ea.	.400	66.00	14.60	80.60
3/4″ tubing	1.000	Ea.	.400	66.00	14.60	80.60

80 Rooftop Systems

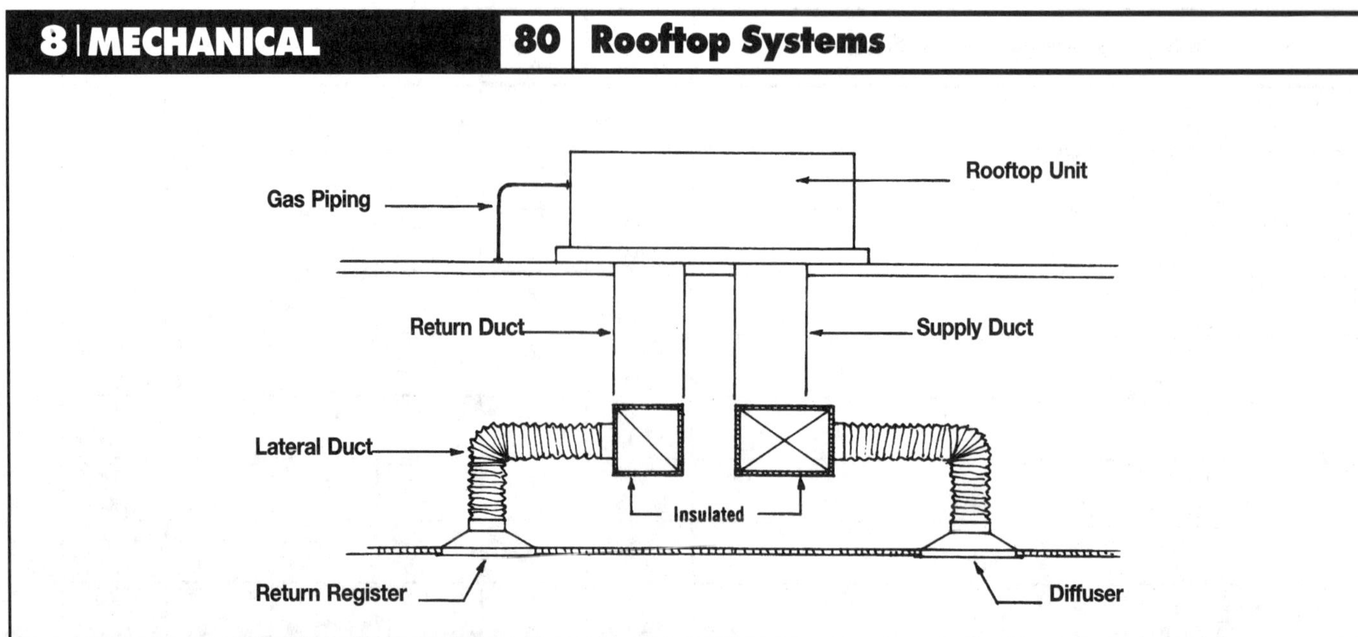

System Description	QUAN.	UNIT	LABOR HOURS	COST EACH		
				MAT.	INST.	TOTAL
ROOFTOP HEATING/COOLING UNIT, AREA TO 2000 S.F.						
Rooftop unit, single zone, electric cool, gas heat, to 2000 s.f.	1.000	Ea.	28.520	5150.00	940.00	6090.00
Gas piping	34.500	L.F.	5.208	69.35	189.75	259.10
Duct, supply and return, galvanized steel	38.000	Lb.	3.881	43.70	131.48	175.18
Insulation, ductwork	33.000	S.F.	1.509	14.19	48.84	63.03
Lateral duct, flexible duct 12" diameter, insulated	72.000	L.F.	11.520	315.36	378.00	693.36
Diffusers	4.000	Ea.	4.571	1040.00	166.00	1206.00
Return registers	1.000	Ea.	.727	113.00	26.50	139.50
TOTAL			55.936	6745.60	1880.57	8626.17
ROOFTOP HEATING/COOLING UNIT, AREA TO 5000 S.F.						
Rooftop unit, single zone, electric cool, gas heat, to 5000 s.f.	1.000	Ea.	42.032	13800.00	1325.00	15125.00
Gas piping	86.250	L.F.	13.019	173.36	474.38	647.74
Duct supply and return, galvanized steel	95.000	Lb.	9.702	109.25	328.70	437.95
Insulation, ductwork	82.000	S.F.	3.749	35.26	121.36	156.62
Lateral duct, flexible duct, 12" diameter, insulated	180.000	L.F.	28.800	788.40	945.00	1733.40
Diffusers	10.000	Ea.	11.429	2600.00	415.00	3015.00
Return registers	3.000	Ea.	2.182	339.00	79.50	418.50
TOTAL			110.913	17845.27	3688.94	21534.21

Description	QUAN.	UNIT	LABOR HOURS	COST EACH		
				MAT.	INST.	TOTAL

Important: See the Reference Section for critical supporting data - Reference Nos., Crews & Location Factors

Rooftop Price Sheet	QUAN.	UNIT	LABOR HOURS	COST EACH		
				MAT.	INST.	TOTAL
Rooftop unit, single zone, electric cool, gas heat to 2000 S.F.	1.000	Ea.	28.520	5150.00	940.00	6090.00
Area to 3000 S.F.	1.000	Ea.	35.982	10400.00	1150.00	11550.00
Area to 5000 S.F.	1.000	Ea.	42.032	13800.00	1325.00	15125.00
Area to 10000 S.F.	1.000	Ea.	68.376	32000.00	2250.00	34250.00
Gas piping, area 2000 through 4000 S.F.	34.500	L.F.	5.208	69.50	190.00	259.50
Area 5000 to 10000 S.F.	86.250	L.F.	13.019	173.00	475.00	648.00
Duct, supply and return, galvanized steel, to 2000 S.F.	38.000	Lb.	3.881	43.50	131.00	174.50
Area to 3000 S.F.	57.000	Lb.	5.821	65.50	197.00	262.50
Area to 5000 S.F.	95.000	Lb.	9.702	109.00	330.00	439.00
Area to 10000 S.F.	190.000	Lb.	19.404	219.00	655.00	874.00
Rigid fiberglass, area to 2000 S.F.	33.000	S.F.	2.263	29.50	76.50	106.00
Area to 3000 S.F.	49.000	S.F.	3.360	43.50	114.00	157.50
Area to 5000 S.F.	82.000	S.F.	5.623	73.00	190.00	263.00
Area to 10000 S.F.	164.000	S.F.	11.246	146.00	380.00	526.00
Insulation, supply and return, blanket type, area to 2000 S.F.	33.000	S.F.	1.509	14.20	49.00	63.20
Area to 3000 S.F.	49.000	S.F.	2.240	21.00	72.50	93.50
Area to 5000 S.F.	82.000	S.F.	3.749	35.50	121.00	156.50
Area to 10000 S.F.	164.000	S.F.	7.497	70.50	243.00	313.50
Lateral ducts, flexible round, 12" insulated, to 2000 S.F.	72.000	L.F.	11.520	315.00	380.00	695.00
Area to 3000 S.F.	108.000	L.F.	17.280	475.00	565.00	1040.00
Area to 5000 S.F.	180.000	L.F.	28.800	790.00	945.00	1735.00
Area to 10000 S.F.	360.000	L.F.	57.600	1575.00	1900.00	3475.00
Rectangular, galvanized steel, to 2000 S.F.	239.000	Lb.	24.409	275.00	825.00	1100.00
Area to 3000 S.F.	360.000	Lb.	36.766	415.00	1250.00	1665.00
Area to 5000 S.F.	599.000	Lb.	61.174	690.00	2075.00	2765.00
Area to 10000 S.F.	998.000	Lb.	101.923	1150.00	3450.00	4600.00
Diffusers, ceiling, 1 to 4 way blow, 24" x 24", to 2000 S.F.	4.000	Ea.	4.571	1050.00	166.00	1216.00
Area to 3000 S.F.	6.000	Ea.	6.857	1550.00	249.00	1799.00
Area to 5000 S.F.	10.000	Ea.	11.429	2600.00	415.00	3015.00
Area to 10000 S.F.	20.000	Ea.	22.857	5200.00	830.00	6030.00
Return grilles, 24" x 24", to 2000 S.F.	1.000	Ea.	.727	113.00	26.50	139.50
Area to 3000 S.F.	2.000	Ea.	1.455	226.00	53.00	279.00
Area to 5000 S.F.	3.000	Ea.	2.182	340.00	79.50	419.50
Area to 10000 S.F.	5.000	Ea.	3.636	565.00	133.00	698.00

Division 9
Electrical

10 Electric Service Systems

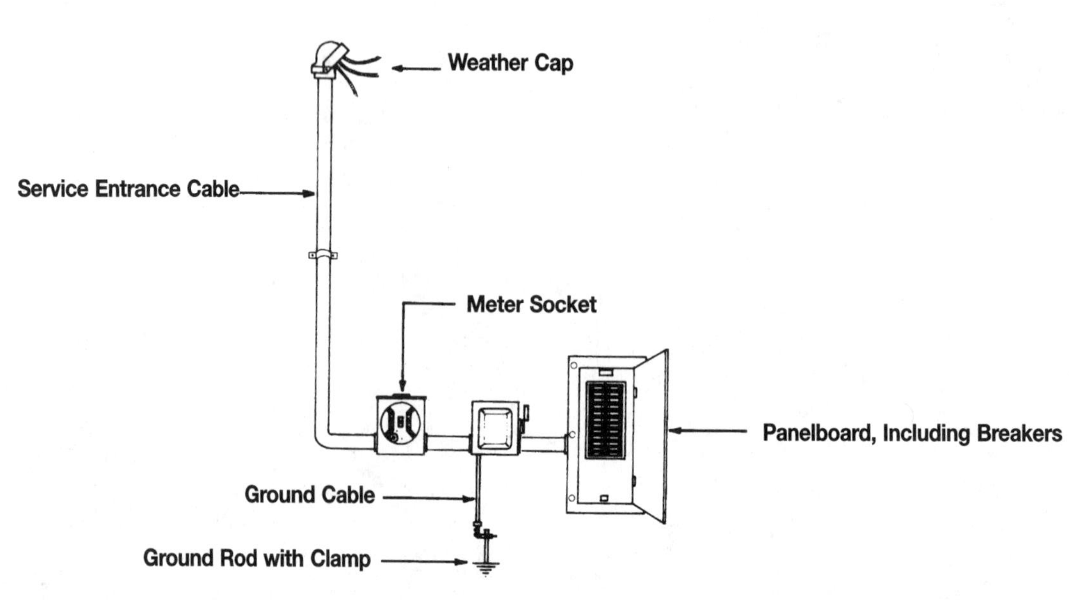

System Description	QUAN.	UNIT	LABOR HOURS	COST EACH		
				MAT.	INST.	TOTAL
100 AMP SERVICE						
Weather cap	1.000	Ea.	.667	7.60	24.00	31.60
Service entrance cable	10.000	L.F.	.762	22.30	27.50	49.80
Meter socket	1.000	Ea.	2.500	30.50	90.50	121.00
Ground rod with clamp	1.000	Ea.	1.455	17.45	52.50	69.95
Ground cable	5.000	L.F.	.250	6.50	9.05	15.55
Panel board, 12 circuit	1.000	Ea.	6.667	173.00	241.00	414.00
TOTAL			12.301	257.35	444.55	701.90
200 AMP SERVICE						
Weather cap	1.000	Ea.	1.000	20.00	36.00	56.00
Service entrance cable	10.000	L.F.	1.143	50.00	41.30	91.30
Meter socket	1.000	Ea.	4.211	44.00	152.00	196.00
Ground rod with clamp	1.000	Ea.	1.818	32.50	65.50	98.00
Ground cable	10.000	L.F.	.500	13.00	18.10	31.10
3/4" EMT	5.000	L.F.	.308	3.00	11.10	14.10
Panel board, 24 circuit	1.000	Ea.	12.308	475.00	380.00	855.00
TOTAL			21.288	637.50	704.00	1341.50
400 AMP SERVICE						
Weather cap	2.000	Ea.	12.509	341.40	453.00	794.40
Service entrance cable	180.000	L.F.	5.760	286.20	208.80	495.00
Meter socket	1.000	Ea.	4.211	44.00	152.00	196.00
Ground rod with clamp	1.000	Ea.	2.068	38.35	74.55	112.90
Ground cable	20.000	L.F.	.485	20.40	17.60	38.00
3/4" greenfield	20.000	L.F.	1.000	9.00	36.20	45.20
Current transformer cabinet	1.000	Ea.	6.154	132.00	222.00	354.00
Panel board, 42 circuit	1.000	Ea.	33.333	2500.00	1200.00	3700.00
TOTAL			65.520	3371.35	2364.15	5735.50

Important: See the Reference Section for critical supporting data - Reference Nos., Crews & Location Factors

Electric Perimeter Heating Systems

Thermostat
Electric Baseboard

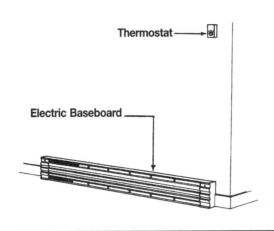

System Description	QUAN.	UNIT	LABOR HOURS	COST EACH MAT.	COST EACH INST.	COST EACH TOTAL
4′ BASEBOARD HEATER						
Electric baseboard heater, 4′ long	1.000	Ea.	1.194	53.00	43.00	96.00
Thermostat, integral	1.000	Ea.	.500	27.50	18.10	45.60
Romex, 12-3 with ground	40.000	L.F.	1.600	14.00	58.00	72.00
Panel board breaker, 20 Amp	1.000	Ea.	.300	7.20	10.80	18.00
TOTAL			3.594	101.70	129.90	231.60
6′ BASEBOARD HEATER						
Electric baseboard heater, 6′ long	1.000	Ea.	1.600	71.50	58.00	129.50
Thermostat, integral	1.000	Ea.	.500	27.50	18.10	45.60
Romex, 12-3 with ground	40.000	L.F.	1.600	14.00	58.00	72.00
Panel board breaker, 20 Amp	1.000	Ea.	.400	9.60	14.40	24.00
TOTAL			4.100	122.60	148.50	271.10
8′ BASEBOARD HEATER						
Electric baseboard heater, 8′ long	1.000	Ea.	2.000	89.00	72.50	161.50
Thermostat, integral	1.000	Ea.	.500	27.50	18.10	45.60
Romex, 12-3 with ground	40.000	L.F.	1.600	14.00	58.00	72.00
Panel board breaker, 20 Amp	1.000	Ea.	.500	12.00	18.00	30.00
TOTAL			4.600	142.50	166.60	309.10
10′ BASEBOARD HEATER						
Electric baseboard heater, 10′ long	1.000	Ea.	2.424	147.00	87.50	234.50
Thermostat, integral	1.000	Ea.	.500	27.50	18.10	45.60
Romex, 12-3 with ground	40.000	L.F.	1.600	14.00	58.00	72.00
Panel board breaker, 20 Amp	1.000	Ea.	.750	18.00	27.00	45.00
TOTAL			5.274	206.50	190.60	397.10

The costs in this system are on a cost each basis and include all necessary conduit fittings.

Description	QUAN.	UNIT	LABOR HOURS	COST EACH MAT.	COST EACH INST.	COST EACH TOTAL

System Description		QUAN.	UNIT	LABOR HOURS	COST EACH MAT.	COST EACH INST.	COST EACH TOTAL
Air conditioning receptacles							
	Using non-metallic sheathed cable	1.000	Ea.	.800	10.45	29.00	39.45
	Using BX cable	1.000	Ea.	.964	21.00	35.00	56.00
	Using EMT conduit	1.000	Ea.	1.194	23.00	43.00	66.00
Disposal wiring							
	Using non-metallic sheathed cable	1.000	Ea.	.889	10.50	32.00	42.50
	Using BX cable	1.000	Ea.	1.067	20.50	38.50	59.00
	Using EMT conduit	1.000	Ea.	1.333	23.50	48.00	71.50
Dryer circuit							
	Using non-metallic sheathed cable	1.000	Ea.	1.455	33.00	52.50	85.50
	Using BX cable	1.000	Ea.	1.739	43.50	63.00	106.50
	Using EMT conduit	1.000	Ea.	2.162	39.50	78.00	117.50
Duplex receptacles							
	Using non-metallic sheathed cable	1.000	Ea.	.615	10.45	22.50	32.95
	Using BX cable	1.000	Ea.	.741	21.00	27.00	48.00
	Using EMT conduit	1.000	Ea.	.920	23.00	33.00	56.00
Exhaust fan wiring							
	Using non-metallic sheathed cable	1.000	Ea.	.800	12.80	29.00	41.80
	Using BX cable	1.000	Ea.	.964	23.00	35.00	58.00
	Using EMT conduit	1.000	Ea.	1.194	25.00	43.00	68.00
Furnace circuit & switch							
	Using non-metallic sheathed cable	1.000	Ea.	1.333	20.00	48.00	68.00
	Using BX cable	1.000	Ea.	1.600	31.00	58.00	89.00
	Using EMT conduit	1.000	Ea.	2.000	32.00	72.50	104.50
Ground fault							
	Using non-metallic sheathed cable	1.000	Ea.	1.000	42.00	36.00	78.00
	Using BX cable	1.000	Ea.	1.212	52.00	44.00	96.00
	Using EMT conduit	1.000	Ea.	1.481	65.00	53.50	118.50
Heater circuits							
	Using non-metallic sheathed cable	1.000	Ea.	1.000	12.20	36.00	48.20
	Using BX cable	1.000	Ea.	1.212	19.75	44.00	63.75
	Using EMT conduit	1.000	Ea.	1.481	22.00	53.50	75.50
Lighting wiring							
	Using non-metallic sheathed cable	1.000	Ea.	.500	12.90	18.10	31.00
	Using BX cable	1.000	Ea.	.602	20.50	21.50	42.00
	Using EMT conduit	1.000	Ea.	.748	21.00	27.00	48.00
Range circuits							
	Using non-metallic sheathed cable	1.000	Ea.	2.000	65.00	72.50	137.50
	Using BX cable	1.000	Ea.	2.424	88.00	87.50	175.50
	Using EMT conduit	1.000	Ea.	2.963	64.50	107.00	171.50
Switches, single pole							
	Using non-metallic sheathed cable	1.000	Ea.	.500	12.80	18.10	30.90
	Using BX cable	1.000	Ea.	.602	23.00	21.50	44.50
	Using EMT conduit	1.000	Ea.	.748	25.00	27.00	52.00
Switches, 3-way							
	Using non-metallic sheathed cable	1.000	Ea.	.667	16.90	24.00	40.90
	Using BX cable	1.000	Ea.	.800	26.00	29.00	55.00
	Using EMT conduit	1.000	Ea.	1.333	34.00	48.00	82.00
Water heater							
	Using non-metallic sheathed cable	1.000	Ea.	1.600	18.10	58.00	76.10
	Using BX cable	1.000	Ea.	1.905	33.50	69.00	102.50
	Using EMT conduit	1.000	Ea.	2.353	27.50	85.00	112.50
Weatherproof receptacle							
	Using non-metallic sheathed cable	1.000	Ea.	1.333	96.50	48.00	144.50
	Using BX cable	1.000	Ea.	1.600	104.00	58.00	162.00
	Using EMT conduit	1.000	Ea.	2.000	106.00	72.50	178.50

DESCRIPTION	QUAN.	UNIT	LABOR HOURS	COST EACH		
				MAT.	INST.	TOTAL
luorescent strip, 4′ long, 1 light, average	1.000	Ea.	.941	28.50	34.00	62.50
Deluxe	1.000	Ea.	1.129	34.00	41.00	75.00
2 lights, average	1.000	Ea.	1.000	31.00	36.00	67.00
Deluxe	1.000	Ea.	1.200	37.00	43.00	80.00
8′ long, 1 light, average	1.000	Ea.	1.194	43.00	43.00	86.00
Deluxe	1.000	Ea.	1.433	51.50	51.50	103.00
2 lights, average	1.000	Ea.	1.290	51.50	46.50	98.00
Deluxe	1.000	Ea.	1.548	62.00	56.00	118.00
Surface mounted, 4′ x 1′, economy	1.000	Ea.	.914	61.50	33.00	94.50
Average	1.000	Ea.	1.143	77.00	41.50	118.50
Deluxe	1.000	Ea.	1.371	92.50	50.00	142.50
4′ x 2′, economy	1.000	Ea.	1.208	79.00	43.50	122.50
Average	1.000	Ea.	1.509	99.00	54.50	153.50
Deluxe	1.000	Ea.	1.811	119.00	65.50	184.50
Recessed, 4′x 1′, 2 lamps, economy	1.000	Ea.	1.123	40.50	40.50	81.00
Average	1.000	Ea.	1.404	50.50	50.50	101.00
Deluxe	1.000	Ea.	1.684	60.50	60.50	121.00
4′ x 2′, 4′ lamps, economy	1.000	Ea.	1.362	49.00	49.00	98.00
Average	1.000	Ea.	1.702	61.50	61.50	123.00
Deluxe	1.000	Ea.	2.043	74.00	74.00	148.00
ncandescent, exterior, 150W, single spot	1.000	Ea.	.500	15.40	18.10	33.50
Double spot	1.000	Ea.	1.167	64.00	42.00	106.00
Recessed, 100W, economy	1.000	Ea.	.800	47.00	29.00	76.00
Average	1.000	Ea.	1.000	58.50	36.00	94.50
Deluxe	1.000	Ea.	1.200	70.00	43.00	113.00
150W, economy	1.000	Ea.	.800	67.50	29.00	96.50
Average	1.000	Ea.	1.000	84.50	36.00	120.50
Deluxe	1.000	Ea.	1.200	101.00	43.00	144.00
Surface mounted, 60W, economy	1.000	Ea.	.800	35.00	29.00	64.00
Average	1.000	Ea.	1.000	39.50	36.00	75.50
Deluxe	1.000	Ea.	1.194	55.00	43.00	98.00
Mercury vapor, recessed, 2′ x 2′ with 250W DX lamp	1.000	Ea.	2.500	275.00	90.50	365.50
2′ x 2′ with 400W DX lamp	1.000	Ea.	2.759	281.00	99.50	380.50
Surface mounted, 2′ x 2′ with 250W DX lamp	1.000	Ea.	2.963	275.00	107.00	382.00
2′ x 2′ with 400W DX lamp	1.000	Ea.	3.333	310.00	121.00	431.00
High bay, single unit, 400W DX lamp	1.000	Ea.	3.478	273.00	126.00	399.00
Twin unit, 400W DX lamp	1.000	Ea.	5.000	460.00	181.00	641.00
Low bay, 250W DX lamp	1.000	Ea.	2.500	281.00	90.50	371.50
Metal halide, recessed 2′ x 2′ 250W	1.000	Ea.	2.500	270.00	90.50	360.50
2′ x 2′, 400W	1.000	Ea.	2.759	315.00	99.50	414.50
Surface mounted, 2′ x 2′, 250W	1.000	Ea.	2.963	270.00	107.00	377.00
2′ x 2′, 400W	1.000	Ea.	3.333	320.00	121.00	441.00
High bay, single, unit, 400W	1.000	Ea.	3.478	375.00	126.00	501.00
Twin unit, 400W	1.000	Ea.	5.000	745.00	181.00	926.00
Low bay, 250W	1.000	Ea.	2.500	360.00	90.50	450.50

Unit Price Section

Table of Contents

Div No.		Page
	General Requirements	**287**
1100	Summary	288
1200	Price & Payment Procedures	288
1300	Administrative Requirements	288
1500	Temporary Facilities & Controls	289
1590	Material & Equipment	292
1700	Execution Requirements	296
	Site Construction	**297**
2050	Basic Site Materials & Methods	298
2200	Site Preparation	298
2300	Earthwork	305
2400	Tunneling, Boring & Jacking	307
2500	Utility Services	308
2600	Drainage & Containment	309
2700	Bases, Ballasts, Pavements, & Appurtenances	310
2800	Site Improvements & Amenities	312
2900	Planting	315
	Concrete	**321**
3050	Basic Conc. Mat. & Methods	322
3100	Concrete Forms & Accessories	322
3200	Concrete Reinforcement	324
3300	Cast-In-Place Concrete	325
3400	Precast Concrete	327
3900	Concrete Restor. & Cleaning	327
	Masonry	**329**
4050	Basic Masonry Mat. & Methods	330
4200	Masonry Units	331
4400	Stone	333
4500	Refractories	335
4800	Masonry Assemblies	335
4900	Masonry Restoration & Cleaning	336
	Metals	**337**
5050	Basic Materials & Methods	338
5100	Structural Metal Framing	340
5300	Metal Decking	341
5400	Cold Formed Metal Framing	342
5500	Metal Fabrications	349
	Wood & Plastics	**351**
6050	Basic Wd. & Plastic Mat. & Meth.	352
6100	Rough Carpentry	356

Div. No.		Page
06200	Finish Carpentry	367
06400	Architectural Woodwork	371
06600	Plastic Fabrications	379
	Thermal & Moisture Protection	**381**
07100	Dampproofing & Waterproofing	382
07200	Thermal Protection	382
07300	Shingles, Roof Tiles & Roof Cov.	386
07400	Roofing & Siding Panels	388
07500	Membrane Roofing	391
07600	Flashing & Sheet Metal	393
07700	Roof Specialties & Accessories	394
07900	Joint Sealers	397
	Doors & Windows	**399**
08100	Metal Doors & Frames	400
08200	Wood & Plastic Doors	401
08300	Specialty Doors	407
08500	Windows	408
08600	Skylights	417
08700	Hardware	418
08800	Glazing	420
	Finishes	**423**
09100	Metal Support Assemblies	424
09200	Plaster & Gypsum Board	424
09300	Tile	428
09400	Terrazzo	430
09500	Ceilings	431
09600	Flooring	432
09700	Wall Finishes	435
09900	Paints & Coatings	436
	Specialties	**449**
10150	Compartments & Cubicles	450
10200	Louvers & Vents	450
10300	Fireplaces & Stoves	451
10340	Manufactured Exterior Specialties	452
10350	Flagpoles	453
10520	Fire Protection Specialties	453
10530	Protective Covers	453
10550	Postal Specialties	454
10670	Storage Shelving	455
10800	Toilet, Bath, & Laundry Access.	455

Div. No.		Page
	Equipment	**457**
11010	Maintenance Equipment	458
11400	Food Service Equipment	458
11450	Residential Equipment	458
	Furnishings	**461**
12300	Manufactured Casework	462
12400	Furnishings & Access.	462
	Special Construction	**465**
13100	Lightning Protection	466
13120	Pre-Engineered Structures	466
13150	Swimming Pools	466
13200	Storage Tanks	467
13280	Hazardous Mat. Remediation	468
13800	Building Automation & Control	469
13850	Detection & Alarm	470
	Conveying Systems	**471**
14200	Elevators	472
	Mechanical	**473**
15050	Basic Materials & Methods	474
15100	Building Services Piping	475
15400	Plumbing Fixtures & Equipment	483
15500	Heat Generation Equipment	486
15700	Heating, Ventilation, & A/C Equip.	489
15800	Air Distribution	492
	Electrical	**495**
16050	Basic Elect. Mat. & Methods	496
16100	Wiring Methods	497
16200	Electrical Power	505
16400	Low-Voltage Distribution	506
16500	Lighting	507
16800	Sound & Video	509

How to Use the Unit Price Pages

The following is a detailed explanation of a sample entry in the Unit Price Section. Next to each bold number below is the item being described with appropriate component of the sample entry following in parenthesis. Some prices are listed as bare costs, others as costs that include overhead and profit of the installing contractor. In most cases, if the work is to be subcontracted, the general contractor will need to add an additional markup (R.S. Means suggests using 10%) to the figures in the column "Total Incl. O&P."

1 Division Number/Title (06100/Rough Carpentry)

Use the Unit Price Section Table of Contents to locate specific items. The sections are classified according to the CSI MasterFormat.

2 Line Numbers (06170 980 5050)

Each unit price line item has been assigned a unique 12-digit code based on the 5-digit CSI MasterFormat classification.

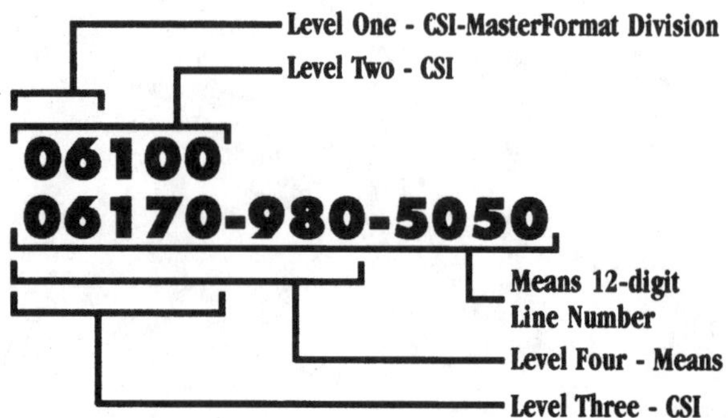

3 Description (Roof Trusses, Etc.)

Each line item is described in detail. Sub-items and additional sizes are indented beneath the appropriate line items. The first line or two after the main item (in boldface) may contain descriptive information that pertains to all line items beneath this boldface listing.

4 Reference Number Information

R06170 -100 You'll see reference numbers shown in bold rectangles at the beginning of some sections. These refer to related items in the Reference Section, visually identified by a vertical gray bar on the edge of pages.

The relation may be: (1) an estimating procedure that should be read before estimating, (2) an alternate pricing method, or (3) technical information.

The "R" designates the Reference Section. The numbers refer to the MasterFormat classification system.

It is strongly recommended that you review all reference numbers that appear within the section in which you are working.

Example: The rectangle number above is directing you to refer to the reference number R06170-100. This particular reference number shows how the unit price lines were formulated and costs derived.

06100 | Rough Carpentry

	06170 | Prefabricated Structural Wood		CREW	DAILY OUTPUT	LABOR-HOURS	UNIT	2000 BARE COSTS MAT.	LABOR	EQUIP.	TOTAL	TOTAL INCL O&P
980	0010	ROOF TRUSSES									
	0020	For timber connectors, see div. 06090-800									
	5000	Com[illegible]n wood, 2" x 4" metal plate connected, 24" O.C. 4[illegible]									
	5010	[illegible]verhang, 12' span	F-5	55	.582	Ea.	27.50	10.05		37.55	47.50
	5050	20' span	F-6	62	.645		36	11.50	7.30	54.80	67
	5100	24' span		60	.667		42.50	11.85	7.55	61.90	75.50
	5150	26' span		57	.702		65	12.50	7.95	85.45	101
	5200	28' span		53	.75[illegible]		50.50	13.45	8.55	72.50	88
	5240	30' span		51	.784		76	13.95	8.90	98.85	117
	5250	32' span		50	.80[illegible]		71	14.25	9.05	94.30	113
	5280	34' span		48	[illegible]		88	14.80	9.45	112.25	132
	5350	8/12 pitch, 1' overhang, 20' span		57	.702		60	12.50	7.95	80.45	96
	5400	24' span		55	.727		71	12.95	8.25	92.20	109
	5450	26' span		52	.769		77	13.70	8.70	99.40	118

R06170 -100

5

Crew (F-6)

The "Crew" column designates the typical trade or crew used to install the item. If an installation can be accomplished by one trade and requires no power equipment, that trade and the number of workers are listed (for example, "2 Carp"). If an installation requires a composite crew, a crew code designation is listed (for example, "F-6"). You'll find full details on all composite crews in the Crew Listings.

- For a complete list of all trades utilized in this book and their abbreviations, see the inside back cover.

Crews

Crew No.	Bare Costs		Incl. Subs O & P		Cost Per Labor-Hour	
Crew F-6	Hr.	Daily	Hr.	Daily	Bare Costs	Incl. O&P
2 Carpenters	$19.70	$315.20	$33.75	$540.00	$17.79	$30.25
2 Building Laborers	14.45	231.20	24.75	396.00		
1 Equip. Oper. (crane)	20.65	165.20	34.25	274.00		
1 Hyd. Crane, 12 Ton		453.45		498.80	11.34	12.47
40 L.H., Daily Totals		$1165.05		$1708.80	$29.13	$42.72

Productivity: Daily Output (62)/ Labor-Hours (.645)

The "Daily Output" represents the typical number of units the designated crew will install in a normal 8-hour day. To find out the number of days the given crew would require to complete the installation, divide your quantity by the daily output. For example:

Quantity	÷	Daily Output	=	Duration
248 Ea.	÷	62 Ea./ Crew Day	=	4 Crew Days

The "Labor-Hours" figure represents the number of labor-hours required to install one unit of work. To find out the number of labor-hours required for your particular task, multiply the quantity of the item times the number of labor-hours shown. For example:

Quantity	x	Productivity Rate	=	Duration
248 Ea.	x	.645 Labor-Hours/ LF	=	159.96 Labor-Hours

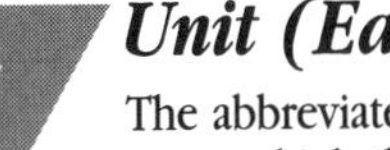

Unit (Ea.)

The abbreviated designation indicates the unit of measure upon which the price, production, and crew are based (Ea. = Each). For a complete listing of abbreviations refer to the Abbreviations Listing in the Reference Section of this book.

8

Bare Costs:

Mat. (Bare Material Cost) (36.00)

The unit material cost is the "bare" material cost with no overhead and profit included. *Costs shown reflect national average material prices for January of the current year and include delivery to the job site. No sales taxes are included.*

Labor (11.50)

The unit labor cost is derived by multiplying bare labor-hour costs for Crew F-6 by labor-hour units. The bare labor-hour cost is found in the Crew Section under F-6. (If a trade is listed, the hourly labor cost—the wage rate—is found on the inside back cover.)

Labor-Hour Cost Crew F-6	x	Labor-Hour Units	=	Labor
$17.79	x	.645	=	$11.50

Equip. (Equipment) (7.30)

Equipment costs for each crew are listed in the description of each crew. Tools or equipment whose value justifies purchase or ownership by a contractor are considered overhead as shown on the inside back cover. The unit equipment cost is derived by multiplying the bare equipment hourly cost by the labor-hour units.

Equipment Cost Crew F-6	x	Labor-Hour Units	=	Equip.
$11.34	x	.645	=	$7.30

Total (54.80)

The total of the bare costs is the arithmetic total of the three previous columns: mat., labor, and equip.

Material	+	Labor	+	Equip.	=	Total
$36.00	+	$11.50	+	$7.30	=	$54.80

Total Costs Including O&P

This figure is the sum of the bare material cost plus 10% for profit; the bare labor cost plus total overhead and profit (per the inside back cover or, if a crew is listed, from the crew listings); and the bare equipment cost plus 10% for profit.

Material is Bare Material Cost + 10% = $36.00 + $3.60	=	$39.60
Labor for Crew F-6 = Labor-Hour Cost ($30.25) x Labor-Hour Units (.645)	=	$19.51
Equip. is Bare Equip. Cost + 10% = $7.30 + $.73	=	$7.73
Total (Rounded)	=	$67.00

Division 1
General Requirements

Estimating Tips

The General Requirements of any contract are very important to both the bidder and the owner. These lay the ground rules under which the contract will be executed and have a significant influence on the cost of operations. Therefore, it is extremely important to thoroughly read and understand the General Requirements both before preparing an estimate and when the estimate is complete, to ascertain that nothing in the contract is overlooked. Caution should be exercised when applying items listed in Division 1 to an estimate. Many of the items are included in the unit prices listed in the other divisions such as mark-ups on labor and company overhead.

01300 Administrative Requirements

- Before determining a final cost estimate, it is a good practice to review all the items listed in subdivision 01300 to make final adjustments for items that may need customizing to specific job conditions.

01330 Submittal Procedures

- Requirements for initial and periodic submittals can represent a significant cost to the General Requirements of a job. Thoroughly check the submittal specifications when estimating a project to determine any costs that should be included.

01400 Quality Requirements

- All projects will require some degree of Quality Control. This cost is not included in the unit cost of construction listed in each division. Depending upon the terms of the contract, the various costs of inspection and testing can be the responsibility of either the owner or the contractor. Be sure to include the required costs in your estimate.

01500 Temporary Facilities & Controls

- Barricades, access roads, safety nets, scaffolding, security and many more requirements for the execution of a safe project are elements of direct cost. These costs can easily be overlooked when preparing an estimate. When looking through the major classifications of this subdivision, determine which items apply to each division in your estimate.

01559 Equipment Rental

- This subdivision contains transportation, handling, storage, protection and product options and substitutions. Listed in this cost manual are average equipment rental rates for all types of equipment. This is useful information when estimating the time and materials requirement of any particular operation in order to establish a unit or total cost.
- A good rule of thumb is that weekly rental is 3 times daily rental and that monthly rental is 3 times weekly rental.
- The figures in the column for Crew Equipment Cost represent the rental rate used in determining the daily cost of equipment in a crew. It is calculated by dividing the weekly rate by 5 days and adding the hourly operating cost times 8 hours.

01770 Closeout Procedures

- When preparing an estimate, read the specifications to determine the requirements for Contract Closeout thoroughly. Final cleaning, record documentation, operation and maintenance data, warranties and bonds, and spare parts and maintenance materials can all be elements of cost for the completion of a contract. Do not overlook these in your estimate.

01830 Operations & Maintenance

- If maintenance and repair are included in your contract, they require special attention. To estimate the cost to remove and replace any unit usually requires a site visit to determine the accessibility and the specific difficulty at that location. Obstructions, dust control, safety, and often overtime hours must be considered when preparing your estimate.

Reference Numbers

Reference numbers are shown in bold squares at the beginning of some major classifications. These numbers refer to related items in the Reference Section. The reference information may be an estimating procedure, an alternate pricing method or technical information.

Note: Not all subdivisions listed here necessarily appear in this publication.

01100 | Summary

		01103 | Models & Renderings	CREW	DAILY OUTPUT	LABOR-HOURS	UNIT	2000 BARE COSTS MAT.	LABOR	EQUIP.	TOTAL	TOTAL INCL O&P
500	0010	**RENDERINGS** Color, matted, 20″ x 30″, eye level,									
	0050	Average				Ea.	2,550			2,550	2,800
		01107 | Professional Consultant									
100	0011	**ARCHITECTURAL FEES** R01107-010									
	0020	For new construction									
	0060	Minimum				Project					4.90%
	0090	Maximum									16%
	0100	For alteration work, to $500,000, add to fee									50%
	0150	Over $500,000, add to fee				↓					25%
200	0011	**CONSTRUCTION MANAGEMENT FEES**									
	0060	For work to $10,000				Project					10%
	0070	To $25,000									9%
	0090	To $100,000				↓					6%
700	0010	**SURVEYING** Conventional, topographical, minimum	A-7	3.30	7.273	Acre	16	148		164	271
	0100	Maximum	A-8	.60	53.333		48	1,075		1,123	1,875
	0300	Lot location and lines, minimum, for large quantities	A-7	2	12		25	245		270	445
	0320	Average	″	1.25	19.200		45	390		435	715
	0400	Maximum, for small quantities	A-8	1	32	↓	72	645		717	1,175
	0600	Monuments, 3′ long	A-7	10	2.400	Ea.	19	49		68	105
	0800	Property lines, perimeter, cleared land	″	1,000	.024	L.F.	.03	.49		.52	.86
	0900	Wooded land	A-8	875	.037	″	.05	.74		.79	1.31
	1100	Crew for building layout, 2 person crew	A-6	1	16	Day		335		335	575
	1200	3 person crew	A-7	1	24	″		490		490	835

01200 | Price & Payment Procedures

		01250 | Contract Modification Procedures	CREW	DAILY OUTPUT	LABOR-HOURS	UNIT	2000 BARE COSTS MAT.	LABOR	EQUIP.	TOTAL	TOTAL INCL O&P
200	0010	**CONTINGENCIES** for estimate at conceptual stage				Project					20%
	0150	Final working drawing stage				″					3%
		01290 | Payment Procedures									
800	0010	**TAXES** Sales tax, State, average R01100-090				%	4.71%				
	0050	Maximum					7.25%				
	0200	Social Security, on first $72,600 of wages R01100-100						7.65%			
	0300	Unemployment, MA, combined Federal and State, minimum						2.10%			
	0350	Average						7%			
	0400	Maximum				↓		8%			

01300 | Administrative Requirements

		01310 | Project Management/Coordination	CREW	DAILY OUTPUT	LABOR-HOURS	UNIT	2000 BARE COSTS MAT.	LABOR	EQUIP.	TOTAL	TOTAL INCL O&P
150	0010	**PERMITS** Rule of thumb, most cities, minimum				Job					.50%
	0100	Maximum				″					2%

Important: See the Reference Section for critical supporting data - Reference Nos., Crews, & Location Factors

01300 | Administrative Requirements

	01310 | Project Management/Coordination		CREW	DAILY OUTPUT	LABOR-HOURS	UNIT	2000 BARE COSTS MAT.	LABOR	EQUIP.	TOTAL	TOTAL INCL O&P	
50	0010	**INSURANCE** Builders risk, standard, minimum R01100 -040				Job					.22%	350
	0050	Maximum									.59%	
	0200	All-risk type, minimum R01100 -060									.25%	
	0250	Maximum									.62%	
	0400	Contractor's equipment floater, minimum				Value					.50%	
	0450	Maximum				"					1.50%	
	0600	Public liability, average				Job					1.55%	
	0800	Workers' compensation & employer's liability, average										
	0850	by trade, carpentry, general				Payroll		19.85%				
	0900	Clerical						.53%				
	0950	Concrete						18.81%				
	1000	Electrical						7.03%				
	1050	Excavation						11.41%				
	1100	Glazing						14.43%				
	1150	Insulation						18.52%				
	1200	Lathing						12.26%				
	1250	Masonry						17.75%				
	1300	Painting & decorating						15.27%				
	1350	Pile driving						29.88%				
	1400	Plastering						15.72%				
	1450	Plumbing						8.82%				
	1500	Roofing						34.62%				
	1550	Sheet metal work (HVAC)						12.27%				
	1600	Steel erection, structural						42.56%				
	1650	Tile work, interior ceramic						10.48%				
	1700	Waterproofing, brush or hand caulking						8.32%				
	1800	Wrecking						43.48%				
	2000	Range of 35 trades in 50 states, excl. wrecking, minimum						2%				
	2100	Average						18.10%				
	2200	Maximum						132.92%				

01500 | Temporary Facilities & Controls

	01540 | Construction Aids		CREW	DAILY OUTPUT	LABOR-HOURS	UNIT	2000 BARE COSTS MAT.	LABOR	EQUIP.	TOTAL	TOTAL INCL O&P	
550	0010	**PUMP STAGING**, Aluminum										550
	1300	System in place, 50' working height, per use based on 50 uses	2 Carp	84.80	.189	C.S.F.	4.95	3.72		8.67	11.80	
	1400	100 uses R01540 -200		84.80	.189		2.48	3.72		6.20	9.05	
	1500	150 uses		84.80	.189		1.66	3.72		5.38	8.15	
750	0014	**SCAFFOLDING** R01540 -100										750
	0015	Steel tubular, rent, 1 use/mo, no plank, erect & dismantle										
	0090	Building exterior, wall face, 1 to 5 stories	3 Carp	24	1	C.S.F.	24.50	19.70		44.20	60.50	
	0200	6 to 12 stories	4 Carp	21.20	1.509		24.50	29.50		54	77.50	
	0310	13 to 20 stories	5 Carp	20	2		24.50	39.50		64	94	
	0460	Building interior, wall face area, up to 16' high	3 Carp	25	.960		24.50	18.90		43.40	59	
	0560	16' to 40' high		23	1.043		24.50	20.50		45	61.50	
	0800	Building interior floor area, up to 30' high		312	.077	C.C.F.	1.87	1.52		3.39	4.66	
	0900	Over 30' high	4 Carp	275	.116	"	1.87	2.29		4.16	6	
	0910	Steel tubular, heavy duty shoring, buy										
	0920	Frames 5' high 2' wide				Ea.	75			75	82.50	
	0925	5' high 4' wide					85			85	93.50	

	01540 | Construction Aids	CREW	DAILY OUTPUT	LABOR-HOURS	UNIT	2000 BARE COSTS MAT.	LABOR	EQUIP.	TOTAL	TOTAL INCL O&P
750 0930	6′ high 2′ wide R01100-100				Ea.	86			86	94.50
0935	6′ high 4′ wide					101			101	111
0940	Accessories									
0945	Cross braces				Ea.	16			16	17.60
0950	U-head, 8″ x 8″					16			16	17.60
0955	J-head, 4″ x 8″					12			12	13.20
0960	Base plate, 8″ x 8″					13			13	14.30
0965	Leveling jack					30.50			30.50	33.50
1000	Steel tubular, regular, buy									
1100	Frames 3′ high 5′ wide				Ea.	58			58	64
1150	5′ high 5′ wide					67			67	73.50
1200	6′-4″ high 5′ wide					84			84	92.50
1350	7′-6″ high 6′ wide					145			145	160
1500	Accessories cross braces					15			15	16.50
1550	Guardrail post					15			15	16.50
1600	Guardrail 7′ section					7.25			7.25	8
1650	Screw jacks & plates					24			24	26.50
1700	Sidearm brackets					28			28	31
1750	8″ casters					33			33	36.50
1800	Plank 2″ x 10″ x 16′-0″					42.50			42.50	47
1900	Stairway section					235			235	259
1910	Stairway starter bar					21			21	23.50
1920	Stairway inside handrail					53			53	58.50
1930	Stairway outside handrail					73			73	80.50
1940	Walk-thru frame guardrail					28			28	31
2000	Steel tubular, regular, rent/mo.									
2100	Frames 3′ high 5′ wide				Ea.	3.75			3.75	4.13
2150	5′ high 5′ wide					3.75			3.75	4.13
2200	6′-4″ high 5′ wide					3.75			3.75	4.13
2250	7′-6″ high 6′ wide					7			7	7.70
2500	Accessories, cross braces					.60			.60	.66
2550	Guardrail post					1			1	1.10
2600	Guardrail 7′ section					.75			.75	.83
2650	Screw jacks & plates					1.50			1.50	1.65
2700	Sidearm brackets					1.50			1.50	1.65
2750	8″ casters					6			6	6.60
2800	Outrigger for rolling tower					3			3	3.30
2850	Plank 2″ x 10″ x 16′-0″					5			5	5.50
2900	Stairway section					10			10	11
2910	Stairway starter bar					.10			.10	.11
2920	Stairway inside handrail					5			5	5.50
2930	Stairway outside handrail					5			5	5.50
2940	Walk-thru frame guardrail					2			2	2.20
3000	Steel tubular, heavy duty shoring, rent/mo.									
3250	5′ high 2′ & 4′ wide				Ea.	5			5	5.50
3300	6′ high 2′ & 4′ wide					5			5	5.50
3500	Accessories, cross braces					1			1	1.10
3600	U - head, 8″ x 8″					1			1	1.10
3650	J - head, 4″ x 8″					1			1	1.10
3700	Base plate, 8″ x 8″					1			1	1.10
3750	Leveling jack					2			2	2.20
5700	Planks, 2″x10″x16′-0″ in place, up to 50′ high	3 Carp	144	.167			3.28		3.28	5.65
5800	In place over 50′ high	4 Carp	160	.200			3.94		3.94	6.75
6820	Erect and dismantle frames, 1st tier	4 Clab	45	.711			10.30		10.30	17.60
6830	Erect and dismantle frames, 2nd tier		93	.344			4.97		4.97	8.50
6840	Erect and dismantle frames, 3rd tier		87	.368			5.30		5.30	9.10

	01540 | Construction Aids		CREW	DAILY OUTPUT	LABOR-HOURS	UNIT	2000 BARE COSTS MAT.	LABOR	EQUIP.	TOTAL	TOTAL INCL O&P	
50	6850	Erect and dismantle frames, 4th tier R01540-100	4 Clab	75	.427	Ea.		6.15		6.15	10.55	750
60	0010	**STAGING AIDS** and fall protection equipment										760
	0100	Sidewall staging bracket, tubular, buy				Ea.	29			29	31.50	
	0110	Cost each per day, based on 250 days use				Day	.11			.11	.13	
	0200	Guard post, buy				Ea.	15			15	16.50	
	0210	Cost each per day, based on 250 days use				Day	.06			.06	.07	
	0300	End guard chains, buy per set				Ea.	25			25	27.50	
	0310	Cost per set per day, based on 250 days use				Day	.11			.11	.13	
	1010	Cost each per day, based on 250 days use				″	.02			.02	.03	
	1100	Wood bracket, buy				Ea.	10.60			10.60	11.65	
	1110	Cost each per day, based on 250 days use				Day	.04			.04	.05	
	2010	Cost per pair per day, based on 250 days use				″	.31			.31	.34	
	2100	Steel siderail jack, buy per pair				Pair	69			69	76	
	2110	Cost per pair per day, based on 250 days use				Day	.28			.28	.30	
	3010	Cost each per day, based on 250 days use				″	.17			.17	.19	
	3100	Aluminum scaffolding plank, 20″ wide x 24′ long, buy				Ea.	660			660	730	
	3110	Cost each per day, based on 250 days use				Day	2.65			2.65	2.91	
	4010	Cost each per day, based on 250 days use				″	.90			.90	.99	
	4100	Rope for safety line, 5/8″ x 100′ nylon, buy				Ea.	35			35	38.50	
	4110	Cost each per day, based on 250 days use				Day	.14			.14	.15	
	4200	Permanent U-Bolt roof anchor, buy				Ea.	29.50			29.50	32.50	
	4300	Temporary (one use) roof ridge anchor, buy				″	18.60			18.60	20.50	
	5000	Installation (setup and removal) of staging aids										
	5010	Sidewall staging bracket	2 Carp	64	.250	Ea.		4.93		4.93	8.45	
	5020	Guard post with 2 wood rails	″	64	.250			4.93		4.93	8.45	
	5030	End guard chains, set	1 Carp	64	.125			2.46		2.46	4.22	
	5100	Roof shingling bracket		96	.083			1.64		1.64	2.81	
	5200	Ladder jack		64	.125			2.46		2.46	4.22	
	5300	Wood plank, 2x10x16′	2 Carp	80	.200			3.94		3.94	6.75	
	5310	Aluminum scaffold plank, 20″ x 24′	″	40	.400			7.90		7.90	13.50	
	5410	Safety rope	1 Carp	40	.200			3.94		3.94	6.75	
	5420	Permanent U-Bolt roof anchor (install only)	2 Carp	40	.400			7.90		7.90	13.50	
	5430	Temporary roof ridge anchor (install only)	1 Carp	64	.125			2.46		2.46	4.22	
00	0010	**TARPAULINS** Cotton duck, 10 oz. to 13.13 oz. per S.Y., minimum				S.F.	.46			.46	.51	800
	0050	Maximum					.55			.55	.61	
	0200	Reinforced polyethylene 3 mils thick, white					.10			.10	.11	
	0300	4 mils thick, white, clear or black					.12			.12	.13	
	0730	Polyester reinforced w/ integral fastening system 11 mils thick					1			1	1.10	
20	0010	**SMALL TOOLS** As % of contractor's work, minimum				Total					.50%	820
	0100	Maximum	″			″					2%	

GENERAL REQUIREMENTS 1

01590 | Equipment Rental

		01590 | Equipment Rental	UNIT	HOURLY OPER. COST	RENT PER DAY	RENT PER WEEK	RENT PER MONTH	CREW EQUIPMENT COST/DAY
100	0010	**CONCRETE EQUIPMENT RENTAL**						
	0100	without operators						
	0600	Cart, concrete, self propelled, operator walking, 10 C.F.	Ea.	1.25	70	210	630	52
	0700	Operator riding, 18 C.F.		2.40	107	320	960	83.20
	0800	Conveyer for concrete, portable, gas, 16" wide, 26' long		3.60	183	550	1,650	138.80
	0900	46' long		3.70	237	710	2,125	171.60
	1100	Core drill, electric, 2-1/2 H.P., 1" to 8" bit diameter		.50	73.50	220	660	48
	1200	Finisher, concrete floor, gas, riding trowel, 48" diameter		2.70	117	350	1,050	91.60
	1300	Gas, manual, 3 blade, 36" trowel		2	53.50	160	480	48
	1400	4 blade, 48" trowel		2.25	58.50	175	525	53
	1500	Float, hand-operated (Bull float) 48" wide		.10	8.35	25	75	5.80
	1570	Curb builder, 14 H.P., gas, single screw		1.75	66.50	200	600	54
	1590	Double screw		2.35	66.50	200	600	58.80
	1600	Grinder, concrete and terrazzo, electric, floor		1.25	66.50	200	600	50
	1700	Wall grinder		.60	33.50	100	300	24.80
	1800	Mixer, powered, mortar and concrete, gas, 6 C.F., 18 H.P.		2.44	58.50	175	525	54.50
	1900	10 C.F., 25 H.P.		3.18	53.50	160	480	57.45
	2000	16 C.F.		3.65	75	225	675	74.20
	2120	Pump, concrete, truck mounted, 4" line, 80' boom		12	935	2,800	8,400	656
	2600	Saw, concrete, manual, gas, 18 H.P.		2.75	80	240	720	70
	2650	Self-propelled, gas, 30 H.P.		5.50	110	330	990	110
	2700	Vibrators, concrete, electric, 60 cycle, 2 H.P.		.25	28.50	85	255	19
	2800	3 H.P.		.26	40	120	360	26.10
	2900	Gas engine, 5 H.P.		.65	40	120	360	29.20
	3000	8 H.P.	↓	1	50	150	450	38
200	0010	**EARTHWORK EQUIPMENT RENTAL** Without operators						
	0050	Augers for truck or trailer mounting, vertical drilling						
	0055	Fence post auger, truck mounted	Ea.	10	485	1,450	4,350	370
	0060	4" to 36" diam., 54 H.P., gas, 10' spindle travel		16	590	1,775	5,325	483
	0070	14' spindle travel		16.40	665	2,000	6,000	531.20
	0100	Excavator, diesel hydraulic, crawler mounted, 1/2 C.Y. cap.		11.65	415	1,248	3,750	342.80
	0120	5/8 C.Y. capacity		15.50	480	1,438	4,325	411.60
	0140	3/4 C.Y. capacity		17.40	530	1,583	4,750	455.80
	0341	Attachments						
	0342	Bucket thumbs		.45	292	875	2,625	178.60
	0345	Grapples		.35	300	900	2,700	182.80
	0350	Gradall type, truck mounted, 3 ton @ 15' radius, 5/8 C.Y.		25.40	690	2,075	6,225	618.20
	0400	Backhoe-loader, 40 to 45 H.P., 5/8 C.Y. capacity		7.15	217	650	1,950	187.20
	0450	45 H.P. to 60 H.P., 3/4 C.Y. capacity		8.50	225	675	2,025	203
	0460	80 H.P., 1-1/4 C.Y. capacity		12.22	292	877	2,625	273.15
	0470	112 H.P., 1-1/2 C.Y. capacity		15.50	395	1,185	3,550	361
	0480	Attachments						
	0482	Compactor, 20,000 lb		1.35	192	575	1,725	125.80
	0485	Hydraulic hammer, 750 ft-lbs		1.15	300	900	2,700	189.20
	0486	Hydraulic hammer, 1200 ft-lbs		1.70	335	1,000	3,000	213.60
	0500	Brush chipper, gas engine, 6" cutter head, 35 H.P.		3.85	142	425	1,275	115.80
	0550	12" cutter head, 130 H.P.		11.88	193	580	1,750	211.05
	0600	15" cutter head, 165 H.P.		15.48	267	800	2,400	283.85
	1200	Compactor, roller, 2 drum, 2000 lb., operator walking		1.78	133	400	1,200	94.25
	1250	Rammer compactor, gas, 1000 lb. blow		.75	50	150	450	36
	1300	Vibratory plate, gas, 13" plate, 1000 lb. blow		.75	43.50	130	390	32
	1350	24" plate, 5000 lb. blow		1.95	75	225	675	60.60
	1860	Grader, self-propelled, 25,000 lb.		12.85	555	1,658	4,975	434.40
	1910	30,000 lb.		14.50	665	2,000	6,000	516
	3710	Screening plant 110 hp. w / 5' x 10'screen		16.28	450	1,350	4,050	400.25
	3720	5' x 16' screen		17.65	520	1,560	4,675	453.20
	4110	Tractor, crawler, with bulldozer, torque converter, diesel 75 H.P.	↓	9.95	335	1,000	3,000	279.60

01590 | Equipment Rental

			UNIT	HOURLY OPER. COST	RENT PER DAY	RENT PER WEEK	RENT PER MONTH	CREW EQUIPMENT COST/DAY	
200	4150	105 H.P.	Ea.	10.75	510	1,535	4,600	393	200
	4200	140 H.P.		17.55	650	1,950	5,850	530.40	
	4260	200 H.P.		25.85	1,000	3,000	9,000	806.80	
	4400	Loader, crawler, torque conv., diesel, 1-1/2 C.Y., 80 H.P.		9.87	420	1,255	3,775	329.95	
	4450	1-1/2 to 1-3/4 C.Y., 95 H.P.		10.23	475	1,425	4,275	366.85	
	4510	1-3/4 to 2-1/4 C.Y., 130 H.P.		12.50	585	1,750	5,250	450	
	4530	2-1/2 to 3-1/4 C.Y., 190 H.P.		19.25	1,050	3,150	9,450	784	
	4610	Tractor loader, wheel, torque conv., 4 x 4, 1 to 1-1/4 C.Y., 65 H.P.		8.45	277	830	2,500	233.60	
	4620	1-1/2 to 1-3/4 C.Y., 80 H.P.		9.75	375	1,125	3,375	303	
	4650	1-3/4 to 2 C.Y., 100 H.P.		11.35	405	1,215	3,650	333.80	
	4710	2-1/2 to 3-1/2 C.Y., 130 H.P.		12.60	500	1,500	4,500	400.80	
	4730	3 to 4-1/2 C.Y., 170 H.P.		15.30	700	2,100	6,300	542.40	
	4880	Wheeled, skid steer, 10 C.F., 30 H.P. gas		4.55	133	400	1,200	116.40	
	4890	1 C.Y., 78 H.P., diesel	↓	6.35	292	875	2,625	225.80	
	4891	Attachments for all skid steer loaders							
	4892	Auger	Ea.	.12	83.50	250	750	50.95	
	4893	Backhoe		.15	110	330	990	67.20	
	4894	Broom		.16	107	320	960	65.30	
	4895	Forks		.08	36.50	110	330	22.65	
	4896	Grapple		.12	83.50	250	750	50.95	
	4897	Concrete hammer		.25	180	540	1,625	110	
	4898	Tree spade		.36	153	460	1,375	94.90	
	4899	Trencher		.41	122	365	1,100	76.30	
	4900	Trencher, chain, boom type, gas, operator walking, 12 H.P.		2.16	142	426	1,275	102.50	
	4910	Operator riding, 40 H.P.		6.36	267	800	2,400	210.90	
	5000	Wheel type, diesel, 4' deep, 12" wide		12.42	515	1,550	4,650	409.35	
	5100	Diesel, 6' deep, 20" wide		15.65	735	2,200	6,600	565.20	
	5150	Ladder type, diesel, 5' deep, 8" wide		9.40	385	1,150	3,450	305.20	
	5200	Diesel, 8' deep, 16" wide		17.61	665	2,000	6,000	540.90	
	5250	Truck, dump, tandem, 12 ton payload		17.41	375	1,130	3,400	365.30	
	5300	Three axle dump, 16 ton payload		20.28	475	1,425	4,275	447.25	
	5350	Dump trailer only, rear dump, 16-1/2 C.Y.		3.45	177	530	1,600	133.60	
	5400	20 C.Y.		3.51	180	540	1,625	136.10	
	5450	Flatbed, single axle, 1-1/2 ton rating		10.87	143	430	1,300	172.95	
	5500	3 ton rating	↓	11.43	148	445	1,325	180.45	
400	0010	**GENERAL EQUIPMENT RENTAL** Without operators							400
	0150	Aerial lift, scissor type, to 15' high, 1000 lb. cap., electric	Ea.	1.18	83.50	250	750	59.45	
	0160	To 25' high, 2000 lb. capacity		1.80	125	375	1,125	89.40	
	0170	Telescoping boom to 40' high, 750 lb. capacity, gas		6.42	335	1,000	3,000	251.35	
	0180	1000 lb. capacity		7.98	435	1,300	3,900	323.85	
	0190	To 60' high, 750 lb. capacity		8.47	500	1,500	4,500	367.75	
	0195	Air compressor, portable, 6.5 CFM, electric		.12	33.50	100	300	20.95	
	0196	gasoline		.13	43.50	130	390	27.05	
	0200	Air compressor, portable, gas engine, 60 C.F.M.		5.41	58.50	175	525	78.30	
	0300	160 C.F.M.		6.80	73.50	220	660	98.40	
	0400	Diesel engine, rotary screw, 250 C.F.M.		6.50	125	375	1,125	127	
	0500	365 C.F.M.		9.30	167	500	1,500	174.40	
	0800	For silenced models, small sizes, add	↓	3%	5%	5%	5%		
	0920	Air tools and accessories							
	0930	Breaker, pavement, 60 lb.	Ea.	.19	30	90	270	19.50	
	0940	80 lb.		.21	40	120	360	25.70	
	0950	Drills, hand (jackhammer) 65 lb.		.23	28.50	85	255	18.85	
	0960	Track or wagon, swing boom, 4" drifter		10.57	380	1,140	3,425	312.55	
	0970	5" drifter		11.38	660	1,975	5,925	486.05	
	0980	Dust control per drill		2.11	12.65	38	114	24.50	
	0990	Hammer, chipping, 12 lb.		.12	26.50	80	240	16.95	
	1000	Hose, air with couplings, 50' long, 3/4" diameter	↓	.15	5	15	45	4.20	

01590 | Equipment Rental

			UNIT	HOURLY OPER. COST	RENT PER DAY	RENT PER WEEK	RENT PER MONTH	CREW EQUIPMENT COST/DAY
400	1100	1" diameter	Ea.	.15	6.65	20	60	5.20
	1200	1-1/2" diameter		.15	10	30	90	7.20
	1300	2" diameter		.21	16.65	50	150	11.70
	1400	2-1/2" diameter		.23	20	60	180	13.85
	1450	Drill, steel, 7/8" x 2'			4	12	36	2.40
	1520	Moil points		.82	3.33	10	30	8.55
	1525	Pneumatic nailer w/accessories		.12	30	90	270	18.95
	1530	Sheeting driver for 60 lb. breaker		.15	11.65	35	105	8.20
	1540	For 90 lb. breaker		.15	18.35	55	165	12.20
	1550	Spade, 25 lb.		.08	8.35	25	75	5.65
	1560	Tamper, single, 35 lb.		.10	25	75	225	15.80
	1580	Wrenches, impact, air powered, up to 3/4" bolt		.25	22	66	198	15.20
	1590	Up to 1-1/4" bolt		.35	41.50	125	375	27.80
	1600	Barricades, barrels, reflectorized, 1 to 50 barrels			2.33	7	21	1.40
	1620	Barrels with flashers, 1 to 50 barrels			3.33	10	30	2
	1640	Barrels with steady burn type C lights			4	12	36	2.40
	1690	Butt fusion machine, electric		1.50	293	880	2,650	188
	1695	Electro fusion machine		1.25	127	380	1,150	86
	1850	Drill, rotary hammer, electric, 1-1/2" diameter		.15	25	75	225	16.20
	1860	Carbide bit for above			6	18	54	3.60
	1870	Emulsion sprayer, 65 gal., 5 H.P. gas engine		.63	56.50	170	510	39.05
	1880	200 gal., 5 H.P. engine		.67	60	180	540	41.35
	1920	Floodlight, mercury, vapor or quartz, on tripod						
	1930	1000 watt	Ea.	.12	25	75	225	15.95
	1940	2000 watt		.13	41.50	125	375	26.05
	2020	Forklift, wheeled, for brick, 18', 3000 lb., 2 wheel drive, gas		9.28	167	500	1,500	174.25
	2100	Generator, electric, gas engine, 1.5 KW to 3 KW		1.13	40	120	360	33.05
	2200	5 KW		1.49	58.50	175	525	46.90
	2300	10 KW		2.40	133	400	1,200	99.20
	2500	Diesel engine, 20 KW		4.35	117	350	1,050	104.80
	2850	Hammer, hydraulic, for mounting on boom, to 500 ft.-lb.		.99	133	400	1,200	87.90
	2860	500 to 1200 ft.-lb.		2.50	258	775	2,325	175
	2900	Heaters, space, oil or electric, 50 MBH		.11	26.50	80	240	16.90
	3000	100 MBH		.12	33.50	100	300	20.95
	3100	300 MBH		.13	50	150	450	31.05
	3150	500 MBH		.15	66.50	200	600	41.20
	3200	Hose, water, suction with coupling, 20' long, 2" diameter		.06	10	30	90	6.50
	3210	3" diameter		.06	15	45	135	9.50
	3220	4" diameter		.06	20	60	180	12.50
	3250	Discharge hose with coupling, 50' long, 2" diameter		.05	8.35	25	75	5.40
	3260	3" diameter		.05	10	30	90	6.40
	3270	4" diameter		.06	13.35	40	120	8.50
	3300	Ladders, extension type, 16' to 36' long			16.65	50	150	10
	3400	40' to 60' long			30	90	270	18
	3460	Builders level with tripod and rod			26.50	80	240	16
	3700	Mixer, powered, plaster and mortar, 6 C.F., 7 H.P.		.85	56.50	170	510	40.80
	3800	10 C.F., 9 H.P.		1.25	78.50	235	705	57
	3850	Nailer, pneumatic		.12	28.50	85	255	17.95
	3900	Paint sprayers complete, 8 CFM		.08	45	135	405	27.65
	4000	17 CFM		.08	61.50	185	555	37.65
	4100	Pump, centrifugal gas pump, 1-1/2", 4 MGPH		.45	35	105	315	24.60
	4200	2", 8 MGPH		.55	36	108	325	26
	4300	3", 15 MGPH		1.25	46.50	140	420	38
	4500	Submersible electric pump, 1-1/4", 55 GPM		.35	36.50	110	330	24.80
	4600	1-1/2", 83 GPM		.41	40	120	360	27.30
	4700	2", 120 GPM		.42	46.50	140	420	31.35
	4800	3", 300 GPM		.75	58.50	175	525	41
	4900	4", 560 GPM		1.21	70	210	630	51.70

	01590 | Equipment Rental	UNIT	HOURLY OPER. COST	RENT PER DAY	RENT PER WEEK	RENT PER MONTH	CREW EQUIPMENT COST/DAY	
5100	Diaphragm pump, gas, single, 1-1/2" diameter	Ea.	.50	26.50	80	240	20	400
5200	2" diameter		.55	41.50	125	375	29.40	
5300	3" diameter		.75	47.50	143	430	34.60	
5400	Double, 4" diameter		1.58	88.50	265	795	65.65	
5500	Trash pump, self-priming, gas, 2" diameter		1.26	40	120	360	34.10	
5600	Diesel, 4" diameter	↓	1.99	91.50	275	825	70.90	
5660	Rollers, see division 01590-200							
5700	Salamanders, L.P. gas fired, 100,000 B.T.U.	Ea.	.75	21.50	65	195	19	
5720	Sandblaster, portable, open top, 3 C.F. capacity		.20	50	150	450	31.60	
5740	Accessories for above		.07	16.65	50	150	10.55	
5750	Sander, floor		.15	40.50	121	365	25.40	
5760	Edger		.10	25	75	225	15.80	
5800	Saw, chain, gas engine, 18" long		.55	40	120	360	28.40	
5900	36" long		1.15	66.50	200	600	49.20	
6000	Masonry, table mounted, 14" diameter, 5 H.P.		1.80	53.50	160	480	46.40	
6100	Circular, hand held, electric, 7-1/4" diameter		.18	21.50	65	195	14.45	
6200	12" diameter		.28	33.50	100	300	22.25	
6275	Shot blaster, walk behind, 20" wide		1.50	217	650	1,950	142	
6300	Steam cleaner, 100 gallons per hour		.45	56.50	170	510	37.60	
6310	200 gallons per hour		.75	66.50	200	600	46	
6340	Tar Kettle/Pot, 400 gallon		.50	75	225	675	49	
6350	Torch, cutting, acetylene-oxygen, 150' hose		7.50	33.50	100	300	80	
6360	Hourly operating cost includes tips and gas		7.65				61.20	
6420	Recycle flush type			20	60	180	12	
7020	Transit with tripod			33.50	100	300	20	
7100	Truck, pickup, 3/4 ton, 2 wheel drive		10.65	75	225	675	130.20	
7200	4 wheel drive		12.08	80	240	720	144.65	
7300	Tractor, 4 x 2, 30 ton capacity, 195 H.P.		11.20	395	1,180	3,550	325.60	
7410	250 H.P.		15.25	435	1,300	3,900	382	
7620	Vacuum truck, hazardous material, 2500 gallon		13.95	305	910	2,725	293.60	
7625	5,000 gallon		15	335	1,000	3,000	320	
7700	Welder, electric, 200 amp		.81	39	117	350	29.90	
7800	300 amp		1.10	66.50	200	600	48.80	
7900	Gas engine, 200 amp		4.40	55	165	495	68.20	
8000	300 amp		5.25	70	210	630	84	
8100	Wheelbarrow, any size	↓		8.35	25	75	5	
0010	**LIFTING AND HOISTING EQUIPMENT RENTAL**							600
0100	without operators							
2400	Truck mounted, hydraulic, 12 ton capacity	Ea.	22.93	450	1,350	4,050	453.45	
2500	25 ton capacity		23.65	615	1,850	5,550	559.20	
2550	33 ton capacity		24.37	835	2,500	7,500	694.95	
2600	55 ton capacity		34.15	890	2,675	8,025	808.20	
2800	Self-propelled, 4 x 4, with telescoping boom, 5 ton		10.18	325	975	2,925	276.45	
2900	12-1/2 ton capacity		16.42	450	1,350	4,050	401.35	
3000	15 ton capacity		18.25	490	1,465	4,400	439	
3100	25 ton capacity		20.99	660	1,980	5,950	563.90	
3600	Hoists, chain type, overhead, manual, 3/4 ton		.06	6.65	20	60	4.50	
3900	10 ton		.25	28.50	85	255	19	
5200	Jacks, hydraulic, 20 ton		.15	14.65	44	132	10	
6350	For each added 10' jackrod section, add			4	12	36	2.40	
6500	125 ton capacity			415	1,250	3,750	250	
6550	For each added 10' jackrod section, add			28.50	85	255	17	
6600	Cable jack, 10 ton capacity with 200' cable			83.50	250	750	50	
6650	For each added 50' of cable, add	↓		8.35	25	75	5	

	01740 | Cleaning	CREW	DAILY OUTPUT	LABOR-HOURS	UNIT	2000 BARE COSTS MAT.	LABOR	EQUIP.	TOTAL	TOTAL INCL O&P	
500	0010 **CLEANING UP** After job completion, allow, minimum				Job					.30%	50
	0040 Maximum				"					1%	

For information about Means Estimating Seminars, see yellow pages 11 and 12 in back of book

Division 2
Site Construction

Estimating Tips

2200 Site Preparation

If possible visit the site and take an inventory of the type, quantity and size of the trees. Certain trees may have a landscape resale value or firewood value. Stump disposal can be very expensive, particularly if they cannot be buried at the site. Consider using a bulldozer in lieu of hand cutting trees.

Estimators should visit the site to determine the need for haul road, access, storage of materials, and security considerations. When estimating for access roads on unstable soil, consider using a geotextile stabilization fabric. It can greatly reduce the quantity of crushed stone or gravel. Sites of limited size and access can cause cost overruns due to lost productivity. Theft and damage is another consideration if the location is isolated. A temporary fence or security guards may be required. Investigate the site thoroughly.

2210 Subsurface Investigation

n preparing estimates on structures involving arthwork or foundations, all information oncerning soil characteristics should be btained. Look particularly for hazardous vaste, evidence of prior dumping of debris, nd previous stream beds.

The costs shown for selective demolition do not include rubbish handling or disposal. These items should be estimated separately using Means data or other sources.

02300 Earthwork

- Estimating the actual cost of performing earthwork requires careful consideration of the variables involved. This includes items such as type of soil, whether or not water will be encountered, dewatering, whether or not banks need bracing, disposal of excavated earth, length of haul to fill or spoil sites, etc. If the project has large quantities of cut or fill, consider raising or lowering the site to reduce costs while paying close attention to the effect on site drainage and utilities if doing this.
- If the project has large quantities of fill, creating a borrow pit on the site can significantly lower the costs. It is very important to consider what time of year the project is scheduled for completion. Bad weather can create large cost overruns from dewatering, site repair and lost productivity from cold weather.

02700 Bases, Ballasts, Pavements/Appurtenances

- When estimating paving, keep in mind the project schedule. If an asphaltic paving project is in a colder climate and runs through to the spring, consider placing the base course in the autumn, then topping it in the spring just prior to completion. This could save considerable costs in spring repair. Keep in mind that prices for asphalt and concrete are generally higher in the cold seasons.

02500 Utility Services
02600 Drainage & Containment

- Never assume that the water, sewer and drainage lines will go in at the early stages of the project. Consider the site access needs before dividing the site in half with open trenches, loose pipe, and machinery obstructions. Always inspect the site to establish that the site drawings are complete. Check off all existing utilities on your drawing as you locate them. If you find any discrepancies, mark up the site plan for further research. Differing site conditions can be very costly if discovered later in the project.

02900 Planting

- The timing of planting and guarantee specifications often dictate the costs for establishing tree and shrub growth and a stand of grass or ground cover. Establish the work performance schedule to coincide with the local planting season. Maintenance and growth guarantees can add from 20% to 100% to the total landscaping cost. The cost to replace trees and shrubs can be as high as 5% of the total cost depending on the planting zone, soil conditions and time of year.

Reference Numbers

Reference numbers are shown in bold squares at the beginning of some major classifications. These numbers refer to related items in the Reference Section. The reference information may be an estimating procedure, an alternate pricing method or technical information.

Note: Not all subdivisions listed here necessarily appear in this publication.

02050 | Basic Site Materials & Methods

	02060 Aggregate		CREW	DAILY OUTPUT	LABOR-HOURS	UNIT	2000 BARE COSTS MAT.	LABOR	EQUIP.	TOTAL	TOTAL INCL O&P
150	0010	**BORROW**									
	0020	and spread, with 200 H.P. dozer, no compaction									
	0100	Bank run gravel	B-15	600	.047	C.Y.	5.20	.79	2.84	8.83	10.15
	0200	Common borrow		600	.047		4.77	.79	2.84	8.40	9.70
	0300	Crushed stone, (1.40 tons per CY), 1-1/2"		600	.047		17.05	.79	2.84	20.68	23.50
	0320	3/4"		600	.047		18.80	.79	2.84	22.43	25
	0340	1/2"		600	.047		18.25	.79	2.84	21.88	24.50
	0360	3/8"		600	.047		15.75	.79	2.84	19.38	22
	0400	Sand, washed, concrete		600	.047		10.65	.79	2.84	14.28	16.15
	0500	Dead or bank sand		600	.047		3.50	.79	2.84	7.13	8.30
	02080	**Utility Materials**									
800	0010	**UTILITY VAULTS** Precast concrete, 6" thick									
	0050	5' x 10' x 6' high, I.D.	B-13	2	24	Ea.	1,400	380	280	2,060	2,475
	0350	Hand hole, precast concrete, 1-1/2" thick									
	0400	1'-0" x 2'-0" x 1'-9", I.D., light duty	B-1	4	6	Ea.	355	90.50		445.50	545
	0450	4'-6" x 3'-2" x 2'-0", O.D., heavy duty	B-6	3	8	"	880	128	67.50	1,075.50	1,250

02200 | Site Preparation

	02210 Subsurface Investigation		CREW	DAILY OUTPUT	LABOR-HOURS	UNIT	2000 BARE COSTS MAT.	LABOR	EQUIP.	TOTAL	TOTAL INCL O&P
310	0010	**BORINGS** Initial field stake out and determination of elevations	A-6	1	16	Day		335		335	575
	0100	Drawings showing boring details				Total		107		107	170
	0200	Report and recommendations from P.E.						235		235	380
	0300	Mobilization and demobilization, minimum	B-55	4	4			60.50	166	226.50	285
	0350	For over 100 miles, per added mile		450	.036	Mile		.54	1.47	2.01	2.53
	0600	Auger holes in earth, no samples, 2-1/2" diameter		78.60	.204	L.F.		3.08	8.45	11.53	14.50
	0800	Cased borings in earth, with samples, 2-1/2" diameter		55.50	.288	"	12.35	4.37	11.95	28.67	34
	1400	Drill rig and crew with truck mounted auger		1	16	Day		242	665	907	1,150
	1500	For inner city borings add, minimum									10%
	1510	Maximum									20%
	02220	**Site Demolition**									
100	0011	**BUILDING DEMOLITION**									
	0500	Small bldgs, or single bldgs, no salvage included, steel	B-3	14,800	.003	C.F.		.05	.11	.16	.21
	0600	Concrete	"	11,300	.004	"		.07	.15	.22	.28
	0605	Concrete, plain	B-5	33	1.212	C.Y.		19.30	29	48.30	65
	0610	Reinforced		25	1.600			25.50	38.50	64	86
	0615	Concrete walls		34	1.176			18.75	28.50	47.25	63
	0620	Elevated slabs		26	1.538			24.50	37	61.50	82.50
	0650	Masonry	B-3	14,800	.003	C.F.		.05	.11	.16	.21
	0700	Wood	"	14,800	.003	"		.05	.11	.16	.21
	1000	Single family, one story house, wood, minimum				Ea.				2,300	2,700
	1020	Maximum								4,000	4,800
	1200	Two family, two story house, wood, minimum								3,000	3,600
	1220	Maximum								5,800	7,000
	1300	Three family, three story house, wood, minimum								4,000	4,800
	1320	Maximum								7,000	8,400
	1400	Gutting building, see division 02225-400									

Important: See the Reference Section for critical supporting data - Reference Nos., Crews, & Location Factor

02220 | Site Demolition

		Crew	Daily Output	Labor-Hours	Unit	2000 Bare Costs Mat.	2000 Bare Costs Labor	2000 Bare Costs Equip.	2000 Bare Costs Total	Total Incl O&P	
0010	**FOOTINGS AND FOUNDATIONS DEMOLITION**										550
0200	Floors, concrete slab on grade,										
0240	4″ thick, plain concrete	B-9C	500	.080	S.F.		1.19	.36	1.55	2.44	
0280	Reinforced, wire mesh		470	.085			1.26	.38	1.64	2.59	
0300	Rods		400	.100			1.49	.45	1.94	3.04	
0400	6″ thick, plain concrete		375	.107			1.58	.48	2.06	3.24	
0420	Reinforced, wire mesh		340	.118			1.75	.53	2.28	3.57	
0440	Rods		300	.133			1.98	.60	2.58	4.05	
1000	Footings, concrete, 1′ thick, 2′ wide	B-5	300	.133	L.F.		2.13	3.21	5.34	7.15	
1080	1′-6″ thick, 2′ wide		250	.160			2.55	3.86	6.41	8.55	
1120	3′ wide		200	.200			3.19	4.82	8.01	10.70	
1200	Average reinforcing, add								10%	10%	
2000	Walls, block, 4″ thick	1 Clab	180	.044	S.F.		.64		.64	1.10	
2040	6″ thick		170	.047			.68		.68	1.16	
2080	8″ thick		150	.053			.77		.77	1.32	
2100	12″ thick		150	.053			.77		.77	1.32	
2400	Concrete, plain concrete, 6″ thick	B-9	160	.250			3.71	1.13	4.84	7.60	
2420	8″ thick		140	.286			4.24	1.29	5.53	8.65	
2440	10″ thick		120	.333			4.95	1.50	6.45	10.15	
2500	12″ thick		100	.400			5.95	1.80	7.75	12.20	
2600	For average reinforcing, add								10%	10%	
4000	For congested sites or small quantities, add up to								200%	200%	
4200	Add for disposal, on site	B-11A	232	.069	C.Y.		1.18	3.48	4.66	5.80	
4250	To five miles	B-30	220	.109	″		1.90	7.30	9.20	11.20	
0010	**SITE DEMOLITION** No hauling, abandon catch basin or manhole	B-6	7	3.429	Ea.		55	29	84	125	875
0020	Remove existing catch basin or manhole, masonry	″	4	6			96	51	147	218	
0030	Catch basin or manhole frames and covers, stored	B-6	13	1.846			29.50	15.60	45.10	67	
0040	Remove and reset	″	7	3.429			55	29	84	125	
0600	Fencing, barbed wire, 3 strand	2 Clab	430	.037	L.F.		.54		.54	.92	
0650	5 strand	″	280	.057			.83		.83	1.41	
0700	Chain link, posts & fabric, remove only, 8′ to 10′ high	B-6	445	.054			.86	.46	1.32	1.96	
0750	Remove and reset	″	70	.343			5.50	2.90	8.40	12.50	
1000	Masonry walls, block or tile, solid, remove	B-5	1,800	.022	C.F.		.35	.54	.89	1.19	
1100	Cavity wall		2,200	.018			.29	.44	.73	.97	
1200	Brick, solid		900	.044			.71	1.07	1.78	2.38	
1300	With block back-up		1,130	.035			.56	.85	1.41	1.90	
1400	Stone, with mortar		900	.044			.71	1.07	1.78	2.38	
1500	Dry set		1,500	.027			.43	.64	1.07	1.43	
1710	Pavement removal, bituminous roads, 3″ thick	B-38	690	.035	S.Y.		.56	.60	1.16	1.60	
1750	4″ to 6″ thick		420	.057			.91	.99	1.90	2.64	
1800	Bituminous driveways		640	.038			.60	.65	1.25	1.74	
1900	Concrete to 6″ thick, hydraulic hammer, mesh reinforced		255	.094			1.51	1.63	3.14	4.35	
2000	Rod reinforced		200	.120			1.92	2.08	4	5.55	
2300	With hand held air equipment, bituminous, to 6″ thick	B-39	1,900	.025	S.F.		.37	.09	.46	.74	
2320	Concrete to 6″ thick, no reinforcing		1,200	.040			.59	.15	.74	1.18	
2340	Mesh reinforced		1,400	.034			.51	.13	.64	1.01	
2360	Rod reinforced		765	.063			.93	.24	1.17	1.85	
2400	Curbs, concrete, plain	B-6	360	.067	L.F.		1.07	.56	1.63	2.42	
2500	Reinforced		275	.087			1.40	.74	2.14	3.17	
2600	Granite		360	.067			1.07	.56	1.63	2.42	
2700	Bituminous		528	.045			.73	.38	1.11	1.65	
2900	Pipe removal, sewer/water, no excavation, 12″ diameter		175	.137			2.19	1.16	3.35	4.99	
2960	24″ diameter		120	.200			3.20	1.69	4.89	7.25	
4000	Sidewalk removal, bituminous, 2-1/2″ thick		325	.074	S.Y.		1.18	.62	1.80	2.69	
4050	Brick, set in mortar		185	.130			2.08	1.10	3.18	4.72	
4100	Concrete, plain, 4″		160	.150			2.40	1.27	3.67	5.45	

02200 | Site Preparation

	02220 | Site Demolition		CREW	DAILY OUTPUT	LABOR-HOURS	UNIT	MAT.	LABOR	EQUIP.	TOTAL	TOTAL INCL O&P
							2000 BARE COSTS				
875	4200	Mesh reinforced	B-6	150	.160	S.Y.		2.56	1.35	3.91	5.80
	5000	Slab on grade removal, plain	B-5	45	.889	C.Y.		14.15	21.50	35.65	47.50
	5100	Mesh reinforced		33	1.212			19.30	29	48.30	65
	5200	Rod reinforced		25	1.600			25.50	38.50	64	86
	5500	For congested sites or small quantities, add up to								200%	200%
	5550	For disposal on site, add	B-11A	232	.069			1.18	3.48	4.66	5.80
	5600	To 5 miles, add	B-34D	76	.105			1.71	7.40	9.11	11
	02225 | Selective Demolition										
310	0010	**CEILING DEMOLITION**									
	0200	Drywall, furred and nailed	2 Clab	800	.020	S.F.		.29		.29	.50
	1000	Plaster, lime and horse hair, on wood lath, incl. lath		700	.023			.33		.33	.57
	1200	Suspended ceiling, mineral fiber, 2′x2′ or 2′x4′		1,500	.011			.15		.15	.26
	1250	On suspension system, incl. system		1,200	.013			.19		.19	.33
	1500	Tile, wood fiber, 12″ x 12″, glued		900	.018			.26		.26	.44
	1540	Stapled		1,500	.011			.15		.15	.26
	2000	Wood, tongue and groove, 1″ x 4″		1,000	.016			.23		.23	.40
	2040	1″ x 8″		1,100	.015			.21		.21	.36
	2400	Plywood or wood fiberboard, 4′ x 8′ sheets		1,200	.013			.19		.19	.33
320	0010	**CUTOUT DEMOLITION** Conc., elev. slab, light reinf., under 6 C.F.	B-9C	65	.615	C.F.		9.15	2.78	11.93	18.70
	0050	Light reinforcing, over 6 C.F.	"	75	.533	"		7.90	2.41	10.31	16.20
	0200	Slab on grade to 6″ thick, not reinforced, under 8 S.F.	B-9	85	.471	S.F.		7	2.12	9.12	14.30
	0250	Not reinforced, over 8 S.F.		175	.229	"		3.39	1.03	4.42	6.95
	0600	Walls, not reinforced, under 6 C.F.		60	.667	C.F.		9.90	3.01	12.91	20.50
	0650	Not reinforced, over 6 C.F.		65	.615			9.15	2.78	11.93	18.70
	1000	Concrete, elevated slab, bar reinforced, under 6 C.F.	B-9C	45	.889			13.20	4.01	17.21	27
	1050	Bar reinforced, over 6 C.F.	"	50	.800			11.90	3.61	15.51	24.50
	1200	Slab on grade to 6″ thick, bar reinforced, under 8 S.F.	B-9	75	.533	S.F.		7.90	2.41	10.31	16.20
	1250	Bar reinforced, over 8 S.F.	"	105	.381	"		5.65	1.72	7.37	11.60
	1400	Walls, bar reinforced, under 6 C.F.	B-9C	50	.800	C.F.		11.90	3.61	15.51	24.50
	1450	Bar reinforced, over 6 C.F.	"	55	.727	"		10.80	3.28	14.08	22
	2000	Brick, to 4 S.F. opening, not including toothing									
	2040	4″ thick	B-9C	30	1.333	Ea.		19.80	6	25.80	40.50
	2060	8″ thick		18	2.222			33	10	43	67.50
	2080	12″ thick		10	4			59.50	18.05	77.55	122
	2400	Concrete block, to 4 S.F. opening, 2″ thick		35	1.143			16.95	5.15	22.10	34.50
	2420	4″ thick		30	1.333			19.80	6	25.80	40.50
	2440	8″ thick		27	1.481			22	6.70	28.70	45
	2460	12″ thick		24	1.667			25	7.50	32.50	51
	2600	Gypsum block, to 4 S.F. opening, 2″ thick	B-9	80	.500			7.45	2.26	9.71	15.20
	2620	4″ thick		70	.571			8.50	2.58	11.08	17.40
	2640	8″ thick		55	.727			10.80	3.28	14.08	22
	2800	Terra cotta, to 4 S.F. opening, 4″ thick		70	.571			8.50	2.58	11.08	17.40
	2840	8″ thick		65	.615			9.15	2.78	11.93	18.70
	2880	12″ thick		50	.800			11.90	3.61	15.51	24.50
	3000	Toothing masonry cutouts, brick, soft old mortar	1 Brhe	40	.200	V.L.F.		3.14		3.14	5.30
	3100	Hard mortar		30	.267			4.19		4.19	7.10
	3200	Block, soft old mortar		70	.114			1.79		1.79	3.04
	3400	Hard mortar		50	.160			2.51		2.51	4.26
	6000	Walls, interior, not including re-framing,									
	6010	openings to 5 S.F.									
	6100	Drywall to 5/8″ thick	A-1	24	.333	Ea.		4.82	2.92	7.74	11.45
	6200	Paneling to 3/4″ thick		20	.400			5.80	3.50	9.30	13.75
	6300	Plaster, on gypsum lath		20	.400			5.80	3.50	9.30	13.75
	6340	On wire lath		14	.571			8.25	5	13.25	19.65

02225 | Selective Demolition

		02225 Selective Demolition	CREW	DAILY OUTPUT	LABOR-HOURS	UNIT	2000 BARE COSTS MAT.	LABOR	EQUIP.	TOTAL	TOTAL INCL O&P	
320	7000	Wood frame, not including re-framing, openings to 5 S.F.										320
	7200	Floors, sheathing and flooring to 2″ thick	A-1	5	1.600	Ea.		23	14	37	55	
	7310	Roofs, sheathing to 1″ thick, not including roofing		6	1.333			19.25	11.65	30.90	46	
	7410	Walls, sheathing to 1″ thick, not including siding		7	1.143			16.50	10	26.50	39.50	
340	0010	DOOR DEMOLITION										340
	0200	Doors, exterior, 1-3/4″ thick, single, 3′ x 7′ high	1 Clab	16	.500	Ea.		7.20		7.20	12.40	
	0220	Double, 6′ x 7′ high		12	.667			9.65		9.65	16.50	
	0500	Interior, 1-3/8″ thick, single, 3′ x 7′ high		20	.400			5.80		5.80	9.90	
	0520	Double, 6′ x 7′ high		16	.500			7.20		7.20	12.40	
	0700	Bi-folding, 3′ x 6′-8″ high		20	.400			5.80		5.80	9.90	
	0720	6′ x 6′-8″ high		18	.444			6.40		6.40	11	
	0900	Bi-passing, 3′ x 6′-8″ high		16	.500			7.20		7.20	12.40	
	0940	6′ x 6′-8″ high		14	.571			8.25		8.25	14.15	
	1500	Remove and reset, minimum	1 Carp	8	1			19.70		19.70	34	
	1520	Maximum	″	6	1.333			26.50		26.50	45	
	2000	Frames, including trim, metal	A-1	8	1			14.45	8.75	23.20	34.50	
	2200	Wood	2 Carp	32	.500			9.85		9.85	16.90	
	2201	Alternate pricing method	A-1	200	.040	L.F.		.58	.35	.93	1.38	
	3000	Special doors, counter doors	2 Carp	6	2.667	Ea.		52.50		52.50	90	
	3300	Glass, sliding, including frames		12	1.333			26.50		26.50	45	
	3400	Overhead, commercial, 12′ x 12′ high		4	4			79		79	135	
	3500	Residential, 9′ x 7′ high		8	2			39.50		39.50	67.50	
	3540	16′ x 7′ high		7	2.286			45		45	77	
	3600	Remove and reset, minimum		4	4			79		79	135	
	3620	Maximum		2.50	6.400			126		126	216	
	3660	Remove and reset elec. garage door opener	1 Carp	8	1			19.70		19.70	34	
	4000	Residential lockset, exterior		30	.267			5.25		5.25	9	
	4200	Deadbolt lock		32	.250			4.93		4.93	8.45	
	9000	Minimum labor/equipment charge		4	2	Job		39.50		39.50	67.50	
380	0010	FLOORING DEMOLITION										380
	0200	Brick with mortar	2 Clab	475	.034	S.F.		.49		.49	.83	
	0400	Carpet, bonded, including surface scraping		2,000	.008			.12		.12	.20	
	0480	Tackless		9,000	.002			.03		.03	.04	
	0800	Resilient, sheet goods		1,400	.011			.17		.17	.28	
	0900	Vinyl composition tile, 12″ x 12″		1,000	.016			.23		.23	.40	
	2000	Tile, ceramic, thin set		675	.024			.34		.34	.59	
	2020	Mud set		625	.026			.37		.37	.63	
	3000	Wood, block, on end	1 Carp	400	.020			.39		.39	.68	
	3200	Parquet		450	.018			.35		.35	.60	
	3400	Strip flooring, interior, 2-1/4″ x 25/32″ thick		325	.025			.48		.48	.83	
	3500	Exterior, porch flooring, 1″ x 4″		220	.036			.72		.72	1.23	
	3800	Subfloor, tongue and groove, 1″ x 6″		325	.025			.48		.48	.83	
	3820	1″ x 8″		430	.019			.37		.37	.63	
	3840	1″ x 10″		520	.015			.30		.30	.52	
	4000	Plywood, nailed		600	.013			.26		.26	.45	
	4100	Glued and nailed		400	.020			.39		.39	.68	
390	0010	FRAMING DEMOLITION										390
	3000	Wood framing, beams, 6″ x 8″	B-2	275	.145	L.F.		2.16		2.16	3.70	
	3040	6″ x 10″		220	.182			2.70		2.70	4.63	
	3080	6″ x 12″		185	.216			3.21		3.21	5.50	
	3120	8″ x 12″		140	.286			4.24		4.24	7.25	
	3160	10″ x 12″		110	.364			5.40		5.40	9.25	
	3400	Fascia boards, 1″ x 6″	1 Clab	500	.016			.23		.23	.40	
	3440	1″ x 8″		450	.018			.26		.26	.44	

2 SITE CONSTRUCTION

	02225 | Selective Demolition		CREW	DAILY OUTPUT	LABOR-HOURS	UNIT	2000 BARE COSTS MAT.	LABOR	EQUIP.	TOTAL	TOTAL INCL O&P
390	3480	1" x 10"	1 Clab	400	.020	L.F.		.29		.29	.50
	3800	Headers over openings, 2 @ 2" x 6"		110	.073			1.05		1.05	1.80
	3840	2 @ 2" x 8"		100	.080			1.16		1.16	1.98
	3880	2 @ 2" x 10"		90	.089			1.28		1.28	2.20
	4230	Joists, 2" x 6"	2 Clab	970	.016			.24		.24	.41
	4240	2" x 8"		940	.017			.25		.25	.42
	4250	2" x 10"		910	.018			.25		.25	.44
	4280	2" x 12"		880	.018			.26		.26	.45
	5400	Posts, 4" x 4"		800	.020			.29		.29	.50
	5440	6" x 6"		400	.040			.58		.58	.99
	5480	8" x 8"		300	.053			.77		.77	1.32
	5500	10" x 10"		240	.067			.96		.96	1.65
	5800	Rafters, ordinary, 2" x 6"		850	.019			.27		.27	.47
	5840	2" x 8"		837	.019			.28		.28	.47
	6200	Stairs and stringers, minimum		40	.400	Riser		5.80		5.80	9.90
	6240	Maximum		26	.615	"		8.90		8.90	15.25
	6600	Studs, 2" x 4"		2,000	.008	L.F.		.12		.12	.20
	6640	2" x 6"		1,600	.010	"		.14		.14	.25
	7000	Trusses, 2" x 4" flat wood construction									
	7050	12' span	2 Clab	74	.216	Ea.		3.12		3.12	5.35
	7150	24' span		66	.242			3.50		3.50	6
	7200	26' span		64	.250			3.61		3.61	6.20
	7250	28' span		62	.258			3.73		3.73	6.40
	7301	30' span		56	.286			4.13		4.13	7.05
	7350	32' span		56	.286			4.13		4.13	7.05
	7400	34' span		54	.296			4.28		4.28	7.35
	7450	36' span		52	.308			4.45		4.45	7.60
	9000	Minimum labor/equipment charge	1 Clab	4	2	Job		29		29	49.50
	9500	See Div. 02225-730 for rubbish handling									
400	0010	**GUTTING** Building interior, including disposal, dumpster fees not included									
	0500	Residential building									
	0560	Minimum	B-16	400	.080	SF Flr.		1.23	1.12	2.35	3.33
	0580	Maximum	"	360	.089	"		1.37	1.24	2.61	3.70
	0900	Commercial building									
	1000	Minimum	B-16	350	.091	SF Flr.		1.41	1.28	2.69	3.80
	1020	Maximum	"	250	.128	"		1.97	1.79	3.76	5.30
610	0010	**MASONRY DEMOLITION**									
	1000	Chimney, 16" x 16", soft old mortar	A-1	24	.333	V.L.F.		4.82	2.92	7.74	11.45
	1020	Hard mortar		18	.444			6.40	3.89	10.29	15.30
	1080	20" x 20", soft old mortar		12	.667			9.65	5.85	15.50	23
	1100	Hard mortar		10	.800			11.55	7	18.55	27.50
	1140	20" x 32", soft old mortar		10	.800			11.55	7	18.55	27.50
	1160	Hard mortar		8	1			14.45	8.75	23.20	34.50
	1200	48" x 48", soft old mortar		5	1.600			23	14	37	55
	1220	Hard mortar		4	2			29	17.50	46.50	69
	2000	Columns, 8" x 8", soft old mortar		48	.167			2.41	1.46	3.87	5.75
	2020	Hard mortar		40	.200			2.89	1.75	4.64	6.90
	2060	16" x 16", soft old mortar		16	.500			7.20	4.38	11.58	17.20
	2100	Hard mortar		14	.571			8.25	5	13.25	19.65
	2140	24" x 24", soft old mortar		8	1			14.45	8.75	23.20	34.50
	2160	Hard mortar		6	1.333			19.25	11.65	30.90	46
	2200	36" x 36", soft old mortar		4	2			29	17.50	46.50	69
	2220	Hard mortar		3	2.667			38.50	23.50	62	91.50
	3000	Copings, precast or masonry, to 8" wide									

	02225 | Selective Demolition		CREW	DAILY OUTPUT	LABOR-HOURS	UNIT	2000 BARE COSTS MAT.	LABOR	EQUIP.	TOTAL	TOTAL INCL O&P	
10	3020	Soft old mortar	A-1	180	.044	L.F.		.64	.39	1.03	1.53	610
	3040	Hard mortar	"	160	.050	"		.72	.44	1.16	1.72	
	3100	To 12" wide										
	3120	Soft old mortar	A-1	160	.050	L.F.		.72	.44	1.16	1.72	
	3140	Hard mortar	"	140	.057	"		.83	.50	1.33	1.96	
	4000	Fireplace, brick, 30" x 24" opening										
	4020	Soft old mortar	A-1	2	4	Ea.		58	35	93	138	
	4040	Hard mortar		1.25	6.400			92.50	56	148.50	220	
	4100	Stone, soft old mortar		1.50	5.333			77	46.50	123.50	184	
	4120	Hard mortar		1	8			116	70	186	275	
	5000	Veneers, brick, soft old mortar		140	.057	S.F.		.83	.50	1.33	1.96	
	5020	Hard mortar		125	.064			.92	.56	1.48	2.20	
	5100	Granite and marble, 2" thick		180	.044			.64	.39	1.03	1.53	
	5120	4" thick		170	.047			.68	.41	1.09	1.61	
	5140	Stone, 4" thick		180	.044			.64	.39	1.03	1.53	
	5160	8" thick		175	.046			.66	.40	1.06	1.57	
	5400	Alternate pricing method, stone, 4" thick		60	.133	C.F.		1.93	1.17	3.10	4.58	
	5420	8" thick		85	.094	"		1.36	.82	2.18	3.24	
20	0010	**MILLWORK AND TRIM DEMOLITION**										620
	1000	Cabinets, wood, base cabinets	2 Clab	80	.200	L.F.		2.89		2.89	4.95	
	1020	Wall cabinets		80	.200			2.89		2.89	4.95	
	1100	Steel, painted, base cabinets		60	.267			3.85		3.85	6.60	
	1500	Counter top, minimum		200	.080			1.16		1.16	1.98	
	1510	Maximum		120	.133			1.93		1.93	3.30	
	2000	Paneling, 4' x 8' sheets, 1/4" thick		2,000	.008	S.F.		.12		.12	.20	
	2100	Boards, 1" x 4"		700	.023			.33		.33	.57	
	2120	1" x 6"		750	.021			.31		.31	.53	
	2140	1" x 8"		800	.020			.29		.29	.50	
	3000	Trim, baseboard, to 6" wide		1,200	.013	L.F.		.19		.19	.33	
	3040	12" wide		1,000	.016			.23		.23	.40	
	3100	Ceiling trim		1,000	.016			.23		.23	.40	
	3120	Chair rail		1,200	.013			.19		.19	.33	
	3140	Railings with balusters		240	.067			.96		.96	1.65	
	3160	Wainscoting		700	.023	S.F.		.33		.33	.57	
	4000	Curtain rod	1 Clab	80	.100	L.F.		1.44		1.44	2.48	
	9000	Minimum labor/equipment charge	"	4	2	Job		29		29	49.50	
90	0010	**ROOFING AND SIDING DEMOLITION**										690
	1200	Wood, boards, tongue and groove, 2" x 6"	2 Clab	960	.017	S.F.		.24		.24	.41	
	1220	2" x 10"		1,040	.015			.22		.22	.38	
	1280	Standard planks, 1" x 6"		1,080	.015			.21		.21	.37	
	1320	1" x 8"		1,160	.014			.20		.20	.34	
	1340	1" x 12"		1,200	.013			.19		.19	.33	
	1350	Plywood, to 1" thick		2,000	.008			.12		.12	.20	
	1360	Flashing, aluminum	1 Clab	290	.028			.40		.40	.68	
	2000	Gutters, aluminum or wood, edge hung	"	240	.033	L.F.		.48		.48	.83	
	2010	Remove and reset, aluminum	1 Shee	125	.064			1.38		1.38	2.32	
	2020	Remove and reset, vinyl	1 Carp	125	.064			1.26		1.26	2.16	
	2100	Built-in	1 Clab	100	.080			1.16		1.16	1.98	
	2500	Roof accessories, plumbing vent flashing		14	.571	Ea.		8.25		8.25	14.15	
	2600	Adjustable metal chimney flashing		9	.889	"		12.85		12.85	22	
	2650	Coping, sheet metal, up to 12" wide		240	.033	L.F.		.48		.48	.83	
	2660	Concrete, up to 12" wide	2 Clab	160	.100	"		1.44		1.44	2.48	
	3000	Roofing, built-up, 5 ply roof, no gravel	B-2	1,600	.025	S.F.		.37		.37	.64	
	3100	Gravel removal, minimum		5,000	.008			.12		.12	.20	
	3120	Maximum		2,000	.020			.30		.30	.51	
	3400	Roof insulation board, up to 2" thick		3,900	.010			.15		.15	.26	

	02225 | Selective Demolition		CREW	DAILY OUTPUT	LABOR-HOURS	UNIT	2000 BARE COSTS MAT.	LABOR	EQUIP.	TOTAL	TOTAL INCL O&P
690	3450	Roll roofing, cold adhesive	1 Clab	12	.667	Sq.		9.65		9.65	16.50
	4000	Shingles, asphalt strip, 1 layer	B-2	3,500	.011	S.F.		.17		.17	.29
	4100	Slate		2,500	.016			.24		.24	.41
	4300	Wood		2,200	.018			.27		.27	.46
	4500	Skylight to 10 S.F.	1 Clab	8	1	Ea.		14.45		14.45	25
	5000	Siding, metal, horizontal		444	.018	S.F.		.26		.26	.45
	5020	Vertical		400	.020			.29		.29	.50
	5200	Wood, boards, vertical		400	.020			.29		.29	.50
	5220	Clapboards, horizontal		380	.021			.30		.30	.52
	5240	Shingles		350	.023			.33		.33	.57
	5260	Textured plywood		725	.011			.16		.16	.27
730	0010	**RUBBISH HANDLING** The following are to be added to the									
	0020	demolition prices									
	0400	Chute, circular, prefabricated steel, 18" diameter	B-1	40	.600	L.F.	19.35	9.05		28.40	37
	0440	30" diameter	"	30	.800	"	25	12.10		37.10	48
	0600	Dumpster, weekly rental, 1 dump/week, 6 C.Y. capacity (2 Tons)				Ea.					300
	0700	10 C.Y. capacity (4 Tons)									375
	0800	30 C.Y. capacity (10 Tons)									640
	0840	40 C.Y. capacity (13 Tons)									775
	1000	Dust partition, 6 mil polyethylene, 4' x 8' panels, 1" x 3" frame	2 Carp	2,000	.008	S.F.	.40	.16		.56	.71
	1080	2" x 4" frame	"	2,000	.008	"	.50	.16		.66	.82
	2000	Load, haul to chute & dumping into chute, 50' haul	2 Clab	24	.667	C.Y.		9.65		9.65	16.50
	2040	100' haul		16.50	.970			14		14	24
	2080	Over 100' haul, add per 100 L.F.		35.50	.451			6.50		6.50	11.15
	2120	In elevators, per 10 floors, add		140	.114			1.65		1.65	2.83
	3000	Loading & trucking, including 2 mile haul, chute loaded	B-16	45	.711			10.95	9.95	20.90	29.50
	3040	Hand loading truck, 50' haul	"	48	.667			10.25	9.30	19.55	27.50
	3080	Machine loading truck	B-17	120	.267			4.28	4.74	9.02	12.40
	5000	Haul, per mile, up to 8 C.Y. truck	B-34B	1,165	.007			.11	.38	.49	.61
	5100	Over 8 C.Y. truck	"	1,550	.005			.08	.29	.37	.46
740	0010	**DUMP CHARGES** Typical urban city, tipping fees only									
	0100	Building construction materials				Ton					55
	0200	Trees, brush, lumber									45
	0300	Rubbish only									50
	0500	Reclamation station, usual charge									80
840	0010	**WALLS AND PARTITIONS DEMOLITION**									
	1000	Drywall, nailed	1 Clab	1,000	.008	S.F.		.12		.12	.20
	1500	Fiberboard, nailed	"	900	.009			.13		.13	.22
	2200	Metal or wood studs, finish 2 sides, fiberboard	B-1	520	.046			.70		.70	1.20
	2250	Lath and plaster		260	.092			1.40		1.40	2.39
	2300	Plasterboard (drywall)		520	.046			.70		.70	1.20
	2350	Plywood		450	.053			.81		.81	1.38
	3000	Plaster, lime and horsehair, on wood lath	1 Clab	400	.020			.29		.29	.50
	3020	On metal lath	"	335	.024			.35		.35	.59
850	0010	**WINDOW DEMOLITION**									
	0200	Aluminum, including trim, to 12 S.F.	1 Clab	16	.500	Ea.		7.20		7.20	12.40
	0240	To 25 S.F.		11	.727			10.50		10.50	18
	0280	To 50 S.F.		5	1.600			23		23	39.50
	0320	Storm windows, to 12 S.F.		27	.296			4.28		4.28	7.35
	0360	To 25 S.F.		21	.381			5.50		5.50	9.45
	0400	To 50 S.F.		16	.500			7.20		7.20	12.40
	0500	Screens, incl. aluminum frame, small		20	.400			5.80		5.80	9.90

Important: See the Reference Section for critical supporting data - Reference Nos., Crews, & Location Factors

02200 | Site Preparation

02225 | Selective Demolition

		CREW	DAILY OUTPUT	LABOR-HOURS	UNIT	2000 BARE COSTS MAT.	LABOR	EQUIP.	TOTAL	TOTAL INCL O&P	
0510	Large	1 Clab	16	.500	Ea.		7.20		7.20	12.40	850
0600	Glass, minimum		200	.040	S.F.		.58		.58	.99	
0620	Maximum		150	.053	"		.77		.77	1.32	
2000	Wood, including trim, to 12 S.F.		22	.364	Ea.		5.25		5.25	9	
2020	To 25 S.F.		18	.444			6.40		6.40	11	
2060	To 50 S.F.		13	.615			8.90		8.90	15.25	
5020	Remove and reset window, minimum	1 Carp	6	1.333			26.50		26.50	45	
5040	Average		4	2			39.50		39.50	67.50	
5080	Maximum		2	4			79		79	135	
9100	Window awning, residential	1 Clab	80	.100	L.F.		1.44		1.44	2.48	

02230 | Site Clearing

		CREW	DAILY OUTPUT	LABOR-HOURS	UNIT	MAT.	LABOR	EQUIP.	TOTAL	TOTAL INCL O&P	
0010	**CLEAR AND GRUB** Cut & chip light, trees to 6" diam.	B-7	1	48	Acre		755	1,100	1,855	2,475	200
0150	Grub stumps and remove	B-30	2	12			209	805	1,014	1,225	
0200	Cut & chip medium, trees to 12" diam.	B-7	.70	68.571			1,075	1,550	2,625	3,550	
0250	Grub stumps and remove	B-30	1	24			420	1,600	2,020	2,475	
0300	Cut & chip heavy, trees to 24" diam.	B-7	.30	160			2,500	3,650	6,150	8,275	
0350	Grub stumps and remove	B-30	.50	48			835	3,225	4,060	4,925	
0400	If burning is allowed, reduce cut & chip									40%	
0010	**CLEARING** Brush with brush saw	A-1	.25	32	Acre		460	280	740	1,100	220
0100	By hand	"	.12	66.667			965	585	1,550	2,300	
0300	With dozer, ball and chain, light clearing	B-11A	2	8			137	405	542	675	
0400	Medium clearing	"	1.50	10.667			183	540	723	900	

02300 | Earthwork

02305 | Equipment

		CREW	DAILY OUTPUT	LABOR-HOURS	UNIT	2000 BARE COSTS MAT.	LABOR	EQUIP.	TOTAL	TOTAL INCL O&P	
0010	**MOBILIZATION OR DEMOBILIZATION** Up to 50 miles										250
0020	Dozer, loader, backhoe or excavator, 70 H.P.- 250 H.P.	B-34K	6	1.333	Ea.		21.50	153	174.50	204	
0900	Shovel or dragline, 3/4 C.Y.	"	3.60	2.222			36	254	290	340	
1100	Delivery charge for small equipment on flatbed trailer, minimum									40	
1150	Maximum									100	

02310 | Grading

		CREW	DAILY OUTPUT	LABOR-HOURS	UNIT	MAT.	LABOR	EQUIP.	TOTAL	TOTAL INCL O&P	
0010	**FINE GRADE** Area to be paved with grader, small area	B-11L	400	.040	S.Y.		.69	1.29	1.98	2.58	440
0100	Large area		2,000	.008			.14	.26	.40	.51	
0200	Grade subgrade for base course, roadways		3,500	.005			.08	.15	.23	.29	
0300	Fine grade, base course for paving, see div. 02720-200										
1020	For large parking lots	B-32C	5,000	.010	S.Y.		.17	.31	.48	.62	
1050	For small irregular areas	"	2,000	.024			.42	.77	1.19	1.56	
1100	Fine grade for slab on grade, machine	B-11L	1,040	.015			.26	.50	.76	.99	
1150	Hand grading	B-18	700	.034			.52	.09	.61	.99	
1200	Fine grade granular base for sidewalks and bikeways	B-62	1,200	.020			.32	.10	.42	.65	
2550	Hand grade select gravel	2 Clab	60	.267	C.S.F.		3.85		3.85	6.60	
3000	Hand grade select gravel, including compaction, 4" deep	B-18	555	.043	S.Y.		.65	.11	.76	1.24	
3100	6" deep		400	.060			.91	.15	1.06	1.72	
3120	8" deep		300	.080			1.21	.20	1.41	2.29	
3300	Finishing grading slopes, gentle	B-11L	8,900	.002			.03	.06	.09	.11	

02300 | Earthwork

	02310 | Grading		CREW	DAILY OUTPUT	LABOR-HOURS	UNIT	2000 BARE COSTS MAT.	LABOR	EQUIP.	TOTAL	TOTAL INCL O&P
440	3310	Steep slopes	B-11L	7,100	.002	S.Y.		.04	.07	.11	.15
460	0010	**LOAM OR TOPSOIL** Remove and stockpile on site									
	0020	6″ deep, 200′ haul	B-10B	865	.009	C.Y.		.18	.93	1.11	1.34
	0100	300′ haul		520	.015			.31	1.55	1.86	2.22
	0150	500′ haul		225	.036			.71	3.59	4.30	5.10
	0200	Alternate method: 6″ deep, 200′ haul		5,090	.002	S.Y.		.03	.16	.19	.22
	0250	500′ haul		1,325	.006	″		.12	.61	.73	.87
	0701	Furnish and place, truck dumped, unscreened, 4″ deep	B-10S	12,000	.001	S.F.	.22	.01	.03	.26	.29
	0801	6″ deep	″	7,400	.001	″	.31	.02	.04	.37	.43
	0900	Fine grading and seeding, incl. lime, fertilizer & seed,									
	1001	With equipment	B-14	9,000	.005	S.F.	.05	.08	.02	.15	.21
		02315 | Excavation and Fill									
100	0010	**BACKFILL** By hand, no compaction, light soil	1 Clab	14	.571	C.Y.		8.25		8.25	14.15
	0100	Heavy soil		11	.727			10.50		10.50	18
	0300	Compaction in 6″ layers, hand tamp, add to above		20.60	.388			5.60		5.60	9.60
	0500	Air tamp, add	B-9C	190	.211			3.13	.95	4.08	6.40
	0600	Vibrating plate, add	A-1	60	.133			1.93	1.17	3.10	4.58
	0800	Compaction in 12″ layers, hand tamp, add to above	1 Clab	34	.235			3.40		3.40	5.80
	1300	Dozer backfilling, bulk, up to 300′ haul, no compaction	B-10B	1,200	.007			.13	.67	.80	.96
	1400	Air tamped	B-11B	240	.067			1.14	4.21	5.35	6.55
130	0010	**BEDDING** For pipe and conduit, not incl. compaction									
	0050	Crushed or screened bank run gravel	B-6	150	.160	C.Y.	6.50	2.56	1.35	10.41	12.90
	0100	Crushed stone 3/4″ to 1/2″		150	.160		17.05	2.56	1.35	20.96	24.50
	0200	Sand, dead or bank		150	.160		3.50	2.56	1.35	7.41	9.65
	0500	Compacting bedding in trench	A-1	90	.089			1.28	.78	2.06	3.06
320	0010	**COMPACTION, STRUCTURAL** Steel wheel tandem roller, 5 tons	B-10E	8	1	Hr.		19.90	16.65	36.55	51.50
	0050	Air tamp, 6″ to 8″ lifts, common fill	B-9	250	.160	C.Y.		2.38	.72	3.10	4.86
	0060	Select fill	″	300	.133			1.98	.60	2.58	4.05
	0600	Vibratory plate, 8″ lifts, common fill	A-1	200	.040			.58	.35	.93	1.38
	0700	Select fill	″	216	.037			.54	.32	.86	1.28
400	0010	**EXCAVATING, BULK BANK MEASURE** Common earth piled									
	0020	For loading onto trucks, add								15%	15%
	0200	Backhoe, hydraulic, crawler mtd., 1 C.Y. cap. = 75 C.Y./hr.	B-12A	600	.013	C.Y.		.28	.90	1.18	1.45
	0310	Wheel mounted, 1/2 C.Y. cap. = 30 C.Y./hr.	B-12E	240	.033			.69	1.43	2.12	2.71
	1200	Front end loader, track mtd., 1-1/2 C.Y. cap. = 70 C.Y./hr.	B-10N	560	.014			.28	.59	.87	1.12
	1500	Wheel mounted, 3/4 C.Y. cap. = 45 C.Y./hr.	B-10R	360	.022			.44	.65	1.09	1.44
	8000	For hauling excavated material, see div. 02320-200									
440	0010	**EXCAVATING, STRUCTURAL** Hand, pits to 6′ deep, sandy soil	1 Clab	8	1	C.Y.		14.45		14.45	25
	0100	Heavy soil or clay		4	2			29		29	49.50
	1100	Hand loading trucks from stock pile, sandy soil		12	.667			9.65		9.65	16.50
	1300	Heavy soil or clay		8	1			14.45		14.45	25
	1500	For wet or muck hand excavation, add to above				%				50%	50%
505	0010	**FILL** Spread dumped material, by dozer, no compaction	B-10B	1,000	.008	C.Y.		.16	.81	.97	1.15
	0100	By hand	1 Clab	12	.667	″		9.65		9.65	16.50
	0500	Gravel fill, compacted, under floor slabs, 4″ deep	B-37	10,000	.005	S.F.	.15	.07	.01	.23	.31
	0600	6″ deep		8,600	.006		.23	.09	.02	.34	.42
	0700	9″ deep		7,200	.007		.38	.10	.02	.50	.61
	0800	12″ deep		6,000	.008		.52	.12	.02	.66	.81
	1000	Alternate pricing method, 4″ deep		120	.400	C.Y.	11.25	6.20	1.11	18.56	24
	1100	6″ deep		160	.300		11.25	4.67	.83	16.75	21.50

02300 | Earthwork

02315 | Excavation and Fill

Line	Description	Crew	Daily Output	Labor-Hours	Unit	2000 Bare Costs Mat.	Labor	Equip.	Total	Total Incl O&P	
1200	9″ deep	B-37	200	.240	C.Y.	11.25	3.73	.67	15.65	19.50	505
1300	12″ deep		220	.218		11.25	3.39	.61	15.25	18.85	
0010	**EXCAVATING, TRENCH** or continuous footing, common earth										900
0050	1′ to 4′ deep, 3/8 C.Y. tractor loader/backhoe	B-11C	150	.107	C.Y.		1.83	1.35	3.18	4.57	
0060	1/2 C.Y. tractor loader/backhoe	B-11M	200	.080			1.37	1.37	2.74	3.81	
0090	4′ to 6′ deep, 1/2 C.Y. tractor loader/backhoe	″	200	.080			1.37	1.37	2.74	3.81	
0100	5/8 C.Y. hydraulic backhoe	B-12Q	250	.032			.66	1.65	2.31	2.91	
0300	1/2 C.Y. hydraulic excavator, truck mounted	B-12J	200	.040			.83	3.09	3.92	4.77	
1400	By hand with pick and shovel 2′ to 6′ deep, light soil	1 Clab	8	1			14.45		14.45	25	
1500	Heavy soil	″	4	2			29		29	49.50	
0010	**EXCAVATING, UTILITY TRENCH** Common earth										940
0050	Trenching with chain trencher, 12 H.P., operator walking										
0100	4″ wide trench, 12″ deep	B-53	800	.010	L.F.		.14	.13	.27	.39	
1000	Backfill by hand including compaction, add										
1050	4″ wide trench, 12″ deep	A-1	800	.010	L.F.		.14	.09	.23	.35	

02320 | Hauling

Line	Description	Crew	Daily Output	Labor-Hours	Unit	Mat.	Labor	Equip.	Total	Total Incl O&P	
0011	**HAULING** Excavated or borrow material, loose cubic yards										200
0015	no loading included, highway haulers										
0020	6 C.Y. dump truck, 1/4 mile round trip, 5.0 loads/hr.	B-34A	195	.041	C.Y.		.66	1.87	2.53	3.17	
0200	4 mile round trip, 1.8 loads/hr.	″	70	.114			1.85	5.20	7.05	8.85	
0310	12 C.Y. dump truck, 1/4 mile round trip 3.7 loads/hr.	B-34B	288	.028			.45	1.55	2	2.46	
0500	4 mile round trip, 1.6 loads/hr.	″	125	.064			1.04	3.58	4.62	5.65	

02360 | Soil Treatment

Line	Description	Crew	Daily Output	Labor-Hours	Unit	Mat.	Labor	Equip.	Total	Total Incl O&P	
0010	**TERMITE PRETREATMENT**										800
0020	Slab and walls, residential	1 Skwk	1,200	.007	SF Flr.	.22	.13		.35	.47	
0400	Insecticides for termite control, minimum		14.20	.563	Gal.	10	11.15		21.15	30	
0500	Maximum		11	.727	″	17.10	14.35		31.45	43.50	

02370 | Erosion & Sedimentation Control

Line	Description	Crew	Daily Output	Labor-Hours	Unit	Mat.	Labor	Equip.	Total	Total Incl O&P	
0010	**EROSION CONTROL** Jute mesh, 100 S.Y. per roll, 4′ wide, stapled	B-80A	2,400	.010	S.Y.	.62	.14	.08	.84	1.01	550
0100	Plastic netting, stapled, 2″ x 1″ mesh, 20 mil	B-1	2,500	.010		.40	.15		.55	.69	
0200	Polypropylene mesh, stapled, 6.5 oz./S.Y.		2,500	.010		1	.15		1.15	1.35	
0300	Tobacco netting, or jute mesh #2, stapled		2,500	.010		.07	.15		.22	.33	
1000	Silt fence, polypropylene, 3′ high, ideal conditions	2 Clab	1,600	.010	L.F.	.30	.14		.44	.58	
1100	Adverse conditions	″	950	.017	″	.30	.24		.54	.75	

02400 | Tunneling, Boring & Jacking

02441 | Microtunneling

Line	Description	Crew	Daily Output	Labor-Hours	Unit	2000 Bare Costs Mat.	Labor	Equip.	Total	Total Incl O&P	
0010	**MICROTUNNELING** Not including excavation, backfill, shoring,										400
0020	or dewatering, average 50′/day, slurry method										
0100	24″ to 48″ outside diameter, minimum				L.F.					600	
0110	Adverse conditions, add				%					50%	
1000	Rent microtunneling machine, average monthly lease				Month					80,000	
1010	Operating technician				Day					600	
1100	Mobilization and demobilization, minimum				Job					40,000	
1110	Maximum				″					400,000	

		02510 | Water Distribution	CREW	DAILY OUTPUT	LABOR-HOURS	UNIT	2000 BARE COSTS MAT.	LABOR	EQUIP.	TOTAL	TOTAL INCL O&P
800	0010	**PIPING, WATER DISTRIBUTION SYSTEMS** Pipe laid in trench, (R15100-050)									
	0020	excavation and backfill not included									
	1400	Ductile Iron, cement lined, class 50 water pipe, 18′ lengths									
	1410	Mechanical joint, 4″ diameter	B-20	144	.167	L.F.	7.15	2.52		9.67	12.20
	2650	Polyvinyl chloride pipe, class 160, S.D.R.-26, 1-1/2″ diameter		300	.080		.27	1.21		1.48	2.37
	2700	2″ diameter		250	.096		.38	1.45		1.83	2.91
	2750	2-1/2″ diameter		250	.096		.50	1.45		1.95	3.04
	2800	3″ diameter		200	.120		.75	1.81		2.56	3.94
	2850	4″ diameter		200	.120		1.18	1.81		2.99	4.41
		02520 | Wells									
900	0010	**WELLS** Domestic water									
	0100	Drilled, 4″ to 6″ diameter	B-23	120	.333	L.F.		4.95	16.45	21.40	26.50
	1500	Pumps, installed in wells to 100′ deep, 4″ submersible									
	1520	3/4 H.P.	Q-1	2.66	6.015	Ea.	475	119		594	720
	1600	1 H.P.	″	2.29	6.987	″	525	138		663	810
		02530 | Sanitary Sewerage									
730	0010	**PIPING, DRAINAGE & SEWAGE, CONCRETE**									
	0020	Not including excavation or backfill									
	1020	8″ diameter	B-14	224	.214	L.F.	3.82	3.33	.91	8.06	10.85
	1030	10″ diameter	″	216	.222	″	4.23	3.46	.94	8.63	11.60
	3780	Concrete slotted pipe, class 4 mortar joint									
	3800	12″ diameter	B-21	168	.167	L.F.	10.80	2.65	.82	14.27	17.35
	3840	18″ diameter	″	152	.184	″	16.75	2.93	.91	20.59	24.50
	3900	Class 4 O-ring									
	3940	12″ diameter	B-21	168	.167	L.F.	12.35	2.65	.82	15.82	19.05
	3960	18″ diameter	″	152	.184	″	18.55	2.93	.91	22.39	26.50
780	0010	**PIPING, DRAINAGE & SEWAGE, POLYVINYL CHLORIDE**									
	0020	Not including excavation or backfill									
	2000	10′ lengths, S.D.R. 35, B&S, 4″ diameter	B-20	375	.064	L.F.	2.21	.97		3.18	4.09
	2040	6″ diameter		350	.069		2.87	1.04		3.91	4.94
	2080	8″ diameter		335	.072		4.40	1.08		5.48	6.70
	2120	10″ diameter	B-21	330	.085		4.84	1.35	.42	6.61	8.05
790	0010	**PIPING, DRAINAGE & SEWAGE, VITRIFIED CLAY** C700									
	0020	Not including excavation or backfill,									
	4030	Extra strength, compression joints, C425									
	5000	4″ diameter x 4′ long	B-20	265	.091	L.F.	1.79	1.37		3.16	4.32
	5020	6″ diameter x 5′ long	″	200	.120		2.93	1.81		4.74	6.35
	5040	8″ diameter x 5′ long	B-21	200	.140		4.14	2.23	.69	7.06	9.10
	5060	10″ diameter x 5′ long	″	190	.147		6.80	2.34	.73	9.87	12.25
		02540 | Septic Tank Systems									
700	0010	**SEPTIC TANKS** Not incl. excav. or piping, precast, 1,000 gallon	B-21	8	3.500	Ea.	490	55.50	17.30	562.80	650
	0100	2,000 gallon		5	5.600		1,000	89	27.50	1,116.50	1,275
	0600	High density polyethylene, 1,000 gallon		6	4.667		800	74.50	23	897.50	1,025
	0700	1,500 gallon		4	7		1,000	111	34.50	1,145.50	1,325
	1000	Distribution boxes, concrete, 7 outlets	2 Clab	16	1		86.50	14.45		100.95	120
	1100	9 outlets	″	8	2		225	29		254	298
	1150	Leaching field chambers, 13′ x 3′-7″ x 1′-4″, standard	B-13	16	3		665	47.50	35	747.50	850
	1420	Leaching pit, 6′, dia, 3′ deep complete					500			500	550
	2200	Excavation for septic tank, 3/4 C.Y. backhoe	B-12F	145	.055	C.Y.		1.14	3.14	4.28	5.35
	2400	4′ trench for disposal field, 3/4 C.Y. backhoe	″	335	.024	L.F.		.49	1.36	1.85	2.32

02500 | Utility Services

	02540 Septic Tank Systems		CREW	DAILY OUTPUT	LABOR-HOURS	UNIT	2000 BARE COSTS MAT.	LABOR	EQUIP.	TOTAL	TOTAL INCL O&P	
00	2600	Gravel fill, run of bank	B-6	150	.160	C.Y.	5.20	2.56	1.35	9.11	11.50	700
	2800	Crushed stone, 3/4"	"	150	.160	"	21	2.56	1.35	24.91	29.50	

	02550 Piped Energy Distribution		CREW	DAILY OUTPUT	LABOR-HOURS	UNIT	MAT.	LABOR	EQUIP.	TOTAL	TOTAL INCL O&P	
64	0010	PIPING, GAS SERVICE & DISTRIBUTION, POLYETHYLENE										464
	0020	not including excavation or backfill										
	1000	60 psi coils, comp cplg @ 100', 1/2" diameter, SDR 9.3	B-20A	608	.053	L.F.	.36	.93		1.29	1.96	
	1040	1-1/4" diameter, SDR 11		544	.059		.59	1.04		1.63	2.38	
	1100	2" diameter, SDR 11		488	.066		.74	1.15		1.89	2.75	
	1160	3" diameter, SDR 11		408	.078		1.55	1.38		2.93	4.02	
	1500	60 PSI 40' joints with coupling, 3" diameter, SDR 11	B-21A	408	.098		1.55	1.79	.98	4.32	5.75	
	1540	4" diameter, SDR 11		352	.114		3.44	2.07	1.14	6.65	8.50	
	1600	6" diameter, SDR 11		328	.122		11.25	2.22	1.22	14.69	17.45	
	1640	8" diameter, SDR 11		272	.147		15	2.68	1.48	19.16	22.50	
56	0010	PIPING, GAS SERVICE & DISTRIBUTION, STEEL										466
	0020	not including excavation or backfill, tar coated and wrapped										
	4000	Schedule 40, plain end										
	4040	1" diameter	Q-4	300	.107	L.F.	2.39	2.22	.16	4.77	6.50	
	4080	2" diameter	"	280	.114	"	3.75	2.38	.17	6.30	8.25	

02600 | Drainage & Containment

	02620 Subdrainage		CREW	DAILY OUTPUT	LABOR-HOURS	UNIT	2000 BARE COSTS MAT.	LABOR	EQUIP.	TOTAL	TOTAL INCL O&P	
10	0010	PIPING, SUBDRAINAGE, CONCRETE										210
	0021	Not including excavation and backfill										
	3000	Porous wall concrete underdrain, std. strength, 4" diameter	B-20	335	.072	L.F.	1.78	1.08		2.86	3.82	
	3020	6" diameter	"	315	.076		2.31	1.15		3.46	4.52	
	3040	8" diameter	B-21	310	.090		2.86	1.44	.45	4.75	6.10	
40	0010	PIPING, SUBDRAINAGE, CORRUGATED METAL										240
	0021	Not including excavation and backfill										
	2010	Aluminum, perforated										
	2020	6" diameter, 18 ga.	B-14	380	.126	L.F.	2.57	1.97	.53	5.07	6.75	
	2200	8" diameter, 16 ga.		370	.130		3.73	2.02	.55	6.30	8.15	
	2220	10" diameter, 16 ga.		360	.133		4.66	2.07	.56	7.29	9.30	
	3000	Uncoated galvanized, perforated										
	3020	6" diameter, 18 ga.	B-20	380	.063	L.F.	4	.95		4.95	6.05	
	3200	8" diameter, 16 ga.	"	370	.065		5.50	.98		6.48	7.75	
	3220	10" diameter, 16 ga.	B-21	360	.078		8.25	1.24	.38	9.87	11.60	
	3240	12" diameter, 16 ga.	"	285	.098		8.65	1.56	.49	10.70	12.70	
	4000	Steel, perforated, asphalt coated										
	4020	6" diameter 18 ga.	B-20	380	.063	L.F.	3.20	.95		4.15	5.15	
	4030	8" diameter 18 ga	"	370	.065		5	.98		5.98	7.20	
	4040	10" diameter 16 ga	B-21	360	.078		5.75	1.24	.38	7.37	8.90	
	4050	12" diameter 16 ga		285	.098		6.60	1.56	.49	8.65	10.45	
	4060	18" diameter 16 ga		205	.137		9	2.17	.67	11.84	14.35	
30	0010	PIPING, SUBDRAINAGE, VITRIFIED CLAY										280
	4000	Channel pipe, 4" diameter	B-20	430	.056	L.F.	2	.84		2.84	3.65	

02600 | Drainage & Containment

	02620 | Subdrainage		CREW	DAILY OUTPUT	LABOR-HOURS	UNIT	2000 BARE COSTS MAT.	LABOR	EQUIP.	TOTAL	TOTAL INCL O&P	
280	4060	8″ diameter	B-20	295	.081	L.F.	4.50	1.23		5.73	7.05	280

	02630 | Storm Drainage		CREW	DAILY OUTPUT	LABOR-HOURS	UNIT	MAT.	LABOR	EQUIP.	TOTAL	TOTAL INCL O&P	
200	0010	**CATCH BASINS OR MANHOLES** not including footing, excavation,										200
	0020	backfill, frame and cover										
	0050	Brick, 4′ inside diameter, 4′ deep	D-1	1	16	Ea.	264	286		550	775	
	1110	Precast, 4′ I.D., 4′ deep	B-22	4.10	7.317		315	119	50.50	484.50	605	
	1600	Frames & covers, C.I., 24″ square, 500 lb.	B-6	7.80	3.077		213	49	26	288	345	

02700 | Bases, Ballasts, Pavements & Appurtenances

	02720 | Unbound Base Courses & Ballasts		CREW	DAILY OUTPUT	LABOR-HOURS	UNIT	2000 BARE COSTS MAT.	LABOR	EQUIP.	TOTAL	TOTAL INCL O&P	
200	0010	**BASE COURSE** For roadways and large paved areas										200
	0051	3/4″ stone compacted to 3″ deep	B-36	36,000	.001	S.F.	.47	.02	.03	.52	.58	
	0101	6″ deep		35,100	.001		.95	.02	.03	1	1.10	
	0201	9″ deep		25,875	.002		1.42	.03	.04	1.49	1.65	
	0305	12″ deep		21,150	.002		1.51	.03	.05	1.59	1.77	
	0306	Crushed 1-1/2″ stone base, compacted to 4″ deep		47,000	.001		.34	.01	.02	.37	.43	
	0307	6″ deep		35,100	.001		.75	.02	.03	.80	.89	
	0308	8″ deep		27,000	.001		1.01	.03	.04	1.08	1.20	
	0309	12″ deep		16,200	.002		1.51	.04	.07	1.62	1.81	
	0350	Bank run gravel, spread and compacted										
	0371	6″ deep	B-32	54,000	.001	S.F.	.29	.01	.03	.33	.37	
	0391	9″ deep		39,600	.001		.43	.01	.04	.48	.54	
	0401	12″ deep		32,400	.001		.58	.02	.05	.65	.71	
	8900	For small and irregular areas, add						50%	50%			
215	0011	**BASE** Prepare and roll sub-base, small areas to 2500 S.Y.	B-32A	13,500	.002	S.F.		.03	.07	.10	.13	215

	02740 | Flexible Pavement		CREW	DAILY OUTPUT	LABOR-HOURS	UNIT	MAT.	LABOR	EQUIP.	TOTAL	TOTAL INCL O&P	
315	0010	**PAVING** Asphaltic concrete										315
	0020	6″ stone base, 2″ binder course, 1″ topping (R02065-300)	B-25C	9,000	.005	S.F.	1.07	.09	.17	1.33	1.52	
	0300	Binder course, 1-1/2″ thick		35,000	.001		.27	.02	.04	.33	.39	
	0400	2″ thick		25,000	.002		.35	.03	.06	.44	.51	
	0500	3″ thick		15,000	.003		.54	.05	.10	.69	.80	
	0600	4″ thick		10,800	.004		.71	.07	.14	.92	1.06	
	0800	Sand finish course, 3/4″ thick		41,000	.001		.15	.02	.04	.21	.24	
	0900	1″ thick		34,000	.001		.19	.02	.05	.26	.30	
	1000	Fill pot holes, hot mix, 2″ thick	B-16	4,200	.008		.39	.12	.11	.62	.74	
	1100	4″ thick		3,500	.009		.56	.14	.13	.83	1	
	1120	6″ thick		3,100	.010		.76	.16	.14	1.06	1.26	
	1140	Cold patch, 2″ thick	B-51	3,000	.016		.45	.24	.06	.75	.97	
	1160	4″ thick		2,700	.018		.86	.27	.06	1.19	1.47	
	1180	6″ thick		1,900	.025		1.33	.38	.09	1.80	2.22	

	02750 | Rigid Pavement		CREW	DAILY OUTPUT	LABOR-HOURS	UNIT	MAT.	LABOR	EQUIP.	TOTAL	TOTAL INCL O&P	
100	0010	**CONCRETE PAVEMENT** Including joints, finishing, and curing										100
	0021	Fixed form, 12′ pass, unreinforced, 6″ thick	B-26	18,000	.005	S.F.	1.99	.08	.10	2.17	2.44	

02700 | Bases, Ballasts, Pavements & Appurtenances

	02750 | Rigid Pavement	CREW	DAILY OUTPUT	LABOR-HOURS	UNIT	2000 BARE COSTS MAT.	LABOR	EQUIP.	TOTAL	TOTAL INCL O&P	
0101	8″ thick	B-26	13,500	.007	S.F.	2.68	.11	.14	2.93	3.29	100
0701	Finishing, broom finish small areas	2 Cefi	1,215	.013			.25		.25	.41	
	02770 | Curbs and Gutters										
0010	**CURBS** Asphaltic, machine formed, 8″ wide, 6″ high, 40 L.F./ton	B-27	1,000	.032	L.F.	.54	.48	.07	1.09	1.50	225
0150	Asphaltic berm, 12″ W, 3″-6″ H, 35 L.F./ton, before pavement	″	700	.046		.80	.68	.10	1.58	2.17	
0200	12″ W, 1-1/2″ to 4″ H, 60 L.F. per ton, laid with pavement	B-2	1,050	.038		.49	.57		1.06	1.51	
0300	Concrete, wood forms, 6″ x 18″, straight	C-2A	500	.096		2.15	1.83		3.98	5.45	
0400	6″ x 18″, radius	″	200	.240		2.26	4.56		6.82	10.25	
0550	Precast, 6″ x 18″, straight	B-29	700	.069		6.25	1.08	.88	8.21	9.70	
0600	6″ x 18″, radius	″	325	.148		7.75	2.34	1.90	11.99	14.55	
1000	Granite, split face, straight, 5″ x 16″	D-13	500	.096		14.10	1.80	.80	16.70	19.40	
1100	6″ x 18″	″	450	.107		18.55	2	.89	21.44	25	
1300	Radius curbing, 6″ x 18″, over 10′ radius	B-29	260	.185		22.50	2.92	2.38	27.80	32.50	
1400	Corners, 2′ radius		80	.600	Ea.	76	9.50	7.75	93.25	109	
1600	Edging, 4-1/2″ x 12″, straight		300	.160	L.F.	7.05	2.53	2.06	11.64	14.35	
1800	Curb inlets, (guttermouth) straight		41	1.171	Ea.	169	18.50	15.10	202.60	234	
2000	Indian granite (belgian block)										
2100	Jumbo, 10-1/2″ x 7-1/2″ x 4″, grey	D-1	150	.107	L.F.	1.75	1.90		3.65	5.15	
2150	Pink		150	.107		2.15	1.90		4.05	5.60	
2200	Regular, 9″ x 4-1/2″ x 4-1/2″, grey		160	.100		1.70	1.79		3.49	4.89	
2250	Pink		160	.100		2	1.79		3.79	5.20	
2300	Cubes, 4″ x 4″ x 4″, grey		175	.091		1.65	1.63		3.28	4.58	
2350	Pink		175	.091		1.75	1.63		3.38	4.69	
2400	6″ x 6″ x 6″, pink		155	.103		3.60	1.84		5.44	7.10	
2500	Alternate pricing method for indian granite										
2550	Jumbo, 10-1/2″ x 7-1/2″ x 4″ (30lb), grey				Ton	100			100	110	
2600	Pink					125			125	138	
2650	Regular, 9″ x 4-1/2″ x 4-1/2″ (20lb), grey					120			120	132	
2700	Pink					140			140	154	
2750	Cubes, 4″ x 4″ x 4″ (5lb), grey					200			200	220	
2800	Pink					225			225	248	
2850	6″ x 6″ x 6″ (25lb), pink					140			140	154	
2900	For pallets, add					15			15	16.50	
	02775 | Sidewalks										
0010	**SIDEWALKS, DRIVEWAYS, & PATIOS** No base										275
0021	Asphaltic concrete, 2″ thick	B-37	6,480	.007	S.F.	.35	.12	.02	.49	.61	
0101	2-1/2″ thick	″	5,950	.008	″	.44	.13	.02	.59	.72	
0300	Concrete, 3000 psi, CIP, 6 x 6 - W1.4 x W1.4 mesh,										
0310	broomed finish, no base, 4″ thick	B-24	600	.040	S.F.	1.05	.71		1.76	2.34	
0350	5″ thick		545	.044		1.40	.78		2.18	2.85	
0400	6″ thick		510	.047		1.63	.83		2.46	3.19	
0450	For bank run gravel base, 4″ thick, add	B-18	2,500	.010		.13	.15	.02	.30	.43	
0520	8″ thick, add	″	1,600	.015		.27	.23	.04	.54	.72	
1000	Crushed stone, 1″ thick, white marble	2 Clab	1,700	.009		.23	.14		.37	.48	
1050	Bluestone	″	1,700	.009		.18	.14		.32	.43	
1700	Redwood, prefabricated, 4′ x 4′ sections	2 Carp	316	.051		7.15	1		8.15	9.55	
1750	Redwood planks, 1″ thick, on sleepers	″	240	.067		4.99	1.31		6.30	7.75	
	02778 | Steps										
0010	**STEPS** Incl. excav., borrow & concrete base, where applicable										280
0100	Brick steps	B-24	35	.686	LF Riser	7.60	12.10		19.70	29	
0200	Railroad ties	2 Clab	25	.640		2.68	9.25		11.93	18.80	
0300	Bluestone treads, 12″ x 2″ or 12″ x 1-1/2″	B-24	30	.800		19.30	14.15		33.45	45	

2 SITE CONSTRUCTION

02700 | Bases, Ballasts, Pavements & Appurtenances

	02778 Steps		CREW	DAILY OUTPUT	LABOR-HOURS	UNIT	2000 BARE COSTS MAT.	LABOR	EQUIP.	TOTAL	TOTAL INCL O&P
280	0600	Precast concrete, see division 03480-800									

	02780 Unit Pavers		CREW	DAILY OUTPUT	LABOR-HOURS	UNIT	2000 BARE COSTS MAT.	LABOR	EQUIP.	TOTAL	TOTAL INCL O&P
100	0010	**ASPHALT BLOCKS**, 6"x12"x1-1/4", w/bed & neopr. adhesive	D-1	135	.119	S.F.	2.84	2.12		4.96	6.70
	0100	3" thick		130	.123		3.98	2.20		6.18	8.10
	0300	Hexagonal tile, 8" wide, 1-1/4" thick		135	.119		2.91	2.12		5.03	6.80
	0400	2" thick		130	.123		4.07	2.20		6.27	8.20
	0500	Square, 8" x 8", 1-1/4" thick		135	.119		2.89	2.12		5.01	6.75
	0600	2" thick		130	.123		4.05	2.20		6.25	8.15
200	0010	**BRICK PAVING** 4" x 8" x 1-1/2", without joints (4.5 brick/S.F.)	D-1	110	.145	S.F.	2.04	2.60		4.64	6.65
	0100	Grouted, 3/8" joint (3.9 brick/S.F.)		90	.178		2.39	3.17		5.56	8
	0200	4" x 8" x 2-1/4", without joints (4.5 bricks/S.F.)		110	.145		2.64	2.60		5.24	7.30
	0300	Grouted, 3/8" joint (3.9 brick/S.F.)		90	.178		2.44	3.17		5.61	8.05
	0500	Bedding, asphalt, 3/4" thick	B-25	5,130	.017		.30	.28	.35	.93	1.18
	0540	Course washed sand bed, 1" thick	B-18	5,000	.005		.15	.07	.01	.23	.29
	0580	Mortar, 1" thick	D-1	300	.053		.36	.95		1.31	2
	0620	2" thick		200	.080		.36	1.43		1.79	2.81
	1500	Brick on 1" thick sand bed laid flat, 4.5 per S.F.		100	.160		3.02	2.86		5.88	8.15
	2000	Brick pavers, laid on edge, 7.2 per S.F.		70	.229		1.99	4.08		6.07	9.10
800	0010	**STONE PAVERS**									
	1100	Flagging, bluestone, irregular, 1" thick,	D-1	81	.198	S.F.	4	3.53		7.53	10.35
	1150	Snapped random rectangular, 1" thick		92	.174		6.05	3.10		9.15	11.90
	1200	1-1/2" thick		85	.188		7.30	3.36		10.66	13.70
	1250	2" thick		83	.193		8.50	3.44		11.94	15.20
	1300	Slate, natural cleft, irregular, 3/4" thick		92	.174		1.80	3.10		4.90	7.25
	1310	1" thick		85	.188		2.10	3.36		5.46	8
	1351	Random rectangular, gauged, 1/2" thick		105	.152		3.90	2.72		6.62	8.90
	1400	Random rectangular, butt joint, gauged, 1/4" thick		150	.107		4.19	1.90		6.09	7.85
	1450	For sand rubbed finish, add					2.50			2.50	2.75
	1500	For interior setting, add								25%	25%
	1550	Granite blocks, 3-1/2" x 3-1/2" x 3-1/2"	D-1	92	.174	S.F.	5.25	3.10		8.35	11.05

	02785 Flexible Pavement Coating		CREW	DAILY OUTPUT	LABOR-HOURS	UNIT	2000 BARE COSTS MAT.	LABOR	EQUIP.	TOTAL	TOTAL INCL O&P
800	0010	**SEALCOATING** 2 coat coal tar pitch emulsion over 10,000 S.Y.	B-45	5,000	.003	S.Y.	.43	.05	.07	.55	.63
	0030	1000 to 10,000 S.Y.	"	3,000	.005		.43	.08	.12	.63	.75
	0100	Under 1000 S.Y.	B-1	1,050	.023		.43	.35		.78	1.06
	0300	Petroleum resistant, over 10,000 S.Y.	B-45	5,000	.003		.50	.05	.07	.62	.71
	0320	1000 to 10,000 S.Y.	"	3,000	.005		.50	.08	.12	.70	.83
	0400	Under 1000 S.Y.	B-1	1,050	.023		.50	.35		.85	1.14

02800 | Site Improvements and Amenities

	02810 Irrigation System		CREW	DAILY OUTPUT	LABOR-HOURS	UNIT	2000 BARE COSTS MAT.	LABOR	EQUIP.	TOTAL	TOTAL INCL O&P
800	0010	**SPRINKLER IRRIGATION SYSTEM** For lawns									
	0800	Residential system, custom, 1" supply	B-20	2,000	.012	S.F.	.25	.18		.43	.59
	0900	1-1/2" supply	"	1,800	.013	"	.28	.20		.48	.66

02820 | Fences & Gates		CREW	DAILY OUTPUT	LABOR-HOURS	UNIT	2000 BARE COSTS				TOTAL	
						MAT.	LABOR	EQUIP.	TOTAL	INCL O&P	
0010	**FENCE, MISC. METAL** Chicken wire, posts @ 4′, 1″ mesh, 4′ high	B-80	410	.059	L.F.	1.10	.89	1.34	3.33	4.21	500
0100	2″ mesh, 6′ high		350	.069		1	1.04	1.57	3.61	4.61	
0200	Galv. steel, 12 ga., 2″ x 4″ mesh, posts 5′ O.C., 3′ high		300	.080		1.50	1.21	1.84	4.55	5.75	
0300	5′ high		300	.080		2	1.21	1.84	5.05	6.30	
0400	14 ga., 1″ x 2″ mesh, 3′ high		300	.080		1.60	1.21	1.84	4.65	5.85	
0500	5′ high		300	.080		2.20	1.21	1.84	5.25	6.50	
1000	Kennel fencing, 1-1/2″ mesh, 6′ long, 3′-6″ wide, 6′-2″ high	2 Clab	4	4	Ea.	250	58		308	375	
1050	12′ long		4	4		300	58		358	430	
1200	Top covers, 1-1/2″ mesh, 6′ long		15	1.067		51	15.40		66.40	82.50	
1250	12′ long		12	1.333		81	19.25		100.25	123	
1300	For kennel doors, see division 08344-350										
0010	**CHAIN LINK FENCE** 11 ga. wire										525
0020	1-5/8″ post 10′O.C.,1-3/8″ top rail,2″ corner post galv. stl., 3′ high	B-1	185	.130	L.F.	3.86	1.96		5.82	7.60	
0050	4′ high		170	.141		5.90	2.13		8.03	10.15	
0100	6′ high		115	.209		6.65	3.16		9.81	12.75	
0150	Add for gate 3′ wide, 1-3/8″ frame 3′ high		12	2	Ea.	39	30		69	95	
0170	4′ high		10	2.400		51.50	36.50		88	119	
0190	6′ high		10	2.400		77.50	36.50		114	147	
0200	Add for gate 4′ wide, 1-3/8″ frame 3′ high		9	2.667		49.50	40.50		90	124	
0220	4′ high		9	2.667		67	40.50		107.50	143	
0240	6′ high		8	3		98	45.50		143.50	186	
0350	Aluminized steel, 9 ga. wire, 3′ high		185	.130	L.F.	4.64	1.96		6.60	8.45	
0380	4′ high		170	.141		5.30	2.13		7.43	9.50	
0400	6′ high		115	.209		6.80	3.16		9.96	12.90	
0450	Add for gate 3′ wide, 1-3/8″ frame 3′ high		12	2	Ea.	45.50	30		75.50	102	
0470	4′ high		10	2.400		62	36.50		98.50	130	
0490	6′ high		10	2.400		92.50	36.50		129	164	
0500	Add for gate 4′ wide, 1-3/8″ frame 3′ high		10	2.400		62	36.50		98.50	130	
0520	4′ high		9	2.667		82.50	40.50		123	160	
0540	6′ high		8	3		129	45.50		174.50	220	
0620	Vinyl covered 9 ga. wire, 3′ high		185	.130	L.F.	4.12	1.96		6.08	7.90	
0640	4′ high		170	.141		6.80	2.13		8.93	11.15	
0660	6′ high		115	.209		7.85	3.16		11.01	14.05	
0720	Add for gate 3′ wide, 1-3/8″ frame 3′ high		12	2	Ea.	51.50	30		81.50	109	
0740	4′ high		10	2.400		67	36.50		103.50	136	
0760	6′ high		10	2.400		103	36.50		139.50	175	
0780	Add for gate 4′ wide, 1-3/8″ frame 3′ high		10	2.400		70	36.50		106.50	139	
0800	4′ high		9	2.667		92.50	40.50		133	171	
0820	6′ high		8	3		134	45.50		179.50	225	
0860	Tennis courts, 11 ga. wire, 2 1/2″ post 10′ O.C., 1-5/8″ top rail										
0900	2-1/2″ corner post, 10′ high	B-1	95	.253	L.F.	10.30	3.82		14.12	17.90	
0920	12′ high		80	.300	"	12.35	4.54		16.89	21.50	
1000	Add for gate 3′ wide, 1-5/8″ frame 10′ high		10	2.400	Ea.	129	36.50		165.50	204	
1040	Aluminized, 11 ga. wire 10′ high		95	.253	L.F.	14.40	3.82		18.22	22.50	
1100	12′ high		80	.300	"	14.95	4.54		19.49	24	
1140	Add for gate 3′ wide, 1-5/8″ frame, 10′ high		10	2.400	Ea.	165	36.50		201.50	243	
1250	Vinyl covered 11 ga. wire, 10′ high		95	.253	L.F.	12.35	3.82		16.17	20	
1300	12′ high		80	.300	"	14.40	4.54		18.94	23.50	
1400	Add for gate 3′ wide, 1-3/8″ frame, 10′ high		10	2.400	Ea.	185	36.50		221.50	266	
0010	**FENCE, WOOD** Basket weave, 3/8″ x 4″ boards, 2″ x 4″										900
0020	stringers on spreaders, 4″ x 4″ posts										
0050	No. 1 cedar, 6′ high	B-1	160	.150	L.F.	7.15	2.27		9.42	11.75	
0070	Treated pine, 6′ high		150	.160		8.55	2.42		10.97	13.60	
0090	Vertical weave 6′ high		145	.166		10.20	2.50		12.70	15.50	
0200	Board fence, 1″ x 4″ boards, 2″ x 4″ rails, 4″ x 4″ post										

or expanded coverage of these items see *Means Site Work and Landscape Cost Data 2000*

	02820 | Fences & Gates		CREW	DAILY OUTPUT	LABOR-HOURS	UNIT	2000 BARE COSTS MAT.	LABOR	EQUIP.	TOTAL	TOTAL INCL O&P
900	0220	Preservative treated, 2 rail, 3′ high	B-1	145	.166	L.F.	5.25	2.50		7.75	10.05
	0240	4′ high		135	.178		5.75	2.69		8.44	10.90
	0260	3 rail, 5′ high		130	.185		6.50	2.79		9.29	11.95
	0300	6′ high		125	.192		7.40	2.90		10.30	13.10
	0320	No. 2 grade western cedar, 2 rail, 3′ high		145	.166		5.70	2.50		8.20	10.60
	0340	4′ high		135	.178		6.75	2.69		9.44	12.05
	0360	3 rail, 5′ high		130	.185		7.80	2.79		10.59	13.35
	0400	6′ high		125	.192		8.55	2.90		11.45	14.35
	0420	No. 1 grade cedar, 2 rail, 3′ high		145	.166		8.40	2.50		10.90	13.55
	0440	4′ high		135	.178		9.55	2.69		12.24	15.10
	0460	3 rail, 5′ high		130	.185		11.10	2.79		13.89	17
	0500	6′ high		125	.192		12.35	2.90		15.25	18.55
	0540	Shadow box, 1″ x 6″ board, 2″ x 4″ rail, 4″ x 4″post									
	0560	Pine, pressure treated, 3 rail, 6′ high	B-1	150	.160	L.F.	9.60	2.42		12.02	14.70
	0600	Gate, 3′-6″ wide		8	3	Ea.	55	45.50		100.50	138
	0620	No. 1 cedar, 3 rail, 4′ high		130	.185	L.F.	11.60	2.79		14.39	17.55
	0640	6′ high		125	.192		14.35	2.90		17.25	20.50
	0860	Open rail fence, split rails, 2 rail 3′ high, no. 1 cedar		160	.150		4.33	2.27		6.60	8.65
	0870	No. 2 cedar		160	.150		3.54	2.27		5.81	7.80
	0880	3 rail, 4′ high, no. 1 cedar		150	.160		6	2.42		8.42	10.75
	0890	No. 2 cedar		150	.160		4.04	2.42		6.46	8.60
	0920	Rustic rails, 2 rail 3′ high, no. 1 cedar		160	.150		2.92	2.27		5.19	7.10
	0930	No. 2 cedar		160	.150		2.84	2.27		5.11	7
	0940	3 rail, 4′ high		150	.160		3.88	2.42		6.30	8.40
	0950	No. 2 cedar		150	.160		3	2.42		5.42	7.45
	0960	Picket fence, gothic, pressure treated pine									
	1000	2 rail, 3′ high	B-1	140	.171	L.F.	3.60	2.59		6.19	8.40
	1020	3 rail, 4′ high		130	.185	″	4.25	2.79		7.04	9.45
	1040	Gate, 3′-6″ wide		9	2.667	Ea.	38	40.50		78.50	111
	1060	No. 2 cedar, 2 rail, 3′ high		140	.171	L.F.	4.50	2.59		7.09	9.40
	1100	3 rail, 4′ high		130	.185	″	4.60	2.79		7.39	9.85
	1120	Gate, 3′-6″ wide		9	2.667	Ea.	45	40.50		85.50	119
	1140	No. 1 cedar, 2 rail 3′ high		140	.171	L.F.	9	2.59		11.59	14.35
	1160	3 rail, 4′ high		130	.185		10.50	2.79		13.29	16.35
	1200	Rustic picket, molded pine, 2 rail, 3′ high		140	.171		4.18	2.59		6.77	9.05
	1220	No. 1 cedar, 2 rail, 3′ high		140	.171		5.55	2.59		8.14	10.55
	1240	Stockade fence, no. 1 cedar, 3-1/4″ rails, 6′ high		160	.150		8.50	2.27		10.77	13.25
	1260	8′ high		155	.155		11	2.34		13.34	16.10
	1300	No. 2 cedar, treated wood rails, 6′ high		160	.150		8.50	2.27		10.77	13.25
	1320	Gate, 3′-6″ wide		8	3	Ea.	50	45.50		95.50	133
	1360	Treated pine, treated rails, 6′ high		160	.150	L.F.	8.30	2.27		10.57	13.05
	1400	8′ high		150	.160	″	12.50	2.42		14.92	17.90
925	0010	**FENCE, RAIL** Picket, No. 2 cedar, Gothic, 2 rail, 3′ high	B-1	160	.150	L.F.	4.47	2.27		6.74	8.80
	0050	Gate, 3′-6″ wide		9	2.667	Ea.	38.50	40.50		79	112
	0400	3 rail, 4′ high		150	.160	L.F.	5.15	2.42		7.57	9.80
	0500	Gate, 3′-6″ wide		9	2.667	Ea.	46	40.50		86.50	120
	1200	Stockade, No. 2 cedar, treated wood rails, 6′ high		160	.150	L.F.	5.55	2.27		7.82	10.05
	1250	Gate, 3′ wide		9	2.667	Ea.	45.50	40.50		86	120
	1300	No. 1 cedar, 3-1/4″ cedar rails, 6′ high		160	.150	L.F.	13.55	2.27		15.82	18.85
	1500	Gate, 3′ wide		9	2.667	Ea.	108	40.50		148.50	188
	2700	Prefabricated redwood or cedar, 4′ high		160	.150	L.F.	10.60	2.27		12.87	15.55
	2800	6′ high		150	.160		14.05	2.42		16.47	19.60
	3300	Board, shadow box, 1″ x 6″, treated pine, 6′ high		160	.150		8.25	2.27		10.52	12.95
	3400	No. 1 cedar, 6′ high		150	.160		16.25	2.42		18.67	22
	3900	Basket weave, No. 1 cedar, 6′ high		160	.150		16.15	2.27		18.42	21.50
	4200	Gate, 3′-6″ wide		9	2.667	Ea.	48	40.50		88.50	122

02800 | Site Improvements and Amenities

02820 Fences & Gates		CREW	DAILY OUTPUT	LABOR-HOURS	UNIT	MAT.	LABOR	EQUIP.	TOTAL	TOTAL INCL O&P	
5000	Fence rail, redwood, 2″ x 4″, merch grade 8′	B-1	2,400	.010	L.F.	.87	.15		1.02	1.22	925

02830 Retaining Walls		CREW	DAILY OUTPUT	LABOR-HOURS	UNIT	MAT.	LABOR	EQUIP.	TOTAL	TOTAL INCL O&P	
0010	**RETAINING WALLS** Aluminized steel bin, excavation										100
0020	and backfill not included, 10′ wide										
0100	4′ high, 5.5′ deep	B-13	650	.074	S.F.	14.70	1.17	.86	16.73	19.10	
0200	8′ high, 5.5′ deep		615	.078		16.90	1.23	.91	19.04	21.50	
0300	10′ high, 7.7′ deep		580	.083		17.75	1.31	.96	20.02	23	
0400	12′ high, 7.7′ deep		530	.091		19.15	1.43	1.06	21.64	24.50	
0500	16′ high, 7.7′ deep		515	.093		20	1.47	1.09	22.56	26	
1800	Concrete gravity wall with vertical face including excavation & backfill										
1850	No reinforcing										
1900	6′ high, level embankment	C-17C	36	2.306	L.F.	45.50	46.50	11.45	103.45	143	
2000	33° slope embankment	"	32	2.594	"	42	52.50	12.85	107.35	150	
2800	Reinforced concrete cantilever, incl. excavation, backfill & reinf.										
2900	6′ high, 33° slope embankment	C-17C	35	2.371	L.F.	42	48	11.75	101.75	141	
0010	**STONE WALL** Including excavation, concrete footing and										400
0020	stone 3′ below grade. Price is exposed face area.										
0200	Decorative random stone, to 6′ high, 1′-6″ thick, dry set	D-1	35	.457	S.F.	7.60	8.15		15.75	22	
0300	Mortar set		40	.400		9.25	7.15		16.40	22.50	
0500	Cut stone, to 6′ high, 1′-6″ thick, dry set		35	.457		11.50	8.15		19.65	26.50	
0600	Mortar set		40	.400		13.50	7.15		20.65	27	
0800	Retaining wall, random stone, 6′ to 10′ high, 2′ thick, dry set		45	.356		9.50	6.35		15.85	21	
0900	Mortar set		50	.320		11.50	5.70		17.20	22.50	
1100	Cut stone, 6′ to 10′ high, 2′ thick, dry set		45	.356		14.75	6.35		21.10	27	
1200	Mortar set		50	.320		15.75	5.70		21.45	27	

02870 Site Furnishings		CREW	DAILY OUTPUT	LABOR-HOURS	UNIT	MAT.	LABOR	EQUIP.	TOTAL	TOTAL INCL O&P	
0010	**BENCHES** Park, precast concrete, w/backs,wood rails, 4′ long	2 Clab	5	3.200	Ea.	370	46		416	485	610
0100	8′ long		4	4		765	58		823	940	
0500	Steel barstock pedestals w/backs, 2″ x 3″ wood rails, 4′ long		10	1.600		755	23		778	870	
0510	8′ long		7	2.286		895	33		928	1,050	
0800	Cast iron pedestals, back & arms, wood slats, 4′ long		8	2		460	29		489	555	
0820	8′ long		5	3.200		765	46		811	920	
1700	Steel frame, fir seat, 10′ long		10	1.600		150	23		173	205	

02900 | Planting

02905 Transplanting		CREW	DAILY OUTPUT	LABOR-HOURS	UNIT	MAT.	LABOR	EQUIP.	TOTAL	TOTAL INCL O&P	
0010	**PLANTING** Moving shrubs on site, 12″ ball	B-62	28	.857	Ea.		13.70	4.16	17.86	27.50	725
0100	24″ ball	"	22	1.091			17.45	5.30	22.75	35.50	
0300	Moving trees on site, 36″ ball	B-6	3.75	6.400			102	54	156	233	
0400	60″ ball	"	1	24			385	203	588	875	

02910 Plant Preparation		CREW	DAILY OUTPUT	LABOR-HOURS	UNIT	MAT.	LABOR	EQUIP.	TOTAL	TOTAL INCL O&P	
0010	**MULCH**										500
0100	Aged barks, 3″ deep, hand spread	1 Clab	100	.080	S.Y.	2	1.16		3.16	4.18	

02900 | Planting

02910 | Plant Preparation

		02910 Plant Preparation	CREW	DAILY OUTPUT	LABOR-HOURS	UNIT	2000 BARE COSTS MAT.	LABOR	EQUIP.	TOTAL	TOTAL INCL O&P
500	0150	Skid steer loader	B-63	13.50	2.963	M.S.F.	220	43	8.60	271.60	325
	0200	Hay, 1" deep, hand spread	1 Clab	475	.017	S.Y.	.25	.24		.49	.70
	0250	Power mulcher, small	B-64	180	.089	M.S.F.	18.50	1.35	1.60	21.45	24.50
	0350	Large	B-65	530	.030	"	18.50	.46	.86	19.82	22
	0400	Humus peat, 1" deep, hand spread	1 Clab	700	.011	S.Y.	1.11	.17		1.28	1.50
	0450	Push spreader	A-1	2,500	.003	"	1.39	.05	.03	1.47	1.64
	0550	Tractor spreader	B-66	700	.011	M.S.F.	104	.22	.27	104.49	115
	0600	Oat straw, 1" deep, hand spread	1 Clab	475	.017	S.Y.	.28	.24		.52	.73
	0650	Power mulcher, small	B-64	180	.089	M.S.F.	25	1.35	1.60	27.95	31.50
	0700	Large	B-65	530	.030	"	25	.46	.86	26.32	29
	0750	Add for asphaltic emulsion	B-45	1,770	.009	Gal.	1.60	.14	.21	1.95	2.22
	0800	Peat moss, 1" deep, hand spread	1 Clab	900	.009	S.Y.	1.55	.13		1.68	1.92
	0850	Push spreader	A-1	2,500	.003	"	1.60	.05	.03	1.68	1.87
	0950	Tractor spreader	B-66	700	.011	M.S.F.	175	.22	.27	175.49	194
	1000	Polyethylene film, 6 mil.	2 Clab	2,000	.008	S.Y.	.15	.12		.27	.37
	1100	Redwood nuggets, 3" deep, hand spread	1 Clab	150	.053	"	6	.77		6.77	7.90
	1150	Skid steer loader	B-63	13.50	2.963	M.S.F.	600	43	8.60	651.60	745
	1200	Stone mulch, hand spread, ceramic chips, economy	1 Clab	125	.064	S.Y.	5.75	.92		6.67	7.95
	1250	Deluxe	"	95	.084	"	8.60	1.22		9.82	11.55
	1300	Granite chips	B-1	10	2.400	C.Y.	28	36.50		64.50	93
	1400	Marble chips		10	2.400		105	36.50		141.50	178
	1500	Onyx gemstone		10	2.400		310	36.50		346.50	400
	1600	Pea gravel		28	.857		61.50	12.95		74.45	89.50
	1700	Quartz	↓	10	2.400	↓	135	36.50		171.50	211
	1800	Tar paper, 15 Lb. felt	1 Clab	800	.010	S.Y.	.40	.14		.54	.69
	1900	Wood chips, 2" deep, hand spread	"	220	.036	"	1.65	.53		2.18	2.72
	1950	Skid steer loader	B-63	20.30	1.970	M.S.F.	108	28.50	5.75	142.25	174

02912 | General Planting

		02912 General Planting	CREW	DAILY OUTPUT	LABOR-HOURS	UNIT	MAT.	LABOR	EQUIP.	TOTAL	TOTAL INCL O&P
275	0010	**GROUND COVER** Plants, pachysandra, in prepared beds	B-1	15	1.600	C	15	24		39	58
	0200	Vinca minor, 1 yr, bare root		12	2	"	19	30		49	73
	0600	Stone chips, in 50 lb. bags, Georgia marble		520	.046	Bag	3.36	.70		4.06	4.90
	0700	Onyx gemstone		260	.092		13.10	1.40		14.50	16.85
	0800	Quartz		260	.092	↓	5.50	1.40		6.90	8.45
	0900	Pea gravel, truckload lots	↓	28	.857	Ton	20.50	12.95		33.45	44.50

02920 | Lawns & Grasses

		02920 Lawns & Grasses	CREW	DAILY OUTPUT	LABOR-HOURS	UNIT	MAT.	LABOR	EQUIP.	TOTAL	TOTAL INCL O&P
500	0010	**SEEDING** Mechanical seeding, 215 lb./acre (R02920-500)	B-66	1.50	5.333	Acre	485	102	125	712	840
	0101	$2.00/lb., 44 lb./M.S.Y.	A-1	13,950	.001	S.F.	.02	.01	.01	.04	.04
	0300	Fine grading and seeding incl. lime, fertilizer & seed,									
	0310	with equipment	B-14	1,000	.048	S.Y.	.16	.75	.20	1.11	1.67
	0600	Limestone hand push spreader, 50 lbs. per M.S.F.	1 Clab	180	.044	M.S.F.	3.25	.64		3.89	4.68
	0800	Grass seed hand push spreader, 4.5 lbs. per M.S.F.	"	180	.044	"	10	.64		10.64	12.10
600	0010	**SODDING** 1" deep, bluegrass sod, on level ground, over 8 M.S.F.	B-63	22	1.818	M.S.F.	165	26.50	5.30	196.80	233
	0200	4 M.S.F.		17	2.353		230	34	6.85	270.85	320
	0300	1000 S.F.		3.50	11.429		230	165	33.50	428.50	575
	0500	Sloped ground, over 8 M.S.F.		6	6.667		165	96.50	19.40	280.90	370
	0600	4 M.S.F.		5	8		230	116	23.50	369.50	475
	0700	1000 S.F.		4	10		230	145	29	404	535
	1000	Bent grass sod, on level ground, over 6 M.S.F.		20	2		315	29	5.80	349.80	400
	1100	3 M.S.F.		18	2.222		230	32	6.45	268.45	315
	1200	Sodding 1000 S.F. or less		14	2.857		345	41.50	8.30	394.80	460
	1500	Sloped ground, over 6 M.S.F.		15	2.667		315	38.50	7.75	361.25	420
	1600	3 M.S.F.		13.50	2.963		230	43	8.60	281.60	335
	1700	1000 S.F.	↓	12	3.333	↓	188	48	9.70	245.70	300

Important: See the Reference Section for critical supporting data - Reference Nos., Crews, & Location Factor

02930 | Exterior Plants

		02930 Exterior Plants	CREW	DAILY OUTPUT	LABOR-HOURS	UNIT	2000 BARE COSTS MAT.	LABOR	EQUIP.	TOTAL	TOTAL INCL O&P	
050	0010	**SHRUBS AND TREES** Evergreen, in prepared beds, B & B										050
	0100	Arborvitae pyramidal, 4'-5'	B-17	30	1.067	Ea.	35	17.10	18.95	71.05	88.50	
	0150	Globe, 12"-15"	B-1	96	.250		10.05	3.78		13.83	17.50	
	0300	Cedar, blue, 8'-10'	B-17	18	1.778		146	28.50	31.50	206	244	
	0500	Hemlock, canadian, 2-1/2'-3'	B-1	36	.667		14.25	10.10		24.35	33	
	0550	Holly, Savannah, 8' - 10' H		9.68	2.479		500	37.50		537.50	615	
	0600	Juniper, andorra, 18"-24"		80	.300		14	4.54		18.54	23	
	0620	Wiltoni, 15"-18"		80	.300		14.50	4.54		19.04	23.50	
	0640	Skyrocket, 4-1/2'-5'	B-17	55	.582		38.50	9.35	10.35	58.20	69.50	
	0660	Blue pfitzer, 2'-2-1/2'	B-1	44	.545		16	8.25		24.25	32	
	0680	Ketleerie, 2-1/2'-3'		50	.480		28	7.25		35.25	43.50	
	0700	Pine, black, 2-1/2'-3'		50	.480		29.50	7.25		36.75	45	
	0720	Mugo, 18"-24"		60	.400		30	6.05		36.05	43.50	
	0740	White, 4'-5'	B-17	75	.427		45.50	6.85	7.60	59.95	70	
	0800	Spruce, blue, 18"-24"	B-1	60	.400		28	6.05		34.05	41.50	
	0840	Norway, 4'-5'	B-17	75	.427		56	6.85	7.60	70.45	81.50	
	0900	Yew, denisforma, 12"-15"	B-1	60	.400		22.50	6.05		28.55	35.50	
	1000	Capitata, 18"-24"		30	.800		21	12.10		33.10	43.50	
	1100	Hicksi, 2'-2-1/2'		30	.800		26	12.10		38.10	49	
410	0010	**SHRUBS** Broadleaf evergreen, planted in prepared beds										410
	0100	Andromeda, 15"-18", container	B-1	96	.250	Ea.	14	3.78		17.78	22	
	0200	Azalea, 15" - 18", container		96	.250		18.50	3.78		22.28	27	
	0300	Barberry, 9"-12", container		130	.185		11	2.79		13.79	16.90	
	0400	Boxwood, 15"-18", B & B		96	.250		17.50	3.78		21.28	25.50	
	0500	Euonymus, emerald gaiety, 12" to 15", container		115	.209		12.25	3.16		15.41	18.90	
	0600	Holly, 15"-18", B & B		96	.250		16.75	3.78		20.53	25	
	0900	Mount laurel, 18" - 24", B & B		80	.300		75	4.54		79.54	90.50	
	1000	Paxistema, 9 - 12" high		130	.185		15	2.79		17.79	21.50	
	1100	Rhododendron, 18"-24", container		48	.500		37.50	7.55		45.05	54.50	
	1200	Rosemary, 1 gal container		600	.040		35	.60		35.60	39.50	
	2000	Deciduous, amelanchier, 2'-3', B & B		57	.421		55	6.35		61.35	71.50	
	2100	Azalea, 15"-18", B & B		96	.250		23	3.78		26.78	32	
	2300	Bayberry, 2'-3', B & B		57	.421		22.50	6.35		28.85	35.50	
	2600	Cotoneaster, 15"-18", B & B		80	.300		21	4.54		25.54	31	
	2800	Dogwood, 3'-4', B & B	B-17	40	.800		30	12.85	14.20	57.05	70	
	2900	Euonymus, alatus compacta, 15" to 18", container	B-1	80	.300		33	4.54		37.54	44.50	
	3200	Forsythia, 2'-3', container	″	60	.400		24	6.05		30.05	37	
	3300	Hibiscus, 3'-4', B & B	B-17	75	.427		36	6.85	7.60	50.45	59.50	
	3400	Honeysuckle, 3'-4', B & B	B-1	60	.400		25.50	6.05		31.55	38.50	
	3500	Hydrangea, 2'-3', B & B	″	57	.421		36	6.35		42.35	50.50	
	3600	Lilac, 3'-4', B & B	B-17	40	.800		39	12.85	14.20	66.05	80	
	3900	Privet, bare root, 18"-24"	B-1	80	.300		7.50	4.54		12.04	16	
	4100	Quince, 2'-3', B & B	″	57	.421		19.50	6.35		25.85	32.50	
	4200	Russian olive, 3'-4', B & B	B-17	75	.427		17	6.85	7.60	31.45	38.50	
	4400	Spirea, 3'-4', B & B	B-1	70	.343		27	5.20		32.20	38.50	
	4500	Viburnum, 3'-4', B & B	B-17	40	.800		24	12.85	14.20	51.05	63.50	
680	0010	**PLANT BED PREPARATION**										680
	0100	Backfill planting pit, by hand, on site topsoil	2 Clab	18	.889	C.Y.		12.85		12.85	22	
	0200	Prepared planting mix	″	24	.667			9.65		9.65	16.50	
	0300	Skid steer loader, on site topsoil	B-62	340	.071			1.13	.34	1.47	2.29	
	0400	Prepared planting mix	″	410	.059			.94	.28	1.22	1.89	
	1000	Excavate planting pit, by hand, sandy soil	2 Clab	16	1			14.45		14.45	25	
	1100	Heavy soil or clay	″	8	2			29		29	49.50	
	1200	1/2 C.Y. backhoe, sandy soil	B-11C	150	.107			1.83	1.35	3.18	4.57	

For expanded coverage of these items see *Means Site Work and Landscape Cost Data 2000*

	02930 | Exterior Plants		CREW	DAILY OUTPUT	LABOR-HOURS	UNIT	2000 BARE COSTS MAT.	LABOR	EQUIP.	TOTAL	TOTAL INCL O&P
680	1300	Heavy soil or clay	B-11C	115	.139	C.Y.		2.39	1.77	4.16	5.95
	2000	Mix planting soil, incl. loam, manure, peat, by hand	2 Clab	60	.267		23	3.85		26.85	31.50
	2100	Skid steer loader	B-62	150	.160		23	2.56	.78	26.34	30
	3000	Pile sod, skid steer loader	″	2,800	.009	S.Y.		.14	.04	.18	.28
	3100	By hand	2 Clab	400	.040			.58		.58	.99
	4000	Remove sod, F.E. loader	B-10S	2,000	.004			.08	.15	.23	.30
	4100	Sod cutter	B-12K	3,200	.002			.05	.26	.31	.37
	4200	By hand	2 Clab	240	.067			.96		.96	1.65
900	0010	**TREES** Deciduous, in prep. beds, balled & burlapped (B&B)									
	0100	Ash, 2″ caliper	B-17	8	4	Ea.	100	64	71	235	296
	0200	Beech, 5′-6′		50	.640		200	10.25	11.35	221.60	250
	0300	Birch, 6′-8′, 3 stems		20	1.600		113	25.50	28.50	167	199
	0500	Crabapple, 6′-8′		20	1.600		150	25.50	28.50	204	240
	0600	Dogwood, 4′-5′		40	.800		58	12.85	14.20	85.05	101
	0700	Eastern redbud 4′-5′		40	.800		115	12.85	14.20	142.05	164
	0800	Elm, 8′-10′		20	1.600		85	25.50	28.50	139	169
	0900	Ginkgo, 6′-7′		24	1.333		100	21.50	23.50	145	172
	1000	Hawthorn, 8′-10′, 1″ caliper		20	1.600		120	25.50	28.50	174	207
	1100	Honeylocust, 10′-12′, 1-1/2″ caliper		10	3.200		120	51.50	57	228.50	281
	1300	Larch, 8′		32	1		75	16.05	17.75	108.80	129
	1400	Linden, 8′-10′, 1″ caliper		20	1.600		120	25.50	28.50	174	207
	1500	Magnolia, 4′-5′		20	1.600		55	25.50	28.50	109	135
	1600	Maple, red, 8′-10′, 1-1/2″ caliper		10	3.200		122	51.50	57	230.50	283
	1700	Mountain ash, 8′-10′, 1″ caliper		16	2		150	32	35.50	217.50	258
	1800	Oak, 2-1/2″-3″ caliper		3	10.667		163	171	189	523	675
	2100	Planetree, 9′-11′, 1-1/4″ caliper		10	3.200		90	51.50	57	198.50	248
	2200	Plum, 6′-8′, 1″ caliper		20	1.600		75	25.50	28.50	129	158
	2300	Poplar, 9′-11′, 1-1/4″ caliper		10	3.200		51.50	51.50	57	160	206
	2500	Sumac, 2′-3′		75	.427		24	6.85	7.60	38.45	46.50
	2700	Tulip, 5′-6′		40	.800		29.50	12.85	14.20	56.55	69.50
	2800	Willow, 6′-8′, 1″ caliper		20	1.600		63	25.50	28.50	117	144

	02945 | Planting Accessories		CREW	DAILY OUTPUT	LABOR-HOURS	UNIT	MAT.	LABOR	EQUIP.	TOTAL	TOTAL INCL O&P
310	0010	**EDGING**									
	0050	Aluminum alloy, including stakes, 1/8″ x 4″, mill finish	B-1	390	.062	L.F.	1.75	.93		2.68	3.52
	0051	Black paint		390	.062		2.03	.93		2.96	3.82
	0052	Black anodized		390	.062		2.34	.93		3.27	4.17
	0100	Brick, set horizontally, 1-1/2 bricks per L.F.	D-1	370	.043		.63	.77		1.40	2
	0150	Set vertically, 3 bricks per L.F.	″	135	.119		1.96	2.12		4.08	5.75
	0200	Corrugated aluminum, roll, 4″ wide	1 Carp	650	.012		.28	.24		.52	.73
	0250	6″ wide	″	550	.015		.35	.29		.64	.88
	0600	Railroad ties, 6″ x 8″	2 Carp	170	.094		2.30	1.85		4.15	5.70
	0650	7″ x 9″	″	136	.118		2.55	2.32		4.87	6.80
	0700	Redwood									
	0750	2″ x 4″	2 Carp	330	.048	L.F.	2.67	.96		3.63	4.57
	0800	Steel edge strips, incl. stakes, 1/4″ x 5″	B-1	390	.062		2.75	.93		3.68	4.62
	0850	3/16″ x 4″	″	390	.062		2.17	.93		3.10	3.98
500	0010	**PLANTERS** Concrete, sandblasted, precast, 48″ diameter, 24″ high	2 Clab	15	1.067	Ea.	425	15.40		440.40	495
	0300	Fiberglass, circular, 36″ diameter, 24″ high		15	1.067		335	15.40		350.40	395
	1200	Wood, square, 48″ side, 24″ high		15	1.067		755	15.40		770.40	855
	1300	Circular, 48″ diameter, 30″ high		10	1.600		675	23		698	785
	1600	Planter/bench, 72″		5	3.200		2,100	46		2,146	2,400
775	0010	**TREE GUYING** Including stakes, guy wire and wrap									
	0100	Less than 3″ caliper, 2 stakes	2 Clab	35	.457	Ea.	15	6.60		21.60	28

02900 | Planting

02945 | Planting Accessories

		Description	Crew	Daily Output	Labor-Hours	Unit	2000 Bare Costs Mat.	Labor	Equip.	Total	Total Incl O&P	
5	0200	3″ to 4″ caliper, 3 stakes	2 Clab	21	.762	Ea.	17.60	11		28.60	38	775
	1000	Including arrowhead anchor, cable, turnbuckles and wrap										
	1100	Less than 3″ caliper, 3″ anchors	2 Clab	20	.800	Ea.	45	11.55		56.55	69.50	
	1200	3″ to 6″ caliper, 4″ anchors		15	1.067		65	15.40		80.40	98	
	1300	6″ caliper, 6″ anchors		12	1.333		80	19.25		99.25	121	
	1400	8″ caliper, 8″ anchors	↓	9	1.778	↓	14.90	25.50		40.40	60.50	

or information about Means Estimating Seminars, see yellow pages 11 and 12 in back of book

Division Notes

		CREW	DAILY OUTPUT	LABOR-HOURS	UNIT	2000 BARE COSTS MAT.	LABOR	EQUIP.	TOTAL	TOTAL INCL O&P

Division 3 Concrete

Estimating Tips

General

- Carefully check all the plans and specifications. Concrete often appears on drawings other than structural drawings, including mechanical and electrical drawings for equipment pads. The cost of cutting and patching is often difficult to estimate. See Subdivision 02225 for demolition costs.
- Always obtain concrete prices from suppliers near the job site. A volume discount can often be negotiated depending upon competition in the area. Remember to add for waste, particularly for slabs and footings on grade.

03100 Concrete Forms & Accessories

- A primary cost for concrete construction is forming. Most jobs today are constructed with prefabricated forms. The selection of the forms best suited for the job and the total square feet of forms required for efficient concrete forming and placing are key elements in estimating concrete construction. Enough forms must be available for erection to make efficient use of the concrete placing equipment and crew.
- Concrete accessories for forming and placing depend upon the systems used. Study the plans and specifications to assure that all special accessory requirements have been included in the cost estimate such as anchor bolts, inserts and hangers.

03200 Concrete Reinforcement

- Ascertain that the reinforcing steel supplier has included all accessories, cutting, bending and an allowance for lapping, splicing and waste. A good rule of thumb is 10% for lapping, splicing and waste. Also, 10% waste should be allowed for welded wire fabric.

03300 Cast-in-Place Concrete

- When estimating structural concrete, pay particular attention to requirements for concrete additives, curing methods and surface treatments. Special consideration for climate, hot or cold, must be included in your estimate. Be sure to include requirements for concrete placing equipment and concrete finishing.

03400 Precast Concrete
03500 Cementitious Decks & Toppings

- The cost of hauling precast concrete structural members is often an important factor. For this reason, it is important to get a quote from the nearest supplier. It may become economically feasible to set up precasting beds on the site if the hauling costs are prohibitive.

Reference Numbers

Reference numbers are shown in bold squares at the beginning of some major classifications. These numbers refer to related items in the Reference Section. The reference information may be an estimating procedure, an alternate pricing method or technical information.

Note: Not all subdivisions listed here necessarily appear in this publication.

03050 | Basic Concrete Materials & Methods

	03060 | Basic Concrete Materials		CREW	DAILY OUTPUT	LABOR-HOURS	UNIT	2000 BARE COSTS MAT.	LABOR	EQUIP.	TOTAL	TOTAL INCL O&P
870	0010	**WINTER PROTECTION** For heated ready mix, add, minimum				C.Y.	3.90			3.90	4.29
	0050	Maximum				″	5			5	5.50
	0100	Protecting concrete and temporary heat, add, minimum	2 Clab	6,000	.003	S.F.	.16	.04		.20	.25
	0200	Temporary shelter for slab on grade, wood frame and polyethylene									
	0201	sheeting, minimum	2 Carp	10	1.600	M.S.F.	299	31.50		330.50	385
	0210	Maximum	″	3	5.333	″	360	105		465	575
	0300	See also Division 03390-200									

03100 | Concrete Forms & Accessories

	03110 | Structural C.I.P. Forms		CREW	DAILY OUTPUT	LABOR-HOURS	UNIT	2000 BARE COSTS MAT.	LABOR	EQUIP.	TOTAL	TOTAL INCL O&P
410	0010	**FORMS IN PLACE, COLUMNS**									
	1500	Round fiber tube, 1 use, 8″ diameter	C-1	155	.206	L.F.	1.72	3.54		5.26	7.95
	1550	10″ diameter		155	.206		2.35	3.54		5.89	8.65
	1600	12″ diameter		150	.213		3.15	3.66		6.81	9.70
	1700	16″ diameter		140	.229		4.80	3.92		8.72	12
	5000	Plywood, 8″ x 8″ columns, 1 use		165	.194	SFCA	1.65	3.33		4.98	7.50
	5500	12″ x 12″ columns, 1 use		180	.178		1.66	3.05		4.71	7.05
	7500	Steel framed plywood, 4 use per mo., rent, 8″ x 8″		340	.094		3.92	1.62		5.54	7.10
	7550	10″ x 10″		350	.091		2.36	1.57		3.93	5.30
	7600	12″ x 12″		370	.086		3.05	1.48		4.53	5.90
430	0010	**FORMS IN PLACE, FOOTINGS** Continuous wall, plywood, 1 use	C-1	375	.085	SFCA	2.16	1.46		3.62	4.88
	0150	4 use	″	485	.066	″	.70	1.13		1.83	2.72
	1500	Keyway, 4 use, tapered wood, 2″ x 4″	1 Carp	530	.015	L.F.	.21	.30		.51	.74
	1550	2″ x 6″	″	500	.016	″	.29	.32		.61	.86
	5000	Spread footings, plywood, 1 use	C-1	305	.105	SFCA	1.69	1.80		3.49	4.94
	5150	4 use	″	414	.077	″	.55	1.33		1.88	2.88
435	0010	**FORMS IN PLACE, GRADE BEAM** Plywood, 1 use	C-2	530	.091	SFCA	1.53	1.59		3.12	4.41
	0150	4 use	″	605	.079	″	.50	1.39		1.89	2.93
445	0010	**FORMS IN PLACE, SLAB ON GRADE**									
	1000	Bulkhead forms with keyway, wood, 1 use, 2 piece	C-1	510	.063	L.F.	.70	1.08		1.78	2.61
	1400	Bulkhead forms w/keyway, 1 piece expanded metal, left in place									
	1410	In lieu of 2 piece form	C-1	1,375	.023	L.F.	1.15	.40		1.55	1.95
	1420	In lieu of 3 piece form		1,200	.027		1.15	.46		1.61	2.05
	1430	In lieu of 4 piece form		1,050	.030		1.15	.52		1.67	2.17
	2000	Curb forms, wood, 6″ to 12″ high, on grade, 1 use		215	.149	SFCA	1.39	2.55		3.94	5.90
	2150	4 use		275	.116	″	.45	2		2.45	3.92
	3000	Edge forms, wood, 4 use, on grade, to 6″ high		600	.053	L.F.	.32	.92		1.24	1.93
	3050	7″ to 12″ high		435	.074	SFCA	.75	1.26		2.01	2.99
	4000	For slab blockouts, to 12″ high, 1 use		200	.160	L.F.	.70	2.75		3.45	5.45
	4100	Plastic (extruded), to 6″ high, multiple use, on grade		800	.040	″	.23	.69		.92	1.43
450	0010	**FORMS IN PLACE, STAIRS** (Slant length x width), 1 use	C-2	165	.291	S.F.	3.68	5.10		8.78	12.80
	0150	4 use		190	.253		1.03	4.43		5.46	8.75
	2000	Stairs, cast on sloping ground (length x width), 1 use		220	.218		2.45	3.82		6.27	9.25
	2100	4 use		240	.200		.81	3.50		4.31	6.90
455	0010	**FORMS IN PLACE, WALLS**									
	0100	Box out for wall openings, to 16″ thick, to 10 S.F.	C-2	24	2	Ea.	24	35		59	86.50
	0150	Over 10 S.F. (use perimeter)	″	280	.171	L.F.	2.09	3		5.09	7.45
	0250	Brick shelf, 4″ w, add to wall forms, use wall area abv shelf									

03100 | Concrete Forms & Accessories

03110 | Structural C.I.P. Forms

			CREW	DAILY OUTPUT	LABOR-HOURS	UNIT	2000 BARE COSTS MAT.	LABOR	EQUIP.	TOTAL	TOTAL INCL O&P	
55	0260	1 use	C-2	240	.200	SFCA	2.29	3.50		5.79	8.50	455
	0350	4 use		300	.160	"	.92	2.80		3.72	5.80	
	0500	Bulkhead forms, with keyway, 1 use, 2 piece		265	.181	L.F.	2.50	3.17		5.67	8.20	
	0550	3 piece		175	.274	"	3.10	4.81		7.91	11.65	
	0600	Bulkhead forms w/keyway, 1 piece expanded metal, left in place										
	0610	In lieu of 2 piece form	C-1	800	.040	L.F.	1.15	.69		1.84	2.45	
	2000	Job built plywood wall forms, to 8′ high, 1 use, below grade	C-2	300	.160	SFCA	1.90	2.80		4.70	6.90	
	2150	4 use, below grade		435	.110		.63	1.93		2.56	4	
	2400	Over 8′ to 16′ high, 1 use		280	.171		3.68	3		6.68	9.20	
	2550	4 use		395	.122		.69	2.13		2.82	4.40	
	3000	For architectural finish, add		1,820	.026		.60	.46		1.06	1.45	
	3500	Polystyrene (expanded) wall forms										
	3510	To 8′ high, 1 use, left in place	1 Carp	295	.027	SFCA	1.60	.53		2.13	2.68	
	7800	Modular prefabricated plywood, to 8′ high, 1 use per month	C-2	1,180	.041		1.13	.71		1.84	2.46	
	7860	4 use per month		1,260	.038		.37	.67		1.04	1.55	
	8000	To 16′ high, 1 use per month		715	.067		1.38	1.18		2.56	3.54	
	8060	4 use per month		790	.061		.46	1.06		1.52	2.32	
	8100	Over 16′ high, 1 use per month		715	.067		1.66	1.18		2.84	3.84	
	8160	4 use per month		790	.061		.55	1.06		1.61	2.43	
00	0010	**SCAFFOLDING** See division 01540-750										800

03150 | Concrete Accessories

			CREW	DAILY OUTPUT	LABOR-HOURS	UNIT	MAT.	LABOR	EQUIP.	TOTAL	TOTAL INCL O&P	
80	0010	**ACCESSORIES, ANCHOR BOLTS** J-type, incl. nut and washer										080
	0020	1/2″ diameter, 6″ long	1 Carp	90	.089	Ea.	1.11	1.75		2.86	4.22	
	0050	10″ long		85	.094		1.22	1.85		3.07	4.52	
	0100	12″ long		85	.094		1.27	1.85		3.12	4.58	
	0200	5/8″ diameter, 12″ long		80	.100		1.40	1.97		3.37	4.92	
	0250	18″ long		70	.114		1.65	2.25		3.90	5.70	
	0300	24″ long		60	.133		1.90	2.63		4.53	6.60	
	0350	3/4″ diameter, 8″ long		80	.100		1.71	1.97		3.68	5.25	
	0400	12″ long		70	.114		2.11	2.25		4.36	6.20	
	0450	18″ long		60	.133		2.58	2.63		5.21	7.35	
	0500	24″ long		50	.160		3.43	3.15		6.58	9.15	
60	0010	**ACCESSORIES, CHAMFER STRIPS**										160
	5000	Wood, 1/2″ wide	1 Carp	535	.015	L.F.	.09	.29		.38	.60	
	5200	3/4″ wide		525	.015		.24	.30		.54	.77	
	5400	1″ wide		515	.016		.31	.31		.62	.86	
70	0010	**ACCESSORIES, COLUMN FORM**										170
	1000	Column clamps, adjustable to 24″ x 24″, buy				Set	73			73	80.50	
	1400	Rent per month				"	6.50			6.50	7.15	
50	0010	**EXPANSION JOINT** Keyed, cold, 24 ga, incl. stakes, 3-1/2″ high	1 Carp	200	.040	L.F.	.50	.79		1.29	1.90	250
	0050	4-1/2″ high		200	.040		.62	.79		1.41	2.03	
	0100	5-1/2″ high		195	.041		.69	.81		1.50	2.14	
	2000	Premolded, bituminous fiber, 1/2″ x 6″		375	.021		.37	.42		.79	1.13	
	2050	1″ x 12″		300	.027		1.58	.53		2.11	2.64	
	2500	Neoprene sponge, closed cell, 1/2″ x 6″		375	.021		1.26	.42		1.68	2.11	
	2550	1″ x 12″		300	.027		5.65	.53		6.18	7.15	
	5000	For installation in walls, add						75%				
	5250	For installation in boxouts, add						25%				
00	0010	**ACCESSORIES, INSERTS**										400
	1000	All size nut insert, 5/8″ & 3/4″, incl. nut	1 Carp	84	.095	Ea.	3	1.88		4.88	6.50	

03100 | Concrete Forms & Accessories

03150 | Concrete Accessories

		03150 Concrete Accessories	CREW	DAILY OUTPUT	LABOR-HOURS	UNIT	2000 BARE COSTS MAT.	LABOR	EQUIP.	TOTAL	TOTAL INCL O&P
400	2000	Continuous slotted, 1-5/8" x 1-3/8"									
	2100	3" long, 12 ga.	1 Carp	65	.123	Ea.	2.75	2.42		5.17	7.20
	2150	6" long, 12 ga.		65	.123		3.60	2.42		6.02	8.10
	2200	12" long, 8 ga.		65	.123		8.80	2.42		11.22	13.85
	2300	36" long, 8 ga.		60	.133		19.50	2.63		22.13	26
620	0010	**ACCESSORIES, SLEEVES AND CHASES**									
	0100	Plastic, 1 use, 9" long, 2" diameter	1 Carp	100	.080	Ea.	.55	1.58		2.13	3.31
	0150	4" diameter		90	.089		1.56	1.75		3.31	4.72
	0200	6" diameter		75	.107		2.75	2.10		4.85	6.65
640	0010	**ACCESSORIES, SNAP TIES, FLAT WASHER**									
	0100	3000 lb., to 8"				C	70			70	77
	0250	16"					90			90	99
	0300	18"					88.50			88.50	97.50
	0500	With plastic cone, to 8"					62.50			62.50	68.50
	0600	11" & 12"					85.50			85.50	94
	0650	16"					89.50			89.50	98.50
	0700	18"					93			93	102
850	0010	**ACCESSORIES, WALL AND FOUNDATION**									
	0020	Coil tie system									
	0700	1-1/4", 36,000 lb., to 8"				C	610			610	670
	1200	1-1/4" diameter x 3" long				"	1,200			1,200	1,300
	4200	30" long				Ea.	2.85			2.85	3.14
	4250	36" long				"	3.10			3.10	3.41
860	0010	**WATERSTOP** PVC, ribbed 3/16" thick, 4" wide	1 Carp	155	.052	L.F.	.56	1.02		1.58	2.36
	0050	6" wide		145	.055		1.21	1.09		2.30	3.19
	0500	Ribbed, PVC, with center bulb, 9" wide, 3/16" thick		135	.059		1.70	1.17		2.87	3.87
	0550	3/8" thick		130	.062		1.94	1.21		3.15	4.21

03200 | Concrete Reinforcement

03210 | Reinforcing Steel

		03210 Reinforcing Steel	CREW	DAILY OUTPUT	LABOR-HOURS	UNIT	2000 BARE COSTS MAT.	LABOR	EQUIP.	TOTAL	TOTAL INCL O&P
600	0010	**REINFORCING IN PLACE** A615 Grade 60									
	0502	Footings, #4 to #7	4 Rodm	4,200	.008	Lb.	.29	.16		.45	.62
	0550	#8 to #18		3.60	8.889	Ton	500	188		688	900
	0702	Walls, #3 to #7		6,000	.005	Lb.	.29	.11		.40	.53
	0750	#8 to #18		4	8	Ton	530	169		699	900
	2400	Dowels, 2 feet long, deformed, #3	2 Rodm	520	.031	Ea.	.23	.65		.88	1.46
	2410	#4		480	.033		.41	.70		1.11	1.76
	2420	#5		435	.037		.65	.78		1.43	2.15
	2430	#6		360	.044		.92	.94		1.86	2.76

03220 | Welded Wire Fabric

		03220 Welded Wire Fabric	CREW	DAILY OUTPUT	LABOR-HOURS	UNIT	MAT.	LABOR	EQUIP.	TOTAL	TOTAL INCL O&P
200	0011	**WELDED WIRE FABRIC** Sheets, 6 x 6 - W1.4 x W1.4 (10 x 10)	2 Rodm	3,500	.005	S.F.	.07	.10		.17	.26
	0301	6 x 6 - W2.9 x W2.9 (6 x 6) 42 lb. per C.S.F.		2,900	.006		.15	.12		.27	.39
	0501	4 x 4 - W1.4 x W1.4 (10 x 10) 31 lb. per C.S.F.		3,100	.005		.11	.11		.22	.32
	0750	Rolls									
	0901	2 x 2 - #12 galv. for gunite reinforcing	2 Rodm	650	.025	S.F.	.21	.52		.73	1.20
	0950	Material prices for above include 10% lap									

Important: See the Reference Section for critical supporting data - Reference Nos., Crews, & Location Factor

03200 | Concrete Reinforcement

03240 Fibrous Reinforcing		CREW	DAILY OUTPUT	LABOR-HOURS	UNIT	2000 BARE COSTS MAT.	LABOR	EQUIP.	TOTAL	TOTAL INCL O&P	
0010	**FIBROUS REINFORCING**										300
0100	Synthetic fibers				Lb.	3.79			3.79	4.17	
0110	1-1/2 lb. per C.Y., add to concrete				C.Y.	5.85			5.85	6.45	
0150	Steel fibers				Lb.	.48			.48	.53	
0155	25 lb. per C.Y., add to concrete				C.Y.	12			12	13.20	
0160	50 lb. per C.Y., add to concrete				↓	24			24	26.50	
0170	75 lb. per C.Y., add to concrete				↓	37			37	40.50	
0180	100 lb. per C.Y., add to concrete				↓	48			48	53	

03300 | Cast-In-Place Concrete

03310 Structural Concrete		CREW	DAILY OUTPUT	LABOR-HOURS	UNIT	2000 BARE COSTS MAT.	LABOR	EQUIP.	TOTAL	TOTAL INCL O&P	
0010	**CONCRETE, READY MIX** Regular weight										220
0020	2000 psi				C.Y.	60.50			60.50	66.50	
0100	2500 psi				↓	61			61	67	
0150	3000 psi					63			63	69	
0200	3500 psi					64.50			64.50	70.50	
0300	4000 psi					67			67	73.50	
0350	4500 psi					69			69	76	
0400	5000 psi					71.50			71.50	78.50	
0411	6000 psi					81.50			81.50	90	
0412	8000 psi					133			133	146	
0413	10,000 psi					189			189	208	
0414	12,000 psi					228			228	251	
1000	For high early strength cement, add					10%					
2000	For all lightweight aggregate, add				↓	45%					
0010	**CONCRETE IN PLACE** Including forms (4 uses), reinforcing										240
0050	steel, including finishing unless otherwise indicated										
0500	Chimney foundations, industrial, minimum	C-14C	32.22	3.476	C.Y.	130	64.50	1.18	195.68	254	
0510	Maximum	↓	23.71	4.724	↓	151	87.50	1.61	240.11	320	
3800	Footings, spread under 1 C.Y.		38.07	2.942		92.50	54.50	1	148	197	
3850	Over 5 C.Y.		81.04	1.382		85	25.50	.47	110.97	138	
3900	Footings, strip, 18″ x 9″, plain		41.04	2.729		84	50.50	.93	135.43	181	
3950	36″ x 12″, reinforced		61.55	1.820		85.50	33.50	.62	119.62	153	
4000	Foundation mat, under 10 C.Y.		38.67	2.896		116	53.50	.98	170.48	222	
4050	Over 20 C.Y.	↓	56.40	1.986	↓	103	36.50	.68	140.18	177	
4520	Handicap access ramp, railing both sides, 3′ wide	C-14H	14.58	3.292	L.F.	97	63.50	2.60	163.10	219	
4525	5′ wide	↓	12.22	3.928	↓	111	75.50	3.10	189.60	256	
4530	With cheek walls and rails both sides, 3′ wide		8.55	5.614		99.50	108	4.44	211.94	300	
4535	5′ wide	↓	7.31	6.566	↓	100	126	5.20	231.20	335	
4650	Slab on grade, not including finish, 4″ thick	C-14E	60.75	1.449	C.Y.	77	27.50	.62	105.12	134	
4700	6″ thick	″	92	.957	″	74	18.05	.41	92.46	114	
4751	Slab on grade, incl. troweled finish, not incl. forms										
4760	or reinforcing, over 10,000 S.F., 4″ thick slab	C-14F	3,425	.021	S.F.	.84	.37	.01	1.22	1.54	
4820	6″ thick slab	″	3,350	.021	″	1.22	.38	.01	1.61	1.98	
5000	Slab on grade, incl. textured finish, not incl. forms										
5001	or reinforcing, 4″ thick slab	C-14G	2,873	.019	S.F.	.82	.34	.01	1.17	1.47	
5010	6″ thick	↓	2,590	.022	↓	1.28	.37	.01	1.66	2.04	
5020	8″ thick	↓	2,320	.024	↓	1.66	.42	.02	2.10	2.54	
6203	Retaining walls, gravity, 4′ high				C.Y.	79			79	87	

03300 | Cast-In-Place Concrete

		03310 | Structural Concrete	CREW	DAILY OUTPUT	LABOR-HOURS	UNIT	2000 BARE COSTS MAT.	LABOR	EQUIP.	TOTAL	TOTAL INCL O&P
240	6800	Stairs, not including safety treads, free standing, 3′-6″ wide	C-14H	83	.578	LF Nose	5.90	11.15	.46	17.51	26.50
	6850	Cast on ground		125	.384	″	4.10	7.40	.30	11.80	17.60
	7000	Stair landings, free standing		200	.240	S.F.	2.30	4.62	.19	7.11	10.75
	7050	Cast on ground		475	.101	″	1.35	1.95	.08	3.38	4.94
700	0010	**PLACING CONCRETE** and vibrating, including labor & equipment									
	1900	Footings, continuous, shallow, direct chute	C-6	120	.400	C.Y.		6.20	.63	6.83	11.25
	1950	Pumped	C-20	150	.427			6.80	4.88	11.68	16.85
	2000	With crane and bucket	C-7	90	.800			12.85	10.15	23	32.50
	2400	Footings, spread, under 1 C.Y., direct chute	C-6	55	.873			13.55	1.38	14.93	24.50
	2600	Footings, spread, over 5 C.Y., direct chute		120	.400			6.20	.63	6.83	11.25
	2900	Foundation mats, over 20 C.Y., direct chute		350	.137			2.13	.22	2.35	3.85
	4300	Slab on grade, 4″ thick, direct chute		110	.436			6.80	.69	7.49	12.25
	4350	Pumped	C-20	130	.492			7.85	5.65	13.50	19.50
	4400	With crane and bucket	C-7	110	.655			10.50	8.30	18.80	27
	4900	Walls, 8″ thick, direct chute	C-6	90	.533			8.30	.84	9.14	15
	4950	Pumped	C-20	100	.640			10.20	7.30	17.50	25.50
	5000	With crane and bucket	C-7	80	.900			14.45	11.40	25.85	37
	5050	12″ thick, direct chute	C-6	100	.480			7.45	.76	8.21	13.50
	5100	Pumped	C-20	110	.582			9.25	6.65	15.90	23
	5200	With crane and bucket	C-7	90	.800			12.85	10.15	23	32.50
	5600	Wheeled concrete dumping, add to placing costs above									
	5610	Walking cart, 50′ haul, add	C-18	32	.281	C.Y.		4.13	1.63	5.76	8.85
	5620	150′ haul, add		24	.375			5.50	2.17	7.67	11.80
	5700	250′ haul, add		18	.500			7.35	2.89	10.24	15.75
	5800	Riding cart, 50′ haul, add	C-19	80	.112			1.65	1.04	2.69	3.97
	5810	150′ haul, add		60	.150			2.20	1.39	3.59	5.30
	5900	250′ haul, add		45	.200			2.93	1.85	4.78	7.10
		03350 | Concrete Finishing									
300	0010	**FINISHING FLOORS** Monolithic, screed finish	1 Cefi	900	.009	S.F.		.17		.17	.27
	0050	Darby finish		750	.011			.20		.20	.33
	0100	Screed and float finish		725	.011			.21		.21	.34
	0150	Screed, float, and broom finish		630	.013			.24		.24	.39
	0200	Screed, float, and hand trowel		600	.013			.25		.25	.41
	0250	Machine trowel		550	.015			.27		.27	.45
	1600	Exposed local aggregate finish, minimum		625	.013		.37	.24		.61	.80
	1650	Maximum		465	.017		1.18	.33		1.51	1.83
350	0010	**FINISHING WALLS** Break ties and patch voids	1 Cefi	540	.015	S.F.	.05	.28		.33	.52
	0050	Burlap rub with grout	″	450	.018		.05	.34		.39	.61
	0300	Bush hammer, green concrete	B-39	1,000	.048		.05	.71	.18	.94	1.48
	0350	Cured concrete	″	650	.074		.05	1.09	.28	1.42	2.23
600	0010	**SLAB TEXTURE STAMPING,** buy									
	0020	Approx. 3 S.F.- 5 S.F. each, minimum				Ea.	40			40	44
	0030	Average				″	44			44	48.50
	0120	Per S.F. of tool, average				S.F.	48			48	53
	0200	Commonly used chemicals for texture systems									
	0210	Hardeners w/colors average				S.F.	.40			.40	.44
	0220	Release agents w/colors, average					.15			.15	.17
	0225	Clear, average					.10			.10	.11
	0230	Sealers, clear, average					.10			.10	.11
	0240	Colors, average					.12			.12	.13
		03390 | Concrete Curing									
200	0011	**CURING** With burlap, 4 uses assumed, 7.5 oz.	2 Clab	5,500	.003	S.F.	.03	.04		.07	.11
	0101	12 oz.		5,500	.003		.06	.04		.10	.13

03300 | Cast-In-Place Concrete

03390 | Concrete Curing

			CREW	DAILY OUTPUT	LABOR-HOURS	UNIT	2000 BARE COSTS MAT.	LABOR	EQUIP.	TOTAL	TOTAL INCL O&P	
00	0201	With waterproof curing paper, 2 ply, reinforced	2 Clab	7,000	.002	S.F.	.05	.03		.08	.11	200
	0301	With sprayed membrane curing compound	↓	9,500	.002		.04	.02		.06	.08	
	0710	Electrically, heated pads, 15 watts/S.F., 20 uses, minimum					.16			.16	.18	
	0800	Maximum				↓	.27			.27	.29	

03400 | Precast Concrete

03480 | Precast Specialties

			CREW	DAILY OUTPUT	LABOR-HOURS	UNIT	2000 BARE COSTS MAT.	LABOR	EQUIP.	TOTAL	TOTAL INCL O&P	
00	0010	**LINTELS**										400
	0800	Precast concrete, 4″ wide, 8″ high, to 5′ long	D-1	175	.091	L.F.	4.72	1.63		6.35	7.95	
	0850	5′-12′ long	D-4	190	.211		4.93	3.43	.57	8.93	11.90	
	1000	6″ wide, 8″ high, to 5′ long		185	.216		6.65	3.53	.59	10.77	14	
	1050	5′-12′ long	↓	190	.211	↓	6.45	3.43	.57	10.45	13.60	
00	0010	**STAIRS**, Precast concrete treads on steel stringers, 3′ wide	C-12	75	.640	Riser	53.50	12.35	6.05	71.90	86.50	800
	0300	Front entrance, 5′ wide with 48″ platform, 2 risers		16	3	Flight	276	58	28.50	362.50	435	
	0350	5 risers		12	4		315	77.50	38	430.50	520	
	0500	6′ wide, 2 risers	↓	15	3.200		315	62	30	407	485	
	1200	Basement entrance stairs, steel bulkhead doors, minimum	B-51	22	2.182		460	33	7.85	500.85	570	
	1250	Maximum	″	11	4.364	↓	680	65.50	15.70	761.20	880	

03900 | Concrete Restoration & Cleaning

03920 | Concrete Resurfacing

			CREW	DAILY OUTPUT	LABOR-HOURS	UNIT	2000 BARE COSTS MAT.	LABOR	EQUIP.	TOTAL	TOTAL INCL O&P	
00	0012	**FLOOR PATCHING** 1/4″ thick, small areas, regular	1 Cefi	170	.047	S.F.	.71	.89		1.60	2.23	400
	0100	Epoxy	″	100	.080	″	1.44	1.51		2.95	4.04	

Division Notes

	CREW	DAILY OUTPUT	LABOR-HOURS	UNIT	2000 BARE COSTS MAT.	LABOR	EQUIP.	TOTAL	TOTAL INCL O&P

Division 4 Masonry

Estimating Tips

04050 Basic Masonry Materials Methods

The terms *mortar* and *grout* are often used interchangeably, and incorrectly. Mortar is used to bed masonry units, seal the entry of air and moisture, provide architectural appearance, and allow for size variations in the units. Grout is used primarily in reinforced masonry construction and is used to bond the masonry to the reinforcing steel. Common mortar types are M(2500 psi), S(1800 psi), N(750 psi), and O(350 psi), and conform to ASTM C270. Grout is either fine or coarse, conforms to ASTM C476, and in-place strengths generally exceed 2500 psi. Mortar and grout are different components of masonry construction and are placed by entirely different methods. An estimator should be aware of their unique uses and costs.

04200 Masonry Units

- The most common types of unit masonry are brick and concrete masonry. The major classifications of brick are building brick (ASTM C62), facing brick (ASTM C216) and glazed brick, fire brick and pavers. Many varieties of texture and appearance can exist within these classifications, and the estimator would be wise to check local custom and availability within the project area. On repair and remodeling jobs, matching the existing brick may be the most important criteria.
- Brick and concrete block are priced by the piece and then converted into a price per square foot of wall. Openings less than two square feet are generally ignored by the estimator because any savings in units used is offset by the cutting and trimming required.
- All masonry walls, whether interior or exterior, require bracing. The cost of bracing walls during construction should be included by the estimator and this bracing must remain in place until permanent bracing is complete. Permanent bracing of masonry walls is accomplished by masonry itself, in the form of pilasters or abutting wall corners, or by anchoring the walls to the structural frame. Accessories in the form of anchors, anchor slots and ties are used, but their supply and installation can be by different trades. For instance, anchor slots on spandrel beams and columns are supplied and welded in place by the steel fabricator, but the ties from the slots into the masonry are installed by the bricklayer. Regardless of the installation method the estimator must be certain that these accessories are accounted for in pricing.

Reference Numbers

Reference numbers are shown in bold squares at the beginning of some major classifications. These numbers refer to related items in the Reference Section. The reference information may be an estimating procedure, an alternate pricing method or technical information.

Note: Not all subdivisions listed here necessarily appear in this publication.

		04060 | Masonry Mortar	CREW	DAILY OUTPUT	LABOR-HOURS	UNIT	2000 BARE COSTS MAT.	LABOR	EQUIP.	TOTAL	TOTAL INCL O&P
200	0010	**CEMENT** Gypsum 80 lb. bag, T.L. lots (R04060-100)				Bag	11.40			11.40	12.55
	0050	L.T.L. lots					11.90			11.90	13.10
	0100	Masonry, 70 lb. bag, T.L. lots					5.90			5.90	6.50
	0150	L.T.L. lots					5.70			5.70	6.25
	0200	White, 70 lb. bag, T.L. lots					16.05			16.05	17.65
	0250	L.T.L. lots					16.60			16.60	18.25
		04070 | Masonry Grout									
420	0010	**GROUTING** Bond bms. & lintels, 8" dp., pumped, not incl. block									
	0200	Concrete block cores, solid, 4" thk., by hand, 0.067 C.F./S.F.	D-8	1,100	.036	S.F.	.25	.66		.91	1.40
	0210	6" thick, pumped, 0.175 C.F. per S.F.	D-4	720	.056		.60	.91	.15	1.66	2.37
	0250	8" thick, pumped, 0.258 C.F. per S.F.		680	.059		.91	.96	.16	2.03	2.81
	0300	10" thick, pumped, 0.340 C.F. per S.F.		660	.061		1.19	.99	.17	2.35	3.17
	0350	12" thick, pumped, 0.422 C.F. per S.F.		640	.063		1.47	1.02	.17	2.66	3.54
		04080 | Masonry Anchor & Reinforcement									
070	0010	**ANCHOR BOLTS** Hooked type with nut and washer, 1/2" diam., 8" long	1 Bric	200	.040	Ea.	.47	.80		1.27	1.87
	0030	12" long		190	.042		1.22	.84		2.06	2.77
	0060	3/4" diameter, 8" long		160	.050		1.71	1		2.71	3.57
	0070	12" long		150	.053		2.11	1.07		3.18	4.13
200	0010	**REINFORCING** Steel bars A615, placed horiz., #3 & #4 bars (R04080-500)	1 Bric	450	.018	Lb.	.28	.36		.64	.91
	0050	Placed vertical, #3 & #4 bars		350	.023		.28	.46		.74	1.08
	0060	#5 & #6 bars		650	.012		.28	.25		.53	.73
	0200	Joint reinforcing, regular truss, to 6" wide, mill std galvanized		30	.267	C.L.F.	10.20	5.35		15.55	20.50
	0250	12" wide		20	.400		11.75	8		19.75	26.50
	0400	Cavity truss with drip section, to 6" wide		30	.267		9.70	5.35		15.05	19.70
	0450	12" wide		20	.400		11.05	8		19.05	25.50
650	0010	**WALL TIES** To brick veneer, galv., corrugated, 7/8" x 7", 22 Ga.	1 Bric	10.50	.762	C	4.47	15.25		19.72	31
	0100	24 Ga.		10.50	.762		3.97	15.25		19.22	30.50
	0150	16 Ga.		10.50	.762		13.45	15.25		28.70	41
	0200	Buck anchors, galv., corrugated, 16 gauge, 2" bend. 8" x 2"		10.50	.762		114	15.25		129.25	151
	0250	8" x 3"		10.50	.762		99	15.25		114.25	135
	0600	Cavity wall, Z type, galvanized, 6" long, 1/4" diameter		10.50	.762		20	15.25		35.25	48
	0650	3/16" diameter		10.50	.762		9.50	15.25		24.75	36.50
	0800	8" long, 1/4" diameter		10.50	.762		24	15.25		39.25	52.50
	0850	3/16" diameter		10.50	.762		10.45	15.25		25.70	37.50
	1000	Rectangular type, galvanized, 1/4" diameter, 2" x 6"		10.50	.762		25.50	15.25		40.75	54.50
	1050	2" x 8" or 4" x 6"		10.50	.762		29	15.25		44.25	57.50
	1100	3/16" diameter, 2" x 6"		10.50	.762		16.95	15.25		32.20	44.50
	1150	2" x 8" or 4" x 6"		10.50	.762		19.40	15.25		34.65	47.50
	1500	Rigid partition anchors, plain, 8" long, 1" x 1/8"		10.50	.762		48	15.25		63.25	79
	1550	1" x 1/4"		10.50	.762		94	15.25		109.25	129
	1580	1-1/2" x 1/8"		10.50	.762		66.50	15.25		81.75	99
	1600	1-1/2" x 1/4"		10.50	.762		158	15.25		173.25	199
	1650	2" x 1/8"		10.50	.762		83	15.25		98.25	118
	1700	2" x 1/4"		10.50	.762		225	15.25		240.25	274
		04090 | Masonry Accessories									
420	0010	**INSULATION** See also division 07210-550									
	0100	Inserts, styrofoam, plant installed, add to block prices									
	0200	8" x 16" units, 6" thick				S.F.	.75			.75	.83
	0250	8" thick					.75			.75	.83
	0300	10" thick					.90			.90	.99
	0350	12" thick					.95			.95	1.05

04050 | Basic Masonry Materials & Methods

04090 | Masonry Accessories		CREW	DAILY OUTPUT	LABOR-HOURS	UNIT	2000 BARE COSTS MAT.	LABOR	EQUIP.	TOTAL	TOTAL INCL O&P	
0010	**SCAFFOLDING & SWING STAGING** See division 01540										700

04200 | Masonry Units

04210 | Clay Masonry Units		CREW	DAILY OUTPUT	LABOR-HOURS	UNIT	2000 BARE COSTS MAT.	LABOR	EQUIP.	TOTAL	TOTAL INCL O&P	
0010	**COMMON BUILDING BRICK** C62, TL lots, material only R04210-120										100
0020	Standard, minimum				M	250			250	275	
0050	Average (select)				"	305			305	335	
0010	**BRICK VENEER** Scaffolding not included, truck load lots										120
0015	Material costs incl. 3% brick and 25% mortar waste										
2000	Standard, sel. common, 4" x 2-2/3" x 8", (6.75/S.F.) R04210-100	D-8	230	.174	S.F.	2.68	3.18		5.86	8.35	
2020	Standard, red, 4" x 2-2/3" x 8", running bond (6.75/SF)		220	.182		2.68	3.32		6	8.60	
2050	Full header every 6th course (7.88/S.F.) R04210-120		185	.216		3.13	3.95		7.08	10.15	
2100	English, full header every 2nd course (10.13/S.F.)		140	.286		4.01	5.20		9.21	13.25	
2150	Flemish, alternate header every course (9.00/S.F.) R04210-180		150	.267		3.57	4.87		8.44	12.15	
2200	Flemish, alt. header every 6th course (7.13/S.F.)		205	.195		2.83	3.57		6.40	9.15	
2250	Full headers throughout (13.50/S.F.) R04210-500		105	.381		5.35	6.95		12.30	17.65	
2300	Rowlock course (13.50/S.F.)		100	.400		5.35	7.30		12.65	18.25	
2350	Rowlock stretcher (4.50/S.F.) R04210-550		310	.129		1.80	2.36		4.16	5.95	
2400	Soldier course (6.75/S.F.)		200	.200		2.68	3.66		6.34	9.15	
2450	Sailor course (4.50/S.F.)		290	.138		1.80	2.52		4.32	6.25	
2600	Buff or gray face, running bond, (6.75/S.F.)		220	.182		2.85	3.32		6.17	8.80	
2700	Glazed face brick, running bond		210	.190		10	3.48		13.48	16.90	
2750	Full header every 6th course (7.88/S.F.)		170	.235		11.65	4.30		15.95	20	
3000	Jumbo, 6" x 4" x 12" running bond (3.00/S.F.)		435	.092		3.53	1.68		5.21	6.75	
3050	Norman, 4" x 2-2/3" x 12" running bond, (4.5/S.F.)		320	.125		3.19	2.28		5.47	7.40	
3100	Norwegian, 4" x 3-1/5" x 12" (3.75/S.F.)		375	.107		2.08	1.95		4.03	5.60	
3150	Economy, 4" x 4" x 8" (4.50/S.F.)		310	.129		2.12	2.36		4.48	6.35	
3200	Engineer, 4" x 3-1/5" x 8" (5.63/S.F.)		260	.154		1.88	2.81		4.69	6.85	
3250	Roman, 4" x 2" x 12" (6.00/S.F.)		250	.160		4.52	2.92		7.44	9.90	
3300	SCR, 6" x 2-2/3" x 12" (4.50/S.F.)		310	.129		4.14	2.36		6.50	8.55	
3350	Utility, 4" x 4" x 12" (3.00/S.F.)		450	.089		2.96	1.62		4.58	6	
3400	For cavity wall construction, add						15%				
3450	For stacked bond, add						10%				
3500	For interior veneer construction, add						15%				
3550	For curved walls, add						30%				
0010	**FACE BRICK** C216, TL lots, material only R04210-120										300
0300	Standard modular, 4" x 2-2/3" x 8", minimum				M	350			350	385	
0350	Maximum				"	315			315	345	
2160	Fire Brick, 2" x 2-2/3" x 9", minimum										
2165	Maximum				M	900			900	990	
2170	For less than truck load lots, add					10					
2180	For buff or gray brick, add					15					

04220 | Concrete Masonry Units		CREW	DAILY OUTPUT	LABOR-HOURS	UNIT	MAT.	LABOR	EQUIP.	TOTAL	TOTAL INCL O&P	
0010	**CONCRETE BLOCK, BACK-UP** Scaffolding not included R04220-200										220
0020	Sand aggregate, 8" x 16" units, tooled joint 1 side										

04220 | Concrete Masonry Units

		04220 Concrete Masonry Units	CREW	DAILY OUTPUT	LABOR-HOURS	UNIT	2000 BARE COSTS MAT.	LABOR	EQUIP.	TOTAL	TOTAL INCL O&P
220	1000	Reinforced, alternate courses, 4" thick R04220-200	D-8	435	.092	S.F.	.88	1.68		2.56	3.81
	1100	6" thick		415	.096		1.18	1.76		2.94	4.27
	1150	8" thick		395	.101		1.38	1.85		3.23	4.65
	1200	10" thick		385	.104		1.92	1.90		3.82	5.35
	1250	12" thick	D-9	365	.132		2.05	2.35		4.40	6.25
230	0010	**CONCRETE BLOCK BOND BEAM** Scaffolding not included									
	0020	Not including grout or reinforcing									
	0100	Regular block, 8" high, 8" thick	D-8	565	.071	L.F.	1.62	1.29		2.91	3.98
	0150	12" thick	D-9	510	.094	"	2.54	1.68		4.22	5.65
240	0010	**CONCRETE BLOCK, DECORATIVE** Scaffolding not included									
	5000	Split rib profile units, 1" deep ribs, 8 ribs									
	5100	8" x 16" x 4" thick	D-8	345	.116	S.F.	1.89	2.12		4.01	5.65
	5150	6" thick		325	.123		2.29	2.25		4.54	6.30
	5200	8" thick		305	.131		2.74	2.40		5.14	7.10
	5250	12" thick	D-9	275	.175		3.35	3.12		6.47	9
	5400	For special deeper colors, 4" thick, add					.17			.17	.19
	5450	12" thick, add					.34			.34	.37
	5600	For white, 4" thick, add					.70			.70	.77
	5650	6" thick, add					.93			.93	1.02
	5700	8" thick, add					1.13			1.13	1.24
	5750	12" thick, add					1.58			1.58	1.74
250	0010	**CONCRETE BLOCK, EXTERIOR** Not including scaffolding									
	0020	Reinforced alt courses, tooled joints 2 sides, foam inserts									
	0100	Regular, 8" x 16" x 6" thick	D-8	390	.103	S.F.	1.17	1.87		3.04	4.45
	0200	8" thick		365	.110		1.36	2		3.36	4.88
	0250	10" thick		355	.113		1.60	2.06		3.66	5.25
	0300	12" thick	D-9	330	.145		2.47	2.60		5.07	7.10
260	0010	**CONCRETE BLOCK FOUNDATION WALL** Scaffolding not included									
	0050	Normal-weight, trowel cut joints, parged 1/2" thick, no reinforcing									
	0200	Hollow, 8" x 16" x 6" thick	D-8	450	.089	S.F.	1.30	1.62		2.92	4.18
	0250	8" thick		430	.093		1.50	1.70		3.20	4.53
	0300	10" thick		420	.095		2.04	1.74		3.78	5.20
	0350	12" thick	D-9	395	.122		2.18	2.17		4.35	6.05
	0500	Solid, 8" x 16" block, 6" thick	D-8	440	.091		1.65	1.66		3.31	4.62
	0550	8" thick	"	415	.096		2.08	1.76		3.84	5.25
	0600	12" thick	D-9	380	.126		3.24	2.25		5.49	7.40
280	0010	**CONCRETE BLOCK, LINTELS** Scaffolding not included									
	0100	Including grout and horizontal reinforcing									
	0200	8" x 8" x 8", 1 #4 bar	D-4	300	.133	L.F.	3.38	2.17	.36	5.91	7.80
	0250	2 #4 bars		295	.136		3.50	2.21	.37	6.08	8
	1000	12" x 8" x 8", 1 #4 bar		275	.145		4.79	2.37	.40	7.56	9.70
	1150	2 #5 bars		270	.148		5.05	2.42	.40	7.87	10.10
340	0010	**COPING** Stock units									
	0050	Precast concrete, 10" wide, 4" tapers to 3-1/2", 8" wall	D-1	75	.213	L.F.	10	3.81		13.81	17.45
	0100	12" wide, 3-1/2" tapers to 3", 10" wall		70	.229		10.10	4.08		14.18	18
	0150	16" wide, 4" tapers to 3-1/2", 14" wall		60	.267		13.90	4.76		18.66	23.50
	0300	Limestone for 12" wall, 4" thick		90	.178		13	3.17		16.17	19.65
	0350	6" thick		80	.200		15.25	3.57		18.82	23
	0500	Marble, to 4" thick, no wash, 9" wide		90	.178		18.90	3.17		22.07	26.50
	0550	12" wide		80	.200		28	3.57		31.57	37
	0700	Terra cotta, 9" wide		90	.178		4.55	3.17		7.72	10.35
	0800	Aluminum, for 12" wall		80	.200		10.75	3.57		14.32	17.85
360	0010	**CORNICES** Brick cornice on existing building									
	0020	Not including scaffolding									

04200 | Masonry Units

	04220 Concrete Masonry Units		CREW	DAILY OUTPUT	LABOR-HOURS	UNIT	2000 BARE COSTS MAT.	LABOR	EQUIP.	TOTAL	TOTAL INCL O&P
360	0110	Face bricks, 12 brick/S.F., minimum	D-1	30	.533	SF Face	4.70	9.50		14.20	21.50
	0150	15 brick/S.F., maximum	"	23	.696	"	5.60	12.40		18	27

	04270 Glass Masonry Units		CREW	DAILY OUTPUT	LABOR-HOURS	UNIT	2000 BARE COSTS MAT.	LABOR	EQUIP.	TOTAL	TOTAL INCL O&P
200	0010	**GLASS BLOCK** Scaffolding not included R04060-200									
	0150	8" x 8" block	D-8	160	.250	S.F.	9.25	4.57		13.82	17.95
	0200	12" x 12" block	"	175	.229	"	11.10	4.18		15.28	19.25
	0700	For solar reflective blocks, add					100%				
	0800	Under 1,000 S.F., 4" x 8" blocks	D-8	145	.276	S.F.	12.60	5.05		17.65	22.50
	1000	Thinline, plain, 3-1/8" thick, under 1,000 S.F., 6" x 6" block		115	.348		9.35	6.35		15.70	21
	1050	8" x 8" block		160	.250		5.80	4.57		10.37	14.10
	1400	For cleaning block after installation (both sides), add	↓	1,000	.040	↓	.10	.73		.83	1.35

	04290 Adobe Masonry Units		CREW	DAILY OUTPUT	LABOR-HOURS	UNIT	2000 BARE COSTS MAT.	LABOR	EQUIP.	TOTAL	TOTAL INCL O&P
100	0010	**ADOBE BRICK** Unstabilized, with adobe mortar (Southwestern States)									
	0260	Brick, 4" x 3" x 8" (6.0 per S.F.)	D-8	266	.150	S.F.	1.80	2.75		4.55	6.65
	0280	4" x 4" x 8" (4.5 per S.F.)		333	.120		1.83	2.20		4.03	5.75
	0300	4" x 4" x 14" (2.5 per S.F.)		540	.074		1.13	1.35		2.48	3.53
	0320	8" x 3" x 16" (3.0 per S.F.)		466	.086		1.52	1.57		3.09	4.33
	0340	4" x 3" x 12" (4.0 per S.F.)		362	.110		1.86	2.02		3.88	5.45
	0360	6" x 3" x 12" (4.0 per S.F.)		362	.110		1.67	2.02		3.69	5.25
	0380	4" x 5" x 16" (1.80 per S.F.)		600	.067		.89	1.22		2.11	3.04
	0400	8" x 4" x 16" (2.25 per S.F.)		577	.069		1.46	1.27		2.73	3.76
	0420	10" x 4" x 14" (2.57 per S.F.)	↓	555	.072		1.54	1.32		2.86	3.93
	0440	Adobe, partially stabilized, add					10%				
	0480	Fully stabilized, add				↓	25%				

04400 | Stone

	04412 Bluestone		CREW	DAILY OUTPUT	LABOR-HOURS	UNIT	2000 BARE COSTS MAT.	LABOR	EQUIP.	TOTAL	TOTAL INCL O&P
800	0010	**WINDOW SILL** Bluestone, thermal top, 10" wide, 1-1/2" thick	D-1	85	.188	S.F.	12.75	3.36		16.11	19.75
	0050	2" thick		75	.213	"	14.75	3.81		18.56	22.50
	0100	Cut stone, 5" x 8" plain		48	.333	L.F.	10	5.95		15.95	21
	0200	Face brick on edge, brick, 8" wide		80	.200		1.90	3.57		5.47	8.15
	0400	Marble, 9" wide, 1" thick		85	.188		7.10	3.36		10.46	13.50
	0600	Precast concrete, 4" tapers to 3", 9" wide		70	.229		9.05	4.08		13.13	16.85
	0650	11" wide		60	.267		12.25	4.76		17.01	21.50
	0700	13" wide, 3 1/2" tapers to 2 1/2", 12" wall		50	.320		12.15	5.70		17.85	23
	0900	Slate, colored, unfading, honed, 12" wide, 1" thick		85	.188		15	3.36		18.36	22
	0950	2" thick	↓	70	.229	↓	21	4.08		25.08	30

	04413 Granite		CREW	DAILY OUTPUT	LABOR-HOURS	UNIT	2000 BARE COSTS MAT.	LABOR	EQUIP.	TOTAL	TOTAL INCL O&P
300	0010	**GRANITE** Cut to size									
	2450	For radius under 5', add				L.F.	100%				
	2500	Steps, copings, etc., finished on more than one surface									
	2550	Minimum	D-10	50	.800	C.F.	73	14.90	8	95.90	114
	2600	Maximum	"	50	.800	"	109	14.90	8	131.90	154
	2800	Pavers, 4" x 4" x 4" blocks, split face and joints									
	2850	Minimum	D-11	80	.300	S.F.	10.40	5.45		15.85	20.50
	2900	Maximum	"	80	.300	"	20.50	5.45		25.95	32

		04413 | Granite	CREW	DAILY OUTPUT	LABOR-HOURS	UNIT	2000 BARE COSTS MAT.	LABOR	EQUIP.	TOTAL	TOTAL INCL O&P
300	3500	Curbing, city street type, See Division 02770-225									
		04414 | Limestone									
400	0012	**LIMESTONE,** Cut to size R04430-500									
	0020	Veneer facing panels									
	0500	Texture finish, light stick, 4-1/2" thick, 5'x 12'	D-4	300	.133	S.F.	18.95	2.17	.36	21.48	25
	0750	5" thick, 5' x 14' panels	D-10	275	.145		22.50	2.71	1.46	26.67	31
	1000	Sugarcube finish, 2" Thick, 3' x 5' panels		275	.145		11.25	2.71	1.46	15.42	18.55
	1050	3" Thick, 4' x 9' panels		275	.145		11.70	2.71	1.46	15.87	19.05
	1200	4" Thick, 5' x 11' panels		275	.145		15.60	2.71	1.46	19.77	23.50
	1400	Sugarcube, textured finish, 4-1/2" thick, 5' x 12'		275	.145		20	2.71	1.46	24.17	28
	1450	5" thick, 5' x 14' panels		275	.145		23.50	2.71	1.46	27.67	31.50
	2000	Coping, sugarcube finish, top & 2 sides		30	1.333	C.F.	52	25	13.35	90.35	114
	2100	Sills, lintels, jambs, trim, stops, sugarcube finish, average		20	2		47	37	20	104	136
	2150	Detailed		20	2		61.50	37	20	118.50	153
	2300	Steps, extra hard, 14" wide, 6" rise		50	.800	L.F.	42.50	14.90	8	65.40	80.50
	3000	Quoins, plain finish, 6"x12"x12"	D-12	25	1.280	Ea.	102	22.50		124.50	150
	3050	6"x16"x24"	"	25	1.280	"	136	22.50		158.50	188
		04415 | Marble									
500	0010	**MARBLE** Base, 3/4" thick, polished, group A, 4" thick	D-1	60	.267	L.F.	9.30	4.76		14.06	18.30
	1000	Facing, polished finish, cut to size, 3/4" to 7/8" thick									
	1050	Average	D-10	130	.308	S.F.	17.95	5.75	3.09	26.79	33
	1100	Maximum	"	130	.308	"	41.50	5.75	3.09	50.34	59
	2200	Window sills, 6" x 3/4" thick	D-1	85	.188	L.F.	6.75	3.36		10.11	13.15
	2500	Flooring, polished tiles, 12" x 12" x 3/8" thick									
	2510	Thin set, average	D-11	90	.267	S.F.	8.30	4.86		13.16	17.40
	2600	Maximum		90	.267		30	4.86		34.86	41.50
	2700	Mortar bed, average		65	.369		8.45	6.75		15.20	20.50
	2740	Maximum		65	.369		27.50	6.75		34.25	42
	2780	Travertine, 3/8" thick, average	D-10	130	.308		11.50	5.75	3.09	20.34	25.50
	2790	Maximum	"	130	.308		28	5.75	3.09	36.84	44
	3500	Thresholds, 3' long, 7/8" thick, 4" to 5" wide, plain	D-12	24	1.333	Ea.	12.95	23.50		36.45	54
	3550	Beveled		24	1.333	"	15.10	23.50		38.60	56
	3700	Window stools, polished, 7/8" thick, 5" wide		85	.376	L.F.	12.10	6.65		18.75	24.50
		04417 | Sandstone									
700	0011	**SANDSTONE OR BROWNSTONE**									
	0100	Sawed face veneer, 2-1/2" thick, to 2' x 4' panels	D-10	130	.308	S.F.	15.35	5.75	3.09	24.19	30
	0150	4' thick, to 3'-6" x 8'panels		100	.400		15.35	7.45	4.01	26.81	34
	0300	Split face, random sizes		100	.400		9.05	7.45	4.01	20.51	27
	0350	Cut stone trim (limestone)									
	0360	Ribbon stone, 4" thick, 5' pieces	D-8	120	.333	Ea.	111	6.10		117.10	132
	0370	Cove stone, 4" thick, 5' pieces		105	.381		111	6.95		117.95	134
	0380	Cornice stone, 10" to 12" wide		90	.444		137	8.10		145.10	165
	0390	Band stone, 4" thick, 5' pieces		145	.276		71	5.05		76.05	86.50
	0410	Window and door trim, 3" to 4" wide		160	.250		60	4.57		64.57	74
	0420	Key stone, 18" long		60	.667		63.50	12.20		75.70	90.50
		04418 | Slate									
800	0010	**SLATE** Pennsylvania, blue gray to gray black; Vermont,									
	3500	Stair treads, sand finish, 1" thick x 12" wide									

04400 | Stone

	04418 | Slate	CREW	DAILY OUTPUT	LABOR-HOURS	UNIT	2000 BARE COSTS MAT.	LABOR	EQUIP.	TOTAL	TOTAL INCL O&P	
3600	3 L.F. to 6 L.F.	D-10	120	.333	L.F.	14.60	6.20	3.34	24.14	30	800
3700	Ribbon, sand finish, 1″ thick x 12″ wide										
3750	To 6 L.F.	D-10	120	.333	L.F.	9.75	6.20	3.34	19.29	25	

	04420 | Collected Stone	CREW	DAILY OUTPUT	LABOR-HOURS	UNIT	MAT.	LABOR	EQUIP.	TOTAL	TOTAL INCL O&P	
0011	**ROUGH STONE WALL**, Dry										750
0100	Random fieldstone, under 18″ thick	D-12	60	.533	C.F.	9.05	9.40		18.45	26	
0150	Over 18″ thick	″	63	.508	″	12.50	8.95		21.45	29	

04500 | Refractories

	04550 | Flue Liners	CREW	DAILY OUTPUT	LABOR-HOURS	UNIT	2000 BARE COSTS MAT.	LABOR	EQUIP.	TOTAL	TOTAL INCL O&P	
0010	**FLUE LINING** Including mortar joints, 8″ x 8″	D-1	125	.128	V.L.F.	2.92	2.28		5.20	7.10	250
0100	8″ x 12″		103	.155		3.80	2.77		6.57	8.90	
0200	12″ x 12″		93	.172		5.70	3.07		8.77	11.45	
0300	12″ x 18″		84	.190		8.75	3.40		12.15	15.35	
0400	18″ x 18″		75	.213		12.05	3.81		15.86	19.70	
0500	20″ x 20″		66	.242		12.65	4.33		16.98	21.50	
0600	24″ x 24″		56	.286		14.85	5.10		19.95	25	
1000	Round, 18″ diameter		66	.242		14.25	4.33		18.58	23	
1100	24″ diameter	↓	47	.340	↓	46	6.10		52.10	61	

	04580 | Refractory Brick	CREW	DAILY OUTPUT	LABOR-HOURS	UNIT	MAT.	LABOR	EQUIP.	TOTAL	TOTAL INCL O&P	
0010	**FIREPLACE** For prefabricated fireplace, see div. 10305-100										270
0100	Brick fireplace, not incl. foundations or chimneys										
0110	30″ x 29″ opening, incl. chamber, plain brickwork	D-1	.40	40	Ea.	355	715		1,070	1,600	
0200	Fireplace box only (110 brick)	″	2	8	″	115	143		258	370	
0300	For elaborate brickwork and details, add					35%	35%				
0400	For hearth, brick & stone, add	D-1	2	8	Ea.	132	143		275	385	
0410	For steel angle, damper, cleanouts, add		4	4		95	71.50		166.50	226	
0600	Plain brickwork, incl. metal circulator		.50	32	↓	700	570		1,270	1,725	
0800	Face brick only, standard size, 8″ x 2-2/3″ x 4″	↓	.30	53.333	M	375	950		1,325	2,025	
0900	Stone fireplace, fieldstone, add				SF Face	9.50			9.50	10.45	
1000	Cut stone, add				″	10			10	11	

04800 | Masonry Assemblies

	04810 | Unit Masonry Assemblies	CREW	DAILY OUTPUT	LABOR-HOURS	UNIT	2000 BARE COSTS MAT.	LABOR	EQUIP.	TOTAL	TOTAL INCL O&P	
0010	**CHIMNEY** See Div. 03310 for foundation, add to prices below [R04210-050]										160
0100	Brick, 16″ x 16″, 8″ flue, scaff. not incl.	D-1	18.20	.879	V.L.F.	16.25	15.70		31.95	44.50	

04800 | Masonry Assemblies

		04810 | Unit Masonry Assemblies	CREW	DAILY OUTPUT	LABOR-HOURS	UNIT	2000 BARE COSTS MAT.	LABOR	EQUIP.	TOTAL	TOTAL INCL O&P
160	0150	16″ x 20″ with one 8″ x 12″ flue R04210-050	D-1	16	1	V.L.F.	19.70	17.85		37.55	51.50
	0200	16″ x 24″ with two 8″ x 8″ flues		14	1.143		28	20.50		48.50	65.50
	0250	20″ x 20″ with one 12″ x 12″ flue		13.70	1.168		19.75	21		40.75	57
	0300	20″ x 24″ with two 8″ x 12″ flues		12	1.333		31.50	24		55.50	75.50
	0350	20″ x 32″ with two 12″ x 12″ flues		10	1.600		33	28.50		61.50	84.50
170	0010	**COLUMNS** Brick, scaffolding not included									
	0050	8″ x 8″, 9 brick	D-1	56	.286	V.L.F.	3.50	5.10		8.60	12.50
	0100	12″ x 8″, 13.5 brick		37	.432		5.25	7.70		12.95	18.80
	0200	12″ x 12″, 20 brick		25	.640		7.80	11.40		19.20	28
	0300	16″ x 12″, 27 brick		19	.842		10.50	15.05		25.55	37
	0400	16″ x 16″, 36 brick		14	1.143		14	20.50		34.50	50
	0500	20″ x 16″, 45 brick		11	1.455		17.50	26		43.50	63.50
	0600	20″ x 20″, 56 brick		9	1.778		22	31.50		53.50	77.50
210	0010	**CONCRETE BLOCK, PARTITIONS** Scaffolding not included									
	1000	Lightweight block, tooled joints, 2 sides, hollow									
	1100	Not reinforced, 8″ x 16″ x 4″ thick	D-8	440	.091	S.F.	.91	1.66		2.57	3.81
	1150	6″ thick		410	.098		1.21	1.78		2.99	4.35
	1200	8″ thick		385	.104		1.52	1.90		3.42	4.90
	1250	10″ thick		370	.108		1.98	1.98		3.96	5.55
	1300	12″ thick	D-9	350	.137		2.12	2.45		4.57	6.50
	4000	Regular block, tooled joints, 2 sides, hollow									
	4100	Not reinforced, 8″ x 16″ x 4″ thick	D-8	430	.093	S.F.	.77	1.70		2.47	3.72
	4150	6″ thick		400	.100		1.06	1.83		2.89	4.27
	4200	8″ thick		375	.107		1.26	1.95		3.21	4.69
	4250	10″ thick		360	.111		1.80	2.03		3.83	5.40
	4300	12″ thick	D-9	340	.141		1.93	2.52		4.45	6.40
400	0010	**LINTELS** See division 05120-480									
650	0016	**WALLS** R04210-500									
	0800	4″ wall, face, 4″ x 2-2/3″ x 8″	D-8	215	.186	S.F.	2.65	3.40		6.05	8.65
	0850	4″ thick, as back up, 6.75 bricks per S.F.		240	.167		1.95	3.05		5	7.30
	0900	8″ thick wall, 13.50 brick per S.F.		135	.296		4	5.40		9.40	13.55
	1000	12″ thick wall, 20.25 bricks per S.F.		95	.421		6	7.70		13.70	19.65
	1050	16″ thick wall, 27.00 bricks per S.F.		75	.533		8.10	9.75		17.85	25.50
	1200	Reinforced, 4″ x 2-2/3″ x 8″ , 4″ wall		205	.195		1.97	3.57		5.54	8.20
	1250	8″ thick wall, 13.50 brick per S.F.		130	.308		4.01	5.60		9.61	13.90
	1300	12″ thick wall, 20.25 bricks per S.F.		90	.444		6.05	8.10		14.15	20.50
	1350	16″ thick wall, 27.00 bricks per S.F.		70	.571		8.15	10.45		18.60	26.50

04900 | Masonry Restoration & Cleaning

		04930 | Unit Masonry Cleaning	CREW	DAILY OUTPUT	LABOR-HOURS	UNIT	2000 BARE COSTS MAT.	LABOR	EQUIP.	TOTAL	TOTAL INCL O&P
200	0010	**CLEAN AND POINT** Smooth brick	1 Bric	300	.027	S.F.	.20	.53		.73	1.12
	0100	Rough brick	″	265	.030	″	.21	.60		.81	1.25
900	0010	**WASHING BRICK** Acid wash, smooth brick R04930-100	1 Bric	560	.014	S.F.	.23	.29		.52	.73
	0050	Rough brick		400	.020		.23	.40		.63	.93
	0060	Stone, acid wash		600	.013		.23	.27		.50	.70
	1000	Muriatic acid, price per gallon in 5 gallon lots				Gal.	4			4	4.40

Division 5 Metals

Estimating Tips

05050 Basic Metal Materials & Methods

- Nuts, bolts, washers, connection angles and plates can add a significant amount to both the tonnage of a structural steel job as well as the estimated cost. As a rule of thumb add 10% to the total weight to account for these accessories.
- Type 2 steel construction, commonly referred to as "simple construction," consists generally of field bolted connections with lateral bracing supplied by other elements of the building, such as masonry walls or x-bracing. The estimator should be aware, however, that shop connections may be accomplished by welding or bolting. The method may be particular to the fabrication shop and may have an impact on the estimated cost.

05200 Metal Joists

- In any given project the total weight of open web steel joists is determined by the loads to be supported and the design. However, economies can be realized in minimizing the amount of labor used to place the joists. This is done by maximizing the joist spacing and therefore minimizing the number of joists required to be installed on the job. Certain spacings and locations may be required by the design, but in other cases maximizing the spacing and keeping it as uniform as possible will keep the costs down.

05300 Metal Deck

- The takeoff and estimating of metal deck involves more than simply the area of the floor or roof and the type of deck specified or shown on the drawings. Many different sizes and types of openings may exist. Small openings for individual pipes or conduits may be drilled after the floor/roof is installed, but larger openings may require special deck lengths as well as reinforcing or structural support. The estimator should determine who will be supplying this reinforcing. Additionally, some deck terminations are part of the deck package, such as screed angles and pour stops, and others will be part of the steel contract, such as angles attached to structural members and cast-in-place angles and plates. The estimator must ensure that all pieces are accounted for in the complete estimate.

05500 Metal Fabrications

- The most economical steel stairs are those that use common materials, standard details and most importantly, a uniform and relatively simple method of field assembly. Commonly available A36 channels and plates are very good choices for the main stringers of the stairs, as are angles and tees for the carrier members. Risers and treads are usually made by specialty shops, and it is most economical to use a typical detail in as many places as possible. The stairs should be pre-assembled and shipped directly to the site. The field connections should be simple and straightforward to be accomplished efficiently and with a minimum of equipment and labor.

Reference Numbers

Reference numbers are shown in bold squares at the beginning of some major classifications. These numbers refer to related items in the Reference Section. The reference information may be an estimating procedure, an alternate pricing method or technical information.

Note: Not all subdivisions listed here necessarily appear in this publication.

	05090 | Metal Fastenings		CREW	DAILY OUTPUT	LABOR-HOURS	UNIT	2000 BARE COSTS MAT.	LABOR	EQUIP.	TOTAL	TOTAL INCL O&P
150	0010	**BOLTS & HEX NUTS** Steel, A307									
	0100	1/4" diameter, 1/2" long				Ea.	.05			.05	.06
	0200	1" long					.06			.06	.07
	0300	2" long					.07			.07	.08
	0400	3" long					.10			.10	.11
	0500	4" long					.15			.15	.17
	0600	3/8" diameter, 1" long					.10			.10	.11
	0700	2" long					.14			.14	.15
	0800	3" long					.17			.17	.19
	0900	4" long					.22			.22	.24
	1000	5" long					.26			.26	.29
	1100	1/2" diameter, 1-1/2" long					.22			.22	.24
	1200	2" long					.25			.25	.28
	1300	4" long					.37			.37	.41
	1400	6" long					.49			.49	.54
	1500	8" long					.66			.66	.73
	1600	5/8" diameter, 1-1/2" long					.38			.38	.42
	1700	2" long					.42			.42	.46
	1800	4" long					.59			.59	.65
	1900	6" long					.80			.80	.88
	2000	8" long					1.06			1.06	1.17
	2100	10" long					1.26			1.26	1.39
	2200	3/4" diameter, 2" long					.66			.66	.73
	2300	4" long					.97			.97	1.07
	2400	6" long					1.23			1.23	1.35
	2500	8" long					1.54			1.54	1.69
	2600	10" long					2.07			2.07	2.28
	2700	12" long					2.36			2.36	2.60
	2800	1" diameter, 3" long					1.61			1.61	1.77
	2900	6" long					2.36			2.36	2.60
	3000	12" long					4.74			4.74	5.20
	3100	For galvanized, add					75%				
	3200	For stainless, add				▼	350%				
340	0010	**DRILLING** For anchors, up to 4" deep, incl. bit and layout									
	0050	in concrete or brick walls and floors, no anchor									
	0100	Holes, 1/4" diameter	1 Carp	75	.107	Ea.	.09	2.10		2.19	3.69
	0150	For each additional inch of depth, add		430	.019		.02	.37		.39	.65
	0200	3/8" diameter		63	.127		.08	2.50		2.58	4.37
	0250	For each additional inch of depth, add		340	.024		.02	.46		.48	.81
	0300	1/2" diameter		50	.160		.07	3.15		3.22	5.50
	0350	For each additional inch of depth, add		250	.032		.02	.63		.65	1.10
	0400	5/8" diameter		48	.167		.14	3.28		3.42	5.80
	0450	For each additional inch of depth, add		240	.033		.03	.66		.69	1.17
	0500	3/4" diameter		45	.178		.14	3.50		3.64	6.15
	0550	For each additional inch of depth, add		220	.036		.04	.72		.76	1.27
	0600	7/8" diameter		43	.186		.17	3.67		3.84	6.50
	0650	For each additional inch of depth, add		210	.038		.04	.75		.79	1.34
	0700	1" diameter		40	.200		.20	3.94		4.14	6.95
	0750	For each additional inch of depth, add		190	.042		.05	.83		.88	1.48
	0800	1-1/4" diameter		38	.211		.30	4.15		4.45	7.45
	0850	For each additional inch of depth, add		180	.044		.07	.88		.95	1.58
	0900	1-1/2" diameter		35	.229		.48	4.50		4.98	8.20
	0950	For each additional inch of depth, add	▼	165	.048	▼	.12	.96		1.08	1.77
	1000	For ceiling installations, add						40%			
	1100	Drilling & layout for drywall or plaster walls, no anchor									

	05090 | Metal Fastenings		CREW	DAILY OUTPUT	LABOR-HOURS	UNIT	2000 BARE COSTS MAT.	LABOR	EQUIP.	TOTAL	TOTAL INCL O&P	
40	1200	Holes, 1/4″ diameter	1 Carp	150	.053	Ea.	.01	1.05		1.06	1.81	340
	1300	3/8″ diameter		140	.057		.01	1.13		1.14	1.94	
	1400	1/2″ diameter		130	.062		.01	1.21		1.22	2.09	
	1500	3/4″ diameter		120	.067		.02	1.31		1.33	2.27	
	1600	1″ diameter		110	.073		.03	1.43		1.46	2.48	
	1700	1-1/4″ diameter		100	.080		.04	1.58		1.62	2.74	
	1800	1-1/2″ diameter		90	.089		.06	1.75		1.81	3.07	
	1900	For ceiling installations, add						40%				
80	0010	**EXPANSION ANCHORS** & shields										380
	0100	Bolt anchors for concrete, brick or stone, no layout and drilling										
	0200	Expansion shields, zinc, 1/4″ diameter, 1″ long, single	1 Carp	90	.089	Ea.	.90	1.75		2.65	3.99	
	0300	1-3/8″ long, double		85	.094		.99	1.85		2.84	4.27	
	0500	2″ long, double		80	.100		1.84	1.97		3.81	5.40	
	0700	2-1/2″ long, double		75	.107		2.38	2.10		4.48	6.20	
	0900	3″ long, double		70	.114		3.52	2.25		5.77	7.75	
	1100	4″ long, double		65	.123		7	2.42		9.42	11.85	
	1410	Concrete anchor, w/rod & epoxy cartridge, 1-3/4″ diameter x 15″ long	E-22	20	1.200		75.50	24.50		100	125	
	1415	18″ long		17	1.412		91	29		120	150	
	1420	2″ diameter x 18″ long		16	1.500		116	30.50		146.50	180	
	1425	24″ long		15	1.600		151	32.50		183.50	222	
	1430	Chemical anchor, w/rod & epoxy cartridge, 3/4″ diam. x 9-1/2″ long		27	.889		11.10	18.15		29.25	43	
	1435	1″ diameter x 11-3/4″ long		24	1		21	20.50		41.50	58	
	1440	1-1/4″ diameter x 14″ long		21	1.143		40	23.50		63.50	84	
	2100	Hollow wall anchors for gypsum wall board, plaster or tile										
	2500	3/16″ diameter, short				Ea.	.59			.59	.65	
	3000	Toggle bolts, bright steel, 1/8″ diameter, 2″ long	1 Carp	85	.094		.26	1.85		2.11	3.47	
	3100	4″ long		80	.100		.33	1.97		2.30	3.74	
	3200	3/16″ diameter, 3″ long		80	.100		.37	1.97		2.34	3.79	
	3300	6″ long		75	.107		.56	2.10		2.66	4.22	
	3400	1/4″ diameter, 3″ long		75	.107		.41	2.10		2.51	4.05	
	3500	6″ long		70	.114		.61	2.25		2.86	4.53	
	3600	3/8″ diameter, 3″ long		70	.114		.90	2.25		3.15	4.85	
	3700	6″ long		60	.133		1.32	2.63		3.95	5.95	
	3800	1/2″ diameter, 4″ long		60	.133		2.86	2.63		5.49	7.65	
	3900	6″ long		50	.160		3.81	3.15		6.96	9.60	
	4000	Nailing anchors										
	4100	Nylon nailing anchor, 1/4″ diameter, 1″ long				C	15.90			15.90	17.50	
	4200	1-1/2″ long					20.50			20.50	22.50	
	4300	2″ long					34			34	37.50	
	4400	Metal nailing anchor, 1/4″ diameter, 1″ long					22			22	24	
	4500	1-1/2″ long					30			30	33	
	4600	2″ long					38.50			38.50	42	
	5000	Screw anchors for concrete, masonry,										
	5100	stone & tile, no layout or drilling included										
	5200	Jute fiber, #6, #8, & #10, 1″ long				Ea.	.19			.19	.21	
	5300	#12, 1-1/2″ long					.28			.28	.31	
	5400	#14, 2″ long					.44			.44	.48	
	5500	#16, 2″ long					.46			.46	.51	
	5600	#20, 2″ long					.74			.74	.81	
	5700	Lag screw shields, 1/4″ diameter, short					.43			.43	.47	
	5800	Long					.50			.50	.55	
	5900	3/8″ diameter, short					.74			.74	.81	
	6000	Long					.84			.84	.92	
	6100	1/2″ diameter, short					1.11			1.11	1.22	
	6200	Long					1.31			1.31	1.44	
	6300	3/4″ diameter, short					2.42			2.42	2.66	

05050 | Basic Materials & Methods

	05090 | Metal Fastenings		CREW	DAILY OUTPUT	LABOR-HOURS	UNIT	2000 BARE COSTS MAT.	LABOR	EQUIP.	TOTAL	TOTAL INCL O&P
380	6400	Long				Ea.	3.09			3.09	3.40
	6600	Lead, #6 & #8, 3/4" long					.18			.18	.20
	6700	#10 - #14, 1-1/2" long					.27			.27	.30
	6800	#16 & #18, 1-1/2" long					.37			.37	.41
	6900	Plastic, #6 & #8, 3/4" long					.04			.04	.04
	7000	#8 & #10, 7/8" long					.04			.04	.04
	7100	#10 & #12, 1" long					.05			.05	.06
	7200	#14 & #16, 1-1/2" long					.06			.06	.07
460	0005	**LAG SCREWS**									
	0010	Steel, 1/4" diameter, 2" long	1 Carp	200	.040	Ea.	.07	.79		.86	1.43
	0100	3/8" diameter, 3" long		150	.053		.18	1.05		1.23	2
	0200	1/2" diameter, 3" long		130	.062		.30	1.21		1.51	2.41
	0300	5/8" diameter, 3" long		120	.067		.59	1.31		1.90	2.90
580	0005	**POWDER ACTUATED** Tools & fasteners									
	0010	Stud driver, .22 caliber, buy, minimum				Ea.	289			289	320
	0100	Maximum				"	505			505	555
	0300	Powder charges for above, low velocity				C	15.15			15.15	16.65
	0400	Standard velocity					25.50			25.50	28
	0600	Drive pins & studs, 1/4" & 3/8" diam., to 3" long, minimum					22			22	24
	0700	Maximum					57.50			57.50	63
600	0010	**RIVETS**									
	0100	Aluminum rivet & mandrel, 1/2" grip length x 1/8" diameter				C	4.53			4.53	4.98
	0200	3/16" diameter					7			7	7.70
	0300	Aluminum rivet, steel mandrel, 1/8" diameter					4.23			4.23	4.65
	0400	3/16" diameter					6.35			6.35	7
	0500	Copper rivet, steel mandrel, 1/8" diameter					5.50			5.50	6.05
	0600	Monel rivet, steel mandrel, 1/8" diameter					19.30			19.30	21
	0700	3/16" diameter					40.50			40.50	44.50
	0800	Stainless rivet & mandrel, 1/8" diameter					12			12	13.20
	0900	3/16" diameter					18.20			18.20	20
	1000	Stainless rivet, steel mandrel, 1/8" diameter					7.60			7.60	8.35
	1100	3/16" diameter					13.75			13.75	15.10
	1200	Steel rivet and mandrel, 1/8" diameter					4.74			4.74	5.20
	1300	3/16" diameter					7.15			7.15	7.85
	1400	Hand riveting tool, minimum				Ea.	76.50			76.50	84
	1500	Maximum					460			460	505
	1600	Power riveting tool, minimum					660			660	730
	1700	Maximum					1,750			1,750	1,925

05100 | Structural Metal Framing

	05120 | Structural Steel		CREW	DAILY OUTPUT	LABOR-HOURS	UNIT	2000 BARE COSTS MAT.	LABOR	EQUIP.	TOTAL	TOTAL INCL O&P
220	0010	**CEILING SUPPORTS**									
	1000	Entrance door/folding partition supports	E-4	60	.533	L.F.	12	11.60	1.40	25	37.50
	1100	Linear accelerator door supports		14	2.286		54.50	49.50	6	110	165
	1200	Lintels or shelf angles, hung, exterior hot dipped galv.		267	.120		8.20	2.61	.32	11.13	14.50
	1250	Two coats primer paint instead of galv.		267	.120		7.10	2.61	.32	10.03	13.30
	1400	Monitor support, ceiling hung, expansion bolted		4	8	Ea.	190	174	21	385	575

05100 | Structural Metal Framing

05120 | Structural Steel		CREW	DAILY OUTPUT	LABOR-HOURS	UNIT	2000 BARE COSTS MAT.	LABOR	EQUIP.	TOTAL	TOTAL INCL O&P	
1450	Hung from pre-set inserts	E-4	6	5.333	Ea.	205	116	14.05	335.05	470	220
1600	Motor supports for overhead doors		4	8		96.50	174	21	291.50	475	
1700	Partition support for heavy folding partitions, without pocket		24	1.333	L.F.	27.50	29	3.51	60.01	91	
1750	Supports at pocket only		12	2.667		54.50	58	7	119.50	182	
2000	Rolling grilles & fire door supports		34	.941		23.50	20.50	2.48	46.48	68.50	
2100	Spider-leg light supports, expansion bolted to ceiling slab		8	4	Ea.	78	87	10.50	175.50	270	
2150	Hung from pre-set inserts		12	2.667	″	84	58	7	149	214	
2400	Toilet partition support		36	.889	L.F.	27.50	19.35	2.34	49.19	70.50	
2500	X-ray travel gantry support		12	2.667	″	93.50	58	7	158.50	225	
0005	**COLUMNS**										260
0800	Steel (lally), concrete filled, extra strong pipe, 3-1/2″ diameter	E-2	660	.073	L.F.	18.35	1.56	1.55	21.46	24.50	
0830	4″ diameter		780	.062		20	1.32	1.31	22.63	26	
0890	5″ diameter		1,020	.047		25.50	1.01	1	27.51	31	
0930	6″ diameter		1,200	.040		34.50	.86	.85	36.21	40.50	
1000	Lightweight units, 3-1/2″ diameter		780	.062		2.33	1.32	1.31	4.96	6.55	
1050	4″ diameter		900	.053		3.26	1.15	1.14	5.55	7.05	
1100	For galvanizing, add				Lb.	.38			.38	.42	
1300	For web ties, angles, etc., add per added lb.	1 Sswk	945	.008		.60	.18		.78	1.01	
1500	Steel pipe, extra strong, no concrete, 3″ to 5″ diameter	E-2	16,000	.003		.69	.06	.06	.81	.95	
1600	6″ to 12″ diameter		14,000	.003		.70	.07	.07	.84	.99	
2400	Structural tubing, rect, 5″ to 6″ wide, light section		11,200	.004		.63	.09	.09	.81	.97	
2700	12″ x 8″ x 1/2″ thk wall		24,000	.002		.61	.04	.04	.69	.80	
2800	Heavy section		32,000	.002		.61	.03	.03	.67	.77	
8000	Lally columns, to 8′, 3-1/2″ diameter	2 Carp	24	.667	Ea.	18.65	13.15		31.80	43	
8080	4″ diameter	″	20	.800	″	26	15.75		41.75	55.50	
0005	**LINTELS**										480
0010	Plain steel angles, under 500 lb.	1 Bric	550	.015	Lb.	.46	.29		.75	1	
0100	500 to 1000 lb.		640	.013	″	.45	.25		.70	.92	
2000	Steel angles, 3-1/2″ x 3″, 1/4″ thick, 2′-6″ long		47	.170	Ea.	6.50	3.40		9.90	12.90	
2100	4′-6″ long		26	.308		11.65	6.15		17.80	23.50	
2600	4″ x 3-1/2″, 1/4″ thick, 5′-0″ long		21	.381		14.90	7.60		22.50	29.50	
2700	9′-0″ long		12	.667		27	13.35		40.35	52	
3500	For precast concrete lintels, see div. 03480-400										
0010	**STRUCTURAL STEEL** Bolted, incl. fabrication										720
0050	Beams, W 6 x 9	E-2	720	.067	L.F.	6.50	1.43	1.42	9.35	11.45	
0100	W 8 x 10		720	.067		7.20	1.43	1.42	10.05	12.20	
0200	Columns, W 6 x 15		540	.089		11.70	1.91	1.89	15.50	18.60	
0250	W 8 x 31		540	.089		24	1.91	1.89	27.80	32.50	

05300 | Metal Decking

05310 | Steel Deck		CREW	DAILY OUTPUT	LABOR-HOURS	UNIT	2000 BARE COSTS MAT.	LABOR	EQUIP.	TOTAL	TOTAL INCL O&P	
0010	**METAL DECKING** Steel decking										300
1900	For multi-story or congested site, add						50%				
2100	Open type, galv., 1-1/2″ deep wide rib, 22 gauge, under 50 squares	E-4	4,500	.007	S.F.	.80	.15	.02	.97	1.20	
2600	20 gauge, under 50 squares		3,865	.008		.94	.18	.02	1.14	1.41	
2900	18 gauge, under 50 squares		3,800	.008		1.22	.18	.02	1.42	1.72	
3050	16 gauge, under 50 squares		3,700	.009		1.51	.19	.02	1.72	2.06	

5 METALS

05300 | Metal Decking

	05310 | Steel Deck		CREW	DAILY OUTPUT	LABOR-HOURS	UNIT	2000 BARE COSTS MAT.	LABOR	EQUIP.	TOTAL	TOTAL INCL O&P
300	3700	4-1/2" deep, long span roof, over 50 squares, 20 gauge	E-4	2,700	.012	S.F.	2.18	.26	.03	2.47	2.94
	6100	Slab form, steel, 28 gauge, 9/16" deep, uncoated		4,000	.008		.46	.17	.02	.65	.87
	6200	Galvanized		4,000	.008		.52	.17	.02	.71	.93
	6220	24 gauge, 1" deep, uncoated		3,900	.008		.45	.18	.02	.65	.87
	6240	Galvanized		3,900	.008		.56	.18	.02	.76	.99
	6300	24 gauge, 1-5/16" deep, uncoated		3,800	.008		.65	.18	.02	.85	1.09
	6400	Galvanized		3,800	.008		.71	.18	.02	.91	1.16
	6500	22 gauge, 1-5/16" deep, uncoated		3,700	.009		.57	.19	.02	.78	1.02
	6600	Galvanized		3,700	.009		.80	.19	.02	1.01	1.27
	6700	22 gauge, 3" deep uncoated		3,600	.009		.66	.19	.02	.87	1.14
	6800	Galvanized		3,600	.009		.77	.19	.02	.98	1.26

05400 | Cold Formed Metal Framing

	05410 | Load-Bearing Metal Studs		CREW	DAILY OUTPUT	LABOR-HOURS	UNIT	2000 BARE COSTS MAT.	LABOR	EQUIP.	TOTAL	TOTAL INCL O&P
100	0010	**BRACING**, shear wall X-bracing, per 10' x 10' bay, one face									
	0120	Metal strap, 20 ga x 4" wide	2 Carp	18	.889	Ea.	13.95	17.50		31.45	45.50
	0130	6" wide		18	.889		21.50	17.50		39	54
	0160	18 ga x 4" wide		16	1		23.50	19.70		43.20	59.50
	0170	6" wide		16	1		34	19.70		53.70	71.50
	0410	Continuous strap bracing, per horizontal row on both faces									
	0420	Metal strap, 20 ga x 2" wide, studs 12" O.C.	1 Carp	7	1.143	C.L.F.	35	22.50		57.50	77
	0430	16" O.C.		8	1		35	19.70		54.70	72.50
	0440	24" O.C.		10	.800		35	15.75		50.75	65.50
	0450	18 ga x 2" wide, studs 12" O.C.		6	1.333		57	26.50		83.50	108
	0460	16" O.C.		7	1.143		57	22.50		79.50	102
	0470	24" O.C.		8	1		57	19.70		76.70	97
120	0010	**BRIDGING**, solid between studs w/ 1-1/4" leg track, per stud bay									
	0200	Studs 12" O.C., 18 ga x 2-1/2" wide	1 Carp	125	.064	Ea.	.55	1.26		1.81	2.77
	0210	3-5/8" wide		120	.067		.63	1.31		1.94	2.94
	0220	4" wide		120	.067		.69	1.31		2	3.01
	0230	6" wide		115	.070		.89	1.37		2.26	3.33
	0240	8" wide		110	.073		1.18	1.43		2.61	3.75
	0300	16 ga x 2-1/2" wide		115	.070		.68	1.37		2.05	3.10
	0310	3-5/8" wide		110	.073		.80	1.43		2.23	3.33
	0320	4" wide		110	.073		.86	1.43		2.29	3.40
	0330	6" wide		105	.076		1.09	1.50		2.59	3.77
	0340	8" wide		100	.080		1.48	1.58		3.06	4.33
	1200	Studs 16" O.C., 18 ga x 2-1/2" wide		125	.064		.71	1.26		1.97	2.94
	1210	3-5/8" wide		120	.067		.81	1.31		2.12	3.14
	1220	4" wide		120	.067		.89	1.31		2.20	3.23
	1230	6" wide		115	.070		1.14	1.37		2.51	3.60
	1240	8" wide		110	.073		1.52	1.43		2.95	4.12
	1300	16 ga x 2-1/2" wide		115	.070		.87	1.37		2.24	3.31
	1310	3-5/8" wide		110	.073		1.02	1.43		2.45	3.58
	1320	4" wide		110	.073		1.11	1.43		2.54	3.67
	1330	6" wide		105	.076		1.40	1.50		2.90	4.11
	1340	8" wide		100	.080		1.90	1.58		3.48	4.79
	2200	Studs 24" O.C., 18 ga x 2-1/2" wide		125	.064		1.03	1.26		2.29	3.29

	05410 | Load-Bearing Metal Studs	CREW	DAILY OUTPUT	LABOR-HOURS	UNIT	2000 BARE COSTS MAT.	LABOR	EQUIP.	TOTAL	TOTAL INCL O&P	
2210	3-5/8" wide	1 Carp	120	.067	Ea.	1.17	1.31		2.48	3.54	120
2220	4" wide		120	.067		1.29	1.31		2.60	3.67	
2230	6" wide		115	.070		1.65	1.37		3.02	4.16	
2240	8" wide		110	.073		2.20	1.43		3.63	4.87	
2300	16 ga x 2-1/2" wide		115	.070		1.27	1.37		2.64	3.74	
2310	3-5/8" wide		110	.073		1.48	1.43		2.91	4.08	
2320	4" wide		110	.073		1.60	1.43		3.03	4.21	
2330	6" wide		105	.076		2.03	1.50		3.53	4.80	
2340	8" wide		100	.080		2.75	1.58		4.33	5.70	
3000	Continuous bridging, per row										
3100	16 ga x 1-1/2" channel thru studs 12" O.C.	1 Carp	6	1.333	C.L.F.	29	26.50		55.50	76.50	
3110	16" O.C.		7	1.143		29	22.50		51.50	70	
3120	24" O.C.		8.80	.909		29	17.90		46.90	62	
4100	2" x 2" angle x 18 ga, studs 12" O.C.		7	1.143		59.50	22.50		82	104	
4110	16" O.C.		9	.889		59.50	17.50		77	95.50	
4120	24" O.C.		12	.667		59.50	13.15		72.65	88	
4200	16 ga, studs 12" O.C.		5	1.600		75	31.50		106.50	137	
4210	16" O.C.		7	1.143		75	22.50		97.50	121	
4220	24" O.C.		10	.800		75	15.75		90.75	110	
0010	**FRAMING,** boxed headers/beams										300
0200	Double, 18 ga x 6" deep	2 Carp	220	.073	L.F.	3.12	1.43		4.55	5.90	
0210	8" deep		210	.076		3.58	1.50		5.08	6.50	
0220	10" deep		200	.080		4.29	1.58		5.87	7.40	
0230	12 " deep		190	.084		4.93	1.66		6.59	8.25	
0300	16 ga x 8" deep		180	.089		4.12	1.75		5.87	7.55	
0310	10" deep		170	.094		4.93	1.85		6.78	8.60	
0320	12 " deep		160	.100		5.40	1.97		7.37	9.35	
0400	14 ga x 10" deep		140	.114		5.85	2.25		8.10	10.30	
0410	12 " deep		130	.123		6.50	2.42		8.92	11.30	
1210	Triple, 18 ga x 8" deep		170	.094		5.15	1.85		7	8.85	
1220	10" deep		165	.097		6.15	1.91		8.06	10	
1230	12 " deep		160	.100		7.10	1.97		9.07	11.20	
1300	16 ga x 8" deep		145	.110		5.95	2.17		8.12	10.25	
1310	10" deep		140	.114		7.10	2.25		9.35	11.65	
1320	12 " deep		135	.119		7.80	2.33		10.13	12.55	
1400	14 ga x 10" deep		115	.139		8.05	2.74		10.79	13.55	
1410	12 " deep		110	.145		9	2.87		11.87	14.80	
0010	**FRAMING, STUD WALLS** w/ top & bottom track, no openings,										400
0020	headers, beams, bridging or bracing										
4100	8' high walls, 18 ga x 2-1/2" wide, studs 12" O.C.	2 Carp	54	.296	L.F.	5.35	5.85		11.20	15.90	
4110	16" O.C.		77	.208		4.29	4.09		8.38	11.70	
4120	24" O.C.		107	.150		3.21	2.95		6.16	8.60	
4130	3-5/8" wide, studs 12" O.C.		53	.302		6	5.95		11.95	16.80	
4140	16" O.C.		76	.211		4.80	4.15		8.95	12.40	
4150	24" O.C.		105	.152		3.60	3		6.60	9.10	
4160	4" wide, studs 12" O.C.		52	.308		6.50	6.05		12.55	17.60	
4170	16" O.C.		74	.216		5.20	4.26		9.46	13.05	
4180	24" O.C.		103	.155		3.92	3.06		6.98	9.55	
4190	6" wide, studs 12" O.C.		51	.314		8.10	6.20		14.30	19.50	
4200	16" O.C.		73	.219		6.50	4.32		10.82	14.55	
4210	24" O.C.		101	.158		4.89	3.12		8.01	10.75	
4220	8" wide, studs 12" O.C.		50	.320		10.40	6.30		16.70	22.50	
4230	16" O.C.		72	.222		8.35	4.38		12.73	16.70	
4240	24" O.C.		100	.160		6.35	3.15		9.50	12.35	
4300	16 ga x 2-1/2" wide, studs 12" O.C.		47	.340		6.15	6.70		12.85	18.30	

	05410 | Load-Bearing Metal Studs	CREW	DAILY OUTPUT	LABOR-HOURS	UNIT	2000 BARE COSTS MAT.	LABOR	EQUIP.	TOTAL	TOTAL INCL O&P
400	4310 16" O.C.	2 Carp	68	.235	L.F.	4.89	4.64		9.53	13.35
	4320 24" O.C.		94	.170		3.61	3.35		6.96	9.70
	4330 3-5/8" wide, studs 12" O.C.		46	.348		7.10	6.85		13.95	19.60
	4340 16" O.C.		66	.242		5.65	4.78		10.43	14.40
	4350 24" O.C.		92	.174		4.16	3.43		7.59	10.45
	4360 4" wide, studs 12" O.C.		45	.356		7.70	7		14.70	20.50
	4370 16" O.C.		65	.246		6.10	4.85		10.95	15.05
	4380 24" O.C.		90	.178		4.52	3.50		8.02	11
	4390 6" wide, studs 12" O.C.		44	.364		9.55	7.15		16.70	23
	4400 16" O.C.		64	.250		7.55	4.93		12.48	16.80
	4410 24" O.C.		88	.182		5.60	3.58		9.18	12.30
	4420 8" wide, studs 12" O.C.		43	.372		12.40	7.35		19.75	26
	4430 16" O.C.		63	.254		9.85	5		14.85	19.40
	4440 24" O.C.		86	.186		7.35	3.67		11.02	14.35
	5100 10' high walls, 18 ga x 2-1/2" wide, studs 12" O.C.		54	.296		6.45	5.85		12.30	17.10
	5110 16" O.C.		77	.208		5.10	4.09		9.19	12.60
	5120 24" O.C.		107	.150		3.75	2.95		6.70	9.20
	5130 3-5/8" wide, studs 12" O.C.		53	.302		7.20	5.95		13.15	18.10
	5140 16" O.C.		76	.211		5.70	4.15		9.85	13.35
	5150 24" O.C.		105	.152		4.20	3		7.20	9.75
	5160 4" wide, studs 12" O.C.		52	.308		7.80	6.05		13.85	19
	5170 16" O.C.		74	.216		6.20	4.26		10.46	14.10
	5180 24" O.C.		103	.155		4.57	3.06		7.63	10.30
	5190 6" wide, studs 12" O.C.		51	.314		9.70	6.20		15.90	21.50
	5200 16" O.C.		73	.219		7.70	4.32		12.02	15.85
	5210 24" O.C.		101	.158		5.70	3.12		8.82	11.60
	5220 8" wide, studs 12" O.C.		50	.320		12.45	6.30		18.75	24.50
	5230 16" O.C.		72	.222		9.90	4.38		14.28	18.40
	5240 24" O.C.		100	.160		7.35	3.15		10.50	13.50
	5300 16 ga x 2-1/2" wide, studs 12" O.C.		47	.340		7.45	6.70		14.15	19.70
	5310 16" O.C.		68	.235		5.85	4.64		10.49	14.40
	5320 24" O.C.		94	.170		4.25	3.35		7.60	10.45
	5330 3-5/8" wide, studs 12" O.C.		46	.348		8.60	6.85		15.45	21
	5340 16" O.C.		66	.242		6.75	4.78		11.53	15.65
	5350 24" O.C.		92	.174		4.90	3.43		8.33	11.25
	5360 4" wide, studs 12" O.C.		45	.356		9.30	7		16.30	22.50
	5370 16" O.C.		65	.246		7.30	4.85		12.15	16.35
	5380 24" O.C.		90	.178		5.30	3.50		8.80	11.85
	5390 6" wide, studs 12" O.C.		44	.364		11.50	7.15		18.65	25
	5400 16" O.C.		64	.250		9.05	4.93		13.98	18.40
	5410 24" O.C.		88	.182		6.60	3.58		10.18	13.40
	5420 8" wide, studs 12" O.C.		43	.372		14.95	7.35		22.30	29
	5430 16" O.C.		63	.254		11.80	5		16.80	21.50
	5440 24" O.C.		86	.186		8.60	3.67		12.27	15.75
	6190 12' high walls, 18 ga x 6" wide, studs 12" O.C.		41	.390		11.30	7.70		19	25.50
	6200 16" O.C.		58	.276		8.90	5.45		14.35	19.10
	6210 24" O.C.		81	.198		6.50	3.89		10.39	13.80
	6220 8" wide, studs 12" O.C.		40	.400		14.50	7.90		22.40	29.50
	6230 16" O.C.		57	.281		11.45	5.55		17	22
	6240 24" O.C.		80	.200		8.35	3.94		12.29	15.95
	6390 16 ga x 6" wide, studs 12" O.C.		35	.457		13.45	9		22.45	30.50
	6400 16" O.C.		51	.314		10.50	6.20		16.70	22
	6410 24" O.C.		70	.229		7.55	4.50		12.05	16.05
	6420 8" wide, studs 12" O.C.		34	.471		17.50	9.25		26.75	35
	6430 16" O.C.		50	.320		13.70	6.30		20	26
	6440 24" O.C.	↓	69	.232	↓	9.85	4.57		14.42	18.70

Important: See the Reference Section for critical supporting data - Reference Nos., Crews, & Location Factor

05410 | Load-Bearing Metal Studs		CREW	DAILY OUTPUT	LABOR-HOURS	UNIT	2000 BARE COSTS MAT.	LABOR	EQUIP.	TOTAL	TOTAL INCL O&P	
6530	14 ga x 3-5/8" wide, studs 12" O.C.	2 Carp	34	.471	L.F.	13.30	9.25		22.55	30.50	400
6540	16" O.C.		48	.333		10.35	6.55		16.90	22.50	
6550	24" O.C.		65	.246		7.40	4.85		12.25	16.45	
6560	4" wide, studs 12" O.C.		33	.485		14.10	9.55		23.65	32	
6570	16" O.C.		47	.340		11	6.70		17.70	23.50	
6580	24" O.C.		64	.250		7.90	4.93		12.83	17.10	
6730	12 ga x 3-5/8" wide, studs 12" O.C.		31	.516		19.75	10.15		29.90	39	
6740	16" O.C.		43	.372		15.20	7.35		22.55	29.50	
6750	24" O.C.		59	.271		10.65	5.35		16	21	
6760	4" wide, studs 12" O.C.		30	.533		21.50	10.50		32	41.50	
6770	16" O.C.		42	.381		16.60	7.50		24.10	31	
6780	24" O.C.		58	.276		11.60	5.45		17.05	22	
7390	16' high walls, 16 ga x 6" wide, studs 12" O.C.		33	.485		17.35	9.55		26.90	35.50	
7400	16" O.C.		48	.333		13.45	6.55		20	26	
7410	24" O.C.		67	.239		9.55	4.70		14.25	18.55	
7420	8" wide, studs 12" O.C.		32	.500		22.50	9.85		32.35	42	
7430	16" O.C.		47	.340		17.50	6.70		24.20	31	
7440	24" O.C.		66	.242		12.40	4.78		17.18	22	
7560	14 ga x 4" wide, studs 12" O.C.		31	.516		18.30	10.15		28.45	37.50	
7570	16" O.C.		45	.356		14.10	7		21.10	27.50	
7580	24" O.C.		61	.262		9.95	5.15		15.10	19.80	
7590	6" wide, studs 12" O.C.		30	.533		23	10.50		33.50	43.50	
7600	16" O.C.		44	.364		17.90	7.15		25.05	32	
7610	24" O.C.		60	.267		12.65	5.25		17.90	23	
7760	12 ga x 4" wide, studs 12" O.C.		29	.552		28	10.85		38.85	49.50	
7770	16" O.C.		40	.400		21.50	7.90		29.40	37	
7780	24" O.C.		55	.291		14.90	5.75		20.65	26	
7790	6" wide, studs 12" O.C.		28	.571		35	11.25		46.25	58	
7800	16" O.C.		39	.410		27	8.10		35.10	43.50	
7810	24" O.C.		54	.296		18.55	5.85		24.40	30.50	
8590	20' high walls, 14 ga x 6" wide, studs 12" O.C.		29	.552		28.50	10.85		39.35	50	
8600	16" O.C.		42	.381		22	7.50		29.50	37	
8610	24" O.C.		57	.281		15.30	5.55		20.85	26.50	
8620	8" wide, studs 12" O.C.		28	.571		35	11.25		46.25	58	
8630	16" O.C.		41	.390		27	7.70		34.70	43	
8640	24" O.C.		56	.286		19	5.65		24.65	30.50	
8790	12 ga x 6" wide, studs 12" O.C.		27	.593		43.50	11.65		55.15	67.50	
8800	16" O.C.		37	.432		33	8.50		41.50	51	
8810	24" O.C.		51	.314		22.50	6.20		28.70	35.50	
8820	8" wide, studs 12" O.C.		26	.615		53	12.10		65.10	79	
8830	16" O.C.		36	.444		40.50	8.75		49.25	59.50	
8840	24" O.C.	↓	50	.320	↓	28	6.30		34.30	41.50	
05420 | Cold-Formed Metal Joists											
0010	**BRACING**, continuous, per row, top & bottom										100
0120	Flat strap, 20 ga x 2" wide, joists at 12" O.C.	1 Carp	4.67	1.713	C.L.F.	37	34		71	98.50	
0130	16" O.C.		5.33	1.501		35.50	29.50		65	89.50	
0140	24" O.C.		6.66	1.201		34	23.50		57.50	78	
0150	18 ga x 2" wide, joists at 12" O.C.		4	2		56.50	39.50		96	130	
0160	16" O.C.		4.67	1.713		55.50	34		89.50	119	
0170	24" O.C.	↓	5.33	1.501	↓	54.50	29.50		84	111	
0010	**BRIDGING**, solid between joists w/ 1-1/4" leg track, per joist bay										120
0230	Joists 12" O.C., 18 ga track x 6" wide	1 Carp	80	.100	Ea.	.89	1.97		2.86	4.36	
0240	8" wide		75	.107		1.18	2.10		3.28	4.90	
0250	10" wide	↓	70	.114	↓	1.52	2.25		3.77	5.55	

5 METALS

	05420 | Cold-Formed Metal Joists		CREW	DAILY OUTPUT	LABOR-HOURS	UNIT	2000 BARE COSTS MAT.	LABOR	EQUIP.	TOTAL	TOTAL INCL O&P	
120	0260	12" wide	1 Carp	65	.123	Ea.	1.75	2.42		4.17	6.10	1
	0330	16 ga track x 6" wide		70	.114		1.09	2.25		3.34	5.05	
	0340	8" wide		65	.123		1.48	2.42		3.90	5.80	
	0350	10" wide		60	.133		1.81	2.63		4.44	6.50	
	0360	12" wide		55	.145		2.15	2.87		5.02	7.25	
	0440	14 ga track x 8" wide		60	.133		1.90	2.63		4.53	6.60	
	0450	10" wide		55	.145		2.34	2.87		5.21	7.50	
	0460	12" wide		50	.160		2.78	3.15		5.93	8.45	
	0550	12 ga track x 10" wide		45	.178		3.55	3.50		7.05	9.90	
	0560	12" wide		40	.200		4.18	3.94		8.12	11.35	
	1230	16" O.C., 18 ga track x 6" wide		80	.100		1.14	1.97		3.11	4.63	
	1240	8" wide		75	.107		1.52	2.10		3.62	5.25	
	1250	10" wide		70	.114		1.95	2.25		4.20	6	
	1260	12" wide		65	.123		2.24	2.42		4.66	6.60	
	1330	16 ga track x 6" wide		70	.114		1.40	2.25		3.65	5.40	
	1340	8" wide		65	.123		1.90	2.42		4.32	6.25	
	1350	10" wide		60	.133		2.33	2.63		4.96	7.05	
	1360	12" wide		55	.145		2.76	2.87		5.63	7.95	
	1440	14 ga track x 8" wide		60	.133		2.44	2.63		5.07	7.20	
	1450	10" wide		55	.145		3	2.87		5.87	8.20	
	1460	12" wide		50	.160		3.56	3.15		6.71	9.30	
	1550	12 ga track x 10" wide		45	.178		4.55	3.50		8.05	11	
	1560	12" wide		40	.200		5.35	3.94		9.29	12.65	
	2230	24" O.C., 18 ga track x 6" wide		80	.100		1.65	1.97		3.62	5.20	
	2240	8" wide		75	.107		2.20	2.10		4.30	6	
	2250	10" wide		70	.114		2.82	2.25		5.07	6.95	
	2260	12" wide		65	.123		3.25	2.42		5.67	7.70	
	2330	16 ga track x 6" wide		70	.114		2.03	2.25		4.28	6.10	
	2340	8" wide		65	.123		2.75	2.42		5.17	7.15	
	2350	10" wide		60	.133		3.37	2.63		6	8.20	
	2360	12" wide		55	.145		3.99	2.87		6.86	9.30	
	2440	14 ga track x 8" wide		60	.133		3.53	2.63		6.16	8.40	
	2450	10" wide		55	.145		4.34	2.87		7.21	9.70	
	2460	12" wide		50	.160		5.15	3.15		8.30	11.05	
	2550	12 ga track x 10" wide		45	.178		6.60	3.50		10.10	13.25	
	2560	12" wide	↓	40	.200	↓	7.75	3.94		11.69	15.30	
200	0010	**FRAMING, BAND JOIST** (track) fastened to bearing wall										2
	0220	18 ga track x 6" deep	2 Carp	1,000	.016	L.F.	.72	.32		1.04	1.34	
	0230	8" deep		920	.017		.97	.34		1.31	1.65	
	0240	10" deep		860	.019		1.24	.37		1.61	1.99	
	0320	16 ga track x 6" deep		900	.018		.89	.35		1.24	1.58	
	0330	8" deep		840	.019		1.21	.38		1.59	1.97	
	0340	10" deep		780	.021		1.48	.40		1.88	2.32	
	0350	12" deep		740	.022		1.75	.43		2.18	2.66	
	0430	14 ga track x 8" deep		750	.021		1.55	.42		1.97	2.43	
	0440	10" deep		720	.022		1.91	.44		2.35	2.85	
	0450	12" deep		700	.023		2.27	.45		2.72	3.26	
	0540	12 ga track x 10" deep		670	.024		2.90	.47		3.37	4	
	0550	12" deep	↓	650	.025	↓	3.41	.48		3.89	4.58	
300	0010	**FRAMING, BOXED HEADERS/BEAMS**										3
	0200	Double, 18 ga x 6" deep	2 Carp	220	.073	L.F.	3.12	1.43		4.55	5.90	
	0210	8" deep		210	.076		3.58	1.50		5.08	6.50	
	0220	10" deep		200	.080		4.29	1.58		5.87	7.40	
	0230	12" deep		190	.084		4.93	1.66		6.59	8.25	
	0300	16 ga x 8" deep	↓	180	.089	↓	4.12	1.75		5.87	7.55	

Important: See the Reference Section for critical supporting data - Reference Nos., Crews, & Location Factors

05420 | Cold-Formed Metal Joists

	05420 Cold-Formed Metal Joists	CREW	DAILY OUTPUT	LABOR-HOURS	UNIT	2000 BARE COSTS MAT.	LABOR	EQUIP.	TOTAL	TOTAL INCL O&P	
0310	10″ deep	2 Carp	170	.094	L.F.	4.93	1.85		6.78	8.60	300
0320	12″ deep		160	.100		5.40	1.97		7.37	9.35	
0400	14 ga x 10″ deep		140	.114		5.85	2.25		8.10	10.30	
0410	12″ deep		130	.123		6.50	2.42		8.92	11.30	
0500	12 ga x 10″ deep		110	.145		7.85	2.87		10.72	13.55	
0510	12″ deep		100	.160		8.75	3.15		11.90	15	
1210	Triple, 18 ga x 8″ deep		170	.094		5.15	1.85		7	8.85	
1220	10″ deep		165	.097		6.15	1.91		8.06	10	
1230	12″ deep		160	.100		7.10	1.97		9.07	11.20	
1300	16 ga x 8″ deep		145	.110		5.95	2.17		8.12	10.25	
1310	10″ deep		140	.114		7.10	2.25		9.35	11.65	
1320	12″ deep		135	.119		7.80	2.33		10.13	12.55	
1400	14 ga x 10″ deep		115	.139		8.50	2.74		11.24	14.05	
1410	12″ deep		110	.145		9.45	2.87		12.32	15.30	
1500	12 ga x 10″ deep		90	.178		11.50	3.50		15	18.65	
1510	12″ deep		85	.188		12.85	3.71		16.56	20.50	
0010	**FRAMING, JOISTS**, no band joists (track), web stiffeners, headers,										410
0020	beams, bridging or bracing										
0030	Joists (2″ flange) and fasteners, materials only										
0220	18 ga x 6″ deep				L.F.	.96			.96	1.05	
0230	8″ deep					1.20			1.20	1.32	
0240	10″ deep					1.41			1.41	1.55	
0320	16 ga x 6″ deep					1.23			1.23	1.35	
0330	8″ deep					1.48			1.48	1.63	
0340	10″ deep					1.74			1.74	1.92	
0350	12″ deep					1.98			1.98	2.18	
0430	14 ga x 8″ deep					1.88			1.88	2.07	
0440	10″ deep					2.24			2.24	2.46	
0450	12″ deep					2.56			2.56	2.82	
0540	12 ga x 10″ deep					3.29			3.29	3.62	
0550	12″ deep					3.75			3.75	4.12	
1010	Installation of joists to band joists, beams & headers, labor only										
1220	18 ga x 6″ deep	2 Carp	110	.145	Ea.		2.87		2.87	4.91	
1230	8″ deep		90	.178			3.50		3.50	6	
1240	10″ deep		80	.200			3.94		3.94	6.75	
1320	16 ga x 6″ deep		95	.168			3.32		3.32	5.70	
1330	8″ deep		70	.229			4.50		4.50	7.70	
1340	10″ deep		60	.267			5.25		5.25	9	
1350	12″ deep		55	.291			5.75		5.75	9.80	
1430	14 ga x 8″ deep		65	.246			4.85		4.85	8.30	
1440	10″ deep		45	.356			7		7	12	
1450	12″ deep		35	.457			9		9	15.45	
1540	12 ga x 10″ deep		40	.400			7.90		7.90	13.50	
1550	12″ deep		30	.533			10.50		10.50	18	
0010	**FRAMING, WEB STIFFENERS** at joist bearing, fabricated from										500
0020	stud piece (1-5/8″ flange) to stiffen joist (2″ flange)										
2120	For 6″ deep joist, with 18 ga x 2-1/2″ stud	1 Carp	120	.067	Ea.	1.19	1.31		2.50	3.56	
2130	3-5/8″ stud		110	.073		1.23	1.43		2.66	3.80	
2140	4″ stud		105	.076		1.23	1.50		2.73	3.93	
2150	6″ stud		100	.080		1.32	1.58		2.90	4.15	
2160	8″ stud		95	.084		1.43	1.66		3.09	4.41	
2220	8″ deep joist, with 2-1/2″ stud		120	.067		1.30	1.31		2.61	3.68	
2230	3-5/8″ stud		110	.073		1.33	1.43		2.76	3.91	
2240	4″ stud		105	.076		1.35	1.50		2.85	4.06	
2250	6″ stud		100	.080		1.45	1.58		3.03	4.29	
2260	8″ stud		95	.084		1.64	1.66		3.30	4.64	

05420 | Cold-Formed Metal Joists		CREW	DAILY OUTPUT	LABOR-HOURS	UNIT	2000 BARE COSTS MAT.	LABOR	EQUIP.	TOTAL	TOTAL INCL O&P
500 2320	10" deep joist, with 2-1/2" stud	1 Carp	110	.073	Ea.	1.84	1.43		3.27	4.47
2330	3-5/8" stud		100	.080		1.89	1.58		3.47	4.78
2340	4" stud		95	.084		1.94	1.66		3.60	4.98
2350	6" stud		90	.089		2.06	1.75		3.81	5.25
2360	8" stud		85	.094		2.20	1.85		4.05	5.60
2420	12" deep joist, with 2-1/2" stud		110	.073		1.94	1.43		3.37	4.59
2430	3-5/8" stud		100	.080		1.98	1.58		3.56	4.88
2440	4" stud		95	.084		2.02	1.66		3.68	5.05
2450	6" stud		90	.089		2.16	1.75		3.91	5.40
2460	8" stud		85	.094		2.45	1.85		4.30	5.85
3130	For 6" deep joist, with 16 ga x 3-5/8" stud		100	.080		1.30	1.58		2.88	4.12
3140	4" stud		95	.084		1.32	1.66		2.98	4.29
3150	6" stud		90	.089		1.42	1.75		3.17	4.56
3160	8" stud		85	.094		1.59	1.85		3.44	4.93
3230	8" deep joist, with 3-5/8" stud		100	.080		1.44	1.58		3.02	4.28
3240	4" stud		95	.084		1.45	1.66		3.11	4.43
3250	6" stud		90	.089		1.58	1.75		3.33	4.73
3260	8" stud		85	.094		1.79	1.85		3.64	5.15
3330	10" deep joist, with 3-5/8" stud		85	.094		1.97	1.85		3.82	5.35
3340	4" stud		80	.100		2.06	1.97		4.03	5.65
3350	6" stud		75	.107		2.20	2.10		4.30	6
3360	8" stud		70	.114		2.42	2.25		4.67	6.55
3430	12" deep joist, with 3-5/8" stud		85	.094		2.15	1.85		4	5.55
3440	4" stud		80	.100		2.16	1.97		4.13	5.75
3450	6" stud		75	.107		2.35	2.10		4.45	6.20
3460	8" stud		70	.114		2.67	2.25		4.92	6.80
4230	For 8" deep joist, with 14 ga x 3-5/8" stud		90	.089		1.97	1.75		3.72	5.15
4240	4" stud		85	.094		2.02	1.85		3.87	5.40
4250	6" stud		80	.100		2.21	1.97		4.18	5.80
4260	8" stud		75	.107		2.39	2.10		4.49	6.25
4330	10" deep joist, with 3-5/8" stud		75	.107		2.77	2.10		4.87	6.65
4340	4" stud		70	.114		2.76	2.25		5.01	6.90
4350	6" stud		65	.123		3.07	2.42		5.49	7.50
4360	8" stud		60	.133		3.23	2.63		5.86	8.05
4430	12" deep joist, with 3-5/8" stud		75	.107		2.94	2.10		5.04	6.85
4440	4" stud		70	.114		3.02	2.25		5.27	7.20
4450	6" stud		65	.123		3.30	2.42		5.72	7.80
4460	8" stud		60	.133		3.56	2.63		6.19	8.40
5330	For 10" deep joist, with 12 ga x 3-5/8" stud		65	.123		3.15	2.42		5.57	7.60
5340	4" stud		60	.133		3.31	2.63		5.94	8.15
5350	6" stud		55	.145		3.59	2.87		6.46	8.85
5360	8" stud		50	.160		3.94	3.15		7.09	9.75
5430	12" deep joist, with 3-5/8" stud		65	.123		3.50	2.42		5.92	8
5440	4" stud		60	.133		3.49	2.63		6.12	8.35
5450	6" stud		55	.145		3.91	2.87		6.78	9.20
5460	8" stud	↓	50	.160	↓	4.50	3.15		7.65	10.35

05500 | Metal Fabrications

05517 | Metal Stairs		CREW	DAILY OUTPUT	LABOR-HOURS	UNIT	2000 BARE COSTS MAT.	LABOR	EQUIP.	TOTAL	TOTAL INCL O&P	
0005	**FIRE ESCAPE STAIRS**										350
0010	One story, disappearing, stainless steel	2 Sswk	20	.800	V.L.F.	142	17		159	190	
0100	Portable ladder				Ea.	50			50	55	
0010	**STAIR** Steel, safety nosing, steel stringers										700
1700	Pre-erected, steel pan tread, 3′-6″ wide, 2 line pipe rail	E-2	87	.552	Riser	165	11.85	11.75	188.60	218	
1800	With flat bar picket rail	″	87	.552		185	11.85	11.75	208.60	240	
1810	Spiral aluminum, 5′-0″ diameter, stock units	E-4	45	.711		185	15.45	1.87	202.32	237	
1820	Custom units		45	.711		350	15.45	1.87	367.32	420	
1900	Spiral, cast iron, 4′-0″ diameter, ornamental, minimum		45	.711		165	15.45	1.87	182.32	215	
1920	Maximum	↓	25	1.280	↓	225	28	3.37	256.37	305	

05520 | Handrails & Railings		CREW	DAILY OUTPUT	LABOR-HOURS	UNIT	MAT.	LABOR	EQUIP.	TOTAL	TOTAL INCL O&P	
0005	**RAILING, PIPE**										700
0010	Aluminum, 2 rail, satin finish, 1-1/4″ diameter	E-4	160	.200	L.F.	13	4.35	.53	17.88	23.50	
0030	Clear anodized		160	.200		16.10	4.35	.53	20.98	27	
0040	Dark anodized		160	.200		18.15	4.35	.53	23.03	29	
0080	1-1/2″ diameter, satin finish		160	.200		15.55	4.35	.53	20.43	26.50	
0090	Clear anodized		160	.200		17.35	4.35	.53	22.23	28.50	
0100	Dark anodized		160	.200		19.25	4.35	.53	24.13	30	
0140	Aluminum, 3 rail, 1-1/4″ diam., satin finish		137	.234		19.90	5.10	.61	25.61	32.50	
0150	Clear anodized		137	.234		25	5.10	.61	30.71	38	
0160	Dark anodized		137	.234		27.50	5.10	.61	33.21	41	
0200	1-1/2″ diameter, satin finish		137	.234		24	5.10	.61	29.71	37	
0210	Clear anodized		137	.234		27	5.10	.61	32.71	40.50	
0220	Dark anodized		137	.234		29.50	5.10	.61	35.21	43	
0500	Steel, 2 rail, on stairs, primed, 1-1/4″ diameter		160	.200		9.35	4.35	.53	14.23	19.50	
0520	1-1/2″ diameter		160	.200		10.30	4.35	.53	15.18	20.50	
0540	Galvanized, 1-1/4″ diameter		160	.200		13	4.35	.53	17.88	23.50	
0560	1-1/2″ diameter		160	.200		14.55	4.35	.53	19.43	25	
0580	Steel, 3 rail, primed, 1-1/4″ diameter		137	.234		13.95	5.10	.61	19.66	26	
0600	1-1/2″ diameter		137	.234		14.80	5.10	.61	20.51	27	
0620	Galvanized, 1-1/4″ diameter		137	.234		19.60	5.10	.61	25.31	32	
0640	1-1/2″ diameter		137	.234		23	5.10	.61	28.71	36	
0700	Stainless steel, 2 rail, 1-1/4″ diam. #4 finish		137	.234		32	5.10	.61	37.71	45.50	
0720	High polish		137	.234		51.50	5.10	.61	57.21	67	
0740	Mirror polish		137	.234		64.50	5.10	.61	70.21	81.50	
0760	Stainless steel, 3 rail, 1-1/2″ diam., #4 finish		120	.267		48	5.80	.70	54.50	65	
0770	High polish		120	.267		79.50	5.80	.70	86	99.50	
0780	Mirror finish		120	.267		97	5.80	.70	103.50	119	
0900	Wall rail, alum. pipe, 1-1/4″ diam., satin finish		213	.150		7.45	3.27	.40	11.12	15.05	
0905	Clear anodized		213	.150		9.05	3.27	.40	12.72	16.90	
0910	Dark anodized		213	.150		11	3.27	.40	14.67	19	
0915	1-1/2″ diameter, satin finish		213	.150		8.25	3.27	.40	11.92	15.95	
0920	Clear anodized		213	.150		10.35	3.27	.40	14.02	18.30	
0925	Dark anodized		213	.150		12.80	3.27	.40	16.47	21	
0930	Steel pipe, 1-1/4″ diameter, primed		213	.150		5.70	3.27	.40	9.37	13.15	
0935	Galvanized		213	.150		8.25	3.27	.40	11.92	15.95	
0940	1-1/2″ diameter		176	.182		5.85	3.95	.48	10.28	14.75	
0945	Galvanized		213	.150		8.30	3.27	.40	11.97	16	
0955	Stainless steel pipe, 1-1/2″ diam., #4 finish		107	.299		25.50	6.50	.79	32.79	41.50	
0960	High polish		107	.299		52	6.50	.79	59.29	70.50	
0965	Mirror polish	↓	107	.299	↓	61	6.50	.79	68.29	81	

METALS 5

		05580 | Formed Metal Fabrications	CREW	DAILY OUTPUT	LABOR-HOURS	UNIT	2000 BARE COSTS MAT.	LABOR	EQUIP.	TOTAL	TOTAL INCL O&P	
600	0005	**LAMP POSTS**										6
	0010	Aluminum, 7' high, stock units, post only	1 Carp	16	.500	Ea.	65	9.85		74.85	88.50	
	0100	Mild steel, plain	"	16	.500	"	39	9.85		48.85	60	

For information about Means Estimating Seminars, see yellow pages 11 and 12 in back of book

Division 6 Wood & Plastics

Estimating Tips

06050 Basic Wood & Plastic Materials & Methods

- Common to any wood framed structure are the accessory connector items such as screws, nails, adhesives, hangers, connector plates, straps, angles and holdowns. For typical wood framed buildings, such as residential projects, the aggregate total for these items can be significant, especially in areas where seismic loading is a concern. For floor and wall framing, nail quantities can be figured on a "pounds per thousand board feet basis", with 10 to 25 lbs. per MBF the range. Holdowns, hangers and other connectors should be taken off by the piece.

06100 Rough Carpentry

- Lumber is a traded commodity and therefore sensitive to supply and demand in the marketplace. Even in "budgetary" estimating of wood framed projects, it is advisable to call local suppliers for the latest market pricing.
- Common quantity units for wood framed projects are "thousand board feet" (MBF). A board foot is a volume of wood, 1″ x 1′ x 1′, or 144 cubic inches. Board foot quantities are generally calculated using nominal material dimensions—dressed sizes are ignored. Board foot per lineal foot of any stick of lumber can be calculated by dividing the nominal cross sectional area by 12. As an example, 2,000 lineal feet of 2 x 12 equates to 4 MBF by dividing the nominal area, 2 x 12, by 12, which equals 2, and multiplying by 2,000 to give 4,000 board feet. This simple rule applies to all nominal dimensioned lumber.
- Waste is an issue of concern at the quantity takeoff for any area of construction. Framing lumber is sold in even foot lengths, i.e., 10′, 12′, 14′, 16′, and depending on spans, wall heights and the grade of lumber, waste is inevitable. A rule of thumb for lumber waste is 5% to 10% depending on material quality and the complexity of the framing.
- Wood in various forms and shapes is used in many projects, even where the main structural framing is steel, concrete or masonry. Plywood as a back-up partition material and 2x boards used as blocking and cant strips around roof edges are two common examples. The estimator should ensure that the costs of all wood materials are included in the final estimate.

06200 Finish Carpentry

- It is necessary to consider the grade of workmanship when estimating labor costs for erecting millwork and interior finish. In practice, there are three grades: premium, custom and economy. The Means daily output for base and case moldings is in the range of 200 to 250 L.F. per carpenter per day. This is appropriate for most average custom grade projects. For premium projects an adjustment to productivity of 25% to 50% should be made depending on the complexity of the job.

Reference Numbers

Reference numbers are shown in bold squares at the beginning of some major classifications. These numbers refer to related items in the Reference Section. The reference information may be an estimating procedure, an alternate pricing method or technical information.

Note: Not all subdivisions listed here necessarily appear in this publication.

	06055 | Wood & Plastic Laminate		CREW	DAILY OUTPUT	LABOR-HOURS	UNIT	2000 BARE COSTS MAT.	LABOR	EQUIP.	TOTAL	TOTAL INCL O&P
740	0010	**COUNTER TOP** Stock, plastic lam., 24" wide w/backsplash, min.	1 Carp	30	.267	L.F.	4.89	5.25		10.14	14.40
	0100	Maximum		25	.320		14.25	6.30		20.55	26.50
	0300	Custom plastic, 7/8" thick, aluminum molding, no splash		30	.267		15.70	5.25		20.95	26.50
	0400	Cove splash		30	.267		20.50	5.25		25.75	31.50
	0600	1-1/4" thick, no splash		28	.286		18.50	5.65		24.15	30
	0700	Square splash		28	.286		23	5.65		28.65	35
	0900	Square edge, plastic face, 7/8" thick, no splash		30	.267		19.80	5.25		25.05	31
	1000	With splash		30	.267		26	5.25		31.25	37.50
	1200	For stainless channel edge, 7/8" thick, add					2.21			2.21	2.43
	1300	1-1/4" thick, add					2.58			2.58	2.84
	1500	For solid color suede finish, add					1.96			1.96	2.16
	1700	For end splash, add				Ea.	12.35			12.35	13.60
	1901	For cut outs, standard, add, minimum	1 Carp	32	.250			4.93		4.93	8.45
	2000	Maximum		8	1		3.14	19.70		22.84	37.50
	2010	Cut out in blacksplash for elec. wall outlet		38	.211			4.15		4.15	7.10
	2020	Cut out for sink		20	.400			7.90		7.90	13.50
	2030	Cut out for stove top		18	.444			8.75		8.75	15
	2100	Postformed, including backsplash and front edge		30	.267	L.F.	8.55	5.25		13.80	18.40
	2110	Mitred, add		12	.667	Ea.		13.15		13.15	22.50
	2200	Built-in place, 25" wide, plastic laminate		25	.320	L.F.	10.80	6.30		17.10	22.50
	2300	Ceramic tile mosaic		25	.320		24.50	6.30		30.80	38
	2500	Marble, stock, with splash, 1/2" thick, minimum	1 Bric	17	.471		30	9.40		39.40	49
	2700	3/4" thick, maximum	"	13	.615		76	12.30		88.30	105
	2900	Maple, solid, laminated, 1-1/2" thick, no splash	1 Carp	28	.286		31	5.65		36.65	43.50
	3000	With square splash		28	.286		35	5.65		40.65	48
	3200	Stainless steel		24	.333	S.F.	73	6.55		79.55	92
	3400	Recessed cutting block with trim, 16" x 20" x 1"		8	1	Ea.	41	19.70		60.70	79.50
	3411	Replace cutting block only		16	.500	"	29	9.85		38.85	48.50
	3600	Table tops, plastic laminate, square edge, 7/8" thick		45	.178	S.F.	7	3.50		10.50	13.70
	3700	1-1/8" thick		40	.200	"	7.20	3.94		11.14	14.70

	06073 | Fire Retardant Treatment		CREW	DAILY OUTPUT	LABOR-HOURS	UNIT	MAT.	LABOR	EQUIP.	TOTAL	TOTAL INCL O&P
400	0011	**LUMBER TREATMENT**									
	0402	Fire retardant, wet				M.B.F.	245			245	270
	0502	KDAT					281			281	310
	0702	Salt treated, water borne, .40 lb. retention					158			158	174
	0802	Oil borne, 8 lb. retention					170			170	187
	1002	Kiln dried lumber, 1" & 2" thick, soft woods					102			102	112
	1102	Hard woods					107			107	118
	1500	For small size 1" stock, add					11			11	12.10
	1700	For full size rough lumber, add					20%				
600	0012	**PLYWOOD TREATMENT** Fire retardant, 1/4" thick				M.S.F.	229			229	252
	0032	3/8" thick					245			245	270
	0052	1/2" thick					260			260	286
	0072	5/8" thick					275			275	305
	0102	3/4" thick					305			305	335
	0200	For KDAT, add					61			61	67
	0502	Salt treated water borne, .25 lb., wet, 1/4" thick					122			122	134
	0532	3/8" thick					127			127	140
	0552	1/2" thick					128			128	140
	0572	5/8" thick					143			143	157
	0602	3/4" thick					148			148	163
	0800	For KDAT add					61			61	67
	0900	For .40 lb., per C.F. retention, add					51			51	56
	1000	For certification stamp, add					30			30	33

	06090 | Wood & Plastic Fastenings	CREW	DAILY OUTPUT	LABOR-HOURS	UNIT	2000 BARE COSTS MAT.	LABOR	EQUIP.	TOTAL	TOTAL INCL O&P	
0010	**NAILS** Prices of material only, based on 50# box purchase, copper, plain				Lb.	4.10			4.10	4.51	600
0400	Stainless steel, plain					5.40			5.40	5.95	
0500	Box, 3d to 20d, bright					1.13			1.13	1.24	
0520	Galvanized					1.31			1.31	1.44	
0600	Common, 3d to 60d, plain					.77			.77	.85	
0700	Galvanized					.99			.99	1.09	
0800	Aluminum					3.31			3.31	3.64	
1000	Annular or spiral thread, 4d to 60d, plain					.66			.66	.73	
1200	Galvanized					.83			.83	.91	
1400	Drywall nails, plain					.77			.77	.85	
1600	Galvanized					1.10			1.10	1.21	
1800	Finish nails, 4d to 10d, plain					.90			.90	.99	
2000	Galvanized					1.04			1.04	1.14	
2100	Aluminum					4.85			4.85	5.35	
2300	Flooring nails, hardened steel, 2d to 10d, plain					1.22			1.22	1.34	
2400	Galvanized					1.34			1.34	1.47	
2500	Gypsum lath nails, 1-1/8″, 13 ga. flathead, blued					1.33			1.33	1.46	
2600	Masonry nails, hardened steel, 3/4″ to 3″ long, plain					1.61			1.61	1.77	
2700	Galvanized					1.44			1.44	1.58	
2900	Roofing nails, threaded, galvanized					1.30			1.30	1.43	
3100	Aluminum					4.70			4.70	5.15	
3300	Compressed lead head, threaded, galvanized					1.44			1.44	1.58	
3600	Siding nails, plain shank, galvanized					1.33			1.33	1.46	
3800	Aluminum					4			4	4.40	
5000	Add to prices above for cement coating					.07			.07	.08	
5200	Zinc or tin plating					.12			.12	.13	
5500	Vinyl coated sinkers, 8d to 16d				▼	.55			.55	.61	
0010	**NAILS** mat. only, for pneumatic tools, framing, per carton of 5000, 2″				Ea.	36.50			36.50	40	650
0100	2-3/8″					41.50			41.50	45.50	
0200	Per carton of 4000, 3″					37			37	40.50	
0300	3-1/4″					39			39	43	
0400	Per carton of 5000, 2-3/8″, galv.					56.50			56.50	62	
0500	Per carton of 4000, 3″, galv.					63.50			63.50	70	
0600	3-1/4″, galv.					79			79	87	
0700	Roofing, per carton of 7200, 1″					34.50			34.50	38	
0800	1-1/4″					32			32	35.50	
0900	1-1/2″					37			37	40.50	
1000	1-3/4″				▼	44.50			44.50	49	
0010	**SHEET METAL SCREWS** Steel, standard, #8 x 3/4″, plain				C	2.69			2.69	2.96	700
0100	Galvanized					3.37			3.37	3.71	
0300	#10 x 1″, plain					3.69			3.69	4.06	
0400	Galvanized					4.26			4.26	4.69	
1500	Self-drilling, with washers, (pinch point) #8 x 3/4″, plain					4.68			4.68	5.15	
1600	Galvanized					7			7	7.70	
1800	#10 x 3/4″, plain					6.80			6.80	7.50	
1900	Galvanized					7.75			7.75	8.55	
3000	Stainless steel w/aluminum or neoprene washers, #14 x 1″, plain					18			18	19.80	
3100	#14 x 2″, plain				▼	24.50			24.50	27	
0010	**WOOD SCREWS** #8, 1″ long, steel				C	3.29			3.29	3.62	750
0100	Brass					11			11	12.10	
0200	#8, 2″ long, steel					3.69			3.69	4.06	
0300	Brass					11.50			11.50	12.65	
0400	#10, 1″ long, steel					4.22			4.22	4.64	
0500	Brass				▼	22.50			22.50	24.50	

		06090 | Wood & Plastic Fastenings	CREW	DAILY OUTPUT	LABOR-HOURS	UNIT	2000 BARE COSTS MAT.	LABOR	EQUIP.	TOTAL	TOTAL INCL O&P
750	0600	#10, 2" long, steel				C	7.50			7.50	8.25
	0700	Brass					39.50			39.50	43.50
	0800	#10, 3" long, steel					13.75			13.75	15.10
	1000	#12, 2" long, steel					4.79			4.79	5.25
	1100	Brass					16			16	17.60
	1500	#12, 3" long, steel					15.90			15.90	17.50
	2000	#12, 4" long, steel					28.50			28.50	31
800	0010	**TIMBER CONNECTORS** Add up cost of each part for total									
	0020	cost of connection									
	0100	Connector plates, steel, with bolts, straight	2 Carp	75	.213	Ea.	16.35	4.20		20.55	25
	0110	Tee	"	50	.320		24	6.30		30.30	37.50
	0200	Bolts, machine, sq. hd. with nut & washer, 1/2" diameter, 4" long	1 Carp	140	.057		.74	1.13		1.87	2.74
	0300	7-1/2" long		130	.062		.94	1.21		2.15	3.11
	0500	3/4" diameter, 7-1/2" long		130	.062		1.69	1.21		2.90	3.94
	0600	15" long		95	.084		2.16	1.66		3.82	5.20
	0720	Machine bolts, sq. hd. w/nut & wash		150	.053	Lb.	1.88	1.05		2.93	3.87
	0800	Drilling bolt holes in timber, 1/2" diameter		450	.018	Inch		.35		.35	.60
	0900	1" diameter		350	.023	"		.45		.45	.77
	1100	Framing anchors, 2 or 3 dimensional, 10 gauge, no nails incl.		175	.046	Ea.	.39	.90		1.29	1.97
	1250	Holdowns, 3 gauge base, 10 gauge body		8	1		13.60	19.70		33.30	49
	1300	Joist and beam hangers, 18 ga. galv., for 2" x 4" joist		175	.046		.49	.90		1.39	2.08
	1400	2" x 6" to 2" x 10" joist		165	.048		.42	.96		1.38	2.10
	1600	16 ga. galv., 3" x 6" to 3" x 10" joist		160	.050		2.29	.99		3.28	4.21
	1700	3" x 10" to 3" x 14" joist		160	.050		2.65	.99		3.64	4.61
	1800	4" x 6" to 4" x 10" joist		155	.052		2	1.02		3.02	3.94
	1900	4" x 10" to 4" x 14" joist		155	.052		2.72	1.02		3.74	4.73
	2000	Two-2" x 6" to two-2" x 10" joists		150	.053		2.19	1.05		3.24	4.21
	2100	Two-2" x 10" to two-2" x 14" joists		150	.053		2.19	1.05		3.24	4.21
	2300	3/16" thick, 6" x 8" joist		145	.055		4.80	1.09		5.89	7.15
	2400	6" x 10" joist		140	.057		5.65	1.13		6.78	8.20
	2500	6" x 12" joist		135	.059		6.80	1.17		7.97	9.50
	2700	1/4" thick, 6" x 14" joist		130	.062		8.45	1.21		9.66	11.40
	2800	Joist anchors, 1/4" x 1-1/4" x 18"		140	.057		3.29	1.13		4.42	5.55
	2900	Plywood clips, extruded aluminum H clip, for 3/4" panels					.12			.12	.13
	3000	Galvanized 18 ga. back-up clip					.11			.11	.12
	3200	Post framing, 16 ga. galv. for 4" x 4" base, 2 piece	1 Carp	130	.062		4.74	1.21		5.95	7.30
	3300	Cap		130	.062		2.33	1.21		3.54	4.64
	3500	Rafter anchors, 18 ga. galv., 1-1/2" wide, 5-1/4" long		145	.055		.38	1.09		1.47	2.28
	3600	10-3/4" long		145	.055		.78	1.09		1.87	2.72
	3800	Shear plates, 2-5/8" diameter		120	.067		1.41	1.31		2.72	3.80
	3900	4" diameter		115	.070		3.26	1.37		4.63	5.95
	4000	Sill anchors, embedded in concrete or block, 18-5/8" long		115	.070		.95	1.37		2.32	3.40
	4100	Spike grids, 4" x 4", flat or curved		120	.067		.34	1.31		1.65	2.62
	4400	Split rings, 2-1/2" diameter		120	.067		1.15	1.31		2.46	3.52
	4500	4" diameter		110	.073		1.77	1.43		3.20	4.40
	4700	Strap ties, 16 ga., 1-3/8" wide, 12" long		180	.044		.92	.88		1.80	2.51
	4800	24" long		160	.050		1.42	.99		2.41	3.25
	5000	Toothed rings, 2-5/8" or 4" diameter		90	.089		.97	1.75		2.72	4.07
	5200	Truss plates, nailed, 20 gauge, up to 32' span		17	.471	Truss	7	9.25		16.25	23.50
	5400	Washers, 2" x 2" x 1/8"				Ea.	.22			.22	.24
	5500	3" x 3" x 3/16"				"	.57			.57	.63
	6101	Beam hangers, polymer painted									
	6102	Bolted, 3 ga., (W x H x L)									
	6104	3-1/4" x 9" x 12" top flange	1 Carp	1	8	C	5,575	158		5,733	6,400
	6106	5-1/4" x 9" x 12" top flange		1	8		5,800	158		5,958	6,650

06090 | Wood & Plastic Fastenings		CREW	DAILY OUTPUT	LABOR-HOURS	UNIT	2000 BARE COSTS MAT.	LABOR	EQUIP.	TOTAL	TOTAL INCL O&P	
6108	5-1/4" x 11" x 11-3/4" top flange	1 Carp	1	8	C	13,000	158		13,158	14,600	800
6110	6-7/8" x 9" x 12" top flange		1	8		5,975	158		6,133	6,850	
6112	6-7/8" x 11" x 13-1/2" top flange		1	8		13,700	158		13,858	15,400	
6114	8-7/8" x 11" x 15-1/2" top flange		1	8		14,600	158		14,758	16,400	
6116	Nailed, 3 ga., (W x H x L)										
6118	3-1/4" x 10-1/2" x 10" top flange	1 Carp	1.80	4.444	C	3,850	87.50		3,937.50	4,375	
6120	3-1/4" x 10-1/2" x 12" top flange		1.80	4.444		4,450	87.50		4,537.50	5,050	
6122	5-1/4" x 9-1/2" x 10" top flange		1.80	4.444		4,150	87.50		4,237.50	4,700	
6124	5-1/4" x 9-1/2" x 12" top flange		1.80	4.444		4,700	87.50		4,787.50	5,325	
6126	5-1/2" x 9-1/2" x 12" top flange		1.80	4.444		4,150	87.50		4,237.50	4,700	
6128	6-7/8" x 8-1/2" x 12" top flange		1.80	4.444		4,275	87.50		4,362.50	4,875	
6130	7-1/2" x 8-1/2" x 12" top flange		1.80	4.444		4,350	87.50		4,437.50	4,925	
6132	8-7/8" x 7-1/2" x 14" top flange		1.80	4.444		4,800	87.50		4,887.50	5,425	
6201	Beam and purlin hangers, galvanized, 12 ga.										
6202	Purlin or joist size, 3" x 8"	1 Carp	1.70	4.706	C	705	92.50		797.50	935	
6204	3" x 10"		1.70	4.706		790	92.50		882.50	1,025	
6206	3" x 12"		1.65	4.848		935	95.50		1,030.50	1,200	
6208	3" x 14"		1.65	4.848		1,100	95.50		1,195.50	1,375	
6210	3" x 16"		1.65	4.848		1,250	95.50		1,345.50	1,550	
6212	4" x 8"		1.65	4.848		705	95.50		800.50	940	
6214	4" x 10"		1.65	4.848		820	95.50		915.50	1,075	
6216	4" x 12"		1.60	5		880	98.50		978.50	1,150	
6218	4" x 14"		1.60	5		955	98.50		1,053.50	1,225	
6220	4" x 16"		1.60	5		1,100	98.50		1,198.50	1,375	
6224	6" x 10"		1.55	5.161		1,200	102		1,302	1,500	
6226	6" x 12"		1.55	5.161		1,275	102		1,377	1,600	
6228	6" x 14"		1.50	5.333		1,400	105		1,505	1,700	
6230	6" x 16"		1.50	5.333		1,575	105		1,680	1,925	
6300	Column bases										
6302	4 x 4, 16 ga.	1 Carp	1.80	4.444	C	980	87.50		1,067.50	1,225	
6306	7 ga.		1.80	4.444		1,825	87.50		1,912.50	2,150	
6314	6 x 6, 16 ga.		1.75	4.571		1,250	90		1,340	1,525	
6318	7 ga.		1.75	4.571		2,550	90		2,640	2,950	
6326	8 x 8, 7 ga.		1.65	4.848		4,250	95.50		4,345.50	4,850	
6330	8 x 10, 7 ga.		1.65	4.848		4,550	95.50		4,645.50	5,175	
6590	Joist hangers, heavy duty 12 ga., galvanized										
6592	2" x 4"	1 Carp	1.75	4.571	C	800	90		890	1,025	
6594	2" x 6"		1.65	4.848		855	95.50		950.50	1,100	
6596	2" x 8"		1.65	4.848		910	95.50		1,005.50	1,175	
6598	2" x 10"		1.65	4.848		975	95.50		1,070.50	1,250	
6600	2" x 12"		1.65	4.848		1,100	95.50		1,195.50	1,375	
6622	(2) 2" x 6"		1.60	5		1,075	98.50		1,173.50	1,350	
6624	(2) 2" x 8"		1.60	5		1,150	98.50		1,248.50	1,450	
6626	(2) 2" x 10"		1.55	5.161		1,300	102		1,402	1,600	
6628	(2) 2" x 12"		1.55	5.161		1,575	102		1,677	1,900	
6890	Purlin hangers, painted										
6892	12 ga., 2" x 6"	1 Carp	1.80	4.444	C	975	87.50		1,062.50	1,225	
6894	2" x 8"		1.80	4.444		1,025	87.50		1,112.50	1,275	
6896	2" x 10"		1.80	4.444		1,050	87.50		1,137.50	1,300	
6898	2" x 12"		1.75	4.571		1,150	90		1,240	1,400	
6934	(2) 2" x 6"		1.70	4.706		985	92.50		1,077.50	1,225	
6936	(2) 2" x 8"		1.70	4.706		1,075	92.50		1,167.50	1,350	
6938	(2) 2" x 10"		1.70	4.706		1,200	92.50		1,292.50	1,475	
6940	(2) 2" x 12"		1.65	4.848		1,275	95.50		1,370.50	1,575	
0010	**ROUGH HARDWARE** Average % of carpentry material, minimum					.50%					825
0200	Maximum					1.50%					

WOOD & PLASTICS 6

06050 | Basic Wood / Plastic Materials / Methods

	06090 | Wood & Plastic Fastenings		CREW	DAILY OUTPUT	LABOR-HOURS	UNIT	2000 BARE COSTS MAT.	LABOR	EQUIP.	TOTAL	TOTAL INCL O&P
850	0010	**BRACING**									
	0302	Let-in, "T" shaped, 22 ga. galv. steel, studs at 16″ O.C.	1 Carp	580	.014	L.F.	.38	.27		.65	.88
	0402	Studs at 24″ O.C.		600	.013		.38	.26		.64	.86
	0502	16 ga. galv. steel straps, studs at 16″ O.C.		600	.013		.49	.26		.75	.99
	0602	Studs at 24″ O.C.		620	.013		.49	.25		.74	.98

6 WOOD & PLASTICS

06100 | Rough Carpentry

	06110 | Wood Framing		CREW	DAILY OUTPUT	LABOR-HOURS	UNIT	2000 BARE COSTS MAT.	LABOR	EQUIP.	TOTAL	TOTAL INCL O&P
100	0010	**BLOCKING** R06100-010									
	1950	Miscellaneous, to wood construction									
	2000	2″ x 4″	1 Carp	250	.032	L.F.	.39	.63		1.02	1.51
	2005	Pneumatic nailed		305	.026		.39	.52		.91	1.32
	2050	2″ x 6″		222	.036		.59	.71		1.30	1.87
	2055	Pneumatic nailed		271	.030		.59	.58		1.17	1.65
	2100	2″ x 8″		200	.040		.83	.79		1.62	2.26
	2105	Pneumatic nailed		244	.033		.83	.65		1.48	2.02
	2150	2″ x 10″		178	.045		1.22	.89		2.11	2.86
	2155	Pneumatic nailed		217	.037		1.22	.73		1.95	2.58
	2200	2″ x 12″		151	.053		1.48	1.04		2.52	3.41
	2205	Pneumatic nailed		185	.043		1.48	.85		2.33	3.08
	2300	To steel construction									
	2320	2″ x 4″	1 Carp	208	.038	L.F.	.39	.76		1.15	1.73
	2340	2″ x 6″		180	.044		.59	.88		1.47	2.15
	2360	2″ x 8″		158	.051		.83	1		1.83	2.62
	2380	2″ x 10″		136	.059		1.22	1.16		2.38	3.33
	2400	2″ x 12″		109	.073		1.48	1.45		2.93	4.10
150	0012	**BRACING** Let-in, with 1″ x 6″ boards, studs @ 16″ O.C.	1 Carp	150	.053	L.F.	.49	1.05		1.54	2.34
	0202	Studs @ 24″ O.C.	″	230	.035	″	.49	.69		1.18	1.71
200	0012	**BRIDGING** Wood, for joists 16″ O.C., 1″ x 3″	1 Carp	130	.062	Pr.	.27	1.21		1.48	2.37
	0017	Pneumatic nailed		170	.047		.27	.93		1.20	1.88
	0102	2″ x 3″ bridging		130	.062		.46	1.21		1.67	2.58
	0107	Pneumatic nailed		170	.047		.46	.93		1.39	2.09
	0302	Steel, galvanized, 18 ga., for 2″ x 10″ joists at 12″ O.C.		130	.062		.59	1.21		1.80	2.72
	0402	24″ O.C.		140	.057		2.20	1.13		3.33	4.35
	0602	For 2″ x 14″ joists at 16″ O.C.		130	.062		.54	1.21		1.75	2.68
	0902	Compression type, 16″ O.C., 2″ x 8″ joists		200	.040		1.26	.79		2.05	2.74
	1002	2″ x 12″ joists		200	.040		1.41	.79		2.20	2.90
505	0010	**FRAMING, BEAMS & GIRDERS** R06100-010									
	1002	Single, 2″ x 6″	2 Carp	700	.023	L.F.	.59	.45		1.04	1.42
	1007	Pneumatic nailed R06110-030		812	.020		.59	.39		.98	1.32
	1022	2″ x 8″		650	.025		.83	.48		1.31	1.74
	1027	Pneumatic nailed		754	.021		.83	.42		1.25	1.63
	1042	2″ x 10″		600	.027		1.22	.53		1.75	2.24
	1047	Pneumatic nailed		696	.023		1.22	.45		1.67	2.12
	1062	2″ x 12″		550	.029		1.48	.57		2.05	2.60
	1067	Pneumatic nailed		638	.025		1.48	.49		1.97	2.47
	1082	2″ x 14″		500	.032		2.06	.63		2.69	3.35

06110 | Wood Framing

			CREW	DAILY OUTPUT	LABOR-HOURS	UNIT	2000 BARE COSTS MAT.	LABOR	EQUIP.	TOTAL	TOTAL INCL O&P
505	1087	Pneumatic nailed (R06100-010)	2 Carp	580	.028	L.F.	2.06	.54		2.60	3.20
	1102	3″ x 8″		550	.029		2.33	.57		2.90	3.54
	1122	3″ x 10″ (R06110-030)		500	.032		2.91	.63		3.54	4.28
	1142	3″ x 12″		450	.036		3.49	.70		4.19	5.05
	1162	3″ x 14″	↓	400	.040		4.07	.79		4.86	5.85
	1170	4″ x 6″	F-3	1,100	.036		2.65	.65	.41	3.71	4.48
	1182	4″ x 8″		1,000	.040		3.54	.72	.45	4.71	5.60
	1202	4″ x 10″		950	.042		4.42	.75	.48	5.65	6.65
	1222	4″ x 12″		900	.044		5.30	.80	.50	6.60	7.75
	1242	4″ x 14″	↓	850	.047		6.20	.84	.53	7.57	8.80
	2002	Double, 2″ x 6″	2 Carp	625	.026		1.18	.50		1.68	2.16
	2007	Pneumatic nailed		725	.022		1.18	.43		1.61	2.04
	2022	2″ x 8″		575	.028		1.66	.55		2.21	2.76
	2027	Pneumatic nailed		667	.024		1.66	.47		2.13	2.63
	2042	2″ x 10″		550	.029		2.44	.57		3.01	3.66
	2047	Pneumatic nailed		638	.025		2.44	.49		2.93	3.53
	2062	2″ x 12″		525	.030		2.95	.60		3.55	4.28
	2067	Pneumatic nailed		610	.026		2.95	.52		3.47	4.14
	2082	2″ x 14″		475	.034		4.12	.66		4.78	5.65
	2087	Pneumatic nailed		551	.029		4.12	.57		4.69	5.50
	3002	Triple, 2″ x 6″		550	.029		1.77	.57		2.34	2.93
	3007	Pneumatic nailed		638	.025		1.77	.49		2.26	2.80
	3022	2″ x 8″		525	.030		2.49	.60		3.09	3.77
	3027	Pneumatic nailed		609	.026		2.49	.52		3.01	3.63
	3042	2″ x 10″		500	.032		3.65	.63		4.28	5.10
	3047	Pneumatic nailed		580	.028		3.65	.54		4.19	4.95
	3062	2″ x 12″		475	.034		4.43	.66		5.09	6
	3067	Pneumatic nailed		551	.029		4.43	.57		5	5.85
	3082	2″ x 14″		450	.036		6.20	.70		6.90	8
	3087	Pneumatic nailed	↓	522	.031	↓	6.20	.60		6.80	7.85
510	0010	**FRAMING, CEILINGS**									
	6002	Suspended, 2″ x 3″	2 Carp	1,000	.016	L.F.	.31	.32		.63	.88
	6052	2″ x 4″		900	.018		.39	.35		.74	1.03
	6102	2″ x 6″		800	.020		.59	.39		.98	1.33
	6152	2″ x 8″	↓	650	.025	↓	.83	.48		1.31	1.74
515	0010	**FRAMING, COLUMNS**									
	0101	4″ x 4″	2 Carp	390	.041	L.F.	1.11	.81		1.92	2.60
	0151	4″ x 6″		275	.058		2.65	1.15		3.80	4.88
	0201	4″ x 8″		220	.073		3.54	1.43		4.97	6.35
	0251	6″ x 6″		215	.074		5.50	1.47		6.97	8.50
	0301	6″ x 8″		175	.091		5.75	1.80		7.55	9.40
	0351	6″ x 10″	↓	150	.107	↓	7.15	2.10		9.25	11.45
520	0010	**FRAMING, HEAVY** Mill timber, beams, single 6″ x 10″	2 Carp	1.10	14.545	M.B.F.	1,425	287		1,712	2,075
	0100	Single 8″ x 16″		1.20	13.333	″	1,625	263		1,888	2,250
	0202	Built from 2″ lumber, multiple 2″ x 14″		900	.018	B.F.	.88	.35		1.23	1.57
	0212	Built from 3″ lumber, multiple 3″ x 6″		700	.023		1.16	.45		1.61	2.05
	0222	Multiple 3″ x 8″		800	.020		1.16	.39		1.55	1.96
	0232	Multiple 3″ x 10″		900	.018		1.16	.35		1.51	1.88
	0242	Multiple 3″ x 12″		1,000	.016		1.16	.32		1.48	1.82
	0252	Built from 4″ lumber, multiple 4″ x 6″		800	.020		1.33	.39		1.72	2.14
	0262	Multiple 4″ x 8″		900	.018		1.33	.35		1.68	2.06
	0272	Multiple 4″ x 10″		1,000	.016		1.33	.32		1.65	2
	0282	Multiple 4″ x 12″		1,100	.015	↓	1.33	.29		1.62	1.95
	0292	Columns, structural grade, 1500f, 4″ x 4″	↓	450	.036	L.F.	1.74	.70		2.44	3.12

	06110	Wood Framing	CREW	DAILY OUTPUT	LABOR-HOURS	UNIT	2000 BARE COSTS MAT.	LABOR	EQUIP.	TOTAL	TOTAL INCL O&P
520	0302	6″ x 6″	2 Carp	225	.071	L.F.	5.40	1.40		6.80	8.35
	0402	8″ x 8″		240	.067		10.40	1.31		11.71	13.70
	0502	10″ x 10″		90	.178		16.85	3.50		20.35	24.50
	0602	12″ x 12″		70	.229		24.50	4.50		29	34.50
	0802	Floor planks, 2″ thick, T & G, 2″ x 6″		1,050	.015	B.F.	1.60	.30		1.90	2.27
	0902	2″ x 10″		1,100	.015		.76	.29		1.05	1.33
	1102	3″ thick, 3″ x 6″		1,050	.015		1.15	.30		1.45	1.78
	1202	3″ x 10″		1,100	.015		1.15	.29		1.44	1.76
	1402	Girders, structural grade, 12″ x 12″		800	.020		1.65	.39		2.04	2.49
	1502	10″ x 16″		1,000	.016		1.61	.32		1.93	2.31
	2050	Roof planks, see division 06150-600									
	2302	Roof purlins, 4″ thick, structural grade	2 Carp	1,050	.015	B.F.	1.13	.30		1.43	1.75
	2502	Roof trusses, add timber connectors, division 06090-800	″	450	.036	″	1.10	.70		1.80	2.41
530	0010	FRAMING, JOISTS									
	2002	Joists, 2″ x 4″	2 Carp	1,250	.013	L.F.	.39	.25		.64	.86
	2007	Pneumatic nailed		1,438	.011		.39	.22		.61	.81
	2100	2″ x 6″		1,250	.013		.59	.25		.84	1.08
	2105	Pneumatic nailed		1,438	.011		.59	.22		.81	1.03
	2152	2″ x 8″		1,100	.015		.83	.29		1.12	1.40
	2157	Pneumatic nailed		1,265	.013		.83	.25		1.08	1.34
	2202	2″ x 10″		900	.018		1.22	.35		1.57	1.94
	2207	Pneumatic nailed		1,035	.015		1.22	.30		1.52	1.86
	2252	2″ x 12″		875	.018		1.48	.36		1.84	2.24
	2257	Pneumatic nailed		1,006	.016		1.48	.31		1.79	2.16
	2302	2″ x 14″		770	.021		2.06	.41		2.47	2.97
	2307	Pneumatic nailed		886	.018		2.06	.36		2.42	2.88
	2352	3″ x 6″		925	.017		1.75	.34		2.09	2.50
	2402	3″ x 10″		780	.021		2.91	.40		3.31	3.89
	2452	3″ x 12″		600	.027		3.49	.53		4.02	4.74
	2502	4″ x 6″		800	.020		2.65	.39		3.04	3.60
	2552	4″ x 10″		600	.027		4.42	.53		4.95	5.75
	2602	4″ x 12″		450	.036		5.30	.70		6	7.05
	2607	Sister joist, 2″ x 6″		800	.020		.59	.39		.98	1.33
	2608	Pneumatic nailed		960	.017		.59	.33		.92	1.21
	3000	Composite wood joist 9-1/2″ deep		.90	17.778	M.L.F.	1,400	350		1,750	2,150
	3010	11-1/2″ deep		.88	18.182		1,525	360		1,885	2,300
	3020	14″ deep		.82	19.512		1,700	385		2,085	2,525
	3030	16″ deep		.78	20.513		2,225	405		2,630	3,150
	4000	Open web joist 12″ deep		.88	18.182		1,575	360		1,935	2,350
	4010	14″ deep		.82	19.512		1,825	385		2,210	2,650
	4020	16″ deep		.78	20.513		1,900	405		2,305	2,800
	4030	18″ deep		.74	21.622		1,975	425		2,400	2,900
	6000	Composite rim joist, 1-1/4″ x 9-1/2″		90	.178		1,700	3.50		1,703.50	1,850
	6010	1-1/4″ x 11-1/2″		.88	18.182		1,850	360		2,210	2,650
	6020	1-1/4″ x 14-1/2″		.82	19.512		2,225	385		2,610	3,075
	6030	1-1/4″ x 16-1/2″		.78	20.513		2,350	405		2,755	3,275
545	0010	FRAMING, MISCELLANEOUS									
	2002	Firestops, 2″ x 4″	2 Carp	780	.021	L.F.	.39	.40		.79	1.12
	2007	Pneumatic nailed		952	.017		.39	.33		.72	1
	2102	2″ x 6″		600	.027		.59	.53		1.12	1.55
	2107	Pneumatic nailed		732	.022		.59	.43		1.02	1.39
	5002	Nailers, treated, wood construction, 2″ x 4″		800	.020		.66	.39		1.05	1.40
	5007	Pneumatic nailed		960	.017		.66	.33		.99	1.28
	5102	2″ x 6″		750	.021		1.05	.42		1.47	1.88

	06110 | Wood Framing	CREW	DAILY OUTPUT	LABOR-HOURS	UNIT	2000 BARE COSTS MAT.	LABOR	EQUIP.	TOTAL	TOTAL INCL O&P	
545	5107 Pneumatic nailed	2 Carp	900	.018	L.F.	1.05	.35		1.40	1.76	545
	5122 2" x 8"		700	.023		1.47	.45		1.92	2.39	
	5127 Pneumatic nailed		840	.019		1.47	.38		1.85	2.26	
	5202 Steel construction, 2" x 4"		750	.021		.66	.42		1.08	1.44	
	5222 2" x 6"		700	.023		1.05	.45		1.50	1.93	
	5242 2" x 8"		650	.025		1.47	.48		1.95	2.45	
	7002 Rough bucks, treated, for doors or windows, 2" x 6"		400	.040		1.05	.79		1.84	2.51	
	7007 Pneumatic nailed		480	.033		1.05	.66		1.71	2.29	
	7102 2" x 8"		380	.042		1.47	.83		2.30	3.04	
	7107 Pneumatic nailed		456	.035		1.47	.69		2.16	2.80	
	8001 Stair stringers, 2" x 10"		130	.123		1.22	2.42		3.64	5.50	
	8101 2" x 12"		130	.123		1.48	2.42		3.90	5.75	
	8151 3" x 10"		125	.128		2.91	2.52		5.43	7.50	
	8201 3" x 12"		125	.128		3.49	2.52		6.01	8.15	
	8870 Composite LSL, 1-1/4" x 11-1/2"		130	.123		1.84	2.42		4.26	6.15	
	8880 1-1/4" x 14-1/2"	↓	130	.123	↓	2.22	2.42		4.64	6.60	
550	0010 **PARTITIONS** Wood stud with single bottom plate and										550
	0020 double top plate, no waste, std. & better lumber										
	0182 2" x 4" studs, 8' high, studs 12" O.C.	2 Carp	80	.200	L.F.	4.78	3.94		8.72	12	
	0187 12" O.C., pneumatic nailed		96	.167		4.78	3.28		8.06	10.90	
	0202 16" O.C.		100	.160		3.91	3.15		7.06	9.70	
	0207 16" O.C., pneumatic nailed		120	.133		3.91	2.63		6.54	8.80	
	0302 24" O.C.		125	.128		3.04	2.52		5.56	7.65	
	0307 24" O.C., pneumatic nailed		150	.107		3.04	2.10		5.14	6.95	
	0382 10' high, studs 12" O.C.		80	.200		5.65	3.94		9.59	12.95	
	0387 12" O.C., pneumatic nailed		96	.167		5.65	3.28		8.93	11.85	
	0402 16" O.C.		100	.160		4.56	3.15		7.71	10.40	
	0407 16" O.C., pneumatic nailed		120	.133		4.56	2.63		7.19	9.50	
	0502 24" O.C.		125	.128		3.47	2.52		5.99	8.15	
	0507 24" O.C., pneumatic nailed		150	.107		3.47	2.10		5.57	7.40	
	0582 12' high, studs 12" O.C.		65	.246		6.50	4.85		11.35	15.45	
	0587 12" O.C., pneumatic nailed		78	.205		6.50	4.04		10.54	14.05	
	0602 16" O.C.		80	.200		5.20	3.94		9.14	12.50	
	0607 16" O.C., pneumatic nailed		96	.167		5.20	3.28		8.48	11.40	
	0700 24" O.C.		100	.160		3.55	3.15		6.70	9.30	
	0705 24" O.C., pneumatic nailed		120	.133		3.55	2.63		6.18	8.40	
	0782 2" x 6" studs, 8' high, studs 12" O.C.		70	.229		7.15	4.50		11.65	15.55	
	0787 12" O.C., pneumatic nailed		84	.190		7.15	3.75		10.90	14.30	
	0802 16" O.C.		90	.178		5.85	3.50		9.35	12.45	
	0807 16" O.C., pneumatic nailed		108	.148		5.85	2.92		8.77	11.45	
	0902 24" O.C.		115	.139		4.54	2.74		7.28	9.70	
	0907 24" O.C., pneumatic nailed		138	.116		4.54	2.28		6.82	8.90	
	0982 10' high, studs 12" O.C.		70	.229		8.45	4.50		12.95	17	
	0987 12" O.C., pneumatic nailed		84	.190		8.45	3.75		12.20	15.75	
	1002 16" O.C.		90	.178		6.80	3.50		10.30	13.50	
	1007 16" O.C., pneumatic nailed		108	.148		6.80	2.92		9.72	12.50	
	1102 24" O.C.		115	.139		5.20	2.74		7.94	10.40	
	1107 24" O.C., pneumatic nailed		138	.116		5.20	2.28		7.48	9.60	
	1182 12' high, studs 12" O.C.		55	.291		9.75	5.75		15.50	20.50	
	1187 12" O.C., pneumatic nailed		66	.242		9.75	4.78		14.53	18.90	
	1202 16" O.C.		70	.229		7.80	4.50		12.30	16.25	
	1207 16" O.C., pneumatic nailed		84	.190		7.80	3.75		11.55	15	
	1302 24" O.C.		90	.178		5.85	3.50		9.35	12.45	
	1307 24" O.C., pneumatic nailed		108	.148		5.85	2.92		8.77	11.45	
	1402 For horizontal blocking, 2" x 4", add		600	.027		.43	.53		.96	1.38	
	1502 2" x 6", add	↓	600	.027	↓	.65	.53		1.18	1.61	

	06110	Wood Framing	CREW	DAILY OUTPUT	LABOR-HOURS	UNIT	2000 BARE COSTS MAT.	LABOR	EQUIP.	TOTAL	TOTAL INCL O&P
550	1600	For openings, add	2 Carp	250	.064	L.F.		1.26		1.26	2.16
	1702	Headers for above openings, material only, add				B.F.	.68			.68	.75
555	0010	**FRAMING, ROOFS**									
	2001	Fascia boards, 2" x 8"	2 Carp	225	.071	L.F.	.83	1.40		2.23	3.31
	2101	2" x 10"		180	.089		1.22	1.75		2.97	4.34
	5002	Rafters, to 4 in 12 pitch, 2" x 6", ordinary		1,000	.016		.59	.32		.91	1.19
	5021	On steep roofs		800	.020		.59	.39		.98	1.33
	5041	On dormers or complex roofs		590	.027		.59	.53		1.12	1.57
	5062	2" x 8", ordinary		950	.017		.83	.33		1.16	1.48
	5081	On steep roofs		750	.021		.83	.42		1.25	1.63
	5101	On dormers or complex roofs		540	.030		.83	.58		1.41	1.91
	5122	2" x 10", ordinary		630	.025		1.22	.50		1.72	2.20
	5141	On steep roofs		495	.032		1.22	.64		1.86	2.43
	5161	On dormers or complex roofs		425	.038		1.22	.74		1.96	2.61
	5182	2" x 12", ordinary		575	.028		1.48	.55		2.03	2.56
	5201	On steep roofs		455	.035		1.48	.69		2.17	2.81
	5221	On dormers or complex roofs		395	.041		1.48	.80		2.28	2.99
	5250	Composite rafter, 9-1/2" deep		575	.028		1.52	.55		2.07	2.61
	5260	11-1/2" deep		575	.028		1.65	.55		2.20	2.76
	5301	Hip and valley rafters, 2" x 6", ordinary		760	.021		.59	.41		1	1.36
	5321	On steep roofs		585	.027		.59	.54		1.13	1.57
	5341	On dormers or complex roofs		510	.031		.59	.62		1.21	1.71
	5361	2" x 8", ordinary		720	.022		.83	.44		1.27	1.66
	5381	On steep roofs		545	.029		.83	.58		1.41	1.90
	5401	On dormers or complex roofs		470	.034		.83	.67		1.50	2.06
	5421	2" x 10", ordinary		570	.028		1.22	.55		1.77	2.29
	5441	On steep roofs		440	.036		1.22	.72		1.94	2.57
	5461	On dormers or complex roofs		380	.042		1.22	.83		2.05	2.76
	5470										
	5481	Hip and valley rafters, 2" x 12", ordinary	2 Carp	525	.030	L.F.	1.48	.60		2.08	2.65
	5501	On steep roofs		410	.039		1.48	.77		2.25	2.94
	5521	On dormers or complex roofs		355	.045		1.48	.89		2.37	3.14
	5541	Hip and valley jacks, 2" x 6", ordinary		600	.027		.59	.53		1.12	1.55
	5561	On steep roofs		475	.034		.59	.66		1.25	1.79
	5581	On dormers or complex roofs		410	.039		.59	.77		1.36	1.97
	5601	2" x 8", ordinary		490	.033		.83	.64		1.47	2.01
	5621	On steep roofs		385	.042		.83	.82		1.65	2.31
	5641	On dormers or complex roofs		335	.048		.83	.94		1.77	2.52
	5661	2" x 10", ordinary		450	.036		1.22	.70		1.92	2.54
	5681	On steep roofs		350	.046		1.22	.90		2.12	2.88
	5701	On dormers or complex roofs		305	.052		1.22	1.03		2.25	3.11
	5721	2" x 12", ordinary		375	.043		1.48	.84		2.32	3.06
	5741	On steep roofs		295	.054		1.48	1.07		2.55	3.45
	5762	On dormers or complex roofs		255	.063		1.48	1.24		2.72	3.74
	5781	Rafter tie, 1" x 4", #3		800	.020		.34	.39		.73	1.06
	5791	2" x 4", #3		800	.020		.39	.39		.78	1.11
	5801	Ridge board, #2 or better, 1" x 6"		600	.027		.90	.53		1.43	1.89
	5821	1" x 8"		550	.029		1.20	.57		1.77	2.30
	5841	1" x 10"		500	.032		1.50	.63		2.13	2.73
	5861	2" x 6"		500	.032		.59	.63		1.22	1.73
	5881	2" x 8"		450	.036		.83	.70		1.53	2.11
	5901	2" x 10"		400	.040		1.22	.79		2.01	2.69
	5921	Roof cants, split, 4" x 4"		650	.025		1.11	.48		1.59	2.05
	5941	6" x 6"		600	.027		5.50	.53		6.03	6.90
	5961	Roof curbs, untreated, 2" x 6"		520	.031		.59	.61		1.20	1.69
	5981	2" x 12"		400	.040		1.48	.79		2.27	2.97

06100 | Rough Carpentry

06110 | Wood Framing		CREW	DAILY OUTPUT	LABOR-HOURS	UNIT	2000 BARE COSTS MAT.	LABOR	EQUIP.	TOTAL	TOTAL INCL O&P	
6001	Sister rafters, 2″ x 6″	2 Carp	800	.020	L.F.	.59	.39		.98	1.33	555
6021	2″ x 8″		640	.025		.83	.49		1.32	1.75	
6041	2″ x 10″		535	.030		1.22	.59		1.81	2.35	
6061	2″ x 12″		455	.035		1.48	.69		2.17	2.81	
0010	**FRAMING, SILLS**										560
2002	Ledgers, nailed, 2″ x 4″	2 Carp	755	.021	L.F.	.39	.42		.81	1.15	
2052	2″ x 6″		600	.027		.59	.53		1.12	1.55	
2102	Bolted, not including bolts, 3″ x 6″		325	.049		1.75	.97		2.72	3.58	
2152	3″ x 12″		233	.069		3.49	1.35		4.84	6.15	
2602	Mud sills, redwood, construction grade, 2″ x 4″		895	.018		2.67	.35		3.02	3.53	
2622	2″ x 6″		780	.021		4	.40		4.40	5.10	
4002	Sills, 2″ x 4″		600	.027		.39	.53		.92	1.33	
4052	2″ x 6″		550	.029		.59	.57		1.16	1.63	
4082	2″ x 8″		500	.032		.83	.63		1.46	1.99	
4101	2″ x 10″		450	.036		1.22	.70		1.92	2.54	
4121	2″ x 12″		400	.040		1.48	.79		2.27	2.97	
4202	Treated, 2″ x 4″		550	.029		.66	.57		1.23	1.70	
4222	2″ x 6″		500	.032		1.05	.63		1.68	2.24	
4242	2″ x 8″		450	.036		1.47	.70		2.17	2.82	
4261	2″ x 10″		400	.040		1.80	.79		2.59	3.33	
4281	2″ x 12″		350	.046		2.33	.90		3.23	4.11	
4402	4″ x 4″		450	.036		1.78	.70		2.48	3.16	
4422	4″ x 6″		350	.046		2.53	.90		3.43	4.32	
4462	4″ x 8″		300	.053		3.37	1.05		4.42	5.50	
4480	4″ x 10″		260	.062		4.59	1.21		5.80	7.15	
0010	**FRAMING, SLEEPERS**										565
0100	On concrete, treated, 1″ x 2″	2 Carp	2,350	.007	L.F.	.20	.13		.33	.45	
0150	1″ x 3″		2,000	.008		.20	.16		.36	.49	
0200	2″ x 4″		1,500	.011		.66	.21		.87	1.08	
0250	2″ x 6″		1,300	.012		1.05	.24		1.29	1.58	
0010	**FRAMING, SOFFITS & CANOPIES**										570
1002	Canopy or soffit framing , 1″ x 4″	2 Carp	900	.018	L.F.	.60	.35		.95	1.26	
1021	1″ x 6″		850	.019		.90	.37		1.27	1.63	
1042	1″ x 8″		750	.021		1.20	.42		1.62	2.04	
1102	2″ x 4″		620	.026		.39	.51		.90	1.30	
1121	2″ x 6″		560	.029		.59	.56		1.15	1.61	
1142	2″ x 8″		500	.032		.83	.63		1.46	1.99	
1202	3″ x 4″		500	.032		1.07	.63		1.70	2.26	
1221	3″ x 6″		400	.040		1.75	.79		2.54	3.27	
1242	3″ x 10″		300	.053		2.91	1.05		3.96	5	
1250											
0010	**FRAMING, TREATED LUMBER**										575
0020	Water-borne salt, C.C.A., A.C.A., wet, .40 P.C.F. retention										
0100	2″ x 4″				M.B.F.	985			985	1,075	
0110	2″ x 6″					1,050			1,050	1,150	
0120	2″ x 8″					1,100			1,100	1,225	
0130	2″ x 10″					1,075			1,075	1,175	
0140	2″ x 12″					1,175			1,175	1,275	
0200	4″ x 4″					1,325			1,325	1,475	
0210	4″ x 6″					1,275			1,275	1,400	
0220	4″ x 8″					1,275			1,275	1,400	
0250	Add for .60 P.C.F. retention					40%					
0260	Add for 2.5 P.C.F. retention					200%					

6 WOOD & PLASTICS

	06110 | Wood Framing		CREW	DAILY OUTPUT	LABOR-HOURS	UNIT	2000 BARE COSTS MAT.	LABOR	EQUIP.	TOTAL	TOTAL INCL O&P
575	0270	Add for K.D.A.T.					20%				
590	0010	**FRAMING, WALLS** R06100-010									
	2002	Headers over openings, 2″ x 6″	2 Carp	360	.044	L.F.	.59	.88		1.47	2.15
	2007	2″ x 6″, pneumatic nailed R06110-030		432	.037		.59	.73		1.32	1.90
	2052	2″ x 8″		340	.047		.83	.93		1.76	2.50
	2057	2″ x 8″, pneumatic nailed		408	.039		.83	.77		1.60	2.23
	2101	2″ x 10″		320	.050		1.22	.99		2.21	3.03
	2106	2″ x 10″, pneumatic nailed		384	.042		1.22	.82		2.04	2.75
	2152	2″ x 12″		300	.053		1.48	1.05		2.53	3.42
	2157	2″ x 12″, pneumatic nailed		360	.044		1.48	.88		2.36	3.12
	2191	4″ x 10″		240	.067		4.42	1.31		5.73	7.10
	2196	4″ x 10″, pneumatic nailed		288	.056		4.42	1.09		5.51	6.75
	2202	4″ x 12″		190	.084		5.30	1.66		6.96	8.70
	2207	4″ x 12″, pneumatic nailed		228	.070		5.30	1.38		6.68	8.20
	2241	6″ x 10″		165	.097		7.15	1.91		9.06	11.10
	2246	6″ x 10″, pneumatic nailed		198	.081		7.15	1.59		8.74	10.60
	2251	6″ x 12″		140	.114		11.20	2.25		13.45	16.20
	2256	6″ x 12″, pneumatic nailed		168	.095		11.20	1.88		13.08	15.55
	5002	Plates, untreated, 2″ x 3″		850	.019		.31	.37		.68	.98
	5007	2″ x 3″, pneumatic nailed		1,020	.016		.31	.31		.62	.87
	5022	2″ x 4″		800	.020		.39	.39		.78	1.11
	5027	2″ x 4″, pneumatic nailed		960	.017		.39	.33		.72	.99
	5040	2″ x 6″		750	.021		.59	.42		1.01	1.37
	5045	2″ x 6″, pneumatic nailed		900	.018		.59	.35		.94	1.25
	5061	Treated, 2″ x 3″		850	.019		.55	.37		.92	1.24
	5066	2″ x 3″, treated, pneumatic nailed		1,020	.016		.55	.31		.86	1.13
	5081	2″ x 4″		800	.020		.66	.39		1.05	1.40
	5086	2″ x 4″, treated, pneumatic nailed		960	.017		.66	.33		.99	1.28
	5101	2″ x 6″		750	.021		1.05	.42		1.47	1.88
	5106	2″ x 6″, treated, pneumatic nailed		900	.018		1.05	.35		1.40	1.76
	5122	Studs, 8′ high wall, 2″ x 3″		1,200	.013		.31	.26		.57	.79
	5127	2″ x 3″, pneumatic nailed		1,440	.011		.31	.22		.53	.72
	5142	2″ x 4″		1,100	.015		.39	.29		.68	.92
	5147	2″ x 4″, pneumatic nailed		1,320	.012		.39	.24		.63	.84
	5162	2″ x 6″		1,000	.016		.59	.32		.91	1.19
	5167	2″ x 6″, pneumatic nailed		1,200	.013		.59	.26		.85	1.10
	5182	3″ x 4″		800	.020		1.07	.39		1.46	1.86
	5187	3″ x 4″, pneumatic nailed		960	.017		1.07	.33		1.40	1.74
	5201	Installed on second story, 2″ x 3″		1,170	.014		.31	.27		.58	.80
	5206	2″ x 3″, pneumatic nailed		1,200	.013		.31	.26		.57	.79
	5221	2″ x 4″		1,015	.016		.39	.31		.70	.96
	5226	2″ x 4″, pneumatic nailed		1,080	.015		.39	.29		.68	.93
	5241	2″ x 6″		890	.018		.59	.35		.94	1.26
	5246	2″ x 6″, pneumatic nailed		1,020	.016		.59	.31		.90	1.18
	5261	3″ x 4″		800	.020		1.07	.39		1.46	1.86
	5266	3″ x 4″, pneumatic nailed		960	.017		1.07	.33		1.40	1.74
	5281	Installed on dormer or gable, 2″ x 3″		1,045	.015		.31	.30		.61	.86
	5286	2″ x 3″, pneumatic nailed		1,254	.013		.31	.25		.56	.77
	5301	2″ x 4″		905	.018		.39	.35		.74	1.03
	5306	2″ x 4″, pneumatic nailed		1,086	.015		.39	.29		.68	.93
	5321	2″ x 6″		800	.020		.59	.39		.98	1.33
	5326	2″ x 6″, pneumatic nailed		960	.017		.59	.33		.92	1.21
	5341	3″ x 4″		700	.023		1.07	.45		1.52	1.95
	5346	3″ x 4″, pneumatic nailed		840	.019		1.07	.38		1.45	1.82
	5361	6′ high wall, 2″ x 3″		970	.016		.31	.32		.63	.90

	06110 | Wood Framing	CREW	DAILY OUTPUT	LABOR-HOURS	UNIT	2000 BARE COSTS MAT.	LABOR	EQUIP.	TOTAL	TOTAL INCL O&P
590	5366 2" x 3", pneumatic nailed R06100-010	2 Carp	1,164	.014	L.F.	.31	.27		.58	.80
	5381 2" x 4"		850	.019		.39	.37		.76	1.07
	5386 2" x 4", pneumatic nailed R06110-030		1,020	.016		.39	.31		.70	.96
	5401 2" x 6"		740	.022		.59	.43		1.02	1.38
	5406 2" x 6", pneumatic nailed		888	.018		.59	.35		.94	1.26
	5421 3" x 4"		600	.027		1.07	.53		1.60	2.08
	5426 3" x 4", pneumatic nailed		720	.022		1.07	.44		1.51	1.93
	5441 Installed on second story, 2" x 3"		950	.017		.31	.33		.64	.91
	5446 2" x 3", pneumatic nailed		1,140	.014		.31	.28		.59	.81
	5461 2" x 4"		810	.020		.39	.39		.78	1.10
	5466 2" x 4", pneumatic nailed		972	.016		.39	.32		.71	.99
	5481 2" x 6"		700	.023		.59	.45		1.04	1.42
	5486 2" x 6", pneumatic nailed		840	.019		.59	.38		.97	1.29
	5501 3" x 4"		550	.029		1.07	.57		1.64	2.16
	5506 3" x 4", pneumatic nailed		660	.024		1.07	.48		1.55	2
	5521 Installed on dormer or gable, 2" x 3"		850	.019		.31	.37		.68	.98
	5526 2" x 3", pneumatic nailed		1,020	.016		.31	.31		.62	.87
	5541 2" x 4"		720	.022		.39	.44		.83	1.18
	5546 2" x 4", pneumatic nailed		864	.019		.39	.36		.75	1.06
	5561 2" x 6"		620	.026		.59	.51		1.10	1.52
	5566 2" x 6", pneumatic nailed		744	.022		.59	.42		1.01	1.38
	5581 3" x 4"		480	.033		1.07	.66		1.73	2.31
	5586 3" x 4", pneumatic nailed		576	.028		1.07	.55		1.62	2.12
	5601 3' high wall, 2" x 3"		740	.022		.31	.43		.74	1.07
	5606 2" x 3", pneumatic nailed		888	.018		.31	.35		.66	.95
	5621 2" x 4"		640	.025		.39	.49		.88	1.27
	5626 2" x 4", pneumatic nailed		768	.021		.39	.41		.80	1.13
	5641 2" x 6"		550	.029		.59	.57		1.16	1.63
	5646 2" x 6", pneumatic nailed		660	.024		.59	.48		1.07	1.47
	5661 3" x 4"		440	.036		1.07	.72		1.79	2.41
	5666 3" x 4", pneumatic nailed		528	.030		1.07	.60		1.67	2.20
	5681 Installed on second story, 2" x 3"		700	.023		.31	.45		.76	1.11
	5686 2" x 3", pneumatic nailed		840	.019		.31	.38		.69	.98
	5701 2" x 4"		610	.026		.39	.52		.91	1.32
	5706 2" x 4", pneumatic nailed		732	.022		.39	.43		.82	1.17
	5721 2" x 6"		520	.031		.59	.61		1.20	1.69
	5726 2" x 6", pneumatic nailed		624	.026		.59	.51		1.10	1.52
	5741 3" x 4"		430	.037		1.07	.73		1.80	2.44
	5746 3" x 4", pneumatic nailed		516	.031		1.07	.61		1.68	2.23
	5761 Installed on dormer or gable, 2" x 3"		625	.026		.31	.50		.81	1.20
	5766 2" x 3", pneumatic nailed		750	.021		.31	.42		.73	1.06
	5781 2" x 4"		545	.029		.39	.58		.97	1.42
	5786 2" x 4", pneumatic nailed		654	.024		.39	.48		.87	1.26
	5801 2" x 6"		465	.034		.59	.68		1.27	1.81
	5806 2" x 6", pneumatic nailed		558	.029		.59	.56		1.15	1.62
	5821 3" x 4"		380	.042		1.07	.83		1.90	2.60
	5826 3" x 4", pneumatic nailed	↓	456	.035	↓	1.07	.69		1.76	2.36
	8250 For second story & above, add						5%			
	8300 For dormer & gable, add						15%			
600	0012 **FURRING** Wood strips, 1" x 2", on walls, on wood	1 Carp	550	.015	L.F.	.18	.29		.47	.69
	0017 Pneumatic nailed		710	.011		.18	.22		.40	.58
	0302 On masonry		495	.016		.18	.32		.50	.75
	0402 On concrete		260	.031		.18	.61		.79	1.24
	0602 1" x 3", on walls, on wood		550	.015		.18	.29		.47	.69
	0607 Pneumatic nailed	↓	710	.011	↓	.18	.22		.40	.58

06100 | Rough Carpentry

6 WOOD & PLASTICS

		06110 | Wood Framing	CREW	DAILY OUTPUT	LABOR-HOURS	UNIT	2000 BARE COSTS MAT.	LABOR	EQUIP.	TOTAL	TOTAL INCL O&P
600	0702	On masonry	1 Carp	495	.016	L.F.	.18	.32		.50	.75
	0802	On concrete		260	.031		.18	.61		.79	1.24
	0852	On ceilings, on wood		350	.023		.18	.45		.63	.97
	0857	Pneumatic nailed		450	.018		.18	.35		.53	.80
	0902	On masonry		320	.025		.18	.49		.67	1.04
	0952	On concrete		210	.038		.18	.75		.93	1.49
700	0012	**GROUNDS** For casework, 1" x 2" wood strips, on wood	1 Carp	330	.024	L.F.	.18	.48		.66	1.02
	0102	On masonry		285	.028		.18	.55		.73	1.15
	0202	On concrete		250	.032		.18	.63		.81	1.28
	0402	For plaster, 3/4" deep, on wood		450	.018		.18	.35		.53	.80
	0502	On masonry		225	.036		.18	.70		.88	1.40
	0602	On concrete		175	.046		.18	.90		1.08	1.74
	0702	On metal lath		200	.040		.18	.79		.97	1.55
		06150 | Wood Decking									
600	0010	**ROOF DECKS**									
	0400	Cedar planks, 3" thick	2 Carp	320	.050	S.F.	5.65	.99		6.64	7.95
	0500	4" thick		250	.064		7.65	1.26		8.91	10.55
	0702	Douglas fir, 3" thick		320	.050		2.12	.99		3.11	4.02
	0802	4" thick		250	.064		2.84	1.26		4.10	5.30
	1002	Hemlock, 3" thick		320	.050		2.12	.99		3.11	4.02
	1102	4" thick		250	.064		2.83	1.26		4.09	5.25
	1302	Western white spruce, 3" thick		320	.050		2.05	.99		3.04	3.95
	1402	4" thick		250	.064		2.73	1.26		3.99	5.15
		06160 | Sheathing									
800	0010	**SHEATHING** Plywood on roof, CDX (R06160-020)									
	0032	5/16" thick	2 Carp	1,600	.010	S.F.	.36	.20		.56	.74
	0037	Pneumatic nailed		1,952	.008		.36	.16		.52	.68
	0052	3/8" thick		1,525	.010		.40	.21		.61	.80
	0057	Pneumatic nailed		1,860	.009		.40	.17		.57	.74
	0102	1/2" thick		1,400	.011		.53	.23		.76	.97
	0103	Pneumatic nailed		1,708	.009		.53	.18		.71	.90
	0202	5/8" thick		1,300	.012		.67	.24		.91	1.15
	0207	Pneumatic nailed		1,586	.010		.67	.20		.87	1.07
	0302	3/4" thick		1,200	.013		.81	.26		1.07	1.34
	0307	Pneumatic nailed		1,464	.011		.81	.22		1.03	1.26
	0502	Plywood on walls with exterior CDX, 3/8" thick		1,200	.013		.40	.26		.66	.90
	0507	Pneumatic nailed		1,488	.011		.40	.21		.61	.81
	0603	1/2" thick		1,125	.014		.53	.28		.81	1.06
	0608	Pneumatic nailed		1,395	.011		.53	.23		.76	.97
	0702	5/8" thick		1,050	.015		.67	.30		.97	1.24
	0707	Pneumatic nailed		1,302	.012		.67	.24		.91	1.14
	0803	3/4" thick		975	.016		.81	.32		1.13	1.44
	0808	Pneumatic nailed		1,209	.013		.81	.26		1.07	1.34
	1000	For shear wall construction, add						20%			
	1200	For structural 1 exterior plywood, add				S.F.	10%				
	1402	With boards, on roof 1" x 6" boards, laid horizontal	2 Carp	725	.022		1.30	.43		1.73	2.17
	1502	Laid diagonal		650	.025		1.30	.48		1.78	2.26
	1702	1" x 8" boards, laid horizontal		875	.018		1.30	.36		1.66	2.05
	1802	Laid diagonal		725	.022		1.30	.43		1.73	2.17
	2000	For steep roofs, add						40%			
	2200	For dormers, hips and valleys, add					5%	50%			
	2402	Boards on walls, 1" x 6" boards, laid regular	2 Carp	650	.025		1.30	.48		1.78	2.26

06160 | Sheathing		CREW	DAILY OUTPUT	LABOR-HOURS	UNIT	2000 BARE COSTS MAT.	LABOR	EQUIP.	TOTAL	TOTAL INCL O&P	
2502	Laid diagonal R06160-020	2 Carp	585	.027	S.F.	1.30	.54		1.84	2.35	800
2702	1" x 8" boards, laid regular		765	.021		1.30	.41		1.71	2.14	
2802	Laid diagonal		650	.025		1.30	.48		1.78	2.26	
2852	Gypsum, weatherproof, 1/2" thick		1,050	.015		.28	.30		.58	.82	
2902	Sealed, 4/10" thick		1,100	.015		.44	.29		.73	.97	
3000	Wood fiber, regular, no vapor barrier, 1/2" thick		1,200	.013		.53	.26		.79	1.03	
3100	5/8" thick		1,200	.013		.71	.26		.97	1.23	
3300	No vapor barrier, in colors, 1/2" thick		1,200	.013		.77	.26		1.03	1.30	
3400	5/8" thick		1,200	.013		.95	.26		1.21	1.50	
3600	With vapor barrier one side, white, 1/2" thick		1,200	.013		.54	.26		.80	1.04	
3700	Vapor barrier 2 sides, 1/2" thick		1,200	.013		.82	.26		1.08	1.35	
3800	Asphalt impregnated, 25/32" thick		1,200	.013		.35	.26		.61	.84	
3850	Intermediate, 1/2" thick		1,200	.013		.29	.26		.55	.77	
4000	Wafer board on roof, 1/2" thick		1,455	.011		.40	.22		.62	.81	
4100	5/8" thick		1,330	.012		.46	.24		.70	.92	
0012	**SUBFLOOR** Plywood, CDX, 1/2" thick R06160-020	2 Carp	1,500	.011	SF Flr.	.53	.21		.74	.94	850
0017	Pneumatic nailed		1,860	.009		.53	.17		.70	.87	
0102	5/8" thick		1,350	.012		.67	.23		.90	1.13	
0107	Pneumatic nailed		1,674	.010		.67	.19		.86	1.05	
0202	3/4" thick		1,250	.013		.81	.25		1.06	1.32	
0207	Pneumatic nailed		1,550	.010		.81	.20		1.01	1.24	
0302	1-1/8" thick, 2-4-1 including underlayment		1,050	.015		2	.30		2.30	2.71	
0502	With boards, 1" x 10" S4S, laid regular		1,100	.015		1.01	.29		1.30	1.60	
0602	Laid diagonal		900	.018		1.02	.35		1.37	1.72	
0802	1" x 8" S4S, laid regular		1,000	.016		.91	.32		1.23	1.54	
0902	Laid diagonal		850	.019		.91	.37		1.28	1.64	
1100	Wood fiber, T&G, 2' x 8' planks, 1" thick		1,000	.016		1.26	.32		1.58	1.93	
1200	1-3/8" thick		900	.018		1.55	.35		1.90	2.30	
1500	Wafer board, 5/8" thick		1,330	.012	S.F.	.41	.24		.65	.86	
1600	3/4" thick		1,230	.013	"	.45	.26		.71	.94	
0010	**UNDERLAYMENT** Plywood, underlayment grade, 3/8" thick	2 Carp	1,500	.011	SF Flr.	.62	.21		.83	1.04	900
0016	Pneumatic nailed		1,860	.009		.62	.17		.79	.97	
0102	1/2" thick		1,450	.011		.80	.22		1.02	1.25	
0107	Pneumatic nailed		1,798	.009		.80	.18		.98	1.18	
0202	5/8" thick		1,400	.011		.76	.23		.99	1.23	
0207	Pneumatic nailed		1,736	.009		.76	.18		.94	1.15	
0302	3/4" thick		1,300	.012		.97	.24		1.21	1.49	
0306	Pneumatic nailed		1,612	.010		.97	.20		1.17	1.40	
0502	Particle board, 3/8" thick		1,500	.011		.38	.21		.59	.78	
0507	Pneumatic nailed		1,860	.009		.38	.17		.55	.71	
0602	1/2" thick		1,450	.011		.40	.22		.62	.81	
0607	Pneumatic nailed		1,798	.009		.40	.18		.58	.74	
0802	5/8" thick		1,400	.011		.46	.23		.69	.90	
0807	Pneumatic nailed		1,736	.009		.46	.18		.64	.82	
0902	3/4" thick		1,300	.012		.57	.24		.81	1.05	
0907	Pneumatic nailed		1,612	.010		.57	.20		.77	.96	
1102	Hardboard, underlayment grade, 4' x 4', .215" thick		1,500	.011		.41	.21		.62	.81	

06170 | Prefabricated Structural Wood		CREW	DAILY OUTPUT	LABOR-HOURS	UNIT	MAT.	LABOR	EQUIP.	TOTAL	TOTAL INCL O&P	
0010	**LAMINATED BEAMS** Fb 2400 psi										200
0050	3" x 15"	F-3	480	.083	L.F.	8.25	1.49	.95	10.69	12.70	
0100	3" x 18"		450	.089		9.90	1.59	1.01	12.50	14.70	
0150	5" x 15"		360	.111		12.40	1.99	1.26	15.65	18.45	

06170 | Prefabricated Structural Wood

		06170 Prefabricated Structural Wood	CREW	DAILY OUTPUT	LABOR-HOURS	UNIT	2000 BARE COSTS MAT.	LABOR	EQUIP.	TOTAL	TOTAL INCL O&P
200	0200	5″ x 18″	F-3	290	.138	L.F.	15.45	2.47	1.56	19.48	23
	0250	5″ x 22-1/2″		220	.182		19.85	3.26	2.06	25.17	30
	0300	6-3/4″ x 18″		320	.125		15.75	2.24	1.42	19.41	22.50
	0350	6-3/4″ x 25-1/2″		260	.154		20.50	2.76	1.74	25	29
	0400	6-3/4″ x 33″		210	.190		26.50	3.42	2.16	32.08	37.50
	0450	8-3/4″ x 34-1/2″		160	.250		37	4.48	2.84	44.32	51
	0500	For premium appearance, add to S.F. prices					5%				
	0550	For industrial type, deduct					15%				
	0600	For stain and varnish, add					5%				
	0650	For 3/4″ laminations, add					25%				
550	0010	**LAMINATED ROOF DECK** Pine or hemlock, 3″ thick	2 Carp	425	.038	S.F.	2.84	.74		3.58	4.39
	0100	4″ thick		325	.049		3.78	.97		4.75	5.80
	0300	Cedar, 3″ thick		425	.038		3.47	.74		4.21	5.10
	0400	4″ thick		325	.049		4.42	.97		5.39	6.50
	0600	Fir, 3″ thick		425	.038		2.89	.74		3.63	4.45
	0700	4″ thick		325	.049		3.63	.97		4.60	5.65
600	0010	**STRUCTURAL JOISTS** Fabricated "I" joists with wood flanges,									
	0100	Plywood webs, incl. bridging & blocking, panels 24″ O.C.									
	1200	15′ to 24′ span, 50 psf live load	F-5	2,400	.013	SF Flr.	1.44	.23		1.67	1.97
	1300	55 psf live load		2,250	.014		1.55	.25		1.80	2.12
	1400	24′ to 30′ span, 45 psf live load		2,600	.012		1.80	.21		2.01	2.34
	1500	55 psf live load		2,400	.013		1.80	.23		2.03	2.37
	1600	Tubular steel open webs, 45 psf, 24″ O.C., 40′ span	F-3	6,250	.006		1.80	.11	.07	1.98	2.26
	1700	55′ span		7,750	.005		1.75	.09	.06	1.90	2.15
	1800	70′ span		9,250	.004		2.27	.08	.05	2.40	2.68
	1900	85 psf live load, 26′ span		2,300	.017		2.11	.31	.20	2.62	3.07
980	0010	**ROOF TRUSSES**									
	0020	For timber connectors, see div. 06090-800									
	5000	Common wood, 2″ x 4″ metal plate connected, 24″ O.C., 4/12 slope									
	5010	1′ overhang, 12′ span	F-5	55	.582	Ea.	27.50	10.05		37.55	47.50
	5050	20′ span R06170-100	F-6	62	.645		36	11.50	7.30	54.80	67
	5100	24′ span		60	.667		42.50	11.85	7.55	61.90	75.50
	5150	26′ span		57	.702		65	12.50	7.95	85.45	101
	5200	28′ span		53	.755		50.50	13.45	8.55	72.50	88
	5240	30′ span		51	.784		76	13.95	8.90	98.85	117
	5250	32′ span		50	.800		71	14.25	9.05	94.30	113
	5280	34′ span		48	.833		88	14.80	9.45	112.25	132
	5350	8/12 pitch, 1′ overhang, 20′ span		57	.702		60	12.50	7.95	80.45	96
	5400	24′ span		55	.727		71	12.95	8.25	92.20	109
	5450	26′ span		52	.769		77	13.70	8.70	99.40	118
	5500	28′ span		49	.816		83	14.50	9.25	106.75	126
	5550	32′ span		45	.889		93	15.80	10.10	118.90	140
	5600	36′ span		41	.976		105	17.35	11.05	133.40	158
	5650	38′ span		40	1		114	17.80	11.35	143.15	168
	5700	40′ span		40	1		130	17.80	11.35	159.15	186

06180 | Glued-Laminated Construction

		06180 Glued-Laminated Construction	CREW	DAILY OUTPUT	LABOR-HOURS	UNIT	MAT.	LABOR	EQUIP.	TOTAL	TOTAL INCL O&P
400	0010	**LAMINATED FRAMING** Not including decking									
	0020	30 lb., short term live load, 15 lb. dead load									
	0200	Straight roof beams, 20′ clear span, beams 8′ O.C.	F-3	2,560	.016	SF Flr.	1.45	.28	.18	1.91	2.27
	0300	Beams 16′ O.C.		3,200	.013		1.06	.22	.14	1.42	1.71
	0500	40′ clear span, beams 8′ O.C.		3,200	.013		2.79	.22	.14	3.15	3.61
	0600	Beams 16′ O.C.		3,840	.010		2.28	.19	.12	2.59	2.96
	0800	60′ clear span, beams 8′ O.C.	F-4	2,880	.014		4.79	.25	.28	5.32	6
	0900	Beams 16′ O.C.	"	3,840	.010		3.57	.19	.21	3.97	4.48

06100 | Rough Carpentry

06180 | Glued-Laminated Construction

	06180 Glued-Laminated Construction	CREW	DAILY OUTPUT	LABOR-HOURS	UNIT	2000 BARE COSTS MAT.	LABOR	EQUIP.	TOTAL	TOTAL INCL O&P	
1100	Tudor arches, 30′ to 40′ clear span, frames 8′ O.C.	F-3	1,680	.024	SF Flr.	6.25	.43	.27	6.95	7.90	400
1200	Frames 16′ O.C.	"	2,240	.018		4.89	.32	.20	5.41	6.15	
1400	50′ to 60′ clear span, frames 8′ O.C.	F-4	2,200	.018		6.75	.33	.37	7.45	8.35	
1500	Frames 16′ O.C.		2,640	.015		5.75	.27	.31	6.33	7.10	
1700	Radial arches, 60′ clear span, frames 8′ O.C.		1,920	.021		6.30	.37	.42	7.09	8.05	
1800	Frames 16′ O.C.		2,880	.014		4.84	.25	.28	5.37	6.05	
2000	100′ clear span, frames 8′ O.C.		1,600	.025		6.50	.45	.51	7.46	8.45	
2100	Frames 16′ O.C.		2,400	.017		5.75	.30	.34	6.39	7.20	
2300	120′ clear span, frames 8′ O.C.		1,440	.028		8.70	.50	.56	9.76	11	
2400	Frames 16′ O.C.		1,920	.021		7.90	.37	.42	8.69	9.80	
2600	Bowstring trusses, 20′ O.C., 40′ clear span	F-3	2,400	.017		3.91	.30	.19	4.40	5	
2700	60′ clear span	F-4	3,600	.011		3.51	.20	.22	3.93	4.45	
2800	100′ clear span		4,000	.010		4.97	.18	.20	5.35	5.95	
2900	120′ clear span		3,600	.011		5.35	.20	.22	5.77	6.45	
3100	For premium appearance, add to S.F. prices					5%					
3300	For industrial type, deduct					15%					
3500	For stain and varnish, add					5%					
3900	For 3/4″ laminations, add to straight					25%					
4100	Add to curved					15%					
4300	Alternate pricing method: (use nominal footage of										
4310	components). Straight beams, camber less than 6″	F-3	3.50	11.429	M.B.F.	2,175	205	130	2,510	2,875	
4400	Columns, including hardware		2	20		2,325	360	227	2,912	3,400	
4600	Curved members, radius over 32′		2.50	16		2,375	287	181	2,843	3,325	
4700	Radius 10′ to 32′		3	13.333		2,350	239	151	2,740	3,175	
4900	For complicated shapes, add maximum					100%					
5100	For pressure treating, add to straight					35%					
5200	Add to curved					45%					
6000	Laminated veneer members, southern pine or western species										
6050	1-3/4″ wide x 5-1/2″ deep	2 Carp	480	.033	L.F.	3.69	.66		4.35	5.20	
6100	9-1/2″ deep		480	.033		3.86	.66		4.52	5.40	
6150	14″ deep		450	.036		5.50	.70		6.20	7.25	
6200	18″ deep		450	.036		7.15	.70		7.85	9.10	
6300	Parallel strand members, southern pine or western species										
6350	1-3/4″ wide x 9-1/4″ deep	2 Carp	480	.033	L.F.	4.42	.66		5.08	6	
6400	11-1/4″ deep		450	.036		5.45	.70		6.15	7.15	
6450	14″ deep		400	.040		6.45	.79		7.24	8.45	
6500	3-1/2″ wide x 9-1/4″ deep		480	.033		8.60	.66		9.26	10.60	
6550	11-1/4″ deep		450	.036		10.60	.70		11.30	12.85	
6600	14″ deep		400	.040		12.65	.79		13.44	15.25	
6650	7″ wide x 9-1/4″ deep		450	.036		16.95	.70		17.65	19.85	
6700	11-1/4″ deep		420	.038		21	.75		21.75	24.50	
6750	14″ deep		400	.040		25	.79		25.79	29	

06200 | Finish Carpentry

06220 | Millwork

	06220 Millwork	CREW	DAILY OUTPUT	LABOR-HOURS	UNIT	2000 BARE COSTS MAT.	LABOR	EQUIP.	TOTAL	TOTAL INCL O&P	
0010	**MILLWORK** Rule of thumb: Milled material cost										100
0020	equals three times cost of lumber										
1000	Typical finish hardwood milled material										
1020	1″ x 12″, custom birch				L.F.	6.10			6.10	6.70	

	06220 | Millwork		CREW	DAILY OUTPUT	LABOR-HOURS	UNIT	2000 BARE COSTS MAT.	LABOR	EQUIP.	TOTAL	TOTAL INCL O&P
100	1040	Cedar				L.F.	3.60			3.60	3.96
	1060	Oak					5.70			5.70	6.30
	1080	Redwood					5.90			5.90	6.50
	1100	Southern yellow pine					1.39			1.39	1.53
	1120	Sugar pine					2.51			2.51	2.76
	1140	Teak					13.15			13.15	14.45
	1160	Walnut					9.35			9.35	10.30
	1180	White pine					3.09			3.09	3.40
200	0010	**MOLDINGS, BASE**									
	0500	Base, stock pine, 9/16″ x 3-1/2″	1 Carp	240	.033	L.F.	1	.66		1.66	2.23
	0501	Oak or birch, 9/16″ x 3-1/2″		240	.033		2.30	.66		2.96	3.66
	0550	9/16″ x 4-1/2″		200	.040		1.28	.79		2.07	2.76
	0561	Base shoe, oak, 3/4″ x 1″		240	.033		.76	.66		1.42	1.97
	0570	Base, prefinished, 2-1/2″ x 9/16″		242	.033		1	.65		1.65	2.22
	0580	Shoe, prefinished, 3/8″ x 5/8″		266	.030		.42	.59		1.01	1.48
	0585	Flooring cant strip, 3/4″ x 1/2″		500	.016		.22	.32		.54	.78
400	0010	**MOLDINGS, CASINGS**									
	0090	Apron, stock pine, 5/8″ x 2″	1 Carp	250	.032	L.F.	.96	.63		1.59	2.14
	0110	5/8″ x 3-1/2″		220	.036		1.40	.72		2.12	2.77
	0300	Band, stock pine, 11/16″ x 1-1/8″		270	.030		.38	.58		.96	1.42
	0350	11/16″ x 1-3/4″		250	.032		.60	.63		1.23	1.74
	0700	Casing, stock pine, 11/16″ x 2-1/2″		240	.033		.77	.66		1.43	1.98
	0701	Oak or birch		240	.033		1.99	.66		2.65	3.32
	0750	11/16″ x 3-1/2″		215	.037		1.48	.73		2.21	2.89
	0760	Door & window casing, exterior, 1-1/4″ x 2″		200	.040		1.40	.79		2.19	2.89
	0770	Finger jointed, 1-1/4″ x 2″		200	.040		1.26	.79		2.05	2.74
	4600	Mullion casing, stock pine, 5/16″ x 2″		200	.040		.62	.79		1.41	2.03
	4601	Oak or birch		200	.040		.72	.79		1.51	2.14
	4700	Teak, custom, nominal 1″ x 1″		215	.037		1.11	.73		1.84	2.48
	4800	Nominal 1″ x 3″		200	.040		3.26	.79		4.05	4.94
450	0010	**MOLDINGS, CEILINGS**									
	0600	Bed, stock pine, 9/16″ x 1-3/4″	1 Carp	270	.030	L.F.	.60	.58		1.18	1.66
	0650	9/16″ x 2″		240	.033		.58	.66		1.24	1.77
	1200	Cornice molding, stock pine, 9/16″ x 1-3/4″		330	.024		.62	.48		1.10	1.50
	1300	9/16″ x 2-1/4″		300	.027		.85	.53		1.38	1.84
	2400	Cove scotia, stock pine, 9/16″ x 1-3/4″		270	.030		.54	.58		1.12	1.59
	2401	Oak or birch, 9/16″ x 1-3/4″		270	.030		.81	.58		1.39	1.89
	2500	11/16″ x 2-3/4″		255	.031		1.15	.62		1.77	2.33
	2600	Crown, stock pine, 9/16″ x 3-5/8″		250	.032		1.51	.63		2.14	2.74
	2700	11/16″ x 4-5/8″		220	.036		1.91	.72		2.63	3.33
500	0010	**MOLDINGS, EXTERIOR**									
	1500	Cornice, boards, pine, 1″ x 2″	1 Carp	330	.024	L.F.	.24	.48		.72	1.08
	1600	1″ x 4″		250	.032		.60	.63		1.23	1.74
	1700	1″ x 6″		250	.032		.86	.63		1.49	2.03
	1800	1″ x 8″		200	.040		.85	.79		1.64	2.29
	1900	1″ x 10″		180	.044		1.12	.88		2	2.73
	2000	1″ x 12″		180	.044		1.14	.88		2.02	2.75
	2200	Three piece, built-up, pine, minimum		80	.100		1.47	1.97		3.44	5
	2300	Maximum		65	.123		5.05	2.42		7.47	9.70
	3000	Corner board, sterling pine, 1″ x 4″		200	.040		.66	.79		1.45	2.08
	3100	1″ x 6″		200	.040		.96	.79		1.75	2.41
	3200	2″ x 6″		165	.048		2	.96		2.96	3.84
	3300	2″ x 8″		165	.048		2.66	.96		3.62	4.57
	3350	Fascia, sterling pine, 1″ x 6″		250	.032		.96	.63		1.59	2.14
	3370	1″ x 8″		225	.036		1.50	.70		2.20	2.85
	3372	2″ x 6″		225	.036		2	.70		2.70	3.40

06220	Millwork	CREW	DAILY OUTPUT	LABOR-HOURS	UNIT	2000 BARE COSTS MAT.	LABOR	EQUIP.	TOTAL	TOTAL INCL O&P	
3374	2″ x 8″	1 Carp	200	.040	L.F.	2.66	.79		3.45	4.28	500
3376	2″ x 10″		180	.044		3.33	.88		4.21	5.15	
3395	Grounds, 1″ x 1″ redwood		300	.027		.18	.53		.71	1.10	
3400	Trim, exterior, sterling pine, back band		250	.032		.57	.63		1.20	1.71	
3500	Casing		250	.032		.60	.63		1.23	1.74	
3600	Crown		250	.032		1.20	.63		1.83	2.40	
3700	Porch rail with balusters		22	.364		9	7.15		16.15	22	
3800	Screen		395	.020		.32	.40		.72	1.03	
4100	Verge board, sterling pine, 1″ x 4″		200	.040		.47	.79		1.26	1.87	
4200	1″ x 6″		200	.040		.72	.79		1.51	2.14	
4300	2″ x 6″		165	.048		1.40	.96		2.36	3.18	
4400	2″ x 8″		165	.048		1.86	.96		2.82	3.69	
4700	For redwood trim, add					200%					
5000	Casing/fascia, rough-sawn cedar										
5100	1″ x 2″	1 Carp	275	.029	L.F.	.23	.57		.80	1.23	
5200	1″ x 6″		250	.032		.70	.63		1.33	1.85	
5300	1″ x 8″		230	.035		.90	.69		1.59	2.16	
5400	2″ x 4″		220	.036		.90	.72		1.62	2.22	
5500	2″ x 6″		220	.036		1.37	.72		2.09	2.74	
5600	2″ x 8″		200	.040		1.82	.79		2.61	3.35	
5700	2″ x 10″		180	.044		2.27	.88		3.15	4	
5800	2″ x 12″		170	.047		2.72	.93		3.65	4.58	
0010	**MOLDINGS, TRIM**										700
0200	Astragal, stock pine, 11/16″ x 1-3/4″	1 Carp	255	.031	L.F.	.78	.62		1.40	1.92	
0250	1-5/16″ x 2-3/16″		240	.033		2.73	.66		3.39	4.13	
0800	Chair rail, stock pine, 5/8″ x 2-1/2″		270	.030		.84	.58		1.42	1.92	
0900	5/8″ x 3-1/2″		240	.033		1.26	.66		1.92	2.52	
1000	Closet pole, stock pine, 1-1/8″ diameter		200	.040		.80	.79		1.59	2.23	
1100	Fir, 1-5/8″ diameter		200	.040		1.19	.79		1.98	2.66	
1150	Corner, inside, 5/16″ x 1″		225	.036		.38	.70		1.08	1.62	
1160	Outside, 1-1/16″ x 1-1/16″		240	.033		.96	.66		1.62	2.19	
1161	1-5/16″ x 1-5/16″		240	.033		1.07	.66		1.73	2.31	
3300	Half round, stock pine, 1/4″ x 1/2″		270	.030		.19	.58		.77	1.21	
3350	1/2″ x 1″		255	.031		.34	.62		.96	1.43	
3400	Handrail, fir, single piece, stock, hardware not included										
3450	1-1/2″ x 1-3/4″	1 Carp	80	.100	L.F.	1.21	1.97		3.18	4.71	
3470	Pine, 1-1/2″ x 1-3/4″		80	.100		1.06	1.97		3.03	4.55	
3500	1-1/2″ x 2-1/2″		76	.105		1.43	2.07		3.50	5.10	
3600	Lattice, stock pine, 1/4″ x 1-1/8″		270	.030		.32	.58		.90	1.35	
3700	1/4″ x 1-3/4″		250	.032		.35	.63		.98	1.47	
3800	Miscellaneous, custom, pine, 1″ x 1″		270	.030		.98	.58		1.56	2.08	
3900	1″ x 3″		240	.033		.78	.66		1.44	1.99	
4100	Birch or oak, nominal 1″ x 1″		240	.033		.46	.66		1.12	1.64	
4200	Nominal 1″ x 3″		215	.037		1.59	.73		2.32	3.01	
4400	Walnut, nominal 1″ x 1″		215	.037		.76	.73		1.49	2.10	
4500	Nominal 1″ x 3″		200	.040		2.29	.79		3.08	3.87	
4700	Teak, nominal 1″ x 1″		215	.037		1.05	.73		1.78	2.41	
4800	Nominal 1″ x 3″		200	.040		3	.79		3.79	4.65	
4900	Quarter round, stock pine, 1/4″ x 1/4″		275	.029		.20	.57		.77	1.20	
4950	3/4″ x 3/4″		255	.031		.41	.62		1.03	1.51	
5600	Wainscot moldings, 1-1/8″ x 9/16″, 2′ high, minimum		76	.105	S.F.	5.90	2.07		7.97	10.05	
5700	Maximum		65	.123	″	13.45	2.42		15.87	18.95	
0010	**MOLDINGS, WINDOW AND DOOR**										800
2800	Door moldings, stock, decorative, 1-1/8″ wide, plain	1 Carp	17	.471	Set	28.50	9.25		37.75	47	
2900	Detailed		17	.471	″	68	9.25		77.25	91	
3150	Door trim set, 1 head and 2 sides, pine, 2-1/2 wide		5.90	1.356	Opng.	12.70	26.50		39.20	60	

	06220 | Millwork		Crew	Daily Output	Labor-Hours	Unit	2000 Bare Costs Mat.	Labor	Equip.	Total	Total Incl O&P
800	3170	4-1/2″ wide	1 Carp	5.30	1.509	Opng.	24	29.50		53.50	77
	3200	Glass beads, stock pine, 1/4″ x 11/16″		285	.028	L.F.	.30	.55		.85	1.28
	3250	3/8″ x 1/2″		275	.029		.36	.57		.93	1.38
	3270	3/8″ x 7/8″		270	.030		.40	.58		.98	1.44
	4850	Parting bead, stock pine, 3/8″ x 3/4″		275	.029		.29	.57		.86	1.30
	4870	1/2″ x 3/4″		255	.031		.36	.62		.98	1.46
	5000	Stool caps, stock pine, 11/16″ x 3-1/2″		200	.040		1.31	.79		2.10	2.79
	5100	1-1/16″ x 3-1/4″		150	.053		1.97	1.05		3.02	3.97
	5300	Threshold, oak, 3′ long, inside, 5/8″ x 3-5/8″		32	.250	Ea.	6.20	4.93		11.13	15.30
	5400	Outside, 1-1/2″ x 7-5/8″		16	.500	″	25	9.85		34.85	44.50
	5900	Window trim sets, including casings, header, stops,									
	5910	stool and apron, 2-1/2″ wide, minimum	1 Carp	13	.615	Opng.	15.30	12.10		27.40	38
	5950	Average		10	.800		26.50	15.75		42.25	56
	6000	Maximum		6	1.333		38	26.50		64.50	86.50
900	0010	**SOFFITS** Wood fiber, no vapor barrier, 15/32″ thick	2 Carp	525	.030	S.F.	.75	.60		1.35	1.86
	0100	5/8″ thick		525	.030		.81	.60		1.41	1.92
	0300	As above, 5/8″ thick, with factory finish		525	.030		.83	.60		1.43	1.94
	0500	Hardboard, 3/8″ thick, slotted		525	.030		1	.60		1.60	2.13
	1000	Exterior AC plywood, 1/4″ thick		420	.038		.70	.75		1.45	2.06
	1100	1/2″ thick		420	.038		.97	.75		1.72	2.36
	06250 | Prefinished Paneling										
200	0010	**PANELING, HARDBOARD**									
	0050	Not incl. furring or trim, hardboard, tempered, 1/8″ thick	2 Carp	500	.032	S.F.	.30	.63		.93	1.41
	0100	1/4″ thick		500	.032		.38	.63		1.01	1.50
	0300	Tempered pegboard, 1/8″ thick		500	.032		.38	.63		1.01	1.50
	0400	1/4″ thick		500	.032		.41	.63		1.04	1.53
	0600	Untempered hardboard, natural finish, 1/8″ thick		500	.032		.32	.63		.95	1.43
	0700	1/4″ thick		500	.032		.31	.63		.94	1.42
	0900	Untempered pegboard, 1/8″ thick		500	.032		.32	.63		.95	1.43
	1000	1/4″ thick		500	.032		.36	.63		.99	1.48
	1200	Plastic faced hardboard, 1/8″ thick		500	.032		.53	.63		1.16	1.66
	1300	1/4″ thick		500	.032		.71	.63		1.34	1.86
	1500	Plastic faced pegboard, 1/8″ thick		500	.032		.50	.63		1.13	1.63
	1600	1/4″ thick		500	.032		.62	.63		1.25	1.76
	1800	Wood grained, plain or grooved, 1/4″ thick, minimum		500	.032		.47	.63		1.10	1.60
	1900	Maximum		425	.038		.89	.74		1.63	2.25
	2100	Moldings for hardboard, wood or aluminum, minimum		500	.032	L.F.	.32	.63		.95	1.43
	2200	Maximum		425	.038	″	.90	.74		1.64	2.26
500	0010	**PANELING, PLYWOOD**									
	2400	Plywood, prefinished, 1/4″ thick, 4′ x 8′ sheets									
	2410	with vertical grooves. Birch faced, minimum	2 Carp	500	.032	S.F.	.72	.63		1.35	1.87
	2420	Average		420	.038		1.10	.75		1.85	2.50
	2430	Maximum		350	.046		1.62	.90		2.52	3.32
	2600	Mahogany, African		400	.040		2.08	.79		2.87	3.64
	2700	Philippine (Lauan)		500	.032		.88	.63		1.51	2.05
	2900	Oak or Cherry, minimum		500	.032		1.74	.63		2.37	2.99
	3000	Maximum		400	.040		2.68	.79		3.47	4.30
	3200	Rosewood		320	.050		3.81	.99		4.80	5.90
	3400	Teak		400	.040		2.68	.79		3.47	4.30
	3600	Chestnut		375	.043		3.97	.84		4.81	5.80
	3800	Pecan		400	.040		1.70	.79		2.49	3.22
	3900	Walnut, minimum		500	.032		2.27	.63		2.90	3.58
	3950	Maximum		400	.040		4.33	.79		5.12	6.10
	4000	Plywood, prefinished, 3/4″ thick, stock grades, minimum		320	.050		1.03	.99		2.02	2.82

06200 | Finish Carpentry

06250 | Prefinished Paneling		CREW	DAILY OUTPUT	LABOR-HOURS	UNIT	2000 BARE COSTS MAT.	LABOR	EQUIP.	TOTAL	TOTAL INCL O&P	
4100	Maximum	2 Carp	224	.071	S.F.	4.48	1.41		5.89	7.35	500
4300	Architectural grade, minimum		224	.071		3.30	1.41		4.71	6.05	
4400	Maximum		160	.100		5.05	1.97		7.02	8.95	
4600	Plywood, "A" face, birch, V.C., 1/2" thick, natural		450	.036		1.55	.70		2.25	2.90	
4700	Select		450	.036		1.70	.70		2.40	3.07	
4900	Veneer core, 3/4" thick, natural		320	.050		1.65	.99		2.64	3.51	
5000	Select		320	.050		1.85	.99		2.84	3.73	
5200	Lumber core, 3/4" thick, natural		320	.050		2.47	.99		3.46	4.41	
5500	Plywood, knotty pine, 1/4" thick, A2 grade		450	.036		1.34	.70		2.04	2.67	
5600	A3 grade		450	.036		1.70	.70		2.40	3.07	
5800	3/4" thick, veneer core, A2 grade		320	.050		1.75	.99		2.74	3.62	
5900	A3 grade		320	.050		1.96	.99		2.95	3.85	
6100	Aromatic cedar, 1/4" thick, plywood		400	.040		1.72	.79		2.51	3.24	
6200	1/4" thick, particle board	↓	400	.040	↓	.83	.79		1.62	2.26	
06260 | Board Paneling											
0010	**PANELING, BOARDS**										400
6400	Wood board paneling, 3/4" thick, knotty pine	2 Carp	300	.053	S.F.	1.25	1.05		2.30	3.18	
6500	Rough sawn cedar		300	.053		1.60	1.05		2.65	3.56	
6700	Redwood, clear, 1" x 4" boards		300	.053		3.75	1.05		4.80	5.95	
6900	Aromatic cedar, closet lining, boards	↓	275	.058	↓	2.89	1.15		4.04	5.15	
06270 | Closet/Utility Wood Shelving											
0010	**SHELVING** Pine, clear grade, no edge band, 1" x 8"	1 Carp	115	.070	L.F.	1.71	1.37		3.08	4.23	200
0100	1" x 10"		110	.073		2.31	1.43		3.74	4.99	
0200	1" x 12"	↓	105	.076		4.18	1.50		5.68	7.15	
0400	For lumber edge band, by hand, add					1.46			1.46	1.61	
0420	By machine, add					.94			.94	1.03	
0600	Plywood, 3/4" thick with lumber edge, 12" wide	1 Carp	75	.107		1.24	2.10		3.34	4.96	
0700	24" wide		70	.114	↓	2.35	2.25		4.60	6.45	
0900	Bookcase, clear grade pine, shelves 12" O.C., 8" deep		70	.114	S.F.	3.35	2.25		5.60	7.55	
1000	12" deep shelves		65	.123	"	4.19	2.42		6.61	8.75	
1200	Adjustable closet rod and shelf, 12" wide, 3' long		20	.400	Ea.	40.50	7.90		48.40	58	
1300	8' long		15	.533	"	57	10.50		67.50	80.50	
1500	Prefinished shelves with supports, stock, 8" wide		75	.107	L.F.	3.65	2.10		5.75	7.60	
1600	10" wide	↓	70	.114	"	4.07	2.25		6.32	8.35	
1800	Custom, high quality dadoed pine shelving units, minimum				S.F.					28	
1900	Maximum				"					40	

06400 | Architectural Woodwork

06410 | Custom Cabinets		CREW	DAILY OUTPUT	LABOR-HOURS	UNIT	2000 BARE COSTS MAT.	LABOR	EQUIP.	TOTAL	TOTAL INCL O&P	
0010	**CABINETS** Corner china cabinets, stock pine,										100
0020	80" high, unfinished, minimum	2 Carp	6.60	2.424	Ea.	430	48		478	550	
0100	Maximum	"	4.40	3.636	"	950	71.50		1,021.50	1,175	
0700	Kitchen base cabinets, hardwood, not incl. counter tops,										
0710	24" deep, 35" high, prefinished										
0800	One top drawer, one door below, 12" wide	2 Carp	24.80	.645	Ea.	125	12.70		137.70	160	

	06410 | Custom Cabinets		CREW	DAILY OUTPUT	LABOR-HOURS	UNIT	2000 BARE COSTS MAT.	LABOR	EQUIP.	TOTAL	TOTAL INCL O&P
100	0820	15″ wide	2 Carp	24	.667	Ea.	177	13.15		190.15	217
	0840	18″ wide		23.30	.687		188	13.55		201.55	230
	0860	21″ wide		22.70	.705		198	13.90		211.90	242
	0880	24″ wide		22.30	.717		223	14.15		237.15	269
	1000	Four drawers, 12″ wide		24.80	.645		305	12.70		317.70	355
	1020	15″ wide		24	.667		310	13.15		323.15	365
	1040	18″ wide		23.30	.687		330	13.55		343.55	385
	1060	24″ wide		22.30	.717		355	14.15		369.15	415
	1200	Two top drawers, two doors below, 27″ wide		22	.727		266	14.35		280.35	315
	1220	30″ wide		21.40	.748		283	14.75		297.75	335
	1240	33″ wide		20.90	.766		296	15.10		311.10	350
	1260	36″ wide		20.30	.788		305	15.55		320.55	360
	1280	42″ wide		19.80	.808		325	15.90		340.90	385
	1300	48″ wide		18.90	.847		340	16.70		356.70	405
	1500	Range or sink base, two doors below, 30″ wide		21.40	.748		222	14.75		236.75	270
	1520	33″ wide		20.90	.766		239	15.10		254.10	289
	1540	36″ wide		20.30	.788		250	15.55		265.55	300
	1560	42″ wide		19.80	.808		267	15.90		282.90	320
	1580	48″ wide		18.90	.847		281	16.70		297.70	340
	1800	For sink front units, deduct					53			53	58.50
	2000	Corner base cabinets, 36″ wide, standard	2 Carp	18	.889		195	17.50		212.50	245
	2100	Lazy Susan with revolving door	″	16.50	.970		320	19.10		339.10	385
	4000	Kitchen wall cabinets, hardwood, 12″ deep with two doors									
	4050	12″ high, 30″ wide	2 Carp	24.80	.645	Ea.	122	12.70		134.70	157
	4100	36″ wide		24	.667		144	13.15		157.15	182
	4400	15″ high, 30″ wide		24	.667		144	13.15		157.15	181
	4420	33″ wide		23.30	.687		154	13.55		167.55	192
	4440	36″ wide		22.70	.705		157	13.90		170.90	197
	4450	42″ wide		22.70	.705		174	13.90		187.90	215
	4700	24″ high, 30″ wide		23.30	.687		166	13.55		179.55	206
	4720	36″ wide		22.70	.705		183	13.90		196.90	225
	4740	42″ wide		22.30	.717		168	14.15		182.15	209
	5000	30″ high, one door, 12″ wide		22	.727		107	14.35		121.35	143
	5020	15″ wide		21.40	.748		126	14.75		140.75	163
	5040	18″ wide		20.90	.766		139	15.10		154.10	179
	5060	24″ wide		20.30	.788		152	15.55		167.55	194
	5300	Two doors, 27″ wide		19.80	.808		209	15.90		224.90	258
	5320	30″ wide		19.30	.829		185	16.35		201.35	231
	5340	36″ wide		18.80	.851		211	16.75		227.75	262
	5360	42″ wide		18.50	.865		231	17.05		248.05	283
	5380	48″ wide		18.40	.870		260	17.15		277.15	315
	6000	Corner wall, 30″ high, 24″ wide		18	.889		132	17.50		149.50	175
	6050	30″ wide		17.20	.930		165	18.35		183.35	214
	6100	36″ wide		16.50	.970		176	19.10		195.10	227
	6500	Revolving Lazy Susan		15.20	1.053		249	20.50		269.50	310
	7000	Broom cabinet, 84″ high, 24″ deep, 18″ wide		10	1.600		350	31.50		381.50	440
	7500	Oven cabinets, 84″ high, 24″ deep, 27″ wide		8	2		530	39.50		569.50	655
	7750	Valance board trim		396	.040	L.F.	7.60	.80		8.40	9.70
	9000	For deluxe models of all cabinets, add					40%				
	9500	For custom built in place, add					25%	10%			
	9550	Rule of thumb, kitchen cabinets not including									
	9560	appliances & counter top, minimum	2 Carp	30	.533	L.F.	80.50	10.50		91	107
	9600	Maximum	″	25	.640	″	225	12.60		237.60	270
210	0010	**CASEWORK, FRAMES**									
	0050	Base cabinets, counter storage, 36″ high, one bay									

06400 | Architectural Woodwork

06410	Custom Cabinets	CREW	DAILY OUTPUT	LABOR-HOURS	UNIT	2000 BARE COSTS MAT.	LABOR	EQUIP.	TOTAL	TOTAL INCL O&P	
0100	18″ wide	1 Carp	2.70	2.963	Ea.	91	58.50		149.50	200	210
0400	Two bay, 36″ wide		2.20	3.636		139	71.50		210.50	276	
1100	Three bay, 54″ wide		1.50	5.333		165	105		270	360	
2800	Book cases, one bay, 7′ high, 18″ wide		2.40	3.333		107	65.50		172.50	231	
3500	Two bay, 36″ wide		1.60	5		155	98.50		253.50	340	
4100	Three bay, 54″ wide		1.20	6.667		257	131		388	510	
6100	Wall mounted cabinet, one bay, 24″ high, 18″ wide		3.60	2.222		59	44		103	140	
6800	Two bay, 36″ wide		2.20	3.636		86	71.50		157.50	218	
7400	Three bay, 54″ wide		1.70	4.706		107	92.50		199.50	277	
8400	30″ high, one bay, 18″ wide		3.60	2.222		64	44		108	146	
9000	Two bay, 36″ wide		2.15	3.721		85	73.50		158.50	220	
9400	Three bay, 54″ wide		1.60	5		106	98.50		204.50	286	
9800	Wardrobe, 7′ high, single, 24″ wide		2.70	2.963		118	58.50		176.50	230	
9950	Partition, adjustable shelves & drawers, 48″ wide	↓	1.40	5.714	↓	225	113		338	440	
0010	CABINET DOORS										220
2000	Glass panel, hardwood frame										
2200	12″ wide, 18″ high	1 Carp	34	.235	Ea.	14.85	4.64		19.49	24.50	
2400	24″ high		33	.242		19.40	4.78		24.18	29.50	
2600	30″ high		32	.250		24.50	4.93		29.43	35.50	
2800	36″ high		30	.267		29.50	5.25		34.75	41.50	
3000	48″ high		23	.348		39	6.85		45.85	54.50	
3200	60″ high		17	.471		49	9.25		58.25	70	
3400	72″ high		15	.533		59	10.50		69.50	83	
3600	15″ wide x 18″ high		33	.242		15.30	4.78		20.08	25	
3800	24″ high		32	.250		19.40	4.93		24.33	30	
4000	30″ high		30	.267		24.50	5.25		29.75	36	
4250	36″ high		28	.286		29.50	5.65		35.15	42	
4300	48″ high		22	.364		39	7.15		46.15	55	
4350	60″ high		16	.500		49	9.85		58.85	71	
4400	72″ high		14	.571		59	11.25		70.25	84.50	
4450	18″ wide, 18″ high		32	.250		15.30	4.93		20.23	25.50	
4500	24″ high		30	.267		19.40	5.25		24.65	30.50	
4550	30″ high		29	.276		24.50	5.45		29.95	36.50	
4600	36″ high		27	.296		29.50	5.85		35.35	42.50	
4650	48″ high		21	.381		39	7.50		46.50	55.50	
4700	60″ high		15	.533		49	10.50		59.50	72	
4750	72″ high	↓	13	.615	↓	59	12.10		71.10	86	
5000	Hardwood, raised panel										
5100	12″ wide, 18″ high	1 Carp	16	.500	Ea.	20.50	9.85		30.35	39.50	
5150	24″ high		15.50	.516		26.50	10.15		36.65	46.50	
5200	30″ high		15	.533		32.50	10.50		43	54	
5250	36″ high		14	.571		40	11.25		51.25	63.50	
5300	48″ high		11	.727		53	14.35		67.35	83	
5320	60″ high		8	1		66	19.70		85.70	107	
5340	72″ high		7	1.143		79.50	22.50		102	126	
5360	15″ wide x 18″ high		15.50	.516		26.50	10.15		36.65	46.50	
5380	24″ high		15	.533		33	10.50		43.50	54.50	
5400	30″ high		14.50	.552		43	10.85		53.85	66	
5420	36″ high		13.50	.593		50	11.65		61.65	75	
5440	48″ high		10.50	.762		65	15		80	97	
5460	60″ high		7.50	1.067		83	21		104	128	
5480	72″ high		6.50	1.231		99	24.50		123.50	151	
5500	18″ wide, 18″ high		15	.533		29.50	10.50		40	50.50	
5550	24″ high		14.50	.552		39	10.85		49.85	61	
5600	30″ high		14	.571		50	11.25		61.25	74.50	
5650	36″ high	↓	13	.615	↓	59	12.10		71.10	86	

		06410 | Custom Cabinets	CREW	DAILY OUTPUT	LABOR-HOURS	UNIT	2000 BARE COSTS MAT.	LABOR	EQUIP.	TOTAL	TOTAL INCL O&P
220	5700	48" high	1 Carp	10	.800	Ea.	79.50	15.75		95.25	115
	5750	60" high		7	1.143		99	22.50		121.50	148
	5800	72" high	↓	6	1.333	↓	118	26.50		144.50	175
	6000	Plastic laminate on particle board									
	6100	12" wide, 18" high	1 Carp	25	.320	Ea.	13	6.30		19.30	25
	6120	24" high		24	.333		17	6.55		23.55	30
	6140	30" high		23	.348		21	6.85		27.85	35
	6160	36" high		21	.381		26	7.50		33.50	41.50
	6200	48" high		16	.500		34	9.85		43.85	54.50
	6250	60" high		13	.615		43	12.10		55.10	68.50
	6300	72" high		12	.667		52	13.15		65.15	79.50
	6320	15" wide x 18" high		24.50	.327		16	6.45		22.45	28.50
	6340	24" high		23.50	.340		22	6.70		28.70	35.50
	6360	30" high		22.50	.356		27	7		34	41.50
	6380	36" high		20.50	.390		32	7.70		39.70	48
	6400	48" high		15.50	.516		43	10.15		53.15	65
	6450	60" high		12.50	.640		54	12.60		66.60	81
	6480	72" high		11.50	.696		65	13.70		78.70	95
	6500	18" wide, 18" high		24	.333		19	6.55		25.55	32.50
	6550	24" high		23	.348		26	6.85		32.85	40.50
	6600	30" high		22	.364		32	7.15		39.15	47.50
	6650	36" high		20	.400		39	7.90		46.90	56.50
	6700	48" high		15	.533		52	10.50		62.50	75
	6750	60" high		12	.667		65	13.15		78.15	94
	6800	72" high	↓	11	.727	↓	77	14.35		91.35	109
	7000	Plywood, with edge band									
	7010	12" wide, 18" high	1 Carp	27	.296	Ea.	17.50	5.85		23.35	29.50
	7100	24" high		26	.308		23.50	6.05		29.55	36.50
	7120	30" high		25	.320		30	6.30		36.30	44
	7140	36" high		23	.348		35.50	6.85		42.35	51.50
	7180	48" high		18	.444		47	8.75		55.75	66.50
	7200	60" high		15	.533		58.50	10.50		69	82
	7250	72" high		14	.571		70.50	11.25		81.75	97
	7300	15" wide x 18" high		26.50	.302		22.50	5.95		28.45	34.50
	7350	24" high		25.50	.314		29	6.20		35.20	42.50
	7400	30" high		24.50	.327		36	6.45		42.45	50.50
	7450	36" high		22.50	.356		43.50	7		50.50	60
	7500	48" high		17.50	.457		58	9		67	79.50
	7550	60" high		14.50	.552		73	10.85		83.85	99
	7600	72" high		13.50	.593		88	11.65		99.65	117
	7650	18" wide, 18" high		26	.308		25.50	6.05		31.55	38.50
	7700	24" high		25	.320		34.50	6.30		40.80	49
	7750	30" high	↓	24	.333	↓	44	6.55		50.55	60
230	0010	**CABINET HARDWARE**									
	1000	Catches, minimum	1 Carp	235	.034	Ea.	.72	.67		1.39	1.94
	1020	Average		119.40	.067		2.19	1.32		3.51	4.67
	1040	Maximum	↓	80	.100	↓	4.19	1.97		6.16	8
	2000	Door/drawer pulls, handles									
	2200	Handles and pulls, projecting, metal, minimum	1 Carp	160	.050	Ea.	1.54	.99		2.53	3.38
	2220	Average		95.24	.084		3.16	1.65		4.81	6.30
	2240	Maximum		68	.118		7.35	2.32		9.67	12.05
	2300	Wood, minimum		160	.050		1.54	.99		2.53	3.38
	2320	Average		95.24	.084		2.08	1.65		3.73	5.10
	2340	Maximum		68	.118		4.20	2.32		6.52	8.60
	2600	Flush, metal, minimum	↓	160	.050	↓	1.47	.99		2.46	3.31

06410 | Custom Cabinets

	Line	Description	Crew	Daily Output	Labor-Hours	Unit	2000 Bare Costs Mat.	Labor	Equip.	Total	Total Incl O&P	
	2620	Average	1 Carp	95.24	.084	Ea.	3.15	1.65		4.80	6.30	230
	2640	Maximum		68	.118		10.50	2.32		12.82	15.50	
	3000	Drawer tracks/glides, minimum		48	.167	Pr.	5.75	3.28		9.03	12	
	3020	Average		32	.250		9.75	4.93		14.68	19.20	
	3040	Maximum		24	.333		16.80	6.55		23.35	30	
	4000	Cabinet hinges, minimum		160	.050		1.43	.99		2.42	3.26	
	4020	Average		95.24	.084		2.57	1.65		4.22	5.65	
	4040	Maximum		68	.118		6.30	2.32		8.62	10.90	
	0010	**DRAWERS**										240
	0100	Solid hardwood front										
	1000	4″ high, 12″ wide	1 Carp	17	.471	Ea.	19.10	9.25		28.35	37	
	1200	18″ wide		16	.500		25	9.85		34.85	44.50	
	1400	24″ wide		15	.533		32.50	10.50		43	54	
	1600	6″ high, 12″ wide		16	.500		25	9.85		34.85	44.50	
	1800	18″ wide		15	.533		32.50	10.50		43	54	
	2000	24″ wide		14	.571		40	11.25		51.25	63.50	
	2200	9″ high, 12″ wide		15	.533		32.50	10.50		43	54	
	2400	18″ wide		14	.571		40	11.25		51.25	63.50	
	2600	24″ wide		13	.615		49.50	12.10		61.60	75.50	
	2800	Plastic laminate on particle board front										
	3000	4″ high, 12″ wide	1 Carp	17	.471	Ea.	19.95	9.25		29.20	38	
	3200	18″ wide		16	.500		23	9.85		32.85	42.50	
	3600	24″ wide		15	.533		27.50	10.50		38	48	
	3800	6″ high, 12″ wide		16	.500		23	9.85		32.85	42.50	
	4000	18″ wide		15	.533		28.50	10.50		39	49	
	4500	24″ wide		14	.571		34.50	11.25		45.75	57.50	
	4800	9″ high, 12″ wide		15	.533		27	10.50		37.50	47.50	
	5000	18″ wide		14	.571		27	11.25		38.25	49	
	5200	24″ wide		13	.615		43.50	12.10		55.60	68.50	
	5400	Plywood, flush panel front										
	6000	4″ high, 12″ wide	1 Carp	17	.471	Ea.	20.50	9.25		29.75	38.50	
	6200	18″ wide	"	16	.500	"	25	9.85		34.85	44.50	
	0010	**VANITIES**										400
	8000	Vanity bases, 2 doors, 30″ high, 21″ deep, 24″ wide	2 Carp	20	.800	Ea.	124	15.75		139.75	163	
	8050	30″ wide		16	1		130	19.70		149.70	177	
	8100	36″ wide		13.33	1.200		146	23.50		169.50	201	
	8150	48″ wide		11.43	1.400		184	27.50		211.50	249	
	9000	For deluxe models of all vanities, add to above					40%					
	9500	For custom built in place, add to above					25%	10%				

06430 | Stairs & Railings

	Line	Description	Crew	Daily Output	Labor-Hours	Unit	2000 Bare Costs Mat.	Labor	Equip.	Total	Total Incl O&P	
	0010	**RAILING** Custom design, architectural grade, hardwood, minimum	1 Carp	38	.211	L.F.	12	4.15		16.15	20.50	500
	0100	Maximum		30	.267		46	5.25		51.25	60	
	0300	Stock interior railing with spindles 6″ O.C., 4′ long		40	.200		29	3.94		32.94	38.50	
	0400	8′ long		48	.167		27	3.28		30.28	35	
	0010	**DECK, WOOD, PRESSURE TREATED LUMBER**										505
	0100	Railings and trim , 1″ x 4″	1 Carp	300	.027	L.F.	.62	.53		1.15	1.58	
	0150	2″ x 2″		300	.027		.43	.53		.96	1.37	
	0200	2″ x 4″		300	.027		.64	.53		1.17	1.60	
	0300	2″ x 6″		300	.027		.85	.53		1.38	1.84	
	0400	Decking, 1″ x 4″		275	.029	S.F.	1.56	.57		2.13	2.70	
	0500	2″ x 4″		300	.027		1.62	.53		2.15	2.68	
	0600	2″ x 6″		320	.025		1.61	.49		2.10	2.61	
	0650	5/4″ x 6″		320	.025		1.68	.49		2.17	2.69	
	0700	Redwood decking, 1″ x 4″		275	.029		4	.57		4.57	5.40	

06430 | Stairs & Railings

		06430 Stairs & Railings	CREW	DAILY OUTPUT	LABOR-HOURS	UNIT	2000 BARE COSTS MAT.	LABOR	EQUIP.	TOTAL	TOTAL INCL O&P
505	0800	2″ x 6″	1 Carp	340	.024	S.F.	11.40	.46		11.86	13.35
	0900	5/4″ x 6″		320	.025		7.15	.49		7.64	8.70
620	0011	**STAIRS, PREFABRICATED** R06430-100									
	0100	Box stairs, prefabricated, 3′-0″ wide									
	0110	Oak treads, no handrails, 2′ high	2 Carp	5	3.200	Flight	216	63		279	345
	0200	4′ high		4	4		430	79		509	610
	0300	6′ high		3.50	4.571		620	90		710	835
	0400	8′ high		3	5.333		775	105		880	1,025
	0600	With pine treads for carpet, 2′ high		5	3.200		88.50	63		151.50	205
	0700	4′ high		4	4		164	79		243	315
	0800	6′ high		3.50	4.571		240	90		330	420
	0900	8′ high		3	5.333		274	105		379	480
	1100	For 4′ wide stairs, add					25%				
	1500	Prefabricated stair rail with balusters, 5 risers	2 Carp	15	1.067	Ea.	229	21		250	288
	1700	Basement stairs, prefabricated, soft wood,									
	1710	open risers, 3′ wide, 8′ high	2 Carp	4	4	Flight	575	79		654	765
	1900	Open stairs, prefabricated prefinished poplar, metal stringers,									
	1910	treads 3′-6″ wide, no railings									
	2000	3′ high	2 Carp	5	3.200	Flight	229	63		292	360
	2100	4′ high		4	4		485	79		564	670
	2200	6′ high		3.50	4.571		555	90		645	770
	2300	8′ high		3	5.333		730	105		835	980
	2500	For prefab. 3 piece wood railings & balusters, add for									
	2600	3′ high stairs	2 Carp	15	1.067	Ea.	31.50	21		52.50	70.50
	2700	4′ high stairs		14	1.143		51	22.50		73.50	95
	2800	6′ high stairs		13	1.231		63	24.50		87.50	111
	2900	8′ high stairs		12	1.333		96.50	26.50		123	151
	3100	For 3′-6″ x 3′-6″ platform, add		4	4		72.50	79		151.50	215
	3300	Curved stairways, 3′-3″ wide, prefabricated, oak, unfinished,									
	3310	incl. curved balustrade system, open one side									
	3400	9′ high	2 Carp	.70	22.857	Flight	6,375	450		6,825	7,800
	3500	10′ high		.70	22.857		7,200	450		7,650	8,700
	3700	Open two sides, 9′ high		.50	32		10,000	630		10,630	12,100
	3800	10′ high		.50	32		10,800	630		11,430	13,000
	4000	Residential, wood, oak treads, prefabricated		1.50	10.667		930	210		1,140	1,375
	4200	Built in place		.44	36.364		1,325	715		2,040	2,675
	4400	Spiral, oak, 4′-6″ diameter, unfinished, prefabricated,									
	4500	incl. railing, 9′ high	2 Carp	1.50	10.667	Flight	4,000	210		4,210	4,750
630	0010	**STAIR PARTS** Balusters, turned, 30″ high, pine, minimum R06430-100	1 Carp	28	.286	Ea.	4.10	5.65		9.75	14.15
	0100	Maximum		26	.308		9	6.05		15.05	20.50
	0300	30″ high birch balusters, minimum		28	.286		6.30	5.65		11.95	16.60
	0400	Maximum		26	.308		10.20	6.05		16.25	21.50
	0600	42″ high, pine balusters, minimum		27	.296		6	5.85		11.85	16.60
	0700	Maximum		25	.320		13	6.30		19.30	25
	0900	42″ high birch balusters, minimum		27	.296		7.65	5.85		13.50	18.40
	1000	Maximum		25	.320		27	6.30		33.30	40.50
	1050	Baluster, stock pine, 1-1/16″ x 1-1/16″		240	.033	L.F.	1.97	.66		2.63	3.30
	1100	1-5/8″ x 1-5/8″		220	.036	″	2.18	.72		2.90	3.63
	1200	Newels, 3-1/4″ wide, starting, minimum		7	1.143	Ea.	33.50	22.50		56	75.50
	1300	Maximum		6	1.333		126	26.50		152.50	184
	1500	Landing, minimum		5	1.600		73.50	31.50		105	135
	1600	Maximum		4	2		179	39.50		218.50	264
	1800	Railings, oak, built-up, minimum		60	.133	L.F.	5.55	2.63		8.18	10.60
	1900	Maximum		55	.145		15.75	2.87		18.62	22.50
	2100	Add for sub rail		110	.073		4.20	1.43		5.63	7.05
	2300	Risers, beech, 3/4″ x 7-1/2″ high		64	.125		5.75	2.46		8.21	10.55

06430 | Stairs & Railings		CREW	DAILY OUTPUT	LABOR-HOURS	UNIT	2000 BARE COSTS MAT.	LABOR	EQUIP.	TOTAL	TOTAL INCL O&P	
2400	Fir, 3/4" x 7-1/2" high (R06430-100)	1 Carp	64	.125	L.F.	1.58	2.46		4.04	5.95	630
2600	Oak, 3/4" x 7-1/2" high		64	.125		4.99	2.46		7.45	9.70	
2800	Pine, 3/4" x 7-1/2" high		66	.121		1.57	2.39		3.96	5.80	
2850	Skirt board, pine, 1" x 10"		55	.145		1.73	2.87		4.60	6.80	
2900	1" x 12"		52	.154		2.10	3.03		5.13	7.50	
3000	Treads, 1-1/16" x 9-1/2" wide, 3' long, oak		18	.444	Ea.	23	8.75		31.75	40.50	
3100	4' long, oak		17	.471		29	9.25		38.25	48	
3300	1-1/16" x 11-1/2" wide, 3' long, oak		18	.444		23.50	8.75		32.25	40.50	
3400	6' long, oak		14	.571		56	11.25		67.25	81	
3600	Beech treads, add					40%					
3800	For mitered return nosings, add				L.F.	8.40			8.40	9.25	

06440 | Wood Ornaments		CREW	DAILY OUTPUT	LABOR-HOURS	UNIT	MAT.	LABOR	EQUIP.	TOTAL	TOTAL INCL O&P	
0010	**BEAMS, DECORATIVE** Rough sawn cedar, non-load bearing, 4" x 4"	2 Carp	180	.089	L.F.	1.30	1.75		3.05	4.43	150
0100	4" x 6"		170	.094		2.50	1.85		4.35	5.95	
0200	4" x 8"		160	.100		3.21	1.97		5.18	6.90	
0300	4" x 10"		150	.107		4.46	2.10		6.56	8.50	
0400	4" x 12"		140	.114		5.40	2.25		7.65	9.80	
0500	8" x 8"		130	.123		7.55	2.42		9.97	12.45	
0600	Plastic beam, "hewn finish", 6" x 2"		240	.067		2.78	1.31		4.09	5.30	
0601	6" x 4"		220	.073		3.24	1.43		4.67	6	
1100	Beam connector plates see div. 06090-800										
0010	**GRILLES** and panels, hardwood, sanded										350
0020	2' x 4' to 4' x 8', custom designs, unfinished, minimum	1 Carp	38	.211	S.F.	12	4.15		16.15	20.50	
0050	Average		30	.267		26	5.25		31.25	37.50	
0100	Maximum		19	.421		40	8.30		48.30	58	
0300	As above, but prefinished, minimum		38	.211		12	4.15		16.15	20.50	
0400	Maximum		19	.421		45	8.30		53.30	63.50	
0010	**LOUVERS** Redwood, 2'-0" diameter, full circle	1 Carp	16	.500	Ea.	105	9.85		114.85	133	400
0100	Half circle		16	.500		104	9.85		113.85	131	
0200	Octagonal		16	.500		86	9.85		95.85	111	
0300	Triangular, 5/12 pitch, 5'-0" at base		16	.500		180	9.85		189.85	215	
0010	**FIREPLACE MANTELS** 6" molding, 6' x 3'-6" opening, minimum	1 Carp	5	1.600	Opng.	130	31.50		161.50	196	500
0100	Maximum		5	1.600		157	31.50		188.50	227	
0300	Prefabricated pine, colonial type, stock, deluxe		2	4		760	79		839	970	
0400	Economy		3	2.667		256	52.50		308.50	370	
0010	**FIREPLACE MANTEL BEAMS** Rough texture wood, 4" x 8"	1 Carp	36	.222	L.F.	4.24	4.38		8.62	12.15	550
0100	4" x 10"		35	.229	"	5.30	4.50		9.80	13.55	
0300	Laminated hardwood, 2-1/4" x 10-1/2" wide, 6' long		5	1.600	Ea.	95.50	31.50		127	159	
0400	8' long		5	1.600	"	133	31.50		164.50	200	
0600	Brackets for above, rough sawn		12	.667	Pr.	8.75	13.15		21.90	32	
0700	Laminated		12	.667	"	13.25	13.15		26.40	37	
0011	**COLUMNS**										700
0050	Aluminum, round colonial, 6" diameter	2 Carp	80	.200	V.L.F.	16	3.94		19.94	24.50	
0100	8" diameter		62.25	.257		19	5.05		24.05	29.50	
0200	10" diameter		55	.291		23.50	5.75		29.25	36	
0250	Fir, stock units, hollow round, 6" diameter		80	.200		13	3.94		16.94	21	
0300	8" diameter		80	.200		15	3.94		18.94	23.50	
0350	10" diameter		70	.229		19	4.50		23.50	28.50	
0400	Solid turned, to 8' high, 3-1/2" diameter		80	.200		7	3.94		10.94	14.45	
0500	4-1/2" diameter		75	.213		10	4.20		14.20	18.20	
0600	5-1/2" diameter		70	.229		14	4.50		18.50	23	
0800	Square columns, built-up, 5" x 5"		65	.246		13	4.85		17.85	22.50	
0900	Solid, 3-1/2" x 3-1/2"		130	.123		6	2.42		8.42	10.75	

06400 | Architectural Woodwork

		06440 | Wood Ornaments	CREW	DAILY OUTPUT	LABOR-HOURS	UNIT	2000 BARE COSTS MAT.	LABOR	EQUIP.	TOTAL	TOTAL INCL O&P
700	1600	Hemlock, tapered, T & G, 12″ diam, 10′ high	2 Carp	100	.160	V.L.F.	30	3.15		33.15	38.50
	1700	16′ high		65	.246		53	4.85		57.85	67
	1900	10′ high, 14″ diameter		100	.160		77	3.15		80.15	90
	2000	18′ high		65	.246		73	4.85		77.85	89
	2200	18″ diameter, 12′ high		65	.246		103	4.85		107.85	121
	2300	20′ high		50	.320		99	6.30		105.30	120
	2500	20″ diameter, 14′ high		40	.400		122	7.90		129.90	148
	2600	20′ high		35	.457		126	9		135	154
	2800	For flat pilasters, deduct					33%				
	3000	For splitting into halves, add				Ea.	60			60	66
	4000	Rough sawn cedar posts, 4″ x 4″	2 Carp	250	.064	V.L.F.	2.43	1.26		3.69	4.83
	4100	4″ x 6″		235	.068		3.62	1.34		4.96	6.30
	4200	6″ x 6″		220	.073		5.45	1.43		6.88	8.45
	4300	8″ x 8″		200	.080		5.60	1.58		7.18	8.85

		06445 | Simulated Wood Ornaments	CREW	DAILY OUTPUT	LABOR-HOURS	UNIT	MAT.	LABOR	EQUIP.	TOTAL	TOTAL INCL O&P
100	0010	**MILLWORK, HIGH DENSITY POLYMER**									
	0100	Base, 9/16″ x 3-3/16″	1 Carp	230	.035	L.F.	1.22	.69		1.91	2.51
	0200	Casing, fluted, 5/8″ x 3-1/4″		215	.037		1.22	.73		1.95	2.60
	0300	Chair rail, 9/16″ x 2-1/4″		260	.031		.62	.61		1.23	1.72
	0400	5/8″ x 3-1/8″		230	.035		1.17	.69		1.86	2.46
	0500	Corner, inside, 1/2″ x 1-1/8″		220	.036		.61	.72		1.33	1.90
	0600	Cove, 13/16″ x 3-3/4″		260	.031		1.22	.61		1.83	2.38
	0700	Crown, 3/4″ x 3-13/16″		260	.031		1.22	.61		1.83	2.38
	0800	Half round, 15/16″ x 2″		240	.033		.67	.66		1.33	1.87

		06470 | Screen, Blinds & Shutters	CREW	DAILY OUTPUT	LABOR-HOURS	UNIT	MAT.	LABOR	EQUIP.	TOTAL	TOTAL INCL O&P
100	0010	**SHUTTERS, EXTERIOR** Aluminum, louvered, 1′-4″ wide, 3′-0″ long	1 Carp	10	.800	Pr.	29.50	15.75		45.25	59.50
	0200	4′-0″ long		10	.800		32.50	15.75		48.25	63
	0300	5′-4″ long		10	.800		37.50	15.75		53.25	68.50
	0400	6′-8″ long		9	.889		48	17.50		65.50	82.50
	1000	Pine, louvered, primed, each 1′-2″ wide, 3′-3″ long		10	.800		41	15.75		56.75	72
	1001	Pine, louvered, primed, each 1′-2″ wide, 3′-3″ long		20	.400	Ea.	24	7.90		31.90	40
	1100	4′-7″ long		10	.800	Pr.	49	15.75		64.75	81
	1101	4′-7″ long		20	.400	Ea.	34.50	7.90		42.40	51.50
	1250	Each 1′-4″ wide, 3′-0″ long		10	.800	Pr.	37.50	15.75		53.25	68.50
	1251	Each 1′-4″ wide, 3′-0″ long		20	.400	Ea.	25.50	7.90		33.40	41.50
	1350	5′-3″ long		10	.800	Pr.	53	15.75		68.75	85.50
	1351	5′-3″ long		20	.400	Ea.	31	7.90		38.90	47.50
	1500	Each 1′-6″ wide, 3′-3″ long		10	.800	Pr.	45	15.75		60.75	76.50
	1600	4′-7″ long		10	.800	″	61	15.75		76.75	94
	1601	4′-7″ long		20	.400	Ea.	38	7.90		45.90	55.50
	1620	Hemlock, louvered, 1′-2″ wide, 5′-7″ long		10	.800	Pr.	60.50	15.75		76.25	93.50
	1630	Each 1′-4″ wide, 2′-2″ long		10	.800		38	15.75		53.75	68.50
	1640	3′-0″ long		10	.800		40	15.75		55.75	71
	1650	3′-3″ long		10	.800		43.50	15.75		59.25	74.50
	1660	3′-11″ long		10	.800		47.50	15.75		63.25	79.50
	1670	4′-3″ long		10	.800		50	15.75		65.75	81.50
	1680	5′-3″ long		10	.800		56.50	15.75		72.25	89
	1690	5′-11″ long		10	.800		61.50	15.75		77.25	95
	1700	Door blinds, 6′-9″ long, each 1′-3″ wide		9	.889		71.50	17.50		89	109
	1710	1′-6″ wide		9	.889		90	17.50		107.50	129
	1720	Hemlock, solid raised panel, each 1′-4″ wide, 3′-3″ long		10	.800		61.50	15.75		77.25	95
	1730	3′-11″ long		10	.800		74	15.75		89.75	108
	1740	4′-3″ long		10	.800		78	15.75		93.75	113

Important: See the Reference Section for critical supporting data - Reference Nos., Crews, & Location Factors

06400 | Architectural Woodwork

06470 | Screen, Blinds & Shutters

	06470 Screen, Blinds & Shutters	CREW	DAILY OUTPUT	LABOR-HOURS	UNIT	2000 BARE COSTS MAT.	LABOR	EQUIP.	TOTAL	TOTAL INCL O&P	
1750	4′-7″ long	1 Carp	10	.800	Pr.	81	15.75		96.75	116	100
1760	4′-11″ long		10	.800		88.50	15.75		104.25	125	
1770	5′-11″ long		10	.800		111	15.75		126.75	149	
1800	Door blinds, 6′-9″ long, each 1′-3″ wide		9	.889		119	17.50		136.50	161	
1900	1′-6″ wide		9	.889		124	17.50		141.50	167	
2500	Polystyrene, solid raised panel, each 1′-4″ wide, 3′-3″ long		10	.800		52	15.75		67.75	84	
2600	3′-11″ long		10	.800		59.50	15.75		75.25	92.50	
2700	4′-7″ long		10	.800		62.50	15.75		78.25	96	
2800	5′-3″ long		10	.800		70.50	15.75		86.25	105	
2900	6′-8″ long		9	.889		97.50	17.50		115	137	
3500	Polystyrene, solid raised panel, each 3′-3″ wide, 3′-0″ long		10	.800		126	15.75		141.75	165	
3600	3′-11″ long		10	.800		156	15.75		171.75	199	
3700	4′-7″ long		10	.800		178	15.75		193.75	223	
3800	5′-3″ long		10	.800		198	15.75		213.75	244	
3900	6′-8″ long		9	.889		222	17.50		239.50	275	
4500	Polystyrene, louvered, each 1′-2″ wide, 3′-3″ long		10	.800		40	15.75		55.75	71	
4600	4′-7″ long		10	.800		50	15.75		65.75	81.50	
4750	5′-3″ long		10	.800		53	15.75		68.75	85.50	
4850	6′-8″ long		9	.889		87	17.50		104.50	126	
6000	Vinyl, louvered, each 1′-2″ x 4′-7″ long		10	.800		52	15.75		67.75	84	
6200	Each 1′-4″ x 6′-8″ long		9	.889		79	17.50		96.50	117	
8000	PVC exterior rolling shutters										
8100	including crank control	1 Carp	8	1	Ea.	360	19.70		379.70	430	
8500	Insulative - 6′ x 6′8″ stock unit	″	8	1	″	515	19.70		534.70	600	

06600 | Plastic Fabrications

06620 | Non-Structural Plastics

	06620 Non-Structural Plastics	CREW	DAILY OUTPUT	LABOR-HOURS	UNIT	2000 BARE COSTS MAT.	LABOR	EQUIP.	TOTAL	TOTAL INCL O&P	
0010	**SOLID SURFACE COUNTERTOPS**, Acrylic polymer										810
2000	Pricing for order of 1 - 50 L.F.										
2100	25″ wide, solid colors	2 Carp	20	.800	L.F.	56.50	15.75		72.25	89	
2200	Patterned colors		20	.800		71.50	15.75		87.25	106	
2300	Premium patterned colors		20	.800		89.50	15.75		105.25	126	
2400	With silicone attached 4″ backsplash, solid colors		19	.842		62	16.60		78.60	96.50	
2500	Patterned colors		19	.842		78.50	16.60		95.10	115	
2600	Premium patterned colors		19	.842		98	16.60		114.60	137	
2700	With hard seam attached 4″ backsplash, solid colors		4	4		62	79		141	203	
2800	Patterned colors		15	1.067		78.50	21		99.50	123	
2900	Premium patterned colors		15	1.067		98	21		119	144	
3800	Sinks, pricing for order of 1 - 50 units										
3900	Single bowl, hard seamed, solid colors, 13″ x 17″	1 Carp	2	4	Ea.	380	79		459	555	
4000	10″ x 15″		4.55	1.758		176	34.50		210.50	254	
4100	Cutouts for sinks		5.25	1.524			30		30	51.50	
0010	**VANITY TOPS**										850
0015	Solid surface, center bowl, 17″ x 19″	1 Carp	12	.667	Ea.	168	13.15		181.15	208	
0020	19″ x 25″		12	.667		203	13.15		216.15	246	
0030	19″ x 31″		12	.667		247	13.15		260.15	295	
0040	19″ x 37″		12	.667		287	13.15		300.15	340	
0050	22″ x 25″		10	.800		229	15.75		244.75	278	

	06620 | Non-Structural Plastics		CREW	DAILY OUTPUT	LABOR-HOURS	UNIT	2000 BARE COSTS MAT.	LABOR	EQUIP.	TOTAL	TOTAL INCL O&P
850	0060	22″ x 31″	1 Carp	10	.800	Ea.	267	15.75		282.75	320
	0070	22″ x 37″		10	.800		310	15.75		325.75	365
	0080	22″ x 43″		10	.800		355	15.75		370.75	415
	0090	22″ x 49″		10	.800		390	15.75		405.75	455
	0110	22″ x 55″		8	1		445	19.70		464.70	525
	0120	22″ x 61″		8	1		510	19.70		529.70	595
	0130	22″ x 68″		8	1		650	19.70		669.70	750
	0140	22″ x 73″		8	1		735	19.70		754.70	845
	0150	22″ x 85″		8	1		850	19.70		869.70	970
	0160	Offset bowl, left or right, 22″ x 49″		10	.800		470	15.75		485.75	540
	0170	22″ x 55″		8	1		535	19.70		554.70	620
	0180	22″ x 61″		8	1		605	19.70		624.70	700
	0190	22″ x 68″		8	1		710	19.70		729.70	815
	0200	22″ x 73″		8	1		875	19.70		894.70	995
	0210	22″ x 85″		8	1		1,025	19.70		1,044.70	1,150
	0220	Double bowl, 22″ x 61″		8	1		555	19.70		574.70	650
	0230	Corner top/bowl, 22″ x 49″	▼	8	1	▼	435	19.70		454.70	510
	0240	For aggregate colors, add					35%				
	0250	For faucets and fittings see 15410-300									

For information about Means Estimating Seminars, see yellow pages 11 and 12 in back of book

Division 7
Thermal & Moisture Protection

stimating Tips

7100 Dampproofing & aterproofing

Be sure of the job specifications before pricing this subdivision. The difference in cost between waterproofing and dampproofing can be great. Waterproofing will hold back standing water. Dampproofing prevents the transmission of water vapor. Also included in this section are vapor retarding membranes.

7200 Thermal Protection

Insulation and fireproofing products are measured by area, thickness, volume or R value. Specifications may only give what the specific R value should be in a certain situation. The estimator may need to choose the type of insulation to meet that R value.

7300 Shingles, Roof Tiles & oof Coverings
7400 Roofing & Siding Panels

Many roofing and siding products are bought and sold by the square. One square is equal to an area that measures 100 square feet.

This simple change in unit of measure could create a large error if the estimator is not observant. Accessories and fasteners necessary for a complete installation must be figured into any calculations for both material and labor.

07500 Membrane Roofing
07600 Flashing & Sheet Metal
07700 Roof Specialties & Accessories

- The items in these subdivisions compose a roofing system. No one component completes the installation and all must be estimated. Built-up or single ply membrane roofing systems are made up of many products and installation trades. Wood blocking at roof perimeters or penetrations, parapet coverings, reglets, roof drains, gutters, downspouts, sheet metal flashing, skylights, smoke vents or roof hatches all need to be considered along with the roofing material. Several different installation trades will need to work together on the roofing system. Inherent difficulties in the scheduling and coordination of various trades must be accounted for when estimating labor costs.

07900 Joint Sealers

- To complete the weather-tight shell the sealants and caulkings must be estimated. Where different materials meet—at expansion joints, at flashing penetrations, and at hundreds of other locations throughout a construction project—they provide another line of defense against water penetration. Often, an entire system is based on the proper location and placement of caulking or sealants. The detail drawings that are included as part of a set of architectural plans, show typical locations for these materials. When caulking or sealants are shown at typical locations, this means the estimator must include them for all the locations where this detail is applicable. Be careful to keep different types of sealants separate, and remember to consider backer rods and primers if necessary.

Reference Numbers

Reference numbers are shown in bold squares at the beginning of some major classifications. These numbers refer to related items in the Reference Section. The reference information may be an estimating procedure, an alternate pricing method or technical information.

Note: Not all subdivisions listed here necessarily appear in this publication.

07100 | Dampproofing & Waterproofing

	07110 | Dampproofing		CREW	DAILY OUTPUT	LABOR-HOURS	UNIT	2000 BARE COSTS MAT.	LABOR	EQUIP.	TOTAL	TOTAL INCL O&P
100	0010	**BITUMINOUS ASPHALT COATING** For foundation									
	0030	Brushed on, below grade, 1 coat	1 Rofc	665	.012	S.F.	.06	.21		.27	.45
	0100	2 coat		500	.016		.10	.27		.37	.62
	0300	Sprayed on, below grade, 1 coat, 25.6 S.F./gal.		830	.010		.07	.16		.23	.39
	0400	2 coat, 20.5 S.F./gal.		500	.016		.14	.27		.41	.67
	0600	Troweled on, asphalt with fibers, 1/16" thick		500	.016		.15	.27		.42	.68
	0700	1/8" thick		400	.020		.28	.34		.62	.94
	1000	1/2" thick		350	.023		.92	.39		1.31	1.74
200	0010	**CEMENT PARGING** 2 coats, 1/2" thick, regular P.C. R07110-010	D-1	250	.064	S.F.	.15	1.14		1.29	2.10
	0100	Waterproofed Portland cement	"	250	.064	"	.17	1.14		1.31	2.12
	07190 | Water Repellents										
700	0010	**RUBBER COATING** Water base liquid, roller applied	2 Rofc	7,000	.002	S.F.	.55	.04		.59	.68
	0200	Silicone or stearate, sprayed on CMU, 1 coat	1 Rofc	4,000	.002		.27	.03		.30	.36
	0300	2 coats	"	3,000	.003		.54	.05		.59	.68

07200 | Thermal Protection

	07210 | Building Insulation		CREW	DAILY OUTPUT	LABOR-HOURS	UNIT	2000 BARE COSTS MAT.	LABOR	EQUIP.	TOTAL	TOTAL INCL O&P
150	0010	**BLOWN-IN INSULATION** Ceilings, with open access									
	0020	Cellulose, 3-1/2" thick, R13	G-4	5,000	.005	S.F.	.13	.07	.05	.25	.32
	0030	5-3/16" thick, R19		3,800	.006		.19	.10	.07	.36	.45
	0050	6-1/2" thick, R22		3,000	.008		.24	.12	.09	.45	.57
	1000	Fiberglass, 5" thick, R11		3,800	.006		.16	.10	.07	.33	.42
	1050	6" thick, R13		3,000	.008		.17	.12	.09	.38	.50
	1100	8-1/2" thick, R19		2,200	.011		.23	.16	.12	.51	.67
	1300	12" thick, R26		1,500	.016		.33	.24	.18	.75	.97
	2000	Mineral wool, 4" thick, R12		3,500	.007		.16	.10	.08	.34	.45
	2050	6" thick, R17		2,500	.010		.18	.15	.11	.44	.57
	2100	9" thick, R23		1,750	.014		.27	.21	.16	.64	.83
	2500	Wall installation, incl. drilling & patching from outside, two 1"									
	2510	diam. holes @ 16" O.C., top & mid-point of wall, add to above									
	2700	For masonry	G-4	415	.058	S.F.	.06	.87	.65	1.58	2.29
	2800	For wood siding		840	.029		.06	.43	.32	.81	1.17
	2900	For stucco/plaster		665	.036		.06	.55	.41	1.02	1.45
350	0010	**FLOOR INSULATION, NONRIGID** Including									
	0020	spring type wire fasteners									
	2000	Fiberglass, blankets or batts, paper or foil backing									
	2100	1 side, 3-1/2" thick, R11	1 Carp	700	.011	S.F.	.27	.23		.50	.69
	2150	6" thick, R19		600	.013		.36	.26		.62	.85
	2200	8-1/2" thick, R30		550	.015		.61	.29		.90	1.16
500	0010	**POURED INSULATION** Cellulose fiber, R3.8 per inch	1 Carp	200	.040	C.F.	.45	.79		1.24	1.85
	0080	Fiberglass wool, R4 per inch		200	.040		.33	.79		1.12	1.71
	0100	Mineral wool, R3 per inch		200	.040		.30	.79		1.09	1.68
	0300	Polystyrene, R4 per inch		200	.040		1.93	.79		2.72	3.47
	0400	Vermiculite or perlite, R2.7 per inch		200	.040		1.45	.79		2.24	2.95
550	0010	**MASONRY INSULATION** Vermiculite or perlite, poured									
	0100	In cores of concrete block, 4" thick wall, .115 CF/SF	D-1	4,800	.003	S.F.	.17	.06		.23	.28

07210	Building Insulation	CREW	DAILY OUTPUT	LABOR-HOURS	UNIT	2000 BARE COSTS MAT.	LABOR	EQUIP.	TOTAL	TOTAL INCL O&P	
0700	Foamed in place, urethane in 2-5/8″ cavity	G-2	1,035	.023	S.F.	.38	.38	.24	1	1.32	550
0800	For each 1″ added thickness, add	″	2,372	.010	↓	.12	.16	.11	.39	.53	
0600	**PERIMETER INSULATION**, polystyrene, expanded, 1″ thick, R4	1 Carp	680	.012	S.F.	.17	.23		.40	.59	600
0700	2″ thick, R8		675	.012	″	.32	.23		.55	.75	
0011	**REFLECTIVE INSULATION**, aluminum foil on reinforced scrim		1,900	.004	S.F.	.14	.08		.22	.29	700
0101	Reinforced with woven polyolefin		1,900	.004		.17	.08		.25	.33	
0501	With single bubble air space, R8.8		1,500	.005		.26	.11		.37	.47	
0601	With double bubble air space, R9.8	↓	1,500	.005	↓	.28	.11		.39	.49	
0010	**WALL INSULATION, RIGID**										900
0040	Fiberglass, 1.5#/CF, unfaced, 1″ thick, R4.1	1 Carp	1,000	.008	S.F.	.25	.16		.41	.55	
0060	1-1/2″ thick, R6.2		1,000	.008		.34	.16		.50	.64	
0080	2″ thick, R8.3		1,000	.008		.41	.16		.57	.72	
0120	3″ thick, R12.4		800	.010		.49	.20		.69	.88	
0370	3#/CF, unfaced, 1″ thick, R4.3		1,000	.008		.32	.16		.48	.62	
0390	1-1/2″ thick, R6.5		1,000	.008		.63	.16		.79	.96	
0400	2″ thick, R8.7		890	.009		.77	.18		.95	1.15	
0420	2-1/2″ thick, R10.9		800	.010		.96	.20		1.16	1.40	
0440	3″ thick, R13		800	.010		1.14	.20		1.34	1.59	
0520	Foil faced, 1″ thick, R4.3		1,000	.008		.76	.16		.92	1.11	
0540	1-1/2″ thick, R6.5		1,000	.008		1.02	.16		1.18	1.39	
0560	2″ thick, R8.7		890	.009		1.27	.18		1.45	1.70	
0580	2-1/2″ thick, R10.9		800	.010		1.50	.20		1.70	1.99	
0600	3″ thick, R13		800	.010		1.64	.20		1.84	2.14	
0670	6#/CF, unfaced, 1″ thick, R4.3		1,000	.008		.73	.16		.89	1.07	
0690	1-1/2″ thick, R6.5		890	.009		1.13	.18		1.31	1.54	
0700	2″ thick, R8.7		800	.010		1.58	.20		1.78	2.08	
0721	2-1/2″ thick, R10.9		800	.010		1.74	.20		1.94	2.25	
0741	3″ thick, R13		730	.011		2.08	.22		2.30	2.66	
0821	Foil faced, 1″ thick, R4.3		1,000	.008		1.03	.16		1.19	1.40	
0840	1-1/2″ thick, R6.5		890	.009		1.48	.18		1.66	1.93	
0850	2″ thick, R8.7		800	.010		1.93	.20		2.13	2.46	
0880	2-1/2″ thick, R10.9		800	.010		2.32	.20		2.52	2.89	
0900	3″ thick, R13		730	.011		2.77	.22		2.99	3.42	
1500	Foamglass, 1-1/2″ thick, R4.5		800	.010		1.45	.20		1.65	1.94	
1550	3″ thick, R9	↓	730	.011	↓	2.71	.22		2.93	3.35	
1600	Isocyanurate, 4′ x 8′ sheet, foil faced, both sides										
1610	1/2″ thick, R3.9	1 Carp	800	.010	S.F.	.27	.20		.47	.64	
1620	5/8″ thick, R4.5		800	.010		.28	.20		.48	.65	
1630	3/4″ thick, R5.4		800	.010		.29	.20		.49	.66	
1640	1″ thick, R7.2		800	.010		.32	.20		.52	.69	
1650	1-1/2″ thick, R10.8		730	.011		.35	.22		.57	.76	
1660	2″ thick, R14.4		730	.011		.44	.22		.66	.85	
1670	3″ thick, R21.6		730	.011		1.05	.22		1.27	1.52	
1680	4″ thick, R28.8		730	.011		1.29	.22		1.51	1.79	
1700	Perlite, 1″ thick, R2.77		800	.010		.24	.20		.44	.60	
1750	2″ thick, R5.55		730	.011		.48	.22		.70	.90	
1900	Extruded polystyrene, 25 PSI compressive strength, 1″ thick, R5		800	.010		.30	.20		.50	.67	
1940	2″ thick R10		730	.011		.59	.22		.81	1.02	
1960	3″ thick, R15		730	.011		.86	.22		1.08	1.32	
2100	Expanded polystyrene, 1″ thick, R3.85		800	.010		.12	.20		.32	.47	
2120	2″ thick, R7.69		730	.011		.32	.22		.54	.72	
2140	3″ thick, R11.49	↓	730	.011	↓	.49	.22		.71	.91	
0010	**WALL OR CEILING INSUL., NON-RIGID**										950
0040	Fiberglass, kraft faced, batts or blankets										
0061	3-1/2″ thick, R11, 11″ wide	1 Carp	1,600	.005	S.F.	.22	.10		.32	.41	
0080	15″ wide	↓	1,600	.005	↓	.22	.10		.32	.41	

	07210 | Building Insulation		CREW	DAILY OUTPUT	LABOR-HOURS	UNIT	2000 BARE COSTS MAT.	LABOR	EQUIP.	TOTAL	TOTAL INCL O&P
950	0141	6″ thick, R19, 11″ wide	1 Carp	1,350	.006	S.F.	.31	.12		.43	.54
	0201	9″ thick, R30, 15″ wide		1,350	.006		.56	.12		.68	.82
	0241	12″ thick, R38, 15″ wide		1,350	.006		.72	.12		.84	.99
	0400	Fiberglass, foil faced, batts or blankets									
	0420	3-1/2″ thick, R11, 15″ wide	1 Carp	1,600	.005	S.F.	.32	.10		.42	.52
	0461	6″ thick, R19, 15″ wide		1,600	.005		.39	.10		.49	.60
	0501	9″ thick, R30, 15″ wide		1,350	.006		.67	.12		.79	.94
	0800	Fiberglass, unfaced, batts or blankets									
	0821	3-1/2″ thick, R11, 15″ wide	1 Carp	1,600	.005	S.F.	.19	.10		.29	.38
	0861	6″ thick, R19, 15″ wide		1,350	.006		.32	.12		.44	.55
	0901	9″ thick, R30, 15″ wide		1,150	.007		.56	.14		.70	.85
	0941	12″ thick, R38, 15″ wide		1,150	.007		.72	.14		.86	1.02
	1300	Mineral fiber batts, kraft faced									
	1320	3-1/2″ thick, R12	1 Carp	1,600	.005	S.F.	.24	.10		.34	.43
	1340	6″ thick, R19		1,600	.005		.37	.10		.47	.58
	1380	10″ thick, R30		1,350	.006		.58	.12		.70	.84
	1850	Friction fit wire insulation supports, 16″ O.C.		960	.008	Ea.	.05	.16		.21	.34
	1900	For foil backing, add				S.F.	.04			.04	.04

	07220 | Roof & Deck Insulation		CREW	DAILY OUTPUT	LABOR-HOURS	UNIT	MAT.	LABOR	EQUIP.	TOTAL	TOTAL INCL O&P
700	0010	ROOF DECK INSULATION									
	0020	Fiberboard low density, 1/2″ thick R1.39	1 Rofc	1,000	.008	S.F.	.17	.14		.31	.44
	0030	1″ thick R2.78		800	.010		.32	.17		.49	.67
	0080	1 1/2″ thick R4.17		800	.010		.48	.17		.65	.85
	0100	2″ thick R5.56		800	.010		.64	.17		.81	1.02
	0110	Fiberboard high density, 1/2″ thick R1.3		1,000	.008		.18	.14		.32	.45
	0120	1″ thick R2.5		800	.010		.34	.17		.51	.69
	0130	1-1/2″ thick R3.8		800	.010		.55	.17		.72	.93
	0200	Fiberglass, 3/4″ thick R2.78		1,000	.008		.44	.14		.58	.73
	0400	15/16″ thick R3.70		1,000	.008		.57	.14		.71	.88
	0460	1-1/16″ thick R4.17		1,000	.008		.72	.14		.86	1.04
	0600	1-5/16″ thick R5.26		1,000	.008		.99	.14		1.13	1.34
	0650	2-1/16″ thick R8.33		800	.010		1.06	.17		1.23	1.49
	0700	2-7/16″ thick R10		800	.010		1.20	.17		1.37	1.64
	1650	Perlite, 1/2″ thick R1.32		1,050	.008		.25	.13		.38	.52
	1655	3/4″ thick R2.08		800	.010		.30	.17		.47	.65
	1660	1″ thick R2.78		800	.010		.24	.17		.41	.58
	1670	1-1/2″ thick R4.17		800	.010		.36	.17		.53	.72
	1680	2″ thick R5.56		700	.011		.48	.19		.67	.89
	1685	2-1/2″ thick R6.67		700	.011		.71	.19		.90	1.14
	1700	Polyisocyanurate, 2#/CF density, 3/4″ thick, R5.1		1,500	.005		.29	.09		.38	.49
	1705	1″ thick R7.14		1,400	.006		.30	.10		.40	.51
	1715	1-1/2″ thick R10.87		1,250	.006		.34	.11		.45	.57
	1725	2″ thick R14.29		1,100	.007		.42	.12		.54	.69
	1735	2-1/2″ thick R16.67		1,050	.008		.48	.13		.61	.77
	1745	3″ thick R21.74		1,000	.008		.59	.14		.73	.90
	1755	3-1/2″ thick R25		1,000	.008		.70	.14		.84	1.02
	1765	Tapered for drainage		1,400	.006	B.F.	.36	.10		.46	.58
	1900	Extruded Polystyrene									
	1910	15 PSI compressive strength, 1″ thick, R5	1 Rofc	1,500	.005	S.F.	.21	.09		.30	.40
	1920	2″ thick, R10		1,250	.006		.33	.11		.44	.56
	1930	3″ thick R15		1,000	.008		.73	.14		.87	1.05
	1932	4″ thick R20		1,000	.008		1.01	.14		1.15	1.36
	1934	Tapered for drainage		1,500	.005	B.F.	.33	.09		.42	.53
	1940	25 PSI compressive strength, 1″ thick R5		1,500	.005	S.F.	.31	.09		.40	.51
	1942	2″ thick R10		1,250	.006		.63	.11		.74	.89

07200 | Thermal Protection

07220 | Roof & Deck Insulation		CREW	DAILY OUTPUT	LABOR-HOURS	UNIT	2000 BARE COSTS MAT.	LABOR	EQUIP.	TOTAL	TOTAL INCL O&P	
1944	3″ thick R15	1 Rofc	1,000	.008	S.F.	.94	.14		1.08	1.28	700
1946	4″ thick R20		1,000	.008		1.05	.14		1.19	1.40	
1948	Tapered for drainage		1,500	.005	B.F.	.38	.09		.47	.59	
1950	40 psi compressive strength, 1″ thick R5		1,500	.005	S.F.	.33	.09		.42	.53	
1952	2″ thick R10		1,250	.006		.65	.11		.76	.91	
1954	3″ thick R15		1,000	.008		.95	.14		1.09	1.30	
1956	4″ thick R20		1,000	.008		1.27	.14		1.41	1.65	
1958	Tapered for drainage		1,400	.006	B.F.	.48	.10		.58	.71	
1960	60 PSI compressive strength, 1″ thick R5		1,450	.006	S.F.	.40	.09		.49	.62	
1962	2″ thick R10		1,200	.007		.71	.11		.82	.99	
1964	3″ thick R15		975	.008		1.06	.14		1.20	1.43	
1966	4″ thick R20		950	.008		1.47	.14		1.61	1.89	
1968	Tapered for drainage		1,400	.006	B.F.	.57	.10		.67	.81	
2010	Expanded polystyrene, 1#/CF density, 3/4″ thick R2.89		1,500	.005	S.F.	.17	.09		.26	.36	
2020	1″ thick R3.85		1,500	.005		.17	.09		.26	.36	
2100	2″ thick R7.69		1,250	.006		.32	.11		.43	.55	
2110	3″ thick R11.49		1,250	.006		.48	.11		.59	.73	
2120	4″ thick R15.38		1,200	.007		.53	.11		.64	.79	
2130	5″ thick R19.23		1,150	.007		.67	.12		.79	.96	
2140	6″ thick R23.26		1,150	.007		.79	.12		.91	1.09	
2150	Tapered for drainage		1,500	.005	B.F.	.32	.09		.41	.52	
2400	Composites with 2″ EPS										
2410	1″ fiberboard	1 Rofc	950	.008	S.F.	.74	.14		.88	1.08	
2420	7/16″ oriented strand board		800	.010		.87	.17		1.04	1.28	
2430	1/2″ plywood		800	.010		.94	.17		1.11	1.35	
2440	1″ perlite		800	.010		.78	.17		.95	1.18	
2450	Composites with 1 1/2″ polyisocyanurate										
2460	1″ fiberboard	1 Rofc	800	.010	S.F.	.80	.17		.97	1.20	
2470	1″ perlite		850	.009		.83	.16		.99	1.21	
2480	7/16″ oriented strand board		800	.010		.95	.17		1.12	1.37	

07240 | Ext. Insulation/Finish Systems		CREW	DAILY OUTPUT	LABOR-HOURS	UNIT	MAT.	LABOR	EQUIP.	TOTAL	TOTAL INCL O&P	
0010	**EXTERIOR INSULATION FINISH SYSTEM**										100
0095	Field applied, 1″ EPS insulation	J-1	295	.136	S.F.	1.70	2.37	.18	4.25	6.05	
0100	With 1/2″ cement board sheathing		220	.182		2.74	3.17	.25	6.16	8.60	
0105	2″ EPS insulation		295	.136		1.90	2.37	.18	4.45	6.25	
0110	With 1/2″ cement board sheathing		220	.182		2.94	3.17	.25	6.36	8.80	
0115	3″ EPS insulation		295	.136		2.07	2.37	.18	4.62	6.45	
0120	With 1/2″ cement board sheathing		220	.182		3.11	3.17	.25	6.53	9	
0125	4″ EPS insulation		295	.136		2.22	2.37	.18	4.77	6.60	
0130	With 1/2″ cement board sheathing		220	.182		4.30	3.17	.25	7.72	10.30	
0140	Premium finish add		1,265	.032		.27	.55	.04	.86	1.27	
0150	Heavy duty reinforcement add		914	.044		1.65	.76	.06	2.47	3.17	
0160	2.5#/S.Y. metal lath substrate add	1 Lath	75	.107	S.Y.	2.12	2.05		4.17	5.70	
0170	3.4#/S.Y. metal lath substrate add	"	75	.107	"	2.21	2.05		4.26	5.80	
0180	Color or texture change,	J-1	1,265	.032	S.F.	.72	.55	.04	1.31	1.76	
0190	With substrate leveling base coat	1 Plas	530	.015		.72	.28		1	1.26	
0210	With substrate sealing base coat	1 Pord	1,224	.007		.07	.12		.19	.28	
0370	V groove shape in panel face				L.F.	.52			.52	.57	
0380	U groove shape in panel face				"	.69			.69	.76	
0390	Architectural features,										
0410	Crown moulding 8″ high x 4″ wide	1 Plas	150	.053	L.F.	1.15	.99		2.14	2.93	
0420	Crown moulding 12″ high x 6″ wide		150	.053		2.15	.99		3.14	4.03	
0430	Crown moulding 16″ high x 12″ wide		150	.053		5.65	.99		6.64	7.90	

07200 | Thermal Protection

07260 | Vapor Retarders

		07260 Vapor Retarders	CREW	DAILY OUTPUT	LABOR-HOURS	UNIT	2000 BARE COSTS MAT.	LABOR	EQUIP.	TOTAL	TOTAL INCL O&P
100	0011	**BUILDING PAPER** Aluminum and kraft laminated, foil 1 side	1 Carp	3,700	.002	S.F.	.04	.04		.08	.11
	0101	Foil 2 sides		3,700	.002		.06	.04		.10	.13
	0301	Asphalt, two ply, 30#, for subfloors		1,900	.004		.11	.08		.19	.26
	0401	Asphalt felt sheathing paper, 15#		3,700	.002		.03	.04		.07	.10
	0450	Housewrap, exterior, spun bonded polypropylene									
	0470	Small roll	1 Carp	3,800	.002	S.F.	.10	.04		.14	.18
	0480	Large roll	"	4,000	.002	"	.09	.04		.13	.17
	0500	Material only, 3′ x 111.1′ roll				Ea.	33			33	36.50
	0520	9′ x 111.1′ roll				"	94			94	103
	0601	Polyethylene vapor barrier, standard, .002″ thick	1 Carp	3,700	.002	S.F.	.01	.04		.05	.08
	0701	.004″ thick		3,700	.002		.02	.04		.06	.10
	0901	.006″ thick		3,700	.002		.03	.04		.07	.10
	1201	.010″ thick		3,700	.002		.06	.04		.10	.14
	1501	Red rosin paper, 5 sq rolls, 4 lb per square		3,700	.002		.02	.04		.06	.09
	1601	5 lbs. per square		3,700	.002		.02	.04		.06	.09
	1801	Reinf. waterproof, .002″ polyethylene backing, 1 side		3,700	.002		.05	.04		.09	.12
	1901	2 sides		3,700	.002		.07	.04		.11	.14
	3000	Building wrap, spunbonded polyethylene	2 Carp	8,000	.002		.10	.04		.14	.18

07300 | Shingles, Roof Tiles & Roof Coverings

07310 | Shingles

		07310 Shingles	CREW	DAILY OUTPUT	LABOR-HOURS	UNIT	2000 BARE COSTS MAT.	LABOR	EQUIP.	TOTAL	TOTAL INCL O&P
100	0010	**ASPHALT SHINGLES**									
	0100	Standard strip shingles									
	0150	Inorganic, class A, 210-235 lb/sq	1 Rofc	5.50	1.455	Sq.	26	25		51	74.50
	0155	Pneumatic nailed		7	1.143		26	19.50		45.50	65
	0200	Organic, class C, 235-240 lb/sq		5	1.600		35.50	27.50		63	90
	0205	Pneumatic nailed		6.25	1.280		35.50	22		57.50	79.50
	0250	Standard, laminated multi-layered shingles									
	0300	Class A, 240-260 lb/sq	1 Rofc	4.50	1.778	Sq.	34	30.50		64.50	94
	0305	Pneumatic nailed		5.63	1.422		34	24.50		58.50	82.50
	0350	Class C, 260-300 lb/square, 4 bundles/square		4	2		49	34		83	118
	0355	Pneumatic nailed		5	1.600		49	27.50		76.50	105
	0400	Premium, laminated multi-layered shingles									
	0450	Class A, 260-300 lb, 4 bundles/sq	1 Rofc	3.50	2.286	Sq.	43	39		82	120
	0455	Pneumatic nailed		4.37	1.831		43	31		74	106
	0500	Class C, 300-385 lb/square, 5 bundles/square		3	2.667		65.50	45.50		111	157
	0505	Pneumatic nailed		3.75	2.133		65.50	36.50		102	140
	0800	#15 felt underlayment		64	.125		2.59	2.13		4.72	6.80
	0825	#30 felt underlayment		58	.138		5.45	2.35		7.80	10.40
	0850	Self adhering polyethylene and rubberized asphalt underlayment		22	.364		37	6.20		43.20	52.50
	0900	Ridge shingles		330	.024	L.F.	.72	.41		1.13	1.56
	0905	Pneumatic nailed		412.50	.019	"	.72	.33		1.05	1.41
	1000	For steep roofs (7 to 12 pitch or greater), add						50%			
500	0010	**FIBER CEMENT** shingles, 16″ x 9.35″, 500 lb per square	1 Carp	4	2	Sq.	244	39.50		283.50	335
	0200	Shakes, 16″ x 9.35″, 550 lb per square		2.20	3.636	"	221	71.50		292.50	365
	0301	Hip & ridge, 4.75 x 14″		100	.080	L.F.	6	1.58		7.58	9.30
	0400	Hexagonal, 16″ x 16″		3	2.667	Sq.	165	52.50		217.50	272
	0500	Square, 16″ x 16″		3	2.667		148	52.50		200.50	253
	2000	For steep roofs (7/12 pitch or greater), add						50%			

07300 | Shingles, Roof Tiles & Roof Coverings

	07310 Shingles		CREW	DAILY OUTPUT	LABOR-HOURS	UNIT	2000 BARE COSTS MAT.	LABOR	EQUIP.	TOTAL	TOTAL INCL O&P	
00	0010	**SLATE**, Buckingham, Virginia, black R07310 -020										800
	0100	3/16" - 1/4" thick	1 Rots	1.75	4.571	Sq.	540	78		618	740	
	0200	1/4" thick		1.75	4.571		720	78		798	935	
	0900	Pennsylvania black, Bangor, #1 clear		1.75	4.571		435	78		513	625	
	1200	Vermont, unfading, green, mottled green		1.75	4.571		395	78		473	580	
	1300	Semi-weathering green & gray		1.75	4.571		296	78		374	470	
	1400	Purple		1.75	4.571		390	78		468	575	
	1500	Black or gray		1.75	4.571		355	78		433	535	
	2700	Ridge shingles, slate		200	.040	L.F.	8.50	.68		9.18	10.60	
80	0010	**WOOD** 16" No. 1 red cedar shingles, 5" exposure, on roof	1 Carp	2.50	3.200	Sq.	151	63		214	274	980
	0015	Pneumatic nailed		3.25	2.462	"	151	48.50		199.50	249	
	0200	7-1/2" exposure, on walls		2.05	3.902	Sq.	101	77		178	243	
	0205	Pneumatic nailed		2.67	2.996		101	59		160	212	
	0300	18" No. 1 red cedar perfections, 5-1/2" exposure, on roof		2.75	2.909		164	57.50		221.50	279	
	0305	Pneumatic nailed		3.57	2.241		164	44		208	257	
	0500	7-1/2" exposure, on walls		2.25	3.556		121	70		191	253	
	0505	Pneumatic nailed		2.92	2.740		121	54		175	226	
	0600	Resquared, and rebutted, 5-1/2" exposure, on roof		3	2.667		199	52.50		251.50	310	
	0605	Pneumatic nailed		3.90	2.051		199	40.50		239.50	288	
	0900	7-1/2" exposure, on walls		2.45	3.265		146	64.50		210.50	271	
	0905	Pneumatic nailed		3.18	2.516		146	49.50		195.50	246	
	1000	Add to above for fire retardant shingles, 16" long					30			30	33	
	1050	18" long					28.50			28.50	31.50	
	1060	Preformed ridge shingles	1 Carp	400	.020	L.F.	1.65	.39		2.04	2.50	
	1100	Hand-split red cedar shakes, 1/2" thick x 24" long, 10" exp. on roof		2.50	3.200	Sq.	138	63		201	260	
	1105	Pneumatic nailed		3.25	2.462		138	48.50		186.50	235	
	1110	3/4" thick x 24" long, 10" exp. on roof		2.25	3.556		138	70		208	272	
	1115	Pneumatic nailed		2.92	2.740		138	54		192	245	
	1200	1/2" thick, 18" long, 8-1/2" exp. on roof		2	4		97.50	79		176.50	242	
	1205	Pneumatic nailed		2.60	3.077		97.50	60.50		158	211	
	1210	3/4" thick x 18" long, 8 1/2" exp. on roof		1.80	4.444		97.50	87.50		185	257	
	1215	Pneumatic nailed		2.34	3.419		97.50	67.50		165	222	
	1255	10" exp. on walls		2	4		110	79		189	256	
	1260	10" exposure on walls, pneumatic nailed		2.60	3.077		110	60.50		170.50	225	
	1700	Add to above for fire retardant shakes, 24" long					30			30	33	
	1800	18" long					30			30	33	
	1810	Ridge shakes	1 Carp	350	.023	L.F.	2.35	.45		2.80	3.35	
	2000	White cedar shingles, 16" long, extras, 5" exposure, on roof		2.40	3.333	Sq.	112	65.50		177.50	237	
	2005	Pneumatic nailed		3.12	2.564		112	50.50		162.50	211	
	2050	5" exposure on walls		2	4		112	79		191	259	
	2055	Pneumatic nailed		2.60	3.077		112	60.50		172.50	228	
	2100	7-1/2" exposure, on walls		2	4		80	79		159	224	
	2105	Pneumatic nailed		2.60	3.077		80	60.50		140.50	193	
	2150	"B" grade, 5" exposure on walls		2	4		117	79		196	264	
	2155	Pneumatic nailed		2.60	3.077		117	60.50		177.50	233	
	2300	For 15# organic felt underlayment on roof, 1 layer, add		64	.125		2.59	2.46		5.05	7.05	
	2400	2 layers, add		32	.250		5.20	4.93		10.13	14.15	
	2600	For steep roofs (7/12 pitch or greater), add to above						50%				
	3000	Ridge shakes or shingle wood	1 Carp	280	.029	L.F.	2.40	.56		2.96	3.60	
	07320	**Roof Tiles**										
00	0010	**CLAY TILE** ASTM C1167, GR 1, severe weathering, acces. incl.										200
	0200	Lanai tile or Classic tile, 158 pc per sq	1 Rots	1.65	4.848	Sq.	490	83		573	695	
	0300	Americana, 158 pc per sq, most colors		1.65	4.848		495	83		578	700	
	0350	Green, gray or brown		1.65	4.848		490	83		573	695	

07300 | Shingles, Roof Tiles & Roof Coverings

	07320 | Roof Tiles		CREW	DAILY OUTPUT	LABOR-HOURS	UNIT	2000 BARE COSTS MAT.	LABOR	EQUIP.	TOTAL	TOTAL INCL O&P
200	0400	Blue	1 Rots	1.65	4.848	Sq.	490	83		573	695
	0600	Spanish tile, 171 pc per sq, red		1.80	4.444		305	76		381	480
	0800	Blend		1.80	4.444		420	76		496	600
	0900	Glazed white		1.80	4.444		500	76		576	690
	1100	Mission tile, 192 pc per sq, machine scored finish, red		1.15	6.957		635	119		754	915
	1700	French tile, 133 pc per sq, smooth finish, red		1.35	5.926		575	101		676	820
	1750	Blue or green		1.35	5.926		685	101		786	945
	1800	Norman black 317 pc per sq		1	8		805	137		942	1,150
	2200	Williamsburg tile, 158 pc per sq, aged cedar		1.35	5.926		490	101		591	730
	2250	Gray or green		1.35	5.926		490	101		591	730
	2350	Ridge shingles, clay tile		200	.040	L.F.	8.65	.68		9.33	10.75
	2510	One piece mission tile, natural red, 75 pc per square		1.65	4.848	Sq.	133	83		216	300
	2530	Mission Tile, 134 pc per square		1.15	6.957		178	119		297	415
	3000	For steep roofs (7/12 pitch or greater), add to above						50%			
300	0010	**CONCRETE TILE** Including installation of accessories									
	0020	Corrugated, 13" x 16-1/2", 90 per sq, 950 lb per sq									
	0050	Earthtone colors, nailed to wood deck	1 Rots	1.35	5.926	Sq.	103	101		204	300
	0150	Blues		1.35	5.926		114	101		215	315
	0200	Greens		1.35	5.926		114	101		215	315
	0250	Premium colors		1.35	5.926		153	101		254	355
	0500	Shakes, 13" x 16-1/2", 90 per sq, 950 lb per sq									
	0600	All colors, nailed to wood deck	1 Rots	1.50	5.333	Sq.	185	91		276	375
	1500	Accessory pieces, ridge & hip, 10" x 16-1/2", 8 lbs. each				Ea.	2.25			2.25	2.48
	1700	Rake, 6-1/2" x 16-3/4", 9 lbs. each					2.25			2.25	2.48
	1800	Mansard hip, 10" x 16-1/2", 9.2 lbs. each					2.25			2.25	2.48
	1900	Hip starter, 10" x 16-1/2", 10.5 lbs. each					9.50			9.50	10.45
	2000	3 or 4 way apex, 10" each side, 11.5 lbs. each					10.25			10.25	11.30

7 THERMAL & MOISTURE PROTECTION

07400 | Roofing & Siding Panels

	07410 | Metal Roof & Wall Panels		CREW	DAILY OUTPUT	LABOR-HOURS	UNIT	2000 BARE COSTS MAT.	LABOR	EQUIP.	TOTAL	TOTAL INCL O&P
100	0010	**ALUMINUM ROOFING** Corrugated or ribbed, .0155" thick, natural	G-3	1,200	.027	S.F.	.61	.48		1.09	1.48
	0300	Painted	"	1,200	.027	"	.87	.48		1.35	1.77
	07420 | Plastic Roof & Wall Panels										
770	0010	**FIBERGLASS** Corrugated panels, roofing, 8 oz per SF	G-3	1,000	.032	S.F.	2.17	.58		2.75	3.37
	0100	12 oz per SF		1,000	.032		3.76	.58		4.34	5.10
	0300	Corrugated siding, 6 oz per SF		880	.036		1.88	.65		2.53	3.18
	0400	8 oz per SF		880	.036		2.17	.65		2.82	3.50
	0600	12 oz. siding, textured		880	.036		3.68	.65		4.33	5.15
	0900	Flat panels, 6 oz per SF, clear or colors		880	.036		1.68	.65		2.33	2.96
	1300	8 oz per SF, clear or colors		880	.036		2.17	.65		2.82	3.50
	07460 | Siding										
100	0011	**ALUMINUM SIDING**									
	6040	.024 thick smooth white single 8" wide	2 Carp	515	.031	S.F.	1.23	.61		1.84	2.40
	6060	Double 4" pattern		515	.031		1.17	.61		1.78	2.34
	6080	Double 5" pattern		550	.029		1.17	.57		1.74	2.27

07460 | Siding

	07460 Siding	CREW	DAILY OUTPUT	LABOR-HOURS	UNIT	2000 BARE COSTS MAT.	LABOR	EQUIP.	TOTAL	TOTAL INCL O&P	
6120	Embossed white, 8″ wide	2 Carp	515	.031	S.F.	1.45	.61		2.06	2.65	100
6140	Double 4″ pattern		515	.031		1.32	.61		1.93	2.50	
6160	Double 5″ pattern		550	.029		1.32	.57		1.89	2.43	
6170	Vertical, embossed white, 12″ wide		590	.027		1.32	.53		1.85	2.37	
6320	.019 thick, insulated, smooth white, 8″ wide		515	.031		1.18	.61		1.79	2.35	
6340	Double 4″ pattern		515	.031		1.16	.61		1.77	2.33	
6360	Double 5″ pattern		550	.029		1.16	.57		1.73	2.26	
6400	Embossed white, 8″ wide		515	.031		1.35	.61		1.96	2.54	
6420	Double 4″ pattern		515	.031		1.37	.61		1.98	2.56	
6440	Double 5″ pattern		550	.029		1.37	.57		1.94	2.49	
6500	Shake finish 10″ wide white		550	.029		1.45	.57		2.02	2.58	
6600	Vertical pattern, 12″ wide, white		590	.027		1.37	.53		1.90	2.43	
6640	For colors add					.08			.08	.09	
6700	Accessories, white										
6720	Starter strip 2-1/8″	2 Carp	610	.026	L.F.	.19	.52		.71	1.10	
6740	Sill trim		450	.036		.30	.70		1	1.53	
6760	Inside corner		610	.026		.92	.52		1.44	1.90	
6780	Outside corner post		610	.026		1.57	.52		2.09	2.62	
6800	Door & window trim		440	.036		.29	.72		1.01	1.55	
6820	For colors add					.08			.08	.09	
6900	Soffit & fascia 1′ overhang solid	2 Carp	110	.145		2.12	2.87		4.99	7.25	
6920	Vented		110	.145		2.12	2.87		4.99	7.25	
6940	2′ overhang solid		100	.160		3.11	3.15		6.26	8.80	
6960	Vented		100	.160		3.11	3.15		6.26	8.80	
0010	**FASCIA** Aluminum, reverse board and batten,										300
0100	.032″ thick, colored, no furring included	1 Shee	145	.055	S.F.	2.10	1.19		3.29	4.31	
0200	Residential type, aluminum	1 Carp	200	.040	L.F.	1.15	.79		1.94	2.62	
0300	Steel, galv and enameled, stock, no furring, long panels	1 Shee	145	.055	S.F.	2.19	1.19		3.38	4.41	
0600	Short panels	″	115	.070	″	3.31	1.50		4.81	6.15	
0010	**FIBER CEMENT SIDING**										500
0020	Lap siding, 5/16″ thick, 6″ wide, smooth texture	2 Carp	415	.039	S.F.	.88	.76		1.64	2.27	
0025	Woodgrain texture		415	.039		.88	.76		1.64	2.27	
0030	7-1/2″ wide, smooth texture		425	.038		.88	.74		1.62	2.24	
0035	Woodgrain texture		425	.038		.88	.74		1.62	2.24	
0040	8″ wide, smooth texture		425	.038		.87	.74		1.61	2.23	
0045	Roughsawn texture		425	.038		.87	.74		1.61	2.23	
0050	9-1/2″ wide, smooth texture		440	.036		.84	.72		1.56	2.16	
0055	Woodgrain texture		440	.036		.84	.72		1.56	2.16	
0060	12″ wide, smooth texture		455	.035		.81	.69		1.50	2.09	
0065	Woodgrain texture		455	.035		.81	.69		1.50	2.09	
0070	Panel siding, 5/16″ thick, smooth texture		750	.021		.73	.42		1.15	1.52	
0075	Stucco texture		750	.021		.73	.42		1.15	1.52	
0080	Grooved woodgrain texture		750	.021		.73	.42		1.15	1.52	
0085	V - grooved woodgrain texture		750	.021		.73	.42		1.15	1.52	
0090	Wood starter strip		400	.040	L.F.	.20	.79		.99	1.57	
0010	**VINYL SIDING** Solid PVC panels, 8″ to 10″ wide, plain	1 Carp	255	.031	S.F.	.62	.62		1.24	1.74	600
2000	Smooth, white, single, 8″ wide	2 Carp	495	.032		.62	.64		1.26	1.77	
2020	Dutch lap, 10″ wide		550	.029		.64	.57		1.21	1.68	
2100	Double 4″ pattern, 8″ wide		495	.032		.55	.64		1.19	1.70	
2120	Double 5″ pattern, 10″ wide		550	.029		.52	.57		1.09	1.55	
2200	Embossed, white, single, 8″ wide		495	.032		.66	.64		1.30	1.82	
2220	10 ″ wide		550	.029		.67	.57		1.24	1.72	
2300	Double 4″ pattern, 8″ wide		495	.032		.57	.64		1.21	1.72	
2320	5″ pattern, 10″ wide		550	.029		.58	.57		1.15	1.62	
2400	Shake finish, 10″ wide, white		550	.029		1.90	.57		2.47	3.07	

	07460 | Siding		CREW	DAILY OUTPUT	LABOR-HOURS	UNIT	2000 BARE COSTS MAT.	LABOR	EQUIP.	TOTAL	TOTAL INCL O&P
600	2600	Vertical pattern, double 5″, 10″ wide, white	2 Carp	550	.029	S.F.	1.30	.57		1.87	2.41
	2620										
	2700	For colors, add				S.F.	.08			.08	.09
	2720	1/4″ extruded polystyrene fan folded insulation	2 Carp	2,000	.008	″	.14	.16		.30	.42
	3000	Accessories, starter strip		700	.023	L.F.	.23	.45		.68	1.02
	3100	"J" channel, 1/2″		700	.023		.23	.45		.68	1.02
	3120	5/8″		700	.023		.24	.45		.69	1.03
	3140	3/4″		695	.023		.26	.45		.71	1.07
	3160	1″		690	.023		.28	.46		.74	1.09
	3180	1-1/8″		685	.023		.27	.46		.73	1.09
	3190	1-1/4″		680	.024		.29	.46		.75	1.11
	3200	Under sill trim		500	.032		.26	.63		.89	1.37
	3300	Outside corner post, 3″ face, pocket 5/8″		700	.023		1.05	.45		1.50	1.92
	3320	7/8″		690	.023		1.05	.46		1.51	1.93
	3340	1-1/4″		680	.024		1.07	.46		1.53	1.97
	3400	Inside corner post, pocket 5/8″		700	.023		.56	.45		1.01	1.39
	3420	7/8″		690	.023		.62	.46		1.08	1.46
	3440	1-1/4″		680	.024		.66	.46		1.12	1.52
	3500	Door & window trim, 2-1/2″ face, pocket 5/8″		510	.031		.54	.62		1.16	1.65
	3520	7/8″		500	.032		.53	.63		1.16	1.66
	3540	1-1/4″		490	.033		.61	.64		1.25	1.77
	3600	Soffit & fascia, 1′ overhang, solid		120	.133		1.33	2.63		3.96	5.95
	3620	Vented		120	.133		1.35	2.63		3.98	6
	3700	2′ overhang, solid		110	.145		1.97	2.87		4.84	7.10
	3720	Vented		110	.145		1.97	2.87		4.84	7.10
750	0010	**SOFFIT** Aluminum, residential, stock units, .020″ thick	1 Carp	210	.038	S.F.	.99	.75		1.74	2.38
	0100	Baked enamel on steel, 16 or 18 gauge		105	.076		3.74	1.50		5.24	6.70
	0300	Polyvinyl chloride, white, solid		230	.035		.65	.69		1.34	1.88
	0400	Perforated		230	.035		.65	.69		1.34	1.88
	0500	For colors, add					.06			.06	.07
800	0010	**STEEL SIDING**, Beveled, vinyl coated, 8″ wide, including fasteners	1 Carp	265	.030	S.F.	1.12	.59		1.71	2.25
	0050	10″ wide	″	275	.029		1.19	.57		1.76	2.29
	0081	Galv., corrugated or ribbed, on steel frame, 30 gauge	G-3	775	.041		.75	.74		1.49	2.09
	0101	28 gauge		775	.041		.79	.74		1.53	2.13
	0301	26 gauge		775	.041		.86	.74		1.60	2.21
	0401	24 gauge		775	.041		1.03	.74		1.77	2.39
	0601	22 gauge		775	.041		1.17	.74		1.91	2.55
	0701	Colored, corrugated/ribbed, on steel frame, 10 yr fnsh, 28 ga.		775	.041		1.08	.74		1.82	2.45
	0901	26 gauge		775	.041		.94	.74		1.68	2.29
	1001	24 gauge		775	.041		1.11	.74		1.85	2.48
900	0010	**WOOD SIDING, BOARDS**									
	2000	Board & batten, cedar, "B" grade, 1″ x 10″	1 Carp	400	.020	S.F.	1.98	.39		2.37	2.86
	2200	Redwood, clear, vertical grain, 1″ x 10″		400	.020		3.59	.39		3.98	4.63
	2400	White pine, #2 & better, 1″ x 10″		400	.020		.70	.39		1.09	1.45
	2410	Board & batten siding, white pine #2, 1″ x 12″		450	.018		.80	.35		1.15	1.48
	3200	Wood, cedar bevel, A grade, 1/2″ x 6″		250	.032		1.90	.63		2.53	3.17
	3300	1/2″ x 8″		275	.029		1.56	.57		2.13	2.70
	3500	3/4″ x 10″, clear grade		300	.027		2.92	.53		3.45	4.11
	3600	"B" grade		300	.027		2.89	.53		3.42	4.08
	3800	Cedar, rough sawn, 1″ x 4″, A grade, natural		240	.033		2.75	.66		3.41	4.16
	3900	Stained		240	.033		3.10	.66		3.76	4.54
	4100	1″ x 12″, board & batten, #3 & Btr., natural		260	.031		2.08	.61		2.69	3.33
	4200	Stained		260	.031		2.42	.61		3.03	3.70
	4400	1″ x 8″ channel siding, #3 & Btr., natural		250	.032		2.02	.63		2.65	3.30

07400 | Roofing & Siding Panels

07460 | Siding

	07460 Siding	CREW	DAILY OUTPUT	LABOR-HOURS	UNIT	2000 BARE COSTS MAT.	LABOR	EQUIP.	TOTAL	TOTAL INCL O&P	
4500	Stained	1 Carp	250	.032	S.F.	2.30	.63		2.93	3.61	900
4700	Redwood, clear, beveled, vertical grain, 1/2″ x 4″		200	.040		3.22	.79		4.01	4.89	
4750	1/2″ x 6″		225	.036		2.70	.70		3.40	4.17	
4800	1/2″ x 8″		250	.032		2.19	.63		2.82	3.49	
5000	3/4″ x 10″		300	.027		3.57	.53		4.10	4.83	
5200	Channel siding, 1″ x 10″, B grade		285	.028		2.30	.55		2.85	3.48	
5250	Redwood, T&G boards, B grade, 1″ x 4″	2 Carp	300	.053		2.74	1.05		3.79	4.81	
5270	1″ x 8″	″	375	.043		2.36	.84		3.20	4.04	
5400	White pine, rough sawn, 1″ x 8″, natural	1 Carp	275	.029		.69	.57		1.26	1.74	
5500	Stained	″	275	.029		1.02	.57		1.59	2.10	
5600	Tongue and groove, 1″ x 8″, horizontal	2 Carp	375	.043		.63	.84		1.47	2.13	
0010	**WOOD PRODUCT SIDING**										950
0030	Lap siding, hardboard, 7/16″ x 8″, primed										
0050	Wood grain texture finish	2 Carp	650	.025	S.F.	1.04	.48		1.52	1.97	
0100	Panels, 7/16″ thick, smooth, textured or grooved, primed		700	.023		.74	.45		1.19	1.58	
0200	Stained		700	.023		.88	.45		1.33	1.74	
0700	Particle board, overlaid, 3/8″ thick		750	.021		.63	.42		1.05	1.41	
0900	Plywood, medium density overlaid, 3/8″ thick		750	.021		1.02	.42		1.44	1.84	
1000	1/2″ thick		700	.023		1.19	.45		1.64	2.08	
1100	3/4″ thick		650	.025		1.56	.48		2.04	2.55	
1600	Texture 1-11, cedar, 5/8″ thick, natural		675	.024		1.09	.47		1.56	2	
1700	Factory stained		675	.024		1.74	.47		2.21	2.71	
1900	Texture 1-11, fir, 5/8″ thick, natural		675	.024		1.01	.47		1.48	1.91	
2000	Factory stained		675	.024		1.14	.47		1.61	2.05	
2050	Texture 1-11, S.Y.P., 5/8″ thick, natural		675	.024		.85	.47		1.32	1.74	
2100	Factory stained		675	.024		.93	.47		1.40	1.82	
2200	Rough sawn cedar, 3/8″ thick, natural		675	.024		1.14	.47		1.61	2.05	
2300	Factory stained		675	.024		1.26	.47		1.73	2.19	
2500	Rough sawn fir, 3/8″ thick, natural		675	.024		.61	.47		1.08	1.47	
2600	Factory stained		675	.024		.68	.47		1.15	1.55	
2800	Redwood, textured siding, 5/8″ thick		675	.024		1.89	.47		2.36	2.88	

07500 | Membrane Roofing

07510 | Built-Up Bituminous Roofing

	07510 Built-Up Bituminous Roofing	CREW	DAILY OUTPUT	LABOR-HOURS	UNIT	2000 BARE COSTS MAT.	LABOR	EQUIP.	TOTAL	TOTAL INCL O&P	
0010	**ASPHALT** Coated felt, #30, 2 sq per roll, not mopped	1 Rofc	58	.138	Sq.	5.45	2.35		7.80	10.40	050
0200	#15, 4 sq per roll, plain or perforated, not mopped		58	.138		2.59	2.35		4.94	7.25	
0250	Perforated		58	.138		2.59	2.35		4.94	7.25	
0300	Roll roofing, smooth, #65		15	.533		12	9.10		21.10	30	
0500	#90		15	.533		13.65	9.10		22.75	32	
0520	Mineralized		15	.533		14.75	9.10		23.85	33	
0540	D.C. (Double coverage), 19″ selvage edge		10	.800		27.50	13.65		41.15	55.50	
0580	Adhesive (lap cement)				Gal.	3.68			3.68	4.05	
0010	**BUILT-UP ROOFING**										300
0120	Asphalt flood coat with gravel/slag surfacing, not including										
0140	Insulation, flashing or wood nailers										
0200	Asphalt base sheet, 3 plies #15 asphalt felt, mopped	G-1	22	2.545	Sq.	35.50	41	19	95.50	137	
0350	On nailable decks		21	2.667		38.50	43	19.90	101.40	145	
0500	4 plies #15 asphalt felt, mopped		20	2.800		49.50	45	21	115.50	161	

07500 | Membrane Roofing

07510 | Built-Up Bituminous Roofing

			CREW	DAILY OUTPUT	LABOR-HOURS	UNIT	2000 BARE COSTS MAT.	LABOR	EQUIP.	TOTAL	TOTAL INCL O&P
300	0550	On nailable decks	G-1	19	2.947	Sq.	45	47.50	22	114.50	162
	2000	Asphalt flood coat, smooth surface									
	2200	Asphalt base sheet & 3 plies #15 asphalt felt, mopped	G-1	24	2.333	Sq.	35.50	37.50	17.45	90.45	128
	2400	On nailable decks		23	2.435		33.50	39	18.20	90.70	130
	2600	4 plies #15 asphalt felt, mopped		24	2.333		41.50	37.50	17.45	96.45	135
	2700	On nailable decks	↓	23	2.435	↓	39.50	39	18.20	96.70	137
	4500	Coal tar pitch with gravel/slag surfacing									
	4600	4 plies #15 tarred felt, mopped	G-1	21	2.667	Sq.	101	43	19.90	163.90	213
	4800	3 plies glass fiber felt (type IV), mopped	″	19	2.947	″	83	47.50	22	152.50	204
400	0010	**CANTS** 4″ x 4″, treated timber, cut diagonally	1 Rofc	325	.025	L.F.	.80	.42		1.22	1.66
	0100	Foamglass		325	.025		1.92	.42		2.34	2.89
	0300	Mineral or fiber, trapezoidal, 1″x 4″ x 48″		325	.025		.17	.42		.59	.97
	0400	1-1/2″ x 5-5/8″ x 48″	↓	325	.025	↓	.29	.42		.71	1.10

07530 | Elastomeric Membrane Roofing

			CREW	DAILY OUTPUT	LABOR-HOURS	UNIT	MAT.	LABOR	EQUIP.	TOTAL	TOTAL INCL O&P
500	0010	**MODIFIED BITUMEN ROOFING** R07550-030									
	0020	Base sheet, #15 glass fiber felt, nailed to deck	1 Rofc	58	.138	Sq.	3.73	2.35		6.08	8.50
	0030	Spot mopped to deck	G-1	295	.190		4.56	3.06	1.42	9.04	12.25
	0040	Fully mopped to deck	″	192	.292		6.10	4.70	2.18	12.98	17.90
	0050	#15 organic felt, nailed to deck	1 Rofc	58	.138		3.32	2.35		5.67	8.05
	0060	Spot mopped to deck	G-1	295	.190		4.15	3.06	1.42	8.63	11.85
	0070	Fully mopped to deck	″	192	.292	↓	5.70	4.70	2.18	12.58	17.45
	0080	SBS modified, granule surf cap sheet, polyester rein., mopped									
	1500	Glass fiber reinforced, mopped, 160 mils	G-1	2,000	.028	S.F.	.36	.45	.21	1.02	1.47
	1600	Smooth surface cap sheet, mopped, 145 mils		2,100	.027		.36	.43	.20	.99	1.42
	1700	Smooth surface flashing, 145 mils		1,260	.044		.36	.72	.33	1.41	2.09
	1800	150 mils		1,260	.044		.35	.72	.33	1.40	2.08
	1900	Granular surface flashing, 150 mils		1,260	.044		.38	.72	.33	1.43	2.11
	2000	160 mils	↓	1,260	.044		.57	.72	.33	1.62	2.32
	2150	170 mils	G-5	2,100	.019		.39	.30	.08	.77	1.08
	2200	Granule surface cap sheet, poly. reinf., torched, 180 mils		2,000	.020		.45	.31	.09	.85	1.18
	2250	Smooth surface flashing, torched, 160 mils		1,260	.032		.35	.50	.14	.99	1.47
	2300	170 mils		1,260	.032		.39	.50	.14	1.03	1.51
	2350	Granule surface flashing, torched, 180 mils	↓	1,260	.032		.45	.50	.14	1.09	1.58
	2400	Fibrated aluminum coating	1 Rofc	3,800	.002	↓	.09	.04		.13	.17

07580 | Roll Roofing

			CREW	DAILY OUTPUT	LABOR-HOURS	UNIT	MAT.	LABOR	EQUIP.	TOTAL	TOTAL INCL O&P
200	0010	**ROLL ROOFING**									
	0100	Asphalt, mineral surface									
	0200	1 ply #15 organic felt, 1 ply mineral surfaced									
	0300	Selvage roofing, lap 19″, nailed & mopped	G-1	27	2.074	Sq.	33.50	33.50	15.50	82.50	116
	0400	3 plies glass fiber felt (type IV), 1 ply mineral surfaced									
	0500	Selvage roofing, lapped 19″, mopped	G-1	25	2.240	Sq.	48.50	36	16.75	101.25	139
	0600	Coated glass fiber base sheet, 2 plies of glass fiber									
	0700	Felt (type IV), 1 ply mineral surfaced selvage									
	0800	Roofing, lapped 19″, mopped	G-1	25	2.240	Sq.	53.50	36	16.75	106.25	144
	0900	On nailable decks	″	24	2.333	″	50.50	37.50	17.45	105.45	145
	1000	3 plies glass fiber felt (type III), 1 ply mineral surfaced									
	1100	Selvage roofing, lapped 19″, mopped	G-1	25	2.240	Sq.	48.50	36	16.75	101.25	139

07590 | Roof Maintenance & Repairs

			CREW	DAILY OUTPUT	LABOR-HOURS	UNIT	MAT.	LABOR	EQUIP.	TOTAL	TOTAL INCL O&P
300	0010	**ROOF COATINGS** Asphalt				Gal.	2.95			2.95	3.25
	0800	Glass fibered roof & patching cement, 5 gallon					3.50			3.50	3.85
	1100	Roof patch & flashing cement, 5 gallon				↓	17.45			17.45	19.20

07610 | Sheet Metal Roofing		CREW	DAILY OUTPUT	LABOR-HOURS	UNIT	2000 BARE COSTS MAT.	LABOR	EQUIP.	TOTAL	TOTAL INCL O&P	
0010	**COPPER ROOFING** Batten seam, over 10 sq, 16 oz, 130 lb/sq	1 Shee	1.10	7.273	Sq.	395	156		551	700	300
0200	18 oz, 145 lb per sq		1	8		440	172		612	775	
0400	Standing seam, over 10 squares, 16 oz, 125 lb per sq		1.30	6.154		380	132		512	645	
0600	18 oz, 140 lb per sq		1.20	6.667		425	143		568	710	
0900	Flat seam, over 10 squares, 16 oz, 115 lb per sq		1.20	6.667		350	143		493	625	
1200	For abnormal conditions or small areas, add					25%	100%				
1300	For lead-coated copper, add					25%					
0010	**ZINC** Copper alloy roofing, batten seam, .020" thick	1 Shee	1.20	6.667	Sq.	510	143		653	800	900
0100	.027" thick		1.15	6.957		615	150		765	935	
0300	.032" thick		1.10	7.273		695	156		851	1,025	
0400	.040" thick		1.05	7.619		820	164		984	1,175	
0600	For standing seam construction, deduct					2%					
0700	For flat seam construction, deduct					3%					
07620 | Sheet Metal Flash & Trim											
0010	**SHEET METAL CLADDING**										100
0100	Aluminum, up to 6 bends, .032" thick, window casing	1 Carp	180	.044	S.F.	.58	.88		1.46	2.14	
0200	Window sill		72	.111	L.F.	.58	2.19		2.77	4.39	
0300	Door casing		180	.044	S.F.	.58	.88		1.46	2.14	
0400	Fascia		250	.032		.58	.63		1.21	1.72	
0500	Rake trim		225	.036		.58	.70		1.28	1.84	
0600	Add for colors					.04			.04	.04	
0700	.024" thick, window casing	1 Carp	180	.044		.90	.88		1.78	2.49	
0800	Window sill		72	.111	L.F.	.90	2.19		3.09	4.74	
0900	Door casing		180	.044	S.F.	.90	.88		1.78	2.49	
1000	Fascia		250	.032		.90	.63		1.53	2.07	
1100	Rake trim		225	.036		.90	.70		1.60	2.19	
1200	Vinyl coated aluminum, up to 6 bends, window casing		180	.044		.61	.88		1.49	2.17	
1300	Window sill		72	.111	L.F.	.61	2.19		2.80	4.42	
1400	Door casing		180	.044	S.F.	.61	.88		1.49	2.17	
1500	Fascia		250	.032		.61	.63		1.24	1.75	
1600	Rake trim		225	.036		.61	.70		1.31	1.87	
07650 | Flexible Flashing											
0010	**FLASHING** Aluminum, mill finish, .013" thick	1 Shee	145	.055	S.F.	.34	1.19		1.53	2.37	600
0030	.016" thick		145	.055		.50	1.19		1.69	2.55	
0060	.019" thick		145	.055		.75	1.19		1.94	2.83	
0100	.032" thick		145	.055		.99	1.19		2.18	3.09	
0200	.040" thick		145	.055		1.36	1.19		2.55	3.50	
0300	.050" thick		145	.055		1.80	1.19		2.99	3.98	
0325	Mill finish 5" x 7" step flashing, .016" thick		1,920	.004	Ea.	.10	.09		.19	.26	
0350	Mill finish 12" x 12" step flashing, .016" thick		1,600	.005	"	.40	.11		.51	.62	
0400	Painted finish, add				S.F.	.24			.24	.26	
1600	Copper, 16 oz, sheets, under 1000 lbs.	1 Shee	115	.070		3.05	1.50		4.55	5.90	
1900	20 oz sheets, under 1000 lbs.		110	.073		3.82	1.56		5.38	6.85	
2200	24 oz sheets, under 1000 lbs.		105	.076		4.60	1.64		6.24	7.80	
2500	32 oz sheets, under 1000 lbs.		100	.080		6.10	1.72		7.82	9.60	
2700	W shape for valleys, 16 oz, 24" wide		100	.080	L.F.	5.90	1.72		7.62	9.40	
2800	Copper, paperbacked 1 side, 2 oz		330	.024	S.F.	.86	.52		1.38	1.83	
2900	3 oz		330	.024		1.12	.52		1.64	2.11	
3100	Paperbacked 2 sides, 2 oz		330	.024		.88	.52		1.40	1.85	
3150	3 oz		330	.024		1.11	.52		1.63	2.10	
3200	5 oz		330	.024		1.70	.52		2.22	2.75	
5800	Lead, 2.5 lb. per SF, up to 12" wide	1 Rofc	135	.059		3.25	1.01		4.26	5.45	

07600 | Flashing & Sheet Metal

07650 | Flexible Flashing			CREW	DAILY OUTPUT	LABOR-HOURS	UNIT	2000 BARE COSTS MAT.	LABOR	EQUIP.	TOTAL	TOTAL INCL O&P
600	5900	Over 12" wide	1 Rofc	135	.059	S.F.	3.25	1.01		4.26	5.45
	6100	Lead-coated copper, fabric-backed, 2 oz	1 Shee	330	.024		1.41	.52		1.93	2.43
	6200	5 oz		330	.024		1.66	.52		2.18	2.71
	6400	Mastic-backed 2 sides, 2 oz		330	.024		1.10	.52		1.62	2.09
	6500	5 oz		330	.024		1.39	.52		1.91	2.41
	6700	Paperbacked 1 side, 2 oz		330	.024		.95	.52		1.47	1.93
	6800	3 oz		330	.024		1.12	.52		1.64	2.11
	7000	Paperbacked 2 sides, 2 oz		330	.024		.98	.52		1.50	1.96
	7100	5 oz		330	.024		1.58	.52		2.10	2.62
	8500	Shower pan, bituminous membrane, 7 oz		155	.052		1.08	1.11		2.19	3.06
	8550	3 ply copper and fabric, 3 oz		155	.052		1.60	1.11		2.71	3.63
	8600	7 oz		155	.052		3.30	1.11		4.41	5.50
	8650	Copper, 16 oz		100	.080		3.05	1.72		4.77	6.25
	8700	Lead on copper and fabric, 5 oz		155	.052		1.66	1.11		2.77	3.70
	8800	7 oz		155	.052		2.87	1.11		3.98	5.05
	8900	Stainless steel sheets, 32 ga, .010" thick		155	.052		2.16	1.11		3.27	4.25
	9000	28 ga, .015" thick		155	.052		2.55	1.11		3.66	4.67
	9100	26 ga, .018" thick		155	.052		3.16	1.11		4.27	5.35
	9200	24 ga, .025" thick		155	.052		4.10	1.11		5.21	6.40
	9290	For mechanically keyed flashing, add					40%				
	9300	Stainless steel, paperbacked 2 sides, .005" thick	1 Shee	330	.024	S.F.	1.97	.52		2.49	3.05
	9320	Steel sheets, galvanized, 20 gauge		130	.062		.72	1.32		2.04	3.02
	9340	30 gauge		160	.050		.30	1.08		1.38	2.14
	9400	Terne coated stainless steel, .015" thick, 28 ga		155	.052		3.97	1.11		5.08	6.25
	9500	.018" thick, 26 ga		155	.052		4.48	1.11		5.59	6.80
	9600	Zinc and copper alloy (brass), .020" thick		155	.052		3.20	1.11		4.31	5.40
	9700	.027" thick		155	.052		4.28	1.11		5.39	6.60
	9800	.032" thick		155	.052		5	1.11		6.11	7.35
	9900	.040" thick		155	.052		6.10	1.11		7.21	8.55

07700 | Roof Specialties & Accessories

07710 | Manufactured Roof Specialties			CREW	DAILY OUTPUT	LABOR-HOURS	UNIT	2000 BARE COSTS MAT.	LABOR	EQUIP.	TOTAL	TOTAL INCL O&P
400	0010	**DOWNSPOUTS** Aluminum 2" x 3", .020" thick, embossed	1 Shee	190	.042	L.F.	.71	.91		1.62	2.31
	0100	Enameled		190	.042		.69	.91		1.60	2.29
	0300	Enameled, .024" thick, 2" x 3"		180	.044		1.09	.96		2.05	2.81
	0400	3" x 4"		140	.057		1.56	1.23		2.79	3.79
	0600	Round, corrugated aluminum, 3" diameter, .020" thick		190	.042		.85	.91		1.76	2.47
	0700	4" diameter, .025" thick		140	.057		1.41	1.23		2.64	3.62
	0900	Wire strainer, round, 2" diameter		155	.052	Ea.	1.75	1.11		2.86	3.80
	1000	4" diameter		155	.052		1.82	1.11		2.93	3.87
	1200	Rectangular, perforated, 2" x 3"		145	.055		2.15	1.19		3.34	4.37
	1300	3" x 4"		145	.055		3.10	1.19		4.29	5.40
	1500	Copper, round, 16 oz., stock, 2" diameter		190	.042	L.F.	4.94	.91		5.85	7
	1600	3" diameter		190	.042		3.79	.91		4.70	5.70
	1800	4" diameter		145	.055		4.60	1.19		5.79	7.05
	1900	5" diameter		130	.062		6.55	1.32		7.87	9.45
	2100	Rectangular, corrugated copper, stock, 2" x 3"		190	.042		3.50	.91		4.41	5.40
	2200	3" x 4"		145	.055		4.36	1.19		5.55	6.80

07710 | Manufactured Roof Specialties		CREW	DAILY OUTPUT	LABOR-HOURS	UNIT	2000 BARE COSTS MAT.	LABOR	EQUIP.	TOTAL	TOTAL INCL O&P	
2400	Rectangular, plain copper, stock, 2″ x 3″	1 Shee	190	.042	L.F.	4.71	.91		5.62	6.75	400
2500	3″ x 4″		145	.055		6.20	1.19		7.39	8.80	
2700	Wire strainers, rectangular, 2″ x 3″		145	.055	Ea.	2.76	1.19		3.95	5.05	
2800	3″ x 4″		145	.055		4.37	1.19		5.56	6.80	
3000	Round, 2″ diameter		145	.055		2.59	1.19		3.78	4.85	
3100	3″ diameter		145	.055		3.62	1.19		4.81	6	
3300	4″ diameter		145	.055		5.60	1.19		6.79	8.15	
3400	5″ diameter		115	.070		8.10	1.50		9.60	11.45	
3600	Lead-coated copper, round, stock, 2″ diameter		190	.042	L.F.	4.95	.91		5.86	7	
3700	3″ diameter		190	.042		4.51	.91		5.42	6.50	
3900	4″ diameter		145	.055		5.65	1.19		6.84	8.25	
4300	Rectangular, corrugated, stock, 2″ x 3″		190	.042		4.09	.91		5	6.05	
4500	Plain, stock, 2″ x 3″		190	.042		6.95	.91		7.86	9.20	
4600	3″ x 4″		145	.055		7.55	1.19		8.74	10.30	
4800	Steel, galvanized, round, corrugated, 2″ or 3″ diam, 28 ga		190	.042		.68	.91		1.59	2.28	
4900	4″ diameter, 28 gauge		145	.055		.90	1.19		2.09	2.99	
5700	Rectangular, corrugated, 28 gauge, 2″ x 3″		190	.042		.53	.91		1.44	2.11	
5800	3″ x 4″		145	.055		1.48	1.19		2.67	3.63	
6000	Rectangular, plain, 28 gauge, galvanized, 2″ x 3″		190	.042		.80	.91		1.71	2.41	
6100	3″ x 4″		145	.055		1.23	1.19		2.42	3.35	
6300	Epoxy painted, 24 gauge, corrugated, 2″ x 3″		190	.042		1.01	.91		1.92	2.64	
6400	3″ x 4″		145	.055		1.95	1.19		3.14	4.15	
6600	Wire strainers, rectangular, 2″ x 3″		145	.055	Ea.	1.62	1.19		2.81	3.78	
6700	3″ x 4″		145	.055		2.61	1.19		3.80	4.87	
6900	Round strainers, 2″ or 3″ diameter		145	.055		1.20	1.19		2.39	3.32	
7000	4″ diameter		145	.055		1.42	1.19		2.61	3.56	
7200	5″ diameter		145	.055		2.20	1.19		3.39	4.42	
7300	6″ diameter		115	.070		2.63	1.50		4.13	5.40	
8200	Vinyl, rectangular, 2″ x 3″		210	.038	L.F.	.72	.82		1.54	2.17	
8300	Round, 2-1/2″		220	.036	″	.72	.78		1.50	2.11	
0010	**DRIP EDGE**, aluminum, .016″ thick, 5″ wide, mill finish	1 Carp	400	.020	L.F.	.20	.39		.59	.90	450
0100	White finish		400	.020		.22	.39		.61	.92	
0200	8″ wide, mill finish		400	.020		.30	.39		.69	1.01	
0300	Ice belt, 28″ wide, mill finish		100	.080		3.42	1.58		5	6.45	
0310	Vented, mill finish		400	.020		1.38	.39		1.77	2.20	
0320	Painted finish		400	.020		1.50	.39		1.89	2.33	
0400	Galvanized, 5″ wide		400	.020		.22	.39		.61	.92	
0500	8″ wide, mill finish		400	.020		.33	.39		.72	1.04	
0510	Rake edge, aluminum, 1-1/2″ x 1-1/2″		400	.020		.13	.39		.52	.82	
0520	3-1/2″ x 1-1/2″		400	.020		.19	.39		.58	.89	
0010	**ELBOWS** Aluminum, 2″ x 3″, embossed	1 Shee	100	.080	Ea.	.90	1.72		2.62	3.89	500
0100	Enameled		100	.080		1.63	1.72		3.35	4.69	
0200	3″ x 4″, .025″ thick, embossed		100	.080		3.15	1.72		4.87	6.35	
0300	Enameled		100	.080		3.15	1.72		4.87	6.35	
0400	Round corrugated, 3″, embossed, .020″ thick		100	.080		1.95	1.72		3.67	5.05	
0500	4″, .025″ thick		100	.080		2.95	1.72		4.67	6.15	
0600	Copper, 16 oz. round, 2″ diameter		100	.080		11	1.72		12.72	15	
0700	3″ diameter		100	.080		5.05	1.72		6.77	8.45	
0800	4″ diameter		100	.080		9.50	1.72		11.22	13.35	
1000	2″ x 3″ corrugated		100	.080		5.05	1.72		6.77	8.50	
1100	3″ x 4″ corrugated		100	.080		9.75	1.72		11.47	13.65	
1300	Vinyl, 2-1/2″ diameter, 45° or 75°		100	.080		2	1.72		3.72	5.10	
1400	Tee Y junction		75	.107		8.50	2.29		10.79	13.20	
0010	**GRAVEL STOP** Aluminum, .050″ thick, 4″ face height, mill finish	1 Shee	145	.055	L.F.	2.72	1.19		3.91	4.99	550
0080	Duranodic finish		145	.055		3.69	1.19		4.88	6.05	

07700 | Roof Specialties & Accessories

	07710 | Manufactured Roof Specialties		CREW	DAILY OUTPUT	LABOR-HOURS	UNIT	2000 BARE COSTS MAT.	LABOR	EQUIP.	TOTAL	TOTAL INCL O&P
550	0100	Painted	1 Shee	145	.055	L.F.	4.26	1.19		5.45	6.70
	1350	Galv steel, 24 ga., 4″ leg, plain, with continuous cleat, 4″ face		145	.055		1.50	1.19		2.69	3.65
	1500	Polyvinyl chloride, 6″ face height		135	.059		3.28	1.27		4.55	5.75
	1800	Stainless steel, 24 ga., 6″ face height		135	.059		7.15	1.27		8.42	10
650	0010	**GUTTERS** Aluminum, stock units, 5″ box, .027″ thick, plain	1 Shee	120	.067	L.F.	.97	1.43		2.40	3.49
	0020	Inside corner		25	.320	Ea.	5.15	6.90		12.05	17.25
	0030	Outside corner		25	.320	″	5.15	6.90		12.05	17.25
	0100	Enameled		120	.067	L.F.	1.07	1.43		2.50	3.60
	0110	Inside corner		25	.320	Ea.	5.20	6.90		12.10	17.30
	0120	Outside corner		25	.320	″	5.20	6.90		12.10	17.30
	0300	5″ box type, .032″ thick, plain		120	.067	L.F.	1.08	1.43		2.51	3.61
	0310	Inside corner		25	.320	Ea.	4.94	6.90		11.84	17.05
	0320	Outside corner		25	.320	″	5.40	6.90		12.30	17.50
	0400	Enameled		120	.067	L.F.	1.20	1.43		2.63	3.74
	0410	Inside corner		25	.320	Ea.	5.50	6.90		12.40	17.65
	0420	Outside corner		25	.320	″	5.50	6.90		12.40	17.65
	0600	5″ x 6″ combination fascia & gutter, .032″ thick, enameled		60	.133	L.F.	3.45	2.87		6.32	8.65
	0700	Copper, half round, 16 oz, stock units, 4″ wide		120	.067		3.18	1.43		4.61	5.90
	0900	5″ wide		120	.067		3.87	1.43		5.30	6.70
	1000	6″ wide		115	.070		4.91	1.50		6.41	7.95
	1200	K type, 16 oz, stock, 4″ wide		120	.067		3.59	1.43		5.02	6.35
	1300	5″ wide		120	.067		3.61	1.43		5.04	6.40
	1500	Lead coated copper, half round, stock, 4″ wide		120	.067		4.58	1.43		6.01	7.45
	1600	6″ wide		115	.070		5.60	1.50		7.10	8.70
	1800	K type, stock, 4″ wide		120	.067		6.95	1.43		8.38	10.05
	1900	5″ wide		120	.067		6.20	1.43		7.63	9.25
	2100	Stainless steel, half round or box, stock, 4″ wide		120	.067		4.65	1.43		6.08	7.50
	2200	5″ wide		120	.067		5	1.43		6.43	7.90
	2400	Steel, galv, half round or box, 28 ga, 5″ wide, plain		120	.067		.90	1.43		2.33	3.41
	2500	Enameled		120	.067		.95	1.43		2.38	3.47
	2700	26 ga, stock, 5″ wide		120	.067		.95	1.43		2.38	3.47
	2800	6″ wide		120	.067		1.61	1.43		3.04	4.19
	3000	Vinyl, O.G., 4″ wide	1 Carp	110	.073		.85	1.43		2.28	3.39
	3100	5″ wide		110	.073		1	1.43		2.43	3.55
	3200	4″ half round, stock units		110	.073		.68	1.43		2.11	3.20
	3250	Joint connectors				Ea.	1.36			1.36	1.50
	3300	Wood, clear treated cedar, fir or hemlock, 3″ x 4″	1 Carp	100	.080	L.F.	6.30	1.58		7.88	9.60
	3400	4″ x 5″	″	100	.080	″	7.30	1.58		8.88	10.70
700	0010	**GUTTER GUARD** 6″ wide strip, aluminum mesh	1 Carp	500	.016	L.F.	.37	.32		.69	.95
	0100	Vinyl mesh	″	500	.016	″	.22	.32		.54	.78
	07720 | Roof Accessories										
550	0010	**RIDGE VENT**									
	0100	Aluminum strips, mill finish	1 Rofc	160	.050	L.F.	1.19	.85		2.04	2.90
	0150	Painted finish		160	.050	″	2.12	.85		2.97	3.93
	0200	Connectors		48	.167	Ea.	1.81	2.84		4.65	7.30
	0300	End caps		48	.167	″	.76	2.84		3.60	6.15
	0400	Galvanized strips, with damper and bird screen		160	.050	L.F.	20.50	.85		21.35	24.50
	0430	Molded polyethylene, shingles not included		160	.050	″	2.50	.85		3.35	4.34
	0440	End plugs		48	.167	Ea.	.76	2.84		3.60	6.15
	0450	Flexible roll, shingles not included		160	.050	L.F.	1.95	.85		2.80	3.73

07900 | Joint Sealers

07920 | Joint Sealants

	07920 Joint Sealants	Crew	Daily Output	Labor-Hours	Unit	2000 Bare Costs Mat.	Labor	Equip.	Total	Total Incl O&P	
0010	CAULKING AND SEALANTS										800
0020	Acoustical sealant, elastomeric, cartridges				Ea.	2.05			2.05	2.26	
0100	Acrylic latex caulk, white										
0200	11 fl. oz cartridge				Ea.	1.99			1.99	2.19	
0500	1/4" x 1/2"	1 Bric	248	.032	L.F.	.16	.65		.81	1.27	
0600	1/2" x 1/2"		250	.032		.32	.64		.96	1.44	
0800	3/4" x 3/4"		230	.035		.73	.70		1.43	1.98	
0900	3/4" x 1"		200	.040		.98	.80		1.78	2.42	
1000	1" x 1"		180	.044		1.22	.89		2.11	2.84	
1400	Butyl based, bulk				Gal.	22			22	24	
1500	Cartridges				"	26.50			26.50	29.50	
1700	Bulk, in place 1/4" x 1/2", 154 L.F./gal.	1 Bric	230	.035	L.F.	.14	.70		.84	1.34	
1800	1/2" x 1/2", 77 L.F./gal.	"	180	.044	"	.29	.89		1.18	1.81	
2000	Latex acrylic based, bulk				Gal.	23			23	25.50	
2100	Cartridges				"	26			26	28.50	
2200	Bulk in place, 1/4" x 1/2", 154 L.F./gal.	1 Bric	230	.035	L.F.	.15	.70		.85	1.34	
2300	Polysulfide compounds, 1 component, bulk				Gal.	43.50			43.50	47.50	
2400	Cartridges				"	46			46	51	
2600	1 or 2 component, in place, 1/4" x 1/4", 308 L.F./gal.	1 Bric	145	.055	L.F.	.14	1.10		1.24	2.02	
2700	1/2" x 1/4", 154 L.F./gal.		135	.059		.28	1.19		1.47	2.32	
2900	3/4" x 3/8", 68 L.F./gal.		130	.062		.64	1.23		1.87	2.78	
3000	1" x 1/2", 38 L.F./gal.		130	.062		1.14	1.23		2.37	3.33	
3200	Polyurethane, 1 or 2 component				Gal.	49			49	54	
3300	Cartridges				"	46.50			46.50	51	
3500	Bulk, in place, 1/4" x 1/4"	1 Bric	150	.053	L.F.	.16	1.07		1.23	1.98	
3600	1/2" x 1/4"		145	.055		.32	1.10		1.42	2.22	
3800	3/4" x 3/8", 68 L.F./gal.		130	.062		.72	1.23		1.95	2.87	
3900	1" x 1/2"		110	.073		1.27	1.45		2.72	3.86	
4100	Silicone rubber, bulk				Gal.	34			34	37.50	
4200	Cartridges				"	40			40	44	

Division Notes

		CREW	DAILY OUTPUT	LABOR-HOURS	UNIT	2000 BARE COSTS MAT.	LABOR	EQUIP.	TOTAL	TOTAL INCL O&P

Division 8 Doors & Windows

Estimating Tips

08100 Metal Doors & Frames

Most metal doors and frames look alike, but there may be significant differences among them. When estimating these items be sure to choose the line item that most closely compares to the specification or door schedule requirements regarding:

- type of metal
- metal gauge
- door core material
- fire rating
- finish

08200 Wood & Plastic Doors

Wood and plastic doors vary considerably in price. The primary determinant is the veneer material. Lauan, birch and oak are the most common veneers. Other variables include the following:

- hollow or solid core
- fire rating
- flush or raised panel
- finish

If the specifications require compliance with AWI (Architectural Woodwork Institute) standards or acoustical standards, the cost of the door may increase substantially. All wood doors are priced pre-mortised for hinges and predrilled for cylindrical locksets.

08300 Specialty Doors

- There are many varieties of special doors, and they are usually priced per each. Add frames, hardware or operators required for a complete installation.

08510 Steel Windows

- Most metal windows are delivered preglazed. However, some metal windows are priced without glass. Refer to 08800 Glazing for glass pricing. The grade C indicates commercial grade windows, usually ASTM C-35.

08550 Wood Windows

- All wood windows are priced preglazed. The two glazing options priced are single pane float glass and insulating glass 1/2″ thick. Add the cost of screens and grills if required.

08700 Hardware

- Hardware costs add considerably to the cost of a door. The most efficient method to determine the hardware requirements for a project is to review the door schedule. This schedule, in conjunction with the specifications, is all you should need to take off the door hardware.
- Door hinges are priced by the pair, with most doors requiring 1-1/2 pairs per door. The hinge prices do not include installation labor because it is included in door installation. Hinges are classified according to the frequency of use.

08800 Glazing

- Different openings require different types of glass. The three most common types are:
  - float
  - tempered
  - insulating
- Most exterior windows are glazed with insulating glass. Entrance doors and window walls, where the glass is less than 18″ from the floor, are generally glazed with tempered glass. Interior windows and some residential windows are glazed with float glass.

08900 Glazed Curtain Wall

- Glazed curtain walls consist of the metal tube framing and the glazing material. The cost data in this subdivision is presented for the metal tube framing alone or the composite wall. If your estimate requires a detailed takeoff of the framing, be sure to add the glazing cost.

Reference Numbers

Reference numbers are shown in bold squares at the beginning of some major classifications. These numbers refer to related items in the Reference Section. The reference information may be an estimating procedure, an alternate pricing method or technical information.

Note: Not all subdivisions listed here necessarily appear in this publication.

	08110 | Steel Doors & Frames	CREW	DAILY OUTPUT	LABOR-HOURS	UNIT	2000 BARE COSTS MAT.	LABOR	EQUIP.	TOTAL	TOTAL INCL O&P
300	0010 FIRE DOOR R08110-120									
	0015 Steel, flush, "B" label, 90 minute									
	0020 Full panel, 20 ga., 2'-0" x 6'-8"	2 Carp	20	.800	Ea.	173	15.75		188.75	217
	0040 2'-8" x 6'-8"		18	.889		180	17.50		197.50	228
	0060 3'-0" x 6'-8"		17	.941		180	18.55		198.55	230
	0080 3'-0" x 7'-0"		17	.941		187	18.55		205.55	238
	0140 18 ga., 3'-0" x 6'-8"		16	1		200	19.70		219.70	253
	0160 2'-8" x 7'-0"		17	.941		213	18.55		231.55	266
	0180 3'-0" x 7'-0"		16	1		207	19.70		226.70	261
	0200 4'-0" x 7'-0"		15	1.067		261	21		282	325
	0210									
	0220 For "A" label, 3 hour, 18 ga., use same price as "B" label									
	0240 For vision lite, add				Ea.	35			35	38.50
	0520 Flush, "B" label 90 min., composite, 20 ga., 2'-0" x 6'-8"	2 Carp	18	.889		233	17.50		250.50	286
	0540 2'-8" x 6'-8"		17	.941		237	18.55		255.55	293
	0560 3'-0" x 6'-8"		16	1		240	19.70		259.70	298
	0580 3'-0" x 7'-0"		16	1		248	19.70		267.70	305
	0640 Flush, "A" label 3 hour, composite, 18 ga., 3'-0" x 6'-8"		15	1.067		260	21		281	320
	0660 2'-8" x 7'-0"		16	1		270	19.70		289.70	330
	0680 3'-0" x 7'-0"		15	1.067		267	21		288	330
	0700 4'-0" x 7'-0"		14	1.143		320	22.50		342.50	395
600	0010 RESIDENTIAL STEEL DOOR R08110-120									
	0020 Prehung, insulated, exterior									
	0030 Embossed, full panel, 2'-8" x 6'-8"	2 Carp	17	.941	Ea.	178	18.55		196.55	228
	0040 3'-0" x 6'-8"		15	1.067		179	21		200	233
	0060 3'-0" x 7'-0"		15	1.067		233	21		254	292
	0070 5'-4" x 6'-8", double		8	2		315	39.50		354.50	420
	0220 Half glass, 2'-8" x 6'-8"		17	.941		226	18.55		244.55	280
	0240 3'-0" x 6'-8"		16	1		229	19.70		248.70	286
	0260 3'-0" x 7'-0"		16	1		277	19.70		296.70	340
	0270 5'-4" x 6'-8", double		8	2		485	39.50		524.50	600
	0720 Raised plastic face, full panel, 2'-8" x 6'-8"		16	1		221	19.70		240.70	277
	0740 3'-0" x 6'-8"		15	1.067		225	21		246	283
	0760 3'-0" x 7'-0"		15	1.067		239	21		260	298
	0780 5'-4" x 6'-8", double		8	2		440	39.50		479.50	555
	0820 Half glass, 2'-8" x 6'-8"		17	.941		263	18.55		281.55	320
	0840 3'-0" x 6'-8"		16	1		266	19.70		285.70	325
	0860 3'-0" x 7'-0"		16	1		291	19.70		310.70	355
	0880 5'-4" x 6'-8", double		8	2		550	39.50		589.50	675
	1320 Flush face, full panel, 2'-6" x 6'-8"		16	1		200	19.70		219.70	254
	1340 3'-0" x 6'-8"		15	1.067		202	21		223	259
	1360 3'-0" x 7'-0"		15	1.067		256	21		277	320
	1380 5'-4" x 6'-8", double		8	2		400	39.50		439.50	510
	1420 Half glass, 2'-8" x 6'-8"		17	.941		247	18.55		265.55	305
	1440 3'-0" x 6'-8"		16	1		250	19.70		269.70	310
	1460 3'-0" x 7'-0"		16	1		305	19.70		324.70	370
	1480 5'-4" x 6'-8", double		8	2		495	39.50		534.50	615
	2300 Interior, residential, closet, bi-fold, 6'-8" x 2'-0" wide		16	1		128	19.70		147.70	175
	2330 3'-0" wide		16	1		144	19.70		163.70	192
	2360 4'-0" wide		15	1.067		218	21		239	276
	2400 5'-0" wide		14	1.143		253	22.50		275.50	315
	2420 6'-0" wide		13	1.231		283	24.50		307.50	350
	2510 Bi-passing closet, incl. hardware, no frame or trim incl.									
	2511 Mirrored, metal frame, 6'-8" x 4'-0" wide	2 Carp	10	1.600	Opng.	164	31.50		195.50	235
	2512 5'-0" wide		10	1.600		296	31.50		327.50	380

08100 | Metal Doors & Frames

08110 | Steel Doors & Frames

	08110 Steel Doors & Frames	CREW	DAILY OUTPUT	LABOR-HOURS	UNIT	2000 BARE COSTS MAT.	LABOR	EQUIP.	TOTAL	TOTAL INCL O&P	
2513	6'-0" wide R08110-120	2 Carp	10	1.600	Opng.	325	31.50		356.50	415	600
2514	7'-0" wide		9	1.778		380	35		415	475	
2515	8'-0" wide		9	1.778		415	35		450	520	
2611	Mirrored, metal, 8'-0" x 4'-0" wide		10	1.600		284	31.50		315.50	370	
2612	5'-0" wide		10	1.600		320	31.50		351.50	405	
2613	6'-0" wide		10	1.600		350	31.50		381.50	440	
2614	7'-0" wide		9	1.778		380	35		415	475	
2615	8'-0" wide		9	1.778		415	35		450	520	
0010	**STEEL FRAMES, KNOCK DOWN** R08110-100										820
0020	18 ga., up to 5-3/4" deep										
0025	6'-8" high, 3'-0" wide, single	2 Carp	16	1	Ea.	61.50	19.70		81.20	102	
0040	6'-0" wide, double		14	1.143		74.50	22.50		97	121	
0100	7'-0" high, 3'-0" wide, single		16	1		63.50	19.70		83.20	104	
0140	6'-0" wide, double		14	1.143		77.50	22.50		100	124	
1000	18 ga., up to 4-7/8" deep, 7'-0" H, 3'-0" W, single		16	1		65.50	19.70		85.20	106	
1140	6'-0" wide, double		14	1.143		79	22.50		101.50	126	
2800	16 ga., up to 3-7/8" deep, 7'-0" high, 3'-0" wide, single		16	1		70	19.70		89.70	111	
2840	6'-0" wide, double		14	1.143		86	22.50		108.50	133	
3600	5-3/4" deep, 7'-0" high, 4'-0" wide, single		15	1.067		73	21		94	117	
3640	8'-0" wide, double		12	1.333		93.50	26.50		120	148	
3700	8'-0" high, 4'-0" wide, single		15	1.067		91	21		112	136	
3740	8'-0" wide, double		12	1.333		92.50	26.50		119	147	
4000	6-3/4" deep, 7'-0" high, 4'-0" wide, single		15	1.067		82	21		103	127	
4040	8'-0"		12	1.333		101	26.50		127.50	156	
4100	8'-0" high, 4'-0" wide, single		15	1.067		95	21		116	141	
4140	8'-0" wide, double		12	1.333		113	26.50		139.50	169	
4400	8-3/4" deep, 7'-0" high, 4'-0" wide, single		15	1.067		93	21		114	138	
4440	8'-0" wide, double		12	1.333		120	26.50		146.50	177	
4500	8'-0" high, 4'-0" wide, single		15	1.067		104	21		125	150	
4540	8'-0" wide, double		12	1.333		113	26.50		139.50	170	
4900	For welded frames, add					32			32	35	
5400	16 ga., "B" label, up to 5-3/4" deep, 7'-0" high, 4'-0" wide, single	2 Carp	15	1.067		80.50	21		101.50	125	
5440	8'-0" wide, double		12	1.333		122	26.50		148.50	180	
5800	6-3/4" deep, 7'-0" high, 4'-0" wide, single		15	1.067		86.50	21		107.50	132	
5840	8'-0" wide, double		12	1.333		116	26.50		142.50	172	
6200	8-3/4" deep, 7'-0" high, 4'-0" wide, single		15	1.067		96.50	21		117.50	142	
6240	8'-0" wide, double		12	1.333		138	26.50		164.50	197	
6300	For "A" label use same price as "B" label										
6400	For baked enamel finish, add					30%	15%				
6500	For galvanizing, add					15%					
7900	Transom lite frames, fixed, add	2 Carp	155	.103	S.F.	25	2.03		27.03	31	
8000	Movable, add	"	130	.123	"	30	2.42		32.42	37	

08200 | Wood & Plastic Doors

08210 | Wood Doors

	08210 Wood Doors	CREW	DAILY OUTPUT	LABOR-HOURS	UNIT	2000 BARE COSTS MAT.	LABOR	EQUIP.	TOTAL	TOTAL INCL O&P	
0010	**PRE-HUNG DOORS**										720
0300	Exterior, wood, comb. storm & screen, 6'-9" x 2'-6" wide	2 Carp	15	1.067	Ea.	208	21		229	265	
0320	2'-8" wide		15	1.067		208	21		229	265	
0340	3'-0" wide		15	1.067		218	21		239	276	

	08210 | Wood Doors		CREW	DAILY OUTPUT	LABOR-HOURS	UNIT	2000 BARE COSTS MAT.	LABOR	EQUIP.	TOTAL	TOTAL INCL O&P
720	0360	For 7′-0″ high door, add				Ea.	28			28	30.50
	0370	For aluminum storm doors, see division 08280-800									
	1600	Entrance door, flush, birch, solid core									
	1620	4-5/8″ solid jamb, 1-3/4″ x 6′-8″ x 2′-8″ wide	2 Carp	16	1	Ea.	191	19.70		210.70	244
	1640	3′-0″ wide	″	16	1		199	19.70		218.70	252
	1680	For 7′-0″ high door, add					13.25			13.25	14.60
	2000	Entrance door, colonial, 6 panel pine									
	2020	4-5/8″ solid jamb, 1-3/4″ x 6′-8″ x 2′-8″ wide	2 Carp	16	1	Ea.	440	19.70		459.70	520
	2040	3′-0″ wide	″	16	1		465	19.70		484.70	545
	2060	For 7′-0″ high door, add					13.25			13.25	14.60
	2200	For 5-5/8″ solid jamb, add					13.25			13.25	14.60
	2250	French door, 6′-8″ x 4′-0″ wide, 1/2″ insul. glass and grille	2 Carp	7	2.286	Pr.	870	45		915	1,025
	2260	5′-0″ wide	″	7	2.286	″	1,050	45		1,095	1,225
	2500	Exterior, metal face, insulated, incl. jamb, brickmold and									
	2520	threshold, flush, 2′-8″ x 6′-8″	2 Carp	16	1	Ea.	139	19.70		158.70	187
	2550	3′-0″ x 6′-8″	″	16	1	″	139	19.70		158.70	187
	2990										
	3500	Embossed, 6 panel, 2′-8″ x 6′-8″	2 Carp	16	1	Ea.	188	19.70		207.70	240
	3550	3′-0″ x 6′-8″		16	1		190	19.70		209.70	243
	3600	2 narrow lites, 2′-8″ x 6′-8″		16	1		297	19.70		316.70	360
	3650	3′-0″ x 6′-8″		16	1		305	19.70		324.70	370
	3700	Half glass, 2′-8″ x 6′-8″		16	1		246	19.70		265.70	305
	3750	3′-0″ x 6′-8″		16	1		250	19.70		269.70	310
	3800	2 top lites, 2′-8″ x 6′-8″		16	1		239	19.70		258.70	297
	3850	3′-0″ x 6′-8″		16	1		242	19.70		261.70	300
	4000	Interior, passage door, 4-5/8″ solid jamb									
	4400	Lauan, flush, solid core, 1-3/8″ x 6′-8″ x 2′-6″ wide	2 Carp	20	.800	Ea.	135	15.75		150.75	175
	4420	2′-8″ wide		20	.800		138	15.75		153.75	179
	4440	3′-0″ wide		19	.842		161	16.60		177.60	206
	4600	Hollow core, 1-3/8″ x 6′-8″ x 2′-6″ wide		20	.800		102	15.75		117.75	139
	4620	2′-8″ wide		20	.800		102	15.75		117.75	139
	4640	3′-0″ wide		19	.842		104	16.60		120.60	143
	4700	For 7′-0″ high door, add					8.95			8.95	9.80
	5000	Birch, flush, solid core, 1-3/8″ x 6′-8″ x 2′-6″ wide	2 Carp	20	.800		153	15.75		168.75	195
	5020	2′-8″ wide		20	.800		156	15.75		171.75	198
	5040	3′-0″ wide		19	.842		162	16.60		178.60	208
	5200	Hollow core, 1-3/8″ x 6′-8″ x 2′-6″ wide		20	.800		122	15.75		137.75	161
	5220	2′-8″ wide		20	.800		146	15.75		161.75	188
	5240	3′-0″ wide		19	.842		127	16.60		143.60	169
	5280	For 7′-0″ high door, add					9.40			9.40	10.35
	5500	Hardboard paneled, 1-3/8″ x 6′-8″ x 2′-6″ wide	2 Carp	20	.800		123	15.75		138.75	162
	5520	2′-8″ wide		20	.800		125	15.75		140.75	164
	5540	3′-0″ wide		19	.842		127	16.60		143.60	169
	6000	Pine paneled, 1-3/8″ x 6′-8″ x 2′-6″ wide		20	.800		210	15.75		225.75	258
	6020	2′-8″ wide		20	.800		221	15.75		236.75	270
	6040	3′-0″ wide		19	.842		229	16.60		245.60	281
	6500	For 5-5/8″ solid jamb, add					21.50			21.50	24
	6520	For split jamb, deduct					12.05			12.05	13.25
910	0010	**WOOD DOORS, DECORATOR** R08210-100									
	3000	Solid wood, 1-3/4″ thick stile and rail									
	3020	Mahogany, 3′-0″ x 7′-0″, minimum	2 Carp	14	1.143	Ea.	440	22.50		462.50	525
	3030	Maximum		10	1.600		640	31.50		671.50	760
	3040	3′-6″ x 8′-0″, minimum		10	1.600		645	31.50		676.50	765
	3050	Maximum		8	2		750	39.50		789.50	895
	3100	Pine, 3′-0″ x 7′-0″, minimum		14	1.143		289	22.50		311.50	355
	3110	Maximum		10	1.600		520	31.50		551.50	625

08200 | Wood & Plastic Doors

08210 | Wood Doors		CREW	DAILY OUTPUT	LABOR-HOURS	UNIT	2000 BARE COSTS MAT.	LABOR	EQUIP.	TOTAL	TOTAL INCL O&P	
3120	3'-6" x 8'-0", minimum R08210 -100	2 Carp	10	1.600	Ea.	570	31.50		601.50	685	910
3130	Maximum		8	2		1,550	39.50		1,589.50	1,775	
3200	Red oak, 3'-0" x 7'-0", minimum		14	1.143		650	22.50		672.50	755	
3210	Maximum		10	1.600		1,150	31.50		1,181.50	1,325	
3220	3'-6" x 8'-0", minimum		10	1.600		825	31.50		856.50	960	
3230	Maximum		8	2		1,350	39.50		1,389.50	1,575	
4000	Hand carved door, mahogany										
4020	3'-0" x 7'-0", minimum	2 Carp	14	1.143	Ea.	645	22.50		667.50	750	
4030	Maximum		11	1.455		1,550	28.50		1,578.50	1,750	
4040	3'-6" x 8'-0", minimum		10	1.600		1,050	31.50		1,081.50	1,200	
4050	Maximum		8	2		2,200	39.50		2,239.50	2,500	
4200	Red oak, 3'-0" x 7'-0", minimum		14	1.143		1,275	22.50		1,297.50	1,450	
4210	Maximum		11	1.455		2,975	28.50		3,003.50	3,325	
4220	3'-6" x 8'-0", minimum		10	1.600		2,550	31.50		2,581.50	2,850	
4280	For 6'-8" high door, deduct from 7'-0" door					23			23	25.50	
4400	For custom finish, add					101			101	112	
4600	Side light, mahogany, 7'-0" x 1'-6" wide, minimum	2 Carp	18	.889		255	17.50		272.50	310	
4610	Maximum		14	1.143		695	22.50		717.50	800	
4620	8'-0" x 1'-6" wide, minimum		14	1.143		315	22.50		337.50	385	
4630	Maximum		10	1.600		805	31.50		836.50	940	
4640	Side light, oak, 7'-0" x 1'-6" wide, minimum		18	.889		345	17.50		362.50	410	
4650	Maximum		14	1.143		800	22.50		822.50	920	
4660	8'-0" x 1-6" wide, minimum		14	1.143		415	22.50		437.50	500	
4670	Maximum		10	1.600		935	31.50		966.50	1,075	
6520	Interior cafe doors, 2'-6" opening, stock, panel pine		16	1		146	19.70		165.70	194	
6540	3'-0" opening		16	1		151	19.70		170.70	200	
6550	Louvered pine										
6560	2'-6" opening	2 Carp	16	1	Ea.	127	19.70		146.70	173	
8000	3'-0" opening		16	1		135	19.70		154.70	183	
8010	2'-6" opening, hardwood		16	1		171	19.70		190.70	222	
8020	3'-0" opening		16	1		201	19.70		220.70	255	
8800	Pre-hung doors, see division 08210-720										
0010	**WOOD DOORS, PANELED** R08210 -100										920
0020	Interior, six panel, hollow core, 1-3/8" thick										
0040	Molded hardboard, 2'-0" x 6'-8"	2 Carp	17	.941	Ea.	41	18.55		59.55	77	
0060	2'-6" x 6'-8"		17	.941		44.50	18.55		63.05	80.50	
0080	3'-0" x 6'-8"		17	.941		49	18.55		67.55	85.50	
0140	Embossed print, molded hardboard, 2'-0" x 6'-8"		17	.941		44.50	18.55		63.05	80.50	
0160	2'-6" x 6'-8"		17	.941		44.50	18.55		63.05	80.50	
0180	3'-0" x 6'-8"		17	.941		49	18.55		67.55	85.50	
0540	Six panel, solid, 1-3/8" thick, pine, 2'-0" x 6'-8"		15	1.067		107	21		128	154	
0560	2'-6" x 6'-8"		14	1.143		120	22.50		142.50	171	
0580	3'-0" x 6'-8"		13	1.231		138	24.50		162.50	194	
1020	Two panel, bored rail, solid, 1-3/8" thick, pine, 1'-6" x 6'-8"		16	1		195	19.70		214.70	249	
1040	2'-0" x 6'-8"		15	1.067		257	21		278	320	
1060	2'-6" x 6'-8"		14	1.143		293	22.50		315.50	365	
1340	Two panel, solid, 1-3/8" thick, fir, 2'-0" x 6'-8"		15	1.067		107	21		128	154	
1360	2'-6" x 6'-8"		14	1.143		120	22.50		142.50	171	
1380	3'-0" x 6'-8"		13	1.231		293	24.50		317.50	365	
1740	Five panel, solid, 1-3/8" thick, fir, 2'-0" x 6'-8"		15	1.067		192	21		213	247	
1760	2'-6" x 6'-8"		14	1.143		293	22.50		315.50	365	
1780	3'-0" x 6'-8"		13	1.231		294	24.50		318.50	365	
0010	**WOOD DOORS, RESIDENTIAL**										930
0200	Exterior, combination storm & screen, pine										
0260	2'-8" wide	2 Carp	10	1.600	Ea.	199	31.50		230.50	273	
0280	3'-0" wide		9	1.778		206	35		241	286	

	08210 | Wood Doors		CREW	DAILY OUTPUT	LABOR-HOURS	UNIT	2000 BARE COSTS MAT.	LABOR	EQUIP.	TOTAL	TOTAL INCL O&P
930	0300	7'-1" x 3'-0" wide	2 Carp	9	1.778	Ea.	209	35		244	290
	0400	Full lite, 6'-9" x 2'-6" wide		11	1.455		204	28.50		232.50	274
	0420	2'-8" wide		10	1.600		204	31.50		235.50	279
	0440	3'-0" wide		9	1.778		211	35		246	292
	0500	7'-1" x 3'-0" wide		9	1.778		230	35		265	315
	0700	Dutch door, pine, 1-3/4" x 6'-8" x 2'-8" wide, minimum		12	1.333		550	26.50		576.50	650
	0720	Maximum		10	1.600		585	31.50		616.50	695
	0800	3'-0" wide, minimum		12	1.333		575	26.50		601.50	680
	0820	Maximum		10	1.600		625	31.50		656.50	740
	1000	Entrance door, colonial, 1-3/4" x 6'-8" x 2'-8" wide		16	1		298	19.70		317.70	365
	1020	6 panel pine, 3'-0" wide		15	1.067		325	21		346	390
	1100	8 panel pine, 2'-8" wide		16	1		445	19.70		464.70	525
	1120	3'-0" wide		15	1.067		480	21		501	565
	1200	For tempered safety glass lites, add					23.50			23.50	26
	1300	Flush, birch, solid core, 1-3/4" x 6'-8" x 2'-8" wide	2 Carp	16	1		80	19.70		99.70	123
	1320	3'-0" wide		15	1.067		76	21		97	120
	1350	7'-0" x 2'-8" wide		16	1		77.50	19.70		97.20	119
	1360	3'-0" wide		15	1.067		91.50	21		112.50	137
	1740	Mahogany, 2'-8" x 6'-8"		15	1.067		675	21		696	775
	1760	3'-0"x 6'-8"		15	1.067		700	21		721	805
	2700	Interior, closet, bi-fold, w/hardware, no frame or trim incl.									
	2720	Flush, birch, 6'-6" or 6'-8" x 2'-6" wide	2 Carp	13	1.231	Ea.	43	24.50		67.50	89
	2740	3'-0" wide		13	1.231		46.50	24.50		71	93
	2760	4'-0" wide		12	1.333		78.50	26.50		105	132
	2780	5'-0" wide		11	1.455		85.50	28.50		114	143
	2800	6'-0" wide		10	1.600		93	31.50		124.50	156
	3000	Raised panel pine, 6'-6" or 6'-8" x 2'-6" wide		13	1.231		119	24.50		143.50	173
	3020	3'-0" wide		13	1.231		134	24.50		158.50	189
	3040	4'-0" wide		12	1.333		203	26.50		229.50	268
	3060	5'-0" wide		11	1.455		234	28.50		262.50	305
	3080	6'-0" wide		10	1.600		263	31.50		294.50	345
	3200	Louvered, pine 6'-6" or 6'-8" x 2'-6" wide		13	1.231		98	24.50		122.50	150
	3220	3'-0" wide		13	1.231		108	24.50		132.50	160
	3240	4'-0" wide		12	1.333		169	26.50		195.50	231
	3260	5'-0" wide		11	1.455		153	28.50		181.50	217
	3280	6'-0" wide		10	1.600		208	31.50		239.50	283
	4400	Bi-passing closet, incl. hardware and frame, no trim incl.									
	4420	Flush, lauan, 6'-8" x 4'-0" wide	2 Carp	12	1.333	Opng.	150	26.50		176.50	210
	4440	5'-0" wide		11	1.455		157	28.50		185.50	222
	4460	6'-0" wide		10	1.600		163	31.50		194.50	233
	4600	Flush, birch, 6'-8" x 4'-0" wide		12	1.333		165	26.50		191.50	227
	4620	5'-0" wide		11	1.455		183	28.50		211.50	250
	4640	6'-0" wide		10	1.600		198	31.50		229.50	272
	4800	Louvered, pine, 6'-8" x 4'-0" wide		12	1.333		330	26.50		356.50	410
	4820	5'-0" wide		11	1.455		360	28.50		388.50	450
	4840	6'-0" wide		10	1.600		405	31.50		436.50	500
	4900	Mirrored, 6'-8" x 4'-0" wide		12	1.333	Ea.	291	26.50		317.50	365
	5000	Paneled, pine, 6'-8" x 4'-0" wide		12	1.333	Opng.	315	26.50		341.50	395
	5020	5'-0" wide		11	1.455		350	28.50		378.50	435
	5040	6'-0" wide		10	1.600		390	31.50		421.50	485
	5061	Hardboard, 6'-8" x 4'-0" wide		10	1.600		169	31.50		200.50	240
	5062	5'-0" wide		10	1.600		169	31.50		200.50	240
	5063	6'-0" wide		10	1.600		197	31.50		228.50	271
	6100	Folding accordion, closet, including track and frame									
	6121	Vinyl, 2 layer, stock	2 Carp	400	.040	S.F.	2.76	.79		3.55	4.39
	6140	Woven mahogany and vinyl, stock		400	.040		1.41	.79		2.20	2.90

08210	Wood Doors	CREW	DAILY OUTPUT	LABOR-HOURS	UNIT	2000 BARE COSTS MAT.	LABOR	EQUIP.	TOTAL	TOTAL INCL O&P	
6160	Wood slats with vinyl overlay, stock	2 Carp	400	.040	S.F.	9.65	.79		10.44	11.95	930
6180	Economy vinyl, stock		400	.040		1.48	.79		2.27	2.98	
6200	Rigid PVC		400	.040		4.17	.79		4.96	5.95	
6220	For custom partition, add					25%	10%				
7310	Passage doors, flush, no frame included										
7320	Hardboard, hollow core, 1-3/8" x 6'-8" x 1'-6" wide	2 Carp	18	.889	Ea.	36	17.50		53.50	69.50	
7330	2'-0" wide		18	.889		37	17.50		54.50	70.50	
7340	2'-6" wide		18	.889		41	17.50		58.50	75	
7350	2'-8" wide		18	.889		42	17.50		59.50	76	
7360	3'-0" wide		17	.941		34.50	18.55		53.05	70	
7420	Lauan, hollow core, 1-3/8" x 6'-8" x 1'-6" wide		18	.889		21.50	17.50		39	54	
7440	2'-0" wide		18	.889		23	17.50		40.50	55.50	
7450	2'-4" wide		18	.889		26.50	17.50		44	59	
7460	2'-6" wide		18	.889		26.50	17.50		44	59	
7480	2'-8" wide		18	.889		27.50	17.50		45	60	
7500	3'-0" wide		17	.941		29.50	18.55		48.05	64.50	
7700	Birch, hollow core, 1-3/8" x 6'-8" x 1'-6" wide		18	.889		31	17.50		48.50	64	
7720	2'-0" wide		18	.889		33.50	17.50		51	67	
7740	2'-6" wide		18	.889		39	17.50		56.50	73	
7760	2'-8" wide		18	.889		40.50	17.50		58	75	
7780	3'-0" wide		17	.941		42.50	18.55		61.05	79	
8000	Pine louvered, 1-3/8" x 6'-8" x 1'-6" wide		19	.842		99.50	16.60		116.10	138	
8020	2'-0" wide		18	.889		115	17.50		132.50	157	
8040	2'-6" wide		18	.889		128	17.50		145.50	171	
8060	2'-8" wide		18	.889		139	17.50		156.50	183	
8080	3'-0" wide		17	.941		147	18.55		165.55	194	
8300	Pine paneled, 1-3/8" x 6'-8" x 1'-6" wide		19	.842		99.50	16.60		116.10	138	
8320	2'-0" wide		18	.889		115	17.50		132.50	157	
8330	2'-4" wide		18	.889		122	17.50		139.50	164	
8340	2'-6" wide		18	.889		128	17.50		145.50	171	
8360	2'-8" wide		18	.889		139	17.50		156.50	183	
8380	3'-0" wide		17	.941		147	18.55		165.55	194	
8450	French door, pine, 15 lites, 1-3/8"x6'-8"x2'-6" wide		18	.889		175	17.50		192.50	223	
8470	2'-8" wide		18	.889		175	17.50		192.50	223	
8490	3'-0" wide		17	.941		183	18.55		201.55	233	
8550	For over 20 doors, deduct					15%					
0010	**WOOD FRAMES**										960
0400	Exterior frame, incl. ext. trim, pine, 5/4 x 4-9/16" deep	2 Carp	375	.043	L.F.	3.89	.84		4.73	5.70	
0420	5-3/16" deep		375	.043		6.60	.84		7.44	8.75	
0440	6-9/16" deep		375	.043		5.70	.84		6.54	7.70	
0600	Oak, 5/4 x 4-9/16" deep		350	.046		8.35	.90		9.25	10.70	
0620	5-3/16" deep		350	.046		9.40	.90		10.30	11.90	
0640	6-9/16" deep		350	.046		10.45	.90		11.35	13.05	
0800	Walnut, 5/4 x 4-9/16" deep		350	.046		9.80	.90		10.70	12.35	
0820	5-3/16" deep		350	.046		14.25	.90		15.15	17.20	
0840	6-9/16" deep		350	.046		16.80	.90		17.70	20	
1000	Sills, 8/4 x 8" deep, oak, no horns		100	.160		11.85	3.15		15	18.40	
1020	2" horns		100	.160		13.20	3.15		16.35	19.90	
1040	3" horns		100	.160		15.20	3.15		18.35	22	
1100	8/4 x 10" deep, oak, no horns		90	.178		15.95	3.50		19.45	23.50	
1120	2" horns		90	.178		17.75	3.50		21.25	25.50	
1140	3" horns		90	.178		19.35	3.50		22.85	27.50	
2000	Exterior, colonial, frame & trim, 3' opng., in-swing, minimum		22	.727	Ea.	276	14.35		290.35	330	
2010	Average		21	.762		235	15		250	284	
2020	Maximum		20	.800		930	15.75		945.75	1,050	
2100	5'-4" opening, in-swing, minimum		17	.941		310	18.55		328.55	375	

08200 | Wood & Plastic Doors

	08210 | Wood Doors		Crew	Daily Output	Labor-Hours	Unit	2000 Bare Costs Mat.	Labor	Equip.	Total	Total Incl O&P
960	2120	Maximum	2 Carp	15	1.067	Ea.	930	21		951	1,050
	2140	Out-swing, minimum		17	.941		320	18.55		338.55	380
	2160	Maximum		15	1.067		965	21		986	1,075
	2400	6′-0″ opening, in-swing, minimum		16	1		299	19.70		318.70	365
	2420	Maximum		10	1.600		965	31.50		996.50	1,100
	2460	Out-swing, minimum		16	1		320	19.70		339.70	385
	2480	Maximum		10	1.600		1,200	31.50		1,231.50	1,375
	2600	For two sidelights, add, minimum		30	.533	Opng.	310	10.50		320.50	360
	2620	Maximum		20	.800	"	985	15.75		1,000.75	1,100
	2700	Custom birch frame, 3′-0″ opening		16	1	Ea.	177	19.70		196.70	229
	2750	6′-0″ opening		16	1		252	19.70		271.70	310
	2900	Exterior, modern, plain trim, 3′ opng., in-swing, minimum		26	.615		26.50	12.10		38.60	50.50
	2920	Average		24	.667		34	13.15		47.15	59.50
	2940	Maximum		22	.727		38.50	14.35		52.85	67
	3000	Interior frame, pine, 11/16″ x 3-5/8″ deep		375	.043	L.F.	3.81	.84		4.65	5.65
	3020	4-9/16″ deep		375	.043		5.10	.84		5.94	7.10
	3040	5-3/16″ deep		375	.043		4.80	.84		5.64	6.75
	3200	Oak, 11/16″ x 3-5/8″ deep		350	.046		4.73	.90		5.63	6.75
	3220	4-9/16″ deep		350	.046		5.10	.90		6	7.15
	3240	5-3/16″ deep		350	.046		5.25	.90		6.15	7.35
	3400	Walnut, 11/16″ x 3-5/8″ deep		350	.046		8.40	.90		9.30	10.80
	3420	4-9/16″ deep		350	.046		8.40	.90		9.30	10.80
	3440	5-3/16″ deep		350	.046		8.75	.90		9.65	11.15
	3600	Pocket door frame		16	1	Ea.	71	19.70		90.70	112
	3800	Threshold, oak, 5/8″ x 3-5/8″ deep		200	.080	L.F.	3.07	1.58		4.65	6.10
	3820	4-5/8″ deep		190	.084		3.91	1.66		5.57	7.15
	3840	5-5/8″ deep		180	.089		4.69	1.75		6.44	8.15
	4000	For casing see division 06220-400									
	08260 | Sliding Wood & Plastic Doors										
700	0010	**GLASS, SLIDING**									
	0012	Vinyl clad, 1″ insul. glass, 6′-0″ x 6′-10″ high	2 Carp	4	4	Opng.	750	79		829	955
	0030	6′-0″ x 8′-0″ high		4	4	Ea.	1,575	79		1,654	1,850
	0100	8′-0″ x 6′-10″ high		4	4	Opng.	1,675	79		1,754	1,975
	0500	3 leaf, 9′-0″ x 6′-10″ high		3	5.333		1,525	105		1,630	1,850
	0600	12′-0″ x 6′-10″ high		3	5.333		1,875	105		1,980	2,250
900	0010	**GLASS, SLIDING**									
	0020	Wood, 5/8″ tempered insul. glass, 6′ wide, premium	2 Carp	4	4	Ea.	930	79		1,009	1,150
	0100	Economy		4	4		680	79		759	880
	0150	8′ wide, wood, premium		3	5.333		1,075	105		1,180	1,350
	0200	Economy		3	5.333		795	105		900	1,050
	0250	12′ wide, wood, premium		2.50	6.400		1,725	126		1,851	2,125
	0300	Economy		2.50	6.400		1,125	126		1,251	1,475
	0350	Aluminum, 5/8″ tempered insulated glass, 6′ wide									
	0400	Premium	2 Carp	4	4	Ea.	1,225	79		1,304	1,475
	0450	Economy		4	4		505	79		584	695
	0500	8′ wide, premium		3	5.333		1,225	105		1,330	1,525
	0550	Economy		3	5.333		1,025	105		1,130	1,300
	0600	12′ wide, premium		2.50	6.400		2,025	126		2,151	2,475
	0650	Economy		2.50	6.400		1,275	126		1,401	1,625
	1000	Replacement doors, wood									
	1050	6′ wide, premium	2 Carp	4	4	Ea.	565	79		644	760
	08280 | Wood/Plastic Storm/Screen Doors										
800	0010	**STORM DOORS & FRAMES** Aluminum, residential,									
	0020	combination storm and screen									

08200 | Wood & Plastic Doors

08280	Wood/Plastic Storm/Screen Doors	CREW	DAILY OUTPUT	LABOR-HOURS	UNIT	2000 BARE COSTS MAT.	LABOR	EQUIP.	TOTAL	TOTAL INCL O&P	
0400	Clear anodic coating, 6′-8″ x 2′-6″ wide	2 Carp	15	1.067	Ea.	138	21		159	187	800
0420	2′-8″ wide		14	1.143		165	22.50		187.50	220	
0440	3′-0″ wide	↓	14	1.143	↓	165	22.50		187.50	220	
0500	For 7′ door height, add					5%					
1000	Mill finish, 6′-8″ x 2′-6″ wide	2 Carp	15	1.067	Ea.	194	21		215	249	
1020	2′-8″ wide		14	1.143		194	22.50		216.50	252	
1040	3′-0″ wide	↓	14	1.143		210	22.50		232.50	270	
1100	For 7′-0″ door, add					5%					
1500	White painted, 6′-8″ x 2′-6″ wide	2 Carp	15	1.067		189	21		210	244	
1520	2′-8″ wide		14	1.143		177	22.50		199.50	233	
1540	3′-0″ wide		14	1.143		196	22.50		218.50	255	
1541	Storm door, painted, alum., insul., 6′-8″ x 2′-6″ wide		14	1.143		210	22.50		232.50	270	
1545	2′-8″ wide	↓	14	1.143		210	22.50		232.50	270	
1600	For 7′-0″ door, add					5%					
1800	Aluminum screen door, minimum, 6′-8″ x 2′-8″ wide	2 Carp	14	1.143		92	22.50		114.50	140	
1810	3′-0″ wide		14	1.143		194	22.50		216.50	252	
1820	Average, 6′-8″ x 2′-8″ wide		14	1.143		153	22.50		175.50	207	
1830	3′-0″ wide		14	1.143		194	22.50		216.50	252	
1840	Maximum, 6′-8″ x 2′-8″ wide		14	1.143		305	22.50		327.50	375	
1850	3′-0″ wide	↓	14	1.143	↓	210	22.50		232.50	270	
2000	Wood door & screen, see division 08210-930										
2020											

08300 | Specialty Doors

08310	Access Doors & Panels	CREW	DAILY OUTPUT	LABOR-HOURS	UNIT	2000 BARE COSTS MAT.	LABOR	EQUIP.	TOTAL	TOTAL INCL O&P	
0010	**BULKHEAD CELLAR DOORS**										150
0020	Steel, not incl. sides, 44″ x 62″	1 Carp	5.50	1.455	Ea.	187	28.50		215.50	255	
0100	52″ x 73″		5.10	1.569		208	31		239	282	
0500	With sides and foundation plates, 57″ x 45″ x 24″		4.70	1.702		244	33.50		277.50	325	
0600	42″ x 49″ x 51″	↓	4.30	1.860	↓	294	36.50		330.50	390	

08360	Overhead Doors	CREW	DAILY OUTPUT	LABOR-HOURS	UNIT	MAT.	LABOR	EQUIP.	TOTAL	TOTAL INCL O&P	
0010	**RESIDENTIAL GARAGE DOORS** Including hardware, no frame										600
0050	Hinged, wood, custom, double door, 9′ x 7′	2 Carp	4	4	Ea.	310	79		389	475	
0070	16′ x 7′		3	5.333		525	105		630	760	
0200	Overhead, sectional, incl. hardware, fiberglass, 9′ x 7′, standard		5.28	3.030		510	59.50		569.50	660	
0220	Deluxe		5.28	3.030		595	59.50		654.50	755	
0300	16′ x 7′, standard		6	2.667		790	52.50		842.50	955	
0320	Deluxe		6	2.667		985	52.50		1,037.50	1,175	
0500	Hardboard, 9′ x 7′, standard		8	2		325	39.50		364.50	425	
0520	Deluxe		8	2		390	39.50		429.50	500	
0600	16′ x 7′, standard		6	2.667		570	52.50		622.50	720	
0620	Deluxe		6	2.667		665	52.50		717.50	825	
0700	Metal, 9′ x 7′, standard		5.28	3.030		257	59.50		316.50	385	
0720	Deluxe		8	2		510	39.50		549.50	630	
0800	16′ x 7′, standard		3	5.333		505	105		610	735	
0820	Deluxe		6	2.667		700	52.50		752.50	860	
0900	Wood, 9′ x 7′, standard	↓	8	2	↓	360	39.50		399.50	470	

08300 | Specialty Doors

	08360 | Overhead Doors		CREW	DAILY OUTPUT	LABOR-HOURS	UNIT	2000 BARE COSTS MAT.	LABOR	EQUIP.	TOTAL	TOTAL INCL O&P
600	0920	Deluxe	2 Carp	8	2	Ea.	1,025	39.50		1,064.50	1,200
	1000	16′ x 7′, standard		6	2.667		725	52.50		777.50	890
	1020	Deluxe		6	2.667		1,525	52.50		1,577.50	1,775
	1800	Door hardware, sectional	1 Carp	4	2		159	39.50		198.50	243
	1810	Door tracks only		4	2		74.50	39.50		114	150
	1820	One side only		7	1.143		51.50	22.50		74	95
	3000	Swing-up, including hardware, fiberglass, 9′ x 7′, standard	2 Carp	8	2		500	39.50		539.50	620
	3020	Deluxe		8	2		590	39.50		629.50	720
	3100	16′ x 7′, standard		6	2.667		630	52.50		682.50	780
	3120	Deluxe		6	2.667		735	52.50		787.50	895
	3200	Hardboard, 9′ x 7′, standard		8	2		258	39.50		297.50	350
	3220	Deluxe		8	2		340	39.50		379.50	445
	3300	16′ x 7′, standard		6	2.667		360	52.50		412.50	485
	3320	Deluxe		6	2.667		535	52.50		587.50	680
	3400	Metal, 9′ x 7′, standard		8	2		283	39.50		322.50	380
	3420	Deluxe		8	2		445	39.50		484.50	560
	3500	16′ x 7′, standard		6	2.667		445	52.50		497.50	580
	3520	Deluxe		6	2.667		715	52.50		767.50	875
	3600	Wood, 9′ x 7′, standard		8	2		310	39.50		349.50	410
	3620	Deluxe		8	2		495	39.50		534.50	615
	3700	16′ x 7′, standard		6	2.667		540	52.50		592.50	680
	3720	Deluxe		6	2.667		760	52.50		812.50	925
	3900	Door hardware only, swing up	1 Carp	4	2		78.50	39.50		118	154
	3920	One side only		7	1.143		50.50	22.50		73	94
	4000	For electric operator, economy, add		8	1		208	19.70		227.70	262
	4100	Deluxe, including remote control		8	1		298	19.70		317.70	365
	4500	For transmitter/receiver control , add to operator				Total	69.50			69.50	76.50
	4600	Transmitters, additional				"	22.50			22.50	24.50
	6000	Replace section, on sectional door, fiberglass, 9′ x 7′	1 Carp	4	2	Ea.	119	39.50		158.50	199
	6020	16′ x 7′		3.50	2.286		221	45		266	320
	6200	Hardboard, 9′ x 7′		4	2		76	39.50		115.50	151
	6220	16′ x 7′		3.50	2.286		135	45		180	226
	6300	Metal, 9′ x 7′		4	2		76	39.50		115.50	151
	6320	16′ x 7′		3.50	2.286		132	45		177	222
	6500	Wood, 9′ x 7′		4	2		76	39.50		115.50	151
	6520	16′ x 7′		3.50	2.286		148	45		193	239

08500 | Windows

	08510 | Steel Windows		CREW	DAILY OUTPUT	LABOR-HOURS	UNIT	2000 BARE COSTS MAT.	LABOR	EQUIP.	TOTAL	TOTAL INCL O&P
700	0010	SCREENS									
	0020	For metal sash, aluminum or bronze mesh, flat screen	2 Sswk	1,200	.013	S.F.	2.94	.28		3.22	3.79
	0500	Wicket screen, inside window	"	1,000	.016	"	4.49	.34		4.83	5.60
	0600	Residential, aluminum mesh and frame, 2′ x 3′	2 Carp	32	.500	Ea.	10.85	9.85		20.70	29
	0610	Rescreen		50	.320		8.35	6.30		14.65	19.95
	0620	3′ x 5′		32	.500		23.50	9.85		33.35	43
	0630	Rescreen		45	.356		20.50	7		27.50	34.50
	0640	4′ x 8′		25	.640		44.50	12.60		57.10	70.50
	0650	Rescreen		40	.400		33	7.90		40.90	49.50
	0660	Patio door		25	.640		126	12.60		138.60	161

08500 | Windows

	08510 | Steel Windows	CREW	DAILY OUTPUT	LABOR-HOURS	UNIT	2000 BARE COSTS MAT.	LABOR	EQUIP.	TOTAL	TOTAL INCL O&P	
0680	Rescreening	2 Carp	1,600	.010	S.F.	1.06	.20		1.26	1.51	700
1000	For solar louvers, add	2 Sswk	160	.100	″	16.40	2.13		18.53	22	
	08520 | Aluminum Windows										
0010	**ALUMINUM WINDOWS** Incl. frame and glazing, grade C										120
1000	Stock units, casement, 3′-1″ x 3′-2″ opening	2 Sswk	10	1.600	Ea.	258	34		292	350	
1040	Insulating glass	″	10	1.600		258	34		292	350	
1050	Add for storms					52			52	57	
1600	Projected, with screen, 3′-1″ x 3′-2″ opening	2 Sswk	10	1.600		184	34		218	270	
1650	Insulating glass	″	10	1.600		172	34		206	257	
1700	Add for storms					49			49	54	
2000	4′-5″ x 5′-3″ opening	2 Sswk	8	2		260	42.50		302.50	370	
2050	Insulating glass	″	8	2		299	42.50		341.50	415	
2100	Add for storms					68			68	74.50	
2500	Enamel finish windows, 3′-1″ x 3′-2″	2 Sswk	10	1.600		165	34		199	249	
2550	Insulating glass		10	1.600		180	34		214	265	
2600	4′-5″ x 5′-3″		8	2		248	42.50		290.50	355	
2700	Insulating glass		8	2		325	42.50		367.50	440	
3000	Single hung, 2′ x 3′ opening, enameled, standard glazed		10	1.600		124	34		158	203	
3100	Insulating glass		10	1.600		150	34		184	232	
3300	2′-8″ x 6′-8″ opening, standard glazed		8	2		263	42.50		305.50	375	
3400	Insulating glass		8	2		340	42.50		382.50	455	
3700	3′-4″ x 5′-0″ opening, standard glazed		9	1.778		170	38		208	262	
3800	Insulating glass		9	1.778		238	38		276	335	
4000	Sliding aluminum, 3′ x 2′ opening, standard glazed		10	1.600		139	34		173	220	
4100	Insulating glass		10	1.600		154	34		188	237	
4300	5′ x 3′ opening, standard glazed		9	1.778		177	38		215	270	
4400	Insulating glass		9	1.778		247	38		285	345	
4600	8′ x 4′ opening, standard glazed		6	2.667		255	56.50		311.50	390	
4700	Insulating glass		6	2.667		410	56.50		466.50	560	
5000	9′ x 5′ opening, standard glazed		4	4		385	85		470	595	
5100	Insulating glass		4	4		615	85		700	850	
5500	Sliding, with thermal barrier and screen, 6′ x 4′, 2 track		8	2		525	42.50		567.50	660	
5700	4 track		8	2		640	42.50		682.50	790	
6000	For above units with bronze finish, add					12%					
6200	For installation in concrete openings, add					5%					
0010	**JALOUSIES**										500
0020	Aluminum incl. glazing & screens, stock, 1′-7″ x 3′-2″	2 Sswk	10	1.600	Ea.	113	34		147	191	
0100	2′-3″ x 4′-0″		10	1.600		162	34		196	245	
0200	3′-1″ x 2′-0″		10	1.600		111	34		145	189	
0300	3′-1″ x 5′-3″		10	1.600		231	34		265	320	
1000	Mullions for above, 2′-0″ long		80	.200		8.70	4.25		12.95	17.95	
1100	5′-3″ long		80	.200		14.90	4.25		19.15	25	
0010	**LOUVERS** See division 06440 & 10210										600
	08550 | Wood Windows										
0010	**AWNING WINDOW** Including frame, screen, and exterior trim										100
0100	Average quality, builders model, 34″ x 22″, double insulated glass	1 Carp	10	.800	Ea.	220	15.75		235.75	269	
0200	Low E glass		10	.800		220	15.75		235.75	269	
0300	40″ x 28″, double insulated glass		9	.889		370	17.50		387.50	440	
0400	Low E Glass		9	.889		370	17.50		387.50	440	
0500	48″ x 36″, double insulated glass		8	1		685	19.70		704.70	790	
0600	Low E glass		8	1		685	19.70		704.70	790	
0800	Vinyl clad, premium, double insulated glass, 24″ x 17″		12	.667		167	13.15		180.15	206	

08550 | Wood Windows

			CREW	DAILY OUTPUT	LABOR-HOURS	UNIT	2000 BARE COSTS MAT.	LABOR	EQUIP.	TOTAL	TOTAL INCL O&P
100	0840	24″ x 28″	1 Carp	11	.727	Ea.	194	14.35		208.35	238
	0860	36″ x 17″		11	.727		191	14.35		205.35	235
	0900	36″ x 40″		10	.800		300	15.75		315.75	360
	1100	40″ x 22″		10	.800		213	15.75		228.75	261
	1200	36″ x 28″		9	.889		227	17.50		244.50	280
	1300	36″ x 36″		9	.889		286	17.50		303.50	345
	1400	48″ x 28″		8	1		270	19.70		289.70	330
	1500	60″ x 36″		8	1		370	19.70		389.70	445
	2000	Metal clad, deluxe, double insulated glass, 34″ x 22″		10	.800		184	15.75		199.75	230
	2100	40″ x 22″		10	.800		201	15.75		216.75	249
	2200	36″ x 25″		9	.889		205	17.50		222.50	256
	2300	40″ x 30″		9	.889		267	17.50		284.50	325
	2400	48″ x 28″		8	1		241	19.70		260.70	299
	2500	60″ x 36″		8	1		237	19.70		256.70	295
150	0010	**BOW-BAY WINDOW** Including frame, screens and grills,									
	0020	end panels operable									
	1000	Bow type, casement, wood, bldrs mdl, 8′ x 5′ dbl insltd glass, 4 panel	2 Carp	10	1.600	Ea.	850	31.50		881.50	990
	1050	Low E glass		10	1.600		1,100	31.50		1,131.50	1,250
	1100	10′-0″ x 5′-0″ , double insulated glass, 6 panels		6	2.667		1,125	52.50		1,177.50	1,350
	1200	Low E glass, 6 panels		6	2.667		1,200	52.50		1,252.50	1,425
	1300	Vinyl clad, bldrs model, double insulated glass, 6′-0″ x 4′-0″, 3 panel		10	1.600		925	31.50		956.50	1,075
	1340	9′-0″ x 4′-0″, 4 panel		8	2		1,225	39.50		1,264.50	1,425
	1380	10′-0″ x 6′-0″, 5 panels		7	2.286		2,075	45		2,120	2,350
	1420	12′-0″ x 6′-0″, 6 panels		6	2.667		2,650	52.50		2,702.50	3,000
	1600	Metal clad, casement, bldrs mdl, 6′-0″ x 4′-0″, dbl insltd gls, 3 panels		10	1.600		770	31.50		801.50	900
	1640	9′-0″ x 4′-0″ , 4 panels		8	2		1,075	39.50		1,114.50	1,275
	1680	10′-0″ x 5′-0″, 5 panels		7	2.286		1,475	45		1,520	1,700
	1720	12′-0″ x 6′-0″, 6 panels		6	2.667		2,075	52.50		2,127.50	2,375
	2000	Bay window, casement, builders model, 8′ x 5′ dbl insul glass, 4 panels		10	1.600		1,300	31.50		1,331.50	1,475
	2050	Low E glass,		10	1.600		1,575	31.50		1,606.50	1,775
	2100	12′-0″ x 6′-0″ , double insulated glass, 6 panels		6	2.667		1,625	52.50		1,677.50	1,875
	2200	Low E glass		6	2.667		1,675	52.50		1,727.50	1,950
	2280	6′-0″ x 4′-0″		11	1.455		955	28.50		983.50	1,100
	2300	Vinyl clad, premium, insulating glass, 8′-0″ x 5′-0″		10	1.600		1,125	31.50		1,156.50	1,275
	2340	10′-0″ x 5′-0″		8	2		1,600	39.50		1,639.50	1,825
	2380	10′-0″ x 6′-0″		7	2.286		1,650	45		1,695	1,900
	2420	12′-0″ x 6′-0″		6	2.667		1,975	52.50		2,027.50	2,275
	2430	14′-0″ x 3′-0″		7	2.286		1,350	45		1,395	1,575
	2440	14′-0″ x 6′-0″		5	3.200		2,175	63		2,238	2,475
	2600	Metal clad, deluxe, insul. glass, 8′-0″ x 5′-0″ high, 4 panels		10	1.600		1,150	31.50		1,181.50	1,300
	2640	10′-0″ x 5′-0″ high, 5 panels		8	2		1,225	39.50		1,264.50	1,425
	2680	10′-0″ x 6′-0″ high, 5 panels		7	2.286		1,450	45		1,495	1,675
	2720	12′-0″ x 6′-0″ high, 6 panels		6	2.667		2,025	52.50		2,077.50	2,325
	3000	Double hung, bldrs. model, bay, 8′ x 4′ high, std. glazed		10	1.600		910	31.50		941.50	1,050
	3050	Insulating glass		10	1.600		975	31.50		1,006.50	1,125
	3100	9′-0″ x 5′-0″ high, standard glazed		6	2.667		980	52.50		1,032.50	1,175
	3200	Insulating glass		6	2.667		1,025	52.50		1,077.50	1,225
	3300	Vinyl clad, premium, insulating glass, 7′-0″ x 4′-6″		10	1.600		945	31.50		976.50	1,075
	3340	8′-0″ x 4′-6″		8	2		960	39.50		999.50	1,125
	3380	8′-0″ x 5′-0″		7	2.286		1,000	45		1,045	1,175
	3420	9′-0″ x 5′-0″		6	2.667		1,025	52.50		1,077.50	1,225
	3600	Metal clad, deluxe, insul. glass, 7′-0″ x 4′-0″ high		10	1.600		870	31.50		901.50	1,025
	3640	8′-0″ x 4′-0″ high		8	2		900	39.50		939.50	1,075
	3680	8′-0″ x 5′-0″ high		7	2.286		935	45		980	1,100
	3720	9′-0″ x 5′-0″ high		6	2.667		990	52.50		1,042.50	1,200
	7000	Drip cap, premolded vinyl, 8′ long		30	.533		67.50	10.50		78	92

Important: See the Reference Section for critical supporting data - Reference Nos., Crews, & Location Facto

08550 | Wood Windows		CREW	DAILY OUTPUT	LABOR-HOURS	UNIT	2000 BARE COSTS MAT.	LABOR	EQUIP.	TOTAL	TOTAL INCL O&P	
7040	12′ long	2 Carp	26	.615	Ea.	73.50	12.10		85.60	102	150
0010	**CASEMENT WINDOW** Including frame, screen, and exterior trim R08550-010										200
0100	Avg. quality, bldrs. model, 2′-0″ x 3′-0″ H, standard glazed	1 Carp	10	.800	Ea.	159	15.75		174.75	202	
0150	Insulating glass		10	.800		196	15.75		211.75	242	
0200	2′-0″ x 4′-6″ high, standard glazed		9	.889		207	17.50		224.50	257	
0250	Insulating glass		9	.889		266	17.50		283.50	325	
0300	2′-3″ x 6′-0″ high, standard glazed		8	1		226	19.70		245.70	282	
0350	Insulating glass		8	1		299	19.70		318.70	365	
0522	Vinyl clad, premium, insulating glass, 2′-0″ x 3′-0″		10	.800		194	15.75		209.75	241	
0524	2′-0″ x 4′-0″		9	.889		231	17.50		248.50	284	
0525	2′-0″ x 5′-0″		8	1		266	19.70		285.70	325	
0528	2′-0″ x 6′-0″		8	1		315	19.70		334.70	385	
3020	Vinyl clad, premium, insulating glass, multiple leaf units										
3080	2 units, 1′-6″ x 5′-0″	2 Carp	20	.800	Ea.	320	15.75		335.75	375	
3100	2′-0″ x 2′-0″		20	.800		335	15.75		350.75	395	
3140	2′-0″ x 2′-6″		20	.800		194	15.75		209.75	241	
3220	2′-0″ x 3′-6″		20	.800		360	15.75		375.75	420	
3260	2′-0″ x 4′-0″		19	.842		231	16.60		247.60	283	
3300	2′-0″ x 4′-6″		19	.842		460	16.60		476.60	535	
3340	2′-0″ x 5′-0″		18	.889		266	17.50		283.50	320	
3460	2′-4″ x 3′-0″		20	.800		325	15.75		340.75	385	
3500	2′-4″ x 4′-0″		19	.842		415	16.60		431.60	485	
3540	2′-4″ x 5′-0″		18	.889		445	17.50		462.50	515	
3700	2′-8″ x 5′-0″		18	.889		345	17.50		362.50	410	
3740	2′-8″ x 6′-0″		17	.941		365	18.55		383.55	435	
3840	3′-0″ x 4′-6″		18	.889		415	17.50		432.50	490	
3860	3′-0″ x 5′-0″		17	.941		515	18.55		533.55	595	
3880	3′-0″ x 6′-0″		17	.941		460	18.55		478.55	535	
3980	3′-4″ x 2′-6″		19	.842		500	16.60		516.60	580	
4000	3′-4″ x 3′-0″		12	1.333		500	26.50		526.50	595	
4030	3′-4″ x 4′-0″		18	.889		415	17.50		432.50	490	
4050	3′-4″ x 5′-0″		12	1.333		515	26.50		541.50	610	
4100	3′-4″ x 6′-0″		11	1.455		460	28.50		488.50	555	
4200	3′-6″ x 3′-0″		18	.889		555	17.50		572.50	640	
4340	4′-0″ x 3′-0″		18	.889		375	17.50		392.50	440	
4380	4′-0″ x 3′-6″		17	.941		405	18.55		423.55	475	
4420	4′-0″ x 4′-0″		16	1		460	19.70		479.70	545	
4460	4′-0″ x 4′-4″		16	1		460	19.70		479.70	540	
4540	4′-0″ x 5′-0″		16	1		510	19.70		529.70	595	
4580	4′-0″ x 6′-0″		15	1.067		590	21		611	685	
4740	4′-8″ x 3′-0″		18	.889		405	17.50		422.50	475	
4780	4′-8″ x 3′-6″		17	.941		435	18.55		453.55	510	
4820	4′-8″ x 4′-0″		16	1		505	19.70		524.70	590	
4860	4′-8″ x 5′-0″		15	1.067		600	21		621	695	
4900	4′-8″ x 6′-0″		15	1.067		690	21		711	795	
5060	5′-0″ x 5′-0″		15	1.067		540	21		561	630	
5100	5′-6″ x 3′-0″		17	.941		585	18.55		603.55	675	
5140	5′-6″ x 3′-6″		16	1		630	19.70		649.70	725	
5180	5′-6″ x 4′-6″		15	1.067		670	21		691	770	
5220	5′-6″ x 5′-6″		15	1.067		845	21		866	965	
5300	6′-0″ x 4′-6″		15	1.067		900	21		921	1,025	
5850	3 units, 5′-0″ x 3′-0″		12	1.333		620	26.50		646.50	725	
5900	5′-0″ x 4′-0″		11	1.455		655	28.50		683.50	770	
6000	5′-0″ x 5′-0″		10	1.600		785	31.50		816.50	915	
6100	5′-0″ x 5′-6″		10	1.600		870	31.50		901.50	1,025	

08550 | Wood Windows

		08550 Wood Windows	CREW	DAILY OUTPUT	LABOR-HOURS	UNIT	2000 BARE COSTS MAT.	LABOR	EQUIP.	TOTAL	TOTAL INCL O&P
200	6150	5'-0" x 6'-0" R08550-010	2 Carp	10	1.600	Ea.	875	31.50		906.50	1,025
	6200	6'-0" x 3'-0"		12	1.333		515	26.50		541.50	610
	6250	6'-0" x 3'-4"		12	1.333		565	26.50		591.50	670
	6300	6'-0" x 4'-0"		11	1.455		615	28.50		643.50	725
	6350	6'-0" x 5'-0"		10	1.600		700	31.50		731.50	825
	6400	6'-0" x 6'-0"		10	1.600		885	31.50		916.50	1,025
	6500	7'-0" x 4'-0"		9	1.778		775	35		810	910
	6700	8'-0" x 4'-6"		9	1.778		1,000	35		1,035	1,150
	6950	4 units, 6'-8" x 4'-0"		10	1.600		835	31.50		866.50	970
	7000	6'-8" x 6'-0"		10	1.600		920	31.50		951.50	1,050
	8000	Solid vinyl, premium, insulating glass, 2'-0" x 3'-0" high	1 Carp	10	.800		162	15.75		177.75	206
	8020	2'-0" x 4'-0" high		9	.889		199	17.50		216.50	249
	8040	2'-0" x 5'-0" high		8	1		229	19.70		248.70	286
	8100	Metal clad, deluxe, insulating glass, 2'-0" x 3'-0" high		10	.800		160	15.75		175.75	203
	8120	2'-0" x 4'-0" high		9	.889		193	17.50		210.50	242
	8140	2'-0" x 5'-0" high		8	1		219	19.70		238.70	275
	8160	2'-0" x 6'-0" high		8	1		252	19.70		271.70	310
	8200	For multiple leaf units, deduct for stationary sash									
	8220	2' high				Ea.	16.65			16.65	18.30
	8240	4'-6" high					19.20			19.20	21
	8260	6' high					25.50			25.50	28
	8300	For installation, add per leaf						15%			
250	0010	**DOUBLE HUNG** Including frame, screen, and exterior trim									
	0100	Avg. quality, bldrs. model, 2'-0" x 3'-0" high, standard glazed	1 Carp	10	.800	Ea.	107	15.75		122.75	144
	0150	Insulating glass		10	.800		151	15.75		166.75	194
	0200	3'-0" x 4'-0" high, standard glazed		9	.889		140	17.50		157.50	184
	0250	Insulating glass		9	.889		185	17.50		202.50	233
	0300	4'-0" x 4'-6" high, standard glazed		8	1		172	19.70		191.70	223
	0350	Insulating glass		8	1		227	19.70		246.70	284
	1000	Vinyl clad, premium, insulating glass, 2'-6" x 3'-0"		10	.800		240	15.75		255.75	291
	1100	3'-0" x 3'-6"		10	.800		291	15.75		306.75	345
	1200	3'-0" x 4'-0"		9	.889		315	17.50		332.50	375
	1300	3'-0" x 4'-6"		9	.889		330	17.50		347.50	390
	1400	3'-0" x 5'-0"		8	1		340	19.70		359.70	410
	1500	3'-6" x 6'-0"		8	1		570	19.70		589.70	660
	1540	Solid vinyl, average quality, insulated glass, 2'-0" x 3'-0"		10	.800		121	15.75		136.75	160
	1542	3'-0" x 4'-0"		9	.889		146	17.50		163.50	191
	1544	4'-0" x 4'-6"		8	1		175	19.70		194.70	227
	1560	Premium, insulating glass, 2'-6" x 3'-0"		10	.800		137	15.75		152.75	178
	1562	3'-0" x 3'-6"		9	.889		159	17.50		176.50	205
	1564	3'-0" x 4"-0"		9	.889		169	17.50		186.50	216
	1566	3'-0" x 4'-6"		9	.889		173	17.50		190.50	221
	1568	3'-0" x 5'-0"		8	1		178	19.70		197.70	230
	1570	3'-6" x 6"-0"		8	1		195	19.70		214.70	249
	2000	Metal clad, deluxe, insulating glass, 2'-6" x 3'-0" high		10	.800		170	15.75		185.75	214
	2100	3'-0" x 3'-6" high		10	.800		201	15.75		216.75	248
	2200	3'-0" x 4'-0" high		9	.889		214	17.50		231.50	265
	2300	3'-0" x 4'-6" high		9	.889		231	17.50		248.50	284
	2400	3'-0" x 5'-0" high		8	1		247	19.70		266.70	305
	2500	3'-6" x 6'-0" high		8	1		300	19.70		319.70	365
260	0010	**HALF ROUND WINDOW**, Vinyl clad, including grill									
	0800	8" radius x 1'-4" base	2 Carp	9	1.778	Ea.	355	35		390	450
	1040	1'-4" radius x 2'-8" base		8	2		400	39.50		439.50	510
	1060	1'-6" radius x 3'-0" base		7	2.286		425	45		470	545
	1080	1'-8" radius x 3'-4" base		7	2.286		470	45		515	590
	2000	2'-0" radius x 4'-0" base	1 Carp	6	1.333		600	26.50		626.50	705

Important: See the Reference Section for critical supporting data - Reference Nos., Crews, & Location Factor

	Line	08550 | Wood Windows	Crew	Daily Output	Labor-Hours	Unit	2000 Bare Costs Mat.	Labor	Equip.	Total	Total Incl O&P
260	2100	2'-4" radius x 4'-8" base	1 Carp	6	1.333	Ea.	700	26.50		726.50	815
	2200	2'-6" radius x 5'-0" base		6	1.333		800	26.50		826.50	925
	2250	2'-8" radius x 5'-4" base	2 Carp	6	2.667		1,100	52.50		1,152.50	1,325
	2300	2'-10" radius x 5'-8" base		6	2.667		1,125	52.50		1,177.50	1,325
	2350	2'-11" radius x 5'-10" base		6	2.667		1,150	52.50		1,202.50	1,350
	3000	3'-0" radius x 6'-0"base	1 Carp	4	2		1,075	39.50		1,114.50	1,250
	3040	3'-4" radius x 6'-8" base	2 Carp	5	3.200		1,400	63		1,463	1,650
	3050	3'-5" radius x 6'-10" base		5	3.200		1,500	63		1,563	1,750
	3060	3'-6" radius x 7'-0" base		5	3.200		1,575	63		1,638	1,825
	4000	4'-0" radius x 8'-0" base		6	2.667		1,950	52.50		2,002.50	2,250
	5000	Elliptical, 3'-2" x 1'-4"	1 Carp	11	.727		740	14.35		754.35	840
	5100	5'-0" x 1'-0"		10	.800		740	15.75		755.75	840
	5200	7'-0" x 1'-0"		10	.800		1,050	15.75		1,065.75	1,175
650	0010	**PALLADIAN WINDOWS**									
	0020	Aluminum clad, including frame and exterior trim									
	0040	3'-2" x 2'-0" high	2 Carp	11	1.455	Ea.	905	28.50		933.50	1,050
	0060	3'-2" x 4'-10"		11	1.455		1,025	28.50		1,053.50	1,175
	0080	3'-2" x 6'-4"		10	1.600		1,225	31.50		1,256.50	1,400
	0100	4'-0" x 4'-0"		10	1.600		1,000	31.50		1,031.50	1,150
	0120	4'-0" x 5'-4"	3 Carp	10	2.400		1,150	47.50		1,197.50	1,350
	0140	4'-0" x 6'-0"		9	2.667		1,200	52.50		1,252.50	1,425
	0160	4'-0" x 7'-4"		9	2.667		1,325	52.50		1,377.50	1,550
	0180	5'-5" x 4'-10"		9	2.667		1,300	52.50		1,352.50	1,525
	0200	5'-5" x 6'-10"		9	2.667		1,550	52.50		1,602.50	1,800
	0220	5'-5" x 7'-9"		9	2.667		1,675	52.50		1,727.50	1,950
	0240	6'-0" x 7'-11"		8	3		2,225	59		2,284	2,550
	0260	8'-0" x 6'-0"		8	3		1,900	59		1,959	2,200
670	0010	**PICTURE WINDOW** Including frame and exterior trim									
	0100	Average quality, bldrs. model, 3'-6" x 4'-0" high, standard glazed	2 Carp	12	1.333	Ea.	238	26.50		264.50	305
	0150	Insulating glass		12	1.333		200	26.50		226.50	265
	0200	4'-0" x 4'-6" high, standard glazed		11	1.455		198	28.50		226.50	267
	0250	Insulating glass		11	1.455		202	28.50		230.50	271
	0300	5'-0" x 4'-0" high, standard glazed		11	1.455		200	28.50		228.50	269
	0350	Insulating glass		11	1.455		250	28.50		278.50	325
	0400	6'-0" x 4'-6" high, standard glazed		10	1.600		280	31.50		311.50	365
	0450	Insulating glass		10	1.600		345	31.50		376.50	435
	1000	Vinyl clad, premium, insulating glass, 4'-0" x 4'-0"		12	1.333		420	26.50		446.50	510
	1100	4'-0" x 6'-0"		11	1.455		575	28.50		603.50	685
	1200	5'-0" x 6'-0"		10	1.600		750	31.50		781.50	880
	1300	6'-0" x 6'-0"		10	1.600		765	31.50		796.50	895
	2000	Metal clad, deluxe, insulating glass, 4'-0" x 4'-0" high		12	1.333		252	26.50		278.50	320
	2100	4'-0" x 6'-0" high		11	1.455		370	28.50		398.50	460
	2200	5'-0" x 6'-0" high		10	1.600		410	31.50		441.50	505
	2300	6'-0" x 6'-0" high		10	1.600		470	31.50		501.50	570
750	0010	**SLIDING WINDOW** Including frame, screen, and exterior trim									
	0100	Average quality, bldrs. model, 3'-0" x 3'-0" high, standard glazed	1 Carp	10	.800	Ea.	128	15.75		143.75	168
	0120	Insulating glass		10	.800		161	15.75		176.75	205
	0200	4'-0" x 3'-6" high, standard glazed		9	.889		152	17.50		169.50	197
	0220	Insulating glass		9	.889		190	17.50		207.50	239
	0300	6'-0" x 5'-0" high, standard glazed		8	1		279	19.70		298.70	340
	0320	Insulating glass		8	1		335	19.70		354.70	400
	1000	Vinyl clad, premium, insulating glass, 3'-0" x 3'-0"		10	.800		465	15.75		480.75	540
	1020	4'-0" x 1'-11"		11	.727		430	14.35		444.35	495
	1040	4'-0" x 3'-0"		10	.800		515	15.75		530.75	590

	08550 | Wood Windows		CREW	DAILY OUTPUT	LABOR-HOURS	UNIT	2000 BARE COSTS MAT.	LABOR	EQUIP.	TOTAL	TOTAL INCL O&P
750	1050	4'-0" x 3'-6"	1 Carp	9	.889	Ea.	560	17.50		577.50	645
	1090	4'-0" x 5'-0"		9	.889		665	17.50		682.50	760
	1100	5'-0" x 4'-0"		9	.889		645	17.50		662.50	740
	1120	5'-0" x 5'-0"		8	1		720	19.70		739.70	830
	1140	6'-0" x 4'-0"		8	1		705	19.70		724.70	810
	1150	6'-0" x 5'-0"		8	1		805	19.70		824.70	920
	2000	Metal clad, deluxe, insulating glass, 3'-0" x 3'-0" high		10	.800		260	15.75		275.75	315
	2050	4'-0" x 3'-6" high		9	.889		320	17.50		337.50	380
	2100	5'-0" x 4'-0" high		9	.889		385	17.50		402.50	450
	2150	6'-0" x 5'-0" high		8	1		470	19.70		489.70	550
760	0010	**TRANSOM WINDOWS**									
	0050	Vinyl clad, premium, insulating glass, 32" x 8"	1 Carp	16	.500	Ea.	182	9.85		191.85	217
	0100	36" x 8"		16	.500		182	9.85		191.85	217
	0110	36" x 12"		16	.500		221	9.85		230.85	260
	0150	36" x 48"		12	.667		290	13.15		303.15	345
770	0010	**WEATHERSTRIPPING** See division 08720									
780	0010	**TRAPEZOID WINDOWS**									
	0100	Vinyl clad, including frame and exterior trim									
	0900	20" base x 44" leg x 53" leg	2 Carp	13	1.231	Ea.	345	24.50		369.50	420
	1000	24" base x 90" leg x 102" leg		8	2		600	39.50		639.50	730
	3000	36" base x 0" leg x 22" leg		12	1.333		375	26.50		401.50	455
	3010	36" base x 4" leg x 25" leg		13	1.231		400	24.50		424.50	480
	3050	36" base x 26" leg x 48" leg		9	1.778		415	35		450	515
	3100	36" base x 42" legs, 50" peak		9	1.778		475	35		510	585
	3200	36" base x 60" leg x 81" leg		11	1.455		625	28.50		653.50	740
	4320	44" base x 23" leg x 56" leg		11	1.455		480	28.50		508.50	580
	4350	44" base x 59" leg x 92" leg		10	1.600		745	31.50		776.50	875
	4500	46" base x 15" leg x 46" leg		8	2		375	39.50		414.50	480
	4550	46" base x 16" leg x 48" leg		8	2		385	39.50		424.50	490
	4600	46" base x 50" leg x 80" leg		7	2.286		610	45		655	745
	6600	66" base x 12" leg x 42" leg		8	2		490	39.50		529.50	605
	6650	66" base x 12" legs, 28" peak		9	1.778		415	35		450	515
	6700	68" base x 3" legs, 31" peak		8	2		500	39.50		539.50	620
800	0010	**WINDOW GRILLE OR MUNTIN** Snap-in type									
	0020	Colonial or diamond pattern									
	2000	Wood, awning window, glass size 28" x 16" high	1 Carp	30	.267	Ea.	22	5.25		27.25	33
	2060	44" x 24" high		32	.250		28	4.93		32.93	39.50
	2100	Casement, glass size, 20" x 36" high		30	.267		27.50	5.25		32.75	39
	2180	20" x 56" high		32	.250		36.50	4.93		41.43	49
	2200	Double hung, glass size, 16" x 24" high		24	.333	Set	57.50	6.55		64.05	75
	2280	32" x 32" high		34	.235	"	104	4.64		108.64	122
	2500	Picture, glass size, 48" x 48" high		30	.267	Ea.	110	5.25		115.25	130
	2580	60" x 68" high		28	.286	"	122	5.65		127.65	144
	2600	Sliding, glass size, 14" x 36" high		24	.333	Set	31.50	6.55		38.05	46
	2680	36" x 36" high		22	.364	"	49.50	7.15		56.65	67
820	0010	**WOOD SASH** Including glazing but not including trim									
	0050	Custom, 5'-0" x 4'-0", 1" dbl. glazed, 3/16" thick lites	2 Carp	3.20	5	Ea.	187	98.50		285.50	375
	0100	1/4" thick lites		5	3.200		192	63		255	320
	0200	1" thick, triple glazed		5	3.200		440	63		503	595
	0300	7'-0" x 4'-6" high, 1" double glazed, 3/16" thick lites		4.30	3.721		445	73.50		518.50	615
	0400	1/4" thick lites		4.30	3.721		505	73.50		578.50	680
	0500	1" thick, triple glazed		4.30	3.721		575	73.50		648.50	760
	0600	8'-6" x 5'-0" high, 1" double glazed, 3/16" thick lites		3.50	4.571		605	90		695	820

08550 | Wood Windows		CREW	DAILY OUTPUT	LABOR-HOURS	UNIT	2000 BARE COSTS MAT.	LABOR	EQUIP.	TOTAL	TOTAL INCL O&P	
0700	1/4" thick lites	2 Carp	3.50	4.571	Ea.	660	90		750	880	820
0800	1" thick, triple glazed	↓	3.50	4.571	↓	665	90		755	885	
0900	Window frames only, based on perimeter length				L.F.	3.69			3.69	4.06	
3000	Replacement sash, double hung, double glazing, to 12 S.F.	1 Carp	64	.125	S.F.	18	2.46		20.46	24	
3100	12 S.F. to 20 S.F.		94	.085		15.05	1.68		16.73	19.40	
3200	20 S.F. and over	↓	106	.075		13.10	1.49		14.59	17	
3800	Triple glazing for above, add				↓	2.71			2.71	2.98	
7000	Sash, single lite, 2'-0" x 2'-0" high	1 Carp	20	.400	Ea.	35.50	7.90		43.40	52.50	
7050	2'-6" x 2'-0" high		19	.421		38	8.30		46.30	56	
7100	2'-6" x 2'-6" high		18	.444		40.50	8.75		49.25	59.50	
7150	3'-0" x 2'-0" high	↓	17	.471	↓	51	9.25		60.25	72	
0010	**WOOD SCREENS**										840
0020	Over 3 S.F., 3/4" frames	2 Carp	375	.043	S.F.	3.03	.84		3.87	4.77	
0100	1-1/8" frames	"	375	.043	"	2.98	.84		3.82	4.72	

08580 | Special Function Windows		CREW	DAILY OUTPUT	LABOR-HOURS	UNIT	MAT.	LABOR	EQUIP.	TOTAL	TOTAL INCL O&P	
0010	**STORM WINDOWS** Aluminum, residential										900
0300	Basement, mill finish, incl. fiberglass screen										
0320	1'-10" x 1'-0" high	2 Carp	30	.533	Ea.	25	10.50		35.50	45	
0340	2'-9" x 1'-6" high		30	.533		27	10.50		37.50	47.50	
0360	3'-4" x 2'-0" high	↓	30	.533	↓	32.50	10.50		43	54	
1600	Double-hung, combination, storm & screen										
1700	Custom, clear anodic coating, 2'-0" x 3'-5" high	2 Carp	30	.533	Ea.	64	10.50		74.50	88.50	
1720	2'-6" x 5'-0" high		28	.571		86	11.25		97.25	114	
1740	4'-0" x 6'-0" high		25	.640		182	12.60		194.60	222	
1800	White painted, 2'-0" x 3'-5" high		30	.533		76.50	10.50		87	102	
1820	2'-6" x 5'-0" high		28	.571		123	11.25		134.25	154	
1840	4'-0" x 6'-0" high		25	.640		220	12.60		232.60	264	
2000	Average quality, clear anodic coating, 2'-0" x 3'-5" high		30	.533		65	10.50		75.50	89.50	
2020	2'-6" x 5'-0" high		28	.571		82	11.25		93.25	110	
2040	4'-0" x 6'-0" high		25	.640		96.50	12.60		109.10	128	
2400	White painted, 2'-0" x 3'-5" high		30	.533		64	10.50		74.50	88.50	
2420	2'-6" x 5'-0" high		28	.571		71	11.25		82.25	97.50	
2440	4'-0" x 6'-0" high		25	.640		77.50	12.60		90.10	107	
2600	Mill finish, 2'-0" x 3'-5" high		30	.533		58.50	10.50		69	82.50	
2620	2'-6" x 5'-0" high		28	.571		65	11.25		76.25	91.50	
2640	4'-0" x 6-8" high	↓	25	.640	↓	73	12.60		85.60	102	
4000	Picture window, storm, 1 lite, white or bronze finish										
4020	4'-6" x 4'-6" high	2 Carp	25	.640	Ea.	98	12.60		110.60	130	
4040	5'-8" x 4'-6" high		20	.800		111	15.75		126.75	149	
4400	Mill finish, 4'-6" x 4'-6" high		25	.640		98	12.60		110.60	130	
4420	5'-8" x 4'-6" high	↓	20	.800	↓	111	15.75		126.75	149	
4600	3 lite, white or bronze finish										
4620	4'-6" x 4'-6" high	2 Carp	25	.640	Ea.	119	12.60		131.60	153	
4640	5'-8" x 4'-6" high		20	.800		133	15.75		148.75	173	
4800	Mill finish, 4'-6" x 4'-6" high		25	.640		105	12.60		117.60	137	
4820	5'-8" x 4'-6" high	↓	20	.800		111	15.75		126.75	149	
5000	Sliding glass door, storm 6' x 6'-8", standard	1 Glaz	2	4		600	77.50		677.50	790	
5100	Economy	"	2	4	↓	265	77.50		342.50	420	
6000	Sliding window, storm, 2 lite, white or bronze finish										
6020	3'-4" x 2'-7" high	2 Carp	28	.571	Ea.	82.50	11.25		93.75	110	
6040	4'-4" x 3'-3" high		25	.640		112	12.60		124.60	145	
6060	5'-4" x 6'-0" high	↓	20	.800	↓	180	15.75		195.75	225	
6400	3 lite, white or bronze finish										

	08580 | Special Function Windows		Crew	Daily Output	Labor-Hours	Unit	2000 Bare Costs Mat.	Labor	Equip.	Total	Total Incl O&P
900	6420	4′-4″ x 3′-3″ high	2 Carp	25	.640	Ea.	130	12.60		142.60	165
	6440	5′-4″ x 6′-0″ high		20	.800		235	15.75		250.75	285
	6460	6′-0″ x 6′-0″ high		18	.889		236	17.50		253.50	289
	6800	Mill finish, 4′-4″ x 3′-3″ high		25	.640		112	12.60		124.60	145
	6820	5′-4″ x 6′-0″ high		20	.800		235	15.75		250.75	286
	6840	6′-0″ x 6-0″ high		18	.889		243	17.50		260.50	298
	9000	Magnetic interior storm window									
	9100	3/16″ plate glass	1 Glaz	107	.075	S.F.	3.79	1.45		5.24	6.55
	08590 | Window Restoration & Replace										
600	0010	SOLID VINYL REPLACEMENT WINDOWS R08550 -200									
	0020	White, double hung, up to 83 united inches	2 Carp	8	2	Ea.	145	39.50		184.50	227
	0040	84 to 93		8	2		145	39.50		184.50	227
	0060	94 to 101		6	2.667		145	52.50		197.50	249
	0080	102 to 111		6	2.667		167	52.50		219.50	273
	0100	112 to 120		6	2.667		186	52.50		238.50	295
	0120	For each united inch over 120 , add		800	.020	Inch	2.17	.39		2.56	3.07
	0140	Casement windows, one operating sash , 42 to 60 united inches		8	2	Ea.	165	39.50		204.50	249
	0160	61 to 70		8	2		188	39.50		227.50	275
	0180	71 to 80		8	2		204	39.50		243.50	292
	0200	81 to 96		8	2		217	39.50		256.50	305
	0220	Two operating sash, 58 to 78 united inches		8	2		299	39.50		338.50	400
	0240	79 to 88		8	2		335	39.50		374.50	440
	0260	89 to 98		8	2		365	39.50		404.50	470
	0280	99 to 108		6	2.667		385	52.50		437.50	515
	0300	109 to 121		6	2.667		420	52.50		472.50	555
	0320	Three operating sash, 73 to 108 united inches		8	2		450	39.50		489.50	565
	0340	109 to 118		8	2		480	39.50		519.50	600
	0360	119 to 128		6	2.667		525	52.50		577.50	665
	0380	129 to 138		6	2.667		570	52.50		622.50	715
	0400	139 to 156		6	2.667		605	52.50		657.50	755
	0420	Four operating sash, 89 to 98 united inches		8	2		600	39.50		639.50	730
	0440	99 to 108		8	2		640	39.50		679.50	775
	0460	109 to 118		6	2.667		695	52.50		747.50	855
	0480	119 to 128		6	2.667		755	52.50		807.50	920
	0500	129 to 138		6	2.667		810	52.50		862.50	980
	0520	139 to 148		6	2.667		870	52.50		922.50	1,050
	0540	149 to 168		4	4		925	79		1,004	1,150
	0560	Fixed picture window, up to 63 united inches		8	2		125	39.50		164.50	205
	0580	64 to 83		8	2		167	39.50		206.50	252
	0600	84 to 101		8	2		213	39.50		252.50	300
	0620	For each united inch over 101, add		900	.018	Inch	2.32	.35		2.67	3.15
	0640	Picture window opt., low E glazing, up to 101 united inches				Ea.	17.70			17.70	19.45
	0660	102 to 124					22.50			22.50	25
	0680	124 and over					34.50			34.50	38
	0700	Options, low E glazing, up to 101 united inches					8.85			8.85	9.70
	0720	102 to 124					22.50			22.50	25
	0740	124 and over					34.50			34.50	38
	0760	Muntins, between glazing, square, per lite					1.67			1.67	1.84
	0780	Diamond shape, per full or partial diamond					2.78			2.78	3.06
	0800	Celluose fiber insulation, poured into sash balance cavity	1 Carp	36	.222	C.F.	.45	4.38		4.83	8
	0820	Silicone caulking at perimeter	″	800	.010	L.F.	.13	.20		.33	.48

Important: See the Reference Section for critical supporting data - Reference Nos., Crews, & Location Facto

08600 | Skylights

08610 | Roof Windows

	08610 Roof Windows	CREW	DAILY OUTPUT	LABOR-HOURS	UNIT	2000 BARE COSTS MAT.	LABOR	EQUIP.	TOTAL	TOTAL INCL O&P	
0010	**METAL ROOF WINDOW** Fixed, high perf tmpd glazing, 46" x 21-1/2"	1 Carp	8	1	Ea.	210	19.70		229.70	264	600
0100	46" x 28"		8	1		310	19.70		329.70	380	
0125	57" x 44"		6	1.333		400	26.50		426.50	485	
0130	72" x 28"		7	1.143		400	22.50		422.50	480	
0150	Venting, high performance tempered glazing, 46" x 21-1/2"		8	1		320	19.70		339.70	390	
0175	46" x 28"		8	1		430	19.70		449.70	505	
0200	57" x 44"		6	1.333		535	26.50		561.50	635	
0500	Flashing set for shingled roof, 46" x 21-1/2"		5	1.600		39.50	31.50		71	97.50	
0525	46" x 28"		5	1.600		42.50	31.50		74	101	
0550	57" x 44"		5	1.600		49	31.50		80.50	108	
0560	72" x 28"		6	1.333		49	26.50		75.50	99	
0575	Flashing set for low pitched roof, 46" x 21-1/2"		5	1.600		161	31.50		192.50	231	
0600	46" x 28"		5	1.600		164	31.50		195.50	235	
0625	57" x 44"		5	1.600		186	31.50		217.50	259	
0650	Flashing set for tile roof 46" x 21-1/2"		5	1.600		94.50	31.50		126	158	
0675	46" x 28"		5	1.600		96	31.50		127.50	159	
0700	57" x 44"	↓	5	1.600	↓	111	31.50		142.50	176	

08620 | Unit Skylights

	08620 Unit Skylights	CREW	DAILY OUTPUT	LABOR-HOURS	UNIT	MAT.	LABOR	EQUIP.	TOTAL	TOTAL INCL O&P	
0010	**SKYLIGHT** Plastic domes, flush or curb mounted, ten or										800
0100	more units, curb not included										
0300	Nominal size under 10 S.F., double	G-3	130	.246	S.F.	12	4.43		16.43	20.50	
0400	Single		160	.200		8.05	3.60		11.65	14.95	
0600	10 S.F. to 20 S.F., double		315	.102		9.80	1.83		11.63	13.85	
0700	Single		395	.081		7.40	1.46		8.86	10.60	
0900	20 S.F. to 30 S.F., double		395	.081		9.80	1.46		11.26	13.25	
1000	Single		465	.069		8.75	1.24		9.99	11.70	
1200	30 S.F. to 65 S.F., double		465	.069		9.20	1.24		10.44	12.20	
1300	Single	↓	610	.052		13.15	.94		14.09	16.05	
1500	For insulated 4" curbs, double, add					25%					
1600	Single, add				↓	30%					
1800	For integral insulated 9" curbs, double, add					30%					
1900	Single, add					40%					
2120	Ventilating insulated plexiglass dome with										
2130	curb mounting, 36" x 36"	G-3	12	2.667	Ea.	340	48		388	455	
2150	52" x 52"		12	2.667		510	48		558	645	
2160	28" x 52"		10	3.200		400	57.50		457.50	540	
2170	36" x 52"	↓	10	3.200		430	57.50		487.50	575	
2180	For electric opening system, add					256			256	281	
2210	Operating skylight, with thermopane glass, 24" x 48"	G-3	10	3.200		500	57.50		557.50	650	
2220	32" x 48"	"	9	3.556	↓	525	64		589	685	
2310	Non venting insulated plexiglass dome skylight with										
2320	Flush mount 22" x 46"	G-3	15.23	2.101	Ea.	279	38		317	370	
2330	30" x 30"		16	2		256	36		292	340	
2340	46" x 46"		13.91	2.301		470	41.50		511.50	590	
2350	Curb mount 22" x 46"		15.23	2.101		244	38		282	335	
2360	30" x 30"		16	2		233	36		269	315	
2370	46" x 46"		13.91	2.301		440	41.50		481.50	550	
2381	Non-insulated flush mount 22" x 46"		15.23	2.101		188	38		226	270	
2382	30" x 30"		16	2		171	36		207	249	
2383	46" x 46"		13.91	2.301		320	41.50		361.50	420	
2384	Curb mount 22" x 46"		15.23	2.101		159	38		197	239	
2385	30" x 30"	↓	16	2	↓	153	36		189	230	

	08710	Door Hardware	CREW	DAILY OUTPUT	LABOR-HOURS	UNIT	2000 BARE COSTS MAT.	LABOR	EQUIP.	TOTAL	TOTAL INCL O&P
150	0010	**AVERAGE** Percentage for hardware, total job cost, minimum									.60%
	0050	Maximum									3%
	0500	Total hardware for building, average distribution					85%	15%			
	1000	Door hardware, apartment, interior				Door	107			107	118
	2100	Pocket door				Ea.	109			109	120
	4000	Door knocker, bright brass	1 Carp	32	.250		39.50	4.93		44.43	52
	4100	Mail slot, bright brass, 2″ x 11″	″	25	.320		52	6.30		58.30	68.50
	4200	Peep hole, add to price of door					14.40			14.40	15.85
340	0010	**DOORSTOPS** Holder and bumper, floor or wall	1 Carp	32	.250	Ea.	27	4.93		31.93	38.50
	1300	Wall bumper, 4″ diameter, with rubber pad, aluminum		32	.250		6.60	4.93		11.53	15.70
	1600	Door bumper, floor type, aluminum		32	.250		3.77	4.93		8.70	12.60
	1900	Plunger type, door mounted		32	.250		23	4.93		27.93	34
400	0010	**ENTRANCE LOCKS** Cylinder, grip handle, deadlocking latch	1 Carp	9	.889	Ea.	103	17.50		120.50	143
	0020	Deadbolt		8	1		125	19.70		144.70	171
	0100	Push and pull plate, dead bolt		8	1		119	19.70		138.70	165
	0900	For handicapped lever, add					130			130	143
520	0010	**HINGES** Full mortise, avg. freq., steel base, 4-1/2″ x 4-1/2″, USP R08700-100				Pr.	17.90			17.90	19.70
	0100	5″ x 5″, USP					29			29	32
	0200	6″ x 6″, USP					60.50			60.50	67
	0400	Brass base, 4-1/2″ x 4-1/2″, US10					37			37	40.50
	0500	5″ x 5″, US10					52.50			52.50	57.50
	0600	6″ x 6″, US10					87			87	96
	0800	Stainless steel base, 4-1/2″ x 4-1/2″, US32					61			61	67.50
	0900	For non removable pin, add				Ea.	2.14			2.14	2.35
	0910	For floating pin, driven tips, add					3.90			3.90	4.29
	0930	For hospital type tip on pin, add					11.45			11.45	12.60
	0940	For steeple type tip on pin, add					7.65			7.65	8.40
	0950	Full mortise, high frequency, steel base, 3-1/2″ x 3-1/2″, US26D				Pr.	19.40			19.40	21.50
	1000	4-1/2″ x 4-1/2″, USP					37.50			37.50	41.50
	1100	5″ x 5″, USP					40			40	44
	1200	6″ x 6″, USP					97			97	107
	1400	Brass base, 3-1/2″ x 3-1/2″, US4					35			35	38.50
	1430	4-1/2″ x 4-1/2″, US10					41.50			41.50	46
	1500	5″ x 5″, US10					74.50			74.50	82
	1600	6″ x 6″, US10					124			124	136
	1800	Stainless steel base, 4-1/2″ x 4-1/2″, US32					94.50			94.50	104
	1930	For hospital type tip on pin, add				Ea.	15.05			15.05	16.55
	1950	Full mortise, low frequency, steel base, 3-1/2″ x 3-1/2″, US26D				Pr.	12.25			12.25	13.45
	2000	4-1/2″ x 4-1/2″, USP					13.45			13.45	14.80
	2100	5″ x 5″, USP					21.50			21.50	23.50
	2200	6″ x 6″, USP					42			42	46.50
	2300	4-1/2″ x 4-1/2″, US3					9.85			9.85	10.80
	2310	5″ x 5″, US3					30.50			30.50	33.50
	2400	Brass bass, 4-1/2″ x 4-1/2″, US10					30.50			30.50	34
	2500	5″ x 5″, US10					46			46	50.50
	2800	Stainless steel base, 4-1/2″ x 4-1/2″, US32					51			51	56
550	0010	**KICK PLATE** 6″ high, for 3′ door, stainless steel	1 Carp	15	.533	Ea.	14.25	10.50		24.75	33.50
	0500	Bronze	″	15	.533	″	20	10.50		30.50	40
650	0010	**LOCKSET** Standard duty, cylindrical, with sectional trim									
	0020	Non-keyed, passage	1 Carp	12	.667	Ea.	37	13.15		50.15	63.50
	0100	Privacy		12	.667		45	13.15		58.15	72.50
	0400	Keyed, single cylinder function		10	.800		66	15.75		81.75	100
	0500	Lever handled, keyed, single cylinder function		10	.800		91	15.75		106.75	127
	1700	Residential, interior door, minimum		16	.500		12.60	9.85		22.45	31
	1720	Maximum		8	1		33.50	19.70		53.20	71
	1800	Exterior, minimum		14	.571		28	11.25		39.25	50.50

08700 | Hardware

	08710 | Door Hardware	CREW	DAILY OUTPUT	LABOR-HOURS	UNIT	2000 BARE COSTS MAT.	LABOR	EQUIP.	TOTAL	TOTAL INCL O&P	
1810	Average	1 Carp	8	1	Ea.	59.50	19.70		79.20	99.50	650
1820	Maximum	↓	8	1	↓	118	19.70		137.70	164	
	08720 | Weatherstripping & Seals										
0010	**WEATHERSTRIPPING** Window, double hung, 3′ x 5′, zinc	1 Carp	7.20	1.111	Opng.	10.70	22		32.70	49.50	300
0100	Bronze		7.20	1.111		19.45	22		41.45	59	
0200	Vinyl V strip		7	1.143		3.50	22.50		26	42.50	
0500	As above but heavy duty, zinc		4.60	1.739		13	34.50		47.50	73	
0600	Bronze		4.60	1.739		22.50	34.50		57	83.50	
1000	Doors, wood frame, interlocking, for 3′ x 7′ door, zinc		3	2.667		11.65	52.50		64.15	103	
1100	Bronze		3	2.667		18.35	52.50		70.85	110	
1300	6′ x 7′ opening, zinc		2	4		12.75	79		91.75	149	
1400	Bronze		2	4	↓	24	79		103	162	
1500	Vinyl V strip	↓	6.40	1.250	Ea.	6.70	24.50		31.20	49.50	
1700	Wood frame, spring type, bronze										
1800	3′ x 7′ door	1 Carp	7.60	1.053	Opng.	15.45	20.50		35.95	52.50	
1900	6′ x 7′ door		7	1.143		16.45	22.50		38.95	56.50	
1920	Felt, 3′ x 7′ door		14	.571		1.91	11.25		13.16	21.50	
1930	6′ x 7′ door		13	.615		2.07	12.10		14.17	23.50	
1950	Rubber, 3′ x 7′ door		7.60	1.053		4.14	20.50		24.64	40	
1960	6′ x 7′ door	↓	7	1.143	↓	4.72	22.50		27.22	43.50	
2200	Metal frame, spring type, bronze										
2300	3′ x 7′ door	1 Carp	3	2.667	Opng.	25.50	52.50		78	118	
2400	6′ x 7′ door	"	2.50	3.200	"	35	63		98	146	
2500	For stainless steel, spring type, add					133%					
2700	Metal frame, extruded sections, 3′ x 7′ door, aluminum	1 Carp	2	4	Opng.	34	79		113	172	
2800	Bronze		2	4		85.50	79		164.50	229	
3100	6′ x 7′ door, aluminum		1.20	6.667		43	131		174	273	
3200	Bronze	↓	1.20	6.667	↓	101	131		232	335	
3500	Threshold weatherstripping										
3650	Door sweep, flush mounted, aluminum	1 Carp	25	.320	Ea.	10	6.30		16.30	22	
3700	Vinyl		25	.320	"	11.85	6.30		18.15	24	
4000	Astragal for double doors, aluminum		4	2	Opng.	16.95	39.50		56.45	86	
4100	Bronze		4	2	"	27.50	39.50		67	98	
5000	Garage door bottom weatherstrip, 12′ aluminum, clear		14	.571	Ea.	16.05	11.25		27.30	37	
5010	Bronze		14	.571		61	11.25		72.25	87	
5050	Bottom protection, 12′ aluminum, clear		14	.571		19.45	11.25		30.70	41	
5100	Bronze	↓	14	.571	↓	76	11.25		87.25	103	
0010	**THRESHOLD** 3′ long door saddles, aluminum	1 Carp	48	.167	L.F.	3.38	3.28		6.66	9.35	800
0100	Aluminum, 8″ wide, 1/2″ thick		12	.667	Ea.	29	13.15		42.15	54.50	
0500	Bronze		60	.133	L.F.	27.50	2.63		30.13	34.50	
0600	Bronze, panic threshold, 5″ wide, 1/2″ thick		12	.667	Ea.	57	13.15		70.15	85	
0700	Rubber, 1/2″ thick, 5-1/2″ wide		20	.400		28.50	7.90		36.40	45	
0800	2-3/4″ wide	↓	20	.400	↓	13.25	7.90		21.15	28	
	08750 | Window Hardware										
0010	**WINDOW HARDWARE**										400
1000	Handles, surface mounted, aluminum	1 Carp	24	.333	Ea.	1.79	6.55		8.34	13.20	
1020	Brass		24	.333		2.08	6.55		8.63	13.55	
1040	Chrome		24	.333		1.90	6.55		8.45	13.35	
1500	Recessed, aluminum		12	.667		1.04	13.15		14.19	23.50	
1520	Brass		12	.667		1.15	13.15		14.30	24	
1540	Chrome		12	.667		1.09	13.15		14.24	23.50	
2000	Latches, aluminum		20	.400		1.49	7.90		9.39	15.15	
2020	Brass		20	.400		1.79	7.90		9.69	15.45	
2040	Chrome	↓	20	.400	↓	1.68	7.90		9.58	15.35	

08700 | Hardware

	08770 | Door/Window Accessories		CREW	DAILY OUTPUT	LABOR-HOURS	UNIT	2000 BARE COSTS MAT.	LABOR	EQUIP.	TOTAL	TOTAL INCL O&P
550	0010	**DETECTION SYSTEMS** See division 13851-065									
560	0010	**DOOR ACCESSORIES**									
	1000	Knockers, brass, standard	1 Carp	16	.500	Ea.	35	9.85		44.85	55.50
	1100	Deluxe		10	.800		108	15.75		123.75	146
	4000	Security chain, standard		18	.444		6	8.75		14.75	21.50
	4100	Deluxe	↓	18	.444	↓	36	8.75		44.75	54.50

08800 | Glazing

	08810 | Glass		CREW	DAILY OUTPUT	LABOR-HOURS	UNIT	2000 BARE COSTS MAT.	LABOR	EQUIP.	TOTAL	TOTAL INCL O&P
260	0010	**FLOAT GLASS** 3/16" thick, clear, plain	2 Glaz	130	.123	S.F.	3.55	2.38		5.93	7.85
	0200	Tempered, clear		130	.123		4.24	2.38		6.62	8.60
	0300	Tinted		130	.123		5.30	2.38		7.68	9.80
	0600	1/4" thick, clear, plain		120	.133		4.47	2.58		7.05	9.20
	0700	Tinted		120	.133		4.24	2.58		6.82	8.95
	0800	Tempered, clear		120	.133		5.30	2.58		7.88	10.15
	0900	Tinted		120	.133		7.35	2.58		9.93	12.40
	1600	3/8" thick, clear, plain		75	.213		7.05	4.13		11.18	14.60
	1700	Tinted		75	.213		8.50	4.13		12.63	16.20
	1800	Tempered, clear		75	.213		10.60	4.13		14.73	18.50
	1900	Tinted		75	.213		13.20	4.13		17.33	21.50
	2200	1/2" thick, clear, plain		55	.291		13.80	5.65		19.45	24.50
	2300	Tinted		55	.291		14.85	5.65		20.50	25.50
	2400	Tempered, clear		55	.291		15.90	5.65		21.55	27
	2500	Tinted		55	.291		19.85	5.65		25.50	31.50
	2800	5/8" thick, clear, plain		45	.356		14.85	6.90		21.75	27.50
	2900	Tempered, clear	↓	45	.356		16.95	6.90		23.85	30
	8900	For low emissivity coating for 3/16" and 1/4" only, add to above				↓	3				
300	0010	**GLAZING VARIABLES**									
	0600	For glass replacement, add				S.F.		100%			
	0700	For gasket settings, add				L.F.	3.18			3.18	3.50
	0900	For sloped glazing, add				S.F.		25%			
	2000	Fabrication, polished edges, 1/4" thick				Inch	.27			.27	.30
	2100	1/2" thick					.70			.70	.77
	2500	Mitered edges, 1/4" thick					.70			.70	.77
	2600	1/2" thick				↓	1.12			1.12	1.23
460	0010	**INSULATING GLASS** 2 lites 1/8" float, 1/2" thk, under 15 S.F.									
	0100	Tinted	2 Glaz	95	.168	S.F.	9.15	3.26		12.41	15.45
	0280	Double glazed, 5/8" thk unit, 3/16" float, 15-30 S.F., clear		90	.178		7.40	3.44		10.84	13.85
	0400	1" thk, dbl. glazed, 1/4" float, 30-70 S.F., clear		75	.213		10.50	4.13		14.63	18.40
	0500	Tinted		75	.213		12.70	4.13		16.83	21
	2000	Both lites, light & heat reflective		85	.188		16.95	3.64		20.59	24.50
	2500	Heat reflective, film inside, 1" thick unit, clear		85	.188		14.85	3.64		18.49	22.50
	2600	Tinted		85	.188		16	3.64		19.64	23.50
	3000	Film on weatherside, clear, 1/2" thick unit		95	.168		10.60	3.26		13.86	17.05
	3100	5/8" thick unit		90	.178		13.45	3.44		16.89	20.50
	3200	1" thick unit	↓	85	.188	↓	14.60	3.64		18.24	22

08800 | Glazing

08810	Glass	CREW	DAILY OUTPUT	LABOR-HOURS	UNIT	2000 BARE COSTS MAT.	LABOR	EQUIP.	TOTAL	TOTAL INCL O&P	
0010	**WINDOW GLASS** Clear float, stops, putty bed, 1/8" thick	2 Glaz	480	.033	S.F.	2.92	.65		3.57	4.28	850
0500	3/16" thick, clear		480	.033		3.60	.65		4.25	5.05	
0600	Tinted		480	.033		4.08	.65		4.73	5.55	
0700	Tempered	↓	480	.033	↓	4.94	.65		5.59	6.50	

08830	Mirrors	CREW	DAILY OUTPUT	LABOR-HOURS	UNIT	MAT.	LABOR	EQUIP.	TOTAL	TOTAL INCL O&P	
0010	**MIRRORS** No frames, wall type, 1/4" plate glass, polished edge										100
0100	Up to 5 S.F.	2 Glaz	125	.128	S.F.	5.85	2.48		8.33	10.50	
0200	Over 5 S.F.		160	.100		5.65	1.94		7.59	9.40	
0500	Door type, 1/4" plate glass, up to 12 S.F.		160	.100		6.05	1.94		7.99	9.85	
1000	Float glass, up to 10 S.F., 1/8" thick		160	.100		3.57	1.94		5.51	7.15	
1100	3/16" thick		150	.107		4.15	2.06		6.21	8	
1500	12" x 12" wall tiles, square edge, clear		195	.082		1.52	1.59		3.11	4.30	
1600	Veined		195	.082		3.87	1.59		5.46	6.90	
2010	Bathroom, unframed, laminated	↓	160	.100	↓	10.10	1.94		12.04	14.30	

Division Notes

	CREW	DAILY OUTPUT	LABOR-HOURS	UNIT	2000 BARE COSTS MAT.	LABOR	EQUIP.	TOTAL	TOTAL INCL O&P

Division 9 Finishes

Estimating Tips

General

- Room Finish Schedule: A complete set of plans should contain a room finish schedule. If one is not available, it would be well worth the time and effort to put one together. A room finish schedule should contain the room number, room name (for clarity), floor materials, base materials, wainscot materials, wainscot height, wall materials (for each wall), ceiling materials and special instructions.
- Surplus Finishes: Review the specifications to determine if there is any requirement to provide certain amounts of extra materials for the owner's maintenance department. In some cases the owner may require a substantial amount of materials, especially when it is a special order item or long lead time item.

09200 Plaster & Gypsum Board

- Lath is estimated by the square yard for both gypsum and metal lath, plus usually 5% allowance for waste. Furring, channels and accessories are measured by the linear foot. An extra foot should be allowed for each accessory miter or stop.
- Plaster is also estimated by the square yard. Deductions for openings vary by preference, from zero deduction to 50% of all openings over 2 feet in width. Some estimators deduct a percentage of the total yardage for openings. The estimator should allow one extra square foot for each linear foot of horizontal interior or exterior angle located below the ceiling level. Also, double the areas of small radius work.
- Each room should be measured, perimeter times maximum wall height. Ceiling areas are equal to length times width.
- Drywall accessories, studs, track, and acoustical caulking are all measured by the linear foot. Drywall taping is figured by the square foot. Gypsum wallboard is estimated by the square foot. No material deductions should be made for door or window openings under 32 S.F. Coreboard can be obtained in a 1″ thickness for solid wall and shaft work. Additions should be made to price out the inside or outside corners.
- Different types of partition construction should be listed separately on the quantity sheets. There may be walls with studs of various widths, double studded, and similar or dissimilar surface materials. Shaft work is usually different construction from surrounding partitions requiring separate quantities and pricing of the work.

09300 Tile
09400 Terrazzo

- Tile and terrazzo areas are taken off on a square foot basis. Trim and base materials are measured by the linear foot. Accent tiles are listed per each. Two basic methods of installation are used. Mud set is approximately 30% more expensive than the thin set. In terrazzo work, be sure to include the linear footage of embedded decorative strips, grounds, machine rubbing and power cleanup.

09600 Flooring

- Wood flooring is available in strip, parquet, or block configuration. The latter two types are set in adhesives with quantities estimated by the square foot. The laying pattern will influence labor costs and material waste. In addition to the material and labor for laying wood floors, the estimator must make allowances for sanding and finishing these areas unless the flooring is prefinished.
- Most of the various types of flooring are all measured on a square foot basis. Base is measured by the linear foot. If adhesive materials are to be quantified, they are estimated at a specified coverage rate by the gallon depending upon the specified type and the manufacturer's recommendations.
- Sheet flooring is measured by the square yard. Roll widths vary, so consideration should be given to use the most economical width, as waste must be figured into the total quantity. Consider also the installation methods available, direct glue down or stretched .

09700 Wall Finishes

- Wall coverings are estimated by the square foot. The area to be covered is measured, length by height of wall above baseboards, to calculate the square footage of each wall. This figure is divided by the number of square feet in the single roll which is being used. Deduct, in full, the areas of openings such as doors and windows. Where a pattern match is required allow 25%-30% waste. One gallon of paste should be sufficient to hang 12 single rolls of light to medium weight paper.

09800 Acoustical Treatment

- Acoustical systems fall into several categories. The takeoff of these materials is by the square foot of area with a 5% allowance for waste. Do not forget about scaffolding, if applicable, when estimating these systems.

09900 Paints & Coatings

- Painting is one area where bids vary to a greater extent than almost any other section of a project. This arises from the many methods of measuring surfaces to be painted. The estimator should check the plans and specifications carefully to be sure of the required number of coats.
- Protection of adjacent surfaces is not included in painting costs. When considering the method of paint application, an important factor is the amount of protection and masking required. These must be estimated separately and may be the determining factor in choosing the method of application.

Reference Numbers

Reference numbers are shown in bold squares at the beginning of some major classifications. These numbers refer to related items in the Reference Section. The reference information may be an estimating procedure, an alternate pricing method or technical information.

Note: Not all subdivisions listed here necessarily appear in this publication.

09100 | Metal Support Assemblies

09110 | Non-Load Bearing Wall Framing

		09110 Non-Load Bearing Wall Framing	CREW	DAILY OUTPUT	LABOR-HOURS	UNIT	MAT.	LABOR	EQUIP.	TOTAL	TOTAL INCL O&P
100	0010	**METAL STUDS, PARTITIONS**, 10′ high, with runners									
	2000	Non-load bearing, galvanized, 25 ga. 1-5/8″ wide, 16″ O.C.	1 Carp	450	.018	S.F.	.15	.35		.50	.76
	2100	24″ O.C.		520	.015		.11	.30		.41	.64
	2200	2-1/2″ wide, 16″ O.C.		440	.018		.16	.36		.52	.79
	2250	24″ O.C.		510	.016		.12	.31		.43	.66
	2300	3-5/8″ wide, 16″ O.C.		430	.019		.19	.37		.56	.84
	2350	24″ O.C.		500	.016		.14	.32		.46	.69
	2400	4″ wide, 16″ O.C.		420	.019		.23	.38		.61	.89
	2450	24″ O.C.		490	.016		.17	.32		.49	.73
	2500	6″ wide, 16″ O.C.		410	.020		.30	.38		.68	.99
	2550	24″ O.C.		480	.017		.22	.33		.55	.80
	2600	20 ga. studs, 1-5/8″ wide, 16″ O.C.		450	.018		.25	.35		.60	.88
	2650	24″ O.C.		520	.015		.19	.30		.49	.73
	2700	2-1/2″ wide, 16″ O.C.		440	.018		.27	.36		.63	.91
	2750	24″ O.C.		510	.016		.20	.31		.51	.75
	2800	3-5/8″ wide, 16″ O.C.		430	.019		.29	.37		.66	.95
	2850	24″ O.C.		500	.016		.22	.32		.54	.78
	2900	4″ wide, 16″ O.C.		420	.019		.37	.38		.75	1.04
	2950	24″ O.C.		490	.016		.27	.32		.59	.85
	3000	6″ wide, 16″ O.C.		410	.020		.43	.38		.81	1.13
	3050	24″ O.C.		480	.017		.32	.33		.65	.91
	5000	Load bearing studs, see division 05410-400									

09130 | Acoustical Suspension

		09130 Acoustical Suspension	CREW	DAILY OUTPUT	LABOR-HOURS	UNIT	MAT.	LABOR	EQUIP.	TOTAL	TOTAL INCL O&P
100	0010	**CEILING SUSPENSION SYSTEMS** For boards and tile									
	0050	Class A suspension system, 15/16″ T bar, 2′ x 4′ grid	1 Carp	800	.010	S.F.	.32	.20		.52	.69
	0300	2′ x 2′ grid	″	650	.012		.40	.24		.64	.86
	0350	For 9/16″ grid, add					.13			.13	.14
	0360	For fire rated grid, add					.07			.07	.08
	0370	For colored grid, add					.15			.15	.17
	0400	Concealed Z bar suspension system, 12″ module	1 Carp	520	.015		.35	.30		.65	.91
	0600	1-1/2″ carrier channels, 4′ O.C., add		470	.017		.21	.34		.55	.80
	0650	1-1/2″ x 3-1/2″ channels		470	.017		.39	.34		.73	1
	0700	Carrier channels for ceilings with									
	0900	recessed lighting fixtures, add	1 Carp	460	.017	S.F.	.34	.34		.68	.96
	5000	Wire hangers, #12 wire	″	300	.027	Ea.	.38	.53		.91	1.32

09200 | Plaster & Gypsum Board

09205 | Furring & Lathing

		09205 Furring & Lathing	CREW	DAILY OUTPUT	LABOR-HOURS	UNIT	MAT.	LABOR	EQUIP.	TOTAL	TOTAL INCL O&P
530	0010	**FURRING** Beams & columns, 7/8″ galvanized channels,									
	0030	12″ O.C.	1 Lath	155	.052	S.F.	.20	.99		1.19	1.84
	0050	16″ O.C.		170	.047		.16	.90		1.06	1.66
	0070	24″ O.C.		185	.043		.11	.83		.94	1.48
	0100	Ceilings, on steel, 7/8″ channels, galvanized, 12″ O.C.		210	.038		.18	.73		.91	1.40
	0300	16″ O.C.		290	.028		.16	.53		.69	1.05
	0400	24″ O.C.		420	.019		.11	.37		.48	.72
	0600	1-5/8″ channels, galvanized, 12″ O.C.		190	.042		.28	.81		1.09	1.62
	0700	16″ O.C.		260	.031		.25	.59		.84	1.24
	0900	24″ O.C.		390	.021		.17	.39		.56	.83

09200 | Plaster & Gypsum Board

09205 | Furring & Lathing

Line	Description	Crew	Daily Output	Labor-Hours	Unit	2000 Bare Costs Mat.	Labor	Equip.	Total	Total Incl O&P	
1000	Walls, 7/8" channels, galvanized, 12" O.C.	1 Lath	235	.034	S.F.	.18	.65		.83	1.27	530
1200	16" O.C.		265	.030		.16	.58		.74	1.13	
1300	24" O.C.		350	.023		.11	.44		.55	.84	
1500	1-5/8" channels, galvanized, 12" O.C.		210	.038		.28	.73		1.01	1.50	
1600	16" O.C.		240	.033		.25	.64		.89	1.32	
1800	24" O.C.		305	.026		.17	.50		.67	1	
8000	Suspended ceilings, including carriers										
8200	1-1/2" carriers, 24" O.C. with:										
8300	7/8" channels, 16" O.C.	1 Lath	165	.048	S.F.	.58	.93		1.51	2.16	
8320	24" O.C.		200	.040		.53	.77		1.30	1.84	
8400	1-5/8" channels, 16" O.C.		155	.052		.67	.99		1.66	2.35	
8420	24" O.C.		190	.042		.59	.81		1.40	1.96	
8600	2" carriers, 24" O.C. with:										
8700	7/8" channels, 16" O.C.	1 Lath	155	.052	S.F.	.16	.99		1.15	1.80	
8720	24" O.C.		190	.042		.11	.81		.92	1.44	
8800	1-5/8" channels, 16" O.C.		145	.055		.63	1.06		1.69	2.43	
8820	24" O.C.		180	.044		.54	.85		1.39	2	
0011	**GYPSUM LATH** Plain or perforated, nailed, 3/8" thick	1 Lath	765	.010	S.F.	.38	.20		.58	.75	540
0101	1/2" thick, nailed		720	.011		.39	.21		.60	.78	
0301	Clipped to steel studs, 3/8" thick		675	.012		.38	.23		.61	.79	
0401	1/2" thick		630	.013		.40	.24		.64	.84	
0601	Firestop gypsum base, to steel studs, 3/8" thick		630	.013		.39	.24		.63	.83	
0701	1/2" thick		585	.014		.44	.26		.70	.91	
0901	Foil back, to steel studs, 3/8" thick		675	.012		.40	.23		.63	.81	
1001	1/2" thick		630	.013		.44	.24		.68	.88	
1501	For ceiling installations, add		1,950	.004			.08		.08	.13	
1601	For columns and beams, add		1,550	.005			.10		.10	.16	
0011	**METAL LATH**										560
3601	2.5 lb. diamond painted, on wood framing, on walls	1 Lath	765	.010	S.F.	.15	.20		.35	.49	
3701	On ceilings		675	.012		.15	.23		.38	.53	
4201	3.4 lb. diamond painted, wired to steel framing, on walls		675	.012		.23	.23		.46	.62	
4301	On ceilings		540	.015		.23	.28		.51	.72	
5101	Rib lath, painted, wired to steel, on walls, 2.75 lb.		675	.012		.23	.23		.46	.62	
5201	3.4 lb.		630	.013		.32	.24		.56	.75	
5701	Suspended ceiling system, incl. 3.4 lb. diamond lath, painted		135	.059		.96	1.14		2.10	2.92	
5801	Galvanized		135	.059		.99	1.14		2.13	2.95	
0010	**ACCESSORIES, PLASTER** Casing bead, expanded flange, galvanized	1 Lath	2.70	2.963	C.L.F.	27	57		84	123	700
0900	Channels, cold rolled, 16 ga., 3/4" deep, galvanized					18			18	19.80	
1620	Corner bead, expanded bullnose, 3/4" radius, #10, galvanized	1 Lath	2.60	3.077		20	59		79	119	
1650	#1, galvanized		2.55	3.137		35	60		95	137	
1670	Expanded wing, 2-3/4" wide, galv. #1		2.65	3.019		20	58		78	117	
1700	Inside corner, (corner rite) 3" x 3", painted		2.60	3.077		17.65	59		76.65	116	
1750	Strip-ex, 4" wide, painted		2.55	3.137		14.75	60		74.75	115	
1800	Expansion joint, 3/4" grounds, limited expansion, galv., 1 piece		2.70	2.963		65	57		122	165	
2100	Extreme expansion, galvanized, 2 piece		2.60	3.077		120	59		179	229	

09210 | Gypsum Plaster

Line	Description	Crew	Daily Output	Labor-Hours	Unit	2000 Bare Costs Mat.	Labor	Equip.	Total	Total Incl O&P	
0010	**GYPSUM PLASTER** 80# bag, less than 1 ton (R09210-105)				Bag	14.20			14.20	15.65	100
0302	2 coats, no lath included, on walls	J-1	750	.053	S.F.	.40	.93	.07	1.40	2.08	
0402	On ceilings		660	.061		.38	1.06	.08	1.52	2.28	
0903	3 coats, no lath included, on walls		620	.065		.56	1.13	.09	1.78	2.60	
1002	On ceilings		560	.071		.53	1.25	.10	1.88	2.77	
1600	For irregular or curved surfaces, add						30%				

9 FINISHES

	09210 | Gypsum Plaster		CREW	DAILY OUTPUT	LABOR-HOURS	UNIT	2000 BARE COSTS MAT.	LABOR	EQUIP.	TOTAL	TOTAL INCL O&P
100	1800	For columns & beams, add R09210-105						50%			
500	0010	**PERLITE OR VERMICULITE PLASTER** 100 lb. bags Under 200 bags				Bag	13.05			13.05	14.35
	0301	2 coats, no lath included, on walls	J-1	830	.048	S.F.	.36	.84	.07	1.27	1.88
	0401	On ceilings		710	.056		.36	.98	.08	1.42	2.12
	0901	3 coats, no lath included, on walls		665	.060		.59	1.05	.08	1.72	2.50
	1001	On ceilings		565	.071		.59	1.24	.10	1.93	2.83
	1700	For irregular or curved surfaces, add to above						30%			
	1800	For columns and beams, add to above						50%			
	1900	For soffits, add to ceiling prices						40%			
900	0010	**THIN COAT** Plaster, 1 coat veneer, not incl. lath	J-1	3,600	.011	S.F.	.07	.19	.02	.28	.42
	1000	In 50 lb. bags				Bag	8.60			8.60	9.45
	09220 | Portland Cement Plaster										
200	0011	**STUCCO** 3 coats 1" thick, float finish, with mesh, on wood frame R09220-300	J-2	470	.102	S.F.	.37	1.81	.12	2.30	3.55
	0101	On masonry construction	J-1	495	.081	"	.20	1.41	.11	1.72	2.70
	0151	2 coats, 3/4" thick, float finish, no lath incl.	"	980	.041	S.F.	.21	.71	.06	.98	1.48
	0301	For trowel finish, add	1 Plas	1,530	.005	"		.10		.10	.16
	0600	For coloring and special finish, add, minimum	J-1	685	.058	S.Y.	.36	1.02	.08	1.46	2.19
	0700	Maximum		200	.200	"	1.26	3.49	.27	5.02	7.55
	1001	Exterior stucco, with bonding agent, 3 coats, on walls		1,800	.022	S.F.	.33	.39	.03	.75	1.04
	1201	Ceilings		1,620	.025		.33	.43	.03	.79	1.12
	1301	Beams		720	.056		.33	.97	.08	1.38	2.06
	1501	Columns		900	.044		.33	.78	.06	1.17	1.73
	1601	Mesh, painted, nailed to wood, 1.8 lb.	1 Lath	540	.015		.34	.28		.62	.84
	1801	3.6 lb.		495	.016		.18	.31		.49	.71
	1901	Wired to steel, painted, 1.8 lb.		477	.017		.34	.32		.66	.90
	2101	3.6 lb.		450	.018		.18	.34		.52	.76
	09250 | Gypsum Board										
200	0010	**CEMENTITIOUS BACKERBOARD**									
	0070	Cementitious backerboard, on floor, 3' x 4'x 1/2" sheets	2 Carp	525	.030	S.F.	1.04	.60		1.64	2.18
	0080	3' x 5' x 1/2" sheets		525	.030		1.04	.60		1.64	2.18
	0090	3' x 6' x 1/2" sheets		525	.030		1.04	.60		1.64	2.17
	0100	3' x 4'x 5/8" sheets		525	.030		1.08	.60		1.68	2.22
	0110	3' x 5' x 5/8" sheets		525	.030		1.06	.60		1.66	2.20
	0120	3' x 6' x 5/8" sheets		525	.030		1.07	.60		1.67	2.20
	0150	On wall, 3' x 4'x 1/2" sheets		350	.046		1.04	.90		1.94	2.69
	0160	3' x 5' x 1/2" sheets		350	.046		1.04	.90		1.94	2.69
	0170	3' x 6' x 1/2" sheets		350	.046		1.04	.90		1.94	2.68
	0180	3' x 4'x 5/8" sheets		350	.046		1.08	.90		1.98	2.73
	0190	3' x 5' x 5/8" sheets		350	.046		1.06	.90		1.96	2.71
	0200	3' x 6' x 5/8" sheets		350	.046		1.07	.90		1.97	2.71
	0250	On counter, 3' x 4'x 1/2" sheets		180	.089		1.04	1.75		2.79	4.15
	0260	3' x 5' x 1/2" sheets		180	.089		1.04	1.75		2.79	4.15
	0270	3' x 6' x 1/2" sheets		180	.089		1.04	1.75		2.79	4.14
	0300	3' x 4'x 5/8" sheets		180	.089		1.08	1.75		2.83	4.19
	0310	3' x 5' x 5/8" sheets		180	.089		1.06	1.75		2.81	4.17
	0320	3' x 6' x 5/8" sheets		180	.089		1.07	1.75		2.82	4.17
300	0010	**BLUEBOARD** For use with thin coat									
	0100	plaster application (see division 09210-900)									
	1000	3/8" thick, on walls or ceilings, standard, no finish included	2 Carp	1,900	.008	S.F.	.27	.17		.44	.58
	1100	With thin coat plaster finish		875	.018		.34	.36		.70	.99
	1400	On beams, columns, or soffits, standard, no finish included		675	.024		.31	.47		.78	1.14
	1450	With thin coat plaster finish		475	.034		.39	.66		1.05	1.57

09200 | Plaster & Gypsum Board

09250 | Gypsum Board		CREW	DAILY OUTPUT	LABOR-HOURS	UNIT	2000 BARE COSTS MAT.	LABOR	EQUIP.	TOTAL	TOTAL INCL O&P	
3000	1/2″ thick, on walls or ceilings, standard, no finish included	2 Carp	1,900	.008	S.F.	.27	.17		.44	.58	300
3100	With thin coat plaster finish		875	.018		.34	.36		.70	.99	
3300	Fire resistant, no finish included		1,900	.008		.27	.17		.44	.58	
3400	With thin coat plaster finish		875	.018		.34	.36		.70	.99	
3450	On beams, columns, or soffits, standard, no finish included		675	.024		.31	.47		.78	1.14	
3500	With thin coat plaster finish		475	.034		.39	.66		1.05	1.57	
3700	Fire resistant, no finish included		675	.024		.31	.47		.78	1.14	
3800	With thin coat plaster finish		475	.034		.39	.66		1.05	1.57	
5000	5/8″ thick, on walls or ceilings, fire resistant, no finish included		1,900	.008		.30	.17		.47	.61	
5100	With thin coat plaster finish		875	.018		.37	.36		.73	1.03	
5500	On beams, columns, or soffits, no finish included		675	.024		.35	.47		.82	1.18	
5600	With thin coat plaster finish		475	.034		.43	.66		1.09	1.61	
6000	For high ceilings, over 8′ high, add		3,060	.005			.10		.10	.18	
6500	For over 3 stories high, add per story		6,100	.003			.05		.05	.09	
0010	**DRYWALL** Gypsum plasterboard, nailed or screwed to studs										700
0100	unless otherwise noted										
0150	3/8″ thick, on walls, standard, no finish included	2 Carp	2,000	.008	S.F.	.26	.16		.42	.56	
0200	On ceilings, standard, no finish included		1,800	.009		.26	.18		.44	.59	
0250	On beams, columns, or soffits, no finish included		675	.024		.30	.47		.77	1.13	
0300	1/2″ thick, on walls, standard, no finish included		2,000	.008		.26	.16		.42	.56	
0350	Taped and finished (level 4 finish)		965	.017		.30	.33		.63	.89	
0390	With compound skim coat (level 5 finish)		775	.021		.32	.41		.73	1.05	
0400	Fire resistant, no finish included		2,000	.008		.25	.16		.41	.55	
0450	Taped and finished (level 4 finish)		965	.017		.29	.33		.62	.88	
0490	With compound skim coat (level 5 finish)		775	.021		.31	.41		.72	1.04	
0500	Water resistant, no finish included		2,000	.008		.32	.16		.48	.62	
0550	Taped and finished (level 4 finish)		965	.017		.36	.33		.69	.96	
0590	With compound skim coat (level 5 finish)		775	.021		.38	.41		.79	1.12	
0600	Prefinished, vinyl, clipped to studs		900	.018		.56	.35		.91	1.22	
1000	On ceilings, standard, no finish included		1,800	.009		.26	.18		.44	.59	
1050	Taped and finished (level 4 finish)		765	.021		.30	.41		.71	1.04	
1090	With compound skim coat (level 5 finish)		610	.026		.32	.52		.84	1.24	
1100	Fire resistant, no finish included		1,800	.009		.25	.18		.43	.58	
1150	Taped and finished (level 4 finish)		765	.021		.29	.41		.70	1.03	
1195	With compound skim coat (level 5 finish)		610	.026		.31	.52		.83	1.23	
1200	Water resistant, no finish included		1,800	.009		.32	.18		.50	.65	
1250	Taped and finished (level 4 finish)		765	.021		.36	.41		.77	1.11	
1290	With compound skim coat (level 5 finish)		610	.026		.38	.52		.90	1.31	
1500	On beams, columns, or soffits, standard, no finish included		675	.024		.30	.47		.77	1.13	
1550	Taped and finished (level 4 finish)		475	.034		.35	.66		1.01	1.52	
1590	With compound skim coat (level 5 finish)		540	.030		.37	.58		.95	1.40	
1600	Fire resistant, no finish included		675	.024		.29	.47		.76	1.12	
1650	Taped and finished (level 4 finish)		475	.034		.33	.66		.99	1.51	
1690	With compound skim coat (level 5 finish)		540	.030		.36	.58		.94	1.39	
1700	Water resistant, no finish included		675	.024		.37	.47		.84	1.20	
1750	Taped and finished (level 4 finish)		475	.034		.41	.66		1.07	1.60	
1790	With compound skim coat (level 5 finish)		540	.030		.44	.58		1.02	1.48	
2000	5/8″ thick, on walls, standard, no finish included		2,000	.008		.29	.16		.45	.59	
2050	Taped and finished (level 4 finish)		965	.017		.33	.33		.66	.92	
2090	With compound skim coat (level 5 finish)		775	.021		.35	.41		.76	1.09	
2100	Fire resistant, no finish included		2,000	.008		.29	.16		.45	.59	
2150	Taped and finished (level 4 finish)		965	.017		.33	.33		.66	.92	
2195	With compound skim coat (level 5 finish)		775	.021		.35	.41		.76	1.09	
2200	Water resistant, no finish included		2,000	.008		.35	.16		.51	.66	
2250	Taped and finished (level 4 finish)		965	.017		.39	.33		.72	.99	
2290	With compound skim coat (level 5 finish)		775	.021		.41	.41		.82	1.15	

09200 | Plaster & Gypsum Board

	09250 Gypsum Board		CREW	DAILY OUTPUT	LABOR-HOURS	UNIT	2000 BARE COSTS MAT.	LABOR	EQUIP.	TOTAL	TOTAL INCL O&P
700	2300	Prefinished, vinyl, clipped to studs	2 Carp	900	.018	S.F.	.65	.35		1	1.31
	3000	On ceilings, standard, no finish included		1,800	.009		.29	.18		.47	.62
	3050	Taped and finished (level 4 finish)		765	.021		.33	.41		.74	1.07
	3090	With compound skim coat (level 5 finish)		615	.026		.35	.51		.86	1.27
	3100	Fire resistant, no finish included		1,800	.009		.29	.18		.47	.62
	3150	Taped and finished (level 4 finish)		765	.021		.33	.41		.74	1.07
	3190	With compound skim coat (level 5 finish)		615	.026		.35	.51		.86	1.27
	3200	Water resistant, no finish included		1,800	.009		.35	.18		.53	.69
	3250	Taped and finished (level 4 finish)		765	.021		.39	.41		.80	1.14
	3290	With compound skim coat (level 5 finish)		615	.026		.41	.51		.92	1.33
	3500	On beams, columns, or soffits, no finish included		675	.024		.33	.47		.80	1.17
	3550	Taped and finished (level 4 finish)		475	.034		.38	.66		1.04	1.56
	3590	With compound skim coat (level 5 finish)		380	.042		.40	.83		1.23	1.86
	3600	Fire resistant, no finish included		675	.024		.33	.47		.80	1.17
	3650	Taped and finished (level 4 finish)		475	.034		.38	.66		1.04	1.56
	3690	With compound skim coat (level 5 finish)		380	.042		.40	.83		1.23	1.86
	3700	Water resistant, no finish included		675	.024		.40	.47		.87	1.24
	3750	Taped and finished (level 4 finish)		475	.034		.45	.66		1.11	1.63
	3790	With compound skim coat (level 5 finish)		380	.042		.47	.83		1.30	1.94
	4000	Fireproofing, beams or columns, 2 layers, 1/2″ thick, incl finish		330	.048		.54	.96		1.50	2.23
	4050	5/8″ thick		300	.053		.66	1.05		1.71	2.53
	4100	3 layers, 1/2″ thick		225	.071		.87	1.40		2.27	3.36
	4150	5/8″ thick		210	.076		.99	1.50		2.49	3.66
	5200	For high ceilings, over 8′ high, add		3,060	.005			.10		.10	.18
	5270	For textured spray, add	2 Lath	1,600	.010		.05	.19		.24	.37
	5350	For finishing corners, inside or outside, add	2 Carp	1,100	.015	L.F.	.06	.29		.35	.56
	5500	For acoustical sealant, add per bead	1 Carp	500	.016	″	.03	.32		.35	.57
	5550	Sealant, 1 quart tube				Ea.	4.89			4.89	5.40
	5600	Sound deadening board, 1/4″ gypsum	2 Carp	1,800	.009	S.F.	.25	.18		.43	.58
	5650	1/2″ wood fiber	″	1,800	.009	″	.34	.18		.52	.67

	09270 Drywall Accessories		CREW	DAILY OUTPUT	LABOR-HOURS	UNIT	MAT.	LABOR	EQUIP.	TOTAL	TOTAL INCL O&P
100	0011	**ACCESSORIES, DRYWALL** Casing bead, galvanized steel	1 Carp	290	.028	L.F.	.14	.54		.68	1.08
	0101	Vinyl		290	.028		.16	.54		.70	1.11
	0401	Corner bead, galvanized steel, 1-1/4″ x 1-1/4″		350	.023		.09	.45		.54	.87
	0411	1-1/4″ x 1-1/4″, 10′ long		35	.229	Ea.	.90	4.50		5.40	8.70
	0601	Vinyl corner bead		400	.020	L.F.	.21	.39		.60	.91
	0901	Furring channel, galv. steel, 7/8″ deep, standard		260	.031		.17	.61		.78	1.23
	1001	Resilient		260	.031		.19	.61		.80	1.24
	1101	J trim, galvanized steel, 1/2″ wide		300	.027			.53		.53	.90
	1121	5/8″ wide		300	.027		.13	.53		.66	1.04
	1160	Screws #6 x 1″ A				M	6.65			6.65	7.30
	1170	#6 x 1-5/8″ A				″	8.65			8.65	9.55
	1501	Z stud, galvanized steel, 1-1/2″ wide	1 Carp	260	.031	L.F.	.24	.61		.85	1.30

09300 | Tile

	09310 Ceramic Tile		CREW	DAILY OUTPUT	LABOR-HOURS	UNIT	2000 BARE COSTS MAT.	LABOR	EQUIP.	TOTAL	TOTAL INCL O&P
100	0010	**CERAMIC TILE**									
	0050	Base, using 1′ x 4″ high pc. with 1″ x 1″ tiles, mud set	D-7	82	.195	L.F.	3.99	3.38		7.37	9.90

09300 | Tile

09310 | Ceramic Tile		CREW	DAILY OUTPUT	LABOR-HOURS	UNIT	2000 BARE COSTS MAT.	LABOR	EQUIP.	TOTAL	TOTAL INCL O&P	
0100	Thin set	D-7	128	.125	L.F.	3.79	2.17		5.96	7.70	100
0300	For 6″ high base, 1″ x 1″ tile face, add					.63			.63	.69	
0400	For 2″ x 2″ tile face, add to above					.33			.33	.36	
0600	Cove base, 4-1/4″ x 4-1/4″ high, mud set	D-7	91	.176		3.02	3.05		6.07	8.25	
0700	Thin set		128	.125		3.02	2.17		5.19	6.85	
0900	6″ x 4-1/4″ high, mud set		100	.160		2.54	2.77		5.31	7.30	
1000	Thin set		137	.117		2.54	2.02		4.56	6.05	
1200	Sanitary cove base, 6″ x 4-1/4″ high, mud set		93	.172		3.52	2.98		6.50	8.70	
1300	Thin set		124	.129		3.52	2.24		5.76	7.50	
1500	6″ x 6″ high, mud set		84	.190		3	3.30		6.30	8.65	
1600	Thin set		117	.137		3	2.37		5.37	7.15	
1800	Bathroom accessories, average		82	.195	Ea.	9.50	3.38		12.88	15.95	
1900	Bathtub, 5′, rec. 4-1/4″ x 4-1/4″ tile wainscot, adhesive set 6′ high		2.90	5.517		140	95.50		235.50	310	
2100	7′ high wainscot		2.50	6.400		160	111		271	355	
2200	8′ high wainscot		2.20	7.273		170	126		296	390	
2400	Bullnose trim, 4-1/4″ x 4-1/4″, mud set		82	.195	L.F.	2.74	3.38		6.12	8.50	
2500	Thin set		128	.125		2.74	2.17		4.91	6.50	
2700	6″ x 4-1/4″ bullnose trim, mud set		84	.190		2.30	3.30		5.60	7.90	
2800	Thin set		124	.129		2.30	2.24		4.54	6.15	
3000	Floors, natural clay, random or uniform, thin set, color group 1		183	.087	S.F.	3.55	1.52		5.07	6.35	
3100	Color group 2		183	.087		3.82	1.52		5.34	6.65	
3260	Floors, glazed, thin set, 8″ x 8″, color group 1		250	.064		3.02	1.11		4.13	5.10	
3270	12″ x 12″ tile		325	.049		3.78	.85		4.63	5.55	
3280	16″ x 16″ tile		550	.029		5.75	.50		6.25	7.15	
3300	Porcelain type, 1 color, color group 2, 1″ x 1″		183	.087		4.11	1.52		5.63	7	
3400	2″ x 2″ or 2″ x 1″, thin set		190	.084		3.97	1.46		5.43	6.75	
3600	For random blend, 2 colors, add					.75			.75	.83	
3700	4 colors, add					1.08			1.08	1.19	
4300	Specialty tile, 4-1/4″ x 4-1/4″ x 1/2″, decorator finish	D-7	183	.087		8.95	1.52		10.47	12.30	
4500	Add for epoxy grout, 1/16″ joint, 1″ x 1″ tile		800	.020		.53	.35		.88	1.14	
4600	2″ x 2″ tile		820	.020		.50	.34		.84	1.10	
4800	Pregrouted sheets, walls, 4-1/4″ x 4-1/4″, 6″ x 4-1/4″										
4810	and 8-1/2″ x 4-1/4″, 4 S.F. sheets, silicone grout	D-7	240	.067	S.F.	4.07	1.16		5.23	6.35	
5100	Floors, unglazed, 2 S.F. sheets,										
5110	urethane adhesive	D-7	180	.089	S.F.	4.05	1.54		5.59	6.95	
5400	Walls, interior, thin set, 4-1/4″ x 4-1/4″ tile		190	.084		2	1.46		3.46	4.56	
5500	6″ x 4-1/4″ tile		190	.084		2.25	1.46		3.71	4.84	
5700	8-1/2″ x 4-1/4″ tile		190	.084		3.19	1.46		4.65	5.85	
5800	6″ x 6″ tile		200	.080		2.63	1.39		4.02	5.15	
5810	8″ x 8″ tile		225	.071		3.08	1.23		4.31	5.40	
5820	12″ x 12″ tile		300	.053		2.92	.92		3.84	4.71	
5830	16″ x 16″ tile		500	.032		3.17	.55		3.72	4.39	
6000	Decorated wall tile, 4-1/4″ x 4-1/4″, minimum		270	.059		4	1.03		5.03	6.05	
6100	Maximum		180	.089		42	1.54		43.54	48.50	
6600	Crystalline glazed, 4-1/4″ x 4-1/4″, mud set, plain		100	.160		3.15	2.77		5.92	7.95	
6700	4-1/4″ x 4-1/4″, scored tile		100	.160		3.97	2.77		6.74	8.85	
6900	6″x 6″ plain		93	.172		4.02	2.98		7	9.25	
7000	For epoxy grout, 1/16″ joints, 4-1/4″ tile, add		800	.020		.33	.35		.68	.92	
7200	For tile set in dry mortar, add		1,735	.009			.16		.16	.26	
7300	For tile set in portland cement mortar, add		290	.055			.96		.96	1.55	

09330 | Quarry Tile		CREW	DAILY OUTPUT	LABOR-HOURS	UNIT	MAT.	LABOR	EQUIP.	TOTAL	TOTAL INCL O&P	
0010	**QUARRY TILE** Base, cove or sanitary, 2″ or 5″ high, mud set										100
0100	1/2″ thick	D-7	110	.145	L.F.	3.70	2.52		6.22	8.15	
0300	Bullnose trim, red, mud set, 6″ x 6″ x 1/2″ thick		120	.133		3.81	2.31		6.12	7.95	
0400	4″ x 4″ x 1/2″ thick		110	.145		3.83	2.52		6.35	8.30	

09300 | Tile

		09330 | Quarry Tile	CREW	DAILY OUTPUT	LABOR-HOURS	UNIT	2000 BARE COSTS MAT.	LABOR	EQUIP.	TOTAL	TOTAL INCL O&P
100	0600	4″ x 8″ x 1/2″ thick, using 8″ as edge	D-7	130	.123	L.F.	3.63	2.13		5.76	7.45
	0700	Floors, mud set, 1,000 S.F. lots, red, 4″ x 4″ x 1/2″ thick		120	.133	S.F.	3.74	2.31		6.05	7.85
	0900	6″ x 6″ x 1/2″ thick		140	.114		2.70	1.98		4.68	6.20
	1000	4″ x 8″ x 1/2″ thick		130	.123		3.74	2.13		5.87	7.55
	1300	For waxed coating, add					.60			.60	.66
	1500	For colors other than green, add					.35			.35	.39
	1600	For abrasive surface, add					.42			.42	.46
	1800	Brown tile, imported, 6″ x 6″ x 3/4″	D-7	120	.133		4.44	2.31		6.75	8.60
	1900	8″ x 8″ x 1″		110	.145		5.05	2.52		7.57	9.65
	2100	For thin set mortar application, deduct		700	.023			.40		.40	.64
	2700	Stair tread, 6″ x 6″ x 3/4″, plain		50	.320		4.04	5.55		9.59	13.45
	2800	Abrasive		47	.340		4.59	5.90		10.49	14.60
	3000	Wainscot, 6″ x 6″ x 1/2″, thin set, red		105	.152		3.42	2.64		6.06	8.05
	3100	Colors other than green		105	.152		3.81	2.64		6.45	8.45
	3300	Window sill, 6″ wide, 3/4″ thick		90	.178	L.F.	4.41	3.08		7.49	9.85
	3400	Corners		80	.200	Ea.	4.81	3.47		8.28	10.90

		09370 | Metal Tile	CREW	DAILY OUTPUT	LABOR-HOURS	UNIT	MAT.	LABOR	EQUIP.	TOTAL	TOTAL INCL O&P
100	0010	**METAL TILE** 4′ x 4′ sheet, 24 ga., tile pattern, nailed									
	0200	Stainless steel	2 Carp	512	.031	S.F.	21.50	.62		22.12	25
	0400	Aluminized steel	"	512	.031	"	11.70	.62		12.32	13.90

09400 | Terrazzo

		09420 | Precast Terrazzo	CREW	DAILY OUTPUT	LABOR-HOURS	UNIT	2000 BARE COSTS MAT.	LABOR	EQUIP.	TOTAL	TOTAL INCL O&P
900	0010	**TERRAZZO, PRECAST** Base, 6″ high, straight	1 Mstz	35	.229	L.F.	8.35	4.37		12.72	16.25
	0100	Cove		30	.267		9	5.10		14.10	18.15
	0300	8″ high base, straight		30	.267		7.75	5.10		12.85	16.80
	0400	Cove		25	.320		12	6.10		18.10	23
	0600	For white cement, add					.33			.33	.36
	0700	For 16 ga. zinc toe strip, add					.99			.99	1.09
	0900	Curbs, 4″ x 4″ high	1 Mstz	19	.421		23	8.05		31.05	38
	1000	8″ x 8″ high	"	15	.533		28.50	10.20		38.70	47.50
	1200	Floor tiles, non-slip, 1″ thick, 12″ x 12″	D-1	29	.552	S.F.	14.60	9.85		24.45	33
	1300	1-1/4″ thick, 12″ x 12″		29	.552		16.75	9.85		26.60	35
	1500	16″ x 16″		23	.696		18.20	12.40		30.60	41
	1600	1-1/2″ thick, 16″ x 16″		21	.762		16.65	13.60		30.25	41.50
	4800	Wainscot, 12″ x 12″ x 1″ tiles	1 Mstz	12	.667		5	12.75		17.75	26
	4900	16″ x 16″ x 1-1/2″ tiles	"	8	1		11.70	19.10		30.80	44

09510 | Acoustical Ceilings

		09510 Acoustical Ceilings	CREW	DAILY OUTPUT	LABOR-HOURS	UNIT	2000 BARE COSTS MAT.	LABOR	EQUIP.	TOTAL	TOTAL INCL O&P	
700	0010	**SUSPENDED ACOUSTIC CEILING TILES**, Not including										700
	0100	suspension system										
	0300	Fiberglass boards, film faced, 2′ x 2′ or 2′ x 4′, 5/8″ thick	1 Carp	625	.013	S.F.	.51	.25		.76	.99	
	0400	3/4″ thick		600	.013		1.06	.26		1.32	1.62	
	0500	3″ thick, thermal, R11		450	.018		1.25	.35		1.60	1.98	
	0600	Glass cloth faced fiberglass, 3/4″ thick		500	.016		1.65	.32		1.97	2.36	
	0700	1″ thick		485	.016		1.86	.32		2.18	2.61	
	0820	1-1/2″ thick, nubby face		475	.017		2.21	.33		2.54	3	
	1110	Mineral fiber tile, lay-in, 2′ x 2′ or 2′ x 4′, 5/8″ thick, fine texture		625	.013		.41	.25		.66	.88	
	1115	Rough textured		625	.013		1	.25		1.25	1.53	
	1125	3/4″ thick, fine textured		600	.013		1.12	.26		1.38	1.68	
	1130	Rough textured		600	.013		1.40	.26		1.66	1.99	
	1135	Fissured		600	.013		1.65	.26		1.91	2.27	
	1150	Tegular, 5/8″ thick, fine textured		470	.017		.99	.34		1.33	1.66	
	1155	Rough textured		470	.017		1.29	.34		1.63	1.99	
	1165	3/4″ thick, fine textured		450	.018		1.40	.35		1.75	2.14	
	1170	Rough textured		450	.018		1.58	.35		1.93	2.34	
	1175	Fissured	↓	450	.018		2.46	.35		2.81	3.31	
	1180	For aluminum face, add					4.49			4.49	4.94	
	1185	For plastic film face, add					.69			.69	.76	
	1190	For fire rating, add					.34			.34	.37	
	1300	Mirror faced panels, 15/16″ thick, 2′ x 2′	1 Carp	500	.016		9.10	.32		9.42	10.55	
	1900	Eggcrate, acrylic, 1/2″ x 1/2″ x 1/2″ cubes		500	.016		1.36	.32		1.68	2.04	
	2100	Polystyrene eggcrate, 3/8″ x 3/8″ x 1/2″ cubes		510	.016		1.13	.31		1.44	1.77	
	2200	1/2″ x 1/2″ x 1/2″ cubes		500	.016		1.52	.32		1.84	2.21	
	2400	Luminous panels, prismatic, acrylic		400	.020		1.64	.39		2.03	2.48	
	2500	Polystyrene		400	.020		.84	.39		1.23	1.60	
	2700	Flat white acrylic		400	.020		2.87	.39		3.26	3.84	
	2800	Polystyrene		400	.020		1.96	.39		2.35	2.84	
	3000	Drop pan, white, acrylic		400	.020		4.20	.39		4.59	5.30	
	3100	Polystyrene		400	.020		3.51	.39		3.90	4.54	
	3600	Perforated aluminum sheets, .024″ thick, corrugated, painted		490	.016		1.67	.32		1.99	2.39	
	3700	Plain		500	.016		1.50	.32		1.82	2.19	
	3750	Wood fiber in cementitious binder, 2′ x 2′ or 4′, painted, 1″ thick		600	.013		1.14	.26		1.40	1.70	
	3760	2″ thick		550	.015		1.88	.29		2.17	2.56	
	3770	2-1/2″ thick		500	.016		2.58	.32		2.90	3.38	
	3780	3″ thick	↓	450	.018	↓	2.89	.35		3.24	3.78	
760	0010	**SUSPENDED CEILINGS, COMPLETE** Including standard										760
	0100	suspension system but not incl. 1-1/2″ carrier channels										
	0600	Fiberglass ceiling board, 2′ x 4′ x 5/8″, plain faced,	1 Carp	500	.016	S.F.	1.02	.32		1.34	1.66	
	0700	Offices, 2′ x 4′ x 3/4″		380	.021		1.13	.41		1.54	1.95	
	1800	Tile, Z bar suspension, 5/8″ mineral fiber tile		150	.053		1.47	1.05		2.52	3.42	
	1900	3/4″ mineral fiber tile	↓	150	.053	↓	1.57	1.05		2.62	3.53	
900	0010	**CEILING TILE**, Stapled or cemented										900
	0100	12″ x 12″ or 12″ x 24″, not including furring										
	0600	Mineral fiber, vinyl coated, 5/8″ thick	1 Carp	1,000	.008	S.F.	.72	.16		.88	1.06	
	0700	3/4″ thick		1,000	.008		1.17	.16		1.33	1.56	
	0900	Fire rated, 3/4″ thick, plain faced		1,000	.008		1	.16		1.16	1.37	
	1000	Plastic coated face		1,000	.008		1.04	.16		1.20	1.41	
	1200	Aluminum faced, 5/8″ thick, plain	↓	1,000	.008		1.09	.16		1.25	1.47	
	3300	For flameproofing, add					.09			.09	.10	
	3400	For sculptured 3 dimensional, add					.24			.24	.26	
	3900	For ceiling primer, add					.12			.12	.13	
	4000	For ceiling cement, add				↓	.32			.32	.35	

09600 | Flooring

		09631 Brick Flooring	CREW	DAILY OUTPUT	LABOR-HOURS	UNIT	2000 BARE COSTS MAT.	LABOR	EQUIP.	TOTAL	TOTAL INCL O&P
100	0010	**FLOORING**									
	0020	Acid proof shales, red, 8" x 3-3/4" x 1-1/4" thick	D-7	.43	37.209	M	695	645		1,340	1,825
	0050	2-1/4" thick	D-1	.40	40		755	715		1,470	2,025
	0200	Acid proof clay brick, 8" x 3-3/4" x 2-1/4" thick	"	.40	40		755	715		1,470	2,025
	0260	Cast ceramic, pressed, 4" x 8" x 1/2", unglazed	D-7	100	.160	S.F.	4.94	2.77		7.71	9.95
	0270	Glazed		100	.160		6.60	2.77		9.37	11.75
	0280	Hand molded flooring, 4" x 8" x 3/4", unglazed		95	.168		6.55	2.92		9.47	11.95
	0290	Glazed		95	.168		8.20	2.92		11.12	13.75
	0300	8" hexagonal, 3/4" thick, unglazed		85	.188		7.15	3.26		10.41	13.20
	0310	Glazed		85	.188		12.95	3.26		16.21	19.50
	0450	Acid proof joints, 1/4" wide	D-1	65	.246		1.13	4.39		5.52	8.70
	0500	Pavers, 8" x 4", 1" to 1-1/4" thick, red	D-7	95	.168		2.88	2.92		5.80	7.90
	0510	Ironspot	"	95	.168		4.07	2.92		6.99	9.20
	0540	1-3/8" to 1-3/4" thick, red	D-1	95	.168		2.78	3.01		5.79	8.15
	0560	Ironspot		95	.168		4.02	3.01		7.03	9.50
	0580	2-1/4" thick, red		90	.178		2.83	3.17		6	8.45
	0590	Ironspot		90	.178		4.38	3.17		7.55	10.15
	0800	For sidewalks and patios with pavers, see division 02780-200									
	0870	For epoxy joints, add	D-1	600	.027	S.F.	2.15	.48		2.63	3.18
	0880	For Furan underlayment, add	"	600	.027		1.78	.48		2.26	2.77
	0890	For waxed surface, steam cleaned, add	D-5	1,000	.008		.15	.13		.28	.38

		09635 Marble Flooring	CREW	DAILY OUTPUT	LABOR-HOURS	UNIT	MAT.	LABOR	EQUIP.	TOTAL	TOTAL INCL O&P
100	0010	**MARBLE** Thin gauge tile, 12" x 6", 3/8", White Carara	D-7	60	.267	S.F.	8.75	4.62		13.37	17.15
	0100	Travertine		60	.267		9.65	4.62		14.27	18.10
	0200	12" x 12" x 3/8", thin set, floors		60	.267		6.70	4.62		11.32	14.85
	0300	On walls		52	.308		8.75	5.35		14.10	18.30

		09637 Stone Flooring	CREW	DAILY OUTPUT	LABOR-HOURS	UNIT	MAT.	LABOR	EQUIP.	TOTAL	TOTAL INCL O&P
100	0010	**SLATE TILE** Vermont, 6" x 6" x 1/4" thick, thin set	D-7	180	.089	S.F.	4.14	1.54		5.68	7.05
200	0010	**SLATE & STONE FLOORS** See division 02780-800									

		09648 Wood Strip Flooring	CREW	DAILY OUTPUT	LABOR-HOURS	UNIT	MAT.	LABOR	EQUIP.	TOTAL	TOTAL INCL O&P
100	0010	**WOOD** Fir, vertical grain, 1" x 4", not incl. finish, B & better	1 Carp	255	.031	S.F.	2.30	.62		2.92	3.59
	0100	C grade & better		255	.031		2.16	.62		2.78	3.44
	0300	Flat grain, 1" x 4", not incl. finish, B & better		255	.031		2.63	.62		3.25	3.95
	0400	C & better		255	.031		2.53	.62		3.15	3.84
	4000	Maple, strip, 25/32" x 2-1/4", not incl. finish, select		170	.047		3.10	.93		4.03	5
	4100	#2 & better		170	.047		2.65	.93		3.58	4.51
	4300	33/32" x 3-1/4", not incl. finish, #1 grade		170	.047		3.25	.93		4.18	5.15
	4400	#2 & better		170	.047		2.90	.93		3.83	4.78
	4600	Oak, white or red, 25/32" x 2-1/4", not incl. finish									
	4700	#1 common	1 Carp	170	.047	S.F.	2.70	.93		3.63	4.56
	4900	Select quartered, 2-1/4" wide		170	.047		3.02	.93		3.95	4.91
	5000	Clear		170	.047		3.25	.93		4.18	5.15
	5200	Parquetry, standard, 5/16" thick, not incl. finish, oak, minimum		160	.050		3.04	.99		4.03	5.05
	5300	Maximum		100	.080		5.10	1.58		6.68	8.35
	5500	Teak, minimum		160	.050		4.15	.99		5.14	6.25
	5600	Maximum		100	.080		7.25	1.58		8.83	10.70
	5650	13/16" thick, select grade oak, minimum		160	.050		8	.99		8.99	10.50
	5700	Maximum		100	.080		12.15	1.58		13.73	16.05
	5800	Custom parquetry, including finish, minimum		100	.080		13.40	1.58		14.98	17.45
	5900	Maximum		50	.160		17.75	3.15		20.90	25

9 FINISHES

09600 | Flooring

09648 | Wood Strip Flooring

		CREW	DAILY OUTPUT	LABOR-HOURS	UNIT	2000 BARE COSTS MAT.	LABOR	EQUIP.	TOTAL	TOTAL INCL O&P	
6100	Prefinished, white oak, prime grade, 2-1/4″ wide	1 Carp	170	.047	S.F.	5.85	.93		6.78	8.05	100
6200	3-1/4″ wide		185	.043		7.50	.85		8.35	9.70	
6400	Ranch plank		145	.055		7.25	1.09		8.34	9.85	
6500	Hardwood blocks, 9″ x 9″, 25/32″ thick		160	.050		4.85	.99		5.84	7.05	
6700	Parquetry, 5/16″ thick, oak, minimum		160	.050		3.30	.99		4.29	5.30	
6800	Maximum		100	.080		8.10	1.58		9.68	11.60	
7000	Walnut or teak, parquetry, minimum		160	.050		4.50	.99		5.49	6.65	
7100	Maximum		100	.080		7.85	1.58		9.43	11.35	
7200	Acrylic wood parquet blocks, 12″ x 12″ x 5/16″,										
7210	irradiated, set in epoxy	1 Carp	160	.050	S.F.	6.55	.99		7.54	8.90	
7400	Yellow pine, 3/4″ x 3-1/8″, T & G, C & better, not incl. finish	″	200	.040		2.15	.79		2.94	3.72	
7500	Refinish wood floor, sand, 2 cts poly, wax, soft wood, min.	1 Clab	400	.020		.65	.29		.94	1.21	
7600	Hard wood, max		130	.062		.98	.89		1.87	2.60	
7800	Sanding and finishing, 2 coats polyurethane		295	.027		.65	.39		1.04	1.38	
7900	Subfloor and underlayment, see division 06160										
8015	Transition molding, 2 1/4″ wide, 5′ long	1 Carp	19.20	.417	Ea.	10.35	8.20		18.55	25.50	
8300	Floating floor, wood composition strip, complete.	1 Clab	133	.060	S.F.	4.16	.87		5.03	6.05	
8310	Floating floor components, T & G wood composite strips					3.87			3.87	4.26	
8320	Film					.09			.09	.10	
8330	Foam					.19			.19	.21	
8340	Adhesive					.06			.06	.07	
8350	Installation kit					.16			.16	.18	
8360	Trim, 2″ wide x 3′ long				L.F.	2.25			2.25	2.48	
8370	Reducer moulding				″	4.15			4.15	4.57	

09658 | Resilient Tile Flooring

		CREW	DAILY OUTPUT	LABOR-HOURS	UNIT	2000 BARE COSTS MAT.	LABOR	EQUIP.	TOTAL	TOTAL INCL O&P	
0010	**RESILIENT FLOORING** R09658 -600										100
0800	Base, cove, rubber or vinyl, .080″ thick										
1100	Standard colors, 2-1/2″ high	1 Tilf	315	.025	L.F.	.40	.49		.89	1.23	
1150	4″ high		315	.025		.44	.49		.93	1.27	
1200	6″ high		315	.025		.73	.49		1.22	1.59	
1450	1/8″ thick, standard colors, 2-1/2″ high		315	.025		.45	.49		.94	1.29	
1500	4″ high		315	.025		.65	.49		1.14	1.50	
1550	6″ high		315	.025		.81	.49		1.30	1.68	
1600	Corners, 2-1/2″ high		315	.025	Ea.	1.02	.49		1.51	1.91	
1630	4″ high		315	.025		1.07	.49		1.56	1.97	
1660	6″ high		315	.025		1.38	.49		1.87	2.31	
1700	Conductive flooring, rubber tile, 1/8″ thick		315	.025	S.F.	2.50	.49		2.99	3.54	
1800	Homogeneous vinyl tile, 1/8″ thick		315	.025		3.42	.49		3.91	4.55	
2200	Cork tile, standard finish, 1/8″ thick		315	.025		3.41	.49		3.90	4.54	
2250	3/16″ thick		315	.025		3.64	.49		4.13	4.79	
2300	5/16″ thick		315	.025		4.56	.49		5.05	5.80	
2350	1/2″ thick		315	.025		5.45	.49		5.94	6.80	
2500	Urethane finish, 1/8″ thick		315	.025		4.33	.49		4.82	5.55	
2550	3/16″ thick		315	.025		5.15	.49		5.64	6.50	
2600	5/16″ thick		315	.025		6.05	.49		6.54	7.45	
2650	1/2″ thick		315	.025		8.10	.49		8.59	9.70	
3700	Polyethylene, in rolls, no base incl., landscape surfaces		275	.029		2.24	.56		2.80	3.36	
3800	Nylon action surface, 1/8″ thick		275	.029		2.41	.56		2.97	3.55	
3900	1/4″ thick		275	.029		3.47	.56		4.03	4.72	
4000	3/8″ thick		275	.029		4.36	.56		4.92	5.70	
5900	Rubber, sheet goods, 36″ wide, 1/8″ thick		120	.067		3.09	1.28		4.37	5.45	
5950	3/16″ thick		100	.080		4.39	1.54		5.93	7.30	
6000	1/4″ thick		90	.089		5.05	1.71		6.76	8.35	
6050	Tile, marbleized colors, 12″ x 12″, 1/8″ thick		400	.020		3.50	.38		3.88	4.47	
6100	3/16″ thick		400	.020		4.75	.38		5.13	5.80	

09658 | Resilient Tile Flooring

			CREW	DAILY OUTPUT	LABOR-HOURS	UNIT	2000 BARE COSTS MAT.	LABOR	EQUIP.	TOTAL	TOTAL INCL O&P
100	6300	Special tile, plain colors, 1/8″ thick R09658-600	1 Tilf	400	.020	S.F.	3.83	.38		4.21	4.83
	6350	3/16″ thick		400	.020		5.15	.38		5.53	6.25
	7000	Vinyl composition tile, 12″ x 12″, 1/16″ thick		500	.016		.69	.31		1	1.26
	7050	Embossed		500	.016		.86	.31		1.17	1.45
	7100	Marbleized		500	.016		.86	.31		1.17	1.45
	7150	Solid		500	.016		.97	.31		1.28	1.57
	7200	3/32″ thick, embossed		500	.016		.85	.31		1.16	1.44
	7250	Marbleized		500	.016		.98	.31		1.29	1.58
	7300	Solid		500	.016		1.43	.31		1.74	2.07
	7350	1/8″ thick, marbleized		500	.016		.94	.31		1.25	1.53
	7400	Solid		500	.016		2.01	.31		2.32	2.71
	7450	Conductive		500	.016		3.33	.31		3.64	4.16
	7500	Vinyl tile, 12″ x 12″, .050″ thick, minimum		500	.016		1.54	.31		1.85	2.19
	7550	Maximum		500	.016		3.01	.31		3.32	3.81
	7600	1/8″ thick, minimum		500	.016		1.94	.31		2.25	2.63
	7650	Solid colors		500	.016		4.35	.31		4.66	5.30
	7700	Marbleized or Travertine pattern		500	.016		3.11	.31		3.42	3.92
	7750	Florentine pattern		500	.016		3.57	.31		3.88	4.43
	7800	Maximum		500	.016		7.35	.31		7.66	8.55
	8000	Vinyl sheet goods, backed, .065″ thick, minimum		250	.032		1.25	.61		1.86	2.38
	8050	Maximum		200	.040		2.32	.77		3.09	3.79
	8100	.080″ thick, minimum		230	.035		1.50	.67		2.17	2.73
	8150	Maximum		200	.040		2.65	.77		3.42	4.16
	8200	.125″ thick, minimum		230	.035		1.63	.67		2.30	2.87
	8250	Maximum		200	.040		3.57	.77		4.34	5.15
	8700	Adhesive cement, 1 gallon does 200 to 300 S.F.				Gal.	13.85			13.85	15.20
	8800	Asphalt primer, 1 gallon per 300 S.F.					8.40			8.40	9.25
	8900	Emulsion, 1 gallon per 140 S.F.					9			9	9.90
	8950	Latex underlayment, liquid, fortified					27.50			27.50	30

09680 | Carpet

			CREW	DAILY OUTPUT	LABOR-HOURS	UNIT	MAT.	LABOR	EQUIP.	TOTAL	TOTAL INCL O&P
800	0010	**CARPET** Commercial grades, direct cement									
	0701	Nylon, level loop, 26 oz., light to medium traffic	1 Tilf	445	.018	S.F.	1.47	.35		1.82	2.18
	0901	32 oz., medium traffic		445	.018		1.95	.35		2.30	2.71
	1101	40 oz., medium to heavy traffic		445	.018		2.39	.35		2.74	3.19
	2101	Nylon, plush, 20 oz., light traffic		445	.018		.94	.35		1.29	1.59
	2801	24 oz., light to medium traffic		445	.018		1.02	.35		1.37	1.68
	2901	30 oz., medium traffic		445	.018		1.40	.35		1.75	2.10
	3001	36 oz., medium traffic		445	.018		1.80	.35		2.15	2.54
	3101	42 oz., medium to heavy traffic		370	.022		2.05	.42		2.47	2.93
	3201	46 oz., medium to heavy traffic		370	.022		2.45	.42		2.87	3.37
	3301	54 oz., heavy traffic		370	.022		2.81	.42		3.23	3.76
	3501	Olefin, 15 oz., light traffic		445	.018		.51	.35		.86	1.12
	3651	22 oz., light traffic		445	.018		.61	.35		.96	1.23
	4501	50 oz., medium to heavy traffic, level loop		445	.018		4.19	.35		4.54	5.15
	4701	32 oz., medium to heavy traffic, patterned		400	.020		4.13	.38		4.51	5.15
	4901	48 oz., heavy traffic, patterned		400	.020		5.35	.38		5.73	6.45
	5000	For less than full roll, add					25%				
	5100	For small rooms, less than 12′ wide, add						25%			
	5200	For large open areas (no cuts), deduct						25%			
	5600	For bound carpet baseboard, add	1 Tilf	300	.027	L.F.	1.02	.51		1.53	1.95
	5610	For stairs, not incl. price of carpet, add	″	30	.267	Riser		5.10		5.10	8.30
	8950	For tackless, stretched installation, add padding to above									
	9001	Sponge rubber pad, minimum	1 Tilf	1,350	.006	S.F.	.28	.11		.39	.49
	9101	Maximum		1,350	.006		.77	.11		.88	1.03

09600 | Flooring

09680 | Carpet		CREW	DAILY OUTPUT	LABOR-HOURS	UNIT	2000 BARE COSTS MAT.	LABOR	EQUIP.	TOTAL	TOTAL INCL O&P	
9201	Felt pad, minimum	1 Tilf	1,350	.006	S.F.	.35	.11		.46	.57	800
9301	Maximum		1,350	.006		.58	.11		.69	.82	
9401	Bonded urethane pad, minimum		1,350	.006		.36	.11		.47	.58	
9501	Maximum		1,350	.006		.62	.11		.73	.86	
9601	Prime urethane pad, minimum		1,350	.006		.21	.11		.32	.41	
9701	Maximum		1,350	.006		.37	.11		.48	.59	
9850	For "branded" fiber, add				S.Y.	25%					
0010	CARPET TILE										900
0100	Tufted nylon, 18" x 18", hard back, 20 oz.	1 Tilf	150	.053	S.Y.	16.85	1.02		17.87	20	
0110	26 oz.		150	.053		29	1.02		30.02	33	
0200	Cushion back, 20 oz.		150	.053		21	1.02		22.02	25	
0210	26 oz.		150	.053		33	1.02		34.02	38	

09700 | Wall Finishes

09710 | Acoustical Wall Treatment		CREW	DAILY OUTPUT	LABOR-HOURS	UNIT	2000 BARE COSTS MAT.	LABOR	EQUIP.	TOTAL	TOTAL INCL O&P	
0010	WALL COVERING Including sizing, add 10%-30% waste at takeoff R09700-700										100
0050	Aluminum foil	1 Pape	275	.029	S.F.	.78	.53		1.31	1.74	
0100	Copper sheets, .025" thick, vinyl backing		240	.033		4.15	.60		4.75	5.55	
0300	Phenolic backing		240	.033		5.35	.60		5.95	6.90	
0600	Cork tiles, light or dark, 12" x 12" x 3/16"		240	.033		2.72	.60		3.32	3.99	
0700	5/16" thick		235	.034		2.77	.61		3.38	4.07	
0900	1/4" basketweave		240	.033		4.25	.60		4.85	5.70	
1000	1/2" natural, non-directional pattern		240	.033		6.35	.60		6.95	7.95	
1100	3/4" natural, non-directional pattern		240	.033		8.05	.60		8.65	9.85	
1200	Granular surface, 12" x 36", 1/2" thick		385	.021		.92	.38		1.30	1.64	
1300	1" thick		370	.022		1.19	.39		1.58	1.96	
1500	Polyurethane coated, 12" x 12" x 3/16" thick		240	.033		2.87	.60		3.47	4.16	
1600	5/16" thick		235	.034		4.08	.61		4.69	5.50	
1800	Cork wallpaper, paperbacked, natural		480	.017		1.63	.30		1.93	2.29	
1900	Colors		480	.017		2.02	.30		2.32	2.72	
2100	Flexible wood veneer, 1/32" thick, plain woods		100	.080		1.64	1.44		3.08	4.21	
2200	Exotic woods		95	.084		2.49	1.52		4.01	5.25	
2400	Gypsum-based, fabric-backed, fire										
2500	resistant for masonry walls, minimum, 21 oz./S.Y.	1 Pape	800	.010	S.F.	.65	.18		.83	1.01	
2600	Average		720	.011		.90	.20		1.10	1.32	
2700	Maximum, (small quantities)		640	.013		1	.23		1.23	1.48	
2750	Acrylic, modified, semi-rigid PVC, .028" thick	2 Carp	330	.048		.82	.96		1.78	2.54	
2800	.040" thick	"	320	.050		1.08	.99		2.07	2.88	
3000	Vinyl wall covering, fabric-backed, lightweight, (12-15 oz./S.Y.)	1 Pape	640	.013		.53	.23		.76	.96	
3300	Medium weight, type 2, (20-24 oz./S.Y.)		480	.017		.63	.30		.93	1.19	
3400	Heavy weight, type 3, (28 oz./S.Y.)		435	.018		1.11	.33		1.44	1.77	
3600	Adhesive, 5 gal. lots, (18SY/Gal.)				Gal.	8.30			8.30	9.15	
3700	Wallpaper, average workmanship, solid pattern, low cost paper	1 Pape	640	.013	S.F.	.27	.23		.50	.68	
3900	basic patterns (matching required), avg. cost paper		535	.015		.49	.27		.76	.99	
4000	Paper at $50 per double roll, quality workmanship		435	.018		.92	.33		1.25	1.56	
4100	Linen wall covering, paper backed										
4150	Flame treatment, minimum				S.F.	.65			.65	.71	

9 FINISHES

09700 | Wall Finishes

		09710 | Acoustical Wall Treatment	CREW	DAILY OUTPUT	LABOR-HOURS	UNIT	2000 BARE COSTS MAT.	LABOR	EQUIP.	TOTAL	TOTAL INCL O&P
100	4180	Maximum R09700-700				S.F.	1.17			1.17	1.29
	4200	Grass cloths with lining paper, minimum	1 Pape	400	.020		.59	.36		.95	1.25
	4300	Maximum	"	350	.023		1.88	.41		2.29	2.76

		09770 | Special Wall Surfaces	CREW	DAILY OUTPUT	LABOR-HOURS	UNIT	MAT.	LABOR	EQUIP.	TOTAL	TOTAL INCL O&P
700	0010	**RAISED PANEL SYSTEM,** 3/4" MDO									
	0100	Standard, paint grade	2 Carp	300	.053	S.F.	6.75	1.05		7.80	9.25
	0110	Oak veneer		300	.053		10.25	1.05		11.30	13.10
	0120	Maple veneer		300	.053		16.65	1.05		17.70	20
	0130	Cherry veneer		300	.053		20	1.05		21.05	24
	0300	Class I fire rated, paint grade		300	.053		6.75	1.05		7.80	9.25
	0310	Oak veneer		300	.053		10.25	1.05		11.30	13.10
	0320	Maple veneer		300	.053		16.65	1.05		17.70	20
	0330	Cherry veneer		300	.053		20	1.05		21.05	24
	5000	For prefinished paneling, see division 06250-500 & 06250-200									

9 FINISHES

09900 | Paints & Coatings

		09910 | Paints & Coatings	CREW	DAILY OUTPUT	LABOR-HOURS	UNIT	2000 BARE COSTS MAT.	LABOR	EQUIP.	TOTAL	TOTAL INCL O&P
100	0010	**CABINETS AND CASEWORK**									
	1000	Primer coat, oil base, brushwork	1 Pord	650	.012	S.F.	.04	.22		.26	.42
	2000	Paint, oil base, brushwork, 1 coat		650	.012		.05	.22		.27	.43
	2500	2 coats		400	.020		.10	.36		.46	.71
	3000	Stain, brushwork, wipe off		650	.012		.04	.22		.26	.42
	4000	Shellac, 1 coat, brushwork		650	.012		.05	.22		.27	.42
	4500	Varnish, 3 coats, brushwork, sand after 1st coat		325	.025		.16	.44		.60	.92
	5000	For latex paint, deduct					10%				
300	0010	**DOORS AND WINDOWS, EXTERIOR** R09910-100									
	0100	Door frames & trim, only									
	0110	Brushwork, primer R09910-220	1 Pord	512	.016	L.F.	.05	.28		.33	.53
	0120	Finish coat, exterior latex		512	.016		.05	.28		.33	.53
	0130	Primer & 1 coat, exterior latex		300	.027		.10	.48		.58	.91
	0140	Primer & 2 coats, exterior latex		265	.030		.16	.54		.70	1.07
	0150	Doors, flush, both sides, incl. frame & trim									
	0160	Roll & brush, primer	1 Pord	10	.800	Ea.	3.83	14.35		18.18	28
	0170	Finish coat, exterior latex		10	.800		4	14.35		18.35	28.50
	0180	Primer & 1 coat, exterior latex		7	1.143		7.85	20.50		28.35	42.50
	0190	Primer & 2 coats, exterior latex		5	1.600		11.85	28.50		40.35	61
	0200	Brushwork, stain, sealer & 2 coats polyurethane		4	2		14.25	36		50.25	75.50
	0210	Doors, French, both sides, 10-15 lite, incl. frame & trim									
	0220	Brushwork, primer	1 Pord	6	1.333	Ea.	1.91	24		25.91	42
	0230	Finish coat, exterior latex		6	1.333		2	24		26	42
	0240	Primer & 1 coat, exterior latex		3	2.667		3.92	48		51.92	84.50
	0250	Primer & 2 coats, exterior latex		2	4		5.80	72		77.80	126
	0260	Brushwork, stain, sealer & 2 coats polyurethane		2.50	3.200		5.10	57.50		62.60	102
	0270	Doors, louvered, both sides, incl. frame & trim									
	0280	Brushwork, primer	1 Pord	7	1.143	Ea.	3.83	20.50		24.33	38

Important: See the Reference Section for critical supporting data - Reference Nos., Crews, & Location Factor

	09910 | Paints & Coatings		Crew	Daily Output	Labor-Hours	Unit	2000 Bare Costs Mat.	Labor	Equip.	Total	Total Incl O&P	
300	0290	Finish coat, exterior latex (R09910-100)	1 Pord	7	1.143	Ea.	4	20.50		24.50	38.50	300
	0300	Primer & 1 coat, exterior latex		4	2		7.85	36		43.85	68.50	
	0310	Primer & 2 coats, exterior latex (R09910-200)		3	2.667		11.60	48		59.60	93	
	0320	Brushwork, stain, sealer & 2 coats polyurethane		4.50	1.778		14.25	32		46.25	68.50	
	0330	Doors, panel, both sides, incl. frame & trim										
	0340	Roll & brush, primer	1 Pord	6	1.333	Ea.	3.83	24		27.83	44	
	0350	Finish coat, exterior latex		6	1.333		4	24		28	44.50	
	0360	Primer & 1 coat, exterior latex		3	2.667		7.85	48		55.85	88.50	
	0370	Primer & 2 coats, exterior latex		2.50	3.200		11.60	57.50		69.10	109	
	0380	Brushwork, stain, sealer & 2 coats polyurethane		3	2.667		14.25	48		62.25	95.50	
	0400	Windows, per ext. side, based on 15 SF										
	0410	1 to 6 lite										
	0420	Brushwork, primer	1 Pord	13	.615	Ea.	.76	11.05		11.81	19.30	
	0430	Finish coat, exterior latex		13	.615		.79	11.05		11.84	19.30	
	0440	Primer & 1 coat, exterior latex		8	1		1.55	17.95		19.50	31.50	
	0450	Primer & 2 coats, exterior latex		6	1.333		2.29	24		26.29	42.50	
	0460	Stain, sealer & 1 coat varnish		7	1.143		2.01	20.50		22.51	36	
	0470	7 to 10 lite										
	0480	Brushwork, primer	1 Pord	11	.727	Ea.	.76	13.05		13.81	23	
	0490	Finish coat, exterior latex		11	.727		.79	13.05		13.84	23	
	0500	Primer & 1 coat, exterior latex		7	1.143		1.55	20.50		22.05	35.50	
	0510	Primer & 2 coats, exterior latex		5	1.600		2.29	28.50		30.79	50.50	
	0520	Stain, sealer & 1 coat varnish		6	1.333		2.01	24		26.01	42	
	0530	12 lite										
	0540	Brushwork, primer	1 Pord	10	.800	Ea.	.76	14.35		15.11	25	
	0550	Finish coat, exterior latex		10	.800		.79	14.35		15.14	25	
	0560	Primer & 1 coat, exterior latex		6	1.333		1.55	24		25.55	41.50	
	0570	Primer & 2 coats, exterior latex		5	1.600		2.29	28.50		30.79	50.50	
	0580	Stain, sealer & 1 coat varnish		6	1.333		2	24		26	42	
	0590	For oil base paint, add					10%					
310	0010	**DOORS & WINDOWS, INTERIOR LATEX** (R09910-100)										310
	0100	Doors flush, both sides, incl. frame & trim										
	0110	Roll & brush, primer (R09910-220)	1 Pord	10	.800	Ea.	3.32	14.35		17.67	27.50	
	0120	Finish coat, latex		10	.800		3.73	14.35		18.08	28	
	0130	Primer & 1 coat latex		7	1.143		7.05	20.50		27.55	42	
	0140	Primer & 2 coats latex		5	1.600		10.60	28.50		39.10	59.50	
	0160	Spray, both sides, primer		20	.400		3.50	7.20		10.70	15.85	
	0170	Finish coat, latex		20	.400		3.92	7.20		11.12	16.30	
	0180	Primer & 1 coat latex		11	.727		7.45	13.05		20.50	30	
	0190	Primer & 2 coats latex		8	1		11.20	17.95		29.15	42.50	
	0200	Doors, French, both sides, 10-15 lite, incl. frame & trim										
	0210	Roll & brush, primer	1 Pord	6	1.333	Ea.	1.66	24		25.66	42	
	0220	Finish coat, latex		6	1.333		1.87	24		25.87	42	
	0230	Primer & 1 coat latex		3	2.667		3.53	48		51.53	84	
	0240	Primer & 2 coats latex		2	4		5.30	72		77.30	126	
	0260	Doors, louvered, both sides, incl. frame & trim										
	0270	Roll & brush, primer	1 Pord	7	1.143	Ea.	3.32	20.50		23.82	37.50	
	0280	Finish coat, latex		7	1.143		3.73	20.50		24.23	38	
	0290	Primer & 1 coat, latex		4	2		6.85	36		42.85	67.50	
	0300	Primer & 2 coats, latex		3	2.667		10.80	48		58.80	92	
	0320	Spray, both sides, primer		20	.400		3.50	7.20		10.70	15.85	
	0330	Finish coat, latex		20	.400		3.92	7.20		11.12	16.30	
	0340	Primer & 1 coat, latex		11	.727		7.45	13.05		20.50	30	
	0350	Primer & 2 coats, latex		8	1		11.45	17.95		29.40	42.50	
	0360	Doors, panel, both sides, incl. frame & trim										
	0370	Roll & brush, primer	1 Pord	6	1.333	Ea.	3.50	24		27.50	44	

	09910 | Paints & Coatings		CREW	DAILY OUTPUT	LABOR-HOURS	UNIT	2000 BARE COSTS MAT.	LABOR	EQUIP.	TOTAL	TOTAL INCL O&P
310	0380	Finish coat, latex (R09910-100)	1 Pord	6	1.333	Ea.	3.73	24		27.73	44
	0390	Primer & 1 coat, latex		3	2.667		7.05	48		55.05	88
	0400	Primer & 2 coats, latex (R09910-220)		2.50	3.200		10.80	57.50		68.30	108
	0420	Spray, both sides, primer		10	.800		3.50	14.35		17.85	28
	0430	Finish coat, latex		10	.800		3.92	14.35		18.27	28.50
	0440	Primer & 1 coat, latex		5	1.600		7.45	28.50		35.95	56
	0450	Primer & 2 coats, latex	↓	4	2	↓	11.45	36		47.45	72.50
	0460	Windows, per interior side, base on 15 SF									
	0470	1 to 6 lite									
	0480	Brushwork, primer	1 Pord	13	.615	Ea.	.66	11.05		11.71	19.15
	0490	Finish coat, enamel		13	.615		.74	11.05		11.79	19.25
	0500	Primer & 1 coat enamel		8	1		1.39	17.95		19.34	31.50
	0510	Primer & 2 coats enamel	↓	6	1.333	↓	2.13	24		26.13	42.50
	0530	7 to 10 lite									
	0540	Brushwork, primer	1 Pord	11	.727	Ea.	.66	13.05		13.71	22.50
	0550	Finish coat, enamel		11	.727		.74	13.05		13.79	23
	0560	Primer & 1 coat enamel		7	1.143		1.39	20.50		21.89	35.50
	0570	Primer & 2 coats enamel	↓	5	1.600	↓	2.13	28.50		30.63	50.50
	0590	12 lite									
	0600	Brushwork, primer	1 Pord	10	.800	Ea.	.66	14.35		15.01	24.50
	0610	Finish coat, enamel		10	.800		.74	14.35		15.09	25
	0620	Primer & 1 coat enamel		6	1.333		1.39	24		25.39	41.50
	0630	Primer & 2 coats enamel	↓	5	1.600		2.13	28.50		30.63	50.50
	0650	For oil base paint, add				↓	10%				
320	0010	**DOORS AND WINDOWS, INTERIOR ALKYD (OIL BASE)**									
	0500	Flush door & frame, 3′ x 7′, oil, primer, brushwork	1 Pord	10	.800	Ea.	1.94	14.35		16.29	26
	1000	Paint, 1 coat		10	.800		1.86	14.35		16.21	26
	1200	2 coats		7	1.143		3.07	20.50		23.57	37.50
	1400	Stain, brushwork, wipe off		18	.444		.90	8		8.90	14.30
	1600	Shellac, 1 coat, brushwork		25	.320		.96	5.75		6.71	10.65
	1800	Varnish, 3 coats, brushwork, sand after 1st coat		9	.889		3.36	15.95		19.31	30
	2000	Panel door & frame, 3′ x 7′, oil, primer, brushwork		6	1.333		1.63	24		25.63	42
	2200	Paint, 1 coat		6	1.333		1.86	24		25.86	42
	2400	2 coats		3	2.667		5.35	48		53.35	86
	2600	Stain, brushwork, panel door, 3′ x 7′, not incl. frame		16	.500		.90	9		9.90	16
	2800	Shellac, 1 coat, brushwork		22	.364		.96	6.55		7.51	11.95
	3000	Varnish, 3 coats, brushwork, sand after 1st coat	↓	7.50	1.067	↓	3.36	19.15		22.51	35.50
	3020	French door, incl. 3′ x 7′, 6 lites, frame & trim									
	3022	Paint, 1 coat, over existing paint	1 Pord	5	1.600	Ea.	3.72	28.50		32.22	52
	3024	2 coats, over existing paint		5	1.600		7.25	28.50		35.75	56
	3026	Primer & 1 coat		3.50	2.286		7	41		48	76
	3028	Primer & 2 coats		3	2.667		10.70	48		58.70	92
	3032	Varnish or polyurethane, 1 coat		5	1.600		4.09	28.50		32.59	52.50
	3034	2 coats, sanding between	↓	3	2.667	↓	8.20	48		56.20	89
	4400	Windows, including frame and trim, per side									
	4600	Colonial type, 6/6 lites, 2′ x 3′, oil, primer, brushwork	1 Pord	14	.571	Ea.	.26	10.25		10.51	17.40
	5800	Paint, 1 coat		14	.571		.29	10.25		10.54	17.40
	6000	2 coats		9	.889		.57	15.95		16.52	27
	6200	3′ x 5′ opening, 6/6 lites, primer coat, brushwork		12	.667		.64	11.95		12.59	20.50
	6400	Paint, 1 coat		12	.667		.73	11.95		12.68	21
	6600	2 coats		7	1.143		1.43	20.50		21.93	35.50
	6800	4′ x 8′ opening, 6/6 lites, primer coat, brushwork		8	1		1.37	17.95		19.32	31.50
	7000	Paint, 1 coat		8	1		1.57	17.95		19.52	31.50
	7200	2 coats		5	1.600		3.04	28.50		31.54	51.50
	8000	Single lite type, 2′ x 3′, oil base, primer coat, brushwork		33	.242		.26	4.35		4.61	7.55
	8200	Paint, 1 coat	↓	33	.242	↓	.29	4.35		4.64	7.55

09910	Paints & Coatings	CREW	DAILY OUTPUT	LABOR-HOURS	UNIT	2000 BARE COSTS MAT.	LABOR	EQUIP.	TOTAL	TOTAL INCL O&P	
8400	2 coats	1 Pord	20	.400	Ea.	.57	7.20		7.77	12.65	320
8600	3′ x 5′ opening, primer coat, brushwork		20	.400		.64	7.20		7.84	12.70	
8800	Paint, 1 coat		20	.400		.73	7.20		7.93	12.80	
9000	2 coats		13	.615		1.43	11.05		12.48	20	
9200	4′ x 8′ opening, primer coat, brushwork		14	.571		1.37	10.25		11.62	18.60	
9400	Paint, 1 coat		14	.571		1.57	10.25		11.82	18.80	
9600	2 coats		8	1		3.04	17.95		20.99	33.50	
0010	**FENCES** R09910-100										400
0100	Chain link or wire metal, one side, water base										
0110	Roll & brush, first coat R09910-220	1 Pord	960	.008	S.F.	.05	.15		.20	.31	
0120	Second coat		1,280	.006		.05	.11		.16	.24	
0130	Spray, first coat		2,275	.004		.05	.06		.11	.17	
0140	Second coat		2,600	.003		.05	.06		.11	.15	
0150	Picket, water base										
0160	Roll & brush, first coat	1 Pord	865	.009	S.F.	.06	.17		.23	.34	
0170	Second coat		1,050	.008		.06	.14		.20	.29	
0180	Spray, first coat		2,275	.004		.06	.06		.12	.17	
0190	Second coat		2,600	.003		.06	.06		.12	.15	
0200	Stockade, water base										
0210	Roll & brush, first coat	1 Pord	1,040	.008	S.F.	.06	.14		.20	.29	
0220	Second coat		1,200	.007		.06	.12		.18	.26	
0230	Spray, first coat		2,275	.004		.06	.06		.12	.17	
0240	Second coat		2,600	.003		.06	.06		.12	.15	
0010	**FLOORS,** Interior										500
0100	Concrete										
0120	1st coat	1 Pord	975	.008	S.F.	.11	.15		.26	.37	
0130	2nd coat		1,150	.007		.07	.12		.19	.29	
0140	3rd coat		1,300	.006		.06	.11		.17	.24	
0150	Roll, latex, block filler										
0160	1st coat	1 Pord	2,600	.003	S.F.	.14	.06		.20	.25	
0170	2nd coat		3,250	.002		.08	.04		.12	.16	
0180	3rd coat		3,900	.002		.06	.04		.10	.13	
0190	Spray, latex, block filler										
0200	1st coat	1 Pord	2,600	.003	S.F.	.12	.06		.18	.22	
0210	2nd coat		3,250	.002		.07	.04		.11	.14	
0220	3rd coat		3,900	.002		.05	.04		.09	.12	
0010	**MISCELLANEOUS, EXTERIOR** R09910-220										620
0100	Railing, ext., decorative wood, incl. cap & baluster										
0110	newels & spindles @ 12″ O.C.										
0120	Brushwork, stain, sand, seal & varnish										
0130	First coat	1 Pord	90	.089	L.F.	.40	1.60		2	3.10	
0140	Second coat	″	120	.067	″	.40	1.20		1.60	2.44	
0150	Rough sawn wood, 42″ high, 2″x2″ verticals, 6″ O.C.										
0160	Brushwork, stain, each coat	1 Pord	90	.089	L.F.	.13	1.60		1.73	2.80	
0170	Wrought iron, 1″ rail, 1/2″ sq. verticals										
0180	Brushwork, zinc chromate, 60″ high, bars 6″ O.C.										
0190	Primer	1 Pord	130	.062	L.F.	.47	1.10		1.57	2.36	
0200	Finish coat		130	.062		.15	1.10		1.25	2	
0210	Additional coat		190	.042		.17	.76		.93	1.45	
0220	Shutters or blinds, single panel, 2′x4′, paint all sides										
0230	Brushwork, primer	1 Pord	20	.400	Ea.	.55	7.20		7.75	12.60	
0240	Finish coat, exterior latex		20	.400		.42	7.20		7.62	12.45	
0250	Primer & 1 coat, exterior latex		13	.615		.85	11.05		11.90	19.40	
0260	Spray, primer		35	.229		.81	4.10		4.91	7.75	

	09910 | Paints & Coatings		CREW	DAILY OUTPUT	LABOR-HOURS	UNIT	2000 BARE COSTS MAT.	LABOR	EQUIP.	TOTAL	TOTAL INCL O&P
620	0270	Finish coat, exterior latex (R09910-220)	1 Pord	35	.229	Ea.	.90	4.10		5	7.85
	0280	Primer & 1 coat, exterior latex		20	.400		.86	7.20		8.06	12.95
	0290	For louvered shutters, add				S.F.	10%				
	0300	Stair stringers, exterior, metal									
	0310	Roll & brush, zinc chromate, to 14", each coat	1 Pord	320	.025	L.F.	.05	.45		.50	.80
	0320	Rough sawn wood, 4" x 12"									
	0330	Roll & brush, exterior latex, each coat	1 Pord	215	.037	L.F.	.06	.67		.73	1.18
	0340	Trellis/lattice, 2"x2" @ 3" O.C. with 2"x8" supports									
	0350	Spray, latex, per side, each coat	1 Pord	475	.017	S.F.	.06	.30		.36	.57
	0450	Decking, Ext., sealer, alkyd, brushwork, sealer coat		1,140	.007		.05	.13		.18	.26
	0460	1st coat		1,140	.007		.05	.13		.18	.26
	0470	2nd coat		1,300	.006		.04	.11		.15	.22
	0500	Paint, alkyd, brushwork, primer coat		1,140	.007		.07	.13		.20	.28
	0510	1st coat		1,140	.007		.06	.13		.19	.28
	0520	2nd coat		1,300	.006		.05	.11		.16	.23
	0600	Sand paint, alkyd, brushwork, 1 coat		150	.053		.08	.96		1.04	1.69
630	0010	**MISCELLANEOUS, INTERIOR**									
	2400	Floors, conc./wood, oil base, primer/sealer coat, brushwork	2 Pord	1,950	.008	S.F.	.05	.15		.20	.31
	2450	Roller		5,200	.003		.06	.06		.12	.15
	2600	Spray		6,000	.003		.06	.05		.11	.14
	2650	Paint 1 coat, brushwork		1,950	.008		.05	.15		.20	.30
	2800	Roller		5,200	.003		.05	.06		.11	.15
	2850	Spray		6,000	.003		.05	.05		.10	.14
	3000	Stain, wood floor, brushwork, 1 coat		4,550	.004		.04	.06		.10	.16
	3200	Roller		5,200	.003		.05	.06		.11	.14
	3250	Spray		6,000	.003		.04	.05		.09	.13
	3400	Varnish, wood floor, brushwork		4,550	.004		.05	.06		.11	.17
	3450	Roller		5,200	.003		.06	.06		.12	.15
	3600	Spray		6,000	.003		.06	.05		.11	.15
	3800	Grilles, per side, oil base, primer coat, brushwork	1 Pord	520	.015		.09	.28		.37	.55
	3850	Spray		1,140	.007		.09	.13		.22	.31
	3880	Paint 1 coat, brushwork		520	.015		.10	.28		.38	.57
	3900	Spray		1,140	.007		.11	.13		.24	.33
	3920	Paint 2 coats, brushwork		325	.025		.19	.44		.63	.95
	3940	Spray		650	.012		.22	.22		.44	.61
	4250	Paint 1 coat, brushwork	2 Pord	1,300	.012	L.F.	.10	.22		.32	.48
	4500	Louvers, one side, primer, brushwork	1 Pord	524	.015	S.F.	.05	.27		.32	.52
	4520	Paint one coat, brushwork		520	.015		.05	.28		.33	.52
	4530	Spray		1,140	.007		.06	.13		.19	.27
	4540	Paint two coats, brushwork		325	.025		.10	.44		.54	.85
	4550	Spray		650	.012		.11	.22		.33	.49
	4560	Paint three coats, brushwork		270	.030		.14	.53		.67	1.05
	4570	Spray		500	.016		.16	.29		.45	.66
	5000	Pipe, to 4" diameter, primer or sealer coat, oil base, brushwork	2 Pord	1,250	.013	L.F.	.06	.23		.29	.44
	5100	Spray		2,165	.007		.05	.13		.18	.28
	5200	Paint 1 coat, brushwork		1,250	.013		.06	.23		.29	.44
	5300	Spray		2,165	.007		.05	.13		.18	.28
	5350	Paint 2 coats, brushwork		775	.021		.10	.37		.47	.73
	5400	Spray		1,240	.013		.11	.23		.34	.51
	5450	To 8" diameter, primer or sealer coat, brushwork		620	.026		.11	.46		.57	.89
	5500	Spray		1,085	.015		.18	.26		.44	.64
	5550	Paint 1 coat, brushwork		620	.026		.16	.46		.62	.94
	5600	Spray		1,085	.015		.17	.26		.43	.63
	5650	Paint 2 coats, brushwork		385	.042		.20	.75		.95	1.46
	5700	Spray		620	.026		.23	.46		.69	1.02
	6600	Radiators, per side, primer, brushwork	1 Pord	520	.015	S.F.	.05	.28		.33	.52

09910 Paints & Coatings		CREW	DAILY OUTPUT	LABOR-HOURS	UNIT	2000 BARE COSTS MAT.	LABOR	EQUIP.	TOTAL	TOTAL INCL O&P	
6620	Paint one coat, brushwork	1 Pord	520	.015	S.F.	.05	.28		.33	.52	630
6640	Paint two coats, brushwork		340	.024		.10	.42		.52	.81	
6660	Paint three coats, brushwork		283	.028		.14	.51		.65	1.01	
7000	Trim, wood, incl. puttying, under 6″ wide										
7200	Primer coat, oil base, brushwork	1 Pord	650	.012	L.F.	.02	.22		.24	.39	
7250	Paint, 1 coat, brushwork		650	.012		.02	.22		.24	.40	
7400	2 coats		400	.020		.05	.36		.41	.65	
7450	3 coats		325	.025		.07	.44		.51	.82	
7500	Over 6″ wide, primer coat, brushwork		650	.012		.04	.22		.26	.42	
7550	Paint, 1 coat, brushwork		650	.012		.05	.22		.27	.42	
7600	2 coats		400	.020		.10	.36		.46	.70	
7650	3 coats		325	.025		.14	.44		.58	.90	
8000	Cornice, simple design, primer coat, oil base, brushwork		650	.012	S.F.	.04	.22		.26	.42	
8250	Paint, 1 coat		650	.012		.05	.22		.27	.42	
8300	2 coats		400	.020		.10	.36		.46	.70	
8350	Ornate design, primer coat		350	.023		.04	.41		.45	.73	
8400	Paint, 1 coat		350	.023		.05	.41		.46	.73	
8450	2 coats		400	.020		.10	.36		.46	.70	
8600	Balustrades, primer coat, oil base, brushwork		520	.015		.04	.28		.32	.51	
8650	Paint, 1 coat		520	.015		.05	.28		.33	.51	
8700	2 coats		325	.025		.10	.44		.54	.84	
8900	Trusses and wood frames, primer coat, oil base, brushwork		800	.010		.04	.18		.22	.35	
8950	Spray		1,200	.007		.04	.12		.16	.25	
9000	Paint 1 coat, brushwork		750	.011		.05	.19		.24	.37	
9200	Spray		1,200	.007		.05	.12		.17	.26	
9220	Paint 2 coats, brushwork		500	.016		.10	.29		.39	.58	
9240	Spray		600	.013		.11	.24		.35	.52	
9260	Stain, brushwork, wipe off		600	.013		.04	.24		.28	.45	
9280	Varnish, 3 coats, brushwork		275	.029		.16	.52		.68	1.05	
9350	For latex paint, deduct					10%					
0010	**SIDING EXTERIOR**, Alkyd (oil base) R09910-100										700
0500	Spray	2 Pord	4,550	.004	S.F.	.08	.06		.14	.19	
0800	Paint 2 coats, brushwork		1,300	.012		.10	.22		.32	.48	
1000	Spray		4,550	.004		.14	.06		.20	.26	
1200	Stucco, rough, oil base, paint 2 coats, brushwork		1,300	.012		.10	.22		.32	.48	
1400	Roller		1,625	.010		.11	.18		.29	.41	
1600	Spray		2,925	.005		.11	.10		.21	.28	
1800	Texture 1-11 or clapboard, oil base, primer coat, brushwork		1,300	.012		.09	.22		.31	.46	
2000	Spray		4,550	.004		.09	.06		.15	.20	
2100	Paint 1 coat, brushwork		1,300	.012		.07	.22		.29	.45	
2200	Spray		4,550	.004		.07	.06		.13	.19	
2400	Paint 2 coats, brushwork		810	.020		.15	.35		.50	.75	
2600	Spray		2,600	.006		.16	.11		.27	.36	
3000	Stain 1 coat, brushwork		1,520	.011		.04	.19		.23	.37	
3200	Spray		5,320	.003		.05	.05		.10	.14	
3400	Stain 2 coats, brushwork		950	.017		.09	.30		.39	.59	
4000	Spray		3,050	.005		.10	.09		.19	.26	
4200	Wood shingles, oil base primer coat, brushwork		1,300	.012		.08	.22		.30	.46	
4400	Spray		3,900	.004		.07	.07		.14	.20	
4600	Paint 1 coat, brushwork		1,300	.012		.06	.22		.28	.44	
4800	Spray		3,900	.004		.08	.07		.15	.20	
5000	Paint 2 coats, brushwork		810	.020		.12	.35		.47	.72	
5200	Spray		2,275	.007		.12	.13		.25	.34	
5800	Stain 1 coat, brushwork		1,500	.011		.04	.19		.23	.37	
6000	Spray		3,900	.004		.04	.07		.11	.17	
6500	Stain 2 coats, brushwork		950	.017		.09	.30		.39	.59	

09910 | Paints & Coatings

		09910 Paints & Coatings	CREW	DAILY OUTPUT	LABOR-HOURS	UNIT	2000 BARE COSTS MAT.	LABOR	EQUIP.	TOTAL	TOTAL INCL O&P
700	7000	Spray	2 Pord	2,660	.006	S.F.	.12	.11		.23	.31
	8000	For latex paint, deduct					10%				
710	0010	**SIDING, MISC.** R09910 -100									
	0100	Aluminum siding									
	0110	Brushwork, primer R09910 -220	2 Pord	2,275	.007	S.F.	.05	.13		.18	.27
	0120	Finish coat, exterior latex		2,275	.007		.04	.13		.17	.25
	0130	Primer & 1 coat exterior latex		1,300	.012		.10	.22		.32	.48
	0140	Primer & 2 coats exterior latex		975	.016		.14	.29		.43	.64
	0150	Mineral Fiber shingles									
	0160	Brushwork, primer	2 Pord	1,495	.011	S.F.	.09	.19		.28	.41
	0170	Finish coat, industrial enamel		1,495	.011		.09	.19		.28	.42
	0180	Primer & 1 coat enamel		810	.020		.18	.35		.53	.79
	0190	Primer & 2 coats enamel		540	.030		.27	.53		.80	1.19
	0200	Roll, primer		1,625	.010		.09	.18		.27	.39
	0210	Finish coat, industrial enamel		1,625	.010		.10	.18		.28	.40
	0220	Primer & 1 coat enamel		975	.016		.20	.29		.49	.70
	0230	Primer & 2 coats enamel		650	.025		.30	.44		.74	1.06
	0240	Spray, primer		3,900	.004		.07	.07		.14	.20
	0250	Finish coat, industrial enamel		3,900	.004		.08	.07		.15	.21
	0260	Primer & 1 coat enamel		2,275	.007		.16	.13		.29	.38
	0270	Primer & 2 coats enamel		1,625	.010		.24	.18		.42	.55
	0280	Waterproof sealer, first coat		4,485	.004		.06	.06		.12	.18
	0290	Second coat		5,235	.003		.06	.05		.11	.16
	0300	Rough wood incl. shingles, shakes or rough sawn siding									
	0310	Brushwork, primer	2 Pord	1,280	.013	S.F.	.11	.22		.33	.49
	0320	Finish coat, exterior latex		1,280	.013		.07	.22		.29	.45
	0330	Primer & 1 coat exterior latex		960	.017		.18	.30		.48	.70
	0340	Primer & 2 coats exterior latex		700	.023		.25	.41		.66	.95
	0350	Roll, primer		2,925	.005		.15	.10		.25	.32
	0360	Finish coat, exterior latex		2,925	.005		.08	.10		.18	.25
	0370	Primer & 1 coat exterior latex		1,790	.009		.23	.16		.39	.53
	0380	Primer & 2 coats exterior latex		1,300	.012		.31	.22		.53	.72
	0390	Spray, primer		3,900	.004		.13	.07		.20	.26
	0400	Finish coat, exterior latex		3,900	.004		.06	.07		.13	.19
	0410	Primer & 1 coat exterior latex		2,600	.006		.19	.11		.30	.39
	0420	Primer & 2 coats exterior latex		2,080	.008		.25	.14		.39	.51
	0430	Waterproof sealer, first coat		4,485	.004		.11	.06		.17	.23
	0440	Second coat		4,485	.004		.06	.06		.12	.18
	0450	Smooth wood incl. butt, T&G, beveled, drop or B&B siding									
	0460	Brushwork, primer	2 Pord	2,325	.007	S.F.	.08	.12		.20	.30
	0470	Finish coat, exterior latex		1,280	.013		.07	.22		.29	.45
	0480	Primer & 1 coat exterior latex		800	.020		.15	.36		.51	.76
	0490	Primer & 2 coats exterior latex		630	.025		.22	.46		.68	1
	0500	Roll, primer		2,275	.007		.09	.13		.22	.31
	0510	Finish coat, exterior latex		2,275	.007		.07	.13		.20	.29
	0520	Primer & 1 coat exterior latex		1,300	.012		.16	.22		.38	.55
	0530	Primer & 2 coats exterior latex		975	.016		.24	.29		.53	.75
	0540	Spray, primer		4,550	.004		.07	.06		.13	.19
	0550	Finish coat, exterior latex		4,550	.004		.06	.06		.12	.18
	0560	Primer & 1 coat exterior latex		2,600	.006		.13	.11		.24	.33
	0570	Primer & 2 coats exterior latex		1,950	.008		.20	.15		.35	.47
	0580	Waterproof sealer, first coat		5,230	.003		.06	.05		.11	.16
	0590	Second coat		5,980	.003		.06	.05		.11	.15
	0600	For oil base paint, add					10%				
800	0010	**TRIM, EXTERIOR** R09910 -100									
	0100	Door frames & trim (see Doors, interior or exterior)									

09910 | Paints & Coatings

	09910 Paints & Coatings	Ref.	CREW	DAILY OUTPUT	LABOR-HOURS	UNIT	2000 BARE COSTS MAT.	LABOR	EQUIP.	TOTAL	TOTAL INCL O&P	
0110	Fascia, latex paint, one coat coverage	R09910 -100										800
0120	1″ x 4″, brushwork		1 Pord	640	.013	L.F.	.02	.22		.24	.39	
0130	Roll	R09910 -220		1,280	.006		.02	.11		.13	.21	
0140	Spray			2,080	.004		.01	.07		.08	.13	
0150	1″ x 6″ to 1″ x 10″, brushwork			640	.013		.06	.22		.28	.43	
0160	Roll			1,230	.007		.06	.12		.18	.25	
0170	Spray			2,100	.004		.05	.07		.12	.16	
0180	1″ x 12″, brushwork			640	.013		.06	.22		.28	.43	
0190	Roll			1,050	.008		.06	.14		.20	.29	
0200	Spray		↓	2,200	.004	↓	.05	.07		.12	.16	
0210	Gutters & downspouts, metal, zinc chromate paint											
0220	Brushwork, gutters, 5″, first coat		1 Pord	640	.013	L.F.	.05	.22		.27	.43	
0230	Second coat			960	.008		.05	.15		.20	.30	
0240	Third coat			1,280	.006		.04	.11		.15	.23	
0250	Downspouts, 4″, first coat			640	.013		.05	.22		.27	.43	
0260	Second coat			960	.008		.05	.15		.20	.30	
0270	Third coat		↓	1,280	.006	↓	.04	.11		.15	.23	
0280	Gutters & downspouts, wood											
0290	Brushwork, gutters, 5″, primer		1 Pord	640	.013	L.F.	.05	.22		.27	.43	
0300	Finish coat, exterior latex			640	.013		.05	.22		.27	.42	
0310	Primer & 1 coat exterior latex			400	.020		.10	.36		.46	.71	
0320	Primer & 2 coats exterior latex			325	.025		.16	.44		.60	.91	
0330	Downspouts, 4″, primer			640	.013		.05	.22		.27	.43	
0340	Finish coat, exterior latex			640	.013		.05	.22		.27	.42	
0350	Primer & 1 coat exterior latex			400	.020		.10	.36		.46	.71	
0360	Primer & 2 coats exterior latex		↓	325	.025	↓	.08	.44		.52	.83	
0370	Molding, exterior, up to 14″ wide											
0380	Brushwork, primer		1 Pord	640	.013	L.F.	.06	.22		.28	.44	
0390	Finish coat, exterior latex			640	.013		.06	.22		.28	.43	
0400	Primer & 1 coat exterior latex			400	.020		.12	.36		.48	.74	
0410	Primer & 2 coats exterior latex			315	.025		.12	.46		.58	.90	
0420	Stain & fill			1,050	.008		.05	.14		.19	.29	
0430	Shellac			1,850	.004		.05	.08		.13	.19	
0440	Varnish	↓	↓	1,275	.006	↓	.06	.11		.17	.26	
0350	**WALLS, MASONRY (CMU), EXTERIOR**											910
0360	Concrete masonry units (CMU), smooth surface											
0370	Brushwork, latex, first coat		1 Pord	640	.013	S.F.	.03	.22		.25	.41	
0380	Second coat			960	.008		.03	.15		.18	.28	
0390	Waterproof sealer, first coat			736	.011		.06	.20		.26	.40	
0400	Second coat			1,104	.007		.04	.13		.17	.26	
0410	Roll, latex, paint, first coat			1,465	.005		.04	.10		.14	.20	
0420	Second coat			1,790	.004		.03	.08		.11	.16	
0430	Waterproof sealer, first coat			1,680	.005		.06	.09		.15	.21	
0440	Second coat			2,060	.004		.04	.07		.11	.16	
0450	Spray, latex, paint, first coat			1,950	.004		.03	.07		.10	.15	
0460	Second coat			2,600	.003		.02	.06		.08	.12	
0470	Waterproof sealer, first coat			2,245	.004		.08	.06		.14	.19	
0480	Second coat		↓	2,990	.003	↓	.03	.05		.08	.11	
0490	Concrete masonry unit (CMU), porous											
0500	Brushwork, latex, first coat		1 Pord	640	.013	S.F.	.07	.22		.29	.44	
0510	Second coat			960	.008		.03	.15		.18	.29	
0520	Waterproof sealer, first coat			736	.011		.08	.20		.28	.41	
0530	Second coat			1,104	.007		.04	.13		.17	.26	
0540	Roll latex, first coat			1,465	.005		.05	.10		.15	.21	
0550	Second coat			1,790	.004		.03	.08		.11	.17	
0560	Waterproof sealer, first coat		↓	1,680	.005	↓	.08	.09		.17	.22	

	09910 Paints & Coatings		Crew	Daily Output	Labor-Hours	Unit	2000 Bare Costs Mat.	Labor	Equip.	Total	Total Incl O&P
910	0570	Second coat	1 Pord	2,060	.004	S.F.	.04	.07		.11	.17
	0580	Spray latex, first coat		1,950	.004		.04	.07		.11	.16
	0590	Second coat		2,600	.003		.02	.06		.08	.12
	0600	Waterproof sealer, first coat		2,245	.004		.06	.06		.12	.18
	0610	Second coat		2,990	.003		.03	.05		.08	.12
920	0010	**WALLS AND CEILINGS**, Interior									
	0100	Concrete, dry wall or plaster, oil base, primer or sealer coat									
	0200	Smooth finish, brushwork	1 Pord	1,150	.007	S.F.	.04	.12		.16	.26
	0240	Roller		2,040	.004		.04	.07		.11	.16
	0300	Sand finish, brushwork		975	.008		.04	.15		.19	.29
	0340	Roller		1,150	.007		.05	.12		.17	.26
	0380	Spray		2,275	.004		.03	.06		.09	.15
	0400	Paint 1 coat, smooth finish, brushwork		1,200	.007		.04	.12		.16	.25
	0440	Roller		1,300	.006		.04	.11		.15	.23
	0480	Spray		2,275	.004		.04	.06		.10	.15
	0500	Sand finish, brushwork		1,050	.008		.04	.14		.18	.27
	0540	Roller		1,600	.005		.04	.09		.13	.20
	0580	Spray		2,100	.004		.04	.07		.11	.15
	0800	Paint 2 coats, smooth finish, brushwork		680	.012		.08	.21		.29	.44
	0840	Roller		800	.010		.08	.18		.26	.39
	0880	Spray		1,625	.005		.07	.09		.16	.23
	0900	Sand finish, brushwork		605	.013		.08	.24		.32	.49
	0940	Roller		1,020	.008		.08	.14		.22	.32
	0980	Spray		1,700	.005		.07	.08		.15	.22
	1200	Paint 3 coats, smooth finish, brushwork		510	.016		.12	.28		.40	.60
	1240	Roller		650	.012		.13	.22		.35	.51
	1280	Spray		1,625	.005		.11	.09		.20	.27
	1300	Sand finish, brushwork		454	.018		.12	.32		.44	.66
	1340	Roller		680	.012		.13	.21		.34	.49
	1380	Spray		1,133	.007		.11	.13		.24	.33
	1600	Glaze coating, 5 coats, spray, clear		900	.009		.60	.16		.76	.93
	1640	Multicolor		900	.009		.81	.16		.97	1.16
	1700	For latex paint, deduct					10%				
	1800	For ceiling installations, add						25%			
	2000	Masonry or concrete block, oil base, primer or sealer coat									
	2100	Smooth finish, brushwork	1 Pord	1,224	.007	S.F.	.04	.12		.16	.25
	2180	Spray		2,400	.003		.06	.06		.12	.17
	2200	Sand finish, brushwork		1,089	.007		.07	.13		.20	.29
	2280	Spray		2,400	.003		.06	.06		.12	.17
	2400	Paint 1 coat, smooth finish, brushwork		1,100	.007		.07	.13		.20	.29
	2480	Spray		2,400	.003		.06	.06		.12	.17
	2500	Sand finish, brushwork		979	.008		.07	.15		.22	.31
	2580	Spray		2,400	.003		.06	.06		.12	.17
	2800	Paint 2 coats, smooth finish, brushwork		756	.011		.13	.19		.32	.47
	2880	Spray		1,360	.006		.12	.11		.23	.32
	2900	Sand finish, brushwork		672	.012		.13	.21		.34	.51
	2980	Spray		1,360	.006		.12	.11		.23	.32
	3200	Paint 3 coats, smooth finish, brushwork		560	.014		.20	.26		.46	.65
	3280	Spray		1,088	.007		.19	.13		.32	.42
	3300	Sand finish, brushwork		498	.016		.20	.29		.49	.70
	3380	Spray		1,088	.007		.19	.13		.32	.42
	3600	Glaze coating, 5 coats, spray, clear		900	.009		.60	.16		.76	.93
	3620	Multicolor		900	.009		.81	.16		.97	1.16
	4000	Block filler, 1 coat, brushwork		425	.019		.12	.34		.46	.69
	4100	Silicone, water repellent, 2 coats, spray		2,000	.004		.24	.07		.31	.38

09900 | Paints & Coatings

09910 Paints & Coatings		CREW	DAILY OUTPUT	LABOR-HOURS	UNIT	2000 BARE COSTS MAT.	LABOR	EQUIP.	TOTAL	TOTAL INCL O&P	
4120	For latex paint, deduct				S.F.	10%					920

09930 Stains/Transp. Finishes		CREW	DAILY OUTPUT	LABOR-HOURS	UNIT	MAT.	LABOR	EQUIP.	TOTAL	TOTAL INCL O&P	
0010	**VARNISH** 1 coat + sealer, on wood trim, no sanding included	1 Pord	400	.020	S.F.	.06	.36		.42	.67	100
0100	Hardwood floors, 2 coats, no sanding included, roller	"	1,890	.004	"	.12	.08		.20	.26	

09963 Glazed Coatings		CREW	DAILY OUTPUT	LABOR-HOURS	UNIT	MAT.	LABOR	EQUIP.	TOTAL	TOTAL INCL O&P	
0010	**WALL COATINGS** Acrylic glazed coatings, minimum	1 Pord	525	.015	S.F.	.24	.27		.51	.72	200
0100	Maximum		305	.026		.50	.47		.97	1.34	
0300	Epoxy coatings, minimum		525	.015		.31	.27		.58	.80	
0400	Maximum		170	.047		.96	.84		1.80	2.47	
0600	Exposed aggregate, troweled on, 1/16" to 1/4", minimum		235	.034		.48	.61		1.09	1.55	
0700	Maximum (epoxy or polyacrylate)		130	.062		1.03	1.10		2.13	2.97	
0900	1/2" to 5/8" aggregate, minimum		130	.062		.95	1.10		2.05	2.89	
1000	Maximum		80	.100		1.62	1.80		3.42	4.78	
1200	1" aggregate size, minimum		90	.089		1.65	1.60		3.25	4.48	
1300	Maximum		55	.145		2.52	2.61		5.13	7.15	
1500	Exposed aggregate, sprayed on, 1/8" aggregate, minimum		295	.027		.45	.49		.94	1.31	
1600	Maximum	↓	145	.055	↓	.82	.99		1.81	2.55	

09990 Paint Restoration		CREW	DAILY OUTPUT	LABOR-HOURS	UNIT	MAT.	LABOR	EQUIP.	TOTAL	TOTAL INCL O&P	
0010	**SCRAPE AFTER FIRE DAMAGE**										500
0050	Boards, 1" x 4"	1 Pord	336	.024	L.F.		.43		.43	.71	
0060	1" x 6"		260	.031			.55		.55	.92	
0070	1" x 8"		207	.039			.69		.69	1.16	
0080	1" x 10"		174	.046			.83		.83	1.38	
0500	Framing, 2" x 4"		265	.030			.54		.54	.90	
0510	2" x 6"		221	.036			.65		.65	1.08	
0520	2" x 8"		190	.042			.76		.76	1.26	
0530	2" x 10"		165	.048			.87		.87	1.45	
0540	2" x 12"		144	.056			1		1	1.66	
1000	Heavy framing, 3" x 4"		226	.035			.64		.64	1.06	
1010	4" x 4"		210	.038			.68		.68	1.14	
1020	4" x 6"		191	.042			.75		.75	1.25	
1030	4" x 8"		165	.048			.87		.87	1.45	
1040	4" x 10"		144	.056			1		1	1.66	
1060	4" x 12"		131	.061	↓		1.10		1.10	1.83	
2900	For sealing, minimum		825	.010	S.F.	.11	.17		.28	.41	
2920	Maximum	↓	460	.017	"	.24	.31		.55	.78	
0010	**SANDING** and puttying interior trim, compared to										800
0100	Painting 1 coat, on quality work				L.F.		100%				
0300	Medium work						50%				
0400	Industrial grade				↓		25%				
0500	Surface protection, placement and removal										
0510	Basic drop cloths	1 Pord	6,400	.001	S.F.		.02		.02	.04	
0520	Masking with paper		800	.010		.02	.18		.20	.32	
0530	Volume cover up (using plastic sheathing, or building paper)	↓	16,000	.001	↓		.01		.01	.01	
0010	**SURFACE PREPARATION, EXTERIOR**										900
0015	Doors, per side, not incl. frames or trim										
0020	Scrape & sand										
0030	Wood, flush	1 Pord	616	.013	S.F.		.23		.23	.39	
0040	Wood, detail		496	.016			.29		.29	.48	
0050	Wood, louvered		280	.029			.51		.51	.86	
0060	Wood, overhead	↓	616	.013	↓		.23		.23	.39	
0070	Wire brush										

	09990 | Paint Restoration		CREW	DAILY OUTPUT	LABOR-HOURS	UNIT	2000 BARE COSTS MAT.	LABOR	EQUIP.	TOTAL	TOTAL INCL O&P
900	0080	Metal, flush	1 Pord	640	.013	S.F.		.22		.22	.37
	0090	Metal, detail		520	.015			.28		.28	.46
	0100	Metal, louvered		360	.022			.40		.40	.67
	0110	Metal or fibr., overhead		640	.013			.22		.22	.37
	0120	Metal, roll up		560	.014			.26		.26	.43
	0130	Metal, bulkhead		640	.013			.22		.22	.37
	0140	Power wash, based on 2500 lb. operating pressure									
	0150	Metal, flush	B-9	2,240	.018	S.F.		.27	.08	.35	.54
	0160	Metal, detail		2,120	.019			.28	.09	.37	.57
	0170	Metal, louvered		2,000	.020			.30	.09	.39	.61
	0180	Metal or fibr., overhead		2,400	.017			.25	.08	.33	.50
	0190	Metal, roll up		2,400	.017			.25	.08	.33	.50
	0200	Metal, bulkhead		2,200	.018			.27	.08	.35	.55
	0400	Windows, per side, not incl. trim									
	0410	Scrape & sand									
	0420	Wood, 1-2 lite	1 Pord	320	.025	S.F.		.45		.45	.75
	0430	Wood, 3-6 lite		280	.029			.51		.51	.86
	0440	Wood, 7-10 lite		240	.033			.60		.60	1
	0450	Wood, 12 lite		200	.040			.72		.72	1.20
	0460	Wood, Bay / Bow		320	.025			.45		.45	.75
	0470	Wire brush									
	0480	Metal, 1-2 lite	1 Pord	480	.017	S.F.		.30		.30	.50
	0490	Metal, 3-6 lite		400	.020			.36		.36	.60
	0500	Metal, Bay / Bow		480	.017			.30		.30	.50
	0510	Power wash, based on 2500 lb. operating pressure									
	0520	1-2 lite	B-9	4,400	.009	S.F.		.14	.04	.18	.28
	0530	3-6 lite		4,320	.009			.14	.04	.18	.29
	0540	7-10 lite		4,240	.009			.14	.04	.18	.29
	0550	12 lite		4,160	.010			.14	.04	.18	.29
	0560	Bay / Bow		4,400	.009			.14	.04	.18	.28
	0600	Siding, scrape and sand, light=10-30%, med.=30-70%									
	0610	Heavy=70-100%, % of surface to sand									
	0620	For Chemical Washing, see Division 04930-900									
	0650	Texture 1-11, light	1 Pord	480	.017	S.F.		.30		.30	.50
	0660	Med.		440	.018			.33		.33	.54
	0670	Heavy		360	.022			.40		.40	.67
	0680	Wood shingles, shakes, light		440	.018			.33		.33	.54
	0690	Med.		360	.022			.40		.40	.67
	0700	Heavy		280	.029			.51		.51	.86
	0710	Clapboard, light		520	.015			.28		.28	.46
	0720	Med.		480	.017			.30		.30	.50
	0730	Heavy		400	.020			.36		.36	.60
	0740	Wire brush									
	0750	Aluminum, light	1 Pord	600	.013	S.F.		.24		.24	.40
	0760	Med.		520	.015			.28		.28	.46
	0770	Heavy		440	.018			.33		.33	.54
	0780	Pressure wash, based on 2500 lb.. operating pressure									
	0790	Stucco	B-9	3,080	.013	S.F.		.19	.06	.25	.39
	0800	Aluminum or vinyl		3,200	.013			.19	.06	.25	.38
	0810	Siding, masonry, brick & block		2,400	.017			.25	.08	.33	.50
	1300	Miscellaneous, wire brush									
	1310	Metal, pedestrian gate	1 Pord	100	.080	S.F.		1.44		1.44	2.40
910	0010	SURFACE PREPARATION, INTERIOR									
	0020	Doors									
	0030	Scrape & sand									
	0040	Wood, flush	1 Pord	616	.013	S.F.		.23		.23	.39

09990 | Paint Restoration		CREW	DAILY OUTPUT	LABOR-HOURS	UNIT	2000 BARE COSTS MAT.	LABOR	EQUIP.	TOTAL	TOTAL INCL O&P	
0050	Wood, detail	1 Pord	496	.016	S.F.		.29		.29	.48	910
0060	Wood, louvered		280	.029			.51		.51	.86	
0070	Wire brush										
0080	Metal, flush	1 Pord	640	.013	S.F.		.22		.22	.37	
0090	Metal, detail		520	.015			.28		.28	.46	
0100	Metal, louvered		360	.022			.40		.40	.67	
0110	Hand wash										
0120	Wood, flush	1 Pord	2,160	.004	S.F.		.07		.07	.11	
0130	Wood, detailed		2,000	.004			.07		.07	.12	
0140	Wood, louvered		1,360	.006			.11		.11	.18	
0150	Metal, flush		2,160	.004			.07		.07	.11	
0160	Metal, detail		2,000	.004			.07		.07	.12	
0170	Metal, louvered		1,360	.006			.11		.11	.18	
0400	Windows, per side, not incl. trim										
0410	Scrape & sand										
0420	Wood, 1-2 lite	1 Pord	360	.022	S.F.		.40		.40	.67	
0430	Wood, 3-6 lite		320	.025			.45		.45	.75	
0440	Wood, 7-10 lite		280	.029			.51		.51	.86	
0450	Wood, 12 lite		240	.033			.60		.60	1	
0460	Wood, Bay / Bow		360	.022			.40		.40	.67	
0470	Wire brush										
0480	Metal, 1-2 lite	1 Pord	520	.015	S.F.		.28		.28	.46	
0490	Metal, 3-6 lite		440	.018			.33		.33	.54	
0500	Metal, Bay / Bow		520	.015			.28		.28	.46	
0600	Walls, sanding, light=10-30%										
0610	Med.=30-70%, heavy=70-100%, % of surface to sand										
0620	For Chemical Washing, see Division 04930-900										
0650	Walls, sand										
0660	Drywall, gypsum, plaster, light	1 Pord	3,077	.003	S.F.		.05		.05	.08	
0670	Drywall, gypsum, plaster, med.		2,160	.004			.07		.07	.11	
0680	Drywall, gypsum, plaster, heavy		923	.009			.16		.16	.26	
0690	Wood, T&G, light		2,400	.003			.06		.06	.10	
0700	Wood, T&G, med.		1,600	.005			.09		.09	.15	
0710	Wood, T&G, heavy		800	.010			.18		.18	.30	
0720	Walls, wash										
0730	Drywall, gypsum, plaster	1 Pord	3,200	.002	S.F.		.04		.04	.07	
0740	Wood, T&G		3,200	.002			.04		.04	.07	
0750	Masonry, brick & block, smooth		2,800	.003			.05		.05	.09	
0760	Masonry, brick & block, coarse		2,000	.004			.07		.07	.12	

Division Notes

	CREW	DAILY OUTPUT	LABOR-HOURS	UNIT	2000 BARE COSTS MAT.	LABOR	EQUIP.	TOTAL	TOTAL INCL O&P

Division 10 Specialties

stimating Tips

:neral

The items in this division are usually priced per square foot or each.

Many items in Division 10 require some type of support system that is not usually furnished with the item. Examples of these systems include blocking for the attachment of grab bars and support angles for ceiling hung toilet partitions. The required blocking or supports must be added to the estimate in the appropriate division.

Some items in Division 10, such as lockers, may require assembly before installation. Verify the amount of assembly required. Assembly can often exceed installation time.

10100 Visual Display Boards

- Toilet partitions are priced by the stall. A stall consists of a side wall, pilaster and door with hardware. Toilet tissue holders and grab bars are extra.

10600 Partitions

- The required acoustical rating of a folding partition can have a significant impact on costs. Verify the sound transmission coefficient rating of the panel priced to the specification requirements.

Reference Numbers

Reference numbers are shown in bold squares at the beginning of some major classifications. These numbers refer to related items in the Reference Section. The reference information may be an estimating procedure, an alternate pricing method or technical information.

Note: Not all subdivisions listed here necessarily appear in this publication.

10150 | Compartments & Cubicles

10185 | Shower/Dressing Compartments

			CREW	DAILY OUTPUT	LABOR-HOURS	UNIT	2000 BARE COSTS MAT.	LABOR	EQUIP.	TOTAL	TOTAL INCL O&P
100	0010	**PARTITIONS, SHOWER** Floor mounted, no plumbing									
	0100	Cabinet, incl. base, no door, painted steel, 1″ thick walls	2 Shee	5	3.200	Ea.	600	69		669	775
	0300	With door, fiberglass		4.50	3.556		490	76.50		566.50	670
	0600	Galvanized and painted steel, 1″ thick walls		5	3.200		635	69		704	815
	0800	Stall, 1″ thick wall, no base, enameled steel		5	3.200		610	69		679	790
	1500	Circular fiberglass, cabinet 36″ diameter,		4	4		505	86		591	700
	1700	One piece, 36″ diameter, less door		4	4		430	86		516	620
	1800	With door		3.50	4.571		695	98.50		793.50	930
	2400	Glass stalls, with doors, no receptors, chrome on brass		3	5.333		1,050	115		1,165	1,350
	2700	Anodized aluminum	↓	4	4		720	86		806	940
	3200	Receptors, precast terrazzo, 32″ x 32″	2 Marb	14	1.143		266	22.50		288.50	330
	3300	48″ x 34″		9.50	1.684		385	33.50		418.50	475
	3500	Plastic, simulated terrazzo receptor, 32″ x 32″		14	1.143		85.50	22.50		108	133
	3600	32″ x 48″		12	1.333		125	26.50		151.50	183
	3800	Precast concrete, colors, 32″ x 32″		14	1.143		152	22.50		174.50	206
	3900	48″ x 48″	↓	8	2		187	39.50		226.50	273
	4100	Shower doors, economy plastic, 24″ wide	1 Shee	9	.889		90	19.10		109.10	132
	4200	Tempered glass door, economy		8	1		174	21.50		195.50	228
	4400	Folding, tempered glass, aluminum frame		6	1.333		213	28.50		241.50	283
	4700	Deluxe, tempered glass, chrome on brass frame, minimum		8	1		220	21.50		241.50	279
	4800	Maximum		1	8		640	172		812	990
	4850	On anodized aluminum frame, minimum		2	4		103	86		189	258
	4900	Maximum	↓	1	8	↓	350	172		522	675
	5100	Shower enclosure, tempered glass, anodized alum. frame									
	5120	2 panel & door, corner unit, 32″ x 32″	1 Shee	2	4	Ea.	350	86		436	530
	5140	Neo-angle corner unit, 16″ x 24″ x 16″	″	2	4		610	86		696	820
	5200	Shower surround, 3 wall, polypropylene, 32″ x 32″	1 Carp	4	2		258	39.50		297.50	350
	5220	PVC, 32″ x 32″		4	2		226	39.50		265.50	315
	5240	Fiberglass		4	2		271	39.50		310.50	365
	5250	2 wall, polypropylene, 32″ x 32″		4	2		187	39.50		226.50	274
	5270	PVC		4	2		226	39.50		265.50	315
	5290	Fiberglass	↓	4	2		251	39.50		290.50	345
	5300	Tub doors, tempered glass & frame, minimum	1 Shee	8	1		148	21.50		169.50	200
	5400	Maximum		6	1.333		340	28.50		368.50	425
	5600	Chrome plated, brass frame, minimum		8	1		193	21.50		214.50	250
	5700	Maximum		6	1.333		385	28.50		413.50	475
	5900	Tub/shower enclosure, temp. glass, alum. frame, minimum		2	4		264	86		350	435
	6200	Maximum		1.50	5.333		515	115		630	760
	6500	On chrome-plated brass frame, minimum		2	4		365	86		451	550
	6600	Maximum	↓	1.50	5.333		755	115		870	1,025
	6800	Tub surround, 3 wall, polypropylene	1 Carp	4	2		148	39.50		187.50	231
	6900	PVC		4	2		226	39.50		265.50	315
	7000	Fiberglass, minimum		4	2		251	39.50		290.50	345
	7100	Maximum	↓	3	2.667	↓	430	52.50		482.50	565

10200 | Louvers & Vents

10210 | Wall Louvers

			CREW	DAILY OUTPUT	LABOR-HOURS	UNIT	2000 BARE COSTS MAT.	LABOR	EQUIP.	TOTAL	TOTAL INCL O&P
800	0010	**LOUVERS** Aluminum with screen, residential, 8″ x 8″	1 Carp	38	.211	Ea.	6.95	4.15		11.10	14.70
	0100	12″ x 12″	↓	38	.211	↓	8.75	4.15		12.90	16.70

10200 | Louvers & Vents

10210 | Wall Louvers		CREW	DAILY OUTPUT	LABOR-HOURS	UNIT	2000 BARE COSTS MAT.	LABOR	EQUIP.	TOTAL	TOTAL INCL O&P	
0200	12″ x 18″	1 Carp	35	.229	Ea.	12.70	4.50		17.20	21.50	800
0250	14″ x 24″		30	.267		16.35	5.25		21.60	27	
0300	18″ x 24″		27	.296		18.85	5.85		24.70	31	
0500	24″ x 30″		24	.333		27.50	6.55		34.05	42	
0700	Triangle, adjustable, small		20	.400		23	7.90		30.90	38.50	
0800	Large		15	.533		61.50	10.50		72	85.50	
2100	Midget, aluminum, 3/4″ deep, 1″ diameter		85	.094		.56	1.85		2.41	3.80	
2150	3″ diameter		60	.133		1.32	2.63		3.95	5.95	
2200	4″ diameter		50	.160		1.66	3.15		4.81	7.25	
2250	6″ diameter		30	.267		2.25	5.25		7.50	11.50	
2300	Ridge vent strip, mill finish	1 Shee	155	.052	L.F.	1.95	1.11		3.06	4.02	
2400	Under eaves vent, aluminum, mill finish, 16″ x 4″	1 Carp	48	.167	Ea.	1.49	3.28		4.77	7.30	
2500	16″ x 8″		48	.167		1.79	3.28		5.07	7.60	
7000	Vinyl gable vent, 8″ x 8″		38	.211		9.50	4.15		13.65	17.55	
7020	12″ x 12″		38	.211		17	4.15		21.15	26	
7080	12″ x 18″		35	.229		22	4.50		26.50	31.50	
7200	18″ x 24″		30	.267		26	5.25		31.25	37.50	

10300 | Fireplaces & Stoves

10305 | Manufactured Fireplaces		CREW	DAILY OUTPUT	LABOR-HOURS	UNIT	2000 BARE COSTS MAT.	LABOR	EQUIP.	TOTAL	TOTAL INCL O&P	
0010	**FIREPLACE, PREFABRICATED** Free standing or wall hung										100
0100	with hood & screen, minimum	1 Carp	1.30	6.154	Ea.	930	121		1,051	1,225	
0150	Average		1	8		1,125	158		1,283	1,500	
0200	Maximum		.90	8.889		2,675	175		2,850	3,250	
0500	Chimney dbl. wall, all stainless, over 8′-6″, 7″ diam., add		33	.242	V.L.F.	31.50	4.78		36.28	42.50	
0600	10″ diameter, add		32	.250		45	4.93		49.93	58	
0700	12″ diameter, add		31	.258		58.50	5.10		63.60	73	
0800	14″ diameter, add		30	.267		74	5.25		79.25	90.50	
1000	Simulated brick chimney top, 4′ high, 16″ x 16″		10	.800	Ea.	160	15.75		175.75	203	
1100	24″ x 24″		7	1.143	″	298	22.50		320.50	370	
1500	Simulated logs, gas fired, 40,000 BTU, 2′ long, minimum		7	1.143	Set	415	22.50		437.50	500	
1600	Maximum		6	1.333		605	26.50		631.50	715	
1700	Electric, 1,500 BTU, 1′-6″ long, minimum		7	1.143		110	22.50		132.50	161	
1800	11,500 BTU, maximum		6	1.333		239	26.50		265.50	310	
2000	Fireplace, built-in, 36″ hearth, radiant		1.30	6.154	Ea.	480	121		601	735	
2100	Recirculating, small fan		1	8		685	158		843	1,025	
2150	Large fan		.90	8.889		1,275	175		1,450	1,700	
2200	42″ hearth, radiant		1.20	6.667		610	131		741	895	
2300	Recirculating, small fan		.90	8.889		800	175		975	1,175	
2350	Large fan		.80	10		1,525	197		1,722	2,025	
2400	48″ hearth, radiant		1.10	7.273		1,125	143		1,268	1,500	
2500	Recirculating, small fan		.80	10		1,400	197		1,597	1,900	
2550	Large fan		.70	11.429		2,175	225		2,400	2,775	
3000	See through, including doors		.80	10		1,800	197		1,997	2,325	
3200	Corner (2 wall)		1	8		890	158		1,048	1,250	

10300 | Fireplaces & Stoves

10310 | Fireplace Specialties & Accessories

		10310 Fireplace Specialties & Accessories	CREW	DAILY OUTPUT	LABOR-HOURS	UNIT	2000 BARE COSTS MAT.	LABOR	EQUIP.	TOTAL	TOTAL INCL O&P
100	0010	**FIREPLACE ACCESSORIES** Chimney screens, galv., 13" x 13" flue	1 Bric	8	1	Ea.	47.50	20		67.50	86.50
	0050	Galv., 24" x 24" flue		5	1.600		125	32		157	191
	0200	Stainless steel, 13" x 13" flue		8	1		264	20		284	325
	0250	20" x 20" flue		5	1.600		360	32		392	450
	0400	Cleanout doors and frames, cast iron, 8" x 8"		12	.667		17.50	13.35		30.85	42
	0450	12" x 12"		10	.800		33.50	16		49.50	63.50
	0500	18" x 24"		8	1		105	20		125	149
	0550	Cast iron frame, steel door, 24" x 30"		5	1.600		227	32		259	305
	0800	Damper, rotary control, steel, 30" opening		6	1.333		63.50	26.50		90	115
	0850	Cast iron, 30" opening		6	1.333		71	26.50		97.50	123
	1200	Steel plate, poker control, 60" opening		8	1		225	20		245	281
	1250	84" opening, special opening		5	1.600		410	32		442	505
	1400	"Universal" type, chain operated, 32" x 20" opening		8	1		153	20		173	203
	1450	48" x 24" opening		5	1.600		271	32		303	350
	1600	Dutch Oven door and frame, cast iron, 12" x 15" opening		13	.615		92	12.30		104.30	122
	1650	Copper plated, 12" x 15" opening		13	.615		177	12.30		189.30	215
	1800	Fireplace forms, no accessories, 32" opening		3	2.667		465	53.50		518.50	605
	1900	36" opening		2.50	3.200		565	64		629	730
	2000	40" opening		2	4		680	80		760	880
	2100	78" opening		1.50	5.333		970	107		1,077	1,250
	2400	Squirrel and bird screens, galvanized, 8" x 8" flue		16	.500		44.50	10		54.50	66
	2450	13" x 13" flue	↓	12	.667	↓	52	13.35		65.35	80

10320 | Stoves

		10320 Stoves	CREW	DAILY OUTPUT	LABOR-HOURS	UNIT	MAT.	LABOR	EQUIP.	TOTAL	TOTAL INCL O&P
100	0010	**WOODBURNING STOVES** Cast iron, minimum	2 Carp	1.30	12.308	Ea.	765	242		1,007	1,250
	0020	Average		1	16		1,225	315		1,540	1,900
	0030	Maximum	↓	.80	20		2,075	395		2,470	2,950
	0050	For gas log lighter, add				↓	36			36	39.50

10340 | Manufactured Exterior Specialties

10342 | Cupolas

		10342 Cupolas	CREW	DAILY OUTPUT	LABOR-HOURS	UNIT	2000 BARE COSTS MAT.	LABOR	EQUIP.	TOTAL	TOTAL INCL O&P
100	0010	**CUPOLA** Stock units, pine, painted, 18" sq., 28" high, alum. roof	1 Carp	4.10	1.951	Ea.	99	38.50		137.50	175
	0100	Copper roof		3.80	2.105		141	41.50		182.50	226
	0300	23" square, 33" high, aluminum roof		3.70	2.162		232	42.50		274.50	330
	0400	Copper roof		3.30	2.424		233	48		281	340
	0600	30" square, 37" high, aluminum roof		3.70	2.162		355	42.50		397.50	465
	0700	Copper roof		3.30	2.424		365	48		413	480
	0900	Hexagonal, 31" wide, 46" high, copper roof		4	2		470	39.50		509.50	590
	1000	36" wide, 50" high, copper roof	↓	3.50	2.286		560	45		605	695
	1200	For deluxe stock units, add to above					25%				
	1400	For custom built units, add to above					50%	50%			
	1600	Fiberglass, 5'-0" base, 63" high minimum	F-3	6	6.667		2,325	120	75.50	2,520.50	2,825
	1650	Maximum		4	10		2,725	179	113	3,017	3,400
	1700	6'-0" base, 63" high, minimum		5	8		3,625	143	90.50	3,858.50	4,325
	1750	Maximum	↓	3	13.333	↓	3,725	239	151	4,115	4,675

10344 | Weathervanes

		10344 Weathervanes	CREW	DAILY OUTPUT	LABOR-HOURS	UNIT	MAT.	LABOR	EQUIP.	TOTAL	TOTAL INCL O&P
800	0005	**WEATHERVANES**									
	0010	Residential types, minimum	1 Carp	8	1	Ea.	40	19.70		59.70	78

10340 | Manufactured Exterior Specialties

10344	Weathervanes	CREW	DAILY OUTPUT	LABOR-HOURS	UNIT	2000 BARE COSTS MAT.	LABOR	EQUIP.	TOTAL	TOTAL INCL O&P	
0100	Maximum	1 Carp	2	4	Ea.	950	79		1,029	1,175	800

10350 | Flagpoles

10355	Flagpoles	CREW	DAILY OUTPUT	LABOR-HOURS	UNIT	2000 BARE COSTS MAT.	LABOR	EQUIP.	TOTAL	TOTAL INCL O&P	
0010	**FLAGPOLE**, Ground set										400
0050	Not including base or foundation										
0100	Aluminum, tapered, ground set 20′ high	K-1	2	8	Ea.	600	142	90	832	1,000	
0200	25′ high	↓	1.70	9.412	↓	725	167	106	998	1,200	
0300	30′ high	↓	1.50	10.667	↓	995	190	120	1,305	1,550	
0500	40′ high	↓	1.20	13.333	↓	1,775	237	150	2,162	2,550	

10520 | Fire Protection Specialties

10525	Fire Extinguishers	CREW	DAILY OUTPUT	LABOR-HOURS	UNIT	2000 BARE COSTS MAT.	LABOR	EQUIP.	TOTAL	TOTAL INCL O&P	
0010	**FIRE EXTINGUISHERS**										300
0120	CO2, portable with swivel horn, 5 lb.				Ea.	101			101	111	
0140	With hose and "H" horn, 10 lb.				"	150			150	165	
1000	Dry chemical, pressurized										
1040	Standard type, portable, painted, 2-1/2 lb.				Ea.	27.50			27.50	30.50	
1080	10 lb.				↓	67			67	73.50	
1100	20 lb.				↓	90			90	99	
1120	30 lb.				↓	145			145	160	
2000	ABC all purpose type, portable, 2-1/2 lb.				↓	27.50			27.50	30.50	
2080	9-1/2 lb.				↓	65			65	71.50	

10530 | Protective Covers

10535	Awning & Canopies	CREW	DAILY OUTPUT	LABOR-HOURS	UNIT	2000 BARE COSTS MAT.	LABOR	EQUIP.	TOTAL	TOTAL INCL O&P	
0010	**CANOPIES, RESIDENTIAL** Prefabricated										100
0500	Carport, free standing, baked enamel, alum., .032″, 40 psf										
0520	16′ x 8′, 4 posts	2 Carp	3	5.333	Ea.	2,375	105		2,480	2,800	
0600	20′ x 10′, 6 posts	"	2	8	↓	2,475	158		2,633	3,000	
1000	Door canopies, extruded alum., .032″, 42″ projection, 4′ wide	1 Carp	8	1	↓	315	19.70		334.70	385	
1020	6′ wide	"	6	1.333	↓	405	26.50		431.50	490	
1040	8′ wide	2 Carp	9	1.778	↓	530	35		565	645	
1060	10′ wide	↓	7	2.286	↓	630	45		675	770	

10530 | Protective Covers

	10535 | Awning & Canopies		CREW	DAILY OUTPUT	LABOR-HOURS	UNIT	2000 BARE COSTS MAT.	LABOR	EQUIP.	TOTAL	TOTAL INCL O&P
100	1080	12′ wide	2 Carp	5	3.200	Ea.	750	63		813	930
	1200	54″ projection, 4′ wide	1 Carp	8	1		405	19.70		424.70	480
	1220	6′ wide	″	6	1.333		535	26.50		561.50	630
	1240	8′ wide	2 Carp	9	1.778		700	35		735	835
	1260	10′ wide		7	2.286		825	45		870	980
	1280	12′ wide	↓	5	3.200		950	63		1,013	1,150
	1300	Painted, add					20%				
	1310	Bronze anodized, add					50%				
	3000	Window awnings, aluminum, window 3′ high, 4′ wide	1 Carp	10	.800		180	15.75		195.75	225
	3020	6′ wide	″	8	1		238	19.70		257.70	296
	3040	9′ wide	2 Carp	9	1.778		325	35		360	420
	3060	12′ wide	″	5	3.200		435	63		498	590
	3100	Window, 4′ high, 4′ wide	1 Carp	10	.800		231	15.75		246.75	281
	3120	6′ wide	″	8	1		315	19.70		334.70	380
	3140	9′ wide	2 Carp	9	1.778		430	35		465	530
	3160	12′ wide	″	5	3.200		580	63		643	750
	3200	Window, 6′ high, 4′ wide	1 Carp	10	.800		325	15.75		340.75	380
	3220	6′ wide	″	8	1		445	19.70		464.70	525
	3240	9′ wide	2 Carp	9	1.778		590	35		625	710
	3260	12′ wide	″	5	3.200		760	63		823	945
	3400	Roll-up aluminum, 2′-6″ wide	1 Carp	14	.571		87.50	11.25		98.75	115
	3420	3′ wide		12	.667		105	13.15		118.15	138
	3440	4′ wide		10	.800		135	15.75		150.75	176
	3460	6′ wide	↓	8	1		167	19.70		186.70	217
	3480	9′ wide	2 Carp	9	1.778		242	35		277	325
	3500	12′ wide	″	5	3.200	↓	297	63		360	435
	3600	Window awnings, canvas, 24″ drop, 3′ wide	1 Carp	30	.267	L.F.	56	5.25		61.25	70.50
	3620	4′ wide		40	.200		52.50	3.94		56.44	64.50
	3700	30″ drop, 3′ wide		30	.267		65.50	5.25		70.75	81
	3720	4′ wide		40	.200		58.50	3.94		62.44	71
	3740	5′ wide		45	.178		53.50	3.50		57	65
	3760	6′ wide		48	.167		50	3.28		53.28	60.50
	3780	8′ wide		48	.167		41.50	3.28		44.78	51.50
	3800	10′ wide	↓	50	.160	↓	38	3.15		41.15	47.50

10550 | Postal Specialties

	10555 | Mail Delivery Systems		CREW	DAILY OUTPUT	LABOR-HOURS	UNIT	2000 BARE COSTS MAT.	LABOR	EQUIP.	TOTAL	TOTAL INCL O&P
600	0011	MAIL BOXES									
	1900	Letter slot, residential	1 Carp	20	.400	Ea.	50.50	7.90		58.40	69
	2400	Residential, galv. steel, small 20″ x 7″ x 9″	1 Clab	16	.500		91	7.20		98.20	112
	2410	With galv. steel post, 54″ long		6	1.333		217	19.25		236.25	272
	2420	Large, 24″ x 12″ x 15″		16	.500		152	7.20		159.20	179
	2430	With galv. steel post, 54″ long		6	1.333		278	19.25		297.25	340
	2440	Decorative, polyethylene, 22″ x 10″ x 10″		16	.500		31	7.20		38.20	46.50
	2450	With alum. post, decorative, 54″ long	↓	6	1.333	↓	49	19.25		68.25	86.50

Important: See the Reference Section for critical supporting data - Reference Nos., Crews, & Location Factor

10670 | Storage Shelving

10674 Storage Shelving		CREW	DAILY OUTPUT	LABOR-HOURS	UNIT	2000 BARE COSTS MAT.	LABOR	EQUIP.	TOTAL	TOTAL INCL O&P	
0010	**SHELVING** Metal, industrial, cross-braced, 3′ wide, 12″ deep	1 Sswk	175	.046	SF Shlf	8.80	.97		9.77	11.60	500
0100	24″ deep		330	.024		6.85	.52		7.37	8.55	
2200	Wide span, 1600 lb. capacity per shelf, 6′ wide, 24″ deep		380	.021		6.55	.45		7	8.10	
2400	36″ deep	↓	440	.018	↓	5.80	.39		6.19	7.10	
3000	Residential, vinyl covered wire, wardrobe, 12″ deep	1 Carp	195	.041	L.F.	2.72	.81		3.53	4.37	
3100	16″ deep		195	.041		2.85	.81		3.66	4.52	
3200	Standard, 6″ deep		195	.041		2.52	.81		3.33	4.15	
3300	9″ deep		195	.041		2.52	.81		3.33	4.15	
3400	12″ deep		195	.041		2.52	.81		3.33	4.15	
3500	16″ deep		195	.041		2.58	.81		3.39	4.22	
3600	20″ deep		195	.041	↓	2.63	.81		3.44	4.27	
3700	Support bracket	↓	80	.100	Ea.	1.32	1.97		3.29	4.83	

10800 | Toilet/Bath/Laundry Accessories

10820 Bath Accessories		CREW	DAILY OUTPUT	LABOR-HOURS	UNIT	2000 BARE COSTS MAT.	LABOR	EQUIP.	TOTAL	TOTAL INCL O&P	
0010	**BATH ACCESSORIES**										100
0200	Curtain rod, stainless steel, 5′ long, 1″ diameter	1 Carp	13	.615	Ea.	20.50	12.10		32.60	43.50	
0300	1-1/4″ diameter		13	.615		24.50	12.10		36.60	48	
0800	Grab bar, straight, 1-1/4″ diameter, stainless steel, 18″ long ♿		24	.333		35.50	6.55		42.05	50.50	
1100	36″ long		20	.400		45	7.90		52.90	63	
3000	Mirror, with stainless steel 3/4″ square frame, 18″ x 24″		20	.400		61	7.90		68.90	80.50	
3100	36″ x 24″		15	.533		108	10.50		118.50	137	
3300	72″ x 24″		6	1.333		196	26.50		222.50	261	
4300	Robe hook, single, regular		36	.222		9.60	4.38		13.98	18.05	
4400	Heavy duty, concealed mounting		36	.222		14.20	4.38		18.58	23	
6400	Towel bar, stainless steel, 18″ long		23	.348		30	6.85		36.85	45	
6500	30″ long		21	.381		34.50	7.50		42	51	
7400	Tumbler holder, tumbler only		30	.267		20.50	5.25		25.75	31.50	
7500	Soap, tumbler & toothbrush	↓	30	.267	↓	20.50	5.25		25.75	31.50	
0010	**MEDICINE CABINETS** With mirror, st. st. frame, 16″ x 22″, unlighted	1 Carp	14	.571	Ea.	65	11.25		76.25	91	400
0100	Wood frame		14	.571		95	11.25		106.25	124	
0300	Sliding mirror doors, 20″ x 16″ x 4-3/4″, unlighted		7	1.143		85	22.50		107.50	132	
0400	24″ x 19″ x 8-1/2″, lighted		5	1.600		143	31.50		174.50	211	
0600	Triple door, 30″ x 32″, unlighted, plywood body		7	1.143		211	22.50		233.50	271	
0700	Steel body		7	1.143		278	22.50		300.50	345	
0900	Oak door, wood body, beveled mirror, single door		7	1.143		124	22.50		146.50	175	
1000	Double door	↓	6	1.333	↓	315	26.50		341.50	390	

Division Notes

	CREW	DAILY OUTPUT	LABOR-HOURS	UNIT	2000 BARE COSTS MAT.	LABOR	EQUIP.	TOTAL	TOTAL INCL O&P

Division 11 Equipment

stimating Tips

eneral

The items in this division are usually priced per square foot or each. Many of these items are purchased by the owner for installation by the contractor. Check the specifications for responsibilities, and include time for receiving, storage, installation and mechanical and electrical hook-ups in the appropriate divisions.

- Many items in Division 11 require some type of support system that is not usually furnished with the item. Examples of these systems include blocking for the attachment of casework and support angles for ceiling hung projection screens. The required blocking or supports must be added to the estimate in the appropriate division.
- Some items in Division 11 may require assembly or electrical hook-ups. Verify the amount of assembly required or the need for a hard electrical connection and add the appropriate costs.

Reference Numbers

Reference numbers are shown in bold squares at the beginning of some major classifications. These numbers refer to related items in the Reference Section. The reference information may be an estimating procedure, an alternate pricing method or technical information.

Note: Not all subdivisions listed here necessarily appear in this publication.

11010 | Maintenance Equipment

	11013 | Floor/Wall Cleaning Equipment		CREW	DAILY OUTPUT	LABOR-HOURS	UNIT	2000 BARE COSTS MAT.	LABOR	EQUIP.	TOTAL	TOTAL INCL O&P
800	0010	**VACUUM CLEANING**									
	0020	Central, 3 inlet, residential	1 Skwk	.90	8.889	Total	685	176		861	1,050
	0400	5 inlet system, residential	↓	.50	16	↓	840	315		1,155	1,450
	0600	7 inlet system, commercial	↓	.40	20	↓	1,050	395		1,445	1,825
	0800	9 inlet system, residential	↓	.30	26.667	↓	1,200	525		1,725	2,225
	4010	Rule of thumb: First 1200 S.F., installed				↓					1,044
	4020	For each additional S.F., add				S.F.					.17

11400 | Food Service Equipment

	11405 | Food Storage Equipment		CREW	DAILY OUTPUT	LABOR-HOURS	UNIT	2000 BARE COSTS MAT.	LABOR	EQUIP.	TOTAL	TOTAL INCL O&P
800	0010	**WINE CELLAR**, refrigerated, Redwood interior, carpeted, walk-in type									
	0020	6'-8" high, including racks									
	0200	80" W x 48" D for 900 bottles	2 Carp	1.50	10.667	Ea.	2,800	210		3,010	3,425
	0250	80" W x 72" D for 1300 bottles	↓	1.33	12.030	↓	3,700	237		3,937	4,475
	0300	80" W x 94" D for 1900 bottles	↓	1.17	13.675	↓	4,800	269		5,069	5,725

11 EQUIPMENT

11450 | Residential Equipment

	11454 | Residential Appliances		CREW	DAILY OUTPUT	LABOR-HOURS	UNIT	2000 BARE COSTS MAT.	LABOR	EQUIP.	TOTAL	TOTAL INCL O&P
500	0010	**RESIDENTIAL APPLIANCES**									
	0020	Cooking range, 30" free standing, 1 oven, minimum	2 Clab	10	1.600	Ea.	259	23		282	325
	0050	Maximum	↓	4	4	↓	1,175	58		1,233	1,400
	0150	2 oven, minimum	↓	10	1.600	↓	1,125	23		1,148	1,275
	0200	Maximum	↓	10	1.600	↓	1,500	23		1,523	1,700
	0350	Built-in, 30" wide, 1 oven, minimum	1 Elec	6	1.333	↓	259	29.50		288.50	335
	0400	Maximum	2 Carp	2	8	↓	795	158		953	1,150
	0500	2 oven, conventional, minimum	↓	4	4	↓	880	79		959	1,100
	0550	1 conventional, 1 microwave, maximum	↓	2	8	↓	1,550	158		1,708	1,975
	0700	Free-standing, 1 oven, 21" wide range, minimum	2 Clab	10	1.600	↓	281	23		304	350
	0750	21" wide, maximum	"	4	4	↓	465	58		523	615
	0900	Counter top cook tops, 4 burner, standard, minimum	1 Elec	6	1.333	↓	172	29.50		201.50	237
	0950	Maximum	↓	3	2.667	↓	430	59		489	570
	1050	As above, but with grille and griddle attachment, minimum	↓	6	1.333	↓	400	29.50		429.50	490
	1100	Maximum	↓	3	2.667	↓	515	59		574	660
	1250	Microwave oven, minimum	↓	4	2	↓	87.50	44		131.50	169
	1300	Maximum	↓	2	4	↓	365	88.50		453.50	550
	1500	Combination range, refrigerator and sink, 30" wide, minimum	L-1	2	5	↓	635	110		745	880
	1550	Maximum	↓	1	10	↓	1,275	220		1,495	1,775
	1570	60" wide, average	↓	1.40	7.143	↓	2,100	157		2,257	2,575
	1590	72" wide, average	↓	1.20	8.333	↓	2,375	183		2,558	2,925
	1750	Compactor, residential size, 4 to 1 compaction, minimum	1 Carp	5	1.600	↓	335	31.50		366.50	425

Important: See the Reference Section for critical supporting data - Reference Nos., Crews, & Location Facto

11454	Residential Appliances	CREW	DAILY OUTPUT	LABOR-HOURS	UNIT	2000 BARE COSTS MAT.	LABOR	EQUIP.	TOTAL	TOTAL INCL O&P	
1800	Maximum	1 Carp	3	2.667	Ea.	370	52.50		422.50	495	500
2000	Deep freeze, 15 to 23 C.F., minimum	2 Clab	10	1.600		370	23		393	445	
2050	Maximum		5	3.200		510	46		556	640	
2200	30 C.F., minimum		8	2		730	29		759	855	
2250	Maximum		3	5.333		850	77		927	1,075	
2450	Dehumidifier, portable, automatic, 15 pint					185			185	204	
2550	40 pint					220			220	242	
2750	Dishwasher, built-in, 2 cycles, minimum	L-1	4	2.500		247	55		302	360	
2800	Maximum		2	5		258	110		368	465	
2950	4 or more cycles, minimum		4	2.500		239	55		294	355	
2960	Average		4	2.500		320	55		375	440	
3000	Maximum		2	5		495	110		605	725	
3200	Dryer, automatic, minimum	L-2	3	5.333		251	92		343	435	
3250	Maximum	″	2	8		545	138		683	835	
3300	Garbage disposer, sink type, minimum	L-1	10	1		37.50	22		59.50	77.50	
3350	Maximum	″	10	1		174	22		196	228	
3550	Heater, electric, built-in, 1250 watt, ceiling type, minimum	1 Elec	4	2		68	44		112	148	
3600	Maximum		3	2.667		111	59		170	219	
3700	Wall type, minimum		4	2		96.50	44		140.50	179	
3750	Maximum		3	2.667		128	59		187	238	
3900	1500 watt wall type, with blower		4	2		119	44		163	204	
3950	3000 watt		3	2.667		244	59		303	365	
4150	Hood for range, 2 speed, vented, 30″ wide, minimum	L-3	5	2		35	40.50		75.50	107	
4200	Maximum		3	3.333		430	67.50		497.50	590	
4300	42″ wide, minimum		5	2		164	40.50		204.50	250	
4330	Custom		5	2		575	40.50		615.50	700	
4350	Maximum		3	3.333		700	67.50		767.50	885	
4500	For ventless hood, 2 speed, add					13.40			13.40	14.75	
4650	For vented 1 speed, deduct from maximum					26			26	28.50	
4850	Humidifier, portable, 8 gallons per day					92.50			92.50	102	
5000	15 gallons per day					149			149	164	
5200	Icemaker, automatic, 20 lb. per day	1 Plum	7	1.143		505	25		530	595	
5350	51 lb. per day	″	2	4		785	88		873	1,000	
5380	Oven, built in, standard	1 Elec	4	2		360	44		404	470	
5390	Deluxe	″	2	4		815	88.50		903.50	1,050	
5500	Refrigerator, no frost, 10 C.F. to 12 C.F. minimum	2 Clab	10	1.600		420	23		443	505	
5600	Maximum		6	2.667		640	38.50		678.50	770	
5750	14 C.F. to 16 C.F., minimum		9	1.778		440	25.50		465.50	530	
5800	Maximum		5	3.200		520	46		566	650	
5950	18 C.F. to 20 C.F., minimum		8	2		485	29		514	585	
6000	Maximum		4	4		805	58		863	985	
6150	21 C.F. to 29 C.F., minimum		7	2.286		790	33		823	925	
6200	Maximum		3	5.333		2,600	77		2,677	2,975	
6400	Sump pump cellar drainer, pedestal, 1/3 H.P., molded PVC base	1 Plum	3	2.667		86.50	58.50		145	192	
6450	Solid brass	″	2	4		179	88		267	340	
6460	Sump pump, see also division 15440-940										
6650	Washing machine, automatic, minimum	1 Plum	3	2.667	Ea.	320	58.50		378.50	445	
6700	Maximum	″	1	8		620	176		796	970	
6900	Water heater, electric, glass lined, 30 gallon, minimum	L-1	5	2		215	44		259	310	
6950	Maximum		3	3.333		299	73.50		372.50	450	
7100	80 gallon, minimum		2	5		415	110		525	635	
7150	Maximum		1	10		575	220		795	995	
7180	Water heater, gas, glass lined, 30 gallon, minimum	2 Plum	5	3.200		284	70		354	425	
7220	Maximum		3	5.333		395	117		512	630	
7260	50 gallon, minimum		2.50	6.400		515	140		655	800	
7300	Maximum		1.50	10.667		720	234		954	1,175	

	11454 | Residential Appliances		CREW	DAILY OUTPUT	LABOR-HOURS	UNIT	2000 BARE COSTS MAT.	LABOR	EQUIP.	TOTAL	TOTAL INCL O&P
500	7310	Water heater, see also division 15480-200									
	7350	Water softener, automatic, to 30 grains per gallon	2 Plum	5	3.200	Ea.	420	70		490	575
	7400	To 100 grains per gallon	"	4	4		510	88		598	710
	7450	Vent kits for dryers	1 Carp	10	.800	↓	9.55	15.75		25.30	37.50
550	0010	**DISAPPEARING STAIRWAY** No trim included									
	0020	One piece, yellow pine, 8′-0″ ceiling	2 Carp	4	4	Ea.	850	79		929	1,075
	0030	9′-0″ ceiling		4	4		875	79		954	1,100
	0040	10′-0″ ceiling		3	5.333		920	105		1,025	1,175
	0050	11′-0″ ceiling		3	5.333		1,075	105		1,180	1,375
	0060	12′-0″ ceiling	↓	3	5.333		1,125	105		1,230	1,425
	0100	Custom grade, pine, 8′-6″ ceiling, minimum	1 Carp	4	2		100	39.50		139.50	178
	0150	Average		3.50	2.286		105	45		150	193
	0200	Maximum		3	2.667		185	52.50		237.50	294
	0500	Heavy duty, pivoted, from 7′-7″ to 12′-10″ floor to floor		3	2.667		315	52.50		367.50	440
	0600	16′-0″ ceiling		2	4		1,025	79		1,104	1,275
	0800	Economy folding, pine, 8′-6″ ceiling		4	2		83	39.50		122.50	159
	0900	9′-6″ ceiling	↓	4	2		93.50	39.50		133	171
	1000	Fire escape, galvanized steel, 8′-0″ to 10′-4″ ceiling	2 Carp	1	16		1,125	315		1,440	1,775
	1010	10′-6″ to 13′-6″ ceiling		1	16		1,400	315		1,715	2,100
	1100	Automatic electric, aluminum, floor to floor height, 8′ to 9′	↓	1	16	↓	5,500	315		5,815	6,625

Division 12 Furnishings

'stimating Tips

eneral

The items in this division are usually priced per square foot or each. Most of these items are purchased by the owner and placed by the supplier. Do not assume the items in Division 12 will be purchased and installed by the supplier. Check the specifications for responsibilities and include receiving, storage, installation and mechanical and electrical hook-ups in the appropriate divisions.

- Some items in this division require some type of support system that is not usually furnished with the item. Examples of these systems include blocking for the attachment of casework and heavy drapery rods. The required blocking must be added to the estimate in the appropriate division.

Reference Numbers

Reference numbers are shown in bold squares at the beginning of some major classifications. These numbers refer to related items in the Reference Section. The reference information may be an estimating procedure, an alternate pricing method or technical information.

Note: Not all subdivisions listed here necessarily appear in this publication.

12300 | Manufactured Casework

	12310 | Metal Casework		Crew	Daily Output	Labor-Hours	Unit	2000 Bare Costs Mat.	Labor	Equip.	Total	Total Incl O&P
560	0010	IRONING CENTER									
	0020	Including cabinet, board & light, minimum	1 Carp	2	4	Ea.	240	79		319	400

12400 | Furnishings & Accessories

	12492 | Blinds and Shades		Crew	Daily Output	Labor-Hours	Unit	2000 Bare Costs Mat.	Labor	Equip.	Total	Total Incl O&P
100	0010	BLINDS, INTERIOR									
	0020	Horizontal, 1″ aluminum slats, solid color, stock	1 Carp	590	.014	S.F.	2.60	.27		2.87	3.32
	0090	Custom, minimum		590	.014		2.06	.27		2.33	2.73
	0100	Maximum		440	.018		6.50	.36		6.86	7.75
	0450	Stock, minimum		590	.014		2.66	.27		2.93	3.39
	0500	Maximum		440	.018		4.94	.36		5.30	6.05
	3000	Wood folding panels with movable louvers, 7″ x 20″ each		17	.471	Pr.	36	9.25		45.25	55.50
	3300	8″ x 28″ each		17	.471		52	9.25		61.25	73
	3450	9″ x 36″ each		17	.471		62	9.25		71.25	84
	3600	10″ x 40″ each		17	.471		70	9.25		79.25	93
	4000	Fixed louver type, stock units, 8″ x 20″ each		17	.471		54	9.25		63.25	75.50
	4150	10″ x 28″ each		17	.471		74	9.25		83.25	97.50
	4300	12″ x 36″ each		17	.471		94	9.25		103.25	119
	4450	18″ x 40″ each		17	.471		112	9.25		121.25	139
	5000	Insert panel type, stock, 7″ x 20″ each		17	.471		13.90	9.25		23.15	31
	5150	8″ x 28″ each		17	.471		25.50	9.25		34.75	44
	5300	9″ x 36″ each		17	.471		32.50	9.25		41.75	51.50
	5450	10″ x 40″ each		17	.471		35	9.25		44.25	54
	5600	Raised panel type, stock, 10″ x 24″ each		17	.471		93	9.25		102.25	118
	5650	12″ x 26″ each		17	.471		108	9.25		117.25	135
	5700	14″ x 30″ each		17	.471		122	9.25		131.25	150
	5750	16″ x 36″ each		17	.471		137	9.25		146.25	167
	6000	For custom built pine, add					22%				
	6500	For custom built hardwood blinds, add					42%				
600	0011	SHADES Basswood roll-up, stain finish, 3/8″ slats	1 Carp	300	.027	S.F.	9	.53		9.53	10.80
	5011	Insulative shades		125	.064		7.55	1.26		8.81	10.45
	6011	Solar screening, fiberglass		85	.094		3.61	1.85		5.46	7.15
	8011	Interior insulative shutter									
	8111	Stock unit, 15″ x 60″	1 Carp	17	.471	Pr.	7.55	9.25		16.80	24

	12493 | Curtains and Drapes		Crew	Daily Output	Labor-Hours	Unit	Mat.	Labor	Equip.	Total	Total Incl O&P
200	0010	DRAPERY HARDWARE									
	0030	Standard traverse, per foot, minimum	1 Carp	59	.136	L.F.	1.40	2.67		4.07	6.10
	0100	Maximum		51	.157	"	7.50	3.09		10.59	13.55
	0200	Decorative traverse, 28″-48″, minimum		22	.364	Ea.	12	7.15		19.15	25.50
	0220	Maximum		21	.381		28	7.50		35.50	44
	0300	48″-84″, minimum		20	.400		16	7.90		23.90	31
	0320	Maximum		19	.421		45.50	8.30		53.80	64
	0400	66″-120″, minimum		18	.444		18	8.75		26.75	35
	0420	Maximum		17	.471		60	9.25		69.25	82
	0500	84″-156″, minimum		16	.500		20	9.85		29.85	39
	0520	Maximum		15	.533		67.50	10.50		78	92
	0600	130″-240″, minimum		14	.571		24	11.25		35.25	46

12493 | Curtains and Drapes		CREW	DAILY OUTPUT	LABOR-HOURS	UNIT	2000 BARE COSTS MAT.	LABOR	EQUIP.	TOTAL	TOTAL INCL O&P	
0620	Maximum	1 Carp	13	.615	Ea.	93	12.10		105.10	123	200
0700	Slide rings, each, minimum					.65			.65	.71	
0720	Maximum					1.40			1.40	1.54	
3000	Ripplefold, snap-a-pleat system, 3′ or less, minimum	1 Carp	15	.533		39.50	10.50		50	61	
3020	Maximum	"	14	.571	↓	56	11.25		67.25	81	
3200	Each additional foot, add, minimum				L.F.	2			2	2.20	
3220	Maximum				"	5.45			5.45	6	
4000	Traverse rods, adjustable, 28″ to 48″	1 Carp	22	.364	Ea.	15.20	7.15		22.35	29	
4020	48″ to 84″		20	.400		24.50	7.90		32.40	40.50	
4040	66″ to 120″		18	.444		26	8.75		34.75	43.50	
4060	84″ to 156″		16	.500		30	9.85		39.85	49.50	
4080	100″ to 180″		14	.571		29.50	11.25		40.75	51.50	
4100	228″ to 312″		13	.615		45.50	12.10		57.60	71	
4500	Curtain rod, 28″ to 48″, single		22	.364		3.91	7.15		11.06	16.55	
4510	Double		22	.364		6.65	7.15		13.80	19.60	
4520	48″ to 86″, single		20	.400		6.70	7.90		14.60	21	
4530	Double		20	.400		11.15	7.90		19.05	26	
4540	66″ to 120″, single		18	.444		11.20	8.75		19.95	27.50	
4550	Double	↓	18	.444		17.50	8.75		26.25	34.50	
4600	Valance, pinch pleated fabric, 12″ deep, up to 54″ long, minimum					28			28	31	
4610	Maximum					70			70	77	
4620	Up to 77″ long, minimum					43			43	47.50	
4630	Maximum					113			113	124	
5000	Stationary rods, first 2 feet				↓	6.95			6.95	7.65	
0010	**SAUNA** Prefabricated, incl. heater & controls, 7′ high, 6′ x 4′, C/C	L-7	2.20	11.818	Ea.	3,200	199		3,399	3,875	800
0050	6′ x 4′, C/P		2	13		2,525	219		2,744	3,150	
0400	6′ x 5′, C/C		2	13		3,200	219		3,419	3,900	
0450	6′ x 5′, C/P		2	13		2,875	219		3,094	3,525	
0600	6′ x 6′, C/C		1.80	14.444		3,750	244		3,994	4,550	
0650	6′ x 6′, C/P		1.80	14.444		3,000	244		3,244	3,725	
0800	6′ x 9′, C/C		1.60	16.250		4,675	274		4,949	5,625	
0850	6′ x 9′, C/P		1.60	16.250		4,675	274		4,949	5,625	
1000	8′ x 12′, C/C		1.10	23.636		6,300	400		6,700	7,600	
1050	8′ x 12′, C/P		1.10	23.636		5,275	400		5,675	6,475	
1400	8′ x 10′, C/C		1.20	21.667		5,400	365		5,765	6,575	
1450	8′ x 10′, C/P		1.20	21.667		4,725	365		5,090	5,825	
1600	10′ x 12′, C/C		1	26		7,200	440		7,640	8,675	
1650	10′ x 12′, C/P	↓	1	26		6,375	440		6,815	7,775	
1700	Door only, cedar, 2′x6′, with tempered insulated glass window	2 Carp	3.40	4.706		275	92.50		367.50	465	
1800	Prehung, incl. jambs, pulls & hardware	"	12	1.333		325	26.50		351.50	405	
2500	Heaters only (incl. above), wall mounted, to 200 C.F.					440			440	485	
2750	To 300 C.F.					465			465	515	
3000	Floor standing, to 720 C.F., 10,000 watts, w/controls	1 Elec	3	2.667		890	59		949	1,075	
3250	To 1,000 C.F., 16,000 watts	"	3	2.667	↓	1,575	59		1,634	1,825	
0010	**STEAM BATH** Heater, timer & head, single, to 140 C.F.	1 Plum	1.20	6.667	Ea.	740	146		886	1,050	940
0500	To 300 C.F.	"	1.10	7.273		800	160		960	1,150	
2700	Conversion unit for residential tub, including door				↓	2,225			2,225	2,450	

For information about Means Estimating Seminars, see yellow pages 11 and 12 in back of book

Division Notes

	CREW	DAILY OUTPUT	LABOR-HOURS	UNIT	2000 BARE COSTS MAT.	LABOR	EQUIP.	TOTAL	TOTAL INCL O&P

Division 13 Special Construction

Estimating Tips

General

- The items and systems in this division are usually estimated, purchased, supplied and installed as a unit by one or more subcontractors. The estimator must ensure that all parties are operating from the same set of specifications and assumptions and that all necessary items are estimated and will be provided. Many times the complex items and systems are covered but the more common ones such as excavation or a crane are overlooked for the very reason that everyone assumes nobody could miss them. The estimator should be the central focus and be able to ensure that all systems are complete.
- Another area where problems can develop in this division is at the interface between systems. The estimator must ensure, for instance, that anchor bolts, nuts and washers are estimated and included for the air-supported structures and pre-engineered buildings to be bolted to their foundations. Utility supply is a common area where essential items or pieces of equipment can be missed or overlooked due to the fact that each subcontractor may feel it is the others' responsibility. The estimator should also be aware of certain items which may be supplied as part of a package but installed by others, and ensure that the installing contractor's estimate includes the cost of installation. Conversely, the estimator must also ensure that items are not costed by two different subcontractors, resulting in an inflated overall estimate.

13120 Pre-Engineered Structures

- The foundations and floor slab, as well as rough mechanical and electrical, should be estimated, as this work is required for the assembly and erection of the structure. Generally, as noted in the book, the pre-engineered building comes as a shell and additional features must be included by the estimator. Here again, the estimator must have a clear understanding of the scope of each portion of the work and all the necessary interfaces.

13200 Storage Tanks

- The prices in this subdivision for above and below ground storage tanks do not include foundations or hold-down slabs. The estimator should refer to Divisions 2 and 3 for foundation system pricing. In addition to the foundations, required tank accessories such as tank gauges, leak detection devices, and additional manholes and piping must be added to the tank prices.

Reference Numbers

Reference numbers are shown in bold squares at the beginning of some major classifications. These numbers refer to related items in the Reference Section. The reference information may be an estimating procedure, an alternate pricing method or technical information.

Note: Not all subdivisions listed here necessarily appear in this publication.

13100 | Lightning Protection

13101 | Lightning Protection

			CREW	DAILY OUTPUT	LABOR-HOURS	UNIT	2000 BARE COSTS				TOTAL INCL O&P
							MAT.	LABOR	EQUIP.	TOTAL	
055	0010	**LIGHTNING PROTECTION**									
	0200	Air terminals & base, copper									
	0400	3/8″ diameter x 10″ (to 75′ high)	1 Elec	8	1	Ea.	21.50	22		43.50	59.50
	1000	Aluminum, 1/2″ diameter x 12″ (to 75′ high)		8	1	″	17	22		39	54.50
	2000	Cable, copper, 220 lb. per thousand ft. (to 75′ high)		320	.025	L.F.	1.07	.55		1.62	2.08
	2500	Aluminum, 101 lb. per thousand ft. (to 75′ high)		280	.029	″	.46	.63		1.09	1.54
	3000	Arrester, 175 volt AC to ground		8	1	Ea.	36	22		58	75.50

13120 | Pre-Engineered Structures

13128 | Pre-Eng. Structures

			CREW	DAILY OUTPUT	LABOR-HOURS	UNIT	2000 BARE COSTS				TOTAL INCL O&P
							MAT.	LABOR	EQUIP.	TOTAL	
540	0010	**GREENHOUSE** Shell only, stock units, not incl. 2′ stub walls,									
	0020	foundation, floors, heat or compartments									
	0300	Residential type, free standing, 8′-6″ long x 7′-6″ wide	2 Carp	59	.271	SF Flr.	35	5.35		40.35	47.50
	0400	10′-6″ wide		85	.188		27	3.71		30.71	36
	0600	13′-6″ wide		108	.148		24	2.92		26.92	31.50
	0700	17′-0″ wide		160	.100		27	1.97		28.97	33
	0900	Lean-to type, 3′-10″ wide		34	.471		31	9.25		40.25	50
	1000	6′-10″ wide		58	.276		24	5.45		29.45	36
	1100	Wall mounted, to existing window, 3′ x 3′	1 Carp	4	2	Ea.	335	39.50		374.50	440
	1120	4′ x 5′	″	3	2.667	″	500	52.50		552.50	640
	1200	Deluxe quality, free standing, 7′-6″ wide	2 Carp	55	.291	SF Flr.	69	5.75		74.75	86
	1220	10′-6″ wide		81	.198		64	3.89		67.89	77
	1240	13′-6″ wide		104	.154		60	3.03		63.03	71
	1260	17′-0″ wide		150	.107		51	2.10		53.10	59.50
	1400	Lean-to type, 3′-10″ wide		31	.516		80	10.15		90.15	105
	1420	6′-10″ wide		55	.291		75	5.75		80.75	92.50
	1440	8′-0″ wide		97	.165		70	3.25		73.25	82.50
880	0010	**SWIMMING POOL ENCLOSURE** Translucent, free standing,									
	0020	not including foundations, heat or light									
	0200	Economy, minimum	2 Carp	200	.080	SF Hor.	10	1.58		11.58	13.70
	0300	Maximum		100	.160		21	3.15		24.15	28.50
	0400	Deluxe, minimum		100	.160		23	3.15		26.15	31
	0600	Maximum		70	.229		250	4.50		254.50	283

13150 | Swimming Pools

13151 | Swimming Pools

			CREW	DAILY OUTPUT	LABOR-HOURS	UNIT	2000 BARE COSTS				TOTAL INCL O&P
							MAT.	LABOR	EQUIP.	TOTAL	
200	0010	**SWIMMING POOLS** Residential in-ground, vinyl lined, concrete sides									
	0020	Sides including equipment, sand bottom	B-52	300	.187	SF Surf	10.40	3.05	1.31	14.76	18.15
	0100	Metal or polystyrene sides R13128 -520	B-14	410	.117		8.70	1.82	.50	11.02	13.20
	0200	Add for vermiculite bottom					.67			.67	.74

Important: See the Reference Section for critical supporting data - Reference Nos., Crews, & Location Factor

13150 | Swimming Pools

13151 | Swimming Pools

	13151 Swimming Pools	CREW	DAILY OUTPUT	LABOR-HOURS	UNIT	2000 BARE COSTS MAT.	LABOR	EQUIP.	TOTAL	TOTAL INCL O&P	
0500	Gunite bottom and sides, white plaster finish R13128-520										200
0600	12' x 30' pool	B-52	145	.386	SF Surf	17.25	6.30	2.70	26.25	33	
0720	16' x 32' pool		155	.361		15.50	5.90	2.53	23.93	30	
0750	20' x 40' pool		250	.224		13.85	3.66	1.57	19.08	23.50	
0810	Concrete bottom and sides, tile finish										
0820	12' x 30' pool	B-52	80	.700	SF Surf	17.40	11.45	4.90	33.75	44	
0830	16' x 32' pool		95	.589		14.40	9.65	4.13	28.18	37	
0840	20' x 40' pool		130	.431		11.45	7.05	3.02	21.52	28	
1600	For water heating system, see division 15510-880										
1700	Filtration and deck equipment only, as % of total				Total				20%	20%	
1800	Deck equipment, rule of thumb, 20' x 40' pool				SF Pool					1.30	
3000	Painting pools, preparation + 3 coats, 20' x 40' pool, epoxy	2 Pord	.33	48.485	Total	595	870		1,465	2,100	
3100	Rubber base paint, 18 gallons	"	.33	48.485	"	450	870		1,320	1,950	
0010	**SWIMMING POOL EQUIPMENT** Diving stand, stainless steel, 3 meter	2 Carp	.40	40	Ea.	4,725	790		5,515	6,525	700
0600	Diving boards, 16' long, aluminum		2.70	5.926		2,100	117		2,217	2,525	
0700	Fiberglass		2.70	5.926		1,575	117		1,692	1,925	
0900	Filter system, sand or diatomite type, incl. pump, 6,000 gal./hr.	2 Plum	1.80	8.889	Total	900	195		1,095	1,325	
1020	Add for chlorination system, 800 S.F. pool	"	3	5.333	Ea.	196	117		313	410	
1200	Ladders, heavy duty, stainless steel, 2 tread	2 Carp	7	2.286		790	45		835	940	
1500	4 tread	"	6	2.667		1,350	52.50		1,402.50	1,600	
2100	Lights, underwater, 12 volt, with transformer, 300 watt	1 Elec	.40	20		150	440		590	890	
2200	110 volt, 500 watt, standard	"	.40	20		148	440		588	890	
3000	Pool covers, reinforced vinyl	3 Clab	1,800	.013	S.F.	.35	.19		.54	.72	
3100	Vinyl water tube		1,800	.013		.25	.19		.44	.61	
3200	Maximum		1,600	.015		.45	.22		.67	.87	
3300	Slides, tubular, fiberglass, aluminum handrails & ladder, 5'-0", straight	2 Carp	1.60	10	Ea.	1,025	197		1,222	1,475	
3320	8'-0", curved	"	3	5.333	"	11,000	105		11,105	12,300	

SPECIAL CONSTRUCTION 13

13200 | Storage Tanks

13201 | Storage Tanks

	13201 Storage Tanks	CREW	DAILY OUTPUT	LABOR-HOURS	UNIT	2000 BARE COSTS MAT.	LABOR	EQUIP.	TOTAL	TOTAL INCL O&P	
3001	**STEEL,** storage, above ground, including supports, coating										300
3020	fittings, not including fdn, pumps or piping										
3040	Single wall, interior, 275 gallon	Q-5	5	3.200	Ea.	231	63.50		294.50	360	
3060	550 gallon	"	2.70	5.926		815	118		933	1,100	
3080	1,000 gallon	Q-7	5	6.400		1,025	127		1,152	1,325	
3320	Double wall, 500 gallon capacity	Q-5	2.40	6.667		2,200	133		2,333	2,650	
3330	2000 gallon capacity	Q-7	4.15	7.711		4,575	153		4,728	5,275	
3340	4000 gallon capacity		3.60	8.889		8,125	177		8,302	9,225	
3350	6000 gallon capacity		2.40	13.333		9,675	265		9,940	11,000	
3360	8000 gallon capacity		2	16		11,100	320		11,420	12,700	
3370	10000 gallon capacity		1.80	17.778		12,200	355		12,555	14,000	
3380	15000 gallon capacity		1.50	21.333		18,200	425		18,625	20,700	
3390	20000 gallon capacity		1.30	24.615		20,700	490		21,190	23,600	
3400	25000 gallon capacity		1.15	27.826		25,100	555		25,655	28,600	
3410	30000 gallon capacity		1	32		27,500	635		28,135	31,400	
0010	**UNDERGROUND STORAGE TANKS**										800
0210	Fiberglass, underground, single wall, U.L. listed, not including										

13200 | Storage Tanks

	13201 Storage Tanks	CREW	DAILY OUTPUT	LABOR-HOURS	UNIT	2000 BARE COSTS MAT.	LABOR	EQUIP.	TOTAL	TOTAL INCL O&P
800	0220 manway or hold-down strap									
	0230 1,000 gallon capacity	Q-5	2.46	6.504	Ea.	1,825	129		1,954	2,225
	0240 2,000 gallon capacity	Q-7	4.57	7.002		2,375	139		2,514	2,825
	0500 For manway, fittings and hold-downs, add					20%	15%			
	2210 Fiberglass, underground, single wall, U.L. listed, including									
	2220 hold-down straps, no manways									
	2230 1,000 gallon capacity	Q-5	1.88	8.511	Ea.	1,975	169		2,144	2,450
	2240 2,000 gallon capacity	Q-7	3.55	9.014	"	2,525	179		2,704	3,075
	5000 Steel underground, sti-P3, set in place, not incl. hold-down bars.									
	5500 Excavation, pad, pumps and piping not included									
	5510 Single wall, 500 gallon capacity, 7 gauge shell	Q-5	2.70	5.926	Ea.	820	118		938	1,100
	5520 1,000 gallon capacity, 7 gauge shell	"	2.50	6.400		1,350	127		1,477	1,700
	5530 2,000 gallon capacity, 1/4" thick shell	Q-7	4.60	6.957		2,175	138		2,313	2,625
	5535 2,500 gallon capacity, 7 gauge shell	Q-5	3	5.333		2,375	106		2,481	2,800
	5610 25,000 gallon capacity, 3/8" thick shell	Q-7	1.30	24.615		17,600	490		18,090	20,200
	5630 40,000 gallon capacity, 3/8" thick shell		.90	35.556		30,700	710		31,410	34,900
	5640 50,000 gallon capacity, 3/8" thick shell		.80	40		38,300	795		39,095	43,500

13280 | Hazardous Material Remediation

	13281 Hazardous Material Remediation	CREW	DAILY OUTPUT	LABOR-HOURS	UNIT	2000 BARE COSTS MAT.	LABOR	EQUIP.	TOTAL	TOTAL INCL O&P
440	0010 **REMOVAL** Existing lead paint, by chemicals, per application									
	0020 See also, Div. 13280, Haz. Mat'l. Abatement									
	0050 Baseboard, to 6" wide	1 Pord	64	.125	L.F.	1.42	2.24		3.66	5.30
	0070 To 12" wide		32	.250	"	2.78	4.49		7.27	10.55
	0200 Balustrades, one side		28	.286	S.F.	3.16	5.15		8.31	12.05
	1400 Cabinets, simple design		32	.250		2.76	4.49		7.25	10.55
	1420 Ornate design		25	.320		3.55	5.75		9.30	13.50
	1600 Cornice, simple design		60	.133		1.48	2.39		3.87	5.60
	1620 Ornate design		20	.400		4.37	7.20		11.57	16.80
	2800 Doors, one side, flush		84	.095		1.07	1.71		2.78	4.03
	2820 Two panel		80	.100		1.11	1.80		2.91	4.22
	2840 Four panel		45	.178		1.96	3.19		5.15	7.45
	2880 For trim, one side, add		64	.125	L.F.	1.42	2.24		3.66	5.30
	3000 Fence, picket, one side		30	.267	S.F.	2.97	4.79		7.76	11.25
	3200 Grilles, one side, simple design		30	.267		2.97	4.79		7.76	11.25
	3220 Ornate design		25	.320		3.55	5.75		9.30	13.50
	4400 Pipes, to 4" diameter		90	.089	L.F.	1.01	1.60		2.61	3.77
	4420 To 8" diameter		50	.160		1.76	2.87		4.63	6.75
	4440 To 12" diameter		36	.222		2.46	3.99		6.45	9.35
	4460 To 16" diameter		20	.400		4.40	7.20		11.60	16.85
	4500 For hangers, add		40	.200	Ea.	2.21	3.59		5.80	8.45
	4800 Siding		90	.089	S.F.	1.01	1.60		2.61	3.77
	5000 Trusses, open		55	.145	SF Face	1.61	2.61		4.22	6.15
	6200 Windows, one side only, double hung, 1/1 light, 24" x 48" high		4	2	Ea.	22	36		58	84.50
	6220 30" x 60" high		3	2.667		29.50	48		77.50	113
	6240 36" x 72" high		2.50	3.200		35.50	57.50		93	135
	6280 40" x 80" high		2	4		44.50	72		116.50	169
	6400 Colonial window, 6/6 light, 24" x 48" high		2	4		44.50	72		116.50	169

13280 | Hazardous Material Remediation

13281 | Hazardous Material Remediation		CREW	DAILY OUTPUT	LABOR-HOURS	UNIT	2000 BARE COSTS MAT.	LABOR	EQUIP.	TOTAL	TOTAL INCL O&P	
6420	30″ x 60″ high	1 Pord	1.50	5.333	Ea.	59.50	95.50		155	225	440
6440	36″ x 72″ high		1	8		89	144		233	340	
6480	40″ x 80″ high		1	8		89	144		233	340	
6600	8/8 light, 24″ x 48″ high		2	4		44.50	72		116.50	169	
6620	40″ x 80″ high		1	8		89	144		233	340	
6800	12/12 light, 24″ x 48″ high		1	8		89	144		233	340	
6820	40″ x 80″ high		.75	10.667		119	191		310	450	
6840	Window frame & trim items, included in pricing above										
0010	**LEAD PAINT ENCAPSULATION**, water based polymer coating,14 mil DFT										460
0020	Interior, brushwork, trim, under 6″	1 Pord	240	.033	L.F.	2.09	.60		2.69	3.30	
0030	6″ to 12″ wide		180	.044		2.77	.80		3.57	4.38	
0040	Balustrades		300	.027		1.67	.48		2.15	2.64	
0050	Pipe to 4″ diameter		500	.016		1	.29		1.29	1.58	
0060	To 8″ diameter		375	.021		1.33	.38		1.71	2.10	
0070	To 12″ diameter		250	.032		2	.57		2.57	3.16	
0080	To 16″ diameter		170	.047		2.94	.84		3.78	4.64	
0090	Cabinets, ornate design		200	.040	S.F.	2.50	.72		3.22	3.95	
0100	Simple design		250	.032	″	2	.57		2.57	3.16	
0110	Doors, 3′x 7′, both sides, incl. frame & trim										
0120	Flush	1 Pord	6	1.333	Ea.	25.50	24		49.50	68	
0130	French, 10-15 lite		3	2.667		5.10	48		53.10	85.50	
0140	Panel		4	2		30.50	36		66.50	94	
0150	Louvered		2.75	2.909		28	52		80	118	
0160	Windows, per interior side, per 15 S.F.										
0170	1 to 6 lite	1 Pord	14	.571	Ea.	17.70	10.25		27.95	36.50	
0180	7 to 10 lite		7.50	1.067		19.50	19.15		38.65	53.50	
0190	12 lite		5.75	1.391		26	25		51	70.50	
0200	Radiators		8	1		62.50	17.95		80.45	98.50	
0210	Grilles, vents		275	.029	S.F.	1.82	.52		2.34	2.87	
0220	Walls, roller, drywall or plaster		1,000	.008		.49	.14		.63	.78	
0230	With spunbonded reinforcing fabric		720	.011		.56	.20		.76	.95	
0240	Wood		800	.010		.62	.18		.80	.98	
0250	Ceilings, roller, drywall or plaster		900	.009		.56	.16		.72	.89	
0260	Wood		700	.011		.71	.21		.92	1.12	
0270	Exterior, brushwork, gutters and downspouts		300	.027	L.F.	1.67	.48		2.15	2.64	
0280	Columns		400	.020	S.F.	1.25	.36		1.61	1.98	
0290	Spray, siding		600	.013	″	.83	.24		1.07	1.31	
0300	Miscellaneous										
0310	Electrical conduit, brushwork, to 2″ diameter	1 Pord	500	.016	L.F.	1	.29		1.29	1.58	
0320	Brick, block or concrete, spray		500	.016	S.F.	1	.29		1.29	1.58	
0330	Steel, flat surfaces and tanks to 12″		500	.016		1	.29		1.29	1.58	
0340	Beams, brushwork		400	.020		1.25	.36		1.61	1.98	
0350	Trusses		400	.020		1.25	.36		1.61	1.98	

13800 | Building Automation & Control

13838 | Pneumatic/Electric Controls		CREW	DAILY OUTPUT	LABOR-HOURS	UNIT	2000 BARE COSTS MAT.	LABOR	EQUIP.	TOTAL	TOTAL INCL O&P	
0010	CONTROL COMPONENTS										200
5000	Thermostats										

13800 | Building Automation & Control

	13838	Pneumatic/Electric Controls	CREW	DAILY OUTPUT	LABOR-HOURS	UNIT	2000 BARE COSTS MAT.	LABOR	EQUIP.	TOTAL	TOTAL INCL O&P
200	5030	Manual	1 Shee	8	1	Ea.	24.50	21.50		46	63.50
	5040	1 set back, electric, timed		8	1		79	21.50		100.50	124
	5050	2 set back, electric, timed	↓	8	1	↓	106	21.50		127.50	153

13850 | Detection & Alarm

	13851	Detection & Alarm	CREW	DAILY OUTPUT	LABOR-HOURS	UNIT	2000 BARE COSTS MAT.	LABOR	EQUIP.	TOTAL	TOTAL INCL O&P
065	0010	**DETECTION SYSTEMS**, not including wires & conduits									
	0100	Burglar alarm, battery operated, mechanical trigger	1 Elec	4	2	Ea.	249	44		293	345
	0200	Electrical trigger		4	2		297	44		341	400
	0400	For outside key control, add		8	1		70.50	22		92.50	114
	0600	For remote signaling circuitry, add		8	1		112	22		134	159
	0800	Card reader, flush type, standard		2.70	2.963		835	65.50		900.50	1,025
	1000	Multi-code		2.70	2.963		1,075	65.50		1,140.50	1,275
	1200	Door switches, hinge switch		5.30	1.509		52.50	33.50		86	113
	1400	Magnetic switch		5.30	1.509		62	33.50		95.50	123
	2800	Ultrasonic motion detector, 12 volt		2.30	3.478		206	77		283	355
	3000	Infrared photoelectric detector		2.30	3.478		170	77		247	315
	3200	Passive infrared detector		2.30	3.478		254	77		331	405
	3420	Switchmats, 30″ x 5′		5.30	1.509		76	33.50		109.50	138
	3440	30″ x 25′		4	2		182	44		226	273
	3460	Police connect panel		4	2		219	44		263	315
	3480	Telephone dialer		5.30	1.509		345	33.50		378.50	435
	3500	Alarm bell		4	2		69.50	44		113.50	149
	3520	Siren		4	2		131	44		175	217
	5200	Smoke detector, ceiling type		6.20	1.290		75	28.50		103.50	129
	5600	Strobe and horn		5.30	1.509		95	33.50		128.50	160
	5800	Fire alarm horn		6.70	1.194		36.50	26.50		63	83
	6600	Drill switch		8	1		86.50	22		108.50	131
	6800	Master box		2.70	2.963		3,100	65.50		3,165.50	3,500
	7800	Remote annunciator, 8 zone lamp	↓	1.80	4.444		175	98		273	355
	8000	12 zone lamp	2 Elec	2.60	6.154		300	136		436	550
	8200	16 zone lamp	"	2.20	7.273		300	161		461	595
	8400	Standpipe or sprinkler alarm, alarm device	1 Elec	8	1		125	22		147	174
	8600	Actuating device	"	8	1	↓	290	22		312	355

For information about Means Estimating Seminars, see yellow pages 11 and 12 in back of book

Division 14 Conveying Systems

[E]stimating Tips

[G]eneral

Many products in Division 14 will require some type of support or blocking for installation not included with the item itself. Examples are supports for conveyors or tube systems, attachment points for lifts, and footings for hoists or cranes. Add these supports in the appropriate division.

[1]4100 Dumbwaiters
[1]4200 Elevators

Dumbwaiters and elevators are estimated and purchased in a method similar to buying a car. The manufacturer has a base unit with standard features. Added to this base unit price will be whatever options the owner or specifications require. Increased load capacity, additional vertical travel, additional stops, higher speed, and cab finish options are items to be considered. When developing an estimate for dumbwaiters and elevators, remember that some items needed by the installers may have to be included as part of the general contract.

Examples are:

- shaftway
- rail support brackets
- machine room
- electrical supply
- sill angles
- electrical connections
- pits
- roof penthouses
- pit ladders

Check the job specifications and drawings before pricing.

14300 Escalators & Moving Walks

- Escalators and moving walks are specialty items installed by specialty contractors. There are numerous options associated with these items. For specific options contact a manufacturer or contractor. In a method similar to estimating dumbwaiters and elevators, you should verify the extent of general contract work and add items as necessary.

14400 Lifts
14500 Material Handling
14600 Hoists & Cranes

- Products such as correspondence lifts, conveyors, chutes, pneumatic tube systems, material handling cranes and hoists as well as other items specified in this subdivision may require trained installers. The general contractor might not have any choice as to who will perform the installation or when it will be performed. Long lead times are often required for these products, making early decisions in scheduling necessary.

Reference Numbers

Reference numbers are shown in bold squares at the beginning of some major classifications. These numbers refer to related items in the Reference Section. The reference information may be an estimating procedure, an alternate pricing method or technical information.

Note: Not all subdivisions listed here necessarily appear in this publication.

		14210 | Electric Traction Elevators	CREW	DAILY OUTPUT	LABOR-HOURS	UNIT	2000 BARE COSTS MAT.	LABOR	EQUIP.	TOTAL	TOTAL INCL O&P
100	0012	ELEVATORS OR LIFTS									
	7000	Residential, cab type, 1 floor, 2 stop, minimum	2 Elev	.20	80	Ea.	7,850	1,825		9,675	11,700
	7100	Maximum		.10	160		13,300	3,675		16,975	20,700
	7200	2 floor, 3 stop, minimum		.12	133		11,700	3,050		14,750	17,900
	7300	Maximum		.06	266		19,000	6,100		25,100	31,000
	7700	Stair climber (chair lift), single seat, minimum		1	16		3,800	365		4,165	4,775
	7800	Maximum	↓	.20	80	↓	5,225	1,825		7,050	8,775

For information about Means Estimating Seminars, see yellow pages 11 and 12 in back of book

Division 15 Mechanical

stimating Tips

100 Building Services Piping

is subdivision is primarily basic pipe and ated materials. The pipe may be used by any the mechanical disciplines, i.e., plumbing, : protection, heating, and air conditioning.

The piping section lists the add to labor for elevated pipe installation. These adds apply to all elevated pipe, fittings, valves, insulation, etc., that are placed above 10′ high. CAUTION: the correct percentage may vary for the same pipe. For example, the percentage add for the basic pipe installation should be based on the maximum height that the craftsman must install for that particular section. If the pipe is to be located 14′ above the floor but it is suspended on threaded rod from beams, the bottom flange of which is 18′ high (4′ rods), then the height is actually 18′ and the add is 20%. The pipe coverer, however, does not have to go above the 14′ and so his add should be 10%.

Most pipe is priced first as straight pipe with a joint (coupling, weld, etc.) every 10′ and a hanger usually every 10′. There are exceptions with hanger spacing such as: for cast iron pipe (5′) and plastic pipe (3 per 10′). Following each type of pipe there are several lines listing sizes and the amount to be subtracted to delete couplings and hangers. This is for pipe that is to be buried or supported together on trapeze hangers. The reason that the couplings are deleted is that these runs are usually long and frequently longer lengths of pipe are used. By deleting the couplings the estimator is expected to look up and add back the correct reduced number of couplings.

- When preparing an estimate it may be necessary to approximate the fittings. Fittings usually run between 25% and 50% of the cost of the pipe. The lower percentage is for simpler runs, and the higher number is for complex areas like mechanical rooms.

15400 Plumbing Fixtures & Equipment

- Plumbing fixture costs usually require two lines, the fixture itself and its "rough-in, supply and waste".
- Remember that gas and oil fired units need venting.

15500 Heat Generation Equipment

- When estimating the cost of an HVAC system, check to see who is responsible for providing and installing the temperature control system. It is possible to overlook controls, assuming that they would be included in the electrical estimate.
- When looking up a boiler be careful on specified capacity. Some manufacturers rate their products on output while others use input.
- Include HVAC insulation for pipe, boiler and duct (wrap and liner).
- Be careful when looking up mechanical items to get the correct pressure rating and connection type (thread, weld, flange).

15700 Heating/Ventilation/ Air Conditioning Equipment

- Combination heating and cooling units are sized by the air conditioning requirements. (See Reference No. R15710-020 for preliminary sizing guide.)
- A ton of air conditioning is nominally 400 CFM.
- Rectangular duct is taken off by the linear foot for each size, but its cost is usually estimated by the pound. Remember that SMACNA standards now base duct on internal pressure.
- Prefabricated duct is estimated and purchased like pipe: straight sections and fittings.
- Note that cranes or other lifting equipment are not included on any lines in Division 15. For example, if a crane is required to lift a heavy piece of pipe into place high above a gym floor, or to put a rooftop unit on the roof of a four-story building, etc., it must be added. Due to the potential for extreme variation—from nothing additional required, to a major crane or helicopter—we feel that including a nominal amount for "lifting contingency" would be useless and detract from the accuracy of the estimate. When using equipment rental from Means do not forget to include the cost of the operator(s).

Reference Numbers

Reference numbers are shown in bold squares at the beginning of some major classifications. These numbers refer to related items in the Reference Section. The reference information may be an estimating procedure, an alternate pricing method or technical information.

Note: Not all subdivisions listed here necessarily appear in this publication.

15055 | Selective Mech Demolition

		15055 Selective Mech Demolition	CREW	DAILY OUTPUT	LABOR-HOURS	UNIT	2000 BARE COSTS MAT.	LABOR	EQUIP.	TOTAL	TOTAL INCL O&P
300	0010	HVAC DEMOLITION									
	0100	Air conditioner, split unit, 3 ton	Q-5	2	8	Ea.		159		159	263
	0150	Package unit, 3 ton	Q-6	3	8	"		153		153	253
	0260	Baseboard, hydronic fin tube, 1/2"	Q-5	117	.137	L.F.		2.72		2.72	4.50
	0300	Boiler, electric	Q-19	2	12	Ea.		248		248	410
	0340	Gas or oil, steel, under 150 MBH	Q-6	3	8	"		153		153	253
	1000	Ductwork, 4" high, 8" wide	1 Clab	200	.040	L.F.		.58		.58	.99
	1100	6" high, 8" wide		165	.048			.70		.70	1.20
	1200	10" high, 12" wide		125	.064			.92		.92	1.58
	1300	12"-14" high, 16"-18" wide		85	.094			1.36		1.36	2.33
	1500	30" high, 36" wide	↓	56	.143	↓		2.06		2.06	3.54
	2200	Furnace, electric	Q-20	2	10	Ea.		199		199	335
	2300	Gas or oil, under 120 MBH	Q-9	4	4			77.50		77.50	131
	2340	Over 120 MBH	"	3	5.333			103		103	174
	2800	Heat pump, package unit, 3 ton	Q-5	2.40	6.667			133		133	219
	2840	Split unit, 3 ton		2	8			159		159	263
	2950	Tank, steel, oil, 275 gal., above ground		10	1.600			32		32	52.50
	2960	Remove and reset	↓	3	5.333	↓		106		106	175
	9000	Minimum labor/equipment charge	Q-6	3	8	Job		153		153	253
600	0010	PLUMBING DEMOLITION									
	1020	Fixtures, including 10' piping									
	1100	Bath tubs, cast iron	1 Plum	4	2	Ea.		44		44	72.50
	1120	Fiberglass		6	1.333			29.50		29.50	48.50
	1140	Steel		5	1.600			35		35	58
	1200	Lavatory, wall hung		10	.800			17.55		17.55	29
	1220	Counter top		8	1			22		22	36.50
	1300	Sink, steel or cast iron, single		8	1			22		22	36.50
	1320	Double		7	1.143			25		25	41.50
	1400	Water closet, floor mounted		8	1			22		22	36.50
	1420	Wall mounted		7	1.143	↓		25		25	41.50
	2000	Piping, metal, to 2" diameter		200	.040	L.F.		.88		.88	1.45
	2050	2" to 4" diameter	↓	150	.053			1.17		1.17	1.94
	2100	4" to 8" diameter	2 Plum	100	.160	↓		3.51		3.51	5.80
	2250	Water heater, 40 gal.	1 Plum	6	1.333	Ea.		29.50		29.50	48.50
	3000	Submersible sump pump		24	.333			7.30		7.30	12.10
	6000	Remove and reset fixtures, minimum		6	1.333			29.50		29.50	48.50
	6100	Maximum		4	2	↓		44		44	72.50
	9000	Minimum labor/equipment charge	↓	2	4	Job		88		88	145

15082 | Duct Insulation

		15082 Duct Insulation	CREW	DAILY OUTPUT	LABOR-HOURS	UNIT	2000 BARE COSTS MAT.	LABOR	EQUIP.	TOTAL	TOTAL INCL O&P
200	0010	INSULATION									
	2900	Domestic water heater wrap kit									
	2920	1-1/2" with vinyl jacket, 20-60 gal.	1 Plum	8	1	Ea.	16.25	22		38.25	54.50
	2930	Insulated protectors, (ADA)									
	2935	For exposed piping under sinks or lavatories.									
	2940	Vinyl coated foam, velcro tabs									
	2945	P Trap, 1-1/4" or 1-1/2"	1 Plum	32	.250	Ea.	12.85	5.50		18.35	23
	2960	Valve and supply cover									
	2965	1/2", 3/8", and 7/16" pipe size	1 Plum	32	.250	Ea.	12.85	5.50		18.35	23
	2970	Extension drain cover									
	2975	1-1/4", or 1-1/2" pipe size	1 Plum	32	.250	Ea.	13.15	5.50		18.65	23.50
	2985	1-1/4" pipe size	"	32	.250	"	13.50	5.50		19	24
	3000	Ductwork									
	3020	Blanket type, fiberglass, flexible									

15082 | Duct Insulation

15082 200		Crew	Daily Output	Labor-Hours	Unit	2000 Bare Costs Mat.	Labor	Equip.	Total	Total Incl O&P
3030	Fire resistant liner, black coating one side									
3050	1/2″ thick, 2 lb. density	Q-14	380	.042	S.F.	.29	.78		1.07	1.68
3060	1″ thick, 1-1/2 lb. density	″	350	.046	″	.39	.84		1.23	1.91
3140	FRK vapor barrier wrap, .75 lb. density									
3160	1″ thick	Q-14	350	.046	S.F.	.27	.84		1.11	1.78
3170	1-1/2″ thick	″	320	.050	″	.24	.92		1.16	1.87
3490	Board type, fiberglass liner, 3 lb. density									
3500	Fire resistant, black pigmented, 1 side									
3520	1″ thick	Q-14	150	.107	S.F.	1.16	1.97		3.13	4.73
3540	1-1/2″ thick	″	130	.123	″	1.43	2.27		3.70	5.55
4000	Pipe covering (price copper tube one size less than IPS)									
6600	Fiberglass, with all service jacket									
6840	1″ wall, 1/2″ iron pipe size	Q-14	240	.067	L.F.	1.04	1.23		2.27	3.29
6860	3/4″ iron pipe size		230	.070		1.19	1.28		2.47	3.56
6870	1″ iron pipe size		220	.073		1.21	1.34		2.55	3.68
6900	2″ iron pipe size		200	.080		1.62	1.48		3.10	4.36
7879	Rubber tubing, flexible closed cell foam									
8100	1/2″ wall, 1/4″ iron pipe size	1 Asbe	90	.089	L.F.	.57	1.82		2.39	3.82
8130	1/2″ iron pipe size		89	.090		.70	1.84		2.54	4
8140	3/4″ iron pipe size		89	.090		.79	1.84		2.63	4.10
8150	1″ iron pipe size		88	.091		.87	1.86		2.73	4.22
8170	1-1/2″ iron pipe size		87	.092		1.27	1.89		3.16	4.70
8180	2″ iron pipe size		86	.093		1.56	1.91		3.47	5.05
8300	3/4″ wall, 1/4″ iron pipe size		90	.089		.86	1.82		2.68	4.14
8330	1/2″ iron pipe size		89	.090		1.12	1.84		2.96	4.46
8340	3/4″ iron pipe size		89	.090		1.38	1.84		3.22	4.75
8350	1″ iron pipe size		88	.091		1.56	1.86		3.42	4.98
8380	2″ iron pipe size		86	.093		2.77	1.91		4.68	6.40
8444	1″ wall, 1/2″ iron pipe size		86	.093		2.18	1.91		4.09	5.75
8445	3/4″ iron pipe size		84	.095		2.64	1.95		4.59	6.30
8446	1″ iron pipe size		84	.095		3.07	1.95		5.02	6.80
8447	1-1/4″ iron pipe size		82	.098		3.47	2		5.47	7.30
8448	1-1/2″ iron pipe size		82	.098		4.03	2		6.03	7.95
8449	2″ iron pipe size		80	.100		5.40	2.05		7.45	9.55
8450	2-1/2″ iron pipe size		80	.100		7	2.05		9.05	11.30
8456	Rubber insulation tape, 1/8″ x 2″ x 30′				Ea.	12.95			12.95	14.25

15100 | Building Services Piping

15107 | Metal Pipe & Fittings

15107 320		Crew	Daily Output	Labor-Hours	Unit	2000 Bare Costs Mat.	Labor	Equip.	Total	Total Incl O&P
0010	**PIPE, CAST IRON** Soil, on hangers 5′ O.C. R15100 -050									
0020	Single hub, service wt., lead & oakum joints 10′ O.C.									
2120	2″ diameter	Q-1	63	.254	L.F.	4.92	5		9.92	13.70
2140	3″ diameter		60	.267		6.35	5.25		11.60	15.65
2160	4″ diameter		55	.291		6.25	5.75		12	16.35
4000	No hub, couplings 10′ O.C.									
4100	1-1/2″ diameter	Q-1	71	.225	L.F.	5.60	4.45		10.05	13.50
4120	2″ diameter		67	.239		5.70	4.72		10.42	14.10
4140	3″ diameter		64	.250		7.15	4.94		12.09	16.05
4160	4″ diameter		58	.276		9.30	5.45		14.75	19.20

	15107 | Metal Pipe & Fittings		CREW	DAILY OUTPUT	LABOR-HOURS	UNIT	2000 BARE COSTS MAT.	LABOR	EQUIP.	TOTAL	TOTAL INCL O&P
360	0010	**PIPE, CAST IRON, FITTINGS** Soil									
	0040	Hub and spigot, service weight, lead & oakum joints									
	0080	1/4 bend, 2"	Q-1	16	1	Ea.	6.10	19.75		25.85	39
	0120	3"		14	1.143		10.60	22.50		33.10	49
	0140	4"		13	1.231		16.55	24.50		41.05	58
	0340	1/8 bend, 2"		16	1		4.88	19.75		24.63	38
	0350	3"		14	1.143		8.90	22.50		31.40	47.50
	0360	4"		13	1.231		12.90	24.50		37.40	54
	0500	Sanitary tee, 2"		10	1.600		9.75	31.50		41.25	62.50
	0540	3"		9	1.778		18.70	35		53.70	78.50
	0620	4"		8	2		22	39.50		61.50	89.50
	5990	No hub									
	6000	Cplg. & labor required at joints not incl. in fitting									
	6010	price. Add 1 coupling per joint for installed price									
	6020	1/4 Bend, 1-1/2"				Ea.	4.47			4.47	4.92
	6060	2"					4.85			4.85	5.35
	6080	3"					6.35			6.35	6.95
	6120	4"					9.70			9.70	10.65
	6184	1/4 Bend, long sweep, 1-1/2"					10.95			10.95	12.05
	6186	2"					10.95			10.95	12.05
	6188	3"					13.05			13.05	14.35
	6189	4"					21			21	23
	6190	5"					38.50			38.50	42
	6191	6"					46.50			46.50	51.50
	6192	8"					114			114	125
	6193	10"					205			205	225
	6200	1/8 Bend, 1-1/2"					3.72			3.72	4.09
	6210	2"					4.13			4.13	4.54
	6212	3"					5.60			5.60	6.15
	6214	4"					7.10			7.10	7.80
	6380	Sanitary Tee, tapped, 1-1/2"					8.55			8.55	9.40
	6382	2" x 1-1/2"					8.55			8.55	9.40
	6384	2"					9.45			9.45	10.40
	6386	3" x 2"					12.10			12.10	13.30
	6388	3"					22.50			22.50	24.50
	6390	4" x 1-1/2"					10.75			10.75	11.85
	6392	4" x 2"					12.10			12.10	13.30
	6394	6" x 1-1/2"					25			25	27.50
	6396	6" x 2"					25.50			25.50	28
	6459	Sanitary Tee, 1-1/2"					6.15			6.15	6.80
	6460	2"					6.75			6.75	7.40
	6470	3"					8.20			8.20	9
	6472	4"					12.70			12.70	13.95
	8000	Coupling, standard (by CISPI Mfrs.)									
	8020	1-1/2"	Q-1	48	.333	Ea.	3.82	6.60		10.42	15.10
	8040	2"		44	.364		3.82	7.20		11.02	16.05
	8080	3"		38	.421		4.54	8.30		12.84	18.75
	8120	4"		33	.485		5.40	9.60		15	22
420	0010	**PIPE, COPPER** Solder joints									
	1000	Type K tubing, couplings & clevis hangers 10' O.C.									
	1180	3/4" diameter	1 Plum	74	.108	L.F.	1.85	2.37		4.22	5.95
	1200	1" diameter	"	66	.121	"	2.37	2.66		5.03	7
	2000	Type L tubing, couplings & hangers 10' O.C.									
	2140	1/2" diameter	1 Plum	81	.099	L.F.	1.14	2.17		3.31	4.84
	2160	5/8" diameter		79	.101		1.39	2.22		3.61	5.20
	2180	3/4" diameter		76	.105		1.50	2.31		3.81	5.45

15107	Metal Pipe & Fittings	CREW	DAILY OUTPUT	LABOR-HOURS	UNIT	2000 BARE COSTS MAT.	LABOR	EQUIP.	TOTAL	TOTAL INCL O&P	
2200	1" diameter	1 Plum	68	.118	L.F.	2.03	2.58		4.61	6.50	420
2220	1-1/4" diameter		58	.138		2.42	3.03		5.45	7.65	
3000	Type M tubing, couplings & hangers 10' O.C.										
3140	1/2" diameter	1 Plum	84	.095	L.F.	.98	2.09		3.07	4.53	
3180	3/4" diameter		78	.103		1.26	2.25		3.51	5.10	
3200	1" diameter		70	.114		1.74	2.51		4.25	6.05	
3220	1-1/4" diameter		60	.133		2.08	2.93		5.01	7.15	
3240	1-1/2" diameter		54	.148		2.67	3.25		5.92	8.35	
3260	2" diameter		44	.182		3.93	3.99		7.92	10.90	
4000	Type DWV tubing, couplings & hangers 10' O.C.										
4100	1-1/4" diameter	1 Plum	60	.133	L.F.	2.06	2.93		4.99	7.10	
4120	1-1/2" diameter		54	.148		2.45	3.25		5.70	8.10	
4140	2" diameter		44	.182		3.04	3.99		7.03	9.95	
4160	3" diameter	Q-1	58	.276		4.74	5.45		10.19	14.20	
4180	4" diameter	"	40	.400		8.25	7.90		16.15	22	
0010	**PIPE, COPPER, FITTINGS** Wrought unless otherwise noted										460
0040	Solder joints, copper x copper										
0100	1/2"	1 Plum	20	.400	Ea.	.18	8.80		8.98	14.70	
0120	3/4"		19	.421		.39	9.25		9.64	15.75	
0250	45° elbow, 1/4"		22	.364		.98	8		8.98	14.30	
0280	1/2"		20	.400		.31	8.80		9.11	14.85	
0290	5/8"		19	.421		1.52	9.25		10.77	16.95	
0300	3/4"		19	.421		.53	9.25		9.78	15.90	
0310	1"		16	.500		1.34	11		12.34	19.60	
0320	1-1/4"		15	.533		1.81	11.70		13.51	21.50	
0450	Tee, 1/4"		14	.571		1.10	12.55		13.65	21.50	
0480	1/2"		13	.615		.30	13.50		13.80	23	
0490	5/8"		12	.667		1.83	14.65		16.48	26	
0500	3/4"		12	.667		.73	14.65		15.38	25	
0510	1"		10	.800		2.12	17.55		19.67	31.50	
0520	1-1/4"		9	.889		3.21	19.50		22.71	36	
0612	Tee, reducing on the outlet, 1/4"		15	.533		1.78	11.70		13.48	21.50	
0613	3/8"		15	.533		1.67	11.70		13.37	21	
0614	1/2"		14	.571		1.46	12.55		14.01	22	
0615	5/8"		13	.615		2.47	13.50		15.97	25	
0616	3/4"		12	.667		1.97	14.65		16.62	26	
0617	1"		11	.727		2.47	15.95		18.42	29	
0618	1-1/4"		10	.800		3.63	17.55		21.18	33	
0619	1-1/2"		9	.889		3.31	19.50		22.81	36	
0620	2"		8	1		5.25	22		27.25	42.50	
0621	2-1/2"	Q-1	9	1.778		16.30	35		51.30	76	
0622	3"		8	2		19.65	39.50		59.15	87	
0623	4"		6	2.667		69	52.50		121.50	163	
0624	5"		5	3.200		161	63		224	281	
0625	6"	Q-2	7	3.429		220	65		285	350	
0626	8"	"	6	4		895	76		971	1,100	
0630	Tee, reducing on the run, 1/4"	1 Plum	15	.533		2.21	11.70		13.91	22	
0631	3/8"		15	.533		3.13	11.70		14.83	23	
0632	1/2"		14	.571		2.64	12.55		15.19	23.50	
0633	5/8"		13	.615		2.97	13.50		16.47	26	
0634	3/4"		12	.667		1.46	14.65		16.11	25.50	
0635	1"		11	.727		2.52	15.95		18.47	29.50	
0636	1-1/4"		10	.800		4.09	17.55		21.64	33.50	
0637	1-1/2"		9	.889		7.05	19.50		26.55	40.50	
0638	2"		8	1		9	22		31	46.50	

15107 | Metal Pipe & Fittings

			CREW	DAILY OUTPUT	LABOR-HOURS	UNIT	2000 BARE COSTS MAT.	LABOR	EQUIP.	TOTAL	TOTAL INCL O&P
460	0639	2-1/2″	Q-1	9	1.778	Ea.	22	35		57	82
	0640	3″		8	2		33	39.50		72.50	102
	0641	4″		6	2.667		69	52.50		121.50	163
	0642	5″		5	3.200		161	63		224	281
	0643	6″	Q-2	7	3.429		220	65		285	350
	0644	8″	″	6	4		895	76		971	1,100
	0650	Coupling, 1/4″	1 Plum	24	.333		.13	7.30		7.43	12.25
	0680	1/2″		22	.364		.13	8		8.13	13.35
	0690	5/8″		21	.381		.38	8.35		8.73	14.25
	0700	3/4″		21	.381		.26	8.35		8.61	14.15
	0710	1″		18	.444		.90	9.75		10.65	17.15
	0715	1-1/4″		17	.471		.90	10.35		11.25	18.10
	2000	DWV, solder joints, copper x copper									
	2030	90° Elbow, 1-1/4″	1 Plum	13	.615	Ea.	1.61	13.50		15.11	24.50
	2050	1-1/2″		12	.667		4.12	14.65		18.77	28.50
	2070	2″		10	.800		5.85	17.55		23.40	35.50
	2090	3″	Q-1	10	1.600		11.45	31.50		42.95	64.50
	2100	4″	″	9	1.778		36.50	35		71.50	98.50
	2250	Tee, Sanitary, 1-1/4″	1 Plum	9	.889		3.17	19.50		22.67	36
	2270	1-1/2″		8	1		3.95	22		25.95	41
	2290	2″		7	1.143		4.61	25		29.61	46.50
	2310	3″	Q-1	7	2.286		16.45	45		61.45	92.50
	2330	4″	″	6	2.667		41.50	52.50		94	133
	2400	Coupling, 1-1/4″	1 Plum	14	.571		.75	12.55		13.30	21.50
	2420	1-1/2″		13	.615		.94	13.50		14.44	23.50
	2440	2″		11	.727		1.29	15.95		17.24	28
	2460	3″	Q-1	11	1.455		2.51	28.50		31.01	50.50
	2480	4″	″	10	1.600		6	31.50		37.50	58.50
620	0010	**PIPE, STEEL**									
	0050	Schedule 40, threaded, with couplings, and clevis type									
	0060	hangers sized for covering, 10′ O.C.									
	0540	Black, 1/4″ diameter	1 Plum	66	.121	L.F.	1.45	2.66		4.11	6
	0570	3/4″ diameter		61	.131		1.54	2.88		4.42	6.45
	0580	1″ diameter		53	.151		1.83	3.31		5.14	7.50
	0590	1-1/4″ diameter	Q-1	89	.180		2.17	3.55		5.72	8.25
	0600	1-1/2″ diameter		80	.200		2.42	3.95		6.37	9.20
	0610	2″ diameter		64	.250		3.28	4.94		8.22	11.75
640	0010	**PIPE, STEEL, FITTINGS** Threaded									
	5000	Malleable iron, 150 lb.									
	5020	Black									
	5040	90° elbow, straight									
	5090	3/4″	1 Plum	14	.571	Ea.	1.28	12.55		13.83	22
	5100	1″	″	13	.615		2.23	13.50		15.73	25
	5120	1-1/2″	Q-1	20	.800		4.81	15.80		20.61	31.50
	5130	2″	″	18	.889		8.30	17.55		25.85	38
	5450	Tee, straight									
	5500	3/4″	1 Plum	9	.889	Ea.	2.04	19.50		21.54	34.50
	5510	1″	″	8	1		3.48	22		25.48	40.50
	5520	1-1/4″	Q-1	14	1.143		5.65	22.50		28.15	43.50
	5530	1-1/2″		13	1.231		7	24.50		31.50	47.50
	5540	2″		11	1.455		11.95	28.50		40.45	60.50
	5650	Coupling									
	5700	3/4″	1 Plum	18	.444	Ea.	1.71	9.75		11.46	18.05
	5710	1″	″	15	.533		2.57	11.70		14.27	22
	5730	1-1/2″	Q-1	24	.667		4.49	13.15		17.64	27

15100 | Building Services Piping

	15107 | Metal Pipe & Fittings	CREW	DAILY OUTPUT	LABOR-HOURS	UNIT	2000 BARE COSTS MAT.	LABOR	EQUIP.	TOTAL	TOTAL INCL O&P	
5740	2"	Q-1	21	.762	Ea.	6.65	15.05		21.70	32.50	640

	15108 | Plastic Pipe & Fittings	CREW	DAILY OUTPUT	LABOR-HOURS	UNIT	MAT.	LABOR	EQUIP.	TOTAL	TOTAL INCL O&P	
0010	**PIPE, PLASTIC**										520
1800	PVC, couplings 10' O.C., hangers 3 per 10'										
1820	Schedule 40										
1860	1/2" diameter	1 Plum	54	.148	L.F.	1.90	3.25		5.15	7.50	
1870	3/4" diameter		51	.157		2.06	3.44		5.50	7.95	
1880	1" diameter		46	.174		2.57	3.82		6.39	9.15	
1890	1-1/4" diameter		42	.190		2.61	4.18		6.79	9.75	
1900	1-1/2" diameter		36	.222		2.81	4.88		7.69	11.15	
1910	2" diameter	Q-1	59	.271		2.83	5.35		8.18	11.95	
1920	2-1/2" diameter		56	.286		3.52	5.65		9.17	13.20	
1930	3" diameter		53	.302		4.05	5.95		10	14.30	
1940	4" diameter		48	.333		5.50	6.60		12.10	16.95	
4100	DWV type, schedule 40, couplings 10' O.C., hangers 3 per 10'										
4120	ABS										
4140	1-1/4" diameter	1 Plum	42	.190	L.F.	2.58	4.18		6.76	9.75	
4150	1-1/2" diameter	"	36	.222		2.56	4.88		7.44	10.85	
4160	2" diameter	Q-1	59	.271		2.66	5.35		8.01	11.75	
4400	PVC										
4410	1-1/4" diameter	1 Plum	42	.190	L.F.	2.66	4.18		6.84	9.80	
4420	1-1/2" diameter	"	36	.222		2.73	4.88		7.61	11.05	
4460	2" diameter	Q-1	59	.271		2.91	5.35		8.26	12.05	
4470	3" diameter		53	.302		3.79	5.95		9.74	14	
4480	4" diameter		48	.333		5.10	6.60		11.70	16.55	
5360	CPVC, couplings 10' O.C., hangers 3 per 10'										
5380	Schedule 40										
5460	1/2" diameter	1 Plum	54	.148	L.F.	2.63	3.25		5.88	8.30	
5470	3/4" diameter		51	.157		3.04	3.44		6.48	9.05	
5480	1" diameter		46	.174		3.98	3.82		7.80	10.70	
5490	1-1/4" diameter		42	.190		4.57	4.18		8.75	11.95	
5500	1-1/2" diameter		36	.222		5.20	4.88		10.08	13.75	
5510	2" diameter	Q-1	59	.271		6.05	5.35		11.40	15.55	
0010	**PIPE, PLASTIC, FITTINGS**										560
2700	PVC (white), schedule 40, socket joints										
2760	90° elbow, 1/2"	1 Plum	22	.364	Ea.	.23	8		8.23	13.45	
2770	3/4"		21	.381		.25	8.35		8.60	14.15	
2780	1"		18	.444		.45	9.75		10.20	16.65	
2790	1-1/4"		17	.471		.80	10.35		11.15	18	
2800	1-1/2"		16	.500		.85	11		11.85	19.10	
2810	2"	Q-1	28	.571		1.34	11.30		12.64	20	
2820	2-1/2"		22	.727		4.06	14.35		18.41	28.50	
2830	3"		17	.941		4.86	18.60		23.46	36	
2840	4"		14	1.143		8.70	22.50		31.20	47	
3180	Tee, 1/2"	1 Plum	14	.571		.28	12.55		12.83	21	
3190	3/4"		13	.615		.32	13.50		13.82	23	
3200	1"		12	.667		.59	14.65		15.24	24.50	
3210	1-1/4"		11	.727		.94	15.95		16.89	27.50	
3220	1-1/2"		10	.800		1.13	17.55		18.68	30	
3230	2"	Q-1	17	.941		1.65	18.60		20.25	32.50	
3240	2-1/2"		14	1.143		5.45	22.50		27.95	43.50	
3250	3"		11	1.455		7.15	28.50		35.65	55.50	
3260	4"		9	1.778		12.90	35		47.90	72	

15108 | Plastic Pipe & Fittings

	15108 Plastic Pipe & Fittings	CREW	DAILY OUTPUT	LABOR-HOURS	UNIT	2000 BARE COSTS MAT.	LABOR	EQUIP.	TOTAL	TOTAL INCL O&P
560 3380	Coupling, 1/2"	1 Plum	22	.364	Ea.	.18	8		8.18	13.40
3390	3/4"		21	.381		.24	8.35		8.59	14.10
3400	1"		18	.444		.41	9.75		10.16	16.60
3410	1-1/4"		17	.471		.57	10.35		10.92	17.75
3420	1-1/2"		16	.500		.61	11		11.61	18.80
3430	2"	Q-1	28	.571		.95	11.30		12.25	19.70
3440	2-1/2"		20	.800		2.08	15.80		17.88	28.50
3450	3"		19	.842		3.27	16.65		19.92	31
3460	4"		16	1		4.72	19.75		24.47	37.50
4500	DWV, ABS, non pressure, socket joints									
4540	1/4 Bend, 1-1/4"	1 Plum	17	.471	Ea.	.99	10.35		11.34	18.20
4560	1-1/2"	"	16	.500		.53	11		11.53	18.75
4570	2"	Q-1	28	.571		.87	11.30		12.17	19.60
4800	Tee, sanitary									
4820	1-1/4"	1 Plum	11	.727	Ea.	1.25	15.95		17.20	28
4830	1-1/2"	"	10	.800		.77	17.55		18.32	30
4840	2"	Q-1	17	.941		1.10	18.60		19.70	31.50
5000	DWV, PVC, schedule 40, socket joints									
5040	1/4 bend, 1-1/4"	1 Plum	17	.471	Ea.	.93	10.35		11.28	18.10
5060	1-1/2"	"	16	.500		.41	11		11.41	18.60
5070	2"	Q-1	28	.571		.63	11.30		11.93	19.35
5080	3"		17	.941		1.57	18.60		20.17	32
5090	4"		14	1.143		2.82	22.50		25.32	40.50
5110	1/4 bend, long sweep, 1-1/2"	1 Plum	16	.500		1.28	11		12.28	19.55
5112	2"	Q-1	28	.571		1.05	11.30		12.35	19.80
5114	3"		17	.941		2.73	18.60		21.33	33.50
5116	4"		14	1.143		5.40	22.50		27.90	43.50
5250	Tee, sanitary 1-1/4"	1 Plum	11	.727		1.45	15.95		17.40	28
5254	1-1/2"	"	10	.800		.61	17.55		18.16	29.50
5255	2"	Q-1	17	.941		.83	18.60		19.43	31.50
5256	3"		11	1.455		2.29	28.50		30.79	50
5257	4"		9	1.778		4.38	35		39.38	63
5259	6"		5	3.200		25.50	63		88.50	132
5261	8"	Q-2	6	4		76	76		152	210
5264	2" x 1-1/2"	Q-1	17	.941		2.05	18.60		20.65	33
5266	3" x 1-1/2"		12	1.333		1.76	26.50		28.26	45.50
5268	4" x 3"		12	1.333		8.05	26.50		34.55	52.50
5271	6" x 4"		8	2		25	39.50		64.50	93
5314	Combination Y & 1/8 bend, 1-1/2"	1 Plum	10	.800		1.88	17.55		19.43	31
5315	2"	Q-1	17	.941		2.50	18.60		21.10	33.50
5317	3"		11	1.455		4.12	28.50		32.62	52
5318	4"		9	1.778		8.10	35		43.10	67
5324	Combination Y & 1/8 bend, reducing									
5325	2" x 2" x 1-1/2"	Q-1	17	.941	Ea.	3.64	18.60		22.24	34.50
5327	3" x 3" x 1-1/2"		13	1.231		4.80	24.50		29.30	45.50
5328	3" x 3" x 2"		12	1.333		3.01	26.50		29.51	47
5329	4" x 4" x 2"		11	1.455		6.70	28.50		35.20	55
5331	Wye, 1-1/4"	1 Plum	11	.727		1.39	15.95		17.34	28
5332	1-1/2"	"	10	.800		1.21	17.55		18.76	30.50
5333	2"	Q-1	17	.941		1.21	18.60		19.81	32
5334	3"		11	1.455		2.79	28.50		31.29	50.50
5335	4"		9	1.778		5.10	35		40.10	63.50
5336	6"		5	3.200		33	63		96	141
5337	8"	Q-2	5	4.800		37.50	91.50		129	193
5341	2" x 1-1/2"	Q-1	17	.941		2.13	18.60		20.73	33
5342	3" x 1-1/2"		12	1.333		2.82	26.50		29.32	46.50

15 MECHANICAL

15108 | Plastic Pipe & Fittings		CREW	DAILY OUTPUT	LABOR-HOURS	UNIT	2000 BARE COSTS MAT.	LABOR	EQUIP.	TOTAL	TOTAL INCL O&P	
5343	4" x 3"	Q-1	10	1.600	Ea.	4.79	31.50		36.29	57.50	560
5344	6" x 4"		6	2.667		18.80	52.50		71.30	108	
5345	8" x 6"	Q-2	8	3		45	57		102	144	
5347	Double wye, 1-1/2"	1 Plum	8	1		2.57	22		24.57	39.50	
5348	2"	Q-1	12	1.333		3.30	26.50		29.80	47	
5349	3"		8	2		8.50	39.50		48	75	
5350	4"		6	2.667		17.25	52.50		69.75	106	
5354	2" x 1-1/2"		11	1.455		3.02	28.50		31.52	51	
5355	3" x 2"		8	2		6.35	39.50		45.85	72.50	
5356	4" x 3"		7	2.286		13.70	45		58.70	89.50	
5357	6" x 4"		5	3.200		28.50	63		91.50	136	
5410	Reducer bushing, 2" x 1-1/4"		31	.516		.46	10.20		10.66	17.35	
5412	3" x 1-1/2"		25	.640		2.01	12.65		14.66	23	
5414	4" x 2"		22	.727		3.51	14.35		17.86	28	
5416	6" x 4"		14	1.143		10.05	22.50		32.55	48.50	
5418	8" x 6"	Q-2	18	1.333		20	25.50		45.50	64	
5500	CPVC, Schedule 80, threaded joints										
5540	90° Elbow, 1/4"	1 Plum	20	.400	Ea.	5.35	8.80		14.15	20.50	
5560	1/2"		18	.444		2.09	9.75		11.84	18.45	
5570	3/4"		17	.471		2.67	10.35		13.02	20	
5580	1"		15	.533		4.24	11.70		15.94	24	
5590	1-1/4"		14	.571		9.20	12.55		21.75	30.50	
5600	1-1/2"		13	.615		10.25	13.50		23.75	34	
5610	2"	Q-1	22	.727		12.40	14.35		26.75	37.50	
6000	Coupling, 1/4"	1 Plum	20	.400		5.70	8.80		14.50	21	
6020	1/2"		18	.444		2.21	9.75		11.96	18.60	
6030	3/4"		17	.471		3.09	10.35		13.44	20.50	
6040	1"		15	.533		4.16	11.70		15.86	24	
6050	1-1/4"		14	.571		6.25	12.55		18.80	27.50	
6060	1-1/2"		13	.615		7.85	13.50		21.35	31	
6070	2"	Q-1	22	.727		9.15	14.35		23.50	34	

15110 | Valves		CREW	DAILY OUTPUT	LABOR-HOURS	UNIT	MAT.	LABOR	EQUIP.	TOTAL	TOTAL INCL O&P	
0010	VALVES, BRONZE										160
1750	Check, swing, class 150, regrinding disc, threaded										
1860	3/4" size	1 Plum	20	.400	Ea.	28.50	8.80		37.30	46	
1870	1" size	"	19	.421	"	43	9.25		52.25	63	
2850	Gate, N.R.S., soldered, 125 psi										
2940	3/4" size	1 Plum	20	.400	Ea.	16.20	8.80		25	32.50	
2950	1" size	"	19	.421	"	23	9.25		32.25	41	
5600	Relief, pressure & temperature, self-closing, ASME, threaded										
5650	1" size	1 Plum	24	.333	Ea.	92	7.30		99.30	113	
5660	1-1/4" size	"	20	.400	"	184	8.80		192.80	217	
6400	Pressure, water, ASME, threaded										
6440	3/4" size	1 Plum	28	.286	Ea.	45	6.25		51.25	60	
6450	1" size	"	24	.333	"	91	7.30		98.30	112	
6900	Reducing, water pressure										
6940	1/2" size	1 Plum	24	.333	Ea.	103	7.30		110.30	126	
6960	1" size	"	19	.421	"	160	9.25		169.25	191	
8350	Tempering, water, sweat connections										
8400	1/2" size	1 Plum	24	.333	Ea.	39	7.30		46.30	55	
8440	3/4" size	"	20	.400	"	48	8.80		56.80	67.50	
8650	Threaded connections										
8700	1/2" size	1 Plum	24	.333	Ea.	48	7.30		55.30	65	
8740	3/4" size	"	20	.400	"	182	8.80		190.80	215	

15100 | Building Services Piping

		15120 | Piping Specialties	CREW	DAILY OUTPUT	LABOR-HOURS	UNIT	2000 BARE COSTS MAT.	LABOR	EQUIP.	TOTAL	TOTAL INCL O&P
320	0010	**EXPANSION TANKS**									
	1505	Fiberglass and steel single / double wall storage, see Div 13201									
	2000	Steel, liquid expansion, ASME, painted, 15 gallon capacity	Q-5	17	.941	Ea.	330	18.75		348.75	395
	2040	30 gallon capacity	↓	12	1.333	↓	370	26.50		396.50	450
	3000	Steel ASME expansion, rubber diaphragm, 19 gal. cap. accept.	↓	12	1.333	↓	1,250	26.50		1,276.50	1,425
	3020	31 gallon capacity	↓	8	2	↓	1,400	40		1,440	1,600
940	0010	**WATER SUPPLY METERS**									
	2000	Domestic/commercial, bronze									
	2020	Threaded									
	2060	5/8″ diameter, to 20 GPM	1 Plum	16	.500	Ea.	66.50	11		77.50	91.50
	2080	3/4″ diameter, to 30 GPM	↓	14	.571	↓	115	12.55		127.55	148
	2100	1″ diameter, to 50 GPM	↓	12	.667	↓	157	14.65		171.65	196
		15140 | Domestic Water Piping									
100	0010	**BACKFLOW PREVENTER** Includes valves									
	0020	and four test cocks, corrosion resistant, automatic operation									
	4100	Threaded, valves are ball									
	4120	3/4″ pipe size	1 Plum	16	.500	Ea.	685	11		696	775
600	0010	**VACUUM BREAKERS** Hot or cold water									
	1030	Anti-siphon, brass									
	1060	1/2″ size	1 Plum	24	.333	Ea.	20.50	7.30		27.80	34.50
	1080	3/4″ size	↓	20	.400	↓	24.50	8.80		33.30	41.50
	1100	1″ size	↓	19	.421	↓	38	9.25		47.25	57.50
800	0010	**WATER HAMMER ARRESTORS / SHOCK ABSORBERS**									
	0490	Copper									
	0500	3/4″ male I.P.S. For 1 to 11 fixtures	1 Plum	12	.667	Ea.	16.05	14.65		30.70	41.50
		15155 | Drainage Specialties									
160	0010	**CLEANOUTS**									
	0080	Round or square, scoriated nickel bronze top									
	0100	2″ pipe size	1 Plum	10	.800	Ea.	84	17.55		101.55	122
	0140	4″ pipe size	″	6	1.333	″	113	29.50		142.50	174
170	0010	**CLEANOUT TEE**									
	0100	Cast iron, B&S, with countersunk plug									
	0220	3″ pipe size	1 Plum	3.60	2.222	Ea.	87	49		136	176
	0240	4″ pipe size	″	3.30	2.424	↓	108	53		161	207
	0500	For round smooth access cover, same price				↓					
	4000	Plastic, tees and adapters. Add plugs									
	4010	ABS, DWV									
	4020	Cleanout tee, 1-1/2″ pipe size	1 Plum	15	.533	Ea.	5.45	11.70		17.15	25.50
300	0010	**DRAINS**									
	2000	Floor, medium duty, C.I., deep flange, 7″ dia top									
	2040	2″ and 3″ pipe size	Q-1	12	1.333	Ea.	60	26.50		86.50	110
	2080	For galvanized body, add				↓	25			25	27.50
	2120	For polished bronze top, add				↓	33			33	36.50
	3860	Roof, flat metal deck, C.I. body, 12″ C.I. dome									
	3890	3″ pipe size	Q-1	14	1.143	Ea.	117	22.50		139.50	167
780	0010	**TRAPS**									
	0030	Cast iron, service weight									
	0050	Running P trap, without vent									
	1100	2″	Q-1	16	1	Ea.	15.95	19.75		35.70	50
	1150	4″	″	13	1.231	↓	50	24.50		74.50	95.50
	1160	6″	Q-2	17	1.412	↓	232	27		259	300

15 MECHANICAL

Important: See the Reference Section for critical supporting data - Reference Nos., Crews, & Location Facto

15100 | Building Services Piping

	15155 | Drainage Specialties		Crew	Daily Output	Labor-Hours	Unit	2000 Bare Costs Mat.	Labor	Equip.	Total	Total Incl O&P
780	3000	P trap, B&S, 2″ pipe size	Q-1	16	1	Ea.	13.20	19.75		32.95	47
	3040	3″ pipe size	″	14	1.143		19.75	22.50		42.25	59.50
	3800	Drum trap, 4″ x 5″, 1-1/2″ tapping	Q-2	17	1.412		13.50	27		40.50	59.50
	4700	Copper, drainage, drum trap									
	4840	3″ x 6″ swivel, 1-1/2″ pipe size	1 Plum	16	.500	Ea.	22	11		33	42.50
	5100	P trap, standard pattern									
	5200	1-1/4″ pipe size	1 Plum	18	.444	Ea.	10.75	9.75		20.50	28
	5240	1-1/2″ pipe size		17	.471		10.75	10.35		21.10	29
	5260	2″ pipe size		15	.533		16.60	11.70		28.30	37.50
	5280	3″ pipe size		11	.727		40	15.95		55.95	70.50
	6710	ABS DWV P trap, solvent weld joint									
	6720	1-1/2″ pipe size	1 Plum	18	.444	Ea.	5.75	9.75		15.50	22.50
	6722	2″ pipe size		17	.471		9.35	10.35		19.70	27.50
	6724	3″ pipe size		15	.533		42.50	11.70		54.20	66.50
	6726	4″ pipe size		14	.571		87.50	12.55		100.05	117
	6860	PVC DWV hub x hub, basin trap, 1-1/4″ pipe size		18	.444		4.22	9.75		13.97	21
	6870	Sink P trap, 1-1/2″ pipe size		18	.444		4.22	9.75		13.97	21
	6880	Tubular S trap, 1-1/2″ pipe size		17	.471		9.70	10.35		20.05	28
	6890	PVC sch. 40 DWV, drum trap									
	6900	1-1/2″ pipe size	1 Plum	16	.500	Ea.	9.90	11		20.90	29
	6910	P trap, 1-1/2″ pipe size		18	.444		1.91	9.75		11.66	18.25
	6920	2″ pipe size		17	.471		2.60	10.35		12.95	19.95
	6930	3″ pipe size		15	.533		12.95	11.70		24.65	33.50
	6940	4″ pipe size		14	.571		31.50	12.55		44.05	55
	6950	P trap w/clean out, 1-1/2″ pipe size		18	.444		4.64	9.75		14.39	21.50
	6960	2″ pipe size		17	.471		7.90	10.35		18.25	26

	15180 | Heating and Cooling Piping		Crew	Daily Output	Labor-Hours	Unit	Mat.	Labor	Equip.	Total	Total Incl O&P
200	0010	**PUMPS, CIRCULATING** Heated or chilled water application									
	0600	Bronze, sweat connections, 1/40 HP, in line									
	0640	3/4″ size	Q-1	16	1	Ea.	110	19.75		129.75	154
	1000	Flange connection, 3/4″ to 1-1/2″ size									
	1040	1/12 HP	Q-1	6	2.667	Ea.	300	52.50		352.50	415
	1060	1/8 HP	″	6	2.667	″	520	52.50		572.50	655

MECHANICAL 15

15400 | Plumbing Fixtures & Equipment

	15410 | Plumbing Fixtures		Crew	Daily Output	Labor-Hours	Unit	2000 Bare Costs Mat.	Labor	Equip.	Total	Total Incl O&P
200	0010	**CARRIERS/SUPPORTS** For plumbing fixtures									
	0600	Plate type with studs, top back plate	1 Plum	7	1.143	Ea.	26.50	25		51.50	70.50
	3000	Lavatory, concealed arm									
	3050	Floor mounted, single									
	3100	High back fixture	1 Plum	6	1.333	Ea.	128	29.50		157.50	190
	3200	Flat slab fixture	″	6	1.333	″	149	29.50		178.50	213
	8200	Water closet, residential									
	8220	Vertical centerline, floor mount									
	8240	Single, 3″ caulk, 2″ or 3″ vent	1 Plum	6	1.333	Ea.	147	29.50		176.50	211
	8260	4″ caulk, 2″ or 4″ vent	″	6	1.333	″	185	29.50		214.50	253
300	0010	**FAUCETS/FITTINGS**									
	0150	Bath, faucets, diverter spout combination, sweat	1 Plum	8	1	Ea.	65	22		87	108

	15410 | Plumbing Fixtures		CREW	DAILY OUTPUT	LABOR-HOURS	UNIT	2000 BARE COSTS MAT.	LABOR	EQUIP.	TOTAL	TOTAL INCL O&P
300	0200	For integral stops, IPS unions, add				Ea.	68.50			68.50	75.50
	0500	Drain, central lift, 1-1/2" IPS male	1 Plum	20	.400		35.50	8.80		44.30	54
	0600	Trip lever, 1-1/2" IPS male		20	.400		36.50	8.80		45.30	55
	1000	Kitchen sink faucets, top mount, cast spout		10	.800		48	17.55		65.55	82
	1100	For spray, add		24	.333		12	7.30		19.30	25.50
	2000	Laundry faucets, shelf type, IPS or copper unions		12	.667		39.50	14.65		54.15	67.50
	2020										
	2100	Lavatory faucet, centerset, without drain	1 Plum	10	.800	Ea.	51	17.55		68.55	85
	2200	For pop-up drain, add		16	.500		14.20	11		25.20	34
	2800	Self-closing, center set		10	.800		102	17.55		119.55	142
	3000	Service sink faucet, cast spout, pail hook, hose end		14	.571		67.50	12.55		80.05	94.50
	4000	Shower by-pass valve with union		18	.444		47.50	9.75		57.25	68
	4200	Shower thermostatic mixing valve, concealed		8	1		226	22		248	286
	4300	For inlet strainer, check, and stops, add					43			43	47.50
	5000	Sillcock, compact, brass, IPS or copper to hose	1 Plum	24	.333		4.94	7.30		12.24	17.55

	15418 | Resi/Comm/Industrial Fixtures		CREW	DAILY OUTPUT	LABOR-HOURS	UNIT	MAT.	LABOR	EQUIP.	TOTAL	TOTAL INCL O&P
100	0010	**BATHS** R15100-420									
	0100	Tubs, recessed porcelain enamel on cast iron, with trim									
	0180	48" x 42"	Q-1	4	4	Ea.	1,025	79		1,104	1,250
	0220	72" x 36"		3	5.333		1,125	105		1,230	1,400
	0300	Mat bottom, 4' long		5.50	2.909		875	57.50		932.50	1,050
	0380	5' long		4.40	3.636		305	72		377	455
	0480	Above floor drain, 5' long		4	4		545	79		624	730
	0560	Corner 48" x 44"		4.40	3.636		1,200	72		1,272	1,450
	2000	Enameled formed steel, 4'-6" long		5.80	2.759		240	54.50		294.50	355
	2200	5' long		5.50	2.909		224	57.50		281.50	340
	4600	Module tub & showerwall surround, molded fiberglass									
	4610	5' long x 34" wide x 76" high	Q-1	4	4	Ea.	575	79		654	760
	6000	Whirlpool, bath with vented overflow, molded fiberglass									
	6100	66" x 48" x 24"	Q-1	1	16	Ea.	1,975	315		2,290	2,700
	6400	72" x 36" x 24"		1	16		1,925	315		2,240	2,650
	6500	60" x 30" x 21"		1	16		1,700	315		2,015	2,375
	6600	72" x 42" x 22"		1	16		2,725	315		3,040	3,525
	6700	83" x 65"		.30	53.333		3,750	1,050		4,800	5,875
	7000	Redwood hot tub system									
	7050	4' diameter x 4' deep	Q-1	1	16	Ea.	1,350	315		1,665	2,000
	7150	6' diameter x 4' deep		.80	20		2,150	395		2,545	3,025
	7200	8' diameter x 4' deep		.80	20		3,175	395		3,570	4,150
	9600	Rough-in, supply, waste and vent, for all above tubs, add		2.07	7.729		102	153		255	365
400	0010	**LAUNDRY SINKS** With trim									
	0020	Porcelain enamel on cast iron, black iron frame									
	0050	24" x 20", single compartment	Q-1	6	2.667	Ea.	340	52.50		392.50	455
	0100	24" x 23", single compartment	"	6	2.667	"	365	52.50		417.50	490
	3000	Plastic, on wall hanger or legs									
	3020	18" x 23", single compartment	Q-1	6.50	2.462	Ea.	82	48.50		130.50	171
	3100	20" x 24", single compartment		6.50	2.462		108	48.50		156.50	200
	3200	36" x 23", double compartment		5.50	2.909		127	57.50		184.50	235
	3300	40" x 24", double compartment		5.50	2.909		187	57.50		244.50	300
	5000	Stainless steel, counter top, 22" x 17" single compartment		6	2.667		253	52.50		305.50	365
	5100	22" x 22", single compartment		6	2.667		320	52.50		372.50	435
	5200	33" x 22", double compartment		5	3.200		295	63		358	430
	9600	Rough-in, supply, waste and vent, for all laundry sinks		2.14	7.477		60	148		208	310

	15418 | Resi/Comm/Industrial Fixtures	CREW	DAILY OUTPUT	LABOR-HOURS	UNIT	2000 BARE COSTS MAT.	LABOR	EQUIP.	TOTAL	TOTAL INCL O&P	
0010	**LAVATORIES** With trim, white unless noted otherwise										450
0500	Vanity top, porcelain enamel on cast iron										
0600	20" x 18"	Q-1	6.40	2.500	Ea.	199	49.50		248.50	300	
0640	33" x 19" oval		6.40	2.500		345	49.50		394.50	455	
0720	18" round		6.40	2.500		180	49.50		229.50	280	
0860	For color, add					25%					
1000	Cultured marble, 19" x 17", single bowl	Q-1	6.40	2.500	Ea.	117	49.50		166.50	211	
1120	25" x 22", single bowl		6.40	2.500		153	49.50		202.50	251	
1160	37" x 22", single bowl		6.40	2.500		184	49.50		233.50	284	
1560											
1900	Stainless steel, self-rimming, 25" x 22", single bowl, ledge	Q-1	6.40	2.500	Ea.	155	49.50		204.50	253	
1960	17" x 22", single bowl		6.40	2.500		128	49.50		177.50	223	
2600	Steel, enameled, 20" x 17", single bowl		5.80	2.759		101	54.50		155.50	201	
2900	Vitreous china, 20" x 16", single bowl		5.40	2.963		184	58.50		242.50	299	
3200	22" x 13", single bowl		5.40	2.963		193	58.50		251.50	310	
3580	Rough-in, supply, waste and vent for all above lavatories		2.30	6.957		52.50	137		189.50	285	
4000	Wall hung										
4040	Porcelain enamel on cast iron, 16" x 14", single bowl	Q-1	8	2	Ea.	345	39.50		384.50	445	
4180	20" x 18", single bowl	"	8	2	"	200	39.50		239.50	286	
4580	For color, add					30%					
6000	Vitreous china, 18" x 15", single bowl with backsplash	Q-1	7	2.286	Ea.	197	45		242	291	
6060	19" x 17", single bowl		7	2.286		146	45		191	235	
6960	Rough-in, supply, waste and vent for above lavatories		1.66	9.639		160	190		350	490	
0010	**SHOWERS**										500
1500	Stall, with drain only. Add for valve and door/curtain										
1520	32" square	Q-1	2	8	Ea.	315	158		473	605	
1530	36" square		2	8		410	158		568	710	
1540	Terrazzo receptor, 32" square		2	8		990	158		1,148	1,350	
1560	36" square		1.80	8.889		820	176		996	1,200	
1580	36" corner angle		1.80	8.889		750	176		926	1,125	
3000	Fiberglass, one piece, with 3 walls, 32" x 32" square		2.40	6.667		290	132		422	540	
3100	36" x 36" square		2.40	6.667		330	132		462	585	
4200	Rough-in, supply, waste and vent for above showers		2.05	7.805		53.50	154		207.50	315	
0010	**SINKS** With faucets and drain										600
2000	Kitchen, counter top style, P.E. on C.I., 24" x 21" single bowl	Q-1	5.60	2.857	Ea.	185	56.50		241.50	298	
2100	31" x 22" single bowl		5.60	2.857		315	56.50		371.50	440	
2200	32" x 21" double bowl		4.80	3.333		276	66		342	415	
3000	Stainless steel, self rimming, 19" x 18" single bowl		5.60	2.857		263	56.50		319.50	385	
3100	25" x 22" single bowl		5.60	2.857		290	56.50		346.50	415	
3200	33" x 22" double bowl		4.80	3.333		415	66		481	565	
3300	43" x 22" double bowl		4.80	3.333		475	66		541	635	
4000	Steel, enameled, with ledge, 24" x 21" single bowl		5.60	2.857		99	56.50		155.50	203	
4100	32" x 21" double bowl		4.80	3.333		116	66		182	237	
4960	For color sinks except stainless steel, add					10%					
4980	For rough-in, supply, waste and vent, counter top sinks	Q-1	2.14	7.477		60	148		208	310	
5000	Kitchen, raised deck, P.E. on C.I.										
5100	32" x 21", dual level, double bowl	Q-1	2.60	6.154	Ea.	350	122		472	585	
5790	For rough-in, supply, waste & vent, sinks		1.85	8.649		60	171		231	350	
6650	Service, floor, corner, P.E. on C.I., 28" x 28"		4.40	3.636		475	72		547	640	
6750	Vinyl coated rim guard, add					61.50			61.50	68	
6760	Mop sink, molded stone, 24" x 36"	1 Plum	3.33	2.402		187	52.50		239.50	293	
6770	Mop sink, molded stone, 24" x 36", w/rim 3 sides	"	3.33	2.402		410	52.50		462.50	535	
6790	For rough-in, supply, waste & vent, floor service sinks	Q-1	1.64	9.756		114	193		307	445	
0010	**WATER CLOSETS**										900
0150	Tank type, vitreous china, incl. seat, supply pipe w/stop										

15400 | Plumbing Fixtures & Equipment

		15418 | Resi/Comm/Industrial Fixtures	CREW	DAILY OUTPUT	LABOR-HOURS	UNIT	2000 BARE COSTS MAT.	LABOR	EQUIP.	TOTAL	TOTAL INCL O&P
900	0200	Wall hung, one piece	Q-1	5.30	3.019	Ea.	445	59.50		504.50	590
	0400	Two piece, close coupled		5.30	3.019		385	59.50		444.50	525
	0960	For rough-in, supply, waste, vent and carrier		2.73	5.861		255	116		371	470
	1000	Floor mounted, one piece		5.30	3.019		435	59.50		494.50	580
	1020	One piece, low profile		5.30	3.019		500	59.50		559.50	650
	1100	Two piece, close coupled, water saver	↓	5.30	3.019	↓	126	59.50		185.50	238
	1960	For color, add					30%				
	1980	For rough-in, supply, waste and vent	Q-1	3.05	5.246	Ea.	125	104		229	310
	3000	Bowl only, with flush valve, seat									
	3100	Wall hung	Q-1	5.80	2.759	Ea.	335	54.50		389.50	455
	3200	For rough-in, supply, waste and vent, single WC		2.56	6.250		261	123		384	490
	3300	Floor mounted		5.80	2.759		299	54.50		353.50	420
	3400	For rough-in, supply, waste and vent, single WC	↓	2.84	5.634	↓	131	111		242	330

		15440 | Plumbing Pumps	CREW	DAILY OUTPUT	LABOR-HOURS	UNIT	MAT.	LABOR	EQUIP.	TOTAL	TOTAL INCL O&P
940	0010	**PUMPS, SUBMERSIBLE** Sump									
	7000	Sump pump, automatic									
	7100	Plastic, 1-1/4" discharge, 1/4 HP	1 Plum	6	1.333	Ea.	105	29.50		134.50	165
	7500	Cast iron, 1-1/4" discharge, 1/4 HP	"	6	1.333	"	125	29.50		154.50	186

		15480 | Domestic Water Heaters	CREW	DAILY OUTPUT	LABOR-HOURS	UNIT	MAT.	LABOR	EQUIP.	TOTAL	TOTAL INCL O&P
200	0010	**WATER HEATERS**									
	1000	Residential, electric, glass lined tank, 5 yr, 10 gal., single element	1 Plum	2.30	3.478	Ea.	188	76.50		264.50	335
	1060	30 gallon, double element		2.20	3.636		239	80		319	395
	1080	40 gallon, double element		2	4		258	88		346	430
	1100	52 gallon, double element		2	4		296	88		384	470
	1120	66 gallon, double element		1.80	4.444		405	97.50		502.50	605
	1140	80 gallon, double element		1.60	5		460	110		570	685
	2000	Gas fired, glass lined tank, 5 yr, vent not incl., 20 gallon		2.10	3.810		265	83.50		348.50	430
	2040	30 gallon		2	4		315	88		403	490
	2100	75 gallon		1.50	5.333		805	117		922	1,075
	3000	Oil fired, glass lined tank, 5 yr, vent not included, 30 gallon		2	4		745	88		833	965
	3040	50 gallon	↓	1.80	4.444	↓	1,000	97.50		1,097.50	1,250

15500 | Heat Generation Equipment

		15510 | Heating Boilers and Accessories	CREW	DAILY OUTPUT	LABOR-HOURS	UNIT	2000 BARE COSTS MAT.	LABOR	EQUIP.	TOTAL	TOTAL INCL O&P
120	0010	**BURNERS**									
	0990	Residential, conversion, gas fired, LP or natural									
	1000	Gun type, atmospheric input 72 to 200 MBH	Q-1	2.50	6.400	Ea.	655	126		781	930
	1020	120 to 360 MBH		2	8		725	158		883	1,050
	1040	280 to 800 MBH	↓	1.70	9.412	↓	1,400	186		1,586	1,850
300	0010	**BOILERS, ELECTRIC, ASME** Standard controls and trim									
	1000	Steam, 6 KW, 20.5 MBH	Q-19	1.20	20	Ea.	7,900	415		8,315	9,375
	1160	60 KW, 205 MBH		1	24		9,775	495		10,270	11,500
	2000	Hot water, 12 KW, 41 MBH		1.30	18.462		3,400	380		3,780	4,350
	2040	24 KW, 82 MBH		1.20	20		3,625	415		4,040	4,675
	2060	30 KW, 103 MBH	↓	1.20	20	↓	3,725	415		4,140	4,775

15510 | Heating Boilers and Accessories

		CREW	DAILY OUTPUT	LABOR-HOURS	UNIT	2000 BARE COSTS MAT.	LABOR	EQUIP.	TOTAL	TOTAL INCL O&P	
0010	**BOILERS, GAS FIRED** Natural or propane, standard controls										400
1000	Cast iron, with insulated jacket										
3000	Hot water, gross output, 80 MBH	Q-7	1.46	21.918	Ea.	1,000	435		1,435	1,850	
3020	100 MBH	"	1.35	23.704	"	1,175	470		1,645	2,075	
4000	Steel, insulating jacket										
6000	Hot water, including burner & one zone valve, gross output										
6010	51.2 MBH	Q-6	2	12	Ea.	1,575	230		1,805	2,100	
6020	72 MBH		2	12		1,750	230		1,980	2,300	
6040	89 MBH		1.90	12.632		1,775	242		2,017	2,375	
6060	105 MBH		1.80	13.333		2,000	256		2,256	2,625	
6080	132 MBH		1.70	14.118		2,275	271		2,546	2,950	
6100	155 MBH	↓	1.50	16	↓	2,650	305		2,955	3,400	
7000	For tankless water heater on smaller gas units, add					10%					
7050	For additional zone valves up to 312 MBH add				Ea.	101			101	111	
0010	**BOILERS, GAS/OIL** Combination with burners and controls										460
1000	Cast iron with insulated jacket										
2000	Steam, gross output, 720 MBH	Q-7	.43	74.074	Ea.	6,375	1,475		7,850	9,450	
2900	Hot water, gross output										
2910	200 MBH	Q-6	.61	39.024	Ea.	4,150	750		4,900	5,800	
2920	300 MBH		.49	49.080		4,150	940		5,090	6,125	
2930	400 MBH		.41	57.971		4,850	1,100		5,950	7,175	
2940	500 MBH	↓	.36	67.039		5,225	1,275		6,500	7,875	
3000	584 MBH	Q-7	.44	72.072	↓	8,100	1,425		9,525	11,300	
4000	Steel, insulated jacket, skid base, tubeless										
4500	Steam, 150 psi gross output, 335 MBH, 10 BHP	Q-6	.54	44.037	Ea.	9,250	845		10,095	11,600	
0010	**BOILERS, OIL FIRED** Standard controls, flame retention burner										500
1000	Cast iron, with insulated flush jacket										
2000	Steam, gross output, 109 MBH	Q-7	1.20	26.667	Ea.	1,325	530		1,855	2,350	
2060	207 MBH	"	.90	35.556	"	1,825	710		2,535	3,175	
3000	Hot water, same price as steam										
7000	Hot water, gross output, 103 MBH	Q-6	1.60	15	Ea.	1,300	288		1,588	1,900	
7020	122 MBH		1.45	16.506		2,100	315		2,415	2,825	
7060	168 MBH		1.30	18.405		3,625	355		3,980	4,575	
7080	225 MBH	↓	1.22	19.704	↓	4,125	380		4,505	5,150	
0010	**SWIMMING POOL HEATERS** Not including wiring, external										880
0020	piping, base or pad,										
0060	Gas fired, input, 115 MBH	Q-6	3	8	Ea.	1,500	153		1,653	1,900	
0100	135 MBH		2	12		1,675	230		1,905	2,225	
0160	155 MBH		1.50	16		1,775	305		2,080	2,450	
0200	190 MBH		1	24		2,300	460		2,760	3,300	
0280	500 MBH	↓	.40	60		5,600	1,150		6,750	8,050	
2000	Electric, 12 KW, 4,800 gallon pool	Q-19	3	8		1,475	165		1,640	1,900	
2020	15 KW, 7,200 gallon pool		2.80	8.571		1,475	177		1,652	1,925	
2040	24 KW, 9,600 gallon pool		2.40	10		1,975	206		2,181	2,525	
2100	55 KW, 24,000 gallon pool	↓	1.20	20	↓	2,825	415		3,240	3,775	

15530 | Furnaces

		CREW	DAILY OUTPUT	LABOR-HOURS	UNIT	MAT.	LABOR	EQUIP.	TOTAL	TOTAL INCL O&P	
0010	**FURNACE COMPONENTS AND COMBINATIONS**										200
0080	Coils, A/C evaporator, for gas or oil furnaces										
0090	Add-on, with holding charge										
0100	Upflow										
0120	1-1/2 ton cooling	Q-5	4	4	Ea.	127	79.50		206.50	272	
0130	2 ton cooling	↓	3.70	4.324	↓	153	86		239	310	

15530 | Furnaces

		15530 Furnaces	CREW	DAILY OUTPUT	LABOR-HOURS	UNIT	2000 BARE COSTS MAT.	LABOR	EQUIP.	TOTAL	TOTAL INCL O&P
200	0140	3 ton cooling	Q-5	3.30	4.848	Ea.	191	96.50		287.50	370
	0150	4 ton cooling		3	5.333		270	106		376	470
	0160	5 ton cooling		2.70	5.926		345	118		463	575
	0300	Downflow									
	0330	2-1/2 ton cooling	Q-5	3	5.333	Ea.	178	106		284	370
	0340	3-1/2 ton cooling		2.60	6.154		239	122		361	465
	0350	5 ton cooling		2.20	7.273		345	145		490	620
	0600	Horizontal									
	0630	2 ton cooling	Q-5	3.90	4.103	Ea.	180	81.50		261.50	335
	0640	3 ton cooling		3.50	4.571		206	91		297	375
	0650	4 ton cooling		3.20	5		259	99.50		358.50	450
	0660	5 ton cooling		2.90	5.517		345	110		455	560
	2000	Cased evaporator coils for air handlers									
	2100	1-1/2 ton cooling	Q-5	4.40	3.636	Ea.	178	72.50		250.50	315
	2110	2 ton cooling		4.10	3.902		180	77.50		257.50	325
	2120	2-1/2 ton cooling		3.90	4.103		219	81.50		300.50	375
	2130	3 ton cooling		3.70	4.324		219	86		305	385
	2140	3-1/2 ton cooling		3.50	4.571		267	91		358	445
	2150	4 ton cooling		3.20	5		268	99.50		367.50	460
	2160	5 ton cooling		2.90	5.517		325	110		435	535
	3010	Air handler, modular									
	3100	With cased evaporator cooling coil									
	3120	1-1/2 ton cooling	Q-5	3.80	4.211	Ea.	470	84		554	655
	3130	2 ton cooling		3.50	4.571		470	91		561	665
	3140	2-1/2 ton cooling		3.30	4.848		530	96.50		626.50	745
	3150	3 ton cooling		3.10	5.161		530	103		633	755
	3160	3-1/2 ton cooling		2.90	5.517		595	110		705	835
	3170	4 ton cooling		2.50	6.400		595	127		722	865
	3180	5 ton cooling		2.10	7.619		715	152		867	1,025
	3500	With no cooling coil									
	3520	1-1/2 ton coil size	Q-5	12	1.333	Ea.	291	26.50		317.50	365
	3530	2 ton coil size		10	1.600		335	32		367	425
	3540	2-1/2 ton coil size		10	1.600		335	32		367	425
	3554	3 ton coil size		9	1.778		345	35.50		380.50	440
	3560	3-1/2 ton coil size		9	1.778		355	35.50		390.50	450
	3570	4 ton coil size		8.50	1.882		325	37.50		362.50	420
	3580	5 ton coil size		8	2		390	40		430	495
	4000	Heater for above handlers									
	4120	5 kW, 17.1 MBH	Q-5	16	1	Ea.	247	19.90		266.90	305
	4130	7.5 kW, 25.6 MBH		15.60	1.026		251	20.50		271.50	310
	4140	10 kW, 34.2 MBH		15.20	1.053		297	21		318	360
	4150	12.5 KW, 42.7 MBH		14.80	1.081		297	21.50		318.50	360
	4160	15 KW, 51.2 MBH		14.40	1.111		370	22		392	445
	4170	25 KW, 85.4 MBH		14	1.143		370	22.50		392.50	450
	4180	30 KW, 102 MBH		13	1.231		505	24.50		529.50	595
400	0010	**FURNACES** Hot air heating, blowers, standard controls (R13600-610)									
	0020	not including gas, oil or flue piping									
	1000	Electric, UL listed									
	1020	10.2 MBH	Q-20	5	4	Ea.	315	79.50		394.50	480
	1100	34.1 MBH	"	4.40	4.545	"	405	90.50		495.50	595
	3000	Gas, AGA certified, direct drive models									
	3020	45 MBH input	Q-9	4	4	Ea.	435	77.50		512.50	610
	3040	60 MBH input		3.80	4.211		580	81.50		661.50	780
	3060	75 MBH input		3.60	4.444		615	86		701	825
	3100	100 MBH input		3.20	5		655	97		752	885

15500 | Heat Generation Equipment

15530 | Furnaces

	15530 Furnaces	CREW	DAILY OUTPUT	LABOR-HOURS	UNIT	2000 BARE COSTS MAT.	LABOR	EQUIP.	TOTAL	TOTAL INCL O&P	
3120	125 MBH input R13600-610	Q-9	3	5.333	Ea.	755	103		858	1,000	400
3130	150 MBH input		2.80	5.714		875	111		986	1,150	
3140	200 MBH input		2.60	6.154		1,850	119		1,969	2,225	
4000	For starter plenum, add		16	1		61.50	19.35		80.85	100	
6000	Oil, UL listed, atomizing gun type burner										
6020	56 MBH output	Q-9	3.60	4.444	Ea.	745	86		831	965	
6030	84 MBH output		3.50	4.571		775	88.50		863.50	1,000	
6040	95 MBH output		3.40	4.706		790	91		881	1,025	
6060	134 MBH output		3.20	5		1,100	97		1,197	1,375	
6080	151 MBH output		3	5.333		1,225	103		1,328	1,500	
6100	200 MBH input		2.60	6.154		1,950	119		2,069	2,350	
0010	**FURNACES, COMBINATION SYSTEMS** Heating, cooling, electric air										440
0020	cleaner, humidification, dehumidification.										
2000	Gas fired, 80 MBH heat output, 24 MBH cooling	Q-9	1.20	13.333	Ea.	2,700	258		2,958	3,400	
2020	80 MBH heat output, 36 MBH cooling		1.20	13.333		2,975	258		3,233	3,675	
2040	100 MBH heat output, 29 MBH cooling		1	16		3,025	310		3,335	3,850	
2060	100 MBH heat output, 36 MBH cooling		1	16		3,175	310		3,485	4,025	
2080	100 MBH heat output, 47 MBH cooling		.90	17.778		3,775	345		4,120	4,725	
2100	120 MBH heat output, 29 MBH cooling	Q-10	1.30	18.462		3,200	370		3,570	4,150	
2120	120 MBH heat output, 42 MBH cooling		1.30	18.462		3,400	370		3,770	4,350	
2140	120 MBH heat output, 47 MBH cooling		1.20	20		3,800	400		4,200	4,850	
2160	120 MBH heat output, 55 MBH cooling		1.10	21.818		4,075	440		4,515	5,225	
2180	144 MBH heat output, 42 MBH cooling		1.20	20		3,525	400		3,925	4,575	
2200	144 MBH heat output, 47 MBH cooling		1.20	20		3,925	400		4,325	5,000	
2220	144 MBH heat output, 58 MBH cooling		1	24		4,125	480		4,605	5,350	
2250	144 MBH heat, 60 MBH cool		.70	34.286		4,125	690		4,815	5,675	
3000	Oil fired, 84 MBH heat output, 24 MBH cooling	Q-9	1.20	13.333		3,025	258		3,283	3,750	
3020	84 MBH heat output, 36 MBH cooling		1.20	13.333		3,325	258		3,583	4,100	
3040	95.2 MBH heat output, 29 MBH cooling		1	16		3,250	310		3,560	4,100	
3060	95.2 MBH heat output, 36 MBH cooling		1	16		3,400	310		3,710	4,250	
3280	184.8 MBH heat, 60 MBH cooling	Q-10	1	24		4,675	480		5,155	5,950	
3500	For precharged tubing with connection, add										
3520	15 feet				Ea.	113			113	124	
3540	25 feet					150			150	165	
3560	35 feet					180			180	198	

15550 | Breechings, Chimneys & Stacks

	15550 Breechings, Chimneys & Stacks	CREW	DAILY OUTPUT	LABOR-HOURS	UNIT	MAT.	LABOR	EQUIP.	TOTAL	TOTAL INCL O&P	
0010	**VENT CHIMNEY** Prefab metal, U.L. listed										200
0020	Gas, double wall, galvanized steel										
0080	3" diameter	Q-9	72	.222	V.L.F.	3.38	4.30		7.68	10.95	
0100	4" diameter		68	.235	"	4.14	4.55		8.69	12.25	
5000	Vent damper bi-metal 6" flue		16	1	Ea.	97.50	19.35		116.85	140	
5100	Gas, auto., electric		8	2	"	156	38.50		194.50	238	

15700 | Heating/Ventilating/Air Conditioning Equipment

15730 | Unitary Air Conditioning Equip

	15730 Unitary Air Conditioning Equip	CREW	DAILY OUTPUT	LABOR-HOURS	UNIT	2000 BARE COSTS MAT.	LABOR	EQUIP.	TOTAL	TOTAL INCL O&P	
0010	**PACKAGED TERMINAL AIR CONDITIONER** Cabinet, wall sleeve,										500
0100	louver, electric heat, thermostat, manual changeover, 208 V										

	15730 | Unitary Air Conditioning Equip		CREW	DAILY OUTPUT	LABOR-HOURS	UNIT	2000 BARE COSTS MAT.	LABOR	EQUIP.	TOTAL	TOTAL INCL O&P
500	0200	6,000 BTUH cooling, 8800 BTU heat	Q-5	6	2.667	Ea.	915	53		968	1,100
	0220	9,000 BTUH cooling, 13,900 BTU heat		5	3.200		935	63.50		998.50	1,125
	0240	12,000 BTUH cooling, 13,900 BTU heat		4	4		970	79.50		1,049.50	1,200
	0260	15,000 BTUH cooling, 13,900 BTU heat	↓	3	5.333	↓	1,050	106		1,156	1,325
600	0010	**ROOF TOP AIR CONDITIONERS** Standard controls, curb, economizer									
	1000	Single zone, electric cool, gas heat									
	1140	5 ton cooling, 112 MBH heating	Q-5	.56	28.520	Ea.	4,675	570		5,245	6,100
	1160	10 ton cooling, 200 MBH heating	Q-6	.67	35.982	"	9,500	690		10,190	11,600
800	0010	**WINDOW UNIT AIR CONDITIONERS**									
	4000	Portable/window, 15 amp 125V grounded receptacle required									
	4060	5000 BTUH	1 Carp	8	1	Ea.	219	19.70		238.70	275
	4340	6000 BTUH		8	1		265	19.70		284.70	325
	4480	8000 BTUH		6	1.333		279	26.50		305.50	350
	4500	10,000 BTUH	↓	6	1.333		380	26.50		406.50	460
	4520	12,000 BTUH	L-2	8	2	↓	470	34.50		504.50	575
	4600	Window/thru-the-wall, 15 amp 230V grounded receptacle required									
	4780	17,000 BTUH	L-2	6	2.667	Ea.	715	46		761	865
	4940	25,000 BTUH		4	4		1,025	69		1,094	1,250
	4960	29,000 BTUH	↓	4	4	↓	1,200	69		1,269	1,450
840	0010	**SELF-CONTAINED SINGLE PACKAGE**									
	0100	Air cooled, for free blow or duct, including remote condenser									
	0200	3 ton cooling	Q-5	1	16	Ea.	4,400	320		4,720	5,350
	0210	4 ton cooling	"	.80	20	"	4,775	400		5,175	5,900
	1000	Water cooled for free blow or duct, not including tower									
	1100	3 ton cooling	Q-6	1	24	Ea.	3,075	460		3,535	4,125
900	0010	**SPLIT DUCTLESS SYSTEM**									
	0100	Cooling only, single zone									
	0110	Wall mount									
	0120	3/4 ton cooling	Q-5	2	8	Ea.	1,375	159		1,534	1,800
	0130	1 ton cooling		1.80	8.889		1,650	177		1,827	2,125
	0140	1-1/2 ton cooling		1.60	10		2,225	199		2,424	2,775
	0150	2 ton cooling	↓	1.40	11.429	↓	2,950	227		3,177	3,625
	1000	Ceiling mount									
	1020	2 ton cooling	Q-5	1.40	11.429	Ea.	3,025	227		3,252	3,700
	1030	3 ton cooling	"	1.20	13.333	"	4,750	265		5,015	5,675
	2000	T-Bar mount									
	2010	2 ton cooling	Q-5	1.40	11.429	Ea.	3,450	227		3,677	4,175
	2020	3 ton cooling		1.20	13.333		4,250	265		4,515	5,125
	2030	3-1/2 ton cooling	↓	1.10	14.545	↓	5,125	289		5,414	6,100
	3000	Multizone									
	3010	Wall mount									
	3020	2 @ 3/4 ton cooling	Q-5	1.80	8.889	Ea.	4,250	177		4,427	4,975
	5000	Cooling / Heating									
	5110	1 ton cooling	Q-5	1.70	9.412	Ea.	2,075	187		2,262	2,600
	5120	1-1/2 ton cooling	"	1.50	10.667	"	2,825	212		3,037	3,450
	5300	Ceiling mount									
	5310	3 ton cooling	Q-5	1	16	Ea.	5,500	320		5,820	6,600
	7000	Accessories for all split ductless systems									
	7010	Add for ambient frost control	Q-5	8	2	Ea.	234	40		274	325
	7020	Add for tube / wiring kit									
	7030	15′ kit	Q-5	32	.500	Ea.	87	9.95		96.95	112
	7040	35′ kit	"	24	.667	"	173	13.25		186.25	213

Important: See the Reference Section for critical supporting data - Reference Nos., Crews, & Location Fact

	15740 | Heat Pumps	CREW	DAILY OUTPUT	LABOR-HOURS	UNIT	2000 BARE COSTS MAT.	LABOR	EQUIP.	TOTAL	TOTAL INCL O&P	
0010	**HEAT PUMPS** (Not including interconnecting tubing)										100
1000	Air to air, split system, not including curbs, pads, or ductwork										
1020	2 ton cooling, 8.5 MBH heat @ 0°F	Q-5	1.20	13.333	Ea.	2,100	265		2,365	2,750	
1054	4 ton cooling, 24 MBH heat @ 0°F	"	.60	26.667	"	3,650	530		4,180	4,900	
1500	Single package, not including curbs, pads, or plenums										
1520	2 ton cooling, 6.5 MBH heat @ 0°F	Q-5	1.50	10.667	Ea.	2,525	212		2,737	3,125	
1580	4 ton cooling, 13 MBH heat @ 0°F	"	.96	16.667	"	3,800	330		4,130	4,725	
2000	Water source to air, single package										
2100	1 ton cooling, 13 MBH heat @ 75°F	Q-5	2	8	Ea.	965	159		1,124	1,350	
2200	4 ton cooling, 31 MBH heat @ 75°F	"	1.20	13.333	"	1,750	265		2,015	2,375	
	15766 | Fin Tube Radiation										
0010	**HYDRONIC HEATING** Terminal units, not incl. main supply pipe										190
1000	Radiation										
1310	Baseboard, pkgd, 1/2" copper tube, alum. fin, 7" high	Q-5	60	.267	L.F.	6.75	5.30		12.05	16.20	
1320	3/4" copper tube, alum. fin, 7" high		58	.276		7.15	5.50		12.65	17	
1340	1" copper tube, alum. fin, 8-7/8" high		56	.286		14.80	5.70		20.50	25.50	
1360	1-1/4" copper tube, alum. fin, 8-7/8" high		54	.296		22	5.90		27.90	34	
3000	Radiators, cast iron										
3100	Free standing or wall hung, 6 tube, 25" high	Q-5	96	.167	Section	21.50	3.32		24.82	29	
	15770 | Floor-Heating & Snow-Melting Eq.										
0010	**ELECTRIC HEATING**, equipments are not including wires & conduits										200
0400	Cable heating, radiant heat plaster, no controls, in South	1 Elec	130	.062	S.F.	7.25	1.36		8.61	10.20	
0600	In North		90	.089		7.25	1.96		9.21	11.20	
0800	Cable on 1/2" board not incl. controls, tract housing		90	.089		6.10	1.96		8.06	9.90	
1000	Custom housing		80	.100		7.10	2.21		9.31	11.40	
1100	Rule of thumb: Baseboard units, including control		4.40	1.818	kW	75	40		115	148	
1300	Baseboard heaters, 2' long, 375 watt		8	1	Ea.	34	22		56	73.50	
1400	3' long, 500 watt		8	1		41	22		63	81	
1600	4' long, 750 watt		6.70	1.194		48	26.50		74.50	96	
1800	5' long, 935 watt		5.70	1.404		57	31		88	113	
2000	6' long, 1125 watt		5	1.600		65	35.50		100.50	130	
2400	8' long, 1500 watt		4	2		81	44		125	162	
2800	10' long, 1875 watt		3.30	2.424		134	53.50		187.50	235	
2950	Wall heaters with fan, 120 to 277 volt										
2970	surface mounted, residential, 750 watt	1 Elec	7	1.143	Ea.	105	25.50		130.50	158	
2980	1000 watt		7	1.143		107	25.50		132.50	160	
2990	1250 watt		6	1.333		120	29.50		149.50	180	
3000	1500 watt		5	1.600		120	35.50		155.50	190	
3010	2000 watt		5	1.600		123	35.50		158.50	193	
3050	2500 watt		4	2		255	44		299	355	
3070	4000 watt		3.50	2.286		260	50.50		310.50	370	
3600	Thermostats, integral		16	.500		25	11.05		36.05	45.50	
3800	Line voltage, 1 pole		8	1		25	22		47	63.50	
5000	Radiant heating ceiling panels, 2' x 4', 500 watt		16	.500		174	11.05		185.05	209	
5050	750 watt		16	.500		192	11.05		203.05	229	
5300	Infra-red quartz heaters, 120 volts, 1000 watts		6.70	1.194		119	26.50		145.50	174	
5350	1500 watt		5	1.600		119	35.50		154.50	189	
5400	240 volts, 1500 watt		5	1.600		119	35.50		154.50	189	
5450	2000 watt		4	2		119	44		163	204	
5500	3000 watt		3	2.667		136	59		195	247	

MECHANICAL 15

	15770 | Floor-Heating & Snow-Melting Eq.		CREW	DAILY OUTPUT	LABOR-HOURS	UNIT	2000 BARE COSTS MAT.	LABOR	EQUIP.	TOTAL	TOTAL INCL O&P
600	0010	DUCTWORK									
	0020	Fabricated rectangular, includes fittings, joints, supports,									
	0030	allowance for flexible connections, no insulation									
	0031	NOTE: Fabrication and installation are combined									
	0040	as LABOR cost. Approx. 25% fittings assumed.									
	0100	Aluminum, alloy 3003-H14, under 100 lb.	Q-10	75	.320	Lb.	2.80	6.40		9.20	13.95
	0110	100 to 500 lb.		80	.300		2.60	6		8.60	13
	0120	500 to 1,000 lb.		95	.253		1.50	5.05		6.55	10.20
	0140	1,000 to 2,000 lb.		120	.200		1.20	4.01		5.21	8.10
	0500	Galvanized steel, under 200 lb.		235	.102		1.05	2.05		3.10	4.61
	0520	200 to 500 lb.		245	.098		.85	1.97		2.82	4.26
	0540	500 to 1,000 lb.	↓	255	.094	↓	.41	1.89		2.30	3.64
	1300	Flexible, coated fiberglass fabric on corr. resist. metal helix									
	1400	pressure to 12″ (WG) UL-181									
	1500	Non-insulated, 3″ diameter	Q-9	400	.040	L.F.	.95	.77		1.72	2.36
	1540	5″ diameter		320	.050		1.11	.97		2.08	2.85
	1560	6″ diameter		280	.057		1.25	1.11		2.36	3.25
	1580	7″ diameter		240	.067		1.64	1.29		2.93	3.98
	1900	Insulated, 1″ thick with 3/4 lb., PE jacket, 3″ diameter		380	.042		1.64	.81		2.45	3.18
	1910	4″ diameter		340	.047		1.64	.91		2.55	3.34
	1920	5″ diameter		300	.053		1.92	1.03		2.95	3.85
	1940	6″ diameter		260	.062		2.15	1.19		3.34	4.38
	1960	7″ diameter		220	.073		2.57	1.41		3.98	5.20
	1980	8″ diameter		180	.089		2.63	1.72		4.35	5.80
	2040	12″ diameter	↓	100	.160	↓	3.98	3.10		7.08	9.65

	15820 | Duct Accessories		CREW	DAILY OUTPUT	LABOR-HOURS	UNIT	MAT.	LABOR	EQUIP.	TOTAL	TOTAL INCL O&P
300	0010	DUCT ACCESSORIES									
	0050	Air extractors, 12″ x 4″	1 Shee	24	.333	Ea.	14.45	7.15		21.60	28
	0100	8″ x 6″		22	.364		14.45	7.80		22.25	29
	3000	Fire damper, curtain type, 1-1/2 hr rated, vertical, 6″ x 6″		24	.333		19.70	7.15		26.85	33.50
	3020	8″ x 6″		22	.364		19.70	7.80		27.50	34.50
	6000	12″ x 12″	↓	21	.381	↓	25	8.20		33.20	41.50
	8000	Multi-blade dampers, parallel blade									
	8100	8″ x 8″	1 Shee	24	.333	Ea.	52.50	7.15		59.65	70
	9200	Plenums, measured by panel surface				S.F.	8.25			8.25	9.10

	15830 | Fans		CREW	DAILY OUTPUT	LABOR-HOURS	UNIT	MAT.	LABOR	EQUIP.	TOTAL	TOTAL INCL O&P
100	0010	FANS									
	8000	Ventilation, residential									
	8020	Attic, roof type									
	8030	Aluminum dome, damper & curb									
	8040	6″ diameter, 300 CFM	1 Elec	16	.500	Ea.	291	11.05		302.05	340
	8050	7″ diameter, 450 CFM		15	.533		320	11.80		331.80	370
	8060	9″ diameter, 900 CFM		14	.571		510	12.65		522.65	580
	8080	12″ diameter, 1000 CFM (gravity)		10	.800		227	17.70		244.70	279
	8090	16″ diameter, 1500 CFM (gravity)		9	.889		274	19.65		293.65	330
	8100	20″ diameter, 2500 CFM (gravity)	↓	8	1	↓	335	22		357	405
	8160	Plastic, ABS dome									
	8180	1050 CFM	1 Elec	14	.571	Ea.	67	12.65		79.65	94
	8200	1600 CFM	″	12	.667	″	100	14.75		114.75	134
	8240	Attic, wall type, with shutter, one speed									
	8250	12″ diameter, 1000 CFM	1 Elec	14	.571	Ea.	194	12.65		206.65	234
	8260	14″ diameter, 1500 CFM	↓	12	.667	↓	233	14.75		247.75	280

15800 | Air Distribution

15830 | Fans		CREW	DAILY OUTPUT	LABOR-HOURS	UNIT	2000 BARE COSTS MAT.	LABOR	EQUIP.	TOTAL	TOTAL INCL O&P	
8270	16″ diameter, 2000 CFM	1 Elec	9	.889	Ea.	325	19.65		344.65	390	100
8290	Whole house, wall type, with shutter, one speed										
8300	30″ diameter, 4800 CFM	1 Elec	7	1.143	Ea.	545	25.50		570.50	635	
8310	36″ diameter, 7000 CFM		6	1.333		595	29.50		624.50	705	
8320	42″ diameter, 10,000 CFM		5	1.600		720	35.50		755.50	850	
8330	48″ diameter, 16,000 CFM		4	2		945	44		989	1,125	
8340	For two speed, add					60			60	66	
8350	Whole house, lay-down type, with shutter, one speed										
8360	30″ diameter, 4500 CFM	1 Elec	8	1	Ea.	590	22		612	685	
8370	36″ diameter, 6500 CFM		7	1.143		650	25.50		675.50	755	
8380	42″ diameter, 9000 CFM		6	1.333		765	29.50		794.50	890	
8390	48″ diameter, 12,000 CFM		5	1.600		1,000	35.50		1,035.50	1,150	
8440	For two speed, add					15			15	16.50	
8450	For 12 hour timer switch, add	1 Elec	32	.250		30	5.55		35.55	42	

15850 | Air Outlets & Inlets		CREW	DAILY OUTPUT	LABOR-HOURS	UNIT	MAT.	LABOR	EQUIP.	TOTAL	TOTAL INCL O&P	
0010	**DIFFUSERS** Aluminum, opposed blade damper unless noted										300
0100	Ceiling, linear, also for sidewall										
0120	2″ wide	1 Shee	32	.250	L.F.	25	5.40		30.40	36.50	
0160	4″ wide		26	.308	″	33.50	6.60		40.10	48	
0500	Perforated, 24″ x 24″ lay-in panel size, 6″ x 6″		16	.500	Ea.	79.50	10.75		90.25	105	
0520	8″ x 8″		15	.533		82	11.45		93.45	109	
0530	9″ x 9″		14	.571		84	12.30		96.30	113	
0590	16″ x 16″		11	.727		107	15.65		122.65	145	
1000	Rectangular, 1 to 4 way blow, 6″ x 6″		16	.500		46	10.75		56.75	69	
1010	8″ x 8″		15	.533		54	11.45		65.45	79	
1014	9″ x 9″		15	.533		56	11.45		67.45	81	
1016	10″ x 10″		15	.533		67	11.45		78.45	93	
1020	12″ x 6″		15	.533		57	11.45		68.45	82	
1040	12″ x 9″		14	.571		69.50	12.30		81.80	97	
1060	12″ x 12″		12	.667		82	14.35		96.35	114	
1070	14″ x 6″		13	.615		61	13.25		74.25	90	
1074	14″ x 14″		12	.667		101	14.35		115.35	135	
1150	18″ x 18″		9	.889		149	19.10		168.10	197	
1170	24″ x 12″		10	.800		134	17.20		151.20	177	
1180	24″ x 24″		7	1.143		236	24.50		260.50	300	
1500	Round, butterfly damper, 6″ diameter		18	.444		15.45	9.55		25	33	
1520	8″ diameter		16	.500		16.35	10.75		27.10	36	
2000	T bar mounting, 24″ x 24″ lay-in frame, 6″ x 6″		16	.500		75.50	10.75		86.25	101	
2020	9″ x 9″		14	.571		84	12.30		96.30	113	
2040	12″ x 12″		12	.667		111	14.35		125.35	146	
2060	15″ x 15″		11	.727		143	15.65		158.65	184	
2080	18″ x 18″		10	.800		149	17.20		166.20	193	
6000	For steel diffusers instead of aluminum, deduct					10%					
0010	**GRILLES**										500
0020	Aluminum										
1000	Air return, 6″ x 6″	1 Shee	26	.308	Ea.	12.35	6.60		18.95	25	
1020	10″ x 6″		24	.333		14.95	7.15		22.10	28.50	
1080	16″ x 8″		22	.364		21.50	7.80		29.30	36.50	
1100	12″ x 12″		22	.364		21.50	7.80		29.30	36.50	
1120	24″ x 12″		18	.444		38.50	9.55		48.05	58	
1180	16″ x 16″		22	.364		30.50	7.80		38.30	46.50	
0010	**REGISTERS**										700
0980	Air supply										
3000	Baseboard, hand adj. damper, enameled steel										
3012	8″ x 6″	1 Shee	26	.308	Ea.	8.30	6.60		14.90	20.50	

15800 | Air Distribution

		15850 | Air Outlets & Inlets	CREW	DAILY OUTPUT	LABOR-HOURS	UNIT	2000 BARE COSTS MAT.	LABOR	EQUIP.	TOTAL	TOTAL INCL O&P
700	3020	10" x 6"	1 Shee	24	.333	Ea.	9.30	7.15		16.45	22.50
	3040	12" x 5"		23	.348		10.65	7.50		18.15	24.50
	3060	12" x 6"	↓	23	.348	↓	9.80	7.50		17.30	23.50
	4000	Floor, toe operated damper, enameled steel									
	4020	4" x 8"	1 Shee	32	.250	Ea.	15.90	5.40		21.30	26.50
	4040	4" x 12"	"	26	.308	"	18.10	6.60		24.70	31
		15854 | Ventilators									
300	0010	**VENTILATORS** Base, damper & bird screen, CFM in 5 MPH wind									
	0500	Rotary syphon, galvanized, 6" neck diameter, 185 CFM	Q-9	16	1	Ea.	50	19.35		69.35	87.50
	0520	8" neck diameter, 215 CFM	"	14	1.143	"	57	22		79	100
		15860 | Air Cleaning Devices									
100	0010	**AIR FILTERS**									
	0050	Activated charcoal type, full flow				MCFM	600			600	660
	2000	Electronic air cleaner, duct mounted									
	2150	400 - 1000 CFM	1 Shee	2.30	3.478	Ea.	545	75		620	720
	2200	1000 - 1400 CFM		2.20	3.636		700	78		778	895
	2250	1400 - 2000 CFM	↓	2.10	3.810	↓	760	82		842	980
	2950	Mechanical media filtration units									
	3000	High efficiency type, with frame, non-supported				MCFM	45			45	49.50
	3100	Supported type					55			55	60.50
	4000	Medium efficiency, extended surface					5			5	5.50
	4500	Permanent washable					20			20	22
	5000	Renewable disposable roll				↓	120			120	132
	5500	Throwaway glass or paper media type				Ea.	4.60			4.60	5.05

For information about Means Estimating Seminars, see yellow pages 11 and 12 in back of book

Division 16 Electrical

stimating Tips

060 Grounding & Bonding

When taking off grounding system, identify eparately the type and size of wire and list ach unique type of ground connection.

100 Wiring Methods

Conduit should be taken off in three main ategories: power distribution, branch power, nd branch lighting, so the estimator can concentrate on systems and components, herefore making it easier to ensure all items have been accounted for.

For cost modifications for elevated conduit nstallation, add the percentages to labor ccording to the height of installation and only the quantities exceeding the different height levels, not to the total conduit quantities.

Remember that aluminum wiring of equal ampacity is larger in diameter than copper and may require larger conduit.

If more than three wires at a time are being pulled, deduct percentages from the labor hours of that grouping of wires.

- The estimator should take the weights of materials into consideration when completing a takeoff. Topics to consider include: How will the materials be supported? What methods of support are available? How high will the support structure have to reach? Will the final support structure be able to withstand the total burden? Is the support material included or separate from the fixture, equipment and material specified?

16200 Electrical Power

- Do not overlook the costs for equipment used in the installation. If scaffolding or highlifts are available in the field, contractors may use them in lieu of the proposed ladders and rolling staging.

16400 Low-Voltage Distribution

- Supports and concrete pads may be shown on drawings for the larger equipment, or the support system may be just a piece of plywood for the back of a panelboard. In either case, it must be included in the costs.

16500 Lighting

- Fixtures should be taken off room by room, using the fixture schedule, specifications, and the ceiling plan. For large concentrations of lighting fixtures in the same area deduct the percentages from labor hours.

16700 Communications
16800 Sound & Video

- When estimating material costs for special systems, it is always prudent to obtain manufacturers' quotations for equipment prices and special installation requirements which will affect the total costs.

Reference Numbers

Reference numbers are shown in bold squares at the beginning of some major classifications. These numbers refer to related items in the Reference Section. The reference information may be an estimating procedure, an alternate pricing method or technical information.

Note: Not all subdivisions listed here necessarily appear in this publication.

	16055 | Selective Demolition	CREW	DAILY OUTPUT	LABOR-HOURS	UNIT	2000 BARE COSTS MAT.	LABOR	EQUIP.	TOTAL	TOTAL INCL O&P
300 0010	ELECTRICAL DEMOLITION									
0020	Conduit to 15′ high, including fittings & hangers									
0100	Rigid galvanized steel, 1/2″ to 1″ diameter	1 Elec	242	.033	L.F.		.73		.73	1.20
0120	1-1/4″ to 2″	″	200	.040	″		.88		.88	1.45
0270	Armored cable, (BX) avg. 50′ runs									
0290	#14, 3 wire	1 Elec	571	.014	L.F.		.31		.31	.51
0300	#12, 2 wire		605	.013			.29		.29	.48
0310	#12, 3 wire		514	.016			.34		.34	.56
0320	#10, 2 wire		514	.016			.34		.34	.56
0330	#10, 3 wire		425	.019			.42		.42	.68
0340	#8, 3 wire		342	.023			.52		.52	.85
0350	Non metallic sheathed cable (Romex)									
0360	#14, 2 wire	1 Elec	720	.011	L.F.		.25		.25	.40
0370	#14, 3 wire		657	.012			.27		.27	.44
0380	#12, 2 wire		629	.013			.28		.28	.46
0390	#10, 3 wire		450	.018			.39		.39	.64
0400	Wiremold raceway, including fittings & hangers									
0420	No. 3000	1 Elec	250	.032	L.F.		.71		.71	1.16
0440	No. 4000		217	.037			.81		.81	1.33
0460	No. 6000		166	.048			1.07		1.07	1.74
0500	Channels, steel, including fittings & hangers									
0520	3/4″ x 1-1/2″	1 Elec	308	.026	L.F.		.57		.57	.94
0540	1-1/2″ x 1-1/2″		269	.030			.66		.66	1.08
0560	1-1/2″ x 1-7/8″		229	.035			.77		.77	1.26
1180	400 amp	2 Elec	6.80	2.353	Ea.		52		52	85
1210	Panel boards, incl. removal of all breakers,									
1220	pipe terminations & wire connections									
1230	3 wire, 120/240V, 100A, to 20 circuits	1 Elec	2.60	3.077	Ea.		68		68	111
1240	200 amps, to 42 circuits		1.30	6.154			136		136	222
1260	4 wire, 120/208V, 125A, to 20 circuits		2.40	3.333			73.50		73.50	121
1270	200 amps, to 42 circuits		1.20	6.667			147		147	241
1720	Junction boxes, 4″ sq. & oct.		80	.100			2.21		2.21	3.62
1760	Switch box		107	.075			1.65		1.65	2.70
1780	Receptacle & switch plates		257	.031			.69		.69	1.13
1800	Wire, THW-THWN-THHN, removed from									
1810	in place conduit, to 15′ high									
1830	#14	1 Elec	65	.123	C.L.F.		2.72		2.72	4.45
1840	#12		55	.145			3.21		3.21	5.25
1850	#10		45.50	.176			3.89		3.89	6.35
2000	Interior fluorescent fixtures, incl. supports									
2010	& whips, to 15′ high									
2100	Recessed drop-in 2′ x 2′, 2 lamp	2 Elec	35	.457	Ea.		10.10		10.10	16.55
2140	2′ x 4′, 4 lamp	″	30	.533	″		11.80		11.80	19.30
2180	Surface mount, acrylic lens & hinged frame									
2220	2′ x 2′, 2 lamp	2 Elec	44	.364	Ea.		8.05		8.05	13.15
2260	2′ x 4′, 4 lamp	″	33	.485	″		10.70		10.70	17.55
2300	Strip fixtures, surface mount									
2320	4′ long, 1 lamp	2 Elec	53	.302	Ea.		6.65		6.65	10.90
2380	8′ long, 2 lamp	″	40	.400	″		8.85		8.85	14.45
2460	Interior incandescent, surface, ceiling									
2470	or wall mount, to 12′ high									
2480	Metal cylinder type, 75 Watt	2 Elec	62	.258	Ea.		5.70		5.70	9.35
2600	Exterior fixtures, incandescent, wall mount									
2620	100 Watt	2 Elec	50	.320	Ea.		7.05		7.05	11.55
3000	Ceiling fan, tear out and remove	1 Elec	18	.444	″		9.80		9.80	16.05
9000	Minimum labor/equipment charge	″	4	2	Job		44		44	72.50

16050 | Basic Electrical Materials & Methods

16060 | Grounding & Bonding		CREW	DAILY OUTPUT	LABOR-HOURS	UNIT	2000 BARE COSTS MAT.	LABOR	EQUIP.	TOTAL	TOTAL INCL O&P	
0010	GROUNDING										800
0030	Rod, copper clad, 8′ long, 1/2″ diameter	1 Elec	5.50	1.455	Ea.	15.85	32		47.85	70	
0050	3/4″ diameter		5.30	1.509		30	33.50		63.50	87.50	
0080	10′ long, 1/2″ diameter		4.80	1.667		17.50	37		54.50	80	
0100	3/4″ diameter		4.40	1.818		29.50	40		69.50	98	
0261	Wire, ground, bare armored, #8-1 conductor		200	.040	L.F.	.63	.88		1.51	2.15	
0271	#6-1 conductor		180	.044	"	.80	.98		1.78	2.49	
0390	Bare copper wire, #8 stranded		11	.727	C.L.F.	14	16.05		30.05	42	
0401	Bare copper, #6 wire		1,000	.008	L.F.	.21	.18		.39	.52	
0601	#2 stranded	2 Elec	1,000	.016	"	.51	.35		.86	1.14	
1800	Water pipe ground clamps, heavy duty										
2000	Bronze, 1/2″ to 1″ diameter	1 Elec	8	1	Ea.	11.70	22		33.70	49	

16100 | Wiring Methods

16120 | Conductors & Cables		CREW	DAILY OUTPUT	LABOR-HOURS	UNIT	2000 BARE COSTS MAT.	LABOR	EQUIP.	TOTAL	TOTAL INCL O&P	
0010	ARMORED CABLE										120
0051	600 volt, copper (BX), #14, 2 conductor, solid	1 Elec	240	.033	L.F.	.47	.74		1.21	1.72	
0101	3 conductor, solid		200	.040		.61	.88		1.49	2.12	
0151	#12, 2 conductor, solid		210	.038		.48	.84		1.32	1.91	
0201	3 conductor, solid		180	.044		.72	.98		1.70	2.41	
0251	#10, 2 conductor, solid		180	.044		.85	.98		1.83	2.55	
0301	3 conductor, solid		150	.053		1.12	1.18		2.30	3.16	
0351	#8, 3 conductor, solid		120	.067		1.83	1.47		3.30	4.42	
0010	NON-METALLIC SHEATHED CABLE 600 volt										550
0100	Copper with ground wire, (Romex)										
0151	#14, 2 wire	1 Elec	250	.032	L.F.	.13	.71		.84	1.31	
0201	3 wire		230	.035		.22	.77		.99	1.51	
0251	#12, 2 wire		220	.036		.19	.80		.99	1.52	
0301	3 wire		200	.040		.32	.88		1.20	1.80	
0351	#10, 2 wire		200	.040		.32	.88		1.20	1.80	
0401	3 wire		140	.057		.48	1.26		1.74	2.60	
0451	#8, 3 conductor		130	.062		.98	1.36		2.34	3.29	
0501	#6, 3 wire		120	.067		1.50	1.47		2.97	4.06	
0550	SE type SER aluminum cable, 3 RHW and										
0601	1 bare neutral, 3 #8 & 1 #8	1 Elec	150	.053	L.F.	.96	1.18		2.14	2.99	
0651	3 #6 & 1 #6	"	130	.062		1.09	1.36		2.45	3.41	
0701	3 #4 & 1 #6	2 Elec	220	.073		1.22	1.61		2.83	3.97	
0751	3 #2 & 1 #4		200	.080		1.79	1.77		3.56	4.86	
0801	3 #1/0 & 1 #2		180	.089		2.71	1.96		4.67	6.20	
0851	3 #2/0 & 1 #1		160	.100		3.19	2.21		5.40	7.10	
0901	3 #4/0 & 1 #2/0		140	.114		4.54	2.53		7.07	9.15	
2401	SEU service entrance cable, copper 2 conductors, #8 + #8 neut.	1 Elec	150	.053		.83	1.18		2.01	2.84	
2601	#6 + #8 neutral		130	.062		1.14	1.36		2.50	3.48	
2801	#6 + #6 neutral		130	.062		1.25	1.36		2.61	3.60	
3001	#4 + #6 neutral	2 Elec	220	.073		1.64	1.61		3.25	4.43	
3201	#4 + #4 neutral		220	.073		1.85	1.61		3.46	4.67	
3401	#3 + #5 neutral		210	.076		2.03	1.68		3.71	4.98	
6500	Service entrance cap for copper SEU										
6600	100 amp	1 Elec	12	.667	Ea.	6.90	14.75		21.65	31.50	

	16120 | Conductors & Cables		CREW	DAILY OUTPUT	LABOR-HOURS	UNIT	2000 BARE COSTS MAT.	LABOR	EQUIP.	TOTAL	TOTAL INCL O&P
550	6700	150 amp	1 Elec	10	.800	Ea.	12.10	17.70		29.80	42.50
	6800	200 amp	↓	8	1	↓	18.25	22		40.25	56
900	0010	**WIRE**									
	0021	600 volt type THW, copper solid, #14	1 Elec	1,300	.006	L.F.	.04	.14		.18	.26
	0031	#12		1,100	.007		.05	.16		.21	.32
	0041	#10		1,000	.008		.08	.18		.26	.38
	0161	#6	↓	650	.012		.25	.27		.52	.72
	0181	#4	2 Elec	1,060	.015		.42	.33		.75	1.01
	0201	#3		1,000	.016		.48	.35		.83	1.10
	0221	#2		900	.018		.59	.39		.98	1.29
	0241	#1		800	.020		.75	.44		1.19	1.55
	0261	1/0		660	.024		.93	.54		1.47	1.90
	0281	2/0		580	.028		1.14	.61		1.75	2.26
	0301	3/0		500	.032		1.45	.71		2.16	2.75
	0351	4/0	↓	440	.036	↓	1.80	.80		2.60	3.29
	16132 | Conduit & Tubing										
205	0010	**CONDUIT** To 15' high, includes 2 terminations, 2 elbows and									
	0020	11 beam clamps per 100 L.F.									
	1750	Rigid galvanized steel, 1/2" diameter	1 Elec	90	.089	L.F.	1.52	1.96		3.48	4.88
	1770	3/4" diameter		80	.100		1.81	2.21		4.02	5.60
	1800	1" diameter		65	.123		2.53	2.72		5.25	7.25
	1830	1-1/4" diameter		60	.133		3.38	2.95		6.33	8.55
	1850	1-1/2" diameter		55	.145		4.02	3.21		7.23	9.65
	1870	2" diameter		45	.178		5.35	3.93		9.28	12.35
	5000	Electric metallic tubing (EMT), 1/2" diameter		170	.047		.36	1.04		1.40	2.10
	5020	3/4" diameter		130	.062		.55	1.36		1.91	2.82
	5040	1" diameter		115	.070		.93	1.54		2.47	3.53
	5060	1-1/4" diameter		100	.080		1.39	1.77		3.16	4.42
	5080	1-1/2" diameter		90	.089		1.76	1.96		3.72	5.15
	9100	PVC, #40, 1/2" diameter		190	.042		.57	.93		1.50	2.15
	9110	3/4" diameter		145	.055		.69	1.22		1.91	2.75
	9120	1" diameter		125	.064		.99	1.41		2.40	3.39
	9130	1-1/4" diameter		110	.073		1.30	1.61		2.91	4.06
	9140	1-1/2" diameter		100	.080		1.55	1.77		3.32	4.60
	9150	2" diameter	↓	90	.089	↓	2.01	1.96		3.97	5.45
230	0010	**CONDUIT IN CONCRETE SLAB** Including terminations,									
	0020	fittings and supports									
	3230	PVC, schedule 40, 1/2" diameter	1 Elec	270	.030	L.F.	.35	.65		1	1.45
	3250	3/4" diameter		230	.035		.42	.77		1.19	1.73
	3270	1" diameter		200	.040		.56	.88		1.44	2.07
	3300	1-1/4" diameter		170	.047		.78	1.04		1.82	2.56
	3330	1-1/2" diameter		140	.057		.96	1.26		2.22	3.13
	3350	2" diameter		120	.067		1.23	1.47		2.70	3.76
	4350	Rigid galvanized steel, 1/2" diameter		200	.040		1.27	.88		2.15	2.84
	4400	3/4" diameter		170	.047		1.56	1.04		2.60	3.41
	4450	1" diameter		130	.062		2.28	1.36		3.64	4.73
	4500	1-1/4" diameter		110	.073		2.96	1.61		4.57	5.90
	4600	1-1/2" diameter		100	.080		3.59	1.77		5.36	6.85
	4800	2" diameter	↓	90	.089	↓	4.79	1.96		6.75	8.45
240	0010	**CONDUIT IN TRENCH** Includes terminations and fittings									
	0200	Rigid galvanized steel, 2" diameter	1 Elec	150	.053	L.F.	4.63	1.18		5.81	7.05
	0400	2-1/2" diameter	"	100	.080		8.05	1.77		9.82	11.75
	0600	3" diameter	2 Elec	160	.100	↓	10.40	2.21		12.61	15.05

16 ELECTRICAL

16100 | Wiring Methods

16132 | Conduit & Tubing

	16132 Conduit & Tubing	CREW	DAILY OUTPUT	LABOR-HOURS	UNIT	2000 BARE COSTS MAT.	LABOR	EQUIP.	TOTAL	TOTAL INCL O&P	
0800	3-1/2" diameter	2 Elec	140	.114	L.F.	13.20	2.53		15.73	18.65	240
0010	CONDUIT FITTINGS FOR RIGID GALVANIZED STEEL										250
2280	LB, LR or LL fittings & covers, 1/2" diameter	1 Elec	16	.500	Ea.	7.90	11.05		18.95	27	
2290	3/4" diameter		13	.615		9.55	13.60		23.15	33	
2300	1" diameter		11	.727		14.10	16.05		30.15	42	
2330	1-1/4" diameter		8	1		21	22		43	59.50	
2350	1-1/2" diameter		6	1.333		26.50	29.50		56	77	
2370	2" diameter		5	1.600		44	35.50		79.50	106	
5280	Service entrance cap, 1/2" diameter		16	.500		6.10	11.05		17.15	25	
5300	3/4" diameter		13	.615		7.05	13.60		20.65	30.50	
5320	1" diameter		10	.800		6.30	17.70		24	36	
5340	1-1/4" diameter		8	1		12.35	22		34.35	49.50	
5360	1-1/2" diameter		6.50	1.231		11.25	27		38.25	57	
5380	2" diameter	↓	5.50	1.455	↓	34	32		66	90	
0010	FLEXIBLE METALLIC CONDUIT										320
0050	Steel, 3/8" diameter	1 Elec	200	.040	L.F.	.23	.88		1.11	1.70	
0100	1/2" diameter		200	.040		.32	.88		1.20	1.80	
0200	3/4" diameter		160	.050		.41	1.11		1.52	2.26	
0250	1" diameter		100	.080		.85	1.77		2.62	3.83	
0300	1-1/4" diameter		70	.114		1.09	2.53		3.62	5.35	
0350	1-1/2" diameter		50	.160		1.40	3.54		4.94	7.35	
0370	2" diameter	↓	40	.200	↓	2.05	4.42		6.47	9.50	

16133 | Multi-outlet Assemblies

	16133 Multi-outlet Assemblies	CREW	DAILY OUTPUT	LABOR-HOURS	UNIT	MAT.	LABOR	EQUIP.	TOTAL	TOTAL INCL O&P	
0010	SURFACE RACEWAY										800
0100	No. 500	1 Elec	100	.080	L.F.	.63	1.77		2.40	3.58	
0110	No. 700		100	.080		.71	1.77		2.48	3.67	
0200	No. 1000		90	.089		1.30	1.96		3.26	4.64	
0400	No. 1500, small pancake		90	.089		1.33	1.96		3.29	4.67	
0600	No. 2000, base & cover, blank		90	.089		1.29	1.96		3.25	4.63	
0800	No. 3000, base & cover, blank		75	.107	↓	2.53	2.36		4.89	6.65	
2400	Fittings, elbows, No. 500		40	.200	Ea.	1.14	4.42		5.56	8.50	
2800	Elbow cover, No. 2000		40	.200		2.30	4.42		6.72	9.80	
2880	Tee, No. 500		42	.190		2.21	4.21		6.42	9.35	
2900	No. 2000		27	.296		7.05	6.55		13.60	18.45	
3000	Switch box, No. 500		16	.500		8.10	11.05		19.15	27	
3400	Telephone outlet, No. 1500		16	.500		8.60	11.05		19.65	27.50	
3600	Junction box, No. 1500	↓	16	.500	↓	6	11.05		17.05	24.50	
3800	Plugmold wired sections, No. 2000										
4000	1 circuit, 6 outlets, 3 ft. long	1 Elec	8	1	Ea.	21.50	22		43.50	59.50	
4100	2 circuits, 8 outlets, 6 ft. long	"	5.30	1.509	"	36	33.50		69.50	94	

16136 | Boxes

	16136 Boxes	CREW	DAILY OUTPUT	LABOR-HOURS	UNIT	MAT.	LABOR	EQUIP.	TOTAL	TOTAL INCL O&P	
0010	OUTLET BOXES										600
0021	Pressed steel, octagon, 4"	1 Elec	18	.444	Ea.	1.46	9.80		11.26	17.65	
0060	Covers, blank		64	.125		.60	2.76		3.36	5.20	
0100	Extension		40	.200		2.33	4.42		6.75	9.80	
0151	Square 4"		18	.444		1.98	9.80		11.78	18.25	
0200	Extension		40	.200		2.44	4.42		6.86	9.95	
0250	Covers, blank		64	.125		.69	2.76		3.45	5.30	
0300	Plaster rings		64	.125		1.11	2.76		3.87	5.75	
0651	Switchbox		24	.333		2.26	7.35		9.61	14.55	
1101	Concrete, floor, 1 gang	↓	4.80	1.667	↓	57.50	37		94.50	124	

		16136 | Boxes	CREW	DAILY OUTPUT	LABOR-HOURS	UNIT	2000 BARE COSTS MAT.	LABOR	EQUIP.	TOTAL	TOTAL INCL O&P
620	0010	OUTLET BOXES, PLASTIC									
	0051	4" diameter, round, with 2 mounting nails	1 Elec	23	.348	Ea.	1.70	7.70		9.40	14.40
	0101	Bar hanger mounted		23	.348		2.90	7.70		10.60	15.75
	0201	Square with 2 mounting nails		23	.348		2.62	7.70		10.32	15.45
	0300	Plaster ring		64	.125		.92	2.76		3.68	5.55
	0401	Switch box with 2 mounting nails, 1 gang		27	.296		1.07	6.55		7.62	11.90
	0501	2 gang		23	.348		2.10	7.70		9.80	14.85
	0601	3 gang		18	.444		3.32	9.80		13.12	19.70
700	0010	PULL BOXES & CABINETS									
	0100	Sheet metal, pull box, NEMA 1, type SC, 6" W x 6" H x 4" D	1 Elec	8	1	Ea.	8.35	22		30.35	45
	0200	8" W x 8" H x 4" D		8	1		11.45	22		33.45	48.50
	0300	10" W x 12" H x 6" D		5.30	1.509		20	33.50		53.50	76.50
		16139 | Residential Wiring									
700	0010	RESIDENTIAL WIRING									
	0020	20' avg. runs and #14/2 wiring incl. unless otherwise noted									
	1000	Service & panel, includes 24' SE-AL cable, service eye, meter,									
	1010	Socket, panel board, main bkr., ground rod, 15 or 20 amp									
	1020	1-pole circuit breakers, and misc. hardware									
	1100	100 amp, with 10 branch breakers	1 Elec	1.19	6.723	Ea.	400	149		549	680
	1110	With PVC conduit and wire		.92	8.696		425	192		617	785
	1120	With RGS conduit and wire		.73	10.959		545	242		787	995
	1150	150 amp, with 14 branch breakers		1.03	7.767		615	172		787	960
	1170	With PVC conduit and wire		.82	9.756		675	216		891	1,100
	1180	With RGS conduit and wire		.67	11.940		885	264		1,149	1,400
	1200	200 amp, with 18 branch breakers	2 Elec	1.80	8.889		810	196		1,006	1,200
	1220	With PVC conduit and wire		1.46	10.959		870	242		1,112	1,350
	1230	With RGS conduit and wire		1.24	12.903		1,150	285		1,435	1,750
	1800	Lightning surge suppressor for above services, add	1 Elec	32	.250		39.50	5.55		45.05	52.50
	2000	Switch devices									
	2100	Single pole, 15 amp, Ivory, with a 1-gang box, cover plate,									
	2110	Type NM (Romex) cable	1 Elec	17.10	.468	Ea.	6.65	10.35		17	24.50
	2120	Type MC (BX) cable		14.30	.559		16.80	12.35		29.15	38.50
	2130	EMT & wire		5.71	1.401		17.45	31		48.45	69.50
	2150	3-way, #14/3, type NM cable		14.55	.550		9.85	12.15		22	30.50
	2170	Type MC cable		12.31	.650		21	14.35		35.35	46.50
	2180	EMT & wire		5	1.600		19.65	35.50		55.15	79.50
	2200	4-way, #14/3, type NM cable		14.55	.550		22.50	12.15		34.65	45
	2220	Type MC cable		12.31	.650		34	14.35		48.35	60.50
	2230	EMT & wire		5	1.600		32.50	35.50		68	93.50
	2250	S.P., 20 amp, #12/2, type NM cable		13.33	.600		11.65	13.25		24.90	34.50
	2270	Type MC cable		11.43	.700		21	15.45		36.45	48.50
	2280	EMT & wire		4.85	1.649		23	36.50		59.50	84.50
	2290	S.P. rotary dimmer, 600W, no wiring		17	.471		15.55	10.40		25.95	34
	2300	S.P. rotary dimmer, 600W, type NM cable		14.55	.550		18.25	12.15		30.40	40
	2320	Type MC cable		12.31	.650		28.50	14.35		42.85	54.50
	2330	EMT & wire		5	1.600		30	35.50		65.50	91
	2350	3-way rotary dimmer, type NM cable		13.33	.600		15.65	13.25		28.90	38.50
	2370	Type MC cable		11.43	.700		26	15.45		41.45	54
	2380	EMT & wire		4.85	1.649		27.50	36.50		64	89.50
	2400	Interval timer wall switch, 20 amp, 1-30 min., #12/2									
	2410	Type NM cable	1 Elec	14.55	.550	Ea.	28.50	12.15		40.65	51
	2420	Type MC cable		12.31	.650		35.50	14.35		49.85	62.50
	2430	EMT & wire		5	1.600		39.50	35.50		75	102
	2500	Decorator style									
	2510	S.P., 15 amp, type NM cable	1 Elec	17.10	.468	Ea.	10.05	10.35		20.40	28

Important: See the Reference Section for critical supporting data - Reference Nos., Crews, & Location Facto

16139 | Residential Wiring		CREW	DAILY OUTPUT	LABOR-HOURS	UNIT	2000 BARE COSTS MAT.	LABOR	EQUIP.	TOTAL	TOTAL INCL O&P	
2520	Type MC cable	1 Elec	14.30	.559	Ea.	20	12.35		32.35	42	700
2530	EMT & wire		5.71	1.401		21	31		52	73.50	
2550	3-way, #14/3, type NM cable		14.55	.550		13.25	12.15		25.40	34.50	
2570	Type MC cable		12.31	.650		24.50	14.35		38.85	50.50	
2580	EMT & wire		5	1.600		23	35.50		58.50	83.50	
2600	4-way, #14/3, type NM cable		14.55	.550		26	12.15		38.15	48.50	
2620	Type MC cable		12.31	.650		37	14.35		51.35	64.50	
2630	EMT & wire		5	1.600		36	35.50		71.50	97.50	
2650	S.P., 20 amp, #12/2, type NM cable		13.33	.600		15.05	13.25		28.30	38	
2670	Type MC cable		11.43	.700		24.50	15.45		39.95	52.50	
2680	EMT & wire		4.85	1.649		26	36.50		62.50	88.50	
2700	S.P., slide dimmer, type NM cable		17.10	.468		25.50	10.35		35.85	45	
2720	Type MC cable		14.30	.559		35.50	12.35		47.85	59	
2730	EMT & wire		5.71	1.401		37	31		68	91.50	
2750	S.P., touch dimmer, type NM cable		17.10	.468		21.50	10.35		31.85	40.50	
2770	Type MC cable		14.30	.559		31.50	12.35		43.85	54.50	
2780	EMT & wire		5.71	1.401		33	31		64	87	
2800	3-way touch dimmer, type NM cable		13.33	.600		39	13.25		52.25	64	
2820	Type MC cable		11.43	.700		49	15.45		64.45	79.50	
2830	EMT & wire		4.85	1.649		50.50	36.50		87	115	
3000	Combination devices										
3100	S.P. switch/15 amp recpt., Ivory, 1-gang box, plate										
3110	Type NM cable	1 Elec	11.43	.700	Ea.	14.45	15.45		29.90	41.50	
3120	Type MC cable		10	.800		24.50	17.70		42.20	56	
3130	EMT & wire		4.40	1.818		26	40		66	94	
3150	S.P. switch/pilot light, type NM cable		11.43	.700		15.05	15.45		30.50	42	
3170	Type MC cable		10	.800		25	17.70		42.70	56.50	
3180	EMT & wire		4.43	1.806		26.50	40		66.50	95	
3190	2-S.P. switches, 2-#14/2, no wiring		14	.571		5.45	12.65		18.10	26.50	
3200	2-S.P. switches, 2-#14/2, type NM cables		10	.800		16.45	17.70		34.15	47	
3220	Type MC cable		8.89	.900		33	19.90		52.90	69	
3230	EMT & wire		4.10	1.951		28	43		71	101	
3250	3-way switch/15 amp recpt., #14/3, type NM cable		10	.800		20.50	17.70		38.20	51.50	
3270	Type MC cable		8.89	.900		31.50	19.90		51.40	67.50	
3280	EMT & wire		4.10	1.951		30.50	43		73.50	104	
3300	2-3 way switches, 2-#14/3, type NM cables		8.89	.900		27	19.90		46.90	62	
3320	Type MC cable		8	1		46	22		68	86.50	
3330	EMT & wire		4	2		35	44		79	111	
3350	S.P. switch/20 amp recpt., #12/2, type NM cable		10	.800		24.50	17.70		42.20	55.50	
3370	Type MC cable		8.89	.900		31	19.90		50.90	67	
3380	EMT & wire		4.10	1.951		35.50	43		78.50	110	
3400	Decorator style										
3410	S.P. switch/15 amp recpt., type NM cable	1 Elec	11.43	.700	Ea.	17.85	15.45		33.30	45	
3420	Type MC cable		10	.800		28	17.70		45.70	60	
3430	EMT & wire		4.40	1.818		29.50	40		69.50	98	
3450	S.P. switch/pilot light, type NM cable		11.43	.700		18.45	15.45		33.90	46	
3470	Type MC cable		10	.800		28.50	17.70		46.20	60.50	
3480	EMT & wire		4.40	1.818		30	40		70	98.50	
3500	2-S.P. switches, 2-#14/2, type NM cables		10	.800		19.85	17.70		37.55	51	
3520	Type MC cable		8.89	.900		36.50	19.90		56.40	73	
3530	EMT & wire		4.10	1.951		31	43		74	105	
3550	3-way/15 amp recpt., #14/3, type NM cable		10	.800		24	17.70		41.70	55.50	
3570	Type MC cable		8.89	.900		35	19.90		54.90	71	
3580	EMT & wire		4.10	1.951		34	43		77	108	
3650	2-3 way switches, 2-#14/3, type NM cables		8.89	.900		30.50	19.90		50.40	66	
3670	Type MC cable		8	1		49.50	22		71.50	90.50	

16139 | Residential Wiring		CREW	DAILY OUTPUT	LABOR-HOURS	UNIT	2000 BARE COSTS MAT.	LABOR	EQUIP.	TOTAL	TOTAL INCL O&P
700 3680	EMT & wire	1 Elec	4	2	Ea.	38	44		82	115
3700	S.P. switch/20 amp recpt., #12/2, type NM cable		10	.800		27.50	17.70		45.20	59.50
3720	Type MC cable		8.89	.900		34.50	19.90		54.40	70.50
3730	EMT & wire		4.10	1.951		39	43		82	114
4000	Receptacle devices									
4010	Duplex outlet, 15 amp recpt., Ivory, 1-gang box, plate									
4015	Type NM cable	1 Elec	14.55	.550	Ea.	5.20	12.15		17.35	25.50
4020	Type MC cable		12.31	.650		15.30	14.35		29.65	40.50
4030	EMT & wire		5.33	1.501		16	33		49	72
4050	With #12/2, type NM cable		12.31	.650		6.30	14.35		20.65	30.50
4070	Type MC cable		10.67	.750		15.65	16.55		32.20	44
4080	EMT & wire		4.71	1.699		17.50	37.50		55	81
4100	20 amp recpt., #12/2, type NM cable		12.31	.650		9.50	14.35		23.85	34
4120	Type MC cable		10.67	.750		18.85	16.55		35.40	48
4130	EMT & wire		4.71	1.699		20.50	37.50		58	84.50
4140	For GFI see line 4300 below									
4150	Decorator style, 15 amp recpt., type NM cable	1 Elec	14.55	.550	Ea.	8.60	12.15		20.75	29.50
4170	Type MC cable		12.31	.650		18.70	14.35		33.05	44
4180	EMT & wire		5.33	1.501		19.40	33		52.40	76
4200	With #12/2, type NM cable		12.31	.650		9.70	14.35		24.05	34
4220	Type MC cable		10.67	.750		19.05	16.55		35.60	48
4230	EMT & wire		4.71	1.699		21	37.50		58.50	84.50
4250	20 amp recpt. #12/2, type NM cable		12.31	.650		12.90	14.35		27.25	37.50
4270	Type MC cable		10.67	.750		22.50	16.55		39.05	51.50
4280	EMT & wire		4.71	1.699		24	37.50		61.50	88
4300	GFI, 15 amp recpt., type NM cable		12.31	.650		33	14.35		47.35	59.50
4320	Type MC cable		10.67	.750		43	16.55		59.55	74.50
4330	EMT & wire		4.71	1.699		43.50	37.50		81	110
4350	GFI with #12/2, type NM cable		10.67	.750		34	16.55		50.55	64.50
4370	Type MC cable		9.20	.870		43.50	19.20		62.70	79
4380	EMT & wire		4.21	1.900		45	42		87	118
4400	20 amp recpt., #12/2 type NM cable		10.67	.750		35.50	16.55		52.05	66
4420	Type MC cable		9.20	.870		45	19.20		64.20	81
4430	EMT & wire		4.21	1.900		47	42		89	120
4500	Weather-proof cover for above receptacles, add		32	.250		4.40	5.55		9.95	13.90
4550	Air conditioner outlet, 20 amp-240 volt recpt.									
4560	30′ of #12/2, 2 pole circuit breaker									
4570	Type NM cable	1 Elec	10	.800	Ea.	39.50	17.70		57.20	72
4580	Type MC cable		9	.889		51.50	19.65		71.15	88.50
4590	EMT & wire		4	2		50	44		94	128
4600	Decorator style, type NM cable		10	.800		43	17.70		60.70	76.50
4620	Type MC cable		9	.889		55.50	19.65		75.15	93
4630	EMT & wire		4	2		54	44		98	132
4650	Dryer outlet, 30 amp-240 volt recpt., 20′ of #10/3									
4660	2 pole circuit breaker									
4670	Type NM cable	1 Elec	6.41	1.248	Ea.	47.50	27.50		75	97.50
4680	Type MC cable		5.71	1.401		57	31		88	114
4690	EMT & wire		3.48	2.299		53	51		104	142
4700	Range outlet, 50 amp-240 volt recpt., 30′ of #8/3									
4710	Type NM cable	1 Elec	4.21	1.900	Ea.	69	42		111	145
4720	Type MC cable		4	2		98.50	44		142.50	181
4730	EMT & wire		2.96	2.703		67.50	59.50		127	172
4750	Central vacuum outlet, Type NM cable		6.40	1.250		41.50	27.50		69	90.50
4770	Type MC cable		5.71	1.401		58.50	31		89.50	115
4780	EMT & wire		3.48	2.299		52.50	51		103.50	141
4800	30 amp-110 volt locking recpt., #10/2 circ. bkr.									

16139 | Residential Wiring

700

		CREW	DAILY OUTPUT	LABOR-HOURS	UNIT	2000 BARE COSTS MAT.	LABOR	EQUIP.	TOTAL	TOTAL INCL O&P
4810	Type NM cable	1 Elec	6.20	1.290	Ea.	48	28.50		76.50	99
4820	Type MC cable		5.40	1.481		70.50	32.50		103	131
4830	EMT & wire		3.20	2.500		61	55.50		116.50	158
4900	Low voltage outlets									
4910	Telephone recpt., 20′ of 4/C phone wire	1 Elec	26	.308	Ea.	6.75	6.80		13.55	18.50
4920	TV recpt., 20′ of RG59U coax wire, F type connector	″	16	.500	″	11.45	11.05		22.50	30.50
4950	Door bell chime, transformer, 2 buttons, 60′ of bellwire									
4970	Economy model	1 Elec	11.50	.696	Ea.	48.50	15.35		63.85	78.50
4980	Custom model		11.50	.696		83.50	15.35		98.85	117
4990	Luxury model, 3 buttons		9.50	.842		228	18.60		246.60	282
6000	Lighting outlets									
6050	Wire only (for fixture), type NM cable	1 Elec	32	.250	Ea.	3.51	5.55		9.06	12.90
6070	Type MC cable		24	.333		11.40	7.35		18.75	24.50
6080	EMT & wire		10	.800		11.15	17.70		28.85	41.50
6100	Box (4″), and wire (for fixture), type NM cable		25	.320		7.50	7.05		14.55	19.80
6120	Type MC cable		20	.400		15.40	8.85		24.25	31.50
6130	EMT & wire		11	.727		15.10	16.05		31.15	43
6200	Fixtures (use with lines 6050 or 6100 above)									
6210	Canopy style, economy grade	1 Elec	40	.200	Ea.	25.50	4.42		29.92	35.50
6220	Custom grade		40	.200		46	4.42		50.42	58
6250	Dining room chandelier, economy grade		19	.421		76	9.30		85.30	98.50
6260	Custom grade		19	.421		225	9.30		234.30	263
6270	Luxury grade		15	.533		495	11.80		506.80	565
6310	Kitchen fixture (fluorescent), economy grade		30	.267		51.50	5.90		57.40	66
6320	Custom grade		25	.320		161	7.05		168.05	189
6350	Outdoor, wall mounted, economy grade		30	.267		27	5.90		32.90	39
6360	Custom grade		30	.267		101	5.90		106.90	121
6370	Luxury grade		25	.320		227	7.05		234.05	262
6410	Outdoor PAR floodlights, 1 lamp, 150 watt		20	.400		20	8.85		28.85	36.50
6420	2 lamp, 150 watt each		20	.400		33.50	8.85		42.35	51.50
6430	For infrared security sensor, add		32	.250		87	5.55		92.55	105
6450	Outdoor, quartz-halogen, 300 watt flood		20	.400		38	8.85		46.85	56.50
6600	Recessed downlight, round, pre-wired, 50 or 75 watt trim		30	.267		35	5.90		40.90	48
6610	With shower light trim		30	.267		42	5.90		47.90	55.50
6620	With wall washer trim		28	.286		52.50	6.30		58.80	68.50
6630	With eye-ball trim		28	.286		49	6.30		55.30	64.50
6640	For direct contact with insulation, add					1.60			1.60	1.76
6700	Porcelain lamp holder	1 Elec	40	.200		3.50	4.42		7.92	11.10
6710	With pull switch		40	.200		3.75	4.42		8.17	11.40
6750	Fluorescent strip, 1-20 watt tube, wrap around diffuser, 24″		24	.333		51.50	7.35		58.85	68.50
6760	1-40 watt tube, 48″		24	.333		65	7.35		72.35	83.50
6770	2-40 watt tubes, 48″		20	.400		78	8.85		86.85	100
6780	With residential ballast		20	.400		89.50	8.85		98.35	113
6800	Bathroom heat lamp, 1-250 watt		28	.286		30.50	6.30		36.80	44
6810	2-250 watt lamps		28	.286		51	6.30		57.30	66.50
6820	For timer switch, see line 2400									
6900	Outdoor post lamp, incl. post, fixture, 35′ of #14/2									
6910	Type NMC cable	1 Elec	3.50	2.286	Ea.	165	50.50		215.50	265
6920	Photo-eye, add		27	.296		29	6.55		35.55	42.50
6950	Clock dial time switch, 24 hr., w/enclosure, type NM cable		11.43	.700		50.50	15.45		65.95	81
6970	Type MC cable		11	.727		60.50	16.05		76.55	93.50
6980	EMT & wire		4.85	1.649		61.50	36.50		98	127
7000	Alarm systems									
7050	Smoke detectors, box, #14/3, type NM cable	1 Elec	14.55	.550	Ea.	28	12.15		40.15	50.50
7070	Type MC cable		12.31	.650		36.50	14.35		50.85	64
7080	EMT & wire		5	1.600		35.50	35.50		71	97

16139 Residential Wiring		CREW	DAILY OUTPUT	LABOR-HOURS	UNIT	2000 BARE COSTS MAT.	LABOR	EQUIP.	TOTAL	TOTAL INCL O&P
700 7090	For relay output to security system, add				Ea.	11.75			11.75	12.95
8000	Residential equipment									
8050	Disposal hook-up, incl. switch, outlet box, 3′ of flex									
8060	20 amp-1 pole circ. bkr., and 25′ of #12/2									
8070	Type NM cable	1 Elec	10	.800	Ea.	19.10	17.70		36.80	50
8080	Type MC cable		8	1		30	22		52	69
8090	EMT & wire		5	1.600		31.50	35.50		67	93
8100	Trash compactor or dishwasher hook-up, incl. outlet box,									
8110	3′ of flex, 15 amp-1 pole circ. bkr., and 25′ of #14/2									
8130	Type MC cable	1 Elec	8	1	Ea.	25.50	22		47.50	64
8140	EMT & wire	″	5	1.600	″	26	35.50		61.50	86.50
8150	Hot water sink dispensor hook-up, use line 8100									
8200	Vent/exhaust fan hook-up, type NM cable	1 Elec	32	.250	Ea.	3.51	5.55		9.06	12.90
8220	Type MC cable		24	.333		11.40	7.35		18.75	24.50
8230	EMT & wire		10	.800		11.15	17.70		28.85	41.50
8250	Bathroom vent fan, 50 CFM (use with above hook-up)									
8260	Economy model	1 Elec	15	.533	Ea.	21.50	11.80		33.30	43
8270	Low noise model		15	.533		30	11.80		41.80	52.50
8280	Custom model		12	.667		111	14.75		125.75	146
8300	Bathroom or kitchen vent fan, 110 CFM									
8310	Economy model	1 Elec	15	.533	Ea.	53	11.80		64.80	78
8320	Low noise model	″	15	.533	″	72	11.80		83.80	98.50
8350	Paddle fan, variable speed (w/o lights)									
8360	Economy model (AC motor)	1 Elec	10	.800	Ea.	94.50	17.70		112.20	133
8370	Custom model (AC motor)		10	.800		165	17.70		182.70	211
8380	Luxury model (DC motor)		8	1		325	22		347	395
8390	Remote speed switch for above, add		12	.667		21	14.75		35.75	47
8500	Whole house exhaust fan, ceiling mount, 36″, variable speed									
8510	Remote switch, incl. shutters, 20 amp-1 pole circ. bkr.									
8520	30′ of #12/2, type NM cable	1 Elec	4	2	Ea.	675	44		719	820
8530	Type MC cable		3.50	2.286		690	50.50		740.50	845
8540	EMT & wire		3	2.667		690	59		749	855
8600	Whirlpool tub hook-up, incl. timer switch, outlet box									
8610	3′ of flex, 20 amp-1 pole GFI circ. bkr.									
8620	30′ of #12/2, type NM cable	1 Elec	5	1.600	Ea.	74.50	35.50		110	140
8630	Type MC cable		4.20	1.905		82	42		124	160
8640	EMT & wire		3.40	2.353		84	52		136	178
8650	Hot water heater hook-up, incl. 1-2 pole circ. bkr., box;									
8660	3′ of flex, 20′ of #10/2, type NM cable	1 Elec	5	1.600	Ea.	19.20	35.50		54.70	79
8670	Type MC cable		4.20	1.905		33.50	42		75.50	106
8680	EMT & wire		3.40	2.353		28	52		80	116
9000	Heating/air conditioning									
9050	Furnace/boiler hook-up, incl. firestat, local on-off switch									
9060	Emergency switch, and 40′ of type NM cable	1 Elec	4	2	Ea.	42	44		86	119
9070	Type MC cable		3.50	2.286		59	50.50		109.50	148
9080	EMT & wire		1.50	5.333		60	118		178	259
9100	Air conditioner hook-up, incl. local 60 amp disc. switch									
9110	3′ sealtite, 40 amp, 2 pole circuit breaker									
9130	40′ of #8/2, type NM cable	1 Elec	3.50	2.286	Ea.	136	50.50		186.50	232
9140	Type MC cable		3	2.667		182	59		241	298
9150	EMT & wire		1.30	6.154		146	136		282	380
9200	Heat pump hook-up, 1-40 & 1-100 amp 2 pole circ. bkr.									
9210	Local disconnect switch, 3′ sealtite									
9220	40′ of #8/2 & 30′ of #3/2									
9230	Type NM cable	1 Elec	1.30	6.154	Ea.	325	136		461	575
9240	Type MC cable		1.08	7.407		445	164		609	760

16100 | Wiring Methods

16139 | Residential Wiring		Crew	Daily Output	Labor-Hours	Unit	2000 Bare Costs Mat.	Labor	Equip.	Total	Total Incl O&P	
9250	EMT & wire	1 Elec	.94	8.511	Ea.	335	188		523	680	700
9500	Thermostat hook-up, using low voltage wire										
9520	Heating only	1 Elec	24	.333	Ea.	5.75	7.35		13.10	18.35	
9530	Heating/cooling	"	20	.400	"	7.10	8.85		15.95	22.50	

16140 | Wiring Devices		Crew	Daily Output	Labor-Hours	Unit	2000 Bare Costs Mat.	Labor	Equip.	Total	Total Incl O&P	
0010	LOW VOLTAGE SWITCHING										500
3600	Relays, 120 V or 277 V standard	1 Elec	12	.667	Ea.	26	14.75		40.75	52.50	
3800	Flush switch, standard		40	.200		9.05	4.42		13.47	17.20	
4000	Interchangeable		40	.200		11.80	4.42		16.22	20.50	
4100	Surface switch, standard		40	.200		6.60	4.42		11.02	14.55	
4200	Transformer 115 V to 25 V		12	.667		93	14.75		107.75	126	
4400	Master control, 12 circuit, manual		4	2		94	44		138	176	
4500	25 circuit, motorized		4	2		102	44		146	185	
4600	Rectifier, silicon		12	.667		30.50	14.75		45.25	57.50	
4800	Switchplates, 1 gang, 1, 2 or 3 switch, plastic		80	.100		3	2.21		5.21	6.90	
5000	Stainless steel		80	.100		8.10	2.21		10.31	12.50	
5400	2 gang, 3 switch, stainless steel		53	.151		15.65	3.34		18.99	22.50	
5500	4 switch, plastic		53	.151		6.70	3.34		10.04	12.80	
5600	2 gang, 4 switch, stainless steel		53	.151		16.35	3.34		19.69	23.50	
5700	6 switch, stainless steel		53	.151		36	3.34		39.34	45	
5800	3 gang, 9 switch, stainless steel		32	.250		50	5.55		55.55	64	
0010	WIRING DEVICES										910
0200	Toggle switch, quiet type, single pole, 15 amp	1 Elec	40	.200	Ea.	4.63	4.42		9.05	12.35	
0600	3 way, 15 amp		23	.348		6.70	7.70		14.40	19.90	
0900	4 way, 15 amp		15	.533		19.40	11.80		31.20	41	
1650	Dimmer switch, 120 volt, incandescent, 600 watt, 1 pole		16	.500		10.80	11.05		21.85	30	
2460	Receptacle, duplex, 120 volt, grounded, 15 amp		40	.200		1.14	4.42		5.56	8.50	
2470	20 amp		27	.296		9.50	6.55		16.05	21	
2490	Dryer, 30 amp		15	.533		11.70	11.80		23.50	32	
2500	Range, 50 amp		11	.727		10.45	16.05		26.50	38	
2600	Wall plates, stainless steel, 1 gang		80	.100		2.05	2.21		4.26	5.90	
2800	2 gang		53	.151		4.10	3.34		7.44	9.95	
3200	Lampholder, keyless		26	.308		8.80	6.80		15.60	21	
3400	Pullchain with receptacle		22	.364		8.90	8.05		16.95	23	

16150 | Wiring Connections		Crew	Daily Output	Labor-Hours	Unit	2000 Bare Costs Mat.	Labor	Equip.	Total	Total Incl O&P	
0010	MOTOR CONNECTIONS										275
0020	Flexible conduit and fittings, 115 volt, 1 phase, up to 1 HP motor	1 Elec	8	1	Ea.	4.30	22		26.30	40.50	

16200 | Electrical Power

16210 | Elect Utility Services		Crew	Daily Output	Labor-Hours	Unit	2000 Bare Costs Mat.	Labor	Equip.	Total	Total Incl O&P	
0010	METER CENTERS AND SOCKETS										600
0100	Sockets, single position, 4 terminal, 100 amp	1 Elec	3.20	2.500	Ea.	27.50	55.50		83	121	
0200	150 amp		2.30	3.478		30	77		107	159	
0300	200 amp		1.90	4.211		40	93		133	196	
0500	Double position, 4 terminal, 100 amp		2.80	2.857		116	63		179	231	
0600	150 amp		2.10	3.810		120	84		204	270	

16200 | Electrical Power

	16210 | Elect Utility Services		CREW	DAILY OUTPUT	LABOR-HOURS	UNIT	2000 BARE COSTS MAT.	LABOR	EQUIP.	TOTAL	TOTAL INCL O&P
600	0700	200 amp	1 Elec	1.70	4.706	Ea.	288	104		392	485
	2590	Basic meter device									
	2600	1P 3W 120/240V 4 jaw 125A sockets, 3 meter	2 Elec	1	16	Ea.	410	355		765	1,025
	2620	5 meter		.80	20		615	440		1,055	1,400
	2640	7 meter		.56	28.571		900	630		1,530	2,025
	2660	10 meter		.48	33.333		1,225	735		1,960	2,550
	2680	Rainproof 1P 3W 120/240V 4 jaw 125A sockets									
	2690	3 meter	2 Elec	1	16	Ea.	410	355		765	1,025
	2710	6 meter		.60	26.667		705	590		1,295	1,750
	2730	8 meter		.52	30.769		985	680		1,665	2,175
	2750	1P 3W 120/240V 4 jaw sockets									
	2760	with 125A circuit breaker, 3 meter	2 Elec	1	16	Ea.	770	355		1,125	1,425
	2780	5 meter		.80	20		1,200	440		1,640	2,050
	2800	7 meter		.56	28.571		1,725	630		2,355	2,925
	2820	10 meter		.48	33.333		2,425	735		3,160	3,875
	2830	Rainproof 1P 3W 120/240V 4 jaw sockets									
	2840	with 125A circuit breaker, 3 meter	2 Elec	1	16	Ea.	770	355		1,125	1,425
	2870	6 meter		.60	26.667		1,425	590		2,015	2,550
	2890	8 meter		.52	30.769		1,925	680		2,605	3,225
	3250	1P 3W 120/240V 4 jaw sockets									
	3260	with 200A circuit breaker, 3 meter	2 Elec	1	16	Ea.	1,150	355		1,505	1,850
	3290	6 meter		.60	26.667		2,300	590		2,890	3,500
	3310	8 meter		.56	28.571		3,125	630		3,755	4,450
	3330	Rainproof 1P 3W 120/240V 4 jaw sockets									
	3350	with 200A circuit breaker, 3 meter	2 Elec	1	16	Ea.	1,150	355		1,505	1,850
	3380	6 meter		.60	26.667		2,300	590		2,890	3,500
	3400	8 meter		.52	30.769		3,125	680		3,805	4,525
	16230 | Generator Assemblies										
450	0010	GENERATOR SET									
	0020	Gas or gasoline operated, includes battery,									
	0050	charger, muffler & transfer switch									
	0200	3 phase 4 wire, 277/480 volt, 7.5 kW	R-3	.83	24.096	Ea.	6,000	530	167	6,697	7,650
	0300	11.5 kW		.71	28.169		8,500	620	195	9,315	10,600
	0400	20 kW		.63	31.746		10,000	700	219	10,919	12,400

16 ELECTRICAL

16400 | Low-Voltage Distribution

	16410 | Encl Switches & Circuit Breakers		CREW	DAILY OUTPUT	LABOR-HOURS	UNIT	2000 BARE COSTS MAT.	LABOR	EQUIP.	TOTAL	TOTAL INCL O&P
200	0010	CIRCUIT BREAKERS (in enclosure)									
	0100	Enclosed (NEMA 1), 600 volt, 3 pole, 30 amp	1 Elec	3.20	2.500	Ea.	365	55.50		420.50	490
	0200	60 amp		2.80	2.857		365	63		428	505
	0400	100 amp		2.30	3.478		415	77		492	585
800	0010	SAFETY SWITCHES									
	0100	General duty 240 volt, 3 pole NEMA 1, fusible, 30 amp	1 Elec	3.20	2.500	Ea.	65.50	55.50		121	163
	0200	60 amp		2.30	3.478		111	77		188	248
	0300	100 amp		1.90	4.211		191	93		284	360
	0400	200 amp		1.30	6.154		410	136		546	670
	0500	400 amp	2 Elec	1.80	8.889		1,025	196		1,221	1,450

16400 | Low-Voltage Distribution

16410 | Encl Switches & Circuit Breakers		Crew	Daily Output	Labor-Hours	Unit	2000 Bare Costs Mat.	Labor	Equip.	Total	Total Incl O&P	
9010	Disc. switch, 600V 3 pole fusible, 30 amp, to 10 HP motor	1 Elec	3.20	2.500	Ea.	184	55.50		239.50	293	800
9050	60 amp, to 30 HP motor		2.30	3.478		420	77		497	590	
9070	100 amp, to 60 HP motor		1.90	4.211		420	93		513	615	
0010	**TIME SWITCHES**										840
0100	Single pole, single throw, 24 hour dial	1 Elec	4	2	Ea.	74.50	44		118.50	154	
0200	24 hour dial with reserve power		3.60	2.222		315	49		364	430	
0300	Astronomic dial		3.60	2.222		128	49		177	222	
0400	Astronomic dial with reserve power		3.30	2.424		405	53.50		458.50	535	
0500	7 day calendar dial		3.30	2.424		115	53.50		168.50	215	
0600	7 day calendar dial with reserve power		3.20	2.500		340	55.50		395.50	465	
0700	Photo cell 2000 watt		8	1		15.15	22		37.15	52.50	

16415 | Transfer Switches		Crew	Daily Output	Labor-Hours	Unit	Mat.	Labor	Equip.	Total	Total Incl O&P	
0010	**AUTOMATIC TRANSFER SWITCHES**										600
0100	Switches, enclosed 480 volt, 3 pole, 30 amp	1 Elec	2.30	3.478	Ea.	2,900	77		2,977	3,325	
0200	60 amp	″	1.90	4.211	″	2,900	93		2,993	3,350	

16440 | Swbds, Panels & Cont Centers		Crew	Daily Output	Labor-Hours	Unit	Mat.	Labor	Equip.	Total	Total Incl O&P	
0010	**LOAD CENTERS** (residential type)										500
0100	3 wire, 120/240V, 1 phase, including 1 pole plug-in breakers										
0200	100 amp main lugs, indoor, 8 circuits	1 Elec	1.40	5.714	Ea.	113	126		239	330	
0300	12 circuits		1.20	6.667		157	147		304	415	
0400	Rainproof, 8 circuits		1.40	5.714		135	126		261	355	
0500	12 circuits		1.20	6.667		190	147		337	450	
0600	200 amp main lugs, indoor, 16 circuits	R-1A	1.80	8.889		262	164		426	560	
0700	20 circuits		1.50	10.667		330	197		527	690	
0800	24 circuits		1.30	12.308		430	227		657	855	
1200	Rainproof, 16 circuits		1.80	8.889		310	164		474	620	
1300	20 circuits		1.50	10.667		375	197		572	740	
1400	24 circuits		1.30	12.308		560	227		787	995	

16500 | Lighting

16510 | Interior Luminaires		Crew	Daily Output	Labor-Hours	Unit	2000 Bare Costs Mat.	Labor	Equip.	Total	Total Incl O&P	
0010	**INTERIOR LIGHTING FIXTURES** Including lamps, mounting										440
0030	hardware and connections										
0100	Fluorescent, C.W. lamps, troffer, recess mounted in grid, RS										
0130	grid ceiling mount										
0200	Acrylic lens, 1′W x 4′L, two 40 watt	1 Elec	5.70	1.404	Ea.	46	31		77	101	
0300	2′W x 2′L, two U40 watt		5.70	1.404		50	31		81	106	
0600	2′W x 4′L, four 40 watt		4.70	1.702		56	37.50		93.50	123	
1000	Surface mounted, RS										
1030	Acrylic lens with hinged & latched door frame										
1100	1′W x 4′L, two 40 watt	1 Elec	7	1.143	Ea.	70	25.50		95.50	119	
1200	2′W x 2′L, two U40 watt		7	1.143		85	25.50		110.50	135	
1500	2′W x 4′L, four 40 watt		5.30	1.509		90	33.50		123.50	154	
2100	Strip fixture										
2200	4′ long, one 40 watt RS	1 Elec	8.50	.941	Ea.	26	21		47	62.50	

16510 | Interior Luminaires

		16510 Interior Luminaires	CREW	DAILY OUTPUT	LABOR-HOURS	UNIT	2000 BARE COSTS MAT.	LABOR	EQUIP.	TOTAL	TOTAL INCL O&P
440	2300	4' long, two 40 watt RS	1 Elec	8	1	Ea.	28	22		50	67
	2600	8' long, one 75 watt, SL	2 Elec	13.40	1.194		39	26.50		65.50	86
	2700	8' long, two 75 watt, SL	"	12.40	1.290		47	28.50		75.50	98
	4450	Incandescent, high hat can, round alzak reflector, prewired									
	4470	100 watt	1 Elec	8	1	Ea.	53	22		75	94.50
	4480	150 watt	"	8	1	"	77	22		99	121
	5200	Ceiling, surface mounted, opal glass drum									
	5300	8", one 60 watt lamp	1 Elec	10	.800	Ea.	32	17.70		49.70	64
	5400	10", two 60 watt lamps		8	1		36	22		58	75.50
	5500	12", four 60 watt lamps		6.70	1.194		50	26.50		76.50	98
	6900	Mirror light, fluorescent, RS, acrylic enclosure, two 40 watt		8	1		77	22		99	121
	6910	One 40 watt		8	1		60	22		82	102
	6920	One 20 watt		12	.667		47.50	14.75		62.25	76.50
800	0010	**RESIDENTIAL FIXTURES**									
	0400	Fluorescent, interior, surface, circline, 32 watt & 40 watt	1 Elec	20	.400	Ea.	68	8.85		76.85	89.50
	0500	2' x 2', two U 40 watt		8	1		77	22		99	121
	0700	Shallow under cabinet, two 20 watt		16	.500		37	11.05		48.05	58.50
	0900	Wall mounted, 4'L, one 40 watt, with baffle		10	.800		90	17.70		107.70	128
	2000	Incandescent, exterior lantern, wall mounted, 60 watt		16	.500		28	11.05		39.05	49
	2100	Post light, 150W, with 7' post		4	2		100	44		144	183
	2500	Lamp holder, weatherproof with 150W PAR		16	.500		14	11.05		25.05	33.50
	2550	With reflector and guard		12	.667		44	14.75		58.75	72.50
	2600	Interior pendent, globe with shade, 150 watt		20	.400		106	8.85		114.85	131

16520 | Exterior Luminaires

		16520 Exterior Luminaires	CREW	DAILY OUTPUT	LABOR-HOURS	UNIT	MAT.	LABOR	EQUIP.	TOTAL	TOTAL INCL O&P
300	0010	**EXTERIOR FIXTURES** With lamps									
	0400	Quartz, 500 watt	1 Elec	5.30	1.509	Ea.	51	33.50		84.50	111
	0800	Wall pack, mercury vapor, 175 watt		4	2		210	44		254	305
	1000	250 watt		4	2		213	44		257	305
	1100	Low pressure sodium, 35 watt		4	2		208	44		252	300
	1150	55 watt		4	2		280	44		324	385
	6420	Wood pole, 4-1/2" x 5-1/8", 8' high		6	1.333		230	29.50		259.50	300
	6440	12' high		5.70	1.404		315	31		346	395
	6460	20' high		4	2		440	44		484	560
	6500	Bollard light, lamp & ballast, 42" high with polycarbonate lens									
	6700	Mercury vapor, 175 watt	1 Elec	3	2.667	Ea.	505	59		564	650
	7200	Incandescent, 150 watt	"	3	2.667	"	390	59		449	525
	7380	Landscape recessed uplight, incl. housing, ballast, transformer									
	7390	& reflector									
	7420	Incandescent, 250 watt	1 Elec	5	1.600	Ea.	400	35.50		435.50	500
	7440	Quartz, 250 watt	"	5	1.600	"	380	35.50		415.50	480

16550 | Special Purpose Ltg.

		16550 Special Purpose Ltg.	CREW	DAILY OUTPUT	LABOR-HOURS	UNIT	MAT.	LABOR	EQUIP.	TOTAL	TOTAL INCL O&P
820	0010	**TRACK LIGHTING**									
	0100	8' section	1 Elec	5.30	1.509	Ea.	55	33.50		88.50	115
	0300	3 circuits, 4' section		6.70	1.194		44	26.50		70.50	91.50
	0400	8' section		5.30	1.509		68.50	33.50		102	130
	0500	12' section		4.40	1.818		136	40		176	216
	1000	Feed kit, surface mounting		16	.500		8	11.05		19.05	27
	1100	End cover		24	.333		3	7.35		10.35	15.35
	1200	Feed kit, stem mounting, 1 circuit		16	.500		22.50	11.05		33.55	43
	1300	3 circuit		16	.500		22.50	11.05		33.55	43
	2000	Electrical joiner, for continuous runs, 1 circuit		32	.250		10.50	5.55		16.05	20.50
	2100	3 circuit		32	.250		24	5.55		29.55	35.50
	2200	Fixtures, spotlight, 75W PAR halogen		16	.500		82.50	11.05		93.55	109

Important: See the Reference Section for critical supporting data - Reference Nos., Crews, & Location Facto

16500 | Lighting

16550	Special Purpose Ltg.	CREW	DAILY OUTPUT	LABOR-HOURS	UNIT	2000 BARE COSTS MAT.	LABOR	EQUIP.	TOTAL	TOTAL INCL O&P	
2210	50W MR16 halogen	1 Elec	16	.500	Ea.	98.50	11.05		109.55	126	820
3000	Wall washer, 250 watt tungsten halogen	↓	16	.500	↓	93.50	11.05		104.55	121	
3100	Low voltage, 25/50 watt, 1 circuit	↓	16	.500	↓	90	11.05		101.05	117	
3120	3 circuit	↓	16	.500	↓	94	11.05		105.05	121	

16585	Lamps	CREW	DAILY OUTPUT	LABOR-HOURS	UNIT	MAT.	LABOR	EQUIP.	TOTAL	TOTAL INCL O&P	
0010	**LAMPS**										600
0081	Fluorescent, rapid start, cool white, 2′ long, 20 watt	1 Elec	100	.080	Ea.	2.74	1.77		4.51	5.90	
0101	4′ long, 40 watt	↓	90	.089	↓	2.53	1.96		4.49	6	
1351	High pressure sodium, 70 watt	↓	30	.267	↓	42	5.90		47.90	56	
1371	150 watt	↓	30	.267	↓	45	5.90		50.90	59	

16800 | Sound & Video

16820	Sound Reinforcement	CREW	DAILY OUTPUT	LABOR-HOURS	UNIT	2000 BARE COSTS MAT.	LABOR	EQUIP.	TOTAL	TOTAL INCL O&P	
0010	**DOORBELL SYSTEM** Incl. transformer, button & signal										300
1000	Door chimes, 2 notes, minimum	1 Elec	16	.500	Ea.	22	11.05		33.05	42	
1020	Maximum	↓	12	.667	↓	112	14.75		126.75	147	
1100	Tube type, 3 tube system	↓	12	.667	↓	163	14.75		177.75	203	
1180	4 tube system	↓	10	.800	↓	264	17.70		281.70	320	
1900	For transformer & button, minimum add	↓	5	1.600	↓	12.50	35.50		48	72	
1960	Maximum, add	↓	4.50	1.778	↓	37	39.50		76.50	105	
3000	For push button only, minimum	↓	24	.333	↓	2.45	7.35		9.80	14.75	
3100	Maximum	↓	20	.400	↓	19.45	8.85		28.30	36	

16850	Television Equipment	CREW	DAILY OUTPUT	LABOR-HOURS	UNIT	MAT.	LABOR	EQUIP.	TOTAL	TOTAL INCL O&P	
0010	**T.V. SYSTEMS** not including rough-in wires, cables & conduits										600
0100	Master TV antenna system										
0200	VHF reception & distribution, 12 outlets	1 Elec	6	1.333	Outlet	152	29.50		181.50	215	
0800	VHF & UHF reception & distribution, 12 outlets	↓	6	1.333	"	151	29.50		180.50	214	
5000	T.V. Antenna only, minimum	↓	6	1.333	Ea.	33	29.50		62.50	84.50	
5100	Maximum	↓	4	2	"	140	44		184	227	

or information about Means Estimating Seminars, see yellow pages 11 and 12 in back of book

Division Notes

	CREW	DAILY OUTPUT	LABOR-HOURS	UNIT	2000 BARE COSTS MAT.	LABOR	EQUIP.	TOTAL	TOTAL INCL O&P

Reference Section

the reference information is in one section aking it easy to find what you need to know . . . d easy to use the book on a daily basis. This ction is visually identified by a vertical gray bar the edge of pages.

the reference number information that lows, you'll see the background that relates the "reference numbers" that appeared in the it Price Sections. You'll find reference tables, planations and estimating information that pport how we arrived at the unit price data. so included are alternate pricing methods, chnical data and estimating procedures along th information on design and economy in nstruction.

so in this Reference Section, we've included ange Orders, information on pricing changes contract documents; Crew Listings, a full ting of all the crews, equipment and their sts; Historical Cost Indexes for cost mparisons over time; City Cost Indexes and cation Factors for adjusting costs to the gion you are in; and an explanation of all breviations used in the book.

Table of Contents

Reference Numbers

R011 Overhead & Misc. Data 512
R012 Contract Modifications Procedures 519
R015 Construction Aids 520
R020 Paving & Surfacing 522
R029 Landscaping 523
R040 Mortar & Masonry Accessories 524
R042 Unit Masonry 525
R044 Stone 529
R049 Restoration & Cleaning 529
R061 Rough Carpentry 530
R064 Architectural Woodwork 533
R071 Waterproofing & Dampproofing 534
R073 Shingles & Roofing Tiles 534

Reference Numbers (cont.)

R075 Membrane Roofing 534
R081 Metal Doors & Frames 535
R082 Wood & Plastic Doors 535
R085 Metal Windows 536
R087 Hardware 536
R092 Plaster & Gypsum Board 537
R096 Flooring 538
R097 Wall Finishes 538
R099 Paints & Coatings 539
R131 Pre-Engineered Structures & Aquatic Facilities 541
R136 Solar & Wind Energy Equip. 541
R151 Plumbing 542

Crew Listings 543

Location Factors 566

Abbreviations 572

R01100-005 Tips for Accurate Estimating

1. Use pre-printed or columnar forms for orderly sequence of dimensions and locations and for recording telephone quotations.
2. Use only the front side of each paper or form except for certain pre-printed summary forms.
3. Be consistent in listing dimensions: For example, length x width x height. This helps in rechecking to ensure that, the total length of partitions is appropriate for the building area.
4. Use printed (rather than measured) dimensions where given.
5. Add up multiple printed dimensions for a single entry where possible.
6. Measure all other dimensions carefully.
7. Use each set of dimensions to calculate multiple related quantities.
8. Convert foot and inch measurements to decimal feet when listing. Memorize decimal equivalents to .01 parts of a foot (1/8″ equals approximately .01′).
9. Do not "round off" quantities until the final summary.
10. Mark drawings with different colors as items are taken off.
11. Keep similar items together, different items separate.
12. Identify location and drawing numbers to aid in future checking for completeness.
13. Measure or list everything on the drawings or mentioned in the specifications.
14. It may be necessary to list items not called for to make the job complete.
15. Be alert for: Notes on plans such as N.T.S. (not to scale); changes in scale throughout the drawings; reduced size drawings; discrepancies between the specifications and the drawings.
16. Develop a consistent pattern of performing an estimate. For example:
    a. Start the quantity takeoff at the lower floor and move to the next higher floor.
    b. Proceed from the main section of the building to the wings.
    c. Proceed from south to north or vice versa, clockwise or counterclockwise.
    d. Take off floor plan quantities first, elevations next, then detail drawings.
17. List all gross dimensions that can be either used again for different quantities, or used as a rough check of other quantities for verification (exterior perimeter, gross floor area, individual floor areas, etc.).
18. Utilize design symmetry or repetition (repetitive floors, repetitive wings, symmetrical design around a center line, similar room layouts, etc.). Note: Extreme caution is needed here so as not to omit or duplicate an area.
19. Do not convert units until the final total is obtained. For instance, when estimating concrete work, keep all units to the nearest cubic foot, then summarize and convert to cubic yards.
20. When figuring alternatives, it is best to total all items involved in the basic system, then total all items involved in the alternates. Therefore you work with positive numbers in all cases. When adds and deducts are used, it is often confusing whether to add or subtract a portion of an item; especially on a complicated or involved alternate.

R01100-040 Builder's Risk Insurance

Builder's Risk Insurance is insurance on a building during construction. Premiums are paid by the owner or the contractor. Blasting, collapse and underground insurance would raise total insurance costs above those listed. Floater policy for materials delivered to the job runs $.75 to $1.25 per $100 value. Contractor equipment insurance runs $.50 to $1.50 per $100 value. Insurance for miscellaneous tools to $1,500 value runs from $3.00 to $7.50 per $100 value.

Tabulated below are New England Builder's Risk insurance rates in dollars per $100 value for $1,000 deductible. For $25,000 deductible, rates can be reduced 13% to 34%. On contracts over $1,000,000, rates may be lower than those tabulated. Policies are written annually for the total completed value in place. For "all risk" insurance (excluding flood, earthquake and certain other perils) add $.025 to total rates below.

Coverage	Frame Construction (Class 1)				Brick Construction (Class 4)				Fire Resistive (Class 6)			
	Range			Average	Range			Average	Range			Average
Fire Insurance	$.300	to	$.420	$.394	$.132	to	$.189	$.174	$.052	to	$.080	$.070
Extended Coverage	.115	to	.150	.144	.080	to	.105	.101	.081	to	.105	.100
Vandalism	.012	to	.016	.015	.008	to	.011	.011	.008	to	.011	.010
Total Annual Rate	$.427	to	$.586	$.553	$.220	to	$.305	$.286	$.141	to	$.196	$.180

GENERAL REQUIREMENTS

REFERENCE NOS.

R01100-050 General Contractor's Overhead

The table below shows a contractor's overhead as a percentage of direct cost in two ways. The figures on the right are for the overhead, markup based on both material and labor. The figures on the left are based on the entire overhead applied only to the labor. This figure would be used if the owner supplied the materials or if a contract is for labor only. Note: Some of these markups are included in the labor rates shown on Reference Table R01100-070.

Items of General Contractor's Indirect Costs	% of Direct Costs	
	As a Markup of Labor Only	As a Markup of Both Material and Labor
Field Supervision	6.0%	2.9%
Main Office Expense (see details below)	16.2	7.7
Tools and Minor Equipment	1.0	0.5
Workers' Compensation & Employers' Liability. See R01100-060	18.1	8.6
Field Office, Sheds, Photos, Etc.	1.5	0.7
Performance and Payment Bond, 0.7% to 1.5%. See R01100-080	2.3	1.1
Unemployment Tax See R01100-100 (Combined Federal and State)	7.0	3.3
Social Security and Medicare, See R01100-100	7.7	3.7
Sales Tax — add if applicable 42/80 x % as markup of total direct costs including both material and labor. See R01100-090		
Sub Total	59.8%	28.5%
*Builder's Risk Insurance ranges from .141% to .586%. See R01100-040	0.6	0.3
*Public Liability Insurance	3.2	1.5
Grand Total	63.6%	30.3%

*Paid by Owner or Contractor

Main Office Expense

A General Contractor's main office expense consists of many items not detailed in the front portion of the book. The percentage of main office expense declines with increased annual volume of the contractor. Typical main office expense ranges from 2% to 20% with the median about 7.2% of total volume. This equals about 7.7% of direct costs. The following are approximate percentages of total overhead for different items usually included in a General Contractor's main office overhead. With different accounting procedures, these percentages may vary.

Item	Typical Range	Average
Managers', clerical and estimators' salaries	40 % to 55 %	48%
Profit sharing, pension and bonus plans	2 to 20	12
Insurance	5 to 8	6
Estimating and project management (not including salaries)	5 to 9	7
Legal, accounting and data processing	0.5 to 5	3
Automobile and light truck expense	2 to 8	5
Depreciation of overhead capital expenditures	2 to 6	4
Maintenance of office equipment	0.1 to 1.5	1
Office rental	3 to 5	4
Utilities including phone and light	1 to 3	2
Miscellaneous	5 to 15	8
Total		100%

R01100-060 Workers' Compensation Insurance Rates by Trade

The table below tabulates the national averages for Workers' Compensation insurance rates by trade and type of building. The average "Insurance Rate" is multiplied by the "% of Building Cost" for each trade. This produces the "Workers' Compensation Cost" by % of total labor cost, to be added for each trade by building type to determine the weighted average Workers' Compensation rate for the building types analyzed.

Trade	Insurance Rate (% Labor Cost)		% of Building Cost			Workers' Compensation		
	Range	Average	Office Bldgs.	Schools & Apts.	Mfg.	Office Bldgs.	Schools & Apts.	Mfg.
Excavation, Grading, etc.	4.0 % to 26.6%	11.4%	4.8%	4.9%	4.5%	.55%	.56%	.51%
Piles & Foundations	8.1 to 80.1	29.9	7.1	5.2	8.7	2.12	1.55	2.60
Concrete	7.5 to 39.4	18.8	5.0	14.8	3.7	.94	2.78	.70
Masonry	5.8 to 48.6	17.8	6.9	7.5	1.9	1.23	1.34	.34
Structural Steel	8.1 to 132.9	42.6	10.7	3.9	17.6	4.56	1.66	7.50
Miscellaneous & Ornamental Metals	5.4 to 34.0	13.8	2.8	4.0	3.6	.39	.55	.50
Carpentry & Millwork	7.0 to 49.9	19.9	3.7	4.0	0.5	.74	.80	.10
Metal or Composition Siding	6.9 to 36.7	18.1	2.3	0.3	4.3	.42	.05	.78
Roofing	8.1 to 88.8	34.6	2.3	2.6	3.1	.80	.90	1.07
Doors & Hardware	3.8 to 28.8	11.7	0.9	1.4	0.4	.11	.16	.05
Sash & Glazing	5.7 to 30.0	14.4	3.5	4.0	1.0	.50	.58	.14
Lath & Plaster	5.3 to 37.4	15.7	3.3	6.9	0.8	.52	1.08	.13
Tile, Marble & Floors	3.5 to 29.5	10.5	2.6	3.0	0.5	.27	.32	.05
Acoustical Ceilings	4.0 to 26.7	12.3	2.4	0.2	0.3	.30	.02	.04
Painting	6.1 to 41.1	15.3	1.5	1.6	1.6	.23	.24	.24
Interior Partitions	7.0 to 49.9	19.9	3.9	4.3	4.4	.78	.86	.88
Miscellaneous Items	2.6 to 112.5	19.0	5.2	3.7	9.7	.99	.70	1.84
Elevators	2.0 to 19.1	8.5	2.1	1.1	2.2	.18	.09	.19
Sprinklers	2.8 to 20.6	9.0	0.5	—	2.0	.05	—	.18
Plumbing	3.0 to 15.9	8.8	4.9	7.2	5.2	.43	.63	.46
Heat., Vent., Air Conditioning	3.9 to 25.8	12.3	13.5	11.0	12.9	1.66	1.35	1.59
Electrical	2.9 to 11.6	7.0	10.1	8.4	11.1	.71	.59	.78
Total	2.0 % to 132.9%	—	100.0%	100.0%	100.0%	18.48%	16.81%	20.67%
		Overall Weighted Average 18.65%						

Workers' Compensation Insurance Rates by States

The table below lists the weighted average Workers' Compensation base rate for each state with a factor comparing this with the national average of 18.1%.

State	Weighted Average	Factor	State	Weighted Average	Factor	State	Weighted Average	Factor
Alabama	30.4%	168	Kentucky	19.8%	109	North Dakota	16.7%	92
Alaska	11.3	62	Louisiana	27.3	151	Ohio	16.2	90
Arizona	14.5	80	Maine	21.8	120	Oklahoma	21.9	121
Arkansas	14.2	78	Maryland	11.9	66	Oregon	19.3	107
California	18.6	103	Massachusetts	26.6	147	Pennsylvania	22.9	127
Colorado	26.5	146	Michigan	20.6	114	Rhode Island	22.4	124
Connecticut	21.1	117	Minnesota	37.7	208	South Carolina	13.8	76
Delaware	12.3	68	Mississippi	23.8	131	South Dakota	17.9	99
District of Columbia	25.2	139	Missouri	15.8	87	Tennessee	13.0	72
Florida	28.0	155	Montana	37.4	207	Texas	24.7	136
Georgia	24.9	138	Nebraska	15.7	87	Utah	13.0	72
Hawaii	16.7	92	Nevada	19.4	107	Vermont	14.9	82
Idaho	11.9	66	New Hampshire	23.2	128	Virginia	11.0	61
Illinois	23.3	129	New Jersey	11.0	61	Washington	12.0	66
Indiana	7.5	41	New Mexico	23.1	128	West Virginia	14.1	78
Iowa	12.1	67	New York	17.0	94	Wisconsin	13.9	77
Kansas	10.1	56	North Carolina	14.1	78	Wyoming	8.9	49
			Weighted Average for U.S. is	18.7% of payroll = 100%				

Rates in the following table are the base or manual costs per $100 of payroll for Workers' Compensation in each state. Rates are usually applied to straight time wages only and not to premium time wages and bonuses.

The weighted average skilled worker rate for 35 trades is 18.1%. For bidding purposes, apply the full value of Workers' Compensation directly to total labor costs, or if labor is 38%, materials 42% and overhead and profit 20% of total cost, carry 38/80 x 18.1% = 8.6% of cost (before overhead and profit) into overhead. Rates vary not only from state to state but also with the experience rating of the contractor.

Rates are the most current available at the time of publication.

01100-060 Workers' Compensation Insurance Rates by Trade and State (cont.)

State	Carpentry — 3 stories or less 5651	Carpentry — interior cab. work 5437	Carpentry — general 5403	Concrete Work — NOC 5213	Concrete Work — flat (flr., sdwk.) 5221	Electrical Wiring — inside 5190	Excavation — earth NOC 6217	Excavation — rock 6217	Glaziers 5462	Insulation Work 5479	Lathing 5443	Masonry 5022	Painting & Decorating 5474	Pile Driving 6003	Plastering 5480	Plumbing 5183	Roofing 5551	Sheet Metal Work (HVAC) 5538	Steel Erection — door & sash 5102	Steel Erection — inter., ornam. 5102	Steel Erection — structure 5040	Steel Erection — NOC 5057	Tile Work — (interior ceramic) 5348	Waterproofing 9014	Wrecking 5701
AL	36.41	14.51	34.52	32.46	16.19	9.57	20.98	20.98	24.55	20.09	15.28	29.78	21.80	64.38	33.79	11.96	57.62	25.78	16.41	16.41	54.68	66.67	16.19	6.92	54.68
AK	9.30	6.63	9.58	9.26	5.69	5.31	8.16	8.16	9.44	8.72	6.37	9.33	8.41	36.08	11.46	4.46	16.70	6.03	10.43	10.43	18.69	18.69	6.28	5.07	25.07
AZ	14.89	9.11	26.40	13.69	8.09	6.28	8.66	8.66	12.90	15.63	7.83	13.69	10.23	19.88	12.27	7.77	19.67	10.13	10.02	10.02	36.62	23.31	7.24	4.94	50.31
AR	16.12	10.60	15.38	16.45	5.97	7.93	9.21	9.21	12.58	15.29	12.66	9.32	11.10	19.14	13.01	6.58	30.74	10.28	8.19	8.19	27.11	25.39	7.80	4.58	27.11
CA	28.28	9.07	28.28	13.18	13.18	9.93	7.88	7.88	16.37	23.34	10.90	14.55	19.45	21.18	18.12	11.59	39.79	15.33	13.55	13.55	23.93	21.91	7.74	19.45	21.91
CO	32.22	13.54	20.95	20.96	15.13	9.20	15.99	15.99	15.98	21.31	14.68	31.06	21.85	41.53	37.40	14.50	59.10	12.54	11.83	11.83	73.46	43.49	15.04	12.26	73.46
CT	19.25	13.14	24.07	20.48	14.17	6.68	8.42	8.42	21.25	34.89	13.64	25.45	18.84	24.54	17.60	12.25	33.98	13.93	13.67	13.67	50.86	35.83	13.94	5.02	50.86
DE	13.83	13.83	11.56	9.28	6.29	5.82	8.54	8.54	11.18	11.56	10.51	9.85	13.41	11.64	10.51	5.41	22.89	10.29	10.21	10.21	26.69	10.21	7.54	9.85	25.69
DC	17.51	10.80	18.88	29.27	16.85	8.76	17.96	17.96	28.00	22.14	12.07	29.67	16.85	42.12	17.06	12.83	22.39	15.78	33.66	33.66	66.43	42.52	16.42	4.66	66.43
FL	36.73	23.28	29.09	29.92	16.15	10.51	15.27	15.27	17.81	25.18	24.27	28.72	28.92	49.35	28.79	13.37	53.20	17.12	18.67	18.67	43.71	50.61	12.56	7.89	43.71
GA	31.04	16.02	32.71	20.67	15.12	9.82	19.55	19.55	21.97	19.15	26.66	23.03	21.20	40.88	19.50	11.14	40.72	18.65	13.74	13.74	36.03	53.09	13.08	10.16	36.03
HI	17.38	11.62	28.20	13.74	12.27	7.10	7.73	7.73	20.55	21.19	10.94	17.87	11.33	20.15	15.61	6.06	33.24	8.02	11.11	11.11	31.71	21.70	9.78	11.54	31.71
ID	12.68	6.50	16.29	9.46	9.04	5.37	6.87	6.87	8.58	11.01	8.97	9.38	8.83	18.27	8.94	5.04	21.34	7.76	7.51	7.51	26.68	24.89	5.35	7.44	26.68
IL	20.76	12.31	21.64	35.48	12.83	8.86	8.69	8.69	23.09	24.90	14.24	20.34	13.15	37.63	13.78	11.68	35.88	14.58	21.53	21.53	58.63	52.47	13.87	6.26	58.63
IN	6.86	3.76	6.99	7.50	3.50	2.89	4.01	4.01	5.67	7.82	3.97	5.79	6.06	12.19	5.30	2.96	12.65	4.33	5.55	5.55	27.19	12.61	3.49	3.67	27.19
IA	10.22	5.19	12.01	9.18	6.42	3.84	5.56	5.56	11.62	16.49	5.95	10.93	7.58	16.30	9.26	5.01	17.38	6.09	9.80	9.80	45.02	27.20	5.60	4.16	45.02
KS	11.44	7.93	9.21	9.68	6.17	4.12	4.14	4.14	6.93	16.00	7.06	11.86	8.63	13.22	9.21	5.11	22.98	5.57	5.38	5.38	16.49	22.99	3.90	3.77	16.49
KY	16.47	15.26	24.34	22.12	9.53	8.95	13.67	13.67	15.96	14.68	10.62	22.00	19.83	30.44	13.94	10.38	32.83	15.92	13.21	13.21	45.21	33.73	10.60	8.85	45.21
LA	29.63	22.95	26.01	20.86	14.80	10.48	24.73	24.73	23.88	20.82	14.00	21.27	27.63	54.39	21.42	13.78	49.11	21.35	15.51	15.51	66.41	36.94	12.82	10.26	66.41
ME	13.43	10.42	43.59	24.29	11.12	10.02	14.75	14.75	13.35	15.39	14.03	17.26	16.07	31.44	17.95	11.12	32.07	17.70	15.08	15.08	37.84	61.81	11.66	9.37	37.84
MD	10.55	5.95	10.55	11.35	5.05	5.15	9.25	9.25	13.20	13.05	6.45	11.35	6.75	27.45	6.65	5.55	22.80	7.00	9.15	9.15	26.80	18.50	6.35	3.50	26.80
MA	15.25	11.38	22.82	39.38	16.79	6.14	10.72	10.72	16.78	27.76	14.02	29.03	14.95	29.58	13.75	8.49	67.48	13.20	16.07	16.07	80.76	79.51	16.46	6.62	62.87
MI	16.08	9.61	17.85	24.50	12.37	5.71	10.99	10.99	9.62	20.95	11.61	19.18	17.86	64.20	18.33	8.19	43.08	10.91	14.52	14.52	36.41	31.20	13.08	10.42	41.58
MN	27.66	28.76	49.94	30.20	22.59	8.28	19.73	19.73	30.01	37.20	23.93	29.89	19.91	80.08	23.93	15.88	88.83	13.91	27.21	27.21	132.92	31.83	29.54	8.71	132.92
MS	21.77	16.71	28.09	17.88	11.02	10.84	13.03	13.03	18.18	21.21	14.41	22.67	17.90	77.98	20.06	10.84	36.25	20.16	14.52	14.52	37.42	42.77	12.76	8.68	37.42
MO	18.72	7.52	11.84	13.28	10.66	5.79	11.05	11.05	8.97	19.99	9.67	13.57	11.77	16.88	14.22	7.10	25.74	9.76	13.82	13.82	47.98	31.96	5.83	5.71	47.98
MT	30.63	15.19	47.53	32.78	24.74	11.01	26.63	26.63	25.04	38.54	19.54	48.55	41.06	53.68	27.00	15.03	85.06	16.93	33.95	33.95	86.98	55.25	13.30	12.10	86.98
NE	16.19	9.23	15.94	17.31	12.73	5.43	10.19	10.19	12.36	16.79	9.34	18.07	11.58	22.01	14.62	9.06	33.91	13.86	10.75	10.75	25.03	26.96	7.88	5.47	25.03
NV	19.86	10.88	14.94	14.89	11.46	9.64	9.26	9.26	13.64	22.78	14.60	14.48	16.01	27.25	17.93	10.76	33.53	18.03	16.53	16.53	45.36	32.56	18.03	7.43	45.36
NH	16.38	9.82	19.68	28.99	9.29	7.10	13.87	13.87	12.87	43.98	10.90	25.97	17.94	28.09	17.33	9.74	58.76	12.48	13.27	13.27	55.90	49.67	12.82	7.37	55.90
NJ	10.40	8.85	10.40	10.83	8.37	4.23	7.95	7.95	7.36	12.78	7.88	10.77	11.49	11.32	7.88	5.48	29.41	6.20	8.03	8.03	22.52	12.96	5.74	5.37	29.92
NM	25.35	13.09	20.45	19.66	13.39	6.34	10.95	10.95	14.58	31.49	16.25	26.97	16.87	33.27	15.64	12.10	43.41	16.46	21.26	21.26	67.82	29.91	9.67	10.65	67.82
NY	15.39	6.82	14.50	24.92	15.16	6.89	9.69	9.69	16.92	15.82	13.00	18.61	12.55	27.49	13.03	9.03	32.78	14.13	14.12	14.12	27.11	24.09	10.56	7.48	32.53
NC	14.43	9.91	15.37	16.82	6.37	7.64	7.73	7.73	9.52	12.92	8.00	12.94	10.40	17.95	14.47	6.93	25.39	10.26	13.32	13.32	39.03	16.68	7.68	3.86	39.03
ND	13.89	13.89	13.89	12.17	12.17	6.16	9.91	9.91	9.85	9.38	13.16	12.20	11.42	29.82	13.16	8.05	31.37	8.05	13.89	13.89	29.82	29.82	9.63	31.37	29.82
OH	13.27	13.27	13.27	13.45	13.45	6.02	13.45	13.45	18.28	16.31	16.31	15.62	18.28	13.45	16.31	7.96	28.62	24.12	13.45	13.45	13.45	13.45	13.08	13.45	13.45
OK	26.16	11.79	19.33	15.51	11.50	7.41	18.45	18.45	13.55	18.00	12.05	16.80	17.10	36.26	18.29	8.72	42.26	12.45	11.17	11.17	69.45	46.76	11.42	8.20	69.45
OR	29.51	9.28	20.03	16.50	10.61	6.63	12.94	12.94	14.95	19.68	12.27	12.43	19.83	22.05	13.85	8.38	38.11	11.85	11.01	11.01	61.69	25.54	9.92	12.10	61.69
PA	15.80	15.80	19.85	28.18	11.84	7.80	12.74	12.74	15.69	19.85	18.94	18.96	22.05	32.12	18.94	11.57	43.30	12.58	24.71	24.71	59.05	24.71	11.83	18.96	81.34
RI	21.32	11.02	18.77	22.11	15.55	5.86	12.16	12.16	17.02	20.95	11.89	20.66	19.86	41.76	14.19	7.25	29.48	10.52	13.18	13.18	78.79	49.67	14.34	9.90	78.79
SC	19.10	13.03	19.24	12.88	6.19	7.02	8.32	8.32	12.22	10.69	7.27	9.55	11.19	16.64	18.89	6.88	28.98	11.42	9.49	9.49	21.80	23.66	6.23	4.65	21.80
SD	14.61	9.94	20.92	21.64	9.04	5.72	11.73	11.73	14.77	17.73	10.89	13.30	13.25	32.14	16.96	9.50	29.39	11.35	11.51	11.51	46.57	33.20	9.85	5.59	46.57
TN	13.22	9.52	11.99	12.39	7.53	5.00	8.68	8.68	8.96	15.04	8.51	12.73	10.88	15.62	12.19	5.91	24.81	8.76	7.36	7.36	45.06	13.95	5.89	4.98	45.06
TX	28.29	18.95	28.29	25.59	19.49	11.60	18.16	18.16	14.17	25.84	13.47	23.12	18.29	43.91	20.99	12.88	47.24	22.92	14.29	14.29	50.47	31.01	9.94	11.38	54.81
UT	11.17	11.17	11.17	17.68	7.99	7.05	6.38	6.38	9.55	10.95	12.32	13.75	15.29	17.24	10.60	6.41	26.27	6.30	8.99	8.99	23.51	23.51	6.06	5.83	26.53
VT	11.93	7.61	14.88	34.92	9.35	4.51	7.91	7.91	11.95	11.32	8.84	13.76	7.82	19.81	10.86	6.71	24.32	10.29	10.29	10.29	32.12	35.79	6.45	8.56	32.12
VA	10.00	6.46	8.65	9.29	5.47	3.45	6.13	6.13	10.17	8.77	7.59	8.07	8.11	24.38	8.20	5.60	17.80	6.74	10.42	10.42	19.28	30.25	6.78	2.57	19.28
WA	8.78	8.78	8.78	8.53	8.53	3.68	8.82	8.82	8.71	9.93	8.78	11.54	10.23	20.99	11.78	5.41	23.39	3.94	16.21	16.21	16.21	16.21	9.81	10.23	16.21
WV	15.43	15.43	15.43	20.87	20.87	5.01	9.14	9.14	7.33	7.33	16.88	16.02	16.88	10.56	16.88	7.40	12.04	7.33	14.98	14.98	13.27	14.98	16.02	4.87	13.27
WI	10.47	10.13	19.98	8.91	8.50	5.73	7.00	7.00	9.67	13.88	13.48	15.43	12.10	17.04	11.81	5.95	29.35	8.64	11.52	11.52	36.51	18.72	8.45	3.92	36.51
WY	8.13	8.13	8.13	8.13	8.13	8.13	8.13	8.13	8.13	8.13	8.13	8.13	8.13	8.13	8.13	8.13	8.13	8.13	8.13	8.13	8.13	8.13	8.13	8.13	8.13
AVG.	18.12	11.65	19.85	18.81	11.46	7.03	11.41	11.41	14.43	18.52	12.26	17.75	15.27	29.88	15.72	8.82	34.62	12.27	13.77	13.77	42.56	31.55	10.48	8.32	43.48

1 GENERAL REQUIREMENTS

REFERENCE NOS.

R01100-060 Workers' Compensation (cont.) (Canada in Canadian dollars)

Province		Alberta	British Columbia	Manitoba	Ontario	New Brunswick	Newfndld. & Labrador	Northwest Territories	Nova Scotia	Prince Edward Island	Quebec	Saskat-chewan	Yukon
Carpentry—3 stories or less	Rate	**2.65**	**6.89**	**5.35**	**8.19**	**3.76**	**5.36**	**2.75**	**6.34**	**5.68**	**13.85**	**5.38**	**2.35**
	Code	25401	70600	40102	723	422	403	4-41	4013	401	80110	B12-02	4-042
Carpentry—interior cab. work	Rate	**2.48**	**2.68**	**5.35**	**8.19**	**3.85**	**5.36**	**2.75**	**6.34**	**3.74**	**13.85**	**3.86**	**2.35**
	Code	42133	60412	40102	723	427	403	4-41	4013	402	80110	B11-25	4-042
CARPENTRY—general	Rate	**2.65**	**6.89**	**5.35**	**8.19**	**3.76**	**5.36**	**2.75**	**6.34**	**5.68**	**13.85**	**5.38**	**2.35**
	Code	25401	70600	40102	723	422	403	4-41	4013	401	80110	B12-02	4-042
CONCRETE WORK—NOC	Rate	**3.26**	**6.89**	**6.72**	**14.29**	**3.76**	**5.36**	**2.75**	**7.87**	**5.68**	**16.29**	**7.99**	**3.25**
	Code	42104	70604	40110	745	422	403	4-41	4222	401	80100	B14-04	2-032
CONCRETE WORK—flat (flr. sidewalk)	Rate	**3.26**	**6.89**	**6.72**	**14.29**	**3.76**	**5.36**	**2.75**	**7.87**	**5.68**	**16.29**	**7.99**	**3.25**
	Code	42104	70604	40110	745	422	403	4-41	4222	401	80100	B14-04	2-032
ELECTRICAL Wiring—inside	Rate	**1.61**	**4.61**	**3.50**	**3.61**	**1.49**	**4.34**	**2.00**	**3.19**	**3.74**	**7.41**	**3.86**	**2.35**
	Code	42124	71100	40203	704	426	400	4-46	4261	402	80170	B11-05	4-041
EXCAVATION—earth NOC	Rate	**2.03**	**4.13**	**5.35**	**4.95**	**2.72**	**5.36**	**2.50**	**3.92**	**3.95**	**7.87**	**4.69**	**3.25**
	Code	40604	72607	40706	711	421	403	4-43	4214	404	80030	R11-06	2-016
EXCAVATION—rock	Rate	**2.03**	**4.13**	**5.35**	**4.95**	**2.72**	**5.36**	**2.50**	**3.92**	**3.95**	**7.87**	**4.69**	**3.25**
	Code	40604	72607	40706	711	421	403	4-43	4214	404	80030	R11-06	2-016
GLAZIERS	Rate	**1.93**	**2.03**	**3.50**	**10.27**	**3.85**	**4.09**	**2.75**	**7.75**	**3.74**	**17.29**	**7.99**	**2.35**
	Code	42121	60236	40109	751	423	402	4-41	4233	402	80150	B13-04	4-042
INSULATION WORK	Rate	**2.28**	**6.86**	**5.35**	**10.27**	**3.85**	**4.09**	**2.75**	**7.75**	**5.68**	**15.05**	**5.38**	**3.25**
	Code	42184	70504	40102	751	423	402	4-41	4234	401	80120	B12-07	2-035
LATHING	Rate	**5.05**	**6.86**	**5.35**	**8.19**	**3.85**	**4.09**	**2.75**	**6.14**	**3.74**	**15.05**	**7.99**	**3.25**
	Code	42135	70500	40102	723	427	402	4-41	4271	402	80120	B13-02	2-036
MASONRY	Rate	**3.26**	**6.89**	**5.35**	**16.52**	**3.85**	**5.36**	**2.75**	**7.75**	**5.68**	**16.29**	**7.99**	**3.25**
	Code	42102	70602	40102	741	423	403	4-41	4231	401	80100	B15-01	2-032
PAINTING & DECORATING	Rate	**2.63**	**6.86**	**6.72**	**9.76**	**3.85**	**4.09**	**2.75**	**6.14**	**3.74**	**15.05**	**5.38**	**3.25**
	Code	42111	70501	40105	719	427	402	4-41	4275	402	80120	B12-01	2-036
PILE DRIVING	Rate	**3.26**	**17.50**	**5.35**	**5.76**	**3.76**	**9.60**	**2.50**	**7.87**	**5.68**	**7.87**	**5.38**	**3.25**
	Code	42159	72502	40706	732	422	404	4-43	4221	401	80030	B12-10	2-030
PLASTERING	Rate	**5.05**	**6.86**	**6.72**	**9.76**	**3.85**	**4.09**	**2.75**	**6.14**	**3.74**	**15.05**	**7.99**	**3.25**
	Code	42135	70502	40108	719	427	402	4-41	4271	402	80120	B13-02	2-036
PLUMBING	Rate	**1.61**	**4.07**	**3.50**	**4.42**	**1.88**	**3.57**	**2.00**	**3.70**	**3.74**	**8.03**	**3.86**	**2.35**
	Code	42122	70712	40204	707	424	401	4-46	4241	402	80160	B11-01	4-039
ROOFING	Rate	**6.56**	**6.89**	**6.72**	**12.05**	**7.72**	**5.36**	**2.75**	**9.30**	**5.68**	**22.52**	**7.99**	**3.25**
	Code	42118	70600	40403	728	430	403	4-41	4235	401	80130	B15-02	2-031
SHEET METAL WORK (HVAC)	Rate	**1.61**	**4.07**	**5.35**	**4.42**	**1.88**	**3.57**	**2.00**	**7.75**	**3.74**	**8.03**	**3.86**	**2.35**
	Code	42117	70714	40402	707	424	401	4-46	4236	402	80160	B11-07	4-040
STEEL ERECTION—door & sash	Rate	**2.28**	**17.50**	**5.35**	**20.96**	**3.76**	**9.60**	**2.75**	**7.87**	**5.68**	**32.75**	**7.99**	**3.25**
	Code	42106	72509	40502	748	422	404	4-41	4223	401	80080	B15-03	2-012
STEEL ERECTION—inter., ornam.	Rate	**2.28**	**17.50**	**5.35**	**20.96**	**3.76**	**5.36**	**2.75**	**7.87**	**5.68**	**32.75**	**7.99**	**3.25**
	Code	42106	72509	40502	748	422	403	4-41	4223	401	80080	B15-03	2-012
STEEL ERECTION—structure	Rate	**2.28**	**17.50**	**5.35**	**20.96**	**3.76**	**9.60**	**2.75**	**10.79**	**5.68**	**32.75**	**7.99**	**3.25**
	Code	42106	72509	40502	748	422	404	4-41	4227	401	80080	B15-04	2-012
STEEL ERECTION—NOC	Rate	**2.28**	**17.50**	**5.35**	**20.96**	**3.76**	**9.60**	**2.75**	**10.79**	**5.68**	**32.75**	**7.99**	**3.25**
	Code	42106	72509	40502	748	422	404	4-41	4227	401	80080	B15-03	2-012
TILE WORK—inter. (ceramic)	Rate	**2.53**	**6.86**	**2.31**	**9.76**	**3.85**	**4.09**	**2.75**	**6.14**	**3.74**	**15.05**	**7.99**	**3.25**
	Code	42113	70506	40103	719	427	402	4-41	4276	402	80120	B13-01	2-034
WATERPROOFING	Rate	**2.63**	**6.89**	**5.35**	**8.19**	**3.85**	**5.36**	**2.75**	**7.75**	**3.74**	**22.52**	**3.86**	**3.25**
	Code	42139	70620	40102	723	423	403	4-41	4239	402	80130	B11-17	2-030
WRECKING	Rate	**12.86**	**6.89**	**6.72**	**20.96**	**2.72**	**5.36**	**2.50**	**3.92**	**5.68**	**26.98**	**7.99**	**3.25**
	Code	42108	70600	40106	748	421	403	4-43	4211	401	80220	B14-07	2-030

01100-070 Contractor's Overhead & Profit

elow are the **average** installing contractor's percentage mark-ups applied base labor rates to arrive at typical billing rates.

olumn A: Labor rates are based on average open shop wages for 7 major S. regions. Base rates including fringe benefits are listed hourly and ily. These figures are the sum of the wage rate and employer-paid fringe nefits such as vacation pay, and employer-paid health costs.

olumn B: Workers' Compensation rates are the national average of state tes established for each trade.

olumn C: Column C lists average fixed overhead figures for all trades. cluded are Federal and State Unemployment costs set at 7.0%; Social curity Taxes (FICA) set at 7.65%; Builder's Risk Insurance costs set at 34%; and Public Liability costs set at 1.55%. All the percentages except ose for Social Security Taxes vary from state to state as well as from mpany to company.

Columns D and E: Percentages in Columns D and E are based on the presumption that the installing contractor has annual billing of $1,000,000 and up. Overhead percentages may increase with smaller annual billing. The overhead percentages for any given contractor may vary greatly and depend on a number of factors, such as the contractor's annual volume, engineering and logistical support costs, and staff requirements. The figures for overhead and profit will also vary depending on the type of job, the job location, and the prevailing economic conditions. All factors should be examined very carefully for each job.

Column F: Column F lists the total of Columns B, C, D, and E.

Column G: Column G is Column A (hourly base labor rate) multiplied by the percentage in Column F (O&P percentage).

Column H: Column H is the total of Column A (hourly base labor rate) plus Column G (Total O&P).

Column I: Column I is Column H multiplied by eight hours.

		A		B	C	D	E	F	G	H	I
		Base Rate Incl. Fringes		Work-ers' Comp. Ins.	Average Fixed Over-head	Over-head	Profit	Total Overhead & Profit		Rate with O & P	
Abbr.	Trade	Hourly	Daily					%	Amount	Hourly	Daily
Skwk	Skilled Workers Average (35 trades)	$19.75	$158.00	18.1%	16.5%	27.0%	10.0%	71.6%	$14.15	$33.90	$271.20
	Helpers Average (5 trades)	14.80	118.40	19.7		25.0		71.2	10.55	25.35	202.80
	Foreman Average, Inside ($.50 over trade)	20.25	162.00	18.1		27.0		71.6	14.50	34.75	278.00
	Foreman Average, Outside ($2.00 over trade)	21.75	174.00	18.1		27.0		71.6	15.55	37.30	298.40
Clab	Common Building Laborers	14.45	115.60	19.9		25.0		71.4	10.30	24.75	198.00
Asbe	Asbestos Workers	20.50	164.00	18.5		30.0		75.0	15.40	35.90	287.20
Boil	Boilermakers	22.10	176.80	16.2		30.0		72.7	16.05	38.15	305.20
Bric	Bricklayers	20.00	160.00	17.8		25.0		69.3	13.85	33.85	270.80
Brhe	Bricklayer Helpers	15.70	125.60	17.8		25.0		69.3	10.90	26.60	212.80
Carp	Carpenters	19.70	157.60	19.9		25.0		71.4	14.05	33.75	270.00
Cefi	Cement Finishers	18.90	151.20	11.5		25.0		63.0	11.90	30.80	246.40
Elec	Electricians	22.10	176.80	7.0		30.0		63.5	14.05	36.15	289.20
Elev	Elevator Constructors	22.90	183.20	8.5		30.0		65.0	14.90	37.80	302.40
Eqhv	Equipment Operators, Crane or Shovel	20.65	165.20	11.4		28.0		65.9	13.60	34.25	274.00
Eqmd	Equipment Operators, Medium Equipment	19.90	159.20	11.4		28.0		65.9	13.10	33.00	264.00
Eqlt	Equipment Operators, Light Equipment	19.10	152.80	11.4		28.0		65.9	12.60	31.70	253.60
Eqol	Equipment Operators, Oilers	17.00	136.00	11.4		28.0		65.9	11.20	28.20	225.60
Eqmm	Equipment Operators, Master Mechanics	21.05	168.40	11.4		28.0		65.9	13.85	34.90	279.20
Glaz	Glaziers	19.35	154.80	14.4		25.0		65.9	12.75	32.10	256.80
Lath	Lathers	19.20	153.60	12.3		25.0		63.8	12.25	31.45	251.60
Marb	Marble Setters	19.85	158.80	17.8		25.0		69.3	13.75	33.60	268.80
Mill	Millwrights	20.60	164.80	11.7		25.0		63.2	13.00	33.60	268.80
Mstz	Mosaic and Terrazzo Workers	19.10	152.80	10.5		25.0		62.0	11.85	30.95	247.60
Pord	Painters, Ordinary	17.95	143.60	15.3		25.0		66.8	12.00	29.95	239.60
Psst	Painters, Structural Steel	18.65	149.20	51.1		25.0		102.6	19.15	37.80	302.40
Pape	Paper Hangers	18.05	144.40	15.3		25.0		66.8	12.05	30.10	240.80
Pile	Pile Drivers	19.45	155.60	29.9		30.0		86.4	16.80	36.25	290.00
Plas	Plasterers	18.65	149.20	15.7		25.0		67.2	12.55	31.20	249.60
Plah	Plasterer Helpers	15.65	125.20	15.7		25.0		67.2	10.50	26.15	209.20
Plum	Plumbers	21.95	175.60	8.8		30.0		65.3	14.35	36.30	290.40
Rodm	Rodmen (Reinforcing)	21.10	168.80	31.6		28.0		86.1	18.15	39.25	314.00
Rofc	Roofers, Composition	17.05	136.40	34.6		25.0		86.1	14.70	31.75	254.00
Rots	Roofers, Tile and Slate	17.10	136.80	34.6		25.0		86.1	14.70	31.80	254.40
Rohe	Roofer Helpers (Composition)	12.75	102.00	34.6		25.0		86.1	11.00	23.75	190.00
Shee	Sheet Metal Workers	21.50	172.00	12.3		30.0		68.8	14.80	36.30	290.40
Spri	Sprinkler Installers	22.05	176.40	9.0		30.0		65.5	14.45	36.50	292.00
Stpi	Steamfitters or Pipefitters	22.10	176.80	8.8		30.0		65.3	14.45	36.55	292.40
Ston	Stone Masons	19.50	156.00	17.8		25.0		69.3	13.50	33.00	264.00
Sswk	Structural Steel Workers	21.25	170.00	42.6		28.0		97.1	20.65	41.90	335.20
Tilf	Tile Layers (Floor)	19.20	153.60	10.5		25.0		62.0	11.90	31.10	248.80
Tilh	Tile Layer Helpers	15.45	123.60	10.5		25.0		62.0	9.60	25.05	200.40
Trlt	Truck Drivers, Light	15.85	126.80	15.6		25.0		67.1	10.65	26.50	212.00
Trhv	Truck Drivers, Heavy	16.20	129.60	15.6		25.0		67.1	10.85	27.05	216.40
Sswl	Welders, Structural Steel	21.25	170.00	42.6		28.0		97.1	20.65	41.90	335.20
Wrck	*Wrecking	14.90	119.20	43.5		25.0		95.0	14.15	29.05	232.40

*Not included in Averages.

R01100-090 Sales Tax by State (United States)

State sales tax on materials is tabulated below (5 states have no sales tax). Many states allow local jurisdictions, such as a county or city, to levy additional sales tax.

Some projects may be sales tax exempt, particularly those constructed with public funds.

State	Tax (%)	State	Tax (%)	State	Tax (%)	State	Tax (%)
Alabama	4	Illinois	6.25	Montana	0	Rhode Island	7
Alaska	0	Indiana	5	Nebraska	4.5	South Carolina	5
Arizona	5	Iowa	5	Nevada	6.875	South Dakota	4
Arkansas	4.625	Kansas	4.9	New Hampshire	0	Tennessee	6
California	7.25	Kentucky	6	New Jersey	6	Texas	6.25
Colorado	3	Louisiana	4	New Mexico	5	Utah	4.75
Connecticut	6	Maine	6	New York	4	Vermont	5
Delaware	0	Maryland	5	North Carolina	4	Virginia	4.5
District of Columbia	5.75	Massachusetts	5	North Dakota	5	Washington	6.5
Florida	6	Michigan	6	Ohio	5	West Virginia	6
Georgia	4	Minnesota	6.5	Oklahoma	4.5	Wisconsin	5
Hawaii	4	Mississippi	7	Oregon	0	Wyoming	4
Idaho	5	Missouri	4.225	Pennsylvania	6	Average	4.71 %

Sales Tax by Province (Canada)

GST - a value-added tax, which the federal government imposes on most goods and services provided in or imported into Canada.
PST - a retail sales tax, which five of the provinces impose on the price of most goods and some services.

QST - a value-added tax, similar to the federal GST, which Quebec imposes.
HST - Three provinces have combined their retail sales tax with the federal GST into the Harmonized Sales Tax.

Province	PST (%)	QST (%)	GST (%)	HST (%)
Alberta	0	0	7	0
British Columbia	7	0	7	0
Manitoba	7	0	7	0
New Brunswick	0	0	0	15
Newfoundland	0	0	0	15
Northwest Territories	0	0	7	0
Nova Scotia	0	0	0	15
Ontario	8	0	7	0
Prince Edward Island	10	0	7	0
Quebec	0	6.5	7	0
Saskatchewan	7	0	7	0
Yukon	0	0	7	0

R01100-100 Unemployment Taxes and Social Security Taxes

Mass. State Unemployment tax ranges from 1.325% to 7.225% plus an experience rating assessment the following year, on the first $10,800 of wages. Federal Unemployment tax is 6.2% of the first $7,000 of wages. This is reduced by a credit for payment to the state. The minimum Federal Unemployment tax is .8% after all credits.

Combined rates in Mass. thus vary from 2.125% to 8.025% of the first $10,800 of wages. Combined average U.S. rate is about 7.0% of the first $7,000. Contractors with permanent workers will pay less since the average annual wages for skilled workers is $28.75 x 2,000 hours or about $57,500 per year. The average combined rate for U.S. would thus be 7.0% x $7,000 ÷ $57,500 = 0.9% of total wages for permanent employees.

Rates vary not only from state to state but also with the experience rating of the contractor.

Social Security (FICA) for 2000 is estimated at time of publication to be 7.65% of wages up to $72,600.

R01107-010 Architectural Fees

Tabulated below are typical percentage fees by project size, for good professional architectural service. Fees may vary from those listed depending upon degree of design difficulty and economic conditions in any particular area.

Rates can be interpolated horizontally and vertically. Various portions of the same project requiring different rates should be adjusted proportionately. For alterations, add 50% to the fee for the first $500,000 of project cost and add 25% to the fee for project cost over $500,000.

Architectural fees tabulated below include Structural, Mechanical and Electrical Engineering Fees. They do not include the fees for special consultants such as kitchen planning, security, acoustical, interior design, etc.

Building Types	Total Project Size in Thousands of Dollars						
	100	250	500	1,000	5,000	10,000	50,000
Factories, garages, warehouses, repetitive housing	9.0%	8.0%	7.0%	6.2%	5.3%	4.9%	4.5%
Apartments, banks, schools, libraries, offices, municipal buildings	12.2	12.3	9.2	8.0	7.0	6.6	6.2
Churches, hospitals, homes, laboratories, museums, research	15.0	13.6	12.7	11.9	9.5	8.8	8.0
Memorials, monumental work, decorative furnishings	—	16.0	14.5	13.1	10.0	9.0	8.3

01250-010 Repair and Remodeling

st figures are based on new construction utilizing the most cost-effective mbination of labor, equipment and material with the work scheduled proper sequence to allow the various trades to accomplish their work in efficient manner.

e costs for repair and remodeling work must be modified due to the lowing factors that may be present in any given repair and remodeling oject.

1. Equipment usage curtailment due to the physical limitations of the project, with only hand-operated equipment being used.
2. Increased requirement for shoring and bracing to hold up the building while structural changes are being made and to allow for temporary storage of construction materials on above-grade floors.
3. Material handling becomes more costly due to having to move within the confines of an enclosed building. For multi-story construction, low capacity elevators and stairwells may be the only access to the upper floors.
4. Large amount of cutting and patching and attempting to match the existing construction is required. It is often more economical to remove entire walls rather than create many new door and window openings. This sort of trade-off has to be carefully analyzed.
5. Cost of protection of completed work is increased since the usual sequence of construction usually cannot be accomplished.
6. Economies of scale usually associated with new construction may not be present. If small quantities of components must be custom fabricated due to job requirements, unit costs will naturally increase. Also, if only small work areas are available at a given time, job scheduling between trades becomes difficult and subcontractor quotations may reflect the excessive start-up and shut-down phases of the job.
7. Work may have to be done on other than normal shifts and may have to be done around an existing production facility which has to stay in production during the course of the repair and remodeling.
8. Dust and noise protection of adjoining non-construction areas can involve substantial special protection and alter usual construction methods.
9. Job may be delayed due to unexpected conditions discovered during demolition or removal. These delays ultimately increase construction costs.
10. Piping and ductwork runs may not be as simple as for new construction. Wiring may have to be snaked through walls and floors.
11. Matching "existing construction" may be impossible because materials may no longer be manufactured. Substitutions may be expensive.
12. Weather protection of existing structure requires additional temporary structures to protect building at openings.
13. On small projects, because of local conditions, it may be necessary to pay a tradesman for a minimum of four hours for a task that is completed in one hour.

All of the above areas can contribute to increased costs for a repair and remodeling project. Each of the above factors should be considered in the planning, bidding and construction stage in order to minimize the increased costs associated with repair and remodeling jobs.

R01540-100 Steel Tubular Scaffolding

On new construction, tubular scaffolding is efficient up to 60′ high or five stories. Above this it is usually better to use a hung scaffolding if construction permits. Swing scaffolding operations may interfere with tenants. In this case, the tubular is more practical at all heights.

In repairing or cleaning the front of an existing building the cost of tubular scaffolding per S.F. of building front increases as the height increases above the first tier. The first tier cost is relatively high due to leveling and alignment.

The minimum efficient crew for erection is three workers. For heights over 50′, a crew of four is more efficient. Use two or more on top and two at the bottom for handing up or hoisting. Four workers can erect and dismantle about nine frames per hour up to five stories. From five to eight stories they will average six frames per hour. With 7′ horizontal spacing this will run about 400 S.F. and 265 S.F. of wall surface, respectively. Time for placing planks must be added to the above. On heights above 50′, five planks can be placed per labor-hour.

The cost per 1,000 S.F. of building front in the table below was developed by pricing the materials required for a typical tubular scaffolding system eleven frames long and two frames high. Planks were figured five wide for standing plus two wide for materials.

Frames are 5′ wide and usually spaced 7′ O.C. horizontally. Sidewalk frame are 6′ wide. Rental rates will be lower for jobs over three months duration.

For jobs under twenty-five frames, add 50% to rental cost. These figures d not include accessories which are listed separately below. Large quantities for long periods can reduce rental rates by 20%.

Item	Unit	Monthly Rent	Per 1,000 S. F. of Building Front	
			No. of Pieces	Rental per Month
5′ Wide Standard Frame, 6′-4″ High	Ea.	$ 3.75	24	$ 90.00
Leveling Jack & Plate	↓	1.50	24	36.00
Cross Brace		.60	44	26.40
Side Arm Bracket, 21″		1.50	12	18.00
Guardrail Post		1.00	12	12.00
Guardrail, 7′ section		.75	22	16.50
Stairway Section		10.00	2	20.00
Stairway Starter Bar		.10	1	.10
Stairway Inside Handrail		5.00	2	10.00
Stairway Outside Handrail		5.00	2	10.00
Walk-Thru Frame Guardrail		2.00	2	4.00
			Total	$243.00
			Per C.S.F., 1 Use/Mo.	$ 24.30

Scaffolding is often used as falsework over 15′ high during construction of cast-in-place concrete beams and slabs. Two foot wide scaffolding is generally used for heavy beam construction. The span between frames depends upon the load to be carried with a maximum span of 5′.

Heavy duty shoring frames with a capacity of 10,000#/leg can be spaced up to 10′ O. C. depending upon form support design and loading.

Scaffolding used as horizontal shoring requires less than half the material required with conventional shoring.

On new construction, erection is done by carpenters.

Rolling towers supporting horizontal shores can reduce labor and speed th job. For maintenance work, catwalks with spans up to 70′ can be supported by the rolling towers.

01540-200 Pump Staging

mp staging is generally not available for rent. Purchase prices for individual ems are shown in the chart below.

			Per 2,400 S.F. of Building Front	
Item	Unit	Purchase, Each	No. of Pieces	Cost to Buy
Aluminum pole section, 24′ long	Ea.	$335.00	6	$2,010.00
Aluminum splice joint, 6′ long		68.00	3	204.00
Aluminum foldable brace		48.50	3	145.50
Aluminum pump jack		111.00	3	333.00
Aluminum support for workbench/back safety rail		59.00	3	177.00
Aluminum scaffold plank/workbench, 14″ wide x 24′ long		545.00	4	2,180.00
Safety net, 22′ long		267.00	2	534.00
Aluminum plank end safety rail	↓	183.00	2	366.00
			Total System	$5,949.50
			Per C.S.F., 1 Use	$247.90

his system in place will cover up to 2,400 square feet of wall area. The ost in place will depend on how many uses are realized during the life of the equipment. Several options are given in Division 01540-550. The above prices are bare costs.

R02065-300 Bituminous Paving

City	Bituminous Asphalt per Ton*	Pavement (3") 6.13 S.Y./ton				Sidewalks (2") 9.2 S.Y./ton			
		Cost per S.Y.			Per Ton	Cost per S.Y.			Per Ton
		Material*	Installation	Total	Total	Material*	Installation	Total	Total
Atlanta	$25.00	$4.08	$ .68	$4.76	$29.15	$2.72	$1.11	$3.83	$35.24
Baltimore	30.50	4.98	.72	5.70	34.94	3.32	1.25	4.57	42.04
Boston	36.25	5.91	.89	6.80	41.69	3.94	1.83	5.77	53.08
Buffalo	31.80	5.19	.86	6.05	37.09	3.46	1.73	5.19	47.75
Chicago	32.50	5.30	.91	6.21	38.06	3.53	1.89	5.42	49.86
Cincinnati	33.25	5.42	.76	6.18	37.88	3.61	1.39	5.00	46.00
Cleveland	30.00	4.89	.82	5.71	34.98	3.26	1.58	4.84	44.53
Columbus	26.50	4.32	.76	5.08	31.14	2.88	1.39	4.27	39.28
Dallas	25.00	4.08	.69	4.77	29.23	2.72	1.15	3.87	35.60
Denver	25.75	4.20	.70	4.90	30.03	2.80	1.18	3.98	36.62
Detroit	29.00	4.73	.82	5.55	34.02	3.15	1.59	4.74	43.61
Houston	30.50	4.98	.68	5.66	34.72	3.32	1.13	4.45	40.94
Indianapolis	27.80	4.54	.77	5.31	32.52	3.02	1.41	4.43	40.76
Kansas City	24.50	4.00	.78	4.78	29.29	2.66	1.45	4.11	37.81
Los Angeles	30.00	4.89	.88	5.77	35.38	3.26	1.80	5.06	46.55
Memphis	29.30	4.78	.65	5.43	33.29	3.18	1.02	4.20	38.64
Milwaukee	26.50	4.32	.86	5.18	31.77	2.88	1.74	4.62	42.50
Minneapolis	27.25	4.45	.82	5.27	32.28	2.96	1.58	4.54	41.77
Nashville	28.00	4.57	.68	5.25	32.15	3.04	1.11	4.15	38.18
New Orleans	37.00	6.04	.64	6.68	40.96	4.02	.99	5.01	46.09
New York City	44.00	7.18	1.04	8.22	50.38	4.78	2.33	7.11	65.41
Philadelphia	27.75	4.53	.82	5.35	32.82	3.02	1.60	4.62	42.50
Phoenix	23.75	3.87	.69	4.56	27.95	2.58	1.15	3.73	34.32
Pittsburgh	32.25	5.26	.80	6.06	37.12	3.51	1.51	5.02	46.18
St. Louis	25.50	4.16	.84	5.00	30.64	2.77	1.65	4.42	40.66
San Antonio	27.25	4.45	.63	5.08	31.14	2.96	.95	3.91	35.97
San Diego	28.50	4.65	.88	5.53	33.91	3.10	1.80	4.90	45.08
San Francisco	31.50	5.14	.91	6.05	37.08	3.42	1.89	5.31	48.85
Seattle	32.75	5.34	.84	6.18	37.87	3.56	1.65	5.21	47.93
Washington, D.C.	35.00	5.71	.71	6.42	39.37	3.80	1.23	5.03	46.28
Average	$29.80	$4.87	$ .78	$5.65	$34.63	$3.24	$1.47	$4.71	$43.34

Assumed density is 145 lb. per C.F.
*Includes delivery within 20 miles

Table below shows quantities and bare costs for 1000 S.Y. of Bituminous Paving.

Item	Roads and Parking Areas, 3" Thick (02740-300-0460)		Sidewalks, 2" Thick (02775-275-0010)	
	Quantities	Cost	Quantities	Cost
Bituminous asphalt	163 tons @ $29.80 per ton	$4,857.40	109 tons @ $29.80 per ton	$3,248.20
Installation using	Crew B-25B @ $3,575.40 /4900SY/ day x 1000	729.67	Crew B-37 @ $880.00 /720 SY/day x 1000	1,222.22
Total per 1000 S.Y.		$5,587.07		$4,470.42
Total per S.Y.		$ 5.65		$ 4.47
Total per Ton		$ 34.28		$ 41.01

02920-500 Seeding

e type of grass is determined by light, shade and moisture content of soil us intended use. Fertilizer should be disked 4″ before seeding. For steep opes disk five tons of mulch and lay two tons of hay or straw on surface per re after seeding. Surface mulch can be staked, lightly disked or tar nulsion sprayed. Material for mulch can be wood chips, peat moss, partially rotted hay or straw, wood fibers and sprayed emulsions. Hemp seed blankets with fertilizer are also available. For spring seeding, watering is necessary. Late fall seeding may have to be reseeded in the spring. Hydraulic seeding, power mulching, and aerial seeding can be used on large areas.

02930-900 Cost of Trees: Based on Pin Oak (Quercus palustris)

Tree Diameter	Normal Height	Catalog List Price of Tree	Guying Material	Equipment Charge	Installation Labor	Total
2 to 3 inch	14 feet	$ 140	$15.00	$ 56.83	$ 51.36	$ 263.59
3 to 4 inch	16 feet	231	18.00	94.72	85.60	429.44
4 to 5 inch	18 feet	371	55.00	113.66	102.72	642.36
6 to 7 inch	22 feet	822	75.00	142.08	128.40	1,167.36
8 to 9 inch	26 feet	1053	95.00	189.43	171.20	1,508.63

Installation Time & Cost for Planting Trees, Bare Costs

Ball Size Diam. X Depth	Soil in Ball	Weight of Ball	Hole Diam. Req'd	Hole Excavation	Amount of Soil Displ.	Topsoil Handled	Time Required in Labor-Hours						Cost	
							Dig & Lace	Handle Ball	Dig Hole	Plant & Prune	Water & Guy	Total L.H.	Crew	Total per Tree
Inches	C.F.	Lbs.	Feet	C.F.	C.F.	C.F.								
12 x 12	.70	56.00	2.00	4.00	3.00	11.00	.25	.17	.33	.25	.07	1.10	1 Clab	$ 15.90
18 x 16	2.00	160.00	2.50	8.00	6.00	21.00	.50	.33	.47	.35	.08	1.70	2 Clab	24.57
24 x 18	4.00	320.00	3.00	13.00	9.00	38.00	1.00	.67	1.08	.82	.20	3.80	3 Clab	54.91
30 x 21	7.50	600.00	4.00	27.00	19.50	76.00	.82	.71	.79	1.22	.26	3.80		92.95
36 x 24	12.50	980.00	4.50	38.00	25.50	114.00	1.08	.95	1.11	1.32	.30	4.76		116.43
42 x 27	19.00	1,520.00	5.50	64.00	45.00	185.00	1.90	1.27	1.87	1.43	.34	6.80	B-6	166.33
48 x 30	28.00	2,040.00	6.00	85.00	57.00	254.00	2.41	1.60	2.06	1.55	.39	8.00	@	195.68
54 x 33	38.50	3,060.00	7.00	127.00	88.50	370.00	2.86	1.90	2.39	1.76	.45	9.40	$24.46	229.92
60 x 36	52.00	4,160.00	7.50	159.00	107.00	474.00	3.26	2.17	2.73	2.00	.51	10.70	per	261.72
66 x 39	68.00	5,440.00	8.00	196.00	128.00	596.00	3.61	2.41	3.07	2.26	.58	11.90	Labor-	291.07
72 x 42	87.00	7,160.00	9.00	267.00	180.00	785.00	3.90	2.60	3.71	2.78	.70	13.70	hour	335.10

R04060-100 Cement Mortar (material only)

Type N - 1:1:6 mix by volume. Use everywhere above grade except as noted below.
- 1:3 mix using conventional masonry cement which saves handling two separate bagged materials.

Type M - 1:1/4:3 mix by volume, or 1 part cement, 1/4 (10% by wt.) lime 3 parts sand. Use for heavy loads and where earthquakes or hurricanes m occur. Also for reinforced brick, sewers, manholes and everywhere below grade.

Cost and Mix Proportions of Various Types of Mortar

Components	Type Mortar and Mix Proportions by Volume										
	M		S		N		O		K	PM	PL
	1:1:6	1:1/4:3	1/2:1:4	1:1/2:4	1:3	1:1:6	1:3	1:2:9	1:3:12	1:1:6	1:1/2:4
Portland cement @ $7.10 per bag	$ 7.10	$ 7.10	$ 3.55	$ 7.10	—	$ 7.10	—	$ 7.10	$ 7.10	$ 7.10	$ 7.10
Masonry cement @ $5.90 per bag	5.90	—	5.90	—	5.90	—	$5.90	—	—	$ 5.90	—
Lime @ $5.60 per 50 lb. bag	—	1.40	—	2.80	—	5.60	—	11.20	16.80	—	2.80
Masonry sand @ $17.50 per C.Y.*	3.89	1.94	2.59	2.59	1.94	3.89	1.94	5.83	7.78	3.89	2.59
Mixing machine incl. fuel**	1.72	.86	1.15	1.15	.86	1.72	.86	2.58	3.45	1.72	1.15
Total for Materials	$18.61	$11.30	$13.19	$13.64	$8.70	$18.31	$8.70	$26.71	$35.13	$18.61	$13.64
Total C.F.	6	3	4	4	3	6	3	9	12	6	4
Approximate Cost per C.F.	$ 3.10	$ 3.77	$ 3.30	$ 3.41	$2.90	$ 3.05	$2.90	$ 2.97	$ 2.93	$ 3.10	$ 3.41

*Includes 10 mile haul
**Based on a daily rental, 10 C.F., 25 H.P. mixer, mix 200 C.F./Day

Mix Proportions by Volume, Compressive Strength and Cost of Mortar

Where Used	Allowable Proportions by Volume					Compressive Strength @ 28 days	Cost per Cubic Foot
	Mortar Type	Portland Cement	Masonry Cement	Hydrated Lime	Masonry Sand		
Plain Masonry	M	1	1	—	6	2500 psi	$3.10
		1	—	1/4	3		3.77
	S	1/2	1	—	4	1800 psi	3.30
		1	—	1/4 to 1/2	4		3.41
	N	—	1	—	3	750 psi	2.90
		1	—	1/2 to 1-1/4	6		3.05
	O	—	1	—	3	350 psi	2.90
		1	—	1-1/4 to 2-1/2	9		2.97
	K	1	—	2-1/2 to 4	12	75 psi	2.93
Reinforced Masonry	PM	1	1	—	6	2500 psi	3.10
	PL	1	—	1/4 to 1/2	4	2500 psi	3.41

Note: The total aggregate should be between 2.25 to 3 times the sum of the cement and lime used.

The labor cost to mix the mortar is included in the labor cost on brickwork.

Machine mixing is usually specified on jobs of any size. There is a large price saving over hand mixing and mortar is more uniform.

There are two types of mortar color used. Prices in Section 04060 are for the inert additive type with about 100 lbs. per M brick as the typical quantity required. These colors are also available in smaller batch size bags (1 lb. to 15 lb.) which can be placed directly into the mixer without measuring. The other type is premixed and replaces the masonry cement with ranges price from $5 to $15 per 70 lb. bag. Dark green color has the highest cost.

R04060-200 Miscellaneous Mortar (material only)

Quantities	Glass Block Mortar		Gypsum Cement Mortar	
White Portland cement at $17.75 per bag	7 bags	$124.25		
Gypsum cement at $11.40 per 80 lb. bag			11.25 bags	$128.25
Lime at $5.60 per 50 lb. bag	280 lbs.	31.36		
Sand at $17.50 per C.Y.*	1 C.Y.	17.50	1 C.Y.	17.50
Mixing machine and fuel		7.75		7.75
Total per C.Y.		$180.86		$153.50
Approximate Total per C.F.		$ 6.70		$ 5.69

* Includes 10 mile haul

04080-500 Masonry Reinforcing

rizontal joint reinforcing helps prevent wall cracks where wall movement y occur and in many locations is required by code. Horizontal joint nforcing is generally not considered to be structural reinforcing and an reinforced wall may still contain joint reinforcing.

inforcing strips come in 10′ and 12′ lengths and in truss and ladder apes, with and without drips. Field labor runs between 2.7 to 5.3 hours r 1000 L.F. for wall thicknesses up to 12″.

The wire meets ASTM A82 for cold drawn steel wire and the typical size is 9 ga. sides and ties with 3/16″ diameter also available. Typical finish is mill galvanized with zinc coating at .10 oz. per S.F. Class I (.40 oz. per S.F.) and Class III (.80 oz per S.F.) are also available, as is hot dipped galvanizing at 1.50 oz. per S.F.

04210-050 Brick Chimneys

Quantities	16″ x 16″		20″ x 20″		20″ x 24″		20″ x 32″	
Brick at $300 per M	28 brick	$ 8.40	37 brick	$11.10	42 brick	$12.60	51 brick	$15.30
Type M mortar at $3.70 per C.F.	.5 C.F.	1.89	.6 C.F.	2.26	1.0 C.F.	3.77	1.3 C.F.	4.90
Flue tile (square)(per foot)	8″ x 8″	3.00	12″ x 12″	5.70	2 @ 8″ x 12″	7.54	2 @ 12″ x 12″	11.40
Install tile & brick, crew D-1	.055 day	15.71	.073 day	20.85	.083 day	23.70	.10 day	28.56
Total per L.F. high		$29.00		$39.91		$47.61		$60.16

aterial costs include 3% waste for brick and 25% waste for mortar.

bor costs are bare costs and do not include contractor's O&P.

bor for chimney brick using D-1 crew is 31 hours per thousand ick or about $530 per thousand brick. An 8″ x 12″ flue takes brick and two 8″ x 8″ flues take 37 brick.

04210-100 Economy in Bricklaying

ave adequate supervision. Be sure bricklayers are always supplied with aterials so there is no waiting. Place best bricklayers at corners and enings.

se only screened sand for mortar. Otherwise, labor time will be wasted cking out pebbles. Use seamless metal tubs for mortar as they do not leak catch the trowel. Locate stack and mortar for easy wheeling.

ave brick delivered for stacking. This makes for faster handling, reduces ipping and breakage, and requires less storage space. Many dealers will eliver select common in 2′ x 3′ x 4′ pallets or face brick packaged. This fords quick handling with a crane or forklift and easy tonging in units of n, which reduces waste.

Use wider bricks for one wythe wall construction. Keep scaffolding away from wall to allow mortar to fall clear and not stain wall.

On large jobs develop specialized crews for each type of masonry unit.

Consider designing for prefabricated panel construction on high rise projects.

Avoid excessive corners or openings. Each opening adds about 50% to labor cost for area of opening.

Bolting stone panels and using window frames as stops reduces labor costs and speeds up erection.

R04210-120 Common and Face Brick Prices

Prices are based on truckload lot purchases for Common Brick and carload lots for Face Brick. Prices are per M, (thousand), brick.

	Material			Installation				Total			
	Brick per M Delivered		Mortar	Common in 8″ Wall		Face Brick, 4″ Veneer		Common in 8″ Wall		Face Brick, 4″ Veneer	
City	Face	3/8″ Joint	Bare Costs	Incl. O & P	Bare Costs	Incl. O & P	Bare Costs	Incl. O & P	Bare Costs	Incl. O & P	
Atlanta	$190	$250	$38.10	$278	$ 471	$334	$ 565	$512	$ 728	$ 623	$ 883
Baltimore	235	280	for 8″ Wall	320	542	384	651	600	850	704	1,002
Boston	305	480	and	573	970	687	1,164	925	1,357	1,213	1,742
Buffalo	260	350	$31.40	491	832	590	998	797	1,168	981	1,429
Chicago	260	380	for 4″ Wall	520	881	624	1,057	826	1,218	1,047	1,522
Cincinnati	230	330		384	650	461	781	659	953	832	1,189
Cleveland	235	325		464	785	556	942	744	1,093	923	1,345
Columbus	235	350		379	641	454	769	659	949	846	1,200
Dallas	200	275		273	463	328	555	517	731	643	901
Denver	210	310		318	538	381	645	572	818	732	1,031
Detroit	235	275		487	825	585	990	768	1,133	900	1,336
Houston	250	275		284	481	341	578	580	806	656	924
Indianapolis	255	265		388	657	466	788	689	988	770	1,123
Kansas City	260	300		416	704	499	845	722	1,041	840	1,220
Los Angeles	250	325		483	818	580	982	779	1,144	946	1,385
Memphis	190	235		273	463	328	555	507	720	601	856
Milwaukee	280	365		458	776	550	931	785	1,135	957	1,379
Minneapolis	255	400		473	801	568	961	774	1,132	1,011	1,449
Nashville	185	245		267	452	321	543	496	704	604	855
New Orleans	210	380		236	400	283	480	491	680	706	945
New York City	300	355		649	1,098	778	1,318	996	1,480	1,176	1,755
Philadelphia	245	345		516	873	619	1,048	806	1,193	1,006	1,473
Phoenix	340	350		296	501	355	601	684	928	747	1,032
Pittsburgh	210	275		420	711	504	854	675	991	819	1,200
St. Louis	235	265		249	422	299	506	529	730	603	841
San Antonio	260	325		429	726	514	871	735	1,062	880	1,274
San Diego	285	360		562	951	674	1,141	893	1,316	1,076	1,584
San Francisco	350	435		470	795	563	954	868	1,233	1,043	1,481
Seattle	340	390		473	801	568	961	861	1,228	1,001	1,438
Washington, D.C.	210	240		330	558	396	670	584	838	674	976
Average	$235	$305	↓	$410	$ 690	$490	$ 823	$700	$1,010	$ 850	$1,226

Common building brick manufactured according to ASTM C62 and facing brick manufactured according to ASTM C216 are the two standard bricks available for general building use. Building brick is made in three grades; SW, where high resistance to damage caused by cyclic freezing is required; MW, where moderate resistance to cyclic freezing is needed; and NW, where little resistance to cyclic freezing is needed. Facing brick is made in only the two grades SW and MW. Additionally, facing brick is available in three types; FBS, for general use; FBX, for general use where a higher degree of precision and lower permissible variation in size than FBS is needed; and FBA, for general use to produce characteristic architectural effects resulting from non-uniformity in size and texture of the units.

In figuring above installation costs, a D-8 Crew (with a daily output of 1.5 M) was used for the 4″ veneer. A D-8 Crew (with a daily output of 1.8 M) was used for the 8″ solid wall.

In figuring the total cost including overhead and profit, an allowance of 10% was added to the sum of the cost of the brick and mortar. Also, 3% breakage was included for both the bare costs and the costs with overhead and profit. If bricks are delivered palletized with 280 to 300 per pallet, or packaged, allow only 1-1/2% for breakage. Then add $10 per M to the cost of brick and deduct two hours helper time. The net result is a savings of $30 to $40 per M in place. Packaged or palletized delivery is practical when a job is big enough to have a crane or other equipment available to handle a package of brick. This is so on all industrial work but not always true on small commercial buildings.

There are many types of red face brick. The prices above are for the most usual type used in commercial, apartment house or industrial construction. it is possible to obtain the price of the actual brick to be used, it should be done and substituted in the table. The use of buff and gray face is increasing, and there is a continuing trend to the Norman, Roman, Jumbo and SCR brick.

See R04210-500 for brick quantities per S.F. and mortar quantities per M brick. (Average prices for the various sizes are listed in Division 4)

Common red clay brick for backup is not used that often. Concrete block i the most usual backup material with occasional use of sand lime or cement brick. Sand lime cost about $15 per M less than red clay and ceme brick are about $5 per M less than red clay. These figures may be substituted in the common brick breakdown for the cost of these items in place, as labor is about the same. Occasionally common brick is being used i solid walls for strength and as a fire stop.

Brick panels built on the ground floor and then crane erected to the upper floors have proven to be economical. This allows the work to be done under cover and without scaffolding.

.04210-180 Brick in Place

ıble below is for common bond with 3/8″ concave joints and ıcludes 3% waste for brick and 25% waste for mortar. :ew costs are bare costs.

Item	8″ Common Brick Wall 8″ x 2-2/3″ x 4″		Select Common Face 8″ x 2-2/3″ x 4″		Red Face Brick 8″ x 2-2/3″ x 4″	
1030 brick delivered	$250 per M	$257.50	$300 per M	$309.00	$325 per M	$334.75
Type N mortar @ $2.99 per C.F.	12.5 C.F.	38.13	10.3 C.F.	31.42	10.3 C.F.	31.42
Installation using indicated crew	Crew D-8 @ .556 Days	406.55	Crew D-8 @ .667 Days	487.71	Crew D-8 @ .667 Days	487.71
Total per M in place		$702.18		$828.13		$853.88
Total per S.F. of wall	13.5 bricks/S.F.	$ 9.48	6.75 bricks/S.F.	$ 5.59	6.75 bricks/S.F.	$ 5.76

ᴿ04210-500 Brick, Block & Mortar Quantities

Running Bond						For Other Bonds Standard Size Add to S.F. Quantities in Table to Left		
Number of Brick per S.F. of Wall - Single Wythe with 3/8″ Joints				C.F. of Mortar per M Bricks, Waste Included				
Type Brick	Nominal Size (incl. mortar) L H W	Modular Coursing	Number of Brick per S.F.	3/8″ Joint	1/2″ Joint	Bond Type	Description	Factor
Standard	8 x 2-2/3 x 4	3C=8″	6.75	10.3	12.9	Common	full header every fifth course	+20%
Economy	8 x 4 x 4	1C=4″	4.50	11.4	14.6		full header every sixth course	+16.7%
Engineer	8 x 3-1/5 x 4	5C=16″	5.63	10.6	13.6	English	full header every second course	+50%
Fire	9 x 2-1/2 x 4-1/2	2C=5″	6.40	550 # Fireclay	—	Flemish	alternate headers every course	+33.3%
Jumbo	12 x 4 x 6 or 8	1C=4″	3.00	23.8	30.8		every sixth course	+5.6%
Norman	12 x 2-2/3 x 4	3C=8″	4.50	14.0	17.9	Header = W x H exposed		+100%
Norwegian	12 x 3-1/5 x 4	5C=16″	3.75	14.6	18.6	Rowlock = H x W exposed		+100%
Roman	12 x 2 x 4	2C=4″	6.00	13.4	17.0	Rowlock stretcher = L x W exposed		+33.3%
SCR	12 x 2-2/3 x 6	3C=8″	4.50	21.8	28.0	Soldier = H x L exposed		—
Utility	12 x 4 x 4	1C=4″	3.00	15.4	19.6	Sailor = W x L exposed		-33.3%

Concrete Blocks Nominal Size	Approximate Weight per S.F.		Blocks per 100 S.F.	Mortar per M block	
	Standard	Lightweight		Partitions	Back up
2″ x 8″ x 16″	20 PSF	15 PSF	113	16 C.F.	36 C.F.
4″	30	20	↓	31	51
6″	42	30	↓	46	66
8″	55	38	↓	62	82
10″	70	47	↓	77	97
12″ ↓	85 ↓	55 ↓	↓	92 ↓	112 ↓

ᴿ04210-550 Brick Veneer in Place

'able below is for running bond with 3/8″ concave joints and includes 3% vaste for brick and 25% waste for mortar.

Item	Buff Face Brick 8″ x 2-2/3″ x 4″		Red Norman Face Brick 12″ x 2-2/3″ x 4″		Buff Roman Face Brick 12″ x 2″ x 4″	
1030 brick delivered	$390 per M	$401.70	$675 per M	$ 695.25	$700 per M	$ 721.00
Type N mortar @ $3.05 per C.F.	10.3 C.F.	31.42	14.0 C.F.	42.70	13.4 C.F.	40.87
Installation using indicated crew	Crew D-8 @ .667 days	487.71	Crew D-8 @ .690 days	504.53	Crew D-8 @ .667 days	487.71
Total per M in place		$920.83		$1,242.48		$1,249.58
Total per S.F. of wall	6.75 brick/S.F.	$ 6.22	4.50 brick/S.F.	$ 5.59	6.00 brick/S.F.	$ 7.50

R04220-200 Concrete Block

8″ x 16″ block, sand aggregate blocks with 3/8″ joints for partitions.

	Material					Material			
	Per Block, Delivered		113 Block, Delivered			Per Block, Delivered		113 Block, Delivered	
City	8″ Thick	4″ Thick	8″ Thick	City	4″ Thick	8″ Thick	4″ Thick	8″ Thick	
Atlanta	$.74	$1.00	$83.62	$113.00	Memphis	.68	.94	$76.84	$106.22
Baltimore	.65	.85	73.45	96.05	Milwaukee	.76	1.10	85.88	124.30
Boston	.60	.85	67.80	96.05	Minneapolis	.75	1.01	84.75	114.13
Buffalo	.70	1.15	79.10	129.95	Nashville	.77	1.02	87.01	115.26
Chicago	.60	.92	67.80	103.96	New Orleans	.82	1.15	92.66	129.95
Cincinnati	.60	.85	67.80	96.05	New York City	.77	1.03	87.01	116.39
Cleveland	.66	.93	74.58	105.09	Philadelphia	.70	.80	79.10	90.40
Columbus	.69	.83	77.97	93.79	Phoenix	.62	.86	70.06	97.18
Dallas	.61	1.02	68.93	115.26	Pittsburgh	.72	.92	81.36	103.96
Denver	.73	1.05	82.49	118.65	St. Louis	.76	1.10	85.88	124.30
Detroit	.77	1.05	87.01	118.65	San Antonio	.75	.96	84.75	108.48
Houston	.75	.97	84.75	109.61	San Diego	.63	.86	71.19	97.18
Indianapolis	.70	1.00	79.10	113.00	San Francisco	.88	1.45	99.44	163.85
Kansas City	.72	1.20	81.36	135.60	Seattle	.88	1.27	99.44	143.51
Los Angeles	.81	.95	91.53	107.35	Washington D.C.	.66	1.07	74.58	120.91
					Average	$.72	$1.01	$80.91	$113.60

The cost of sand aggregate block is based on the delivered price of 113 blocks, which is the equivalent of a 100 S.F. wall area. Mortar for a 100 S.F. wall area will average $10.50 for 4″ thick block and $21.10 for 8″ thick block.

Cost for 100 S.F. of 8″ x 16″ Concrete Block Partitions to Four Stories High, Tooled Joint One Side						
8″ x 16″ Sand Aggregate	4″ Thick Block		8″ Thick Block		12″ Thick Block	
113 block delivered	$.72 Ea.	$ 79.10	$1.01 Ea.	$113.00	$1.60 Ea.	$175.15
Type N mortar at $3.05 per C.F.	3.5 C.F.	10.36	7.0 C.F.	20.72	10.4 C.F.	30.78
Installation Crew (Bare Costs)	D-2 @ 9.32 L.H.	164.22	D-2 @ 10.68 L.H.	188.18	D-6 @ 14.70 L.H.	303.41
Total per 100 S.F.		$253.68		$321.90		$509.34
Add for filling cores solid	6.7 C.F.	$ 85.58	25.8 C.F.	$183.19	42.2 C.F.	$251.22

Cost for 100 S.F. of 8″ x 16″ Concrete Block Backup with Trowel Cut Joints						
8″ x 16″ Sand Aggregate	4″ Thick Block		8″ Thick Block		12″ Thick Block	
113 block delivered	$.72 Ea.	$ 79.10	$1.01 Ea.	$113.00	$1.60 Ea.	$175.15
Type N mortar at $3.05 per C.F.	5.8 C.F.	17.17	9.3 C.F.	27.53	12.7 C.F.	37.59
Installation Crew (Bare Costs)	D-2 @ 9.20 L.H.	162.10	D-2 @ 10.10 L.H.	177.96	D-6 @ 13.20 L.H.	272.45
Total per 100 S.F.		$258.37		$318.49		$485.19

Special block: corner, jamb and head block are same price as ordinary block of same size. Tabulated on the next page are national average prices per block. Labor on specials is about the same as equal sized regular block. Bond beam and 16″ high lintel blocks cost 30% more than regular units of equal size. Lintel blocks are 8″ long and 8″ or 16″ high. Costs in individual cities may be factored from the table above.

Use of motorized mortar spreader box will speed construction of continuou walls.

04430-500 Ashlar Veneer

Quantities			Low Price Stone		High Price Stone	
4″ Stone, random ashlar,	2.5	tons average	$275 per ton	$ 687.50	$375 per ton	$ 937.50
Type N mortar @	$3.05	C.F.	20 C.F.	61.00	20 C.F.	61.00
50 - 1″ x 1/4″ x 6″ galv. stone anchors @	$1.10	each		55.00		55.00
Using crew D-8 @	$731.20	per day	.714 days	522.00	.833 days	609.09
Total per 100 S.F.				$1,325.50		$1,662.59

ɔne coverage varies from 30 S.F. to 50 S.F. per ton. Mortar quantities include 1″ backing bed.

04930-100 Cleaning Face Brick

ı smooth brick a person can clean 70 S.F. an hour; on rough brick 50 S.F. ɛr hour. Use one gallon muriatic acid to 20 gallons of water for 1000 F. Do not use acid solution until wall is at least seven days old, but a mild soap solution may be used after two days. Commercial cleaners cost from $9 to $12 per gallon.

Time has been allowed for clean-up in brick prices.

R06100-010 Thirty City Lumber Prices

Prices for boards are for #2 or better or sterling, whichever is in best supply. Dimension lumber is "Standard or Better" either Southern Yellow Pine (S.Y.P.), Spruce-Pine-Fir (S.P.F.), Hem-Fir (H.F.) or Douglas Fir (D.F.). The species of lumber used in a geographic area is listed by city. Plyform is 3/4" BB oil sealed fir or S.Y.P. whichever prevails locally, 3/4" CDX is S.Y.P. or Fir.

These are prices at the time of publication and should be checked against the current market price. Relative differences between cities will stay approximately constant.

		Contractor Purchases per M.B.F. S4S								Contractor Purchases per M.S.F.	
		Dimensions						Boards		3/4" Ext.	3/4" Thick
City	Species	2"x4"	2"x6"	2"x8"	2"x10"	2"x12"	4"x4"	1"x6"	1"x12"	Plyform	CDX T&G
Atlanta	S.P.F.	$603	$577	$644	$819	$915	$ 824	$ 757	$1,289	$ 900	$ 804
Baltimore	S.P.F.	596	528	642	703	681	943	1000	1500	746	789
Boston	S.P.F.	600	568	630	790	830	1042	1920	1550	1032	831
Buffalo	H.F.	513	553	579	642	650	755	1280	1610	635	515
Chicago	S.P.F.	608	586	614	748	685	690	936	1543	779	774
Cincinnati	S.P.F.	599	683	691	795	882	863	1200	1446	824	904
Cleveland	S.P.F.	673	631	650	892	893	787	984	1910	1030	837
Columbus	S.P.F.	566	722	696	894	908	744	1419	1947	1186	775
Dallas	S.P.F.	771	656	609	690	731	668	728	1029	718	718
Denver	H.F.	631	631	646	766	658	958	771	1131	1196	900
Detroit	S.P.F.	550	589	614	851	886	594	1147	1859	804	889
Houston	S.Y.P.	528	536	536	592	641	802	707	855	985	703
Indianapolis	S.Y.F.	633	615	705	795	835	888	917	1177	825	840
Kansas City	D.F.	590	570	570	645	670	864	1050	1300	785	830
Los Angeles	D.F.	525	536	625	642	654	696	671	1385	690	790
Memphis	S.P.F.	698	654	693	834	770	1068	909	1314	954	1045
Milwaukee	S.P.F.	785	750	862	900	995	842	850	1237	828	859
Minneapolis	S.P.F.	559	545	536	675	754	674	1197	1409	1138	753
Nashville	S.P.F.	609	625	642	804	753	864	831	1543	910	754
New Orleans	S.Y.P.	577	577	561	612	570	732	591	1547	870	717
New York City	S.P.F.	539	522	605	666	577	1044	800	1101	1064	796
Philadelphia	H.F.	576	561	599	704	676	804	1223	1468	749	796
Phoenix	D.F.	614	658	715	838	768	1134	1178	1455	1308	924
Pittsburgh	H.F.	584	624	679	749	780	799	1300	1490	780	830
St. Louis	S.P.F.	488	470	412	438	506	858	910	654	742	691
San Antonio	S.Y.P.	527	531	618	662	737	616	743	712	753	802
San Diego	D.F.	522	522	550	618	588	768	1059	1178	1284	904
San Francisco	D.F.	490	490	487	630	625	840	760	1355	906	937
Seattle	S.P.F.	638	627	673	794	801	817	666	910	800	792
Washington, DC	S.P.F.	567	548	586	752	716	908	819	1210	1126	820
Average		$592	$590	$622	$731	$738	$ 830	$ 977	$1,337	$ 912	$ 811

To convert square feet of surface to board feet, 4% waste included.

S4S Size	Multiply S.F. by	T & G Size	Multiply S.F. by	Flooring Size	Multiply S.F. by
1 x 4	1.18	1 x 4	1.27	25/32" x 2-1/4"	1.37
1 x 6	1.13	1 x 6	1.18	25/32" x 3-1/4"	1.29
1 x 8	1.11	1 x 8	1.14	15/32" x 1-1/2"	1.54
1 x 10	1.09	2 x 6	2.36	1" x 3"	1.28
				1" x 4"	1.24

06110-030 Lumber Product Material Prices

he price of forest products fluctuates widely from location to location and om season to season depending upon economic conditions. The table low indicates National Average material prices in effect Jan. 1, 2000. The ble shows relative differences between various sizes, grades and species. hese percentage differentials remain fairly constant even though lumber ices in general may change significantly during the year.

Availability of certain items depends upon geographic location and must be checked prior to firm price bidding.

	National Average Contractor Price, Quantity Purchase							
	Dimension Lumber, S4S, #2 & Better, KD						Heavy Timbers, Fir	
	Species	2″x4″	2″x6″	2″x8″	2″x10″	2″x12″		
Framing Lumber per MBF	Douglas Fir	$ 548	$ 555	$ 589	$ 681	$ 661	3″ x 4″ thru 3″ x 12″	$1,164
	Spruce	614	604	635	767	780	4″ x 4″ thru 4″ x 12″	1,326
	Southern Yellow Pine	544	548	571	622	649	6″ x 6″ thru 6″ x 12″	2,000
	Hem-Fir	576	592	625	715	691	8″ x 8″ thru 8″ x 12″	2,055
							10″ x 10″ and 10″ x 12″	2,086

	S4S "D" Quality or Clear, KD						S4S # 2 & Better or Sterling, KD					
	Species	1″x4″	1″x6″	1″x8″	1″x10″	1″x12″	Species	1″x4″	1″x6″	1″x8″	1″x10″	1″x12″
Boards per MBF *See also Cedar Siding	Sugar Pine	$1,491	$1,645	$1,757	$2,079	$2,506	Sugar Pine	$ 602	$ 805	$ 861	$ 987	$1,288
	Idaho Pine	1,288	1,645	1,624	1,834	2,394	Idaho Pine	1,029	1,036	1,022	1,043	1,253
	Engleman Spruce	1,260	1,722	1,750	1,841	2,275	Engleman Spruce	630	756	847	966	1,295
	So. Yellow Pine	1,275	1,498	1,589	1,505	1,695	So. Yellow Pine	504	728	790	616	833
	Ponderosa Pine	1,300	1,624	1,645	1,869	2,394	Ponderosa Pine	553	756	812	945	1,218
	Redwood, CVG	3,940	3,800	3,750	3,770	5,940						

	Item	Price	Item	Price
Flooring per MSF	1″ x 4″ Vertical grain, Fir "B" & better	$2,400	2-1/4″ x 25/32″ Maple, select	$2,970
	2-1/4″ x 25/32″, Oak, clear	3,200	#2 & better	2,700
	Select	3,050	2-1/4″ x 33/32″ Maple, #2 & better	3,132
	#1 common	2,700	3-1/4″ x 33/32″ Maple, #2 & better	2,916
	Oak, prefinished, standard & better	3,750	Parquet, unfinished, 5/16″, minimum	2,700
	Standard	3,200	Maximum	5,000

	Item	Price	Item	Price
Siding per MBF	Clapboard, Cedar, beveled		*Rough sawn, Cedar, T&G, "A" grade 1″ x 4″	$2,434
	1/2″ x 6″ thru 1/2″ x 8″, clear	$1,344	1″ x 6″	3,201
	"A" grade	1,246	"STK" grade, 1″ x 6″	1,614
	"B" grade	933	Board, "STK" grade, 1″ x 8″	1,561
	3/4″ x 10″ "clear"	2,592	Board & Batten 1″ x 12″	1,561
	"A" grade	2,334	Cedar channel siding 1″ x 8″, #3 & better	$1,092
	Redwood, beveled		Factory stained	1,292
	1/2″ x 6″ thru 1/2″ x 8″, vertical grain, clear	1,907	White Pine siding, T&G, rough sawn	$ 637
	3/4″ x 10″ vertical grain, clear	2,889	Factory stained	837

	Item	Price	Item	Price
Shingles per CSF	Red Cedar		White Cedar shingles	
	5X—16″ long #1 regular	$ 151	16″ long, extra grade (East Coast)	$ 125
	#2	170	Clear, 1st grade (East Coast)	115
	18″ long perfections #1	184		
	#2	128	Fire retardant Red Cedar	
	Resquared & Rebutted #1	116	5X — 16″ long	$ 251
			18″ long perfections	284
	Handsplit shakes, resawn		Handsplit & resawn	
	24″ long, 1/2″ to 3/4″	145	24″ long, 1/2″ to 3/4″	245
	18″ long, 1/2″ to 3/4″	125	18″ long, 1/2″ to 3/4″	225

R06160-020 Plywood

There are two types of plywood used in construction: interior, which is moisture resistant but not waterproofed, and exterior, which is waterproofed.

The grade of the exterior surface of the plywood sheets is designated by the first letter: A, for smooth surface with patches allowed; B, for solid surface with patches and plugs allowed; C, which may be surface plugged or may have knot holes up to 1" wide; and D, which is used only for interior type plywood and may have knot holes up to 2-1/2" wide. "Structural Grade" is specifically designed for engineered applications such as box beams. All CC & DD grades have roof and floor spans marked on them.

Underlayment grade plywood runs from 1/4" to 1-1/4" thick. Thicknesses 5/8" and over have optional tongue and groove joints which eliminates the need for blocking the edges. Underlayment 19/32" and over may be referred to as Sturd-i-Floor.

The price of plywood can fluctuate widely due to geographic and economic conditions. When one or two local prices are known, the relative prices for other types and sizes may be found by direct factoring of the prices in the table below.

Typical uses for various plywood grades are as follows:

AA-AD Interior — cupboards, shelving, paneling, furniture
BB Plyform — concrete form plywood
CDX — wall and roof sheathing
Structural — box beams, girders, stressed skin panels
AA-AC Exterior — fences, signs, siding, soffits, etc.
Underlayment — base for resilient floor coverings
Overlaid HDO — high density for concrete forms & highway signs
Overlaid MDO — medium density for painting, siding, soffits & signs
303 Siding — exterior siding, textured, striated, embossed, etc.

Grade	National Average Price in Lots of 10 MSF, per MSF-January 2000				
	Type	4'x8'	Type	4'x8'	4'x10'
Sanded Grade	1/4" Interior AD	$ 609	1/4" Exterior AC	$ 616	$ 696
	3/8"	693	3/8"	707	727
	1/2"	805	1/2"	819	871
	5/8"	959	5/8"	937	1045
	3/4"	1,050	3/4"	1064	1,239
	1"	1,221	1"	1,380	1,427
	1-1/4"	1,464	Exterior AA, add	155	160
	Interior AA, add	155	Exterior AB, add	130	135
Unsanded Grade 4' x 8' Sheets			CD Structural 1		Underlayment
	5/16" CDX	$ 400			
	3/8"	405	5/16", 4'x8' sheets	$365	3/8", 4'x8' sheets $ 572
	1/2"	531	3/8"	459	1/2" 733
	5/8"	666	1/2"	647	5/8"T&G 756
	3/4"	806	5/8"	777	3/4"T&G 1,211
	3/4" T&G	831	3/4"	889	1-1/8" 2-4-1T&G 1,897
Form Plywood	5/8" Exterior, oiled BB, plyform	$ 966	5/8" HDO (overlay 2 sides)		$2,148
	3/4" Exterior, oiled BB, plyform	1,036	3/4" HDO (overlay 2 sides)		2,214
Overlaid 4'x8' Sheets	Overlay 2 Sides MDO		Overlay 1 Side MDO		
	3/8" thick	$1,212	3/8" thick		$1,029
	1/2"	1,418	1/2"		1,186
	5/8"	1,600	5/8"		1,332
	3/4"	1,830	3/4"		1,562
303 Siding	Fir, rough sawn, natural finish, 3/8" thick	$ 707	Texture 1-11	5/8" thick, Fir	$1,093
	Redwood	1,890		Redwood	1,890
	Cedar	1,890		Cedar	1,537
	Southern Yellow Pine	539		Southern Yellow	850
Waferboard/O.S.B.	1/4" sheathing	$ 298	19/32" T&G		$ 407
	7/16" sheathing	435	23/32" T&G		594

For 2 MSF to 10 MSF, add 10%. For less than 2 MSF, add 15%.

R06170-100 Wood Roof Trusses

Loading figures represent live load. An additional load of 10 psf on the top chord and 10 psf on the bottom chord is included in the truss design. Spacing is 24″ O.C.

Span in Feet	Cost per Truss for Different Live Loads and Roof Pitches					
	Flat	4 in 12 Pitch		5 in 12 Pitch		8 in 12 Pitch
	40 psf	30 psf	40 psf	30 psf	40 psf	30 psf
20	$ 66	$ 48	$ 50	$ 48	$ 52	$ 58
22	73	53	55	53	57	63
24	79	58	60	58	62	69
26	86	62	64	63	67	75
28	92	67	69	68	72	81
30	99	72	74	73	77	86
32	123	86	88	87	91	101
34	131	92	94	93	97	107
36	139	97	99	99	103	113
38	146	103	105	104	108	120
40	154	108	110	110	114	126

R06430-100 Wood Stair, Residential

One Flight with 8′-6″ Story Height, 3′-6″ Wide Oak Treads Open One Side, Built in Place				
Item	Quantity	Unit Cost	Bare Costs	Costs Incl. Subs O & P
Treads 10-1/2″ x 1-1/16″ thick	11 Ea.	$ 27.50	$ 302.50	$ 332.80
Landing tread nosing, rabbeted	1 Ea.	12.50	12.50	13.80
Risers 3/4″ thick	12 Ea.	12.40	148.80	163.60
Single end starting step (range $175 to $210)	1 Ea.	193.00	193.00	212.40
Balusters (range $4.50 to $25.00)	22 Ea.	7.75	170.50	187.60
Newels, starting & landing (range $40 to $120)	2 Ea.	73.00	146.00	160.60
Rail starter (range $39 to $190)	1 Ea.	67.00	67.00	73.80
Handrail (range $4.50 to $10.30)	26 L.F.	7.00	182.00	200.20
Cove trim	50 L.F.	.78	39.00	43.00
Rough stringers three - 2 x 12′s, 14′ long	84 B.F.	.74	62.16	68.40
Carpenter′s installation: Bare Cost	36 Hrs.	$ 19.70	$ 709.20	
Incl. Subs O & P		$ 33.75		$1,215.00
	Total per Flight		$2,032.66	$2,671.20

Add for rail return on second floor and for varnishing or other finish. Adjoining walls or landings must be figured separately.

R07110-010 1/2″ Pargeting (rough dampproofing plaster)

1:2-1/2 Mix, 4.5 C.F. Covers 100 S.F., 2 Coats, Waste Included		Regular Portland Cement		Waterproofed Portland Cement	
1.7 lbs. integral waterproofing admixture				$.65 per lb.	$ 1.11
1.7 bags Portland cement		$7.10 per bag	$ 12.07	$7.10 per bag	12.07
4.25 C.F. sand @	$18.70 per C.Y.		2.94		2.94
Labor mix, apply, crew D-1 @	$285.60 per day	.40 days	114.24	.40 days	114.24
Total Bare Cost per 100 S.F.			$129.25		$130.36

R07310-020 Roof Slate

16″, 18″ and 20″ are standard lengths and slate usually comes in random widths. For standard 3/16″ thickness use 1-1/2″ copper nails. Allow for 3% breakage.

Quantities per Square	Unfading Vermont Colored	Weathering Sea Green	Buckingham, Virginia Black
Slate delivered (incl. punching)	$385.00	$286.00	$470.00
# 30 Felt, 2.5 lbs. copper nails	14.49	14.49	14.49
Slate roofer 4.6 hrs. @ $17.10 per hr.	78.66	78.66	78.66
Total Bare Cost per Square	$478.15	$379.15	$563.15

R07550-030 Modified Bitumen Roofing

The cost of modified bitumen roofing is highly dependent on the type of installation that is planned. Installation is based on the type of modifier used in the bitumen. The two most popular modifiers are atactic polypropylene (APP) and styrene butadiene styrene (SBS). The modifiers are added to heated bitumen during the manufacturing process to change its characteristics. A polyethylene, polyester or fiberglass reinforcing sheet is then sandwiched between layers of this bitumen. When completed, the result is a pre-assembled, built-up roof that has increased elasticity and weatherablility. Some manufacturers include a surfacing material such as ceramic or mineral granules, metal particles or sand.

The preferred method of adhering SBS-modified bitumen roofing to the substrate is with hot-mopped asphalt (much the same as built-up roofing). This installation method requires a tar kettle/pot to heat the asphalt, as well as the labor, tools and equipment necessary to distribute and spread the hot asphalt.

The alternative method for applying APP and SBS modified bitumen is as follows. A skilled installer uses a torch to melt a small pool of bitumen off th membrane. This pool must form across the entire roll for proper adhesion. The installer must unroll the roofing at a pace slow enough to melt the bitumen, but fast enough to prevent damage to the rest of the membrane.

Modified bitumen roofing provides the advantages of both built-up and single-ply roofing. Labor costs are reduced over those of built-up roofing because only a single ply is necessary. The elasticity of single-ply roofing is attained with the reinforcing sheet and polymer modifiers. Modifieds have some self-healing characteristics and because of their multi-layer constructio they offer the reliability and safety of built-up roofing.

For information about Means Estimating Seminars, see yellow pages 11 and 12 in back of bo

08110-100 Hollow Metal Frames

ble below lists material prices only.

Frame for Opening Size	16 Ga. Frames		16 Ga. UL Frames		16 Ga. Drywall Frames		16 Ga. UL Drywall Frames	
	6-3/4″ Deep	8-3/4″ Deep	6-3/4″ Deep	8-3/4″ Deep	7-1/8″ Deep	8-1/4″ Deep	7-1/8″ Deep	8-1/4″ Deep
2′-0″ x 6′-8″	$ 69	$ 84	$ 78	$ 93	$ 83	$ 89	$ 92	$ 98
2′-6″ x 6′-8″	69	84	78	93	83	89	92	98
3′-0″ x 7′-0″	75	85	84	94	84	90	93	99
3′-6″ x 7′-0″	80	87	89	96	88	94	97	103
4′-0″ x 7′-0″	85	97	94	104	89	94	98	103
6′-0″ x 7′-0″	90	101	108	126	99	106	117	124
8′-0″ x 8′-0″	107	124	116	134	119	126	137	144

r welded frames, add $26.00.
r galvanized frames, add $22.00.

08110-120 Hollow Metal Doors

ble below lists material prices only, not including hardware or labor.

Door Thickness and Size		Full Flush Doors, 18 Ga.				Flush Fire Doors			
		Hollow Metal		Composite Core		Hollow Metal "B"		Hollow Metal Class "A"	
		Plain	Glazed	Plain	Glazed	20 Ga.	18 Ga.	16 Ga.	18 Ga.
1-3/4″	3′-0″ x 6′-8″	$189	$246	$214	$271	$179	$205	$255	$217
	3′-0″ x 7′-0″	199	255	224	280	192	211	265	222
	3′-6″ x 7′-0″	227	300	252	325	215	244	285	258
	3′-0″ x 8′-0″	247	330	272	355	249	269	310	283
	4′-0″ x 8′-0″	285	380	310	405	320	316	349	321
1-3/8″	2′-0″ x 6′-8″	146	203	—	—	160	—	—	—
	2′-6″ x 6′-8″	155	212	—	—	170	—	—	—
	3′-0″ x 7′-0″	165	222	—	—	180	—	—	—

ndicates 20 gauge doors.

08210-100 Wood Doors

ble below, except where indicated, lists price per door only, not including ame, hardware or labor. For pre-hung exterior door units up to 3′ x 7′, d $125 per door for wood frame and hardware for types not listed under pre-hung. Pricing is for ten or more doors. Doors are factory trimmed for butts and locksets.

Door Thickness and Size		Flush Type Doors (Interior)					Architectural (1-3/4″ Ext., 1-3/8″ Int.)					
		Hollow Core		Solid Particle Core			Pine		Fir		Pre-hung	
		Lauan	Birch	Lauan	Birch	Oak	Panel	Glazed	Panel	Glazed	Pine Panel	Flush Birch S. C.
1-3/4″	2′-6″ x 6′-8″	$36	$56	$ 73	$ 91 *	$ 97	$289	$304	$242	$257	$394	$250
	3′-0″ x 6′-8″	41	64	79	96 *	106	311	326	247	262	417	260
	3′-0″ x 7′-0″	52	71	97	102 *	113	354	369	260	270	464	274
	3′-6″ x 7′-0″	63	82	104	125 *	137	—	—	—	—	—	—
	4′-0″ x 7′-0″	69	97	112	136 *	147	—	—	—	—	—	—
1-3/8″	2′-0″ x 6′-8″	33	47	69	76 *	86	110	—	140	—	177	140
	2′-6″ x 6′-8″	35	49	73	83 *	91	124	139	159	—	190	147
	3′-0″ x 6′-8″	40	51	77	94 *	98	148	163	182	—	220	157
	3′-0″ x 7′-0″	46	56	82	99 *	106	166	—	206	—	—	—

dd to the above for the following birch face door types:

lid wood stave core, add $25
4 hour label door, add $61
1 hour label door, add $72
1-1/2 hour label door, add $82
8′ high door, add 31%

R08550-010 D. H. Wood Window (ready hung)

Ponderosa pine and vinyl clad sash, exterior primed with insulating glass.

Description	2′-0″ x 3′-0″				3′-0″ x 4′-0″			
	Wood		Vinyl Clad		Wood		Vinyl Clad	
Window and frame, 2 lights		$122.50		$177.00		$171.00		$236.00
Removable grilles		13.50		13.50		15.60		15.60
Aluminum storm/screen		57.00		60.00		72.00		72.00
Interior trim set		15.00		15.00		19.00		19.00
Carpenter @ $19.70 per hr.	1.7 hr.	33.49	1.7 hr.	33.49	2 hr.	39.40	2 hr.	39.40
Complete in place		$241.49		$298.99		$317.00		$382.00

Mullions for above windows are about $28 each. Aluminum clad double-hung windows run about $28.60 per S.F. of glass.

R08550-200 Replacement Windows

Replacement windows are typically measured per United Inch.

United Inches are calculated by rounding the width and height of the window opening up to the nearest inch, then adding the two figures.

The labor cost for replacement windows includes removal of sash, existing sash balance or weights, parting bead where necessary and installation of new window.

Debris hauling and dump fees are not included.

Doors & Windows | R087 | Hardware

R08700-100 Hinges

All closer equipped doors should have ball bearing hinges. Lead lined or extremely heavy doors require special strength hinges. Usually 1-1/2 pair of hinges are used per door up to 7′-6″ high openings. Table below shows typical hinge requirements.

Use Frequency	Type Hinge Required	Type of Opening	Type of Structure
High	Heavy weight ball bearing	Entrances Toilet Rooms	Banks, Office buildings, Schools, Stores & Theaters Office buildings and Schools
Average	Standard weight ball bearing	Entrances Corridors Toilet Rooms	Dwellings Office buildings and Schools Stores
Low	Plain bearing	Interior	Dwellings

Door Thickness	Weight of Doors in Pounds per Square Foot				
	White Pine	Oak	Hollow Core	Solid Core	Hollow Metal
1-3/8″	3 psf	6 psf	1-1/2 psf	3-1/2 — 4 psf	6-1/2 psf
1-3/4″	3-1/2	7	2-1/2	4-1/2 — 5-1/4	6-1/2
2-1/4″	4-1/2	9	—	5-1/2 — 6-3/4	6-1/2

09210-105 Gypsum Plaster

Quantities for 100 S.Y.	2 Coat, 5/8" Thick		3 Coat, 3/4" Thick		
	Base 1:3 Mix	Finish 2:1 Mix	Scratch 1:2 Mix	Brown 1:3 Mix	Finish 2:1 Mix
Gypsum plaster	1300 lb.		1350 lb.	650 lb.	
Sand	1.75 C.Y.		1.16 C.Y.	.84 C.Y.	
Finish hydrated lime		340 lb.			340 lb.
Gauging plaster		170 lb.			170 lb.

Total, in Place for 100 S.Y. on Walls					2 Coat, 5/8" Thick			3 Coat, 3/4" Thick		
					Quantities	Bare Cost	Incl. O & P	Quantities	Bare Cost	Incl. O & P
Gypsum plaster @	$13.02	per	80	lb. bag	1300 lb.	$211.58	$ 232.74	2000 lb.	$ 325.50	$ 358.05
Finish hydrated lime @	$ 5.60	per	50	lb. bag	340 lb.	38.08	41.89	340 lb.	38.08	41.89
Gauging plaster @	$19.02	per	100	lb. bag	170 lb.	32.33	35.56	170 lb.	32.33	35.56
Sand @	$17.25	per	C.Y.		1.7 C.Y.	29.33	32.26	2.0 C.Y.	34.50	37.95
J-1 crew @	$17.45	&	$29.18	per L.H.	34.8 L.H.	607.26	1,015.46	42.0 L.H.	732.90	1,225.56
Cleaning, staging, handling, patching					3.6 L.H.	62.82	105.05	4.0 L.H.	69.80	116.72
Total per 100 S.Y. in place						$981.40	$1,462.96		$1,233.11	$1,815.73

09210-115 Vermiculite or Perlite Plaster

oportions: Over lath, scratch coat and brown coat, 0# gypsum plaster to 2 C.F. aggregate; over masonry, 0# gypsum plaster to 3 C.F. aggregate.

Quantities for 100 S.Y.					2 Coat, 5/8" Thick			3 Coat, 3/4" Thick		
					Quantities	Bare Cost	Incl. O & P	Quantities	Bare Cost	Incl. O & P
Gypsum plaster @	$13.02	per	80	lb. bag	1250 lb.	$ 203.44	$ 223.78	2250 lb.	$ 366.19	$ 402.81
Vermiculite or perlite @	$13.03	per		bag	7.8 bags	101.63	111.79	11.3 bags	147.24	161.96
Finish hydrated lime @	$ 5.60	per	50	lb. bag	340 lb.	38.08	41.89	340 lb.	38.08	41.89
Gauging plaster @	$19.02	per	100	lb. bag	170 lb.	32.33	35.56	170 lb.	32.33	35.56
J-1 crew @	$18.81	&	$30.68	per L.H.	40.0 L.H.	752.50	1,227.15	50.0 L.H.	940.63	1,533.94
Cleaning, staging, handling, patching					3.6 L.H.	67.73	110.44	4.0 L.H.	75.25	122.72
Total per 100 S.Y. in place						$1,195.71	$1,750.61		$1,599.72	$2,298.88

09220-300 Stucco

Quantities for 100 S.Y. 3 Coats, 1" Thick		On Wood Frame			On Masonry		
		Quantities	Bare Cost	Incl. O & P	Quantities	Bare Cost	Incl. O & P
Portland cement @	$ 7.10 per bag	29 bags	$ 205.90	$ 226.49	21 bags	$149.10	$164.01
Sand @	$15.85 per C.Y.	2.6 C.Y.	41.21	45.33	2 C.Y.	31.70	34.87
Hydrated lime @	$ 5.60 per 50 lb. bag	180 lb.	20.16	22.18	120 lb.	13.44	14.78
Painted stucco mesh @	$ 2.32 per S.Y., 3.6#	105 S.Y.	243.60	267.96	—	—	—
Furring nails and jute fiber			13.84	15.22	—	—	—
Mix and install with crew indicated		J-2 @ .74 days	670.51	1,094.28	J-1 @ 0.5 days	376.25	613.58
Cleaning, staging, handling, patching		.10 days	85.16	143.63	.10 days	85.16	143.63
Total per 100 S.Y. in place			$1,280.38	$1,815.09		$655.65	$970.87

R09658-600 Resilient Flooring and Base

Description	12″ x 12″ x 3/32″ V.C. Tile			4″ x 1/8″ Vinyl Base		
	Quantities	Bare Cost	Incl. O & P	Quantities	Bare Cost	Incl. O & P
Vinyl composition, tile, standard line	100 S.F.	$112.00	$123.20	100 L.F.	$ 63.00	$ 69.30
Vinyl cement @ $10.95 per gallon	0.8 gallon	8.76	9.64	0.5 gallon	5.48	6.03
Tile layer @ $19.20 and $31.10 per L.H.	1.5 L.H.	28.80	46.65	2.7 L.H.	51.84	83.97
Total per 100 S.F. in place		$149.56	$179.49	100 L.F.	$120.32	$159.30

R09700-700 Wall Covering

Quantities for 100 S.F.	Medium Price Paper			Expensive Paper		
	Quantities	Bare Cost	Incl. O & P	Quantities	Bare Cost	Incl. O & P
Paper @ $30.00 and $57.00 per double roll	1.6 dbl. rolls	$48.00	$ 52.80	1.6 dbl. rolls	$ 91.20	$100.32
Wall sizing @ $19.60 per gallon	.25 gallon	4.90	5.39	.25 gallon	4.90	5.39
Vinyl wall paste @ $ 9.00 per gallon	0.4 gallon	3.60	3.96	0.4 gallon	3.60	3.96
Apply sizing @ $17.95 and $29.95 per hour	0.3 hour	5.39	8.99	0.3 hour	5.39	8.99
Apply paper @ $17.95 and $29.95 per hour	1.2 hours	21.54	35.94	1.5 hours	26.93	44.93
Total including waste allowance per 100 S.F.		$83.43	$107.08		$132.02	$163.59

This is equivalent to about $66.95 and $102.25 per double roll complete in place. Most wallpapers now come in double rolls only. To remove old paper allow 1.3 hours per 100 S.F.

R09910-100 Paint

Material prices per gallon in 5 gallon lots, up to 25 gallons. For 100 gallons, deduct 10%.

Exterior, Alkyd (oil base)	
Flat	$23.00
Gloss	22.25
Primer	22.00
Exterior, Latex (water base)	
Acrylic stain	19.00
Gloss enamel	23.00
Flat	19.50
Primer	21.75
Semi-gloss	22.50
Interior, Alkyd (oil base)	
Enamel undercoater	20.75
Flat	18.50
Gloss	23.00
Primer sealer	17.25
Semi-gloss	22.25
Interior, Latex (water base)	
Enamel undercoater	15.50
Flat	15.75
Floor and deck	21.00
Gloss	22.50
Primer sealer	18.00
Semi-gloss	21.50
Masonry, Exterior	
Alkali resistant primer	$19.25
Block filler, epoxy	20.25
Block filler, latex	10.00
Latex, flat or semi-gloss	17.00
Masonry, Interior	
Alkali resistant primer	15.75
Block filler, epoxy	23.25
Block filler, latex	10.25
Floor, alkyd	21.00
Floor, latex	21.25
Latex, flat acrylic	11.25
Latex, flat emulsion	11.00
Latex, sealer	12.25
Latex, semi-gloss	18.75
Varnish and Stain	
Alkyd clear	21.25
Polyurethane, clear	21.25
Primer sealer	15.00
Semi-transparent stain	15.75
Solid color stain	15.75
Metal Coatings	
Galvanized	24.50
High heat	48.50
Machinery enamel, alkyd	25.00
Normal heat	25.50
Rust inhibitor ferrous metal	$23.50
Zinc chromate	20.00
Heavy Duty Coatings	
Acrylic urethane	51.50
Chlorinated rubber	29.75
Coal tar epoxy	29.25
Metal pretreatment (polyvinyl butyral)	23.50
Polyamide epoxy finish	31.50
Polyamide epoxy primer	31.50
Silicone alkyd	32.00
2 component solvent based acrylic epoxy	59.50
2 component solvent based polyester epoxy	88.50
Vinyl	27.00
Zinc rich primer	94.50
Special Coatings/Miscellaneous	
Aluminum	22.75
Dry fall out, flat	12.25
Fire retardant, intumescent	36.50
Linseed oil	10.25
Shellac	18.25
Swimming pool, epoxy or urethane base	36.00
Swimming pool, rubber base	27.75
Texture paint	11.25
Turpentine	10.75
Water repellent 5% silicone	17.75

R09910-220 Painting

Item	Coat	One Gallon Covers			In 8 Hours a Laborer Covers			Labor-Hours per 100 S.F.		
		Brush	Roller	Spray	Brush	Roller	Spray	Brush	Roller	Spray
Paint wood siding	prime	250 S.F.	225 S.F.	290 S.F.	1150 S.F.	1300 S.F.	2275 S.F.	.695	.615	.351
	others	270	250	290	1300	1625	2600	.615	.492	.307
Paint exterior trim	prime	400	—	—	650	—	—	1.230	—	—
	1st	475	—	—	800	—	—	1.000	—	—
	2nd	520	—	—	975	—	—	.820	—	—
Paint shingle siding	prime	270	255	300	650	975	1950	1.230	.820	.410
	others	360	340	380	800	1150	2275	1.000	.695	.351
Stain shingle siding	1st	180	170	200	750	1125	2250	1.068	.711	.355
	2nd	270	250	290	900	1325	2600	.888	.603	.307
Paint brick masonry	prime	180	135	160	750	800	1800	1.066	1.000	.444
	1st	270	225	290	815	975	2275	.981	.820	.351
	2nd	340	305	360	815	1150	2925	.981	.695	.273
Paint interior plaster or drywall	prime	400	380	495	1150	2000	3250	.695	.400	.246
	others	450	425	495	1300	2300	4000	.615	.347	.200
Paint interior doors and windows	prime	400	—	—	650	—	—	1.230	—	—
	1st	425	—	—	800	—	—	1.000	—	—
	2nd	450	—	—	975	—	—	.820	—	—

13128-520 Swimming Pools

ol prices given per square foot of surface area include pool structure, er and chlorination equipment where required, pumps, related piping, ing boards, ladders, maintenance kit, skimmer and vacuum system. Decks d electrical service to equipment are not included.

sidential in-ground pool construction can be divided into two categories: nyl lined and gunite. Vinyl lined pool walls are constructed of different aterials including wood, concrete, plastic or metal. The bottom is often aded with sand over which the vinyl liner is installed. Costs are generally the $16 to $23 per S.F. range. Vermiculite or soil cement bottoms may be bstituted for an added cost.

unite pool construction is used both in residential and municipal stallations. These structures are steel reinforced for strength and finished with a white cement limestone plaster. Residential costs run from $26 to $38 per S.F. surface. Municipal costs vary from $50 to $75 because plumbing codes require more expensive materials, chlorination equipment and higher filtration rates.

Municipal pools greater than 1,800 S.F. require gutter systems to control waves. This gutter may be formed into the concrete wall. Often a vinyl, stainless steel gutter or gutter and wall system is specified, which will raise the pool cost an additional $51 to $150 per L.F. of gutter installed and up to $290 per L.F. if a gutter and wall system is installed.

Competition pools usually require tile bottoms and sides with contrasting lane striping. Add $12 per S.F. of wall or bottom to be tiled.

13600-610 Solar Heating (Space and Hot Water)

ollectors should face as close to due South as possible, however, variations up to 20 degrees on either side of true South are acceptable. Local imate and collector type may influence the choice between east or west eviations. Obviously they should be located so they are not shaded from the n's rays. Incline collectors at a slope of latitude minus 5 degrees for omestic hot water and latitude plus 15 degrees for space heating.

at plate collectors consist of a number of components as follows: Insulation reduce heat loss through the bottom and sides of the collector. The nclosure which contains all the components in this assembly is usually eatherproof and prevents dust, wind and water from coming in contact ith the absorber plate. The cover plate usually consists of one or more yers of a variety of glass or plastic and reduces the reradiation by creating air space which traps the heat between the cover and the absorber ates.

ne absorber plate must have a good thermal bond with the fluid passages. ne absorber plate is usually metallic and treated with a surface coating hich improves absorptivity. Black or dark paints or selective coatings are sed for this purpose, and the design of this passage and plate combination elps determine a solar system's effectiveness.

Heat transfer fluid passage tubes are attached above and below or integral with an absorber plate for the purpose of transferring thermal energy from the absorber plate to a heat transfer medium. The heat exchanger is a device for transferring thermal energy from one fluid to another.

Piping and storage tanks should be well insulated to minimize heat losses.

Size domestic water heating storage tanks to hold 20 gallons of water per user, minimum, plus 10 gallons per dishwasher or washing machine. For domestic water heating an optimum collector size is approximately 3/4 square foot of area per gallon of water storage. For space heating of residences and small commercial applications the collector is commonly sized between 30% and 50% of the internal floor area. For space heating of large commercial applications, collector areas less than 30% of the internal floor area can still provide significant heat reductions.

A supplementary heat source is recommended for Northern states for December through February.

The solar energy transmission per square foot of collector surface varies greatly with the material used. Initial cost, heat transmittance and useful life are obviously interrelated.

R15100-050 Pipe Material Considerations

1. Malleable fittings should be used for gas service.
2. Malleable fittings are used where there are stresses/strains due to expansion and vibration.
3. Cast fittings may be broken as an aid to disassembling of heating lines frozen by long use, temperature and minerals.
4. Cast iron pipe is extensively used for underground and submerged service.
5. Type M (light wall) copper tubing is available in hard temper only and is used for nonpressure and less severe applications than K and L.
6. Type L (medium wall) copper tubing, available hard or soft for interio service.
7. Type K (heavy wall) copper tubing, available in hard or soft temper fo use where conditions are severe. For underground and interior service
8. Hard drawn tubing requires fewer hangers or supports but should not be bent. Silver brazed fittings are recommended, however soft solder i normally used.
9. Type DMV (very light wall) copper tubing designed for drainage, wast and vent plus other non-critical pressure services.

Domestic/Imported Pipe and Fittings Cost

The prices shown in this publication for steel/cast iron pipe and steel, cast iron, malleable iron fittings are based on domestic production sold at the normal trade discounts. The above listed items of foreign manufacture may be available at prices of 1/3 to 1/2 those shown. Some imported items after minor machining or finishing operations are being sold as domestic to further complicate the system.

Caution: Most pipe prices in this book also include a coupling and pipe hangers which for the larger sizes can add significantly to the per meter c and should be taken into account when comparing "book cost" with quoted supplier's cost.

R15100-420 Plumbing Fixture Installation Time

Item	Rough-In	Set	Total Hours	Item	Rough-In	Set	Total Hours
Bathtub	5	5	10	Shower head only	2	1	3
Bathtub and shower, cast iron	6	6	12	Shower drain	3	1	4
Fire hose reel and cabinet	4	2	6	Shower stall, slate		15	15
Floor drain to 4 inch diameter	3	1	4	Slop sink	5	3	8
Grease trap, single, cast iron	5	3	8	Test 6 fixtures			14
Kitchen gas range		4	4	Urinal, wall	6	2	8
Kitchen sink, single	4	4	8	Urinal, pedestal or floor	6	4	10
Kitchen sink, double	6	6	12	Water closet and tank	4	3	7
Laundry tubs	4	2	6	Water closet and tank, wall hung	5	3	8
Lavatory wall hung	5	3	8	Water heater, 45 gals. gas, automatic	5	2	7
Lavatory pedestal	5	3	8	Water heaters, 65 gals. gas, automatic	5	2	7
Shower and stall	6	4	10	Water heaters, electric, plumbing only	4	2	6

Fixture prices in front of book are based on the cost per fixture set in place. The rough-in cost, which must be added for each fixture, includes carrier, if required, some supply, waste and vent pipe connecting fittings and stops. The lengths of rough-in pipe are nominal runs which would connect to the larger runs and stacks. The supply runs and DWV runs and stacks must be accounted for in separate entries. In the eastern half of the United States it is common for the plumber to carry these to a point 5′ outside the building.

Crews

Crew No.	Bare Costs		Incl. Subs O & P		Cost Per Labor-Hour	
Crew A-1	**Hr.**	**Daily**	**Hr.**	**Daily**	**Bare Costs**	**Incl. O&P**
ilding Laborer	$14.45	$115.60	$24.75	$198.00	$14.45	$24.75
as Eng. Power Tool		70.00		77.00	8.75	9.63
H., Daily Totals		$185.60		$275.00	$23.20	$34.38

Crew A-1A	Hr.	Daily	Hr.	Daily	Bare Costs	Incl. O&P
illed Worker	$19.75	$158.00	$33.90	$271.20	$19.75	$33.90
ot Blaster, 20"		142.00		156.20	17.75	19.53
H., Daily Totals		$300.00		$427.40	$37.50	$53.43

Crew A-2	Hr.	Daily	Hr.	Daily	Bare Costs	Incl. O&P
aborers	$14.45	$231.20	$24.75	$396.00	$14.92	$25.33
uck Driver (light)	15.85	126.80	26.50	212.00		
ght Truck, 1.5 Ton		172.95		190.25	7.21	7.93
.H., Daily Totals		$530.95		$798.25	$22.13	$33.26

Crew A-2A	Hr.	Daily	Hr.	Daily	Bare Costs	Incl. O&P
aborers	$14.45	$231.20	$24.75	$396.00	$14.92	$25.33
uck Driver (light)	15.85	126.80	26.50	212.00		
ght Truck, 1.5 ton		172.95		190.25		
oncrete Saw		110.00		121.00	11.79	12.97
L.H., Daily Totals		$640.95		$919.25	$26.71	$38.30

Crew A-3	Hr.	Daily	Hr.	Daily	Bare Costs	Incl. O&P
uck Driver (heavy)	$16.20	$129.60	$27.05	$216.40	$16.20	$27.05
ump Truck, 12 Ton		365.30		401.85	45.66	50.23
.H., Daily Totals		$494.90		$618.25	$61.86	$77.28

Crew A-3A	Hr.	Daily	Hr.	Daily	Bare Costs	Incl. O&P
ruck Driver (light)	$15.85	$126.80	$26.50	$212.00	$15.85	$26.50
ickup truck (4x4)		144.65		159.10	18.08	19.89
.H., Daily Totals		$271.45		$371.10	$33.93	$46.39

Crew A-3B	Hr.	Daily	Hr.	Daily	Bare Costs	Incl. O&P
quip. Oper. (medium)	$19.90	$159.20	$33.00	$264.00	$18.05	$30.02
ruck Driver (heavy)	16.20	129.60	27.05	216.40		
Dump Truck, 16 Ton		447.25		492.00		
.E. Loader, 3 C.Y.		400.80		440.90	53.00	58.30
L.H., Daily Totals		$1136.85		$1413.30	$71.05	$88.32

Crew A-3C	Hr.	Daily	Hr.	Daily	Bare Costs	Incl. O&P
quip. Oper. (light)	$19.10	$152.80	$31.70	$253.60	$19.10	$31.70
Wheeled Skid Steer Loader		225.80		248.40	28.23	31.05
.H., Daily Totals		$378.60		$502.00	$47.33	$62.75

Crew A-4	Hr.	Daily	Hr.	Daily	Bare Costs	Incl. O&P
Carpenters	$19.70	$315.20	$33.75	$540.00	$19.12	$32.48
Painter, Ordinary	17.95	143.60	29.95	239.60		
L.H., Daily Totals		$458.80		$779.60	$19.12	$32.48

Crew A-5	Hr.	Daily	Hr.	Daily	Bare Costs	Incl. O&P
Laborers	$14.45	$231.20	$24.75	$396.00	$14.61	$24.94
5 Truck Driver (light)	15.85	31.70	26.50	53.00		
5 Light Truck, 1.5 Ton		43.24		47.55	2.40	2.64
3 L.H., Daily Totals		$306.14		$496.55	$17.01	$27.58

Crew A-6	Hr.	Daily	Hr.	Daily	Bare Costs	Incl. O&P
Chief Of Party	$22.10	$176.80	$38.15	$305.20	$20.93	$36.03
Instrument Man	19.75	158.00	33.90	271.20		
5 L.H., Daily Totals		$334.80		$576.40	$20.93	$36.03

Crew No.	Bare Costs		Incl. Subs O & P		Cost Per Labor-Hour	
Crew A-7	**Hr.**	**Daily**	**Hr.**	**Daily**	**Bare Costs**	**Incl. O&P**
1 Chief Of Party	$22.10	$176.80	$38.15	$305.20	$20.40	$34.72
1 Instrument Man	19.75	158.00	33.90	271.20		
1 Rodman/Chainman	19.35	154.80	32.10	256.80		
24 L.H., Daily Totals		$489.60		$833.20	$20.40	$34.72

Crew A-8	Hr.	Daily	Hr.	Daily	Bare Costs	Incl. O&P
1 Chief Of Party	$22.10	$176.80	$38.15	$305.20	$20.14	$34.06
1 Instrument Man	19.75	158.00	33.90	271.20		
2 Rodmen/Chainmen	19.35	309.60	32.10	513.60		
32 L.H., Daily Totals		$644.40		$1090.00	$20.14	$34.06

Crew A-9	Hr.	Daily	Hr.	Daily	Bare Costs	Incl. O&P
1 Asbestos Foreman	$21.00	$168.00	$36.75	$294.00	$20.56	$36.01
7 Asbestos Workers	20.50	1148.00	35.90	2010.40		
64 L.H., Daily Totals		$1316.00		$2304.40	$20.56	$36.01

Crew A-10	Hr.	Daily	Hr.	Daily	Bare Costs	Incl. O&P
1 Asbestos Foreman	$21.00	$168.00	$36.75	$294.00	$20.56	$36.01
7 Asbestos Workers	20.50	1148.00	35.90	2010.40		
64 L.H., Daily Totals		$1316.00		$2304.40	$20.56	$36.01

Crew A-10A	Hr.	Daily	Hr.	Daily	Bare Costs	Incl. O&P
1 Asbestos Foreman	$21.00	$168.00	$36.75	$294.00	$20.67	$36.18
2 Asbestos Workers	20.50	328.00	35.90	574.40		
24 L.H., Daily Totals		$496.00		$868.40	$20.67	$36.18

Crew A-10B	Hr.	Daily	Hr.	Daily	Bare Costs	Incl. O&P
1 Asbestos Foreman	$21.00	$168.00	$36.75	$294.00	$20.63	$36.11
3 Asbestos Workers	20.50	492.00	35.90	861.60		
32 L.H., Daily Totals		$660.00		$1155.60	$20.63	$36.11

Crew A-10C	Hr.	Daily	Hr.	Daily	Bare Costs	Incl. O&P
3 Asbestos Workers	$20.50	$492.00	$35.90	$861.60	$20.50	$35.90
1 Flatbed truck		172.95		190.25	7.21	7.93
24 L.H., Daily Totals		$664.95		$1051.85	$27.71	$43.83

Crew A-10D	Hr.	Daily	Hr.	Daily	Bare Costs	Incl. O&P
2 Asbestos Workers	$20.50	$328.00	$35.90	$574.40	$19.66	$33.56
1 Equip. Oper. (crane)	20.65	165.20	34.25	274.00		
1 Equip. Oper. Oiler	17.00	136.00	28.20	225.60		
1 Hydraulic crane, 33 ton		694.95		764.45	21.72	23.89
32 L.H., Daily Totals		$1324.15		$1838.45	$41.38	$57.45

Crew A-11	Hr.	Daily	Hr.	Daily	Bare Costs	Incl. O&P
1 Asbestos Foreman	$21.00	$168.00	$36.75	$294.00	$20.56	$36.01
7 Asbestos Workers	20.50	1148.00	35.90	2010.40		
2 Chipping Hammers		33.90		37.30	.53	.58
64 L.H., Daily Totals		$1349.90		$2341.70	$21.09	$36.59

Crew A-12	Hr.	Daily	Hr.	Daily	Bare Costs	Incl. O&P
1 Asbestos Foreman	$21.00	$168.00	$36.75	$294.00	$20.56	$36.01
7 Asbestos Workers	20.50	1148.00	35.90	2010.40		
1 Large Prod. Vac. Loader		529.80		582.80	8.28	9.11
64 L.H., Daily Totals		$1845.80		$2887.20	$28.84	$45.12

Crew A-13	Hr.	Daily	Hr.	Daily	Bare Costs	Incl. O&P
1 Equip. Oper. (light)	$19.10	$152.80	$31.70	$253.60	$19.10	$31.70
1 Large Prod. Vac. Loader		529.80		582.80	66.23	72.85
8 L.H., Daily Totals		$682.60		$836.40	$85.33	$104.55

Crews

Crew No.	Bare Costs		Incl. Subs O & P		Cost Per Labor-Hour	

Crew B-1	Hr.	Daily	Hr.	Daily	Bare Costs	Incl. O&P
1 Labor Foreman (outside)	$16.45	$131.60	$28.20	$225.60	$15.12	$25.90
2 Laborers	14.45	231.20	24.75	396.00		
24 L.H., Daily Totals		$362.80		$621.60	$15.12	$25.90

Crew B-2	Hr.	Daily	Hr.	Daily	Bare Costs	Incl. O&P
1 Labor Foreman (outside)	$16.45	$131.60	$28.20	$225.60	$14.85	$25.44
4 Laborers	14.45	462.40	24.75	792.00		
40 L.H., Daily Totals		$594.00		$1017.60	$14.85	$25.44

Crew B-3	Hr.	Daily	Hr.	Daily	Bare Costs	Incl. O&P
1 Labor Foreman (outside)	$16.45	$131.60	$28.20	$225.60	$16.27	$27.47
2 Laborers	14.45	231.20	24.75	396.00		
1 Equip. Oper. (med.)	19.90	159.20	33.00	264.00		
2 Truck Drivers (heavy)	16.20	259.20	27.05	432.80		
1 F.E. Loader, T.M., 2.5 C.Y.		784.00		862.40		
2 Dump Trucks, 16 Ton		894.50		983.95	34.97	38.47
48 L.H., Daily Totals		$2459.70		$3164.75	$51.24	$65.94

Crew B-3A	Hr.	Daily	Hr.	Daily	Bare Costs	Incl. O&P
4 Laborers	$14.45	$462.40	$24.75	$792.00	$15.54	$26.40
1 Equip. Oper. (med.)	19.90	159.20	33.00	264.00		
1 Hyd. Excavator, 1.5 C.Y.		713.20		784.50	17.83	19.61
40 L.H., Daily Totals		$1334.80		$1840.50	$33.37	$46.01

Crew B-3B	Hr.	Daily	Hr.	Daily	Bare Costs	Incl. O&P
2 Laborers	$14.45	$231.20	$24.75	$396.00	$16.25	$27.39
1 Equip. Oper. (med.)	19.90	159.20	33.00	264.00		
1 Truck Drivers (heavy)	16.20	129.60	27.05	216.40		
1 Backhoe Loader, 80 H.P.		273.15		300.45		
1 Dump Trucks, 16 Ton		447.25		492.00	22.51	24.76
32 L.H., Daily Totals		$1240.40		$1668.85	$38.76	$52.15

Crew B-3C	Hr.	Daily	Hr.	Daily	Bare Costs	Incl. O&P
3 Laborers	$14.45	$346.80	$24.75	$594.00	$15.81	$26.81
1 Equip. Oper. (med.)	19.90	159.20	33.00	264.00		
1 F.E. Loader, 3.75 C.Y.		542.40		596.65	16.95	18.65
32 L.H., Daily Totals		$1048.40		$1454.65	$32.76	$45.46

Crew B-4	Hr.	Daily	Hr.	Daily	Bare Costs	Incl. O&P
1 Labor Foreman (outside)	$16.45	$131.60	$28.20	$225.60	$15.08	$25.71
4 Laborers	14.45	462.40	24.75	792.00		
1 Truck Driver (heavy)	16.20	129.60	27.05	216.40		
1 Tractor, 4 x 2, 195 H.P.		325.60		358.15		
1 Platform Trailer		152.00		167.20	9.95	10.95
48 L.H., Daily Totals		$1201.20		$1759.35	$25.03	$36.66

Crew B-5	Hr.	Daily	Hr.	Daily	Bare Costs	Incl. O&P
1 Labor Foreman (outside)	$16.45	$131.60	$28.20	$225.60	$15.94	$27.09
3 Laborers	14.45	346.80	24.75	594.00		
1 Equip. Oper. (med.)	19.90	159.20	33.00	264.00		
1 Air Compr., 250 C.F.M.		127.00		139.70		
2 Air Tools & Accessories		39.00		42.90		
2-50 Ft. Air Hoses, 1.5″ Dia.		14.40		15.85		
1 F.E. Loader, T.M., 2.5 C.Y.		784.00		862.40	24.11	26.52
40 L.H., Daily Totals		$1602.00		$2144.45	$40.05	$53.61

Crew B-5A	Hr.	Daily	Hr.	Daily	Bare Costs	Incl. O&P
1 Foreman	$16.45	$131.60	$28.20	$225.60	$16.20	$27.3
6 Laborers	14.45	693.60	24.75	1188.00		
2 Equip. Oper. (med.)	19.90	318.40	33.00	528.00		
1 Equip. Oper. (light)	19.10	152.80	31.70	253.60		
2 Truck Drivers (heavy)	16.20	259.20	27.05	432.80		
1 Air Compr. 365 C.F.M.		174.40		191.85		
2 Pavement Breakers		39.00		42.90		
8 Air Hoses w/Coup.,1″		41.60		45.75		
2 Dump Trucks, 12 Ton		730.60		803.65	10.27	11.2
96 L.H., Daily Totals		$2541.20		$3712.15	$26.47	$38.6

Crew B-5B	Hr.	Daily	Hr.	Daily	Bare Costs	Incl. O&P
1 Powderman	$19.75	$158.00	$33.90	$271.20	$18.02	$30.1
2 Equip. Oper. (med.)	19.90	318.40	33.00	528.00		
3 Truck Drivers (heavy)	16.20	388.80	27.05	649.20		
1 F.E. Ldr. 2-1/2 CY		400.80		440.90		
3 Dump Trucks, 16 Ton		1341.75		1475.95		
1 Air Compr. 365 C.F.M.		174.40		191.85	39.94	43.9
48 L.H., Daily Totals		$2782.15		$3557.10	$57.96	$74.1

Crew B-5C	Hr.	Daily	Hr.	Daily	Bare Costs	Incl. O&P
3 Laborers	$14.45	$346.80	$24.75	$594.00	$16.66	$27.9
1 Equip. Oper. (medium)	19.90	159.20	33.00	264.00		
2 Truck Drivers (heavy)	16.20	259.20	27.05	432.80		
1 Equip. Oper. (crane)	20.65	165.20	34.25	274.00		
1 Equip. Oper. Oiler	17.00	136.00	28.20	225.60		
2 Dump Truck, 16 Ton		894.50		983.95		
1 F.E. Lder, 3.75 C.Y.		1035.00		1138.50		
1 Hyd. Crane, 25 Ton		563.90		620.30	38.96	42.8
64 L.H., Daily Totals		$3559.80		$4533.15	$55.62	$70.8

Crew B-6	Hr.	Daily	Hr.	Daily	Bare Costs	Incl. O&P
2 Laborers	$14.45	$231.20	$24.75	$396.00	$16.00	$27.07
1 Equip. Oper. (light)	19.10	152.80	31.70	253.60		
1 Backhoe Loader, 48 H.P.		203.00		223.30	8.46	9.30
24 L.H., Daily Totals		$587.00		$872.90	$24.46	$36.37

Crew B-7	Hr.	Daily	Hr.	Daily	Bare Costs	Incl. O&P
1 Labor Foreman (outside)	$16.45	$131.60	$28.20	$225.60	$15.69	$26.70
4 Laborers	14.45	462.40	24.75	792.00		
1 Equip. Oper. (med.)	19.90	159.20	33.00	264.00		
1 Chipping Machine		211.05		232.15		
1 F.E. Loader, T.M., 2.5 C.Y.		784.00		862.40		
2 Chain Saws, 36″		98.40		108.25	22.78	25.06
48 L.H., Daily Totals		$1846.65		$2484.40	$38.47	$51.76

Crew B-7A	Hr.	Daily	Hr.	Daily	Bare Costs	Incl. O&P
2 Laborers	$14.45	$231.20	$24.75	$396.00	$16.00	$27.07
1 Equip. Oper. (light)	19.10	152.80	31.70	253.60		
1 Rake w/Tractor		219.05		240.95		
2 Chain Saws, 18″		56.80		62.50	11.49	12.64
24 L.H., Daily Totals		$659.85		$953.05	$27.49	$39.71

Crew B-8	Bare Costs Hr.	Bare Costs Daily	Incl. Subs O & P Hr.	Incl. Subs O & P Daily	Cost Per Labor-Hour Bare Costs	Cost Per Labor-Hour Incl. O&P
abor Foreman (outside)	$16.45	$131.60	$28.20	$225.60	$16.79	$28.26
aborers	14.45	231.20	24.75	396.00		
quip. Oper. (med.)	19.90	318.40	33.00	528.00		
ruck Drivers (heavy)	16.20	259.20	27.05	432.80		
yd. Crane, 25 Ton		559.20		615.10		
.E. Loader, T.M., 2.5 C.Y.		784.00		862.40		
ump Trucks, 16 Ton		894.50		983.95	39.96	43.95
L.H., Daily Totals		$3178.10		$4043.85	$56.75	$72.21

Crew B-9	Hr.	Daily	Hr.	Daily	Bare Costs	Incl. O&P
abor Foreman (outside)	$16.45	$131.60	$28.20	$225.60	$14.85	$25.44
aborers	14.45	462.40	24.75	792.00		
ir Compr., 250 C.F.M.		127.00		139.70		
ir Tools & Accessories		39.00		42.90		
0 Ft. Air Hoses, 1.5″ Dia.		14.40		15.85	4.51	4.96
L.H., Daily Totals		$774.40		$1216.05	$19.36	$30.40

Crew B-9A	Hr.	Daily	Hr.	Daily	Bare Costs	Incl. O&P
aborers	$14.45	$231.20	$24.75	$396.00	$15.03	$25.52
ruck Driver (heavy)	16.20	129.60	27.05	216.40		
Vater Tanker		205.90		226.50		
ractor		325.60		358.15		
0 Ft. Disch. Hoses		12.80		14.10	22.68	24.95
L.H., Daily Totals		$905.10		$1211.15	$37.71	$50.47

Crew B-9B	Hr.	Daily	Hr.	Daily	Bare Costs	Incl. O&P
Laborers	$14.45	$231.20	$24.75	$396.00	$15.03	$25.52
Truck Driver (heavy)	16.20	129.60	27.05	216.40		
50 Ft. Disch. Hoses		12.80		14.10		
Water Tanker		205.90		226.50		
Tractor		325.60		358.15		
Pressure Washer		60.40		66.45	25.20	27.72
L.H., Daily Totals		$965.50		$1277.60	$40.23	$53.24

Crew B-9C	Hr.	Daily	Hr.	Daily	Bare Costs	Incl. O&P
Labor Foreman (outside)	$16.45	$131.60	$28.20	$225.60	$14.85	$25.44
Laborers	14.45	462.40	24.75	792.00		
Air Compr., 250 C.F.M.		127.00		139.70		
50 Ft. Air Hoses, 1.5″ Dia.		14.40		15.85		
Breaker, Pavement, 60 lb.		39.00		42.90	4.51	4.96
) L.H., Daily Totals		$774.40		$1216.05	$19.36	$30.40

Crew B-10	Hr.	Daily	Hr.	Daily	Bare Costs	Incl. O&P
Equip. Oper. (med.)	$19.90	$159.20	$33.00	$264.00	$19.90	$33.00
L.H., Daily Totals		$159.20		$264.00	$19.90	$33.00

Crew B-10A	Hr.	Daily	Hr.	Daily	Bare Costs	Incl. O&P
Equip. Oper. (med.)	$19.90	$159.20	$33.00	$264.00	$19.90	$33.00
Roll. Compact., 2K Lbs.		94.25		103.70	11.78	12.96
L.H., Daily Totals		$253.45		$367.70	$31.68	$45.96

Crew B-10B	Hr.	Daily	Hr.	Daily	Bare Costs	Incl. O&P
Equip. Oper. (med.)	$19.90	$159.20	$33.00	$264.00	$19.90	$33.00
Dozer, 200 H.P.		806.80		887.50	100.85	110.94
L.H., Daily Totals		$966.00		$1151.50	$120.75	$143.94

Crew B-10C	Bare Costs Hr.	Bare Costs Daily	Incl. Subs O & P Hr.	Incl. Subs O & P Daily	Cost Per Labor-Hour Bare Costs	Cost Per Labor-Hour Incl. O&P
1 Equip. Oper. (med.)	$19.90	$159.20	$33.00	$264.00	$19.90	$33.00
1 Dozer, 200 H.P.		806.80		887.50		
1 Vibratory Roller, Towed		110.00		121.00	114.60	126.06
8 L.H., Daily Totals		$1076.00		$1272.50	$134.50	$159.06

Crew B-10D	Hr.	Daily	Hr.	Daily	Bare Costs	Incl. O&P
1 Equip. Oper. (med.)	$19.90	$159.20	$33.00	$264.00	$19.90	$33.00
1 Dozer, 200 H.P		806.80		887.50		
1 Sheepsft. Roller, Towed		126.40		139.05	116.65	128.32
8 L.H., Daily Totals		$1092.40		$1290.55	$136.55	$161.32

Crew B-10E	Hr.	Daily	Hr.	Daily	Bare Costs	Incl. O&P
1 Equip. Oper. (med.)	$19.90	$159.20	$33.00	$264.00	$19.90	$33.00
1 Tandem Roller, 5 Ton		133.20		146.50	16.65	18.32
8 L.H., Daily Totals		$292.40		$410.50	$36.55	$51.32

Crew B-10F	Hr.	Daily	Hr.	Daily	Bare Costs	Incl. O&P
1 Equip. Oper. (med.)	$19.90	$159.20	$33.00	$264.00	$19.90	$33.00
1 Tandem Roller, 10 Ton		224.00		246.40	28.00	30.80
8 L.H., Daily Totals		$383.20		$510.40	$47.90	$63.80

Crew B-10G	Hr.	Daily	Hr.	Daily	Bare Costs	Incl. O&P
1 Equip. Oper. (med.)	$19.90	$159.20	$33.00	$264.00	$19.90	$33.00
1 Sheepsft. Roll., 130 H.P.		567.60		624.35	70.95	78.05
8 L.H., Daily Totals		$726.80		$888.35	$90.85	$111.05

Crew B-10H	Hr.	Daily	Hr.	Daily	Bare Costs	Incl. O&P
1 Equip. Oper. (med.)	$19.90	$159.20	$33.00	$264.00	$19.90	$33.00
1 Diaphr. Water Pump, 2″		29.40		32.35		
1-20 Ft. Suction Hose, 2″		6.50		7.15		
2-50 Ft. Disch. Hoses, 2″		10.80		11.90	5.84	6.42
8 L.H., Daily Totals		$205.90		$315.40	$25.74	$39.42

Crew B-10I	Hr.	Daily	Hr.	Daily	Bare Costs	Incl. O&P
1 Equip. Oper. (med.)	$19.90	$159.20	$33.00	$264.00	$19.90	$33.00
1 Diaphr. Water Pump, 4″		65.65		72.20		
1-20 Ft. Suction Hose, 4″		12.50		13.75		
2-50 Ft. Disch. Hoses, 4″		17.00		18.70	11.89	13.08
8 L.H., Daily Totals		$254.35		$368.65	$31.79	$46.08

Crew B-10J	Hr.	Daily	Hr.	Daily	Bare Costs	Incl. O&P
1 Equip. Oper. (med.)	$19.90	$159.20	$33.00	$264.00	$19.90	$33.00
1 Centr. Water Pump, 3″		38.00		41.80		
1-20 Ft. Suction Hose, 3″		9.50		10.45		
2-50 Ft. Disch. Hoses, 3″		12.80		14.10	7.54	8.29
8 L.H., Daily Totals		$219.50		$330.35	$27.44	$41.29

Crew B-10K	Hr.	Daily	Hr.	Daily	Bare Costs	Incl. O&P
1 Equip. Oper. (med.)	$19.90	$159.20	$33.00	$264.00	$19.90	$33.00
1 Centr. Water Pump, 6″		157.20		172.90		
1-20 Ft. Suction Hose, 6″		22.50		24.75		
2-50 Ft. Disch. Hoses, 6″		41.10		45.20	27.60	30.36
8 L.H., Daily Totals		$380.00		$506.85	$47.50	$63.36

Crew B-10L	Hr.	Daily	Hr.	Daily	Bare Costs	Incl. O&P
1 Equip. Oper. (med.)	$19.90	$159.20	$33.00	$264.00	$19.90	$33.00
1 Dozer, 75 H.P.		279.60		307.55	34.95	38.45
8 L.H., Daily Totals		$438.80		$571.55	$54.85	$71.45

Crews

Crew No.	Bare Costs		Incl. Subs O & P		Cost Per Labor-Hour	
Crew B-10M	**Hr.**	**Daily**	**Hr.**	**Daily**	**Bare Costs**	**Incl. O&P**
1 Equip. Oper. (med.)	$19.90	$159.20	$33.00	$264.00	$19.90	$33.00
1 Dozer, 300 H.P.		1108.00		1218.80	138.50	152.35
8 L.H., Daily Totals		$1267.20		$1482.80	$158.40	$185.35
Crew B-10N	**Hr.**	**Daily**	**Hr.**	**Daily**	**Bare Costs**	**Incl. O&P**
1 Equip. Oper. (med.)	$19.90	$159.20	$33.00	$264.00	$19.90	$33.00
1 F.E. Loader, T.M., 1.5 C.Y		329.95		362.95	41.24	45.37
8 L.H., Daily Totals		$489.15		$626.95	$61.14	$78.37
Crew B-10O	**Hr.**	**Daily**	**Hr.**	**Daily**	**Bare Costs**	**Incl. O&P**
1 Equip. Oper. (med.)	$19.90	$159.20	$33.00	$264.00	$19.90	$33.00
1 F.E. Loader, T.M., 2.25 C.Y.		450.00		495.00	56.25	61.88
8 L.H., Daily Totals		$609.20		$759.00	$76.15	$94.88
Crew B-10P	**Hr.**	**Daily**	**Hr.**	**Daily**	**Bare Costs**	**Incl. O&P**
1 Equip. Oper. (med.)	$19.90	$159.20	$33.00	$264.00	$19.90	$33.00
1 F.E. Loader, T.M., 2.5 C.Y.		784.00		862.40	98.00	107.80
8 L.H., Daily Totals		$943.20		$1126.40	$117.90	$140.80
Crew B-10Q	**Hr.**	**Daily**	**Hr.**	**Daily**	**Bare Costs**	**Incl. O&P**
1 Equip. Oper. (med.)	$19.90	$159.20	$33.00	$264.00	$19.90	$33.00
1 F.E. Loader, T.M., 5 C.Y.		1035.00		1138.50	129.38	142.31
8 L.H., Daily Totals		$1194.20		$1402.50	$149.28	$175.31
Crew B-10R	**Hr.**	**Daily**	**Hr.**	**Daily**	**Bare Costs**	**Incl. O&P**
1 Equip. Oper. (med.)	$19.90	$159.20	$33.00	$264.00	$19.90	$33.00
1 F.E. Loader, W.M., 1 C.Y.		233.60		256.95	29.20	32.12
8 L.H., Daily Totals		$392.80		$520.95	$49.10	$65.12
Crew B-10S	**Hr.**	**Daily**	**Hr.**	**Daily**	**Bare Costs**	**Incl. O&P**
1 Equip. Oper. (med.)	$19.90	$159.20	$33.00	$264.00	$19.90	$33.00
1 F.E. Loader, W.M., 1.5 C.Y.		303.00		333.30	37.88	41.66
8 L.H., Daily Totals		$462.20		$597.30	$57.78	$74.66
Crew B-10T	**Hr.**	**Daily**	**Hr.**	**Daily**	**Bare Costs**	**Incl. O&P**
1 Equip. Oper. (med.)	$19.90	$159.20	$33.00	$264.00	$19.90	$33.00
1 F.E. Ldr, W.M.,2.5CY		400.80		440.90	50.10	55.11
8 L.H., Daily Totals		$560.00		$704.90	$70.00	$88.11
Crew B-10U	**Hr.**	**Daily**	**Hr.**	**Daily**	**Bare Costs**	**Incl. O&P**
1 Equip. Oper. (med.)	$19.90	$159.20	$33.00	$264.00	$19.90	$33.00
1 F.E. Loader, W.M., 5.5 C.Y.		861.20		947.30	107.65	118.42
8 L.H., Daily Totals		$1020.40		$1211.30	$127.55	$151.42
Crew B-10V	**Hr.**	**Daily**	**Hr.**	**Daily**	**Bare Costs**	**Incl. O&P**
1 Equip. Oper. (med.)	$19.90	$159.20	$33.00	$264.00	$19.90	$33.00
1 Dozer, 700 H.P.		2752.00		3027.20	344.00	378.40
8 L.H., Daily Totals		$2911.20		$3291.20	$363.90	$411.40
Crew B-10W	**Hr.**	**Daily**	**Hr.**	**Daily**	**Bare Costs**	**Incl. O&P**
1 Equip. Oper. (med.)	$19.90	$159.20	$33.00	$264.00	$19.90	$33.00
1 Dozer, 105 H.P.		393.00		432.30	49.13	54.04
8 L.H., Daily Totals		$552.20		$696.30	$69.03	$87.04
Crew B-10X	**Hr.**	**Daily**	**Hr.**	**Daily**	**Bare Costs**	**Incl. O&P**
1 Equip. Oper. (med.)	$19.90	$159.20	$33.00	$264.00	$19.90	$33.00
1 Dozer, 410 H.P.		1397.00		1536.70	174.63	192.09
8 L.H., Daily Totals		$1556.20		$1800.70	$194.53	$225.09
Crew B-10Y	**Hr.**	**Daily**	**Hr.**	**Daily**	**Bare Costs**	**Incl. O&P**
1 Equip. Oper. (med.)	$19.90	$159.20	$33.00	$264.00	$19.90	$33.00
1 Vibratory Drum Roller		372.00		409.20	46.50	51.15
8 L.H., Daily Totals		$531.20		$673.20	$66.40	$84.15
Crew B-11	**Hr.**	**Daily**	**Hr.**	**Daily**	**Bare Costs**	**Incl. O&P**
1 Equipment Oper. (med.)	$19.90	$159.20	$33.00	$264.00	$17.17	$28.88
1 Laborer	14.45	115.60	24.75	198.00		
16 L.H., Daily Totals		$274.80		$462.00	$17.17	$28.88
Crew B-11A	**Hr.**	**Daily**	**Hr.**	**Daily**	**Bare Costs**	**Incl. O&P**
1 Equipment Oper. (med.)	$19.90	$159.20	$33.00	$264.00	$17.17	$28.88
1 Laborer	14.45	115.60	24.75	198.00		
1 Dozer, 200 H.P.		806.80		887.50	50.43	55.47
16 L.H., Daily Totals		$1081.60		$1349.50	$67.60	$84.35
Crew B-11B	**Hr.**	**Daily**	**Hr.**	**Daily**	**Bare Costs**	**Incl. O&P**
1 Equipment Oper. (med.)	$19.90	$159.20	$33.00	$264.00	$17.17	$28.88
1 Laborer	14.45	115.60	24.75	198.00		
1 Dozer, 200 H.P.		806.80		887.50		
1 Air Powered Tamper		15.80		17.40		
1 Air Compr. 365 C.F.M.		174.40		191.85		
2-50 Ft. Air Hoses, 1.5" Dia.		14.40		15.85	63.21	69.53
16 L.H., Daily Totals		$1286.20		$1574.60	$80.38	$98.41
Crew B-11C	**Hr.**	**Daily**	**Hr.**	**Daily**	**Bare Costs**	**Incl. O&P**
1 Equipment Oper. (med.)	$19.90	$159.20	$33.00	$264.00	$17.17	$28.88
1 Laborer	14.45	115.60	24.75	198.00		
1 Backhoe Loader, 48 H.P.		203.00		223.30	12.69	13.96
16 L.H., Daily Totals		$477.80		$685.30	$29.86	$42.84
Crew B-11K	**Hr.**	**Daily**	**Hr.**	**Daily**	**Bare Costs**	**Incl. O&P**
1 Equipment Oper. (med.)	$19.90	$159.20	$33.00	$264.00	$17.17	$28.88
1 Laborer	14.45	115.60	24.75	198.00		
1 Trencher, 8' D., 16" W.		540.90		595.00	33.81	37.19
16 L.H., Daily Totals		$815.70		$1057.00	$50.98	$66.07
Crew B-11L	**Hr.**	**Daily**	**Hr.**	**Daily**	**Bare Costs**	**Incl. O&P**
1 Equipment Oper. (med.)	$19.90	$159.20	$33.00	$264.00	$17.17	$28.88
1 Laborer	14.45	115.60	24.75	198.00		
1 Grader, 30,000 Lbs.		516.00		567.60	32.25	35.48
16 L.H., Daily Totals		$790.80		$1029.60	$49.42	$64.36
Crew B-11M	**Hr.**	**Daily**	**Hr.**	**Daily**	**Bare Costs**	**Incl. O&P**
1 Equipment Oper. (med.)	$19.90	$159.20	$33.00	$264.00	$17.17	$28.88
1 Laborer	14.45	115.60	24.75	198.00		
1 Backhoe Loader, 80 H.P.		273.15		300.45	17.07	18.78
16 L.H., Daily Totals		$547.95		$762.45	$34.24	$47.66
Crew B-12	**Hr.**	**Daily**	**Hr.**	**Daily**	**Bare Costs**	**Incl. O&P**
1 Equip. Oper. (crane)	$20.65	$165.20	$34.25	$274.00	$20.65	$34.25
8 L.H., Daily Totals		$165.20		$274.00	$20.65	$34.25
Crew B-12A	**Hr.**	**Daily**	**Hr.**	**Daily**	**Bare Costs**	**Incl. O&P**
1 Equip. Oper. (crane)	$20.65	$165.20	$34.25	$274.00	$20.65	$34.25
1 Hyd. Excavator, 1 C.Y.		538.00		591.80	67.25	73.98
8 L.H., Daily Totals		$703.20		$865.80	$87.90	$108.23

CREWS

Crew No.	Bare Costs		Incl. Subs O & P		Cost Per Labor-Hour	
Crew B-12B	**Hr.**	**Daily**	**Hr.**	**Daily**	**Bare Costs**	**Incl. O&P**
quip. Oper. (crane)	$20.65	$165.20	$34.25	$274.00	$20.65	$34.25
yd. Excavator, 1.5 C.Y.		713.20		784.50	89.15	98.07
.H., Daily Totals		$878.40		$1058.50	$109.80	$132.32

Crew B-12C	Hr.	Daily	Hr.	Daily	Bare Costs	Incl. O&P
quip. Oper. (crane)	$20.65	$165.20	$34.25	$274.00	$20.65	$34.25
yd. Excavator, 2 C.Y.		1017.00		1118.70	127.13	139.84
.H., Daily Totals		$1182.20		$1392.70	$147.78	$174.09

Crew B-12D	Hr.	Daily	Hr.	Daily	Bare Costs	Incl. O&P
quip. Oper. (crane)	$20.65	$165.20	$34.25	$274.00	$20.65	$34.25
yd. Excavator, 3.5 C.Y.		2126.00		2338.60	265.75	292.33
.H., Daily Totals		$2291.20		$2612.60	$286.40	$326.58

Crew B-12E	Hr.	Daily	Hr.	Daily	Bare Costs	Incl. O&P
quip. Oper. (crane)	$20.65	$165.20	$34.25	$274.00	$20.65	$34.25
yd. Excavator, .5 C.Y.		342.80		377.10	42.85	47.14
.H., Daily Totals		$508.00		$651.10	$63.50	$81.39

Crew B-12F	Hr.	Daily	Hr.	Daily	Bare Costs	Incl. O&P
Equip. Oper. (crane)	$20.65	$165.20	$34.25	$274.00	$20.65	$34.25
Hyd. Excavator, .75 C.Y.		455.80		501.40	56.98	62.67
.H., Daily Totals		$621.00		$775.40	$77.63	$96.92

Crew B-12G	Hr.	Daily	Hr.	Daily	Bare Costs	Incl. O&P
Equip. Oper. (crane)	$20.65	$165.20	$34.25	$274.00	$20.65	$34.25
Power Shovel, .5 C.Y.		477.00		524.70		
Clamshell Bucket, .5 C.Y		48.80		53.70	65.73	72.30
L.H., Daily Totals		$691.00		$852.40	$86.38	$106.55

Crew B-12H	Hr.	Daily	Hr.	Daily	Bare Costs	Incl. O&P
Equip. Oper. (crane)	$20.65	$165.20	$34.25	$274.00	$20.65	$34.25
Power Shovel, 1 C.Y.		516.80		568.50		
Clamshell Bucket, 1 C.Y.		70.00		77.00	73.35	80.69
L.H., Daily Totals		$752.00		$919.50	$94.00	$114.94

Crew B-12I	Hr.	Daily	Hr.	Daily	Bare Costs	Incl. O&P
Equip. Oper. (crane)	$20.65	$165.20	$34.25	$274.00	$20.65	$34.25
Power Shovel, .75 C.Y.		492.60		541.85		
Dragline Bucket, .75 C.Y.		33.20		36.50	65.73	72.30
L.H., Daily Totals		$691.00		$852.35	$86.38	$106.55

Crew B-12J	Hr.	Daily	Hr.	Daily	Bare Costs	Incl. O&P
Equip. Oper. (crane)	$20.65	$165.20	$34.25	$274.00	$20.65	$34.25
Gradall, 3 Ton, .5 C.Y.		618.20		680.00	77.28	85.00
L.H., Daily Totals		$783.40		$954.00	$97.93	$119.25

Crew B-12K	Hr.	Daily	Hr.	Daily	Bare Costs	Incl. O&P
Equip. Oper. (crane)	$20.65	$165.20	$34.25	$274.00	$20.65	$34.25
Gradall, 3 Ton, 1 C.Y.		817.85		899.65	102.23	112.45
L.H., Daily Totals		$983.05		$1173.65	$122.88	$146.70

Crew B-12L	Hr.	Daily	Hr.	Daily	Bare Costs	Incl. O&P
Equip. Oper. (crane)	$20.65	$165.20	$34.25	$274.00	$20.65	$34.25
Power Shovel, .5 C.Y.		477.00		524.70		
F.E. Attachment, .5 C.Y.		56.20		61.80	66.65	73.32
L.H., Daily Totals		$698.40		$860.50	$87.30	$107.57

Crew No.	Bare Costs		Incl. Subs O & P		Cost Per Labor-Hour	
Crew B-12M	**Hr.**	**Daily**	**Hr.**	**Daily**	**Bare Costs**	**Incl. O&P**
1 Equip. Oper. (crane)	$20.65	$165.20	$34.25	$274.00	$20.65	$34.25
1 Power Shovel, .75 C.Y		492.60		541.85		
1 F.E. Attachment, .75 C.Y.		101.20		111.30	74.23	81.65
8 L.H., Daily Totals		$759.00		$927.15	$94.88	$115.90

Crew B-12N	Hr.	Daily	Hr.	Daily	Bare Costs	Incl. O&P
1 Equip. Oper. (crane)	$20.65	$165.20	$34.25	$274.00	$20.65	$34.25
1 Power Shovel, 1 C.Y.		516.80		568.50		
1 F.E. Attachment, 1 C.Y.		133.20		146.50	81.25	89.38
8 L.H., Daily Totals		$815.20		$989.00	$101.90	$123.63

Crew B-12O	Hr.	Daily	Hr.	Daily	Bare Costs	Incl. O&P
1 Equip. Oper. (crane)	$20.65	$165.20	$34.25	$274.00	$20.65	$34.25
1 Power Shovel, 1.5 C.Y.		714.30		785.75		
1 F.E. Attachment, 1.5 C.Y.		161.00		177.10	109.41	120.35
8 L.H., Daily Totals		$1040.50		$1236.85	$130.06	$154.60

Crew B-12P	Hr.	Daily	Hr.	Daily	Bare Costs	Incl. O&P
1 Equip. Oper. (crane)	$20.65	$165.20	$34.25	$274.00	$20.65	$34.25
1 Crawler Crane, 40 Ton		714.30		785.75		
1 Dragline Bucket, 1.5 C.Y.		48.95		53.85	95.41	104.95
8 L.H., Daily Totals		$928.45		$1113.60	$116.06	$139.20

Crew B-12Q	Hr.	Daily	Hr.	Daily	Bare Costs	Incl. O&P
1 Equip. Oper. (crane)	$20.65	$165.20	$34.25	$274.00	$20.65	$34.25
1 Hyd. Excavator, 5/8 C.Y.		411.60		452.75	51.45	56.60
8 L.H., Daily Totals		$576.80		$726.75	$72.10	$90.85

Crew B-12R	Hr.	Daily	Hr.	Daily	Bare Costs	Incl. O&P
1 Equip. Oper. (crane)	$20.65	$165.20	$34.25	$274.00	$20.65	$34.25
1 Hyd. Excavator, 1.5 C.Y.		713.20		784.50	89.15	98.07
8 L.H., Daily Totals		$878.40		$1058.50	$109.80	$132.32

Crew B-12S	Hr.	Daily	Hr.	Daily	Bare Costs	Incl. O&P
1 Equip. Oper. (crane)	$20.65	$165.20	$34.25	$274.00	$20.65	$34.25
1 Hyd. Excavator, 2.5 C.Y.		1673.00		1840.30	209.13	230.04
8 L.H., Daily Totals		$1838.20		$2114.30	$229.78	$264.29

Crew B-12T	Hr.	Daily	Hr.	Daily	Bare Costs	Incl. O&P
1 Equip. Oper. (crane)	$20.65	$165.20	$34.25	$274.00	$20.65	$34.25
1 Crawler Crane, 75 Ton		939.20		1033.10		
1 F.E. Attachment, 3 C.Y.		300.40		330.45	154.95	170.45
8 L.H., Daily Totals		$1404.80		$1637.55	$175.60	$204.70

Crew B-12V	Hr.	Daily	Hr.	Daily	Bare Costs	Incl. O&P
1 Equip. Oper. (crane)	$20.65	$165.20	$34.25	$274.00	$20.65	$34.25
1 Crawler Crane, 75 Ton		939.20		1033.10		
1 Dragline Bucket, 3 C.Y.		81.60		89.75	127.60	140.36
8 L.H., Daily Totals		$1186.00		$1396.85	$148.25	$174.61

Crew B-13	Hr.	Daily	Hr.	Daily	Bare Costs	Incl. O&P
1 Labor Foreman (outside)	$16.45	$131.60	$28.20	$225.60	$15.82	$26.91
4 Laborers	14.45	462.40	24.75	792.00		
1 Equip. Oper. (crane)	20.65	165.20	34.25	274.00		
1 Hyd. Crane, 25 Ton		559.20		615.10	11.65	12.82
48 L.H., Daily Totals		$1318.40		$1906.70	$27.47	$39.73

Crew No.	Bare Costs		Incl. Subs O & P		Cost Per Labor-Hour	
Crew B-13A	**Hr.**	**Daily**	**Hr.**	**Daily**	**Bare Costs**	**Incl. O&P**
1 Foreman	$16.45	$131.60	$28.20	$225.60	$16.79	$28.26
2 Laborers	14.45	231.20	24.75	396.00		
2 Equipment Operator	19.90	318.40	33.00	528.00		
2 Truck Drivers (heavy)	16.20	259.20	27.05	432.80		
1 Crane, 75 Ton		939.20		1033.10		
1 F.E. Lder, 3.75 C.Y.		1035.00		1138.50		
2 Dump Trucks, 12 Ton		730.60		803.65	48.30	53.13
56 L.H., Daily Totals		$3645.20		$4557.65	$65.09	$81.39
Crew B-13B	**Hr.**	**Daily**	**Hr.**	**Daily**	**Bare Costs**	**Incl. O&P**
1 Labor Foreman (outside)	$16.45	$131.60	$28.20	$225.60	$15.99	$27.09
4 Laborers	14.45	462.40	24.75	792.00		
1 Equip. Oper. (crane)	20.65	165.20	34.25	274.00		
1 Equip. Oper. Oiler	17.00	136.00	28.20	225.60		
1 Hyd. Crane, 55 Ton		808.20		889.00	14.43	15.88
56 L.H., Daily Totals		$1703.40		$2406.20	$30.42	$42.97
Crew B-13C	**Hr.**	**Daily**	**Hr.**	**Daily**	**Bare Costs**	**Incl. O&P**
1 Labor Foreman (outside)	$16.45	$131.60	$28.20	$225.60	$15.99	$27.09
4 Laborers	14.45	462.40	24.75	792.00		
1 Equip. Oper. (crane)	20.65	165.20	34.25	274.00		
1 Equip. Oper. Oiler	17.00	136.00	28.20	225.60		
1 Crawler Crane, 100 Ton		1139.00		1252.90	20.34	22.37
56 L.H., Daily Totals		$2034.20		$2770.10	$36.33	$49.46
Crew B-14	**Hr.**	**Daily**	**Hr.**	**Daily**	**Bare Costs**	**Incl. O&P**
1 Labor Foreman (outside)	$16.45	$131.60	$28.20	$225.60	$15.56	$26.48
4 Laborers	14.45	462.40	24.75	792.00		
1 Equip. Oper. (light)	19.10	152.80	31.70	253.60		
1 Backhoe Loader, 48 H.P.		203.00		223.30	4.23	4.65
48 L.H., Daily Totals		$949.80		$1494.50	$19.79	$31.13
Crew B-15	**Hr.**	**Daily**	**Hr.**	**Daily**	**Bare Costs**	**Incl. O&P**
1 Equipment Oper. (med)	$19.90	$159.20	$33.00	$264.00	$17.01	$28.42
.5 Laborer	14.45	57.80	24.75	99.00		
2 Truck Drivers (heavy)	16.20	259.20	27.05	432.80		
2 Dump Trucks, 16 Ton		894.50		983.95		
1 Dozer, 200 H.P.		806.80		887.50	60.76	66.84
28 L.H., Daily Totals		$2177.50		$2667.25	$77.77	$95.26
Crew B-16	**Hr.**	**Daily**	**Hr.**	**Daily**	**Bare Costs**	**Incl. O&P**
1 Labor Foreman (outside)	$16.45	$131.60	$28.20	$225.60	$15.39	$26.19
2 Laborers	14.45	231.20	24.75	396.00		
1 Truck Driver (heavy)	16.20	129.60	27.05	216.40		
1 Dump Truck, 16 Ton		447.25		492.00	13.98	15.37
32 L.H., Daily Totals		$939.65		$1330.00	$29.37	$41.56
Crew B-17	**Hr.**	**Daily**	**Hr.**	**Daily**	**Bare Costs**	**Incl. O&P**
2 Laborers	$14.45	$231.20	$24.75	$396.00	$16.05	$27.06
1 Equip. Oper. (light)	19.10	152.80	31.70	253.60		
1 Truck Driver (heavy)	16.20	129.60	27.05	216.40		
1 Backhoe Loader, 48 H.P.		203.00		223.30		
1 Dump Truck, 12 Ton		365.30		401.85	17.76	19.54
32 L.H., Daily Totals		$1081.90		$1491.15	$33.81	$46.60
Crew B-18	**Hr.**	**Daily**	**Hr.**	**Daily**	**Bare Costs**	**Incl. O&P**
1 Labor Foreman (outside)	$16.45	$131.60	$28.20	$225.60	$15.12	$25.90
2 Laborers	14.45	231.20	24.75	396.00		
1 Vibrating Compactor		60.60		66.65	2.53	2.78
24 L.H., Daily Totals		$423.40		$688.25	$17.65	$28.68

Crew No.	Bare Costs		Incl. Subs O & P		Cost Per Labor-Ho	
Crew B-19	**Hr.**	**Daily**	**Hr.**	**Daily**	**Bare Costs**	**Incl. O&P**
1 Pile Driver Foreman	$21.45	$171.60	$40.00	$320.00	$19.19	$34.8
4 Pile Drivers	19.45	622.40	36.25	1160.00		
1 Equip. Oper. (crane)	20.65	165.20	34.25	274.00		
1 Building Laborer	14.45	115.60	24.75	198.00		
1 Crane, 40 Ton & Access.		714.30		785.75		
60 L.F. Leads, 15K Ft. Lbs.		264.00		290.40		
1 Hammer, 15K Ft. Lbs.		311.40		342.55		
1 Air Compr., 600 C.F.M.		278.00		305.80		
2-50 Ft. Air Hoses, 3" Dia.		33.80		37.20	28.60	31.4
56 L.H., Daily Totals		$2676.30		$3713.70	$47.79	$66.3
Crew B-19A	**Hr.**	**Daily**	**Hr.**	**Daily**	**Bare Costs**	**Incl. O&P**
1 Pile Driver Foreman	$21.45	$171.60	$40.00	$320.00	$19.69	$35.2
4 Pile Drivers	19.45	622.40	36.25	1160.00		
2 Equip. Oper. (crane)	20.65	330.40	34.25	548.00		
1 Equip. Oper. Oiler	17.00	136.00	28.20	225.60		
1 Crawler Crane, 75 Ton		939.20		1033.10		
60 L.F. Leads, 25K Ft. Lbs.		315.00		346.50		
1 Air Compressor, 750 CFM		300.80		330.90		
4-50 Ft. Air Hose, 3" Dia.		67.60		74.35	25.35	27.8
64 L.H., Daily Totals		$2883.00		$4038.45	$45.04	$63.1
Crew B-20	**Hr.**	**Daily**	**Hr.**	**Daily**	**Bare Costs**	**Incl. O&P**
1 Labor Foreman (out)	$16.45	$131.60	$28.20	$225.60	$15.12	$25.90
2 Laborer	14.45	231.20	24.75	396.00		
24 L.H., Daily Totals		$362.80		$621.60	$15.12	$25.90
Crew B-20A	**Hr.**	**Daily**	**Hr.**	**Daily**	**Bare Costs**	**Incl. O&P**
1 Labor Foreman	$16.45	$131.60	$28.20	$225.60	$17.60	$29.56
1 Laborer	14.45	115.60	24.75	198.00		
1 Plumber	21.95	175.60	36.30	290.40		
1 Plumber Apprentice	17.55	140.40	29.00	232.00		
32 L.H., Daily Totals		$563.20		$946.00	$17.60	$29.56
Crew B-21	**Hr.**	**Daily**	**Hr.**	**Daily**	**Bare Costs**	**Incl. O&P**
1 Labor Foreman (out)	$16.45	$131.60	$28.20	$225.60	$15.91	$27.09
2 Laborer	14.45	231.20	24.75	396.00		
.5 Equip. Oper. (crane)	20.65	82.60	34.25	137.00		
.5 S.P. Crane, 5 Ton		138.23		152.05	4.94	5.43
28 L.H., Daily Totals		$583.63		$910.65	$20.85	$32.52
Crew B-21A	**Hr.**	**Daily**	**Hr.**	**Daily**	**Bare Costs**	**Incl. O&P**
1 Labor Foreman	$16.45	$131.60	$28.20	$225.60	$18.21	$30.50
1 Laborer	14.45	115.60	24.75	198.00		
1 Plumber	21.95	175.60	36.30	290.40		
1 Plumber Apprentice	17.55	140.40	29.00	232.00		
1 Equip. Oper. (crane)	20.65	165.20	34.25	274.00		
1 S.P. Crane, 12 Ton		401.35		441.50	10.03	11.04
40 L.H., Daily Totals		$1129.75		$1661.50	$28.24	$41.54
Crew B-22	**Hr.**	**Daily**	**Hr.**	**Daily**	**Bare Costs**	**Incl. O&P**
1 Labor Foreman (out)	$16.45	$131.60	$28.20	$225.60	$16.22	$27.57
2 Laborer	14.45	231.20	24.75	396.00		
.75 Equip. Oper. (crane)	20.65	123.90	34.25	205.50		
.75 S.P. Crane, 5 Ton		207.34		228.05	6.91	7.60
30 L.H., Daily Totals		$694.04		$1055.15	$23.13	$35.17

Crew No.	Bare Costs		Incl. Subs O & P		Cost Per Labor-Hour	
Crew B-22A	**Hr.**	**Daily**	**Hr.**	**Daily**	**Bare Costs**	**Incl. O&P**
abor Foreman (out)	$16.45	$131.60	$28.20	$225.60	$16.97	$28.90
killed Worker	19.75	158.00	33.90	271.20		
aborers	14.45	231.20	24.75	396.00		
Equipment Oper. (crane)	20.65	123.90	34.25	205.50		
Crane, 5 Ton		207.34		228.05		
enerator, 5 KW		46.90		51.60		
utt Fusion Machine		188.00		206.80	11.64	12.80
L.H., Daily Totals		$1086.94		$1584.75	$28.61	$41.70
Crew B-22B	**Hr.**	**Daily**	**Hr.**	**Daily**	**Bare Costs**	**Incl. O&P**
killed Worker	$19.75	$158.00	$33.90	$271.20	$17.10	$29.33
aborer	14.45	115.60	24.75	198.00		
lectro Fusion Machine		86.00		94.60	5.38	5.91
L.H., Daily Totals		$359.60		$563.80	$22.48	$35.24
Crew B-23	**Hr.**	**Daily**	**Hr.**	**Daily**	**Bare Costs**	**Incl. O&P**
abor Foreman (outside)	$16.45	$131.60	$28.20	$225.60	$14.85	$25.44
aborers	14.45	462.40	24.75	792.00		
rill Rig, Wells		1796.00		1975.60		
ight Truck, 3 Ton		180.45		198.50	49.41	54.35
L.H., Daily Totals		$2570.45		$3191.70	$64.26	$79.79
Crew B-23A	**Hr.**	**Daily**	**Hr.**	**Daily**	**Bare Costs**	**Incl. O&P**
abor Foreman (outside)	$16.45	$131.60	$28.20	$225.60	$16.93	$28.65
aborers	14.45	115.60	24.75	198.00		
quip. Operator, medium	19.90	159.20	33.00	264.00		
Drill Rig, Wells		1796.00		1975.60		
Pickup Truck, 3/4 Ton		130.20		143.20	80.26	88.28
L.H., Daily Totals		$2332.60		$2806.40	$97.19	$116.93
Crew B-23B	**Hr.**	**Daily**	**Hr.**	**Daily**	**Bare Costs**	**Incl. O&P**
Labor Foreman (outside)	$16.45	$131.60	$28.20	$225.60	$16.93	$28.65
Laborer	14.45	115.60	24.75	198.00		
Equip. Operator, medium	19.90	159.20	33.00	264.00		
Drill Rig, Wells		1796.00		1975.60		
Pickup Truck, 3/4 Ton		130.20		143.20		
Pump, Cntfgl, 6"		157.20		172.90	86.81	95.49
L.H., Daily Totals		$2489.80		$2979.30	$103.74	$124.14
Crew B-24	**Hr.**	**Daily**	**Hr.**	**Daily**	**Bare Costs**	**Incl. O&P**
Cement Finisher	$18.90	$151.20	$30.80	$246.40	$17.68	$29.77
Laborer	14.45	115.60	24.75	198.00		
Carpenter	19.70	157.60	33.75	270.00		
L.H., Daily Totals		$424.40		$714.40	$17.68	$29.77
Crew B-25	**Hr.**	**Daily**	**Hr.**	**Daily**	**Bare Costs**	**Incl. O&P**
Labor Foreman	$16.45	$131.60	$28.20	$225.60	$16.12	$27.31
Laborers	14.45	809.20	24.75	1386.00		
Equip. Oper. (med.)	19.90	477.60	33.00	792.00		
Asphalt Paver, 130 H.P		1309.00		1439.90		
Tandem Roller, 10 Ton		224.00		246.40		
Roller, Pneumatic Wheel		240.80		264.90	20.16	22.17
8 L.H., Daily Totals		$3192.20		$4354.80	$36.28	$49.48

Crew No.	Bare Costs		Incl. Subs O & P		Cost Per Labor-Hour	
Crew B-25B	**Hr.**	**Daily**	**Hr.**	**Daily**	**Bare Costs**	**Incl. O&P**
1 Labor Foreman	$16.45	$131.60	$28.20	$225.60	$16.43	$27.79
7 Laborers	14.45	809.20	24.75	1386.00		
4 Equip. Oper. (medium)	19.90	636.80	33.00	1056.00		
1 Asphalt Paver, 130 H.P.		1309.00		1439.90		
2 Rollers, Steel Wheel		448.00		492.80		
1 Roller, Pneumatic Wheel		240.80		264.90	20.81	22.89
96 L.H., Daily Totals		$3575.40		$4865.20	$37.24	$50.68
Crew B-26	**Hr.**	**Daily**	**Hr.**	**Daily**	**Bare Costs**	**Incl. O&P**
1 Labor Foreman (outside)	$16.45	$131.60	$28.20	$225.60	$16.63	$28.43
6 Laborers	14.45	693.60	24.75	1188.00		
2 Equip. Oper. (med.)	19.90	318.40	33.00	528.00		
1 Rodman (reinf.)	21.10	168.80	39.25	314.00		
1 Cement Finisher	18.90	151.20	30.80	246.40		
1 Grader, 30,000 Lbs.		516.00		567.60		
1 Paving Mach. & Equip.		1324.00		1456.40	20.91	23.00
88 L.H., Daily Totals		$3303.60		$4526.00	$37.54	$51.43
Crew B-27	**Hr.**	**Daily**	**Hr.**	**Daily**	**Bare Costs**	**Incl. O&P**
1 Labor Foreman (outside)	$16.45	$131.60	$28.20	$225.60	$14.95	$25.61
3 Laborers	14.45	346.80	24.75	594.00		
1 Berm Machine		73.00		80.30	2.28	2.51
32 L.H., Daily Totals		$551.40		$899.90	$17.23	$28.12
Crew B-28	**Hr.**	**Daily**	**Hr.**	**Daily**	**Bare Costs**	**Incl. O&P**
2 Carpenters	$19.70	$315.20	$33.75	$540.00	$17.95	$30.75
1 Laborer	14.45	115.60	24.75	198.00		
24 L.H., Daily Totals		$430.80		$738.00	$17.95	$30.75
Crew B-29	**Hr.**	**Daily**	**Hr.**	**Daily**	**Bare Costs**	**Incl. O&P**
1 Labor Foreman (outside)	$16.45	$131.60	$28.20	$225.60	$15.82	$26.91
4 Laborers	14.45	462.40	24.75	792.00		
1 Equip. Oper. (crane)	20.65	165.20	34.25	274.00		
1 Gradall, 3 Ton, 1/2 C.Y.		618.20		680.00	12.88	14.17
48 L.H., Daily Totals		$1377.40		$1971.60	$28.70	$41.08
Crew B-30	**Hr.**	**Daily**	**Hr.**	**Daily**	**Bare Costs**	**Incl. O&P**
1 Equip. Oper. (med.)	$19.90	$159.20	$33.00	$264.00	$17.43	$29.03
2 Truck Drivers (heavy)	16.20	259.20	27.05	432.80		
1 Hyd. Excavator, 1.5 C.Y.		713.20		784.50		
2 Dump Trucks, 16 Ton		894.50		983.95	66.99	73.69
24 L.H., Daily Totals		$2026.10		$2465.25	$84.42	$102.72
Crew B-31	**Hr.**	**Daily**	**Hr.**	**Daily**	**Bare Costs**	**Incl. O&P**
1 Labor Foreman (outside)	$16.45	$131.60	$28.20	$225.60	$14.85	$25.44
4 Laborers	14.45	462.40	24.75	792.00		
1 Air Compr., 250 C.F.M.		127.00		139.70		
1 Sheeting Driver		12.20		13.40		
2-50 Ft. Air Hoses, 1.5" Dia.		14.40		15.85	3.84	4.22
40 L.H., Daily Totals		$747.60		$1186.55	$18.69	$29.66
Crew B-32	**Hr.**	**Daily**	**Hr.**	**Daily**	**Bare Costs**	**Incl. O&P**
1 Laborer	$14.45	$115.60	$24.75	$198.00	$18.54	$30.94
3 Equip. Oper. (med.)	19.90	477.60	33.00	792.00		
1 Grader, 30,000 Lbs.		516.00		567.60		
1 Tandem Roller, 10 Ton		224.00		246.40		
1 Dozer, 200 H.P.		806.80		887.50	48.34	53.17
32 L.H., Daily Totals		$2140.00		$2691.50	$66.88	$84.11

Crew No.	Bare Costs		Incl. Subs O & P		Cost Per Labor-Hour	
Crew B-32A	**Hr.**	**Daily**	**Hr.**	**Daily**	**Bare Costs**	**Incl. O&P**
1 Laborer	$14.45	$115.60	$24.75	$198.00	$18.08	$30.25
2 Equip. Oper. (medium)	19.90	318.40	33.00	528.00		
1 Grader, 30,000 Lbs.		516.00		567.60		
1 Roller, Vibratory, 29,000 Lbs.		432.40		475.65	39.52	43.47
24 L.H., Daily Totals		$1382.40		$1769.25	$57.60	$73.72
Crew B-32B	**Hr.**	**Daily**	**Hr.**	**Daily**	**Bare Costs**	**Incl. O&P**
1 Laborer	$14.45	$115.60	$24.75	$198.00	$18.08	$30.25
2 Equip. Oper. (medium)	19.90	318.40	33.00	528.00		
1 Dozer, 200 H.P.		806.80		887.50		
1 Roller, Vibratory, 29,000 Lbs.		432.40		475.65	51.63	56.80
24 L.H., Daily Totals		$1673.20		$2089.15	$69.71	$87.05
Crew B-32C	**Hr.**	**Daily**	**Hr.**	**Daily**	**Bare Costs**	**Incl. O&P**
1 Labor Foreman	$16.45	$131.60	$28.20	$225.60	$17.51	$29.45
2 Laborers	14.45	231.20	24.75	396.00		
3 Equip. Oper. (medium)	19.90	477.60	33.00	792.00		
1 Grader, 30,000 Lbs.		516.00		567.60		
1 Roller, Steel Wheel		224.00		246.40		
1 Dozer, 200 H.P.		806.80		887.50	32.23	35.45
48 L.H., Daily Totals		$2387.20		$3115.10	$49.74	$64.90
Crew B-33	**Hr.**	**Daily**	**Hr.**	**Daily**	**Bare Costs**	**Incl. O&P**
1 Equip. Oper. (med.)	$19.90	$159.20	$33.00	$264.00	$19.90	$33.00
.25 Equip. Oper. (med.)	19.90	39.80	33.00	66.00		
10 L.H., Daily Totals		$199.00		$330.00	$19.90	$33.00
Crew B-33A	**Hr.**	**Daily**	**Hr.**	**Daily**	**Bare Costs**	**Incl. O&P**
1 Equip. Oper. (med.)	$19.90	$159.20	$33.00	$264.00	$19.90	$33.00
.25 Equip. Oper. (med.)	19.90	39.80	33.00	66.00		
1 Scraper, Towed, 7 C.Y.		81.00		89.10		
1.25 Dozer, 300 H.P.		1385.00		1523.50	146.60	161.26
10 L.H., Daily Totals		$1665.00		$1942.60	$166.50	$194.26
Crew B-33B	**Hr.**	**Daily**	**Hr.**	**Daily**	**Bare Costs**	**Incl. O&P**
1 Equip. Oper. (med.)	$19.90	$159.20	$33.00	$264.00	$19.90	$33.00
.25 Equip. Oper. (med.)	19.90	39.80	33.00	66.00		
1 Scraper, Towed, 10 C.Y.		204.80		225.30		
1.25 Dozer, 300 H.P.		1385.00		1523.50	158.98	174.88
10 L.H., Daily Totals		$1788.80		$2078.80	$178.88	$207.88
Crew B-33C	**Hr.**	**Daily**	**Hr.**	**Daily**	**Bare Costs**	**Incl. O&P**
1 Equip. Oper. (med.)	$19.90	$159.20	$33.00	$264.00	$19.90	$33.00
.25 Equip. Oper. (med.)	19.90	39.80	33.00	66.00		
1 Scraper, Towed, 12 C.Y.		204.80		225.30		
1.25 Dozer, 300 H.P.		1385.00		1523.50	158.98	174.88
10 L.H., Daily Totals		$1788.80		$2078.80	$178.88	$207.88
Crew B-33D	**Hr.**	**Daily**	**Hr.**	**Daily**	**Bare Costs**	**Incl. O&P**
1 Equip. Oper. (med.)	$19.90	$159.20	$33.00	$264.00	$19.90	$33.00
.25 Equip. Oper. (med.)	19.90	39.80	33.00	66.00		
1 S.P. Scraper, 14 C.Y.		1626.00		1788.60		
.25 Dozer, 300 H.P.		277.00		304.70	190.30	209.33
10 L.H., Daily Totals		$2102.00		$2423.30	$210.20	$242.33

Crew No.	Bare Costs		Incl. Subs O & P		Cost Per Labor-Ho	
Crew B-33E	**Hr.**	**Daily**	**Hr.**	**Daily**	**Bare Costs**	**Incl. O&P**
1 Equip. Oper. (med.)	$19.90	$159.20	$33.00	$264.00	$19.90	$33.0
.25 Equip. Oper. (med.)	19.90	39.80	33.00	66.00		
1 S.P. Scraper, 24 C.Y.		1913.00		2104.30		
.25 Dozer, 300 H.P.		277.00		304.70	219.00	240.9
10 L.H., Daily Totals		$2389.00		$2739.00	$238.90	$273.9
Crew B-33F	**Hr.**	**Daily**	**Hr.**	**Daily**	**Bare Costs**	**Incl. O&P**
1 Equip. Oper. (med.)	$19.90	$159.20	$33.00	$264.00	$19.90	$33.0
.25 Equip. Oper. (med.)	19.90	39.80	33.00	66.00		
1 Elev. Scraper, 11 C.Y.		698.55		768.40		
.25 Dozer, 300 H.P.		277.00		304.70	97.56	107.3
10 L.H., Daily Totals		$1174.55		$1403.10	$117.46	$140.3
Crew B-33G	**Hr.**	**Daily**	**Hr.**	**Daily**	**Bare Costs**	**Incl. O&P**
1 Equip. Oper. (med.)	$19.90	$159.20	$33.00	$264.00	$19.90	$33.0
.25 Equip. Oper. (med.)	19.90	39.80	33.00	66.00		
1 Elev. Scraper, 20 C.Y.		984.80		1083.30		
.25 Dozer, 300 H.P.		277.00		304.70	126.18	138.8
10 L.H., Daily Totals		$1460.80		$1718.00	$146.08	$171.8
Crew B-34A	**Hr.**	**Daily**	**Hr.**	**Daily**	**Bare Costs**	**Incl. O&P**
1 Truck Driver (heavy)	$16.20	$129.60	$27.05	$216.40	$16.20	$27.05
1 Dump Truck, 12 Ton		365.30		401.85	45.66	50.23
8 L.H., Daily Totals		$494.90		$618.25	$61.86	$77.28
Crew B-34B	**Hr.**	**Daily**	**Hr.**	**Daily**	**Bare Costs**	**Incl. O&P**
1 Truck Driver (heavy)	$16.20	$129.60	$27.05	$216.40	$16.20	$27.05
1 Dump Truck, 16 Ton		447.25		492.00	55.91	61.50
8 L.H., Daily Totals		$576.85		$708.40	$72.11	$88.55
Crew B-34C	**Hr.**	**Daily**	**Hr.**	**Daily**	**Bare Costs**	**Incl. O&P**
1 Truck Driver (heavy)	$16.20	$129.60	$27.05	$216.40	$16.20	$27.05
1 Truck Tractor, 40 Ton		427.60		470.35		
1 Dump Trailer, 16.5 C.Y.		133.60		146.95	70.15	77.17
8 L.H., Daily Totals		$690.80		$833.70	$86.35	$104.22
Crew B-34D	**Hr.**	**Daily**	**Hr.**	**Daily**	**Bare Costs**	**Incl. O&P**
1 Truck Driver (heavy)	$16.20	$129.60	$27.05	$216.40	$16.20	$27.05
1 Truck Tractor, 40 Ton		427.60		470.35		
1 Dump Trailer, 20 C.Y.		136.10		149.70	70.46	77.51
8 L.H., Daily Totals		$693.30		$836.45	$86.66	$104.56
Crew B-34E	**Hr.**	**Daily**	**Hr.**	**Daily**	**Bare Costs**	**Incl. O&P**
1 Truck Driver (heavy)	$16.20	$129.60	$27.05	$216.40	$16.20	$27.05
1 Truck, Off Hwy., 25 Ton		700.00		770.00	87.50	96.25
8 L.H., Daily Totals		$829.60		$986.40	$103.70	$123.30
Crew B-34F	**Hr.**	**Daily**	**Hr.**	**Daily**	**Bare Costs**	**Incl. O&P**
1 Truck Driver (heavy)	$16.20	$129.60	$27.05	$216.40	$16.20	$27.05
1 Truck, Off Hwy., 22 C.Y.		1058.00		1163.80	132.25	145.48
8 L.H., Daily Totals		$1187.60		$1380.20	$148.45	$172.53
Crew B-34G	**Hr.**	**Daily**	**Hr.**	**Daily**	**Bare Costs**	**Incl. O&P**
1 Truck Driver (heavy)	$16.20	$129.60	$27.05	$216.40	$16.20	$27.05
1 Truck, Off Hwy., 34 C.Y.		1405.00		1545.50	175.63	193.19
8 L.H., Daily Totals		$1534.60		$1761.90	$191.83	$220.24

Crew B-34H	Bare Costs Hr.	Bare Costs Daily	Incl. Subs O & P Hr.	Incl. Subs O & P Daily	Cost Per Labor-Hour Bare Costs	Cost Per Labor-Hour Incl. O&P
ruck Driver (heavy)	$16.20	$129.60	$27.05	$216.40	$16.20	$27.05
ruck, Off Hwy., 42 C.Y.		1599.00		1758.90	199.88	219.86
.H., Daily Totals		$1728.60		$1975.30	$216.08	$246.91

Crew B-34J	Hr.	Daily	Hr.	Daily	Bare Costs	Incl. O&P
ruck Driver (heavy)	$16.20	$129.60	$27.05	$216.40	$16.20	$27.05
ruck, Off Hwy., 60 C.Y.		2240.00		2464.00	280.00	308.00
.H., Daily Totals		$2369.60		$2680.40	$296.20	$335.05

Crew B-34K	Hr.	Daily	Hr.	Daily	Bare Costs	Incl. O&P
ruck Driver (heavy)	$16.20	$129.60	$27.05	$216.40	$16.20	$27.05
ruck Tractor, 240 H.P.		531.20		584.30		
ow Bed Trailer		384.75		423.25	114.49	125.94
.H., Daily Totals		$1045.55		$1223.95	$130.69	$152.99

Crew B-35	Hr.	Daily	Hr.	Daily	Bare Costs	Incl. O&P
aborer Foreman (out)	$16.45	$131.60	$28.20	$225.60	$18.65	$31.48
Skilled Worker	19.75	158.00	33.90	271.20		
Velder (plumber)	21.95	175.60	36.30	290.40		
.aborer	14.45	115.60	24.75	198.00		
quip. Oper. (crane)	20.65	165.20	34.25	274.00		
Electric Welding Mach.		48.80		53.70		
Hyd. Excavator, .75 C.Y.		455.80		501.40	12.62	13.88
L.H., Daily Totals		$1250.60		$1814.30	$31.27	$45.36

Crew B-35A	Hr.	Daily	Hr.	Daily	Bare Costs	Incl. O&P
aborer Foreman (out)	$16.45	$131.60	$28.20	$225.60	$17.81	$30.05
aborers	14.45	231.20	24.75	396.00		
Skilled Worker	19.75	158.00	33.90	271.20		
Welder (plumber)	21.95	175.60	36.30	290.40		
Equip. Oper. (crane)	20.65	165.20	34.25	274.00		
Equip. Oper. Oiler	17.00	136.00	28.20	225.60		
Welder, 300 amp		84.00		92.40		
Crane, 75 Ton		939.20		1033.10	18.27	20.10
L.H., Daily Totals		$2020.80		$2808.30	$36.08	$50.15

Crew B-36	Hr.	Daily	Hr.	Daily	Bare Costs	Incl. O&P
Labor Foreman (outside)	$16.45	$131.60	$28.20	$225.60	$17.03	$28.74
Laborers	14.45	231.20	24.75	396.00		
Equip. Oper. (med.)	19.90	318.40	33.00	528.00		
Dozer, 200 H.P.		806.80		887.50		
Aggregate Spreader		73.80		81.20		
Tandem Roller, 10 Ton		224.00		246.40	27.62	30.38
) L.H., Daily Totals		$1785.80		$2364.70	$44.65	$59.12

Crew B-36A	Hr.	Daily	Hr.	Daily	Bare Costs	Incl. O&P
Labor Foreman (outside)	$16.45	$131.60	$28.20	$225.60	$17.85	$29.96
Laborers	14.45	231.20	24.75	396.00		
Equip. Oper. (med.)	19.90	636.80	33.00	1056.00		
Dozer, 200 H.P.		806.80		887.50		
Aggregate Spreader		73.80		81.20		
Roller, Steel Wheel		224.00		246.40		
Roller, Pneumatic Wheel		240.80		264.90	24.03	26.43
5 L.H., Daily Totals		$2345.00		$3157.60	$41.88	$56.39

Crew B-36B	Hr.	Daily	Hr.	Daily	Bare Costs	Incl. O&P
1 Labor Foreman (outside)	$16.45	$131.60	$28.20	$225.60	$17.64	$29.59
2 Laborers	14.45	231.20	24.75	396.00		
4 Equip. Oper. (medium)	19.90	636.80	33.00	1056.00		
1 Truck Driver, Heavy	16.20	129.60	27.05	216.40		
1 Grader, 30,000 Lbs.		516.00		567.60		
1 F.E. Loader, crl, 1.5 C.Y.		366.85		403.55		
1 Dozer, 300 H.P.		1108.00		1218.80		
1 Roller, Vibratory		432.40		475.65		
1 Truck, Tractor, 240 H.P.		531.20		584.30		
1 Water Tanker, 5000 Gal.		205.90		226.50	49.38	54.32
64 L.H., Daily Totals		$4289.55		$5370.40	$67.02	$83.91

Crew B-37	Hr.	Daily	Hr.	Daily	Bare Costs	Incl. O&P
1 Labor Foreman (outside)	$16.45	$131.60	$28.20	$225.60	$15.56	$26.48
4 Laborers	14.45	462.40	24.75	792.00		
1 Equip. Oper. (light)	19.10	152.80	31.70	253.60		
1 Tandem Roller, 5 Ton		133.20		146.50	2.78	3.05
48 L.H., Daily Totals		$880.00		$1417.70	$18.34	$29.53

Crew B-38	Hr.	Daily	Hr.	Daily	Bare Costs	Incl. O&P
2 Laborers	$14.45	$231.20	$24.75	$396.00	$16.00	$27.07
1 Equip. Oper. (light)	19.10	152.80	31.70	253.60		
1 Backhoe Loader, 48 H.P.		203.00		223.30		
1 Hyd.Hammer,(1200 lb)		213.60		234.95	17.36	19.09
24 L.H., Daily Totals		$800.60		$1107.85	$33.36	$46.16

Crew B-39	Hr.	Daily	Hr.	Daily	Bare Costs	Incl. O&P
1 Labor Foreman (outside)	$16.45	$131.60	$28.20	$225.60	$14.78	$25.32
5 Laborers	14.45	578.00	24.75	990.00		
1 Air Compr., 250 C.F.M.		127.00		139.70		
2 Air Tools & Accessories		39.00		42.90		
2-50 Ft. Air Hoses, 1.5" Dia.		14.40		15.85	3.76	4.13
48 L.H., Daily Totals		$890.00		$1414.05	$18.54	$29.45

Crew B-40	Hr.	Daily	Hr.	Daily	Bare Costs	Incl. O&P
1 Pile Driver Foreman (out)	$21.45	$171.60	$40.00	$320.00	$19.19	$34.86
4 Pile Drivers	19.45	622.40	36.25	1160.00		
1 Building Laborer	14.45	115.60	24.75	198.00		
1 Equip. Oper. (crane)	20.65	165.20	34.25	274.00		
1 Crane, 40 Ton		714.30		785.75		
1 Vibratory Hammer & Gen.		1218.00		1339.80	34.51	37.96
56 L.H., Daily Totals		$3007.10		$4077.55	$53.70	$72.82

Crew B-41	Hr.	Daily	Hr.	Daily	Bare Costs	Incl. O&P
1 Labor Foreman (outside)	$16.45	$131.60	$28.20	$225.60	$15.21	$25.97
4 Laborers	14.45	462.40	24.75	792.00		
.25 Equip. Oper. (crane)	20.65	41.30	34.25	68.50		
.25 Equip. Oper. Oiler	17.00	34.00	28.20	56.40		
.25 Crawler Crane, 40 Ton		178.57		196.45	4.06	4.46
44 L.H., Daily Totals		$847.87		$1338.95	$19.27	$30.43

Crew B-42	Hr.	Daily	Hr.	Daily	Bare Costs	Incl. O&P
1 Labor Foreman (outside)	$16.45	$131.60	$28.20	$225.60	$16.69	$28.25
4 Laborers	14.45	462.40	24.75	792.00		
1 Equip. Oper. (crane)	20.65	165.20	34.25	274.00		
1 Welder	21.95	175.60	36.30	290.40		
1 Hyd. Crane, 25 Ton		559.20		615.10		
1 Gas Welding Machine		84.00		92.40		
1 Horz. Boring Csg. Mch.		498.00		547.80	20.38	22.42
56 L.H., Daily Totals		$2076.00		$2837.30	$37.07	$50.67

Crew No.	Bare Costs		Incl. Subs O & P		Cost Per Labor-Hour	
Crew B-43	**Hr.**	**Daily**	**Hr.**	**Daily**	**Bare Costs**	**Incl. O&P**
1 Labor Foreman (outside)	$16.45	$131.60	$28.20	$225.60	$14.85	$25.44
4 Laborers	14.45	462.40	24.75	792.00		
1 Drill Rig & Augers		1796.00		1975.60	44.90	49.39
40 L.H., Daily Totals		$2390.00		$2993.20	$59.75	$74.83
Crew B-44	**Hr.**	**Daily**	**Hr.**	**Daily**	**Bare Costs**	**Incl. O&P**
1 Pile Driver Foreman	$21.45	$171.60	$40.00	$320.00	$18.60	$33.59
4 Pile Drivers	19.45	622.40	36.25	1160.00		
1 Equip. Oper. (crane)	20.65	165.20	34.25	274.00		
2 Laborer	14.45	231.20	24.75	396.00		
1 Crane, 40 Ton, & Access.		714.30		785.75		
45 L.F. Leads, 15K Ft. Lbs.		198.00		217.80	14.25	15.68
64 L.H., Daily Totals		$2102.70		$3153.55	$32.85	$49.27
Crew B-45	**Hr.**	**Daily**	**Hr.**	**Daily**	**Bare Costs**	**Incl. O&P**
1 Building Laborer	$14.45	$115.60	$24.75	$198.00	$15.33	$25.90
1 Truck Driver (heavy)	16.20	129.60	27.05	216.40		
1 Dist. Tank Truck, 3K Gal.		374.95		412.45	23.43	25.78
16 L.H., Daily Totals		$620.15		$826.85	$38.76	$51.68
Crew B-46	**Hr.**	**Daily**	**Hr.**	**Daily**	**Bare Costs**	**Incl. O&P**
1 Pile Driver Foreman	$21.45	$171.60	$40.00	$320.00	$17.28	$31.13
2 Pile Drivers	19.45	311.20	36.25	580.00		
3 Laborers	14.45	346.80	24.75	594.00		
1 Chain Saw, 36" Long		49.20		54.10	1.03	1.13
48 L.H., Daily Totals		$878.80		$1548.10	$18.31	$32.26
Crew B-47	**Hr.**	**Daily**	**Hr.**	**Daily**	**Bare Costs**	**Incl. O&P**
1 Blast Foreman	$16.45	$131.60	$28.20	$225.60	$15.45	$26.48
1 Driller	14.45	115.60	24.75	198.00		
1 Crawler Type Drill, 4"		312.55		343.80		
1 Air Compr., 600 C.F.M.		278.00		305.80		
2-50 Ft. Air Hoses, 3" Dia.		33.80		37.20	39.02	42.92
16 L.H., Daily Totals		$871.55		$1110.40	$54.47	$69.40
Crew B-47A	**Hr.**	**Daily**	**Hr.**	**Daily**	**Bare Costs**	**Incl. O&P**
1 Drilling Foreman	$16.45	$131.60	$28.20	$225.60	$18.03	$30.22
1 Equip. Oper. (heavy)	20.65	165.20	34.25	274.00		
1 Oiler	17.00	136.00	28.20	225.60		
1 Quarry Drill		486.05		534.65	20.25	22.28
24 L.H., Daily Totals		$918.85		$1259.85	$38.28	$52.50
Crew B-47C	**Hr.**	**Daily**	**Hr.**	**Daily**	**Bare Costs**	**Incl. O&P**
1 Laborer	$14.45	$115.60	$24.75	$198.00	$16.77	$28.23
1 Equip. Oper. (light)	19.10	152.80	31.70	253.60		
1 Air Compressor, 750 CFM		300.80		330.90		
2-50' Air Hose, 3"		33.80		37.20		
1 Air Track Drill, 4"		312.55		343.80	40.45	44.49
16 L.H., Daily Totals		$915.55		$1163.50	$57.22	$72.72
Crew B-47E	**Hr.**	**Daily**	**Hr.**	**Daily**	**Bare Costs**	**Incl. O&P**
1 Laborer Forman	$16.45	$131.60	$28.20	$225.60	$14.95	$25.61
3 Laborers	14.45	346.80	24.75	594.00		
1 Truck, Flatbed, 3 ton		180.45		198.50	5.64	6.20
32 L.H., Daily Totals		$658.85		$1018.10	$20.59	$31.81

Crew No.	Bare Costs		Incl. Subs O & P		Cost Per Labor-Ho	
Crew B-48	**Hr.**	**Daily**	**Hr.**	**Daily**	**Bare Costs**	**Incl. O&P**
1 Labor Foreman (outside)	$16.45	$131.60	$28.20	$225.60	$15.82	$26.9
4 Laborers	14.45	462.40	24.75	792.00		
1 Equip. Oper. (crane)	20.65	165.20	34.25	274.00		
1 Centr. Water Pump, 6"		157.20		172.90		
1-20 Ft. Suction Hose, 6"		22.50		24.75		
1-50 Ft. Disch. Hose, 6"		20.55		22.60		
1 Drill Rig & Augers		1796.00		1975.60	41.59	45.75
48 L.H., Daily Totals		$2755.45		$3487.45	$57.41	$72.66
Crew B-49	**Hr.**	**Daily**	**Hr.**	**Daily**	**Bare Costs**	**Incl. O&P**
1 Labor Foreman (outside)	$16.45	$131.60	$28.20	$225.60	$16.47	$28.74
5 Laborers	14.45	578.00	24.75	990.00		
1 Equip. Oper. (crane)	20.65	165.20	34.25	274.00		
2 Pile Drivers	19.45	311.20	36.25	580.00		
1 Hyd. Crane, 25 Ton		559.20		615.10		
1 Centr. Water Pump, 6"		157.20		172.90		
1-20 Ft. Suction Hose, 6"		22.50		24.75		
1-50 Ft. Disch. Hose, 6"		20.55		22.60		
1 Drill Rig & Augers		1796.00		1975.60	35.49	39.04
72 L.H., Daily Totals		$3741.45		$4880.55	$51.96	$67.78
Crew B-50	**Hr.**	**Daily**	**Hr.**	**Daily**	**Bare Costs**	**Incl. O&P**
1 Pile Driver Foremen	$21.45	$171.60	$40.00	$320.00	$17.77	$31.96
6 Pile Drivers	19.45	933.60	36.25	1740.00		
1 Equip. Oper. (crane)	20.65	165.20	34.25	274.00		
5 Laborers	14.45	578.00	24.75	990.00		
1 Crane, 40 Ton		714.30		785.75		
60 L.F. Leads, 15K Ft. Lbs.		264.00		290.40		
1 Hammer, 15K Ft. Lbs.		311.40		342.55		
1 Air Compr., 600 C.F.M.		278.00		305.80		
2-50 Ft. Air Hoses, 3" Dia.		33.80		37.20		
1 Chain Saw, 36" Long		49.20		54.10	15.87	17.46
104 L.H., Daily Totals		$3499.10		$5139.80	$33.64	$49.42
Crew B-51	**Hr.**	**Daily**	**Hr.**	**Daily**	**Bare Costs**	**Incl. O&P**
1 Labor Foreman (outside)	$16.45	$131.60	$28.20	$225.60	$15.02	$25.62
4 Laborers	14.45	462.40	24.75	792.00		
1 Truck Driver (light)	15.85	126.80	26.50	212.00		
1 Light Truck, 1.5 Ton		172.95		190.25	3.60	3.96
48 L.H., Daily Totals		$893.75		$1419.85	$18.62	$29.58
Crew B-52	**Hr.**	**Daily**	**Hr.**	**Daily**	**Bare Costs**	**Incl. O&P**
1 Labor Foreman	$16.45	$131.60	$28.20	$225.60	$16.35	$28.15
1 Carpenter	19.70	157.60	33.75	270.00		
4 Laborers	14.45	462.40	24.75	792.00		
.5 Rodman (reinf.)	21.10	84.40	39.25	157.00		
.5 Equip. Oper. (med.)	19.90	79.60	33.00	132.00		
.5 F.E. Ldr., T.M., 2.5 C.Y.		392.00		431.20	7.00	7.70
56 L.H., Daily Totals		$1307.60		$2007.80	$23.35	$35.85
Crew B-53	**Hr.**	**Daily**	**Hr.**	**Daily**	**Bare Costs**	**Incl. O&P**
1 Building Laborer	$14.45	$115.60	$24.75	$198.00	$14.45	$24.75
1 Trencher, Chain, 12 H.P.		102.50		112.75	12.81	14.09
8 L.H., Daily Totals		$218.10		$310.75	$27.26	$38.84
Crew B-54	**Hr.**	**Daily**	**Hr.**	**Daily**	**Bare Costs**	**Incl. O&P**
1 Equip. Oper. (light)	$19.10	$152.80	$31.70	$253.60	$19.10	$31.70
1 Trencher, Chain, 40 H.P.		210.90		232.00	26.36	29.00
8 L.H., Daily Totals		$363.70		$485.60	$45.46	$60.70

CREWS

Crew No.	Bare Costs		Incl. Subs O & P		Cost Per Labor-Hour	
Crew B-54A	**Hr.**	**Daily**	**Hr.**	**Daily**	**Bare Costs**	**Incl. O&P**
.abor Foreman (outside)	$16.45	$22.37	$28.20	$38.35	$19.40	$32.30
uipment Operator (med.)	19.90	159.20	33.00	264.00		
eel Trencher, 67HP		409.35		450.30	43.73	48.11
L.H., Daily Totals		$590.92		$752.65	$63.13	$80.41
Crew B-54B	**Hr.**	**Daily**	**Hr.**	**Daily**	**Bare Costs**	**Incl. O&P**
.abor Foreman (outsidc)	$16.45	$32.90	$28.20	$56.40	$19.21	$32.04
uipment Operator (med.)	19.90	159.20	33.00	264.00		
eel Trencher,150HP		565.20		621.70	56.52	62.17
.H., Daily Totals		$757.30		$942.10	$75.73	$94.21
Crew B-55	**Hr.**	**Daily**	**Hr.**	**Daily**	**Bare Costs**	**Incl. O&P**
borers	$14.45	$115.60	$24.75	$198.00	$15.15	$25.63
ıck Driver (light)	15.85	126.80	26.50	212.00		
ger, 4" to 36" Dia		483.00		531.30		
tbed 3 Ton Truck		180.45		198.50	41.47	45.61
.H., Daily Totals		$905.85		$1139.80	$56.62	$71.24
Crew B-56	**Hr.**	**Daily**	**Hr.**	**Daily**	**Bare Costs**	**Incl. O&P**
borer	$14.45	$231.20	$24.75	$396.00	$14.45	$24.75
awler Type Drill, 4"		312.55		343.80		
Compr., 600 C.F.M.		278.00		305.80		
Ft. Air Hose, 3" Dia.		16.90		18.60	37.97	41.76
.H., Daily Totals		$838.65		$1064.20	$52.42	$66.51
Crew B-57	**Hr.**	**Daily**	**Hr.**	**Daily**	**Bare Costs**	**Incl. O&P**
bor Foreman (outside)	$16.45	$131.60	$28.20	$225.60	$16.09	$27.34
borers	14.45	346.80	24.75	594.00		
uip. Oper. (crane)	20.65	165.20	34.25	274.00		
rge, 400 Ton		453.65		499.00		
wer Shovel, 1 C.Y.		516.80		568.50		
amshell Bucket, 1 C.Y.		70.00		77.00		
ntr. Water Pump, 6"		157.20		172.90		
Ft. Suction Hose, 6"		22.50		24.75		
0 Ft. Disch. Hoses, 6"		411.00		452.10	40.78	44.86
.H., Daily Totals		$2274.75		$2887.85	$56.87	$72.20
Crew B-58	**Hr.**	**Daily**	**Hr.**	**Daily**	**Bare Costs**	**Incl. O&P**
borers	$14.45	$231.20	$24.75	$396.00	$16.00	$27.07
uip. Oper. (light)	19.10	152.80	31.70	253.60		
ackhoe Loader, 48 H.P.		203.00		223.30		
nall Helicopter, w/pilot		3540.00		3894.00	155.96	171.55
.H., Daily Totals		$4127.00		$4766.90	$171.96	$198.62
Crew B-59	**Hr.**	**Daily**	**Hr.**	**Daily**	**Bare Costs**	**Incl. O&P**
uck Driver (heavy)	$16.20	$129.60	$27.05	$216.40	$16.20	$27.05
uck, 30 Ton		325.60		358.15		
ater tank, 5000 Gal.		205.90		226.50	66.44	73.08
.H., Daily Totals		$661.10		$801.05	$82.64	$100.13
Crew B-60	**Hr.**	**Daily**	**Hr.**	**Daily**	**Bare Costs**	**Incl. O&P**
abor Foreman (outside)	$16.45	$131.60	$28.20	$225.60	$16.59	$28.07
aborers	14.45	346.80	24.75	594.00		
quip. Oper. (crane)	20.65	165.20	34.25	274.00		
quip. Oper. (light)	19.10	152.80	31.70	253.60		
rawler Crane, 40 Ton		714.30		785.75		
L.F. Leads, 15K Ft. Lbs.		198.00		217.80		
ackhoe Loader, 48 H.P.		203.00		223.30	23.24	25.56
L.H., Daily Totals		$1911.70		$2574.05	$39.83	$53.63

Crew No.	Bare Costs		Incl. Subs O & P		Cost Per Labor-Hour	
Crew B-61	**Hr.**	**Daily**	**Hr.**	**Daily**	**Bare Costs**	**Incl. O&P**
1 Labor Foreman (outside)	$16.45	$131.60	$28.20	$225.60	$14.85	$25.44
4 Laborers	14.45	462.40	24.75	792.00		
1 Cement Mixer, 2 C.Y.		222.00		244.20		
1 Air Compr., 160 C.F.M.		98.40		108.25	8.01	8.81
40 L.H., Daily Totals		$914.40		$1370.05	$22.86	$34.25
Crew B-62	**Hr.**	**Daily**	**Hr.**	**Daily**	**Bare Costs**	**Incl. O&P**
2 Laborers	$14.45	$231.20	$24.75	$396.00	$16.00	$27.07
1 Equip. Oper. (light)	19.10	152.80	31.70	253.60		
1 Loader, Skid Steer		116.40		128.05	4.85	5.34
24 L.H., Daily Totals		$500.40		$777.65	$20.85	$32.41
Crew B-63	**Hr.**	**Daily**	**Hr.**	**Daily**	**Bare Costs**	**Incl. O&P**
5 Laborers	$14.45	$578.00	$24.75	$990.00	$14.45	$24.75
1 Loader, Skid Steer		116.40		128.05	2.91	3.20
40 L.H., Daily Totals		$694.40		$1118.05	$17.36	$27.95
Crew B-64	**Hr.**	**Daily**	**Hr.**	**Daily**	**Bare Costs**	**Incl. O&P**
1 Laborer	$14.45	$115.60	$24.75	$198.00	$15.15	$25.63
1 Truck Driver (light)	15.85	126.80	26.50	212.00		
1 Power Mulcher (small)		115.80		127.40		
1 Light Truck, 1.5 Ton		172.95		190.25	18.05	19.85
16 L.H., Daily Totals		$531.15		$727.65	$33.20	$45.48
Crew B-65	**Hr.**	**Daily**	**Hr.**	**Daily**	**Bare Costs**	**Incl. O&P**
1 Laborer	$14.45	$115.60	$24.75	$198.00	$15.15	$25.63
1 Truck Driver (light)	15.85	126.80	26.50	212.00		
1 Power Mulcher (large)		283.85		312.25		
1 Light Truck, 1.5 Ton		172.95		190.25	28.55	31.41
16 L.H., Daily Totals		$699.20		$912.50	$43.70	$57.04
Crew B-66	**Hr.**	**Daily**	**Hr.**	**Daily**	**Bare Costs**	**Incl. O&P**
1 Equip. Oper. (light)	$19.10	$152.80	$31.70	$253.60	$19.10	$31.70
1 Backhoe Ldr. w/Attchmt.		187.20		205.90	23.40	25.74
8 L.H., Daily Totals		$340.00		$459.50	$42.50	$57.44
Crew B-67	**Hr.**	**Daily**	**Hr.**	**Daily**	**Bare Costs**	**Incl. O&P**
1 Millwright	$20.60	$164.80	$33.60	$268.80	$19.85	$32.65
1 Equip. Oper. (light)	19.10	152.80	31.70	253.60		
1 Forklift		171.20		188.30	10.70	11.77
16 L.H., Daily Totals		$488.80		$710.70	$30.55	$44.42
Crew B-68	**Hr.**	**Daily**	**Hr.**	**Daily**	**Bare Costs**	**Incl. O&P**
2 Millwrights	$20.60	$329.60	$33.60	$537.60	$20.10	$32.97
1 Equip. Oper. (light)	19.10	152.80	31.70	253.60		
1 Forklift		171.20		188.30	7.13	7.85
24 L.H., Daily Totals		$653.60		$979.50	$27.23	$40.82
Crew B-69	**Hr.**	**Daily**	**Hr.**	**Daily**	**Bare Costs**	**Incl. O&P**
1 Labor Foreman (outside)	$16.45	$131.60	$28.20	$225.60	$16.24	$27.48
3 Laborers	14.45	346.80	24.75	594.00		
1 Equip Oper. (crane)	20.65	165.20	34.25	274.00		
1 Equip Oper. Oiler	17.00	136.00	28.20	225.60		
1 Truck Crane, 80 Ton		1098.00		1207.80	22.88	25.16
48 L.H., Daily Totals		$1877.60		$2527.00	$39.12	$52.64

CREWS

Crew B-69A	Bare Costs Hr.	Bare Costs Daily	Incl. Subs O & P Hr.	Incl. Subs O & P Daily	Cost Per Labor-Hour Bare Costs	Cost Per Labor-Hour Incl. O&P
1 Labor Foreman	$16.45	$131.60	$28.20	$225.60	$16.43	$27.71
3 Laborers	14.45	346.80	24.75	594.00		
1 Equip. Oper. (medium)	19.90	159.20	33.00	264.00		
1 Concrete Finisher	18.90	151.20	30.80	246.40		
1 Curb Paver		454.00		499.40	9.46	10.40
48 L.H., Daily Totals		$1242.80		$1829.40	$25.89	$38.11

Crew B-69B	Hr.	Daily	Hr.	Daily	Bare Costs	Incl. O&P
1 Labor Foreman	$16.45	$131.60	$28.20	$225.60	$16.43	$27.71
3 Laborers	14.45	346.80	24.75	594.00		
1 Equip. Oper. (medium)	19.90	159.20	33.00	264.00		
1 Cement Finisher	18.90	151.20	30.80	246.40		
1 Curb/Gutter Paver		911.85		1003.05	19.00	20.90
48 L.H., Daily Totals		$1700.65		$2333.05	$35.43	$48.61

Crew B-70	Hr.	Daily	Hr.	Daily	Bare Costs	Incl. O&P
1 Labor Foreman (outside)	$16.45	$131.60	$28.20	$225.60	$17.07	$28.78
3 Laborers	14.45	346.80	24.75	594.00		
3 Equip. Oper. (med.)	19.90	477.60	33.00	792.00		
1 Motor Grader, 30,000 Lb.		516.00		567.60		
1 Grader Attach., Ripper		64.00		70.40		
1 Road Sweeper, S.P.		218.40		240.25		
1 F.E. Loader, 1-3/4 C.Y.		303.00		333.30	19.67	21.63
56 L.H., Daily Totals		$2057.40		$2823.15	$36.74	$50.41

Crew B-71	Hr.	Daily	Hr.	Daily	Bare Costs	Incl. O&P
1 Labor Foreman (outside)	$16.45	$131.60	$28.20	$225.60	$17.07	$28.78
3 Laborers	14.45	346.80	24.75	594.00		
3 Equip. Oper. (med.)	19.90	477.60	33.00	792.00		
1 Pvmt. Profiler, 450 H.P.		3442.00		3786.20		
1 Road Sweeper, S.P.		218.40		240.25		
1 F.E. Loader, 1-3/4 C.Y.		303.00		333.30	70.78	77.85
56 L.H., Daily Totals		$4919.40		$5971.35	$87.85	$106.63

Crew B-72	Hr.	Daily	Hr.	Daily	Bare Costs	Incl. O&P
1 Labor Foreman (outside)	$16.45	$131.60	$28.20	$225.60	$17.42	$29.31
3 Laborers	14.45	346.80	24.75	594.00		
4 Equip. Oper. (med.)	19.90	636.80	33.00	1056.00		
1 Pvmt. Profiler, 450 H.P.		3442.00		3786.20		
1 Hammermill, 250 H.P.		1177.00		1294.70		
1 Windrow Loader		929.85		1022.85		
1 Mix Paver 165 H.P.		1440.00		1584.00		
1 Roller, Pneu. Tire, 12 T.		240.80		264.90	112.96	124.26
64 L.H., Daily Totals		$8344.85		$9828.25	$130.38	$153.57

Crew B-73	Hr.	Daily	Hr.	Daily	Bare Costs	Incl. O&P
1 Labor Foreman (outside)	$16.45	$131.60	$28.20	$225.60	$18.11	$30.34
2 Laborers	14.45	231.20	24.75	396.00		
5 Equip. Oper. (med.)	19.90	796.00	33.00	1320.00		
1 Road Mixer, 310 H.P.		1008.00		1108.80		
1 Roller, Tandem, 12 Ton		224.00		246.40		
1 Hammermill, 250 H.P.		1177.00		1294.70		
1 Motor Grader, 30,000 Lb.		516.00		567.60		
.5 F.E. Loader, 1-3/4 C.Y.		151.50		166.65		
.5 Truck, 30 Ton		162.80		179.10		
.5 Water Tank 5000 Gal.		102.95		113.25	52.22	57.44
64 L.H., Daily Totals		$4501.05		$5618.10	$70.33	$87.78

Crew B-74	Hr.	Daily	Hr.	Daily	Bare Costs	Incl. O&P
1 Labor Foreman (outside)	$16.45	$131.60	$28.20	$225.60	$17.86	$29.8
1 Laborer	14.45	115.60	24.75	198.00		
4 Equip. Oper. (med.)	19.90	636.80	33.00	1056.00		
2 Truck Drivers (heavy)	16.20	259.20	27.05	432.80		
1 Motor Grader, 30,000 Lb.		516.00		567.60		
1 Grader Attach., Ripper		64.00		70.40		
2 Stabilizers, 310 H.P.		1394.10		1533.50		
1 Flatbed Truck, 3 Ton		180.45		198.50		
1 Chem. Spreader, Towed		101.20		111.30		
1 Vibr. Roller, 29,000 Lb.		432.40		475.65		
1 Water Tank 5000 Gal.		205.90		226.50		
1 Truck, 30 Ton		325.60		358.15	50.31	55.3
64 L.H., Daily Totals		$4362.85		$5454.00	$68.17	$85.2

Crew B-75	Hr.	Daily	Hr.	Daily	Bare Costs	Incl. O&P
1 Labor Foreman (outside)	$16.45	$131.60	$28.20	$225.60	$18.10	$30.2
1 Laborer	14.45	115.60	24.75	198.00		
4 Equip. Oper. (med.)	19.90	636.80	33.00	1056.00		
1 Truck Driver (heavy)	16.20	129.60	27.05	216.40		
1 Motor Grader, 30,000 Lb.		516.00		567.60		
1 Grader Attach., Ripper		64.00		70.40		
2 Stabilizers, 310 H.P.		1394.10		1533.50		
1 Dist. Truck, 3000 Gal.		374.95		412.45		
1 Vibr. Roller, 29,000 Lb.		432.40		475.65	49.67	54.6
56 L.H., Daily Totals		$3795.05		$4755.60	$67.77	$84.9

Crew B-76	Hr.	Daily	Hr.	Daily	Bare Costs	Incl. O&P
1 Dock Builder Foreman	$21.45	$171.60	$40.00	$320.00	$19.67	$35.3
5 Dock Builders	19.45	778.00	36.25	1450.00		
2 Equip. Oper. (crane)	20.65	330.40	34.25	548.00		
1 Equip. Oper. Oiler	17.00	136.00	28.20	225.60		
1 Crawler Crane, 50 Ton		850.80		935.90		
1 Barge, 400 Ton		453.65		499.00		
1 Hammer, 15K Ft. Lbs.		311.40		342.55		
60 L.F. Leads, 15K Ft. Lbs.		264.00		290.40		
1 Air Compr., 600 C.F.M.		278.00		305.80		
2-50 Ft. Air Hoses, 3" Dia.		33.80		37.20	30.44	33.4
72 L.H., Daily Totals		$3607.65		$4954.45	$50.11	$68.8

Crew B-77	Hr.	Daily	Hr.	Daily	Bare Costs	Incl. O&P
1 Labor Foreman	$16.45	$131.60	$28.20	$225.60	$15.13	$25.7
3 Laborers	14.45	346.80	24.75	594.00		
1 Truck Driver (light)	15.85	126.80	26.50	212.00		
1 Crack Cleaner, 25 H.P.		76.95		84.65		
1 Crack Filler, Trailer Mtd.		145.20		159.70		
1 Flatbed Truck, 3 Ton		180.45		198.50	10.07	11.0
40 L.H., Daily Totals		$1007.80		$1474.45	$25.20	$36.8

Crew B-78	Hr.	Daily	Hr.	Daily	Bare Costs	Incl. O&P
1 Labor Foreman	$16.45	$131.60	$28.20	$225.60	$14.85	$25.4
4 Laborers	14.45	462.40	24.75	792.00		
1 Paint Striper, S.P.		209.00		229.90		
1 Flatbed Truck, 3 Ton		180.45		198.50		
1 Pickup Truck, 3/4 Ton		130.20		143.20	12.99	14.2
40 L.H., Daily Totals		$1113.65		$1589.20	$27.84	$39.7

rews

Crew B-79	Bare Costs Hr.	Bare Costs Daily	Incl. Subs O & P Hr.	Incl. Subs O & P Daily	Cost Per Labor-Hour Bare Costs	Cost Per Labor-Hour Incl. O&P
oor Foreman	$16.45	$131.60	$28.20	$225.60	$14.95	$25.61
oorers	14.45	346.80	24.75	594.00		
ermo. Striper, T.M.		245.05		269.55		
tbed Truck, 3 Ton		180.45		198.50		
kup Truck, 3/4 Ton		260.40		286.45	21.43	23.58
.H., Daily Totals		$1164.30		$1574.10	$36.38	$49.19

Crew B-80	Hr.	Daily	Hr.	Daily	Bare Costs	Incl. O&P
oor Foreman	$16.45	$131.60	$28.20	$225.60	$15.12	$25.90
oorer	14.45	231.20	24.75	396.00		
tbed Truck, 3 Ton		180.45		198.50		
nce Post Auger, TM		370.00		407.00	22.94	25.23
.H., Daily Totals		$913.25		$1227.10	$38.06	$51.13

Crew B-80A	Hr.	Daily	Hr.	Daily	Bare Costs	Incl. O&P
borers	$14.45	$346.80	$24.75	$594.00	$14.45	$24.75
tbed Truck, 3 Ton		180.45		198.50	7.52	8.27
.H., Daily Totals		$527.25		$792.50	$21.97	$33.02

Crew B-80B	Hr.	Daily	Hr.	Daily	Bare Costs	Incl. O&P
borers	$14.45	$346.80	$24.75	$594.00	$15.61	$26.49
uip. Oper. (light)	19.10	152.80	31.70	253.60		
ane, flatbed mnt		283.20		311.50	8.85	9.74
.H., Daily Totals		$782.80		$1159.10	$24.46	$36.23

Crew B-81	Hr.	Daily	Hr.	Daily	Bare Costs	Incl. O&P
borer	$14.45	$115.60	$24.75	$198.00	$15.33	$25.90
uck Driver (heavy)	16.20	129.60	27.05	216.40		
dromulcher, T.M.		301.75		331.95		
actor Truck, 4x2		325.60		358.15	39.21	43.13
.H., Daily Totals		$872.55		$1104.50	$54.54	$69.03

Crew B-82	Hr.	Daily	Hr.	Daily	Bare Costs	Incl. O&P
aborer	$14.45	$115.60	$24.75	$198.00	$16.77	$28.23
quip. Oper. (light)	19.10	152.80	31.70	253.60		
oriz. Borer, 6 H.P.		49.80		54.80	3.11	3.42
.H., Daily Totals		$318.20		$506.40	$19.88	$31.65

Crew B-83	Hr.	Daily	Hr.	Daily	Bare Costs	Incl. O&P
ugboat Captain	$19.90	$159.20	$33.00	$264.00	$17.17	$28.88
ugboat Hand	14.45	115.60	24.75	198.00		
ugboat, 250 H.P.		444.80		489.30	27.80	30.58
L.H., Daily Totals		$719.60		$951.30	$44.97	$59.46

Crew B-84	Hr.	Daily	Hr.	Daily	Bare Costs	Incl. O&P
quip. Oper. (med.)	$19.90	$159.20	$33.00	$264.00	$19.90	$33.00
otary Mower/Tractor		209.50		230.45	26.19	28.81
.H., Daily Totals		$368.70		$494.45	$46.09	$61.81

Crew B-85	Hr.	Daily	Hr.	Daily	Bare Costs	Incl. O&P
aborers	$14.45	$346.80	$24.75	$594.00	$15.89	$26.86
quip. Oper. (med.)	19.90	159.20	33.00	264.00		
ruck Driver (heavy)	16.20	129.60	27.05	216.40		
erial Lift Truck, 80'		569.60		626.55		
rush Chipper, 130 H.P.		211.05		232.15		
runing Saw, Rotary		22.25		24.50	20.07	22.08
L.H., Daily Totals		$1438.50		$1957.60	$35.96	$48.94

Crew B-86	Hr.	Daily	Hr.	Daily	Bare Costs	Incl. O&P
1 Equip. Oper. (med.)	$19.90	$159.20	$33.00	$264.00	$19.90	$33.00
1 Stump Chipper, S.P.		165.60		182.15	20.70	22.77
8 L.H., Daily Totals		$324.80		$446.15	$40.60	$55.77

Crew B-86A	Hr.	Daily	Hr.	Daily	Bare Costs	Incl. O&P
1 Equip. Oper. (medium)	$19.90	$159.20	$33.00	$264.00	$19.90	$33.00
1 Grader, 30,000 Lbs.		516.00		567.60	64.50	70.95
8 L.H., Daily Totals		$675.20		$831.60	$84.40	$103.95

Crew B-86B	Hr.	Daily	Hr.	Daily	Bare Costs	Incl. O&P
1 Equip. Oper. (medium)	$19.90	$159.20	$33.00	$264.00	$19.90	$33.00
1 Dozer, 200 H.P.		806.80		887.50	100.85	110.94
8 L.H., Daily Totals		$966.00		$1151.50	$120.75	$143.94

Crew B-87	Hr.	Daily	Hr.	Daily	Bare Costs	Incl. O&P
1 Laborer	$14.45	$115.60	$24.75	$198.00	$18.81	$31.35
4 Equip. Oper. (med.)	19.90	636.80	33.00	1056.00		
2 Feller Bunchers, 50 H.P.		868.80		955.70		
1 Log Chipper, 22" Tree		2017.00		2218.70		
1 Dozer, 105 H.P.		279.60		307.55		
1 Chainsaw, Gas, 36" Long		49.20		54.10	80.37	88.40
40 L.H., Daily Totals		$3967.00		$4790.05	$99.18	$119.75

Crew B-88	Hr.	Daily	Hr.	Daily	Bare Costs	Incl. O&P
1 Laborer	$14.45	$115.60	$24.75	$198.00	$19.12	$31.82
6 Equip. Oper. (med.)	19.90	955.20	33.00	1584.00		
2 Feller Bunchers, 50 H.P.		868.80		955.70		
1 Log Chipper, 22" Tree		2017.00		2218.70		
2 Log Skidders, 50 H.P.		784.00		862.40		
1 Dozer, 105 H.P.		279.60		307.55		
1 Chainsaw, Gas, 36" Long		49.20		54.10	71.40	78.54
56 L.H., Daily Totals		$5069.40		$6180.45	$90.52	$110.36

Crew B-89	Hr.	Daily	Hr.	Daily	Bare Costs	Incl. O&P
1 Skilled Worker	$19.75	$158.00	$33.90	$271.20	$17.10	$29.33
1 Building Laborer	14.45	115.60	24.75	198.00		
1 Cutting Machine		48.00		52.80	3.00	3.30
16 L.H., Daily Totals		$321.60		$522.00	$20.10	$32.63

Crew B-89A	Hr.	Daily	Hr.	Daily	Bare Costs	Incl. O&P
1 Skilled Worker	$19.75	$158.00	$33.90	$271.20	$17.10	$29.33
1 Laborer	14.45	115.60	24.75	198.00		
1 Core Drill (large)		60.60		66.65	3.79	4.17
16 L.H., Daily Totals		$334.20		$535.85	$20.89	$33.50

Crew B-89B	Hr.	Daily	Hr.	Daily	Bare Costs	Incl. O&P
1 Equip. Oper. (light)	$19.10	$152.80	$31.70	$253.60	$17.48	$29.10
1 Truck Driver, Light	15.85	126.80	26.50	212.00		
1 Wall Saw, Hydraulic, 10 H.P.		133.20		146.50		
1 Generator, Diesel, 100 KW		207.05		227.75		
1 Water Tank, 65 Gal.		13.00		14.30		
1 Flatbed Truck, 3 Ton		180.45		198.50	33.36	36.69
16 L.H., Daily Totals		$813.30		$1052.65	$50.84	$65.79

Crew No.	Bare Costs		Incl. Subs O & P		Cost Per Labor-Hour	
Crew B-90	**Hr.**	**Daily**	**Hr.**	**Daily**	**Bare Costs**	**Incl. O&P**
1 Labor Foreman (outside)	$16.45	$131.60	$28.20	$225.60	$16.30	$27.49
3 Laborers	14.45	346.80	24.75	594.00		
2 Equip. Oper. (light)	19.10	305.60	31.70	507.20		
2 Truck Drivers (heavy)	16.20	259.20	27.05	432.80		
1 Road Mixer, 310 H.P.		1008.00		1108.80		
1 Dist. Truck, 2000 Gal.		350.30		385.35	21.22	23.35
64 L.H., Daily Totals		$2401.50		$3253.75	$37.52	$50.84

Crew B-90A	Hr.	Daily	Hr.	Daily	Bare Costs	Incl. O&P
1 Labor Foreman	$16.45	$131.60	$28.20	$225.60	$17.85	$29.96
2 Laborers	14.45	231.20	24.75	396.00		
4 Equip. Oper. (medium)	19.90	636.80	33.00	1056.00		
2 Graders, 30,000 Lbs.		1032.00		1135.20		
1 Roller, Steel Wheel		224.00		246.40		
1 Roller, Pneumatic Wheel		240.80		264.90	26.73	29.40
56 L.H., Daily Totals		$2496.40		$3324.10	$44.58	$59.36

Crew B-90B	Hr.	Daily	Hr.	Daily	Bare Costs	Incl. O&P
1 Labor Foreman	$16.45	$131.60	$28.20	$225.60	$17.51	$29.45
2 Laborers	14.45	231.20	24.75	396.00		
3 Equip. Oper. (medium)	19.90	477.60	33.00	792.00		
1 Roller, Steel Wheel		224.00		246.40		
1 Roller, Pneumatic Wheel		240.80		264.90		
1 Road Mixer, 310 H.P.		1008.00		1108.80	30.68	33.75
48 L.H., Daily Totals		$2313.20		$3033.70	$48.19	$63.20

Crew B-91	Hr.	Daily	Hr.	Daily	Bare Costs	Incl. O&P
1 Labor Foreman (outside)	$16.45	$131.60	$28.20	$225.60	$17.64	$29.59
2 Laborers	14.45	231.20	24.75	396.00		
4 Equip. Oper. (med.)	19.90	636.80	33.00	1056.00		
1 Truck Driver (heavy)	16.20	129.60	27.05	216.40		
1 Dist. Truck, 3000 Gal.		374.95		412.45		
1 Aggreg. Spreader, S.P.		640.40		704.45		
1 Roller, Pneu. Tire, 12 Ton		240.80		264.90		
1 Roller, Steel, 10 Ton		224.00		246.40	23.13	25.44
64 L.H., Daily Totals		$2609.35		$3522.20	$40.77	$55.03

Crew B-92	Hr.	Daily	Hr.	Daily	Bare Costs	Incl. O&P
1 Labor Foreman (outside)	$16.45	$131.60	$28.20	$225.60	$14.95	$25.61
3 Laborers	14.45	346.80	24.75	594.00		
1 Crack Cleaner, 25 H.P.		76.95		84.65		
1 Air Compressor		78.30		86.15		
1 Tar Kettle, T.M.		23.20		25.50		
1 Flatbed Truck, 3 Ton		180.45		198.50	11.22	12.34
32 L.H., Daily Totals		$837.30		$1214.40	$26.17	$37.95

Crew B-93	Hr.	Daily	Hr.	Daily	Bare Costs	Incl. O&P
1 Equip. Oper. (med.)	$19.90	$159.20	$33.00	$264.00	$19.90	$33.00
1 Feller Buncher, 50 H.P.		434.40		477.85	54.30	59.73
8 L.H., Daily Totals		$593.60		$741.85	$74.20	$92.73

Crew B-94A	Hr.	Daily	Hr.	Daily	Bare Costs	Incl. O&P
1 Laborer	$14.45	$115.60	$24.75	$198.00	$14.45	$24.75
1 Diaph. Water Pump, 2"		29.40		32.35		
1-20 Ft. Suction Hose, 2"		6.50		7.15		
2-50 Ft. Disch. Hoses, 2"		10.80		11.90	5.84	6.42
8 L.H., Daily Totals		$162.30		$249.40	$20.29	$31.17

Crew No.	Bare Costs		Incl. Subs O & P		Cost Per Labor-Hour	
Crew B-94B	**Hr.**	**Daily**	**Hr.**	**Daily**	**Bare Costs**	**Incl. O&P**
1 Laborer	$14.45	$115.60	$24.75	$198.00	$14.45	$24.7
1 Diaph. Water Pump, 4"		65.65		72.20		
1-20 Ft. Suction Hose, 4"		12.50		13.75		
2-50 Ft. Disch. Hoses, 4"		17.00		18.70	11.89	13.0
8 L.H., Daily Totals		$210.75		$302.65	$26.34	$37.8

Crew B-94C	Hr.	Daily	Hr.	Daily	Bare Costs	Incl. O&P
1 Laborer	$14.45	$115.60	$24.75	$198.00	$14.45	$24.7
1 Centr. Water Pump, 3"		38.00		41.80		
1-20 Ft. Suction Hose, 3"		9.50		10.45		
2-50 Ft. Disch. Hoses, 3"		12.80		14.10	7.54	8.2
8 L.H., Daily Totals		$175.90		$264.35	$21.99	$33.0

Crew B-94D	Hr.	Daily	Hr.	Daily	Bare Costs	Incl. O&P
1 Laborer	$14.45	$115.60	$24.75	$198.00	$14.45	$24.7
1 Centr. Water Pump, 6"		157.20		172.90		
1-20 Ft. Suction Hose, 6"		22.50		24.75		
2-50 Ft. Disch. Hoses, 6"		41.10		45.20	27.60	30.3
8 L.H., Daily Totals		$336.40		$440.85	$42.05	$55.1

Crew B-95	Hr.	Daily	Hr.	Daily	Bare Costs	Incl. O&P
1 Equip. Oper. (crane)	$20.65	$165.20	$34.25	$274.00	$17.55	$29.5
1 Laborer	14.45	115.60	24.75	198.00		
16 L.H., Daily Totals		$280.80		$472.00	$17.55	$29.5

Crew B-95A	Hr.	Daily	Hr.	Daily	Bare Costs	Incl. O&P
1 Equip. Oper. (crane)	$20.65	$165.20	$34.25	$274.00	$17.55	$29.5
1 Laborer	14.45	115.60	24.75	198.00		
1 Hyd. Excavator, 5/8 C.Y.		411.60		452.75	25.73	28.3
16 L.H., Daily Totals		$692.40		$924.75	$43.28	$57.8

Crew B-95B	Hr.	Daily	Hr.	Daily	Bare Costs	Incl. O&P
1 Equip. Oper. (crane)	$20.65	$165.20	$34.25	$274.00	$17.55	$29.5
1 Laborer	14.45	115.60	24.75	198.00		
1 Hyd. Excavator, 1.5 C.Y.		713.20		784.50	44.58	49.0
16 L.H., Daily Totals		$994.00		$1256.50	$62.13	$78.5

Crew B-95C	Hr.	Daily	Hr.	Daily	Bare Costs	Incl. O&P
1 Equip. Oper. (crane)	$20.65	$165.20	$34.25	$274.00	$17.55	$29.5
1 Laborer	14.45	115.60	24.75	198.00		
1 Hyd. Excavator, 2.5 C.Y.		1673.00		1840.30	104.56	115.0
16 L.H., Daily Totals		$1953.80		$2312.30	$122.11	$144.5

Crew C-1	Hr.	Daily	Hr.	Daily	Bare Costs	Incl. O&P
2 Carpenters	$19.70	$315.20	$33.75	$540.00	$17.16	$29.40
1 Carpenter Helper	14.80	118.40	25.35	202.80		
1 Laborer	14.45	115.60	24.75	198.00		
32 L.H., Daily Totals		$549.20		$940.80	$17.16	$29.40

Crew C-2	Hr.	Daily	Hr.	Daily	Bare Costs	Incl. O&P
1 Carpenter Foreman (out)	$21.70	$173.60	$37.20	$297.60	$17.52	$30.03
2 Carpenters	19.70	315.20	33.75	540.00		
2 Carpenter Helpers	14.80	236.80	25.35	405.60		
1 Laborer	14.45	115.60	24.75	198.00		
48 L.H., Daily Totals		$841.20		$1441.20	$17.52	$30.03

rews

Crew C-2A	Bare Costs Hr.	Bare Costs Daily	Incl. Subs O & P Hr.	Incl. Subs O & P Daily	Cost Per Labor-Hour Bare Costs	Cost Per Labor-Hour Incl. O&P
arpenter Foreman (out)	$21.70	$173.60	$37.20	$297.60	$19.02	$32.33
arpenters	19.70	472.80	33.75	810.00		
ement Finisher	18.90	151.20	30.80	246.40		
borer	14.45	115.60	24.75	198.00		
.H., Daily Totals		$913.20		$1552.00	$19.02	$32.33

Crew C-3	Hr.	Daily	Hr.	Daily	Bare Costs	Incl. O&P
odman Foreman	$23.10	$184.80	$43.00	$344.00	$18.61	$33.34
odmen (reinf.)	21.10	506.40	39.25	942.00		
uip. Oper. (light)	19.10	152.80	31.70	253.60		
aborers	14.45	346.80	24.75	594.00		
ressing Equipment		78.00		85.80		
routing Equipment		125.00		137.50	3.17	3.49
.H., Daily Totals		$1393.80		$2356.90	$21.78	$36.83

Crew C-4	Hr.	Daily	Hr.	Daily	Bare Costs	Incl. O&P
odman Foreman	$23.10	$184.80	$43.00	$344.00	$19.94	$36.56
odmen (reinf.)	21.10	337.60	39.25	628.00		
uilding Laborer	14.45	115.60	24.75	198.00		
tressing Equipment		78.00		85.80	2.44	2.68
L.H., Daily Totals		$716.00		$1255.80	$22.38	$39.24

Crew C-5	Hr.	Daily	Hr.	Daily	Bare Costs	Incl. O&P
odman Foreman	$23.10	$184.80	$43.00	$344.00	$19.14	$34.21
odmen (reinf.)	21.10	337.60	39.25	628.00		
quip. Oper. (crane)	20.65	165.20	34.25	274.00		
uilding Laborers	14.45	231.20	24.75	396.00		
yd. Crane, 25 Ton		559.20		615.10	11.65	12.82
L.H., Daily Totals		$1478.00		$2257.10	$30.79	$47.03

Crew C-6	Hr.	Daily	Hr.	Daily	Bare Costs	Incl. O&P
abor Foreman (outside)	$16.45	$131.60	$28.20	$225.60	$15.53	$26.33
aborers	14.45	462.40	24.75	792.00		
ement Finisher	18.90	151.20	30.80	246.40		
as Engine Vibrators		76.00		83.60	1.58	1.74
L.H., Daily Totals		$821.20		$1347.60	$17.11	$28.07

Crew C-7	Hr.	Daily	Hr.	Daily	Bare Costs	Incl. O&P
abor Foreman (outside)	$16.45	$131.60	$28.20	$225.60	$16.06	$27.11
aborers	14.45	578.00	24.75	990.00		
ement Finisher	18.90	151.20	30.80	246.40		
quip. Oper. (med.)	19.90	159.20	33.00	264.00		
quip. Oper. (oiler)	17.00	136.00	28.20	225.60		
Gas Engine Vibrators		76.00		83.60		
Concrete Bucket, 1 C.Y.		27.20		29.90		
Hyd. Crane, 55 Ton		808.20		889.00	12.66	13.92
L.H., Daily Totals		$2067.40		$2954.10	$28.72	$41.03

Crew C-8	Hr.	Daily	Hr.	Daily	Bare Costs	Incl. O&P
Labor Foreman (outside)	$16.45	$131.60	$28.20	$225.60	$16.79	$28.15
Laborers	14.45	346.80	24.75	594.00		
Cement Finishers	18.90	302.40	30.80	492.80		
Equip. Oper. (med.)	19.90	159.20	33.00	264.00		
Concrete Pump (small)		656.00		721.60	11.71	12.89
5 L.H., Daily Totals		$1596.00		$2298.00	$28.50	$41.04

Crew C-8A	Hr.	Daily	Hr.	Daily	Bare Costs	Incl. O&P
Labor Foreman (outside)	$16.45	$131.60	$28.20	$225.60	$16.27	$27.34
Laborers	14.45	346.80	24.75	594.00		
Cement Finishers	18.90	302.40	30.80	492.80		
8 L.H., Daily Totals		$780.80		$1312.40	$16.27	$27.34

Crew C-8B	Bare Costs Hr.	Bare Costs Daily	Incl. Subs O & P Hr.	Incl. Subs O & P Daily	Cost Per Labor-Hour Bare Costs	Cost Per Labor-Hour Incl. O&P
1 Labor Foreman (outside)	$16.45	$131.60	$28.20	$225.60	$15.94	$27.09
3 Laborers	14.45	346.80	24.75	594.00		
1 Equipment Operator	19.90	159.20	33.00	264.00		
1 Vibrating Screed		49.20		54.10		
1 Vibratory Roller		432.40		475.65		
1 Dozer, 200 HP		806.80		887.50	32.21	35.43
40 L.H., Daily Totals		$1926.00		$2500.85	$48.15	$62.52

Crew C-8C	Hr.	Daily	Hr.	Daily	Bare Costs	Incl. O&P
1 Labor Foreman (outside)	$16.45	$131.60	$28.20	$225.60	$16.43	$27.71
3 Laborers	14.45	346.80	24.75	594.00		
1 Cement Finisher	18.90	151.20	30.80	246.40		
1 Equipment Operator (med.)	19.90	159.20	33.00	264.00		
1 Shotcrete Rig, 12 CY/hr		346.00		380.60	7.21	7.93
48 L.H., Daily Totals		$1134.80		$1710.60	$23.64	$35.64

Crew C-8D	Hr.	Daily	Hr.	Daily	Bare Costs	Incl. O&P
1 Labor Foreman (outside)	$16.45	$131.60	$28.20	$225.60	$17.23	$28.86
1 Laborers	14.45	115.60	24.75	198.00		
1 Cement Finisher	18.90	151.20	30.80	246.40		
1 Equipment Operator (light)	19.10	152.80	31.70	253.60		
1 Compressor, 250 CFM		127.00		139.70		
2 Hoses, 1", 50'		10.40		11.45	4.29	4.72
32 L.H., Daily Totals		$688.60		$1074.75	$21.52	$33.58

Crew C-8E	Hr.	Daily	Hr.	Daily	Bare Costs	Incl. O&P
1 Labor Foreman (outside)	$16.45	$131.60	$28.20	$225.60	$17.23	$28.86
1 Laborers	14.45	115.60	24.75	198.00		
1 Cement Finisher	18.90	151.20	30.80	246.40		
1 Equipment Operator (light)	19.10	152.80	31.70	253.60		
1 Compressor, 250 CFM		127.00		139.70		
2 Hoses, 1", 50'		10.40		11.45		
1 Concrete Pump (small)		656.00		721.60	24.79	27.27
32 L.H., Daily Totals		$1344.60		$1796.35	$42.02	$56.13

Crew C-10	Hr.	Daily	Hr.	Daily	Bare Costs	Incl. O&P
1 Laborer	$14.45	$115.60	$24.75	$198.00	$17.42	$28.78
2 Cement Finishers	18.90	302.40	30.80	492.80		
24 L.H., Daily Totals		$418.00		$690.80	$17.42	$28.78

Crew C-11	Hr.	Daily	Hr.	Daily	Bare Costs	Incl. O&P
1 Skilled Worker Foreman	$21.75	$174.00	$37.30	$298.40	$20.16	$34.44
5 Skilled Worker	19.75	790.00	33.90	1356.00		
1 Equip. Oper. (crane)	20.65	165.20	34.25	274.00		
1 Truck Crane, 150 Ton		1563.00		1719.30	27.91	30.70
56 L.H., Daily Totals		$2692.20		$3647.70	$48.07	$65.14

Crew C-12	Hr.	Daily	Hr.	Daily	Bare Costs	Incl. O&P
1 Carpenter Foreman (out)	$21.70	$173.60	$37.20	$297.60	$19.32	$32.91
3 Carpenters	19.70	472.80	33.75	810.00		
1 Laborer	14.45	115.60	24.75	198.00		
1 Equip. Oper. (crane)	20.65	165.20	34.25	274.00		
1 Hyd. Crane, 12 Ton		453.45		498.80	9.45	10.39
48 L.H., Daily Totals		$1380.65		$2078.40	$28.77	$43.30

Crew C-13	Hr.	Daily	Hr.	Daily	Bare Costs	Incl. O&P
2 Struc. Steel Worker	$21.25	$340.00	$41.90	$670.40	$20.73	$39.18
1 Carpenter	19.70	157.60	33.75	270.00		
1 Gas Welding Machine		84.00		92.40	3.50	3.85
24 L.H., Daily Totals		$581.60		$1032.80	$24.23	$43.03

Crew No.	Bare Costs		Incl. Subs O & P		Cost Per Labor-Hour	
Crew C-14	**Hr.**	**Daily**	**Hr.**	**Daily**	**Bare Costs**	**Incl. O&P**
1 Carpenter Foreman (out)	$21.70	$173.60	$37.20	$297.60	$17.56	$30.19
3 Carpenters	19.70	472.80	33.75	810.00		
2 Carpenter Helpers	14.80	236.80	25.35	405.60		
4 Laborers	14.45	462.40	24.75	792.00		
2 Rodmen (reinf.)	21.10	337.60	39.25	628.00		
2 Rodman Helpers	14.80	236.80	25.35	405.60		
2 Cement Finishers	18.90	302.40	30.80	492.80		
1 Equip. Oper. (crane)	20.65	165.20	34.25	274.00		
1 Crane, 80 Ton, & Tools		1098.00		1207.80	8.07	8.88
136 L.H., Daily Totals		$3485.60		$5313.40	$25.63	$39.07

Crew C-14A	Hr.	Daily	Hr.	Daily	Bare Costs	Incl. O&P
1 Carpenter Foreman (out)	$21.70	$173.60	$37.20	$297.60	$19.56	$33.90
16 Carpenters	19.70	2521.60	33.75	4320.00		
4 Rodmen (reinf.)	21.10	675.20	39.25	1256.00		
2 Laborers	14.45	231.20	24.75	396.00		
1 Cement Finisher	18.90	151.20	30.80	246.40		
1 Equip. Oper. (med)	19.90	159.20	33.00	264.00		
1 Gas Engine Vibrator		38.00		41.80		
1 Concrete Pump (small)		656.00		721.60	3.47	3.82
200 L.H., Daily Totals		$4606.00		$7543.40	$23.03	$37.72

Crew C-14B	Hr.	Daily	Hr.	Daily	Bare Costs	Incl. O&P
1 Carpenter Foreman (out)	$21.70	$173.60	$37.20	$297.60	$19.53	$33.78
16 Carpenters	19.70	2521.60	33.75	4320.00		
4 Rodmen (reinf.)	21.10	675.20	39.25	1256.00		
2 Laborers	14.45	231.20	24.75	396.00		
2 Cement Finishers	18.90	302.40	30.80	492.80		
1 Equip. Oper. (med)	19.90	159.20	33.00	264.00		
1 Gas Engine Vibrator		38.00		41.80		
1 Concrete Pump (small)		656.00		721.60	3.34	3.67
208 L.H., Daily Totals		$4757.20		$7789.80	$22.87	$37.45

Crew C-14C	Hr.	Daily	Hr.	Daily	Bare Costs	Incl. O&P
1 Carpenter Foreman (out)	$21.70	$173.60	$37.20	$297.60	$18.49	$32.00
6 Carpenters	19.70	945.60	33.75	1620.00		
2 Rodmen (reinf)	21.10	337.60	39.25	628.00		
4 Laborers	14.45	462.40	24.75	792.00		
1 Cement Finisher	18.90	151.20	30.80	246.40		
1 Gas Engine Vibrator		38.00		41.80	.34	.37
112 L.H., Daily Totals		$2108.40		$3625.80	$18.83	$32.37

Crew C-14D	Hr.	Daily	Hr.	Daily	Bare Costs	Incl. O&P
1 Carpenter Foreman (out)	$21.70	$173.60	$37.20	$297.60	$19.45	$33.46
18 Carpenters	19.70	2836.80	33.75	4860.00		
2 Rodmen (reinf.)	21.10	337.60	39.25	628.00		
2 Laborers	14.45	231.20	24.75	396.00		
1 Cement Finisher	18.90	151.20	30.80	246.40		
1 Equip. Oper. (med.)	19.90	159.20	33.00	264.00		
1 Gas Engine Vibrator		38.00		41.80		
1 Concrete Pump (small)		656.00		721.60	3.47	3.82
200 L.H., Daily Totals		$4583.60		$7455.40	$22.92	$37.28

Crew C-14E	Hr.	Daily	Hr.	Daily	Bare Costs	Incl. O&P
1 Carpenter Foreman (out)	$21.70	$173.60	$37.20	$297.60	$18.89	$33.34
2 Carpenters	19.70	315.20	33.75	540.00		
4 Rodmen (reinf.)	21.10	675.20	39.25	1256.00		
3 Laborers	14.45	346.80	24.75	594.00		
1 Cement Finisher	18.90	151.20	30.80	246.40		
1 Gas Engine Vibrator		38.00		41.80	.43	.48
88 L.H., Daily Totals		$1700.00		$2975.80	$19.32	$33.82

Crew No.	Bare Costs		Incl. Subs O & P		Cost Per Labor-H	
Crew C-14F	**Hr.**	**Daily**	**Hr.**	**Daily**	**Bare Costs**	**Inc O&**
1 Laborer Foreman (out)	$16.45	$131.60	$28.20	$225.60	$17.64	$29.1
2 Laborers	14.45	231.20	24.75	396.00		
6 Cement Finishers	18.90	907.20	30.80	1478.40		
1 Gas Engine Vibrator		38.00		41.80	.53	.5
72 L.H., Daily Totals		$1308.00		$2141.80	$18.17	$29.7

Crew C-14G	Hr.	Daily	Hr.	Daily	Bare Costs	Inc O&
1 Laborer Foreman (out)	$16.45	$131.60	$28.20	$225.60	$17.28	$28.7
2 Laborers	14.45	231.20	24.75	396.00		
4 Cement Finishers	18.90	604.80	30.80	985.60		
1 Gas Engine Vibrator		38.00		41.80	.68	.7
56 L.H., Daily Totals		$1005.60		$1649.00	$17.96	$29.4

Crew C-14H	Hr.	Daily	Hr.	Daily	Bare Costs	Incl O&
1 Carpenter Foreman (out)	$21.70	$173.60	$37.20	$297.60	$19.26	$33.2
2 Carpenters	19.70	315.20	33.75	540.00		
1 Rodman (reinf.)	21.10	168.80	39.25	314.00		
1 Laborer	14.45	115.60	24.75	198.00		
1 Cement Finisher	18.90	151.20	30.80	246.40		
1 Gas Engine Vibrator		38.00		41.80	.79	.8
48 L.H., Daily Totals		$962.40		$1637.80	$20.05	$34.1

Crew C-15	Hr.	Daily	Hr.	Daily	Bare Costs	Incl O&P
1 Carpenter Foreman (out)	$21.70	$173.60	$37.20	$297.60	$18.15	$31.0
2 Carpenters	19.70	315.20	33.75	540.00		
3 Laborers	14.45	346.80	24.75	594.00		
2 Cement Finishers	18.90	302.40	30.80	492.80		
1 Rodman (reinf.)	21.10	168.80	39.25	314.00		
72 L.H., Daily Totals		$1306.80		$2238.40	$18.15	$31.0

Crew C-16	Hr.	Daily	Hr.	Daily	Bare Costs	Incl. O&P
1 Labor Foreman (outside)	$16.45	$131.60	$28.20	$225.60	$17.74	$30.6
3 Laborers	14.45	346.80	24.75	594.00		
2 Cement Finishers	18.90	302.40	30.80	492.80		
1 Equip. Oper. (med.)	19.90	159.20	33.00	264.00		
2 Rodmen (reinf.)	21.10	337.60	39.25	628.00		
1 Concrete Pump (small)		656.00		721.60	9.11	10.0
72 L.H., Daily Totals		$1933.60		$2926.00	$26.85	$40.64

Crew C-17	Hr.	Daily	Hr.	Daily	Bare Costs	Incl. O&P
2 Skilled Worker Foremen	$21.75	$348.00	$37.30	$596.80	$20.15	$34.58
8 Skilled Workers	19.75	1264.00	33.90	2169.60		
80 L.H., Daily Totals		$1612.00		$2766.40	$20.15	$34.58

Crew C-17A	Hr.	Daily	Hr.	Daily	Bare Costs	Incl. O&P
2 Skilled Worker Foremen	$21.75	$348.00	$37.30	$596.80	$20.16	$34.58
8 Skilled Workers	19.75	1264.00	33.90	2169.60		
.125 Equip. Oper. (crane)	20.65	20.65	34.25	34.25		
.125 Crane, 80 Ton, & Tools		137.25		151.00	1.69	1.86
81 L.H., Daily Totals		$1769.90		$2951.65	$21.85	$36.44

Crew C-17B	Hr.	Daily	Hr.	Daily	Bare Costs	Incl. O&P
2 Skilled Worker Foremen	$21.75	$348.00	$37.30	$596.80	$20.16	$34.57
8 Skilled Workers	19.75	1264.00	33.90	2169.60		
.25 Equip. Oper. (crane)	20.65	41.30	34.25	68.50		
.25 Crane, 80 Ton, & Tools		274.50		301.95		
.25 Walk Behind Power Tools		13.25		14.60	3.51	3.86
82 L.H., Daily Totals		$1941.05		$3151.45	$23.67	$38.43

Crew No.	Bare Costs		Incl. Subs O & P		Cost Per Labor-Hour	
Crew C-17C	**Hr.**	**Daily**	**Hr.**	**Daily**	**Bare Costs**	**Incl. O&P**
killed Worker Foremen	$21.75	$348.00	$37.30	$596.80	$20.17	$34.57
killed Workers	19.75	1264.00	33.90	2169.60		
5 Equip. Oper. (crane)	20.65	61.95	34.25	102.75		
5 Crane, 80 Ton & Tools		411.75		452.95	4.96	5.46
.H., Daily Totals		$2085.70		$3322.10	$25.13	$40.03

Crew C-17D	Hr.	Daily	Hr.	Daily	Bare Costs	Incl. O&P
killed Worker Foremen	$21.75	$348.00	$37.30	$596.80	$20.17	$34.56
killed Workers	19.75	1264.00	33.90	2169.60		
quip. Oper. (crane)	20.65	82.60	34.25	137.00		
rane, 80 Ton & Tools		549.00		603.90	6.54	7.19
L.H., Daily Totals		$2243.60		$3507.30	$26.71	$41.75

Crew C-17E	Hr.	Daily	Hr.	Daily	Bare Costs	Incl. O&P
killed Worker Foremen	$21.75	$348.00	$37.30	$596.80	$20.15	$34.58
killed Workers	19.75	1264.00	33.90	2169.60		
yd. Jack with Rods		60.50		66.55	.76	.83
L.H., Daily Totals		$1672.50		$2832.95	$20.91	$35.41

Crew C-18	Hr.	Daily	Hr.	Daily	Bare Costs	Incl. O&P
5 Labor Foreman (out)	$16.45	$16.45	$28.20	$28.20	$14.67	$25.13
aborer	14.45	115.60	24.75	198.00		
oncrete Cart, 10 C.F.		52.00		57.20	5.78	6.36
.H., Daily Totals		$184.05		$283.40	$20.45	$31.49

Crew C-19	Hr.	Daily	Hr.	Daily	Bare Costs	Incl. O&P
5 Labor Foreman (out)	$16.45	$16.45	$28.20	$28.20	$14.67	$25.13
aborer	14.45	115.60	24.75	198.00		
oncrete Cart, 18 C.F.		83.20		91.50	9.24	10.17
.H., Daily Totals		$215.25		$317.70	$23.91	$35.30

Crew C-20	Hr.	Daily	Hr.	Daily	Bare Costs	Incl. O&P
abor Foreman (outside)	$16.45	$131.60	$28.20	$225.60	$15.94	$26.97
aborers	14.45	578.00	24.75	990.00		
ement Finisher	18.90	151.20	30.80	246.40		
quip. Oper. (med.)	19.90	159.20	33.00	264.00		
as Engine Vibrators		76.00		83.60		
oncrete Pump (small)		656.00		721.60	11.44	12.58
L.H., Daily Totals		$1752.00		$2531.20	$27.38	$39.55

Crew C-21	Hr.	Daily	Hr.	Daily	Bare Costs	Incl. O&P
abor Foreman (outside)	$16.45	$131.60	$28.20	$225.60	$15.94	$26.97
Laborers	14.45	578.00	24.75	990.00		
Cement Finisher	18.90	151.20	30.80	246.40		
Equip. Oper. (med.)	19.90	159.20	33.00	264.00		
Gas Engine Vibrators		76.00		83.60		
Concrete Conveyer		182.00		200.20	4.03	4.43
L.H., Daily Totals		$1278.00		$2009.80	$19.97	$31.40

Crew C-22	Hr.	Daily	Hr.	Daily	Bare Costs	Incl. O&P
Rodman Foreman	$23.10	$184.80	$43.00	$344.00	$21.37	$39.58
Rodmen (reinf.)	21.10	675.20	39.25	1256.00		
25 Equip. Oper. (crane)	20.65	20.65	34.25	34.25		
25 Equip. Oper. Oiler	17.00	17.00	28.20	28.20		
25 Hyd. Crane, 25 Ton		69.90		76.90	1.66	1.83
2 L.H., Daily Totals		$967.55		$1739.35	$23.03	$41.41

Crew No.	Bare Costs		Incl. Subs O & P		Cost Per Labor-Hour	
Crew C-23	**Hr.**	**Daily**	**Hr.**	**Daily**	**Bare Costs**	**Incl. O&P**
2 Skilled Worker Foremen	$21.75	$348.00	$37.30	$596.80	$19.97	$34.05
6 Skilled Workers	19.75	948.00	33.90	1627.20		
1 Equip. Oper. (crane)	20.65	165.20	34.25	274.00		
1 Equip. Oper. Oiler	17.00	136.00	28.20	225.60		
1 Crane, 90 Ton		1022.00		1124.20	12.78	14.05
80 L.H., Daily Totals		$2619.20		$3847.80	$32.75	$48.10

Crew C-24	Hr.	Daily	Hr.	Daily	Bare Costs	Incl. O&P
2 Skilled Worker Foremen	$21.75	$348.00	$37.30	$596.80	$19.97	$34.05
6 Skilled Workers	19.75	948.00	33.90	1627.20		
1 Equip. Oper. (crane)	20.65	165.20	34.25	274.00		
1 Equip. Oper. Oiler	17.00	136.00	28.20	225.60		
1 Truck Crane, 150 Ton		1563.00		1719.30	19.54	21.49
80 L.H., Daily Totals		$3160.20		$4442.90	$39.51	$55.54

Crew C-25	Hr.	Daily	Hr.	Daily	Bare Costs	Incl. O&P
2 Rodmen (reinf.)	$21.10	$337.60	$39.25	$628.00	$16.93	$31.50
2 Rodmen Helpers	12.75	204.00	23.75	380.00		
32 L.H., Daily Totals		$541.60		$1008.00	$16.93	$31.50

Crew D-1	Hr.	Daily	Hr.	Daily	Bare Costs	Incl. O&P
1 Bricklayer	$20.00	$160.00	$33.85	$270.80	$17.85	$30.23
1 Bricklayer Helper	15.70	125.60	26.60	212.80		
16 L.H., Daily Totals		$285.60		$483.60	$17.85	$30.23

Crew D-2	Hr.	Daily	Hr.	Daily	Bare Costs	Incl. O&P
3 Bricklayers	$20.00	$480.00	$33.85	$812.40	$18.28	$30.95
2 Bricklayer Helpers	15.70	251.20	26.60	425.60		
40 L.H., Daily Totals		$731.20		$1238.00	$18.28	$30.95

Crew D-3	Hr.	Daily	Hr.	Daily	Bare Costs	Incl. O&P
3 Bricklayers	$20.00	$480.00	$33.85	$812.40	$18.35	$31.08
2 Bricklayer Helpers	15.70	251.20	26.60	425.60		
.25 Carpenter	19.70	39.40	33.75	67.50		
42 L.H., Daily Totals		$770.60		$1305.50	$18.35	$31.08

Crew D-4	Hr.	Daily	Hr.	Daily	Bare Costs	Incl. O&P
1 Bricklayer	$20.00	$160.00	$33.85	$270.80	$16.31	$27.68
3 Bricklayer Helpers	15.70	376.80	26.60	638.40		
1 Building Laborer	14.45	115.60	24.75	198.00		
1 Grout Pump, 50 C.F./hr		67.20		73.90		
1 Hoses & Hopper		31.60		34.75		
1 Accessories		10.55		11.60	2.73	3.01
40 L.H., Daily Totals		$761.75		$1227.45	$19.04	$30.69

Crew D-5	Hr.	Daily	Hr.	Daily	Bare Costs	Incl. O&P
1 Block Mason Helper	$15.70	$125.60	$26.60	$212.80	$15.70	$26.60
8 L.H., Daily Totals		$125.60		$212.80	$15.70	$26.60

Crew D-6	Hr.	Daily	Hr.	Daily	Bare Costs	Incl. O&P
3 Bricklayers	$20.00	$480.00	$33.85	$812.40	$17.85	$30.23
3 Bricklayer Helpers	15.70	376.80	26.60	638.40		
48 L.H., Daily Totals		$856.80		$1450.80	$17.85	$30.23

Crew D-7	Hr.	Daily	Hr.	Daily	Bare Costs	Incl. O&P
1 Tile Layer	$19.20	$153.60	$31.10	$248.80	$17.33	$28.08
1 Tile Layer Helper	15.45	123.60	25.05	200.40		
16 L.H., Daily Totals		$277.20		$449.20	$17.33	$28.08

Crew No.	Bare Costs		Incl. Subs O & P		Cost Per Labor-Hour	
Crew D-8	**Hr.**	**Daily**	**Hr.**	**Daily**	**Bare Costs**	**Incl. O&P**
3 Bricklayers	$20.00	$480.00	$33.85	$812.40	$18.28	$30.95
2 Bricklayer Helpers	15.70	251.20	26.60	425.60		
40 L.H., Daily Totals		$731.20		$1238.00	$18.28	$30.95
Crew D-9	**Hr.**	**Daily**	**Hr.**	**Daily**	**Bare Costs**	**Incl. O&P**
3 Block Masons	$20.00	$480.00	$33.85	$812.40	$17.85	$30.23
3 Block Mason Helpers	15.70	376.80	26.60	638.40		
48 L.H., Daily Totals		$856.80		$1450.80	$17.85	$30.23
Crew D-10	**Hr.**	**Daily**	**Hr.**	**Daily**	**Bare Costs**	**Incl. O&P**
1 Stone Mason Foreman	$21.50	$172.00	$36.40	$291.20	$18.61	$31.37
1 Stone Mason	19.50	156.00	33.00	264.00		
2 Stone Mason Helpers	15.70	251.20	26.60	425.60		
1 Equip. Oper. (crane)	20.65	165.20	34.25	274.00		
1 Truck Crane, 12.5 Ton		401.35		441.50	10.03	11.04
40 L.H., Daily Totals		$1145.75		$1696.30	$28.64	$42.41
Crew D-11	**Hr.**	**Daily**	**Hr.**	**Daily**	**Bare Costs**	**Incl. O&P**
2 Stone Masons	$19.50	$312.00	$33.00	$528.00	$18.23	$30.87
1 Stone Mason Helper	15.70	125.60	26.60	212.80		
24 L.H., Daily Totals		$437.60		$740.80	$18.23	$30.87
Crew D-12	**Hr.**	**Daily**	**Hr.**	**Daily**	**Bare Costs**	**Incl. O&P**
2 Stone Masons	$19.50	$312.00	$33.00	$528.00	$17.60	$29.80
2 Stone Mason Helpers	15.70	251.20	26.60	425.60		
32 L.H., Daily Totals		$563.20		$953.60	$17.60	$29.80
Crew D-13	**Hr.**	**Daily**	**Hr.**	**Daily**	**Bare Costs**	**Incl. O&P**
1 Stone Mason Foreman	$21.50	$172.00	$36.40	$291.20	$18.76	$31.64
2 Stone Masons	19.50	312.00	33.00	528.00		
2 Stone Mason Helpers	15.70	251.20	26.60	425.60		
1 Equip. Oper. (crane)	20.65	165.20	34.25	274.00		
1 Truck Crane, 12.5 Ton		401.35		441.50	8.36	9.20
48 L.H., Daily Totals		$1301.75		$1960.30	$27.12	$40.84
Crew E-1	**Hr.**	**Daily**	**Hr.**	**Daily**	**Bare Costs**	**Incl. O&P**
2 Struc. Steel Workers	$21.25	$340.00	$41.90	$670.40	$21.25	$41.90
1 Gas Welding Machine		84.00		92.40	5.25	5.78
16 L.H., Daily Totals		$424.00		$762.80	$26.50	$47.68
Crew E-2	**Hr.**	**Daily**	**Hr.**	**Daily**	**Bare Costs**	**Incl. O&P**
1 Struc. Steel Foreman	$23.25	$186.00	$45.85	$366.80	$21.48	$41.28
4 Struc. Steel Workers	21.25	680.00	41.90	1340.80		
1 Equip. Oper. (crane)	20.65	165.20	34.25	274.00		
1 Crane, 90 Ton		1022.00		1124.20	21.29	23.42
48 L.H., Daily Totals		$2053.20		$3105.80	$42.77	$64.70
Crew E-3	**Hr.**	**Daily**	**Hr.**	**Daily**	**Bare Costs**	**Incl. O&P**
1 Struc. Steel Foreman	$23.25	$186.00	$45.85	$366.80	$21.92	$43.22
2 Struc. Steel Worker	21.25	340.00	41.90	670.40		
1 Gas Welding Machine		84.00		92.40	3.50	3.85
24 L.H., Daily Totals		$610.00		$1129.60	$25.42	$47.07
Crew E-4	**Hr.**	**Daily**	**Hr.**	**Daily**	**Bare Costs**	**Incl. O&P**
1 Struc. Steel Foreman	$23.25	$186.00	$45.85	$366.80	$21.75	$42.89
3 Struc. Steel Workers	21.25	510.00	41.90	1005.60		
1 Gas Welding Machine		84.00		92.40	2.63	2.89
32 L.H., Daily Totals		$780.00		$1464.80	$24.38	$45.78

Crew No.	Bare Costs		Incl. Subs O & P		Cost Per Labor-H	
Crew E-5	**Hr.**	**Daily**	**Hr.**	**Daily**	**Bare Costs**	**Inc O&**
1 Struc. Steel Foremen	$23.25	$186.00	$45.85	$366.80	$21.41	$41.4
7 Struc. Steel Workers	21.25	1190.00	41.90	2346.40		
1 Equip. Oper. (crane)	20.65	165.20	34.25	274.00		
1 Crane, 90 Ton		1022.00		1124.20		
1 Gas Welding Machine		84.00		92.40	15.36	16.9
72 L.H., Daily Totals		$2647.20		$4203.80	$36.77	$58.3
Crew E-6	**Hr.**	**Daily**	**Hr.**	**Daily**	**Bare Costs**	**Inc O&**
1 Struc. Steel Foreman	$23.25	$186.00	$45.85	$366.80	$21.20	$40.9
12 Struc. Steel Workers	21.25	2040.00	41.90	4022.40		
1 Equip. Oper. (crane)	20.65	165.20	34.25	274.00		
1 Equip. Oper. (light)	19.10	152.80	31.70	253.60		
1 Crane, 90 Ton		1022.00		1124.20		
1 Gas Welding Machine		84.00		92.40		
1 Air Compr., 160 C.F.M.		98.40		108.25		
2 Impact Wrenches		55.60		61.15	10.50	11.5
120 L.H., Daily Totals		$3804.00		$6302.80	$31.70	$52.5
Crew E-7	**Hr.**	**Daily**	**Hr.**	**Daily**	**Bare Costs**	**Incl O&**
1 Struc. Steel Foreman	$23.25	$186.00	$45.85	$366.80	$21.41	$41.4
7 Struc. Steel Workers	21.25	1190.00	41.90	2346.40		
1 Equip. Oper. (crane)	20.65	165.20	34.25	274.00		
1 Crane, 90 Ton		1022.00		1124.20		
2 Gas Welding Machines		168.00		184.80	16.53	18.1
72 L.H., Daily Totals		$2731.20		$4296.20	$37.94	$59.6
Crew E-8	**Hr.**	**Daily**	**Hr.**	**Daily**	**Bare Costs**	**Incl O&P**
1 Struc. Steel Foreman	$23.25	$186.00	$45.85	$366.80	$21.38	$41.5
9 Struc. Steel Workers	21.25	1530.00	41.90	3016.80		
1 Equip. Oper. (crane)	20.65	165.20	34.25	274.00		
1 Crane, 90 Ton		1022.00		1124.20		
4 Gas Welding Machines		336.00		369.60	15.43	16.9
88 L.H., Daily Totals		$3239.20		$5151.40	$36.81	$58.5
Crew E-9	**Hr.**	**Daily**	**Hr.**	**Daily**	**Bare Costs**	**Incl. O&P**
2 Struc. Steel Foremen	$23.25	$372.00	$45.85	$733.60	$21.19	$40.6
5 Struc. Steel Workers	21.25	850.00	41.90	1676.00		
1 Welder Foreman	23.25	186.00	45.85	366.80		
5 Welders	21.25	850.00	41.90	1676.00		
1 Equip. Oper. (crane)	20.65	165.20	34.25	274.00		
1 Equip. Oper. Oiler	17.00	136.00	28.20	225.60		
1 Equip. Oper. (light)	19.10	152.80	31.70	253.60		
1 Crane, 90 Ton		1022.00		1124.20		
5 Gas Welding Machines		420.00		462.00	11.27	12.3
128 L.H., Daily Totals		$4154.00		$6791.80	$32.46	$53.0
Crew E-10	**Hr.**	**Daily**	**Hr.**	**Daily**	**Bare Costs**	**Incl. O&P**
1 Struc. Steel Foreman	$23.25	$186.00	$45.85	$366.80	$21.92	$43.2
2 Struc. Steel Workers	21.25	340.00	41.90	670.40		
4 Gas Welding Machines		336.00		369.60		
1 Truck, 3 Ton		180.45		198.50	21.52	23.6
24 L.H., Daily Totals		$1042.45		$1605.30	$43.44	$66.8
Crew E-11	**Hr.**	**Daily**	**Hr.**	**Daily**	**Bare Costs**	**Incl. O&P**
2 Painters, Struc. Steel	$18.65	$298.40	$37.80	$604.80	$17.25	$33.45
1 Building Laborer	14.45	115.60	24.75	198.00		
1 Air Compressor 250 C.F.M.		127.00		139.70		
1 Sand Blaster		31.60		34.75		
1 Sand Blasting Accessories		10.55		11.60	7.05	7.75
24 L.H., Daily Totals		$583.15		$988.85	$24.30	$41.20

Crew No.	Bare Costs		Incl. Subs O & P		Cost Per Labor-Hour	
Crew E-12	**Hr.**	**Daily**	**Hr.**	**Daily**	**Bare Costs**	**Incl. O&P**
elder Foreman	$23.25	$186.00	$45.85	$366.80	$21.18	$38.78
quip. Oper. (light)	19.10	152.80	31.70	253.60		
as Welding Machine		84.00		92.40	5.25	5.78
L.H., Daily Totals		$422.80		$712.80	$26.43	$44.56
Crew E-13	**Hr.**	**Daily**	**Hr.**	**Daily**	**Bare Costs**	**Incl. O&P**
elder Foreman	$23.25	$186.00	$45.85	$366.80	$21.87	$41.13
quip. Oper. (light)	19.10	76.40	31.70	126.80		
as Welding Machine		84.00		92.40	7.00	7.70
L.H., Daily Totals		$346.40		$586.00	$28.87	$48.83
Crew E-14	**Hr.**	**Daily**	**Hr.**	**Daily**	**Bare Costs**	**Incl. O&P**
truc. Steel Worker	$21.25	$170.00	$41.90	$335.20	$21.25	$41.90
as Welding Machine		84.00		92.40	10.50	11.55
.H., Daily Totals		$254.00		$427.60	$31.75	$53.45
Crew E-16	**Hr.**	**Daily**	**Hr.**	**Daily**	**Bare Costs**	**Incl. O&P**
elder Foreman	$23.25	$186.00	$45.85	$366.80	$22.25	$43.88
elder	21.25	170.00	41.90	335.20		
as Welding Machine		84.00		92.40	5.25	5.78
L.H., Daily Totals		$440.00		$794.40	$27.50	$49.66
Crew E-17	**Hr.**	**Daily**	**Hr.**	**Daily**	**Bare Costs**	**Incl. O&P**
tructural Steel Foreman	$23.25	$186.00	$45.85	$366.80	$22.25	$43.88
tructural Steel Worker	21.25	170.00	41.90	335.20		
ower Tool		2.40		2.65	.15	.17
L.H., Daily Totals		$358.40		$704.65	$22.40	$44.05
Crew E-18	**Hr.**	**Daily**	**Hr.**	**Daily**	**Bare Costs**	**Incl. O&P**
tructural Steel Foreman	$23.25	$186.00	$45.85	$366.80	$21.38	$40.91
tructural Steel Workers	21.25	510.00	41.90	1005.60		
quipment Operator (med.)	19.90	159.20	33.00	264.00		
rane, 20 Ton		532.55		585.80	13.31	14.65
L.H., Daily Totals		$1387.75		$2222.20	$34.69	$55.56
Crew E-19	**Hr.**	**Daily**	**Hr.**	**Daily**	**Bare Costs**	**Incl. O&P**
Structural Steel Worker	$21.25	$170.00	$41.90	$335.20	$21.20	$39.82
Structural Steel Foreman	23.25	186.00	45.85	366.80		
Equip. Oper. (light)	19.10	152.80	31.70	253.60		
Power Tool		2.40		2.65		
Crane, 20 ton		532.55		585.80	22.29	24.52
L.H., Daily Totals		$1043.75		$1544.05	$43.49	$64.34
Crew E-20	**Hr.**	**Daily**	**Hr.**	**Daily**	**Bare Costs**	**Incl. O&P**
Structural Steel Foreman	$23.25	$186.00	$45.85	$366.80	$20.89	$39.73
Structural Steel Workers	21.25	850.00	41.90	1676.00		
Equip. Oper. (crane)	20.65	165.20	34.25	274.00		
Oiler	17.00	136.00	28.20	225.60		
Power Tool		2.40		2.65		
Crane, 40 ton		704.80		775.30	11.05	12.16
4 L.H., Daily Totals		$2044.40		$3320.35	$31.94	$51.89
Crew E-22	**Hr.**	**Daily**	**Hr.**	**Daily**	**Bare Costs**	**Incl. O&P**
Skilled Worker Foreman	$21.75	$174.00	$37.30	$298.40	$20.42	$35.03
Skilled Worker	19.75	316.00	33.90	542.40		
4 L.H., Daily Totals		$490.00		$840.80	$20.42	$35.03

Crew No.	Bare Costs		Incl. Subs O & P		Cost Per Labor-Hour	
Crew E-24	**Hr.**	**Daily**	**Hr.**	**Daily**	**Bare Costs**	**Incl. O&P**
3 Structural Steel Worker	$21.25	$510.00	$41.90	$1005.60	$20.91	$39.68
1 Equipment Operator (medium)	19.90	159.20	33.00	264.00		
1-25 ton crane		559.20		615.10	17.48	19.22
32 L.H., Daily Totals		$1228.40		$1884.70	$38.39	$58.90
Crew F-3	**Hr.**	**Daily**	**Hr.**	**Daily**	**Bare Costs**	**Incl. O&P**
2 Carpenters	$19.70	$315.20	$33.75	$540.00	$17.93	$30.49
2 Carpenter Helpers	14.80	236.80	25.35	405.60		
1 Equip. Oper. (crane)	20.65	165.20	34.25	274.00		
1 Hyd. Crane, 12 Ton		453.45		498.80	11.34	12.47
40 L.H., Daily Totals		$1170.65		$1718.40	$29.27	$42.96
Crew F-4	**Hr.**	**Daily**	**Hr.**	**Daily**	**Bare Costs**	**Incl. O&P**
2 Carpenters	$19.70	$315.20	$33.75	$540.00	$17.93	$30.49
2 Carpenter Helpers	14.80	236.80	25.35	405.60		
1 Equip. Oper. (crane)	20.65	165.20	34.25	274.00		
1 Hyd. Crane, 55 Ton		808.20		889.00	20.21	22.23
40 L.H., Daily Totals		$1525.40		$2108.60	$38.14	$52.72
Crew F-5	**Hr.**	**Daily**	**Hr.**	**Daily**	**Bare Costs**	**Incl. O&P**
2 Carpenters	$19.70	$315.20	$33.75	$540.00	$17.25	$29.55
2 Carpenter Helpers	14.80	236.80	25.35	405.60		
32 L.H., Daily Totals		$552.00		$945.60	$17.25	$29.55
Crew F-6	**Hr.**	**Daily**	**Hr.**	**Daily**	**Bare Costs**	**Incl. O&P**
2 Carpenters	$19.70	$315.20	$33.75	$540.00	$17.79	$30.25
2 Building Laborers	14.45	231.20	24.75	396.00		
1 Equip. Oper. (crane)	20.65	165.20	34.25	274.00		
1 Hyd. Crane, 12 Ton		453.45		498.80	11.34	12.47
40 L.H., Daily Totals		$1165.05		$1708.80	$29.13	$42.72
Crew F-7	**Hr.**	**Daily**	**Hr.**	**Daily**	**Bare Costs**	**Incl. O&P**
2 Carpenters	$19.70	$315.20	$33.75	$540.00	$17.08	$29.25
2 Building Laborers	14.45	231.20	24.75	396.00		
32 L.H., Daily Totals		$546.40		$936.00	$17.08	$29.25
Crew G-1	**Hr.**	**Daily**	**Hr.**	**Daily**	**Bare Costs**	**Incl. O&P**
1 Roofer Foreman	$19.05	$152.40	$35.45	$283.60	$16.11	$29.99
4 Roofers, Composition	17.05	545.60	31.75	1016.00		
2 Roofer Helpers	12.75	204.00	23.75	380.00		
1 Application Equipment		171.60		188.75		
1 Tar Kettle/Pot		49.00		53.90		
1 Crew Truck		197.60		217.35	7.47	8.21
56 L.H., Daily Totals		$1320.20		$2139.60	$23.58	$38.20
Crew G-2	**Hr.**	**Daily**	**Hr.**	**Daily**	**Bare Costs**	**Incl. O&P**
1 Plasterer	$18.65	$149.20	$31.20	$249.60	$16.25	$27.37
1 Plasterer Helper	15.65	125.20	26.15	209.20		
1 Building Laborer	14.45	115.60	24.75	198.00		
1 Grouting Equipment		250.00		275.00	10.42	11.46
24 L.H., Daily Totals		$640.00		$931.80	$26.67	$38.83
Crew G-3	**Hr.**	**Daily**	**Hr.**	**Daily**	**Bare Costs**	**Incl. O&P**
2 Sheet Metal Workers	$21.50	$344.00	$36.30	$580.80	$17.98	$30.52
2 Building Laborers	14.45	231.20	24.75	396.00		
32 L.H., Daily Totals		$575.20		$976.80	$17.98	$30.52

Crews

Crew G-4	Bare Costs Hr.	Bare Costs Daily	Incl. Subs O & P Hr.	Incl. Subs O & P Daily	Cost Per Labor-Hour Bare Costs	Cost Per Labor-Hour Incl. O&P
1 Labor Foreman (outside)	$16.45	$131.60	$28.20	$225.60	$15.12	$25.90
2 Building Laborers	14.45	231.20	24.75	396.00		
1 Light Truck, 1.5 Ton		172.95		190.25		
1 Air Compr., 160 C.F.M.		98.40		108.25	11.31	12.44
24 L.H., Daily Totals		$634.15		$920.10	$26.43	$38.34

Crew G-5	Hr.	Daily	Hr.	Daily	Bare Costs	Incl. O&P
1 Roofer Foreman	$19.05	$152.40	$35.45	$283.60	$15.73	$29.29
2 Roofers, Composition	17.05	272.80	31.75	508.00		
2 Roofer Helpers	12.75	204.00	23.75	380.00		
1 Application Equipment		171.60		188.75	4.29	4.72
40 L.H., Daily Totals		$800.80		$1360.35	$20.02	$34.01

Crew G-6A	Hr.	Daily	Hr.	Daily	Bare Costs	Incl. O&P
2 Roofers Composition	$17.05	$272.80	$31.75	$508.00	$17.05	$31.75
1 Small Compressor		20.95		23.05		
2 Pneumatic Nailers		37.90		41.70	3.68	4.05
16 L.H., Daily Totals		$331.65		$572.75	$20.73	$35.80

Crew G-7	Hr.	Daily	Hr.	Daily	Bare Costs	Incl. O&P
1 Carpenter	$19.70	$157.60	$33.75	$270.00	$19.70	$33.75
1 Small Compressor		20.95		23.05		
1 Pneumatic Nailer		18.95		20.85	4.99	5.49
8 L.H., Daily Totals		$197.50		$313.90	$24.69	$39.24

Crew H-1	Hr.	Daily	Hr.	Daily	Bare Costs	Incl. O&P
2 Glaziers	$19.35	$309.60	$32.10	$513.60	$20.30	$37.00
2 Struc. Steel Workers	21.25	340.00	41.90	670.40		
32 L.H., Daily Totals		$649.60		$1184.00	$20.30	$37.00

Crew H-2	Hr.	Daily	Hr.	Daily	Bare Costs	Incl. O&P
2 Glaziers	$19.35	$309.60	$32.10	$513.60	$17.72	$29.65
1 Building Laborer	14.45	115.60	24.75	198.00		
24 L.H., Daily Totals		$425.20		$711.60	$17.72	$29.65

Crew H-3	Hr.	Daily	Hr.	Daily	Bare Costs	Incl. O&P
1 Glazier	$19.35	$154.80	$32.10	$256.80	$17.08	$28.73
1 Helper	14.80	118.40	25.35	202.80		
16 L.H., Daily Totals		$273.20		$459.60	$17.08	$28.73

Crew J-1	Hr.	Daily	Hr.	Daily	Bare Costs	Incl. O&P
3 Plasterers	$18.65	$447.60	$31.20	$748.80	$17.45	$29.18
2 Plasterer Helpers	15.65	250.40	26.15	418.40		
1 Mixing Machine, 6 C.F.		54.50		59.95	1.36	1.50
40 L.H., Daily Totals		$752.50		$1227.15	$18.81	$30.68

Crew J-2	Hr.	Daily	Hr.	Daily	Bare Costs	Incl. O&P
3 Plasterers	$18.65	$447.60	$31.20	$748.80	$17.74	$29.56
2 Plasterer Helpers	15.65	250.40	26.15	418.40		
1 Lather	19.20	153.60	31.45	251.60		
1 Mixing Machine, 6 C.F.		54.50		59.95	1.14	1.25
48 L.H., Daily Totals		$906.10		$1478.75	$18.88	$30.81

Crew J-3	Hr.	Daily	Hr.	Daily	Bare Costs	Incl. O&P
1 Terrazzo Worker	$19.10	$152.80	$30.95	$247.60	$17.33	$28.08
1 Terrazzo Helper	15.55	124.40	25.20	201.60		
1 Terrazzo Grinder, Electric		50.00		55.00		
1 Terrazzo Mixer		74.20		81.60	7.76	8.54
16 L.H., Daily Totals		$401.40		$585.80	$25.09	$36.62

Crew J-4	Hr.	Daily	Hr.	Daily	Bare Costs	Incl. O&P
1 Tile Layer	$19.20	$153.60	$31.10	$248.80	$17.33	$28.0
1 Tile Layer Helper	15.45	123.60	25.05	200.40		
16 L.H., Daily Totals		$277.20		$449.20	$17.33	$28.0

Crew K-1	Hr.	Daily	Hr.	Daily	Bare Costs	Incl. O&P
1 Carpenter	$19.70	$157.60	$33.75	$270.00	$17.78	$30.1
1 Truck Driver (light)	15.85	126.80	26.50	212.00		
1 Truck w/Power Equip.		180.45		198.50	11.28	12.4
16 L.H., Daily Totals		$464.85		$680.50	$29.06	$42.5

Crew K-2	Hr.	Daily	Hr.	Daily	Bare Costs	Incl. O&P
1 Struc. Steel Foreman	$23.25	$186.00	$45.85	$366.80	$20.12	$38.0
1 Struc. Steel Worker	21.25	170.00	41.90	335.20		
1 Truck Driver (light)	15.85	126.80	26.50	212.00		
1 Truck w/Power Equip.		180.45		198.50	7.52	8.2
24 L.H., Daily Totals		$663.25		$1112.50	$27.64	$46.3

Crew L-1	Hr.	Daily	Hr.	Daily	Bare Costs	Incl. O&P
.25 Electrician	$22.10	$44.20	$36.15	$72.30	$21.98	$36.2
1 Plumber	21.95	175.60	36.30	290.40		
10 L.H., Daily Totals		$219.80		$362.70	$21.98	$36.2

Crew L-2	Hr.	Daily	Hr.	Daily	Bare Costs	Incl. O&P
1 Carpenter	$19.70	$157.60	$33.75	$270.00	$17.25	$29.5
1 Carpenter Helper	14.80	118.40	25.35	202.80		
16 L.H., Daily Totals		$276.00		$472.80	$17.25	$29.5

Crew L-3	Hr.	Daily	Hr.	Daily	Bare Costs	Incl. O&P
1 Carpenter	$19.70	$157.60	$33.75	$270.00	$20.18	$34.2
.25 Electrician	22.10	44.20	36.15	72.30		
10 L.H., Daily Totals		$201.80		$342.30	$20.18	$34.2

Crew L-3A	Hr.	Daily	Hr.	Daily	Bare Costs	Incl. O&P
1 Carpenter Foreman (outside)	$21.70	$173.60	$37.20	$297.60	$21.63	$36.9
.5 Sheet Metal Worker	21.50	86.00	36.30	145.20		
12 L.H., Daily Totals		$259.60		$442.80	$21.63	$36.9

Crew L-4	Hr.	Daily	Hr.	Daily	Bare Costs	Incl. O&P
1 Skilled Workers	$19.75	$158.00	$33.90	$271.20	$17.27	$29.6
1 Helper	14.80	118.40	25.35	202.80		
16 L.H., Daily Totals		$276.40		$474.00	$17.27	$29.6

Crew L-5	Hr.	Daily	Hr.	Daily	Bare Costs	Incl. O&P
1 Struc. Steel Foreman	$23.25	$186.00	$45.85	$366.80	$21.45	$41.37
5 Struc. Steel Workers	21.25	850.00	41.90	1676.00		
1 Equip. Oper. (crane)	20.65	165.20	34.25	274.00		
1 Hyd. Crane, 25 Ton		559.20		615.10	9.99	10.98
56 L.H., Daily Totals		$1760.40		$2931.90	$31.44	$52.35

Crew L-5A	Hr.	Daily	Hr.	Daily	Bare Costs	Incl. O&P
1 Structural Steel Foreman	$23.25	$186.00	$45.85	$366.80	$21.60	$40.98
2 Structural Steel Workers	21.25	340.00	41.90	670.40		
1 Equip. Oper. (crane)	20.65	165.20	34.25	274.00		
1 Crane,SP, 25 Ton		563.90		620.30	17.62	19.38
32 L.H., Daily Totals		$1255.10		$1931.50	$39.22	$60.36

Crew No.	Bare Costs Hr.	Bare Costs Daily	Incl. Subs O & P Hr.	Incl. Subs O & P Daily	Cost Per Labor-Hour Bare Costs	Cost Per Labor-Hour Incl. O&P
Crew L-6	**Hr.**	**Daily**	**Hr.**	**Daily**	**Bare Costs**	**Incl. O&P**
umber	$21.95	$175.60	$36.30	$290.40	$22.00	$36.25
lectrician	22.10	88.40	36.15	144.60		
.H., Daily Totals		$264.00		$435.00	$22.00	$36.25
Crew L-7	**Hr.**	**Daily**	**Hr.**	**Daily**	**Bare Costs**	**Incl. O&P**
arpenters	$19.70	$157.60	$33.75	$270.00	$16.87	$28.77
arpenter Helpers	14.80	236.80	25.35	405.60		
Electrician	22.10	44.20	36.15	72.30		
.H., Daily Totals		$438.60		$747.90	$16.87	$28.77
Crew L-8	**Hr.**	**Daily**	**Hr.**	**Daily**	**Bare Costs**	**Incl. O&P**
arpenters	$19.70	$157.60	$33.75	$270.00	$18.19	$30.90
arpenter Helper	14.80	118.40	25.35	202.80		
lumber	21.95	87.80	36.30	145.20		
.H., Daily Totals		$363.80		$618.00	$18.19	$30.90
Crew L-9	**Hr.**	**Daily**	**Hr.**	**Daily**	**Bare Costs**	**Incl. O&P**
killed Worker Foreman	$21.75	$174.00	$37.30	$298.40	$18.26	$31.11
killed Worker	19.75	158.00	33.90	271.20		
elpers	14.80	236.80	25.35	405.60		
lectrician	22.10	88.40	36.15	144.60		
L.H., Daily Totals		$657.20		$1119.80	$18.26	$31.11
Crew L-10	**Hr.**	**Daily**	**Hr.**	**Daily**	**Bare Costs**	**Incl. O&P**
tructural Steel Foreman	$23.25	$186.00	$45.85	$366.80	$21.72	$40.67
tructural Steel Worker	21.25	170.00	41.90	335.20		
quip. Oper. (crane)	20.65	165.20	34.25	274.00		
yd. Crane, 12 Ton		453.45		498.80	18.89	20.78
L.H., Daily Totals		$974.65		$1474.80	$40.61	$61.45
Crew M-1	**Hr.**	**Daily**	**Hr.**	**Daily**	**Bare Costs**	**Incl. O&P**
levator Constructors	$22.90	$549.60	$37.80	$907.20	$21.75	$35.90
levator Apprentice	18.30	146.40	30.20	241.60		
and Tools		95.00		104.50	2.97	3.27
L.H., Daily Totals		$791.00		$1253.30	$24.72	$39.17
Crew M-3	**Hr.**	**Daily**	**Hr.**	**Daily**	**Bare Costs**	**Incl. O&P**
lectrician Foreman (out)	$24.10	$192.80	$39.40	$315.20	$19.94	$33.04
ommon Laborer	14.45	115.60	24.75	198.00		
5 Equipment Operator, Medium	19.90	39.80	33.00	66.00		
levator Constructor	22.90	183.20	37.80	302.40		
levator Apprentice	18.30	146.40	30.20	241.60		
5 Crane, SP, 4 x 4, 20 ton		118.83		130.70	3.49	3.84
L.H., Daily Totals		$796.63		$1253.90	$23.43	$36.88
Crew M-4	**Hr.**	**Daily**	**Hr.**	**Daily**	**Bare Costs**	**Incl. O&P**
Electrician Foreman (out)	$24.10	$192.80	$39.40	$315.20	$19.81	$32.84
Common Laborer	14.45	115.60	24.75	198.00		
5 Equipment Operator, Crane	20.65	41.30	34.25	68.50		
5 Equipment Operator, Oiler	17.00	34.00	28.20	56.40		
Elevator Constructor	22.90	183.20	37.80	302.40		
Elevator Apprentice	18.30	146.40	30.20	241.60		
5 Crane, hyd, SP, 4WD, 40 ton		195.95		215.55	5.44	5.99
1 L.H., Daily Totals		$909.25		$1397.65	$25.25	$38.83
Crew Q-1	**Hr.**	**Daily**	**Hr.**	**Daily**	**Bare Costs**	**Incl. O&P**
Plumber	$21.95	$175.60	$36.30	$290.40	$19.75	$32.65
Plumber Apprentice	17.55	140.40	29.00	232.00		
5 L.H., Daily Totals		$316.00		$522.40	$19.75	$32.65

Crew No.	Bare Costs Hr.	Bare Costs Daily	Incl. Subs O & P Hr.	Incl. Subs O & P Daily	Cost Per Labor-Hour Bare Costs	Cost Per Labor-Hour Incl. O&P
Crew Q-1C	**Hr.**	**Daily**	**Hr.**	**Daily**	**Bare Costs**	**Incl. O&P**
1 Plumber	$21.95	$175.60	$36.30	$290.40	$19.80	$32.77
1 Plumber Apprentice	17.55	140.40	29.00	232.00		
1 Equip. Oper. (medium)	19.90	159.20	33.00	264.00		
1 Trencher, Chain		540.90		595.00	22.54	24.79
24 L.H., Daily Totals		$1016.10		$1381.40	$42.34	$57.56
Crew Q-2	**Hr.**	**Daily**	**Hr.**	**Daily**	**Bare Costs**	**Incl. O&P**
1 Plumber	$21.95	$175.60	$36.30	$290.40	$19.02	$31.43
2 Plumber Apprentices	17.55	280.80	29.00	464.00		
24 L.H., Daily Totals		$456.40		$754.40	$19.02	$31.43
Crew Q-3	**Hr.**	**Daily**	**Hr.**	**Daily**	**Bare Costs**	**Incl. O&P**
2 Plumbers	$21.95	$351.20	$36.30	$580.80	$19.75	$32.65
2 Plumber Apprentices	17.55	280.80	29.00	464.00		
32 L.H., Daily Totals		$632.00		$1044.80	$19.75	$32.65
Crew Q-4	**Hr.**	**Daily**	**Hr.**	**Daily**	**Bare Costs**	**Incl. O&P**
2 Plumbers	$21.95	$351.20	$36.30	$580.80	$20.85	$34.47
1 Welder (plumber)	21.95	175.60	36.30	290.40		
1 Plumber Apprentice	17.55	140.40	29.00	232.00		
1 Electric Welding Mach.		48.80		53.70	1.53	1.68
32 L.H., Daily Totals		$716.00		$1156.90	$22.38	$36.15
Crew Q-5	**Hr.**	**Daily**	**Hr.**	**Daily**	**Bare Costs**	**Incl. O&P**
1 Steamfitter	$22.10	$176.80	$36.55	$292.40	$19.90	$32.90
1 Steamfitter Apprentice	17.70	141.60	29.25	234.00		
16 L.H., Daily Totals		$318.40		$526.40	$19.90	$32.90
Crew Q-6	**Hr.**	**Daily**	**Hr.**	**Daily**	**Bare Costs**	**Incl. O&P**
1 Steamfitters	$22.10	$176.80	$36.55	$292.40	$19.17	$31.68
2 Steamfitter Apprentices	17.70	283.20	29.25	468.00		
24 L.H., Daily Totals		$460.00		$760.40	$19.17	$31.68
Crew Q-7	**Hr.**	**Daily**	**Hr.**	**Daily**	**Bare Costs**	**Incl. O&P**
2 Steamfitters	$22.10	$353.60	$36.55	$584.80	$19.90	$32.90
2 Steamfitter Apprentices	17.70	283.20	29.25	468.00		
32 L.H., Daily Totals		$636.80		$1052.80	$19.90	$32.90
Crew Q-8	**Hr.**	**Daily**	**Hr.**	**Daily**	**Bare Costs**	**Incl. O&P**
2 Steamfitters	$22.10	$353.60	$36.55	$584.80	$21.00	$34.72
1 Welder (steamfitter)	22.10	176.80	36.55	292.40		
1 Steamfitter Apprentice	17.70	141.60	29.25	234.00		
1 Electric Welding Mach.		48.80		53.70	1.53	1.68
32 L.H., Daily Totals		$720.80		$1164.90	$22.53	$36.40
Crew Q-9	**Hr.**	**Daily**	**Hr.**	**Daily**	**Bare Costs**	**Incl. O&P**
1 Sheet Metal Worker	$21.50	$172.00	$36.30	$290.40	$19.35	$32.67
1 Sheet Metal Apprentice	17.20	137.60	29.05	232.40		
16 L.H., Daily Totals		$309.60		$522.80	$19.35	$32.67
Crew Q-10	**Hr.**	**Daily**	**Hr.**	**Daily**	**Bare Costs**	**Incl. O&P**
2 Sheet Metal Workers	$21.50	$344.00	$36.30	$580.80	$20.07	$33.88
1 Sheet Metal Apprentice	17.20	137.60	29.05	232.40		
24 L.H., Daily Totals		$481.60		$813.20	$20.07	$33.88
Crew Q-11	**Hr.**	**Daily**	**Hr.**	**Daily**	**Bare Costs**	**Incl. O&P**
2 Sheet Metal Workers	$21.50	$344.00	$36.30	$580.80	$19.35	$32.67
2 Sheet Metal Apprentices	17.20	275.20	29.05	464.80		
32 L.H., Daily Totals		$619.20		$1045.60	$19.35	$32.67

Crew Q-12	Bare Costs Hr.	Bare Costs Daily	Incl. Subs O & P Hr.	Incl. Subs O & P Daily	Cost Per Labor-Hour Bare Costs	Cost Per Labor-Hour Incl. O&P
1 Sprinkler Installer	$22.05	$176.40	$36.50	$292.00	$19.85	$32.85
1 Sprinkler Apprentice	17.65	141.20	29.20	233.60		
16 L.H., Daily Totals		$317.60		$525.60	$19.85	$32.85

Crew Q-13	Hr.	Daily	Hr.	Daily	Bare Costs	Incl. O&P
2 Sprinkler Installers	$22.05	$352.80	$36.50	$584.00	$19.85	$32.85
2 Sprinkler Apprentices	17.65	282.40	29.20	467.20		
32 L.H., Daily Totals		$635.20		$1051.20	$19.85	$32.85

Crew Q-14	Hr.	Daily	Hr.	Daily	Bare Costs	Incl. O&P
1 Asbestos Worker	$20.50	$164.00	$35.90	$287.20	$18.45	$32.30
1 Asbestos Apprentice	16.40	131.20	28.70	229.60		
16 L.H., Daily Totals		$295.20		$516.80	$18.45	$32.30

Crew Q-15	Hr.	Daily	Hr.	Daily	Bare Costs	Incl. O&P
1 Plumber	$21.95	$175.60	$36.30	$290.40	$19.75	$32.65
1 Plumber Apprentice	17.55	140.40	29.00	232.00		
1 Electric Welding Mach.		48.80		53.70	3.05	3.36
16 L.H., Daily Totals		$364.80		$576.10	$22.80	$36.01

Crew Q-16	Hr.	Daily	Hr.	Daily	Bare Costs	Incl. O&P
2 Plumbers	$21.95	$351.20	$36.30	$580.80	$20.48	$33.87
1 Plumber Apprentice	17.55	140.40	29.00	232.00		
1 Electric Welding Mach.		48.80		53.70	2.03	2.24
24 L.H., Daily Totals		$540.40		$866.50	$22.51	$36.11

Crew Q-17	Hr.	Daily	Hr.	Daily	Bare Costs	Incl. O&P
1 Steamfitter	$22.10	$176.80	$36.55	$292.40	$19.90	$32.90
1 Steamfitter Apprentice	17.70	141.60	29.25	234.00		
1 Electric Welding Mach.		48.80		53.70	3.05	3.36
16 L.H., Daily Totals		$367.20		$580.10	$22.95	$36.26

Crew Q-17A	Hr.	Daily	Hr.	Daily	Bare Costs	Incl. O&P
1 Steamfitter	$22.10	$176.80	$36.55	$292.40	$20.15	$33.35
1 Steamfitter Apprentice	17.70	141.60	29.25	234.00		
1 Equip. Oper. (crane)	20.65	165.20	34.25	274.00		
1 Truck Crane, 12 Ton		453.45		498.80		
1 Electric Welding Mach.		48.80		53.70	20.93	23.02
24 L.H., Daily Totals		$985.85		$1352.90	$41.08	$56.37

Crew Q-18	Hr.	Daily	Hr.	Daily	Bare Costs	Incl. O&P
2 Steamfitters	$22.10	$353.60	$36.55	$584.80	$20.63	$34.12
1 Steamfitter Apprentice	17.70	141.60	29.25	234.00		
1 Electric Welding Mach.		48.80		53.70	2.03	2.24
24 L.H., Daily Totals		$544.00		$872.50	$22.66	$36.36

Crew Q-19	Hr.	Daily	Hr.	Daily	Bare Costs	Incl. O&P
1 Steamfitter	$22.10	$176.80	$36.55	$292.40	$20.63	$33.98
1 Steamfitter Apprentice	17.70	141.60	29.25	234.00		
1 Electrician	22.10	176.80	36.15	289.20		
24 L.H., Daily Totals		$495.20		$815.60	$20.63	$33.98

Crew Q-20	Hr.	Daily	Hr.	Daily	Bare Costs	Incl. O&P
1 Sheet Metal Worker	$21.50	$172.00	$36.30	$290.40	$19.90	$33.37
1 Sheet Metal Apprentice	17.20	137.60	29.05	232.40		
.5 Electrician	22.10	88.40	36.15	144.60		
20 L.H., Daily Totals		$398.00		$667.40	$19.90	$33.37

Crew Q-21	Bare Costs Hr.	Bare Costs Daily	Incl. Subs O & P Hr.	Incl. Subs O & P Daily	Cost Per Labor-Hour Bare Costs	Cost Per Labor-Hour Incl. O&P
2 Steamfitters	$22.10	$353.60	$36.55	$584.80	$21.00	$34.6
1 Steamfitter Apprentice	17.70	141.60	29.25	234.00		
1 Electrician	22.10	176.80	36.15	289.20		
32 L.H., Daily Totals		$672.00		$1108.00	$21.00	$34.6

Crew Q-22	Hr.	Daily	Hr.	Daily	Bare Costs	Incl. O&P
1 Plumber	$21.95	$175.60	$36.30	$290.40	$19.75	$32.6
1 Plumber Apprentice	17.55	140.40	29.00	232.00		
1 Truck Crane, 12 Ton		453.45		498.80	28.34	31.1
16 L.H., Daily Totals		$769.45		$1021.20	$48.09	$63.8

Crew Q-22A	Hr.	Daily	Hr.	Daily	Bare Costs	Incl. O&P
1 Plumber	$21.95	$175.60	$36.30	$290.40	$18.65	$31.0
1 Plumber Apprentice	17.55	140.40	29.00	232.00		
1 Laborer	14.45	115.60	24.75	198.00		
1 Equip. Oper. (crane)	20.65	165.20	34.25	274.00		
1 Truck Crane, 12 Ton		453.45		498.80	14.17	15.5
32 L.H., Daily Totals		$1050.25		$1493.20	$32.82	$46.6

Crew Q-23	Hr.	Daily	Hr.	Daily	Bare Costs	Incl. O&P
1 Plumber Foreman	$23.95	$191.60	$39.60	$316.80	$21.93	$36.3
1 Plumber	21.95	175.60	36.30	290.40		
1 Equip. Oper. (medium)	19.90	159.20	33.00	264.00		
1 Power Tools		2.40		2.65		
1 Crane, 20 Ton		532.55		585.80	22.29	24.5
24 L.H., Daily Totals		$1061.35		$1459.65	$44.22	$60.8

Crew R-1	Hr.	Daily	Hr.	Daily	Bare Costs	Incl. O&P
1 Electrician Foreman	$22.60	$180.80	$36.95	$295.60	$19.75	$32.6
3 Electricians	22.10	530.40	36.15	867.60		
2 Helpers	14.80	236.80	25.35	405.60		
48 L.H., Daily Totals		$948.00		$1568.80	$19.75	$32.6

Crew R-1A	Hr.	Daily	Hr.	Daily	Bare Costs	Incl. O&P
1 Electrician	$22.10	$176.80	$36.15	$289.20	$18.45	$30.7
1 Helper	14.80	118.40	25.35	202.80		
16 L.H., Daily Totals		$295.20		$492.00	$18.45	$30.7

Crew R-2	Hr.	Daily	Hr.	Daily	Bare Costs	Incl. O&P
1 Electrician Foreman	$22.60	$180.80	$36.95	$295.60	$19.88	$32.9
3 Electricians	22.10	530.40	36.15	867.60		
2 Helpers	14.80	236.80	25.35	405.60		
1 Equip. Oper. (crane)	20.65	165.20	34.25	274.00		
1 S.P. Crane, 5 Ton		276.45		304.10	4.94	5.4
56 L.H., Daily Totals		$1389.65		$2146.90	$24.82	$38.3

Crew R-3	Hr.	Daily	Hr.	Daily	Bare Costs	Incl. O&P
1 Electrician Foreman	$22.60	$180.80	$36.95	$295.60	$22.01	$36.0
1 Electrician	22.10	176.80	36.15	289.20		
.5 Equip. Oper. (crane)	20.65	82.60	34.25	137.00		
.5 S.P. Crane, 5 Ton		138.23		152.05	6.91	7.6
20 L.H., Daily Totals		$578.43		$873.85	$28.92	$43.6

Crew R-4	Hr.	Daily	Hr.	Daily	Bare Costs	Incl. O&P
1 Struc. Steel Foreman	$23.25	$186.00	$45.85	$366.80	$21.82	$41.54
3 Struc. Steel Workers	21.25	510.00	41.90	1005.60		
1 Electrician	22.10	176.80	36.15	289.20		
1 Gas Welding Machine		84.00		92.40	2.10	2.31
40 L.H., Daily Totals		$956.80		$1754.00	$23.92	$43.85

Crew No.	Bare Costs		Incl. Subs O & P		Cost Per Labor-Hour	
Crew R-5	**Hr.**	**Daily**	**Hr.**	**Daily**	**Bare Costs**	**Incl. O&P**
ectrician Foreman	$22.60	$180.80	$36.95	$295.60	$19.49	$32.30
ectrician Linemen	22.10	707.20	36.15	1156.80		
ectrician Operators	22.10	353.60	36.15	578.40		
ectrician Groundmen	14.80	473.60	25.35	811.20		
ew Truck		197.60		217.35		
ol Van		214.40		235.85		
ckup Truck, 3/4 Ton		130.20		143.20		
rane, 55 Ton		161.64		177.80		
rane, 12 Ton		90.69		99.75		
uger, Truck Mtd.		359.20		395.10		
actor w/Winch		298.80		328.70	16.51	18.16
.H., Daily Totals		$3167.73		$4439.75	$36.00	$50.46
Crew R-6	**Hr.**	**Daily**	**Hr.**	**Daily**	**Bare Costs**	**Incl. O&P**
ectrician Foreman	$22.60	$180.80	$36.95	$295.60	$19.49	$32.30
ectrician Linemen	22.10	707.20	36.15	1156.80		
ectrician Operators	22.10	353.60	36.15	578.40		
ectrician Groundmen	14.80	473.60	25.35	811.20		
ew Truck		197.60		217.35		
ol Van		214.40		235.85		
ckup Truck, 3/4 Ton		130.20		143.20		
rane, 55 Ton		161.64		177.80		
rane, 12 Ton		90.69		99.75		
uger, Truck Mtd.		359.20		395.10		
actor w/Winch		298.80		328.70		
able Trailers		422.40		464.65		
ensioning Rig		141.20		155.30		
able Pulling Rig		819.50		901.45	32.22	35.45
.H., Daily Totals		$4550.83		$5961.15	$51.71	$67.75
Crew R-7	**Hr.**	**Daily**	**Hr.**	**Daily**	**Bare Costs**	**Incl. O&P**
lectrician Foreman	$22.60	$180.80	$36.95	$295.60	$16.10	$27.28
lectrician Groundmen	14.80	592.00	25.35	1014.00		
rew Truck		197.60		217.35	4.12	4.53
L.H., Daily Totals		$970.40		$1526.95	$20.22	$31.81
Crew R-8	**Hr.**	**Daily**	**Hr.**	**Daily**	**Bare Costs**	**Incl. O&P**
lectrician Foreman	$22.60	$180.80	$36.95	$295.60	$19.75	$32.68
lectrician Linemen	22.10	530.40	36.15	867.60		
lectrician Groundmen	14.80	236.80	25.35	405.60		
ickup Truck, 3/4 Ton		130.20		143.20		
rew Truck		197.60		217.35	6.83	7.51
L.H., Daily Totals		$1275.80		$1929.35	$26.58	$40.19
Crew R-9	**Hr.**	**Daily**	**Hr.**	**Daily**	**Bare Costs**	**Incl. O&P**
lectrician Foreman	$22.60	$180.80	$36.95	$295.60	$18.51	$30.85
lectrician Lineman	22.10	176.80	36.15	289.20		
lectrician Operators	22.10	353.60	36.15	578.40		
lectrician Groundmen	14.80	473.60	25.35	811.20		
ickup Truck, 3/4 Ton		130.20		143.20		
rew Truck		197.60		217.35	5.12	5.63
L.H., Daily Totals		$1512.60		$2334.95	$23.63	$36.48
Crew R-10	**Hr.**	**Daily**	**Hr.**	**Daily**	**Bare Costs**	**Incl. O&P**
lectrician Foreman	$22.60	$180.80	$36.95	$295.60	$20.97	$34.48
lectrician Linemen	22.10	707.20	36.15	1156.80		
lectrician Groundman	14.80	118.40	25.35	202.80		
rew Truck		197.60		217.35		
Tram Cars		541.20		595.30	15.39	16.93
L.H., Daily Totals		$1745.20		$2467.85	$36.36	$51.41

Crew No.	Bare Costs		Incl. Subs O & P		Cost Per Labor-Hour	
Crew R-11	**Hr.**	**Daily**	**Hr.**	**Daily**	**Bare Costs**	**Incl. O&P**
1 Electrician Foreman	$22.60	$180.80	$36.95	$295.60	$20.04	$33.09
4 Electricians	22.10	707.20	36.15	1156.80		
1 Helper	14.80	118.40	25.35	202.80		
1 Common Laborer	14.45	115.60	24.75	198.00		
1 Crew Truck		197.60		217.35		
1 Crane, 12 Ton		453.45		498.80	11.63	12.79
56 L.H., Daily Totals		$1773.05		$2569.35	$31.67	$45.88
Crew R-12	**Hr.**	**Daily**	**Hr.**	**Daily**	**Bare Costs**	**Incl. O&P**
1 Carpenter Foreman	$20.20	$161.60	$34.60	$276.80	$18.00	$31.23
4 Carpenters	19.70	630.40	33.75	1080.00		
4 Common Laborers	14.45	462.40	24.75	792.00		
1 Equip. Oper. (med.)	19.90	159.20	33.00	264.00		
1 Steel Worker	21.25	170.00	41.90	335.20		
1 Dozer, 200 H.P.		806.80		887.50		
1 Pickup Truck, 3/4 Ton		130.20		143.20	10.65	11.71
88 L.H., Daily Totals		$2520.60		$3778.70	$28.65	$42.94
Crew R-15	**Hr.**	**Daily**	**Hr.**	**Daily**	**Bare Costs**	**Incl. O&P**
1 Electrician Foreman	$22.60	$180.80	$36.95	$295.60	$21.68	$35.54
4 Electricians	22.10	707.20	36.15	1156.80		
1 Equipment Operator	19.10	152.80	31.70	253.60		
1 Aerial Lift Truck		251.35		276.50	5.24	5.76
48 L.H., Daily Totals		$1292.15		$1982.50	$26.92	$41.30
Crew R-18	**Hr.**	**Daily**	**Hr.**	**Daily**	**Bare Costs**	**Incl. O&P**
.25 Electrician Foreman	$22.60	$45.20	$36.95	$73.90	$17.65	$29.57
1 Electrician	22.10	176.80	36.15	289.20		
2 Helpers	14.80	236.80	25.35	405.60		
26 L.H., Daily Totals		$458.80		$768.70	$17.65	$29.57
Crew R-19	**Hr.**	**Daily**	**Hr.**	**Daily**	**Bare Costs**	**Incl. O&P**
.5 Electrician Foreman	$22.60	$90.40	$36.95	$147.80	$22.20	$36.31
2 Electricians	22.10	353.60	36.15	578.40		
20 L.H., Daily Totals		$444.00		$726.20	$22.20	$36.31
Crew R-21	**Hr.**	**Daily**	**Hr.**	**Daily**	**Bare Costs**	**Incl. O&P**
1 Electrician Foreman	$22.60	$180.80	$36.95	$295.60	$22.17	$36.27
3 Electricians	22.10	530.40	36.15	867.60		
.1 Equip. Oper. (med.)	19.90	15.92	33.00	26.40		
.1 Hyd. Crane 25 Ton		56.39		62.05	1.72	1.89
32. L.H., Daily Totals		$783.51		$1251.65	$23.89	$38.16
Crew R-22	**Hr.**	**Daily**	**Hr.**	**Daily**	**Bare Costs**	**Incl. O&P**
.66 Electrician Foreman	$22.60	$119.33	$36.95	$195.10	$19.04	$31.63
2 Helpers	14.80	236.80	25.35	405.60		
2 Electricians	22.10	353.60	36.15	578.40		
37.28 L.H., Daily Totals		$709.73		$1179.10	$19.04	$31.63
Crew R-30	**Hr.**	**Daily**	**Hr.**	**Daily**	**Bare Costs**	**Incl. O&P**
.25 Electricians	$24.10	$48.20	$39.40	$78.80	$17.55	$29.38
1 Electricians	22.10	176.80	36.15	289.20		
2 Laborers, (Semi-Skilled)	14.45	231.20	24.75	396.00		
26 L.H., Daily Totals		$456.20		$764.00	$17.55	$29.38

Location Factors

Costs shown in *Means cost data publications* are based on National Averages for materials and installation. To adjust these costs to a specific location, simply multiply the base cost by the factor for that city. The data is arranged alphabetically by state and postal zip code numbers. For a city not listed, use the factor for a nearby city with similar economic characteristics.

STATE/ZIP	CITY	Residential	Commercial
ALABAMA			
350-352	Birmingham	.86	.87
354	Tuscaloosa	.82	.80
355	Jasper	.75	.76
356	Decatur	.81	.82
357-358	Huntsville	.83	.84
359	Gadsden	.81	.82
360-361	Montgomery	.81	.79
362	Anniston	.74	.75
363	Dothan	.81	.79
364	Evergreen	.81	.79
365-366	Mobile	.83	.84
367	Selma	.80	.78
368	Phenix City	.81	.79
369	Butler	.80	.78
ALASKA			
995-996	Anchorage	1.26	1.25
997	Fairbanks	1.26	1.25
998	Juneau	1.26	1.25
999	Ketchikan	1.31	1.30
ARIZONA			
850,853	Phoenix	.93	.90
852	Mesa/Tempe	.88	.85
855	Globe	.90	.87
856-857	Tucson	.91	.88
859	Show Low	.92	.88
860	Flagstaff	.94	.90
863	Prescott	.91	.87
864	Kingman	.91	.87
865	Chambers	.90	.86
ARKANSAS			
716	Pine Bluff	.80	.80
717	Camden	.71	.71
718	Texarkana	.75	.74
719	Hot Springs	.70	.70
720-722	Little Rock	.81	.81
723	West Memphis	.80	.80
724	Jonesboro	.80	.80
725	Batesville	.77	.77
726	Harrison	.77	.77
727	Fayetteville	.70	.67
728	Russellville	.79	.76
729	Fort Smith	.83	.80
CALIFORNIA			
900-902	Los Angeles	1.10	1.10
903-905	Inglewood	1.07	1.07
906-908	Long Beach	1.08	1.08
910-912	Pasadena	1.07	1.07
913-916	Van Nuys	1.09	1.09
917-918	Alhambra	1.09	1.09
919-921	San Diego	1.11	1.07
922	Palm Springs	1.10	1.07
923-924	San Bernardino	1.10	1.06
925	Riverside	1.12	1.08
926-927	Santa Ana	1.10	1.07
928	Anaheim	1.11	1.09
930	Oxnard	1.14	1.09
931	Santa Barbara	1.11	1.08
932-933	Bakersfield	1.11	1.06
934	San Luis Obispo	1.22	1.09
935	Mojave	1.10	1.06
936-938	Fresno	1.12	1.08
939	Salinas	1.12	1.12
940-941	San Francisco	1.21	1.24
942,956-958	Sacramento	1.11	1.10
943	Palo Alto	1.14	1.17
944	San Mateo	1.16	1.18
945	Vallejo	1.13	1.15
946	Oakland	1.16	1.19
947	Berkeley	1.16	1.18
948	Richmond	1.14	1.17
949	San Rafael	1.24	1.18
950	Santa Cruz	1.16	1.14
951	San Jose	1.21	1.19
952	Stockton	1.14	1.10
953	Modesto	1.14	1.10

STATE/ZIP	CITY	Residential	Commercial
CALIFORNIA (CONT'D)			
954	Santa Rosa	1.14	1.17
955	Eureka	1.10	1.09
959	Marysville	1.11	1.10
960	Redding	1.11	1.10
961	Susanville	1.11	1.10
COLORADO			
800-802	Denver	.98	.94
803	Boulder	.89	.85
804	Golden	.96	.92
805	Fort Collins	.98	.92
806	Greeley	.91	.85
807	Fort Morgan	.97	.91
808-809	Colorado Springs	.94	.91
810	Pueblo	.94	.92
811	Alamosa	.90	.88
812	Salida	.90	.88
813	Durango	.88	.86
814	Montrose	.86	.84
815	Grand Junction	.91	.86
816	Glenwood Springs	.96	.91
CONNECTICUT			
060	New Britain	1.03	1.04
061	Hartford	1.03	1.04
062	Willimantic	1.03	1.04
063	New London	1.03	1.02
064	Meriden	1.03	1.04
065	New Haven	1.03	1.04
066	Bridgeport	1.01	1.04
067	Waterbury	1.04	1.04
068	Norwalk	.99	1.03
069	Stamford	1.03	1.07
D.C.			
200-205	Washington	.94	.96
DELAWARE			
197	Newark	.99	1.00
198	Wilmington	.98	.99
199	Dover	.99	1.00
FLORIDA			
320,322	Jacksonville	.85	.84
321	Daytona Beach	.88	.87
323	Tallahassee	.77	.79
324	Panama City	.71	.73
325	Pensacola	.86	.84
326,344	Gainesville	.85	.82
327-328,347	Orlando	.88	.86
329	Melbourne	.89	.88
330-332,340	Miami	.84	.86
333	Fort Lauderdale	.85	.87
334,349	West Palm Beach	.86	.83
335-336,346	Tampa	.82	.84
337	St. Petersburg	.82	.84
338	Lakeland	.81	.83
339,341	Fort Myers	.81	.81
342	Sarasota	.80	.82
GEORGIA			
300-303,399	Atlanta	.84	.89
304	Statesboro	.66	.68
305	Gainesville	.70	.74
306	Athens	.77	.81
307	Dalton	.69	.68
308-309	Augusta	.78	.79
310-312	Macon	.82	.82
313-314	Savannah	.81	.82
315	Waycross	.75	.75
316	Valdosta	.77	.77
317	Albany	.78	.80
318-319	Columbus	.79	.79
HAWAII			
967	Hilo	1.26	1.22
968	Honolulu	1.26	1.22

STATE/ZIP	CITY	Residential	Commercial
ATES & POSS.			
9	Guam	.85	.82
AHO			
2	Pocatello	.95	.94
3	Twin Falls	.81	.80
4	Idaho Falls	.85	.84
5	Lewiston	1.11	1.03
6-837	Boise	.95	.94
8	Coeur d'Alene	.98	.91
INOIS			
0-603	North Suburban	1.08	1.07
4	Joliet	1.06	1.05
5	South Suburban	1.08	1.07
6	Chicago	1.12	1.11
9	Kankakee	.99	.99
0-611	Rockford	1.04	1.03
2	Rock Island	1.05	.97
3	La Salle	1.06	.98
4	Galesburg	1.05	.98
5-616	Peoria	1.07	1.01
7	Bloomington	1.04	.99
8-619	Champaign	1.03	1.00
0-622	East St. Louis	.99	.99
3	Quincy	.97	.95
4	Effingham	1.00	.97
5	Decatur	1.00	.97
6-627	Springfield	1.01	.98
8	Centralia	.98	.98
9	Carbondale	.96	.96
DIANA			
0	Anderson	.94	.93
1-462	Indianapolis	.96	.95
3-464	Gary	1.02	1.00
5-466	South Bend	.92	.90
7-468	Fort Wayne	.91	.92
9	Kokomo	.91	.90
0	Lawrenceburg	.91	.88
1	New Albany	.92	.88
2	Columbus	.95	.92
3	Muncie	.93	.92
4	Bloomington	.95	.92
5	Washington	.93	.93
6-477	Evansville	.94	.94
8	Terre Haute	.95	.94
9	Lafayette	.90	.90
WA			
0-503,509	Des Moines	.96	.92
4	Mason City	.86	.80
5	Fort Dodge	.83	.78
6-507	Waterloo	.88	.83
8	Creston	.90	.85
0-511	Sioux City	.89	.83
2	Sibley	.80	.78
3	Spencer	.81	.79
4	Carroll	.85	.81
5	Council Bluffs	.94	.88
6	Shenandoah	.82	.77
20	Dubuque	.97	.87
21	Decorah	.90	.80
22-524	Cedar Rapids	1.00	.92
25	Ottumwa	.94	.86
26	Burlington	.92	.86
27-528	Davenport	.96	.94
ANSAS			
60-662	Kansas City	.96	.95
64-666	Topeka	.86	.85
67	Fort Scott	.86	.84
68	Emporia	.82	.81
69	Belleville	.88	.82
70-672	Wichita	.88	.85
73	Independence	.82	.80
74	Salina	.85	.81
75	Hutchinson	.80	.77
76	Hays	.86	.82
77	Colby	.86	.82
78	Dodge City	.86	.83
79	Liberal	.79	.76
ENTUCKY			
00-402	Louisville	.94	.91
03-405	Lexington	.88	.85

STATE/ZIP	CITY	Residential	Commercial
KENTUCKY (CONT'D)			
406	Frankfort	.94	.87
407-409	Corbin	.79	.74
410	Covington	.96	.93
411-412	Ashland	.95	.96
413-414	Campton	.78	.74
415-416	Pikeville	.82	.83
417-418	Hazard	.77	.74
420	Paducah	.95	.90
421-422	Bowling Green	.94	.89
423	Owensboro	.91	.89
424	Henderson	.94	.92
425-426	Somerset	.76	.73
427	Elizabethtown	.92	.88
LOUISIANA			
700-701	New Orleans	.86	.85
703	Thibodaux	.85	.85
704	Hammond	.83	.82
705	Lafayette	.85	.82
706	Lake Charles	.83	.83
707-708	Baton Rouge	.83	.82
710-711	Shreveport	.81	.81
712	Monroe	.79	.79
713-714	Alexandria	.78	.78
MAINE			
039	Kittery	.78	.79
040-041	Portland	.87	.89
042	Lewiston	.88	.89
043	Augusta	.78	.78
044	Bangor	.92	.92
045	Bath	.78	.78
046	Machias	.83	.83
047	Houlton	.80	.80
048	Rockland	.83	.83
049	Waterville	.79	.78
MARYLAND			
206	Waldorf	.89	.89
207-208	College Park	.90	.90
209	Silver Spring	.90	.90
210-212	Baltimore	.91	.91
214	Annapolis	.88	.89
215	Cumberland	.88	.89
216	Easton	.74	.75
217	Hagerstown	.90	.88
218	Salisbury	.78	.79
219	Elkton	.84	.85
MASSACHUSETTS			
010-011	Springfield	1.05	1.03
012	Pittsfield	.99	.99
013	Greenfield	1.02	1.00
014	Fitchburg	1.09	1.05
015-016	Worcester	1.11	1.07
017	Framingham	1.07	1.08
018	Lowell	1.10	1.10
019	Lawrence	1.10	1.10
020-022, 024	Boston	1.16	1.17
023	Brockton	1.07	1.09
025	Buzzards Bay	1.04	1.06
026	Hyannis	1.06	1.07
027	New Bedford	1.07	1.08
MICHIGAN			
480,483	Royal Oak	1.03	1.02
481	Ann Arbor	1.05	1.04
482	Detroit	1.06	1.05
484-485	Flint	.98	.99
486	Saginaw	.96	.97
487	Bay City	.96	.97
488-489	Lansing	1.02	.99
490	Battle Creek	.99	.93
491	Kalamazoo	1.00	.94
492	Jackson	1.00	.97
493,495	Grand Rapids	.90	.87
494	Muskegan	.96	.93
496	Traverse City	.89	.86
497	Gaylord	.89	.90
498-499	Iron mountain	.98	.95
MINNESOTA			
550-551	Saint Paul	1.11	1.09
553-555	Minneapolis	1.14	1.10

STATE/ZIP	CITY	Residential	Commercial
556-558	Duluth	1.03	1.05
559	Rochester	1.04	1.01
560	Mankato	1.01	1.00
561	Windom	.90	.89
562	Willmar	.93	.92
563	St. Cloud	1.09	1.02
564	Brainerd	1.06	.99
565	Detroit Lakes	.88	.95
566	Bemidji	.91	.98
567	Thief River Falls	.87	.94
MISSISSIPPI			
386	Clarksdale	.71	.67
387	Greenville	.82	.78
388	Tupelo	.72	.73
389	Greenwood	.73	.69
390-392	Jackson	.82	.78
393	Meridian	.77	.76
394	Laurel	.73	.70
395	Biloxi	.86	.82
396	Mccomb	.69	.68
397	Columbus	.72	.73
MISSOURI			
630-631	St. Louis	1.00	1.03
633	Bowling Green	.92	.94
634	Hannibal	1.00	.93
635	Kirksville	.85	.89
636	Flat River	.94	.97
637	Cape Girardeau	.93	.96
638	Sikeston	.89	.91
639	Poplar Bluff	.89	.91
640-641	Kansas City	1.01	.98
644-645	St. Joseph	.86	.90
646	Chillicothe	.80	.83
647	Harrisonville	.93	.91
648	Joplin	.83	.85
650-651	Jefferson City	.96	.90
652	Columbia	.94	.88
653	Sedalia	.94	.88
654-655	Rolla	.87	.82
656-658	Springfield	.84	.86
MONTANA			
590-591	Billings	.98	.96
592	Wolf Point	.98	.96
593	Miles City	.97	.95
594	Great Falls	.97	.96
595	Havre	.96	.95
596	Helena	.96	.95
597	Butte	.94	.93
598	Missoula	.95	.94
599	Kalispell	.94	.93
NEBRASKA			
680-681	Omaha	.91	.90
683-685	Lincoln	.88	.83
686	Columbus	.75	.74
687	Norfolk	.85	.84
688	Grand Island	.88	.84
689	Hastings	.83	.79
690	Mccook	.77	.74
691	North Platte	.84	.80
692	Valentine	.78	.75
693	Alliance	.76	.72
NEVADA			
889-891	Las Vegas	1.06	1.05
893	Ely	.96	.97
894-895	Reno	.95	1.00
897	Carson City	.97	.99
898	Elko	.93	.96
NEW HAMPSHIRE			
030	Nashua	.93	.94
031	Manchester	.93	.94
032-033	Concord	.92	.93
034	Keene	.79	.80
035	Littleton	.81	.82
036	Charleston	.77	.78
037	Claremont	.77	.78
038	Portsmouth	.92	.91

STATE/ZIP	CITY	Residential	Commercial
NEW JERSEY			
070-071	Newark	1.14	1.12
072	Elizabeth	1.09	1.07
073	Jersey City	1.12	1.11
074-075	Paterson	1.12	1.12
076	Hackensack	1.10	1.10
077	Long Branch	1.11	1.09
078	Dover	1.12	1.10
079	Summit	1.08	1.06
080,083	Vineland	1.11	1.07
081	Camden	1.12	1.09
082,084	Atlantic City	1.11	1.09
085-086	Trenton	1.13	1.11
087	Point Pleasant	1.11	1.09
088-089	New Brunswick	1.12	1.10
NEW MEXICO			
870-872	Albuquerque	.89	.91
873	Gallup	.89	.91
874	Farmington	.89	.91
875	Santa Fe	.89	.91
877	Las Vegas	.88	.90
878	Socorro	.87	.89
879	Truth/Consequences	.87	.87
880	Las Cruces	.85	.85
881	Clovis	.90	.90
882	Roswell	.90	.90
883	Carrizozo	.91	.91
884	Tucumcari	.90	.90
NEW YORK			
100-102	New York	1.34	1.34
103	Staten Island	1.28	1.28
104	Bronx	1.28	1.28
105	Mount Vernon	1.18	1.18
106	White Plains	1.17	1.17
107	Yonkers	1.20	1.20
108	New Rochelle	1.18	1.18
109	Suffern	1.12	1.12
110	Queens	1.29	1.29
111	Long Island City	1.30	1.30
112	Brooklyn	1.30	1.30
113	Flushing	1.30	1.30
114	Jamaica	1.29	1.29
115,117,118	Hicksville	1.27	1.27
116	Far Rockaway	1.30	1.30
119	Riverhead	1.27	1.27
120-122	Albany	.98	.98
123	Schenectady	.98	.98
124	Kingston	1.11	1.09
125-126	Poughkeepsie	1.13	1.12
127	Monticello	1.09	1.08
128	Glens Falls	.95	.93
129	Plattsburgh	.96	.94
130-132	Syracuse	.99	.97
133-135	Utica	.91	.94
136	Watertown	.94	.97
137-139	Binghamton	.94	.94
140-142	Buffalo	1.07	1.03
143	Niagara Falls	1.06	1.02
144-146	Rochester	.99	1.00
147	Jamestown	.97	.94
148-149	Elmira	.95	.93
NORTH CAROLINA			
270,272-274	Greensboro	.76	.77
271	Winston-Salem	.76	.77
275-276	Raleigh	.77	.77
277	Durham	.76	.77
278	Rocky Mount	.69	.69
279	Elizabeth City	.71	.71
280	Gastonia	.75	.76
281-282	Charlotte	.75	.76
283	Fayetteville	.76	.76
284	Wilmington	.74	.76
285	Kinston	.68	.68
286	Hickory	.67	.68
287-288	Asheville	.74	.76
289	Murphy	.67	.68
NORTH DAKOTA			
580-581	Fargo	.78	.83
582	Grand Forks	.78	.82
583	Devils Lake	.78	.82
584	Jamestown	.78	.82
585	Bismarck	.81	.85

STATE/ZIP	CITY	Residential	Commercial
RTH DAKOTA ONT'D)			
56	Dickinson	.79	.83
57	Minot	.81	.85
58	Williston	.78	.82
HIO			
30-432	Columbus	.96	.94
33	Marion	.91	.92
34-436	Toledo	.99	.98
37-438	Zanesville	.91	.90
39	Steubenville	.97	.97
40	Lorain	1.03	.97
41	Cleveland	1.09	1.02
2-443	Akron	1.00	.99
44-445	Youngstown	.99	.96
46-447	Canton	.96	.95
48-449	Mansfield	.95	.94
50	Hamilton	.98	.92
51-452	Cincinnati	.99	.93
53-454	Dayton	.93	.92
55	Springfield	.94	.92
56	Chillicothe	1.01	.95
57	Athens	.88	.87
58	Lima	.95	.94
KLAHOMA			
30-731	Oklahoma City	.82	.84
34	Ardmore	.83	.82
35	Lawton	.84	.83
36	Clinton	.80	.82
37	Enid	.83	.82
38	Woodward	.82	.81
39	Guymon	.70	.69
40-741	Tulsa	.85	.82
43	Miami	.85	.83
44	Muskogee	.76	.74
45	Mcalester	.76	.78
46	Ponca City	.82	.81
47	Durant	.79	.81
48	Shawnee	.79	.81
49	Poteau	.85	.81
REGON			
70-972	Portland	1.08	1.06
73	Salem	1.07	1.06
74	Eugene	1.06	1.05
75	Medford	1.06	1.05
76	Klamath Falls	1.07	1.06
77	Bend	1.07	1.06
78	Pendleton	1.04	1.02
79	Vale	.99	.97
ENNSYLVANIA			
50-152	Pittsburgh	1.04	1.02
53	Washington	1.03	1.01
54	Uniontown	1.02	1.00
55	Bedford	1.04	.97
56	Greensburg	1.03	1.01
57	Indiana	1.06	.99
58	Dubois	1.04	.97
59	Johnstown	1.05	.98
60	Butler	1.00	.98
61	New Castle	1.00	.97
62	Kittanning	1.03	1.00
63	Oil City	.90	.95
64-165	Erie	.98	.97
66	Altoona	1.05	.96
67	Bradford	.99	.97
68	State College	.98	.98
69	Wellsboro	.94	.95
70-171	Harrisburg	.97	.96
72	Chambersburg	.96	.95
73-174	York	.97	.95
75-176	Lancaster	.96	.94
77	Williamsport	.92	.92
78	Sunbury	.96	.95
79	Pottsville	.96	.95
80	Lehigh Valley	1.02	1.01
81	Allentown	1.02	1.01
82	Hazleton	.97	.96
83	Stroudsburg	.96	.95
84-185	Scranton	.97	1.00
86-187	Wilkes-Barre	.94	.97
88	Montrose	.94	.97
89	Doylestown	.94	1.05
PENNSYLVANIA (CONT'D)			
190-191	Philadelphia	1.14	1.12
193	Westchester	1.07	1.05
194	Norristown	1.09	1.07
195-196	Reading	.97	.98
RHODE ISLAND			
028	Newport	1.02	1.04
029	Providence	1.02	1.04
SOUTH CAROLINA			
290-292	Columbia	.72	.75
293	Spartanburg	.72	.75
294	Charleston	.74	.76
295	Florence	.71	.73
296	Greenville	.72	.75
297	Rock Hill	.65	.68
298	Aiken	.66	.68
299	Beaufort	.69	.71
SOUTH DAKOTA			
570-571	Sioux Falls	.88	.82
572	Watertown	.85	.79
573	Mitchell	.84	.79
574	Aberdeen	.86	.80
575	Pierre	.86	.80
576	Mobridge	.86	.79
577	Rapid City	.85	.79
TENNESSEE			
370-372	Nashville	.84	.84
373-374	Chattanooga	.83	.82
375,380-381	Memphis	.85	.85
376	Johnson City	.80	.79
377-379	Knoxville	.80	.80
382	Mckenzie	.69	.69
383	Jackson	.68	.75
384	Columbia	.76	.76
385	Cookeville	.69	.69
TEXAS			
750	Mckinney	.90	.83
751	Waxahackie	.83	.83
752-753	Dallas	.91	.87
754	Greenville	.80	.75
755	Texarkana	.90	.79
756	Longview	.85	.75
757	Tyler	.92	.81
758	Palestine	.75	.75
759	Lufkin	.79	.79
760-761	Fort Worth	.85	.84
762	Denton	.88	.80
763	Wichita Falls	.82	.82
764	Eastland	.76	.75
765	Temple	.78	.77
766-767	Waco	.82	.81
768	Brownwood	.74	.73
769	San Angelo	.80	.77
770-772	Houston	.88	.89
773	Huntsville	.75	.75
774	Wharton	.77	.78
775	Galveston	.86	.87
776-777	Beaumont	.83	.85
778	Bryan	.82	.83
779	Victoria	.81	.81
780	Laredo	.79	.79
781-782	San Antonio	.83	.84
783-784	Corpus Christi	.81	.80
785	Mc Allen	.80	.78
786-787	Austin	.79	.82
788	Del Rio	.70	.70
789	Giddings	.75	.74
790-791	Amarillo	.81	.81
792	Childress	.76	.79
793-794	Lubbock	.80	.82
795-796	Abilene	.78	.78
797	Midland	.80	.81
798-799,885	El Paso	.78	.77
UTAH			
840-841	Salt Lake City	.90	.89
842,844	Ogden	.90	.88
843	Logan	.91	.89
845	Price	.82	.81
846-847	Provo	.90	.89

Location Factors

STATE/ZIP	CITY	Residential	Commercial
VERMONT			
050	White River Jct.	.73	.72
051	Bellows Falls	.74	.73
052	Bennington	.70	.69
053	Brattleboro	.74	.73
054	Burlington	.83	.84
056	Montpelier	.81	.82
057	Rutland	.84	.83
058	St. Johnsbury	.74	.75
059	Guildhall	.73	.74
VIRGINIA			
220-221	Fairfax	.89	.90
222	Arlington	.90	.90
223	Alexandria	.90	.91
224-225	Fredericksburg	.85	.85
226	Winchester	.79	.80
227	Culpeper	.80	.80
228	Harrisonburg	.76	.77
229	Charlottesville	.84	.82
230-232	Richmond	.87	.85
233-235	Norfolk	.82	.82
236	Newport News	.83	.82
237	Portsmouth	.81	.81
238	Petersburg	.86	.84
239	Farmville	.75	.74
240-241	Roanoke	.78	.77
242	Bristol	.80	.75
243	Pulaski	.71	.70
244	Staunton	.78	.76
245	Lynchburg	.81	.77
246	Grundy	.70	.70
WASHINGTON			
980-981,987	Seattle	1.01	1.06
982	Everett	.98	1.04
983-984	Tacoma	1.06	1.04
985	Olympia	1.06	1.04
986	Vancouver	1.11	1.05
988	Wenatchee	.94	.98
989	Yakima	1.04	1.02
990-992	Spokane	1.01	1.00
993	Richland	1.02	1.01
994	Clarkston	1.01	1.00
WEST VIRGINIA			
247-248	Bluefield	.88	.88
249	Lewisburg	.90	.90
250-253	Charleston	.94	.94
254	Martinsburg	.77	.77
255-257	Huntington	.95	.97
258-259	Beckley	.92	.92
260	Wheeling	.94	.96
261	Parkersburg	.94	.96
262	Buckhannon	.97	.94
263-264	Clarksburg	.98	.95
265	Morgantown	.98	.95
266	Gassaway	.94	.94
267	Romney	.91	.91
268	Petersburg	.95	.92
WISCONSIN			
530,532	Milwaukee	1.02	1.01
531	Kenosha	1.02	1.01
534	Racine	1.05	1.00
535	Beloit	1.00	.98
537	Madison	1.01	.99
538	Lancaster	.93	.91
539	Portage	.99	.96
540	New Richmond	1.02	.94
541-543	Green Bay	1.00	.97
544	Wausau	.97	.93
545	Rhinelander	.97	.93
546	La Crosse	.98	.95
547	Eau Claire	1.04	.96
548	Superior	1.03	.97
549	Oshkosh	.97	.94
WYOMING			
820	Cheyenne	.88	.83
821	Yellowstone Nat. Pk.	.82	.79
822	Wheatland	.84	.80
823	Rawlins	.83	.79
824	Worland	.80	.77
825	Riverton	.84	.80
826	Casper	.87	.83

STATE/ZIP	CITY	Residential	Commercial
WYOMING (CONT'D)			
827	Newcastle	.81	.77
828	Sheridan	.85	.82
829-831	Rock Springs	.84	.80
CANADIAN FACTORS (reflect Canadian currency)			
ALBERTA			
	Calgary	1.01	.98
	Edmonton	1.01	.98
	Fort McMurray	1.00	.97
	Lethbridge	1.00	.97
	Lloydminster	1.00	.97
	Medicine Hat	1.00	.97
	Red Deer	1.00	.97
BRITISH COLUMBIA			
	Kamloops	1.04	1.05
	Prince George	1.06	1.07
	Vancouver	1.07	1.08
	Victoria	1.06	1.07
MANITOBA			
	Brandon	.99	.98
	Portage la Prairie	.99	.98
	Winnipeg	.99	.98
NEW BRUNSWICK			
	Bathurst	.94	.92
	Dalhousie	.94	.92
	Fredericton	.97	.95
	Moncton	.94	.92
	Newcastle	.94	.92
	Saint John	.97	.95
NEWFOUNDLAND			
	Corner Brook	.96	.95
	St. John's	.96	.95
NORTHWEST TERRITORIES			
	Yellowknife	.93	.92
NOVA SCOTIA			
	Dartmouth	.98	.97
	Halifax	.98	.97
	New Glasgow	.98	.97
	Sydney	.96	.95
	Yarmouth	.98	.97
ONTARIO			
	Barrie	1.10	1.08
	Brantford	1.12	1.10
	Cornwall	1.10	1.08
	Hamilton	1.14	1.10
	Kingston	1.10	1.08
	Kitchener	1.06	1.04
	London	1.10	1.08
	North Bay	1.09	1.07
	Oshawa	1.10	1.08
	Ottawa	1.11	1.09
	Owen Sound	1.10	1.08
	Peterborough	1.10	1.08
	Sarnia	1.12	1.10
	St. Catharines	1.06	1.04
	Sudbury	1.05	1.03
	Thunder Bay	1.06	1.04
	Toronto	1.13	1.12
	Windsor	1.07	1.05
PRINCE EDWARD ISLAND			
	Charlottetown	.93	.91
	Summerside	.93	.92
QUEBEC			
	Cap-de-la-Madeleine	1.05	1.04
	Charlesbourg	1.05	1.04
	Chicoutimi	1.04	1.03
	Gatineau	1.04	1.03
	Laval	1.05	1.04
	Montreal	1.11	1.04
	Quebec	1.13	1.05
	Sherbrooke	1.05	1.04
	Trois Rivieres	1.05	1.04

STATE/ZIP	CITY	Residential	Commercial
SKATCHEWAN			
	Moose Jaw	.93	.93
	Prince Albert	.93	.93
	Regina	.93	.93
	Saskatoon	.93	.93
KON			
	Whitehorse	.93	.92

Abbreviations

A	Area Square Feet; Ampere
ABS	Acrylonitrile Butadiene Stryrene; Asbestos Bonded Steel
A.C.	Alternating Current; Air-Conditioning; Asbestos Cement; Plywood Grade A & C
A.C.I.	American Concrete Institute
AD	Plywood, Grade A & D
Addit.	Additional
Adj.	Adjustable
af	Audio-frequency
A.G.A.	American Gas Association
Agg.	Aggregate
A.H.	Ampere Hours
A hr.	Ampere-hour
A.H.U.	Air Handling Unit
A.I.A.	American Institute of Architects
AIC	Ampere Interrupting Capacity
Allow.	Allowance
alt.	Altitude
Alum.	Aluminum
a.m.	Ante Meridiem
Amp.	Ampere
Anod.	Anodized
Approx.	Approximate
Apt.	Apartment
Asb.	Asbestos
A.S.B.C.	American Standard Building Code
Asbe.	Asbestos Worker
A.S.H.R.A.E.	American Society of Heating, Refrig. & AC Engineers
A.S.M.E.	American Society of Mechanical Engineers
A.S.T.M.	American Society for Testing and Materials
Attchmt.	Attachment
Avg.	Average
A.W.G.	American Wire Gauge
AWWA	American Water Works Assoc.
Bbl.	Barrel
B. & B.	Grade B and Better; Balled & Burlapped
B. & S.	Bell and Spigot
B. & W.	Black and White
b.c.c.	Body-centered Cubic
B.C.Y.	Bank Cubic Yards
BE	Bevel End
B.F.	Board Feet
Bg. cem.	Bag of Cement
BHP	Boiler Horsepower; Brake Horsepower
B.I.	Black Iron
Bit.; Bitum.	Bituminous
Bk.	Backed
Bkrs.	Breakers
Bldg.	Building
Blk.	Block
Bm.	Beam
Boil.	Boilermaker
B.P.M.	Blows per Minute
BR	Bedroom
Brg.	Bearing
Brhe.	Bricklayer Helper
Bric.	Bricklayer
Brk.	Brick
Brng.	Bearing
Brs.	Brass
Brz.	Bronze
Bsn.	Basin
Btr.	Better
BTU	British Thermal Unit
BTUH	BTU per Hour
B.U.R.	Built-up Roofing
BX	Interlocked Armored Cable
c	Conductivity, Copper Sweat
C	Hundred; Centigrade
C/C	Center to Center, Cedar on Cedar
Cab.	Cabinet
Cair.	Air Tool Laborer
Calc	Calculated
Cap.	Capacity
Carp.	Carpenter
C.B.	Circuit Breaker
C.C.A.	Chromate Copper Arsenate
C.C.F.	Hundred Cubic Feet
cd	Candela
cd/sf	Candela per Square Foot
CD	Grade of Plywood Face & Back
CDX	Plywood, Grade C & D, exterior glue
Cefi.	Cement Finisher
Cem.	Cement
CF	Hundred Feet
C.F.	Cubic Feet
CFM	Cubic Feet per Minute
c.g.	Center of Gravity
CHW	Chilled Water; Commercial Hot Water
C.I.	Cast Iron
C.I.P.	Cast in Place
Circ.	Circuit
C.L.	Carload Lot
Clab.	Common Laborer
C.L.F.	Hundred Linear Feet
CLF	Current Limiting Fuse
CLP	Cross Linked Polyethylene
cm	Centimeter
CMP	Corr. Metal Pipe
C.M.U.	Concrete Masonry Unit
CN	Change Notice
Col.	Column
CO_2	Carbon Dioxide
Comb.	Combination
Compr.	Compressor
Conc.	Concrete
Cont.	Continuous; Continued
Corr.	Corrugated
Cos	Cosine
Cot	Cotangent
Cov.	Cover
C/P	Cedar on Paneling
CPA	Control Point Adjustment
Cplg.	Coupling
C.P.M.	Critical Path Method
CPVC	Chlorinated Polyvinyl Chloride
C.Pr.	Hundred Pair
CRC	Cold Rolled Channel
Creos.	Creosote
Crpt.	Carpet & Linoleum Layer
CRT	Cathode-ray Tube
CS	Carbon Steel, Constant Shear Bar Joist
Csc	Cosecant
C.S.F.	Hundred Square Feet
CSI	Construction Specifications Institute
C.T.	Current Transformer
CTS	Copper Tube Size
Cu	Copper, Cubic
Cu. Ft.	Cubic Foot
cw	Continuous Wave
C.W.	Cool White; Cold Water
Cwt.	100 Pounds
C.W.X.	Cool White Deluxe
C.Y.	Cubic Yard (27 cubic feet)
C.Y./Hr.	Cubic Yard per Hour
Cyl.	Cylinder
d	Penny (nail size)
D	Deep; Depth; Discharge
Dis.;Disch.	Discharge
Db.	Decibel
Dbl.	Double
DC	Direct Current
DDC	Direct Digital Control
Demob.	Demobilization
d.f.u.	Drainage Fixture Units
D.H.	Double Hung
DHW	Domestic Hot Water
Diag.	Diagonal
Diam.	Diameter
Distrib.	Distribution
Dk.	Deck
D.L.	Dead Load; Diesel
DLH	Deep Long Span Bar Joist
Do.	Ditto
Dp.	Depth
D.P.S.T.	Double Pole, Single Throw
Dr.	Driver
Drink.	Drinking
D.S.	Double Strength
D.S.A.	Double Strength A Grade
D.S.B.	Double Strength B Grade
Dty.	Duty
DWV	Drain Waste Vent
DX	Deluxe White, Direct Expansio
dyn	Dyne
e	Eccentricity
E	Equipment Only; East
Ea.	Each
E.B.	Encased Burial
Econ.	Economy
EDP	Electronic Data Processing
EIFS	Exterior Insulation Finish Syste
E.D.R.	Equiv. Direct Radiation
Eq.	Equation
Elec.	Electrician; Electrical
Elev.	Elevator; Elevating
EMT	Electrical Metallic Conduit; Thin Wall Conduit
Eng.	Engine, Engineered
EPDM	Ethylene Propylene Diene Monomer
EPS	Expanded Polystyrene
Eqhv.	Equip. Oper., Heavy
Eqlt.	Equip. Oper., Light
Eqmd.	Equip. Oper., Medium
Eqmm.	Equip. Oper., Master Mechanic
Eqol.	Equip. Oper., Oilers
Equip.	Equipment
ERW	Electric Resistance Welded
E.S.	Energy Saver
Est.	Estimated
esu	Electrostatic Units
E.W.	Each Way
EWT	Entering Water Temperature
Excav.	Excavation
Exp.	Expansion, Exposure
Ext.	Exterior
Extru.	Extrusion
f.	Fiber stress
F	Fahrenheit; Female; Fill
Fab.	Fabricated
FBGS	Fiberglass
F.C.	Footcandles
f.c.c.	Face-centered Cubic
f'c.	Compressive Stress in Concrete; Extreme Compressive Stress
F.E.	Front End
FEP	Fluorinated Ethylene Propylene (Teflon)
F.G.	Flat Grain
F.H.A.	Federal Housing Administration
Fig.	Figure
Fin.	Finished
Fixt.	Fixture
Fl. Oz.	Fluid Ounces
Flr.	Floor
F.M.	Frequency Modulation; Factory Mutual
Fmg.	Framing
Fndtn.	Foundation
Fori.	Foreman, Inside
Foro.	Foreman, Outside

nt.	Fountain
	Feet per Minute
	Female Pipe Thread
	Frame
	Fire Rating
	Foil Reinforced Kraft
	Fiberglass Reinforced Plastic
	Forged Steel
	Cast Body; Cast Switch Box
	Foot; Feet
g.	Fitting
	Footing
Lb.	Foot Pound
n.	Furniture
IR	Full Voltage Non-Reversing
A	Female by Male
	Minimum Yield Stress of Steel
	Gram
	Gauss
	Gauge
	Gallon
/Min.	Gallon per Minute
v.	Galvanized
n.	General
I.	Ground Fault Interrupter
z.	Glazier
D	Gallons per Day
H	Gallons per Hour
M	Gallons per Minute
	Grade
n.	Granular
d.	Ground
	High; High Strength Bar Joist; Henry
C.	High Capacity
D.	Heavy Duty; High Density
D.O.	High Density Overlaid
r.	Header
we.	Hardware
p.	Helpers Average
PA	High Efficiency Particulate Air Filter
	Mercury
C	High Interrupting Capacity
A	Hollow Metal
O.	High Output
riz.	Horizontal
P.	Horsepower; High Pressure
P.F.	High Power Factor
	Hour
s./Day	Hours per Day
C	High Short Circuit
	Height
g.	Heating
rs.	Heaters
AC	Heating, Ventilation & Air-Conditioning
y.	Heavy
W	Hot Water
d.;Hydr.	Hydraulic
z.	Hertz (cycles)
	Moment of Inertia
C.	Interrupting Capacity
	Inside Diameter
D.	Inside Dimension; Identification
	Inside Frosted
M.C.	Intermediate Metal Conduit
	Inch
can.	Incandescent
cl.	Included; Including
t.	Interior
st.	Installation
sul.	Insulation/Insulated
P.	Iron Pipe
P.S.	Iron Pipe Size
P.T.	Iron Pipe Threaded
W.	Indirect Waste
	Joule

J.I.C.	Joint Industrial Council
K	Thousand; Thousand Pounds; Heavy Wall Copper Tubing, Kelvin
K.A.H.	Thousand Amp. Hours
KCMIL	Thousand Circular Mils
KD	Knock Down
K.D.A.T.	Kiln Dried After Treatment
kg	Kilogram
kG	Kilogauss
kgf	Kilogram Force
kHz	Kilohertz
Kip.	1000 Pounds
KJ	Kiljoule
K.L.	Effective Length Factor
K.L.F.	Kips per Linear Foot
Km	Kilometer
K.S.F.	Kips per Square Foot
K.S.I.	Kips per Square Inch
kV	Kilovolt
kVA	Kilovolt Ampere
K.V.A.R.	Kilovar (Reactance)
KW	Kilowatt
KWh	Kilowatt-hour
L	Labor Only; Length; Long; Medium Wall Copper Tubing
Lab.	Labor
lat	Latitude
Lath.	Lather
Lav.	Lavatory
lb.; #	Pound
L.B.	Load Bearing; L Conduit Body
L. & E.	Labor & Equipment
lb./hr.	Pounds per Hour
lb./L.F.	Pounds per Linear Foot
lbf/sq.in.	Pound-force per Square Inch
L.C.L.	Less than Carload Lot
Ld.	Load
LE	Lead Equivalent
LED	Light Emitting Diode
L.F.	Linear Foot
Lg.	Long; Length; Large
L & H	Light and Heat
LH	Long Span Bar Joist
L.H.	Labor Hours
L.L.	Live Load
L.L.D.	Lamp Lumen Depreciation
L-O-L	Lateralolet
lm	Lumen
lm/sf	Lumen per Square Foot
lm/W	Lumen per Watt
L.O.A.	Length Over All
log	Logarithm
L.P.	Liquefied Petroleum; Low Pressure
L.P.F.	Low Power Factor
LR	Long Radius
L.S.	Lump Sum
Lt.	Light
Lt. Ga.	Light Gauge
L.T.L.	Less than Truckload Lot
Lt. Wt.	Lightweight
L.V.	Low Voltage
M	Thousand; Material; Male; Light Wall Copper Tubing
M^2CA	Meters Squared Contact Area
m/hr; M.H.	Man-hour
mA	Milliampere
Mach.	Machine
Mag. Str.	Magnetic Starter
Maint.	Maintenance
Marb.	Marble Setter
Mat; Mat'l.	Material
Max.	Maximum
MBF	Thousand Board Feet
MBH	Thousand BTU's per hr.
MC	Metal Clad Cable
M.C.F.	Thousand Cubic Feet
M.C.F.M.	Thousand Cubic Feet per Minute
M.C.M.	Thousand Circular Mils

M.C.P.	Motor Circuit Protector
MD	Medium Duty
M.D.O.	Medium Density Overlaid
Med.	Medium
MF	Thousand Feet
M.F.B.M.	Thousand Feet Board Measure
Mfg.	Manufacturing
Mfrs.	Manufacturers
mg	Milligram
MGD	Million Gallons per Day
MGPH	Thousand Gallons per Hour
MH, M.H.	Manhole; Metal Halide; Man-Hour
MHz	Megahertz
Mi.	Mile
MI	Malleable Iron; Mineral Insulated
mm	Millimeter
Mill.	Millwright
Min., min.	Minimum, minute
Misc.	Miscellaneous
ml	Milliliter, Mainline
M.L.F.	Thousand Linear Feet
Mo.	Month
Mobil.	Mobilization
Mog.	Mogul Base
MPH	Miles per Hour
MPT	Male Pipe Thread
MRT	Mile Round Trip
ms	Millisecond
M.S.F.	Thousand Square Feet
Mstz.	Mosaic & Terrazzo Worker
M.S.Y.	Thousand Square Yards
Mtd.	Mounted
Mthe.	Mosaic & Terrazzo Helper
Mtng.	Mounting
Mult.	Multi; Multiply
M.V.A.	Million Volt Amperes
M.V.A.R.	Million Volt Amperes Reactance
MV	Megavolt
MW	Megawatt
MXM	Male by Male
MYD	Thousand Yards
N	Natural; North
nA	Nanoampere
NA	Not Available; Not Applicable
N.B.C.	National Building Code
NC	Normally Closed
N.E.M.A.	National Electrical Manufacturers Assoc.
NEHB	Bolted Circuit Breaker to 600V.
N.L.B.	Non-Load-Bearing
NM	Non-Metallic Cable
nm	Nanometer
No.	Number
NO	Normally Open
N.O.C.	Not Otherwise Classified
Nose.	Nosing
N.P.T.	National Pipe Thread
NQOD	Combination Plug-on/Bolt on Circuit Breaker to 240V.
N.R.C.	Noise Reduction Coefficient
N.R.S.	Non Rising Stem
ns	Nanosecond
nW	Nanowatt
OB	Opposing Blade
OC	On Center
OD	Outside Diameter
O.D.	Outside Dimension
ODS	Overhead Distribution System
O.G.	Ogee
O.H.	Overhead
O & P	Overhead and Profit
Oper.	Operator
Opng.	Opening
Orna.	Ornamental
OSB	Oriented Strand Board
O. S. & Y.	Outside Screw and Yoke
Ovhd.	Overhead
OWG	Oil, Water or Gas

Abbreviations

Oz.	Ounce
P.	Pole; Applied Load; Projection
p.	Page
Pape.	Paperhanger
P.A.P.R.	Powered Air Purifying Respirator
PAR	Weatherproof Reflector
Pc., Pcs.	Piece, Pieces
P.C.	Portland Cement; Power Connector
P.C.F.	Pounds per Cubic Foot
P.C.M.	Phase Contract Microscopy
P.E.	Professional Engineer; Porcelain Enamel; Polyethylene; Plain End
Perf.	Perforated
Ph.	Phase
P.I.	Pressure Injected
Pile.	Pile Driver
Pkg.	Package
Pl.	Plate
Plah.	Plasterer Helper
Plas.	Plasterer
Pluh.	Plumbers Helper
Plum.	Plumber
Ply.	Plywood
p.m.	Post Meridiem
Pntd.	Painted
Pord.	Painter, Ordinary
pp	Pages
PP; PPL	Polypropylene
P.P.M.	Parts per Million
Pr.	Pair
P.E.S.B.	Pre-engineered Steel Building
Prefab.	Prefabricated
Prefin.	Prefinished
Prop.	Propelled
PSF; psf	Pounds per Square Foot
PSI; psi	Pounds per Square Inch
PSIG	Pounds per Square Inch Gauge
PSP	Plastic Sewer Pipe
Pspr.	Painter, Spray
Psst.	Painter, Structural Steel
P.T.	Potential Transformer
P. & T.	Pressure & Temperature
Ptd.	Painted
Ptns.	Partitions
Pu	Ultimate Load
PVC	Polyvinyl Chloride
Pvmt.	Pavement
Pwr.	Power
Q	Quantity Heat Flow
Quan.;Qty.	Quantity
Q.C.	Quick Coupling
r	Radius of Gyration
R	Resistance
R.C.P.	Reinforced Concrete Pipe
Rect.	Rectangle
Reg.	Regular
Reinf.	Reinforced
Req'd.	Required
Res.	Resistant
Resi.	Residential
Rgh.	Rough
RGS	Rigid Galvanized Steel
R.H.W.	Rubber, Heat & Water Resistant; Residential Hot Water
rms	Root Mean Square
Rnd.	Round
Rodm.	Rodman
Rofc.	Roofer, Composition
Rofp.	Roofer, Precast
Rohe.	Roofer Helpers (Composition)
Rots.	Roofer, Tile & Slate
R.O.W.	Right of Way
RPM	Revolutions per Minute
R.R.	Direct Burial Feeder Conduit
R.S.	Rapid Start
Rsr	Riser
RT	Round Trip
S.	Suction; Single Entrance; South
SCFM	Standard Cubic Feet per Minute
Scaf.	Scaffold
Sch.; Sched.	Schedule
S.C.R.	Modular Brick
S.D.	Sound Deadening
S.D.R.	Standard Dimension Ratio
S.E.	Surfaced Edge
Sel.	Select
S.E.R.;	Service Entrance Cable
S.E.U.	Service Entrance Cable
S.F.	Square Foot
S.F.C.A.	Square Foot Contact Area
S.F.G.	Square Foot of Ground
S.F. Hor.	Square Foot Horizontal
S.F.R.	Square Feet of Radiation
S.F. Shlf.	Square Foot of Shelf
S4S	Surface 4 Sides
Shee.	Sheet Metal Worker
Sin.	Sine
Skwk.	Skilled Worker
SL	Saran Lined
S.L.	Slimline
Sldr.	Solder
SLH	Super Long Span Bar Joist
S.N.	Solid Neutral
S-O-L	Socketolet
sp	Standpipe
S.P.	Static Pressure; Single Pole; Self-Propelled
Spri.	Sprinkler Installer
spwg	Static Pressure Water Gauge
Sq.	Square; 100 Square Feet
S.P.D.T.	Single Pole, Double Throw
SPF	Spruce Pine Fir
S.P.S.T.	Single Pole, Single Throw
SPT	Standard Pipe Thread
Sq. Hd.	Square Head
Sq. In.	Square Inch
S.S.	Single Strength; Stainless Steel
S.S.B.	Single Strength B Grade
sst	Stainless Steel
Sswk.	Structural Steel Worker
Sswl.	Structural Steel Welder
St.;Stl.	Steel
S.T.C.	Sound Transmission Coefficient
Std.	Standard
STK	Select Tight Knot
STP	Standard Temperature & Pressure
Stpi.	Steamfitter, Pipefitter
Str.	Strength; Starter; Straight
Strd.	Stranded
Struct.	Structural
Sty.	Story
Subj.	Subject
Subs.	Subcontractors
Surf.	Surface
Sw.	Switch
Swbd.	Switchboard
S.Y.	Square Yard
Syn.	Synthetic
S.Y.P.	Southern Yellow Pine
Sys.	System
t.	Thickness
T	Temperature; Ton
Tan	Tangent
T.C.	Terra Cotta
T & C	Threaded and Coupled
T.D.	Temperature Difference
T.E.M.	Transmission Electron Microscopy
TFE	Tetrafluoroethylene (Teflon)
T. & G.	Tongue & Groove; Tar & Gravel
Th.;Thk.	Thick
Thn.	Thin
Thrded	Threaded
Tilf.	Tile Layer, Floor
Tilh.	Tile Layer, Helper
THHN	Nylon Jacketed Wire
THW.	Insulated Strand Wire
THWN;	Nylon Jacketed Wire
T.L.	Truckload
T.M.	Track Mounted
Tot.	Total
T-O-L	Threadolet
T.S.	Trigger Start
Tr.	Trade
Transf.	Transformer
Trhv.	Truck Driver, Heavy
Trlr	Trailer
Trlt.	Truck Driver, Light
TV	Television
T.W.	Thermoplastic Water Resistant Wire
UCI	Uniform Construction Index
UF	Underground Feeder
UGND	Underground Feeder
U.H.F.	Ultra High Frequency
U.L.	Underwriters Laboratory
Unfin.	Unfinished
URD	Underground Residential Distribution
US	United States
USP	United States Primed
UTP	Unshielded Twisted Pair
V	Volt
V.A.	Volt Amperes
V.C.T.	Vinyl Composition Tile
VAV	Variable Air Volume
VC	Veneer Core
Vent.	Ventilation
Vert.	Vertical
V.F.	Vinyl Faced
V.G.	Vertical Grain
V.H.F.	Very High Frequency
VHO	Very High Output
Vib.	Vibrating
V.L.F.	Vertical Linear Foot
Vol.	Volume
VRP	Vinyl Reinforced Polyester
W	Wire; Watt; Wide; West
w/	With
W.C.	Water Column; Water Closet
W.F.	Wide Flange
W.G.	Water Gauge
Wldg.	Welding
W. Mile	Wire Mile
W-O-L	Weldolet
W.R.	Water Resistant
Wrck.	Wrecker
W.S.P.	Water, Steam, Petroleum
WT., Wt.	Weight
WWF	Welded Wire Fabric
XFER	Transfer
XFMR	Transformer
XHD	Extra Heavy Duty
XHHW; XLPE	Cross-Linked Polyethylene Wire Insulation
XLP	Cross-linked Polyethylene
Y	Wye
yd	Yard
yr	Year
Δ	Delta
%	Percent
~	Approximately
Ø	Phase
@	At
#	Pound; Number
<	Less Than
>	Greater Than

RESIDENTIAL COST ESTIMATE

OWNER'S NAME: ______ APPRAISER: ______

RESIDENCE ADDRESS: ______ PROJECT: ______

CITY, STATE, ZIP CODE: ______ DATE: ______

CLASS OF CONSTRUCTION
- ☐ ECONOMY
- ☐ AVERAGE
- ☐ CUSTOM
- ☐ LUXURY

RESIDENCE TYPE
- ☐ 1 STORY
- ☐ 1-1/2 STORY
- ☐ 2 STORY
- ☐ 2-1/2 STORY
- ☐ 3 STORY
- ☐ BI-LEVEL
- ☐ TRI-LEVEL

CONFIGURATION
- ☐ DETACHED
- ☐ TOWN/ROW HOUSE
- ☐ SEMI-DETACHED

OCCUPANCY
- ☐ ONE STORY
- ☐ TWO FAMILY
- ☐ THREE FAMILY
- ☐ OTHER ______

EXTERIOR WALL SYSTEM
- ☐ WOOD SIDING—WOOD FRAME
- ☐ BRICK VENEER—WOOD FRAME
- ☐ STUCCO ON WOOD FRAME
- ☐ PAINTED CONCRETE BLOCK
- ☐ SOLID MASONRY (AVERAGE & CUSTOM)
- ☐ STONE VENEER—WOOD FRAME
- ☐ SOLID BRICK (LUXURY)
- ☐ SOLID STONE (LUXURY)

*LIVING AREA (Main Building)		*LIVING AREA (Wing or Ell) ( )		*LIVING AREA (Wing or Ell) ( )	
First Level	______ S.F.	First Level	______ S.F.	First Level	______ S.F.
Second Level	______ S.F.	Second Level	______ S.F.	Second Level	______ S.F.
Third Level	______ S.F.	Third Level	______ S.F.	Third Level	______ S.F.
Total	______ S.F.	Total	______ S.F.	Total	______ S.F.

*Basement Area is not part of living area.

MAIN BUILDING	COSTS PER S.F. LIVING AREA
Cost per Square Foot of Living Area, from Page ______	$
Basement Addition: ______ % Finished, ______ % Unfinished	+
Roof Cover Adjustment: ______ Type, Page ______ (Add or Deduct)	( )
Central Air Conditioning: ☐ Separate Ducts ☐ Heating Ducts, Page ______	+
Heating System Adjustment: ______ Type, Page ______ (Add or Deduct)	( )
Main Building: Adjusted Cost per S.F. of Living Area	$

MAIN BUILDING TOTAL COST: $ ______ /S.F. (Cost per S.F. Living Area) x ______ S.F. (Living Area) x ______ (Town/Row House Multiplier (Use 1 for Detached)) = $ ______ (TOTAL COST)

WING OR ELL ( ) ______ STORY	COSTS PER S.F. LIVING AREA
Cost per Square Foot of Living Area, from Page ______	$
Basement Addition: ______ % Finished, ______ % Unfinished	+
Roof Cover Adjustment: ______ Type, Page ______ (Add or Deduct)	( )
Central Air Conditioning: ☐ Separate Ducts ☐ Heating Ducts, Page ______	+
Heating System Adjustment: ______ Type, Page ______ (Add or Deduct)	( )
Wing or Ell ( ): Adjusted Cost per S.F. of Living Area	$

WING OR ELL ( ) TOTAL COST: $ ______ /S.F. (Cost per S.F. Living Area) x ______ S.F. (Living Area) = $ ______ (TOTAL COST)

WING OR ELL ( ) ______ STORY	COSTS PER S.F. LIVING AREA
Cost per Square Foot of Living Area, from Page ______	$
Basement Addition: ______ % Finished, ______ % Unfinished	+
Roof Cover Adjustment: ______ Type, Page ______ (Add or Deduct)	( )
Central Air Conditioning: ☐ Separate Ducts ☐ Heating Ducts, Page ______	+
Heating System Adjustment: ______ Type, Page ______ (Add or Deduct)	( )
Wing or Ell ( ): Adjusted Cost per S.F. of Living Area	$

WING OR ELL ( ) TOTAL COST: $ ______ /S.F. (Cost per S.F. Living Area) x ______ S.F. (Living Area) = $ ______ (TOTAL COST)

TOTAL THIS PAGE ______

Page 1 of 2

RESIDENTIAL COST ESTIMATE

Total Page 1			$
	QUANTITY	UNIT COST	
Additional Bathrooms: _______ Full, _______ Half			
Finished Attic: _______ Ft. x _______ Ft.	S.F.		+
Breezeway: ☐ Open ☐ Enclosed _______ Ft. x _______ Ft.	S.F.		+
Covered Porch: ☐ Open ☐ Enclosed _______ Ft. x _______ Ft.	S.F.		+
Fireplace: ☐ Interior Chimney ☐ Exterior Chimney ☐ No. of Flues ☐ Additional Fireplaces			+
Appliances:			+
Kitchen Cabinets Adjustment: (+/–)			
☐ Garage ☐ Carport: _______ Car(s) Description _______ (+/–)			
Miscellaneous:			+

ADJUSTED TOTAL BUILDING COST $

REPLACEMENT COST	
ADJUSTED TOTAL BUILDING COST	$ _______
Site Improvements	
(A) Paving & Sidewalks	$ _______
(B) Landscaping	$ _______
(C) Fences	$ _______
(D) Swimming Pool	$ _______
(E) Miscellaneous	$ _______
TOTAL	$ _______
Location Factor	x _______
Location Replacement Cost	$ _______
Depreciation	–$ _______
LOCAL DEPRECIATED COST	$ _______

INSURANCE COST	
ADJUSTED TOTAL BUILDING COST	$ _______
Insurance Exclusions	
(A) Footings, Site Work, Underground Piping	–$ _______
(B) Architects' Fees	–$ _______
Total Building Cost Less Exclusion	$ _______
Location Factor	x _______
LOCAL INSURABLE REPLACEMENT COST	$ _______

SKETCH AND ADDITIONAL CALCULATIONS

Page 2 of 2

A

andon catch basin 299
BC extinguisher 453
rasive floor tile 430
tread 430
S DWV pipe 479
C coils furnace 487
cess door & panels 407
door basement 327
door duct 492
cessories bath 455
bathroom 429, 455
door 420
door and window 420
drywall 428
duct 492
plaster 425
cessory drainage 395
fireplace 452
formwork 323, 324
masonry 330
cordion door 404
id proof floor 432
oustic ceiling board 431
oustical ceiling 431
sealant 397, 428
wallboard 428
rylic carpet 228
ceiling 431
wall coating 445
wallcovering 435
wood block 433
t. tools & fast. powder 340
lhesive cement 434
roof 391
wallpaper 435
dobe brick 333
masonry 333
erial lift 293
ggregate exposed 326
stone 316
r cleaner electronic 494
compressor 293
conditioner cooling & heating 490
conditioner packaged terminal 489
conditioner portable 490
conditioner receptacle 502
conditioner removal 474
conditioner rooftop 490
conditioner self-contained 490
conditioner thru-wall 490
conditioner window 490
conditioner wiring 504
conditioning 268, 270
conditioning ventilating 490, 493
extractor 492
filter 494
filter roll type 494
handler heater 488
handler modular 488
hose 293
return grille 493
spade 294
supply register 493
tool 293
larm burglar 470
detection 470
residential 503
lteration fee 288
luminum ceiling tile 431
chain link 313
column 377
coping 332
diffuser perforated 493
downspout 394
drip edge 395
ductwork 492
edging 318
flagpole 453
flashing 211, 393
foil 383, 386
gravel stop 396
grille 493
gutter 396
louver 450
nail 353
rivet 340
roof 388
service entrance cable 497
sheet metal 394
siding 166, 388
siding accessories 389
siding paint 442
sliding door 406
stair 349
storm door 406
tile 430
window 409
window demolition 304
Anchor bolt 323, 330
brick 330
buck 330
chemical 339
expansion 339
framing 354
hollow wall 339
joist 354
masonry 330
nailing 339
partition 330
rafter 354
rigid 330
screw 339
sill 354
steel 330
stone 330
wall 338
Antenna system 509
T.V. 509
Appliance 236, 458
residential 458, 504
Apron wood 368
Arch laminated 367
radial 367
Architectural equipment 458
fee 288, 518
woodwork 371
Armored cable 497
Arrester lightning 466
Ashlar veneer 529
Asphalt base sheet 391
block 312
block floor 312
coating 382
curb 311
driveway 110
felt 391
flood coat 391
paper 386
primer 434
sheathing 365
shingle 386
sidewalk 108, 311
tile 229
Asphaltic emulsion 316
paving 522
Astragal 419
molding 369
Attic stair 460
ventilation fan 492
Auger 292
hole 298
Automatic transfer switch 507
washing machine 459
Awning canvas 454
window 174, 409

B

Backerboard 426
Backfill 306
planting pit 317
trench 307
Backflow preventer 482
Backhoe 306
excavation 307
Backsplash countertop 352
Backup block 332
Baked enamel frame 401
Baluster 231, 376
birch 231
pine 231
Balustrade painting 441
Band joist framing 346
molding 368
Bankrun gravel 298
Bar grab 455
towel 455
Zee 428
Bark mulch 315
mulch redwood 316
Barrel 294
Barricade 294
Base cabinet 371, 373
carpet 434
ceramic tile 429
column 354
course 310
cove 433
gravel 310
molding 368
quarry tile 429
resilient 433
road 310
sheet 391
sink 372
stone 310
terrazzo 430
vanity 375
wood 368, 378
Baseboard demolition 303
heat 491
heat electric 491
heating 279
register 493
Basement stair 231, 327
Basic meter device 506
Basketweave fence 313
Bath accessories 455
accessory 237
communal 484
steam 463
whirlpool 484
Bathroom 248, 250, 252, 254, 256, 258, 260, 262, 264, 266, 484
accessories 429, 455
faucet 484
fixture 484, 485
Bathtub 484
enclosure 450
removal 474
Batt insulation 168, 383
Bay window 178
Bead casing 428
corner 425
parting 370
Beam & girder framing 356
bondcrete 426
ceiling 377
drywall 427
hanger 354
laminated 365, 366
mantel 377
plaster 426
steel 341
wood 356, 357
Bed molding 368
Bedding brick 312
pipe 306
Beech tread 377
Belgian block 311
Bell & spigot pipe 475
Bench park 315
players 315
Berm pavement 311
Bevel siding 390
Bi-fold door 226, 400, 404
Bi-passing closet door 404
door 226
Birch door 402, 404, 405
molding 369
paneling 370
stair 376
wood frame 406
Bituminous block 312
coating 382
expansion joint 323
paving 522
Blanket insulation 383, 474
Blaster shot 295
Blind 191
exterior 378
venetian 462
window 462
Block asphalt 312
asphalt floor 312
backup 332
belgian 311
bituminous 312
Block, brick and mortar 527
Block concrete 331, 332, 336
concrete bond beam 332
concrete exterior 332
decorative concrete 332
filler 444
floor 433
glass 333
insulation 330
lightweight 336
lintel 332
partition 336
profile 332
reflective 333
split rib 332
wall 158
wall foundation 120
wall removal 299
Blocking 356
carpentry 356
wood 359
Blockout slab 322
Blown in cellulose 382
in fiberglass 382
in insulation 382
Blueboard 426, 427
Bluegrass sod 316
Bluestone sidewalk 311
sill 333
step 311
Board & batten fence 314
& batten siding 390
ceiling 431

  - fence 313
  - gypsum 427
  - insulation 383
  - paneling 371
  - sheathing 364
  - siding 162
  - valance 372
  - verge 369
- Boiler 486
  - demolition 474
  - electric 486
  - electric steam 486
  - gas fired 487
  - gas/oil combination 487
  - hot water 487
  - oil fired 487
  - steam 487
- Bollard 341
  - light 508
- Bolt 338
  - anchor 330
  - expansion 339
  - steel 338, 354
  - toggle 339
- Bondcrete 426
- Bookcase 371, 373
- Boom lift 293
- Boring 298
  - cased 298
- Borrow 298
- Bow window 178, 410
- Bowstring truss 367
- Box 499
  - distribution 308
  - fence shadow 314
  - out 322
  - pull 500
  - stair 376
- Boxed headers & beams 343, 346
- Boxes & wiring device 500
- Bracing 356
  - let-in 356
  - metal joists 345
  - stud walls 342
- Brass hinge 418
  - screw 353
- Breaker circuit 506
  - vacuum 482
- Brick 527
  - adobe 333
  - anchor 330
  - bedding 312
- Brick, block and mortar 527
- Brick chimney simulated 451
  - chimneys 525
  - cleaning 336
  - demolition 301, 302
  - driveway 110
  - edging 318
  - face 331
  - forklift 294
  - molding 368, 369
  - paving 312
  - removal 299
  - shelf 322
  - sidewalk 108, 312
  - sill 333
  - step 311
  - veneer 160, 331, 527
  - veneer demolition 303
  - wall 160
  - wash 336
- Bricklaying 525
- Bridging 356
  - metal joists 345
  - stud walls 342
- Bronze valve 481
- Broom cabinet 372
  - finish concrete 326
- Brownstone 334
- Brush clearing 305
  - cutter 292
- Buck anchor 330
  - rough 359
- Buggy concrete 292, 326
- Builder's risk insurance 512
- Building demolition 298
  - excavation 102, 104
  - greenhouse 466
  - hardware 418
  - insulation 382
  - paper 386
  - permit 288
  - prefabricated 466
- Built-in range 458
- Built-up roof 391
  - roofing 210
- Bulk bank measure excavating 306
- Bulkhead door 407
  - formwork 323
- Bulldozer 292, 293, 306
- Bumper door 418
  - wall 418
- Burglar alarm 470
- Burlap curing 326
  - rubbing 326
- Burner gas conversion 486
  - gun type 486
  - residential 486
- Bush hammer 326
  - hammer concrete 326
- Butt fusion machine 294
- Butyl caulking 397

C

- Cabinet 234
  - base 371, 373
  - broom 372
  - casework 373
  - corner base 372
  - corner wall 372
  - demolition 303
  - door 373, 374
  - electrical 500
  - hardboard 371
  - hardware 374, 375
  - hinge 375
  - kitchen 371
  - medicine 455
  - oven 372
  - shower 450
  - stain 436
  - varnish 436
  - wall 372
- Cable armored 497
  - electric 497, 498
  - heating 491
  - jack 295
  - sheathed nonmetallic 497
  - sheathed romex 497
- Cafe door 403
- Canopy door 453
  - framing 361
- Cant roof 360
- Cantilever retaining wall 315
- Canvas awning 454
- Cap post 354
  - service entrance 497
- Carbon dioxide extinguisher 453
- Carpentry finish 367
  - rough 356
- Carpet 228, 434
  - base 434
  - felt pad 435
  - floor 434
  - nylon 434
  - olefin 434
  - padding 434
  - removal 301
  - stair 434
  - urethane pad 435
  - wool 434
- Carrier ceiling 425
  - channel 424
  - fixture 483
- Cart concrete 292, 326
- Case work 372
- Cased boring 298
  - evaporator coils 488
- Casement window 172, 410, 411
  - windows 411
- Casework cabinet 373
  - custom 371
  - ground 364
  - metal 462
  - painting 436
  - stain 436
  - varnish 436
- Casing bead 428
  - wood 368
- Cast in place concrete 325
  - iron bench 315
  - iron damper 452
  - iron drain 482
  - iron fitting 476
  - iron pipe 475
  - iron pipe fitting 476
  - iron radiator 491
  - iron stair 349
  - iron trap 482
- Catch basin 310
  - basin precast 310
  - basin removal 299
  - door 374
- Caulking 397
  - polyurethane 397
  - sealant 397
- Cavity wall insulation 382
  - wall reinforcing 330
- Cedar closet 371
  - fence 314
  - paneling 371
  - post 378
  - roof deck 366
  - roof plank 364
  - shake siding 164
  - shingle 387
  - siding 390
  - stair 376
- Ceiling 216, 220
  - acoustical 431
  - beam 377
  - board 431
  - board acoustic 431
  - board fiberglass 431
  - bondcrete 426
  - carrier 425
  - demolition 300
  - diffuser 493
  - drill 338
  - drywall 427
  - eggcrate 431
  - framing 357
  - furring 364, 424, 425
  - heater 459
  - insulation 382
  - lath 42
  - luminous 431, 50
  - molding 36
  - painting 44
  - plaster 425, 42
  - stair 46
  - support 34
  - suspended 425, 43
  - tile 43
- Cellar door 40
  - wine 45
- Cellulose blown in 38
  - insulation 38
- Cement adhesive 43
  - flashing 39
  - gypsum 33
  - masonry 33
  - masonry unit 332, 33
  - mortar 429, 52
  - parging 38
- Central vacuum 45
- Centrifugal pump 29
- Ceramic mulch 31
  - tile 229, 428, 42
  - tile countertop 35
  - tile demolition 30
  - tile floor 42
- Chain hoist 29
  - link fence 114, 31
  - link fence paint 43
  - link fence removal 29
  - saw 29
  - trencher 29
- Chair molding 36
  - rail demolition 30
- Chamfer strip 32
  - strip wood 32
- Channel carrier 42
  - furring 424, 42
  - siding 39
  - steel 42
- Chase formwork 32
- Chemical anchor 33
  - dry extinguisher 45
- Chime door 50
- Chimney 33
  - accessories 45
  - brick 33
  - demolition 30
  - flue 33
  - foundation 32
  - metal 45
  - screen 45
  - simulated brick 45
  - vent 48
- China cabinet 371
- Chipper brush 292
- Chipping hammer 293
- Chips wood 316
- Chlorination system 467
- Chute rubbish 304
- C.I.P. concrete 325
- Circline fixture 508
- Circuit wiring 280
- Circuit-breaker 506
- Circular saw 295
- Circulating pump 483
- Cladding 388
  - sheet metal 393
- Clamp water pipe ground 497
- Clapboard painting 441
- Clay masonry 331
  - roofing tile 387
  - tile 387
- Cleaner steam 295
- Cleaning brick 336

INDEX

face brick 529
masonry 336
up 296
eanout door 452
floor 482
pipe 482
tee 482
ear & grub 305
earing brush 305
site 305
ip plywood 354
ock timer 503
oset cedar 371
door 226, 400, 404, 405
pole 369
rod 371
water 485
othes dryer residential 459
MU 331
oal tar pitch 392
oat glaze 444
oating bituminous 382
glazed 445
roof 392
rubber 382
silicone 382
spray 382
trowel 382
wall 445
water repellent 382
waterproofing 382
oil tie 324
oldformed framing 342
joists 345
metal joists 347
ollected stone 335
olonial door 402, 404
wood frame 405, 406
olumn 341
base 354
bondcrete 426
brick 336
demolition 302
drywall 427
lally 341
laminated wood 367
lath 425
pipe 341
plaster 425, 426
removal 302
wood 357, 377
olumns formwork 322
ombination device 501
storm door 402, 404
ommercial gutting 302
ommon brick 331
nail 353
rafter roof framing 138
ommunal bath 484
ompact fill 306
ompaction 306
soil 306
ompactor 236
earth 292
residential 458
ompartments shower 450
ompensation workers' 289
omponents furnace 487
omposite insulation 385
rafter 360
ompressor air 293
oncrete block 331, 332, 336, 528
block back-up 331
block bond beam 332
block decorative 332
block demolition 300
block exterior 332
block foundation 332
block grout 330
block insulation 382
block painting 444
block wall 158
broom finish 326
buggy 326
bush hammer 326
cart 292, 326
cast in place 325
C.I.P. 325
conveyer 292
coping 332
curb 311
curing 326
cutout 300
darby finish 326
demolition 300
drill 338
driveway 110
finish 311, 326
float 292
float finish 326
footing 325, 326
formwork 322
foundation 122, 325
furring 363
hand trowel finish 326
lintel 327
mixer 292
monolithic finish 326
paving 310
pipe 308, 309
placing 326
precast 327
protection 322
pump 292
ready mix 325
reinforcement 324
removal 299
retaining wall 315
saw 292
septic tank 308
shingle 388
sidewalk 108, 311
sill 333
slab 126, 326
stair 326
stamping 326
structural 325
tile 388
trowel 292
utility vault 298
vibrator 292
wall 326
wheeling 326
Conditioner portable air 490
Conditioning ventilating air 490
Conductive floor 433
Conductor 497
& grounding 497
wire 498
Conduit & fitting flexible 505
electrical 498
fitting 499
flexible metallic 499
greenfield 499
in slab 498
in slab PVC 498
in trench electrical 498
in trench steel 498
PVC 498
rigid in slab 498
rigid steel 498
Connection motor 505
Connector joist 354
stud 340
timber 354
Construction aids 289
management fee 288
time 288
Contractor equipment 292
overhead 289
pump 294
Control component 469
Conveyor 292
Cooking equipment 458
range 458
Cooling 268, 270
& heating A/C 490
electric furnace 489
Coping 332, 334
aluminum 332
concrete 332
removal 303
terra cotta 332
Copper cable 497
downspout 394
drum trap 483
DWV tubing 477
fitting 477
flashing 393
gutter 396
pipe 476, 477
rivet 340
roof 393
wall covering 435
wire 498
Core drill 292
Cork floor 433
tile 433
wall tile 435
Corner base cabinet 372
bead 425
wall cabinet 372
Cornice 332
molding 368
painting 441
Corrugated metal pipe 309
roof tile 388
siding 388, 390
Costs of trees 523
Counter top 352
top demolition 303
Countertop backsplash 352
sink 485
Course base 310
Cove base 433
base ceramic tile 429
base terrazzo 430
molding 368
molding scotia 368
Cover ground 316
pool 467
CPVC pipe 479
Crane hydraulic 295
Crew survey 288
Crown molding 368, 369
Crushed stone sidewalk 311
Cubicle shower 485
Cupola 452
Curb 311
asphalt 311
concrete 311
edging 311
formwork 322
granite 311
inlet 311
precast 311
removal 299
roof 360
terrazzo 430
Curing concrete 326
paper 386
Curtain damper fire 492
rod 455
Curved stair 376
stairway 231
Custom casework 371
Cut stone trim 334
Cutout 300
counter 352
slab 300
Cutter brush 292
Cutting block 352
torch 295
Cylinder lockset 418

D

Damper fireplace 452
multi-blade 492
Darby finish concrete 326
Deadbolt 418
Deciduous shrub 317
tree 318
Deciduous tree 318
Deck metal 341
roof 341, 364
steel 341
wood 244, 364, 366
Decorator device 500
switch 500
Deep freeze 459
Dehumidifier 459
Delivery charge 305
Demolition 298, 299, 301, 303, 474
baseboard 303
boiler 474
brick 301, 302
building 298
cabinet 303
ceiling 300
ceramic tile 301
chimney 302
column 302
concrete 300
concrete block 300
door 301
ductwork 474
electric 496
fireplace 303
flooring 301
framing 301
furnace 474
glass 305
granite 303
gutter 303
house 298
HVAC 474
joist 302
masonry 300, 302
metal stud 304
millwork 303
paneling 303
partition 304
pavement 299
plaster 300, 304
plumbing 474
plywood 300, 303, 304
post 302
rafter 302
railing 303
roofing 303
selective 300
siding 303

site 298, 299
tile 300
truss 302
walls and partitions 304
window 304
wood 300
Detection alarm 470
system 470
Detector infra-red 470
motion 470
smoke 470
Device combination 501
decorator 500
GFI 502
receptacle 502
residential 500
wiring 505
Diaphragm pump 295
Diffuser ceiling 493
linear 493
opposed blade damper 493
perforated aluminum 493
rectangular 493
steel 493
T-bar mount 493
Dimmer switch 500, 505
Disappearing stairs 460
Discharge hose 294
Disconnect safety switch 507
Dishwasher 236, 459
Disposal 299
field 308
garbage 459
Disposer garbage 236
Distribution box 308
system gas 309
Diving board 467
Door & panels access 407
accessories 420
accordion 404
bell residential 503
bell system 509
bi-fold 400, 404
bi-passing closet 404
birch 405
blind 379
bulkhead 407
bumper 418
cabinet 373, 374
cafe 403
canopy 453
catch 374
chime 509
cleanout 452
closet 400, 405
combination storm 404
demolition 301
double 400
dutch 404
dutch oven 452
entrance 404
exterior 182
fire 400
flush 404
folding accordion 404
frame 401, 405
frame interior 406
french 402
garage 186, 407
glass 406
handle 374
hardware 418
kick plate 418
knocker 420
labeled 400
metal 400
mirror 421
molding 369
moulded 403
opener 408
overhead 407
panel 403
paneled 400
passage 402, 405
plastic 401
pre-hung 401
removal 301
residential 400, 403, 406
residential garage 407
rough buck 359
sauna 463
sectional overhead 407
shower 450
sill 370, 405, 419
sliding 406
special 407
stain 438
steel 400
stop 418
storm 190, 406
switch burglar alarm 470
threshold 406
trim vinyl 390
varnish 438
weatherstrip 419
weatherstrip garage 419
wood 224, 226, 401
Doors & windows interior paint 437
Dormer framing 150, 152
gable 363
roofing 204, 206
Double hung window 170, 410, 412
Dowels reinforcing 324
Downspout 394, 395
aluminum 394
copper 394
elbow 395
lead coated copper 395
painting 440
steel 395
strainer 394
Dozer 293, 306
Drain cast iron 482
floor 482
Drainage accessory 395
pipe 308
site 310
trap 482
Drains roof 482
Drapery hardware 462, 463
rings 463
Drawer 375
track 375
wood 375
Drill ceiling 338
concrete 338
core 292
drywall 339
hammer 294
plaster 339
rig 298
steel 294
track 293
wall 338
wood 354
Drinking fountain support 483
Drip edge 395
edge aluminum 395
Drive pin 340
Driver sheeting 294
Driveway 110, 311
removal 299
Drop pan ceiling 431
Drum trap 483
trap copper 483
Dryer clothes 459
receptacle 502
vent 460
wiring 280
Drywall 214, 216, 427
accessories 428
column 427
cutout 300
demolition 304
drill 339
frame 401
gypsum 427
nail 353
painting 444
prefinished 428
screw 428
Duck tarpaulin 291
Duct access door 492
accessories 492
flexible insulated 492
flexible noninsulated 492
HVAC 492
insulation 474
liner 475
mechanical 492
Ductile iron pipe cement lined 308
Ductless split system 490
Ductwork 492
aluminum 492
demolition 474
fabric coated flexible 492
fabricated 492
galvanized 492
rectangular 492
rigid 492
Dump charge 304
truck 293
Dumpster 304
Duplex receptacle 280, 505
Dust partition 304
Dutch door 182, 404
oven door 452
DWV pipe ABS 479
PVC pipe 479
tubing copper 477

E

Earth compactor 292
vibrator 306
Earthwork 305
equipment 292, 306
Edge drip 395
form 322
Edging 318
aluminum 318
curb 311
Eggcrate ceiling 431
EIFS 385
Elbow downspout 395
Electric appliance 458
baseboard heater 491
boiler 486
cable 497, 498
demolition 496
fixture 507, 508
furnace 488
generator 294
generator set 506
heater 459
heating 279, 491
lamp 509
log 451
metallic tubing 498
pool heater 487
service 278, 506
stair 460
switch 505-507
unit heater 491
water heater 480
wire 498
Electrical cabinet 500
conduit 498
conduit greenfield 499
Electronic air cleaner 494
Elevator 472
Embossed metal door 402
print door 403
Employer liability 289
EMT 498
Emulsion adhesive 434
asphaltic 316
pavement 312
sprayer 294
Enclosure bathtub 450
shower 450
swimming pool 466
Entrance cable service 497
door 182, 402, 404
frame 405
lock 418
weather cap 499
Epoxy grout 429
wall coating 445
Equipment 292
earthwork 292, 306
insurance 289
rental 292, 295
Erosion control 307
Estimate electrical heating 491
Estimating 512
Evaporator coils cased 488
Evergreen shrub 317
tree 317
Excavating bulk bank measure 306
Excavation 306
backhoe 307
footing 102
foundation 104
hand 306, 307
hauling 307
planting pit 317
septic tank 308
structural 306
trench 307
utility 106
Excavator rental 292
Exhaust hood 459
vent 494
Expansion anchor 339
bolt 339
joint 323, 425
joint bituminous 323
joint neoprene 323
joint premolded 323
shield 339
tank 482
tank steel 482
Exposed aggregate 326
aggregate coating 445
Exterior blind 378
concrete block 332
door 182
door frame 405
insulation 385
insulation finish system 385
light 281
lighting fixture 508

molding ... 368
plaster ... 426
pre-hung door ... 401
residential door ... 403
wall framing ... 136
wood frame ... 405
tinguisher ABC ... 453
carbon dioxide ... 453
chemical dry ... 453
fire ... 453
standard ... 453
tractor air ... 492

F

bric welded wire ... 324
bricated ductwork ... 492
brication metal ... 349
brications plastic ... 379
ce brick ... 331, 526
brick cleaning ... 529
cing stone ... 334
n ... 492
attic ventilation ... 492
house ventilation ... 493
paddle ... 504
residential ... 504
ventilation ... 504
wiring ... 280, 504
scia aluminum ... 389
board ... 360
board demolition ... 301
metal ... 389
vinyl ... 390
wood ... 368
astener timber ... 353
wood ... 353
astenings metal ... 338
aucet & fitting ... 483, 484
bathroom ... 484
ee architectural ... 288
elt asphalt ... 391
carpet pad ... 435
ence and gate ... 313
auger, truck mounted ... 292
basketweave ... 313
board ... 313
board & batten ... 314
cedar ... 314
metal ... 313
open rail ... 314
picket paint ... 439
removal ... 299
residential ... 313
wire ... 313
wood ... 313
ences ... 439
iber cement shingle ... 386
cement siding ... 389
reinforcing ... 325
tube formwork ... 322
iberboard insulation ... 384
ibergalss tank ... 467
iberglass blown in ... 382
ceiling board ... 431
cupola ... 452
insulation ... 382, 383
panel ... 388
planter ... 318
shower stall ... 450
tank ... 468
wool ... 382
ield disposal ... 308
septic ... 112
ieldstone ... 335

Fill ... 306
gravel ... 298, 306
Filler block ... 444
Film polyethylene ... 316
Filter air ... 494
mechanical media ... 494
swimming pool ... 467
Filtration equipment ... 467
Fine grade ... 305, 316
Finish carpentry ... 367
concrete ... 311, 326
floor ... 433, 445
nail ... 353
wall ... 326
Finisher floor ... 292
Finishes wall ... 435
Fir column ... 377
floor ... 432
molding ... 369
roof deck ... 366
roof plank ... 364
Fire call pullbox ... 470
damage repair ... 445
damper curtain type ... 492
door ... 400
door frame ... 401
escape disappearing ... 460
extinguisher ... 453
extinguishers ... 453
horn ... 470
resistant drywall ... 427, 428
retardant lumber ... 352
retardant plywood ... 352
Firebrick ... 335
Fireplace accessory ... 452
box ... 335
built-in ... 451
damper ... 452
demolition ... 303
form ... 452
free standing ... 451
mantel ... 377
masonry ... 238, 335
prefabricated ... 240, 451
Firestop gypsum ... 425
wood ... 358
Fitting cast iron ... 476
conduit ... 499
copper ... 477
malleable iron ... 478
plastic ... 479
PVC ... 479
steel ... 478
Fixed window ... 180
Fixture bathroom ... 484, 485
carrier ... 483
electric ... 507
exterior mercury vapor ... 508
fluorescent ... 507, 508
incandescent ... 508
interior light ... 507
lantern ... 508
mirror light ... 508
plumbing ... 483, 484, 486
removal ... 474
residential ... 503, 508
Flagging ... 312, 432
slate ... 312
Flagpole ... 453
aluminum ... 453
Flashing ... 393
aluminum ... 393
cement ... 392
copper ... 393
lead coated ... 394
masonry ... 393

stainless ... 394
Flatbed truck ... 293
Flexible conduit & fitting ... 505
ductwork fabric coated ... 492
insulated duct ... 492
metallic conduit ... 499
Float concrete ... 292
finish concrete ... 326
glass ... 420
Floater equipment ... 289
Floating floor ... 433
pin ... 418
Floodlight ... 294
Floor ... 432
acid proof ... 432
asphalt block ... 312
brick ... 432
carpet ... 434
ceramic tile ... 429
cleanout ... 482
concrete ... 126
conductive ... 433
cork ... 433
drain ... 482
finish ... 445
finisher ... 292
flagging ... 312
framing ... 130, 132, 134
framing removal ... 301
insulation ... 382
marble ... 334
nail ... 353
oak ... 432
paint ... 440
parquet ... 432
plank ... 358
plywood ... 365
polyethylene ... 433
quarry tile ... 430
register ... 494
removal ... 299
rubber ... 433
sander ... 295
stain ... 440
subfloor ... 365
tile terrazzo ... 430
underlayment ... 365
varnish ... 440
vinyl ... 434
wood ... 432
Flooring ... 229, 432
demolition ... 301
Flue chimney ... 335
chimney metal ... 489
liner ... 335, 336
screen ... 452
tile ... 335
Fluorescent fixture ... 507, 508
lamp ... 509
Flush door ... 404
Foam glass insulation ... 383
insulation ... 383
Foil aluminum ... 383, 386
back insulation ... 384
Folding accordion door ... 404
door shower ... 450
Footing ... 118
concrete ... 325, 326
excavation ... 102
reinforcing ... 324
removal ... 299
spread ... 322
Forklift brick ... 294
Form edge ... 322
fireplace ... 452
polystyrene ... 323

slab ... 342
Formwork ... 322
accessory ... 323, 324
bulkhead ... 323
chase ... 324
columns ... 322
concrete ... 322
curb ... 322
fiber tube ... 322
insert ... 323
plywood ... 323
sleeve ... 324
snap tie ... 324
steel frame ... 322
structural ... 322
wall ... 322
Foundation ... 118
chimney ... 325
concrete ... 122, 325
concrete block ... 332
excavation ... 104
masonry ... 120
mat ... 325, 326
removal ... 298
wall ... 332
wood ... 124
Frame baked enamel ... 401
door ... 401, 405
drywall ... 401
entrance ... 405
fire door ... 401
labeled ... 401
metal ... 401
metal butt ... 401
steel ... 401
welded ... 401
window ... 415
wood ... 373, 405
Framing anchor ... 354
band joist ... 346
beam & girder ... 356
canopy ... 361
coldformed ... 342
demolition ... 301
dormer ... 150, 152
floor ... 130, 132, 134
heavy ... 357
joists ... 347
laminated ... 366
metal ... 340
metal joists ... 347
metal partitions ... 343
metal studs ... 343
partition ... 154, 424
removal ... 301
roof ... 138, 140, 142, 144, 146, 148
roof rafters ... 360
sleepers ... 361
timber ... 357
wall ... 362
wood ... 356-358
Freeze deep ... 459
French door ... 402
Front end loader ... 306
Furnace A/C coils ... 487
components ... 487
demolition ... 474
electric ... 488
gas fired ... 488, 489
heating & cooling ... 489
hot air ... 488
oil fired ... 489
wiring ... 280
Furnishings site ... 315
Furring and lathing ... 424
ceiling ... 364, 424, 425

channel 424, 428
metal 424
steel 424
wall 363, 425
Fusion machine butt 294

G

Gable dormer 363
dormer framing 150
dormer roofing 204
roof framing 138
Galvanized ductwork 492
Gambrel roof framing 144
Garage door 186, 407
door weatherstrip 419
Garbage disposal 459
Gas conversion burner 486
distribution system 309
fired boiler 487
fired furnace 488, 489
fired pool heater 487
generator 506
heat air conditioner 490
heating 268, 272
log 451
pipe 309
vent 489
Gasoline generator 506
Gate metal 313
General contractor's overhead . . 513
Generator construction 294
emergency 506
gas 506
gasoline 506
set 506
GFI receptacle 502
Girder wood 356, 358
Glass 420, 421
bead molding 370
block 333
demolition 305
door 406
door shower 450
float 420
heat reflective 420
lined water heater 459
low emissivity 420
masonry 333
mirror 421
shower stall 450
tempered 420
tile 421
tinted 420
window 421
Glaze coat 444
Glazed ceramic tile 429
coating 445
wall coating 445
Glazing 420
Glued laminated 365, 366
laminated construction 366
Grab bar 455
Gradall 292
Grade fine 305, 316
Grader motorized 292
Grading 305, 306
slope 305
Granite 333
building 333
chips 316
curb 311
demolition 303
indian 311
paver 333
sidewalk 312
Grass cloth wallpaper 436
lawn 316
seed 316
sprinkler 312
Grating 349
Gravel bankrun 298
base 310
fill 298, 306
pea 316
stop 395
Gravity retaining wall 315
Greenfield conduit 499
Greenhouse 242, 466
Grid spike 354
Grille air return 493
aluminum 493
decorative wood 377
painting 440
window 414
Grinder terrazzo 292
Ground 364
clamp water pipe 497
cover 316
fault 280
rod 497
wire 497
Grounding 497
& conductor 497
Grout 330
concrete block 330
epoxy 429
tile 429
wall 330
Guard gutter 396
Gunite pool 243
Gutter 396
aluminum 396
copper 396
demolition 303
guard 396
lead coated copper 396
stainless 396
steel 396
strainer 396
vinyl 396
wood 396
Gutting 302
Guying tree 318
Gypsum block demolition 300
board 427
board accessories 428
cement 330
drywall 427
fabric wallcovering 435
firestop 425
lath 425
lath nail 353
plaster 425, 537
sheathing 365
weatherproof 365

H

Half round molding 369
Hammer bush 326
chipping 293
drill 294
hydraulic 294
Hand carved door 403
clearing 305
excavation 306, 307
hole 298
rail 349
split shake 387
trowel finish concrete 326
Handicap lever 418
ramp 325
Handle door 374
Handling rubbish 304
Handrail and railing 349
oak 231
wood 369, 375
Hand-split shake 164
Hanger beam 354
joist 354
Hardboard cabinet 371
overhead door 407
paneling 370
siding 391
soffit 370
tempered 370
underlayment 365
Hardware 418
cabinet 374, 375
door 418
drapery 462, 463
finish 418
ooor 418
rough 355
window 418
Hardwood floor 432
grille 377
Hauling excavation 307
truck 307
Hay 316
Header wood 362
Headers & beams boxed . . . 343, 346
Hearth 335
Heat baseboard 491
electric baseboard 491
pump 491
pump residential 504
radiant 491
reflective glass 420
Heater air handler 488
contractor 294
electric 459
electric wall 491
infra-red quartz 491
sauna 463
swimming pool 487
water 459, 486
wiring 280
Heating 268, 270, 272, 475, 487
cable 491
electric 279, 491
estimate electrical 491
hot air 488
hydronic 487
panel radiant 491
Heavy framing 357
Hemlock column 378
Hex nut 338
nuts steel 338
Highway paver 312
Hinge brass 418
cabinet 375
residential 418
steel 418
Hinges 536
Hip rafter 360
roof framing 142
Hoist contractor 295
lift equipment 295
Holdown 354
Hole drilling 338
Hollow core door 405
metal 401
metal frames 535
Hood range 459
Hook robe 455
Horizontal aluminum siding 388
vinyl siding 389
Horn fire 470
Hose air 293
discharge 294
suction 294
water 294
Hospital tip pin 418
Hot air furnace 488
air heating 488
tub 484
water boiler 487
water heating 486, 491
Hot-air heating 268, 270
Hot-water heating 272
House demolition 298
ventilation fan 493
Housewrap 386
Humidifier 459
Humus peat 316
HVAC demolition 474
duct 492
Hydraulic crane 295
hammer 294
jack 295
Hydronic heating 487

I

Icemaker 459
Impact wrench 294
In insulation blown 382
slab conduit 498
Incandescent fixture 508
light 281
Indian granite 311
Infra-red detector 470
quartz heater 491
Inlet curb 311
Insecticide 307
Insert formwork 323
Insulated panels 385
protectors ADA 474
Insulation 168, 382, 383, 474
batt 383
blanket 383, 474
board 211, 383
building 382
cavity wall 382
cellulose 382
composite 385
duct 474
exterior 385
fiberglass 382, 383
foam 383
foam glass 383
insert 330
isocyanurate 383
masonry 382
mineral fiber 384
pipe 475
polystyrene 382, 383
roof 384
roof deck 384
vapor barrier 386
vermiculite 382
wall 330, 383
water heater 474
Insurance 289, 512
builder risk 289
equipment 289
public liability 289
Interior door 224, 226
door frame 406

:ght 281
-ght fixture 507
artition framing 154
re-hung door 402
esidential door 404
vood frame 406
erval timer 500
rusion system 470
estigation subsurface 298
ning center 462
nspot brick 432
gation system 312
cyanurate insulation 383

J

k cable 295
hydraulic 295
rafter 360
khammer 293
ousie 409
nt expansion 323, 425
reinforcing 330
sealer 397
st anchor 354
connector 354
demolition 302
hanger 354
sister 358
web stiffeners 347
wood 358, 366
sts coldformed 345
framing 347
e mesh 307

K

nnel fence 313
ttle/pot tar 295
yway 322
ck plate 418
plate door 418
n dried lumber 352
tchen 234
appliance 458
cabinet 371
sink 485
sink faucet 484
nocker door 420

L

beled door 400
frame 401
dder 294
swimming pool 467
g screw 339
screws 340
lly column 341
minated beam 365, 366
construction glued 366
countertop 352
countertop plastic 352
framing 366
glued 365
roof deck 366
veneer members 367
wood 365, 366
amp fluorescent 509
post 350
ampholder 505
andfill fees 304
anding newel 376

Landscape light 508
surface 433
Lantern fixture 508
Latch set 418
Latex caulking 397
underlayment 434
Lath gypsum 425
metal 425
Lattice molding 369
Lauan door 405
Laundry faucet 484
sink 484
tray 484
Lavatory 248, 485
faucet 484
removal 474
support 483
vanity top 485
wall hung 485
Lawn grass 316
Lazy susan 372
Leaching pit 308
Lead coated copper downspout 395
coated copper gutter 396
coated downspout 395
coated flashing 394
flashing 393
paint encapsulation 469
paint removal 468
Lean-to type greenhouse 466
Leeching field 112
Let-in bracing 356
Letter slot 454
Lever handicap 418
Liability employer 289
insurance 513
Lift aerial 293
Light bollard 508
fixture interior 507
landscape 508
post 508
support 341
track 509
Lighting 280, 508, 509
fixture exterior 508
incandescent 508
outdoor 294
outlet 503
residential 503
strip 507
track 508
Lightning arrester 466
protection 466
suppressor 500
Lightweight block 336
Limestone 316, 334
coping 332
Linear diffuser 493
Linen wallcovering 435
Liner duct 475
flue 335
Lintel 327, 334, 341
block 332
concrete 327
precast 327
Load center indoor 507
center plug-in breaker 507
center rainproof 507
Loadbearing studs 342
Loadcenter residential 500
Loader front end 306
tractor 293
Loam 306
Lock entrance 418
Locking receptacle 502
Lockset cylinder 418

Log electric 451
gas 451
Louver 451
aluminum 450
midget 451
redwood 377
ventilation 377
wall 451
wood 377
Louvered blind 378
door 224, 226, 402, 405
Low-voltage silicon rectifier 505
switching 505
switchplate 505
transformer 505
Lumber 356, 530, 531
core paneling 371
kiln dried 352
treatment 352
Luminous ceiling 431, 509
panel 431

M

Machine trowel finish 326
welding 295
Mahogany door 403
Malleable iron fitting 478
iron pipe fitting 478
Management fee construction 288
Manhole precast 310
removal 299
Mansard roof framing 146
Mantel beam 377
fireplace 377
Maple countertop 352
Marble 334
base 334
chips 316
coping 332
countertop 352
floor 334
sill 333
synthetic 432
tile 432
Masonry accessory 330
adobe 333
anchor 330
brick 336, 432
cement 330
clay 331
cleaning 336
cornice 333
demolition 300, 302
fireplace 238, 335
flashing 393
foundation 120
furring 363
glass 333
insulation 382
nail 353
painting 444
point 336
reinforcing 330, 525
removal 299, 302
saw 295
sill 333
step 311
toothing 300
unit 331
wall 158, 160, 315, 332
wall tie 330
Mat foundation 325, 326
Mechanical duct 492
media filter 494

Medicine cabinet 455
Membrane 211
curing 327
roofing 391
Mercury vapor exterior fixture 508
vapor fixture 281
Metal butt frame 401
casework 462
chimney 451
clad window 170, 172, 174, 176, 178, 180
deck 341
door 400
door residential 400
fabrication 349
faced door 402
fascia 389
fastenings 338
fence 313
fireplace 240
flue chimney 489
frame 401
framing 340
furring 424
gate 313
halide fixture 281
hollow 401
joists bracing 345
joists bridging 345
joists coldformed 347
joists framing 347
lath 425
overhead door 407
partitions framing 343
pipe 308
roof 388
sheet 393
shelf 455
siding 166
soffit 390
stair 349
stud 424
stud demolition 304
stud partitions 343
stud walls 343
studs 343
studs framing 343
support assemblies 424
threshold 419
tile 430
window 409
Metallic foil 383
Meter center rainproof 506
device basic 506
socket 505
water supply 482
water supply domestic 482
Microtunneling 307
Microwave oven 458
Mill construction 357
Millwork 367, 373
demolition 303
Mineral fiber ceiling 431
fiber insulation 384
roof 392
wool blown in 382
Mirror 421
ceiling board 431
door 421
glass 421
light fixture 508
wall 421
Mix planting pit 318
Mixer concrete 292
mortar 292, 294
plaster 294

- Mixing valve 484
- Mobilization 305
- Modified bitumen roofing 534
- Modular air handler 488
- Module tub-shower 484
- Moil point 294
- Molding 368
  - base 368
  - brick 369
  - ceiling 368
  - chair 369
  - cornice 368
  - exterior 368
  - hardboard 370
  - pine 368
  - trim 369
  - wood 376
  - wood transition 433
- Monel rivet 340
- Monitor support 340
- Monolithic finish concrete 326
- Monument survey 288
- Mortar 524
- Mortar, brick and block 527
- Mortar cement 429
  - mixer 292, 294
  - thinset 430
- Moss peat 316
- Motion detector 470
- Motor connection 505
  - generator 294
  - support 341
- Moulded door 403
- Mounting board plywood 364
- Movable louver blind 462
- Moving shrub 315
  - tree 315
- Mulch 315
  - bark 315
  - ceramic 316
  - stone 316
- Multi-blade damper 492
- Muntin window 414

N

- Nail 353
- Nailer 294
  - pneumatic 294
  - steel 359
  - wood 359
- Neoprene expansion joint 323
- Newel 231, 376
- No hub pipe 475
- Non-removable pin 418
- Nut hex 338
- Nylon carpet 228, 434, 436

O

- Oak door frame 405
  - floor 229, 432
  - molding 369
  - paneling 370
  - stair tread 377
  - threshold 406
- Oil fired boiler 487
  - fired furnace 489
  - heater temporary 294
  - heating 270, 272
  - water heater 486
- Olefin carpet 434
- Onyx 316
- Ooor hardware 418
- Open rail fence 314
- Opener door 408
- Operating cost equipment 292
- Outdoor lighting 294
- Outlet box plastic 500
  - box steel 499
  - lighting 503
- Oven 458
  - cabinet 372
  - microwave 458
- Overhaul 304
- Overhead and profit 517
  - contractor 289, 513
  - door 186, 407

P

- P trap 483
  - trap running 482
- Padding 228
  - carpet 434
- Paddle fan 504
- Paint 539
  - aluminum siding 442
  - chain link fence 439
  - doors & windows interior 437
  - encapsulation lead 469
  - fence picket 439
  - floor 440
  - floor concrete 439, 440
  - floor wood 440
  - removal 468
  - siding 442
  - trim, exterior 442
  - walls masonry, exterior 443
- Painting 540
  - balustrade 441
  - casework 436
  - ceiling 444
  - clapboard 441
  - concrete block 444
  - cornice 441
  - decking 440
  - downspout 440
  - drywall 444
  - grille 440
  - masonry 444
  - pipe 440
  - plaster 444
  - siding 441
  - sprayer 294
  - steel siding 441
  - stucco 441
  - swimming pool 467
  - trim 441
  - truss 441
  - wall 444
  - window 438, 439
- Paints & coatings 436
- Palladian windows 413
- Pan form stair 349
  - shower 394
- Panel door 403, 404
  - fiberglass 388
  - luminous 431
  - radiant heat 491
- Paneled door 224, 400
  - pine door 405
- Paneling 370
  - board 371
  - cutout 300
  - demolition 303
  - hardboard 370
  - plywood 370
  - wood 370
- Panels insulated 385
  - prefabricated 385
- Paper building 386
  - curing 327
  - sheathing 386
- Paperhanging 435
- Pargeting 534
- Parging cement 382
- Park bench 315
- Parquet floor 229, 432
- Particle board siding 391
  - board underlayment 365
- Parting bead 370
- Partition 214, 218
  - anchor 330
  - block 336
  - demolition 304
  - dust 304
  - framing 154, 424
  - shower 450
  - support 340, 341
  - wood frame 359
- Passage door 224, 402, 405
- Patch roof 392
- Patching concrete floor 327
- Patio door 406
- Pavement 312
  - berm 311
  - demolition 299
  - emulsion 312
- Paver floor 432
  - highway 312
- Paving asphaltic 522
  - brick 312
  - concrete 310
- Pea gravel 316
- Peastone 316
- Peat humus 316
  - moss 316
- Pegboard 370
- Perforated aluminum pipe 309
  - ceiling 431
  - pipe 309
- Perlite insulation 383
  - plaster 426, 537
- Permit building 288
- Pickett fence 314
- Pickup truck 295
- Picture window 180, 413
- Pier brick 336
- Pilaster wood column 378
- Pile sod 318
- Pin floating 418
  - non-removable 418
- Pine door 404
  - door frame 405
  - fireplace mantel 377
  - floor 433
  - molding 368
  - roof deck 366
  - shelving 371
  - siding 391
  - stair 376
  - stair tread 376
- Pipe & fittings 477-481
  - and fittings 542
  - bedding 306
  - cast iron 475
  - cleanout 482
  - column 341
  - concrete 308, 309
  - copper 476, 477
  - corrugated metal 309
  - covering 475
  - covering fiberglass 475
  - CPVC 479
  - drainage 30
  - DWV ABS 47
  - DWV PVC 47
  - fitting cast iron 47
  - fitting copper 47
  - fitting DWV 48
  - fitting no hub 47
  - fitting plastic 479, 48
  - fitting soil 47
  - gas 30
  - insulation 47
  - no hub 47
  - painting 44
  - perforated aluminum 30
  - plastic 47
  - polyethylene 30
  - PVC 308
  - rail 34
  - removal 299, 47
  - sewage 308
  - single hub 47
  - soil 47
  - stainless 34
  - steel 309
  - subdrainage 309
  - vitrified clay 308, 309
- Piping subdrainage 309
- Pit leaching 308
- Pitch coal tar 392
  - emulsion tar 312
- Placing concrete 320
- Plank floor 358
  - roof 364
  - sheathing 364
- Plant bed preparation 317
  - screening 292
- Planter 318
  - fiberglass 318
- Planting 315
- Plaster 218, 220, 424
  - accessories 425
  - beam 426
  - board 214, 216
  - ceiling 426
  - column 425
  - cutout 300
  - demolition 300, 304
  - drill 339
  - ground 364
  - gypsum 425
  - mixer 294
  - painting 444
  - perlite 426
  - thincoat 426
  - wall 425, 426
- Plasterboard 427
- Plastic blind 191
  - clad window 170, 172, 174, 176, 178 , 180
  - door 401
  - fabrications 379
  - faced hardboard 370
  - fitting 479
  - laminated countertop 352
  - outlet box 500
  - pipe 479
  - pipe fitting 480, 481
  - siding 166
  - skylight 417
  - trap 483
- Plate shear 354
  - wall switch 505
- Plating zinc 353
- Players bench 315
- Plenum silencer 492
- Plugmold raceway 499

umbing 482
demolition 474
fixture 483, 484, 486, 542
fixtures removal 474
wood 532
clip 354
demolition 300, 303, 304
floor 365
formwork 323
joist 366
mounting board 364
paneling 370
sheathing, roof and walls 364
shelving 371
siding 391
soffit 370
subfloor 365
treatment 352
underlayment 365
eumatic nailer 294
cket door frame 406
int masonry 336
moil 294
le closet 369
lice connect panel 470
lyethylene film 316
floor 433
pipe 309
pool cover 467
septic tank 308
tarpaulin 291
lystyrene blind 379
ceiling 431
ceiling panel 431
insulation 382, 383
lysulfide caulking 397
lyurethane caulking 397
lyvinyl soffit 390
ol accessory 467
cover 467
cover polyethylene 467
heater electric 487
heater gas fired 487
swimming 466
rcelain tile 429
rch molding 369
rtable air compressor 293
st cap 354
cedar 378
demolition 302
lamp 350
light 508
wood 357
ured insulation 168
owder act. tools & fast. 340
ower wiring 506
recast catch basin 310
concrete 327
coping 332
curb 311
lintel 327
manhole 310
receptor 450
sill 333
stair 327
terrazzo 430
re-engineered structure 466
refabricated building 466
fireplace 240, 451
panels 385
refinished drywall 428
floor 229, 433
shelving 371
reformed roof panel 388
roofing & siding 388
re-hung door 401, 402
Premolded expansion joint 323
Preparation plant bed 317
surface 445, 446
Pressure valve relief 481
wash 446
Pretreatment termite 307
Primer asphalt 434
Profile block 332
Projected window 409
Property line survey 288
Protection lightning 466
slope 307
termite 307
Protectors ADA insulated 474
P&T relief valve 481
Pull box 500
door 374
Pump 486
circulating 483
concrete 292
diaphragm 295
heat 491
rental 294
staging 289, 521
submersible 294, 486
sump 459
trash 295
water 294, 308
Purlin roof 358
Push button lock 418
Puttying 445
PVC conduit 498
conduit in slab 498
DWV pipe 479
fitting 479
gravel stop 396
pipe 308, 479
siding 389
waterstop 324

Q

Quarry tile 429
Quarter round molding 369
Quartz 316
heater 491
Quoins 334

R

Raceway plugmold 499
surface 499
wiremold 499
Radial arch 367
Radiant heat 491
heating panel 491
Radiator cast iron 491
Rafter 360
anchor 354
composite 360
demolition 302
roof framing 138
wood 360
Rail hand 349
pipe 349
wall 349
Railing aluminum 349
demolition 303
steel 349
wood 369, 375, 376
Railroad tie 318
tie step 311
Rainproof meter center 506
Ramp handicap 325
Ranch plank floor 433
Range circuit 280
cooking 458
hood 236, 459
receptacle 502, 505
Ready mix concrete 325
Receptacle air conditioner 502
device 502
dryer 502
duplex 505
GFI 502
locking 502
range 502, 505
telephone 503
television 503
weatherproof 502
Receptor precast 450
shower 450
terrazzo 450
Rectangular diffuser 493
ductwork 492
Rectifier low-voltage silicon 505
Redwood bark mulch 316
cupola 452
louver 377
paneling 371
siding 162, 391
trim 369
wine cellar 458
Refinish floor 433
Reflective block 333
insulation 383
Refrigerated wine cellar 458
Refrigeration residential 459
Register air supply 493
baseboard 493
Reinforcement concrete 324
Reinforcing dowels 324
fiber 325
footing 324
joint 330
masonry 330
steel 324
wall 324
Removal air conditioner 474
block wall 299
catch basin 299
chain link fence 299
concrete 299
curb 299
driveway 299
fence 299
floor 299
foundation 298
lavatory 474
masonry 299
paint 468
pipe 299, 474
plumbing fixtures 474
shingle 304
sidewalk 299
sod 318
stone 299
tree 305
water closet 474
window 305
Rental equipment 292, 295
equipment rate 289
generator 294
Repair fire damage 445
Repellent water 444
Replacement sash 415
sliding door 406
windows 416, 536
Residential alarm 503
appliance 458, 504
burner 486
closet door 400
device 500
door 400, 403, 406
door bell 503
elevator 472
fan 504
fence 313
fixture 503, 508
garage door 407
greenhouse 466
gutting 302
heat pump 504
hinge 418
lighting 503
load center 500
lock 418
overhead door 407
refrigeration 459
service 500
smoke detector 503
stair 376
storm door 407
switch 500
ventilation 492
water heater 486, 504
wiring 500, 504
Resilient floor 433
flooring 229, 538
Resquared shingle 387
Restoration window 468
Retaining wall 315
Retarder vapor 386
Ribbed waterstop 324
Ridge board 360
shingle 387
shingle clay 388
shingle slate 387
vent 396, 451
Rig drill 298
Rigid anchor 330
conduit in-trench 498
in slab conduit 498
insulation 168, 383
Ring split 354
toothed 354
Rings drapery 463
Riser beech 231
oak 231
wood stair 376
Rivet 340
tool 340
Robe hook 455
Rod closet 371
curtain 455
ground 497
Roll roof 391
roofing 392
type air filter 494
Romex copper 497
Roof adhesive 391
aluminum 388
beam 366
built-up 391
cant 360, 392
clay tile 387
coating 392
copper 393
deck 341, 364
deck insulation 384
deck laminated 366
deck wood 366
drains 482
fiberglass 388
framing 138, 140, 142, 144, 146, 148

framing removal 301
insulation 384
metal 388
mineral 392
modified bitumen 392
nail 353
panel preformed 388
patch 392
purlin 358
rafter 360
rafters framing 360
roll 391
sheathing 364
skylight 417
slate 387, 534
specialties, prefab 394
tile 387
truss 358, 366, 367
zinc 393
Roofing 194, 196, 198, 200, 202
& siding preformed 388
demolition 303
gable end 194
gambrel 198
hip roof 196
mansard 200
membrane 391
roll 392
shed 202
sheet metal 393
single ply 392
Rooftop air conditioner 490
Rosewood door 403
Rosin paper 386
Rotary hammer drill 294
Rough buck 359
carpentry 356
hardware 355
stone wall 335
Rough-in sink countertop 485
sink raised deck 485
sink service floor 485
tub 484
Round diffuser 493
Rubber base 433
coating 382
floor 433
floor tile 433
sheet 433
threshold 419
tile 433
Rubbish chute 304
handling 304

S

Safety switch 506
switch disconnect 507
Salamander 295
Sales tax 288, 518, 519
Salt treatment lumber 352
Sand fill 298
Sandblasting equipment 295
Sander floor 295
Sanding 445
floor 433
Sandstone 334
flagging 312
Sanitary base cove 429
Sash replacement 415
wood 414
Sauna 463
door 463
Saw 295
chain 295
circular 295
concrete 292
masonry 295
Scaffold steel tubular 289, 290
Scaffolding 520
tubular 289
Scrape after damage 445
Screen chimney 452
fence 314
molding 369
security 408, 409
squirrel and bird 452
window 408, 415
wood 415
Screening plant 292
Screw anchor 339
brass 353
drywall 428
lag 339
sheet metal 353
steel 353
wood 353
Screws lag 340
Seal pavement 312
Sealant 382
acoustical 428
caulking 397
Sealcoat 312
Sealer joint 397
Sectional door 408
overhead door 407
Security screen 408, 409
Seeding 316, 523
Selective demolition 300
Self propelled crane 295
Self-contained air conditioner 490
Septic system 112, 308
tank 308
tank concrete 308
Service electric 278, 506
entrance cable aluminum 497
entrance cap 497
residential 500
sink 485
sink faucet 484
Sewage pipe 308
Shadow box fence 314
Shake siding 164
wood 387
Shear plate 354
wall 364
Sheathed nonmetallic cable 497
romex cable 497
Sheathing 364
asphalt 365
gypsum 365
paper 386
roof 364
Sheathing, roof & walls plyw. 364
Shed dormer framing 152
dormer roofing 206
roof framing 148
Sheet metal 393
metal aluminum 394
metal cladding 393
metal roofing 393
metal screw 353
rock 214, 216
Sheeting driver 294
Sheetrock 427
Shelf brick 322
metal 455
Shellac door 438
Shelving 371
pine 371
plywood 371
prefinished 371
storage 455
wood 371
Shield expansion 339
Shingle 194, 196, 198, 200, 202, 386
asphalt 386
concrete 388
removal 304
ridge 387
siding 164
stain 441
strip 386
wood 387
Shock absorber 482
Shot blaster 295
Shower by-pass valve 484
compartments 450
cubicle 485
door 450
enclosure 450
glass door 450
pan 394
partition 450
receptor 450
stall 485
surround 450
Shrub broadleaf evergreen 317
deciduous 317
evergreen 317
moving 315
Shutter 191, 462
Sidelight 406
Sidewalk 108, 311, 432
asphalt 311
brick 312
concrete 311
removal 299
Siding aluminum 388
bevel 390
cedar 390
demolition 303
fiber cement 389
fiberglass 388
hardboard 391
metal 166
nail 353
paint 442
painting 441
plywood 391
PVC 389
redwood 391
removal 304
shingle 164
stain 441
steel 390
vinyl 389
wood 162, 390
wood product 391
Silencer plenum 492
Silicone caulking 397
coating 382
water repellent 382
Sill 334, 361
anchor 354
door 370, 405, 419
masonry 333
precast 333
quarry tile 430
Sillcock 484
Silt fence 307
Single hub pipe 475
hung window 409
ply roofing 392
zone rooftop unit 490
Sink 236, 485
base 372
countertop 48
countertop rough-in 48
kitchen 48
laundry 48
lavatory 48
raised deck rough-in 48
removal 47
service 48
service floor rough-in 48
Siren 47
Sister joist 35
Site clearing 30
demolition 298, 29
drainage 31
furnishings 31
improvement 312, 31
preparation 29
Site work 10
Skim coat 214, 21
Skirtboard 37
Skylight 208, 41
removal 30
roof 41
Skywindow 20
Slab blockout 32
concrete 126, 32
cutout 30
form 34
on grade 322, 32
on grade removal 30
textured 32
Slate 33
flagging 31
removal 30
roof 38
shingle 38
sidewalk 31
sill 33
stair 33
tile 43
Slatwall 436
Sleeper 361
wood 229
Sleepers framing 361
Sleeve formwork 324
Sliding door 184, 185, 406
glass door 406
mirror 455
window 176, 409, 413
Slop sink 484, 485
Slope grading 305
protection 307
Slot letter 454
Slotted pipe 308
Small tools 291
Smoke detector 470
Snap tie formwork 324
Soap holder 455
Socket meter 505
Sod 316
Sodium low pressure fixture 508
Soffit 361
aluminum 389
drywall 427
metal 390
plaster 426
plywood 370
vinyl 390
wood 370
Softener water 460
Soil compaction 306
compactor 292
pipe 475
tamping 306
treatment 307
Solar heating 541

id wood door 403
bath 484
ace heater rental 294
ade air 294
anish roof tile 388
ecial door 407
stairway 231
systems 470, 491
ecialties & accessories, roof ... 394
ike grid 354
ral stair 349, 376
it rib block 332
ring 354
system ductless 490
ray coating 382
rayer emulsion 294
read footing 322, 325
rinkler grass 312
ging aids 291
pump 289
in cabinet 436
casework 436
door 438
floor 440
lumber 367
shingle 441
siding 391, 441
truss 441
inless flashing 394
gutter 396
pipe 349
steel bolt 338
steel gravel stop 396
steel hinge 418
steel rivet 340
ir 376
basement 327, 376
carpet 434
ceiling 460
climber 472
concrete 326
electric 460
metal 349
pan form 349
precast 327
railroad tie 312
removal 302
residential 376
slate 335
spiral 376
stringer 359
stringer wood 359
tread 231
tread tile 430
tread wood 377
wood 376
airs disappearing 460
fire escape 349
airway 230
airwork and handrails 375
all shower 485
amping concrete 326
texture 326
andard extinguisher 453
arting newel 376
eam bath 463
bath residential 463
boiler 487
boiler electric 486
cleaner 295
eel anchor 330
bin wall 315
bolt 338, 354
bridging 356
chain link 313
channel 425
conduit in slab 498
conduit in trench 498
conduit rigid 498
deck 341
diffuser 493
door 400
downspout 395
drill 294
edging 318
expansion tank 482
fireplace 240
fitting 478
flashing 394
frame 401
frame formwork 322
furring 424
gravel stop 396
gutter 396
hex nuts 338
hinge 418
lintel 341
nailer 359
pipe 309
reinforcing 324
scaffold tubular 289, 290
screw 353
siding 390
structural 340
stud 424
window 408
Steeple tip pin 418
Step bluestone 311
brick 311
masonry 311
railroad tie 311
stone 334
Stockade fence 314
Stone 333
aggregate 316
anchor 330
base 310
collected 335
fill 298
floor 334
ground cover 316
mulch 316
paver 312, 333
removal 299
step 334
threshold 333
trim cut 334
wall 160, 315
Stool cap 370
window 333, 334
Stop door 369
gravel 395
Storage shelving 455
tank 467
Storm door 190, 406
window 190, 415
Stove 452
wood burning 452
Strainer downspout 394
gutter 396
roof 395
wire 395
Strap tie 354
Straw 316
Stringer 231
stair 359
Strip chamfer 323
floor 432
footing 325
lighting 507
shingle 386
soil 306
Structural concrete 325
excavation 306
formwork 322
steel 340
Structure pre-engineered 466
Stucco 220, 426, 537
interior 218
painting 441
wall 158
Stud connector 340
demolition 302
metal 424
partition 359
partitions metal 343
steel 424
wall 136, 359, 424
walls bracing 342
walls bridging 342
walls metal 343
Studs loadbearing 342
metal 343
Subdrainage pipe 309
piping 309
system 309
Subfloor 365
plywood 229, 365
Submersible pump 294, 308, 486
Subsurface investigation 298
Suction hose 294
Sump pump 459
Support ceiling 340
drinking fountain 483
lavatory 483
light 341
monitor 340
motor 341
partition 340, 341
Suppressor lightning 500
Surface landscape 433
preparation 445, 446
raceway 499
Surfacing 312
Surround shower 450
tub 450
Survey crew 288
monument 288
property line 288
topographic 288
Suspended ceiling 222, 425, 431
Suspension system ceiling 424
Swimming pool 243, 466, 467
pool enclosure 466
pool heater 487
pool ladder 467
pools 541
Swing check valve 481
Swing-up door 186
overhead door 408
Switch box 499
box plastic 500
decorator 500
dimmer 500, 505
electric 505-507
general duty 506
residential 500
safety 506
time 507
toggle 505
wiring 280
Switching low-voltage 505
Switchplate low-voltage 505
Synthetic marble 432
Syphon ventilator rotary 494
System antenna 509
irrigation 312
septic 308
subdrainage 309
T.V. 509
VHF 509

T

T & G siding 162
Table top 352
Tamper 294
Tamping soil 306
Tank expansion 482
fibergalss 467
oil & gas 482
septic 112, 308
storage 467
Tar kettle/pot 295
paper 316
pitch emulsion 312
Tarpaulin 291
duck 291
polyethylene 291
Tax 288
sales 288
social security 288
unemployment 288
Taxes 518
T-bar mount diffuser 493
Teak floor 229, 432
molding 368, 369
paneling 370
Tee cleanout 482
Telephone receptacle 503
Television receptacle 503
Temperature relief valve 481
Tempered glass 420
hardboard 370
Tempering valve 481
Temporary oil heater 294
Tennis court fence 114, 313
Terminal A/C packaged 489
Termite pretreatment 307
protection 307
Terne coated flashing 394
Terra cotta coping 332
cotta demolition 300
Terrazzo 430
precast 430
receptor 450
wainscot 430
Texture stamping 326
Textured slab 325
Thermostat 469
integral 491
wire 505
Thincoat 214, 216
plaster 426
Thinset ceramic tile 429
mortar 430
Threshold 419
door 406
stone 333, 334
wood 370
Thru-wall air conditioner 490
Tie formwork snap 324
rafter 360
railroad 318
strap 354
wall 330
Tile 428, 433
ceiling 431
ceramic 428, 429
clay 387
concrete 388
cork 433
cork wall 435

- demolition 300
- flue 335
- glass 421
- grout 429
- marble 432
- metal 430
- porcelain 429
- quarry 429
- roof 387
- rubber 433
- slate 432
- stainless steel 430
- stair tread 430
- vinyl 434
- wall 428
- window sill 430

Timber connector 354
- fastener 353
- framing 357
- laminated 367

Time switch 507
Timer clock 503
- interval 500
- switch ventilator 493

Tinted glass 420
Toggle bolt 339
- switch 505

Toilet bowl 486
Tongue and groove siding 162
Tool air 293
- rivet 340

Tools small 291
Toothed ring 354
Toothing masonry 300
Top demolition counter 303
- dressing 306
- table 352

Topographic survey 288
Topsoil 306
Torch cutting 295
Towel bar 455
Track drawer 375
- drill 293
- light 509
- lighting 508
- traverse 462

Tractor 293, 307
- loader 293
- truck 295

Trailer truck 293
Transfer switch automatic 507
Transformer low-voltage 505
Transit 295
Transition molding wood 433
Transom lite frame 401
- windows 414

Transplanting 315
Trap cast iron 482
- drainage 482
- P 483
- plastic 483

Trapezoid windows 414
Trash pump 295
Traverse 462, 463
- track 462

Travertine 334
Tray laundry 484
Tread abrasive 430
- beech 231
- oak 231
- stone 335
- wood 377

Treatment lumber 352
- plywood 352
- wood 352

Tree 318
- deciduous 318

Tree deciduous 318
Tree evergreen 317
- guying 318
- moving 315
- removal 305

Trench backfill 307
- excavation 307
- utility 307

Trencher 293
- chain 293

Trenching 106
Trim exterior 368
Trim, exterior paint 442
Trim molding 369
- painting 441
- redwood 369
- tile 429
- wood 369

Trowel coating 382
- concrete 292

Truck dump 293
- flatbed 293
- hauling 307
- loading 306
- pickup 295
- rental 293
- tractor 295
- trailer 293
- vacuum 295

Truss bowstring 367
- demolition 302
- flat wood 366
- painting 441
- plate 354
- roof 358, 366, 367
- roof framing 140
- stain 441
- varnish 441

Tub hot 484
- redwood 484
- rough-in 484
- surround 450

Tubing copper 476
- electric metallic 498

Tub-shower module 484
Tubular scaffolding 289
- steel joist 366

Tumbler holder 455
Turned column 377
TV antenna 509
- system 509

U

Undereave vent 451
Underlayment 365
- latex 434

Unemployment tax 288
Unit heater electric 491
- masonry 331

Utility 106
- excavation 106
- trench 307
- vault 298

V

Vacuum breaker 482
- central 458
- cleaning 458
- truck 295

Valance board 372
Valley rafter 360
Valve 481
- bronze 481
- mixing 484
- relief pressure 481
- shower by-pass 484
- swing check 481
- tempering 481
- water pressure 481

Vanity base 375
- top lavatory 485

Vapor barrier 386
- barrier sheathing 365
- retarder 386

Varnish cabinet 436
- casework 436
- door 438
- floor 440, 445
- truss 441

Vault utility 298
Vct removal 301
Veneer ashlar 529
- brick 160, 331, 527
- core paneling 371
- members laminated 367
- removal 303

Venetian blind 462
Vent chimney 489
- dryer 460
- exhaust 494
- ridge 396, 451
- ridge strip 451

Ventilating air conditioning 490, 493
Ventilation fan 504
- louver 377
- residential 492

Ventilator rotary syphon 494
- timer switch 493

Verge board 369
Vermiculite insulation 382
Vertical aluminum siding 389
- vinyl siding 390

VHF system 509
Vibrator concrete 292
- earth 306

Vibratory equipment 292
Vinyl blind 191, 379
- chain link 313
- composition floor 434
- door trim 390
- downspout 395
- faced wallboard 427
- fascia 390
- floor 434
- flooring 229
- gutter 396
- shutter 191
- siding 166, 389
- siding accessories 390
- soffit 390
- tile 434
- wallpaper 435
- window trim 390

Vitrified clay pipe 308, 309
Volume control damper 492

W

Wainscot ceramic tile 429
- molding 369
- quarry tile 430
- terrazzo 430

Walk 311
Wall anchor 338
- bumper 418
- cabinet 372
- ceramic tile 42
- coating 44
- concrete 32
- covering 53
- cutout 30
- drill 33
- drywall 42
- finish 32

Wall finishes 43
Wall formwork 32
- foundation 33
- framing 136, 36
- framing removal 30
- furring 363, 42
- grout 33
- heater 45
- heater electric 49
- hung lavatory 48
- insulation 330, 382, 38
- interior 214, 21
- lath 42
- louver 45
- masonry 158, 160, 315, 33
- mirror 42
- painting 44
- paneling 37
- plaster 425, 42
- rail 34
- reinforcing 32
- removal 29
- retaining 31
- shear 36
- sheathing 36
- siding 39
- steel bin 31
- stucco 42
- stud 359, 42
- switch plate 50
- tie 33
- tie masonry 33
- tile 42
- tile cork 43

Wallboard acoustical 428
Wallcovering 43
- acrylic 43
- gypsum fabric 43

Wallpaper 43
- grass cloth 43
- vinyl 43

Walls and partitions demolition 304
Walnut door frame 40
- floor 43

Wardrobe wood 373
Wash bowl 485
- brick 336

Washer 354
- residential 459

Washing machine automatic 459
Water closet 485
- closet removal 474
- closet support 483
- distribution system 308
- heater 459, 486, 504
- heater electric 486
- heater insulation 474
- heater oil 486
- heater removal 474
- heater residential 486
- heater wrap kit 474
- heating hot 486, 491
- heating wiring 280
- hose 294
- pipe ground clamp 497
- pressure relief valve 481
- pressure valve 481
- pump 294, 308, 459, 483

epellent 444
epellent coating 382
epellent silicone 382
oftener 460
supply domestic meter 482
supply meter 482
empering valve 481
well 308
terproofing 382
coating 382
aterstop 324
PVC 324
ribbed 324
eather cap entrance 499
eatherproof receptacle 280, 502
eatherstrip 419
and seals 419
door 419
eathervanes 452
eb stiffeners joist 347
elded frame 401
wire fabric 324
elding machine 295
ell 308
water 308
heelbarrow 295
hirlpool bath 484
indow 417
air conditioner 490
aluminum 409
awning 409
blind 378, 462
casement 410, 411
casing 368
demolition 304
double hung 170, 410, 412
frame 415
glass 421
grille 414
hardware 418
metal 409
muntin 414
painting 438, 439
picture 413
removal 304, 305
restoration 468
screen 408, 415
sill 333
sill marble 334
sill tile 430
sliding 413
steel 408
stool 333, 334
storm 190, 415
trim set 370
trim vinyl 390
wood 409-414
Windows casement 411
transom 414
trapezoid 414
Wine cellar 458
Winter protection 322
Wire copper 498
electric 498
fence 313
ground 497
mesh 426
strainer 395
thermostat 505
THW 498
Wiremold raceway 499
Wiring air conditioner 504
device 505
fan 504
methods 497
power 506
residential 500, 504
Wood base 368, 378
beam 356, 357
blind 191, 378, 462
block floor demolition 301
blocking 211, 359
canopy 361
casing 368
chamfer strip 323
chips 316
column 357, 377
cupola 452
deck 244, 364, 366
demolition 300
door 224, 226, 401, 403, 535
drawer 375
fascia 368
fastener 353
fence 313
fiber ceiling 431
fiber sheathing 365
fiber soffit 370
fiber subfloor 365
fiber underlayment 365
floor 130, 132, 134, 229, 432
floor demolition 301
foundation 124
frame 373, 405
framing 356
furring 363
girder 356
gutter 396
handrail 369
joist 358, 366
laminated 365, 366
louver 377
molding 376
nailer 359
overhead door 407
panel door 403
paneling 370
partition 359
planter 318
product siding 391
rafter 360
railing 375, 376
roof deck 364
roof deck demolition 303
roof trusses 533
sash 414
screen 415
screw 353
shake 387
sheathing 364
shelving 371
shingle 387
shutter 191
sidewalk 311
siding 162, 390
siding demolition 304
sill 361
soffit 370
stair 376, 533
stair stringer 359
storm door 190, 401
storm window 190
subfloor 365
threshold 370
tread 377
treatment 352
trim 369
truss 366
veneer wallpaper 435
wall framing 136
wardrobe 373
window 409-414, 536
window demolition 305
Woodwork architectural 371
Wool fiberglass 382
Workers' compensation 289
Workers' compensation 514-516
Wrench impact 294

Y

Yellow pine floor 433

Z

Z bar suspension 424
Zee bar 428
Zinc plating 353
roof 393
weatherstrip 419

Division Notes

		CREW	DAILY OUTPUT	LABOR-HOURS	UNIT	2000 BARE COSTS MAT.	LABOR	EQUIP.	TOTAL	TOTAL INCL O&P

Division Notes

	CREW	DAILY OUTPUT	LABOR-HOURS	UNIT	2000 BARE COSTS				TOTAL INCL O&P
					MAT.	LABOR	EQUIP.	TOTAL	

Division Notes

		CREW	DAILY OUTPUT	LABOR-HOURS	UNIT	2000 BARE COSTS MAT.	LABOR	EQUIP.	TOTAL	TOTAL INCL O&P

Division Notes

	CREW	DAILY OUTPUT	LABOR-HOURS	UNIT	2000 BARE COSTS MAT.	LABOR	EQUIP.	TOTAL	TOTAL INCL O&P

Division Notes

	CREW	DAILY OUTPUT	LABOR-HOURS	UNIT	2000 BARE COSTS MAT.	LABOR	EQUIP.	TOTAL	TOTAL INCL O&P

Division Notes

	CREW	DAILY OUTPUT	LABOR-HOURS	UNIT	2000 BARE COSTS MAT.	LABOR	EQUIP.	TOTAL	TOTAL INCL O&P

Notes

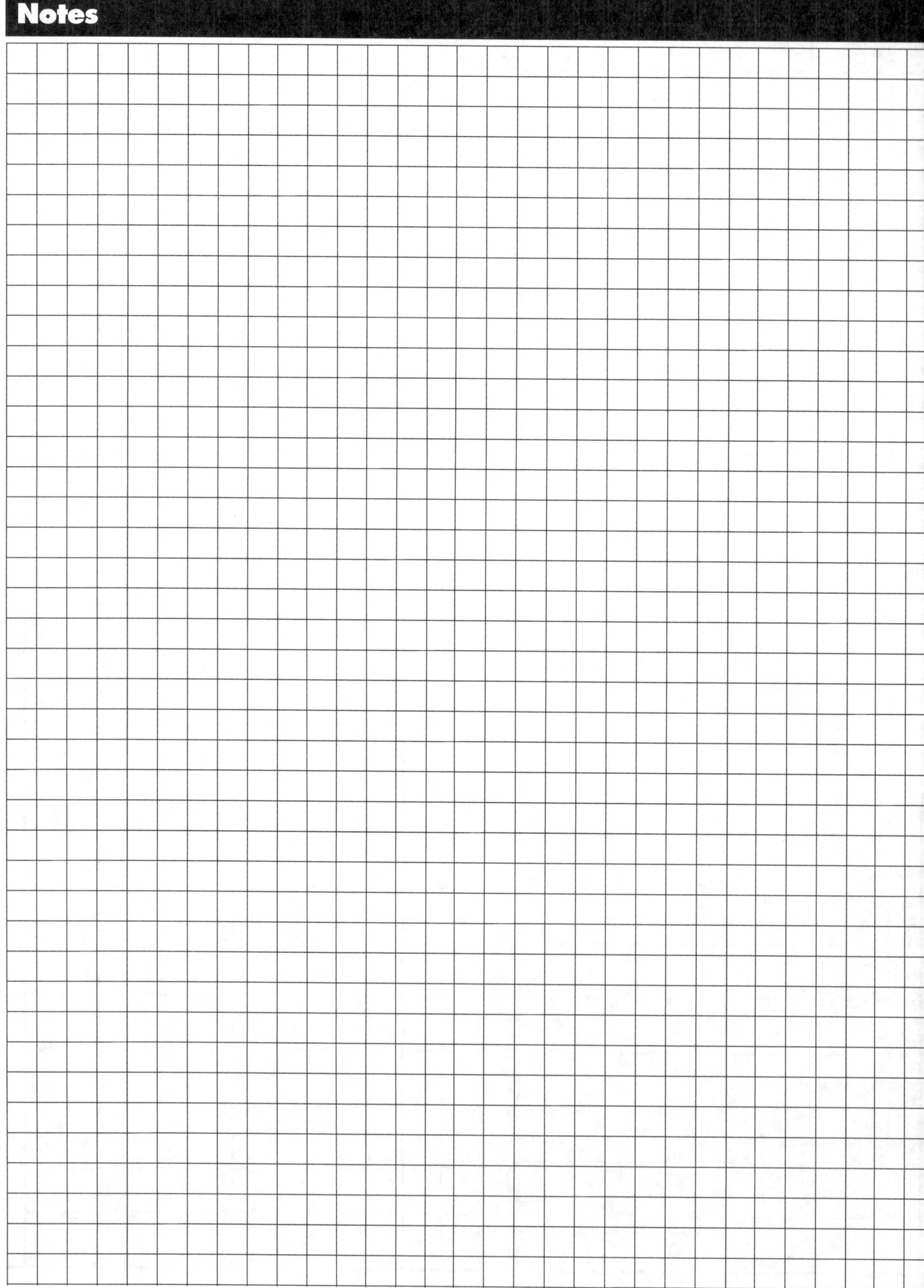

Notes

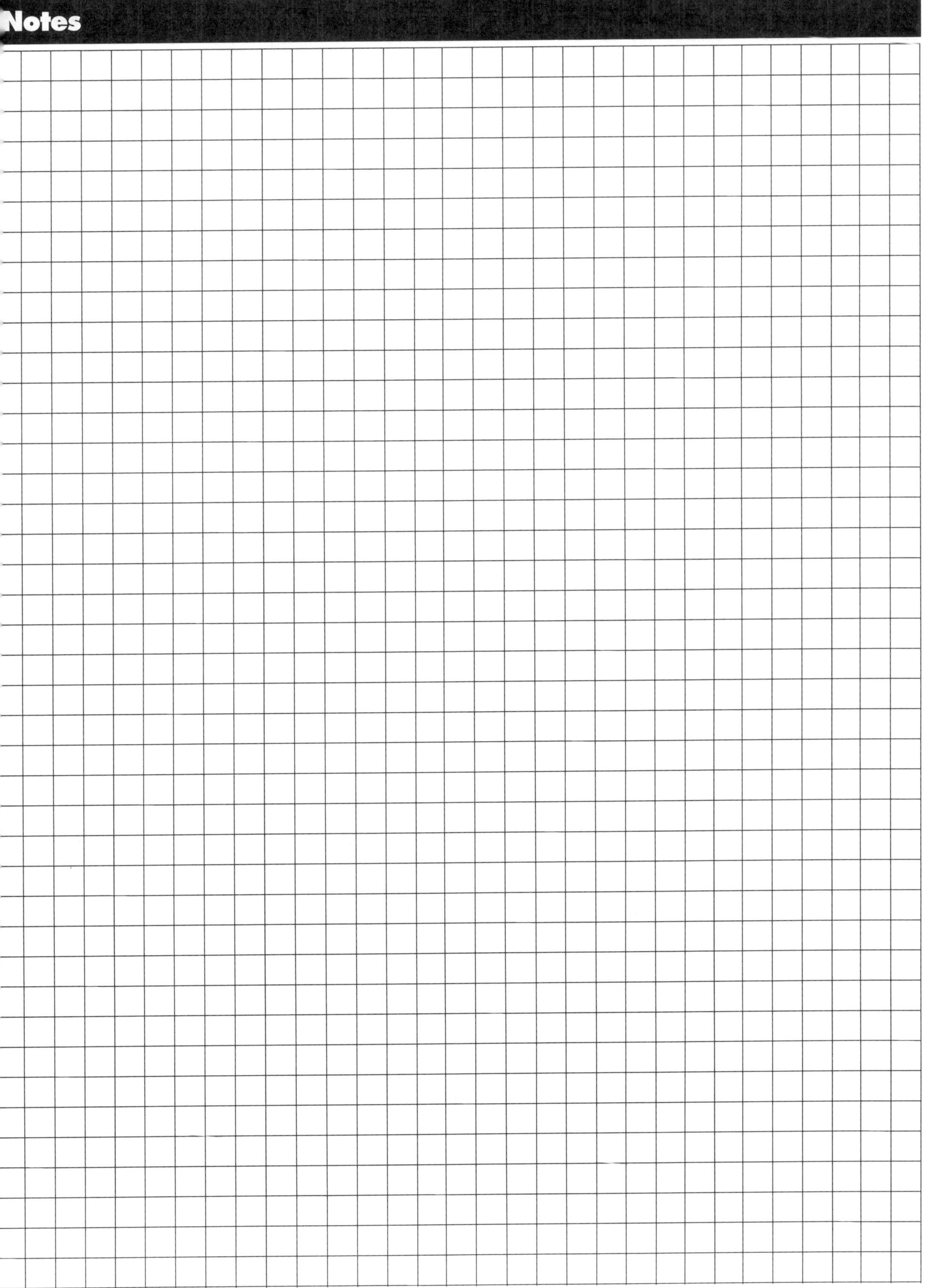

Notes

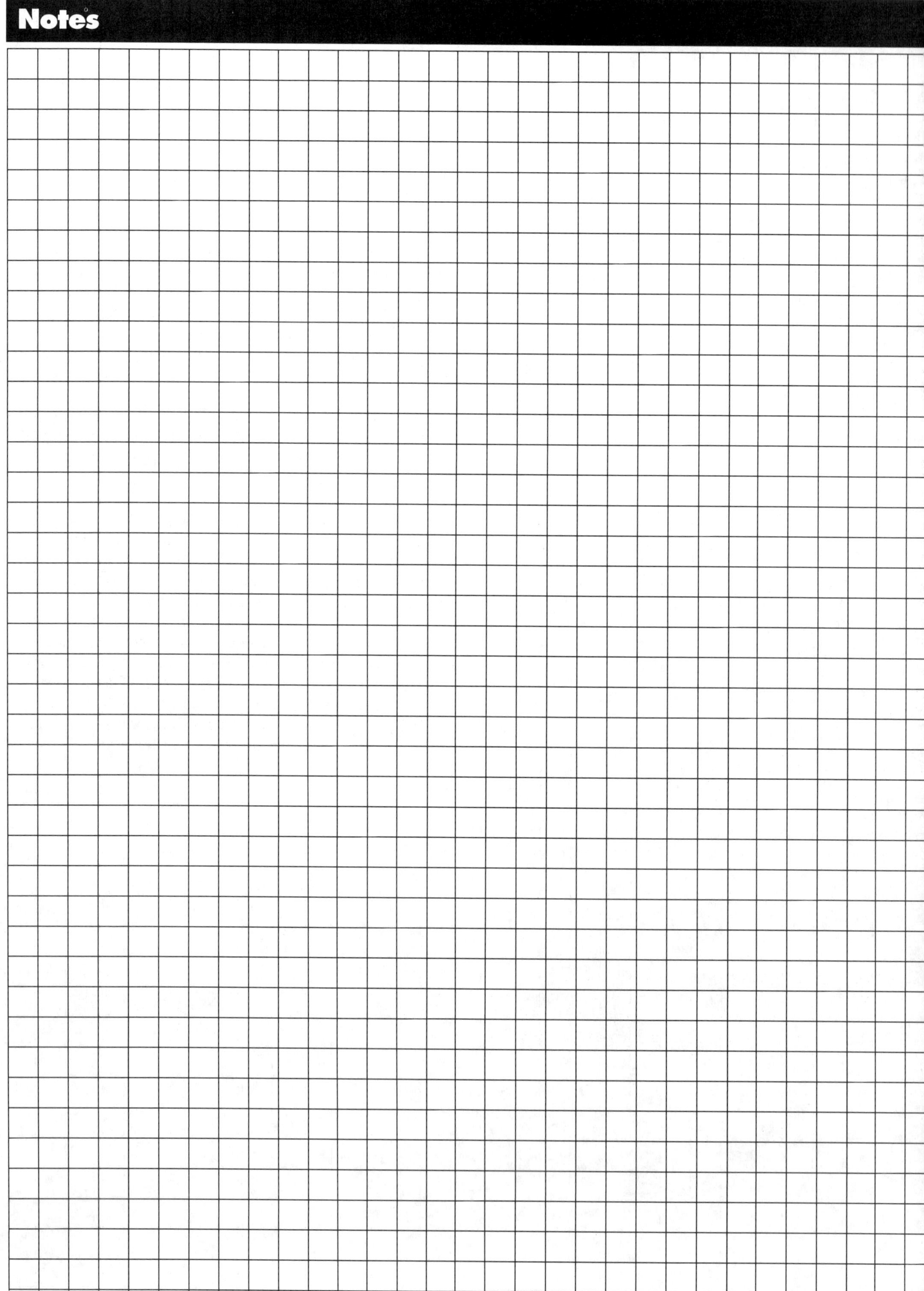

CMD Group...

S. Means Company, Inc., a CMD Group company, the leading provider construction cost data in North America, supplies comprehensive nstruction cost guides, related technical publications and educational vices.

ID Group, a leading worldwide provider of proprietary construction formation, is comprised of three synergistic product groups crafted be the complete resource for reliable, timely and actionable nstruction market data. In North America, CMD Group encompasses:

- Architects' First Source
- Construction Market Data (CMD)
- Associated Construction Publications
- Manufacturer's Survey Associates (MSA)
- R.S. Means
- CMD Canada
- BIMSA/Mexico
- Worldwide, CMD Group includes Byggfakta Scandinavia (Denmark, Estonia, Finland, Norway and Sweden) and Cordell Building Information Services (Australia).

rst Source for Products, available in print, on the Means CostWorks CD, d on the Internet, is a comprehensive product information source. In liance with The Construction Specifications Institute (CSI) and Thomas egister, Architects' First Source also produces CSI's SPEC-DATA® and ANU-SPEC® as well as CADBlocks.[SM] Together, these products offer mmercial building product information for the building team at each age of the construction process.

MD Exchange, a single electronic community where all members of the nstruction industry can communicate, collaborate, and conduct business, tter, faster, easier.

onstruction Market Data provides complete, accurate and timely project formation through all stages of construction. Construction Market Data pplies industry data through productive leads, project reports, contact sts, market penetration analysis and sales evaluation reports. Any of these oducts can pinpoint a county, look at a state, or cover the country. Data is livered via paper, e-mail or the Internet.

onstruction Market Data Canada serves the Canadian construction arket with reliable and comprehensive information services that cover all cets of construction. Core services include: Buildcore Product Source, a reliminary product selection tool available in print and on the Internet; MD Building Reports, a national construction project lead service; anaData, statistical and forecasting information; Daily Commercial News, construction newspaper reporting on news and projects in Ontario; and urnal of Commerce, reporting news in British Columbia and Alberta.

Manufacturers' Survey Associates is a quantity survey and specification service whose experienced estimating staff examines project documents throughout the bidding process and distributes edited information to its clients. Material estimates, edited plans and specifications and distribution of addenda form the heart of the Manufacturers' Survey Associates product line, which is available in print and on CD-ROM.

Associated Construction Publications, founded in 1930, are a group of 14 regional magazines focused on highway and heavy construction and cost data. With one of the world's largest editorial staffs dedicated to construction coverage, the magazine group focuses on local and regional news of projects in all phases of construction, precise project volumes, plus material and labor costs. The magazines have a cumulative ABC and BPA audited circulation of 115,000, an estimated pass-along readership of one-half million, and more than 25,000 pages of annual advertising.

Clark Reports is the premier provider of industrial construction project data, noted for providing earlier, more complete project information in the planning and pre-planning stages for all segments of industrial and institutional construction.

Byggfakta Scandinavia AB, founded in 1936, is the parent company for the leaders of customized construction market data for Denmark, Estonia, Finland, Norway and Sweden. Each company fully covers the local construction market and provides information across several platforms including subscription, ad-hoc basis, electronically and on paper.

Cordell Building Information Services, with its complete range of project and cost and estimating services, is Australia's specialist in the construction information industry. Cordell provides in-depth and historical information on all aspects of construction projects and estimation, including several customized reports, construction and sales leads, and detailed cost information among others.

For more information, please visit our website at www.cmdg.com

CMD Group Corporate Offices
30 Technology Parkway South #100
Norcross, GA 30092-2912
(770) 417-4000
(770) 417-4002 (fax)
www.cmdg.com

Means Project Cost Report

By filling out and returning the Project Description, you can receive a discount of $20.00 off any one of the Means products advertised in the following pages. The cost information required includes all items marked (✓) except those where no costs occurred. The sum of all major items should equal the Total Project Cost.

$20.00 Discount per product for each report you submit.

DISCOUNT PRODUCTS AVAILABLE—FOR U.S. CUSTOMERS ONLY—STRICTLY CONFIDENTIAL

Project Description (No remodeling projects, please.)

✓ Type Building ____________
✓ Location ____________
Capacity ____________
✓ Frame ____________
✓ Exterior ____________
✓ Basement: full ☐ partial ☐ none ☐ crawl ☐
✓ Height in Stories ____________
✓ Total Floor Area ____________
Ground Floor Area ____________
✓ Volume in C.F. ____________
% Air Conditioned ____________ Tons ____________
Comments ____________

Owner ____________
Architect ____________
General Contractor ____________
✓ Bid Date ____________
Typical Bay Size ____________
✓ Labor Force: _____ % Union _____ % Non-Union
✓ Project Description (Circle one number in each line)
1. Economy 2. Average 3. Custom 4. Luxury
1. Square 2. Rectangular 3. Irregular 4. Very Irregular

	✓	**Total Project Cost**		$
A	✓	**General Conditions**		$
B	✓	**Site Work**		$
BS		**Site Clearing & Improvement**		
BE		*Excavation*	(C.Y.)	
BF		*Caissons & Piling*	(L.F.)	
BU		*Site Utilities*		
BP		*Roads & Walks Exterior Paving*	(S.Y.)	
C	✓	**Concrete**		$
C		*Cast in Place*	(C.Y.)	
CP		*Precast*	(S.F.)	
D	✓	**Masonry**		$
DB		*Brick*	(M)	
DC		*Block*	(M)	
DT		*Tile*	(S.F.)	
DS		*Stone*	(S.F.)	
E	✓	**Metals**		$
ES		*Structural Steel*	(Tons)	
EM		*Misc. & Ornamental Metals*		
F	✓	**Wood & Plastics**		$
FR		*Rough Carpentry*	(MBF)	
FF		*Finish Carpentry*		
FM		*Architectural Millwork*		
G	✓	**Thermal & Moisture Protection**		$
GW		*Waterproofing-Dampproofing*	(S.F.)	
GN		*Insulation*	(S.F.)	
GR		*Roofing & Flashing*	(S.F.)	
GM		*Metal Siding/Curtain Wall*	(S.F.)	
H	✓	**Doors and Windows**		$
HD		*Doors*	(Ea.)	
HW		*Windows*	(S.F.)	
HH		*Finish Hardware*		
HG		*Glass & Glazing*	(S.F.)	
HS		*Storefronts*	(S.F.)	

Product Name ____________
Product Number ____________
Your Name ____________
Title ____________
Company ____________
☐ Company
☐ Home Street Address ____________
City, State, Zip ____________
☐ Please send _______ forms.

J	✓	**Finishes**		$
JL		*Lath & Plaster*	(S.Y.)	
JD		*Drywall*	(S.F.)	
JM		*Tile & Marble*	(S.F.)	
JT		*Terrazzo*	(S.F.)	
JA		*Acoustical Treatment*	(S.F.)	
JC		*Carpet*	(S.Y.)	
JF		*Hard Surface Flooring*	(S.F.)	
JP		*Painting & Wall Covering*	(S.F.)	
K	✓	**Specialties**		$
KB		*Bathroom Partitions & Access.*	(S.F.)	
KF		*Other Partitions*	(S.F.)	
KL		*Lockers*	(Ea.)	
L	✓	**Equipment**		$
LK		*Kitchen*		
LS		*School*		
LO		*Other*		
M	✓	**Furnishings**		$
MW		*Window Treatment*		
MS		*Seating*	(Ea.)	
N	✓	**Special Construction**		$
NA		*Acoustical*	(S.F.)	
NB		*Prefab. Bldgs.*	(S.F.)	
NO		*Other*		
P	✓	**Conveying Systems**		$
PE		*Elevators*	(Ea.)	
PS		*Escalators*	(Ea.)	
PM		*Material Handling*		
Q	✓	**Mechanical**		$
QP		*Plumbing*	*(No. of fixtures)*	
QS		*Fire Protection (Sprinklers)*		
QF		*Fire Protection (Hose Standpipes)*		
QB		*Heating, Ventilating & A.C.*		
QH		*Heating & Ventilating*	*(BTU Output)*	
QA		*Air Conditioning*	(Tons)	
R	✓	**Electrical**		$
RL		*Lighting*	(S.F.)	
RP		*Power Service*		
RD		*Power Distribution*		
RA		*Alarms*		
RG		*Special Systems*		
S	✓	**Mech./Elec. Combined**		$

Please specify the Means product you wish to receive. Complet the address information as requested and return this form with your check (product cost less $20.00) to address below.

R.S. Means Company, Inc.,
Square Foot Costs Department
100 Construction Plaza, P.O. Box 800
Kingston, MA 02364-9988

For more information visit Means Web Site at www.rsmeans.com

R.S. Means Company, Inc. . . a tradition of excellence in Construction Cost Information and Services since 1942.

Table of Contents

Annual Cost Guides, Page 2
Reference Books, Page 6
Seminars, Page 11
Consulting Services, Page 13
Electronic Data, Page 14
New Titles, Page 15
Order Form, Page 16

ook Selection uide

The following table provides definitive information on the content of each cost data publication. The number of lines of data provided in each unit price or assemblies division, as well as the number of reference tables and crews is listed for each book. The presence of other elements such as an historical cost index, city cost indexes, square foot models or cross-referenced index is also indicated. You can use the table to help select the Means' book that has the quantity and type of information you most need in your work.

: Cost sions	Building Construction Costs	Mechanical	Electrical	Repair & Remodel.	Square Foot	Site Work Landsc.	Assemblies	Interior	Concrete Masonry	Open Shop	Heavy Construc.	Resi-dential	Light Commercial	Facil. Construc.	Plumbing	Western Construction Costs
1	1108	587	588	780		1035		488	1000	1098	1075	430	625	1524	670	1095
2	3126	1460	506	2292		9120		1150	1636	3339	5584	1152	1217	5043	1819	3100
3	1516	90	68	764		1320		193	1935	1489	1492	263	208	1356	43	1493
4	875	28	0	675		770		642	1193	857	677	320	405	1143	0	853
5	1743	176	173	829		780		837	681	1710	1054	634	661	1733	95	1726
6	1396	95	82	1297		482		1317	334	1364	590	1515	1454	1393	59	1749
7	1365	178	91	1331		492		545	454	1366	353	803	1037	1359	188	1367
8	1898	60	0	1900		335		1818	723	1878	9	1131	1180	2030	0	1898
9	1610	48	0	1466		184		1706	322	1556	113	1250	1350	1805	48	1603
10	978	58	32	551		204		823	197	982	0	253	454	979	251	978
11	1157	434	212	621		152		915	35	1051	96	117	239	1220	369	1050
12	352	0	0	48		226		1536	29	343	0	71	68	1537	0	343
13	1384	1217	493	582		443		1071	106	1355	302	230	1331	965	1080	1335
14	364	43	0	261		36		331	0	365	39	9	18	363	17	363
15	2518	14708	764	2249		1802		1471	77	2518	2088	1065	1549	12986	10913	2547
16	1517	801	10627	1195		1098		1289	60	1529	911	703	1286	10207	708	1453
17	459	369	459	0		0		0	0	460	0	0	0	459	369	459
Totals	**23366**	**20352**	**14095**	**16841**		**18479**		**16132**	**8782**	**23260**	**14383**	**9946**	**13082**	**46102**	**16629**	**23412**

embly isions	Building Construction Costs	Mechanical	Electrical	Repair & Remodel.	Square Foot	Site Work Landsc.	Assemblies	Interior	Concrete Masonry	Open Shop	Heavy Construc.	Resi-dential	Light Commercial	Facil. Construc.	Plumbing	Western Construction Costs
1		0	0	202	131	738	775	0	689		752	844	131	89	0	
2		0	0	40	33	35	48	0	49		0	589	32	0	0	
3		0	0	443	1114	0	3064	228	1245		0	1546	806	174	0	
4		0	0	714	1353	0	3172	255	1273		0	2046	1171	26	0	
5		0	0	288	227	0	459	0	0		0	1082	223	25	0	
6		0	0	1006	845	0	1295	1723	152		0	1082	738	330	0	
7		0	0	42	89	0	179	159	0		0	723	33	71	0	
8		2257	149	998	1691	0	2720	926	0		0	1694	1191	1114	2084	
9		0	1346	355	357	0	1283	306	0		0	241	358	319	0	
10		0	0	0	0	0	0	0	0		0	0	0	0	0	
11		0	0	365	465	0	724	153	0		0	0	466	201	0	
12		501	160	539	84	2392	699	0	698		541	0	84	119	850	
Totals		**2758**	**1655**	**4992**	**6389**	**3165**	**14418**	**3750**	**4106**		**1293**	**9847**	**5233**	**2468**	**2934**	

ference ection	Building Construction Costs	Mechanical	Electrical	Repair & Remodel.	Square Foot	Site Work Landsc.	Assemblies	Interior	Concrete Masonry	Open Shop	Heavy Construc.	Resi-dential	Light Commercial	Facil. Construc.	Plumbing	Western Construction Costs
les	150	46	84	66	4	84	223	59	83	146	55	49	70	86	49	148
dels					102							32	43			
ws	408	408	408	389		408		408	408	391	408	391	391	389	408	408
/ st exes	yes	yes	yes	yes	yes	yes	yes	yes	yes	yes	yes	yes	yes	yes	yes	yes
torical st exes	yes	yes	yes	yes	yes	yes	yes	yes	yes	yes	yes	no	yes	yes	yes	yes
ex	yes	yes	yes	yes	no	yes	yes	yes	yes	yes	yes	yes	yes	yes	yes	yes

For more informati
visit Means Web Si
at www.rsmeans.c

Means Building Construction Cost Data 2000

Available in Both Softbound and Looseleaf Editions

The "Bible" of the industry comes in the standard softcover edition or the looseleaf edition.

Many customers enjoy the convenience and flexibility of the looseleaf binder, which increases the usefulness of *Means Building Construction Cost Data 2000* by making it easy to add and remove pages. You can insert your own cost information pages, so everything is in one place. Copying pages for faxing is easier also. Whichever edition you prefer, softbound or the convenient looseleaf edition, you'll get the *DesignBuildIntelligence* newsletter at no extra cost.

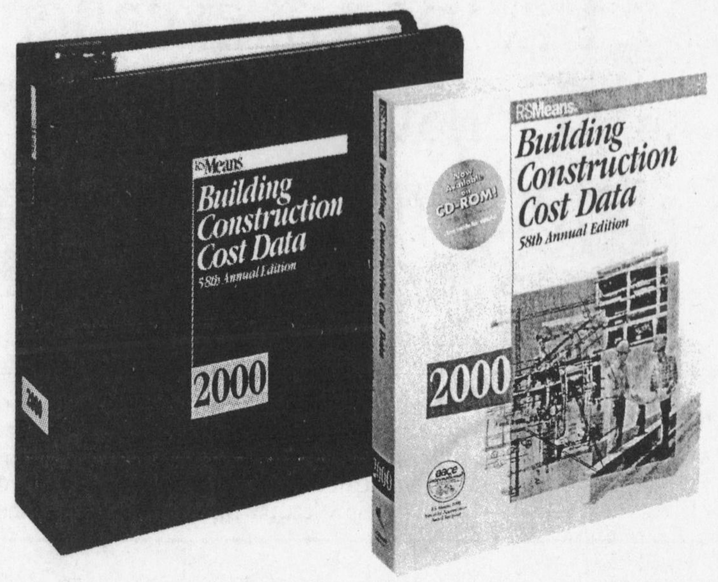

$85.95 per copy, Softbound
Catalog No. 60010

$109.95 per copy, Looseleaf
Catalog No. 61010

Means Building Construction Cost Data 2000

Offers you unchallenged unit price reliability in an easy-to-use arrangement. Whether used for complete, finished estimates or for periodic checks, it supplies more cost facts better and faster than any comparable source. Over 23,000 unit prices for 2000. The City Cost Indexes cover over 930 areas, for indexing to any project location in North America. Order and get the *DesignBuildIntelligence* newsletter sent to you FREE. You'll have year-long access to the Means Estimating **Hotline** FREE with your subscription. Expert assistance when using Means data is just a phone call away.

$85.95 per copy
Over 650 pages, illustrated, available Oct. 1999
Catalog No. 60010

Means Building Construction Cost Data 2000

Metric Version

The Federal Government has stated that all federal construction projects must now use metric documentation. The *Metric Version* of *Means Building Construction Cost Data 2000* is presented in metric measurements covering all construction areas. Don't miss out on these billion dollar opportunities. Make the switch to metric today.

$89.95 per copy
Over 650 pages, illustrated, available Nov. 1999
Catalog No. 63010

more information
it Means Web Site
www.rsmeans.com

Annual Cost Guides

eans Mechanical ost Data 2000

IVAC • Controls

al unit and systems price guidance for chanical construction . . . materials, parts, ings, and complete labor cost information. ludes prices for piping, heating, air conditioning, tilation, and all related construction.

s new 2000 unit costs for:
- ver 2500 installed HVAC/controls assemblies
- On Site" Location Factors for over 930 cities ad towns in the U.S. and Canada
- rews, labor and equipment

.95 per copy
er 600 pages, illustrated, available Oct. 1999
talog No. 60020

Means Plumbing Cost Data 2000

Comprehensive unit prices and assemblies for plumbing, irrigation systems, commercial and residential fire protection, point-of-use water heaters, and the latest approved materials. This publication and its companion, *Means Mechanical Cost Data,* provide full-range cost estimating coverage for all the mechanical trades.

$89.95 per copy
Over 500 pages, illustrated, available Oct. 1999
Catalog No. 60210

eans Electrical ost Data 2000

cing information for every part of electrical cost nning: More than 15,000 unit and systems costs th design tables; clear specifications and awings; engineering guides and illustrated imating procedures; complete labor-hour and terials costs for better scheduling and ocurement; the latest electrical products and nstruction methods.

- Variety of Special Electrical Systems including Cathodic Protection
- Costs for maintenance, demolition, HVAC/ nechanical, specialties, equipment, and more

9.95 per copy
ver 450 pages, illustrated, available Oct. 1999
atalog No. 60030

Means Electrical Change Order Cost Data 2000

You are provided with electrical unit prices exclusively for pricing change orders—based on the recent, direct experience of contractors and suppliers. Analyze and check your own change order estimates against the experience others have had doing the same work. It also covers productivity analysis and change order cost justifications. With useful information for calculating the effects of change orders and dealing with their administration.

$89.95 per copy
Over 450 pages, available Oct. 1999
Catalog No. 60230

leans Facilities laintenance & Repair Cost Data 2000

ablished in a looseleaf format, *Means Facilities aintenance & Repair Cost Data* gives you a mplete system to manage and plan your facility pair and maintenance costs and budget efficiently. uidelines for auditing a facility and developing an nual maintenance plan. Budgeting is included, ong with reference tables on cost and anagement and information on frequency and oductivity of maintenance operations.

he only nationally recognized source of aintenance and repair costs. Developed in operation with the Army Corps of Engineers.

199.95 per copy
ver 600 pages, illustrated, available Dec. 1999
atalog No. 60300

Means Square Foot Costs 2000

It's Accurate and Easy To Use!

- **Updated 2000 price information,** based on nationwide figures from suppliers, estimators, labor experts and contractors.
- "How-to-Use" Sections, with **clear examples** of commercial, residential, industrial, and institutional structures.
- Realistic graphics, offering true-to-life illustrations of building projects.
- Extensive information on using square foot cost data, including **sample estimates** and **alternate pricing methods.**

$99.95 per copy
Over 450 pages, illustrated, available Nov. 1999
Catalog No. 60050

Annual Cost Guides

For more informati
visit Means Web Si
at www.rsmeans.c

Means Repair & Remodeling Cost Data 2000

Commercial/Residential

You can use this valuable tool to estimate commercial and residential renovation and remodeling.

Includes: New costs for hundreds of unique methods, materials and conditions that only come up in repair and remodeling. PLUS:

- Unit costs for over 16,000 construction components
- Installed costs for over 90 assemblies
- Costs for 300+ construction crews
- Over 930 "On Site" localization factors for the U.S. and Canada.

$79.95 per copy
Over 600 pages, illustrated, available Oct. 1999
Catalog No. 60040

Means Facilities Construction Cost Data 2000

For the maintenance and construction of commercial, industrial, municipal, and institution properties. Costs are shown for new and remodeli construction and are broken down into materials, labor, equipment, overhead, and profit. Special emphasis is given to sections on mechanical, electrical, furnishings, site work, building maintenance, finish work, and demolition. More than 45,000 unit costs plus assemblies and reference sections are included.

$209.95 per copy
Over 1150 pages, illustrated, available Nov. 1999
Catalog No. 60200

Means Residential Cost Data 2000

Contains square foot costs for 30 basic home models with the look of today—plus hundreds of custom additions and modifications you can quote right off the page. With costs for the 100 residential systems you're most likely to use in the year ahead. Complete with blank estimating forms, sample estimates and step-by-step instructions.

$74.95 per copy
Over 550 pages, illustrated, available Dec. 1999
Catalog No. 60170

Means Light Commercial Cost Data 2000

Specifically addresses the light commercial market which is an increasingly specialized niche in the industry. Aids you, the owner/designer/contractor, in preparing all types of estimates, from budgets to detailed bids. Includes new advances in methods and materials. Assemblies section allows you to evaluate alternatives in the early stages of design/planning.

Over 13,000 unit costs for 2000 ensure you have the prices you need... when you need them.

$76.95 per copy
Over 600 pages, illustrated, available Dec. 1999
Catalog No. 60180

Means Assemblies Cost Data 2000

Means Assemblies Cost Data 2000 takes the guesswork out of preliminary or conceptual estimates. Now you don't have to try to calculate the assembled cost by working up individual components costs. We've done all the work for you.

Presents detailed illustrations, descriptions, specifications and costs for every conceivable building assembly—240 types in all—arranged in the easy-to-use UniFormat system. Each illustrated "assembled" cost includes a complete grouping of materials and associated installation costs including the installing contractor's overhead and profit.

$139.95 per copy
Over 550 pages, illustrated, available Oct. 1999
Catalog No. 60060

Means Site Work & Landscape Cost Data 2000

Means Site Work & Landscape Cost Data 2000 is organized to assist you in all your estimating needs Hundreds of fact-filled pages help you make accurate cost estimates efficiently.

Updated for 2000!

- Demolition features—including ceilings, doors, electrical, flooring, HVAC, millwork, plumbing, roofing, walls and windows
- State-of-the-art segmental retaining walls
- Flywheel trenching costs and details
- Updated Wells section
- Thousands of landscape materials, flowers, shrubs and trees

$89.95 per copy
Over 600 pages, illustrated, available Nov. 1999
Catalog No. 60280

more information
it Means Web Site
www.rsmeans.com

Annual Cost Guides

eans Open Shop Building onstruction Cost Data 2000

: latest costs for accurate budgeting and estimating of
commercial and residential construction...
ovation work... change orders... cost engineering.
ins Open Shop BCCD will assist you to...
evelop benchmark prices for change orders
ug gaps in preliminary estimates, budgets
timate complex projects
bstantiate invoices on contracts
ice ADA-related renovations

.95 per copy
er 650 pages, illustrated, available Oct. 1999
talog No. 60150

Means Heavy Construction Cost Data 2000

A comprehensive guide to heavy construction costs. Includes costs for highly specialized projects such as tunnels, dams, highways, airports, and waterways. Information on different labor rates, equipment, and material costs is included. Has unit price costs, systems costs, and numerous reference tables for costs and design. Valuable not only to contractors and civil engineers, but also to government agencies and city/town engineers.

$89.95 per copy
Over 450 pages, illustrated, available Nov. 1999
Catalog No. 60160

eans Building Construction ost Data 2000

estern Edition

is regional edition provides more precise cost ormation for western North America. Labor rates are ed on union rates from 13 western states and stern Canada. Included are western practices and terials not found in our national edition: tilt-up ncrete walls, glu-lam structural systems, specialized nber construction, seismic restraints, landscape and gation systems.

9.95 per copy
er 600 pages, illustrated, available Dec. 1999
talog No. 60220

Means Heavy Construction Cost Data 2000

Metric Version

Make sure you have the Means industry standard metric costs for the federal, state, municipal and private marketplace. With thousands of up-to-date metric unit prices in tables by CSI standard divisions. Supplies you with assemblies costs using the metric standard for reliable cost projections in the design stage of your project. Helps you determine sizes, material amounts, and has tips for handling metric estimates.

$89.95 per copy
Over 450 pages, illustrated, available Nov. 1999
Catalog No. 63160

leans Construction Cost ndexes 2000

no knows what 2000 holds? What materials d labor costs will change unexpectedly? how much?

- reakdowns for 305 major cities.
- ational averages for 30 key cities.
- xpanded five major city indexes.
- istorical construction cost indexes.

98.00 per year/$49.50 individual quarters
talog No. 60140

Means Interior Cost Data 2000

Provides you with prices and guidance needed to make accurate interior work estimates. Contains costs on materials, equipment, hardware, custom installations, furnishings, labor costs . . . every cost factor for new and remodel commercial and industrial interior construction, including updated information on office furnishings, plus more than 50 reference tables. For contractors, facility managers, owners.

$89.95 per copy
Over 550 pages, illustrated, available Oct. 1999
Catalog No. 60090

leans Concrete & Masonry Cost Data 2000

ovides you with cost facts for virtually all ncrete/masonry estimating needs, from complicated rmwork to various sizes and face finishes of brick and ock, all in great detail. The comprehensive unit cost ction contains more than 8,500 selected entries. Also ntains an assemblies cost section, and a detailed ference section which supplements the cost data.

79.95 per copy
ver 450 pages, illustrated, available Nov. 1999
atalog No. 60110

Means Labor Rates for the Construction Industry 2000

Complete information for estimating labor costs, making comparisons and negotiating wage rates by trade for over 300 cities (United States and Canada). With 46 construction trades listed by local union number in each city, and historical wage rates included for comparison. No similar book is available through the trade.

Each city chart lists the county and is alphabetically arranged with handy visual flip tabs for quick reference.

$189.95 per copy
Over 300 pages, available Dec. 1999
Catalog No. 60120

Reference Books

Project Scheduling & Management for Construction

New Revised Edition

By David R. Pierce, Jr.

A comprehensive, yet easy-to-follow guide to construction project scheduling and control—from vital project management principles through the latest scheduling, tracking, and controlling techniques. The author is a leading authority on scheduling with years of field and teaching experience at leading academic institutions. Spend a few hours with this book and come away with a solid understanding of this essential management topic.

$64.95 per copy
Over 250 pages; illustrated, Hardcover
Catalog No. 67247A

Means Landscape Estimatir Methods

New 3rd Edition

By Sylvia H. Fee

This revised edition offers expert guidance for preparing accurate estimates for new landscape construction and grounds maintenance. Includes a complete project estimate featuring the latest equipment and methods, and **two totally new chapters—Life Cycle Costing and Landscape Maintenance Estimating.**

$62.95 per copy
Over 300 pages, illustrated, Hardcover
Catalog No. 67295A

Total Productive Facilities Management

By Richard W. Sievert, Jr.

A New Operational Standard for Facilities Management.

The TPFM program incorporates the best of today's cost and project management, quality, and value engineering principles. The book includes:

- Realistic ways to collect and use benchmarking data.
- Value Engineering to find economical answers to the organization's needs.
- Selecting the best scheduling, control, and contracting methods for construction projects.

$79.95 per copy
Over 270 pages, illustrated, Hardcover
Catalog No. 67321

Cyberplaces: The Internet Guide for Architects, Engineers & Contractors

By Paul Doherty

Internet applications for business and project management. Includes Book, CD-ROM and Web Site.

The CD-ROM offers built-in links to web sites, tours of captured sites, FREE browser software and a document workshop, plus a test for Continuing Education Credits. **The Web Site** keeps the book current with updates on new technologies, interactive workshops, links to new tools and sites, and reports from top firms.

$59.95 per copy
Over 700 pages, illustrated, Softcover
Catalog No. 67317

Value Engineering: Practical Applications

...For Design, Construction, Maintenance & Operations

By Alphonse Dell'Isola, P.E., leading authority on VE in construction

A tool for immediate application—for engineers, architects, facility managers, owners and contractors. Includes: Making the Case for VE—The Management Briefing, Integrating VE into Planning and Budgeting, Conducting Life Cycle Costing, Integrating VE into the Design Process, Using VE Methodology in Design Review and Consultant Selection, Case Studies, VE Workbook, and a Life Cycle Costing program on disk.

$79.95 per copy
Over 450 pages, illustrated, Hardcover
Catalog No. 67319

Means Environmental Remediation Estimating Methods

By Richard R. Rast

The first-ever guide to estimating any size environmental remediation project... anywhere in the country.

Field-tested guidelines for estimating 50 standard remediation technologies. This resource will help you: prepare preliminary budgets, develop detailed estimates, compare costs, select solutions, estimate liability, review quotes, and negotiate settlements.

A valuable support tool for ***Means Environmental Remediation Unit Price*** and ***Assemblies books.***

$99.95 per copy
Over 600 pages, illustrated, Hardcover
Catalog No. 64777

r more information
sit Means Web Site
www.rsmeans.com

Reference Books

ost Planning & Estimating for acilities Maintenance

this unique book, a team of facilities management thorities shares their expertise at:
Evaluating and budgeting maintenance operations
Maintaining & repairing key building components
Applying *Means Facilities Maintenance & Repair Cost Data* to your estimating

th the special maintenance requirements of the 10 ajor building types.

2.95 per copy
ver 475 pages, Hardcover
atalog No. 67314

Facilities Planning & Relocation

By David D. Owen

An A-Z working guide—complete with the checklists, schematic diagrams and questionnaires to ensure the success of every office relocation. Complete with a step-by-step manual, 100-page technical reference section, over 50 reproducible forms in hard copy and on computer diskettes.
Winner of the International Facility Managers Assoc. "Distinguished Author of the Year" award.

$109.95 per copy, Textbook-384 pages,
Forms Binder-146 pages, illustrated, Hardcover
Catalog No. 67301

leans Facilities Maintenance tandards

nique features of this one-of-a-kind orking guide for facilities maintenance

working encyclopedia that points the way to lutions to every kind of maintenance and repair lemma. With a labor-hours section to provide oductivity figures for over 180 maintenance tasks. cluded are ready-to-use forms, checklists, worksheets d comparisons, as well as analysis of materials stems and remedies for deterioration and wear.

159.95 per copy, 600 pages, 205 tables, ecklists and diagrams, Hardcover
atalog No. 67246

HVAC: Design Criteria, Options, Selection

Expanded Second Edition

By William H. Rowe III, AIA, PE

Includes Indoor Air Quality, CFC Removal, Energy Efficient Systems and Special Systems by Building Type.
Helps you solve a wide range of HVAC system design and selection problems effectively and economically. Gives you clear explanations of the latest ASHRAE standards.

$84.95 per copy
Over 600 pages, illustrated, Hardcover
Catalog No. 67306

he Facilities Manager's eference

y Harvey H. Kaiser, PhD

he tasks and tools the facility manager needs to ccomplish the organization's objectives, and develop dividual and staff skills. Includes Facilities and roperty Management, Administrative Control, lanning and Operations, Support Services, and a omplete building audit with forms and instructions, idely used by facilities managers nationwide.

86.95 per copy, over 250 pages,
ith prototype forms and graphics, Hardcover
atalog No. 67264

Facilities Maintenance Management

By Gregory H. Magee, PE

Now you can get successful management methods and techniques for all aspects of facilities maintenance. This comprehensive reference explains and demonstrates successful management techniques for all aspects of maintenance, repair and improvements for buildings, machinery, equipment and grounds. Plus, guidance for outsourcing and managing internal staffs.

$86.95 per copy
Over 280 pages with illustrations, Hardcover
Catalog No. 67249

Jnderstanding Building Automation Systems

Direct Digital Control	· Energy Management
Security/Access Control	· Life Safety
	· Lighting

y Reinhold A. Carlson, PE & Robert Di Giandomenico

he authors, leading authorities on the design and nstallation of these systems, describe the major uilding systems in both an overview with estimating nd selection criteria, and in system configuration-level etail.

Now $39.98 per copy, limited quantity
Over 200 pages, illustrated, Hardcover
Catalog No. 67284

The ADA in Practice

By Deborah S. Kearney, PhD

Helps you meet and budget for the requirements of the Americans with Disabilities Act. Shows how to do the job right by understanding what the law requires, allows, and enforces. Includes a "buyers guide" for 70 ADA-compliant products.

*Winner of IFMA's "Distinguished Author of the Year" award.

See page 9 for another ADA resource.

$72.95 per copy
Over 600 pages, illustrated, Softcover
Catalog No. 67147A

Reference Books

For more informat
visit Means Web S
at www.rsmeans.c

Basics for Builders: Plan Reading & Material Takeoff

By Wayne J. DelPico

For Residential and Light Commercial Construction

A valuable tool for understanding plans and specs, and accurately calculating material quantities. Step-by-step instructions and takeoff procedures based on a full set of working drawings.

$35.95 per copy
Over 420 pages, Softcover
Catalog No. 67307

Means Illustrated Construction Dictionary Unabridged Edition

Written in contractor's language, the information is adaptable for report writing, specifications or just intelligent discussion. Contains over 13,000 construction terms, words, phrases, acronyms and abbreviations, slang, regional terminology, and hundr of illustrations.

$99.95 per copy
Over 700 pages, Hardcover
Catalog No. 67292

Superintending for Contractors: How to Bring Jobs in On-time, On-Budget

By Paul J. Cook

This book examines the complex role of the superintendent/field project manager, and provides guidelines for the efficient organization of this job. Includes administration of contracts, change orders, purchase orders, and more.

$35.95 per copy
Over 220 pages, illustrated, Softcover
Catalog No. 67233

Means Forms for Contractors

For general and specialty contractors

Means editors have created and collected the most needed forms as requested by contractors of various-sized firms and specialties. This book covers a project phases. Includes sample correspondence and personnel administration.
Full-size forms on durable paper for photocopying or reprinting.

$79.95 per copy
Over 400 pages, three-ring binder, more than 80 for
Catalog No. 67288

Risk Management for Building Professionals

By Thomas E. Papageorge, RA

Outlines a master plan for phasing a risk management system into each company function. Guides the manager in writing a detailed Risk Management Plan for each project. Contains sample procedures manual.

Now $29.98 per copy, limited quantity
Over 200 pages, illustrated, Hardcover
Catalog No. 67254

Means Heavy Construction Handbook

Informed guidance for planning, estimating and performing today's heavy construction projects. Provid expert advice on every aspect of heavy construction work including hazardous waste remediation and estimating. To assist planning, estimating, performing, or overseeing work.

$74.95 per copy
Over 430 pages, illustrated, Hardcover
Catalog No. 67148

Estimating for Contractors: How to Make Estimates that Win Jobs

By Paul J. Cook

Estimating for Contractors is a reference that will be used over and over, whether to check a specific estimating procedure, or to take a complete course in estimating.

$35.95 per copy
Over 225 pages, illustrated, Softcover
Catalog No. 67160

HVAC Systems Evaluation

By Harold R. Colen, PE

You get direct comparisons of how each type of system works, with the relative costs of installation, operation maintenance and applications by type of building. Wit requirements for hooking up electrical power to HVAC components. Contains experienced advice for repairing operational problems in existing HVAC systems, ductwork, fans, cooling coils, and much more!

$84.95 per copy
Over 500 pages, illustrated, Hardcover
Catalog No. 67281

Quantity Takeoff for Contractors: How to Get Accurate Material Counts

By Paul J. Cook

Contractors who are new to material takeoffs or want to be sure they are using the best techniques will find helpful information in this book, organized by CSI MasterFormat division.

Now $17.98 per copy, limited quantity
Over 250 pages, illustrated, Softcover
Catalog No. 67262

Roofing: Design Criteria, Options, Selection

By R.D. Herbert, III

This book is required reading for those who specify, install or have to maintain roofing systems. It covers al types of roofing technology and systems. You'll get the facts needed to intelligently evaluate and select both traditional and new roofing systems.

Now $31.48 per copy
Over 225 pages with illustrations, Hardcover
Catalog No. 67253

more information
t Means Web Site
www.rsmeans.com

Reference Books

eans Estimating Handbook

s comprehensive reference covers a full spectrum of nical data for estimating, with information on ng, productivity, equipment requirements, codes, gn standards and engineering factors.

ins Estimating Handbook will help you: evaluate itectural plans and specifications, prepare accurate ntity takeoffs, prepare estimates from conceptual to iled, and evaluate change orders.

.95 per copy
er 900 pages, Hardcover
alog No. 67276

Means Repair and Remodeling Estimating Third Edition

By Edward B. Wetherill & R.S. Means

Focuses on the unique problems of estimating renovations of existing structures. It helps you determine the true costs of remodeling through careful evaluation of architectural details and a site visit.
New section on disaster restoration costs.

$69.95 per copy
Over 450 pages, illustrated, Hardcover
Catalog No. 67265A

ccessful Estimating Methods: m Concept to Bid

John D. Bledsoe, PhD, PE

ighly practical, all-in-one guide to the tips and ctices of today's successful estimator. Presents hniques for all types of estimates, *and* advanced ics such as life cycle cost analysis, value engineering, automated estimating.
imate spreadsheets available at Means Web site.

.95 per copy
er 300 pages, illustrated, Hardcover
talog No. 67287

Means Productivity Standards for Construction

Expanded Edition (*Formerly* Man-Hour Standards)

Here is the working encyclopedia of labor productivity information for construction professionals, with labor requirements for thousands of construction functions in CSI MasterFormat.
Completely updated, with over 3,000 new work items.

$159.95 per copy
Over 800 pages, Hardcover
Catalog No. 67236A

eans Electrical Estimating ethods Second Edition

panded version includes sample estimates and cost ormation in keeping with the latest version of the I MasterFormat. Contains new coverage of Fiber tic and Uninterruptible Power Supply electrical tems, broken down by components and explained detail. A practical companion to *Means Electrical st Data.*

4.95 per copy
ver 325 pages, Hardcover
talog No. 67230A

Means Mechanical Estimating Methods Second Edition

This guide assists you in making a review of plans, specs and bid packages with suggestions for takeoff procedures, listings, substitutions and pre-bid scheduling. Includes suggestions for budgeting labor and equipment usage. Compares materials and construction methods to allow you to select the best option.

$64.95 per copy
Over 350 pages, illustrated, Hardcover
Catalog No. 67294

eans Scheduling Manual

ird Edition

F. William Horsley

st, convenient expertise for keeping your scheduling ills right in step with today's cost-conscious times. vers bar charts, PERT, precedence and CPM heduling methods. Now updated to include computer plications.

4.95 per copy
ver 200 pages, spiral-bound, Softcover
atalog No. 67291

Maintenance Management Audit

This annual audit program is essential for every organization in need of a proper assessment of its maintenance operation. The forms in this easy-to-use workbook allow managers to identify and correct problems and enhance productivity. Includes electronic forms on disk.

Now $32.48 per copy, limited quantity
125 pages, spiral bound, illustrated, Hardcover
Catalog No. 67299

leans Square Foot Estimating lethods Second Edition

Billy J. Cox and F. William Horsley

oven techniques for conceptual and design-stage cost anning. Steps you through the square foot cost ocess, demonstrating faster, better ways to relate the sign to the budget. Now updated to the latest version UniFormat.

69.95 per copy
ver 300 pages, illustrated, Hardcover
atalog No. 67145A

Means ADA Compliance Pricing Guide

Gives you detailed cost estimates for budgeting modification projects, including estimates for each of 260 alternates. You get the 75 most commonly needed modifications for ADA compliance, each an assembly estimate with detailed cost breakdown.

$72.95 per copy
Over 350 pages, illustrated, Softcover
Catalog No. 67310

Reference Books

For more informat
visit Means Web S
at www.rsmeans.c

Means Unit Price Estimating Methods
2nd Edition

$59.95 per copy
Catalog No. 67303

Legal Reference for Design & Construction
By Charles R. Heuer, Esq., AIA

Now $54.98 per copy, limited quantity
Catalog No. 67266

Means Plumbing Estimating Methods
New 2nd Edition

By Joseph J. Galeno & Sheldon T. Greene

$59.95 per copy
Catalog No. 67283

Structural Steel Estimating
By S. Paul Bunea, PhD

Now $39.98 per copy, limited quantity
Catalog No. 67241

Business Management for Contractors
How to Make Profits in Today's Market

By Paul J. Cook

Now $17.98 per copy, limited quantity
Catalog No. 67250

Understanding Legal Aspects of Design/Build
By Timothy R. Twomey, Esq., AIA

$79.95 per copy
Catalog No. 67259

Construction Paperwork
An Efficient Management System

By J. Edward Grimes

Now $26.48 per copy, limited quantity
Catalog No. 67268

Contractor's Business Handbook
By Michael S. Milliner

Now $21.48 per copy, limited quantity
Catalog No. 67255

Basics for Builders: How to Survive and Prosper in Construction
By Thomas N. Frisby

$34.95 per copy
Catalog No. 67273

Successful Interior Projects Through Effective Contract Documents
By Joel Downey & Patricia K. Gilbert

Now $34.98 per copy, limited quantity
Catalog No. 67313

The Building Professional's Guide to Contract Documents
By Waller S. Poage, AIA, CSI, CCS

$64.95 per copy
Catalog No. 67261

Illustrated Construction Dictionary, Condensed
$59.95 per copy
Catalog No. 67282

Managing Construction Purchasing
By John G. McConville, CCC, CPE

Now $31.48 per copy, limited quantity
Catalog No. 67302

Hazardous Material & Hazardous Waste
By Francis J. Hopcroft, PE, David L. Vitale, M. Ed., & Donald L. Anglehart, Esq.

Now $44.98 per copy, limited quantity
Catalog No. 67258

Fundamentals of the Construction Process
By Kweku K. Bentil, AIC

Now $34.98 per copy, limited quantity
Catalog No. 67260

Construction Delays
By Theodore J. Trauner, Jr., PE, PP

$59.95 per copy
Catalog No. 67278

Basics for Builders: Framing & Rough Carpentry
By Scot Simpson

$24.95 per copy
Catalog No. 67298

Interior Home Improvement Costs
6th Edition

$19.95 per copy
Catalog No. 67308B

Exterior Home Improvement Costs
6th Edition

$19.95 per copy
Catalog No. 67309B

Concrete Repair and Maintenance Illustrated
By Peter H. Emmons

$69.95 per copy
Catalog No. 67146

How to Estimate with Metric Units
Now $24.98 per copy, limited quantity
Catalog No. 67304

Seminars

eveloping Facility Assessment Programs

is two-day program concentrates on the management process required planning, conducting, and documenting the physical condition and ctional adequacy of buildings and other facilities. Regular facilities ndition inspections are one of the facility department's most important ies. However, gathering reliable data hinges on the design of the rall program used to identify and gauge deferred maintenance uirements. Knowing where to look . . . and reporting results effectively the keys. This seminar is designed to give the facility executive ential steps for conducting facilities inspection programs.

pection Program Requirements • Where is the deficiency? • What he nature of the problem? • How can it be remedied? • How much l it cost in labor, equipment and materials? • When should it be complished? • Who is best suited to do the work?

te: Because of its management focus, this course will not address de practices and procedures.

Repair and Remodeling Estimating

Repair and remodeling work is becoming increasingly competitive as more professionals enter the market. Recycling existing buildings can pose difficult estimating problems. Labor costs, energy use concerns, building codes, and the limitations of working with an existing structure place enormous importance on the development of accurate estimates. Using the exclusive techniques associated with Means' widely acclaimed **Repair & Remodeling Cost Data**, this seminar sorts out and discusses solutions to the problems of building alteration estimating. Attendees will receive two intensive days of eye-opening methods for handling virtually every kind of repair and remodeling situation . . . from demolition and removal to final restoration.

echanical and Electrical Estimating

is seminar is tailored to fit the needs of those seeking to develop or prove their skills and to have a better understanding of how chanical and electrical estimates are prepared during the conceptual, nning, budgeting and bidding stages. Learn how to avoid costly issions and overlaps between these two interrelated specialties by paring complete and thorough cost estimates for both trades. Featured order of magnitude, assemblies, and unit price estimating. In mbination with the use of **Means Mechanical Cost Data, Means umbing Cost Data** and **Means Electrical Cost Data**, this seminar will sure more accurate and complete Mechanical/Electrical estimates for th unit price and preliminary estimating procedures.

Unit Price Estimating

This seminar shows how today's advanced estimating techniques and cost information sources can be used to develop more reliable unit price estimates for projects of any size. It demonstrates how to organize data, use plans efficiently, and avoid embarrassing errors by using better methods of checking.

You'll get down-to-earth help and easy-to-apply guidance for:

- making maximum use of construction cost information sources
- organizing estimating procedures in order to save time and reduce mistakes
- sorting out and identifying unusual job requirements to improve estimating accuracy.

quare Foot Cost Estimating

arn how to make better preliminary estimates with a limited amount budget and design information. You will benefit from examples of a de range of systems estimates with specifications limited to building e requirements, budget, building codes, and type of building. And yet, th minimal information, you will obtain a remarkable degree of curacy.

orkshop sessions will provide you with model square foot estimating oblems and other skill-building exercises. The exclusive Means building semblies square foot cost approach shows how to make very reliable timates using "bare bones" budget and design information.

Scheduling and Project Management

This seminar helps you successfully establish project priorities, develop realistic schedules, and apply today's advanced management techniques to your construction projects. Hands-on exercises familiarize participants with network approaches such as the Critical Path Method. Special emphasis is placed on cost control, including use of computer-based systems. Through this seminar you'll perfect your scheduling and management skills, ensuring completion of your projects *on time* and *within budget.* Includes hands-on application of **Means Scheduling Manual** and **Means Building Construction Cost Data.**

acilities Maintenance and Repair Estimating

ith our Facilities Maintenance and Repair Estimating seminar, you'll arn how to plan, budget, and estimate the cost of ongoing and eventive maintenance and repair for all your buildings and grounds. ased on R.S. Means' groundbreaking cost estimating book, this two-day minar will show you how to decide to either contract out or retain aintenance and repair work in-house. In addition, you'll learn how to epare budgets and schedules that help cut down on unplanned and stly emergency repair projects. Facilities Maintenance and Repair stimating crystallizes what facilities professionals have learned over the ars, but never had time to organize or document. This program covers a ariety of maintenance and repair projects, from underground storage tank moval, roof repair and maintenance, exterior wall renovations, and ergy source conversions, to service upgrades and estimating ergy-saving alternatives.

Managing Facilities Construction and Maintenance

In you're involved in new facility construction, renovation or maintenance projects and are concerned about getting quality work done on time and on or below budget, in Means' seminar **Managing Facilities Construction and Maintenance** you'll learn management techniques needed to effectively plan, organize, control and get the most out of your limited facilities resources.

Learn how to develop budgets, reduce expenditures and check productivity. With the knowledge gained in this course, you'll be better prepared to successfully sell accurate project budgets, timing and manpower needs to senior management . . . plus understand how to evaluate the impact of today's facility decisions on tomorrow's budget.

Call 1-800-448-8182 for more information

Seminars

For more informati
visit Means Web Si
at www.rsmeans.co

2000 Means Seminar Schedule

Location	Dates
Las Vegas, NV	March 13 - 16
Dallas, TX	April 10-13
Washington, DC	April 17-20
Denver, CO	May 22-25
San Francisco, CA	June 12-15
Cape Cod, MA	September 11-14
Washington, DC	September 25-28
San Diego, CA	October TBD
Atlantic City, NJ	November TBD
Orlando, FL	November 13-16

Registration Information

Register Early... Save up to $150! Register 30 days before the start date of a seminar and save $150 off your total fee. ***Note: This discount can be applied only once per order.***

How to Register Register by phone today! Means toll-free number for making reservations is: **1-800-448-8182.**

Individual Seminar Registration Fee $875 To register by mail, complete the registration form and return with your full fee to: Seminar Division, R.S. Means Company, Inc., 63 Smiths Lane, Kingston, MA 02364.

Federal Government Pricing All Federal Government employees save 25% off regular seminar price. Other promotional discounts cannot be combined with Federal Government discount.

Team Discount Program Two to four seminar registrations: $760 per person—Five or more seminar registrations: $710 per person—Ten or more seminar registrations: Call for pricing.

Consecutive Seminar Offer One individual signing up for two separate courses at the same location during the designated time period pays only $1,400. You get the second course for only $525 (**a 40% discount**). Payment must be received at least ten days prior to seminar dates to confirm attendance.

Refunds Cancellations will be accepted up to ten days prior to the seminar start. There are no refunds for cancellations postmarked later than ten working days prior to the first day of the seminar. A $150 processing fee will be charged for all cancellations. Written notice or telegram is required for all cancellations. Substitutions can be made at any time before the session starts. **No-shows are subject to the full seminar fee.**

AACE Approved Courses The R.S. Means Construction Estimatin and Management Seminars described and offered to you here have each been approved for 14 hours (1.4 recertification credits) of credit by the AACE International Certification Board toward meeting the continuing education requirements for re-certification as a Certified Cost Engineer/Certified Cost Consultant.

AIA Continuing Education R.S. Means is registered with the AIA Continuing Education System (AIA/CES) and is committed to developing quality learning activities in accordance with the CES criteria. R.S. Mean seminars meet the AIA/CES criteria for Quality Level 2. AIA members w receive (28) learning units (LUs) for each two day R.S. Means Course.

Daily Course Schedule The first day of each seminar session begi at 8:30 A.M. and ends at 4:30 P.M. The second day is 8:00 A.M.–4:00 P.M Participants are urged to bring a hand-held calculator since many actual problems will be worked out in each session.

Continental Breakfast Your registration includes the cost of a continental breakfast, a morning coffee break, and an afternoon break. These informal segments will allow you to discuss topics of mutual interest with other members of the seminar. (You are free to make your own lunch and dinner arrangements.)

Hotel/Transportation Arrangements R.S. Means has arranged to hold a block of rooms at each hotel hosting a seminar. To take advantage of special group rates when making your reservation be sure to mention that you are attending the Means Seminar. You are of course free to stay at the lodging place of your choice. **(Hotel reservations and transportation arrangements should be made directly by seminar attendees.)**

Important Class sizes are limited, so please register as soon as possibl

Registration Form

Call 1-800-448-8182 to register or FAX 1-800-632-67

Please register the following people for the Means Construction Seminars as shown here. Full payment or deposit is enclosed, and we understand that we must make our own hotel reservations if overnight stays are necessary.

☐ Full payment of $ ____________________ enclosed.

☐ Bill me

Name of Registrant(s)
(To appear on certificate of completion)

____

____

____

P.O. #: ____________________

GOVERNMENT AGENCIES MUST SUPPLY PURCHASE ORDER NUMBER

Firm Name ____________________

Address ____________________

City/State/Zip ____________________

Telephone No. ____________________ Fax No. ____________________

E-Mail Address ____________________

Charge our registration(s) to: ☐ MasterCard ☐ VISA ☐ American Express ☐ Discover

Account No. ____________________ Exp. Date ____________________

Cardholder's Signature ____________________

Seminar Name City Dates

____

____

Please mail check to: R.S. MEANS COMPANY, INC., 63 Smiths Lane, P.O. Box 800, Kingston, MA 02364 USA

acilities Operations & Engineering Reference

all-in-one technical reference for planning & managing facility projects & solving day-to-day operations problems

1 major sections address:

anagement Skills
onomics (budgeting/cost control, financial analysis, VE, etc.)
ivil Engineering & Construction Practices
aintenance (detailed staffing guidance and job descriptions, CMMS, anning/scheduling, training, work orders, preventive/predictive maintenance)
nergy Efficiencies (optimizing energy use, including heating, cooling, ghting, and water)
VAC
echanical Engineering
strumentation & Controls
lectrical Engineering
nvironmental, Health & Safety

Produced jointly by R.S. Means and the Association for Facilities Engineering.

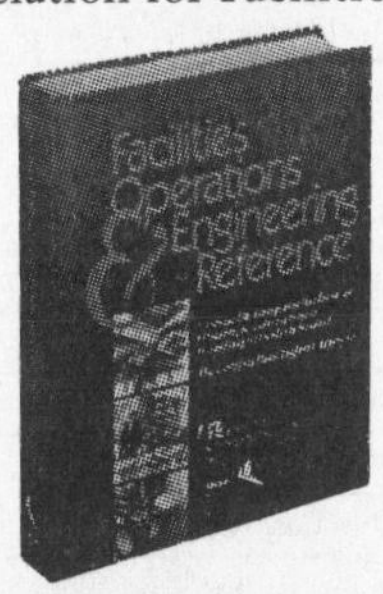

$99.95 per copy
Over 700 pages, Illustrated, Hardcover
Catalog No. 67318

lanning & Managing Interior Projects New Second Edition

Carol E. Farren, CFM

evised edition addresses the major changes in technology and business that have affected all aspects of interiors ojects. Guides you through every step in managing interior design and construction for commercial endeavors, om initial client meetings to post-project administration. Corporate downsizing, company mergers, and advances information technology have revolutionized the workplace. This book addresses all of these changes, showing w to plan effectively for current and future organizational needs. Includes new sections on:

- Evaluating space requirements and selecting the right site
- Alternative work models (telecommuting, hoteling, etc.)
- Records and document management
- Telecommunications and data issues
- Working with consultants
- Environmental considerations

.95 per copy
r 400 pages, Illustrated, Hardcover
alog No. 67245A

esidential & Light Commercial Construction Standards

Unique Collection of Industry Standards That Define Quality in Construction

r Contractors & Subcontractors, Owners, Developers, Architects & Engineers, Attorneys & Insurance Personnel

ompiled from the nation's major building codes, and from scores f publications and reports from professional institutes and other uthorities, this one-of-a-kind resource enables you to:

Set a standard for subcontractors and employees
Protect yourself against defect claims
Substantiate your own claim for inferior workmanship
Resolve disputes
Overview installation methods
Answer client questions with an authoritative reference

$59.95 per copy
Over 500 pages, Illustrated, Softcover
Catalog No. 67322

ORDER TOLL FREE 1-800-334-3509
OR FAX 1-800-632-6732.

2000 Order Form

Qty.	Book No.	COST ESTIMATING BOOKS	Unit Price	Total
	60060	Assemblies Cost Data 2000	$139.95	
	60010	Building Construction Cost Data 2000	85.95	
	61010	Building Const. Cost Data–Looseleaf Ed. 2000	109.95	
	63010	Building Const. Cost Data–Metric Version 2000	89.95	
	60220	Building Const. Cost Data–Western Ed. 2000	89.95	
	60110	Concrete & Masonry Cost Data 2000	79.95	
	50140	Construction Cost Indexes 2000	198.00	
	60140A	Construction Cost Index–January 2000	49.50	
	60140B	Construction Cost Index–April 2000	49.50	
	60140C	Construction Cost Index–July 2000	49.50	
	60140D	Construction Cost Index–October 2000	49.50	
	60310	Contr. Pricing Guide: Framing/Carpentry 2000	36.95	
	60330	Contr. Pricing Guide: Resid. Detailed 2000	36.95	
	60320	Contr. Pricing Guide: Resid. Sq. Ft. 2000	39.95	
	64020	ECHOS Assemblies Cost Book 2000	149.95	
	64010	ECHOS Unit Cost Book 2000	99.95	
	54000	ECHOS (Combo set of both books)	214.95	
	60230	Electrical Change Order Cost Data 2000	89.95	
	60030	Electrical Cost Data 2000	89.95	
	60200	Facilities Construction Cost Data 2000	209.95	
	60300	Facilities Maintenance & Repair Cost Data 2000	199.95	
	60160	Heavy Construction Cost Data 2000	89.95	
	63160	Heavy Const. Cost Data–Metric Version 2000	89.95	
	60090	Interior Cost Data 2000	89.95	
	60120	Labor Rates for the Const. Industry 2000	189.95	
	60180	Light Commercial Cost Data 2000	76.95	
	60020	Mechanical Cost Data 2000	89.95	
	60150	Open Shop Building Const. Cost Data 2000	89.95	
	60210	Plumbing Cost Data 2000	89.95	
	60040	Repair and Remodeling Cost Data 2000	79.95	
	60170	Residential Cost Data 2000	74.95	
	60280	Site Work & Landscape Cost Data 2000	89.95	
	60050	Square Foot Costs 2000	99.95	
		REFERENCE BOOKS		
	67147A	ADA in Practice	72.95	
	67310	ADA Pricing Guide	72.95	
	67298	Basics for Builders: Framing & Rough Carpentry	24.95	
	67273	Basics for Builders: How to Survive and Prosper	34.95	
	67307	Basics for Builders: Plan Reading & Takeoff	35.95	
	67261	Building Prof. Guide to Contract Documents	64.95	
	67312	Building Spec Homes Profitably	29.95	
	67250	Business Management for Contractors	17.98	
	67146	Concrete Repair & Maintenance Illustrated	69.95	
	67278	Construction Delays	59.95	
	67268	Construction Paperwork	26.48	
	67255	Contractor's Business Handbook	21.48	
	67314	Cost Planning & Est. for Facil. Maint.	82.95	
	67317	Cyberplaces: The Internet Guide for A/E/C	59.95	
	67230A	Electrical Estimating Methods–2nd Ed.	64.95	
	64777	Environmental Remediation Est. Methods	99.95	
	67160	Estimating for Contractors	35.95	
	67276	Estimating Handbook	99.95	
	67249	Facilities Maintenance Management	86.95	

Qty.	Book No.	REFERENCE BOOKS (Con't)	Unit Price	Total
	67246	Facilities Maintenance Standards	$159.95	
	67264	Facilities Manager's Reference	86.95	
	67318	Facilities Operations & Engineering Reference	99.95	
	67301	Facilities Planning & Relocation	109.95	
	67231	Forms for Building Const. Profess.	94.95	
	67288	Forms for Contractors	79.95	
	67260	Fundamentals of the Construction Process	34.98	
	67258	Hazardous Material & Hazardous Waste	44.98	
	67148	Heavy Construction Handbook	74.95	
	67308B	Home Improvement Costs–Interior Projects	19.95	
	67309B	Home Improvement Costs–Exterior Projects	19.95	
	67304	How to Estimate with Metric Units	24.98	
	67306	HVAC: Design Criteria, Options, Select.–2nd Ed.	84.95	
	67281	HVAC Systems Evaluation	84.95	
	67282	Illustrated Construction Dictionary, Condensed	59.95	
	67292	Illustrated Construction Dictionary, Unabridged	99.95	
	67295A	Landscape Estimating–3rd Ed.	62.95	
	67266	Legal Reference for Design & Construction	54.98	
	67299	Maintenance Management Audit	32.48	
	67302	Managing Construction Purchasing	31.48	
	67294	Mechanical Estimating–2nd Ed.	64.95	
	67245A	Planning and Managing Interior Projects, 2nd Ed.	69.95	
	67283A	Plumbing Estimating Methods, 2nd Ed.	59.95	
	67236A	Productivity Standards for Constr.–3rd Ed.	159.95	
	67247A	Project Scheduling & Management for Constr.	64.95	
	67262	Quantity Takeoff for Contractors	17.98	
	67265A	Repair & Remodeling Estimating–3rd Ed.	69.95	
	67322	Residential & Light Commercial Const. Stds.	59.95	
	67254	Risk Management for Building Professionals	29.98	
	67253	Roofing: Design Criteria, Options, Selection	31.48	
	67291	Scheduling Manual–3rd Ed.	64.95	
	67145A	Square Foot Estimating Methods–2nd Ed.	69.95	
	67241	Structural Steel Estimating	39.98	
	67287	Successful Estimating Methods	64.95	
	67313	Successful Interior Projects	34.98	
	67233	Superintending for Contractors	35.95	
	67321	Total Productive Facilities Management	79.95	
	67284	Understanding Building Automation Systems	39.98	
	67259	Understanding Legal Aspects of Design/Build	79.95	
	67303	Unit Price Estimating Methods–2nd Ed.	59.95	
	67319	Value Engineering: Practical Applications	79.95	
		MA residents add 5% state sales tax		
		Shipping & Handling**		
		Total (U.S. Funds)*		

Prices are subject to change and are for U.S. delivery only. *Canadian customers may call for current prices. **Shipping & handling charges: Add 7% of total order for check and credit card payments. Add 9% of total order for invoiced orders.

Send Order To: **ADDV-1001**

Name (Please Print) ____________________

Company ____________________

☐ **Company**

☐ **Home** Address ____________________

____

City/State/Zip ____________________

Phone # ____________ P.O. # ____________

(Must accompany all orders being billed)

Mail To: **R.S. Means Company, Inc.,** P.O. Box 800, Kingston, MA 02364-0800